MODERN BIOPOLYMER SCIENCE

BRIDGING THE DIVIDE BETWEEN FUNDAMENTAL TREATISE AND INDUSTRIAL APPLICATION

MODERN BIOPOLYMER SCIENCE

BRIDGING THE DIVIDE BETWEEN FUNDAMENTAL TREATISE AND INDUSTRIAL APPLICATION

Edited by

STEFAN KASAPIS
School of Applied Sciences
RMIT University
Melbourne, Australia

IAN T. NORTON
School of Engineering/Chemical Engineering
The University of Birmingham
Birmingham, UK

JOHAN B. UBBINK
Nestle Research Centre
Switzerland

ELSEVIER

AMSTERDAM • BOSTON • HEIDELBERG • LONDON • NEW YORK • OXFORD
PARIS • SAN DIEGO • SAN FRANCISCO • SINGAPORE • SYDNEY • TOKYO

Academic Press is an imprint of Elsevier

Academic Press is an imprint of Elsevier
32 Jamestown Road, London NW1 7BY, UK
30 Corporate Drive, Suite 400, Burlington, MA 01803, USA
525 B Street, Suite 1900, San Diego, CA 92101-4495, USA

First edition 2009

British Library Cataloguing in Publication Data
A catalogue record for this book is available form the British Library

Library of Congress Cataloging-in-Publication Data
A catalog record for this book is available from the Library of Congress

ISBN: 978-0-12-374195-0

Printed and bound in United States of America

09 10 11 12 13   10 9 8 7 6 5 4 3 2 1

Contents

Contributors

Anthony R. Bird Commonwealth Scientific and Industrial Research Organisation, Food Futures National Research Flagship, and CSIRO Human Nutrition, Adelaide, Australia

Charles Stephen Brennan Hollings Faculty, Manchester Metropolitan University, Manchester, UK

Margaret Anne Brennan Institute of Food, Nutrition and Human Health, Massey University, Palmerston North, New Zealand

Sarah L. Buckley Highton, Australia

Allan H. Clark Pharmaceutical Science Division, King's College London, London, UK

Phil W. Cox School of Engineering-Chemical Engineering, University of Birmingham, Edgbaston, UK

Steve W. Cui Guelph Research Food Centre, Agriculture and Agri-Food Canada, Guelph, Canada

David E. Dunstan Chemical & Biomolecular Engineering, University of Melbourne, Victoria, Australia

E. Allen Foegeding Department of Food Science, North Carolina State University, Raleigh, USA

Michael J. Gidley Centre for Nutrition & Food Sciences, University of Queensland, Brisbane, Australia

Liam M. Grover School of Chemical Engineering, University of Birmingham, Edgbaston, UK

Victoria A. Hughes Chemical & Biomolecular Engineering, University of Melbourne, Victoria, Australia

Stefan Kasapis School of Applied Sciences, RMIT University, Melbourne, Australia

Sandra I. Laneuville Dairy Research Centre STELA and Institute of Nutraceutical and Functional Foods INAF, Laval University, Quebec, Canada

Peter J. Lillford CNAP-Department of Biology, The University of York, York, UK

Erik van der Linden Agrotechnology and Food Sciences Group, Wageningen University, Wageningen, The Netherlands

Amparo Lopez-Rubio Australian Nuclear Science and Technology Organisation, Bragg Institute, Menai, Australia

David Julian McClements Department of Food Science, University of Massachussets Amherst, Amherst, USA

Edwin R. Morris Department of Food & Nutritional Sciences, University College Cork, Ireland

Vic J. Morris Institute of Food Research, Colney, UK

Ian T. Norton School of Engineering-Chemical Engineering, University of Birmingham, Edgbaston, UK

Amos Nussinovitch Faculty of Agricultural, Food and Environmental Quality Sciences, The Hebrew University of Jerusalem, Rehovot, Israel

Kunal Pal Department of Chemistry and Biology, Ryerson University, Toronto, Canada

Allan T. Paulson Department of Chemistry and Biology, Ryerson University, Toronto, Canada

Keisha Roberts Guelph Research Food Centre, Agriculture and Agri-Food Canada, Guelph, Canada

Yrjö H. Roos Department of Food & Nutritional Sciences, University College Cork, Ireland

Simon B. Ross-Murphy Pharmaceutical Science Division, King's College London, London, UK

Dérick Rousseau School of Nutrition, Ryerson University, Toronto, Canada

Ashok K. Shrestha Centre for Nutrition & Food Sciences, University of Queensland, St. Lucia, Australia

Alan M. Smith School of Chemical Engineering, University of Birmingham, Edgbaston, UK

Fotios Spyropoulos School of Engineering-Chemical Engineering, University of Birmingham, Edgbaston, UK

Sylvie L. Turgeon Dairy Research Centre STELA and Institute of Nutraceutical and Functional Foods INAF, Laval University, Quebec, Canada

Johan B. Ubbink Nestle Research Centre Switzerland, Savigny, Switzerland

Preface

It has been a while since a book was put together to address the issues of the physics and chemistry of biopolymers in industrial formulations, including concise treatments of the relation between biopolymer functionality and their conformation, structure, and interactions. In these intervening years, some materials and concepts came to prominence while other ones have changed in their appeal or application. As ever, the industrialist is faced with the challenge of innovation in an increasingly competitive market in terms of ingredient cost, product added-value, expectations of a healthy life-style, improved sensory impact, controlled delivery of bioactive compounds and, last but not least, product stability. Proteins, polysaccharides and their co-solutes remain the basic tools of achieving the required properties in product formulations, and much has been said about the apparent properties of these ingredients in relation to their practical use. There is also an ever increasing literature on the physicochemical behaviour of well-characterised biopolymer systems based on the molecular physics of glassy materials, the fundamentals of gelation, and component interactions in the bulk and at interfaces. It appears, however, that a gap has emerged between the recent advances in fundamental knowledge and the direct application to product situations with a growing need for scientific input.

The above statement does not detract from the pioneering work of the forefathers in the field who developed the origins of biopolymer science. For example, there is no question that the pioneering work on conformational transitions and gelation, the idea of phase separation into water in emulsions, the development of physicochemical understanding that lead to the concept of fluid gels and the application of the glass transition temperature to dehydrated and partially frozen biomaterials has resulted not only in academic progress but in several healthy and novel products in the market place. Thus the first phase of the scientific quest for developing comprehensive knowledge at both the theoretical and applied levels of functional properties in basic preparations and systems has largely been accomplished. It is clear, though, that the future lies in the utilization of this understanding in both established and novel foodstuffs, and non-food materials (e.g. pharmaceuticals) with their multifaceted challenges. A clear pathway for processing, preservation and innovation is developing which is particularly important if progress is to be made in the preparation of indulgent yet healthy foods which are stable, for example, in distribution and storage. This requires a multi-scale engineering approach in which material properties and microstructure, hence the product performance are designed by careful selection of ingredients and processes. Examples of this can be found in the pioneering work on fat replacement and the reliance on the phenomenon of glass transition to rationalise the structural stability and mouthfeel of a complex embodiment.

Within this context of matching science to application, one feels compelled to note that a dividing line has emerged, which is quite rigorous, with researchers in the structure-function relationships of biopolymers opting to address issues largely in either high or low-solid systems. This divide is becoming more and more pronounced, as scientists working in the high-solid regime are increasingly inspired by the apparently "universal" molecular physics of glassy materials, which may or may not consider much of the chemical detail at the vicinity of the glass transition temperature. By comparison, their colleagues working on low-solid systems are shifting their focus from the relatively universal structure-function relationships of biopolymers in solution to the much more specific ones involving multi-scale assembly, complexation and molecular interactions. Sharing the expertise of the two camps under the unified framework of the materials science approach is a prerequisite to ensuring fully "functional solutions" to contemporary needs, spanning the full range of relevant time-, length- and concentration scales. This effort may prove to be the beginning of a modernized biopolymer science that, one the one hand, utilizes and further develops fundamental insights from molecular physics and the advanced synthetic polymer research as a source of inspiration for contemporary bio-related applications. On the other hand, such modernized science should be able to forward novel concepts dealing with the specific and often intricate problems of biopolymer science, such as the strong tendency for macromolecular hydrogen bonding, thus serving as an inspiration for related polymer advances and industrial applications. Sincere thanks are due to all our friends and colleagues whose outstanding contributions within their specialized areas made this a very worthwhile undertaking.

Stefan Kasapis
Ian T. Norton
Johan B. Ubbink

CHAPTER

1

Biopolymer Network Assembly: Measurement and Theory

Allan H. Clark and Simon B. Ross-Murphy

King's College London, Franklin-Wilkins Building, 150 Stamford Street, London, UK

A number of biopolymer systems can self-assemble to form networks and gels and the assembly can occur by a variety of mechanisms. In this chapter we consider the nature of biopolymer gels and networks, the kinetics of assembly, and their characterization by rheological methods. The necessary theory to explain, for example, the complexities of gelation kinetics is then described in some detail. Before reaching this, we discuss the nature of network assembly, and the character of gels and their gelation.

1.1 BIOPOLYMER NETWORKS AND GELS

1.1.1 Gels Versus Thickeners

1.1.1.1 What is a Polymer Network?

Polymer networks are molecular-based systems, whose network structure depends upon covalent or non-covalent interactions between macromolecules. The interactions can be simple covalent cross-links, or more complex junction zone or particulate-type interactions. Figure 1.1 illustrates different types of polymer network. Solvent swollen polymer networks are commonly known as gels – un-swollen networks are important for synthetic polymer systems, but are less relevant for biopolymers. Here, where the solvent is water or electrolyte, we can also introduce the term 'hydrogel'.

1.1.1.2 What is a Gel?

We have already defined a gel above as a swollen polymer network, but unfortunately, one of the major issues in chapters such as the present one is that the term 'gel' means very different things to different audiences. In this respect, the widely cited 1926 definition by Dorothy Jordan Lloyd, that 'the colloidal condition, the gel, is one which is easier to recognize than to define' (Jordan Lloyd, 1926) is quite unhelpful, since it implies that a gel is whatever the observer thinks it is. Consequently we commonly see such products described as shower gels and pain release or topical gels.

Neither of these classes of systems follows a rheological definition such as that of the late John Ferry, in his classic monograph (Ferry, 1980). He suggests that a gel is a swollen polymeric system showing no steady-state flow; in other words if subjected to simple steady shear deformation it will fracture or rupture. Clearly neither shower nor topical gels follows this rule;

Kasapis, Norton, and Ubbink: Modern Biopolymer Science
ISBN: 978-0-12-374195-0

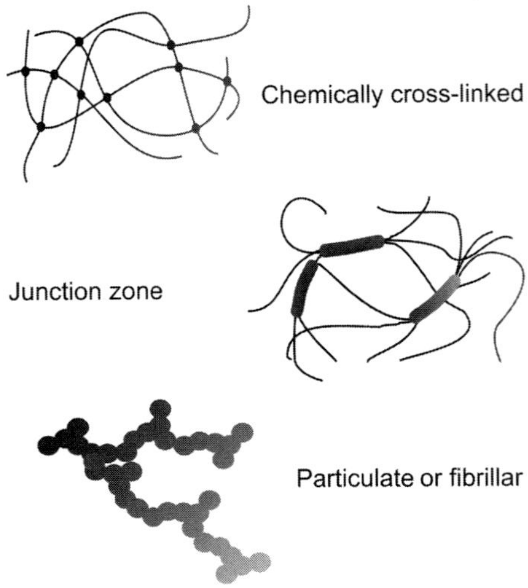

FIGURE 1.1 These diagrams illustrate three different types of polymer network; note that the three figures are not necessarily to scale.

indeed if they did, they would not be useful as products. In fact, commercial shower gels, for example, are simply highly viscous fluids formed by the entanglement of (often rod-like) micelles. For more rigorous definitions, at this stage it is necessary to introduce some common terminology.

Most modern rheological experiments on gelation (see below) employ oscillatory shear. In the simplest form of this, a small sinusoidal strain wave of frequency ω (typically 10^{-3}–10 s^{-1}) is applied to the top surface of a gelling system (most likely constrained between parallel metal discs) and the resultant stress transmitted through the sample is measured. In general the stress and strain waves differ in both phase and amplitude, but using phase resolution, it is easy to extract the in-phase and 90° out-of-phase components. Then G' is the storage modulus given as the ratio of in-phase stress divided by strain, and G'' is the loss modulus, the ratio of 90° out-of-phase stress to strain. There are other relationships between these and common experimentally determined parameters, as we describe later, but for now we are interested only in the storage – sometimes called elastic component – of the modulus, G'. For a perfect, so-called Hookean elastic material, such as a steel rod, G' is effectively independent of the oscillatory frequency. The constancy of G' with respect to frequency is then a useful definition of a solid.

One rheological definition of a gel is therefore a system that shows 'a plateau in the real part of the complex modulus' – G' – 'extending over an appreciable window of frequencies … they are … viscoelastic solids' (Burchard and Ross-Murphy, 1990). A slightly later definition accepts this, but extends it and the Ferry definition by identifying a gel as a soft, solid or solid-like material, which consists of two or more components, one of which is a liquid, present in substantial quantity (Almdal et al., 1993). They therefore follow Ferry in accepting substantially swollen polymer networks as gels. However, according to them, a gel must also show a flat mechanical spectrum in an oscillatory shear experiment. In other words it should show a value of G' which exhibits a pronounced plateau extending to times of the order of seconds, and a G'' which is considerably smaller than the storage modulus in this region.

1.1.1.3 *'Viscosifiers'*

One of the problems in this area follows directly from the overuse of the term gel – as we outlined above, many viscous fluids are also described as gels or hydrogels. These include biopolymer solutions, whose properties are determined all but exclusively by entanglements of long chains, in this area typically represented by solutions of the galactomannan guar. These are analogous to solutions of common synthetic polymers in organic solvents, where entanglements involve reptation of chains (Doi and Edwards, 1986). Rheologically there are also a number of so-called structured liquids – which can suspend particles and appear solid-like – typically formed from liquid crystalline polymers or micellar solutions – and usefully exemplified in the present context by ordered solutions of the microbial polysaccharide xanthan (Richardson and Ross-Murphy, 1987b). To confuse matters, these have been referred to, in the past, including by one of the present authors as 'weak gels' (Ross-Murphy and Shatwell, 1993). We now reject this term totally, both because of its anthropomorphic connotation, and for its lack of precision – since they can show steady-state flow – in terms of the Ferry definition above.

1.1.1.4 Viscoelastic Solids vs. Viscoelastic Liquids

What then is the main difference between solids and liquids? It is the existence of an equilibrium modulus, i.e. a finite value of G' even as the time of measurement becomes very long (or the oscillatory frequency tends to zero), usually referred to simply as the equilibrium shear modulus G. This means that a gel has (at least one) infinite relaxation time. Of course such a definition is partly philosophical, since given infinite time, all systems show flow, and in any case, most biopolymer gels will tend to degrade, not least by microbial action. However, this remains an important distinction, and in subsequent pages we regard biopolymer networks and gels as viscoelastic solids, and non-gelled systems, included pre-gelled solutions, 'sols', as viscoelastic liquids.

1.1.2 Brief History of Gels

1.1.2.1 Flory Types 1–4

Historically the term gel follows from the Latin gelatus 'frozen, immobile', and gelatin, produced by partial hydrolysis of collagen from, e.g. pigs, cattle or fish was probably recognized by early man. Gelatin has certainly been used in photography for almost 150 years, although this is, of course, a shrinking market.

In 1974, Flory (Flory, 1974) proposed a classification of gels based on the following:

1. Well-ordered lamellar structure, including gel mesophases.
2. Covalent polymeric networks; completely disordered.
3. Polymer networks formed through physical aggregation, predominantly disordered, but with regions of local order.
4. Particular, disordered structures.

In the present chapter, although we will not discuss specific systems in much depth, type 3 gels are represented by 'cold set' gelatins, and type 4 gels are represented by denatured protein systems. Type 2 systems are archetypal polymer gels. These are made up, at least formally, by cross-linking simpler linear polymers into networks, and their mechanical properties, such as elasticity, reflect this macroscopic structure.

1.1.2.2 Structural Implications

The structural implications of the above should be clear – gels will be formed whenever a super-molecular structure is formed, and Figure 1.1 illustrates the underlying organization of type 2, 3 and 4 gels. Of course this is highly idealized; for example if the solvent is 'poor', gel collapse is seen. Examples of each of these classes include the rubber-like arterial protein elastin – type 2; many of the gels formed from marine-sourced polysaccharides such as the carrageenans and alginates, as well as gelatin, type 3; and the globular protein gels formed by heating and/or changing pH, without substantial unfolding, type 4.

Of course, Figure 1.1 is highly idealized and the nature of network strands can vary substantially. For example, for the polysaccharide gels, such as the carrageenans, the classic Rees model of partial double helix formation (Morris et al., 1980) has been challenged by both small-angle X-ray scattering (SAXS) and atomic force microscopy (AFM) measurements, and it now seems likely that aggregation of junction zones and intertwining of pre-formed fibrils are additional contributory factors. This is certainly an on-going controversy, but one outside the remit of this chapter, except for its implications for the kinetic processes occurring during gelation. There are similar variations for protein gels too. When heated close to the isoelectric point, a coarse and random coagulate network is commonly formed but heating many globular proteins above their unfolding temperatures under acid conditions – say at pH 2 – results in fibrillar structures (Stading et al., 1992) that, at least at the nano-length scale, resemble the amyloid structures seen in a number of critical diseases such as Alzheimer's (Gosal, 2002; Gosal et al., 2002; Dobson, 2003). This is now a very active area of research, but the subject of a separate chapter in this volume (Hughes and Dunstan, 2009).

1.2 RHEOLOGICAL CHARACTERIZATION OF BIOPOLYMER GELS

1.2.1 Traditional Methods for Gel Characterization

A number of more traditional techniques have been used for gel measurements. They often have a major advantage in their low cost, compared to commercial apparatus. On the debit side, the actual strain deformation is sometimes unknown or, at best, requires calibration. Nowadays these approaches are less commonly employed, as almost all labs possess at least one oscillating rheometer, but they still have some advantages – not least from the financial viewpoint.

1.2.1.1 *Falling Ball*

This is one of the simplest and cheapest methods but, given a few precautions, it can still prove useful. In its simplest form, a magnet is used to raise a small metal sphere within a tube containing gelling material, and then the time taken to fall a fixed distance is registered (Richardson and Ross-Murphy, 1981). Clearly as gelation proceeds from the sol state, the rate of fall decreases, and eventually the sphere does not move any more. For low modulus systems there are potential problems since the sphere may locally rupture the gel and cut a channel through it – so-called 'tunneling' – and in this limit the method is more akin to a large deformation or failure method. The converse method of monitoring the fall of a sphere above a melting gel (or a series of such samples at different concentrations) is very commonly used to determine 'melting temperatures' (Eldridge and Ferry, 1954; Takahashi, 1972), but again care must be taken to ensure that true melting is involved rather than localized pre-melt tunneling.

1.2.1.2 *Oscillatory Microsphere*

The microsphere rheometer is just the oscillatory analogue of a falling ball system. A small magnetic sphere is placed into the sample and using external AC and DC coils, the sphere can be positioned and made to oscillate with the frequency of the AC supply. The maximum deformation can be observed with a traveling microscope, or alternatively tracked, for example, using a position-sensitive detector array. A number of different designs have been published and used for measurements on systems including agarose and gelatin gels, and mucous glycoproteins (King, 1979; Adam et al., 1984). The major limitation is that the measurement is very localized, so that again for some systems local rupture and tunneling can occur and then the modulus determined may not be representative of the whole system.

1.2.1.3 *U-tube Rheometer*

In this very simple assembly, originally designed by Ward and Saunders in the early 1950s for work on gelatin, the gel is allowed to set in a simple U-tube manometer, one arm of which is attached to an air line of known pressure, the other free to the air. Both may be observed with a traveling microscope. The air pressure exerts a compression stress in the sample (stress and pressure both have units of force/area), and the deformation of the sample can be measured from the differential heights of the manometer arms. The static (equilibrium Young's modulus) can be calculated directly using the analogue of Poiseuille's equation for capillary flow (Arenaz and Lozano, 1998).

As well as cheapness, this apparatus has the advantage that it becomes more sensitive for low modulus systems, since the deformation observed will be larger. However, in view of this, great care must be taken that the deformation induced is still in the linear region. The method has recently been extended for use with gels which synerese, by roughening the inner glass surfaces and by using an oscillatory set up (Arenaz et al., 1998; Xu and Raphaelides, 2005).

1.2.2 Modern Experimental Methods Employing Oscillatory Shear

Nowadays the vast majority of physical measurements on gels are made using oscillatory shear rheometry (Ferry, 1980; Ross-Murphy, 1994; Kavanagh and Ross-Murphy, 1998). This is because rheometers are far cheaper and 'user friendly' than used to be the case. However, by the same token, some published data are poor and, just as seriously, the degree of understanding does not always appear to have kept pace with the rate of data collection. One of the major objectives of succeeding sections is to try to modify this situation.

The essential features of a typical rheometer for studying biopolymer systems consists of a vertically mounted motor (which can drive either steadily in one direction or can oscillate). In a controlled stress machine, this is usually attached to the upper fixture. A stress is produced, for example by applying a computer-generated voltage to a DC motor, and the strain induced in the sample can be measured using an optical encoder or radial position transducers attached to the driven member. In a controlled strain instrument, a position-controlled motor, which can be driven from above or below, is attached to one fixture, and opposed to this is a transducer housing with torque and in some cases, normal force transducers. Figure 1.2 represents a typical controlled stress instrument. The sample geometry can be changed from, e.g. Couette, to cone/plate and disc/plate, and the sample temperature controlled. Such a general description covers most of the commercial constant strain rate instruments (e.g. those produced under the names of TA Instruments, ARES series) and controlled stress rheometers (e.g. Malvern Bohlin, TA Instruments Carrimed, Rheologica, Anton Paar). In recent years the latter have begun to dominate the market, since they are intrinsically cheaper to construct, and they can provide good specifications at lower cost. Most claim to be usable in a servo-controlled (feedback) controlled strain mode, and are widely used in this mode. However, there are limitations here, as discussed in detail below.

Controlled stress instruments are ideal for time domain experiments, i.e. measuring creep, whereby a small fixed stress is applied to a gelled sample and the strain ('creep') is monitored over time (Higgs and Ross-Murphy, 1990). The time domain constant strain analogue of the creep experiment is stress relaxation. In this, a fixed deformation is quickly applied to the sample and then held constant. The decrease in induced stress with time is monitored. Few such measurements have been discussed for biopolymer systems and nowadays practically all modern instruments appear to be used predominantly in the oscillatory mode.

1.2.2.1 Mechanical Spectroscopy

We have already introduced the storage and loss moduli, G' and G'', but there are a number of

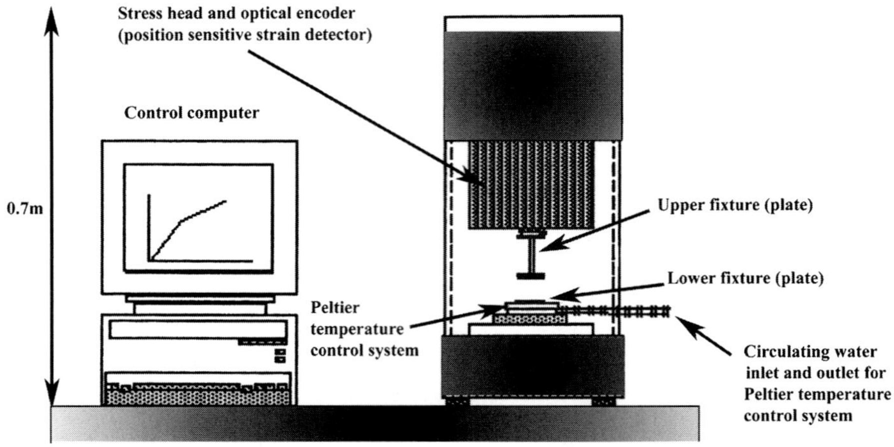

FIGURE 1.2 A typical controlled stress rheometer with parallel plate geometry.

other commonly used rheological parameters, and all are interrelated (Ferry, 1980; Ross-Murphy, 1994).

For example, G*, the complex modulus is given by:

$$G^* = \sqrt{(G')^2 + (G'')^2} \tag{1.1}$$

and the ratio:

$$\frac{G''}{G'} = \tan(\delta) \tag{1.2}$$

In the early days of oscillatory rheometry the phase angle, δ, was an experimentally observed parameter; nowadays instruments tend to hide the experimental measurables, the phase angle and the amplitude ratio, from the user.

Finally the complex viscosity, η^*, is given by:

$$\eta^* = \frac{G^*}{\omega} \tag{1.3}$$

with ω the oscillatory shear (radial) frequency; here ω is just 2π x the frequency in Hertz. Of course, oscillatory measurements can also be made in tension/compression, leading to alternative parameters, such as E′ and E″, etc. However, for biopolymer gels and networks, this is relatively uncommon, and so we do not discuss these further.

1.2.2.1.1 Controlled Strain Versus Controlled Stress We mentioned above that the majority of modern instruments are now of the controlled stress type. However most usually still generate results in the controlled strain form, that is as the modulus components, G′ and G″. Strictly speaking, since stress is applied and the strain is measured, then results should be reported as the components of complex compliance J′ and J″. However, most of the instruments circumvent this by applying a stress, measuring the strain, but in a servo- or feedback mode, so that it appears that they are indeed controlling the strain. For many applications and systems this is acceptable, but for systems very close to gelation, it is certainly not ideal. This is because there is no sure way of controlling the feedback when the system just changes from solution (sol) to gel, and yet at the same time guaranteeing that the strain remains very low. For such systems there is a further advantage in a genuine controlled strain technique, in that the mechanical driving head and the measurement transducer are completely separate assemblies – the only link between them is the test sample and geometry.

1.2.2.1.2 *Time Independent Systems* Below we describe a typical experimental regime to collect the data in a form that is appropriate for an exploration of the kinetic assembly of biopolymer networks. However, since the overall outcome usually involves the conversion of a biopolymer solution (sol) to a viscoelastic solid (gel) it is useful to first understand the so-called mechanical spectra of these two systems, and their dependence on the experimental variables of oscillatory frequency, shear strain deformation (or shear stress, bearing in mind the caveats above) and temperature.

1.2.2.2 *Frequency and Strain Dependence*

1.2.2.2.1 *Biopolymer Solutions* The mechanical spectrum of a liquid has the general form illustrated in Figure 1.3. At low frequencies (note the double log scale) G″ is greater than G′ but as the oscillatory frequency increases, G′ increases more rapidly than G″ (with a slope ~ 2 in the log–log representation, compared to a slope of 1 for G″) and at some frequency there is a 'cross-over'. After this both G′ and G″ become much

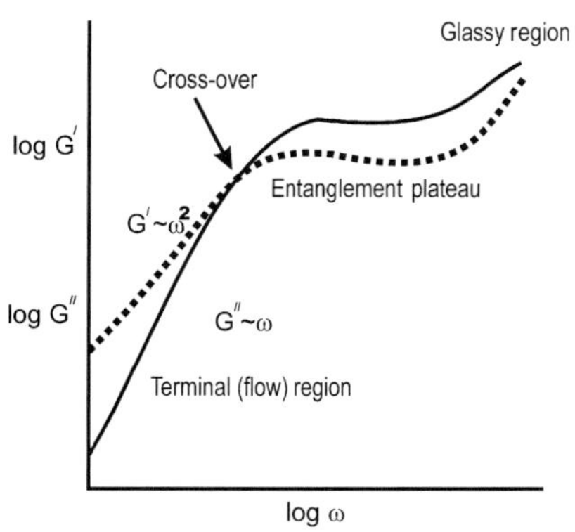

FIGURE 1.3 The mechanical spectrum of a liquid from the terminal zone to the start of the glassy region has the general form illustrated here.

less frequency-dependent – we enter the so-called rubbery plateau region.

Whether or not the cross-over region is reached in the frequency window of conventional oscillatory measurements depends upon the biopolymer concentration, relative molecular mass (MW), and chain flexibility. For example for a typical high MW viscosifier such as guar, the G″– G′ cross-over may occur for concentrations of say 2–3% w/w (Richardson and Ross-Murphy, 1987a), whereas for a more flexible and lower MW biopolymer such as gelatin above its gel melting temperature, the concentration required may be above 25% w/w, and therefore essentially outside the experimentally interesting range.

At the same time, the mechanical spectrum measured will be essentially independent of the amount of shear strain, out to say 100% 'strain units' (i.e. a strain, in terms of the geometry of deformation, of 1). Rheologists may express this by saying that the linear viscoelastic (LV) strain extends out to ca. 100%.

1.2.2.2.2 *Biopolymer Gels* The mechanical spectrum of a viscoelastic solid will, as we already mentioned in the discussion of the equilibrium modulus, have a finite G′, with a value usually well above (say 5–50 x) that of G″, at all frequencies, as illustrated in Figure 1.4 (Clark and Ross-Murphy, 1987; te Nijenhuis, 1997; Kavanagh et al., 1998; Kavanagh, 1998). In this respect it shows some similarities with the plateau region of the solution mentioned above – such a plateau has been referred to, somewhat imprecisely, as gel-like, for exactly this reason.

The strain-dependent behavior for biopolymer gels is more difficult to generalize, although the LV strain is rarely as great as 100% (some gelatin gels may be the exception here), and may be extremely low – say 0.1% as less. At values just greater than the LV strain, G′ and G″ may show an apparent increase with strain. This is, of course, largely an artefact of the experiment, since G′ and G″ are only defined within the LV

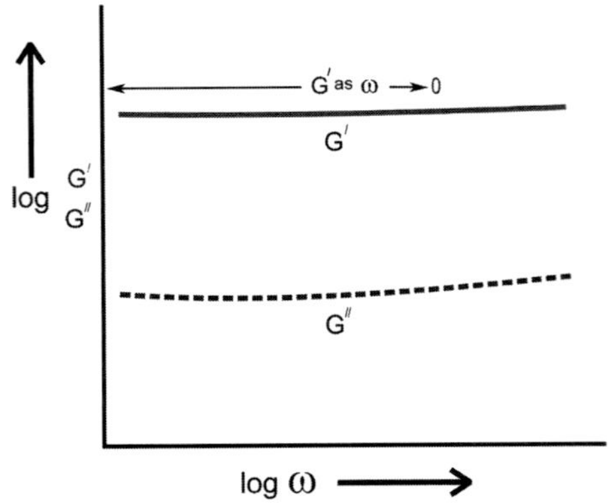

FIGURE 1.4 The mechanical spectrum of a viscoelastic solid has a finite G′, with a value usually well above (say 5–50 x) that of G″, at all frequencies.

region. This is then followed by a dramatic decrease, caused by failure – either by rupture or fracture, sometimes macroscopic – as often failure occurs at the geometry interface, especially if measuring in a disc plate (parallel plate) configuration.

1.2.2.3 *Temperature Dependence*

In this chapter we are not particularly interested in the temperature dependence of time-independent systems, since we are essentially concerned with the processes of self-assembly. However, in the study of synthetic polymer solutions and melts, this is of course of great importance. Again, although it has little to do with the formation of gel networks, many biopolymer gels do show so-called 'glassy' behavior at high enough frequencies or low enough temperatures, and the study of gels under these conditions, perhaps induced by measuring in highly viscous low MW solvents such as saturated sucrose, is a very active area of interest. This is discussed in further detail elsewhere in this book.

What the above does suggest, of course, is the well-known effect in polymer materials science, that high frequencies and low temperatures may be regarded as equivalent. This is the basis of the principle of time–temperature superposition (TTS). This is applied, for example, in the characterization of low-water gels, as mentioned above. Very often it works well, but caution should always be applied. The glass transition itself is related to polymer free volume, and temperature discontinuities in said free volume should make the approach invalid. If we are to follow the principles outlined by Ferry (Ferry, 1980) – one of the co-devisers of the method, and its strongest protagonist – then TTS should never be applied within 50°C of a phase transition within the system. For biopolymer gels, this should eliminate all TTS approaches from –50°C to 150°C – i.e. more than the whole regime of potential interest. In fact TTS can work well within this region, but should not be relied upon.

1.2.2.4 *Time-Dependent Systems*

1.2.2.4.1 *The Kinetic Gelation Experiment*

Clearly if we are, by some physical method (say heating), converting a biopolymer solution to a biopolymer gel, we will change the initial sol mechanical spectrum (Figure 1.3) to the gel spectrum (Figure 1.4). In a typical experiment, following the progress of gelation using mechanical spectroscopy, the oscillatory frequency is kept constant – and ca. 1Hz (6.28 rad s^{-1}) – for convenience many workers use a frequency of 10 rad s^{-1} – and the strain is maintained constant and low – say typically 10% or less. The choice of frequency is always a compromise – we need a high enough value that a single frequency measurement does not take too long – so we can collect enough data – but not so high that instrumental artefacts begin to appear. In our experience these can be seen quite commonly for frequencies > say 30 rad s^{-1}.

The temperature regime employed must also be carefully controlled, whether for heat-set, e.g. globular protein or cold-set, e.g. gelatin, gellan or carrageenan gels. A very common approach, not least because the instrument manufacturers supply it as an option, is to use a temperature ramp – say heating from 25°C to 75°C at 1°C per minute. The problem with this is, of course, that no serious study can be made of the kinetics of assembly, when the time-dependent assembly is convoluted with the change in temperature. Unfortunately many published data do employ such a heating ramp approach. Although an isothermal temperature profile can be difficult to achieve, modern Peltier heating systems are usually very fast to heat, cool and re-equilibrate. Originally these were only available on controlled stress instruments, but that limitation has now been overcome.

1.2.2.4.2 Gelation Time Measurement Before considering the different approaches to the determination of say gelation time, we consider the expected self-assembly time profile. If we consider the equilibrium gel modulus, the ideal profile is seen in Figure 1.5a. Initially there is no response, but then G rises very rapidly, even on a log scale, at or just after the gelation time, before reaching a final asymptotic level, and the behavior illustrated is a simple consequence of the positive order kinetics of self-assembly (cross-linking) and the requirement for a minimum number of cross-links per 'chain' at the gel point. We note that some phenomenological models have neglected the pre-gel behavior, and simply fitted the G (>0) versus t behavior to an n-order kinetic model. From the data-fitting viewpoint, this is quite acceptable, providing it is appreciated that the underlying physics of self-assembly has been perverted.

The above scenario is, of course, complicated by the consideration that what is being evaluated by the instrument is not G, but G′ and G″. Both of these are finite even for a solution, although the respective moduli values may be very low. However, because of the finite frequency effect, and the contribution of non-ideal network assembly contributions, both G′ and G″ will tend to rise before the true gelation point, and something akin to Figure 1.5b is usually seen. The flattening off of G″ is not something predicted from theory, indeed some would expect a pronounced maximum in G″ after gelation, but this is rarely seen, except for some low concentration gelatin gels. This asymptotic level G″ behavior has been associated with the 'stiffness' of the network strands.

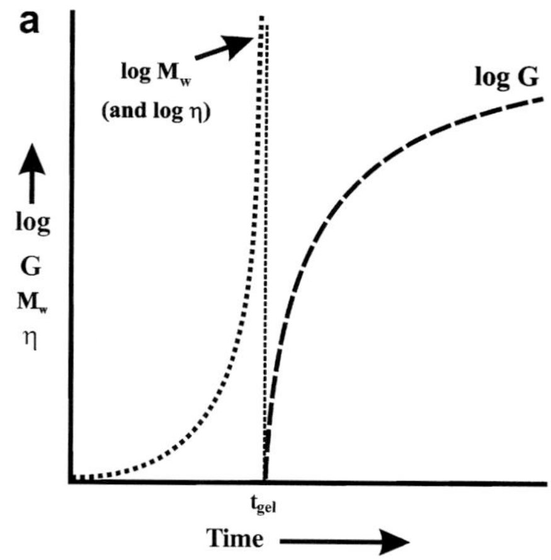

FIGURE 1.5a Idealized profile for a gelation process, showing how M_W and (zero shear) viscosity both become infinite at the gel point, and the equilibrium modulus G begins to increase from zero.

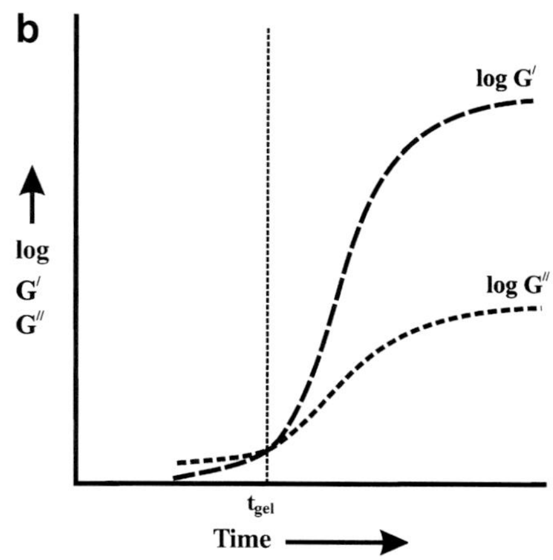

FIGURE 1.5b Experimentally, G′ and G″ tend to rise before the gel point, and close to the latter a cross-over is usually seen (depending on frequency and the nature of the system).

1.2.2.4.3 *Extrapolation Methods for t_{gel}*
Accepting for now that there is a definitive gelation time, t_{gel}, and that it is an important parameter, how best do we establish its value? There are a number of essentially empirical approaches, and we discuss some of these here (Clark et al., 1987; Kavanagh et al., 1998).

1. t_{gel} is the time when G′ becomes greater than some pre-defined threshold value. This is a useful approach, albeit that the choice of the threshold value is obviously arbitrary. It needs to be in the very fast G′ increase region, but above the noise level. This, in turn, will depend upon the instrument and the system being investigated, but is typically in the range 1–10 Pa.
2. By linear back extrapolation of the G′ versus time to a pre-defined level. This can work well, but again is fairly arbitrary; normally the level here will be lower than that employed in method 1.
3. When there is a cross-over in G′, G″ – i.e. where G′ becomes greater than G″. This is a very common empirical approach in the external literature, and has been used for many years, for example when studying the setting of acrylic paints. However, this cross-over time will depend upon frequency, and so is really just as subjective as the previous methods. It also has limitations – for example some globular protein solutions act like charged colloids, so although G′ is low, it is always above G″. This means a cross-over will never be seen.

The different approaches above may appear somewhat capricious, but in practice, they can all be useful even if, typically, they give t_{gel}'s which differ by say 50% or more. As we see later, the real interest is in investigating, and understanding, the concentration dependence of t_{gel}. Here the concentration exponent can be such that (on the required log scale) a 50% difference in t_{gel} is actually not so significant. The important thing is that, somehow or other, a value for t_{gel} can be established, since this is an important parameter in subsequent kinetic modeling.

1.2.2.4.4 *Chambon-Winter Method and Applicability in Biopolymer Self-Assembly*
Some 20 years ago, Winter and Chambon (Chambon and Winter, 1985; Winter and Chambon, 1986) suggested an elegant method, which despite its testing experimental requirements seemed to furnish a more absolute criterion for determining t_{gel}, as that time when, in the mechanical spectrum of a gelling system, both (log) G′ and G″ versus log (ω), first show the same power law exponent. As we mentioned above, for a sol well before gelation, G″ and G′ will be proportional respectively to the oscillatory frequency ω, and to ω^2. Well after the gel point, for a viscoelastic solid (gel), both G′ and G″ become essentially independent of frequency. In other words, the respective slopes of a log (G′, G″) versus log (ω) 'frequency sweep' will change from (2,1) to (0,0).

In the Winter approach (Winter et al., 1986) the time of congruency of slope, n, where

$$G' \propto \omega^n, G'' \propto \omega^n \tag{1.4}$$

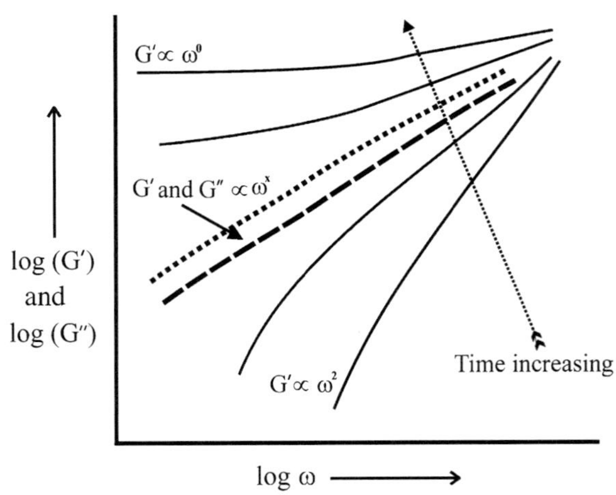

FIGURE 1.6 The Winter-Chambon method. For kinetic gelation, as time increases, the slope of G′ versus frequency ω decreases from ~ 2 to ~ 0, and that of G″ from ~1 to ~0. At some point close to t_{gel}, they will have congruent slopes.

corresponds exactly to the gelation time, t_{gel} (Figure 1.6).

The Winter-Chambon method for determining the gel point has become so popular that it has almost replaced more classical definitions, i.e. that the gel point is the conversion (or the corresponding time) when the average molecular weight (relative molecular mass) M_w becomes infinite (in other words where the system first develops an infinite relaxation time) and it is now assumed, almost without reflection, that the two must be identical. However it might be more reasonable to say that there are so-called dynamic and static gel points, the former measured in a viscoelastic experiment, and the latter in an equilibrium (e.g. light scattering determination of M_W) experiment (Trappe et al., 1992). In practice the two may be very close, but despite much effort it has not been proven (nor may it be possible to prove) that they are actually identical. This reflects the fact that there are always problems in making mechanical measurement on critical gels, connected with the strain dependence, the long relaxation times involved, and also the effect of entanglements, as we discuss below. This may be simplified by new techniques such as particle tracking, although these also tend to measure at high strains and frequencies.

The precise value of the slope n where congruency occurs can be calculated from a number of theories and is usually around 0.7 (Chambon et al., 1985; Winter et al., 1986; te Nijenhuis, 1997). In practice it is found that the experimentally observed congruent slope lies somewhere between 0.1 and 0.9, depending upon the precise system. This range is, unfortunately, very close to the extremes, viz. 0 and 1, given by the G'' slopes. This suggests that congruency of slope may not, of itself, be sufficient to identify the gelation point/time. This is a point that is sometimes misunderstood – if the exponent n is close to 0, the spectrum is that of Figure 1.3 – which is well past, and has little to do with, the critical gelation profile described in the Winter-Chambon method.

1.2.2.4.5 *Range of Viscoelastic Linearity* Yet another aspect of gel time measurement, and arguably one of greater significance, is the effect of finite strain on the tenuous mechanical system close to gelation (Ross-Murphy, 2005). In performing the kinetic gelation experiment it is usual practice to employ the smallest strain consistent with obtaining reliable data. In principle this can be checked to be within the linear viscoelastic region both before and after gelation by stopping the experiment and performing a so-called strain sweep. However, as we discussed above, many experiments are performed using controlled stress instruments in their pseudo-controlled strain mode, and such instruments do have more problems measuring a gelling system where the properties are changing quite rapidly within the oscillatory cycle, than when using the 'controlled' strain mode. This is because one might expect that the linear viscoelastic strain of the gelling system, rather than being constant, would tend to change during the gelation process, and would be a minimum just at the gel point (Rodd et al., 2001). The overall conclusion would appear to be that, even where, for biopolymer self-assembly systems, a definitive Winter-Chambon 'gel point' has been established, it may be impossible to equate this precisely with $t_{gel.}$

1.2.2.4.6 G_{inf} *and the Equilibrium Modulus* For modeling the kinetics of network assembly, another parameter, in addition to t_{gel}, is of value. This is, of course, the gel modulus. However, which gel modulus? As we have already hinted, what is really required is the equilibrium modulus, G; what we have measured is G' (and G''), after a particular time, and at a preset frequency and strain. Assuming for the moment that the strain is sufficiently low, that the system is linearly viscoelastic, we still have the implicit effect of time and frequency. Strictly speaking we need to extrapolate to zero frequency, and very long (nominally infinite) set up (or 'cure') times. Of course, if the system is well into the gel state, G' will be very largely independent of frequency (although this needs to be checked), so all that is required is to somehow extrapolate the values obtained during the cure experiment to infinite time, to obtain the parameter we have called G_{inf} (Kavanagh et al., 1998, 2000).

One approach is simply to appeal to empiricism, and we have found, in practice (Kavanagh et al., 2000) that the form

$$G' \approx G_{inf} \exp(-B/t) \tag{1.5}$$

where t is the time in seconds, B is an empirical parameter, and G_{inf} is the required value of G' at infinite time. As can be seen in Figure 1.7, this form reproduces satisfactorily both the asymptotic limit as $t \rightarrow \infty$ and the required behavior (technically a logarithmic singularity) that $\log G' \rightarrow -\infty$, as $t \rightarrow t_{gel}$. Further thought reduces this to the simple case where we plot $\log(G')$ versus $1/t$, and find the intercept on the $1/t = 0$ axis; this seems to be a valuable aid (Clark et al., 2001).

Two comments are worth noting here. First, very few workers make (or even appreciate the significance of) this extrapolation, and just assume G' (after say 100 minutes of cure) is the same as G_{inf}. This is risky, because only in retrospect can we judge the validity (or otherwise) of this. Second, for some systems, and the archetypal example is gelatin, it is almost impossible to make such an extrapolation anyway, because G' never levels off. Instead gelatin gels enter an apparent 'log phase' of modulus growth, which is assumed to be due to formation of new structures, perhaps via the slow kinetic processes associated with the *cis* proline 'flip' (Djabourov et al., 1985; Busnel et al., 1989). Since this chapter is concerned with generic issues, we mention this, but do not discuss it further.

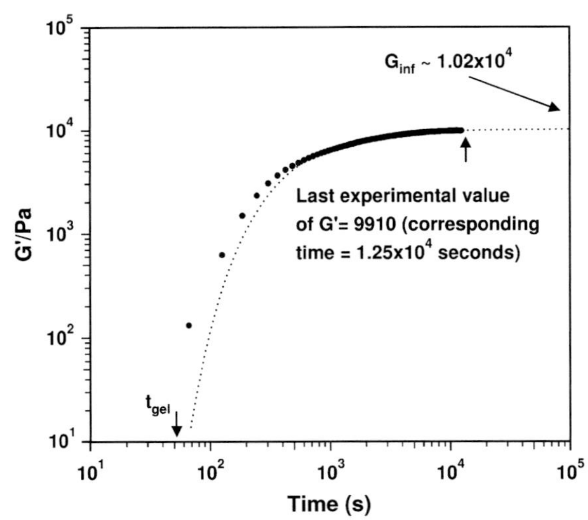

FIGURE 1.7 Extrapolation to G_{inf} using Eq. 1.5 (data from Kavanagh, 1998).

1.3 THEORETICAL ASPECTS

If we take a series of measurements of G_{inf} for different initial biopolymer concentrations, and plot in the form $\log(G_{inf})$ versus $\log(C)$, several features are immediately apparent (Clark and Ross-Murphy, 1985; Clark et al., 1987); in fact, as we shall see, they already reflect the underlying 'percolation' type assembly behavior, and discussion of this behavior leads us naturally into the theoretical part of this chapter. As Figure 1.8 shows, at high concentrations, there is an apparent (power law) dependence of $\log(G_{inf})$ versus $\log(C)$, whereas at lower concentrations, $\log(G_{inf})$ shows increasingly pronounced curvature, and at a particular concentration, appears to vanish. This is the so-called critical gel concentration, here denoted, C_0.

1.3.1 Critical Gel Concentration

That a biopolymer gel does have a critical gel concentration seems, on the surface, obvious. It is well appreciated that there needs to be a certain concentration of the biopolymer present before a contiguous gel can be formed. (Self-evident though this may appear to be, it is not

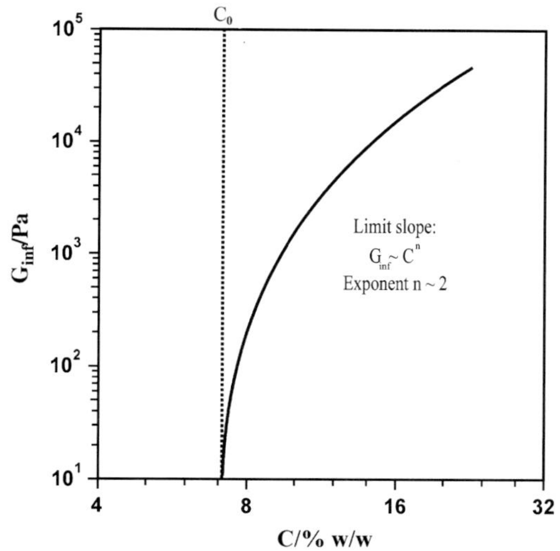

FIGURE 1.8 At high concentration, there is an apparent (power law) dependence of log(G_{inf}) versus log(C), whereas at lower concentrations, log(G_{inf}) shows increasing slope, and at the critical gel concentration, C_0, the slope (on a log–log plot) becomes infinite.

a feature of, for example, the fractal gel model we discuss later.)

What then is the significance of C_0, and what does it tell us about the mechanisms of self-assembly? The answer to this, and the corresponding behavior of the gelation time, is the subject of much of the remaining discussion. In real terms, values of C_0 vary from $< 0.05\%$ for some microbial polysaccharides to $> 10\%$ for certain more particulate gel systems. However, despite this, the form of the scaled log G versus log (C) or better still, from the generality viewpoint, versus log (C/C_0) curve, remains the same.

1.3.2 Gelation Time

We have already discussed methods for the determination of the gelation time, t_{gel}, and its significance in the experimental context. If, for example, we repeat the type of plot seen in Figure 1.8, but instead of plotting log(G_{inf}) we plot log($1/t_{gel}$), the experimental data seem to follow essentially the same behavior, with a power law behavior at high concentrations, and a logarithmic singularity, at the same concentration, and consequently with the same C_0. Many years ago, we conjectured that the shapes of these two curves were essentially congruent (Richardson et al., 1981), and in practice this often works quite well. However like many such practical conjectures, theory shows that this should not really be the case. Indeed, as we discuss in detail below, fitting modulus concentration and (reciprocal) gel time concentration data simultaneously can be extremely testing, and usually requires sophisticated multistage kinetic modeling (Clark et al., 2001).

1.3.3 Kinetic Modeling

1.3.3.1 Flory-Stockmayer (FS) Model

The basic model for gelation is that of non-linear or random step-growth polymerization (or, to use old terminology, polycondensation) which goes back to the classical work of Flory and Stockmayer in the 1940s on covalently formed, irreversible, networks (Flory, 1941, 1942; Stockmayer, 1943; Stockmayer and Zimm, 1984). This model, which in today's terms we would describe as percolation on an infinite dimension, tree-like or Bethe lattice (Gordon and Ross-Murphy, 1975; Stauffer et al., 1982; Stauffer, 1985) has proved of enormous value even though it neglects many features, such as the pre-gel formation of intramolecular links (cycles), which 'wastes' cross-links (Gordon and Scantlebury, 1968). Indeed it is fair to say that gelation in the absence of cross-link wastage reactions is well understood in terms of the Flory-Stockmayer theory and the gel point can be used as a reference point for the consideration of the effects of intramolecular reaction.

In the FS model, at the gel point, the species of infinite molar mass has a tree-like structure permeating through the whole reaction mixture. The critical conversion occurs when there is a non-zero probability that a randomly chosen chain continues to infinity. Given the previously mentioned random reaction (or equal reactivity) of like functional groups or sites, the gel point, and properties relating to the gelling system, may be predicted quite generally, in terms of the parameter α, representing the proportion of reacted groups, and the gel point conversion, α_c. For example many properties can be related back directly to the ratio α/α_c, although very close to α/α_c – in the so-called critical region – critical fluctuations need to be taken into account. (The extent of this critical region is governed by criteria such as that of Ginzburg; evaluation of the extent of the critical domain remains the realm of the theorist but a practical guide is that the upper limit is say $10^{-2} <= (\alpha/\alpha_c) - 1 <= 10^{-1}$. This has relevance later, when we discuss the critical region in more detail.)

The original Flory-Stockmayer model was developed to describe the formation of polymer networks in the absence of a solvent, either through the condensation reaction of monomeric species or the cross-linking of pre-formed polymer chains. As Stockmayer demonstrated, this model can be developed in kinetic terms (Stockmayer, 1943) through a second-order differential equation for the change of α in terms of the fraction of unreacted sites (1 – α), i.e.

$$\frac{d\alpha}{dt} = K(1-\alpha)^2 \tag{1.6}$$

subject to the initial conditions $\alpha = 0$ at $t = 0$. This allows the degree of reaction α to be specified as a function of time and properties of the network calculated from it. For example, one version of the model (Dobson and Gordon, 1965; Gordon and Ross-Murphy, 1975) tells us that the gel modulus is given by:

$$G = \frac{aRTN_e(\alpha)}{V_{mol}} \tag{1.7}$$

Here N_e, which is a function of the degree of conversion α, is the number of elastically active network chains (EANCs) per biopolymer chain. The parameter a is the so-called rubber theory 'front factor', RT is the usual gas constant term, and V_{mol} is volume per mole of biopolymer chains. N_e is zero before the gel point, so $G = 0$ until the gel point, which occurs (Flory, 1941) at:

$$\alpha_c = \frac{1}{f-1} \tag{1.8}$$

where f is the so-called network functionality, the number of functional groups or attachment sites available per primary molecule to form cross-links. By substituting the appropriate expression from Eq. (1.6) into Eq. (1.7), albeit ignoring the various rate and other constants, we obtain G as a function of time.

Figure 1.9 illustrates this result with the dependence of both α and G on time plotted on logarithmic time axes (Ross-Murphy, 2005). At first glance, in the linear time axis plot, both α and G look to have a quite similar pseudo-rectilinear relationship, albeit that there is a small lag time, the gelation time, t_{gel}, in the G (t) behavior.

1.3.4 Random Branching in Solution

The formation of highly solvated networks which is characteristic of biopolymer self-assembly in solution requires development of the Flory-Stockmayer model in a number of directions. A principal change is that the kinetic equations determining the reaction extent must now be written in terms of the concentrations of reacting functionalities and in some cases cross-link reversibility must be anticipated. Where cross-linking is believed to occur irreversibly there is no longer the possibility of ignoring wastage reactions such as the formation of cycles as, in solution, the absence of these would lead to some form of network collapse: i.e. the formation of homogeneous gels with a finite critical concentration

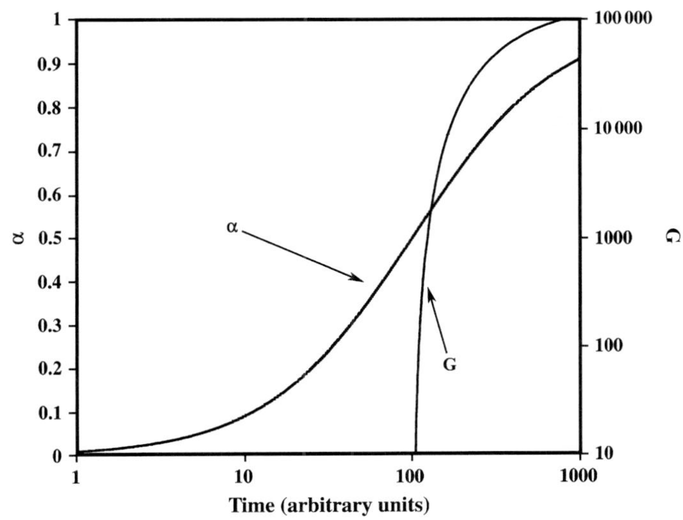

FIGURE 1.9 The dependence of α and G on time plotted versus logarithmic time.

would be impossible. Extension of the Flory-Stockmayer approach along these lines has been described in past literature by Gordon and co-workers (Gordon, 1962; Gordon and Scantlebury, 1964, 1966, 1968; Gordon et al., 1975; Dusek et al., 1978) and their model is described in outline below. In this description their elegant mathematical approach using probability generating functions is adopted, an approach sometimes referred to as branching or cascade theory.

The model starts by assuming the presence in solution of a concentration C of molecular species each of molecular weight M and each bearing f equivalent reactive functional groups or sites available for bonding. Assuming a totally random cross-linking process which includes also the possibility of intramolecular cross-linking within aggregates, i.e. cyclization, the state of cross-linking of the system can be specified at any time by the link probability generating function (lpgf):

$$F_0(\theta_\alpha\theta_\sigma\theta_\omega) = \sum_{i=0,f}\sum_{j=0,f-i} P_{ij}\theta_\alpha^i\theta_\sigma^j\theta_\omega^k \qquad (1.9)$$

The subscript 0 of $F_0(\theta_\alpha\theta_\sigma\theta_\omega)$ indicates the zeroth or root generation of the tree, θ_α, θ_σ and θ_ω are so-called 'dummy variables' and P_{ij} is the probability of a starting unit being subsequently in a state where it has i cross-links to other units through which it participates in the formation of tree-like aggregates and an increase in molecular weight. It also has j bonds which give rise to cycles within the trees and, finally, k unreacted sites ($i + j + k = f$). Since the P_{ij} are normalized, average properties can be obtained over the distribution represented by the P_{ij} in the usual way. Two quantities that are immediately important and can be calculated in this way from Eq. (1.9) are α and σ, the fractions of functionalities present which are engaged in cross-links and cycles, respectively. These are $\alpha = \langle i \rangle / f$ and $\sigma = \langle j \rangle / f$.

Another important generating function which can be derived from F_0 is F_1, the 'offspring' probability generating function (subscript 1 indicating the first and subsequent generations) which gives the probabilities of obtaining various states of cross-linking and cyclization for $f-1$ of the functionalities of a unit, given that at least one of the unit's functionalities is already engaged in an aggregate-forming cross-link. This is:

$$F_1(\theta_\alpha\theta_\sigma\theta_\omega) = \sum_{i=1,f}\sum_{j=0,f-i} iP_{ij}\theta_\alpha^{i-1}\theta_\sigma^j\theta_\omega^k/\sum\sum iP_{ij} \quad (1.10)$$

where the denominator is required to normalize the new distribution iP_{ij}. An important quantity that can be derived from Eq. (1.10) is the average number of cross-links emanating from a unit on one generation to the next. This is:

$$g = (\partial F_1/\partial\theta_\alpha)_{\theta_\alpha\theta_\sigma\theta_\omega=1} = \langle i(i-1)\rangle/\langle i\rangle \quad (1.11)$$

where averages are again taken over the original distribution P_{ij}. When $g = 1$ the aggregating system becomes critical and the infinite gel molecule referred to in the previous section first appears. Where there is no cyclization present this is equivalent to the Flory gel point criterion $\alpha_c = 1/(f-1)$ but this is not generally the case the total conversion at the gel point being $\alpha_c + \sigma_c$ (with $\alpha_c > 1/(f-1)$ when $\sigma_c \neq 0$). After the gel point g becomes greater than unity.

Another fundamentally important quantity associated with the aggregation process which allows statistical calculations to proceed through the critical gel point is the so-called extinction probability v. This is the probability that a bond formed from a given unit gives rise only to a finite number of higher generations, i.e. becomes 'extinct'. It follows from this definition that:

$$v = F_1(v,1,1) \quad (1.12)$$

since $F_1(v,1,1)$ is just the probability that all bonds emanating from a unit into the next generation become extinct and this is clearly just the probability v that the bond to the unit from the previous generation is itself extinct. The extinction probability is equal to unity up to the critical gel point then changes abruptly after that, falling to lower values. The extinction probability allows F_0 to be re-written in the form:

$$G_0(\theta_v\theta_{1-v}\theta_\sigma\theta_\omega) = \sum_{i=0,f}\sum_{j=0,f-i} P_{ij}(v\theta_v + (1-v)\ \theta_{1-v})^i\theta_\sigma^j\theta_\omega^k \quad (1.13)$$

where the i cross-links are now specified more precisely in terms of whether they become extinct (θ_v) or extend to the boundaries of the gel (θ_{1-v}). A similar expression can be obtained from F_1 and written as G_1.

The generating function G_0 is essential for calculating the shear modulus of the gel at any stage of cross-linking beyond the critical point as it allows the average number of elastically active chains (EANCs) per repeat unit in the system (N_e) to be calculated. The concept of the EANC was introduced in the previous section and is now defined more precisely as any sequence of cross-linked units joining two units known as 'ties' where a tie is any unit which has at least three non-extinct cross-links, i.e. is 'tied' into the infinite gel molecule by at least three separate paths to infinity. The quantity N_e is then equal to half the average number of non-extinct cross-links per unit present provided that the average is calculated to exclude contributions from units with only one or two non-extinct cross-links. This average can be derived from Eq. (1.13) using methods described elsewhere (Dusek et al., 1978). The factor of 0.5 arises because each EANC has two ends.

Once N_e is available, the gel modulus at any time during curing and for any starting concentration C, can be calculated using Eq. (1.7) though it is emphasized that for biopolymer gels the front factor 'a' will not necessarily lie close to unity. Early work by the present authors (Clark et al., 1985) using an adapted version of the Flory-Stockmayer model slightly less rigorous than that described here strongly suggested that for biopolymer gels, in general, 'a' could have values much greater than one. This was interpreted in terms of the non-ideality of the EANCs in biopolymer gels in the sense that they could not be convincingly identified with the random flight ('entropic') chains of ideal rubber theory. Their average contribution to the gel modulus was therefore likely to include enthalpic as well as entropic contributions rather than pure configurational entropy alone. This introduction of a modified front factor was bought at the expense, however, of losing the temperature independence of the original ideal 'a'.

It follows from the above that the central problem of modeling the kinetics of network building in solution using the random branching approach is to determine the P_{ij} as functions of time. Everything else follows from these. As has been described elsewhere (Gordon et al., 1968) in more detail differential equations can be written down for these and solved numerically on a computer. Here only an outline of the procedure is given. What is involved is as follows: starting with the simplest case of purely irreversible bonding the time differential of a given P_{ij} is composed of processes by which P_{ij} units are lost through conversion to $P_{i+1,j}$ or $P_{i,j+1}$ units through cross-linking or cyclization involving free functionalities and P_{ij} units are gained from $P_{i-1,j}$ or $P_{i,j-1}$ units by the same processes. For the cross-linking reactions, a rate constant k_α is introduced and a bimolecular second-order reaction is envisaged between free sites on reacting units. This necessarily introduces the overall concentration C of units in the system as the frequency of encounters of a unit with others depends on concentration. For reactions of sites to form cycles the situation is different with a first-order intramolecular reaction being considered, the rate of which is determined by a rate constant k_σ and the number of free functionalities available on units in higher generations, to close rings. For a ring of a given size x there is an additional $x^{-3/2}$ multiplicative factor if one adopts the ideal random flight model of polymer chains. This factor determines the likelihood of two ring-closing functionalities occupying the same region of space and, like the original ideal front factor of unity, it is likely to require modification for biopolymer systems. The change in the probability P_{ij} due to cyclization is summed over all possible ring sizes x the ring size increasing as higher and higher generations are considered. Since an intramolecular process is involved there is no explicit dependence on the concentration C.

Where cyclization processes are concerned it should be noted that cycles are calculated to form only within the sol fraction, or any extinct sub-chains attached to the gel network, or within regions of EANCs between ties. These cross-links produce no increases in the gel modulus and are genuine wastage processes. Quite elaborate calculations are involved to ensure this restriction using the probability generating functions G_0 and G_1 as described in detail elsewhere (Dusek et al., 1978). The gel molecule actually contains many other cycles which do contribute to the modulus but as has been discussed by, for example, Gordon and Ross-Murphy (Gordon et al., 1975) these cycles are considered to form by second-order reactions in the gel and to be part of the normal cross-linking process.

The differential equations defining changes to the P_{ij} can be extended (Gordon et al., 1966) to include bond reversibility by introducing a reverse rate constant k' which is normally assumed to be the same for both cross-links and cycles. The differential equations are then solved numerically on a computer for the P_{ij} as functions of time assuming the starting conditions $P_{00} = 1.0$ and all other $P_{ij} = 0$. Input parameters for the model are the concentration C, the molecular weight M, the functionality f, front factor 'a', the rate constants k_α, k_σ and, if reversibility is included, k'. The P_{ij} are used to calculate α, σ, g, v and the gel shear modulus G as functions of time. Theoretical predictions of gel times t_{gel} are obtained as the time required for g to reach unity. Long time limiting G_{inf} values are determined in the reversible case as equilibrium values, or in the irreversible case by extrapolation to infinite time using log G versus 1/t plots. A critical concentration C_0 emerges naturally in such calculations because, in the irreversible case, intramolecular cyclization overtakes cross-linking as concentration falls and the condition $g = 1$ eventually cannot be realized or, in the reversible case, because bond dissociation has a similar effect. The critical concentration is obtained as the concentration at which the gel time diverges to infinity. G_{inf}/G_{scale} versus C/C_0, t_{gel}/t_{scale} versus C/C_0 and G/G_{inf} versus t/t_{gel} plots are then constructed as master plots from the results to facilitate comparison with experimental data from cure experiments. The reduction factors G_{scale} and t_{scale} can be written explicitly in terms of 'a' and the rate constants even if the master curves themselves are not available in analytical form. Comparison of the model with experiment is discussed in the next section.

1.3.4.1 *Random Branching and Experiment*

Previously, the present authors (Clark et al., 1985, 1987) used the random branching model in more approximate form than just described to explain experimental modulus data for a range of biopolymer gels. The focus was originally on G_{inf} versus C data and, generally, both an irreversible and a reversible form of the model were successful in describing this (Clark, 1993). Where thermoreversible gels were concerned the reversible model was obviously the more appropriate and good fits to modulus-temperature data were obtained for pectin and gelatin gels and sensible values for heats and entropies of cross-linking extracted (Clark et al., 1994; Clark and Farrer, 1996). When fitting was extended to gel time versus concentration data, or to reduced cure data (G/G_{inf} versus t/t_{gel}), neither model was particularly convincing, however, both giving very similar and sometimes equally erroneous descriptions (Clark, 1993, 2000; Clark and Amici, 2003). Initially, it was not clear how much this problem stemmed from aspects of the gelling system being studied as opposed to the approximations adopted in doing the calculations. In the light of this, the model described in the previous section has now been adopted to provide as rigorous as possible an ideal branching model for comparison with experiment. The same basic assumptions are involved as in the earlier models, but cyclization is now calculated more correctly and a unified approach adopted combining both cycle formation and bond reversibility. While it can be shown that this new model includes the early reversible model as a special case (when k'/k_σ is large), in fully irreversible form, it differs significantly from the earlier highly approximate wastage model (Clark, 1993).

Since one can now have more confidence in the new random branching model as a reference against which real system behavior can be assessed it is interesting to see whether deviations between theory and experiment still take the same form and whether, if still present, these can be interpreted and provide better models and understanding. An example of this endeavor is now given based on previously published studies (Kavanagh et al., 2000; Gosal et al., 2004a, 2004b) of pH 2 heat-set β-lactoglobulin gels made by the present authors and co-workers. Cure data from these provide a good test for kinetic models as efforts were made to step-change the temperature to the gelling temperature (rather than a temperature ramp) and modulus data are available at two temperatures and over a range of concentrations. The type of measurements involved and the methods used to extract experimental G_{inf} and t_{gel} values from them have already been described in this chapter and are also made clear in the relevant literature.

In the current exercise using the new model, the published lactoglobulin data were fitted to master curves using log-log plots and data translation. Scale factors such as C_0, G_{scale} and

t_{scale} mentioned above were obtained in this way. Theoretical master curves corresponded to the simplest choice of f = 3 for the functionality as globular proteins are expected to have only a small number of sites available for bonding. Another reasonable choice of f such as five, for example, would not have changed conclusions radically. In the first instance, the master curves considered were for the purely irreversible model and parameters such as the front factor 'a' and the rate constants k_α and k_σ were calculable from the scale factors if required.

A comparison of theory and experiment for a range of β-lactoglobulin data (batches of protein and heating temperatures) appears in Figures 1.10–1.12 for G_{inf} versus C, t_{gel} versus C and G/G_{inf} versus t/t_{gel} data, respectively. While Figure 1.10 shows that the limiting modulus–concentration data can be adequately described by the irreversible model, Figure 1.11 shows that there are serious problems with the corresponding gel time data, the experimental data

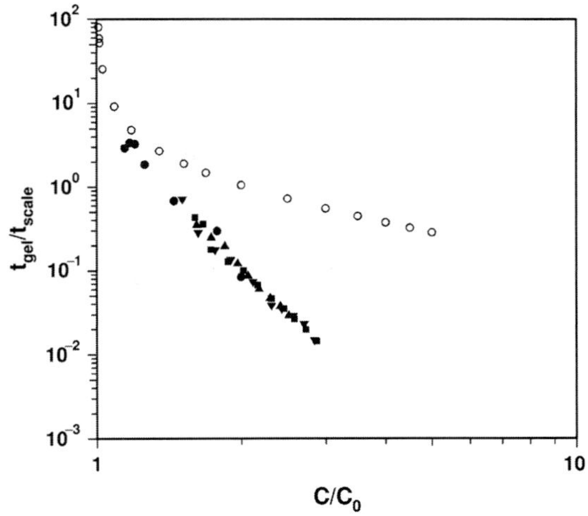

FIGURE 1.11 β-lactoglobulin acid gels: t_{gel} data reduced to master curve form using irreversible branching theory. Solid symbols are experimental, open symbols represent the model. 75°C batch 1, $C_0 = 6.0\%$, upper triangle; 75°C batch 2, $C_0 = 5.1\%$, lower triangle; 80°C batch 1, $C_0 = 4.8\%$, square (Gosal, 2002); 85°C $C_0 = 6.9\%$, filled circle (Kavanagh, 1998). Theory f = 3 model, open circle.

showing a much higher power law dependence on concentration than predicted by the theory. These results are similar to what had been found earlier (Kavanagh et al., 2000; Clark et al., 2001; Gosal et al., 2004a) using the more approximate random branching model but Figure 1.12 shows a considerable improvement in the fit to reduced cure data obtained for one particular batch of lactoglobulin and heating temperature (a typical dataset). The more rounded shape of the data is now much better reproduced by the new, more rigorous, random branching model and there is much more superposition in the theoretical master curves over the relevant C/C_0 range. However, there is still a significant difference between theory and experiment in the slope of the data close to the gel point ($t/t_{gel} = 1$) this being particularly obvious when linear plots are made.

The deviations between the ideal model and experiment for pH 2 heat-set β-lactoglobulin gels are clearly significant and might seem discouraging, but their origin can be explored

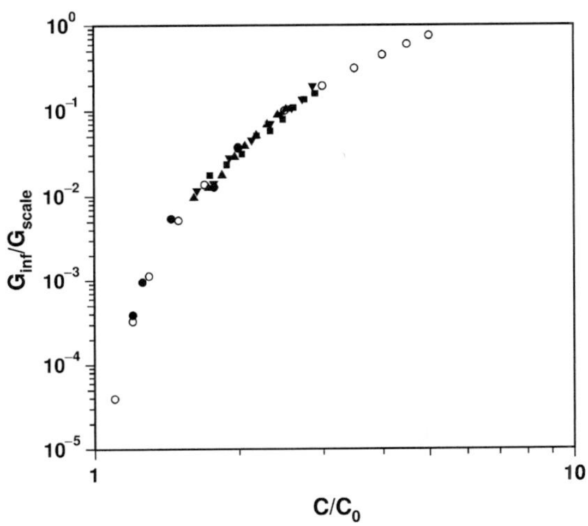

FIGURE 1.10 β-lactoglobulin pH 2 gels: G_{inf} data reduced to master curve form using irreversible branching theory. Solid symbols are experimental, open symbols represent the model. 75°C batch 1, $C_0 = 6.0\%$, upper triangle; 75°C batch 2, $C_0 = 5.1\%$, lower triangle; 80°C batch 1, $C_0 = 4.8\%$, square (Gosal, 2002), 85°C $C_0 = 6.9\%$, filled circle (Kavanagh, 1998). Theory f = 3 model, open circle.

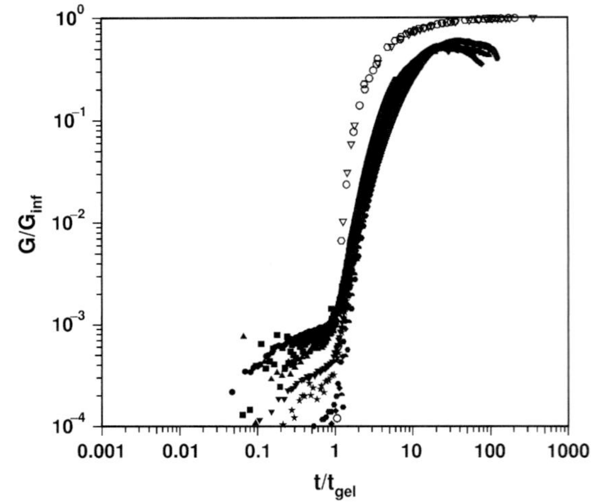

FIGURE 1.12 Irreversible f = 3 random branching model compared with β-lactoglobulin reduced cure data. Solid symbols are experimental, open symbols represent the model. Concentrations are 9.7% w/w: square; 10.4%: circle; 11.1%: triangle (up); 11.8%: triangle (down); 12.4%: star; 13.1%: hexagon; 13.8%: diamond; 14.5%: semicircle (upper); 15.1%: semicircle (lower). (Gosal, 2002) Model fits: $C/C_0 = 1.1$: hexagon; $C/C_0 = 3$: triangle (down); $C/C_0 = 5$: circle.

systematically and valuable insights gained. Manipulation of the idealized ring-closing kinetics through changes to the $x^{-3/2}$ ring-closing factor can be pursued, for example, but this only produces a fine-tuning in the shape and degree of superposition of the reduced cure master-curves. Introduction of increased reversibility in the model through the rate constant k′ has a similar effect altering the exact shape of all master curves but does not relieve the major discrepancies (gel time power law behavior and reduced cure curve slope near $t/t_{gel} = 1$). Introducing a different reactivity for one of the three functionalities in the model to better reproduce a non-random aspect of β-lactoglobulin gel networks (Gosal, 2002; Gosal et al., 2004b) shown by microscopy (that is, their highly fibrillar character – see Figure 1.13) can also be attempted using the first-shell substitution effect (FSSE) described by Gordon and Scantlebury (Gordon et al., 1964, 1966). While this does improve the situation as regards the shape of reduced cure data it does not solve the gel time problem nor does redefining the basic unit of the gel as a high-molecular-weight linear polymer of high functionality which subsequently cross-links randomly. In fact, as has been suggested in an earlier publication (Clark et al., 2001), the most successful approach appears to be to depart radically from the random branching model by making fibril formation the rate determining step in network assembly and to assume

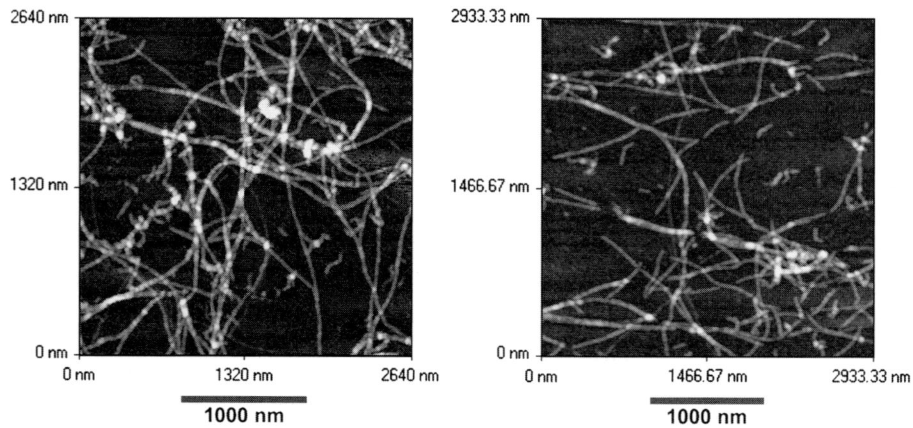

FIGURE 1.13 Fibrillar strands produced by incubating 4.0% w/w β-lactoglobulin pH 2.0 at 80°C for 24 hours, diluting 1 in 50 using distilled water, applying onto mica by pipetting, and imaging by AFM (Topometrix Explorer, TM Microscopes, Veeco Metrology Group, Santa Barbara, CA 93117, USA) under ambient conditions (Gosal, 2002).

higher-order (n = 4–5) kinetics consistent with a nucleation event. All master curves for this 'delivery model' turn out to be in reasonable agreement with experiment, the random branching aspect of the model being relegated to rapid reversible cross-linking of the fibrils as they are assembled. Despite the fact that other explanations of the β-lactoglobulin cure data (particularly the gel time data) may be possible relating, for example, to factors such as pH or salt 'drift' or non-ideal solution behavior, and although the nucleation model is not yet perfected in all aspects, it is a promising development and suggests that the same mechanism may prevail in other types of biopolymer gel where a structured unit is assembled first and then randomly associated into a network: the helical fibers underlying polysaccharide gels, for example.

The delivery model approach will not be pursued further in this section but the above exercise suggests some general conclusions. Most importantly it shows how useful the kinetic random-branching model can be as a reference in efforts to interpret cure data for gelling biopolymer systems. Real systems may never follow this behavior closely but deviations are useful sources of information emphasizing particular features of a system (such as some form of non-random self-assembly) and suggesting new and more realistic models. The above exercise also indicates the type and extent of the cure data that is required for the comparison. Early work focused on G_{inf} values but it emerges that G_{inf} concentration relationships are probably the least sensitive measures of differences between models unless measured very close to the critical concentration. Gel time concentration data and accurate cure measurements close to the gel point seem to be more important. While such data are currently scarce in the literature for gelling biopolymers it is hoped that the situation will soon improve and that exercises of the kind demonstrated here will become more common.

1.3.5 Branching Models and Phase Separation

So far it has been assumed that the biopolymer gels of interest are homogeneous and transparent but phase separation and turbidity in biopolymer gels are common. The introduction of a solvent makes this an inevitable possibility and while never a consideration for the original Flory-Stockmayer approach, phase separation must inevitably pose an issue when this mean field approach is adapted for solvated gels. One solution is to restrict application of the model to uniform transparent gels only and to anticipate different methods for heterogeneous networks such as those that have been applied to the gelation of biopolymer mixtures. While, for reversibly forming networks this seems perfectly feasible without contradiction, there is something of a problem for the fully irreversible model even when it includes cyclization. At long times, when cure is extrapolated, the theory concludes that all functionalities on a unit are used either to form cycles or cross-links. This seems to imply that systems at concentrations near critical will inevitably show network collapse and be heterogeneous. If this conclusion is correct it would seem that for uniform gels to form irreversibly (and there appear to be numerous examples of these) extra processes of steric hindrance must be invoked with an inevitable slowing down of the cross-linking kinetics at long times. This was not included in the random branching model described earlier and suggests that application of the model may have to be restricted to gel time data and the initial parts of cure curves. The problem is not perhaps as significant as it sounds, however, as much real data are restricted to quite short cure times for experimental reasons and G_{inf} values in irreversible cases are often obtained by quite extensive extrapolation if they can be obtained at all.

Where genuinely phase-separated gels are the subjects of mechanical measurement the issue of data interpretation is much less certain. Of course phase separation itself is a broad term, as gels can range from those showing only density fluctuations to those with internal phases and definite phase boundaries. While the random branching model may still be relevant to the former case, one would expect the latter situation to be more amenable to the models developed to describe mixed biopolymer gels (Clark, 1987). Much more systematic study of the mechanical development of heterogeneous gels will be needed to resolve this matter.

1.3.6 Exponents and Critical Exponents

The concept of critical exponents in statistical physics is that close to a (formally continuous) phase transition, a whole range of properties show so-called universal behavior. More simply, we can say that log-log (scaling law) plots of some property, versus the 'distance' from the phase transition, show common slopes regardless of the nature of the particular phase transition, but dependent only on say the dimensionality and range of the system. Such critical behavior is seen, for example, for liquid–gas critical points, and liquid crystal and magnetization critical behavior close to the Curie point.

Following work, particularly by Stauffer and co-workers (Stauffer et al., 1982; Stauffer, 1985) it was suggested that the critical behavior close to the gel point, could also be described by such a treatment. This already begs the question as to whether gelation is a phase transition – a contentious topic in its own right, going back to the early work of Stockmayer (Stockmayer, 1943, 1944) and recent work by Tanaka (Tanaka, 2006), who compared gelation to a Bose-Einstein condensation process – and one which has certainly not been completely settled. (Here we must distinguish clearly between an ordering process, say a disorder to order transition occurring pre-gelation – such as a coil-helix transition – which may indeed be a first-order transition, in the sense of Ehrenfest (Stanley, 1998) and the gelation itself.) However, since a number of papers have been published claiming to show 'universal behavior', particularly in regard to the gel modulus, we discuss this topic here.

In previous discussion we have used the Flory expressions α and α_c to describe the conversion and its critical value, as in Eq. (1.8). We now use the term p (here defined to be identical to α) to show correspondence with other work in this area, where p simply represents some intensive (i.e. sample volume independent) property of the system. The relative distance, ϵ, from the gel point is then given by:

$$\varepsilon \equiv \left(\frac{p - p_c}{p_c}\right) \text{ or } \left[\left(\frac{p}{p_c} - 1\right)\right] \tag{1.14}$$

As we have already noted, the classical exponent(s) should hold when the value of ε is less than the Ginzburg limit – and we have already cited the range of this above (Stauffer et al., 1982) as typically from $\varepsilon = 0$ (practically, since it will be plotted on a log scale say 10^{-5}) up to an upper limit of 10^{-2} to 10^{-1}.

The most usual critical exponent is γ, which relates ε to the gel modulus G is defined as:

$$G \approx \varepsilon^{\gamma} \tag{1.15}$$

The classical value for γ is exactly 3, whereas non-classical ('critical') values lie below this, and are typically 1.8–2.3, depending on dimensionality, etc. Experimental determination of γ is extremely difficult, even for carefully designed chemically cross-linked gels, firstly because the range of ε is very testing to access, and secondly because all exponents are true strictly only in the limit as ε approaches zero. This means even classical theories give a measured value lower than the predicted exponent, 3, for finite ε, and log-log plots even within the Ginzburg region are often slightly curved.

However, in one particular type of application, for physical gels such as those described in this chapter, there are other pitfalls. Several papers have appeared in which a slope of ~ 2 has been reported in plots of G versus say C/C_0, and then related directly to the non-classical critical exponents. As we have pointed out elsewhere (Ross-Murphy, 2007), this is completely incorrect. The reasons for this are several – firstly, all such published work we have seen actually covers a range almost exclusively outside the Ginzburg region. Secondly in the critical region the strain dependence of modulus becomes even more relevant – one of us has suggested for physical gels the linear strain actually tends to zero with ε. Finally the mapping from C/C_0 on to ε is not a trivial or even nominal exercise (as we have seen when performed more formally in preceding pages). In other words it cannot be assumed that $[(C/C_0) - 1]$ is related directly to ε. Indeed exponents of around 2 for γ, have been reported for many years, but without either making the connection or detailed comment; certainly, as we have seen above, values such as this arise quite naturally from branching theory. Whether or not critical exponents have ever been measured in a chemical gelation is open to discussion – what we can assert with confidence is that no such genuine result has been obtained from so far published measurements on biopolymer physical gels.

1.3.7 Fractal Models

Another very physical model, which shows some superficial similarities to that above is the model of fractal geometry. Fractals are well known to many, via beautiful representations of the Mandelbrot set (Mandelbrot, 1983), but also occur in colloid science, as useful descriptors of colloidal aggregate structures. For example, measurement of the scattering behavior of such model colloids via a plot of log scattering intensity versus log scattering vector (or scattering angle) often gives a constant slope over more than two decades of scattering vector, consistent with so-called self-similarity.

Of course the aggregation processes occurring in physical gels usually involve much less random mechanisms including as they do H-bonding, hydrophobic interactions, van der Waals and electrostatic forces (Clark et al., 1987). They also show hierarchical levels of structure, so they are extremely unlikely to show genuine self-similarity.

Despite this, some workers, for example Bremer and co-workers have attempted to model the dependence of modulus and concentration by such a physical fractal model (Bremer et al., 1989). Fractal (Hausdorff) dimensions, D, were calculated using the modulus–concentration relationship (their Eq. 12):

$$G = K'C^{2/(3-D)} \tag{1.16}$$

Values obtained for casein gels using this approach were found to compare favorably with those estimated by direct structural probes. The implicit assumption was made, however, that the limiting critical concentration C_o was effectively zero, which is, of course, contradicted by many experimental results for biopolymer gels, and particularly for the lactoglobulin gels studied here. A more recent approach by Ikeda and co-workers (Ikeda et al., 1999) has applied a fractal model to investigating the effects of ionic strength on fractal structures in heat-induced gels prepared from globular proteins. They obtain fractal dimensions of 2.2 in 25 mM NaCl, with lower values (~ 1.8) being obtained as the salt concentration is increased (Bremer et al., 1989).

When D is, say, 2.0 Eq. (1.16) requires that $G \propto C^2$. Unfortunately, as discussed above, it is observed all but universally that a gel will follow this behavior in some regime of concentration, which, in elementary kinetic treatments, reflects the binary reaction of 'functional groups'. However, as we have pointed out in detail above, the overall concentration dependence of modulus should not, and does not, follow a simple power law, because the exponent n in $G \propto C^n$ is not constant over the entire regime from $C \rightarrow C_0$, but only at higher concentrations.

When measurements are made over a more extensive range, for β-lactoglobulin gels, such as those illustrated in Figure 1.13, the present authors found that plainly unphysical Hausdorff dimensions were obtained (Kavanagh et al., 2000). For example at pH 7, $D = 2.9$, at pH 3, $D = 2.8$, at pH 2.5, $D = 2.8$ and at pH 2, $D = 2.7$. All these values are much larger than would be expected for a diffusion- or reaction-limited cluster–cluster aggregation mechanism. Indeed, they were very close to 3, implying that compact three-dimensional clusters had formed with little room for the solvent, something not supported by microscope images of β-Lg networks (Figure 1.13).

It may be that some protein gels do indeed show real self-similarity over decades of scattering vector, and that equations such as Eq. (1.16) can be applied, but in our experience modulus exponents of ca. 2 do not reflect any such physical model, merely that data have been collected over too limited a range, and (relatively) far from C_0.

1.3.8 Lattice and Off-Lattice Simulation

In addition to the mean field random branching and the fractal models discussed above, simulation approaches have also been widely applied to modeling gelation. These include lattice percolation models and off-lattice simulations. These approaches offer greater realism through their use of a real three-dimensional arena in which molecular interactions can take place. Such work has provided new insights and understanding particularly in the area of colloidal particle gels (Dickinson, 1994) but the simulation methods do not appear so far to have provided predictions (e.g. form of cure curves) of the kind addressed here by the random branching model. For this reason this growing activity will not be addressed further in the present work but it does hold considerable promise for the future.

1.4 CONCLUSIONS

As can be seen from the above, there is much work left to complete from both the experimental and theoretical viewpoints. The availability of excellent modern rheometers makes the kinetic gelation experiment quite straightforward. Indeed much has changed since the authors started on this road some 30 years ago; at that time there were practically no data available. However, many precautions still need to be taken, and unfortunately quite a few datasets are still either incomplete, or frankly poor, because journal authors are unaware of the many pitfalls, and simply repeat them.

As far as theory is concerned, it could be said that, from all that has been repeated above about models and deviations from the theory, there must be easier ways of studying self-assembly in biopolymer systems than mechanical measurements, such as scattering and spectroscopic approaches. However, where mechanical aspects of biopolymer self-assembly are important the approach discussed here comes into its own. It is only by assembling and testing cure models against appropriate data that a link can be made between self-assembly and its mechanical implications. This is important as hydrated biopolymer networks form structural elements of living tissue such as, for example, the self-assembled actin and tubulin networks in animal cells, and the fibrin networks of the blood clotting (fibrin–fibrinogen assembly) mechanism. It is also possible that some mechanical element is involved in the role of self-assembled protein networks in disease, for example, in amyloid structure formation, and as far as the current volume is concerned, there is also the very crucial importance of their role in food processing, pharmacy and other related industries. Such topics are discussed in more detail in later parts of this volume.

ACKNOWLEDGMENTS

The authors are delighted to acknowledge the work and efforts of two of their past joint research students, Gaynor Kavanagh and Walraj Gosal. Some of the data presented here were first given in their respective PhD theses and referenced as such.

References

Adam, M., Delsanti, M., Pieransky, P., and Meyer, R. (1984). Magnetic sphere rheometer. *Revue de Physique Appliquee* **19**, 253–264.

Almdal, M., Dyre, J., Hvidt, S., and Kramer, O. (1993). Towards a phenomenological description of the term "gel". *Polymer Gels and Networks* **1**, 5–17.

Arenaz, M. F. and Lozano, J. E. (1998). Measurement of gelpoint temperature and modulus of pectin gels. *Journal of Food Science* **63**, 979–982.

Bremer, L. G. B., van Vliet, T., and Walstra, P. (1989). Theoretical and experimental-study of the fractal nature of the structure of casein gels. *Journal of the Chemical Society-Faraday Transactions I* **85**, 3359–3372.

Burchard, W. and Ross-Murphy, S. B. (1990). Introduction: physical gels from synthetic and biological macromolecules. *In* Physical networks – polymers and gels, 1–14, Burchard, W. and Ross-Murphy, S. B. (eds.). Elsevier Applied Science, Barking, U.K.

Busnel, J. P., Morris, E. R., and Ross-Murphy, S. B. (1989). Interpretation of the renaturation kinetics of gelatin solutions. *International Journal of Biological Macromolecules* **11**, 119–125.

Chambon, F. and Winter, H. H. (1985). Stopping of cross-linking reaction in a PDMS polymer at the gel point. *Polymer Bulletin* **13**, 499–503.

Clark, A. H. (1987). The application of network theory to food systems. *In* Food structure and behaviour, 13–34, Blanshard, J. M. V. and Lillford, P. J. (eds.). Academic Press, London, U.K.

Clark, A. H. (1993). Biopolymer gelation – comparison of reversible and irreversible cross-link descriptions. *Polymer Gels and Networks* **1**, 139–158.

Clark, A. H. (2000). Biopolymer gelation – The structure-property relationship. *In* Gums and stabilisers for the food industry – 10, 91–110, Williams, P. A. and Phillips, G. O. (eds.). Royal Society of Chemistry, Cambridge, UK.

Clark, A. H. and Amici, E. (2003). The formation and properties of biopolymer gels. *In* Food colloids – biopolymers and materials, 35–48, Dickinson, E. and van Vliet, T. (eds.). Royal Society of Chemistry, Cambridge, UK.

Clark, A. H., Evans, K. T., and Farrer, D. B. (1994). Shear modulus-temperature meltdown profiles of gelatin and pectin gels – A cascade theory description. *International Journal of Biological Macromolecules* **16**, 125–130.

Clark, A. H. and Farrer, D. B. (1996). Shear modulus-concentration relationships for low DE pectin-calcium gels In the temperature range 20-85 degrees C. *Food Hydrocolloids* **10**, 31–39.

Clark, A. H., Kavanagh, G. M., and Ross-Murphy, S. B. (2001). Globular protein gelation – theory and experiment. *Food Hydrocolloids* **15**, 383–400.

Clark, A. H. and Ross-Murphy, S. B. (1985). The concentration-dependence of biopolymer gel modulus. *British Polymer Journal* **17**, 164–168.

Clark, A. H. and Ross-Murphy, S. B. (1987). Structural and mechanical properties of biopolymer gels. *Advances in Polymer Science* **83**, 57–192.

Dickinson, E. (1994). Computer-simulation of particle gel formation. *Journal of the Chemical Society-Faraday Transactions* **90**, 173–180.

Djabourov, M., Maquet, J., Theveneau, H., Leblond, J., and Papon, P. (1985). Kinetics of gelation of aqueous gelatin solutions. *British Polymer Journal* **17**, 169–174.

Dobson, C. M. (2003). Protein folding and misfolding. *Nature* **426**, 884–890.

Dobson, G. R. and Gordon, M. (1965). Theory of branching processes and statistics of rubber elasticity. *Journal of Chemical Physics* **43**, 705–713.

Doi, M. and Edwards, S. F. (1986). The theory of polymer dynamics. Clarendon Press, Oxford, U.K.

Dusek, K., Gordon, M., and Ross-Murphy, S. B. (1978). The Graph-like state of matter X: Cyclisation and concentration of elastically active network chains in polymer networks. *Macromolecules* **11**, 236–245.

Eldridge, J. E. and Ferry, J. D. (1954). Studies of the cross-linking process in gelatin gels. III. Dependence of melting point on concentration and molecular weight. *Journal of Physical Chemistry* **58**, 992–995.

Ferry, J. D. (1980). Viscoelastic properties of polymers. Wiley Interscience, New York, NY, U.S.A.

Flory, P. J. (1941). Molecular size distribution in three dimensional polymers. I. Gelation. *Journal of the American Chemical Society* **63**, 3083–3090.

Flory, P. J. (1942). Constitution of three-dimensional polymers and the theory of gelation. *Journal of Physical Chemistry* **46**, 132–140.

Flory, P. J. (1974). Introduction. *Faraday Discussions of the Chemical Society* **57**, 7–18.

Gordon, M. (1962). Goods theory of cascade processes applied to statistics of polymer distributions. *Proceedings of the Royal Society of London Series A-Mathematical and Physical Sciences* **268**, 240–259.

Gordon, M. and Ross-Murphy, S. B. (1975). The structure and properties of molecular trees and networks. *Pure and Applied Chemistry* **43**, 1–26.

Gordon, M. and Scantlebury, G. R. (1964). Non-random polycondensation – statistical theory of substitution effect. *Transactions of the Faraday Society* **60**, 604–621.

Gordon, M. and Scantlebury, G. R. (1966). Theory of ring-chain equilibria in branched non-random polycondensation systems with applications of $POCl_3/P_2O_5$. *Proceedings of the Royal Society of London Series A-Mathematical and Physical Sciences* **292**, 380–402.

Gordon, M. and Scantlebury, G. R. (1968). Theory of branching processes and kinetically controlled ring-chain competition processes. *Journal of Polymer Science Part C-Polymer Symposium*, 3933–3942.

Gosal, W. S. (2002). Morphological, structural & mechanical properties of self-assembled fibrillar aggregates, networks & gels, derived from non-disease associated globular proteins. PhD Thesis. University of London.

Gosal, W. S., Clark, A. H., Pudney, P. D. A., and Ross-Murphy, S. B. (2002). Novel amyloid fibrillar networks derived from a globular protein: β-lactoglobulin. *Langmuir* **18**, 7174–7181.

Gosal, W. S., Clark, A. H., Ross-Murphy, S. B. (2004a). Fibrillar β-lactoglobulin gels: Part 2-Dynamic mechanical characterisation of heat-set systems. *Biomacromolecules* **5**, 2420–2429.

Gosal, W. S., Clark, A. H., Ross-Murphy, S. B. (2004b). Fibrillar β-lactoglobulin Gels: Part 1–Fibril Formation and Structure. *Biomacromolecules* **5**, 2408–2419.

Higgs, P. G. and Ross-Murphy, S. B. (1990). Creep measurements on gelatin gels. *International Journal of Biological Macromolecules* **12**, 233–240.

Hughes, V. A. and Dunstan, D. (2009). Amyloid Fibrils – Self-Assembling Proteins. *Modern Biopolymer Science*, 561–596, Kasapis, S., Ubbink, J.B., and Norton, I.T. (eds). Elsevier, San Diego, CA.

Ikeda, S., Foegeding, E. A., and Hagiwara, T. (1999). Rheological study on the fractal nature of the protein gel structure. *Langmuir* **15**, 8584–8589.

Jordan Lloyd, D. (1926). The problem of gel structure. *In* Colloid Chemistry, 767–782, Alexander, J. (ed.). Chemical Catalog Company, New York, NY.

Kavanagh, G. M. (1998). Rheological and structural properties of denatured globular protein gels and sols. PhD Thesis. University of London.

Kavanagh, G. M., Clark, A. H., and Ross-Murphy, S. B. (2000). Heat-induced gelation of globular proteins: Part 4 Gelation kinetics of low pH β-lactoglobulin gels. *Langmuir* **16**, 9584–9594.

Kavanagh, G. M. and Ross-Murphy, S. B. (1998). Rheological characterisation of polymer gels. *Progress in Polymer Science* **23**, 533–562.

King, M. (1979). Interrelation between mechanical-properties of mucus and mucociliary transport – effect of pharmacologic interventions. *Biorheology* **16**, 57–68.

Mandelbrot, B. B. (1983). The fractal geometry of nature. WH Freeman and Co., New York, NY.

Morris, E. R., Rees, D. A., and Robinson, G. (1980). The domain model for gelation. *Journal of Molecular Biology* **138**, 349–362.

Richardson, R. K. and Ross-Murphy, S. B. (1981). Mechanical-properties of globular protein gels .1. Incipient gelation behaviour. *International Journal of Biological Macromolecules* **3**, 315–322.

Richardson, R. K., Ross-Murphy, S. B. (1987a). Nonlinear viscoelasticity of polysaccharide solutions.1. Guar galactomannan solutions. *International Journal of Biological Macromolecules* **9**, 250–256.

Richardson, R. K., Ross-Murphy, S. B. (1987b). Nonlinear viscoelasticity of polysaccharide solutions.2. Xanthan polysaccharide solutions. *International Journal of Biological Macromolecules* **9**, 257–264.

Rodd, A. B., Dunstan, D. E., Ross-Murphy, S. B., and Boger, D. V. (2001). Dependence of linear viscoelastic critical strain and stress values on extent of gelation for a thermally activated gelling system. *Rheologica Acta* **40**, 23–29.

Ross-Murphy, S. B. (1994). Rheological methods. *In* Physical Techniques for the Study of Food Biopolymers, 342–392, Ross-Murphy, S. B. (ed.). Blackie Academic and Professional, Glasgow, U.K.

Ross-Murphy, S. B. (2005). Gelation kinetics – problems and prospects. *Journal of Macromolecular Science – Physics Edition* **44**, 1007–1019.

Ross-Murphy, S. B. (2007). Biopolymer gelation – exponents and critical exponents. *Polymer Bulletin* **58**, 119–126.

Ross-Murphy, S. B. and Shatwell, K. P. (1993). Polysaccharide strong and weak gels. *Biorheology* **30**, 217–227.

Stading, M., Langton, M., and Hermansson, A. M. (1992). Inhomogeneous fine-stranded β-lactoglobulin gels. *Food Hydrocolloids* **6**, 455–470.

Stanley, H. E. (1998). Introduction to phase transitions and critical phenomena. Oxford University Press, UK.

Stauffer, D. (1985). Introduction to percolation theory. Taylor and Francis, London, England.

Stauffer, D., Coniglio, A., and Adam, M. (1982). Gelation and critical phenomena. *Advances in Polymer Science* **44**, 103–158.

Stockmayer, W. H. (1943). Theory of molecular size distribution and gel formation in branched-chain polymers. *Journal of Chemical Physics* **11**, 45–55.

Stockmayer, W. H. (1944). Theory of molecular size distribution and gel formation in branched polymers II General cross linking. *Journal of Chemical Physics* **12**, 125–131.

Stockmayer, W. H. and Zimm, B. H. (1984). When polymer science looked easy. *Annual Review of Physical Chemistry* **35**, 1–21.

Takahashi, A. (1972). The melting temperature of thermally reversible gel. I. Poly(vinyl chloride) organic solvent gels. *Polymer Journal* **3**, 207–216.

Tanaka, F. (2006). Thermoreversible gelation is a Bose-Einstein condensation. Physical Review E **73** 061405-1—061405-6.

te Nijenhuis, K. (1997). Thermoreversible networks. Viscoelastic properties and structure of gels. *Advances in Polymer Science* **130**, 1–193.

Trappe, V., Richtering, W., and Burchard, W. (1992). Critical-behaviour of anhydride cured epoxies. *Journal de Physique II* **2**, 1453–1463.

Winter, H. H. and Chambon, F. (1986). Analysis of linear viscoelasticity of a cross-linking polymer at the gel point. *Journal of Rheology* **30**, 367–382.

Xu, Z. M. and Raphaelides, S. N. (2005). A dynamic U-tube rheometer of novel design for the study of weak gels and foams. *Rheologica Acta* **45**, 77–82.

CHAPTER

2

Gelation: Principles, Models and Applications to Proteins

Erik van der Linden[1] *and E. Allen Foegeding*[2]

[1]Agrotechnology and Food Sciences Group, Wageningen University, Bomenweg 2, 6703 HD Wageningen, The Netherlands

[2]Department of Food Bioprocessing and Nutrition Sciences, North Carolina State University, Raleigh, NC 27695, USA

2.1 INTRODUCTION

Gelation is a process that is of importance to a wide range of applications in food, pharmaceutical and material sciences. Gelation is a general way to convert a fluid to a solid and has been used since antiquity to produce a variety of foods with distinctive textures. Foods such as gelatin-based desserts are some of the simpler food gels, consisting of a water–gelatin gel with added sweetener, flavor and color. In contrast, all cheeses start as a simple gel then undergo further processing to remove some compounds and add others. The 'fresh cheese' could be the final step or it may require aging for slow chemical, physical and biological processes to transform it to a desired product over the months or years.

The investigation of food protein gels is complex in that it encompasses everything from the condensed matter physics of forming the gel to the physiological and psychological processes involved in food evaluation. This chapter will focus on the formation of food protein gels and the mechanical and sensory evaluation of protein gels as model foods. We will start with describing and defining the gel state based on key historical developments on the concept of just what exactly is a gel. This broad approach will be extended to the various theoretical models which have been, and are being, used to describe gelation. While emphasis will be given to examples associated with protein gels, other materials will be discussed to better illustrate key concepts. A common feature to all protein gelation reactions is that they require some initial structural transition that can be considered transformation from an unreactive to a reactive structure that increases the probability of intermolecular interactions. Various models will be discussed that describe the denaturation and aggregation processes that occur prior to formation of a gel network. Gel networks trap large quantities of water into a solid mass that behaves elastically on a timescale such that it is perceived as a solid. The solid, elastic structure of gels provides them with textures that are desirable to consumers. When considering

Kasapis, Norton, and Ubbink: Modern Biopolymer Science
ISBN: 978-0-12-374195-0

texture as a property experienced by animals during mastication of foods, it is apparent that fracture properties of gels are important to the overall consumer desirability. However, since mastication is a complex physiological process involving not only the breakdown of food but a constant evaluation of the bolus to determine when swallowing is possible, fracture properties alone may not be enough to predict sensory texture. To that end, gel microstructure and the fracture process at the microstructural level will also be considered. Finally, areas requiring further investigation will be mentioned. This chapter builds on prior excellent reviews on protein gelation that will be acknowledged at appropriate sections in the chapter.

2.2 MODELING GEL NETWORKS AND THEIR RHEOLOGICAL BEHAVIOR

2.2.1 Introduction

From a macroscopic point of view, a gel may be defined as a 'substance that resists deformation when one applies a tractive force to it' (translated from De Gennes, 1976a). The gel is, as a consequence of that, able to store some of the energy that is needed for the deformation. This definition relates well to the known behavior of, for example, a gelatin-based dessert known in the US as 'Jello™'. However, there are numerous other definitions of a gel and they are all practiced at the same time. Illustrative with respect to this large variety of definitions of a gel is a short history on the gel state as given in one of the papers by M. Djabourov of which parts are quoted here. According to this paper (Djabourov, 1988), the gel state was first proposed by Thomas Graham in 1861. Graham differentiated systems by grouping them according to their ability to exhibit diffusion. He distinguished crystalloids such as sugar and salts, which exist in the form of crystals and exhibit the property to quickly dissolve, showing fast diffusion of the components. This type of matter was denoted by Graham as having large 'diffusive power'. Another type of matter was classified by Graham as having a low diffusive power, where the materials seem to be held together by some weak force, as if it were a gluey substance. (It is this gluey character from which this class of materials has obtained its name: colloid. This term stems from the Greek word 'colla' (κολλα), meaning glue.) Examples are substances containing gelatin, starch and gum arabic. According to Graham, the gluey substances can be characterized as homogeneous solutions:

- that are continuous
- the solution may exhibit a thickening modification into an insoluble state
- the resulting insoluble state is soft, elastic, solid-like, but does not show any sharp angles, the latter in contrast to crystal shapes.

The insoluble soft solid-type gluey state, as defined by Graham, serves as the starting point of the definition of the term gel that takes into account structural information. Hermans (1949; Kanai et al., 1992; Trappe and Weitz, 2000) proposed a classification of the different types of gel based on the types of structures present on a scale of micrometers and smaller. This classification takes into account the architecture of the network and the molecular nature of the components.

According to the classification by Hermans, the term gel refers to systems that:

- are coherent, two-component systems formed by a solid substance, finely dispersed or dissolved in a liquid phase (referred to as solvent)
- exhibit solid-type of behavior as mechanical forces are exerted on it
- are continuous in both the dispersed and the solvent phase.

The last criterion could only be formulated as electron microscopy and other ultrafine microscopy techniques became available. Hermans arrived at four different microscopic structures based on the available microstructural information. The possible building blocks leading to the structures were thought to be: (a) spherical particles (leading to a so-called particle gel), (b) rod- or fiber-like particles, (c) polymers, where the polymers are held together by crystalline junction zones, or (d) polymers where the polymers are held together by covalent bonds (like sulfur binds polymers in gelled rubber). Depending on whether the bonds between the building blocks of the network are of a physical or covalent (chemical changes) character, one differentiated between *physical* and *chemical* gels, respectively. We note that the distinction between a particle gel and a polymer gel is based on the view that the primary building block in the former is a particle (which may be a molecule or a larger piece of some sort of matter), while in the latter the primary building block is polymeric, i.e. consisting of a sequence of molecules attached to one another in some fashion. This distinction is somewhat arbitrary, since during one of the earlier steps in the gelation process molecules may first assemble into polymeric aggregates, and subsequently aggregate further. Moreover, it becomes problematic in protein gelation when globular proteins are technically polymers but often viewed as particles.

The third criterion of Hermans (continuous in the dispersed phase and solvent phase) refers to the idea that the building blocks form a (macroscopically) connected network throughout the system, or in other words, form a space-spanning network. Such a space-spanning network then has the size of the system, which on a molecular scale may be considered almost infinite (in fact the network must become infinite if one requires the argument to hold for any system size).

Such an 'infinite' space-spanning network was already addressed by Flory in 1941 for the polymer case (Flory, 1941) since he notes '... gelation is the result of the formation of infinitely large molecules'. Interestingly, Flory also uses the term gel as '... meant to include only those polymeric materials portions of which are insoluble in all solvents that do not destroy the primary polymer chain structure; the term gel will be applied in particular to that portion of such polymers that are insoluble in all solvents.' Hereby he addresses criterion 2 by Graham. He realized that there are other types of gels, as is clear from the following quote from the same paper '... The jelly-like masses which one obtains through the action of a swelling agent on a polymer, which in another solvent would dissolve completely, are not under consideration here.' We note that the insolubility exists by virtue of an attractive interaction between the building blocks. It is also clear that there are other types of gel definitions that include systems where the interactions setting up the gel structure are not necessarily attractive. These will be discussed later. Once immersed in excess solvent, such gels do exhibit dissolution.

Taking the notion of an infinite space-spanning network of polymeric nature, being formed by reactions of units with constant number of functional groups per monomer in solution, Flory derived a critical probability for one polymeric branch to connect to another polymeric branch rather than an end group. This critical probability depended on the functionality (number of reaction sites) of the branches. If the probability of connection is less than the critical value, no infinite network is formed, while if the probability exceeds the critical value, one infinite network will be formed, in conjunction with smaller polymeric chains. This situation of criticality is mentioned in another paper by Flory (1942) and by Stockmayer (1943). The existence of a gel network in conjunction with smaller polymeric chains leads to a 'structural demarcation' of the gel fraction versus a fraction containing the smaller, still soluble, polymeric fragments, referred to as the sol fraction. This demarcation is 'as sharp as that between a liquid and its saturated vapor'. This analogy is substantiated by a remark of JE Mayer given as a footnote that 'combinations of chains through multi-functional units in three dimensional polymers can be regarded as a simplified analogue of molecular clusters in the statistical-mechanical theory of condensing systems'. As an example of the network theory described, Flory mentions gelation of the protein egg albumin in aqueous solution containing acetic acid.

The approach of a network being formed by linear branches as formulated by Flory (1941), and later on by Stockmayer (1943), does not take into account the possibility of cyclic connections, i.e. crosslinks between units of the same branch or multiple crosslinks between two chains. Later more extended approaches involving cascade theory (Gordon, 1962, Gordon and Ross-Murphy, 1975) were developed that do allow for such cyclic connections. Neglecting cyclic connections may be a good approximation for the formation of, for example, rubber, where only a few sulfur bonds are being formed per chain (De Gennes, 1976a), but for the case of gelatin for example, cyclic connections were already thought probable in dilute solution (Eldridge and Ferry, 1954). This shortcoming implies that gel points will be falsely estimated. It was clear from the beginning of the classical network theories that the model for network formation should in general include the possibility of cyclic connections in order to broaden its applicability. The initial idea to view the point in time where an infinite network is formed as a critical point, denoting a transition as depicted by Flory, still remained the attractive starting point for such a more generic model. This idea required connecting gelation phenomena to the area of phase transitions in general, which was intensively studied theoretically and experimentally.

One of the concepts in phase transition models is the so-called percolation phenomenon. Percolation was most notably connected to gelation by Stauffer (1976, 1982). The phenomenon itself was proposed in 1956 by the mathematician Hammersley (1956). Percolation refers to the formation of one infinite cluster in a system where clusters are allowed to grow in a statistically defined way. One such way of letting clusters form and grow statistically is to first divide a system into small blocks (referred to as sites) and to define a chance for having this site being filled by a network building block. Those sites that are filled by building blocks and that are neighbors of one another are defined to be sites that belong to the same cluster, thus defining a cluster (Figure 2.1).

In this way, visiting all sites and statistically determining whether the site is filled or not, one obtains, for a given chance, a set of clusters whose size and number will depend on this predefined chance. If the predefined chance is

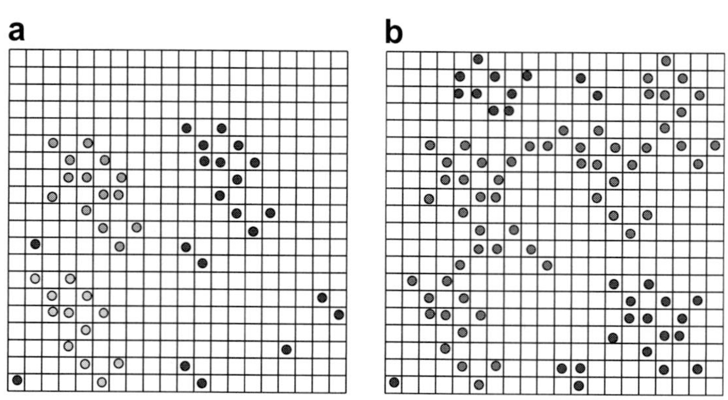

FIGURE 2.1 (a) Differently sized clusters on a lattice. (b) Finite clusters and one infinite cluster, forming a network.

chosen to be higher, more clusters and larger clusters will be formed. It turns out theoretically that there is one, and only one, specific ('critical') chance where you have one space-spanning cluster being formed. Small predefined chances yield separate clusters only, and increasing the predefined chance to higher values can be viewed as approaching a well-defined transition. This transition is obtained for the 'critical' chance. This chance is found to converge in the limit of infinite systems.

Alternatively, one may choose to set predefined chances for filling one site, and then investigate the influence of choice of site size. This is identical to choosing the number of sites to subdivide the system. It is also identical to choosing the concentration of building blocks (i.e. filled sites) at the end of the procedure. Repeating the same statistical procedure of filling sites, again with a predefined and constant chance, but now with a varying end concentration of building blocks, one will observe that there is one and only one concentration below which one has separate clusters, and above which one has one infinite cluster. In this way, one arrives at a critical concentration for formation of an infinite network. Note that according to the above statistical procedure, cyclic connections are allowed, and as such already the percolation approach yields a broader applicability than the single branch theories. We add that one may derive other algorithms for studying percolation but they are similar in many respects to the idea described above (see references by Stauffer et al. on percolation theory (1979) and on the connection to gelation (1982)). We will devote one section to describe, in a non-exhaustive manner, how percolation theory can and has been applied to gelation and how critical exponents describing critical behavior in general (like in phase transitions) can also be found in describing gelation behavior and consequent rheological properties. We refer to that section as 'percolation description of gel networks'. We note that the critical exponents based on classical phase transition theories, to which linear branch theories also belong, do not describe the critical points adequately (De Gennes, 1976a, 1976b; Stauffer, 1976; Stauffer et al., 1982) and in that respect one may favor the percolation approach. We however come back to this point later.

The critical concentration arrived at in a natural way using the percolation approach is an important experimental feature encountered in many biopolymer and food gelling systems according to Clark (1987). It must be said that one can also arrive at a critical concentration for gelation from branch theories. For example, Clark derives such a critical concentration in terms of molecular weight of the building blocks and number of potential reaction sites using a cascade approach. Such a critical concentration follows automatically from the assumption (originally proposed already by Eldridge and Ferry (1954)) that the binding between two functional sites obeys a monomer–dimer equilibrium, as was already originally applied for the weak binding case by Hermans (1965). We gladly refer at this point also to the classic review by Clark and Ross-Murphy (1987) for more details on similar and related approaches regarding the critical concentration using branch theories, to Clark and Ross-Murphy (1985), and to Chapter 1 of this book. These references provide an overview of the applications of branch theories up to 1987. We devote one section to recent advances in applying this approach to gelation, entitled: 'Branch description of gel networks'. Specific linear branch theories neglect cyclic connections. Percolation theories do take these into account, which ensures the right critical exponents.

Both percolation theory and linear branch theory arrive at a critical concentration, an important experimental feature for protein gelation. Both also describe the connectedness within a network, and lead to scaling predictions on rheological behavior in the linear regimen. The critical exponents using the percolation approach do follow experimental critical behavior, while linear branch theories do not show this analogy.

In the review by Clark (1987), another shortcoming of the linear branch (classical) theories is briefly mentioned (in addition to neglecting the cyclic connections): ' ...it is tacitly assumed ...' that during the gelation the system remains homogeneous, i.e. no (local) phase separation is assumed to occur during the process of network formation. This indeed forms an important comment for protein gelation, since one of the most often occurring phenomena during gelation of many proteins is that upon gelation (caused by events like a temperature change, a pH change, or a salt concentration change), the proteins tend to exhibit local densification, suggesting that local phase separation is relevant to gelation in many instances. This is especially relevant for globular proteins under conditions of decreased electrostatic repulsion and increased attractive interaction. It is also possible to a lesser extent to linear polymeric molecules like gelatin, but in some cases it may not be relevant at all.

The realization that densification occurs due to formation of protein aggregates during gelation has given rise to modeling the gelation process according to the aggregation of proteins to one another, leading to clusters. These clusters grow in time as more proteins diffuse onto the cluster parts, and the clusters will ultimately connect to one another, though not necessarily forming a closed packed system of clusters. It is this connecting of clusters that effectively leads to network formation and thus gelation. Depending on whether the growth of the clusters is mainly limited by the diffusion of proteins onto the already existing cluster parts, or that the growth is limited by the 'reaction' needed for the actual 'attachment', one distinguishes diffusion limited aggregation (DLA) and reaction limited aggregation (RLA). Depending on whether the cluster aggregation is limited by diffusion of reaction, one distinguishes DLCA and RLCA, respectively. After this cluster formation rearrangements of the proteins are discarded in this approach. One denotes the clusters being formed in this manner as fractal clusters (Stauffer, 1979) since the amount of proteins being part of the cluster scale in a particular way with the size of the cluster. The dimension that defines this scaling is referred to as the fractal dimension (Stauffer, 1979). This approach of modeling gelation by means of cluster formation is covered in the section entitled 'colloidal cluster description of gel networks'. The gelation by means of such cluster–cluster aggregation has been introduced by Meakin (1983) and Kolb et al. (1983).

We want to note at this point that the percolation approach may seem remote from the colloidal cluster approach, but these two points of view can be united by the realization that if the clusters themselves connect randomly one can envision the colloidal clusters as the building blocks for exhibiting the percolation. In fact, in percolation models, the clusters that are observable in simulations before the gelation point also exhibit a fractal dimension (Stauffer, 1979). It has to be stressed at the same time that the colloidal cluster description of gel networks tries to incorporate the changing spatial distribution of the network-forming building blocks upon aggregation, while the percolation approach refers to the end situation and associated scaling. In fact, the changing spatial distribution incorporated in the fractal model refers to a phase separation type of transition. For the case of the colloidal clusters, it is essential that cluster densification after its initial formation becomes arrested because of the strong attachment of the building blocks to one another. In contrast, if the building blocks would be able to rearrange, the final clusters will be closed packed, much like an oil phase separates from water in the form of clusters that are closed packed arrangements of oil molecules in the form of droplets. This issue of describing the phase separation in relation to percolation, (colloidal) aggregation and gelation has received considerable attention in the last two decades, and this is why we devote a section on 'phase separation and transitions in relation to percolation, aggregation and gelation'. Here we will distinguish between equilibrium network formation and non-equilibrium, arrested, network formation. One issue important to phase separation and clusters forming a space-spanning network in relation to gelation, is the arrestment of aggregation, and suggested similarities with glass transitions. This is also a subject that has received much attention lately (cf. 'Arrested states and gelation').

Finally we note that we choose for this classification of models and include the effects of particle morphology and network morphology within the relevant paragraphs, since the models are not based on the specific building block morphology.

2.2.2 Percolation Description of Gel Networks

Despite the implicit, if not always justified, assumption of homogeneity during gelation in percolation theories, these models have improved our understanding of gelation. This is indeed the case where no distinct phase separation is expected. This has been shown for gelation of polysaccharides that bind to one another by ion binding (calcium ions in pectin and alginate systems, see Axelos and Kolb, 1990). Protein gelation, however, can occur due to a phase separation type of process and therefore gelation by globular proteins usually does not strictly follow the percolation model. We feel it is nonetheless illustrative to cover this type of model because: (1) it is applicable to gelatin for instance (Djabourov et al., 1988a, 1988b; van der Linden and Parker, 2005), (2) it also should be applicable to cluster network formation (applicable to particulate protein gels) and (3) it is an important aspect that is utilized in describing phase behavior in relation to transitions between liquid–liquid phases and gel states (Axelos and Kolb, 1990).

We choose to illustrate the percolation approach by means of examples in the general complex fluid field and therefore not limited to proteins. Advantages reported for the percolation approach, as opposed to models which build on fractal structure aggregation leading to gelation by means of cluster–cluster arrested states, will be discussed in this paragraph. The reason for exploring the larger area of complex fluids regarding gelation is that elegant similarities can be found in the available literature, since modeling gelation received considerable attention during many decades already in the complex fluid field in general, partially because it is applicable in a wide range of industrial areas (Keller, 1995). In the past few decades a considerable amount of work has been undertaken to arrive at a generic description of gel elasticity. Much of this effort has been based on percolation theories (Stauffer et al., 1982) and on the application of what is often called fractal theory (Shih et al., 1990). We quote the following paragraph (van der Linden and Sagis, 2001):

> 'The work based on percolation theories recognizes the existence of a gelation threshold concentration, depending on the strength of interaction between the monomers that form the network (Safran et al., 1985) and on their morphology (Safran et al., 1985), (Haan and Zwanzig, 1977). Elasticity, G′, in this case can be written as a power law according to $G' \sim (c-c_p)^t$, with c the concentration of monomers in the system, c_p the critical percolation threshold concentration, and t a scaling exponent, which depends on the type of Hamiltonian that describes the network elasticity. In contrast, the fractal models predict a power law behaviour written as $G' \sim c^w$, where the exponent w is related to the fractal dimension of so-called flocs that form the network (Bremer et al., 1990; Buscall et al., 1988; Yanez et al., 1999) and where it follows implicitly that one arrives at the fact that c_p can tend to 0, in principle. The fractal dimension depends on the interaction between the monomers. This is described in more detail in a subsequent section. Recently it was reported that fractal models were not able to arrive at a generic explanation for temperature and concentration dependence of the elasticity of alumina particle gels (Yanez et al., 1999). This lack of general applicability also followed from

work by Grant and Russel on silica particle gels (Grant and Russel, 1993). In their case, the use of a fractal model implied a structural change as a function of temperature which was contradicted by light scattering analysis of the structure. In contrast, using a percolation model allows for incorporation of the temperature dependence in c_p. As a consequence of this percolation approach, both the prefactor in the power law (characterising the elasticity of the interparticle bonds) as well as the scaling exponent t (characterising the structure) turned out to be independent of temperature, which was in accordance with experiment. Hence, the percolation model seems to offer a higher level of generality than fractal models in their case.'

In order to further test the general applicability of percolation models, authors have determined elasticity of various protein gels and check for the scaling relation given above for the elasticity G′ (see van der Linden and Sagis, 2001, and references therein). In general, protein gels are excellent candidates for such studies since the critical percolation threshold can differ by orders of magnitude, depending on the type of protein, solvent characteristics and temperature, since these parameters determine the interaction regimen. One of the recently appearing notions is that these differences are at higher concentrations not independent of the total protein concentration (Prinsen and Odijk, 2004), thereby hampering a clear-cut extrapolation toward the eventually present critical concentration. Another reason why protein gels are excellent candidates in principle for investigating percolation models is that a large variety of self-assembling structures is possible, ranging from thin fibers (e.g. actin (MacKintosh et al., 1995)), soy protein (Nakamura et al., 1984) and β-lactoglobulin far from its iso-electric point (Aymard et al., 1999), to particle gels (e.g. whey protein near its iso-electric point (Verheul, 1998)). This large variety of structures in principle allows testing the predictions of the dependence of the scaling exponent on the mesostructure (Mitescu and Musolf, 1983; Pandey and Stauffer, 1983; Kantor and Webman, 1984).

Regarding the values of exponents we quote (van der Linden and Sagis, 2001):

'One of the first theoretical models for the elasticity of a gel, based on the percolation concept, was developed by De Gennes (1976b). He proposed a critical exponent t equal to 11/6 for dimensionality d=3. This scaling relation is based on an analogy between the electrical conductivity and the elastic modulus, which is rigorous for a lattice model of gelation with an isotropic force between nearest neighbours on the lattice. De Gennes (1976b) remarks that the validity of this analogy may be extended to the situation where the connection between the sites that form the network is not through single bonds but through flexible chains. More recent theoretical work (Mitescu and Musolf, 1983), also using an isotropic force between nearest neighbours, resulted in a somewhat higher value for t, i.e. $t = 2.06 \pm 0.16$. This value turned out to be in accordance with subsequent numerical work (Pandey and Stauffer, 1983), giving $t = 2 \pm 0.2$, "…." If one does not consider an isotropic force between neigbouring sites but a central force, the numerical work of Feng and Sen (1984) strongly suggests that the presence of this type of force leads to a different universality class, with a value for $t = 4.4 \pm 0.6$ in the case of a 3d fcc lattice (added comment: a three dimensional face centered cubic lattice). In addition to the two models above (i.e. the isotropic force model and the central force model), Kantor and Webman (1984) have considered a model based on two contributions to the elastic energy, i.e. a central force term and a bending energy term. This approach would apply to macroscopically inhomogeneous materials, consisting of rigid and soft regions. Their approach also leads to a universality class that is different from the isotropic force problem, with a lower bound of a critical exponent τ (equivalent to t) of 2.85 and an upper bound of 3.55. In the case of a homogenous system where percolation theory does not apply, the exponent becomes 3 (Stauffer et al., 1982).'

Experimental values for the exponent t have been reported for a variety of gels (De Gennes, 1979; Kanai et al., 1992; Takigawa et al., 1992; Grant and Russel, 1993; Rueb and Zukoski, 1997; Yanez et al., 1999), with t values lying in the range of 11/6 (De Gennes, 1976b; Takigawa et al., 1992; Rueb and Zukoski, 1997; Yanez et al., 1999), but also cases with large deviations from that value (Kanai et al., 1992; Grant and Russel,

1993). In an attempt to check the validity of the percolation approach for several protein systems, G′ data as a function of concentration were re-analyzed to determine t and c_p (van der Linden and Sagis, 2001). The proteins included ovalbumin, casein, soy-glycinin, and soy β-conglycinin. The accuracy in c_p was claimed to be about 10% while an average value was found for the percolation exponent $t = 1.79 \pm 0.25$. The value of c_p varied by an order of magnitude. This value for t lies in the range of predictions of several models such as that of Mitescu and Musolf (1983), giving $t = 2.06 \pm 0.16$. It is in contrast to experimental t values for systems studied by Grant et al., Kanai et al., and Trappe and Weitz (Kanai et al., 1992; Grant and Russel, 1993; Trappe and Weitz, 2000) who report on systems in a two-phase region (Grant and Russel, 1993) and systems that are pre-sheared (Kanai et al., 1992; Trappe and Weitz, 2000), respectively. However, these systems are non-homogeneous, and, according to Kantor and Webman (1984), indeed are not expected to follow the isotropic percolation value but should be characterized by a lower bound of $t = 2.85$ and an upper bound of $t = 3.55$. Indeed, Grant and Russel (1993) report 3 ± 0.5, Kanai et al. (1992) report 1–4, depending on the initial particle concentration and the strength of the pre-shearing effect, and Trappe and Weitz report 4.1. Initial theoretical work on the deviations from the so-called conductivity exponent put forward by De Gennes was reported by Bergman and Kantor (1984). Different exponents have been reported in silica gelation by Devreux et al. (1993), depending on if measurements were made at times that are close to or further away from the gel time.

The value for c_p will depend on the type of protein, and on parameters that affect molecular interactions like pH and salt concentration (van der Linden and Sagis, 2001). Indeed, c_p decreases upon increasing salt concentration for ovalbumin, i.e. upon increasing protein–protein attraction. Such a decrease in c_p upon increasing attraction between the particles has also been reported for microemulsions (Safran et al., 1985; Hilfiker and Eicke, 1987). In fact, for microemulsions, the value of c_p as a function of experimental parameters has been applied to explain conductivity, Kerr-effect and light-scattering data in a self-consistent way (Hilfiker and Eicke, 1987). We note that in particular, the effect of an external (electric in that case) field on c_p was incorporated successfully (Eicke et al., 1986).

In a similar fashion in relating interaction of particles with c_p and corresponding universal behavior, Trappe and Weitz (2000) have explained visco-elastic behavior of systems containing weakly attractive carbon black particles. In their work, a change in the attractive interaction was also found to shift the percolation threshold while the visco-elastic properties of the network also remain scale invariant. See also the section on glass transitions below.

It was suggested that the so-called isotropic-force percolation model is applicable to some protein gels (van der Linden and Sagis, 2001). Taking a step further, systematic studies of c_p as a function of preparation conditions and/or protein type (i.e. charge distribution and consequent anisotropic interactions, leading to an anisotropic aggregate morphology) were suggested to be fruitful in describing the resulting gel properties of protein systems. The main reason being that such a route would allow one to systematically incorporate effects by external fields, like, e.g. gravitational and shear fields, and effects due to mixing with polysaccharides and other proteins. One successful example of incorporation of shear flow effects on final rheological behavior has been reported by Yanez et al. (1999). These authors were able to rescale all their data, independent of shear history, on a single expression using a critical volume fraction, which was a function of shear history.

Regarding the effects of flow, c_p was determined using flow-birefringence (Veerman et al., 2005) and the resulting flow curves could be fitted by using this c_p as a function of shear for β-lactoglobulin fibrils of micron size. Regarding the effects of anisotropy, the c_p could be determined very accurately by Djabourov (Joly-Duhamel et al., 2002a, 2002b) and related to the anisotropy characterization of the helices, and later on explained in terms of the persistence length of the triple helix formed by the gelatin (van der Linden and Parker, 2005), extending the general models of Philipse (Philipse, 1996; Philipse and Verberkmoes, 1997; Philipse and Wierenga, 1998). The results for critical gel concentrations are based on the assumption that the point at which one finds a caging-induced restriction of fibrillar movement is in fact equivalent to the gel point. However, this generalization may not hold for all conditions. A rheological model for higher concentrations of gelatin (based on thermal form fluctuations of the triple helices) was combined with the scaling in the percolation regimen to explain the rheological profile from the percolation concentration onwards (van der Linden and Parker, 2005). For globular protein (β-lactoglobulin, ovalbumin and BSA) derived fibrillar assemblies, gelation data have also been analyzed in terms of the aggregate anisotropy, and their interaction tuned by electrostatics (Veerman et al., 2002a, 2002b, 2003b, 2003c). Using the fibrillar aggregates, extreme low c_p could be obtained if gelation was being induced by a cold gelation step using Ca-binding (Veerman et al., 2003a). This was possible since fibrils being made at low pH maintained their form while adjusting the pH to higher values, thereby allowing a cold gelation step with Ca (Veerman et al., 2003a). The percolation concentration is found to relate well with the theoretical models by Philipse et al. (Philipse, 1996; Philipse and Verberkmoes, 1997; Philipse and Wierenga, 1998; Philipse and Kluijtmans, 1999), where we want to note that these models actually refer to caging of the particles being equivalent to the gel transition. The caging may be due to repulsion or attraction, since both interaction types are able to freeze out particle motions. See also the section on glass transition in relation to percolation. We note that in the case of fibrils formed from globular proteins it is not clear whether the fibril interactions are of a repulsive or attractive nature. Experimental evidence on this level is lacking still (EvdL is indebted to Dr. Taco Nicolai for discussing this point in the past). It may well be that the gelation of fibrillar systems is in fact not a pure percolation transition: (EvdL is indebted to Dr. Max Kolb for bringing this important point up in discussions on this issue). Instead one may think of it more as a caging phenomena, or related to a glass transition (see also section on glass transitions).

The scaling of G′ with concentration in the case of gelatin gels (te Nijenhuis, 1981a, 1981b; Djabourov et al., 1988a, 1988b) leads to an expected scaling of the low-frequency velocity of sound with volume fraction with an exponent ½, which was indeed also found recently by Bot et al. (1995). The authors also derive an acoustic fractal dimension of around 2 from these measurements, referring to the connectivity of the network. In aging gelatin gels, high-frequency rheological measurements (Bot et al., 1997) reveal much of the power law spectra according to Winter (1994), and discussed by Axelos and Kolb (Axelos and Kolb, 1990; Devreux et al., 1993). From the high-frequency measurements of G′ and G″at the gel point, one deduces a fractal dimension of 1.5 for the network, similar to the fractal dimensions of 1.1 to 2 obtained for fibrillar semi-flexible β-lactoglobulin aggregates, varying with ionic strength during preparation (Aymard et al., 1997).

In order to fully establish whether so-called scalar percolation theory does apply to gelation of a system or not, one should preferably check other scaling for that system. Scalar percolation theory predicts the divergence of the zero-shear viscosity, η according to $(c_p\text{-}c)^{-s}$, and the equal frequency (ω) dependence of both G′ and G″ according to $G' \sim G'' \sim \omega^{\Delta}$, together with the already given scaling of the static elastic modulus $G \sim (c\text{-}c_p)^t$. This was clearly pointed out by Axelos and Kolb (1990). They discuss that in the case of gelatin, t is within the scalar percolation regimen, while s is not (Djabourov et al., 1988a, 1988b). We refer to the work of Axelos and Kolb (Axelos and Kolb, 1990; Devreux et al., 1993) for an extensive explanation on how to analyze the rheological data in order to determine whether scalar percolation applies or not. In their paper, they show its utilization for pectin polymer solutions. In a later paper their approach is utilized by Devreux et al. (1993). Parts of these types of analyses have been performed on alginate gels by Matsumoto et al. (1992), where a relation between Δ and the fractal dimension of aggregates is checked according to predictions of Muthukumar (1985, 1989). The extensive way of analyzing the rheological data according to what Axelos and Kolb have done is seldom applied to protein systems, probably finding its reason in the fact that protein systems usually exhibit phase separation as they gel.

As an application of percolation theory in relation to gelation we mention a paper by De Kruif et al. (1995). For several milk protein systems they investigated the effect of interaction strength on the gel line and compared that to an expression by Chiew and Glandt (1983) that relates the interaction strength with volume fraction necessary for gelation (see also in this respect, Seaton and Glandt (1987)). They also mention the dilute regimen where 'the percolation line seems to follow the spinodal line'. The experimentally accessible interaction parameter plotted against the experimentally determined volume fraction of gelation reveals a good agreement between theory and experiment. The experimental gel line is determined using diffusing wave spectroscopy. The gel was caused by rennetting action and by acidification.

It was reported for protein systems that the point at which the $\tan(\delta)$ sharply decreases coincides with the point where the aggregate size diverges (Tobitani and Ross Murphy, 1997; Le Bon et al., 1999). This may serve as a practical definition of the gel point, although the more rigorously defined criterion of Winter and Chambon (1986) may be more attractive from a theoretical point of view (see however also remarks in the section on percolation about more rigorous cross-checks before deciding whether one actually has a critical point or not). For β-lactoglobulin, BSA, ovalbumin and α-lactalbumin it was reported that the power laws obeying the Winter criterion were not observed. Beyond the gelation point such behavior was observed for β-lactoglobulin gels (though the gels were not strictly speaking critical), and previously such behavior beyond the gel point was also reported for ovalbumin (Koike et al., 1998).

2.2.3 Linear Branch Description of Gel Networks

The critical concentration arrived at in a natural way using the percolation approach is an important experimental feature encountered in many biopolymer and food-gelling systems according to Clark (1987). As already said earlier, one can also arrive at a critical concentration for gelation from branch theories. For example, Clark derives such a critical concentration in terms of molecular weight of the building blocks and number of potential reaction sites by using a cascade approach. Such a critical concentration follows automatically from the assumption (originally proposed by Eldridge and Ferry (1954)) that the binding between two functional sites obeys a monomer–dimer equilibrium, as was already originally applied for the weak binding case by Hermans (1965). We refer at this point also to the classic review by Clark and Ross-Murphy (1987), to Clark and Ross-Murphy (1985) and to Chapter 1 of this book. We also refer to these references for an overview of the applications of branch theories up to 1987 and devote this section to recent advances in applying this approach to gelation. In principle, the irreversible branching theory refers to a possible number of reactive sites on each monomer (f), the proportion of sites reacted (α), and a critical value, α_f, for the required proportion in order to arrive at a gel (Clark, 1987), reflecting the critical concentration for gelation, c_0. This part of the theory refers to the structure of the network. Subsequently one connects the structure to the elastic response by assuming a rubber-like stress response, i.e. $G' = aT/V_{molar}$ with the constant, a, reflecting the deviation from the ideal rubber elasticity, R the gas constant, T the temperature and V_{molar} the molar volume of the primary particle. In order to fit the G′ data versus concentration according to the theory, one has to choose a value for f, and determine by least squares fitting the according values for c_0 and a, and subsequently determine the best overall fit and according best value for f (and accordingly best values for c_0 and a). This model can be elegantly extended to a biopolymer mixed gel (Clark, 1987). The time dependence of the modulus is less adequately described by this irreversible branching theory, predictions namely are a gel time proportional to c_0, but instead larger dependencies on c_0 occur experimentally (Kavanagh et al., 2000). This is due to more complex gelation mechanisms related to pre-aggregation and collective effects (Kavanagh et al., 2000). Alternatively Ross-Murphy has proposed for the time dependence and gel time, t_{gel} the following relationship: $t_{gel} \sim [(c/c_0)^n - 1]^p$ (see Kavanagh et al., 2000). This expression on the one hand allows an alternative determination from experimental data of c_0, which coincides reasonably well with the value as derived from the modulus versus concentration data (Kavanagh et al., 2000). It should be stressed that this expression for t_{gel} leads to fits of G′ versus time that are very satisfactory (Kavanagh et al., 2000). Furthermore, it is clear that the existence of a critical concentration in certain systems is an essential feature which is not present in so-called fractal theories (Kavanagh et al., 2000).

2.2.4 Colloidal Cluster Description of Gel Networks

As has been outlined in Section 2.2.1, one distinguishes that the formation of clusters is rate limited based on diffusion of 'activated' molecules (DCLA) or reaction limited. The first step in the cluster aggregation is of course the formation of the clusters themselves, with their structural features. The second step is the connectedness being formed by the clusters. In this section we treat the descriptions that do not take into account the fact that the formation of clusters and their connecting is in fact related to a phase separation process, and that one can have a multitude of different phases, since we treat that in another section. Here we focus on the work describing cluster structure, and gelation by means of connecting clusters.

At suitable pH, several food proteins form so-called fractal clusters that lead to gelation once the cluster concentration is high enough. For example, heating samples of bovine serum albumin (BSA), ovalbumin (Weijers et al., 2004) and β-lactoglobulin (Pouzot et al., 2004) forms gels if above a critical concentration. The structures that build up the gels are fractal clusters. In the case of BSA, it was recently found by applying light scattering (Donato et al., 2005) that the fractal dimension of the structure was not detectably affected by the concentration of salt (NaCl) in the range of 25–100 mM, and not affected by the type of salt (NaCl versus $CaCl_2$). The formation of particulates instead of fibrils, within a certain pH range (close to the protein's iso-electric point) has been recently proposed to be another generic feature of proteins, in addition to the feature of fibrillization (Krebs et al., 2007). Both features are suggested not to be dependent only on the specific amino-acid sequence but on how all molecular details together give rise to specific physico-chemical properties on a molecular scale, such as hydrophobic surface areas, charge distribution and dipole moments. For other type of systems, one has reported that the formation of clusters exhibits universality (Lin et al., 1989).

Recently, two papers on simulations have appeared by the Durand group (Rottereau et al., 2004a, 2004b) where it is shown that the growth rate of the large clusters in DLCA is exhibiting universal behavior upon scaling the time by the gel time, defined as the point of overlap of clusters. At low-volume fractions of particles, the gel structure is dominated by connectivity, while at high-volume fractions it is dominated by excluded volume interactions between the clusters.

The interactions between the colloidal particles determine the type of fractal structure, and the continuum between strong and weak is described using the Fuchs factor, related to the sticking probability of the particles. Low sticking probability implies the reaction limited case and high sticking probability refers to the diffusion limited case. Interestingly, the fractal parameters turn out (by simulations) to be less dependent on the sticking probability for higher-volume fractions (Mellema et al., 1999b). This was explained in terms of limited formation of depletion zones around the clusters at higher-volume fractions, hinting at transient phase behavior processes, which are more thoroughly described in Section 2.2.5. The detailed dependence on sticking probability is quite relevant for the description of casein gelation in milk (Mellema et al., 1999a).

In experiments on colloidal silica, it was found that there is a concentration-dependent crossover length scale below which there exists a fractal structure and above which the clusters behave as scatterers that are distributed randomly.

In emulsions, Bibette et al. (1992) showed that in the case of DLCA, clusters of nearly equal size are formed. In contrast, RLCA yields polydisperse clusters, similar to what are observed in percolation phenomena (De Gennes, 1976a; Stauffer, 1976). So it is clear that in the case of emulsions, one should deal with a different description for gelation then percolation on its own. Indeed, relevance of phase behavior and spinodal decomposition was alluded to in that emulsion work (see Section 2.2.5).

Relating the rheological properties to the concentration of building blocks can be done in various ways as outlined in the introduction and one popular way is to use the fractal dimensionality of the clusters building up the gel in relation to the volume fraction of building blocks. In fact, the consequent scaling factor then reflects part of the structure in the gel. Brown and Ball (1985) first proposed such scaling factor. Shih et al. (1990) extended these results and incorporated in their work elements of the work by Buscall et al. (1988) and Kantor and Webman (1984). The two cases described by Shih et al. have been connected by one another using a description with two other scaling exponents by Mellema et al. (Mellema, 2000; Mellema et al., 2002). Later on Wu and Morbidelli (2001) extended on the ideas further. The basic idea is that the elastic properties of the clusters are derived from an effective backbone of the cluster. One distinguishes between a regimen where the links between the clusters are stronger than the links within one cluster, and an opposite (weak link) regimen where the links between the clusters are weaker than those within the cluster. In the former case, the elasticity is determined by the links within the cluster (they give rise to the initial response of the system). One has to distinguish between the fractal dimension of the backbone and that of the clusters as two important parameters. Wu and Morbidelli have included one parameter α in addition to the model of Shih. This parameter can take the value between 0 and 1, indicating the relative importance of the inter- and intra-cluster links. Both the elasticity as well as the linear regimen are a function of the fractal dimension of the backbone and of the clusters, and of the parameter α. The backbone exhibits an intrinsic stiffness that will manifest itself upon larger deformations that probe the backbone itself. Thus one expects critical strains above which the backbone stiffness is probed, and this leads to strain hardening. A model relating strain hardening to the fractal geometry of the gel is treated in a paper by Gisler et al. (1999). The RLCA to DLCA transition as a function of interaction between proteins has been demonstrated nicely by Ikeda et al. (1999). In that paper also the critical strain as a function of fractal geometry is addressed, according to predictions by Shih et al. (1990). Vreeker et al. (1992) have shown that fractal dimensions as obtained by rheology are in accordance with light-scattering data, using the predictions by Shih et al. (1990) for a β-lactoglobulin system.

Diffusing wave spectroscopy (DWS) measurements have revealed that in a phase-separated emulsion/polymer system (Meller et al., 1999), which forms a polydispersed cluster phase, the modulus scales linearly with volume fraction particles. This was explained by the idea that the modulus is caused by the number of contact points between the clusters, since all the voids between the large clusters are filled up by the smaller clusters. For gels made by particles that interact strongly, the stress is carried by a strong backbone and therefore gives rise to a different scaling with volume fraction. Regular rheological measurements versus DWS measurements reveal a difference in frequency dependence of the modulus, caused by the different length scales being probed by the two methods.

In gelation by cluster–cluster aggregation, one also has to take into account that larger clusters may themselves show sedimentation or creaming. The criteria for gelation to occur before creaming have been worked out by Allain et al. (1995) and considered by Bremer et al. (1995) who also include effects of flow. The case of RLCA has been addressed for various systems in an excellent paper by Lin et al. (1990), where several different model systems exhibit the same cluster growth kinetics, pointing at universality of RLCA. For protein systems under certain specific circumstances, results by Feder et al. (1984) and Horne (1987) are interpreted by Lin et al. (1990) as belonging to the same RCLA kinetics, but at the same time the higher fractal dimension of 2.5, instead of 2.1, would point towards restructuring within the gels. Fractal dimensions between 2.2 and 2.4 are reported for two different casein gels (Bremer et al., 1990) though in the analysis the clusters are assumed to be monodisperse, in contrast to what is expected for RCLA. Monodispersity is mostly linked to DLCA (Krall and Weitz, 1998).

2.2.5 Phase Separation and Structural Transitions in Relation to Percolation, Aggregation and Gelation

In the sections above, several viewpoints on gelation and network formation have been formulated. One is a percolation approach that assumes that the network is based on links between the co-solutes in the solvent where the distribution of co-solutes is not changed, i.e., an equilibrium condition. In contrast, the colloidal cluster aggregation models do take the changing positioning of the co-solutes into account, leading to clusters, i.e. regions of colloidal particles glued together and clusters touching one another, which is a non-equilibrium situation. The glueing together of the clusters themselves may still exhibit a percolation type of transition. More recently, attempts to combine the two viewpoints have been investigated; not so much directly applied to protein systems, but the work is very illustrative of what one could do with proteins in solution.

In a paper by Poon et al. (1997), a model colloid system (PMMA) is studied with emphasis on the consequences of interactions between and concentration of the PMMA particles for the phase behavior. One distinguishes experimentally a crystal phase and a liquid phase, and their coexistence, which is an equilibrium situation. In addition, there are non-equilibrium phenomena. At higher colloid concentration, Carpineti and Giglio report, based on scattering measurements as a function of time, a spinodal type of behavior for intermediate times that is characterized by a time invariant term Q^{d}_{maxf} S (Q/Q_{max}), where Q denotes the scattering vector and Q_{max} its (time-dependent) peak position (Carpineti and Giglio, 1992), and where the fractal dimension (d_f) is found to vary with interaction strength. $2\pi/Q_{max}$ refers to a lengthscale characteristic of the size of a typical cluster plus their depletion zone (see for a good review on these matters by Poon and Haw (1997)). Such scaling was also found for 2D systems by Robinson and Earnshaw (1993). For strong attractive interactions, one finds that d_f can get as low as 1.7, while d_f increases for less strong interactions up to the value 3, the latter being reminiscent of classical spinodal decomposition, reflecting the dimensionality of three. The value 1.7 can be identified with the fractal dimension related to a diffusion limited aggregation (DLA) phenomenon. It has been reported that it is possible that at first one may consider a reaction limited aggregation with according fractal dimension, but that the behavior yields a cross-over towards diffusion limited behavior as the aggregation process has progressed (Asnaghi et al., 1992). Interestingly, there exists order in the system for the case of DLA, according to the values for Q_{max} but this order seems to be lost for the case of RLA (Carpineti and Giglio, 1993).

One could view the observations above by saying that if the interactions are strong enough, a system will undergo phase separation by means of a spinodal decomposition mechanism, where this mechanism is taking place with an efficiency more than what is the case for dimensionality 3 (where one needs to have the phase separating masses to fill up space completely, i.e. $d_f = 3$). The decomposition is taking place with less mass per volume necessary to effectively fill up the phase separating space, i.e. it is effectively filled with a dimensionality of 1.7 instead of 3. So, there is a bulk phase separating out, where this bulk phase contains less mass per unit volume than it would if the dimensionality is simply 3, i.e. what it would be if one simple fluid is separating out from another one. One may consider the 'fluid' that separates out being indeed a fluid separating out according to spinodal decomposition, but where this fluid is characterized by a dimensionality of 1.7, which refers to a fractal structure, instead of 3 which refers to a bulk and closed packed fluid. For lower concentrations, one observes a nucleation type of phenomenon, i.e. nucleation of clusters that show growth upon time, and at the same time a zone being depleted of colloids around the growing clusters. When the concentration is chosen a bit higher than is necessary for nucleation type of behavior, but lower than necessary for spinodal decomposition, one observes transient gelation, similar to the transition between spinodal decomposition and transient percolation as suggested to exist deep inside the region of two phase coexistence (Hayward et al., 1987; Lironis et al., 1990). Findings such as these are in line with observations on lysozyme systems reported recently based on optical microscopy (Sedgwick et al., 2005) showing bicontinuous structures, droplet-like or bead-like aggregates, larger aggregates and also 'percolating' clusters. Similarly, Bromley et al. (2006) have reported on the effect of temperature changes on aggregation rate and aggregate size for β-lactoglobulin systems which are known to aggregate upon temperature increase. They find spherical and 'well-packed' spheres instead of fractal spherical objects, implying a relatively slow nucleation growth mechanism instead of diffusion limited aggregation. The above non-equililbrium behavior was explained in terms of a so-called hidden gas–liquid binodal within the equilibrium crystal fluid coexistence region (Poon et al., 1995; Poon and Haw, 1997). For a more recent overview on phase behavior in colloidal systems in relation to gelation see Anderson and Lekkerkerker (2002). Interestingly, the idea of network formation by means of aggregated particles arising from spinodal decomposition has been addressed as early as 1985 by Miller and Miller (1985). Considerations such as whether the gel is actually equilibrium or not, and why there are two correlation lengths of which one does not converge at the gel point (in contradiction with 'pure percolation' theory) were already addressed by Martin et al. (1987). The effects of range of interactions have been addressed by Noro et al. (1999). An accessible introduction to the various basic elements of spinodal decomposition, nucleation, phase boundaries and gelation is presented by Verhaegh and Lekkerkerker (1997). Specific work on colloidal aggregation in emulsion systems in relation to spinodal decomposition and gelation has been given extensively in a paper by Poulin et al. (1999). These authors formulate an interesting viewpoint that the observed behavior in emulsion systems in fact suggests that 'DLCA is simply one limit of spinodal decomposition, that is achieved with a deep quench and a very low volume fraction of one of the phase separating phases'.

The above picture clearly relates colloidal aggregation (applicable to proteins) in terms of models yielding fractal clusters to phenomena that are relevant beyond the cluster approach, i.e. including depletion zones around the clusters and cluster–cluster structures. The equilibrium phase behavior is related to non-equilibrium phase behavior, and the structures being formed (followed by e.g. SALS) relate to how far the system is into the unstable region. We also gladly refer for recent developments to the review by Poon (1998) and to an illustrative article (Poon et al., 2002) that crystallization as found in systems of hard spheres and small depletants is present in protein solutions.

A somewhat related topic has been described by Anderson and Jones (2001) on the phase separation of gelatin and dextran, which exhibits also a spinodal decomposition which is yielding an interconnected structure of gelatin when the gelatin phase is allowed to gel at the same time during the phase separation process. Gelation, microphase formation and macroscopic phase separation appearing in associating polymer systems has been addressed theoretically by, for example, Tanaka (1996). For gelatin gels, experimental work was already been reported by Tanaka et al. in 1979 regarding spinodal decomposition and gelation, as induced in gelatin/water systems by alcohol (Tanaka et al., 1979), later followed by theoretical work on the same systems by Sciortino et al. (1993).

Interestingly in the light of the above is the work by Zilman and Safran (2005) who adapt theories from the above work on spherical colloidal phase separation and gelation to systems containing long filaments, such as actin filaments, in terms of a topological phase transition, analogous to the ideas on phase transitions in dipolar fluids by Tlusty and Safran (Tlusty and Safran, 2000; Tlusty et al., 2000), and which also may apply to protein-based fibrils (Aymard et al., 1996; Veerman et al., 2002a). Other work on biopolymeric structures regarding gelation by means of spinodal decomposition phenomena and in relation to other phenomena was reported by San Biagio et al. (1996b). These authors report that gelation at low concentrations is the result of three different processes occurring on different and/or equal timescales. They report that one can have gelation (as the topological phase transition by means of molecular crosslinking percolation) also below the value of random crosslinked percolation if one has correlations (interactions) between the solute molecules leading to what they refer to as 'a break of symmetry in the initially homogeneous sol'. The system under study was BSA. Similar work was performed on agarose gelation (San Biagio et al., 1990, 1996a). In a paper by Manno et al. (1999) the relation with amyloid type of diseases is put forth. In agarose systems, it was shown how the fractal dimensionality of the structures during the spinodal decomposition process increases until it freezes at the gelpoint (Manno and Palma, 1997).

The non-equilibrium transitions described above in relation to phase separation, network formation and percolation can be applied to spherically symmetric clusters but also as is clear from the later examples, to anisotropic structures such as gelatin.

One intermediate system of interest consists of proteins that give rise to formation of anisotropic structures, i.e. fibrils. One example is β-lactoglobulin-derived peptides assembling into fibrils. The network is being formed in two steps: one is assembly into fibrils, the second is the gel transition. The formation of linear aggregates forms an interesting special class of aggregates.

The above-described effects on fibrillization by temperature, pH and ionic strength reveal already three possible necessary molecular properties for describing the interaction between two aggregating proteins. These properties are: hydrophobicity, magnitude of charge and charge distribution, and screening of electrostatic interactions. Apart from the fact that temperature induces a conformational change that will expose hydrophobic regions, it will on itself affect the entropic contribution to the free energy of the system.

Light-scattering experiments during the early stages of assembly of β-lactoglobulin have shown that the assembly is reversible upon cooling (Arnaudov et al., 2003; Akkermans et al., 2008a), while at later stages (after at least several hours) the assembly would not be reversible upon cooling and remains stable under dilution. This irreversibility is also seen with ovalbumin (Veerman et al., 2003b) while for BSA the assembly was reported to be reversible at any time (Veerman et al., 2003c). In these latter two reports the samples (after cooling) have been diluted to establish reversibility (instead of using in-situ light scattering). The stability of the fibrils has also been tested upon pH change and it was found that the fibrils remain intact after changing the pH from 2 to 7 and 8. This was shown using microscopy (Veerman et al., 2003a). The general picture is that somewhat aged (a few hours) β-lactoglobulin fibrils are very stable against disassembly.

It is an important issue to establish whether the structures that are being formed during the redistribution of the proteins as they aggregate are actually equilibrium structures from a thermodynamic point of view or not. The aggregation of proteins into gel networks thus can be described based on (1) a non-equilibrium transient process of phase separation according to nucleation and/or spinodal decomposition type of dynamics, including the fractality of the aggregates, which are again related to the strength of the interaction, and (2) formation of equilibrium interconnected mesostructures. When one studies simple colloidal particles without any surface structure, the clusters they form are usually non-equilibrium clusters, in particular when they have a fractal dimension less than 3, since then the most probable structure is frozen. Once repulsive interactions are introduced, the picture becomes more complicated and there are all sorts of equilibrium situations possible.

Leaving aside for a moment the structural details of a protein, we would like to refer to an analogy with surfactant molecules. Their assembly into micelles (be it spherical or longitudinal) is also attributed to the three parameters: hydrophobicity, charge and charge distribution (Israelachvili et al., 1976), following ideas of Tanford (1980). In describing the assembly into either a spherical micelle, sheet- or rod-like geometry, Israelachvili et al. (1976) used as a first step the chemical potential in the assembly versus the chemical potential in the bulk as two parameters to arrive at important experimentally accessible parameters like size distribution of micelles and critical micelle concentration. The specific geometry allows them to write the chemical potential within the aggregate as a function of the interaction energy between the molecules, and as a function of end effects. The latter refers to the fact that the molecules at the boundary of the assembly have interactions with less molecules of the same kind as their interior colleagues. For a rod, only two molecules are involved in this end effect, while for a sphere, many more are involved. The second step is to give explicit expressions for the chemical potential. One differentiates between bulk and surface terms. The latter are calculated by taking into account repulsive electrostatic interaction and attractive hydrophobic interaction. The surface terms again will depend on the geometry of the assembly. One also finds criteria for the morphology of the assembly as a function of the 'geometry' (balance between hydrophobic and electrostatic contributions) of the surfactant (Israelachvili et al., 1976). Important in the surfactant assembly case is the fact that one has a surface tension contribution and a repulsive contribution.

An important distinction between proteins and surfactants is that surfactants have two distinct regions, a hydrophobic region and a hydrophilic region, for which the 'geometry' can be determined to a satisfactory degree (and thus the morphology of the assembly can be determined to a certain degree), while proteins in general do not allow a straightforward way of incorporating the geometry, since they may have hydrophobic patches distributed over the protein surface, and a complex surface charge distribution (leading to a dipole moment).

The simplest ansatz to try to model protein assembly would be to assume two features (Akkermans et al., 2007) to be of importance. One is the presence of an attractive interaction and the other is the presence of a repulsive interaction, and assume spherically symmetric potentials, thereby discarding any detailed protein information. Taking such an approach, Groenewold and Kegel (2001, 2004) have shown that if one considers spherical particles (and assuming no specific geometry, while the attractive interaction is of short range and the repulsive one of a long-range nature), one ends up having an equilibrium size of clusters of particles. The equilibrium cluster size is again determined by a balance between attractive and repulsive interactions. The features that are used have similarities with features used in describing the stability of a nucleus (Groenewold and Kegel, 2004). The possibility of equilibrium cluster size also was reported by Sciortino et al. (2004). Experimental observations in favor of such an equilibrium cluster size for proteins have been reported by Stradner et al. (2004) using lysozyme. The fact that the system of clusters dispersed in the liquid remains stable is due to: (1) forces acting within the cluster and (2) forces in between clusters. Increasing ionic strength results in a Debije screening length smaller than the cluster size, leading to cluster instability and precipitation (Stradner et al., 2004). The attractive interaction can be tuned by temperature, where it was found that lower temperature increases the attraction (Malfois et al., 1996).

Ultimately one would like to predict not only whether assembly takes place and whether assembly transitions into phase separation but also the form of the assembly (Akkermans et al., 2007). It is the number of degrees of freedom regarding the assembly of proteins that poses difficulties to predict form and size of the assembly accurately, etc. There is no model available as for surfactant molecules. However, it is important to remember that if one knows the form and size, remarkably accurate predictions can be made about temperature dependence and ionic strength dependence and on whether or not assembly will take place or not, as was shown by Kegel and van der Schoot in explaining the assembly behavior of hepatitis B viruses (Kegel and van der Schoot, 2004).

An attempt at describing aggregation of proteins into various forms has been reported, and it yielded criteria for the formation of stable fibrillar aggregates (Akkermans et al., 2007). In this work, a first attempt is made to explain morphology of protein aggregates, ranging from fibrillar to particulates, on the basis of a balance between electrostatic and hydrophobic interactions. That analysis gives an explanation why at high charges fibrils are expected, while at low charges and large hydrophobic interactions particulates are expected. The latter general feature of protein assembly was also put forward qualitatively by Krebs et al. (2007). This was later on more quantified by applying a nucleation growth model for particulates which formed with a narrow size distribution (Bromley et al., 2006). Experiments in this line have been reported, e.g. by Pouzot et al. (2005). Apart from equilibrium aggregate size and form, the non-equilibrium aspects have to be considered, i.e. phase separation, whether local or at a macroscopic level. In particular the local phase separation has to be considered in interpreting the light-scattering measurements (Nicolai et al., 2006).

These works may be applied irrespective of whether the building blocks are proteins or a sequence of peptides stemming from the proteins (see above). Interestingly, it is known that one may end up with equilibrium assemblies in surfactant lamellar phases even if the building blocks are not all having the same chemical fine structure (Sein et al., 1993). The fibrillization derived from β-lactoglobulin is possible also with a range of peptides instead of only one specific sequence (Akkermans et al., 2008c). At the same time, it is not at all clear whether the fibrillar assembly by the peptides in this case is an equilibrium structure or not.

A good example experimentally of how different circumstances give rise to different structures and where the results are compared mutually is by Aymard et al. (1996, 1997). The structures range from fibrillar to particulate fractal structures, and in between different degrees of branching. The different degrees of branching can be tuned by ionic strength. These structures are, however, most likely non-equilibrium. The effect of branching and strand thickness has been demonstrated already some while ago by, for example, Hermanson (1986). The effects of ionic strength have also been addressed by Bromley et al. (2005). They conclude that for zero ionic strength, there are several monomers necessary before they can form a stable fibrillar state, while for 50 mM NaCl the fibrils grow from monomers upwards. Also, shorter and more flexible fibrils are observed, in line with the findings of Veerman et al. (2002a, 2002b). These results are also in line with later published results of Ikeda and Morris (2002). Techniques such as confocal microscopy might serve as a valuable tool to characterize the topology of the network structure, as was demonstrated nicely by Dinsmore et al. (2001), distinguishing loops and strands, and applying the analysis also to determine spatial correlation functions within the network, complementing light-scattering results.

Apparently, some main features of protein assembly can be traced back to a colloidal description and discarding structural details on a protein scale. In order to develop further understanding regarding the influence of structural details, one route would be to investigate protein crystallization, as crystallization and crystal structure are directly related to molecular details. Indeed, as Piazza (2000) remarks in an excellent review on proteins in solution, 'Isotropic potentials are intrinsically unable to describe crystal nucleation and growth, and anisotropic interactions (physically related to hydrophilic–hydrophobic "patches" on the particle surface) might also change the overall nature of the phase diagram'.

We will now discuss in a little more detail some progress on protein crystallization, in particular with the aim of outlining possible connections to phase behavior of proteins in solution more generally, and to show results of work on phases found in the vicinity of the crystalline phase. As mentioned above, crystallization is reported to be a region within the phase diagram which to a large extent can be described while discarding molecular details. Because of the promising results in protein crystallization while discarding molecular details, we want to treat shortly this field, in particular because the approaches may be relevant for understanding the origin of adjacent regions in the phase diagram such as the nucleation and spinodal decomposition regions, that are relevant for protein gelation.

It was found by George and Wilson (1994) that favorable crystallization conditions in dilute protein solutions corresponded to a value of the osmotic second virial coefficient, B_2, which does not change much for different types of protein. They referred to the range of values of B_2 favoring crystallization as the crystallization slot. Vliegenthart and Lekkerkerker (2000) predicted the gas–liquid critical point from the second virial coefficient (calculated B_2) for a variety of short-range attraction potentials and showed that there exists a meta-stable liquid–liquid phase separation line, and that this line is rather flat over a large range of protein concentration. Their value of B_2 is for any interaction found to be in the order of what George and Wilson found. Vliegenthart and Lekkerkerker also refer to an earlier proposal (ten Wolde and Frenkel, 1997), and Haas and Drenth (1999) of the crystallization route occurring according to a two-step process, i.e. formation of regions high in protein concentration and subsequent crystallization near these regions high in protein concentration. Once below the metastable liquid–liquid line, one may enter a spinodal decomposition phase (further away from the liquid–liquid line, leading to localized densification of protein due to fluctuations) or a droplet phase of densified protein regions dispersed in the liquid (nucleation type of phase). Subsequently, the droplets may act as a heterogeneous nucleation site for crystallization, as the likelihood of crystallization is in this case enlarged due to the increased number of proteins available to find the right mutual orientation. The experimentally observed boundaries for optimizing crystallization coincide with simulation data (Poon, 1997; Poon and Haw, 1997).

It may also be that the interface of these droplets exhibits characteristics that enhance the nucleation as suggested by Kuznetsov et al. (2001). By way of an example of how an interface can induce polymerization, actin under conditions where it does not polymerize in bulk is found to polymerize at the gas–liquid interface (Renault et al., 1999). When the virial coefficient is reaching large negative values, the according larger distance to the critical point leads to precipitation, implying less favorable crystallization conditions. Similar observations have been made in colloidal phase diagrams (Anderson and Lekkerkerker, 2002). In addition to the references already mentioned on concentrated protein droplets, such droplets of concentrated protein are also mentioned in studies of protein crystallization (Kuznetsov et al., 2001) and protein-rich droplets are also observed as beads, where in this particular reference the droplets are reported to be of a solid-like nature (Sedgwick et al., 2005). In this latter reference, it is also suggested that these droplets may act as heterogeneous nucleation sites for protein crystallization.

The equilibrium clusters discussed above, composed of colloids due to a delicate balance between longer-range repulsions and shorter-range attractions, are also speculated to be pre-nucleation clusters in crystallization by Stradner et al. (2004). We conclude that the speculation of highly concentrated protein regions in the form of droplets that lead to crystallization is a valuable one, but it remains to be determined whether these highly concentrated protein droplets are in fact an equilibrium cluster or part of a metastable droplet phase. As was argued by Groenewold and Kegel (2001) analogous to surfactant micelles, the size is a balance between (in this case long-range) repulsive and (in this case short-range) attractive interactions. The possibility of equilibrium cluster size was reported by Sciortino (2004). Experimentally these equilibrium configurations have been reported recently by Stradner et al. (2004) for lysozyme. The system equilibrium comes from the suitable droplet interactions as well as the forces within the droplet itself. The attractive interaction between the droplets can be increased by salt addition. The droplet attractive interaction also can be tuned by temperature (Malfois et al., 1996).

In regards to the phase behavior of protein solutions, it should be mentioned that apart from droplet phases as equilibrium phases, and crystals, one may envision other topologies that could be in equilibrium. Some years ago Tlusty and Safran (2000) proposed a model for a dense colloidal phase of a bicontinuous morphology, in equilibrium with a dilute colloidal gas phase. This model is in particular applicable to anisotropic particles, exhibiting a dipolar interaction (originally it was devised to be valid for magnetic dipoles). It therefore is possible that at higher concentrations an equilibrium bicontinuous network phase can be found in protein solutions, though this has not yet been claimed experimentally. Interestingly, experimentally linear chains and interconnected topologies at higher-volume fraction of particles have been observed in systems of two oppositely charged colloidal particles. Their formation is attributed to the dimerization of the oppositely charged particles to yield dipoles, that subsequently form linear chains. The structures that are being formed are very dilute structures (Kim et al., 2003).

Parts of the above described important colloidal aspect of protein crystallization, i.e. the existing strong correlation between osmotic virial coefficient and metastable liquid–liquid phase separation has also been explored in terms of solubility, as reported by Guo et al. (1999). In this way, examples of earlier and later work on effect of type of salt versus solubility (Green, 1931, 1932; Green et al., 1935; Linderstrom-Lang, 1953; Edsall and Wyman, 1958; Foster et al., 1971, 1976; Arakawa and Timasheff, 1985; Iyer and Przybycien, 1994), hydrophobic interactions (Melander and Horváth, 1977), dielectric properties (Ferry and Oncley, 1941; Oncley, 1942), morphology (Edsall, 1943), hydrophobic patch distribution (Lijnzaad et al., 1996a, 1996b; Patro and Przybycien, 1996; Preißner et al., 1998; Dima and Thirumalai, 2002; Levy et al., 2003), surface area involved in association (Carugo and Argos, 1997; Conte et al., 1999; Cudney et al., 1994; Kam et al., 1978; Larsen et al., 1998; McPherson, 1990; Mikol and Giegé, 1992; Mikol et al., 1990; Reiss-Husson, 1992; Rosenbaum and Zukoski, 1996; Rosenberger et al., 1996; Takahashi et al., 1993; Vekilov and Alexander, 2000) could also be linked to the more recent work on equilibrium cluster sizes, pre-nucleation clusters, etc. In this respect, the Hofmeister series for example (Melander and Horváth, 1977) also becomes relevant for protein crystallization as also mentioned recently by Piazza (2000). Its relevance is also clear for the phase behavior of proteins in solution in general, both referring to equilibrium as well as non-equilibrium circumstances. A more recent example of the effect of ionic strength in combination with sucrose on the heat-induced gelation of BSA shows that the final gel structure and its properties can be affected by such factors as sucrose concentration. This is analogous to effects of co-solvents (urea 4–5 M leading to fibrillar structures of β-lactoglobulin) and hydrotropes on protein gelation (alcohol, work of Ross Murphy on beta-lac aggregation into fibrils (Gosal et al., 2002b)). The role of protein hydrophobicity relevant to solvent quality is important to protein assembly (Fernandez-Recio et al., 2005). Relevant to this consideration is how to measure the hydrophobicity of proteins, which was addressed by Urry et al. (1992).

The non-equilibrium considerations and its kinetics have to be regarded in parallel to descriptions of spinodal decomposition in phase separating polymer solutions, such as protein/polysaccharide solutions. Also these systems consist of relatively small concentrations of colloid of polymer and large amounts of solvent. In such systems the kinetics and the driving forces are observed to be of a different magnitude compared to normal liquid/liquid spinodal decomposition. Interesting effects of low interfacial tension and the influence of colloidal concentration (which becomes transient itself) on this tension will complicate the description. These effects have not been incorporated yet, but over the last decade considerable progress has been made regarding the description of such aqueous colloid interfaces (see e.g. Scholten et al., 2005). In a more complete description of the kinetics of spinodal decomposition into transient networks with fractal dimensionality the knowledge on interfacial properties of such phase separated aqueous systems should be incorporated. These works are however beyond the scope of the current review.

Many mixed colloid systems will exhibit the non-equilibrium phase behavior as described above. Differently sized colloids will yield phase separation and if the resulting clustering colloids stick, gelation, transient percolation and phase separation may occur.

In addition, of course, mixed biopolymer systems of which one of the components yield gelation upon an external trigger, will show even more complicated phase behavior and the resulting morphologies can be of a rich variety. The presence of a polysaccharide can induce depletion flocculation and the concentration enables one to tune the interaction potential. Indeed, presence of kappa-carrageenan in β-lactoglobulin systems enhances gelation rate, while at the same time the structure does not seem to be affected (Croguennoc et al., 2001). We leave a review of gelation for these type of systems out of this chapter.

2.2.6 Arrested States and Gelation

Gelation of particle gels has been regarded as the formation of an 'attractive' glass for some years now (Dawson, 2002), since the glass transition laws and ideas also seem to be describing the density correlation for particle gels upon approaching the gel point. This is in particular true for particles that exhibit a short-range attractive interaction. Before that it was already found that dense and repulsive particle systems exhibit a glass transition (van Megen and Underwood, 1993). In such a system of hard spheres, the glassy transition can be characterized by more and more caging of particles by their neighbors, as the glass transition is approached. At a critical density the caging becomes optimal, and the optimal caging stops that caged particle from exhibiting longer-range motions. The system then has become non-ergodic, and glassy.

In his review, Dawson treats colloidal systems with repulsive interactions, like hard sphere fluids, which show a glass transition as they simply cannot move any more above a certain volume fraction. Subsequently he explains the history on systems that exhibit a glass transition and that are composed of hard spheres with a short-range attractive interaction, and which cannot move '... because they stick'. Such sticky hard spheres have even been reported to exhibit two glassy states, due to their sticky character (Pham et al., 2002, 2004; Puertas et al., 2002). In fact, one glassy state is characterized by the fact that the particles are caged by means of excluded volume effects (the hard sphere case) and the other case is characterized by the fact that the particles are stuck, due to the forming of a gel (attraction induced). The aging behavior of the fractal colloidal gel has similarities with the aging characteristics of disordered glassy systems. There have been attempts also to apply these findings to gelation of globular proteins in solution because of the idea that proteins often exhibit a range of interaction that is smaller than their size (see Dawson (2002) and references therein) and therefore can be considered as sticky hard spheres. Another way of relating the glassy state to protein gelation has been outlined by Normand and Parker (2003) where the analogies are pointed out between the dynamics of gelatin gelation and spin glass behavior. Some examples on the usefulness of the glass transition analogue can be found in work by Segrè et al. (2001) where the gelation of polymethyl-metha-acrylate spheres is affiliated with the kinetic arrest of particle clusters forming a fully connected network, implying a non-ergodic network. The kinetic arrest is found to be a function of the interaction depth, and of the volume fraction of particles, and both are interrelated (Segrè et al., 2001). A critical volume fraction of particles is given as an exponential function of the interaction strength at contact. Recent developments of light-scattering techniques in order to characterize these type of transitions are well described in a review by Cipelletti and Ramos (2002). The gelation being linked to the kinetic trapping of particle clusters is related to the fact that the clusters are trapped in between the other clusters: the clusters are being crowded or in other words, they jam. Such a jamming transition has been explored in general terms by Liu and Nagel (1998) to depend on interaction strength between the particles, their volume fraction, and stress being applied onto the system (e.g. flow). In a paper by Trappe et al. (2001), it is beautifully demonstrated how the viscosity, elastic modulus and stress all scale according to critical-like behavior, as also referred to in the section on percolation. The critical volume fractions were also a function of the interaction depth. The results yield a so-called jamming phase diagram for such colloidal systems (see also a paper by Trappe and Weitz (2000)). The gels that are in a jammed state can be rather fragile, as is argued to be the case in general of all jammed systems by Cates et al. (1998) and it is this fragility that also gives rise to spontaneous collapse of these systems, as was investigated by Kilfoil et al. (2003). More recently similar results on critical volume fraction for gelation versus interaction strength has been reported on emulsion gels (Berli et al., 2002). The concepts of jamming, glass transitions and the relation with transient gelation have been nicely summarized by Prasad et al. (2003).

2.2.7 Modeling Gel Networks and their Rheological Behavior: Conclusions

In the sections above, several viewpoints on modeling of network formation and according gelation have been formulated.

The first aspect is the structure one obtains (size and morphology) as a function of the molecular interaction, the second aspect is which gel structures one can distinguish and how they can be obtained (including phenomena such as nucleation, spinodal decomposition, glass transitions, syneresis, and combinations thereof) and the third aspect is how the structures and their mutual interactions lead to rheological responses. Table 2.1 summarizes which structures and morphologies one may distinguish and what the key elements are to obtain them. This is an attempt to present an overview and key elements of the concepts that have been treated in the previous sections. We note that the term equilibrium refers to a minimal free energy of the system as a whole. Furthermore, the term topology refers to the characterization of the surface, i.e. the level of interconnectedness and its two principal curvatures (Jones, 2004).

Table 2.2 outlines which gel structures one can distinguish and how they can be obtained.

During gelation, the structures may grow and lead to thickening of the structure in time. This is clearly non-equilibrium and in contrast to the equilibrium case where structures are being formed which are defining the minimal free energy of the system and subsequently are made to interact by, for example Ca bridges, change in salt concentration, etc.

The structure of a gel is interconnected (meaning exhibiting a density–density correlation), either by means of repulsive interactions between the building blocks, or physical/chemical attachment.

The various models relating structure with rheology relate to various assumptions (see also Table 2.3). One is a percolation approach that assumes that the gelation is associated with the

TABLE 2.1 Morphologies and Details of Aggregate Structures and the Key Elements to Obtain them

Morphology	Key elements	Details of structure
Asymmetric Linear strands Linear tubes Concentric tubes	Specific interactions leading to anisotropic structures, in an equilibrium or non-equilibrium situation. Interactions may be specific due to strong dipoles, specifically situated hydrophobic sites (may act as an intermediate step of aggregation), and beta-sheets	Strands may be curved or not Tubes, closed strands or any other topology determines also partly the free energy of system and thus topology is a parameter that should be accounted for in models
Asymmetric Branches	Either seen as defects (implying still an equilibrium nature, or non-equilibrium aberrations (slow growth onto fibril)	Number of branches per unit strand length may be an equilibrium property but need not be. It also may depend on growth rate under non-equilibrium circumstances, and different growth rate for longitudinal versus transversal growth. These growth rates may also depend on protein concentration.
Asymmetric Multiple strands around one another, and or branches	Minimization of number of hydrophobic sites e.g., per area, may still be equilibrium, but not necessarily	
Asymmetric Multiple strands bundled or branches	Separating out of a bundled phase versus a single fibril phase with or without monomers	
Symmetric Isotropic structures Spherical dense Spherical fractal	Equilibrium (small dense spheres), but usually non-equilibrium, with density a function of interaction strength and nature is fractal due to diffusional growth pattern and non-specific interaction.	

TABLE 2.2 Gel Structures, their Building Blocks and how to Obtain them

Building block morphology	Gel structure	Details and variations
linear	Repulsive or attractive interaction, rods isotropically arranged	Linear may be rod, or flexible Linear may be non-flexible, while being curved or being straight (curved can also be stiff)
linear	Attractive interaction induced, giving sticky rods (mostly fractal aggregates of rods at lower concentration); Higher concentration isotropic sticky rod system	
branched	Branches may act as crosslinkers more than sticking between the strands themselves	Same variations apply as to linear, where the branches just dangle along with the strands that do most of the interaction leading to the gel structure

TABLE 2.2 Gel Structures, their Building Blocks and how to Obtain them—*Cont'd*

Building block morphology	Gel structure	Details and variations
branched	The network may grow by both growing branches and new branches and by linkages between strands	Linkages between strands may look like additional branches
spherical	At high enough concentration the equilibrium clusters may form a colloidal glass The non-equilibrium clusters may grow until they meet and form a connected structure That connected structure may further densify upon time by syneresis or by growth from proteins still in solution	One obtains an interconnected structure that also is inhomogeneous in thickness, since the growth process onto the original structures is random
Non-spherical, bicontinuous phase, one interconnected phase is gel like (glass) and the other may be either fluid or glass like	Densified gel material as continuous phase within fluid phase	Starting at high enough protein concentrations for binodal decomposition
Non-spherical, bicontinuous phase, one interconnected phase is gel like and the other may be either fluid or glass like	Densified gel material as continuous phase within fluid phase	Starting at high enough protein concentrations for binodal decomposition
Asymmetric densified structures stemming from binodal decomposition which is frozen due to gelification of the densified structure (transport and gelification at the same time or short after one another, i.e. high enough concentrations of protein above where there is spherical growth of domains	Glass type of bicontinuous structure	

formation of an infinite network, where the network is based on links between the co-solutes in the solvent, and where the system does not exhibit a phase separation before the transition takes place. The transition is characterized by the fact that there is a critical concentration below which no network, and above which one and only one network exists, no matter what the size of the system. Before transition aggregate clusters are formed; however, they are assumed to be distributed evenly through space and not form a space-spanning cluster. This is all in contrast to the colloidal cluster aggregation models that do take the changing positioning of the co-solutes into account, leading to clusters, i.e. regions of colloidal particles glued together and clusters touching one another, which is a non-equilibrium situation. The glueing together of the clusters themselves may still exhibit a percolation type of transition, but then the clusters form the primary particles in the percolation description.

More recently, attempts to combine the two viewpoints have been investigated. Not so much directly applied to protein systems, but the work

TABLE 2.3 An Overview of Different Models Relating Structure with Rheology, Together with Characteristic Quantities

Theory and scaling	Critical concentration	Structures and scaling	Prefactor and scaling	Experimental
Percolation Visco-elastic quantities $\sim(c\text{-}c_p)^{\alpha}$	Critical concentration exists Depends on interaction strength and morphology of particles. Examples on gelatin, fibrils, model systems. Care should be taken that it concerns a connectivity transition which leads to gelation, i.e. in that case it is a phase transition	Scaling exponent depending on interaction type (isotropic bonds), level of in-homogeneity, and on how the building blocks are connecting together. Structures (infinite cluster and other clusters) exhibit fractal dimension, different whether before or after the gel transition	Prefactor not generally articulated in any of these models. Beyond the critical concentration the prefactor has been estimated correctly by scaling arguments and deflection length of helix for the case of gelatin	Experimentally one should follow the scheme as described by Axelos and Kolb to really establish whether it is percolation or not. Less rigorous is the Winter criterium It is difficult to check all points of the Axelos Kolb procedure experimentally. This hampers progress experimentally regarding the theoretical description Applicable to systems that are built of building blocks that do not exhibit phase separation but instead stay in solution also during the gelation
Fractal theories, phase separation via nucleation Visco-elastic quantities $\sim c^{\alpha}$	No critical concentration This stems from the argument that cluster size depends on concentration and that the more dilute the larger the cluster becomes. This is not completely right since one interchanges the limit to zero concentration and the limit to an infinite large system (connectivity criterium becomes gelation only for an infinite system, strictly speaking). If one takes first the limit to zero concentration, there will never be a gelation concentration in the first place from a phase transition point of view	Refer to systems that are phase separating (cluster being formed due to attractive interaction) and this phase separation cannot be brought in line with a phase transition for interconnectivity	Scaling factor depends on type of cluster, and strengths of bonding within the cluster (which determines the stiffness of the strain baring chains) and between clusters. Also here no prefactors, the relative strengths are reflected in the scaling factors	If one considers the fractal clusters as main objects, gelation may be viewed as a percolation transition of clusters forming an infinite network, i.e. in that case one rescales the argument and the fractal 2-phase theories are linked again to connectivity and a real connectivity transition by means of the percolation idea. This is hard to test experimentally though

Jamming theories	Critical concentration exists and is related to bond strength	Transient structures, flow is included within parameters. Also applies to glassy systems	Bond strength explicit in critical concentration, not in prefactor	Applies under flow conditions, examples throughout different systems
Effective medium theories	Determined by concentration above which one obtains phase separation via spinodal decomposition	Applicable to the bicontinuous and single continuous structures arising from phase separation		Strength of fractal gel network as the continuous phase should be known as input parameter, filler theories apply to inhomogeneous samples
Fracture theories	Far beyond criticality usually	Bond strength needed (elastic and fracture), and permeability (for dissipative processes)		

is very illustrative of what one could do with proteins in solution. If the interactions between particles are strong enough upon the initial transition leading to gelation, a system will undergo phase separation. This will occur by means of a spinodal decomposition mechanism at high enough concentrations. During this decomposition, the particles will stick to one another, and because of this sticking, the resulting network is characterized by a fractal dimension less than 3, i.e. this mechanism ensures a more weight-efficient filling up of space compared to the case of dimensionality 3 (i.e. for spinodal decomposition with ordinary fluid molecules which do not exhibit much sticking, and therefore leading to a fractal dimension $d_f = 3$). In other words, the decomposition is taking place with less mass per volume necessary to effectively fill up the phase-separating space. One may consider the 'fluid' that separates out being a fluid with dimensionality of a fractal structure, with dimension less than 3, which refers to a bulk and closed packed fluid. For lower concentrations one observes a nucleation type of phenomenon, i.e. nucleation of clusters that show growth upon time, and at the same time a zone being depleted of colloids around the growing clusters. When the concentration is chosen a bit higher than is necessary for nucleation type of behavior, but lower than necessary for spinodal decomposition, one observes transient gelation, similar to the transition between spinodal decomposition and transient percolation as suggested to exist deep inside the region of two-phase coexistence. Though speculative at the moment, this statement may bare relevance to the observation that one finds a smooth transition from a fine stranded gel network at low ionic concentration to a particulate gel at higher ionic strength, where in between one observes a maximum in G′. This maximum may well be related to the transition between nucleation and spinodal decomposition.

In summary we have for low concentration and/or low attraction a nucleation type of behavior, and for high concentration and/or higher attraction a spinodal decomposition with fractal dimension related to the strength of the interaction and specific directional dependence. The former (nucleation type) is equivalent to the fractal particle cluster–cluster aggregation. In the former case, the space-spanning network consists of a series of fractal objects glued together. In the latter case (spinodal decomposition) one ends up with one space-spanning network which is continuous and exhibits a fractal dimension, which is determined by the strength of the interaction induced upon gelation.

In using either percolation, fractal clusters that stick (the cluster arising from a nucleation type of phase separation) or spinodal decomposition, the particle motions within the network are tacitly assumed to be zero. In contrast, another possibility would be that short-range motions are still allowed but that the long-range motions are frozen out in the gel state. In other words one could envision the particles to be caged in between their neighbors. The system then has become non-ergodic and glassy. The existence of such caging of particles could also be referred to as a specific class of gel. Such caging is dependent on interaction strength and volume fraction, and one can identify a critical concentration, which is dependent on the interaction strength. This type of gelation seems to be mostly applicable to short-range interactions between proteins, at relatively high protein concentrations in the case of spherical particles but may also be relevant to lower particle concentrations if the particles are anisotropic.

One theory refers to a one-phase system (real phase transition then regarding connectivity, percolation theory applies, and critical concentration depends on morphology and interaction between building blocks) and one refers to a two-phase system (described rheologically at best by the fractal scaling theories), reflecting the structure in a reasonable way. Both of these theories do not incorporate the bond strength. This is relevant to the fracture, but implicitly also relevant to things like bending elasticity of the strands. A theory that relates bond strength between molecules with the strength of the network lacks in both of these theoretical approaches. This might be an interesting avenue to investigate in the future.

Some ways of describing how molecular interactions are related to the type of aggregate structure and subsequent gel structure are available (but are far from being complete, and experimental evidence is not overwhelming). Describing the phase separation window in the phase diagram is available to quite some extent (but not much on how one travels through the phase diagram, i.e. the kinetics is mostly lacking). The models on gel strength as a function of the structure mostly refer to scaling relations and sometimes also to the strength of building blocks. Relationships between the molecular-scale bond strength and the gel strength, via the structure, are usually not made.

Regarding the various types of structures that can be formed, from a molecular scale up towards the gel structure scale, as depicted above, we refer also to Figures 2.2 and 2.3 in which the various structural scales are summarized, and what phenomena are important for a typical structure.

By way of example, we may go through Figures 2.2 and 2.3 for the case of a particle network. One may deduce the following steps:

1. The system starts in equilibrium with proteins evenly distributed in solution with repulsive forces exceeding attractive forces.
2. Something changes (proteins unfold or solvent quality changes), and the attraction becomes greater than the repulsion.
3. Primary flocs are formed.
4. Primary flocs associate into a gel network.
5. The rheological properties of the gel network depend on: interactions among molecules within the flocs, interactions between flocs and concentration of flocs forming a continuous network.

Step 3 may arise via a nucleation mechanism, i.e. at moderate to low concentrations. If

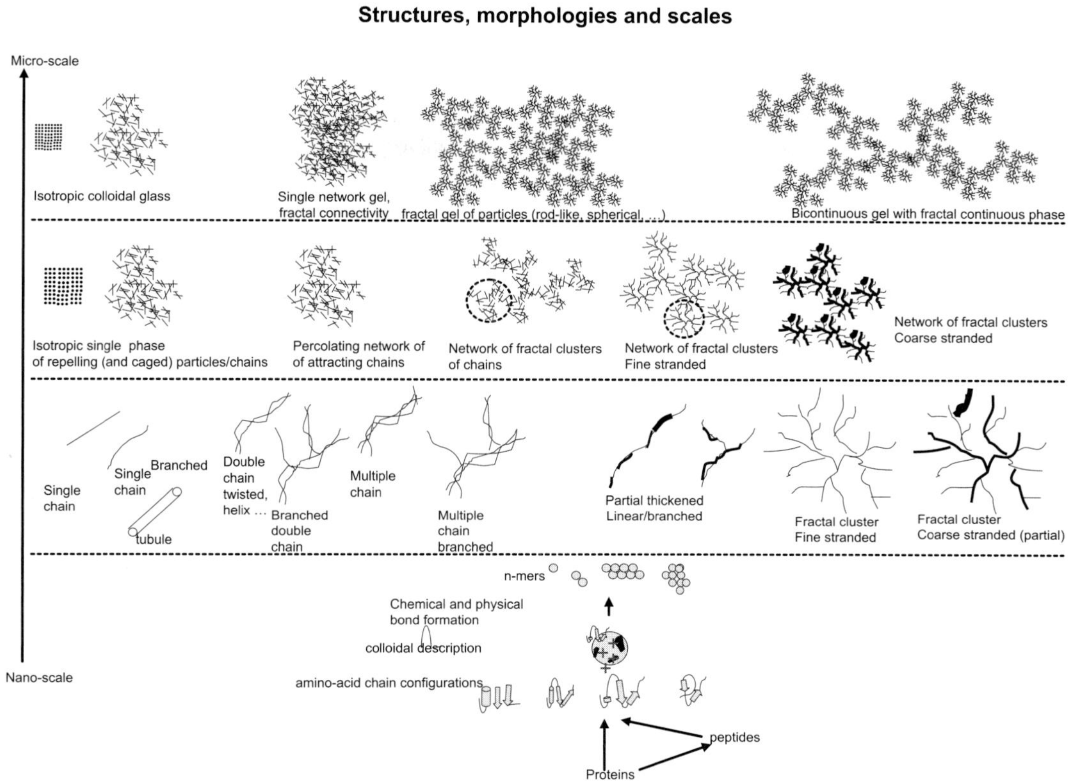

FIGURE 2.2 A schematic overview of the possible protein (assembly) structures at different length scales.

at higher concentrations (and short-range interactions) one finds a spinodal decomposition mechanism by which the system proceeds to separation. This may lead to gelation when the particles stick to one another during the decomposition. If the interactions are non-isotropic, an asymmetric primary floc may arise at step 3. This may occur via a nucleation mechanism, or the assemblies are equilibrium assemblies. The floc may grow linearly, or with branches, either equilibrium or non-equilibrium. The branches may connect and monomers may also assemble onto the primary anisotropic flocs. Whether the strands form independently of the network or not depends on whether:

(a) formation of strands occurs following equilibium arguments, and subsequent network formation by these strands by means of interaction (repulsive or attractive), or
(b) network formation occurs by means of the branches linking the strands, where the branches still result as defects from the strand formation (still an equilibrium possibility following the free energy minimum idea). The network can be an equilibrium situation where the number of branches is determined by the free energy minimum, or
(c) aggregation into asymmetric particulates occurs (tuned by a stronger electrostatic interaction favoring asymmetric particulates) which can either floc together (fractal aggregation of asymmetric particulates) or grow until touching one another and at the same time allowing for

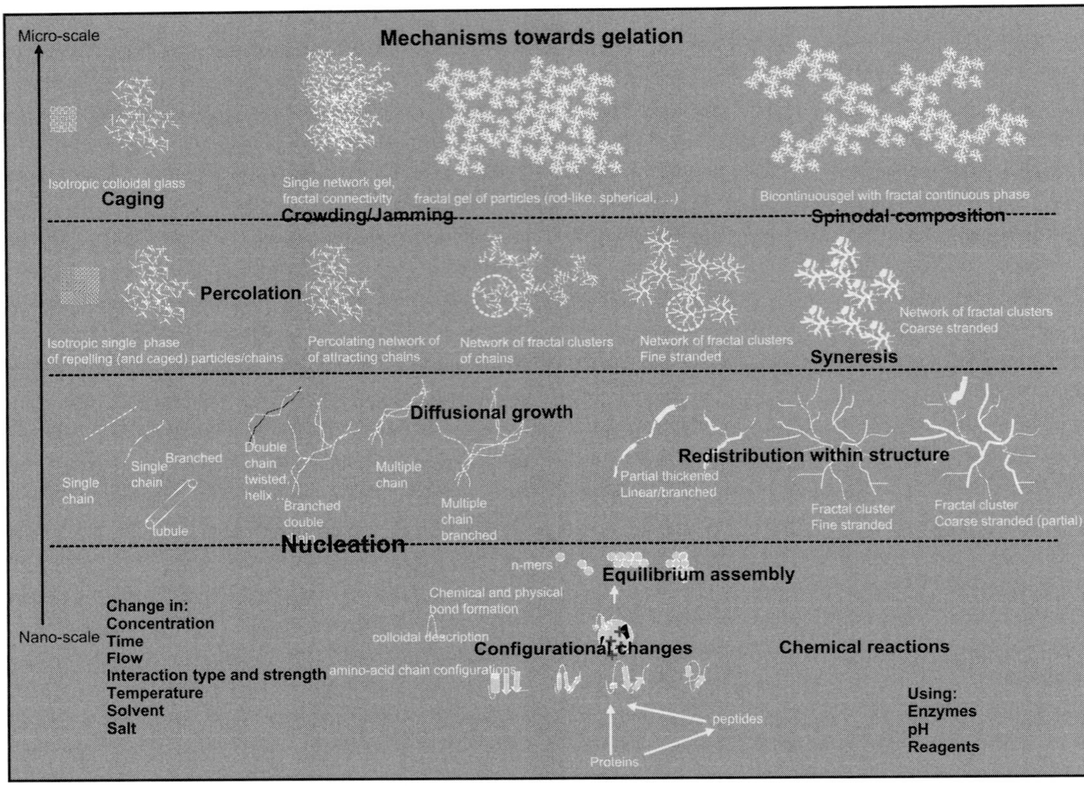

FIGURE 2.3 An overview of the different mechanisms of structure formation.

growth in thickness of the asymmetric particulates or

(d) network formation occurs because of particulates flocking together and where one observes stress-baring strands, etc., but these are the effective structures arising because of the network being formed in the first place.

2.3 MOLECULAR MECHANISMS CAUSING AGGREGATION/GELATION

2.3.1 Introduction

The first step in protein gelation involves a modification of protein structure, or solvent conditions, such that protein–protein interactions are favored. The interactions linking the protein molecules into a gel network can either be covalent (disulfide or iso-amide), non-covalent (hydrogen bonds, hydrophobic interactions or electrostatic interactions) or a combination of both. In many cases, it is a combination of covalent and non-covalent interactions and this prevents the classical categorization of physical (non-covalent) or chemical (covalent) gels (see discussion in Section 2.2.1). Likewise, separating gels into particle and polymer gels becomes difficult when disulfide bonds are present. Protein gels most resemble particulate gels with surface interactions linking individual protein molecules. However, when disulfide bonds are considered, the possibility of forming covalently linked polymer strands exists. One could imagine a chain of protein monomers linked by disulfide bonds interacting with other disulfide-linked chains through non-covalent interactions. Indeed, it has always been the range of interactions that can occur among proteins that makes them more complex when compared with other biopolymers such as polysaccharides. There are, however, exceptions. Casein gels formed by decreasing pH and cleavage of the κ-casein molecule appear to be solely particle gels. In addition, gels formed by extensive hydrolysis of whey proteins can be linked by non-covalent interaction (Doucet et al., 2003a). The general mechanisms that cause or contribute to gelation are denaturation (i.e, unfolding), refolding and enzymatic and non-enzymatic hydrolysis.

2.3.2 Denaturation

The unifying element of the denaturation process is that protein structure is changed without altering the primary sequence (hydrolysis) or modification of amino acids (e.g., deamidation). This is consistent with a definition provided by Neurath et al. (1944) as 'Denaturation is any non-proteolytic modification of the unique structure of native protein, giving rise to definite changes in chemical, physical, or biological properties.' Proteins can be denatured by changes in pH, solvent quality (e.g., ethanol) and temperature, in addition to addition of co-solutes (e.g., urea or guanidine hydrochloride) (Neurath et al., 1944). The prevailing cause of denaturation relevant to food protein gels is heat and therefore that will be the primary mode discussed here.

When globular proteins are progressively increased in temperature, the resistance to denaturation can be viewed as competition between the strengthening of the solvophobic effect stabilizing hydrophobic clusters of amino acids in the folded protein and the conformational entropy gain upon unfolding (Dill et al., 1989). The denatured state has been classically approached as a reversible transformation from a structured, native state to an unfolded random coil. This permits rigorous thermodynamic analysis of the reaction (Shortle, 1996). However, the more common view presently is that the denatured state is a distribution of microstates with varying degrees of structures (Shortle, 1996). This is seen in the simplest form in the model proposed by Dobson to be discussed later (Dobson, 2003; Chiti and Dobson, 2006). The transformation from a native to a denatured state for food globular proteins has been shown experimentally to consist of all or some of the following steps: dissociation of subunits, changes in tertiary structure while maintaining secondary structure (molten globule state) and loss of secondary structure (Qi et al., 1997; Dobson, 2003; Bhattacharjee et al., 2005). The existence of intermediate states is supported by evidence from heat-induced denaturation and molecular refolding. When a single molecule of ribonuclease H is mechanically unfolded using optical tweezers, a distinct molten globule-like intermediate is observed upon refolding (Cecconi et al., 2005). An unfolded protein may or may not be in a state that favors aggregation. Chiti and Dobson (2006) have shown, using human muscle acylphosphatase, that there are areas of the sequence that regulate aggregation and other areas that determine folding.

Bovine β-lactoglobulin serves as a good example for transitions that can occur in globular food proteins at different temperatures. The 162 residue protein has a primary feature of an anti-parallel β-barrel made up of eight β-strands arranged in two groups of four (Sawyer et al., 1999). It has the ability to bind hydrophobic molecules but the precise biological function has yet to be established (Sawyer et al., 1999). The protein exists as a native dimer at neutral pH (Swaisgood, 1982). As temperature is increased, the dimer dissociates into monomers and this is associated with an increased reactivity of the free sulfhydryl (cys_{121}) but no change in secondary structure (Iametti et al., 1996; Panick et al., 1999). This occurs in the temperate range of 40–50°C at neutral pH. As temperature is increased from

50 to 80°C the α-helix and β-sheet structures decrease and random structures increase (Qi et al., 1997; Panick et al., 1999; Bhattacharjee et al., 2005). Under some conditions, the transitions become irreversible at $T > 63$–65° C (Iametti et al., 1996; Prabakaran and Damodaran, 1997). A more detailed description of the unfolding process was obtained using NMR and deuterium exchange to investigate heat denaturation of β-lactoglobulin A at pH 3.0 (Edwards et al., 2002). Changes in the nine β-strands (labeled A–I) and α-helix proceeds as follows: 55–60°C (D–E strand); 60–65°C (α-helix and C–D strand); 75–80°C (A–H, B–C and F–G strand). The involvement of strand E at 55°C was also observed in a previous NMR and deuterium exchange study on β-lactoglobulin at pH 2 (Belloque and Smith, 1998). The transition temperatures for secondary structures observed by spectroscopy techniques coincide with differential scanning calorimetry results, where major transitions are seen at 55, 77, 105 and 125°C (Holt et al., 1998). Since spectroscopy studies generally do not exceed 100°C, the structural transitions occurring at 105 and 125°C are unclear. Further analysis of the structural details of β-lactoglobulin denaturation has been provided by molecular dynamic studies (Euston et al., 2007). A model describing solution properties and denaturation of β-lactoglobulin is shown in Figure 2.4.

There is also the possibility of different denatured states due to intramolecular changes occurring during denaturation. β-Lactoglobulin has two disulfide bonds and one free sulfhydryl that can undergo sulfhydryl–disulfide interchange during heating. Heating at pH 6.6–6.7 causes formation of two different molten globule states, a reversible one with the expected sulfhydryls exposed (Cys_{121}), and a second more stable state with Cys_{119} exposed (Croguennec et al., 2003, 2004). The non-native molten-globule state is irreversible and has a slower aggregation rate, possibly due to a decreased reactivity of Cys_{119} compared to Cys_{121} or an altered structure with less exposed hydrophobic binding sites (Croguennec et al., 2003, 2004). Whatever the mechanism, it illustrates that even with one protein, heat-induced reactions (in this case, sulfhydryl–disulfide exchange) can alter the denatured structure so that more than one species is available for aggregation.

While the previous discussion provides a good reference point (neutral pH, low ionic strength, low concentration of co-solutes) for the

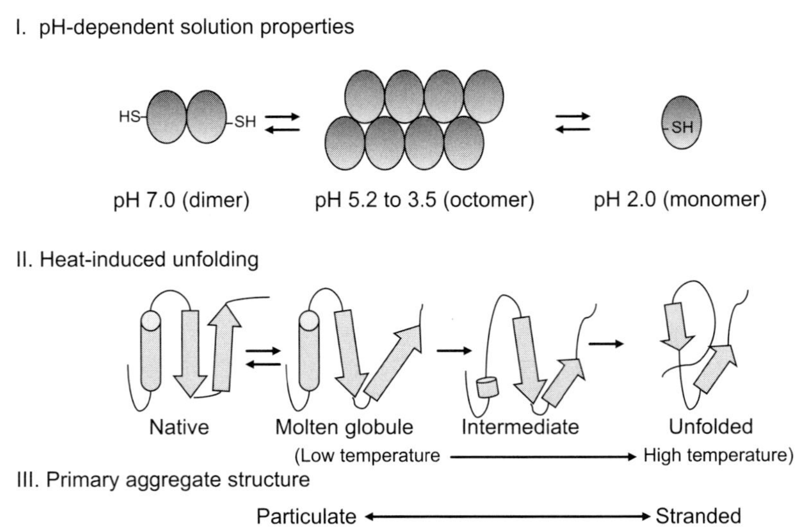

FIGURE 2.4 Model for β-lactoglobulin showing the range of structures in solution, denatured state and aggegated state.

denaturation process, it should be emphasized that denaturation will change also with solvent conditions. For example, addition of palmate, a fatty acid that could be present in foods, increased the thermal stability of β-lactoglobulin (Considine et al., 2005). Just raising the pH to 9.0 increases sulfhydryl reactivity such that β-lactoglobulin polymers are formed (Monahan et al., 1995). Solvent quality can also play a role in structural transitions. The addition of alcohols, which alter the solvent's dielectric constant, changes β-lactoglobulin secondary structure from primary β-sheet (51%) to primarily α-helix (Dufour et al., 1993; Dufour and Haertle, 1993; Dib et al., 1996; Kanjilal et al., 2003). The pH and ionic strength of the solution will alter the denaturation process. Raman and FT-IR spectroscopy have shown that β-lactoglobulin heated under conditions of low electrostatic repulsion (pH close to pI or high salt concentration) have a lower degree of unfolding (i.e., less loss of secondary structures) as compared to proteins heated with high electrostatic repulsion (pH > or < pI and low salt concentration) (Lefevre and Subirade, 2000; Ikeda and Li-Chan, 2004). This suggests that degree of electrostatic repulsion regulates the balance between the extent of structural change and initiation of aggregation. These are just a few examples of how solution properties alter the denaturation process.

Another source of variation is protein isoforms. Three of the genetic variants (A, B and C) of bovine β-lactoglobulin were investigated regarding heat stability using a range of methods (circular dichroism, tryptophane fluorescence or thiol reactivity). Heat stability of variants followed the relative relationship that C > A > B, independent of the method used to detect transitions (Manderson et al., 1999a, 1999b).

There are also specific elements of protein structure that may alter thermal stability depending on solution condition. For example, α-lactalbumin is a 123 amino acid calcium metalloprotein that has historically been considered as very thermally stable because it did not aggregate after denaturation (Bernal and Jelen, 1985; Farrell et al., 2004). The thermal stability depends on whether the protein is in the calcium-bound state (holo-α-lactalbumin) or calcium-free state (apo-α-lactalbumin), with relative transition temperatures of 62–64°C and 22–40°C (Boye et al., 1997; Hendrix et al., 2000; McGuffey et al., 2005). Lowering the pH (<3.5) causes a loss of calcium and the predominate species is the unstable apo-form (Pedersen et al., 2006). This can also be caused at neutral pH by addition of calcium chelating agents (Hendrix et al., 2000). Therefore, in the context of a food gel, the addition of a compound to inhibit lipid oxidation, such as ethylenediamine tetraacetic acid, could have a profound effect on the gelation process. α-Lactalbumin is also unique in that it can be denatured, cooled to refold, and show 24–37% reversibility upon a second heating (McGuffey et al., 2005). However, it should be noted that the protein structure after the first heating and cooling is not the same as the native structure (Fang and Dalgleish, 1998).

2.3.3 Refolding

Most food proteins form gels upon heating. The notable exception to this rule is gelatin, which gels upon cooling; a typical mechanism for polysaccharides such as agar and carrageenans (Ledward, 2000). In many aspects, gelatin behaves more like a polysaccharide than a protein. Gelatin is derived from collagen that has a repeating sequence of glycine as the third residue and this is frequently followed by proline (Ledward, 1986). The manufacturing of gelatin involves releasing the molecule from biological tissues using alkali or acid treatments and this can cause some covalent modification such as deamidation of glutamine and asparagine or hydrolysis of peptide bonds (Ledward, 1986). This produces a range of gelatins that have been slightly modified from their parent collagen structure. Upon cooling, gelatin molecules start to refold into collagen-like helices and these, along with aggregates of helices, serve as junction zones for gel formation (Ledward, 1986; Izmailova et al., 2004). Gelatin gelation had been extensively investigated (for example, see reviews by Ross-Murphy (1997) and te Nijenhuis (1997)) and only the structural changes associated with the initiation of gelation will be discussed. The critical helix concentration required for gelation is independent of the temperature of gelation (Chatellier et al., 1985). Renaturation kinetics has been used to show that the rate-limiting process includes both intramolecular nucleation and bimolecular nucleation. This concept was expanded by proposing that the rate-limiting step is formation of a nucleus consisting of two strands, which could be either intramolecular (first order) or intermolecular (second order). A third strand is then added to the two-strand nucleus, completing the triple helix formation (Guo et al., 2003).

2.3.4 Enzymatic Hydrolysis

Enzymatic hydrolysis changes protein structure and this can increase or decrease the potential for aggregation. The outcome depends on the protease specificity, extent of hydrolysis and specific protein being hydrolyzed. A pre-hydrolysis (3–12% degree of hydrolysis) of whey proteins caused an increase in propensity to form a heat-induced gel when Neutrase® or *Bacillus licheniformis* protease were used but a decrease in gelation with trypsin (Ju et al., 1995). Hydrolyzing soy proteins (soy protein isolate, glycinin and β-conglycinin) with a range of enzymes (Subtilisin Carlsberg, chymotrypsin, bromelain or papain) increased solubility around the pI (~ pH 5) but decreased the solubility at pH 6.0–7.0; showing that hydrolysis can alter the pH where aggregation is favored (Kuipers et al., 2007). Examination of the cleavage sites suggested that hydrolysis in hydrophobic regions increases the probability for aggregation. It should be noted that determining the ability of peptides to aggregate alone may not be indicative of gelation, as peptide aggregation may not always produce gelation (Creusot and Gruppen, 2007). Therefore, investigations on hydrolysis-induced gelation shed some light on how changes in covalent structure can initiate gel formation. The type of gel structure formed (fine stranded or particulate), or formation of fibrils, can be directed by altering the hydrolysis conditions (Otte et al., 1996, 2005; Ju et al., 1997). In addition, starting with native or denatured whey proteins will also alter protease-induced gelation (Otte et al., 1999; Ipsen et al., 2000). Whey protein isolate hydrolyzed with Alcalase 2.4L® form turbid solutions and gels at relatively high degrees of hydrolysis (> 18%) (Doucet et al., 2001). Based on the sequence of 130 peptides and types of interactions forming the gels, it appears that gelation is favored by having a medium degree of hydrophobicity and a net neutral charge (Doucet et al., 2003a, 2003b). A further purification of the major peptides shows that they have a molecular weight range of 265 (dipeptide) to 1485 Da (14 amino acid peptide) and a high proportion of proline residues (Doucet and Foegeding, 2005).

Overall, it is apparent the enzymatic hydrolysis can alter protein gelation by two general mechanisms. In the first, a limited amount of hydrolysis changes the molecular properties such that gelation is altered. This can cause modification in the gelation temperature (Huang et al., 1999) or alter the properties of the gels, as discussed above. In essence, partial hydrolysis is a way to alter the balance of intermolecular repulsive and attractive forces. However, due to the specificity of enzymes, one can utilize a more tailored approach to modifying specific segments of the protein. Alternatively, extensive hydrolysis can create peptides that are prone to aggregate and form gels. The gelation of low-molecular-weight peptides presents a series of considerations that are not as relevant to proteins and therefore will not be discussed further.

2.3.5 Non-enzymatic Hydrolysis

It is possible to modify the protein by hydrolyzing peptide bonds or deamidation of asparagine or glutamine. The susceptibility to deamidation depends on the relative locations of amino acids in the sequence and three-dimensional structure (Wright, 1991). Deamidation can occur in food proteins as the result of a combination of heat, acidic or basic conditions (Riha et al., 1996). This can be caused by a basic hydrolysis reaction where a molecule of water is added to an amide, releasing ammonia, or a more complex β-shift mechanism (Riha et al., 1996). Deamidation may affect protein structure because it converts a polar amide to a potentially charged carboxyl group. This reaction has been used to improve the functional properties of cereal protein. For example, Finley (1975) heated wheat gluten in 0.05–1.0 N HCl at 50–95°C, causing up to 90% deamidation. Protein solubility was maximized when deamidation was between 5 and 20%. More severe reaction conditions (temperature and pH) can cause peptide bond hydrolysis in addition to deamidation (Hamada, 1994). Heating whey protein isolate at pH 1.2–1.3 at 90°C causes deamidation and hydrolysis of peptide bonds (Lupano, 1994). A mild hydrolysis was also observed when heating whey protein isolate at pH 3.5–4.0 for 3 h (Hudson et al., 2000).

Deamidation also occurs in the process of converting collagen to gelatin (Ledward, 1986). The deamidation of asparagine causes a stabilizing of the structure and increase in denaturation temperature. In contrast, deamidation of glutamine can either increase or decrease denaturation temperature, depending on its sequence location (Silva et al., 2005).

2.3.6 General Model for Denaturation and Aggregation

The starting model for structural changes causing aggregation and ultimately gelation is one describing either the denaturation/unfolding process, alterations in specific amino acids (deamidation) or primary sequence. What all three processes have in common is an alteration of the intermolecular attractive and repulsive forces. The operating hypothesis is that some event, be it an increase in temperature, pressure, enzymatic activity or addition of a compound, favors protein structure changes exposing previously inaccessible segments that favor intermolecular interactions. Chiti and Dobson (2006) used natively unfolded proteins as a model to understand what factors regulate aggregation. This class of proteins can be divided into two general groups: ones with hydrodynamic dimensions indicative of a random coil and little to no ordered secondary structure, and ones that are more compact and have some amount of secondary structure (Uversky, 2002). Based on analysis of this group of proteins, protein aggregation is favored by the following factors: increased degree of hydrophobicity, decreased net charge and a high tendency to form β-sheet structures (Chiti and Dobson, 2006). Therefore, with structured proteins, aggregation of denatured proteins will depend on the extent to which the modified state increases the above three factors involved in aggregation. It should always be remembered that solvent quality and pH will modify the above-mentioned factors (see Figure 2.5).

The basic model for protein folding, unfolding and aggregation proposed by Dobson (Dobson, 2003; Chiti and Dobson, 2006), and seen in Figure 2.5, can account for many of the molecular changes that favor aggregation. Based on the model, the protein aggregation/gelation pathway can be viewed as the reverse of protein folding. One starts with a native molecule that can have quaternary structures of functional fibers or oligomers (Figure 2.5). For example, β-lactoglobulin is a dimer at neutral pH. The model shows that the native monomer can interact to form native-like aggregates that ultimately produce β-structured aggregates that can

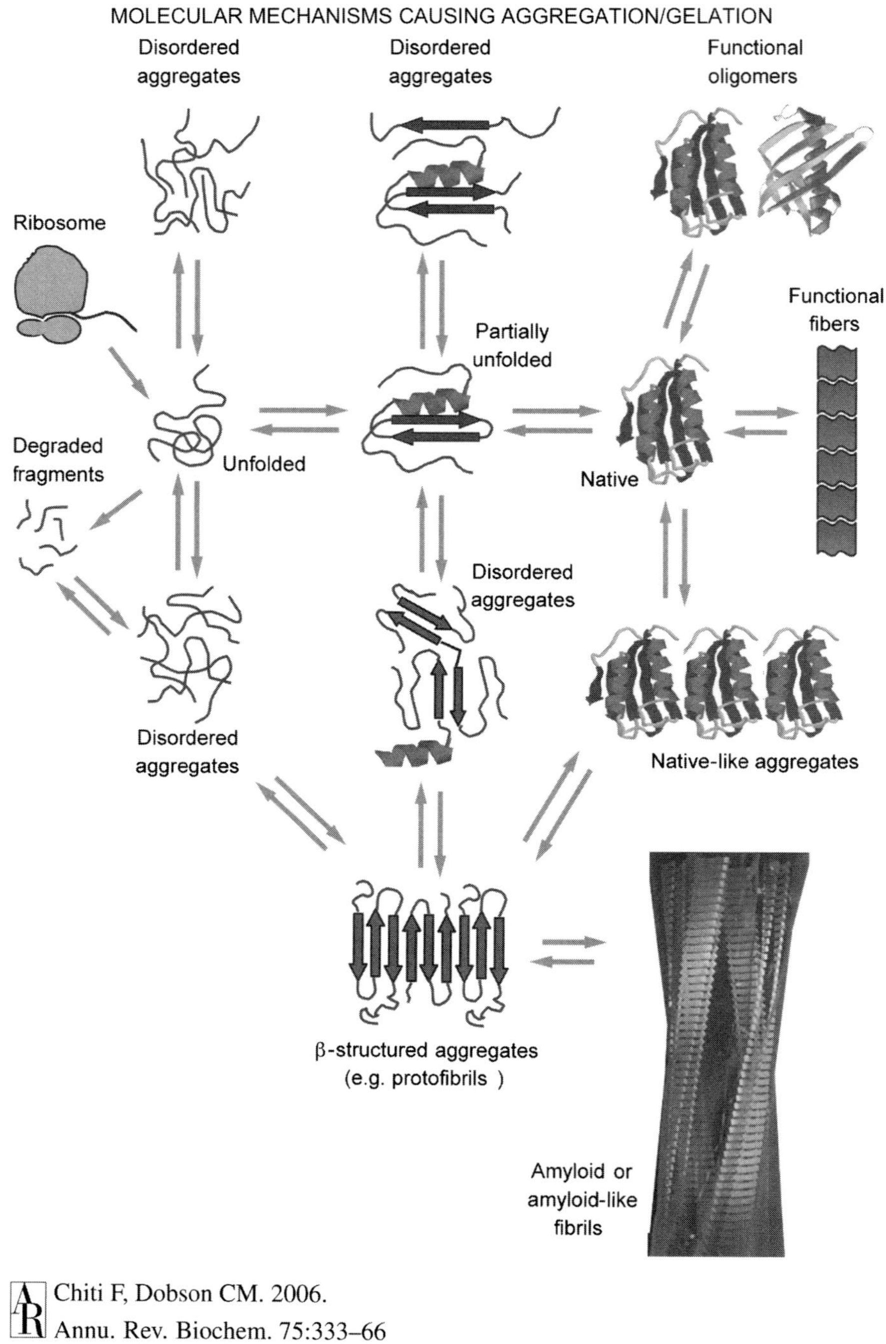

FIGURE 2.5 Model for protein folding, unfolding and aggregation proposed by Chiti and Dobson (2006). Reprinted, with permission, from the *Annual Review of Biochemistry*, Volume 75 ©2006 by Annual Reviews www.annualreviews.org.

form amyloid fibers. It also shows one intermediate unfolded state and an unfolded state that can interact to form disordered aggregates or amyloid fibrils. Another possibility is forming disordered aggregates or amyloid fibrils from degraded fragments that would be the result of peptide bond hydrolysis (as discussed above). This model accounts for much of what is known about protein denaturation and aggregation regarding gelation, including the effect of enzymatic or chemical hydrolysis. However, the number and type of intermediate states may vary, as has been proposed by other researchers. Shortle (2002) proposed a three-state model of expanded denatured, compact denatured (molten-globule) and native states. The molten-globule state was characterized as one 'which displays the characteristic property of severe broadening of NMR lines due to multiple conformations with different chemical shift environment interconverting on the millisecond time scale' (Shortle, 2002). Other groups have proposed a four-state model of: ordered native state, pre-molten globule, molten globule and random coil (possibly equal to expanded denatured or unfolded) (Uversky, 2002; Farrell et al., 2004). While there may still be debate about the precise characteristics of the intermediate and unfolded states, the fact that proteins can have different degrees of unfolding that possess structural elements with different susceptibility to aggregation appears to be well established.

The Dobson model appears to be robust enough to account for all protein gelation seen in food protein systems. Gelation at the iso-electric point or under conditions where protein salting-out is favored is represented by the native-like aggregates (Figure 2.5). Likewise, the combination of pH change and hydrolysis of peptide bonds, the two mechanisms associated with gelation of casein micelles in many cheese-making processes, can also be considered a special case for native-like aggregates where the majority of the structure of the casein micelle remains intact, and aggregation is cased by a decrease in electrostatic repulsion (pH decrease) and removal of a stabilizing factor (glycomacropeptide). What also comes out of the model, and nicely fits with the theoretical concept of morphology (see Table 2.1), is the implication that there are different aggregate geometries. The model shows two disordered aggregates, a native-like aggregate, and a β-structured aggregate. This implies some primary aggregate structure or at least an aggregation mode that is different due to the extent and type of structural change in protein molecules. This concept will be further developed in the following section (see Figure 2.6).

2.4 GEL STRUCTURE TYPE

It must be remembered that the propensity to aggregate is not only related to the changes in protein structure, but also the solvent surrounding the protein. Indeed, the importance of solvent conditions on aggregation and gelation was clearly stated in the pioneering work of Clark et al. (1981) as '....gels can be made from many native globular proteins by heating them in concentrated aqueous solution at suitable pH and ionic strength. Sometimes the products formed are transparent, but often they are turbid, and in extreme cases they can be opaque, water-releasing, materials better described as coagulates or precipitates.' These observations and those of others produced the concept of the gel 'structure' being the three-dimensional arrangement of the network. Therefore, it was only logical that microscopy was used to evaluate the microstructure of gels. Close inspection of gel microstructure showed that transparent gels were composed of strands with diameters between one and two times the diameter of the native proteins and those from turbid or opaque gels having much thicker networks (the coagulate had particles > 0.5 μm). Subsequent investigations referred to gels that were essentially

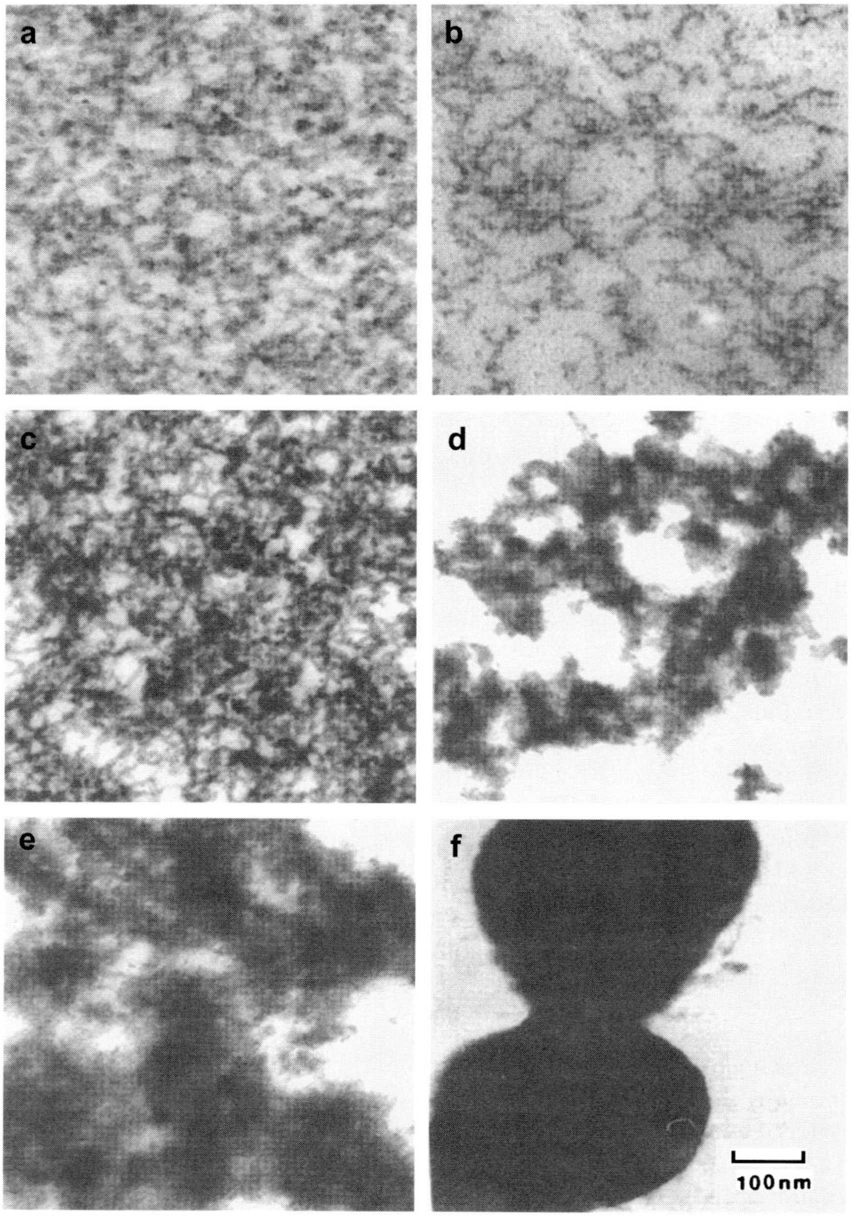

FIGURE 2.6 Transmission electron micrographs of heat-set globular protein gels: (a) 10% β-lactoglobulin pH 7, (b) 15% soy glycinin gel pH 5.3, (c) 10% α-lactalbumin pH 7, (d) 15% α-chymotrypsin pH 3, (e) 10% α-lactalbumin pH 7, 100 mM NaCl and (f) 10% BSA coagulate pH 5.1. Clark, A. 1998. Gelation of Globular Proteins In *Functional Properties of Food Macromolecules*. 120p. [Figure 3.19 Permission granted. Reprinted with kind permission from Springer Science and Business Media.]

clear to the eye and composed of network strands on the order of one or several molecular diameters as 'fine-stranded', and those from turbid/opaque gels with much larger network structures as 'particulate' (Langton and Hermansson, 1992). The type of network structure is clearly linked to the degree of electrostatic repulsion. At pH values above and below a protein's iso-electric point (pI), fine-stranded gels are formed that generally similar in microstructure but differ in rheological properties (Stading and Hermansson, 1991; Langton and Hermansson, 1992; Tang et al., 1995). A transition at pH $>$ pI from a fine-stranded to a particulate gel is also caused by addition of salts to act as counter ions and shield electrostatic charge repulsion. This transformation coincides with an increase then decrease in gel rigidity and a continuous decrease in water-holding properties (Barbut, 1995; Ikeda and Foegeding, 1999a, 1999b; Chantrapornchai and McClements, 2002). These physical transitions are consistent with a progressive coarsening of the gel structure. The Hofmeister series of salts are known to promote increased (salting-in) or decreased (salting-out) solubility of proteins. The general salt-induced transitions in rheological and microstructural properties in whey protein gels are observed with salts from the Hofmeister series (Na_2SO_4, NaH_2PO_4, NaSCN, NaCl); however, magnitude of the changes varies (Mulvihill and Kinsella, 1990; Bowland and Foegeding, 1995; Bowland et al., 1995; Foegeding et al., 1995).

The inherent limitation of using a microscopy-based technique to categorize gels has been eloquently pointed out by Mehalebi et al. (2008), who used light scattering to show that all turbid gels had the same self-similar structure at length scales below that of R_a (z-average radius of gyration), even though their structures seen by confocal light microscopy were different. Also, factors such as macro- or microscopic phase separation, as discussed in Section 2.2 may be of key importance to gel properties but less apparent in microscopy. This suggests that the structures formed from the initial few molecules forming aggregates and proceeding all the way to form a gel network are critical to a precise understanding of gel structure.

2.4.1 Formation of Protein Particulates and Fibers

Bromley et al. (2005), using β-lactoglobulin, proposed that either particulate or fibrillar aggregates are formed, with the latter being able to form spherulites in addition to gel networks. Subsequently, the same group showed that a range of proteins, representing different molecular masses, secondary structures and iso-electric points, all were capable of forming particulate aggregates at pH values close to their iso-electric points (Krebs et al., 2007). It was suggested that formation of particulates was a general property of proteins where molecular charge is minimal. However, it was also shown that adding 0.1 M NaCl increased the particle radius of some proteins but had minimal effect on others. This suggested that charge alone is not enough to explain the effects of pH and salts on formation of protein aggregates.

2.4.2 Fibers

Formation of protein fibers has received more attention than protein aggregates because the former is associated with amyloid fibers and disease states. Here we mainly want to address findings related to fibrils based on globular proteins or peptides derived from globular proteins. The other types of fibril-forming proteins have been extensively reviewed by Clark and Ross-Murphy (Clark and Ross-Murphy, 1987) and will be omitted in this review. Examples of fibrillar protein assembly, most relevant in fact to foods, have been reported by several authors. For instance, a series of publications by Joly and co-workers, from as early as 1949 (Joly, 1949a, 1949b; Joly and Barbu, 1949, 1950). Also see Oosawa and Asakura (1975) for an extensive overview of different possible morphologies, strandness, twist of chains, etc. It took until about two decades ago before the topic of assemblies of food proteins again received much wider attention. A more detailed overview of food-relevant fibril aggregation has recently appeared (van der Linden and Venema, 2007a). The next paragraph is a slightly modified version of part of the reference van der Linden and Venema (2007a):

> 'One example reference describes differences between assembly structure as a function of pH and salt concentration, and where the fibrillar nature of β-lactoglobulin was reported within a region of experimental conditions, including the level of branching (Aymard et al., 1996). This was taken up in another publication later on, where more quantitative measures for e.g. stiffness of the fibril were discussed (Aymard et al., 1999). Effects of protein concentration versus fibril formation were explored by Schokker *et al.* (2000). It is interesting to note that an implicit indication of fibril formation at low pH for β-lactoglobulin was already available in 1985, as Harwalkar and Kalab (1985) observed the formation of clear gels at pH=2.5, while turbid gels were reported for other conditions in their paper. For ovalbumin gels experimental conditions which lead to fibril formation were e.g. examined by Weijers *et al.* (2002). Such fibrils can be utilized as gelling material, clotting material, or for instance elongated and thin templates for metal coating to produce ultra-thin metal wires (as reviewed by Gazit (2007)). An important class of fibrils is the amyloid type of fibrils, defined by their characteristic to bind ThT by means of inter molecular beta-sheets present in the fibril. Nelson and Eisenberg (2006) have reviewed models on arrangements of proteins in such amyloid type of fibrils as proposed in between the years 1999-2005. They distinguish between three cases of fibril formation: 1) native protein, (partially) unfolding and then assembling into the fibril, possibly exhibiting conformational changes along the fibrillisation process, 2) natively disordered proteins (including small peptides) assembling into the fibril, and 3) proteins that locally exhibit a conformational change and assemble into a fibril.'

Regarding class 1 of fibril formation, most food proteins have been thought till recently to belong to this class, but recently it was reported that for the assembly of proteins like lysozyme (Mishra et al., 2007), β-lactoglobulin (Akkermans et al., 2008c), and alpha-lactalbumin (Ipsen et al., 2001), chemical changes in the protein and resulting peptide formation are relevant. Hen egg white lysozyme fibrillization at low pH is found to have so-called protein nicking as a prerequisite for fibrillization (Mishra et al., 2007). This refers to a local conformational change, in this case caused by hydrolysis of the molecule but not yet leading to its disintegration. The intact (non-hydrolyzed) lysozyme was found to slow down fibrillization and was suggested to act as an endcap for fibrillization (Mishra et al., 2007). Once the partially hydrolyzed protein is built hydrolysis reactions occur with the fibril (so-called shaving). The irregularities observed using AFM (Arnaudov and de Vries, 2005) may relate to the results of this shaving process. Recent data on the kinetics of hydrolysis may serve as input parameters for modeling fibrillization including the hydrolysis step, thus enabling extension of the kinetic modeling by Ferrone (1999). We note that similar nicking effects on fibrillization also have been reported on α-lactalbumin (Polverino de Laureto et al., 2005). Effects of hydrolysis were addressed by Bolder et al. for β-lactoglobulin. A hypothesis was put forward that the full protein shows fibrillization, and that the hydrolysis products would stop the fibrillization (Bolder et al., 2007). Hydrolysis had been taken into account by Arnaudov and de Vries (2007) and their kinetic data from dynamic light scattering could be explained using this hydrolysis reaction. The hypothesis of Bolder et al. was recently quantitatively investigated by Akkermans et al. (2008c) and it turned out that in fact the hydrolysis products are actually the building blocks of the fibril and that they do not act in stopping the fibrillization. Interestingly, hydrolysis was utilized to induce gelation by Doucet and Foegeding, where the hydrolysis was induced by an enzyme (Doucet and Foegeding, 2005). In that case both the gelation and hydrolysis took place at pH 8. In another paper, enzymatic hydrolysis was also induced of β-lactoglobulin, but at pH 7, and electron microscopy did not reveal fibrils. Interestingly, once that solution had been brought towards pH 2, fibril formation was observed (Akkermans et al., 2008b). Thus, the pH is important in the aggregation.

The chemical changes during the process that lead to fibrillar assembly will complicate the physico-chemical modeling of the assembly. One needs to measure the fibril composition and the types of molecules present in solution, all as a function of time. Important to note is that it is not at all clear whether the formation is thermodynamically determined or not. A rigorous approach for deriving thermodynamic data was addressed by Wetzel (2006). The effect of molecular crowding due to polyethylene glycol was investigated on polyglutamine aggregation. In other work, PEG modified silk like beta-sheet polypeptides were investigated on fibril formation, and the largest PEG molecular weight influenced the length of the fibril. The PEG acted as a coating agent to prevent fibril–fibril aggregation (Smeenk et al., 2006).

Whether beta-sheet-containing proteins will always form amyloids was addressed by Zandomeneghi et al., (2004). It was found by FTIR that the molecular structure within the amyloid fibrils is different from the beta-sheets in the native proteins. Interestingly, lithostatine, a beta-sheet-containing protein, was even found to form non-beta-sheet-containing based fibrils (Laurine et al., 2003).

For the protein Apomyoglobulin (Fändrich et al., 2006) it was shown that, as a first step towards fibrillar aggregates, spherical aggregates without a clear inner structure are being formed, which in turn act as a seed for protofibrils. The building blocks of the spherical aggregates are suggested to contain many different conformations, and the aggregates are a product of a random nucleation process. Interestingly, it was noted by these authors that this mechanism shows similarities with the extension into nanotubes from ordered arrays of large non-fibrillar particles.

It was shown (Smith et al., 2006) that the existence and size of various oligomers in the early stage of fibrillization were related to whether they were formed according to a nucleation mechanism or transiently. The nucleated mechanism gave rise to rod-like amyloids, while the larger oligomers (up to 16-mers) yielded worm-like fibrils. The approach holds promise to give more general insight into fibrillization of proteins and their peptides. Differences in fibril morphology, though to a lesser extent, were observed for β-lactoglobulin as a function of salt concentration, which at the same time influenced the critical nucleus size (Arnaudov and de Vries, 2006). For a bar star protein (Kumar et al., 2007) several oligomers were found as well as protofibrils, which grow from lateral association of critical aggregates via different pathways. Interestingly, changing the net charge of the protein yielded changes in fibrillization morphology from stiff to flexible (Broersen et al., 2007).

As unfolding and refolding are important steps in the fibrillization process, it is important to study these as a function of pH, salt concentration, as a function of the concentration of urea and/or guanidine hydrochloride, and also pressure. We refer to some recent observations as summarized in van der Linden and Venema (2007b) from which we quote:

> 'Applying high hydrostatic pressure is found to be an interesting perturbation method to test the relevance of non-covalent interactions in regards to formation and stability of fibrils. This was done for hydrophobic interactions in the early stages of fibrillisation as recently addressed by Meersman and Dobson (2006). For the particular protein under

investigation the small aggregates formed before fibrillisation are easily dissociated by high pressure, while this is not the case of mature fibrils, suggesting that the matured fibril exhibits an optimized side chain packing and optimises hydrogen bond formation, while during the early stages, of small aggregates, the hydrophobic and electrostatic interactions are mainly determining the aggregation (Meersman and Dobson, 2006). These two effects on the initial process of fibril formation are in line with findings that NaCl increases amyloid formation initially (Krebs et al., 2005). The modulation of amyloid formation has also been utilized using high pressure by Torrent et al (2006). In older work the significance of hydrophobic interactions have been reported by Nirnajan (2003) in the case of actin filament formation. In line of different interactions and different phenomena playing a role in the assembly of proteins into fibrils the mechanisms for different proteins have been examined as a function of solvent. For instance, ethanol, methanol, propan-2-ol and 2,2,2-trifluorethanol have been used (Gosal et al., 2002a) as well as urea (Hamada and Dobson, 2002).

In order to rapidly screen the effect of modulators on amyloid formation, a microplate assay was developed (Berthelier and Wetzel, 2006) which may serve as an efficient tool to quantify the fibrillisation elongation kinetics as a function of the various parameters.'

2.5 GEL TEXTURE: ORAL PROCESSING, RHEOLOGY/ FRACTURE, MICROSTRUCTURE AND SENSORY ANALYSIS

2.5.1 Introduction

Gelation is used to create texture in food products such as yogurt, cheese, processed meats, cooked egg white and gelatin-based desserts. One standard definition for texture of a food product is 'all the rheological and structural (geometric and surface) attributes of the product perceptible by means of mechanical, tactile, and, when appropriate, visual and auditory receptors' (ISO, 1981). Consumers use a combination of textural, taste and olfactory stimulations to determine if they like or dislike a food product; therefore, it is of great importance to food manufacturers to understand what aspects of gel structure determine sensory perception of texture. In contrast, from a physiological perspective, food texture is what has to be broken-down in order to prepare the food for swallowing. This general process is called 'oral processing' and can be considered what occurs from the point of food entering the mouth until it is swallowed. When evaluating food gel textures, we tend to not consider visual stimuli because they are not a component of oral processing, and since gels are soft solids, auditory signals such as crunchiness found in dry brittle foods are also ignored. If we limit the properties to mechanical (rheological and fracture) and tactile, the challenge becomes determining precisely which physical properties of gels are evaluated during oral processing. This involves not only an understanding of the mechanical properties of the food material, but also how it interacts with saliva during the breakdown process. There is a long-standing interest in understanding how food structure relates to sensory texture (Hutchins and Lillford, 1988; Wilkinson et al., 2000; Lillford, 2001; van Vliet, 2002; Bourne, 2004; Foegeding, 2005; Lucas et al., 2004), and an acknowledgment of the limitations of just looking for simple correlations between instrumental evaluation of mechanical properties and sensory texture terms (Dejmek et al., 2006; Renard et al., 2006). Guinard and Mazzucchelli 1996 astutely pointed out that there was a limited understanding of the role physiology plays in texture perception. This view has been reinforced in several articles pointing to the greater need of incorporating physiological elements into the understanding of how food structure produces sensory texture (Malone et al., 2003; Nishinari, 2004; Foegeding, 2007; Chen, 2008). The array of factors contribution to sensory texture of protein gels will be discussed in this section.

2.5.2 Oral Processing

Oral processing starts with either a 'bite' or piece of food entering the mouth and then being set-up for chewing; this is followed with chewing to reduce particle size, formation into a bolus and finally swallowing (Hiiemae et al., 1996). The general steps in oral processing can be viewed as (de Wijk et al., 2003; Okada et al., 2007):

1. The mouth opens and the tongue moves forward to touch (capture) the food.
2. Food is moved to the anterior third of the tongue.
3. The tongue is lifted, compressing the food against the pallet.
4. The tongue is lowered and this transfers the food to the post-canine teeth.
5. Food is chewed to reduce the particle size and form a bolus.
6. A preswallowing cycle precedes the terminal swallow; and interspersed swallows may occur during the chewing cycles.

The chewing process will not only depend on the physical characteristics of the food material but also on the size of the sample (Hiiemae et al., 1996; Fontijn-Tekamp et al., 2004). Chewing gum of different sizes shows that the opening, closing or total cycle duration do not change with size. However, vertical cycle velocity ranged from 58 to 96 mm/s as bolus (gum) weight increased from 1 to 8 g (Bhatka et al., 2004). This is important when considering rheological properties because it indicates that the deformation rate may vary with sample size. The ability of a person to reduce particle size during chewing is defined as masticatory function and is usually expressed as some measure of particle reduction normalized by the number of chews (van der Bilt et al., 2006; Gambareli et al., 2007). Masticatory function will vary among individuals due to a variety of factors including: occlusal factors, maximum bite force, sensory feedback and manipulation of food, age, saliva, food texture and taste and swallowing threshold (Bourdiol and Mioche, 2000; van der Bilt, 2002; van der Bilt et al., 2006). This means that any particular group of sensory panelists will have a range of oral processing abilities. For example, a group of 87 subjects required between 17 and 110 chewing cycles to process the same amount of peanuts (van der Bilt et al., 2006). Likewise, a group of six panelists required from 52 to 103 total chews when evaluating a gelatin-based gel (Sprunt and Smith, 2002).

The variability in masticatory parameters among individuals is in stark contrast to the fact that the particle size distribution is similar among individuals for a given food (Woda et al., 2006b). In a study involving 87 subjects evaluating carrots, peanuts and cheese, the median particle size ranged from 1 to 5 mm at the time of swallowing (Fontijn-Tekamp et al., 2004). Another investigation used 10 subjects evaluating 10 different foods. The mean particle size ranged from 0.82 mm for peanuts to 3.04 mm for gherkins (Jalabert-Malbos et al., 2007). They suggested that harder foods needed to be reduced to a mean particle size of less than 2 mm for ease of swallowing while softer foods could be swallowed with a larger particle size. The association between particle size and food texture is supported by the observation that chewed nuts (peanut, almond and pistachio) had a peak particle size of 1.4 mm while peak particle size for chewed vegetables (carrot, radish and cauliflower) was 2.5 mm (Peyron et al., 2004). The relative contribution of oral physiology and product characteristics to swallowing threshold was investigated using 266 subjects and seven different food products (Engelen et al., 2005). Lubrication was shown to be important by demonstrating that subjects with higher saliva flow rates required fewer chewing cycles to masticate melba toast and breakfast cake (dry product) and buttering dry products reduced the number of chews. While oral physiology did contribute to determining the urge to swallow, product characteristics were shown to be the major factor (Engelen et al., 2005). Gelatin-based gels were used to determine the amount of saliva and surface area generated during mastication. Six panelists with saliva flow rates of 2.77–5.55 g/min chewed 4.28 g samples of gels to completion (52–103 total chews). This caused the surface area to increase from 1000 mm^2 to 3000 mm^2 and produced 2.13–4.25 g of saliva (Sprunt and Smith, 2002). All these observations show that a combination of lubrication and particle size reduction contributes to forming a bolus capable of swallowing. For a given food, the mastication process may vary among individuals but the particle size distribution in the pre-swallowed bolus is similar among individuals.

The above discussion pertains to oral processing of foods that are sufficiently hard to require jaw movement to fracture samples into small particles in preparation for swallowing. It should be noted that some food gels are sufficiently weak in structure such that they can undergo 'chewing' by simply compressing the food between the tongue and the pallet (de Wijk et al., 2003; Prinz et al., 2007). Both types of food will be discussed.

2.5.3 Large Strain and Fracture Properties

Mechanical properties of protein gels can be generally grouped based on level and rate of deformation into linear visco-elastic, non-linear visco-elastic and fracture. While linear visco-elastic properties are essential to understanding gel formation and physical properties of the formed gel (see Section 2.2), they are generally shown to poorly correlate with sensory texture (Brownsey et al., 1986; Brown et al., 2003; Barrangou et al., 2006). This is logical because, as seen in the above steps of oral processing, non-destructive strains at best only occur during the initial positioning of the food within the mouth. The initial or 'first bite' chew takes into account non-linear deformations and fracture, therefore, instruments that apply an approximately linear force are expected to correlate with information from stress–strain curves, providing it is applied at a strain rate resembling chewing (Shama and Sherman, 1973). In contrast, chewing is detecting the extent of breakdown with each cycle until a bolus is formed that allows for swallowing, and this is not a simple extension of stress–strain curves.

Large strain and fracture properties can be determined by compressing, extending, bending or twisting the sample to the point of fracture. Comparisons among the methods have been discussed in several reviews (Hamann, 1983; Luyten et al., 1992; Truong and Daubert, 2000; Hamann et al., 2006). One of the key elements in testing viscoelastic materials, such as protein gels, involves critical stresses and strains as a function of deformation rate. When conducting large strain and fracture testing on protein gels, the strain rates are limited by a particular instrument's ability to deform the sample and collect data. This tends to restrict the upper limit of deformation rate, which is more reflective of the chewing process. The importance of strain rate was noted in the pioneering work of Shama and Sherman (1973), where it was shown that sensory ranking of cheese hardness was only predicted by compression testing when the compression speed was $\geq$ 3.3 mm/s. This observation is supported by a later investigation showing that for 10 different commercial cheeses, chewing velocities ranged from 16.6 to 39 mm/s (Meullenet et al., 2002). Since cheese is a visco-elastic material, it suggests that the deformation rate during mechanical testing has to be sufficiently high to account for the key viscous components that contribute to texture. Chewing velocity will also vary due to product characteristics. Increasing sample hardness causes a decrease in closing velocity, while closing velocity is increased with sample size and moving from an elastic (gelatin gel) to a plastic (caramel) food structure (Foster et al., 2006; Woda et al., 2006a). This clearly shows that a range of deformation rates is required to fully characterize large strain and fracture properties relevant to food texture. Moreover, it suggests that for visco-elastic materials there is a critical strain rate. Sensory and mechanical tests should correlate above the critical strain rate and show minimal correlation below the critical strain rate. This concept requires further experimental validation.

Determining stress–strain curves at different deformation rates not only shows how the material responds to deformation rates but this information can be used to gain insight into the basic nature of the material. van Vliet et al. (1991) proposed an energy balance for fracture as:

$$W = W' + W''_m + W''_c + W_f$$

This model stipulates that the energy (W) supplied during deformation is either stored (W') or dissipated by various processes ($W''_m + W''_c$), or used to break network structure (W_f). Energy can be dissipated due to friction between components arising from inhomogeneous deformation (W''_c), or by viscous flow (W''_m) (van Vliet et al., 1991, 1993). This equation shows that the molecules composing the gel network, the structure of the gel network, and interactions between molecules, have various ways to influence the fracture process. Moreover, it provides a way to classify visco-elastic materials based on their strain rate dependence. An increase in fracture stress and strain with increasing strain rate is predicted based on energy dissipation due to inhomogeneous deformation (W''_c) being critical to fracture of these materials (van Vliet et al., 1991, 1993, 2007). This type of behavior has been observed in agarose and alginate gels (Barrangou et al., 2006; Zhang et al., 2007). In gels where fracture is based on viscous flow (W''_m), fracture stress increases and fracture strain decreases with an increase in strain rate. This behavior is observed for fine-stranded whey protein isolate gels (Lowe et al., 2003). While it is clearly desirable to conduct fracture experiments over a wide range of strain rates, the increased amount of material and testing time often restricts this approach.

2.5.4 Microstructure

Microscopy has often been used to categorize gel structure. Early work using transmission electron microscopy showed that, depending on the protein and gelation conditions, the gel network consisted of strands with diameters on the level of several molecules or very large aggregates (Clark et al., 1981). This observation coined the terms 'stranded or fine-stranded' and 'particulate' as gel descriptors, as discussed in Section 2.4 (Clark et al., 1981; Stading and Hermansson, 1991; Langton and Hermansson, 1992). Microscopic evaluation of gel networks during fracture has provided new insight into the initial fracture process. Protein gels that have networks visible using confocal laser scanning microscopy, i.e., particulate gels, can be deformed by means of a tension/compression stage so that changes in stress, stain and microstructure can be recorded dynamically (Plucknett et al., 2001). This approach was used to show that the fracture pattern of particulate β-lactoglobulin gels changes with protein concentration (Ohgren et al., 2004). Gels containing 6% protein have lower fracture stress, higher fracture strain, and the pores grow and protein clusters turn during deformation. In contrast, gels containing 12% protein have higher fracture stress, lower fracture strain and the microstructure shows smaller changes with deformation. An additional difference was in the crack growth pattern, with a smooth progression with the stronger gels and a step-wise growth with the weaker gels. This clearly showed that, at the microscopic level, the fracture pattern will depend on overall network concentration. Addition of another protein will alter the fracture process. Brink et al. (2007) showed that crack propagation was different between particulate whey protein isolate gels and whey protein isolate gels filled with gelatin. One of the key observations in this investigation was that fracture starts within the network prior to the point of macroscopic fracture, indicating that it would be occurring in the non-linear region of a stress–strain plot. It was also established that there was a shift in rheological properties occurring at a critical amount of added gelatin, suggesting phase separation or some overall change in gel structure. However, this shift was not reflected in the microstructure observed at the length scale limits of CLSM. Two types of fracture were observed and classified as 'porous' and 'stretched' fracture. With the former, the structure ruptured gradually immediately after applied stress. In contrast, a 'stretched fracture' is when the pores and network are stretched but the number of pores does not increase. In some situations a combination of the two mechanisms was observed. As will be discussed in a subsequent section, one of the challenges in relating large strain and fracture properties to sensory texture is that there is poor correlation with sensory properties measured during the chewing process. It is intriguing to think that observations of fracture on the microscale may be related to sensory properties such as smoothness and adhesiveness.

2.5.5 Sensory Analysis of Protein Gels

Before discussing how mechanical (large deformation and fracture) and microstructural properties relate to sensory properties, a brief discussion of approaches relating mastication and mechanical properties is warranted. It should be noted that models relating mechanical properties to the mastication process are generally concerned with understanding the physiology of oral processing. The goal there is understanding how well chewing breaks down food into particles capable of swallowing, i.e., the physiology of mastication. In contrast, sensory analysis investigations of food texture ultimately want to know how to produce specific textures and what is desirable or undesirable about a specific texture. These different goals influence the experimental measurements and designs one uses to understand the relationships between oral processing and food texture. The model of Lucas, Prinz, Agrawal and Bruce (Agrawal et al., 1997; Lucas et al., 2002, 2004) starts with a piece of food being loaded between upper and lower teeth such that a three-point bend configuration is possible. This can produce fracture between the two points supporting the food (after bending like a beam) or under the tooth cusp. They derive relationships between food toughness (R, defined as energy dissipated in crack propagation) and Young's modulus (E, normal stress divided by normal strain). If the number and size of the fragments are limited by the displacement available, then $(ER)^{0.5}$ is the relationship of food properties that determine crack propagation. Fragmentation is limited by stress. If fragmentation is limited by strain, then $(R/E)^{0.5}$ should be the defining ratio. They tested these models by determining the increase in particle area after one chew for 38 different foods. Particles were created by sealing a fixed sample size into a latex bag (to prevent mixing with saliva), having the subject make one bite with postcanine teeth, then determining total surface area of particles based on image analysis. A particle breakage function [(after chewed area/volume) $^{0.5}$ – (initial area/volume) $^{0.5}$] was scaled with either R, E or $(ER)^{0.5}$ or $(R/E)^{0.5}$ to evaluate which best described the data. The best fit was found with the deformation-limited model, $(R/E)^{0.5}$. One should view this model as linking material properties with the extent of chewing that is required to reduce the food to particle sizes acceptable to swallow; bearing in mind that saliva and other factors are involved (Lucas et al., 2002). This is essentially describing the 'food-crushing' process that occurs during oral processing. In contrast, sensory analysis is concerned with all of the oral stimuli that take place during the food-crushing process. The coinciding processes of oral processing and sensory analysis are shown in Table 2.4.

In contrast to the extensive number of investigations on formation of heat-induced protein gels, there are few investigations on the sensory properties of said gels. This is due in part to heat-induced food protein gels, with the

TABLE 2.4 Coinciding Events in Oral Processing and Sensory Texture Evaluation

Oral processing	Sensory analysis		
	Evaluation phase	Term	Description[a]
Food moved to the anterior third of tongue	Pre-fracture	Smoothness	Sensation between oral surface and food
		Slipperiness	Sensation between oral surface and food
Tongue elevates food and compresses against palate		Firmness	Force required to deform the sample
Tongue is depressed to move the food to post-canine teeth			
Communition	First bite	Moisture release	Extent of fluid released
		Crumbliness (Fracturability)	Degree to which sample fractures into particles
		First bite firmness	Force required to fracture
		First bite deformability	Deformation required to fracture
	Chewdown (chew for a fixed number of chews and evaluate after chewing)	Degree of breakdown (particle breakdown)	Extent of breakdown such as number of particles or loss of structure
		Particle size	Relative size
		Particle size distribution	Relative distribution of particle sizes
		Particle shape	Rounded or irregular shape
		Cohesiveness	Degree to which the chewed mass holds together
		Adhesiveness	Degree to which the chewed mass sticks to mouth surfaces
		Smoothness	Degree to which the chewed mass surface is smooth
Swallowing	Residual	Number of chews (chewiness)	Total number of chews required for swallowing
Clearance		Smoothness of mouth coating	Degree of smoothness in mouth after expectoration or swallowing

[a]Sensory terms were derived from Brown, J.A., Foegeding, E. A., Daubert, C. R., Drake, M. A., Gumperz, M. (2003) Relationship among rheological and sensorial properties of young cheeses. J. Dairy Sci. 86, 3054–3067; Gwartney, E. A., Larick, D. K., Foegeding E. A. (2004) Sensory texture and mechanical properties of stranded and particulate whey protein emulsion gels. J. Food Sci. 69, 333–339; and Barrangou, L. M., Drake, M., Daubert, C. R., Foegeding, E. A. (2006) Textural properties of agarose gels. II. Relationships between rheological properties and sensory texture. Food Hydrocolloids. 20, 196–203.

exception of egg white, containing other ingredients such as lipid and therefore more complex systems are evaluated to be relevant to food products. For example, Gwartney et al. (2004) investigated the sensory texture of whey protein isolate gels containing 0–20% emulsified fat. Two different gel structures were produced by forming gels in the presence of either 25 mM NaCl (stranded) or 25 mM NaCl + 10 mM $CaCl_2$ (particulate). Stranded gels had fracture stresses ranging from 11.1 to 54.5 kPa, and fracture strains from 2.01 to 2.95. They were characterized as having a smooth/slippery surface, a high degree of springiness and compressibility, breaking-down slowly and releasing very little water during mastication. Particulate gels had fracture stresses ranging from 11.9 to 43.4 kPa, and fracture strains from 0.97 to 1.23. The texture of the gels was described as a rough surface with a low level of springiness that broke-down rapidly into small, regular-shaped particles during chewing and released detectable amounts of fluid during mastication. Adding lipid altered the degree of some properties but did not change the overall characteristics of the gel breakdown pattern during mastication. Interestingly, the firmness and number of chews were the same between the two types of gels containing 20% fat.

In a series of articles, van den Berg et al. (2007, 2008a) investigated the texture of gels made from combinations of whey protein isolate and different hydrocolloids. They were able to generate gels with different microstructures by altering the type and amount of hydrocolloid mixed with protein. When viewed under confocal microscopy (160 μm by 160 μm area), gels were classified as being either homogeneous or phase separated. The phase separated gels were further classified into coarse stranded, bicontinuous or protein continuous. This allowed for the generation of one category of microstructure using different types of hydrocolloids. The gels were weak enough that texture was evaluated by pressing between the tongue and palate, essentially stopping at stage 3 in oral processing (see above) and evaluating texture by manipulation between the tongue and palate. The fracture stress and strain of these gels ranged from 3.2 to 16.8 kPa and 0.87 to 1.4, respectively. Protein continuous microstructures showed spontaneous fracture during compression prior to macroscopic fracture; for example, appearing at a strain of 0.3 when fracture strain was 1.04 (van den Berg et al., 2007). Fracture via spontaneous cracks or with notched samples proceeded by a similar mechanism, by fracturing perpendicular through a protein strand in the gel network. Spontaneous cracks were not observed in bicontinuous structure; however, fracture was perpendicular through protein strands, as was observed with the protein continuous structure. For coarse stranded networks, a single fracture plane was not observed and fracture started at various locations, coinciding with an overall coarsening of the structure. Gels were separated based on principal component analysis into two groups with firm and crumbly at one end and spreadable at the other. Gels with the coarse stranded network were perceived as the most spreadable, coinciding with its unique fracture pattern. In contrast, protein continuous and bicontinuous gels were the most firm and crumbly. It is interesting to note that gels containing 0.09% pectin were coarse stranded (spreadable) and increasing the concentration to 0.14% pectin converted the structure to protein continuous (firm and crumbly). Therefore, one can change the microstructure, network fracture pattern and overall textural properties by simply altering the amount of one biopolymer in a two biopolymer system. This suggests that, at least in the case of phase separated systems, the dominant structure determines the main textural properties. Finally, the firmness of the gels correlated ($r = 0.88$–0.94) with physical properties such as Young's modulus, energy at fracture and true fracture stress. These are all associated with the ability of the gel network to resist deformation and fracture and thus logically correlated with sensory firmness ('effort to compress the sample between tongue and palate') (van den Berg et al., 2007). However, spreadability ('the sample spreads between tongue and palate') was not correlated with the measured physical properties but associated with specific gel structure types and microscopically observed fracture patterns. Since fracture stress and strain were not correlated with spreadability, it suggests that some measure of the network altering processes prior to or after macroscopic fracture is needed. That was shown to be the case for the crumbly attribute ('sample falls apart in (small) pieces upon compression between tongue and palate'; (van den Berg et al., 2007)) in the subsequent publication (van den Berg et al., 2008a). Gels with high crumbly scores (53–58) showed a rapid decrease in normalized force after fracture, suggesting a free running crack (Walstra and van Vliet, 2003; van den Berg et al., 2008b). In contrast, gels with a much slower normalized force decrease had lower crumbly scores (26–38). The visco-elastic properties relevant to fracture were determined by examining the compression speed at which 50% of the samples fracture by a propagating crack, termed a 'critical speed for fracture'. This property supported the post-fracture normalized force pattern by showing that gels with high and low crumbly scores had respective critical speeds for fracture of 0.01 and 1 mm s^{-1}, respectively. Finally, a recoverable energy test was used to probe visco-elastic properties at strain levels beyond the linear visco-elastic region but below fracture. It calculated the ratio of areas under the compression and decompression curves to determine the recoverable energy. A pure elastic gel would have a recoverable energy of 100%. Overall, compressing at a rate of 20 mm s^{-1} produced a correlation between crumbly score and recoverable energy ($r^2 = 0.87$), whereas compressing at 1 mm s^{-1} did not produce recoverable energy values showing a correlation. Close inspection of the data showed that compression rate altered recoverable energy for gels with high crumbly scores and had little effect on gels with low crumbly scores. It was also shown that crumbly scores increased as serum release decreased (van den Berg et al., 2008b).

Gelled emulsions is one category of protein gels that has received considerable attention regarding rheological properties (Chen and Dickinson, 1998, 1999a, 1999b; Dickinson and Casanova, 1999; Dickinson and Chen, 1999; Dickinson and Yamamoto, 1996a, 1996b; Dickinson and Yamamoto, 1996c; Kerstens et al., 2006; Langley et al., 1991; van Vliet, 1988). Indeed, the topic of how filler particles alter rheological properties of gels is a subject of scientific interest but will not be covered in detail here as it is considered a separate system. Nonetheless, it will be address when considering the sensory properties of protein gels. As pointed out earlier, Gwartney et al. (2004) investigated the sensory texture of whey protein isolate gels (12% protein) containing 0–20% emulsified fat that had fracture stresses ranging from 11.1 to 54.5 kPa and fracture strains from 0.97 to 2.95. The overall trend was that increased fat content caused an increase in fracture stress and sensory firmness (i.e., first bite properties) but the gel type (microstructure) determined the properties sensed during mastication (e.g., particle size, crumbliness, moisture release, cohesiveness of the mass, smoothness and adhesiveness). This coincides with the findings of van den Berg et al. (2007, 2008b) that gel microstructure determines the breakdown pattern during mastication. Sala et al. (2008) investigated a series of emulsion-filled gels formed with a protein (whey protein isolate or gelatin) or polysaccharide gel network (kappa and iota carrageenan). Fracture stress and strain ranged from 4 to 12 kPa and 0.6 to 1.4, respectively, much less than the investigation of Gwartney et al. (2004). Interestingly, fracture stress did not show a correlation r^2 value above 0.4 with any of the texture terms. Firm and elastic terms were correlated with fracture strain and crumbling effort (effort needed to break the sample into pieces) was correlated with fracture strain and energy to fracture (Sala et al., 2008). The spreadable term was not correlated with any of the rheological properties, similar to what was observed for protein–polysaccharide gels (van den Berg et al., 2007).

What emerges from the few investigations where both sensory and mechanical properties are determined is that the first bite terms, such as firmness, hardness or elastic, are generally correlated with some mechanical terms determined at fracture (fracture stress, fracture strain or fracture energy). This is not surprising in that the oral processing and mechanical testing are essentially doing the same thing, deforming the sample to a point of fracture. In contrast, the sensory terms determined during the chewing process are less likely to correlate with mechanical terms determined at fracture. However, correlations are found with mechanical properties determined at strains before or after fracture and in the microstructure (van den Berg et al., 2007, 2008a). This general trend is also observed in complex soft solid foods such as cheese (Brown et al., 2003). Moreover, again based on sensory texture of cheese, it is the properties that are evaluated during chewing that are the most important in providing a unique characteristic to the product (Brown et al., 2003; Gwartney et al., 2004).

2.6 CONCLUDING REMARKS AND FUTURE CHALLENGES

In this review, the protein gel is discussed in terms of its relation to sensory response, in particular its sensory response in terms of structural and mechanical properties of the gel. Secondly, the mechanical response of protein gels has been discussed in terms of the various structures that exist on a meso-scale. Thirdly, the phenomena that describe how to obtain the

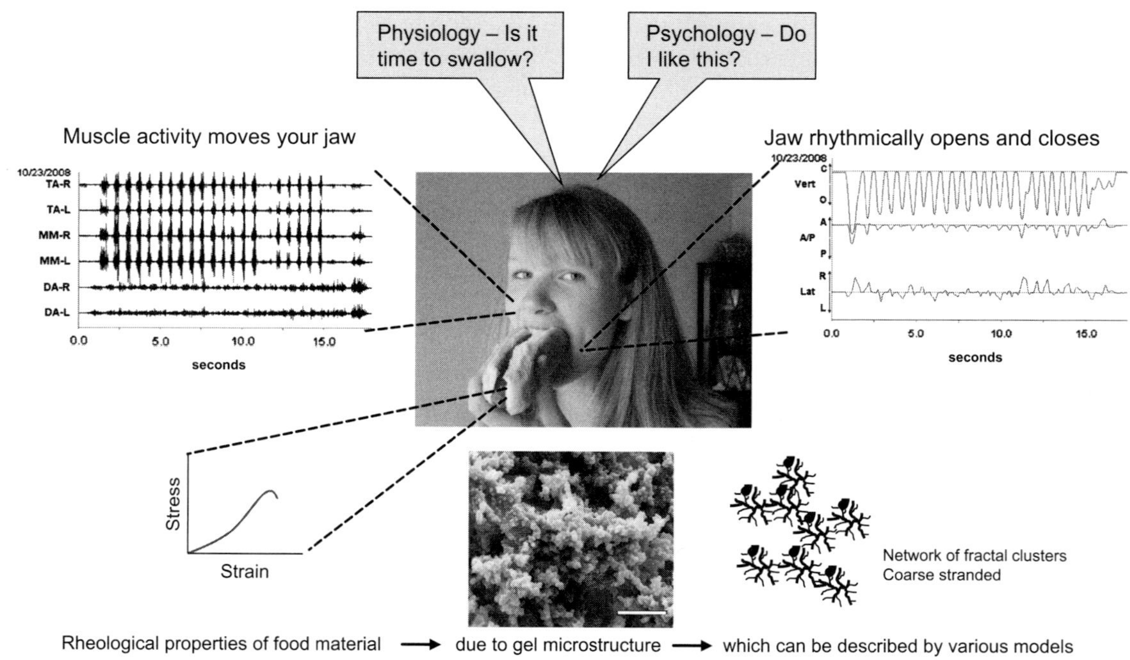

FIGURE 2.7 Overview of the processes involved in understanding food texture. Food structure can be described based on various models and measured using microscopic and rheological techniques. Oral processing provides the physiological function of getting food ready to swallow while simultaneously the brain is evaluating textural properties and desirability.

various meso-structures have been reviewed. This was done on a macroscopic scale, in terms of equilibrium phase diagrams and non-equilibrium paths through the diagram, and was done from the molecular scale upwards, where the chemical details of the protein structure (charge, etc.) were taken into account to try to predict the type of morphologies that the protein meso-structures exhibit (strands, fractal geometries etc.). In this way, an attempt has been made to review protein gels from the macroscopic sensory point of view down to the molecular properties of their constituents.

One of the future challenges is to identify the characteristic and distinct interaction-concentration regimes for the existing regimes in the phase diagram of proteins in solution. This should be done for a large variety of food proteins and relate these to the molecular properties of the specific proteins. This should also include a description of possible equilibrium meso-structures. We also envision the need to develop models that predict the response of food protein systems on a given trajectory through the phase diagram, i.e. non-equilibrium phenomena, leading to various gel structures.

Regarding the relatonships between rheological properties and meso-structures, one would like to extend the exisiting models beyond scaling arguments and to include prefactors to have a quantitative relation between molecular interaction forces, meso-structure and macroscopic elasticity of protein gels. Finally, a description of how gels are deformed and broken down during mastication should be developed in terms of theories that can be connected to the ones that describe the ways the gel structures are produced during gelation, and fracture patterns linked with specific textural properties. This would allow for construction of a specific texture starting with molecular interactions. This overall integration of protein gelation is illustrated in Figure 2.7, which shows all of the associated phenomena.

ACKNOWLEDGMENTS

Evdl thanks Max Kolb and Monique Axelos for continued discussions on percolation theory and gelation, Rolf Hilfiker for his continued interest in the work and helpfull discussions, Alan Parker for discussions on glass transitions, gelation, and percolation, Albert Philipse for discussing caging models and Paul Venema for discussions on protein assembly. We thank Mrs. Els Jansen and Mrs. Paige Luck for their help in editing and 'reference managing'.

References

Agrawal, K. R., et al. (1997). Mechanical properties of foods responsible for resisting food breakdown in the human mouth. *Archives of Oral Biology* **42**, 1–9.

Akkermans, C., et al. (2007). Micrometer-Sized Fibrillar Protein Aggregates from Soy Glycinin and Soy Protein Isolate. *Journal of Agricultural and Food Chemistry* **55**, 9877–9882.

Akkermans, C., et al. (2008a). Formation of fibrillar whey protein aggregates: Influence of heat and shear treatment, and resulting rheology. *Food Hydrocolloids* **22**, 1315–1325.

Akkermans, C., et al. (2008b). Enzyme induced formatin of beta-lactoglobulin fibrils by AspN Endoproteinase. *Food Biophysics* **3**, 390–394.

Akkermans, C., et al. (2008c). Peptides are building blocks of heat-induced fibrillar protein aggregates of β-lactoglobulin formed at pH 2. *Biomacromolecules* **9**, 1474–1479.

Allain, C., et al. (1995). Aggregation and sedimentation in colloidal suspensions. *Physical Review Letters* **74**, 1478–1481.

Anderson, V. J. and Jones, R. A. L. (2001). The influence of gelation on the mechanism of phase separation of a biopolymer mixture. *Polymer* **42**, 9601–9610.

Anderson, V. J. and Lekkerkerker, H. N. W. (2002). Insights into phase transition kinetics from colloid science. *Nature* **416**, 811–815.

Arakawa, T., Timasheff, S. N., et al. (1985). Theory of Protein Solubility. *In* Diffraction Methods for Biological Macromolecules Part A, 114, 49–146, Wyckoff, H. W. et al. (ed.). Academic Press, Orlando.

Arnaudov, L. N. and de Vries, R. (2005). Thermally induced fibrillar aggregation of hen egg white lysozyme. *Biophysical Journal* **88**, 515–526.

Arnaudov, L. N. and de Vries, R. (2006). Strong impact of ionic strength on the kinetics of fibrilar aggregation of bovine ß-lactoglobulin. *Biomacromolecules* **7**, 3490–3498.

Arnaudov, L.N., and de Vries, R. (2007). Theoretical modeling of the kinetics of fibrilar aggregation of bovine-lactoglobulin at pH 2. *Journal of Chemical Physics* 126, 145106-1 - 145106-7.

Arnaudov, L. N., et al. (2003). Multiple steps during the formation of β-lactoglobulin fibrils. *Biomacromolecules* **4**, 1614–1622.

Asnaghi, D., et al. (1992). Coagulation kinetics and aggregate morphology in the intermediate regimes between diffusion-limited and reaction-limited cluster aggregation. *Physical Review A* **45**, 1018–1023.

Axelos, M. A. V. and Kolb, M. (1990). Crosslinked biopolymers: Experimental evidence for scalar percolation theory. *Physical Review Letters* **64**, 1457–1460.

Aymard, P., et al. (1996). A comparison of the structure of β-lactoglobulin aggregates formed at pH 7 and pH 2. *International Journal of Polymer Analysis and Characterization* **2**, 115–119.

Aymard, P., et al. (1997). Fractality of globular protein aggregates: From the molecular to the microscopic level. *Fractals* **5**, 23–43.

Aymard, P., et al. (1999). Static and dynamic scattering of β-lactoglobulin aggregates formed after heat-induced denaturation pH 2. *Macromolecules* **32**, 2542–2552.

Barbut, S. (1995). Effect of sodium level on the microstructure and texture of whey protein isolate gels. *Food Research International* **28**, 437–443.

Barrangou, L. M., et al. (2006). Textural properties of agarose gels. II. Relationships between rheological properties and sensory texture. *Food Hydrocolloids* **20**, 196–203.

Belloque, J. and Smith, G. M. (1998). Thermal denaturation of β-lactoglobulin A H-1 NMR study. *Journal of Agricultural and Food Chemistry* **46**, 1805–1813.

Bergman, D. J. and Kantor, Y. (1984). Critical properties of an elastic fractal. *Physical Review Letters* **53**, 511–514.

Berli, C. L. A., et al. (2002). Modelling the viscosity of depletion flocculated emulsions. *Colloids and Surfaces A: Physicochemical and Engineering Aspects* **203**, 11–20.

Bernal, V. and Jelen, P. (1985). Thermal stability of whey proteins – a calorimetric study. *Journal of Dairy Science* **68**, 2847–2852.

Berthelier, V. and Wetzel, R. (2006). Screening for modulators of aggregation with a microplate elongation assay Methods in Enzymology **413**, pp. 313–325.

Bhatka, R., et al. (2004). Bolus size and unilateral chewing cycle kinematics. *Archives of Oral Biology* **49**, 559–566.

Bhattacharjee, C., et al. (2005). Structural changes of β-lactoglobulin during thermal unfolding and refolding – An FT-IR and circular dichroism study. *The Protein Journal* **24**, 27–35.

Bibette, J., et al. (1992). Kinetically induced ordering in gelation of emulsions. *Physical Review Letters* **69**, 981–984.

Bolder, S. G., et al. (2007). Heat-induced whey protein isolate fibrils: Conversion, hydrolysis, and disulphide bond formation. *International Dairy Journal* **17**, 846–853.

Bot, A., et al. (1995). Brillouin light scattering from a biopolymer gel: hypersonic sound waves in gelatin. *Colloid & Polymer Science* **273**, 252–256.

Bot, A., et al. (1997). The transition zone for gelatin gels as measured by high-frequency rheology. *Imaging Science Journal* **45**, 191–196.

Bourdiol, P. and Mioche, L. (2000). Correlations between functional and occlusal tooth-surface areas and food texture during natural chewing sequences in humans. *Archives of Oral Biology* **45**, 691–699.

Bourne, M. (2004). Relation between texture and mastication. *Journal of Texture Studies* **35**, 125–143.

Bowland, E. L. and Foegeding, E. A. (1995). Effects of anions on thermally induced whey protein isolate gels. *Food Hydrocolloids* **9**, 47–56.

Bowland, E. L., et al. (1995). Rheological analysis of anion-induced matrix transformations in thermally induced whey protein isolate gels. *Food Hydrocolloids* **9**, 57–64.

Boye, J. I., et al. (1997). Use of differential scanning calorimetry and infrared spectroscopy in the study of thermal and structural stability of α-lactalbumin. *Journal of Agricultural Food Chemistry* **45**, 1116–1125.

Bremer, L. G. B., et al. (1990). On the fractal nature of the structure of acid casein gels. *Colloids and Surfaces* **51**, 159–170.

Bremer, L. G. B., et al. (1995). Estimations of the aggregation time of various colloidal systems. *Colloids and Surfaces A: Physicochemical and Engineering Aspects* **99**, 121–127.

Brink, J., et al. (2007). Simultaneous analysis of the structural and mechanical changes during large deformation of whey protein isolate/gelatin gels at the macro and micro levels. *Food Hydrocolloids* **21**, 409–419.

Broersen, K., et al. (2007). Effect of protein charge on the generation of aggregation-prone conformers. *Biomacromolecules* **8**, 1648–1656.

Bromley, E. H. C., et al. (2005). Aggregation across the lengthscales in β-lactoglobulin. *Faraday Discuss* **128**, 13–27.

Bromley, E. H. C., et al. (2006). Mechanisms of structure formation in particulate gels of β-lactoglobulin formed near the isoelectric point. *European Physics Journal E* **21**, 145–152.

Brown, J. A., et al. (2003). Relationships among rheological and sensorial properties of young cheeses. *Journal of Dairy Science* **86**, 3054–3067.

Brown, W. D. and Ball R. C. (1985). *Computer simulation of chemically limited aggregation. Journal Physics A.* 18, L517-L521

Brownsey, G. J., et al. (1986). The influence of food gel structure on texture perception. *In* Gums and Stabilisers for the Food Industry 3, Phillips, G. O., et al. (ed.). Elsevier Applied Science Publisher 103–110.

Buscall, R., et al. (1988). Scaling behavior of the rheology of aggregate networks formed from colloidal particles. *Journal of the Chemical Society-Faraday Transactions I* **84**, 4249–4260.

Carpineti, M. and Giglio, M. (1992). Spinodal-type dynamics in fractal aggregation of colloidal clusters. *Physical Review Letters* **68**, 3327–3330.

Carpineti, M. and Giglio, M. (1993). Transition from semi-order to disorder in the aggregation of dense colloidal solutions. *Physical Review Letters* **70**, 3828–3831.

Carugo, O. and Argos, P. (1997). Protein-protein crystal-packing contacts. *Protein Science* **6**, 2261–2263.

Cates, M. E., et al. (1998). Jamming, force chains, and fragile matter. *Physical Review Letters* **81**, 1841–1844.

Cecconi, C., et al. (2005). Direct observation of the three-state folding of a single protein molecule. *Science* **309**, 2057–2060.

Chantrapornchai, W. and McClements, D. J. (2002). Influence of NaCl on optical properties, large-strain rheology and water holding capacity of heat-induced whey protein isolate gels. *Food Hydrocolloids* **16**, 467–476.

Chatellier, J. Y., et al. (1985). Critical helix content in gelatin gels. *International Journal Biological Macromolecules* **7**, 311–314.

Chen, J. (2009). Food oral processing-A review. *Food Hydrocolloids* **23**, 1–25.

Chen, J. and Dickinson, E. (1998). Viscoelastic properties of heat-set whey protein emulsion gels. *Journal of Texture Studies* **29**, 285–304.

Chen, J. and Dickinson, E. (1999a). Effect of surface character of filler particals on rheology of heat-set whey protein emulsion gels. *Colloids and Surfaces B: Biointerfaces* **12**, 373–381.

Chen, J. and Dickinson, E. (1999b). Interfacial ageing effect on the rheology of a heat-set protein emulsion gel *Food Hydrocolloids* **13**, 363–369.

Chiew, Y. C. and Glandt, E. D. (1983). Percolation behavior of permeable and of adhesive spheres. *Journal of Physics a-Mathematical and General* **16**, 2599–2608.

Chiti, F. and Dobson, C. M. (2006). Protein misfolding, functional amyloid, and human disease. *Annual Reviews of Biochemistry* **75**, 333–366.

Cipelletti, L. and Ramos, L. (2002). Slow dynamics in glasses, gels and foams. *Current Opinion in Colloid & Interface Science* **7**, 228–234.

Clark, A. H. (1987). The application of network theory to food systems. *In* Food structure and behaviour, 13–34, Blanshard, J. M. V. and Lillford, P. (eds.). Academic Press Limited.

Clark, A. H., et al. (1981). Electron microscopy of network structures in thermally-induced globular protein gels. *International Journal of Peptide and Protein Research* **17**, 380–392.

Clark, A. H. (1998). Gelation of globular proteins. *In* Functional Properties of Food Macromolecules, 77–142, Hill, S.E., Ledward, D.A., and Mitchell, J.R. (eds.). Aspen Publishers, Maryland.

Clark, A. H. and Ross-Murphy, S. B. (1985). The concentration dependence of biopolymer gel modulus. *British Polymer Journal* **17**, 164–168.

Clark, A. H. and Ross-Murphy, S. B. (1987). Structural and mechanical properties of biopolymer gels. *Advances in Polymer Science* **83**, 57–192.

Considine, T., et al. (2005). Influence of binding of sodium dodecyl sulfate, all-trans-retinol, palmitate 8-anilino-1-naphthalenesulfonate on the heat-induced unfolding and aggregation of β-lactoglobulin B. *Journal of Agricultural and Food Chemistry* **53**, 3197–3205.

Conte, L. L., et al. (1999). The atomic structure of protein-protein recognition sites. *Journal of Molecular Biology* **285**, 2177–2198.

Creusot, N. and Gruppen, H. (2007). Enzyme-induced aggregation and gelation of proteins. *Biotechnology* **25**, 597–601.

Croguennec, T., et al. (2003). Stable monomeric intermediate with exposed Cys-119 is formed during heat denaturation of β-lactoglobulin. *Biophysical and Biochemical Research Communications* **301**, 465–471.

Croguennec, T., et al. (2004). Spectroscopic characterization of heat-induced nonnative β-lactoglobulin monomers. *Protein Science* **13**, 1340–1346.

Croguennoc, P., et al. (2001). Phase separation and association of globular protein aggregates in the presence of polysaccharides: 2. Heated mixtures of native β-Lactoglobulin and k-Carrageenan. *Langmuir* **17**, 4380–4385.

Cudney, B., et al. (1994). Screening and optimization strategies for macromolecular crystal-growth. *Acta Crystallographica Section D-Biological Crystallography* **50**, 414–423.

Dawson, K. A. (2002). The glass paradigm for colloidal glasses, gels, and other arrested states driven by attractive interactions. *Current Opinion in Colloid & Interface Science* **7**, 218–227.

De Gennes, P. G. (1976a). La percolation: un concept unificateur. *La recherche* 7, 919–927.

De Gennes, P. G. (1976b). On a relation between percolation theory and the elasticity of gels. *Journal of Physique Letter* **37**, L-1, L-2.

De Gennes, P. G. (1979). Scaling properties in polymer physics. Cornell University Press, New York. Ithaka.

de Kruif, K. G., et al. (1995). Gelation of proteins from milk. *Faraday Discussions* **101**, 185–200.

de Wijk, R. A., et al. (2003). The role of intra-oral manipulation in the perception of sensory attributes. *Appetite* **40**, 1–7.

Dejmek, P. and Visschers, R. W. (2006). Discussion session on food gels. *Food Hydrocolloids* **20**, 446–447.

Devreux, F., et al. (1993). Crossover from scalar to vectorial percolation in silica gelation. *Physical Review E* **47**, 2689–2694.

Dib, R., et al. (1996). Secondary structure changes and peptic hydrolysis of β-lactoglobulin induced by diols. *Bioploymers* **39**, 23–30.

Dickinson, E. and Casanova, H. (1999). A thermoreversible emulsion gel based on sodium caseinate. *Food Hydrocolloids* **13**, 285–289.

Dickinson, E. and Chen, J. (1999). Heat-set whey protein emulsion gels: Role of active and inactive filler particles. *Journal of Dispersion Science and Technology* **20**, 197–213.

Dickinson, E. and Yamamoto, Y. (1996a). Effect of lecithin on the visoelastic properties of β-Lactoglobulin stabilized emulsion gels. *Food Hydrocolloids* **10**, 301–307.

Dickinson, E. and Yamamoto, Y. (1996b). Rheology of milk protein gels and protein-stabilized emulsion gels cross-linked with transglutaminase. *Journal of Agricultural and Food Chemistry* **44**, 1371–1377.

Dickinson, E. and Yamamoto, Y. (1996c). Viscoelastic properties of heat-set whey protein-stabilized emulsion gels with added lecithin. *Journal of Food Science* **61**, 811–816.

Dill, K. A., et al. (1989). Thermal stabilities of globular-proteins. *Biochemistry* **28**, 5439–5449.

Dima, R. I. and Thirumalai, D. (2002). Exploring protein aggregation and self-propagation using lattice models: Phase diagram and kinetics. *Protein Science* **11**, 1036–1049.

Dinsmore, A. (2001). Three-dimensional confocal microscopy of colloids. *Applied Optics* **40**, 4152–4159.

Djabourov, M. (1988). Architecture of gelatin gels. *Contemporary Physics* **29**, 273–297.

Djabourov, M., et al. (1988a). Gelation of aqueous gelatin solutions. 1. Structural investigation. *Journal De Physique* **49**, 319–332.

Djabourov, M., et al. (1988b). Gelation of aqueous gelatin solutions. 2. rheology of the sol-gel transition. *Journal De Physique* **49**, 333–343.

Dobson, C. M. (2003). Protein folding and misfolding. *Nature* **426**, 884–890.

Donato, L., et al. (2005). Influence of the NaCl or $CaCl_2$ concentration on the structure of heat-set serum albumin gels at pH 7. *Biomacromolecules* **6**, 2157–2163.

Doucet, D. and Foegeding, E. A. (2005). Gel formation of peptides produced by extensive enzymatic hydrolysis of β-lactoglobulin. *Biomacromolecules* **6**, 1140–1148.

Doucet, D., et al. (2001). Rheological characterizaion of a gel formed during extensive enzymatic hydrolysis. *Journal of Food Science* **66**, 711–715.

Doucet, D., et al. (2003a). Enzyme-induced gelation of extensively hydrolyzed whey proteins by alcalase: comparison with the plastein reaction and characterization of interactions. *Journal of Agricultural and Food Chemistry* **51**, 6036–6042.

Doucet, D., et al. (2003b). Enzyme-induced gelation of extensively hydrolyzed whey proteins by alcalase: peptide identification and determination of enzyme specificity. *Journal of Agricultural and Food Chemistry* **51**, 6300–6308.

Dufour, E., et al. (1993). Reversible effects of medium dielectric-constant on structural transformation of β-lactoglobulin and its retinol binding. *Bioploymers* **33**, 589–598.

Dufour, E. and Haertle, T. (1993). Temperature-induced folding changes of β-lactoglobulin in hydro-methanolic solutions. *International Journal of Biological Macromolecules* **15**, 293–297.

Edsall, J. T. (1943). Rotary Brownian Movement. The Shape of Protein Molecules As Determined from Viscosity and Double Refraction of Flow. *In* Proteins, Amino Acids and Peptides as Ions and Dipolar Ions. 506–542, Cohn, E. J. (ed.). Reinhold Publishing Corporation, New York (USA).

Edsall, J. T. and Wyman, J. (1958). Electrostatics: Its Application to Polar Molecules and Ionic Solutions. *In* Thermodynamics, Electrostatics, and the Biological Significance of the Properties of Matter 1 *Biophysical Chemistry*, Edsall, J. T. and Wyman, J. (eds.). Academic Press, New York.

Edwards, P. J. B., et al. (2002). Heat-resistant structural features of bovine β-lactoglobulin A revealed by NMR H/D exchange observations. *International Dairy Journal* **12**, 331–344.

Eicke, H. F. and Hilfiker, R. (1986). Connectivity-controlled charge transport in water-in-oil microemulsions. *Chemical Physics Letters* **125**, 295–298.

Eldridge, J. E. and Ferry, J. D. (1954). Studies of the cross-linking process in gelatin gels. 3. Dependence of melting point on concentration and molecular weight. *Journal of Physical Chemistry* **58**, 992–995.

Engelen, L., et al. (2005). The influence of product and oral characteristics on swallowing. *Archives of Oral Biology* **50**, 739–746.

Euston, S. R., et al. (2007). Denaturation and aggregation of β-lactoglobulin – a preliminary molecular dynamics study. *Food Hydrocolloids* **21**, 1081–1091.

Fändrich, M., et al. (2006). Apomyoglobin reveals a random-nucleation mechanism in amyloid protofibril formation. *Acta Histochemica* **108**, 215–219.

Fang, Y. and Dalgleish, D. G. (1998). The conformation of α-lactalbumin as a function of pH, heat treatment and adsorption at hydrophobic surfaces studied by FTIR. *Food Hydrocolloids* **12**, 121–126.

Farrell, H. M., et al. (2004). Nomenclature of the proteins of cows' milk – Sixth revision. *Journal of Dairy Science* **87**, 1641–1674.

Feder, J., et al. (1984). Scaling behavior and cluster fractal dimension determined by light scattering from aggregating proteins. *Physical Review Letters* **53**, 1403–1406.

Feng, S. and Sen, P. N. (1984). Percolation on elastic networks – new exponent and threshold. *Physical Review Letters* **52**, 216–219.

Fernandez-Recio, J., et al. (2005). Optimal docking area: A new method for predicting protein-protein interaction sites. *Proteins: Structure, Function, and Bioinformatics* **58**, 134–143.

Ferrone, F. (1999). Analysis of protein aggregation kinetics. *Methods in enzymology* **309**. Academic Press 256–274.

Ferry, J. D. and Oncley, J. L. (1941). Studies of the dielectric properties of protein solutions. III. Lactoglobulin. *Journal American Chemist Society* **63**, 272–278.

Finley, J. W. (1975). Deamidated gluten – a potential fortifier for fruit juices. *Journal of Food Science* **40**, 1283–1285.

Flory, P. J. (1941). Molecular size distribution in three dimensional polymers. I. Gelation. *Journal of the American Chemical Society* **63**, 3083–3090.

Flory, P. J. (1942). Constitution of three-dimensional polymers and the theory of gelation. *Journal of Physical Chemistry* **46**, 132–140.

Foegeding, E. A. (2005). Rheology, structure and texture perception in food protein gels. *In* Food Colloids Interactions, Microstructure and Processing, Dickinson, E. (ed.). Royal Society of Chemistry 3–15.

Foegeding, E. A. (2007). Rheology and sensory texture of biopolymer gels. *Current Opinion in Colloid & Interface Science* **12**, 242–250.

Foegeding, E. A., et al. (1995). Factors that determine the fracture properties and microstructure of globular protein gels. *Food Hydrocolloids* **9**, 237–249.

Fontijn-Tekamp, F. A., et al. (2004). Swallowing threshold and masticatory performance in dentate adults. *Physiology and Behavior* **83**, 431–436.

Foster, K. D., et al. (2006). Effect of texture of plastic and elastic model foods on the parameters of mastication. *Journal of Neurophysiology* **95**, 3469–3479.

Foster, P. R., et al. (1971). Salting-out of enzymes with ammonium sulphate. *Biotechnology and Bioengineering* **13**, 713–718.

Foster, P. R., et al. (1976). The kinetics of protein salting-out: Precipitation of yeast enzymes by ammonium sulfate. *Biotechnology and Bioengineering* **18**, 545–580.

Gambareli, F. R., et al. (2007). Influence of measurement technique, test food, teeth and muscle force interactions in masticatory performance. *Journal of Texture Studies* **38**, 2–20.

Gazit, E. (2007). Use of biomolecular templates for the fabrication of metal nanowires. *FEBS Journal* **274**, 317–322.

George, A. and Wilson, W. W. (1994). Predicting protein crystallization from a dilute-solution property. *Acta Crystallographica Section D-Biological Crystallography* **50**, 361–365.

Gisler, T., et al. (1999). Strain hardening of fractal colloidal gels. *Physical Review Letters* **82**, 1064–1067.

Gordon, M. (1962). Goods theory of cascade processes applied to statistics of polymer distributions. *Proceedings of the Royal Society of London Series a-Mathematical and Physical Sciences* **268**, 240–256.

Gordon, M. and Ross-Murphy, S. B. (1975). Structure and properties of molecular trees and networks. *Pure and Applied Chemistry* **43**, 1–26.

Gosal, W. S., et al., (2002b). Novel amyloid fibrillar networks derived from a globular protein: β-lactoglobulin. *Langmuir.* **18**, 7174–7181.

Grant, M. C. and Russel, W. B. (1993). Volume-fraction dependence of elastic-moduli and transition-temperature for colloidal silica-gels. *Physical Review E* **47**, 2606–2614.

Green, A. A. (1931). Studies in the physical chemistry of the proteins. VIII. The solubility of hemoglobulin in concentrated salt solutions. A study of the salting out of proteins. *Journal of Biological Chemistry* **93**, 495–516.

Green, A. A. (1932). Studies in the physical chemistry of the proteins X. The solubility of hemoglobin in solutions of chlorides and sulfates of varying concentration. *Journal of Biological Chemistry* **95**, 47–66.

Green, A. A., et al. (1935). Studies in the physical chemistry of the proteins. XII. The solubility of human hemoglobin in concentrated salt solutions. *Journal of Biological Chemistry* **109**, 631–634.

Groenewold, J. and Kegel, W. K. (2001). Anomalously large equilibrium clusters of colloids. *Journal of Physical Chemistry B* **105**, 11702–11709.

Groenewold, J. and Kegel, W. K. (2004). Colloidal cluster phases, gelation and nuclear matter. *Journal of Physics-Condensed Matter* **16**, S4877–S4886.

Guinard, J.-X. and Mazzucchelli, R. (1996). The sensory perception of texture and mouthfeel. *Trends in Food Science and Technology* **7**, 213–219.

Guo, B., et al. (1999). Correlation of second virial coefficients and solubilities useful in protein crystal growth. *Journal of Crystal Growth* **196**, 424–433.

Guo, L., et al. (2003). Kinetics of triple helix formation in semidilute gelatin solutions. *Macromolecules* **36**, 9999–10008.

Gwartney, E. A., et al. (2004). Sensory texture and mechanical properties of stranded and particulate whey protein emulsion gels. *Journal of Food Science* **69**, S333–S339.

Haan, S. W. and Zwanzig, R. (1977). Series expansions in a continuum percolation problem. *Journal of Physics a-Mathematical and General* **10**, 1547–1555.

Haas, C. and Drenth, J. (1999). Understanding protein crystallization on the basis of the phase diagram. *Journal of Crystal Growth* **196**, 388–394.

Hamada, D. and Dobson, C. M. (2002). A kinetic study of β-lactoglobulin amyloid fibril formation promoted by urea. *Protein Science* **11**, 2417–2426.

Hamada, J. S. (1994). Deamination of food proteins to improve functionality. *Critical Reviews in Food Science and Technology* **34**, 283–292.

Hamann, D. D. (1983). Structural failure in solid foods. *In* Physical Properties of Foods AVI Publishing Co., 351–383, Bagley, E. B. and Peleg, M. (eds.). Westport, CT.

Hamann, D. D., et al. (2006). Analysis of compression, tension and torsion for testing food gel fracture properties. *Journal of Texture Studies* **37**, 620–639.

Hammersley, J. M. (1956). Conditional Monte Carlo. *Journal ACM* **3**, 73–76.

Harwalkar, V. R. and Kalab, M. (1985). Thermal denaturation and aggregation of β-lactoglobulin at pH 2.5. Effect of ionic strength and protein concentration. *Milchwissenschaft* **40**, 31–34.

Hayward, S., et al. (1987). Dynamic percolation transition induced by phase separation: a Monte Carlo analysis. *Journal of Statistical Physics* **49**, 1053–1081.

Hendrix, T., et al. (2000). A calorimetric study of the influence of calcium on the stability of bovine α-lactalbumin. *Biophysical Chemistry* **84**, 27–34.

Hermans, J. J. (1965). Investigation of the elastic properties of the particle network in gelled solutions of hydrocolloids. I. Carboxymethyl cellulose. *Journal of Polymer Science Part A: General Papers* **3**, 1859–1868.

Hermans, P. (1949). Physics and chemistry of cellulose fibers. Elsevier Publishing Company, New York.

Hermansson, A. M., et al. (1986). Formation of 2 types of gels from bovine myosin. *Journal of the Science of Food and Agriculture* **37**, 69–84.

Hiiemae, K., et al. (1996). Natural bites, food consistency and feeding behaviour in man. *Archives of Oral Biology* **41**, 175–189.

Hilfiker, R. and Eicke, H. F. (1987). Self-consistency of the percolation model as applied to a macrofluid-like water-in-oil microemulsion. *Journal of the Chemical Society-Faraday Transactions I* **83**, 1621–1629.

Holt, C., et al. (1998). Comparison of the effect of heating on the thermal denaturation of nine different β-lactoglobulin preparations of genetic variants A, B or *International Dairy Journal* **8**, 99–104.

Horne, D. S. (1987). Determination of the fractal dimension using turbidimetric techniques – application to aggregating protein systems. *Faraday Discussions*, 259–270.

Huang, X. L., et al. (1999). Modification of rheological properties of whey protein isolates by limites proteolysis. *Nahrung* **43**, 79–85.

Hudson, H. M., et al. (2000). Rheological and physical properties of derivitized whey protein isolate powders. *Journal of Agricultural Food Chemistry* **48**, 3112–3119.

Hutchins, J. B. and Lillford, P. J. (1988). The perception of food texture – The philosophy of the breakdown path. *Journal of Texture Studies* **19**, 103–115.

Iametti, S., et al. (1996). Modifications occur at different structural levels during the heat denaturation of β-lactoglobulin. *European Journal Biochemistry* **237**, 106–112.

Ikeda, S., and Foegeding, E. A. (1999a). Dynamic viscoelastic properties of thermally induced whey protein isolate gels with added lecithin. *Food Hydrocolloids* **13**, 245–254.

Ikeda, S., and Foegeding, E.A. (1999b). Effects of lecithin on thermally induced whey protein isolate gels. *Food Hydrocolloids* **13**, 239–244.

Ikeda, S., et al. (1999). Rheological study on the fractal nature of the protein gel structure. *Langmuir* **15**, 8584–8589.

Ikeda, S. and Li-Chan, E. C. Y. (2004). Raman spectroscopy of heat-induced fine-stranded and particulate β-lactoglobulin gels. *Food Hydrocolloids* **18**, 489–498.

Ikeda, S. and Morris, V. J. (2002). fine-stranded and particulate aggregates of heat-denatured whey proteins visualized by atomic force Microscopy. *Biomacromolecules* **3**, 382–389.

Ipsen, R., et al. (2000). Standardized reaction times used to describe the mechanism of enzyme-induced gelation in whey protein systems. *Journal of Dairy Research* **67**, 403–413.

Ipsen, R., et al. (2001). Molecular self-assembly of partially hydrolysed α-lactalbumin resulting in strong gels with a novel microstructure. *Journal of Dairy Research* **68**, 277–286.

ISO (1981). Sensory analysis vocabulary Part 4. International Organization for Standardization. Switzerland, Geneva.

Israelachvili, J. N., et al. (1976). Theory of self-assembly of hydrocarbon amphiphiles into micelles and bilayers. *Journal of the Chemical Society – Faraday Transactions II.* **72**, 1525–1568.

Iyer, H. V. and Przybycien, T. M. (1994). Protein precipitation: Effects of mixing on protein solubility. *AIChE Journal* **40**, 349–360.

Izmailova, V. N., et al. (2004). Gelation in gelatin and gelatin-containing multicomponent blends. *Polymer Science Series C* **46**, 73–92.

Jalabert-Malbos, M. L., et al. (2007). Particle size distribution in the food bolus after mastication of natural foods. *Food Quality and Preference* **18**, 803–812.

Joly-Duhamel, C., et al. (2002a). All gelatin networks: 2. The master curve for elasticity. *Langmuir* **18**, 7158–7166.

Joly-Duhamel, C., et al. (2002b). All gelatin networks: 1. Biodiversity and physical chemistry. *Langmuir* **18**, 7208–7217.

Joly, M., (1949a). Etude par la birefringence decoulement de la denaturation thermique de la serumal bumine. *Bulletin de la Societe de chimie biologique* **31**, 1642–1655.

Joly, M. (1949b). Etude par la birefringence decoulement de la gelification de la gelatine et de ses interactions avec dautres substances organiques. *Bulletin De La Societe De Chimie Biologique* **31**, 105–107.

Joly, M. and Barbu, E. (1949). Etude par la birefringence decoulement de la denaturation thermique de la serum albumine. *Bulletin De La Societe De Chimie Biologique* **31**, 1642–1655.

Joly, M. and Barbu, E. (1950). L'Action des electrolytes sur lagregation de la eerum albumine – etude par la birefringence decoulement. *Bulletin De La Societe De Chimie Biologique* **32**, 849–850.

Jones, R. A. L. (2004). Soft Condensed Matter. Oxford University Press.

Ju, Z. Y., et al. (1995). Effects of limited proteolysis on gelation and gel properties of whey protein isolate. *Journal of Dairy Science* **78**, 2119–2128.

Ju, Z. Y., et al. (1997). Enzyme-induced gelation of whey proteins: effect of protein denaturation. *International Dairy Journal* **7**, 71–78.

Kam, Z., et al. (1978). On the crystallization of proteins. *Journal of Molecular Biology* **123**, 539–555.

Kanai, H., et al. (1992). Fragile networks and rheology of concentrated suspensions. *Rheologica Acta* **31**, 333–344.

Kanjilal, S., et al. (2003). Ultrasonic studies of alcohol-induced transconformation in β-lactoglobulin: The intermediate state. *Biophysical Journal* **85**, 3928–3934.

Kantor, Y. and Webman, I. (1984). Elastic properties of random percolating systems. *Physical Review Letters* **52**, 1891–1894.

Kavanagh, G. M., et al. (2000). Heat-induced gelation of globular proteins: 4. Gelation kinetics of low pH β-lactoglobulin gels. *Langmuir* **16**, 9584–9594.

Kegel, W. K. and van der Schoot, P. (2004). Competing hydrophobic and screened-coulomb interactions in hepatitis B Virus capsid assembly. *Biophysical Journal* **86**, 3905–3913.

Keller, A. (1995). Introductory lecture – Aspects of polymer gels. *Faraday Discussions*, 1–49.

Kerstens, S., et al. (2006). Influence of ionic surfactants on the microstructure of heat-set β-lactoglobulin-stabilized emulsion gels. *FOBI* **1**, 133–143.

Kilfoil, M. L., et al. (2003). Dynamics of weakly aggregated colloidal particles. philosophical transactions: mathematical. *Physical and Engineering Sciences* **361**, 753–766.

Kim, A. Y., et al. (2003). Linear chains and chain-like fractals from electrostatic heteroaggregation. *Journal of Colloid and Interface Science* **260**, 149–159.

Koike, A., et al. (1998). Structure and dynamics of ovalbumin gels: II. Gels induced by heat treatment at 80°C. *Polymer Gels and Networks* **6**, 257–271.

Kolb, M., et al. (1983). Scaling of kinetically growing clusters. *Physical Review Letters* **51**, 1123–1126.

Krall, A. H. and Weitz, D. A. (1998). Internal dynamics and elasticity of fractal colloidal gels. *Physics Review Letters* **80**, 778–781.

Krebs, M. R. H., et al. (2005). The mechanism of amyloid spherulite formation by bovine insulin. *Biophysical Journal* **88**, 2013–2021.

Krebs, M. R. H., et al. (2007). Protein particulates: another generic form of protein aggregation? *Biophysical Journal* **92**, 1336–1342.

Kuipers, B. J. H., et al. (2007). Comparison of the aggregation behavior of soy and bovine whey protein hydrolysates. *Biotechnology Advances* **25**, 606–610.

Kumar, S., et al. (2007). Mechanism of formation of amyloid protofibrils of barstar from soluble oligomers: Evidence for multiple steps and lateral association coupled to conformational conversion. *Journal of Molecular Biology* **367**, 1186–1204.

Kuznetsov, Y. G., et al. (2001). The liquid protein phase in crystallization: a case study–intact immunoglobulins. *Journal of Crystal Growth* **232**, 30–39.

Langley, K. R., et al. (1991). Mechanical properties and structure of model composite foods. *Food polymers, Gels and Colloids* **82**, 383–391.

Langton, M. and Hermansson, A.-M. (1992). Fine-stranded and particulate gels of β-lactoglobulin and whey protein at varying pH. *Food Hydrocolloids* **6**, 523–539.

Larsen, T. A., et al. (1998). Morphology of protein-protein interfaces. *Structure* **6**, 421–427.

Laurine, E., et al. (2003). Lithostathine quadruple-helical filaments form proteinase k-resistant deposits in creutzfeldt-jakob disease. *Journal Biological Chemistry* **278**, 51770–51778.

Le Bon, C., et al. (1999). Kinetics of aggregation and gelation of globular proteins after heat-induced denaturation. *Macromolecules* **32**, 6120–6127.

Ledward, D. A. (1986). Gelation of gelatin. *In* Functional Properties of Food Macromolecules, 171–202, Mitchell, J. R. and Ledward, D. A. (eds.). Elsevier Applied Science Publishers, New York.

Ledward, D. A. (2000). Gelation. *In* Handbook of hydrocolloids, 450, Phillips, G. O, W. P. A, (eds.). Woodhead Publishing Limited and CRC Press LLC, Cambridge, England.

Lefevre, T. and Subirade, M. (2000). Molecular differences in the formation and structure of fine-stranded and particulate β-lactoglobulin gels. *Biopolymers* **54**, 578–586.

Levy, R. M., et al. (2003). On the nonpolar hydration free energy of proteins: surface area and continuum solvent models for the solute-solvent interaction energy. *Journal American Chemical Society* **125**, 9523–9530.

Lijnzaad, P., et al. (1996a). Hydrophobic patches on the surfaces of protein structures. *Proteins-Structure Function and Genetics* **25**, 389–397.

Lijnzaad, P., et al. (1996b). A method for detecting hydrophobic patches protein. *Proteins-Structure Function and Genetics* **26**, 192–203.

Lillford, P. J. (2001). Mechansims of fracture in foods. *Journal of Texture Studies* **32**, 397–417.

Lin, M. Y., et al. (1989). Universality in colloid aggregation. *Nature* **339**, 360–362.

Lin, M. Y., et al. (1990). Universal reaction-limited colloid aggregation. *Physical Review A* **41**, 2005–2020.

Linderstrom-Lang, K. U. (1953). The activity coefficient of large multipolar ions. *In* Comtes rendus des travaux du Laboratoire Carlsberg, Séries Chimique, 21, 281–316, K. Linderstrom-Lang, (ed.).

Lironis, G., et al. (1990). Finite-size scaling study of non-equilibrium percolation. *Journal of Physics A: Mathematical and General*. **23,** L329–L334.

Liu, A. J. and Nagel, S. R. (1998). Nonlinear dynamics – Jamming is not just cool any more. *Nature* **396**, 21–22.

Lowe, L. L., et al. (2003). Rheological properties of fine-stranded whey protein isolate gels. *Food Hydrocolloids* **17**, 515–522.

Lucas, P. W., et al. (2004). Food texture and its effect on ingestion, mastication and swallowing. *Journal of Texture Studies* **35**, 159–170.

Lucas, P. W., et al. (2002). Food physics and oral physiology. *Food Quality and Preference* **13**, 203–213.

Lupano, C. E. (1994). Effect of heat-treatments in very acidic conditions on whey-protein isolate properties. *Journal of Dairy Science* **77**, 2191–2198.

Luyten, H., et al. (1992). Comparison of various methods to evaluate fracture phenomena in food materials. *Journal of Texture Studies* **23**, 245–266.

MacKintosh, F. C., et al. (1995). Elasticity of semiflexible biopolymer networks. *Physical Review Letters* **75**, 4425–4428.

Malfois, M., et al. (1996). A model of attractive interactions to account for fluid-fluid phase separation of protein solutions. *Journal of Chemical Physics* **105**, 3290–3300.

Malone, M. E., et al. (2003). Oral behaviour of food hydrocolloids and emulsions. Part 1 Lubrication and deposition considerations. *Food Hydrocolloids* **17**, 763–773.

Manderson, G.A., et al. (1999a). Effect of heat treatment on the circular dichroism spectra of bovine β-lactoglobulin A, B, and C. *Journal of Agricultural Food Chemistry* **47**, 4557–4567.

Manderson, G.A., et al. (1999b). Effect of heat treatment on bovine β-lactoglobulin A, B, and C explored using thiol availability and fluorescence. *Journal of Agricultural and Food Chemistry* **47**, 3617–3627.

Manno, M., et al. (1999). Multiple interactions between molecular and supramolecular ordering. *Physical Review E* **59**, 2222–2230.

Manno, M. and Palma, M. U. (1997). Fractal morphogenesis and interacting processes in gelation. *Physical Review Letters* **79**, 4286–4289.

Martin, J. E., et al. (1987). Critical exponents for the sol-gel transition. *Physical Review A* **36**, 1803–1810.

Matsumoto, T., et al. (1992). Viscoelastic and saxs investigation of fractal structure near the gel point in alginate aqueous systems. *Macromolecules* **25**, 5430–5433.

McGuffey, M. K., et al. (2005). Denaturation and aggregation of three α-lactalbumin preparations at neutral pH. *Journal of Agricultural Food Chemistry* **53**, 3182–3190.

McPherson, A. (1990). Current approaches to macromolecular crystallization. *European Journal of Biochemistry* **189**, 1–23.

Meakin, P. (1983). Formation of fractal clusters and networks by irreversible diffusion-limited aggregation. *Physical Review Letters* **51**, 1119–1122.

Meersman, F. and Dobson, C. M. (2006). Probing the pressure-temperature stability of amyloid fibrils provides new insights into their molecular properties. *Biochimica et Biophysica Acta (BBA) – Proteins & Proteomics* **1764**, 452–460.

Mehalebi, S., et al. (2008). The influence of electrostatic interaction on the structure and the shear modulus of heat-set globular protein gels. *Soft Matter* **4**, 893–900.

Melander, W. and Horváth, C. (1977). Salt effects on hydrophobic interactions in precipitation and chromatography of proteins: An interpretation of the lyotropic series. *Archives of Biochemistry and Biophysics* **183**, 200–215.

Mellema, M., Scaling relations between structure and rheology of ageing casein particle gels. Wageningen Universiteit. Promotor: dr.ir.P. Walstra, dr. E van der Linden, co-promotor(en): dr.ir.T. van Vliet. - Wageningen: Michel Mellema, 2000.

Mellema, M., et al. (1999a). Molecular mechanism of the renneting process of casein micelles in skim milk, examined by viscosity and light-scattering experiments and simulated by model SCF calculations. *Langmuir* **15**, 6304–6313.

Mellema, M., et al. (1999b). Relating colloidal particle interactions to gel structure using Brownian dynamics simulations and the Fuchs stability ratio. *Journal of Chemical Physics* **111**, 6129–6135.

Mellema, M., et al. (2002). Categorization of rheological scaling models for particle gels applied to casein gels. *Journal of Rheology* **46**, 11–29.

Meller, A., et al. (1999). Viscoelasticity of depletion-induced gels in emulsion-polymer systems. *Langmuir* **15**, 1918–1922.

Meullenet, J. F., et al. (2002). Measurement of biting velocities, and predetermined and individual crosshead speed instrumental imitative tests for predicting cheese hardness. *Journal of Texture Studies* **33**, 45–58.

Mikol, V., and Giegé, R. (1992). The physical chemistry of protein crystallization. *In* Crystallisation of nucleic acids and proteins, 219–239, Ducruix, A., Giegé, R. (eds.). IRL Press, Oxford, New York, Tokyo.

Mikol, V., et al. (1990). Diagnostic of precipitant for biomacromolecule crystallization by quasi-elastic light-scattering. *Journal of Molecular Biology* **213**, 187–195.

Miller, C. A. and Miller, D. D. (1985). Network formation by spinodal decomposition in colloidal dispersions. *Colloids and Surfaces* **16**, 219–223.

Mishra, R., et al. (2007). Lysozyme amyloidogenesis is accelerated by specific nicking and fragmentation but decelerated by intact protein binding and conversion. *Journal of Molecular Biology* **366**, 1029–1044.

Mitescu, C. D. and Musolf, M. J. (1983). Critical exponent for 3-D percolation conductivity Revisited. *Journal De Physique Lettres* **44**, L679–L683.

Monahan, F. J., et al. (1995). Effect of pH and temperature on protein unfolding and thiol-disculfide interchange reactions during heat-induced gelation of whey proteins. *Journal of Agricultural Food Chemistry* **43**, 46–52.

Mulvihill, D. M. and Kinsella, J. E. (1990). Effects of structuring and destructuring anionic ions on the rheological properties of thermally induced β-lactoglobulin gels. *Food Hydrocolloids* **4**, 267–276.

Muthukumar, M. (1985). Dynamics of polymeric fractals. *The Journal of Chemical Physics* **83**, 3161–3168.

Muthukumar, M. (1989). Screening effect on viscoelasticity near the gel point. *Macromolecules* **22**, 4656–4658.

Nakamura, T., et al. (1984). Network structure formation in thermally induced gelation of glycinin. *Journal of Agricultural and Food Chemistry* **32**, 349–352.

Nelson, R. and Eisenberg, D. (2006). Recent atomic models of amyloid fibril structure. *Current Opinion in Structural Biology* **16**, 260–265.

Neurath, H., et al. (1944). The chemistry of protein denaturation. *Chemical Reviews* **34**, 157–265.

Nicolai, T., et al. (2006). Iso-scattering points during heat-induced aggregation and gelation of globular proteins indicating micro-phase separation. *Europhysics Letters* **73**, 299–305.

Niranjan, P. S., et al. (2003). The polymerization of actin: Thermodynamics near the polymerization line. *The Journal of Chemical Physics* **119**, 4070–4084.

Nishinari, K. (2004). Rheology, food texture and mastication. *Journal of Texture Studies* **35**, 113–124.

Normamnd, V. and Parker, A. (2003). Scaling in the dynamics of gelatin gels. *In* Proceedings of the third International Symposium on Food Rheology and Structure, Fischer, P. et al., (eds.). Kerschensteiner Verlag, Lappersdorf, Germany.

Noro, M. G., et al. (1999). The role of long-range forces in the phase behavior of colloids and proteins. *Europhysics Letters* **48**, 332–338.

Ohgren, C., et al. (2004). Structure-fracture measurements of particulate gels. *Journal of Material Science* **39**, 6473–6482.

Okada, A., et al. (2007). Oral behavior from food intake until terminal swallow. *Physiology and Behavior* **90**, 172–179.

Oncley, J. L. (1942). The investigation of proteins by dielectric measurements. *Chemical Review* **30**, 433–450.

Oosawa, F. and Asakura, S. (1975). Thermodynamics of the Polymerization of Protein. Academic Press, London, New York, San Francisco.

Otte, J., et al. (2005). Formation of amyloid-like fibrils upon limited proteolysis of bovine α-lactalbumin. *International Dairy Journal* **15**, 219–229.

Otte, J., et al. (1996). Protease-induced aggregation and gelation of whey proteins. *Journal of Food Science* **61**, 911–915.

Otte, J., et al. (1999). Protease-induced gelation of unheated and heated whey proteins: effects of pH, temperature, and concentrations of protein, enzyme and salts. *International Dairy Journal* **9**, 801–812.

Pandey, R. B. and Stauffer, D. (1983). Confirmation of dynamical scaling at the percolation-threshold. *Physical Review Letters* **51**, 527–529.

Panick, G., et al. (1999). Differences between the pressure- and temperature-induced denaturation and aggregation of β-lactoglobulin A, B, and AB monitored by FT-IR spectroscopy and small-angle X-ray scattering. *Biochemistry* **38**, 6512–6519.

Patro, S. Y. and Przybycien, T. M. (1996). Simulations of reversible protein aggregate and crystal structure. *Biophysical Journal* **70**, 2888–2902.

Pedersen, J. B., et al. (2006). Towards control of aggregational behaviour of α-lactalbumin at acidic pH. *Journal of Fluorescence* **16**, 611–621.

Peyron, M. A., et al. (2004). Particle size distribution of food boluses after mastication of six natural foods. *Journal of Dental Research* **83**, 578–582.

Pham, K. N., et al. (2004). Glasses in hard spheres with short-range attraction. *Physical Review E* 69, 011503-1-011503-13.

Pham, K. N., et al. (2002). Multiple glassy states in a simple model system. *Science* **296**, 104–106.

Philipse, A. P. (1996). The random contact equation and its implications for (colloidal) rods in packings, suspensions and anisotropic powders. *Langmuir* **12**, 1127–1133.

Philipse, A. P. and Kluijtmans, S. G. J. M. (1999). Sphere caging by a random fibre network. *Physica a-Statistical Mechanics and Its Applications* **274**, 516–524.

Philipse, A. P. and Verberkmoes, A. (1997). Statistical geometry of caging effects in random thin-rod structures. *Physica A: Statistical and Theoretical Physics* **235**, 186–193.

Philipse, A. P. and Wierenga, A. M. (1998). On the density and structure formation in gels and clusters of colloidal rods and fibers. *Langmuir* **14**, 49–54.

Piazza, R. (2000). Interactions and phase transitions in protein solutions. *Current Opinion in Colloid & Interface Science* **5**, 38–43.

Plucknett, K. P., et al. (2001). Dynamic experimentation on the confocal laser scanning microscope: application to soft-solid, composite food materials. *Journal of Microscopy* **201**, 279–290.

Polverino de Laureto, P., et al. (2005). Protein dissection enhances the amyloidogenic properties of α-lactalbumin. *FEBS Journal* **272**, 2176–2188.

Poon, W., et al. (2002). Soft condensed matter: where physics meets biology. *Journal of Physics Education* **37**, 25–33.

Poon, W. C. K. (1997). Crystallization of globular proteins. *Physical Review E* **55**, 3762–3764.

Poon, W. C. K. (1998). Phase separation, aggregation and gelation in colloid-polymer mixtures and related systems. *Current Opinion in Colloid and Interface Science* **3**, 593–599.

Poon, W. C. K. and Haw, M. D. (1997). Mesoscopic structure formation in colloidal aggregation and gelation. *Advances in Colloid and Interface Science* **73**, 71–126.

Poon, W. C. K., et al. (1997). Non-equilibrium behaviour of colloid-polymer mixture. *Physica A* **235**, 110–119.

Poon, W. C. K., et al. (1995). Gelation in colloid-polymer mixtures. *Faraday Discussions* **101**, 65–76.

Poulin, P., et al. (1999). From colloidal aggregation to spinodal decomposition in sticky emulsions. *The European Physical Journal B – Condensed Matter and Complex Systems* **7**, 277–281.

Pouzot, M., et al. (2004). Influence of the ionic strength on the structure of heat-set globular protein gels at pH 7. ß-Lactoglobulin. *Macromolecules* **37**, 8703–8708.

Pouzot, M., et al. (2005). X-ray and light scattering study of the structure of large protein aggregates at neutral pH. *Food Hydrocolloids* **19**, 231–238.

Prabakaran, S. and Damodaran, S. (1997). Thermal unfolding of β-lactoglobulin: Characterization of initial unfolding events responsible for heat-induced aggregation. *Journal of Agricultural Food Chemistry* **45**, 4303–4308.

Prasad, V., et al. (2003). Universal features of the fluid to solid transition for attractive colloidal particles. *Faraday Discussions* **123**, 1–12.

Preißner, R., et al. (1998). Dictionary of interfaces in proteins (DIP). Data bank of complementary molecular surface patches. *Journal of Molecular Biology* **280**, 535–550.

Prinsen, P. and Odijk, T. (2004). Optimized Baxter model of protein solutions: Electrostatics versus adhesion. *The Journal of Chemical Physics* **121**, 6525–6537.

Prinz, J. F., et al. (2007). In vitro simulation of the oral processing of semi-solid foods. *Food Hydrocolloids* **21**, 397–401.

Puertas, A. M., et al. (2002). Comparative simulation study of colloidal gels and glasses. *Physical Review Letters* **88**, 098301-1–098301-4.

Qi, X. L., et al. (1997). Effect of temperature on the secondary structure of β-lactoglobulin at pH 6.7, as determined by CD and IR spectroscopy: A test of the molten globule hypothesis. *Biochemistry Journal* **324**, 341–346.

Reiss-Husson, F. (1992). Crystallization of membrane proteins. *In* Crystallization of nucleic acids and proteins: A practical approach, 175–239, Ducruix, A. and Giegé, R. (eds.). CNRS, Strasburg (F).

Renard, D., et al. (2006). The gap between food gel structure, texture and perception. *Food Hydrocolloids* **20**, 423–431.

Renault, A., et al. (1999). Surface-induced polymerization of actin. *Biophysical Journal* **76**, 1580–1590.

Riha, W. E., et al. (1996). Nonenzymatic deamidation of food proteins. *Critical Reviews in Food Science and Technology* **36**, 225–255.

Robinson, D. J. and Earnshaw, J. C. (1993). Long range order in two dimensional fractal aggregation. *Physical Review Letters* **71**, 715–718.

Rosenbaum, D. F. and Zukoski, C. F. (1996). Protein interactions and crystallization. *Journal of Crystal Growth* **169**, 752–758.

Rosenberger, F., et al. (1996). Nucleation and crystallization of globular proteins – what we know and what is missing. *Journal of Crystal Growth* **168**, 1–27.

Ross-Murphy, S. B. (1997). Structure and rheology of gelatin gels. *The Imaging Science Journal* **45**, 205–209.

Rottereau, M., et al. (2004a). Monte Carlo simulation of particle aggregation and gelation: I. Growth, structure and size distribution of the clusters. *The European Physical Journal E – Soft Matter* **15**, 133–140.

Rottereau, M., et al., (2004b). Monte Carlo simulation of particle aggregation and gelation: II. Pair correlation function and structure factor. *The European Physical Journal E – Soft Matter* **15**, 141–148.

Rueb, C. J. and Zukoski, C. F. (1997). Viscoelastic properties of colloidal gels. *Journal Rheology* **41**, 197–218.

Safran, S. A., et al. (1985). Percolation in interacting colloids. *Physical Review A* **32**, 506–511.

Sala, G., et al. (2008). Matrix properties affect the sensory perception of emulsion-filled gels. *Food Hydrocolloids* **22**, 353–363.

San Biagio, P. L., et al. (1990). Spinodal demixing, percolation and gelation of biostructural polymers. *Makromolekulare Chemie-Macromolecular Symposia* **40**, 33–44.

San Biagio, P.L., et al. (1996a). Spontaneous symmetry-breaking pathways: Time-resolved study of agarose gelation. *Food Hydrocolloids* **10**, 91–97.

San Biagio, P.L., et al. (1996b). Self-assembly of biopolymeric structures below the threshold of random cross-link percolation. *Biophysical Journal* **70**, 494–499.

Sawyer, L., et al. (1999). β-Lactoglobulin – a three-dimensional perspective. *International Journal of Food Science and Technology* **34**, 409–418.

Schokker, E. P., et al. (2000). Heat-induced aggregation of β-lactoglobulin AB at pH 2.5 as influenced by ionic strength and protein concentration. *International Dairy Journal* **10**, 233–240.

Scholten, E., et al. (2005). Coarsening rates of bicontinuous structures in polymer mixtures. *Macromolecules* **38**, 3515–3518.

Sciortino, F., et al. (1993). Interference of phase-separation and gelation – a zeroth-order kinetic-model. *Physical Review E* **47**, 4615–4618.

Sciortino, F., et al. (2004). Equilibrium cluster phases and low-density arrested disordered states: The role of short-range attraction and long-range repulsion. *Physical Review Letters* **93**, 055701-1–055701-4.

Seaton, N. A. and Glandt, E. D. (1987). Aggregation and percolation in a system of adhesive spheres. *The Journal of Chemical Physics* **86**, 4668–4677.

Sedgwick, H., et al. (2005). Non-equilibrium behavior of sticky colloidal particles: beads, clusters and gels. *The European Physical Journal E – Soft Matter* **16**, 77–80.

Segrè, P. N., et al. (2001). Glasslike kinetic arrest at the colloidal-gelation transition. *Physical Review Letters* **86**, 6042–6045.

Sein, A., et al. (1993). Salt-induced transition from a micellar to a lamellar liquid crystalline phase in dilute mixtures of anionic and non-ionic surfactants in aqueous solution. *Langmuir* **9**, 1714–1720.

Shama, F. and Sherman, P. (1973). Evaluation of some textural properties of foods with the Instron universal testing machine. *Journal of Texture Studies* **4**, 344–353.

Shih, W. H., et al. (1990). Scaling behavior of the elastic properties of colloidal gels. *Physical Review A* **42**, 4772–4779.

Shortle, D. (1996). The denatured state (the other half of the folding equation) and its role in protein stability. *FASEB* **10**, 27–34.

Shortle, D. (2002). The expanded denatured state: An ensemble of conformations trapped in a locally encoded topological space. *Unfolded Proteins* **62**, 1–23.

Silva, T., et al. (2005). Effect of deamidation on stability for the collagen to gelatin trasition. *Journal of Agricultural Food Chemistry* **53**, 7802–7806.

Smeenk, J. M., et al. (2006). Fibril formation by triblock copolymers of silklike ß-sheet polypeptides and poly (ethylene glycol). *Macromolecules* **39**, 2989–2997.

Smith, A. M., et al. (2006). Direct observation of oligomeric species formed in the early stages of amyloid fibril formation using electrospray ionisation mass spectrometry. *Journal of Molecular Biology* **364**, 9–19.

Sprunt, J. C. and Smith, A. C. (2002). Measurement of surface area, gel weight and saliva weight in a gelatin-based gel over the course of mastication. *International Journal of Food Sciences and Nutrition* **53**, 261–271.

Stading, M. and Hermansson, A. M. (1991). Large deformation properties of β-lactoglobulin gel structures. *Food Hydrocolloids* **5**, 339–352.

Stauffer, D. (1976). Gelation in concentrated critically branched polymer-solutions – percolation scaling theory of intramolecular bond cycles. *Journal of the Chemical Society-Faraday Transactions II.* **72**, 1354–1364.

Stauffer, D. (1979). Scaling theory of percolation clusters. *Physics Reports-Review Section of Physics Letters* **54**, 1–74.

Stauffer, D., et al. (1982). Gelation and critical phenomena. *Advances in Polymer Science* **44**, 103–158.

Stockmayer, W. H. (1943). Theory of molecular size distribution and gel formation in branched-chain polymers. *The Journal of Chemical Physics* **11**, 45–55.

Stradner, A., et al. (2004). Equilibrium cluster formation in concentrated protein solutions and colloids. *Nature* **432**, 492–495.

Swaisgood, H. E. (1982). Chemistry of milk protein. *In* Developments in Dairy Chemistry – 1, 1–59, Fox, P. F. (ed.). Applied Science Publishers, LTD, Essex, England.

Takahashi, T., et al. (1993). Stabilization of protein crystals by electrostatic interactions as revealed by a numerical approach. *Journal of Molecular Biology* **234**, 421–432.

Takigawa, T., et al. (1992). Comparison of model prediction with experiment for concentration-dependent modulus of poly(vinyl alcohol) (PVA) gels near the gelation point. *Chemical Physics Letters* **195**, 509–512.

Tanaka, F. (1996). Phase formation of associating polymers: gelation, phase separation and microphase formation. *Advances in Colloid and Interface Science* **63**, 23–40.

Tanaka, T., et al. (1979). Phase separation and gelation in gelatin gels. *Physical Review Letters* **42**, 1556–1559.

Tanford, C. (1980). The hydrophobic effect: Formation of micelles and biological membranes. Wiley-Interscience, New York. p.233.

Tang, Q., et al. (1995). Effects of pH on whey protein concentrate gel properties: Comparisons between small deformation (dynamic) and large deformation (failure) testing. *Journal of Texture Studies* **26**, 255–272.

te Nijenhuis, K. (1981a). Investigation into the ageing process in gels of gelatin/water systems by the measurement of their dynamic moduli Part I. *Colloid & Polymer Science* **259**, 522–535.

te Nijenhuis, K. (1981b). Investigation into the ageing process in gels of gelatin/water systems by the measurement of their dynamic moduli Part II. *Colloid & Polymer Science* **259**, 1017–1026.

te Nijenhuis, K. (1997). Thermoreversible Networks: Viscoelastic properties and structures of gels. Springer-Verlag Berlin, Germany.

ten Wolde, P. R. and Frenkel, D. (1997). Enhancement of protein crystal nucleation by critical density fluctuations. *Science* **277**, 1975–1978.

Tlusty, T. and Safran, S. A. (2000). Defect-induced phase separation in dipolar fluids. *Science* **290**, 1328–1331.

Tlusty, T., et al. (2000). Topology, phase instabilities, and wetting of microemulsion networks. *Physical Review Letters* **84**, 1244–1247.

Tobitani, A. and Ross-Murphy, S. B. (1997). Heat-induced gelation of globular proteins .1. Model for the effects of time and temperature on the gelation time of BSA gels. *Macromolecules* **30**, 4845–4854.

Torrent, J., et al. (2006). High pressure modulates amyloid formation. *Protein & Peptide Letters* **13**, 271–277.

Trappe, V., et al. (2001). Jamming phase diagram for attractive particles. *Nature* **411**, 772–775.

Trappe, V. and Weitz, D. A. (2000). Scaling of the viscoelasticity of weakly attractive particles. *Physical Review Letters* **85**, 449–452.

Truong, V. D. and Daubert, C. R. (2000). Comparative study of large strain methods for assessing failure characteristics of selected food gels. *Journal of Texture Studies* **31**, 335–353.

Urry, D. W., et al. (1992). Hydrophobicity scale for proteins based on inverse temperature transitions. *Biopolymers* **32**, 1243–1250.

Uversky, V. N. (2002). Natively unfolded proteins: A point where biology waits for physics. *Protein Science* **11**, 739–756.

van den Berg, L., et al. (2008a). Energy storage controls crumbly perception in whey proteins/polysaccharide mixed gels. *Food Hydrocolloids* **22**, 1404–1417.

van den Berg, L., et al. (2008b). Quantification of a 3D structural evolution of food composites under large deformations using microrheology. *Food Hydrocolloids* **22**, 1574–1583.

van den Berg, L., et al. (2007). Breakdown properties and sensory perception of whey proteins/polysaccharide mixed gels as a function of microstructure. *Food Hydrocolloids* **21**, 961–976.

van der Bilt, A. (2002). Human oral function: a review. *Brazilian Journal of Oral Science* **1**, 7–18.

van der Bilt, A., et al. (2006). Oral physiology and mastication. *Physiology & Behavior* **89**, 22–27.

van der Linden, E. and Parker, A. (2005). Elasticity due to semiflexible protein assemblies near the critical gel concentration and beyond. *Langmuir* **21**, 9792–9794.

van der Linden, E. and Sagis, L. M. C. (2001). Isotropic force percolation in protein gels. *Langmuir* **17**, 5821–5824.

van der Linden, E. and Venema, P. (2007a). Self-assembly and aggregation of proteins. *Current Opinion in Colloid & Interface Science* **12**, 158–165.

van der Linden, E. and Venema, P. (2007b). Similarities in Self-Assembly of Proteins and Surfactants: an Attempt to Bridge the Gap. Food Colloids. Self-Assembly and Material Science. The Royal Society of Chemistry, Cambridge UK, 57–67.

van Megen, W. and Underwood, S. M. (1993). Glass transition in colloidal hard spheres: Mode-coupling theory analysis. *Physical Review Letters* **70**, 2766–2769.

van Vliet, T. (1988). Rheological properties of filled gels. Influence of filler matrix interaction. *Colloid Polymer Science* **266**, 518–524.

van Vliet, T. (2002). On the relation between texture perception and fundamental mechanical parameters for liquids and time dependent solids. *Food Quality and Preference* **13**, 227–236.

van Vliet, T., et al. (1993). Time dependent fracture behaviour of food. *In* Stability and Mechanical Properties, Dickinson, E. and Walstra, P. (eds.). Royal Society of Chemistry, Cambridge Food Colloids and Polymers.

van Vliet, T., et al. (1991). Fracture and yielding of gels. Food Polymers **82** Gels and Colloids (Dickinson E.(ed.)). Royal Society of Chemistry, Cambridge 392.

van Vliet, T., et al. (2007). On the mechanism by which oil uptake decreases crispy/crunchy behaviour of fried products. *Food Research International* **40**, 1122–1128.

Veerman, C., et al. (2003a). A new multistep Ca^{2+}-Induced cold gelation process for β-lactoglobulin. *Journal of Agricultural and Food Chemistry* **51**, 3880–3885.

Veerman, C., et al. (2003b). Irreversible self-assembly of ovalbumin into fibrils and the resulting network rheology. *International Journal of Biological Macromolecules* **33**, 121–127.

Veerman, C., et al. (2002a). Effect of electrostatic interactions on the percolation concentration of fibrillar β-lactoglobulin gels. *Biomacromolecules* **3**, 869–873.

Veerman, C., et al. (2002b). Rheology and mesostructure of β-lactoglobulin gels at pH 2. In: Proceedings of the 6th European Conference on Rheology / H. Munsted, J. Kaschta, A. Merten. – Erlangen (D): Institute of Polymer Materials, 321–322.

Veerman, C., et al. (2003c). Mesostructure of fibrillar bovine serum albumin gels. *International Journal of Biological Macromolecules* **31**, 139–146.

Veerman, C., et al. (2005). The effect of shear flow on the percolation concentration of fibrillar protein assemblies. *Journal of Rheology* **49**, 355–368.

Vekilov, P. G. and Alexander, J. I. D. (2000). Dynamics of layer growth in protein crystallization. *Chemical Review.* **100**, 2061–2090.

Verhaegh, N. A. M. and Lekkerkerker, H. N. W. (1997). Phase transitions in colloidal suspensions. *In* Course CXXXIV, 347–381, Mallamace, F. and Stanley, H. E. (eds.). IOS Press, Amsterdam.

Verheul, M. (1998). Aggregation and gelation of whey proteins. University of Twente, The Netherlands, Enschede, technische natuurkunde pp. 347–381.

Vliegenthart, G. A. and Lekkerkerker, H. N. W. (2000). Predicting the gas-liquid critical point from the second virial coefficient. *Journal of Chemical Physics* **112**, 5364–5369.

Vreeker, R., et al. (1992). Fractal aggregation of whey proteins. *Food Hydrocolloids* **6**, 423–435.

Walstra, P. and van Vliet, T. (2003). Functional properties in industrial applications. *Progress in Biotechnology* **23**, 9–30.

Weijers, M., et al. (2002). Rheology and structure of ovalbumin gels at low pH and low ionic strength. *Food Hydrocolloids* **16**, 269–276.

Weijers, M., et al. (2004). Influence of the ionic strength on the structure of heat-set globular protein gels at pH 7 Ovalbumin. *Macromolecules* **37**, 8709–8714.

Wetzel, R. (2006). Kinetics and thermodynamics of amyloid fibril assembly. *Accounts of Chemical Research* **39**, 671–679.

Wilkinson, C., et al. (2000). From food structure to texture. *Trends in Food Science & Technology* **11**, 442–450.

Winter, H. H. and Chambon, F. (1986). Analysis of linear viscoelasticity of a cross-linking polymer at the gel point. *Journal of Rheology* **30**, 367–382.

Winter, H. H., et al. (1994). Experimental observation of the molecular weight dependence of the critical exponents for the rheology near the gel point. *Polymer Gels and Networks* **2**, 239–245.

Woda, A., et al. (2006a). Adaptation of healthy mastication to factors pertaining to the individual or to the food. *Physiology & Behavior* **89**, 28–35.

Woda, A., et al. (2006b). The regulation of masticatory function and food bolus formation. *Journal of Oral Rehabilitation* **33**, 840–849.

Wright, H. T. (1991). Nonenzymatic deamidation of asparaginyl and glutaminyl redidues in proteins. *Critical Reviews in Biochemistry and Molecular Biology* **26**, 1–52.

Wu, H. and Morbidelli, M. (2001). A model relating structure of colloidal gels to their elastic properties. *Langmuir* **17**, 1030–1036.

Yanez, J. A., et al. (1999). Viscoelastic properties of particle gels. *Journal of Colloid and Interface Science* **209**, 162–172.

Zandomeneghi, G., et al. (2004). FTIR reveals structural differences between native β-sheet proteins and amyloid fibrils. *Protein Science* **13**, 3314–3321.

Zhang, Z., et al. (2007). Molecular dynamics simulations on the oligomer formation process of the GNNQQNY peptide from yeast prion protein Sup35. *Biophysical Journal*.

Zilman, A. G. and Safran, S. A. (2005). Role of cross-links in bundle formation, phase separation and gelation of long filaments. *Europhysics Letters* **71**, 515–516.

CHAPTER

3

Antifreeze Proteins: Their Structure, Binding and Use

Sarah L. Buckley [1] *and Peter J. Lillford* [2]

[1]Highton, Victoria 3216, Australia

[2]CNAP, Dept. of Biology, University of York, Heslington, York YO10 5DG, UK

3.1 ANTIFREEZE PROTEINS

3.1.1 Discovery and Purpose

Antarctica, the southernmost continent was once thought to be uninhabitable but over the last century, the coldest region in the world was found to be alive with a marine ecosystem. Most of the fish in this system is comprised of Notothenioidei, a group of teleost bony fish, which comprises of up to 100 species and accounts for over 90% of the fish in this region (Eastman and DeVries, 1986). In 1957, Scholander et al. observed that even in these conditions, where the temperature of the water was typically –1.9°C, certain fish species were able to survive. This temperature was maintained in deeper waters all year round, and fish whose blood had a freezing point of –0.9°C, were supercooled by almost 1°C. If these fish were put into contact with ice, which is able to enter their body through the gills, they would freeze and die (Scholander and Maggert, 1971). The ice crystallization would rapidly spread through their bodies as all their fluids were in a supercooled state. The absence of ice within their habitat prevented them from being seeded. However, in shallower waters, fish existed in permanent contact with ice and were able to survive. On extraction of the blood plasma from these fish, it was observed that although its melting point was an equilibrium transition, further cooling produced delayed crystallization. These fish had depressed the freezing point of their blood away from the equilibrium melting point and were able to maintain 0.5°C of supercooling. It was suggested that these fish species possessed antifreeze properties, even with the presence of ice crystals in their habitat, as they froze at temperatures as low as –2.2°C.

On further examination it was discovered that this antifreeze effect was due to a group of glycoproteins (DeVries and Wohlschlag, 1969). For most fish, salts make up 85% of the freezing point depressing solutes in the blood. Other molecules make up the remaining 15% such as glucose, urea, calcium and amino acids (Cheng and DeVries, 1991). In the Notothenioidei, these molecules were only able to account for 40–50% of the freezing point depression. The remaining 50–60% is due to the glycoproteins. These proteins have since been characterized and

Kasapis, Norton, and Ubbink: Modern Biopolymer Science
ISBN: 978-0-12-374195-0

isolated by DeVries and co-workers (1970) and in addition to their antifreeze effect, are able to alter the size and shape of the ice crystals within their blood plasma. Since this discovery, many more fish species were found to contain antifreeze glycopeptides (AFGPs) and antifreeze peptides (AFPs) (DeVries and Lin, 1977). Other species that have to survive similar environments were probed and many more types of AFP have been discovered in various animals and plants.

3.1.2 Evolution and Natural Occurrence of AFPs

Antifreeze proteins have now been discovered in various fish, insect and plant species (Wu et al., 1991; Urrutia et al., 1992; Duman and Olsen, 1993; Deng et al., 1997; Worral et al., 1998, Doucet et al., 2000; Sidebottom et al., 2000; Robinson, 2001; Voituron et al., 2002; Marshall et al., 2004; Graham and Davies, 2005). Not all species contain the same antifreeze glycopeptides discovered in the Notothenioidei. A great variety of protein structures are observed to modify ice crystal growth. Even within fish species there are now five known types, AFGPs, AFP I, II, III and IV.

Each of their respective characteristics will be discussed in Section 3.1.3, but whilst they all have different structures, their actions in the presence of ice at subzero temperatures is identical. They all bind to an ice surface. It has become clear that the natural source of the protein is no indication as to its type or structure. Neither the plant, nor insect AFPs has identical structures or behavior to the fish AFPs and characterization of many secondary/tertiary structures is ongoing.

Due to the variety in structure and distribution of AFPs across species, phyla and kingdoms, their evolution remains a topic of research interest (Smaglik, 1998; Ewart et al., 1999). Current theory suggests that AFPs were generated via convergent evolutions. Fish adaptation may have been initiated as Antarctica separated from the southern continents (Eastman and DeVries, 1986). On separation, the water surrounding the region was cooled and separated from the warmer water by a thermal barrier. The temperate fish species then disappeared through natural selection and the Notothenioidei adapted to the colder environment. The Antarctic shelf at the edge of the continent now maintains this lack of diversity in the native species. This cooling occurred recently in evolution and prompted the natural selection of a protein already present in fish blood (Davies and Sykes, 1997). Most recently, it has been proposed that fortuitous expression of 'junk DNA' has given organisms a selective advantage for cold survival (Chen et al., 1997). The nucleic acid sequences that have been co-opted to aid cold survival in grasses have been mapped by Spanenburg et al. (2007). Thus, there are latent proteins in many organisms that could potentially act as AFPs. During adaptation, different AFP molecules were produced in different species, leading to a natural diversity in molecular structure. Indeed, a single species can contain more than one active molecule (Marshall et al., 2004). The common molecular architecture, which is necessary to produce the function of ice crystal modification and antifreeze activity, remains unclear.

3.1.3 Types and Structure of AFPs

Because of their earliest discovery, the molecular structure of AFPs from fish has been studied most extensively.

3.1.3.1 AFGP

Antifreeze glycoproteins exist within the Notothenioidei fish, such as *Trematomas bernacchii*, *Dissostichus mawsoni* or Antarctic cod, and *Gadus ogac* has a group of eight peptides. Each molecule consists of a repeating three residue peptide, with a disaccharide unit attached to each third residue (see Figure 3.1). The eight

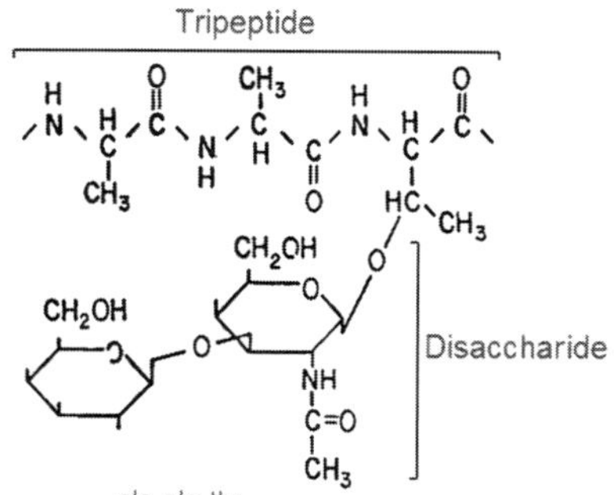

Tripeptide -ala-ala-thr-
Disaccharide β-D-galactopyranosyl-(1- 3)-2-acetamido-2-deoxy-α-D-galactopyranose

FIGURE 3.1 AFGP typical primary structure.

forms differ only by weight and length of their protein chain, and are numbered 1–8 for convenience. The peptides have varying molecular weights between 2.6–33 kDa, with each peptide 1–8 having between 4–50 repeats of the peptide respectively (Feeney and Yeh, 1996). The largest proteins (1–5) use the repeating unit alanine-alanine-threonine in their protein chain, whereas the smallest molecules (6–8) contain some proline residues in place of alanine. Each molecule acts in the same way, with the largest proteins (1–5) being the most active antifreezes, but with the smallest molecules being most abundant in a sample.

The antifreeze properties are dependent on the presence of the glycosylation in the molecule. The disaccharide unit is 3-O-(β-D-galactosyl)-D-N-acetylgalactosamine, with a 1→3 link. Its conformation in solution is less understood, but it is very apparent that there is little α-helical protein (DeVries et al., 1970). The most recent evidence suggests that the secondary structure of these peptides is a left-handed extended three-fold helix, comparable to poly-L-proline (Lillford and Holt, 1994; Yeh and Feeney, 1996). The advantage of this conformation, confirmed by Raman spectroscopy, vacuum CD and proton NMR, is that the disaccharide units are all on one side of the helix, tucked against the protein backbone. The carbohydrate is able to present a hydrophilic face to an ice surface and is proposed to hydrogen bond to water molecules within the ice lattice.

3.1.3.2 AFP I

Duman and DeVries first reported the existence of non-glycosylated antifreeze proteins. These AFPs have since been found in right-eyed flounders, Alaskan plaice and shorthorn sculpin, with the most study being performed on the Winter Flounder, *Pleuronectes americanus*. This AFP is again present in several isoforms but has two main components, HPLC6 and HPLC8 (see Figure 3.2). The major amino acid component in the primary structure is alanine, with two residues difference between the two forms.

This protein again has a repeating sequence of 11 amino acids that is present three times in the native molecule. The 37-residue peptide is 3.6 kDa in weight. In solution, this AFP has been found to exist predominantly as an α-helix, and at

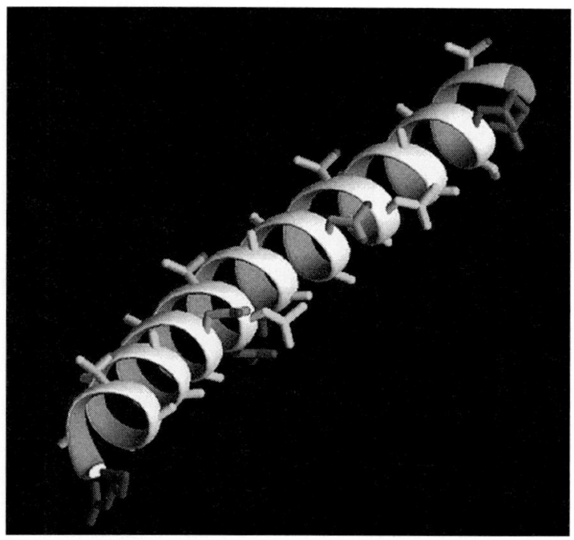

Primary sequence:
DTASDAAAAALTAANAAAAAKLTAD
NAAAAAAATAA

FIGURE 3.2 The main component of AFP I, HPLC6, primary sequence. Secondary structure supplied by Dr Frank Sonnichsen, Case Western Reserve University, Canada.

its functioning temperatures, helicity increases to 85% (Ananthanarayanan et al., 1977). This structure enabled the molecule to be crystallized and X-ray diffraction data collected (Yang et al., 1998). This evidence showed that the non-polar alanine residues were placed entirely along one side of the helix, and the threonine residues, thought to bind to the ice, systematically positioned on the other side in ridged positions. A salt bridge is present in the structure to stabilize the molecule.

3.1.3.3 AFP II

This class of AFPs contains some of the largest AFP molecules discovered to date, the smelt and the herring AFPs, whose molecular weights are 24 and 14.6 kDa respectively. Only the plant AFP from the perennial rye grass *Lolium perenne* is larger (see Section 3.1.3.6). Recent discoveries about the molecules in this group suggest that subtypes exist (Davies and Sykes, 1997). The most studied protein is that from the Sea Raven, *Hemitripterus americanus* (see Figure 3.3). This protein has a molecular weight of 14–16 kDa, with the characteristic type II feature of having a high cysteine content, and as a consequence, disulfide bonds. However, the primary structure still contains many alanine residues as well as asparagine, glutamate and threonine. These AFPs are also non-glycosylated, and their secondary structure is comprised of a mixture of coil, α-helix and β-sheet conformations. The tertiary structure is globular, with the disulfide bonds helping to maintain conformation and expose the hydrophilic residues thought to be involved in ice binding (Ewart et al., 1998; Loewen et al., 1998).

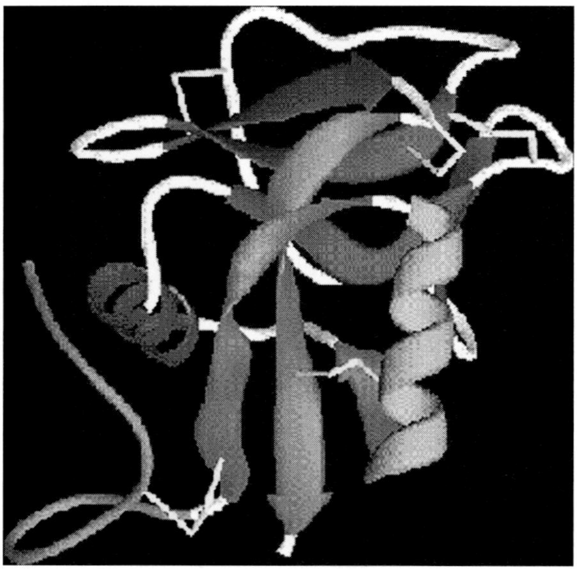

FIGURE 3.3 Secondary structure of AFP II from Sea Raven, supplied by Dr Frank Sonnichsen, Case Western Reserve University, Canada.

3.1.3.4 AFP III

This AFP type is not analogous to any of the other antifreeze proteins. A typical type III protein is found in the Ocean Pout, *Macrozoarces americanus*, and contains fewer alanine residues and no cysteine residues. AFP III molecules lack any particular dominance in their primary structure, and contain no carbohydrate. The molecular weight is typically 6.5 kDa, from 66 residues. The predominant isoform is HPLC 12, though several isoforms exist. The secondary structure is compact and includes β-strands, one turn of an α-helix and a rigid fold (Jia et al., 1996). The fold is unusual in that it contains eight β-strands that form two sheets of three anti-parallel, and one sheet of two anti-parallel, strands. These are then packed into a β-sandwich. This leads to a globular protein with several flat surfaces and hydrophilic areas between the strands. The crystal structure shows a flat amphipathic region, on one side of the protein, which has been suggested as the region binding to the ice lattice (see Figure 3.4).

3.1.3.5 Insect and Plant AFPs

Insect and plant AFPs were discovered during the last decade and structural information on many of the proteins is often limited to primary structure. However, several in each category

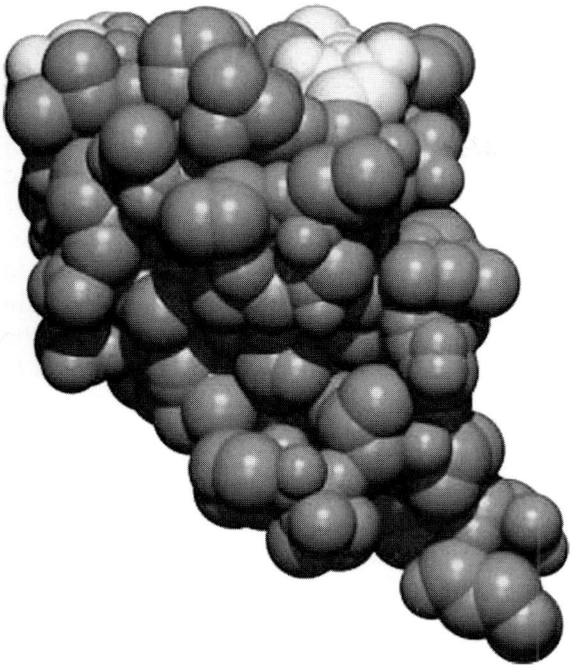

Primary Structure:
NQASVVANQLIPINTALTLVMMRSEVVTPVGI
PAEDIPRLVSMQVNRAVPLGTTLMPDMVKG
YPPA

FIGURE 3.4 HPLC12 primary and tertiary structure, Smith (2000) and Whiteman (personal communication).

have now been described and there seems to be no trend in structure and homology or similarity with the fish AFPs.

Insect AFPs have been isolated and characterized from the beetle larva, the bark beetle, the moth larva spruce budworm, and the snow flea.

The most studied plant AFPs include proteins from Winter Rye or *Secale cereale* (Griffith et al., 1992), Bittersweet nightshade or *Solanum dulcamara* (Duman, 1994), carrot, *Daucus carota*, (Worral et al., 1998) and the perennial rye grass *Lolium perenne* (Sidebottom et al., 2000). Screening the plant kingdom, using ice crystal growth inhibition, showed that most cold-tolerant plants have activity in their apoplast fluids after cold stress, but most of the active agents have not yet been isolated or characterized.

We have selected the most complete structural studies to describe in detail here.

3.1.3.5.1 Spruce Budworm (Duman and Olsen, 1993) The Spruce budworm larvae are freeze-tolerant insects from North America, who suffer temperatures as low as –30°C in the conifer trees on which they live. Not only do they produce AFPs in this seasonal period, but also they increase the concentration of glycerol in their blood ten-fold. The protein is 90 amino acids long, with a molecular weight of 9.1 kDa and exists in several other isoforms. It is threonine rich, 15 out of its 90 residues, and is reported to have an activity as high as 6°C (Tyshenko et al., 1997; Graham et al., 1997; Doucet et al., 2000; Graether et al., 2000). Its low molecular mass is due to its proportion of serine, glycine, and cysteine residues. This composition suggests that it is a hydrophilic protein, in contrast to the predominantly hydrophobic fish proteins. The secondary structure has recently been resolved by NMR analysis and shown to be a β-sheet structure. The eight sheets arrange themselves in a β-helix, with one fold. This conformation means that all but one of the threonine residues are on one side of the protein helix and arranged in a regular array with a periodicity of 7.35, 4.5 and 7.8Å. This motif of residues is proposed as the ice-binding surface. Several disulfide bonds support the structure, which is not homologous to any of the classifications for the fish AFPs (Graether et al., 2000).

3.1.3.5.2 Daucus Carota (Worrall et al., 1998) AFP is found in the native British carrot *Daucus carota*, after cold acclimation for several days. To date, it is the only dicotyledonous crop from which the active molecular species has been extracted and purified. It is present in the crop in one or two isoforms and is a glycoprotein. Its molecular weight is estimated at 34–36 kDa and it is thought to be hydrophilic from its primary amino acid sequence. The protein contains 332 amino acids and is N-glycosylated but is not dependent on the glycan chain for its activity.

There are no dominant amino acid residues, though the protein does contain a region of 11 tandem leucine-rich repeats. This repeating structure is akin to polygalacturanose inhibiting protein (PGIP) and a homology model proposes a similar tertiary conformation (Smith, 2000; Buckley et al., 2001).

3.1.3.5.3 Lolium Perenne *Lolium perenne* is a common grass found in Britain, and although many overwintering grasses show activity within their apoplastic fluid, the active protein from this species has been separated and characterized. There is only one isoform present in the plant (Sidebottom, personal communication) and this has been characterized to homogeneity. The grass AFP is N-glycosylated and the carbohydrate mannose rich. The carbohydrate component makes up over half the molecular weight of the native molecule. The protein is predicted to have a β-sheet secondary structure, with a molecular weight of only 11.8 kDa compared to the total weight of 29 kDa when including the carbohydrate moiety. The protein is very hydrophilic and has a high percentage of asparagine, serine and threonine. The secondary structure has been examined using FT-IR and contains 58% β-sheet, 21% α-helix, 16% unordered and 3% turns (Sidebottom et al., 2000). More recently, this structure has been modeled and found to be a β-roll, with eight loops of 14–15 amino acids (Kuiper et al., 2001). This structure produces two opposite-facing ice-binding sites with good surface complementarity to the prism face of ice. The authors suggest that this dual binding capability is responsible for this protein's phenomenal recrystallization inhibition activity.

3.2 AFP PROPERTIES

The antifreeze proteins differ from each other in structure and source, but share the same physical characteristics. The basic requirement for an antifreeze molecule is the display of thermal hysteresis, a gap between the equilibrium melting point and the freezing point of the solution, as described in Section 3.1.1. However, closely linked to this phenomenon, are the properties of inhibition of recrystallization, shape or morphology modification, implying a change in the ice crystal growth characteristics. Each of these effects is due to the protein's ability to bind to the ice crystal surface, and to recognize certain ice crystal planes. They are therefore not mutually exclusive. Each of the four attributes will now be described in detail.

3.2.1 Thermal Hysteresis

Antifreeze proteins possess a unique ability that has not been observed with any other biopolymer, in that they can depress the freezing point of a solution, whilst maintaining an equilibrium melting point.

In solutions without AFPs, freezing point depression (FPD) is a colligative property, whereby the amount of deviation from pure solvent behavior is proportional to the number of solute particles present, regardless of their nature or identity. This depression is achieved by the solute, due to the increase in entropy of the solution compared to the solvent in isolation. This in turn reduces the tendency of the system to freeze, and a lower temperature is reached before the equilibrium phase transition occurs. By thermodynamic definition, the equilibrium between a solid and a liquid occurs at a specific point or single temperature (Atkins, 1990).

The depression in freezing point for AFP solutions is non-colligative. A typical freezing point depression of an AFP solution can be 0.8°C or more, which, by referral to the thermodynamic relationship for freezing point depression, predicts that 0.43 M of solute should be required. However, the actual concentration required for a depression of freezing point of this magnitude by AFP III is only 10^{-3} M (Feeney and Yeh, 1998).

Thermal hysteresis (ΔT) was first demonstrated by DeVries (1986), who used the concept of 'Nanoliter Osmometry' in his studies of ice crystal shape, and this has since become the most widely used technique in AFP research. This apparatus is used for the measurement of changes in ice crystal morphology, and will be described in Section 3.2.4. The thermal hysteresis measurement involves the generation of a small ice crystal in a solution of known AFP concentration. The sample is housed in a sealed capillary or chamber, which has its temperature controlled by circulated coolant or a peltier device. The temperature is changed at a rate of 0.01°C min^{-1} and the freezing and melting of an ice crystal can be observed. The point at which total melting occurs is taken as the melting point, and when unrestricted growth is observed, this temperature is the freezing point. The melting point can be independently confirmed by use of a vapor pressure osmometer.

In a water or sugar solution, the melting and freezing occurs over a small range, approximately 0.02–0.03°C. In antifreeze protein solutions, the difference is significantly higher. Thermal hysteresis is concentration dependent, in that it increases with concentration of the protein in solution. Whilst antifreeze activity is usually defined as thermal hysteresis, an exact figure cannot be given due to this variance and it is often quoted over a concentration range or as an activity at a specific concentration. Typical melting and freezing point temperatures can be seen in the plot in Figure 3.5. This determination procedure for activity also relies on the antifreeze's ability to totally inhibit ice crystal growth until the freezing point is reached. The exact definition of the freezing point becomes more complex in mixtures of isoforms, dilute solutions, or mutated antifreezes.

Each antifreeze protein has its own characteristic values for thermal hysteresis. This is due to the variety of organisms and environmental conditions each organism has to survive, and the particular requirements for each molecule. For example, the fish proteins display a much higher thermal hysteresis than the plant AFPs and characteristically have less recrystallization inhibition (RI) activity. The differences in thermal hysteresis can be seen in Figure 3.6. Not only is there a variance in activity between phyla of organisms but there is also a difference between species in each group. AFP II has higher activity than AFP I for example. Insect AFP activity is displayed on a second plot for convenience. The relationship between RI and thermal hysteresis has always been assumed, but as more data are collected this correlation is now being questioned and is in doubt (Knight, 2000; Sidebottom et al., 2000).

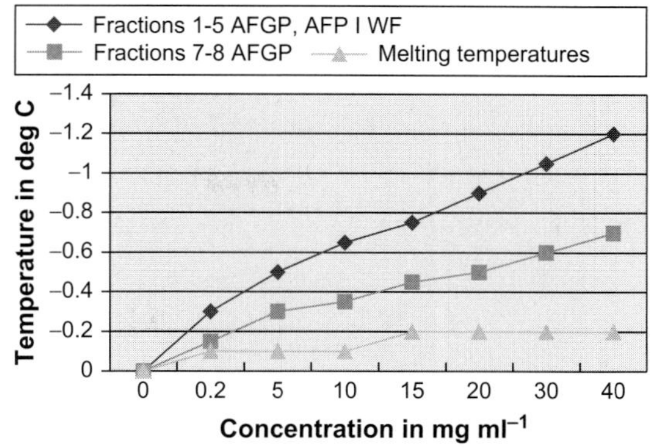

FIGURE 3.5 Typical ΔT for AFGP fractions. Thermal hysteresis is the gap between the melting point and the freezing points.

It is more reasonable to display ΔT data on a molar rather than a weight basis and these data can be seen in Figure 3.7. If the insect antifreeze were added to the chart in Figure 3.7, the obvious superiority of that AFP would become apparent. At 0.2 mM/L, a thermal hysteresis of 3.75°C is observed. The plant antifreeze protein thermal hysteresis data available are those of carrot extract and grass AFP expressed in *Pichia pastoris*, which are displayed in Figure 3.6.

3.2.2 Recrystallization Inhibition

Recrystallization of ice is an established physical property and occurs due to the instability in ice crystals once they are formed. As

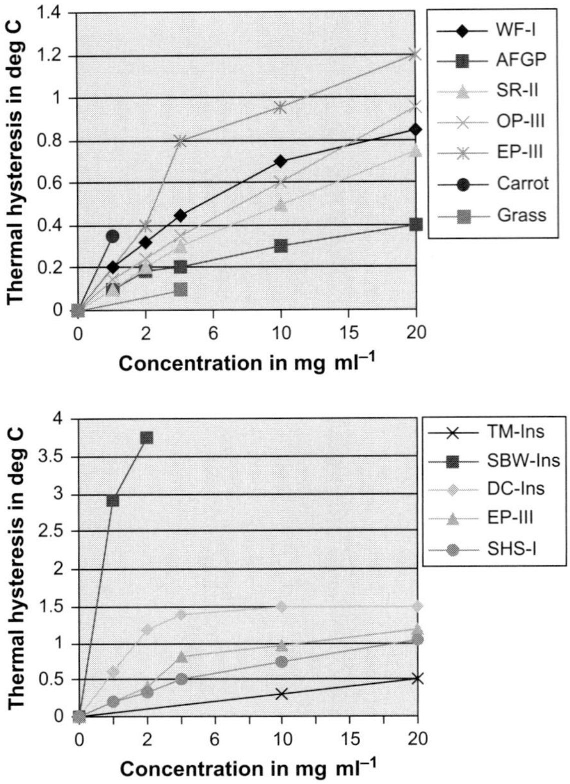

FIGURE 3.6 Thermal hysteresis on a weight scale. (WF = Winter flounder, SHS = Shorthorn sculpin, SR = Sea raven, OP = Ocean pout, AFGP = Atlantic cod, EP = Eel pout, TM = Tenebrio molitor, DC = Dendroides canadensis, SBW = Spruce budworm. Types of peptide are indicated by roman numerals and Ins = Insect AFP). Ewart et al. (1998), Kao et al. (1986), Tyshenko et al. (1997), Duman and Olsen (1993).

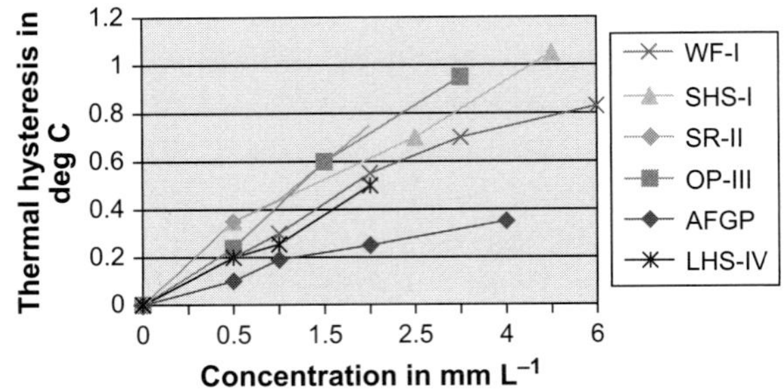

FIGURE 3.7 Thermal hysteresis on a molar scale. Same legend labels apply as in Figure 3.6. LHS = Long horn sculpin. Deng et al. (1997), Kao et al. (1986).

the ice crystals themselves are not at equilibrium, there is a dynamic driving force for all the crystals to become a single mass, where short- and long-range structure is perfect, and there are no boundaries. The term 'recrystallization' therefore refers to any change in the number, size, shape, orientation or perfection in the crystals. This change causes a free energy reduction as the chemical potential of water becomes equal in all phases present in the system (Fennema et al., 1973). There are several classes of recrystallization such as accretive, migratory and ripening.

Accretive recrystallization or accretion, is observed when crystals that are in contact, join together. This has a thermodynamic advantage as the number of crystals decreases and the surface energy of the entire phase is also reduced. Migratory recrystallization is a form of 'Ostwald' ripening and occurs when large crystals grow in a dispersed polycrystalline phase at the expense of the smaller crystals present (Mullin, 1993). The smaller crystals then dissolve to provide a reduction in the number of crystals, a reduction in surface energy and an increase in the average crystal size. This mechanism can occur at a constant temperature, purely due to the surface energy difference in the crystals and the consequential higher melting point of the smaller crystals. Iso-mass recrystallization is recognized as the change in surface or internal structure of a mass of crystalline material as it moves to a lower energy state. This usually occurs to crystals with a high surface area to volume ratio or those in an irregular shape, such as dendritic crystal forms. This process is thought to occur by surface diffusion.

The concentration-dependent inhibitory effect of the AFP manifests itself in the complete halt of the recrystallization mechanisms described above, so that in large volumes of ice, when small quantities of AFP are added, no growth of individual ice crystals is observed. Knight et al. in 1984, and Knight, Hallett and DeVries in 1988, first published the phenomenon of recrystallization inhibition. This was the first time that AFGP had been shown to produce enhanced effects when compared to standard biopolymer or organic compounds. A 'splat' method was developed as an assay for this property and involves a single drop of antifreeze solution that falls onto a cooled aluminum plate over a fixed distance. This produces a wafer of ice, 1cm in diameter and 20 μm thick. After being held at a fixed temperature of –6°C for over 6 hours, the resulting grains are observed and compared to a pure water control. The wafers are photographed between crossed polaroid filters. Typical examples of ice crystals generated in this way are shown in Figure 3.8.

Each AFP sample tested displays recrystallization inhibition at much lower concentrations than those found for its thermal hysteresis activity, or in physiological conditions. The AFP is thought to have this effect on recrystallization due to its ability to bind to the ice (Feeney and Yeh, 1998). As the AFP binds to any ice crystal it comes into contact with, it is a fair assumption that the protein is stopping the recrystallization due to its affiliation with the boundary. Mutant antifreeze proteins, that did not possess thermal hysteresis activity but were the same shape and size, also had no RI activity; thus demonstrating that it is common antifreeze properties that are responsible for this effect (Knight et al., 1995).

3.2.3 AFP Binding Site

3.2.3.1 Crystallographic Principles

A main area of interest in AFP research is the antifreeze's ability to bind to an ice surface. The method of binding is not yet fully understood and several models have been proposed (e.g. Hall and Lips, 1999; Section 3.3). Vital to the development of the current theories, is the determination of the binding site of the protein on an ice crystal.

Ice can be formed in several polymorphic forms. However, without the application of extreme pressure and/or temperature, there is only one form produced. The ice structure (I) is composed of a regular array of water molecules, and there are a total of four molecules in the basic repeating unit. Due to the hydrogen bonding between those molecules and the intermolecular angles formed, the crystal takes on a hexagonal shape.

The standard crystallographic hexagonal unit cell, which is applied to all crystals of this shape, has several crystallographic axes (Barnes, 1929; Megaw, 1934). Three a-axes and one c-axis produce a unit crystal with the dimensions of 7.37 Å along the c-axis and 4.52 Å along the a-axis at 0°C (Fennema et al., 1973). The planes or faces of the crystal unit can be defined by positions along these axes, the international standard definition being Miller-Bravais indices.

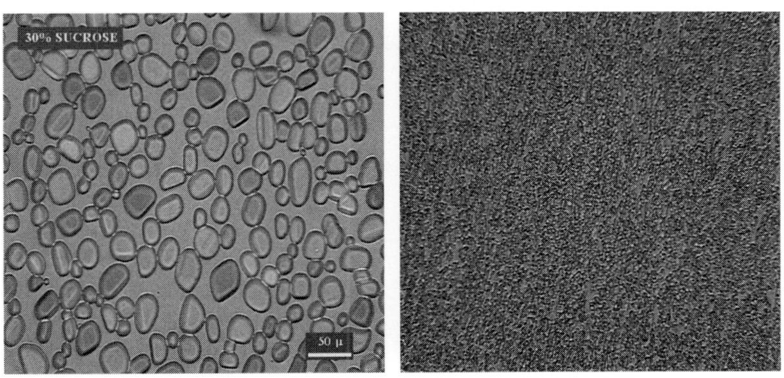

FIGURE 3.8 Examples of RI assay ice crystal sizes: 30% sucrose control (left) and AFP in 30% sucrose (right).

During ice crystal growth, water molecules will be added to these faces and in the absence of any additive or solute, the rate of growth will be much faster along the prism plane direction (Miller-Bravais index $[10\bar{1}0]$) than the basal plane direction $[000\bar{1}]$, shown in Figure 3.9.

3.2.3.2 *Antifreeze Protein-Binding Sites*

The effect of AFP on ice crystal shape has been well reported in the early AFP literature. It was assumed that this was due to the protein adsorption onto the growing ice surface during the crystal growth, thus altering its natural morphology. This effect is further described in Section 3.2.4.5.

The initial experiments to explore the stereospecificity of antifreeze binding to an ice surface were reported by Knight and DeVries (1988) who used a hemispheric single ice crystal. In this most elegant technique, AFPs are allowed to influence a slow-growing single ice crystal and the protein attaches to its preferred crystal plane. Though there was no appreciable inhibition of the growth of the hemispherical crystal, Knight determined that the protein was attaching to specific areas. The protein, once attracted to the ice becomes embedded on the same ice plane throughout the hemisphere. They discovered that they could simply 'etch' back the ice surface by water sublimation and the hemisphere develops *rough patches*, concentrated areas of opaque protein on clear ice. The plane to which the protein attaches is then determined by the crystallographic relationship between the patches and the crystallographic axes. A typical ice hemisphere result is shown in Figure 3.10. A hemisphere measures between 4 and 6cm in diameter, and the patches are visible to the naked eye.

Cheng and DeVries (1991) and Knight, Cheng and DeVries (1991) continued this pioneering work by publishing further AFP binding planes and the experimental technique. They described a series of binding planes from several fish AFPs, including the four types of protein known at that time. Cheng and DeVries also described a second type III Eel pout binding plane determination, *Rhigohila dearborni,* and a new Alaskan Plaice binding site. The original AFGP binding plane

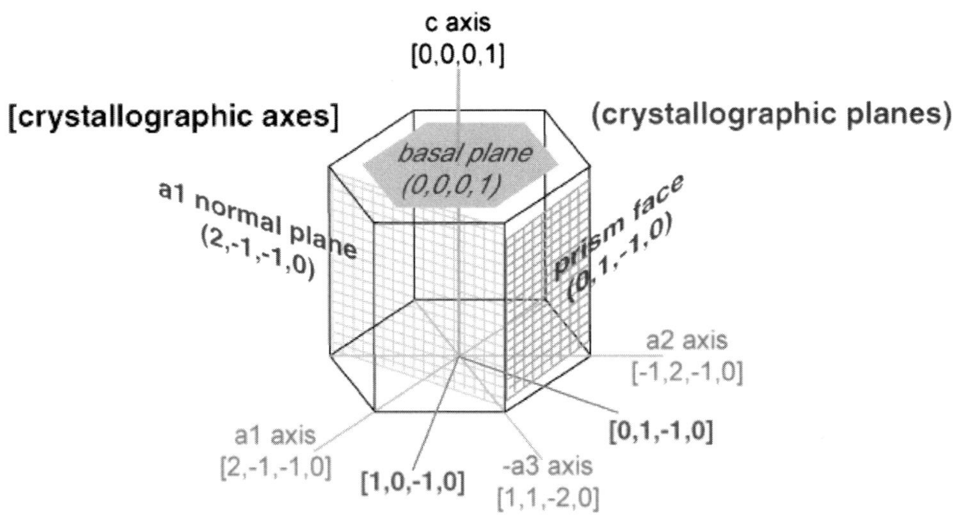

FIGURE 3.9 Unit cell of an ice crystal showing the basal plane, c- and a-axes and Miller-Bravais indices of some hexagonal crystal planes.

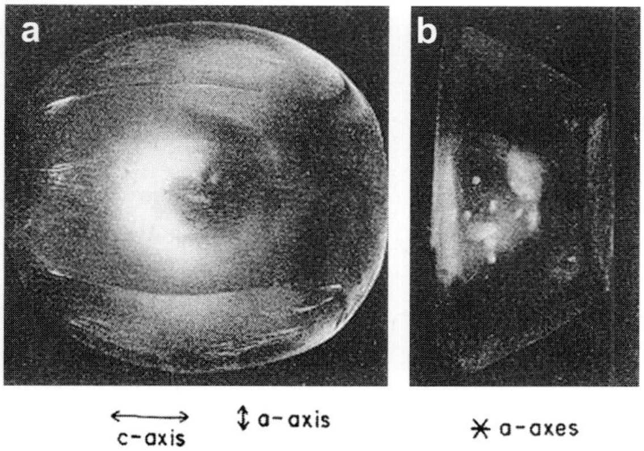

FIGURE 3.10 AFGP ice hemisphere. Reproduced from Knight, Driggers, and DeVries (1993).

determination was expanded and the short horn sculpin was found to bind to a secondary prism plane. These results are summarized in Table 3.1.

The number of possible binding planes on an ice crystal is large. However, Knight and DeVries (1994) discovered that the number of binding planes to which AFP molecules adsorbed is four: the primary and secondary prism planes and the primary and secondary pyramidal planes. These planes are identified in Figure 3.11. In each diagram, one plane is highlighted, but the AFP molecule attaches to the six equivalent planes around the crystal. Thus, when AFP molecules adsorb to any plane, a characteristic six-fold symmetrical pattern is formed on the ice hemisphere.

The experimental publication (Knight et al., 1991) contained the first photographs of binding sites and has become one of the most cited publications in this area. The method outlined the determination of the binding site of type I α-helical proteins from the winter flounder, Alaskan plaice and sculpin. The etch pattern of the Winter Flounder, *Pleuronectes americanus*, is shown in Figure 3.12. The planes were identified by the measurement of the interplaner angle and the alignment of the molecules was discussed. Following optical orientation of the single crystal hemispheres and with reference to the c-axis, the angles between the etched regions or planes were measured on a universal goniometer stage. An average was taken from the four angles generated, by rotating the stage 180° and repeating each measurement. This was completed several times to eliminate statistical error. From crystallographic principles, the interplaner angle is determined and the binding plane taken as the lowest index plane. Higher index planes have increasing molecular complexity and are eliminated unless required.

The authors conclude with speculation as to the fitting of the protein molecules on the ice surface via hydrogen bonding of the amino acid residues to the ice. They observed that the periodicity of these residues along the helix matches the distance between bonding water molecules. The amino acid threonine, a polar residue, appears every 16.5 Å; whilst the ice structure along the alignment direction has a periodicity of 16.7 Å. The flexibility of the peptide backbone may compensate for the 0.2 Å difference in the match, whilst also maintaining the orientation dependence for adsorption.

TABLE 3.1 AFP Binding Planes (From Cheng and DeVries, 1991)

Fish name	Common name	AFP type	Binding plane	Miller-Bravais index
Australocicthys brachycephalus	Eel pout	III	Primary prism plane	$[10\bar{1}0]$
Myoxocephalus scorpius	Atlantic sculpin	I	Secondary prism plane	$[\bar{1}2\bar{1}0]$
Pleuronectes americanus	Atlantic winter flounder	I	Pyramid plane	$[20\bar{2}1]$
Hemitripterus americanus	Sea raven	II	Pyramid plane	$[11\bar{2}1]$
Dissostichus mawsoni	Antarctic cod	AFGP	Primary prism plane	$[10\bar{1}0]$

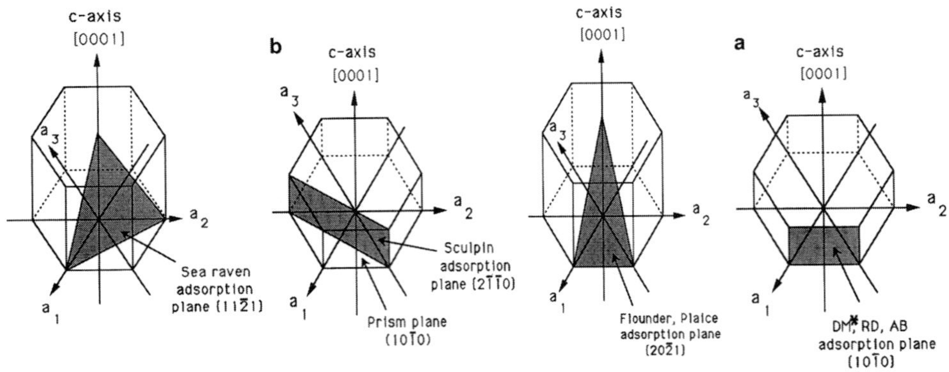

FIGURE 3.11 Binding sites of AFPs (Cheng and DeVries, 1991). From left to right above: Secondary pyramidal binding plane; binding site of the sea raven AFP II; secondary prism plane, binding site of sculpin AFP I; pyramidal plane, binding site of winter flounder and Alaskan plaice AFP I and primary prism plane-binding site of AFGP *Dissostichus mawsoni*.

3.2.3.3 *Native and Synthetic AFP I Ice Binding*

The study of type I AFPs was extended following the initial publication concerning winter flounder (Knight et al., 1991). Synthetic molecules were used to probe the model proposed by Knight and DeVries and ice etching was used as a confirmatory technique to determine the stereospecificity of AFP binding (Laursen et al., 1994; Madura et al., 1994). Modeling suggested that the L-AFP could only bind along the [$\bar{1}$102] direction of the pyramidal plane and not the reflected [0$\bar{1}$12] plane, even though this is structurally equivalent. This argument was reversed for the D-form.

A D-AFP I, from winter flounder, was synthesized to correspond exactly to the naturally occurring L-form. It possessed the same thermal hysteresis activity as the naturally occurring form. A dilute solution was used to grow the hemispheres (0.015 mg/ml) and three etching experiments were completed; hemispheres with the D- and L-forms alone and one with both forms together. The experimental results supported the prediction. The six patches observed in the etched pattern indicate binding

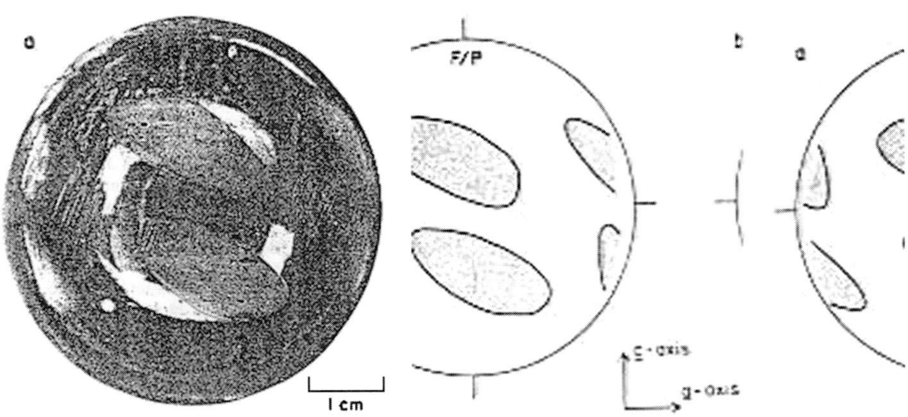

FIGURE 3.12 Etch pattern photograph and diagram for winter flounder AFP I (Knight et al., 1991).

to the pyramidal plane $[20\bar{2}1]$ with the predicted elongation directions for each form. The D- and L-forms are mirror images of each other and produce mirrored etching patterns. The hemisphere containing both forms displays a cross where the etch patterns interact.

A further type I AFP to be investigated by the hemisphere technique is that from the sculpin fish (Wierzbicki et al., 1996; Davies and Sykes, 1997). It is another alpha helical protein whose binding plane is $[2\bar{1}\bar{1}0]$, the secondary prism plane. The initial result appears in the original ice etching paper by Knight et al. (1988), and this was followed by further etching and molecular modeling by Wierzbicki et al. (1996). This paper shows a clear photograph of the etch pattern, as seen in Figure 3.13.

The shorthorn sculpin AFP hemisphere displays vertical bright line segments oriented parallel to the c-axis, as can be seen on the surface of the hemisphere in Figure 3.13. The rough patches provide a distinct flattened surface, though the patches themselves are irregular in surface coverage and area. The protein does not have conventionally spaced residues for ice adsorption like the winter flounder, and only some structural match of the protein to the ice surface is proposed. Each 11th residue falls in an approximate line, 16.9 Å apart. (The ice itself having a regular spacing of 16.7Å.) The protein can align itself along the c-axis grooves by hydrogen bonding with the ice only in the [122] direction. Lysine residues provide a means of ice adsorption for the protein as they contain an NH_3+ group (Wierzbicki et al., 1996).

FIGURE 3.13 Etch pattern of the shorthorn sculpin (Wierzbicki et al., 1996).

3.2.3.3.1 *The AFP I Ice Binding Motif*

AFP I was the most researched AFP in the field's infancy and study of its molecular-level mechanism of binding is therefore well advanced. Work using computer simulations and amino acid substitution continued (Cheng and Merz, 1997; Chao et al., 1997). Many workers suggest that the binding has not been completely explained and claim that the protein does not rely on hydrogen bonding alone for adsorption. This was indicated by the observation that, although four hydrogen bonding amino acid residues (threonine) are potentially available for complete docking onto the surface, in reality, only two are required. The authors speculated that if the growing ice front then comes into contact with the other two threonine binding residues, the adsorption becomes complete. If this does not occur, and the protein molecule desorbs from the growing ice front, then it will have sufficiently restricted the growth to allow an alternative molecule to bind more easily. The binding would then become more permanent as all four hydrogen bonds would need to be broken before removal of the protein could occur.

Baardsnes et al. (1999) and Haymet (Harding et al., 1999; Haymet et al., 1999) have both used mutational studies to probe the binding further. They have challenged the conclusions drawn from the data discussed in this section. Haymet mutated the threonine groups along the AFP I helix, changing the OH groups to CH_3 and found that this does not significantly change the protein's binding characteristics. Using characterization tools such as thermal hysteresis, shape and binding site tests, Haymet's results show that the OH groups are not absolutely required for antifreeze activity and that methyl groups can play a dominant role in binding of the proteins to ice.

Baardsnes' results agree with Haymet's conclusions. The alanine-rich face of the helix may form a critical part of the ice-binding site and long side chains added here by mutation sterically hinder the binding. These authors propose that ice binding involves Thr 13, Ala 17 and Ala 21, and the threonine γ-methyl group is essential for activity, where surface complementarity is the overriding consideration. It is likely that Van der Waals interactions, hydrogen bonds, and generation of the 'best fit' between the protein and ice to prevent displacement of the protein from the surface, are equally as important in the binding.

3.2.3.4 *AFGP Ice Binding*

Antifreeze glycoproteins were the subject of the initial studies using the ice hemisphere technique (Knight et al., 1988) and their binding is discussed in further detail in papers by Knight et al. (1993) and Knight and DeVries (1994). Using AFGP fractions 7 and 8, the adsorption and binding mechanism of the protein was examined.

The alignment direction of the proteins on the primary prism plane $[10\bar{1}0]$, was found to be parallel to an a-axis and normal to the c-axis. The etch pattern is shown in Figure 3.10. Molecular models were used to investigate the method of adsorption of the AFGP to the ice plane. Two models were proposed: the glycoprotein binds by chemical adsorption making it virtually permanent by either (i) hydrogen bonding to the ice or (ii) becoming part of the ice structure itself. This model is shown in Figure 3.14.

Knight and DeVries advanced the AFGP binding model in 1994. They proposed that the binding of the protein takes place via the AFGP sugar residues bonding directly with water molecules to form part of the ice structure. In

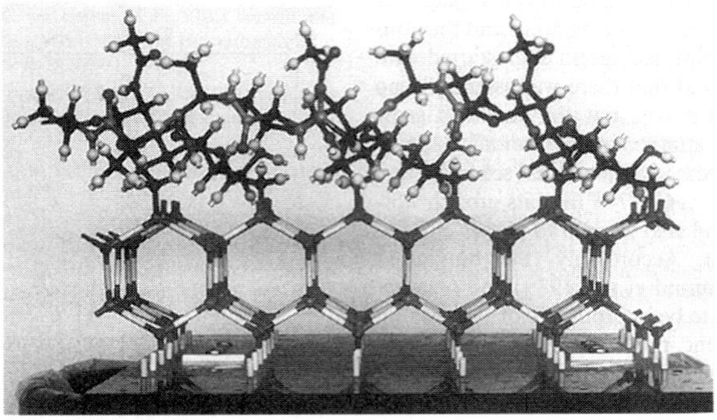

FIGURE 3.14 AFGP model on the primary prism plane of the ice surface (Knight et al., 1993).

AFGP protein chains, OH or NH_2 groups on the saccharides can provide three hydrogen bonding sites each, to make a total of 24 for the smallest AFGP (Isoform 8). A mechanistic description of AFP action was proposed and is now known as the 'button mattress' model. The button mattress model is defined by the firm attachment of AFPs to the ice surface at the ice–water interface. This attachment prevents the migration of the growing ice front and will be further discussed in Section 3.3.

Etching of hemispheres containing AFGP fractions 7 and 8, the shorter fractions containing five and four repeating protein units respectively, provides the three elongated rough patches expected for primary prism plane binding. When the concentration is increased, faceting appears, and hexagonal-sided hemispheres are produced. When similar experiments are performed with the longer AFGP fractions, 1–5, which can have up to 52 protein repeat units, the hemisphere yields further information. In dilute solution, the growing ice interface on the hemisphere displays flat-bottomed sharp-sided grooves. Etching of the interface reveals discrete lines representing concentrations of incorporated AFGP. As the concentration increases from 0.01 mg/ml through to 0.1mg/ml, these grooves and lines move closer together and eventually form a mosaic of tiny rhombi and hexagons.

The author explains the grooves as a second AFGP binding plane. The orientation of the grooves and patches suggests that the adsorption plane changes gradually from $[41\bar{5}0]$ to $[10\bar{1}0]$, the primary prism plane. The explanation proposed is that the prism plane can only accommodate five repeat units of the protein adsorbing onto that region of ice, before the helix structurally prevents further binding. A dislocation in the ice crystal face cannot reasonably accommodate the remainder of the protein molecule, but the AFGP is able to bind to a higher index ice plane, very close to the primary prism plane.

These experiments suggest that the long AFGPs are flexible in solution, allowing adsorption of the appropriate regions of the glycoprotein onto the ice surface whilst leaving the remainder of the molecule in the solution. This is able to explain the shift of adsorption plane with concentration, as the molecule becomes less mobile at higher concentrations. Thus the higher index plane is highlighted as there is no other way to accommodate the protein molecules on the ice surface.

3.2.3.5 *AFP III Ice Binding*

Whilst the investigation into AFGP and AFP I proteins has been extensive, the interest in AFP III has only increased over the last few years since discovery of its solution and crystalline structure (Sonnichsen et al., 1993; Jia et al., 1996; Antson et al., 2001). The *Australocicthys brachycephalus* AFP III binding plane was published initially by Cheng and DeVries as being the primary prism plane, following work by Knight and DeVries. This binding plane was used for modeling of peptides from the AFP III from Antarctic eel pout (*Lycodichthys dearborni* formally *Rhigohila dearborni).* Work by Knight and DeVries (unpublished data) indicated adsorption of the molecule onto the primary prism plane. However, the authors describe the etch pattern as 'non-specific' and the binding mechanism and alignment direction as unclear (Wang et al., 1995).

The most recent developments in the structure and mechanism have been published by Antson et al. (2001). These authors propose that the binding is not confined to the $[10\bar{1}0]$plane as previously implied. The investigation examined AFP III from three different fish sources*: Australocicthys brachycephalus, Macrozoarces americanus* (both expressed in *E.coli*) and HPLC 12, the most active isoform of AFP III, expressed in *Saccromyces cerivisae* yeast. The ice hemisphere experiment for all three sources of AFP III is identical. The etch pattern for AFP III and HPLC 12 consists of one patch which is smeared across the prism planes of the ice hemisphere. The edge of the patch takes the form of three distinct lobes, which are found on the boundary of the prism plane with both basal planes. The lobes on the edge of the AFP III patch, similar to those seen on the AFP II hemisphere, also appear angled. This suggests that the binding is non-specific, and due to the broadness of the patch, cannot be pinpointed to any one plane. However, this result concurs with that of Knight (Cheng and DeVries, 1991; C.A. Knight, personal communication) and is the standard one found with this AFP type. If the edges of the lobes are taken as an indication of selectivity, this AFP could be binding to the primary prism plane. However, the protein covers both the primary [1010] and the pyramidal planes [20-21]. Elongation of the etch pattern suggests AFP III interaction with a variety of crystal planes offset from the main binding planes. In conclusion, these results suggest a very broad coverage of AFP III across an ice crystal and lead to further advancements in modeling the AFP–ice crystal interaction.

As AFP III is a globular protein, the periodic residues found with AFGP and AFP I are not present for ice binding. However, a cluster of polar residues, that could constitute an ice-binding site, were identified by protein mutagenesis, on AFP III from *M. americanus* (ocean pout) (Chao et al 1994; Jia et al., 1996; Madura et al., 1996). The 1.25 Å crystallographic structure confirmed that these five residues all lie on a flat surface (Figure 3.15) of the globular molecule. Due to the flatness of the binding surfaces on the protein, it is possible for it to rotate 180°, and still bind to the same plane (Jia et al., 1996).

The amino acids identified (glutamine 9 and 44, threonine 15 and 18 and asparagine 14) attach the molecule to an ice surface without steric interference. By systematic residue substitution the authors determined that asparagine 14 was the most important residue and recorded a drastic loss in AF activity, indicating that the hydrogen bonding of each side group is complementary. Large side chains substituted next to the identified binding site produced significant steric interference, which also resulted in a significant loss of activity. The ice-binding surface is largely hydrophobic, with the remainder of the protein containing charged and polar residues. This suggests that the back of the molecule also plays a role in AFP action by preferring to be in solution rather than on ice (Haymet et al., 1999).

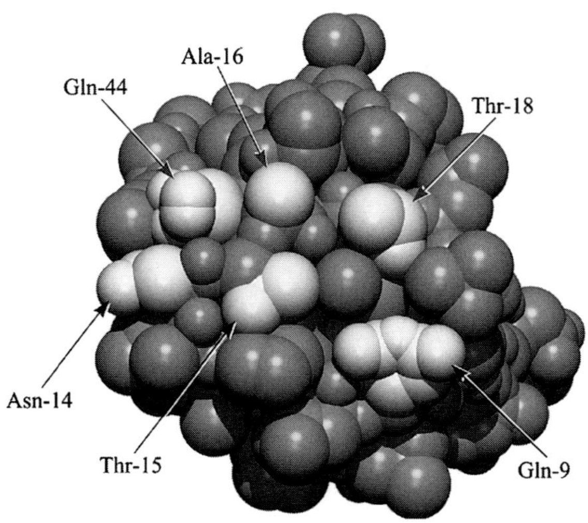

FIGURE 3.15 AFP III binding site model showing important ice-binding residues (Smith, 2000).

The NMR and crystallographic structures of ocean pout AFP III allowed molecular dynamics simulations of the HPLC-12 isoform binding to ice, and all modeling involving this protein has used the $[10\bar{1}0]$ plane (Madura et al., 1996; Jia et al., 1996). However, Madura et al. state that the ice etching studies have not sufficiently identified the adsorption plane for AFP III. Jia et al. (1996) modeled the AFP to be a higher index plane very close to the prism face. This was initially modeled to be reversible, with an asparagine residue completing the docking on the rounded edge of the prism plane, near the basal plane.

The debate regarding the molecular binding interaction of AFP III to an ice surface has continued (DeLuca et al., 1998; Yang et al., 1998). The most recent developments in the structure and mechanism have been published by Antson et al. (2001). These authors propose that the binding is not confined to the $[10\bar{1}0]$ plane. Modeling with a high-resolution structure suggested favorable binding interactions with the $[20\bar{2}1]$ pyramidal ice plane, and several other planes. Manual docking calculations suggested that HPLC 12 could be fitted to almost any ice plane in a number of different ways and still make a substantial number of hydrogen bonds and Van der Waals interactions with the surface of the crystal.

3.2.3.6 *Plant AFPs*

Use of the ice-binding technique has also proved pivotal in the investigation of the plant-based AFPs and has shown the startling similarities between the plant AFPs and their fish counterparts. Sidebottom et al. (2000) and Pudney et al. (2003) investigated the binding of the AFP from *Lolium perenne* expressed in *Pichia pastoris* and *E. coli*. Following the growth of a single ice crystal hemisphere and subsequent surface etching, a distinctive etch pattern emerged. This pattern is similar to that of AFGP (Knight et al., 1988, 1993, 1994) and spruce budworm (Graether et al., 2000), the patches being oblong in shape with their edges well defined, positioned regularly around the equator of the ice hemisphere and having an elongation direction parallel to the c-axis. The authors concluded that the protein is binding to the six equivalent primary prism planes with index $(10\bar{1}0)$.

These data, in addition to the novel study in FT-IR discussed in Pudney et al. (2003), raised interesting comments. This study showed that the protein section of this glycosylated molecule was used for binding and not the carbohydrate as found with AFGP but also that the link between ice binding, thermal hysteresis and recrystallization inhibition needs to be re-examined; *Lolium perenne* having high recrystallization activity and low thermal hysteresis, whilst the AFGP displays the opposite behavior (Knight, 2000). The study also suggested that the grass AFP does not change its conformation on binding to ice, whilst other proteins do.

3.2.3.7 Recent Use of the Ice Hemisphere Binding Site Technique

The increase in crystallographic structures of AF proteins has led to advancements in the binding theories and postulated mechanism of action of these molecules. The hemisphere binding site determination is extremely useful for correlating surface residue positions to periodicity on the ice lattice. This information allows lattice matching exercises to be completed, which indicate sites for hydrogen bonding to take place. This interaction usually forms the basis of molecular binding models and has a very big influence on the development of protein-specific mechanisms. Many examples of the modeling of AFPs binding to different ice planes, can be found in the literature (Wen and Laursen, 1992b; Knight et al., 1993; Lal et al., 1993; Sicheri and Yang, 1995; Jia et al., 1996; Madura et al., 1996; Graether et al., 2000; Antson et al., 2001; Kuiper et al., 2001; Pudney et al., 2003).

Knight has since used the hemisphere technique for confirmatory work to investigate synthetic proteins and AFP mechanisms. For example, experiments include the fluorescent microscope imaging of AFGP in the ice hemisphere to determine more crystallographic details of AFGP adsorption (Yu et al., 1998). However, the ice hemisphere technique has been adapted by several groups in the AFP area and hemisphere binding site determinations have been quoted in investigations of several AFPs (Haymet et al., 1999; Sidebottom et al., 2000; Graether et al., 2000; Antson et al., 2001; Pudney et al., 2003). It is clear that this technique offers a vital insight into AFP binding and has become an integral part of AFP research.

3.2.4 Ice Crystal Shape

3.2.4.1 Ice Crystal Growth: Principles

Crystallization has been studied extensively over the last century. It is often studied as its component parts, namely, nucleation and growth (see Figure 3.16), and these two events are intimately connected. For any crystal to form, a nucleation event must occur. Before crystals can grow, a small group of the crystallizing molecules must come together to form an embryo, nuclei or seed (Mullin, 1993). This event is thermodynamically unfavorable and environmental conditions have to be correct for the event to occur. Nucleation can either be homogeneous or heterogeneous; the crystallizing molecules either self-associate, or collect on an inert body (Knight, 1967). Once nucleation has occurred, either secondary nucleation or crystal growth will occur. This is entirely dependent on the conditions, such as supercooling, in which the substance of interest is crystallizing. As AFPs directly affect the mechanism of crystal growth, this will be discussed in this section.

Crystallization as a process relies on a surface phenomenon, where a growing crystal has contact with its growth solution. Growth of ice

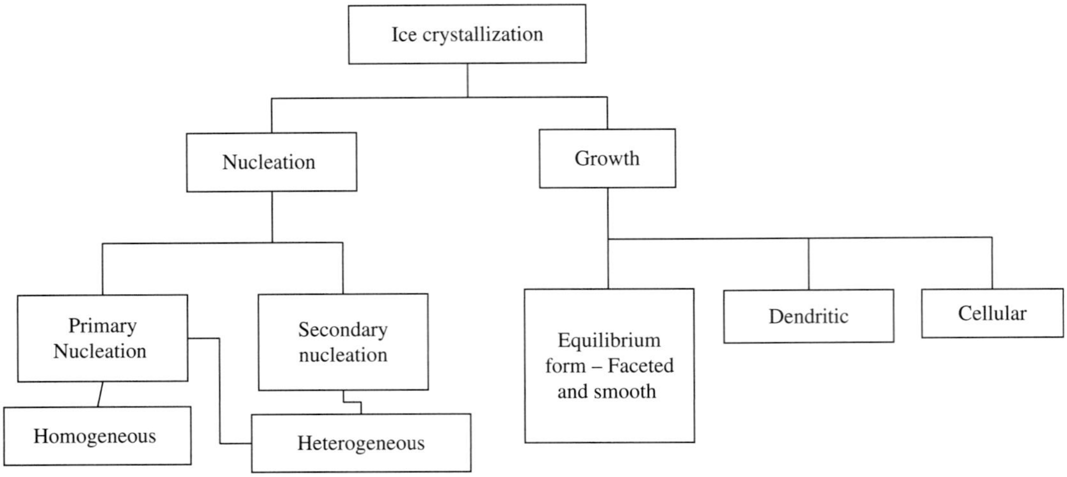

FIGURE 3.16 Crystallization processes (adapted from Mullin, 1993).

crystals in water or vapor is dependent on the same mechanism; the transport of water molecules to the growing interface and their accommodation into the ice crystal lattice. This has to be combined with the removal of latent heat of crystallization and molecular solutes away from the site of growth. The inclusion of solutes creates complications for the ice crystal growth, and for this reason makes ice crystal growth in water less understood than that in vapor (Fletcher, 1970). The effect of additives on crystal shape, or habit, is well known (Buckley, 1961). Anything that alters the rate of deposition of the crystal's component molecules on that surface will alter crystal habit proportionally to its effectiveness on the various planes.

If the shape of any ice crystal was determined thermodynamically, then Wulffs theorem (1901) would apply. The Wulff theorem states that 'in equilibrium, the distance of any face from the center of the crystal is proportional to the free energy per unit area for that face'. As the surface energy (the Gibbs free energy required to increase the surface by one unit area against the force of surface tension) of each crystal face cannot be determined exactly, it can be assumed from the atomic structure of the lattice.

Assuming that water molecules in ice only interact with their nearest neighbors, then the surface energy of any face is proportional to the concentration of nearest neighbor pairs across planes parallel to the face. In ice, the lowest energy planes are the basal and prism planes. Pyramidal planes will grow faster than the basal and prism but they tend to disappear, unless the crystal is growing particularly slowly. The surface energy for the prism face is 6% larger than that for the basal plane, and therefore the equilibrium crystal habit for ice should be a hexagonal prism with an aspect ratio (c/a) of 0.82.

This structure suggests that shapes such as hexagonal prisms, hexagonal bipyramids and prisms with pyramidal ends will be formed the most frequently and in the vapor this is certainly the case. Growth of ice from vapor has been well studied in the areas of atmospheric and meteorological science. Nakaya (1954), Mason (1958) and Magono and Lee (1966) classified naturally occurring snow crystals by their shape or morphology. Crystal habits formed in the vapor include columns, plates, needles and dendritic structures. Each form is influenced by the conditions of its growth, namely the temperature and supersaturation of the vapor.

Ice crystal habits in water are less understood. In solution, the interface between ice and water is molecularly rough, and the transport of water molecules to the growing crystal surface is limited by diffusion. Therefore, circular or hexagonal discs often form due to the roughening of the basal and prism planes or just the basal plane, respectively.

3.2.4.2 *Crystal Growth in Solution*

Crystal growth in solution again exhibits a range of morphology. Water systems have been extensively studied, and continue to be examined due to the complexity of ice growth. However, several common morphologies exist and are of importance to crystal growth with AFP.

The equilibrium form of ice is determined by the Wulff theorem and has been described in the preceding section. This occurs at little or no supercooling, where there is no energy available for crystal growth, and the temperature remains essentially constant. Ice growth in supercooled water utilizes the heat flow from the crystal as ice forms within the liquid. However, if the growing interface becomes unstable, then one region may develop enhanced drivers for growth, and dendrites will form.

Hallett (1960) discovered the relationship of dendrite growth to crystallographic orientation of the crystal and likened it to growth on a supercooled pond. Hallett suggested that two situations can result in this supercooled aqueous environment: if the c-axis of the disc is perpendicular to the surface of the liquid, cells and dendrites grow identically, and a dendritic star with hexagonal symmetry is formed. If the c-axis of the disc is parallel to the surface of the water, the crystal will grow as a needle in the surface, with dendrites protruding into the solution. In general, ice on the surface of any body of water will generally have its c-axis either parallel or perpendicular to the surface, and these orientations produce the crystal growth observed by Hallett; circular disks when the c-axis is vertical, and long needles when the c-axis is horizontal (Fletcher, 1970).

This crystal growth is due to two reasons. Firstly, thermal conductivity is 5% greater in the c-axis direction than in the a-axis direction. And secondly, crystal growth velocity is greater in the a-axis direction. Any nucleating crystal that has its c-axis away from the vertical will grow as a needle. Crystal growth will be slow and in the form of ribs or cells with dendrites growing slowly downwards into the solution. When the c-axis is in the vertical position, hexagonal or circular plates form from rapid growth along the horizontal a-axes and the ice will cover a much wider area, faster.

Since this early work, much research has been devoted to understanding and modeling ice growth in solutions (Tiller, 1991; Trivedi and Kurz, 1994). The studies reported here demonstrate the complexity of ice crystal growth in various environments. The rapid growth, and range of morphologies, that can form in an aqueous system is large and varied. However, this growth becomes significantly more complicated in the presence of solutes or molecules that are able to change the kinetics of the crystallization. Therefore, any molecule that has the ability to alter standard growth forms whilst maintaining hexagonal ice structure is of great interest in this area.

3.2.4.3 *Ice Crystal Growth in the Presence of AFPs*

The effect of AFP on ice crystal shape manifests itself in the change of standard crystal growth, described in the preceding section. Normally, ice growth in slightly supercooled water occurs along the a-axes to form flat hexagonal or circular plates, but in the presence of AFPs, ice growth favors the c-axis direction (Knight and DeVries, 1988). It is assumed that this is due to the protein adsorption. The method of binding and its implications for ice growth are not yet fully understood, and several models have been proposed to account for AFPs effect, discussed in Section 3.3 (Knight et al., 1991; Feeney and Yeh, 1996; Hall and Lips, 1999). This section will summarize the experimental evidence in this area that relates to ice crystal shape and morphology, and a variety of techniques used to measure the AFPs' effect. Thermal hysteresis will be referred to where appropriate, as this is a related property.

3.2.4.4 *The Nanoliter Osmometer Method*

Ice crystal shape characterization is used in various guises by most of the researchers working with antifreeze proteins. It is generally accepted that AFPs alter the standard growth of an ice crystal from solution. In water, a hexagonal or circular ice crystal forms and grows most rapidly in the a-axis direction. When AFPs are added, a thermal hysteresis develops, a gap between the melting and freezing point of the solution. At low AFP concentration, the ice crystals form bi-pyramidal-shaped crystals within this hysteresis gap. At higher concentrations, needles are formed. These are both due to the enhanced c-axis growth that occurs in the presence of AFPs. On reaching the freezing point, the growth becomes explosive and needles are often formed.

The osmometer microscopy method is used for examining AFP effects on ice crystal shape and for measuring thermal hysteresis. DeVries determined the importance of the antifreeze effect on the shapes of ice crystals and characterized AFP activity by the thermal hysteresis measurement (1986). This author measured these two parameters using a constructed apparatus, which consisted of a refrigerated chamber, a binocular microscope, and a sample capillary tube. This method was technically improved by the use of a Clifton Nanolitre Osmometer (Clifton Technical Physics, Hartford, NY; see descriptions in http://www.otago-osmometers.com/nanolitreosmometers.htm). A standard method for looking at the development of shape in the hysteresis gap was defined by Chakrabartty and Hew (1989, 1991).

In nanoliter osmometry, the melting point of an AFP sample is measured by taking 2–5 μl of liquid in a sample capillary. The sample size ensures that the crystal does not interact with the walls of the capillary and that supercooling is uniform. A seed ice crystal, ~50–100 μm in size, is formed, either by using spray freezing (DeVries, 1986), or by crashing the temperature to –40 °C (Chakrabartty and Hew, 1989, 1991), and then applying controlled thawing until only one crystal remains. The capillary is placed on or in the temperature control unit, which is either a refrigerated viewing chamber or a peltier device, and monitored using a microscope. Crossed Polaroids, generating polarized light, are used to view the growth and to determine the crystallographic orientation of the crystal. The temperature of the solution is increased, or decreased, by 0.01°C/min so that both the melting and freezing point are determined accurately. The gap between the two temperature values is defined as the thermal hysteresis, or ΔT gap (see Section 3.2.1).

In the Clifton nanoliter osmometer, the sample capillaries are often filled with mineral oil and a capillary tube is used to suspend a drop of about 10 nl of the sample solution, in the center of the oil column, where it becomes encased in the oil. This ensures that a more uniform temperature change is experienced by the AFP solution. The relevant temperatures are measured in milli-osmols, rather than degrees Celsius. The Celsius scale of temperature and osmolality are related by the following equation:

$$T\ (^{\circ}C) = 0.00186 \times \text{mOs}$$

The crystal habit of the ice crystals in the chosen solution is observed in tandem to the thermal hysteresis measurement. DeVries observed that alternative ice crystal morphologies were formed with differing protein

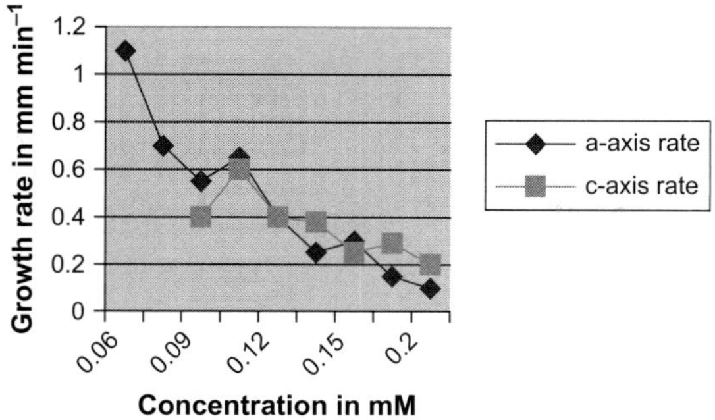

FIGURE 3.17 Mean growth rates of native AFPI winter flounder AFP crystal (Chakrabartty and Hew, 1991).

concentration. At low protein concentration, 0.01%, slow growth aids the formation of hexagonal crystals. At high concentrations, 0.1–4%, narrow spicules form as a result of fast growth along the c-axis.

3.2.4.5 AFP Investigations using the Nanoliter Osmometer

Many studies have used the nanoliter osmometry method to observe the effects of AFP on ice crystal shape. Often, the temperature is lowered to 0.1°C supercooling and the crystal growth and shape recorded for up to 2 minutes, via video. These images are reviewed post experiment and the aspect ratio components, the growth along the a- and c-axes, are measured. Whilst the c-axis growth was relatively easy to measure, the a-axis growth along all a-axes was taken as the average growth of the visible axis (Chakrabartty and Hew, 1989, 1991).

This protocol delivers reproducible measurements, with the error increasing with growth rate. The rate of growth was found to be independent of starting size of the crystals, when starting size was between 50 and 200 μm. However, in the original study, only crystal growth rates were measured (Figure 3.17) and no crystal shape photographs from the individual experiments are shown. This experiment is now routinely used, and ice crystal shapes are usually photographed for publication.

Various researchers in the area quote this method (Wen and Laursen, 1992; Chao et al., 1995; DeLuca et al., 1996, 1998; Deng et al., 1997; Chao Houston et al., 1997; Haymet et al., 1999; Graether et al., 2000). Typically, photographs of the ice crystals are taken after a fixed growth time, and often included in each publication. Occasionally, a series of photographs can be seen which illustrate the growth of the bipyramid crystal (see Figure 3.18).

3.2.4.5.1 AFP Type I Two of the main types of AFP have been studied using the nanoliter osmometer. AFP I analogs of winter flounder were studied by Wen and Laursen (1992) using the HPLC 6 isoform as the native comparison. They state that in AFP solutions, ice crystals exist as hexagonal bipyramids that elongate at the freezing point to form needles. The rates of a- and c-axis growth were measured, and the rate of c-axis growth was observed to be 3.3 times faster than the a-axis growth direction. However, it is interesting to note that the aspect ratio changes with crystal size, being 3.2 for 40 μm crystals and 3.5 for 80 μm crystals.

Wen and Laursen conclude that the crystal that is produced in these experiments, the hexagonal

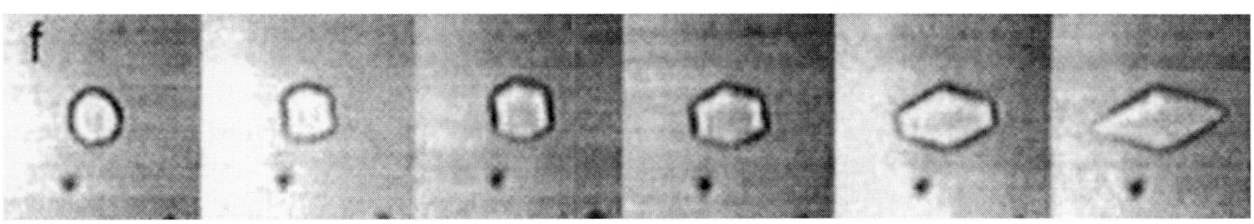

FIGURE 3.18 Growth of a hexagonal bipyramid over the time period of the experiment. Reproduced from Haymet et al. (1999).

bipyramid with an aspect ratio of 3.3/1, would be predicted from the crystallographic plane to which the AFP was attaching. This was determined by Knight et al. (1991) as the primary pyramidal plane $[20\bar{2}1]$, whose theoretical aspect ratio has been calculated as 3.26. The helices of the protein are all aligned down the patch in the $[01\bar{1}2]$ direction, possibly with their tips on the edge of the pyramidal plane in the $[\bar{1}\bar{1}26]$ direction (DeLuca et al., 1998). This arrangement aids the formation of the bipyramid crystal shape.

This characterization of ice crystal shape has also been used to compare AFPs in protein mutation studies (Wen and Laursen, 1992; Chao et al., 1997). Both sets of authors report that mutants have very marked differences from the native behavior of type I AFP. For example, the thermal hysteresis of the native protein at 2 mg/ml is 0.4°C. However, the serine mutant (a threonine on the proposed binding site was mutated to serine) was unable to stop growth at any concentration below 2 mg/ml. At concentrations above this, only a slight thermal hysteresis was measurable (0.1°C). However, within the gap, the ice crystal expressed the bipyramidal shape. In contrast, the valine mutant had 70–80% thermal hysteresis activity of the wild type and ice crystals did not grow when held supercooled in its solution for 10 minutes. This seems to suggest that the binding to ice has not changed, in that the same shape is formed (Figure 3.19). But clearly the differences in thermal hysteresis confirm that the mechanism of crystal growth cessation has been affected.

Mutant analogs of the native winter flounder protein (HPLC6) have been increasingly used to probe its ice binding. In a study by Haymet et al. (1999), hemisphere tests are used alongside the thermal hysteresis and ice crystal morphology measurement to probe the action of the mutants. Where there was a change in the binding site, there was a change in the ice crystal morphology as evident in the pictures presented.

3.2.4.5.2 AFP Type III AFP III ice crystal shape data can be elucidated from two

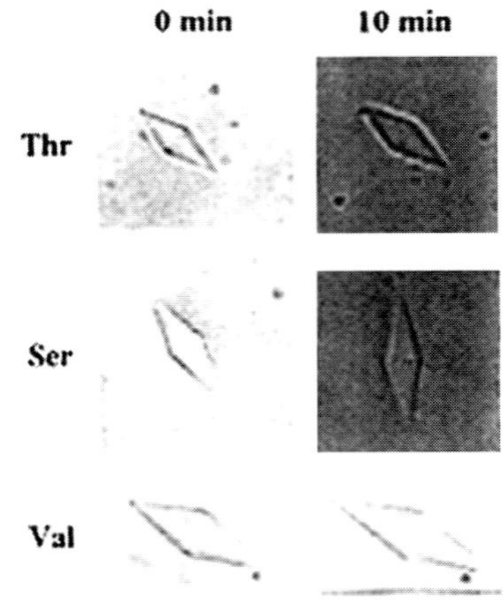

FIGURE 3.19 Mutant AFP I proteins and their ice crystal morphology (Chao et al., 1997).

publications describing mutation studies on this protein. AFP III does display a similar concentration to ice crystal growth relationship as AFP I (Wen and Laursen, 1992). There is a threshold concentration below which the protein cannot inhibit ice crystal growth (0.06 M) and below that growth is proportional to the concentration of AFP. In solutions of AFP III produced in *E. coli*, at a concentration of 1 mg/ml, the ice crystals have a 2:1 ratio of their c- and a-axes, and do not grow. All the mutated proteins, however, produced different ice crystal dimensions, and crystals grew over the experiment time of 10 minutes at a supercooling of 0.2°C. Most still formed a hexagonal bipyramid or a needle-type morphology. The morphology dimensions increased, mainly along the c-axis, with decreasing thermal hysteresis activity of each mutant (DeLuca et al., 1996).

A molar dilution series test was also performed with the native proteins, I and III, and the shape monitored. The authors state that the ratio of the ice crystal axes in type I solutions remained unchanged, while the ice crystals grew perceptively to over 100 times their original size at a protein concentration of 0.25 mg/ml, or below, and an undercooling of 0.1°C. At the equivalent molar concentration of AFP III (0.5 mg/ml), growth was not observed. As this solution was diluted and the test repeated, growth was observed but the c/a axis ratio increased from 2.0/1 to 3.0/1.

3.2.4.5.3 *AFP Mixtures* The shapes produced in mixed AFP solutions were the subject of a study by Chao, DeLuca and Davies (1995), to compare the shapes produced in different solutions and discover if the AFPs worked independently from one another. The experimenters studied the effect of adding together the various types of AFP, I, II and III, and looked at the effect on thermal hysteresis and shape of the growing ice crystal using the nanoliter osmometer. They compared AFP solutions with the same thermal hysteresis activity. Therefore, each AFP solution had a different starting concentration, depending on the type of AFP and its respective dependence of its hysteresis effect on concentration. Still photographs were taken 1 minute after the formation of the bipyramidal ice crystal under 0.1°C undercooling and the crystal dimensions measured. The aspect ratios of crystals in each mixed solution are given in Table 3.2. The three AFPs gave distinct aspect ratios specific for their type, i.e. type I, 2.2; type II, 1.3; and type III, 1.9.

Note that the figure quoted for AFP I is significantly different to that reported in the AFP I literature. When alone in solution, type I produces the largest crystal, with the longest c- and a-axial lengths. Type II has the shortest c-axial length and the lowest ratio. Type II also has an indicative rounding of the crystal faces at the pyramidal junction, whereas type III has the shortest a-axial length and an intermediate aspect ratio (see Figure 3.20).

The mixtures represent a movement from one form to another, depending on which type of AFP is in excess. When either of the globular AFPs is added to type I, then the globular protein dominated the growth, in that the c/a axis ratio was altered. The authors suggest that the crystals formed are 'hybrids' of the two types of AFP, and that both are influencing the crystal growth. They propose that the results correlate to the ice-binding plane, as each AFP used has a completely different structure and assumed binding plane. Additional evidence of AFP action comes from the thermal hysteresis values. The thermal hysteresis values for the mixes are neither increased nor decreased no matter what the mix ratio. (The original solutions were all adjusted so that they had the same thermal hysteresis values.) The authors suggest that each AFP must be acting independently to have these effects.

TABLE 3.2 Ice Crystal Ratios and Thermal Hysteresis Measurements of AFP Mixtures. Taken from Chao et al. (1995)

Sample: AFP type (% of sample solution)	Ratio c/a	Thermal hysteresis
I: 100%	2.2	0.35 ± 0.01
II: 100%	1.3	0.35 ± 0.01
III: 100%	1.9	0.35 ± 0.01
I:II 75:25	1.5	0.35 ± 0.02
I:II 50:50	1.5	0.34 ± 0.02
I:II 25:75	1.4	0.35 ± 0.01
I:III 75:25	2.0	0.35 ± 0.01
I:III 50:50	1.5	0.35 ± 0.02
I:III 25:75	2.0	0.35 ± 0.02
II:III 75:25	1.3	0.35 ± 0.02
II:III 50:50	1.1	0.34 ± 0.02
II:III 25:75	1.7	0.35 ± 0.02

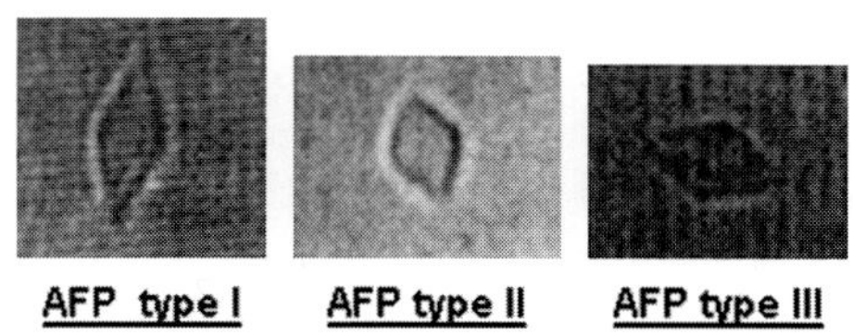

FIGURE 3.20 Examples of the shapes produced by each of the AFPs (Chao et al., 1995).

3.2.4.5.4 *AFP II* Other AFPs have been studied with respect to the ice crystal shapes

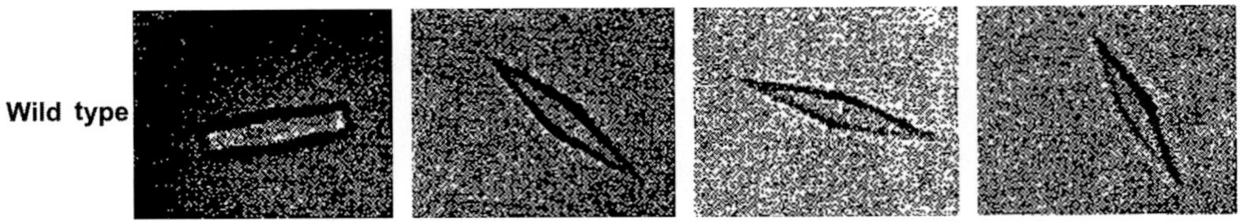

FIGURE 3.21 Ice crystal morphology seen with herring AFP (Ewart et al., 1998). From left to right, concentrations of AFP are 0.10, 0.5, 0.75, 1.5 mg/ml.

they form in solution. For example, AFP II from Atlantic herring (Ewart et al., 1998), where the ice-binding site is assumed to be the calcium-binding carbohydrate domain. The thermal hysteresis was measured, and photographs taken of the ice crystals (see Figure 3.21). These are bipyramids in the presence of the native, or recombinant, protein and circular or hexagonal prisms in the presence of a mutant protein. It is interesting to note that the shape of the crystal in Figure 3.21 is significantly different from that seen with AFP II from the sea raven in Figure 3.20. Unfortunately, there is no scale in this paper to make a judgment about the size. The maximum thermal hysteresis measured was 0.8°C at 1.5 mg/ml recombinant herring AFP.

AFP type II has been further investigated using the protein from the sea raven (Loewen et al., 1998). Growth rate data were taken from photographs captured at 0 and 10 minutes, at a fixed level of supercooling (0.1°C), from tests in the nanoliter osmometer. The crystals formed were all pseudo-bipyramids, but the mutants showed more faceted bipyramid crystals or increased growth than the rounded lemon-shaped crystals seen with the native protein. The authors conclude that the ice-binding site of the sea raven protein is different though near to the binding site of the herring AFP II example.

3.2.4.5.5 *AFP IV* The AFP IV literature reports an interesting shape of ice crystal with this new type of AFP. Deng, Andrews and Laursen (1997) discovered a HPLC extract that exhibited antifreeze activity, and this has since been identified as containing a new type of AFP. This protein, AFP IV, was then subjected to physical characterization, including a binding site determination, thermal hysteresis and a shape measurement (Deng and Laursen, 1998). In the initial publication, an unusual crystal morphology was identified. Though a bipyramid forms, with a c/a-axis ratio of 3.4 which is similar to that for AFP I crystals from winter flounder, the faces are now slightly rotated with respect to one another (see Figure 3.22). This suggests a new binding plane. This shape is further discussed in a subsequent paper where the binding site is determined by the ice hemisphere experiment (Deng and Laursen, 1998). The binding site is unusual, and the explanation is given via the diagrams reproduced in Figure 3.22. The binding plane cannot be identified as a single crystallographic plane, but the diagrams correlate the two aspects of AFP IV binding well.

3.2.4.5.6 *Plant AFP Ice Crystal Shapes* The ice crystal morphology property of AFPs is also used to provide information on ice crystal shape control in the presence of plant and insect AFPs.

The cold acclimated winter rye, *Secale cereal,* crude extract contains seven major peptides, which were found to be distinct from all fish and insect AFPs (Griffith et al., 1992; Hon et al., 1994). Of the seven proteins, the five larger peptides exhibit the greatest antifreeze activity. The antifreeze activity in this case was determined solely on the ability of the protein to alter ice crystal morphology observed in a Clifton nanoliter osmometer. No thermal hysteresis or recrystallization inhibition measurement was made.

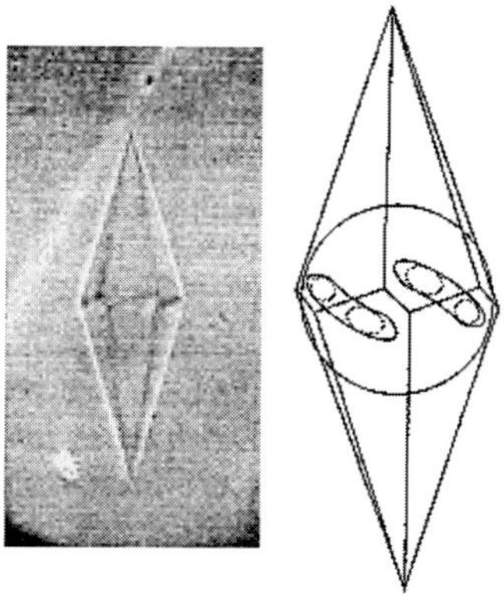

FIGURE 3.22 Ice crystal shape (left) and binding planes (right) found with AFP IV, with reference to the ice hemisphere binding site, suggesting a new AFP binding plane $[3\bar{1}\bar{2}1]$(Deng and Laursen, 1998).

The morphology of the ice crystal present in winter rye extract is photographed as it changes over the time period of the experiment. This series of photographs, taken using a phase contrast microscope, of the ice crystal is shown in Figure 3.23. The ice crystal is viewed normal to the visible a-axis direction and the slow formation of a bipyramid ice crystal continues until growth is halted. The temperature is slowly decreased during, and after this time, and as it continues to decrease, no further growth occurs until the explosive expansion at the thermal hysteretic freezing point.

An AFP from the carrot, *Daucus carota*, has recently been discovered (Worral et al., 1998). Here, the ice crystal shape has been probed by cryomicroscopy (Meyer et al., 1999). The assay is based on the principle nanoliter osmometry and involves one microliter of the AFP in solution. The sample is applied to the center of a cold stage and heated to +20°C. The stage is then crash cooled using a temperature controlling unit to –40°C, at a rate of 99°C per minute. Once frozen the sample is warmed to –5°C at the same rate, and a single crystal generated by controlled thawing. The crystal is then encouraged to grow by slight supercooling and photographs taken. In these conditions, high concentrations of AFP result in multifaceted, bipyramidal or spear-like crystals, while low concentrations produce hexagonal prisms. A pure water sample generates ice crystals that are flat circular discs. The shape assay is again used purely for identification of AFP activity.

3.2.4.5.7 Summary The data provided by the ice crystal shape experiments, completed in the osmometer and described in the preceding sections, are summarized in Table 3.3.

These results need to be examined carefully as many of the researchers in the field use the shape assay as a method to detect antifreeze activity. Therefore, the publications described in this section are mostly concerned with testing mutants and/or specific aspects of the protein binding. Though many of the researchers measure aspect ratio, several of the results do not agree, and the individual authors ignore the physical implications of the osmometer method for generating ice crystals, and the temperature regimen they apply to them. For example, there is confusion about the influences of solutes or buffers on ice crystal shape. Very few of the papers refer to the binding experiments

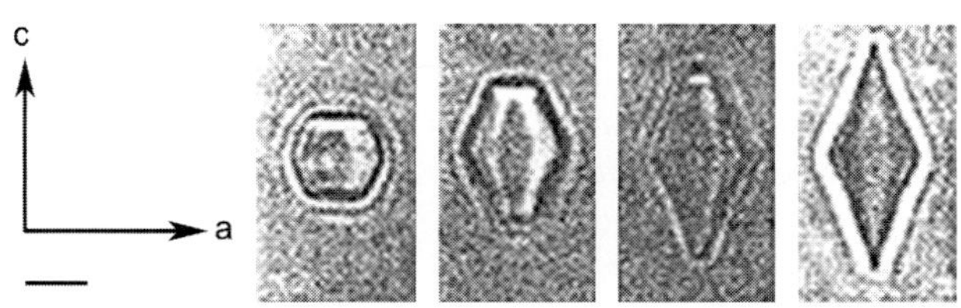

FIGURE 3.23 Series of photographs of ice morphology, taken in the presence of winter rye apoplastic extract 0.66mg/ml (Hon et al., 1994). Growth takes place from left to right.

TABLE 3.3 Summary of Aspect Ratio and Shape Data (From References Given in the Table)

AFP type	Aspect ratio	Shape	Reference
I from winter flounder	3.3/1	Hexagonal Bipyramid	Wen and Laursen (1992)
I from WF and mutants	3.3,3.5/1	Hex. Bipyramid in Hysteresis gap	Chao et al. (1997)
I from WF	2.2/1	Bipyramid in ΔT gap	Chao et al. (1995)
II from sea raven	1.3/1	Rounded Bipyramid in ΔT gap	Chao et al. (1995)
II from sea raven	Not given	Rounded Bipyramid in ΔT gap	Loewen et al. (1998)
II from Atlantic herring	Not given	Sharp Bipyramid in ΔT gap	Ewart et al. (1998)
III from recombinant ocean pout	1.9/1	Small Bipyramid in ΔT gap	Chao et al. (1995)
III from OP from *E. coli*	2.0/1	Bipyramid in ΔT gap	DeLuca et al. (1996)
IV from longhorn sculpin	3.4/1	Hexagonal Trapezohedra	Deng and Laursen (1998)
Plant: Bitter sweet nightshade	Not given	Bipyramid and Needles	Urrutia et al. (1992)
Plant: Winter rye	Not given	Bipyramids with explosive growth along the a-axes	Griffith et al. (1994)

performed by Knight (1993, 1991) and a suitable explanation to describe the effect of AFPs on shape is not forthcoming.

3.3 AFP MECHANISM OF FUNCTION

Several of the publications described in the preceding section have reviewed current opinion about AFP mechanism. There are two main schools of thought, which describe the binding of AFPs as small particles to the ice surface. Various additions to these theories, such as models and experiments that involve particular atomic and charge forces between AFPs and ice, try to substantiate the hypotheses in terms of molecular interactions. The molecular interactions required for AFP action will be discussed where appropriate.

3.3.1 The Adsorption-Inhibition Models

The original model for the antifreeze mechanism was proposed by Raymond and DeVries (1977), and has become known as adsorption-inhibition. Inhibition of crystal growth by impurities is well known. Crystal faces are expressed where an adsorbing molecule has retarded their growth, and the remaining crystal growth takes place on faster-growing faces. The adsorption-inhibition model of AFP action proposes that

> 'the protein adsorbs onto a face in the path of a growing step, the step will be forced to grow between the impurities, increasing the curvature and hence the area of the edge. Because of the surface tension, the increased work will affect the equilibrium state of the system.'

3.3.2 The Kelvin Effect

The Kelvin, or Gibbs-Thomson, effect, considers the energy barrier that arises from the need to increase surface area locally in order for an ice crystal to grow around an AFP molecule (Figure 3.24). This can only occur if the supersaturation is such that the 2D critical radius of the penetrating front is less than the average separation of the particles. By applying the Kelvin equations, an expression can be derived from the lowering of the equilibrium freezing temperature of water due to these effects (Raymond and DeVries, 1977).

Estimating the distribution of the AFP particles on the ice surface as a quadrilateral pattern, the limits of the radius of curvature that the growing step experiences can be estimated. This can be written as the following relationship:

$$\Delta T = \frac{2\sigma M T_o}{L\rho_i l}$$

where ΔT is the thermal hysteresis gap, T_o is the normal freezing point, L is the molar latent heat, l is spacing of the molecules in a random distribution on the crystal face, M is the molecular weight of water, ρ_i is the density of ice, and σ is the surface tension.

This implies that ΔT is proportional to $1/l$, the spacing of the adsorbed impurity molecules. Hence, as the spacing increases, the undercooling required for the step to grow between them will decrease. This equation can be further developed as Raymond and DeVries measured the partition coefficient of AFPs across the solid and liquid phases of a frozen AFP system. Assuming that the values of the partition coefficient relate to the true number of molecules involved in binding, and given information regarding the size of the impurities (molecular weight and dimensions), indirect measures of the amount of AFP on the surface can be gained. This relation presumes that each AFP lies flat on the ice surface and the AFP surface density can be calculated. This measure can therefore be related back to the assumed quadrilateral pattern of the molecules on the surface and the equation becomes:

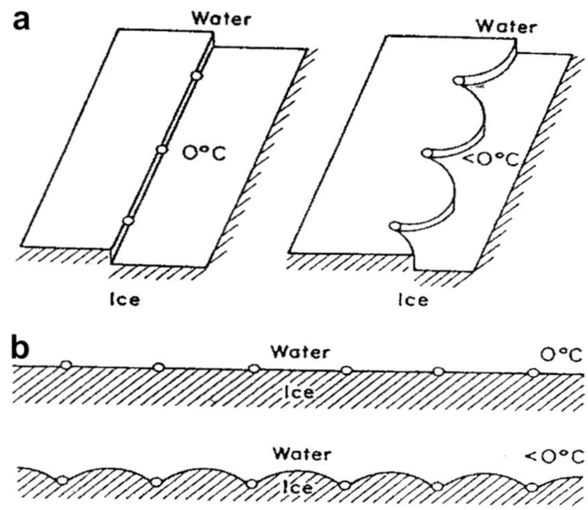

FIGURE 3.24 Adsorption-inhibition (a) and Button Mattress model (b) of AFP adsorption, reproduced from Knight et al. (1991).

$$\Delta T = \frac{2\sigma M T_o}{L\rho_i}\left[\frac{2r\alpha CN}{1000MW}\right]^{1/2}$$

where r is the AFP I or AFGP helix radius, α is the partition coefficient, C the concentration of AFP in molecules per ml, ($2r\alpha C$ = surface density), N is Avagadro's number, and 1000 MW is the molecular weight of the antifreeze in mg.

Inputting experimentally determined values for these parameters ($r = 8$Å, $\sigma = 5.24 \times 10^{-7}$ cal/cm^3, M = 18g, $T_o = 273°$, L = 1436 cal/mol, and $\rho_i = 0.917$ g/cm^3), the relationship can be improved:

$$\Delta T = 27.2[\alpha C/MW]^{1/2}$$

This function, plotted as ΔT against concentration of AFP, correlates very well to experimentally determined values of thermal hysteresis, with respect to concentration, when the molecular weight of the AFP is above 3000 approximately.

3.3.3 The Button Mattress model

Knight developed the adsorption-inhibition model in more physical terms, and proposed that the AFP stops growth of an ice front, not just a 2D step on a surface as proposed by DeVries. This model again describes the maintenance of maximum surface curvature (Figure 3.24) (Knight et al., 1991), by the Kelvin effect. Growth of the ice front will only occur when the supercooling increases and the ice–water interface can overcome this energy barrier to reach the minimum curvature. Essentially, the lobe has to become the size of an ice crystal embryo that has a critical radius above the threshold where growth will occur. The Button Mattress model, therefore, translates the Kelvin effect into three dimensions rather than two.

In this way, Knight reconciles shape development in the hysteresis gap with the hemisphere experiment. The patches on an AFP I ice hemisphere (see Section 3.2.3.2) represent regions where the AFP molecules bind to a particular ice plane (or planes) and stop the ice crystal growth. By considering this in three dimensions, the button mattress surface is generated. The ice lobes on this surface may join together or simply burst through the AFP barrier, as they reach the critical curvature for growth and they can continue to grow until another layer of AFP molecules is able to stop them. This barrier will therefore be dependent on AFP concentration, and correlate with the thermal hysteresis values. The AFP layers may be built up by positive adsorption of the AFP species onto the ice, or a freeze concentration effect may occur, increasing concentration at the interface with the solution, or both. This strategy represents a move from high to low surface energy for the ice–water interface.

3.3.4 Hall and Lips Model of AFP Action

Hall and Lips proposed an alternative model in 1999. This theory suggested several main differences from the Button Mattress model, relating to the degree of permanency of adsorption of the AFP particles. Knight suggests that adsorption of the AFPs is permanent (or nearly so). Ice crystals in an AFP environment, natural or artificial, exist in a supercooled solution. The Kelvin effect requires that the particles are irreversibly adsorbed to prevent any crystal growth. If they are not, and they detach from the surface, attachment of water molecules will occur immediately and the crystal will grow due to the energy available for crystallization.

Hall and Lips suggest that AFPs are adsorbing reversibly to a rough surface and by influencing the roughening transition thermodynamically, make it smooth – hence causing faceting of ice crystals at the beginning of the hysteresis gap as the temperature is decreased from the melting point. Adsorption-inhibition theories, in contrast, propose that AFPs are either pinning an already faceted surface, or as in the Button Mattress model, pinning a rough surface and making it flatter. The fact that in their natural environment, fish encounter ice crystals that are rough and have not undergone a transition to a smooth faceted form seems to imply that either the AFPs need to initiate this transition (as in the Hall and Lips approach), or the Button Mattress is the more correct adsorption-inhibition theory.

In all proposed models, as the AFPs adsorb, they raise the free energy barriers to growth, to such an extent, that further growth (rough or smooth) is inhibited. Hall and Lips envisage a monolayer coverage of AFPs on the crystal surface (terraced areas) with a dynamic exchange of molecules at the ice–water interface. This equilibrium slows growth of a roughened face and eventually converts it to a flat faceted surface that stops growing (within the hysteresis gap) due to an inhibition of 2D surface nucleation or the growth of dislocations. The Hall-Lips model is based on the idea of raising the step free energy via adsorption near a step and so reducing the rate of 2D surface nucleation.

In the Button Mattress model, on the other hand, the growing roughened surface is simply pinned in a roughened form. Very slow growth then leads to generation of what might be described as a roughened facet, and this may stop growing altogether.

The Hall-Lips model has the advantage that it leads to a mathematical prediction of the hysteresis–concentrations relationship, much as does the original adsorption-inhibition theory. Where the models have been tested against hysteresis data, they have tended to be equally successful. Studies of surfaces of the growing crystals by microscopy and measurements of adsorption isotherms, neither of which are currently available, will be needed to clearly discriminate these approaches.

3.4 APPLICATIONS OF AFP

There has been great interest over the recent years regarding the commercial application of AFPs and their use in biotechnology, medicine and frozen foods. Several reviews on this subject have been published such as Griffith and Ewart (1995), Feeney and Yeh (1998), Fletcher, Goddard and Wu (1999) and Litvak (1999), who describe the potential of AFPs in these areas. Due to the various properties of AFPs described in this chapter, and subsequent discovery of their actions in various biological and manufactured environments, they can be applied to enhance the survival of microstructures at subzero temperatures, whether biologically derived or man made.

Cryopreservation of mammalian cells and tissues continues to be an area of great interest and research (Griffith and Ewart, 1995). AFPs are found in various tissues of their native organism, and are extracellular. In fish, the proteins are synthesized in the liver and then secreted into the blood to be distributed to tissues. The AFPs are then thought to interact with the cell membranes to promote survival. In both fish and insects, AFPs have been immuno-localized on cell surfaces to provide a barrier to ice propagation (Urrutia et al., 1992; Duman et al., 1993b). A further discovery that AFPs may enhance hypothermia survival at temperatures above freezing by blocking calcium channels in the cell membrane and interacting with the lipid bi-layer to prevent loss of fluidity, naturally led to the application of cryopreservation of mammalian cells and organs. However, these effects are dependent on the type and concentrations of AFP used and in some cases have been found to be cytotoxic to cells. But successful subzero preservation of rat liver has been reported (Soltys et al., 2001).

The cryoprotective research is inextricably linked to that of cryosurgery, where probes cooled to very low temperatures are used to treat tumors. The development of new imaging technology has allowed the medical profession to increase its use of this technique as a more favorable alternative to laser surgery (Rubinsky, personal communication). However, in its current form, the procedure is not perfected to provide complete cell death regardless of freezing rate, freezing temperatures and number of freeze–thaw cycles. Control of these variables is difficult due to the procedure employed and an improved method is sought. The ability of AFPs to affect the shape and morphology of an ice crystal as it grows is an advantage in this case. It was observed that tumor cells frozen in the presence of AFP were destroyed regardless of freezing regimen, in contrast to the high percentage that survived in the control sample. The reason for this is unclear, but the results of this study have been reproducible in subsequent studies (Fletcher et al., 1999). Research in this area understandably remains a high priority due to its implications and the demand for less invasive surgical techniques (Muldrew et al., 2001).

A further use of AFPs has been in the area of biotechnology. The availability of AFP genes from cold-tolerant fish, plants and insects can be exploited in the protection of other species that do not currently have the genomes to survive cold or subzero temperatures. This application has been successfully applied to aquaculture or fish farming but is not currently in use (Fletcher et al., 1999; Litvak, 1999; Hew and Fletcher, 2001; Melamed et al., 2002). The climate in Eastern Canada is such that freezing of the seawater can occur in winter. This is a risk to the fishing industry located in these areas as they farm Atlantic salmon, a species that does not produce antifreeze proteins. Each year many fish die as they come into contact with ice beyond their tolerant limit of –0.9°C. The AFP I winter flounder gene was successfully introduced to this species and made them increasingly resistant to lower temperatures. The benefit of this research is two-fold; as the number of fish lost to hypothermia decreases and the economics of the farms are improved. Secondly, as more species of fish are made increasingly resistant to lower temperatures, there is less of a barrier to positioning aqua farms in colder regions of the world. This goes part way to improving the commercial output of these farms and it is hoped that in the future, transgenic technology may aid the crisis over the rapidly disappearing fish population by the introduction of transgenic lines. However, due to the recent public concern regarding genetic modification of foodstuffs, the fish that produce antifreeze proteins naturally are becoming targets for exploitation in aquaculture (Litvak, 1999).

The same advantages may be available to the crop farming community who lose revenue each year due to frost damage or death to the many species in production. The discovery of AFPs in plants offers a similar future to this industry and may allow the production of many fruit, vegetable and common crops in non-native environmental conditions in the future (Holmberg et al., 2001).

The application of AFPs to improve the quality of frozen food has been one of the biggest areas of commercial interest in recent years. Ice crystals are a predominant concern in many areas of food production such as the freezing of fruit, vegetables, processed foods and ice cream/desserts. The quality of the final product is dependent on the ice within it, which can be affected by the thermal cycling, or increased temperatures found within distribution chains or when storage conditions are not optimal. The ice in this environment recrystallizes and the ice crystals within the product grow and begin to affect taste and quality (Griffith and Ewart, 1995; Feeney and Yeh, 1998; Fletcher et al., 1999). This effect manifests in a change in texture of ice cream, as the ice crystals become detectable in the mouth, and in frozen vegetables or meat as drip loss, when the internal structure is damaged and water is released, thus removing nutrients. Both these problems affect food quality both during manufacture, throughout the chilled and frozen distribution chain, and once the product reaches the consumer.

AFPs are known to inhibit recrystallization of ice and could be used to overcome these issues. On testing in ice cream or water ices, it was found that on storage between –6 and –8°C for different periods of time, ice cream with AFP added to manufactured products showed significantly different recrystallization behavior. The control samples had larger ice crystals within their microstructure compared to the AFP samples that showed no growth at all (Feeney and Yeh, 1998).

The effect of AFPs in frozen meat is similarly dramatic in its efficiency. Payne et al. (1994) soaked bovine and ovine muscle in AFGP solutions before freezing. The reduced crystal size was evident from the muscle tissue on thawing when observed through optical microscopy and SEM. The control samples had larger intracellular spaces containing large ice crystals, and on treatment with AFP these became smaller and increased in number within the fibers. This effect was directly correlated to the concentration of antifreeze administered and the length of time each portion of muscle was soaked. This effect was also demonstrated when AFGP was injected into the animals prior to slaughter. The resulting meat demonstrated enhanced quality and reduced nutrient deficiency due to reduced drip loss (Fletcher et al., 1999). AFPs provide assistance for the freezing of the notoriously difficult to freeze fruits such as strawberries, raspberries or tomatoes. Again, the reduction in flavor and quality is due to the tissue damage caused by large, growing ice crystals within the tissue, and/or the associated cell dehydration. At low concentration, AFPs could inhibit these effects (Griffith and Ewart, 1995).

Before the addition of AFPs to these foods becomes a reality, there are several economic and safety considerations that need to be examined. For example, the cost of AFP from its native source is currently US $10/mg (AFProtein Inc.). Each liter of fish plasma would supply 2 g of AFP and 100 metric tonnes of fish would be required to produce 1 kg of AFP (Fletcher et al., 1999). To control recrystallization in frozen food, a concentration of the order of 1 μg/ml of AFP would be required. On the above calculations, that would be the equivalent of 150 000 tonnes of fish to supply the current ice cream market. Antifreeze proteins are naturally consumed in the diet in most cultures where subzero temperatures are part of the climatic conditions of primary production, or where freezing occurs in the food chain. Therefore, it is not unreasonable that the addition of these proteins to other foods should be acceptable, provided that levels are not high, and appropriate safety studies are performed. But due to these economics, genetic engineering has been employed to introduce the AFPs into food raw materials directly, or to produce sufficient quantity, at a reasonable cost, so that they can be added during manufacture. From what has been discussed above, it is obvious that whilst the primary function of AFPs in nature is to limit the damage caused by ice formation, the range of AFPs now identified have different activities and modes of action. Even with the demonstrated capability of heterologous expression of AFPs at high levels in various organisms, the GM route of manufacture raises questions about the behavior of each AFP, its activity with respect to the typical AFP properties and the selection of AFP for any application.

Nonetheless, the ability to manipulate ice structure and stability is an important new technology, and is still under investigation by several groups, and a continuous flow of patents are published. In recent years three groups have been particularly active.

Davies et al. have focused on the AFPs from insects, whose activity is particularly high. The uses of the protein from spruce budworm and Tenebrio Molitor are described in CA2257115 (Tyshenko et al., 1997), and US165383 (Graham et al., 2002), both with regard to their heterologous expression and potential to protect farmed fish. Further, a novel AFP from Tenebrio is described in US116506 (Graham et al., 2006).

The National Institute of Advanced Industrial Science (Japan) has been engaged in exploring novel sources of AFPs, and describes low-temperature culture of basiomycetes fungi (Hoshino et al., 2006). Synthetic AFPs have been produced by expression of fragments of ice nucleating proteins (Kobashigawa et al., 2004), and high activity has been reported from protein constructs of ligated AFPs from various sources (Nishimiya and Tsuda, 2003). A particular use of AFPs, not to modify ice, but to restrict the formation of gas clathrates in oil recovery is also disclosed (Uchida et al., 2005).

Unilever is now using AFPs commercially in some of its frozen desserts. The high activity of a particular peptide of AFP III has been demonstrated by cloning into yeast (Chapman et al., 1997) and the benefit obtained by adding disrupted cells to various formulations. The unfortunate reduction of active yield by random glycosylation has been overcome by using mutants low in glycosylating enzymes (Chapman et al., 2007). This allows cost-effective production of active materials. The fortuitous thermal stability of AFP from grasses allows their simple fractionation, and may be of significant value when pasteurization or other heat treatments form part of the downstream product-manufacturing process (Lillford et al., 1999). In a series of application patents, it was shown that by a combination of crystallization processing and the encorporation of AFPs, radically new product structures with remarkable mechanical properties can be formed (Daniel et al., 2001a, 2001b; Daniel and Oldroyd, 2001), requiring very small quantities of AFPs to be present (0.005%). These attributes, which do not relate to the prevention of freezing, but the control of ice crystal size and shape, led to the proposal that AFPs would be better described as ice structuring proteins (Clarke et al., 2002).

Thus there is no doubt that AFPs remain a fascinating and vital part of cold survival in many natural organisms (see 'Coping with the Cold', 2002). Far from being limited to fish in cold waters, it is now evident that biology co-opts appropriate protein sequences to provide freezing protection wherever and whenever necessary. Their technological value has also been widely demonstrated in crop protection, cryobiology, food manufacture and other industrial uses. A wide selection of molecular actives is now identified, but each has subtle differences in the way ice growth is limited, and the structure of ice which subsequently forms. In many cases, the amount of protein necessary to produce an effect are remarkably small, and unrefined natural materials containing these molecules can be identified and used commercially. The power of genetic engineering to produce cost-effective quantities of almost any of the molecular structures identified so far has also been demonstrated, but the acceptance of GM routes remains to be proven, even when the expressed protein occurs naturally. This probably should be argued on a case-by-case basis. There is certainly no need to argue for gene transfer from animals to plants and vice versa, since active molecules can be identified throughout all biological kingdoms. Indeed, the selective breeding of crop species in the absence of cold stress may be the cause of the absence of AFPs in most agricultural species, as the recent identification of an AFP in wild rice may demonstrate (Zhong, 2006).

References

Ananthanarayanan, V. S. and Hew, C. L. (1977). Synthetic polypeptide with antifreeze activity. *Nature* **268**, 560–561.

Antson, A. A., Smith, D. J., Roper, D. I., Lewis, S., Caves, L. S. D., Verma, C. S., Buckley, S. L., Lillford, P. J., and Hubbard, R. E. (2001). Understanding the mechanism of ice binding by type III antifreeze proteins. *Journal of Molecular Biology* **305**, 875–889.

Atkins, P. W. (1990). *Physical Chemistry.* Oxford University Press.

Baardsnes, J., Kondejewski, L. H., Hodges, R. S., Chao, H., Kay, C., and Davies, P. L. (1999). New ice-binding face for type I antifreeze protein. *Febs Letters* **463**, 87–91.

Barnes, W. H. (1929). The crystal structure of ice between 0°C and −183°C. *Proceedings of Royal Society A* **125**, 670.

Bowles, D., Lillford, P., and Rees, D. (2002). Coping with the cold. *Phil. Trans. Roy. Soc.* **357**, 1423.

Buckley, H. (1961). *Crystal Growth*. John Wiley and Sons, U.S.A.

Buckley, S. L., Smith, D. J., Clark, A. H., Lillford, P. J., and Hubbard, R. E. (2001). A Structural Model for the Activity of the Carrot Antifreeze Protein and its Binding to Ice. *Cryobiology* **43**.

Chakrabartty, A. and Hew, C. L. (1991). The Effect Of Enhanced Alpha-Helicity On The Activity Of A Winter Flounder Antifreeze Polypeptide. *European Journal of Biochemistry* **202**, 1057–1063.

Chakrabartty, A., Yang, D. S. C., and Hew, C. L. (1989). Structure-Function Relationship. *In* A Winter Flounder Antifreeze Polypeptide 2. Alteration Of The Component Growth-Rates Of Ice By Synthetic Antifreeze Polypeptides. *Journal of Biological Chemistry* **264**, 11313–11316.

Chao, H., DeLuca, C. I., and Davies, P. L. (1995). Mixing antifreeze protein types changes ice crystal morphology without affecting antifreeze activity. *Febs Letters* **357**, 183–186.

Chao, H., Houston, M. E., Hodges, R. S., Kay, C. M., Sykes, B. D., Loewen, M. C., Davies, P. L., and Sonnichsen, F. D. (1997). A diminished role for hydrogen bonds in antifreeze protein binding to ice. *Biochemistry* **36**, 14652–14660.

Chao, H., Sonnichsen, F. D., DeLuca, C. I., Sykes, B. D., and Davies, P. L. (1994). Structure-function relationship in the globular type III antifreeze protein: Identification of a cluster of surface residues required for binding to ice. *Protein Science* **3**, 1760–1769.

Chapman, J. W., Musters, W., and Van Wassenaar, P. D. (1997). WO Patent No. 02343. World Intellectual Property Organisation, International Bureau.

Chapman, J. W., Van De Laar, T., Lindner, N. M., and Visser, C. (2007). US Patent No. 155956. US Patent and Trademark Office, Washington D.C.

Chen, L., de Vries, A. L., and Cheng C-H, C. (1997). Convergent Evolution of Antifreeze Glycoproteins. *Proc. Nat. Acad. Sci.* **94** (8), 3811–3816.

Cheng, A. and Merz, K. M. (1997). Ice-Binding Mechanism of Winter flounder Antifreeze Proteins. *Biophysical Journal* **73**, 2851–2873.

Cheng, C. C. and DeVries, A. L. (1991). The Role of Antifreeze Glycopeptides and Peptides in the Freezing Avoidance of Cold-Water fish. *In* Life Under Extreme conditions , 1–14, Guido di Prisco (ed.). Springer-Verlag, Berlin.

Clarke, C. J., Buckley, S. L., and Lindner, N. (2002). Ice Structuring Proteins. *CryoLetters* **23**, 89–92.

Daniel, A., Fenn, R.A., Oldroyd, J.R. (2001b). European Patent No. 1158865. European Patent Office.

Daniel, A., Lacy, I., Oldroyd, J.R. (2001a). European Patent No. 1158862. European Patent Office.

Daniel, A. and Oldroyd, J. R. (2001). European Patent No. 1158864. European Patent Office.

Davies, P. L. and Sykes, B. D. (1997). Antifreeze Proteins. *Current Opinion in Structural Biology* **7**, 828–834.

DeLuca, C. I., Chao, H., Sonnichsen, F. D., Sykes, B. D., and Davies, P. L. (1996). Effect of type III antifreeze protein dilution and mutation on the growth inhibition of ice. *Biophysical Journal* **71**, 2346–2355.

DeLuca, C. I., Davies, P. L., Ye, Q., and Jia, Z. (1998). The Effects of Steric Mutations on the Structure of Type III Antifreeze Protein and its Interaction with Ice. *J. Mol. Biol.* **275**, 515–525.

Deng, G. J. and Laursen, R. A. (1998). Isolation and characterization of an antifreeze protein from the longhorn sculpin, Myoxocephalus octodecimspinosis. *Biochimica et Biophysica Acta-Protein Structure and Molecular Enzymology* **1388**, 305–314.

Deng, G. J., Andrews, D. W., and Laursen, R. A. (1997). Amino acid sequence of a new type of antifreeze protein: From the longhorn sculpin Myoxocephalus octodecimspinosis. *Febs Letters* **402**, 17–20.

DeVries, A. L. and Lin, Q. (1977). The role of glycoprotein antifreezes in the survival of antarctic fishes. *In* Adaptations within Antarctic Ecosystems, 439–458.

DeVries, A. L. and Wohlschlag, E. E. (1969). Freezing resistance in some antarctic fishes. *Science* **163**, 1073–1075.

DeVries, A. L. (1986). Antifreeze Glycopeptides And Peptides – Interactions With Ice And Water. *Methods in Enzymology* **127**, 293–303.

DeVries, A. L., Komatsu, S. K., and Feeney, R. E. (1970). Chemical and physical properties of freezing point-depressing Glycoproteins from antarctic fish. *J. Biol. chem.* **245**, 2901–2908.

Doucet, D., Tyshenko, M. G., Kuiper, M. J., Graether, S. P., Sykes, B. D., Daugulis, A. J., Davies, P. L., and Walker, V. K. (2000). Structure-function relationships in spruce budworm antifreeze protein revealed by isoform diversity. *European Journal of Biochemistry* **267**, 6082–6088.

Duman, J. G. and Olsen, T. M. (1993). Thermal Hysteresis Protein Activity in Bacteria, Fungi and Phylogenetically Diverse Plants. *Cryobiology* **30**, 322–328.

Duman, J. G. (1994). Purification And Characterization Of A Thermal Hysteresis Protein From A Plant, The Bittersweet Nightshade Solanum-Dulcamara. *Biochimica et Biophysica Acta-Protein Structure and Molecular Enzymology* **1206**, 129–135.

Eastman, J. T. and DeVries, A. L. (1986). Antarctic Fishes. *Scientific American* **255**, 96–103.

Ewart, K. V., Li, Z. J., Yang, D. S. C., Fletcher, G. L., and Hew, C. L. (1998). The ice-binding site of Atlantic herring antifreeze protein corresponds to the carbohydrate-binding site of C-type lectins. *Biochemistry* **37**, 4080–4085.

Ewart, K. V., Lin, Q., and Hew, C. L. (1999). Structure, function and evolution of antifreeze proteins. *Cellular and Molecular Life Sciences* **55**, 271–283.

Feeney, R. E. and Yeh, Y. (1993). Antifreeze Proteins: Properties, Mechansim of Action and Possible Applications. *Food Technology*, 82–90.

Feeney, R. E. and Yeh, Y. (1998). Antifreeze Proteins: Current Status and possible food uses. *Trends in Food Science and Technology* **9**, 102–106.

Fennema, O. R., Powrie, W. D., and Marth, E. H. (1973). Low Temperature Preservation of Foods and Living Matter. *Marcel Dekker Inc.*

Fletcher, G. L., Goddard, S. V., and Wu, Y. L. (1999). Antifreeze proteins and their genes: From basic research to business opportunity. *Chemtech* **29**, 17–28.

Fletcher, N. H. (1970). *The Chemical Physics of Ice*. Cambridge University Press.

Graether, S. P., Kuiper, M. J., Gagne, S. M., Walker, V. K., Jia, Z. C., Sykes, B. D., and Davies, P. L. (2000). Beta-helix structure and ice-binding properties of a hyperactive antifreeze protein from an insect. *Nature* **406**, 325–328.

Graham, A. L. and Davies, P. L. (2005). Glycine Rich Antifreeze Proteins from Snow Fleas. *Science* **310**, 461–463.

Graham, L. A., Liou, Y. C., Walker, V. K., and Davies, P. L. (1997). Hyperactive antifreeze protein from beetles. *Nature* **388**, 727–728.

Graham, L. A., Liou, Y.-C., Walker, V. K., and Davies, P. L. (2002). US Patent No. 165383. US Patent and Trademark Office, Washington D.C.

Graham, L. A., Liou, Y.-C., Walker, V. K., and Davies, P. L. (2006). US Patent No. 116506. US Patent and Trademark Office, Washington D.C.

Griffith, M. and Ewart, K. V. (1995). Antifreeze Proteins and their potential use in Frozen Foods. *Biotechnology Advances* **13**, 375–402.

Griffith, M., Ala, P., Yang, D. S. C., Hon, W. C., and Moffatt, B. A. (1992). Antifreeze Protein Produced Endogenously In Winter Rye Leaves. *Plant Physiology* **100**, 593–596.

Harding, M. M., Ward, L. G., and Haymet, A. D. J. (1999). Type I 'antifreeze' proteins – Structure-activity studies and mechanisms of ice growth inhibition. *European Journal of Biochemistry* **264**, 653–665.

Haymet, A. D. J., Ward, L. G., and Harding, M. M. (1999). Winter flounder "antifreeze" proteins: Synthesis and ice growth inhibition of analogues that probe the relative importance of hydrophobic and hydrogen-bonding interactions. *Journal of the American Chemical Society* **121**, 941–948.

Holmberg, N., Farres, J., Bailey, J. E., and Kallio, P. T. (2001). Targetted Expression of a synthetic codon optimised Gene. *Gene* **275** (1), 115–124.

Hon, W. C., Griffith, M., Chong, P. L., and Yang, D. S. C. (1994). Extraction and Isolation Of Antifreeze Proteins From Winter Rye (Secale-Cereale L) Leaves. *Plant Physiology* **104**, 971–980.

Hoshino, T., Kiriaki, M., Tsuda, S., Ohgiya, S., Kondo, H., Yokota, Y., and Yumoto, I. (2006). US Patent No. 068470. US Patent and Trademark Office, Washington D.C.

Jia, Z., DeLuca, C. I., Chao, H., and Davies, P. L. (1996). Structural Basis for the binding of a globular antifreeze protein to ice. *Nature* **384**, 285–288.

Kao, M. H., Fletcher, G. L., Wang, N. C., and Hew, C. L. (1986). The Relationship Between Molecular-Weight And Antifreeze Polypeptide Activity In Marine Fish. *Canadian Journal of Zoology-Revue Canadienne de Zoologie* **64**, 578–582.

Knight, C. A. and DeVries, A. L. (1988). The Prevention of Ice Crystal Growth from Water by Antifreeze Proteins. *In* Atmospheric Aerosols and Nucleation, 717–720. Springer, Berlin.

Knight, C. A. and DeVries, A. L. (1994). Effects of a polymeric, nonequilibrium "antifreeze" upon ice growth from water. *Journal of Crystal Growth* **143**, 301–310.

Knight, C. A. (1967). Freezing of Supercooled Fluids. *Van Nostrand Co.*

Knight, C. A. (2000). Structural biology – Adding to the antifreeze agenda. *Nature* **406**, 249–251.

Knight, C. A., Cheng, C. C., and DeVries, A. L. (1991). Adsorption of α-helical antifreeze peptides on specific ice crystal surface planes. *Biophysical Journal* **59**, 409–418.

Knight, C. A., DeVries, A. L., and Oolman, L. D. (1984). Fish Antifreeze Protein and The freezing and recrystallization of ice. *Nature* **308**, 295–296.

Knight, C. A., Driggers, E., and DeVries, A. L. (1993). Adsorption To Ice Of Fish Antifreeze Glycopeptide-7 And Glycopeptide-8. *Biophysical Journal* **64**, 252–259.

Knight, C. A., Hallett, J., and DeVries, A. L. (1988). Solute Effects On Ice Recrystallization – An Assessment Technique. *Cryobiology* **25**, 55–60.

Knight, C. A., Wen, D. Y., and Laursen, R. A. (1995). Nonequilibrium Antifreeze Peptides And The Recrystallization Of Ice. *Cryobiology* **32**, 23–34.

Kobashigawa, Y., Ohgiya, S., Miura, A., and Tsuda, S. (2004). WO Patent No. 072283. Japanese Patent Office.

Kuiper, M. J., Davies, P. L., and Walker, V. K. (2001). A theoretical model of a plant antifreeze protein from Lolium perenne. *Biophysical Journal* **81**, 3560–3565.

Lal, M., Clark, A. H., Lips, A., Ruddock, J. N., and White, D. N. J. (1993). Inhibition Of Ice Crystal-Growth By Preferential Peptide Adsorption – A Molecular Modeling Study. *Faraday Discussions*, 299–306.

Laursen, R. A., Wen, D., and Knight, C. A. (1994). Enantioselective Adsorption of the D- and L-Forms of an α-Helical Antifreeze Polypeptide to the {20-21} Planes of Ice. *J Am Chem Soc.* **116**, 12057–12058.

Lillford, P. J. and Holt, C. B. (1994). Antifreeze Proteins. *Journal of Food Engineering* **22**, 475–482.

Lillford, P. J., Mcarthur, A. J., Sidebottom, C. M., and Wilding, P. (1999). European Patent No. 0918863. European Patent Office.

Litvak, M. K. (1999). The development of winter flounder (Pleuronectes americanus) for aquaculture in Atlantic Canada: current status and future prospects. *Aquaculture* **176**, 55–64.

Loewen, M. C. and Davies, P. L. (1998). Binding of an oligopeptide to a specific plane of ice. *Journal of Biological Chemistry* **273**, 11714–11718.

Madura, J. D., Taylor, M. S., Wierzbicki, A., Harrington, J. P., Sikes, C. S., and Sonnichsen, F. D. (1996). The dynamics and binding of a Type III antifreeze proteins in water and on ice. *Jounal of Molecular Structure* **388**, 65–77.

Madura, J. D., Wierzbicki, A., Harrington, J. P., Maughon, R. H., Raymond, J. A., and Sikes, C. S. (1994). Interactions of the D- and L-Forms of Winter Flounder Antifreeze Peptide with the {201} Planes of Ice. *J Am Chem Soc.* **116**, 417–418.

Magono, C. and Lee, C. (1966). Meterological Classification of Natural Snow Crystals. *J Fac Sci.* **2**, 321–335.

Marshall, C. B., Fletcher, G. L., and Davies, P. L. (2004). Hyperactive Antifreeze Proteins in Fish. *Nature* **429**, 153–158.

Mason, B. J. (1958). The Growth of Ice Crystals from the Vapour and Melt. *Adv. Physics* **7**, 235.

Megaw, H. D. (1934). Cell Dimensions of Ordinary and Heavy Ice. *Nature* **134**, 900.

Melamed, P., Gong, Z. Y., Fletcher, G., and Hew, C. L. (2002). The potential impact of modern biotechnology on fish aquaculture. *Aquaculture* **204**, 255–269.

Meyer, K., Keil, M., and Naldrett, M. J. (1999). A leucine-rich repeat protein of carrot that exhibits antifreeze activity. *Febs Letters* **447**, 171–178.

Muldrew, K., Rewcastle, J., Donnelly, B. J., Saliken, J. C., Liang, S., Goldie, S., Olson, M., Baissalov, R., and Sandison, G. (2001). Flounder antifreeze peptides increase the efficacy of cryosurgery. *Cryobiology* **42**, 182–189.

Mullin, J. (1993). Crystallisation Third Edition, Ispwich Book Co. Ltd., Ipswich, England.

Nakaya, U. (1954). Snow Crystals. Harvard University Press.

Nishimiya, Y. and Tsuda, S. (2003). European Patent No. 1334980. European Patent Office.

Payne, S. R., Sandford, D., Harris, A., and Young, O. A. (1994). The Effects Of Antifreeze Proteins On Chilled And Frozen Meat. *Meat Science* **37**, 429–438.

Raymond, J. A. and DeVries, A. L. (1977). Adsorption Inhibition As A Mechanism Of Freezing Resistance In Polar Fishes. *Proceedings of the National Academy of Sciences of the United States of America*, **74**, 2589–2593.

Robinson, C. H. (2001). Cold adaptation in Arctic and Antarctic fungi. *New Phytologist* **151**, 341–353.

Scholander, P. F. and Maggert, J. E. (1971). Supercooling and Ice Propagation in Blood from Arctic fishes. *Cryobiology* **8**, 371–374.

Sicheri, F. and Yang, D. S. C. (1995). Ice-binding structure and mechanism of an antifreeze protein from winter flounder. *Nature* **375**, 427–431.

Sidebottom, C., Buckley, S., Pudney, P., Twigg, S., Jarman, C., Holt, C., Telford, J., McArthur, A., Worrall, D., Hubbard, R., and Lillford, P. (2000). Phytochemistry – Heat-stable antifreeze protein from grass. *Nature* **406**, 256.

Smaglik, P. (1998). Angling for answers – How do antifreeze proteins work? *Scientist* **12**, 4.

Smith, D. J. (2000). Molecular Modelling of Antifreeze Proteins. PhD Thesis, University of York, UK.

Soltys, K. A., Batta, A. K., and Koneru, B. (2001). Successful nonfreezing, subzero preservation of rat liver with 2,3-butanediol and type I antifreeze protein. *Journal of Surgical Research* **96**, 30–34.

Sonnichsen, F. D., Sykes, B. D., Chao, H., and Davies, P. L. (1993). The Nonhelical Structure of Antifreeze Protein Type III. *Science* **259**, 1154–1157.

Spanenburg, G., Ulrick, P. J., and Polotonianka, R. M. (2007). Ice Recrystallisation Inhibition Proteins from Deschampsia, Lolium and Festuca Species of Grasses. US Patent. 2007/0250946.

Tiller, W. A. (1991). *In* The Scince of Crystallisation, Macroscopic Phenomena and Defect Generation, 231–364. Cambridge University Press.

Trivedi, R. and Kurz, W. (1994). Dendritic Growth. *International Materials Reviews* **39**, 49.

Tyshenko, M. G., Davies, P. L., Walker, V. K., and Rahavard, M. (1997). Canadian Patent No. 2257115. CIPO Canadian Intellectual Property Office.

Tyshenko, M. G., Doucet, D., Davies, P. L., and Walker, V. K. (1997). The antifreeze ptential of the spruce budworm thermal hysteresis protein. *Nature Biotechnology* **15**, 887–890.

Uchida, T., Ikeda, I., Omura, A., Tsuda, S., Miura, K., and Ebinuma, T. (2005). Japanese Patent No. 089353. Japanese Patent Office.

Urrutia, M. E., Duman, J. G., and Knight, C. A. (1992). Plant Thermal Hysteresis Proteins. *Biochimica et Biophysica Acta* **1121**, 199–206.

Voituron, Y., Storey, J. M., Grenot, C., and Storey, K. B. (2002). Freezing survival, body ice content and blood composition of the freeze-tolerant European common lizard, Lacerta vivipara. *Journal of Comparative Physiology B-Biochemical Systemic and Environmental Physiology* **172**, 71–76.

Wang, X., DeVries, A. L., and Cheng, C. C. (1995). Antifreeze peptide hetrogenity in an antarctic eel pout includes an unusually large major variant comprised of two 7 KDa type III AFPs linked in tandem. *Biochimica et Biophysica Acta* **1247**, 163–172.

Wen, D. Y. and Laursen, R. A. (1992b). Structure-Function-Relationships In An Antifreeze Polypeptide – The Role Of Neutral, Polar Amino-Acids. *Journal of Biological Chemistry*, **267**, 14102–14108.

Wierzbicki, A., Taylor, M. S., Knight, C. A., Madura, J. D., Harrington, J. P., and Sikes, C. S. (1996). Analysis of Shorthorn Sculpin Antifreeze Protein Stereospecific Binding of (2-10) Faces of Ice. *Biophysical Journal* **71**, 8–18.

Worrall, D., Elias, L., Ashford, D., Smallwood, M., Sidebottom, C., Lillford, P., Telford, J., Holt, C., and Bowles, D. (1998). A carrot leucine-rich-repeat protein that inhibits ice recrystallization. *Science* **282**, 115–117.

Wu, D. W., Duman, J. G., Cheng, C. H. C., and Castellino, F. J. (1991). Purification And Characterization Of Antifreeze Proteins From Larvae Of The Beetle Dendroides-Canadensis. *Journal of Comparative Physiology B-Biochemical Systemic and Environmental Physiology* **161**, 271–278.

Wulff, G. (1901). Zur Frage der Geshwindigkeit des Wachtums und Auflosung der Krystallflachen. *Z.Krystallogr.* **34**, 449–530.

Yang, D. S. C., Hon, W.-C., Bubanko, S., Xue, Y., Seetharaman, J., Hew, C. L., and Sicheri, F. (1998). Identification of the ice binding surface on a type III AFP with a "Flatness Function" Algorithm. *Biophysical Journal* **74**, 2142–2151.

Yeh, Y. and Feeney, R. E. (1996). Antifreeze Proteins: Structures and Mechanisms of Function. *Chemical Reviews* **96**, 601–617.

Yu, Z.-W., Yu, W., DeVries, A. L., and Cheng, C. C. (1998). Fluorescent Microscopic Imaging of Antifreeze Protein Adsorption to Ice. *Biophysical Journal* **2** (2), A370.

Zhong, K. D. (2006). Chinese Patent No. 1772898. Chinese Patent Office.

CHAPTER

4

Biopolymers in Food Emulsions

David Julian McClements

Department of Food Science, University of Massachusetts, Amherst, Massachusetts, USA

4.1 INTRODUCTION

Biopolymers are commonly used by the food industry to improve the stability and texture of emulsion-based food products (Friberg et al., 2004; McClements, 2005). A wide variety of food products can be classified as emulsions, including milk, cream, whipped cream, ice cream, yogurts, soft drinks, nutritional beverages, infant formula, desserts, sauces, dips, mayonnaise, dressings, cheese, spreads, butter, and margarine (Friberg et al., 2004; McClements, 2005). Some of these products naturally contain functional biopolymers (such as milk), whereas others contain biopolymers that have been added as ingredients because of their unique functional attributes (such as dressings, dips and sauces). The functional biopolymers may be added as specific ingredients (such as emulsifiers, thickening agents, or gelling agents), or they may form part of more complex ingredients (such as eggs, milk, or flour). This chapter will begin by describing the formation, stability and properties of food emulsions, and then it will consider the application of biopolymers as functional ingredients in emulsion-based foods.

4.2 EMULSION SCIENCE AND TECHNOLOGY TERMINOLOGY

It is useful to begin with an overview of the terminology commonly used by those working in the field of emulsion science and technology. An emulsion can be defined as a dispersion of droplets of one liquid in another liquid with which it is (at least partly) immiscible (Figure 4.1).

In foods, the two immiscible liquids are usually an oil phase and a watery phase (although this is not always necessary). Emulsions can be conveniently classified according to the relative organization of the two immiscible phases (McClements, 2005). A system consisting of oil droplets dispersed in a watery phase is called an oil-in-water (O/W) emulsion (e.g., milk, salad dressing and mayonnaise), whereas a system consisting of water droplets dispersed in an oil phase is called a water-in-oil (W/O) emulsion (e.g., butter or margarine). The material within the droplets is usually referred to as the *dispersed* or *internal phase,* whereas the material within the surrounding liquid is called the *continuous* or *external phase*. It is also possible to form more complex arrangements of oil and water within a single system, e.g., W/O/W or

Kasapis, Norton, and Ubbink: Modern Biopolymer Science
ISBN: 978-0-12-374195-0

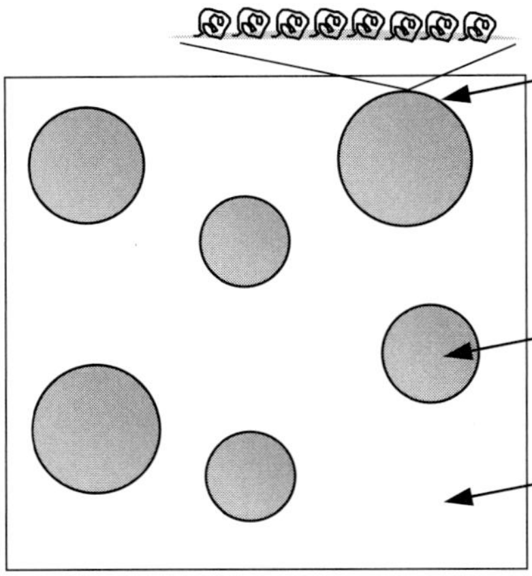

FIGURE 4.1 Emulsions consist of liquid droplets (disperse phase) suspended in an immiscible liquid (continuous phase). The droplets are usually surrounded by a thin interfacial layer that protects them from aggregating. The properties of this interfacial layer have a major impact on emulsion stability and performance.

O/W/O 'double' emulsions (Garti, 1997; Garti and Bisperink, 1998). These double emulsions can be useful for protecting labile ingredients, for controlling the release of specific ingredients, or for formulating low-fat products.

Conventional emulsions are thermodynamically unstable systems because of the positive free energy associated with the contact of oil and water phases, as manifested by a relatively high positive interfacial tension (Friberg et al., 2004). The origin of this positive free energy is the unfavorable interaction between oil and water molecules, which can mainly be attributed to the fact that water molecules are capable of forming strong hydrogen bonds with neighboring water molecules but not with oil molecules, i.e., the hydrophobic effect (McClements, 2005). There is therefore a thermodynamic driving force favoring the reduction of the contact area between the two immiscible liquids, which can be achieved through coalescence and macroscopic phase separation. The thermodynamic instability of conventional 'macroemulsions' clearly distinguishes them from microemulsions, which are thermodynamically stable systems. In addition, the droplets in macroemulsions ($d \approx$ 100 nm to 100 μm) tend to be considerably higher than the droplets in microemulsions ($d \approx 5$ nm to 50 μm). This means that macroemulsions tend to be optically opaque because the droplets have dimensions that are on the same order as light waves (300–800 nm), whereas microemulsions are optically transparent because they are much smaller than light waves. In addition, the relatively large surface area of microemulsions per unit amount of droplets means that a much greater amount of surfactant is required to form them.

The physical instability of an emulsion may occur through a number of different processes, including creaming, sedimentation, flocculation, coalescence, partial coalescence, Ostwald ripening, and phase inversion (Figure 4.2).

The instability mechanism that dominates within a particular food product largely determines its overall shelf life. A major concern of the food industry is therefore how to control these instability mechanisms so as to prolong the shelf life of the final product. For this reason, ingredients known as *stabilizers* are usually added to emulsion-based food products to prevent or arrest these instability mechanisms. Stabilizers can be conveniently divided into two categories depending on the mode of action: *emulsifiers* and *texture modifiers*. Emulsifiers are surface-active molecules that rapidly adsorb to the surfaces of the droplets formed within a homogenizer during the emulsion formation process, where they create a protective coating that helps prevent droplets from coming close together and aggregating. Most food emulsifiers are *amphiphilic* molecules, such as lipid-based emulsifiers (small-molecule surfactants and phospholipids) and biopolymers (proteins and polysaccharides). In addition, some types of small solid particles are also surface active and can act as emulsifiers in foods, e.g., granules from egg or mustard. Texture modifiers are ingredients that are normally added to the continuous phase of an

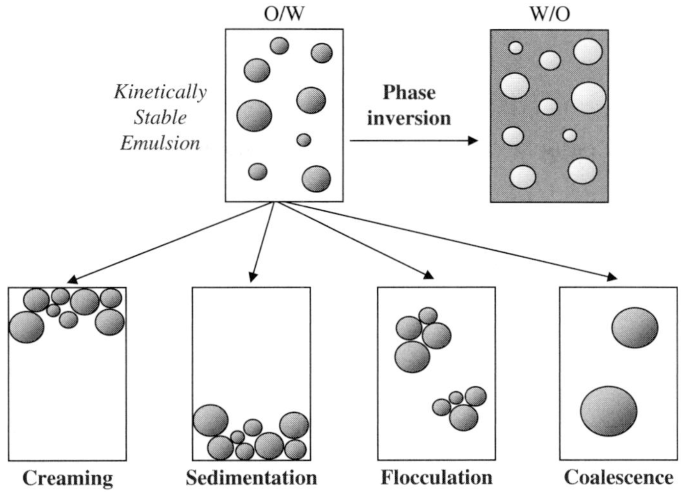

FIGURE 4.2 Summary of the physicochemical mechanisms that typically cause instability in food emulsions. The dominant mechanism in a particular emulsion depends on its composition, structure and environment.

emulsion to increase its viscosity ('thickening agents') or to form a gel ('gelling agents'), thereby reducing the movement of the droplets. In addition, texture modifiers are often added to provide desirable textural and sensory attributes to emulsion-based food products.

4.3 EMULSION DROPLET CHARACTERISTICS

4.3.1 Droplet Concentration

The concentration of droplets within an emulsion is one of the most important factors influencing its overall physicochemical, sensory and nutritional properties (Friberg et al., 2004; McClements, 2005). In emulsions, the droplet concentration is usually expressed as the *disperse phase volume fraction*: $\phi = V_D/V_E$, where V_D and V_E are the volumes of droplets and emulsion, respectively. Nevertheless, it can also be expressed as a *disperse phase mass fraction*, $\phi_m = M_D/M_E$, where M_D and M_E are the masses of droplets and emulsion, respectively. The mass and volume fractions of emulsions can be related to one another using simple expressions:

$$\phi_m = \frac{\phi\rho_2}{\rho_2\phi + (1-\phi)\rho_1} \tag{4.1}$$

$$\phi = \frac{\phi_m\rho_1}{\rho_1\phi_m + (1-\phi_m)\rho_2} \tag{4.2}$$

where ρ_1 and ρ_2 are the densities of the continuous and dispersed phases, respectively. When the densities of the two phases are similar ($\rho_1 = \rho_2$), the mass fraction is equivalent to the volume fraction. The droplet concentration is also commonly presented as a volume percentage ($\%\phi = 100 \times \phi$) or a mass percentage ($\%\phi_m = 100 \times \phi_m$).

4.3.2 Droplet Size Distribution

The droplet size distribution (DSD) of an emulsion-based food product often has a major impact on its physicochemical and sensory properties, e.g., shelf life, appearance, flavor, and texture (McClements, 2005). Many food manufacturers therefore carry out preliminary experiments to establish the most suitable DSD for a particular emulsion-based food product, which will depend on the desired shelf life, optical properties and rheology of the product (McClements, 2005). Product formulation (e.g., emulsifier-to-oil ratio) and homogenization conditions (e.g., valve pressure and number of passes) are then selected to create a product whose DSD falls within the specified range. If the product does not meet these specifications it may be reprocessed or discarded.

The DSD of an emulsion can be broadly characterized in terms of the droplet concentration versus droplet size profile. If an emulsion contains droplets that are all the same size, then it can be referred to as being *'monodisperse'*, whereas if it contains droplets that are a range of different sizes then it is referred to as being *'polydisperse'*. Polydisperse emulsions can be further classified according to the general shape of the DSD as being 'monomodal' (single-peaked), 'bimodal' (double-peaked) or 'multi-modal' (multiple-peaked). The dimensions of the droplets in a monodisperse emulsion are completely characterized by a single number, such as the droplet diameter (d) or radius (r). In practice, monodisperse emulsions are not found in real food products because they cannot currently be economically prepared. Instead, the droplets in most real food products have a range of different sizes and so it is necessary to determine the full *droplet size distribution*, i.e. the concentration of droplets in specified size ranges. Having said this in some situations it is sufficient to have knowledge of the mean droplet size and the width of the distribution. The mean particle size can be calculated in a number of different ways, with the number-weighted (d_{10}), surface-weighted (d_{32}) and volume-weighted (d_{43}) mean diameters being the most common (McClements, 2005). It is particularly important to choose the most appropriate mean diameter to reflect the characteristic of an emulsion of interest when reporting data, and to be aware of the type of mean used by other people when they report their data.

4.3.3 Droplet Charge

The electrical characteristics of the droplets in an emulsion-based food product may play an important role in determining their properties and shelf life (McClements, 2005). The electrical charge on the droplets influences droplet–droplet interactions because it determines the magnitude of the electrostatic repulsion between them. If the droplet charge is sufficiently high, then the emulsion may be stable to droplet aggregation because the electrostatic repulsion is strong enough to prevent them from coming close together (McClements, 2005). The droplet charge also determines how droplets interact with other charged species within an emulsion, such as biopolymers, minerals, antioxidants and flavors. For example, charged biopolymers may promote bridging flocculation when added to an emulsion that contains oppositely charged droplets (Dickinson and Pawlowsky, 1997, 1998; Gu et al., 2004a, 2004b; Guzey and McClements, 2006). The electrical charge on emulsion droplets is usually the result of adsorption of emulsifier molecules that contain ionized or ionizable groups, such as ionic surfactants, phospholipids, proteins and polysaccharides. The magnitude and sign of the electrical charge largely depend on the type and concentration of surface-active molecules present at the interface, as well as the pH and ionic composition of the aqueous phase. The electrical charge on an emulsion droplet can be manipulated by choosing emulsifiers with desirable charge characteristics (e.g., sign, magnitude, isoelectric point) and controlling the aqueous phase properties (e.g., pH and ionic strength).

4.3.4 Droplet Interfacial Properties

The droplets in emulsions are usually surrounded by a thin coating of material (typically 1–50 nm thick) that contains a mixture of various types of molecule, e.g., oil, water, emulsifier, biopolymers, and minerals (Dickinson, 2003). The composition, structure, thickness, rheology and responsiveness of this interfacial layer often play a major role in determining the overall properties of emulsions (McClements, 2005). The composition and structure of the interfacial region are determined by the type and concentration of surface-active species present, as well as by the events that occur both during and after emulsion formation, e.g., molecular rearrangements, competitive adsorption, complex formation. The thickness and rheology of the interfacial region influences the stability of emulsions to gravitational separation, coalescence and flocculation, as well as influencing their physicochemical and sensory properties, such as bulk rheology and mouthfeel.

4.3.5 Droplet–Droplet Interactions

The various attractive and repulsive forces acting between droplets determine their stability to aggregation, and therefore the overall shelf life, texture and appearance of food emulsions. The interactions between two droplets are usually described by an *interaction potential*, $w(h)$, which is the free energy required to bring two droplets from an infinite distance apart to a surface-to-surface separation of h (Figure 4.3).

The overall interaction potential is made up from contributions from various different types of interactions, including van der Waals, steric, electrostatic, depletion, hydrophobic and hydration interactions (Table 4.1) (McClements, 2005).

These individual interactions vary in their sign (attractive or repulsive), magnitude (weak to strong) and range (short to long). Each of the individual interactions usually has a simple dependence on surface-to-surface separation, but the sum of the interactions can exhibit fairly complex behavior, exhibiting maximum or minimum values at certain separations (Figure 4.4).

Droplets tend to aggregate when attractive interactions dominate, but remain as individual entities when repulsion interactions dominate.

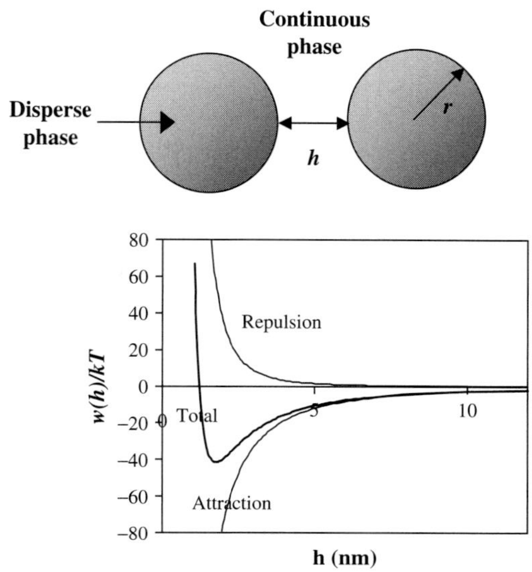

FIGURE 4.3 The interaction between two emulsion droplets is described by the droplet interaction potential ($w(h)$) versus surface-to-surface separation (h) profile. These curves can be used to estimate the relative importance of various attractive and repulsive interactions.

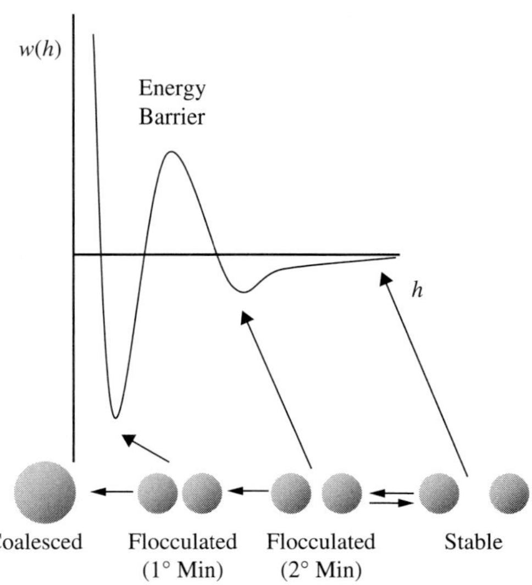

FIGURE 4.4 The interaction potential determines the stability of emulsion droplets to aggregation, such as weak flocculation, strong flocculation (coagulation) or coalescence.

TABLE 4.1 Summary of the Most Important Colloidal Interactions Acting Between Emulsion Droplets

Type	Sign	Magnitude	Range	Major factors influencing
Van der Waals	A	S	L	Refractive index, dielectric constant, ionic strength
Electrostatic	R or A	W – S	S – L	Droplet charge, pH, ionic composition
Steric	R	S	S	Solvent quality, interfacial thickness, interfacial packing
Depletion	A	W – M	M	Excluded species size and concentration
Hydrophobic	A	S	L	Surface hydrophobicity, temperature

Sign: Attractive (A), Repulsive (R); **Magnitude**: Weak (W), Medium (M), Long (L); **Range**: Short (S), Medium (M), Long (L).

4.4 PRODUCTION OF FOOD EMULSIONS

The overall properties of an emulsion-based food product depend on the type and concentration of ingredients that it contains, as well as the processing operations used to manufacture it (Friberg et al., 2004; McClements, 2005). The industrial production of a food emulsion usually involves a number of consecutive steps, which depend on the nature of the starting materials and the desired properties of the end product (Walstra, 1993, 2003). These processes can be divided into pre-homogenization, homogenization and post-homogenization steps.

4.4.1 Major Steps in Homogenization

4.4.1.1 Pre-Homogenization

The pre-homogenization steps typically involve dispersion, mixing and dissolution of the various ingredients into the phase in which they are most soluble. Thus, oil-soluble ingredients (e.g., non-polar vitamins, flavors, coloring agents, antioxidants, and surfactants) are mixed with the oil, while water-soluble ingredients (e.g., polar vitamins, flavors, proteins, polysaccharides, sugars, salts, coloring agents, antioxidants, and surfactants) are mixed with the water. The intensity and duration of the mixing process depends on the time required to solvate and uniformly distribute the various ingredients. Adequate solvation is important for the optimum functionality of a number of functional food ingredients (and particularly food biopolymers) and must be carried out properly in order to produce a high-quality and reliable final product. The composition and properties of the aqueous phase (e.g., pH, ionic strength, mineral type) may also play an important role in determining the solubility and functionality of biopolymers and should usually be controlled. Detailed information about the appropriate dispersion procedures for specific biopolymers and other functional components can usually be obtained from ingredient suppliers. It should be noted that the order of ingredient addition relative to subsequent processing steps is often an important factor in ensuring good dispersion and solubility, and that sometimes certain powdered ingredients are best mixed together prior to dispersion.

If the lipid phase contains any crystalline material, it is usually necessary to heat it before homogenization to a temperature at which all of the fat crystals melt; otherwise it is difficult to create a stable emulsion. Some other functional ingredients may also need careful control of temperature during their dispersion and solvation. For example, it may be necessary to heat a protein or polysaccharide above a critical temperature (e.g., its helix-coil transition temperature), or not to heat a protein above a certain temperature (e.g., its thermal denaturation temperature). Again, ingredient suppliers will often provide detailed information about the conditions required to ensure good dispersion and dissolution of functional biopolymer ingredients.

4.4.1.2 Homogenization

The homogenization step involves converting the two immiscible phases (usually oil and water) into an emulsion using a mechanical device known as a *homogenizer* (Walstra, 1993, 2003). The homogenization process can conveniently be divided into two types depending on the nature of the starting material. The creation of an emulsion directly from two separate liquids can be referred to as *primary* homogenization (e.g., blending of oil and vinegar to form a salad dressing), whereas the reduction in the size of the droplets in an existing emulsion can be referred to as *secondary* homogenization (e.g., size reduction in raw milk). The creation of a food emulsion may use just primary homogenization, just secondary homogenization, or a combination of both. In many food processing operations and laboratory studies the most efficient way of preparing an emulsion is to use a two-step procedure, i.e., primary homogenization followed by secondary homogenization. The separate oil and water phases are converted into a coarse emulsion with one type of homogenizer (e.g., high shear mixer), then the droplet size is reduced by means of another type of homogenizer (e.g., colloid mill, high-pressure valve homogenizer). In reality, many of the physical processes that occur during primary homogenization also occur during secondary homogenization, e.g., droplet disruption, droplet coalescence and emulsifier adsorption. Emulsions that have undergone secondary homogenization usually contain smaller droplets than those that have undergone primary homogenization, although this is not always the case. The main goal of the homogenization process is to produce an emulsion with a stipulated DSD as efficiently and economically as possible.

Efficient homogenization usually requires selection of the most appropriate homogenizer type (Table 4.2) and homogenizer operating conditions (e.g., pressure, flow rate, temperature) for the particular food being produced (Walstra, 1993, 2003; McClements, 2005). In addition, it requires the selection of an appropriate type and amount of emulsifier: there should be sufficient emulsifier present to stabilize the newly formed emulsion droplet surfaces created during homogenization, the emulsifier should rapidly adsorb to the oil–water interface, and the emulsifier should form a coating that protects the droplets against aggregation.

4.4.1.3 Post-Homogenization

Once an emulsion has been formed by homogenization it may be subjected to a number of additional processing steps (McClements, 2005). Probably the most common post-homogenization unit-operations involve temperature treatments, such as heating (pasteurization, sterilization, cooking) or cooling (chilling or freezing) to ensure long-term safety or produce particular quality attributes. As well as inhibiting or deactivating enzymes or microbes, these temperature treatments may also induce physical or chemical changes that alter the properties of the system, e.g., thermal denaturation of proteins, helix-coil transitions of biopolymers, phase inversion of surfactants, and crystallization/melting of fats. Another common post-homogenization unit-operation is dehydration, which is used to convert a liquid emulsion

TABLE 4.2 Comparison Different Types of Homogenizer Used to Prepare Emulsions

Homogenizer type	Throughput	Energy density	Relative energy efficiency	Minimum droplet size	Sample viscosity
High-speed mixer	Batch or continuous	Low–high	Low	≈ 2 μm	Low to medium
Colloid mill	Continuous	Low–high	Medium	≈ 1 μm	Medium to high
High-pressure homogenizer	Continuous	Medium–high	High	≈ 0.1 μm	Low to medium
Microfluidization	Continuous	Medium–high	High	< 0.1 μm	Low to medium
Ultrasonic probe	Batch or continuous	Medium–high	Low	≈ 0.1 μm	Low to medium
Ultrasonic jet homogenizer	Continuous	Medium–high	High	≈ 1 μm	Low to medium
Membrane processing	Batch or continuous	Low–medium	Very high	≈ 0.3 μm	Low to medium

product into a powder, e.g., freeze-drying, spray drying or roller drying.

4.4.2 Physical Principles of Homogenization

The size of the emulsion droplets produced by a homogenizer depends on a balance between two opposing mechanisms: *droplet disruption* and *droplet coalescence* (Walstra, 1993). The tendency for emulsion droplets to break up during homogenization depends on the strength of the *interfacial forces* that hold the droplets together, compared to the strength of the *disruptive forces* in the homogenizer (Walstra, 1993, 2003). In the absence of any applied external forces, emulsion droplets tend to be spherical because this shape minimizes the unfavorable contact area between oil and water phases. Deforming a droplet, or fragmenting it into smaller droplets, increases this contact area and therefore requires the input of free energy. The forces responsible for holding the droplets together are referred to as *interfacial forces* and are proportional to the interfacial tension and the reciprocal of the droplet diameter. Consequently, it becomes more difficult to disrupt droplets as their interfacial tension increases or their size decreases. The nature of the disruptive forces that act on a droplet during homogenization depends on the flow conditions that a droplet experiences within a homogenizer and therefore on the type of homogenizer used to create the emulsion (Walstra, 1993, 2003). To deform and disrupt a droplet during homogenization, it is necessary to generate a disruptive force that is greater than the *interfacial force* and to ensure that this disruptive force is applied to the droplet long enough to allow it to become deformed and disrupted (Schubert et al., 2003; Schubert and Engel, 2004).

Droplet–droplet collisions occur frequently during homogenization because of the intense mechanical agitation of the emulsion (Walstra, 1993, 2003). When the droplets collide with each other in the homogenizer they are likely to coalesce unless they are covered by a protective layer of emulsifier. Immediately after the disruption of an emulsion droplet during homogenization, the newly formed surfaces are not covered by emulsifier, and therefore the droplets are prone to coalesce with their neighbors. To prevent coalescence from occurring, it is necessary for a sufficiently dense emulsifier layer to form around a droplet before it has time to collide with its neighbors. The size of droplets produced during homogenization therefore depends on the time taken for the emulsifier to be adsorbed to the surface of the droplets (τ_A) compared to the time between droplet–droplet collisions (τ_C). If $\tau_A << \tau_C$, then the droplets are rapidly coated with emulsifier and are stable to coalescence; but if $\tau_A >> \tau_C$, the droplets tend to rapidly coalesce because they are not completely coated by emulsifier before colliding with their neighbors. The values of these two times depend on the flow profile the droplets experience during homogenization, as well as the physicochemical properties of the bulk phases and the emulsifier.

4.4.3 Role of Emulsifiers

The preceding discussion has highlighted two of the most important roles of emulsifiers during the homogenization process:

- Their ability to decrease the interfacial tension between oil and water phases and thus reduce the amount of free energy required to deform and disrupt a droplet.
- Their ability to form a protective coating around droplets that prevents them from coalescing with their neighbors during a collision.

The effectiveness of emulsifiers at creating emulsions containing small droplets depends on a number of factors: (a) the concentration of emulsifier present; (b) the time required for the emulsifier to move to the droplet surface and be adsorbed; (c) the amount by which the emulsifier reduces the interfacial tension; and (d) the effectiveness of the adsorbed emulsifier layer at protecting the droplets against aggregation. Emulsifiers used in the food industry vary greatly in their ability to form and stabilize emulsions and it is important for food manufacturers to identify the most appropriate emulsifier for each application. The properties of some common biopolymer emulsifiers are discussed in a later section.

4.4.4 Homogenization Devices

A number of different homogenization technologies are available to the food industry for the production of emulsions, each with its own particular advantages and limitations, and each having a range of foods to which it is most suitably applied. The choice of a particular homogenizer depends on many factors, including the equipment available, the site of the process (i.e., a factory or a laboratory), the physicochemical properties of the starting materials and final product, the volume of material to be homogenized, the desired throughput, the desired droplet size distribution of the final product, and the cost of purchasing and running the equipment (Walstra, 1993, 2003). The most important types of homogenizer used in the food industry are listed in Table 4.2. Once a suitable homogenizer has been identified it is necessary to establish the optimum operating conditions for the particular product being manufactured, e.g., pressure, flow rate, number of passes, temperature, etc. This can often be done in a laboratory scale or pilot plant situation before being scaled up to full-scale production.

4.5 EMULSION STABILITY

As mentioned earlier, conventional emulsions (or 'macroemulsions') are thermodynamically unstable systems that tend to separate back into their individual oil and water phases over time (Figure 4.2). The term 'emulsion stability' refers to the ability of an emulsion to resist changes in its properties: the greater the stability of an emulsion, the slower the change in its properties. Changes in emulsion properties may be the result of physical processes that cause changes in the spatial distribution of its components (e.g., creaming, sedimentation, flocculation, coalescence, Ostwald ripening and phase inversion) or they may be the result of chemical processes that cause alterations in the type of molecules present (e.g., oxidation, hydrolysis, polymerization). It is important for food scientists to elucidate the relative importance of each of these mechanisms in specific food systems, the relationship between them, and the factors that affect them, so that effective means of controlling the properties of specific emulsion-based food products can be established and implemented.

4.5.1 Droplet–Droplet Interactions

As mentioned earlier, one of the most important factors affecting the overall stability and physicochemical properties of food emulsions is the nature of the forces acting between the droplets (McClements, 2005). If there is a relatively strong attraction between the droplets they will tend to aggregate, but if there is a relatively strong repulsion they will tend to remain as separate entities. The overall interaction between droplets depends on the *sign*, *magnitude* and *range* of a number of different types of interactions, e.g., steric, van de Waals, electrostatic, hydrophobic. Knowledge of the origin and nature of these interactions is important because it enables food scientists to predict and control the stability and physicochemical properties of food emulsions in a more rational manner. For example, biopolymers can be added to modulate the electrostatic, steric or depletion interactions acting between emulsion droplets.

Droplet interactions can be characterized by a droplet interaction potential, $w(h)$, which describes the variation of the free energy with droplet separation (Figures 4.3 and 4.4). The overall interaction potential between a pair of emulsion droplets is the sum of various attractive and repulsive contributions (McClements, 2005):

$$w_{total}(h) = w_{VDW}(h) + w_{electrostatic}(h) + w_{SR}(h) + w_{depletion}(h) + w_{hydrophobic}(h) + \ldots \tag{4.3}$$

where w_{VDW}, $w_{electrostatic}$, w_{SR} $w_{depletion}$, and $w_{hydrophobic}$ refer to the interaction potentials associated with van der Waals, electrostatic, short range, depletion and hydrophobic interactions, respectively. The stability of food emulsions to aggregation depends on the overall form of the interaction potential versus droplet separation profile (Figures 4.3 and 4.4), which is governed by the relative contributions of the different types of attractive and repulsive interaction (McClements, 2005). The most important colloidal interactions that operate between the droplets in emulsions are discussed below (Hunter, 1986b, 1989; Israelachvili, 1992).

4.5.1.1 Van Der Waals Interactions

The van der Waals forces act between emulsion droplets of all types and are usually attractive. At close separations, the van der Waals interaction potential between two droplets of equal radius r separated by a surface-to-surface distance h is given approximately by the following equation (McClements, 2005):

$$w_{VDW}(h) = -\frac{Ar}{12h} \tag{4.4}$$

where A is the Hamaker function, which depends on the physical properties of the oil and water phases. In practice, this equation tends to overestimate the attractive forces acting between droplets because it ignores the effects of electrostatic screening, retardation, and the interfacial layer on the Hamaker function (McClements, 2005). It should be noted that the utilization of biopolymers as emulsifiers can considerably improve the stability of emulsions to aggregation by reducing the strength of the van der Waals interactions between them (Guzey and McClements, 2007).

4.5.1.2 Electrostatic Interactions

Electrostatic interactions occur between emulsion droplets with electrically charged surfaces, which may occur in emulsions due to adsorption of ionic or ionizable components, such as surfactants, phospholipids, proteins, or polysaccharides. To a first approximation, the electrostatic interaction between two droplets at close separation is given by the following relationship (McClements, 2005):

$$w_{electrostatic}(h) = -2\pi\varepsilon_0\varepsilon_R r\psi_\delta^2 \ln(1 - \exp(-\kappa h)) \tag{4.5}$$

where

$$\kappa^{-1} = \left(\frac{\varepsilon_0\varepsilon_r kT}{e^2 \Sigma c_i z_i^2}\right)^{1/2} \tag{4.6}$$

Here κ^{-1} is the thickness of the electric double layer, c_i, and z_i, are the molar concentration and valence of ions of species i, ϵ_0 is the dielectric constant of a vacuum, ϵ_r is the relative dielectric constant of the medium surrounding the droplets, e is the elementary electrical charge, ψ_δ is the surface potential of the droplets, k is the Boltzmann constant, and T is the absolute temperature. This equation assumes that the interaction occurs at constant surface charge density, and that the electrostatic interaction between the droplets is not too strong (Hunter, 1986b, 1989).

Usually all the droplets in food emulsions have the same electrical charge so that they repel one another, which is often important for preventing droplet aggregation (e.g., in protein stabilized emulsions). The strength of the electrostatic repulsion increases as the magnitude of the surface potential increases, therefore the greater the number of charges per unit area at a surface, the greater the protection against aggregation. The strength of the electrostatic repulsion decreases as the concentration or valence of ions in the aqueous phase increases because counter-ions 'screen' the charges between droplets. Emulsions stabilized by proteins are particularly sensitive to the pH and ionic strength of the aqueous solution, since altering pH changes ψ_δ and altering ionic strength changes κ^{-1}. In some emulsions, the electrostatic interactions may be attractive and therefore lead to droplet aggregation, for example if droplet–droplet bridging occurs because of the presence of an oppositely charged polymer or colloidal particle. Many biopolymers utilized as emulsifiers are electrically charged, and therefore determine the electrostatic interactions between emulsion droplets. For example, proteins change from positive to negative as the pH moves from below to above their isoelectric point, which has major implications for the stability of protein-coated droplets (Gu et al., 2004b; Guzey et al., 2004).

4.5.1.3 *Hydrophobic Interactions*

The outer surfaces of the emulsion droplets in many oil-in-water emulsions are covered with emulsifier and so they are dominantly hydrophilic in character. Nevertheless, there are certain situations in which the droplet surfaces are not completely covered by emulsifier (e.g., during homogenization) or where the droplet surfaces have some hydrophobicity (e.g., due to exposure of non-polar groups because of protein unfolding) (McClements, 2005). Consequently, an attractive hydrophobic interaction may exist between the emulsion droplets:

$$w_{hydrophobic}(h) = -2\pi r\gamma_i\phi_H\lambda_0 e^{-h/\lambda_0} \tag{4.7}$$

where, γ_i is the interfacial tension between the non-polar groups and water (typically between 10 to 50 mJ m^{-2} for food oils), λ_0 is the decay length of the interaction (typically between 1 to 2 nm), and ϕ_H is a measure of the surface hydrophobicity of the droplets. The surface hydrophobicity varies from 0 (for a fully polar surface) to 1 (for a fully non-polar surface). Thus, the magnitude of the hydrophobic interaction increases as the surface hydrophobicity increases. The hydrophobic attraction between droplets with non-polar surfaces is fairly strong and relatively long range. Hydrophobic interactions are believed to play an important role in determining the stability of a number of emulsions stabilized by biopolymers, e.g., globular protein-stabilized emulsions due to surface or thermal denaturation of the adsorbed protein molecules (Kim et al., 2002a, 2002b).

4.5.1.4 Short-Range Forces

When two emulsion droplets closely approach each other their adsorbed interfacial layers will start to interact. A number of short-range forces result from these interactions, including steric, hydration, protrusion, and undulation forces (Israelachvili, 1992; McClements, 2005). Some progress has been made in developing theories to predict the magnitude and range of short-range forces associated with interfacial layers of fairly simple geometry. Nevertheless, both the magnitude and the range of these forces are highly sensitive to the size, shape, conformation, packing, interactions, mobility, and hydration of the molecules in the adsorbed layer, and so it is difficult to predict their contribution to the overall interaction potential with any certainty. Even so, they are usually repulsive and tend to increase strongly as the interfacial layers overlap.

To a first approximation the short range interactions can be described by the following empirical equation:

$$w_{SR}(h) = \left(\frac{2\delta}{h}\right)^{12} \quad \text{for } h < 2\delta \tag{4.8a}$$

$$w_{SR}(h) = 0 \quad \text{for } h \geq 2\delta \tag{4.8b}$$

where δ is the thickness of the adsorbed layer of emulsifier molecules surrounding the droplets. At separations greater than 2δ the short-range repulsion is negligible, but at separations less than this value there is a steep increase in the interaction potential, which means that the droplets strongly repel one another.

4.5.1.5 Overall Interaction Potential

To a first approximation, the overall droplet interaction potential can be modeled by summing together the major interactions operating in a particular system (Eq. 4.3). To demonstrate the utility of this approach we will consider the behavior of an emulsion in which the dominant droplet interactions are van der Waals attraction, electrostatic repulsion, and short-range repulsion. An emulsion stabilized by a globular protein may demonstrate this kind of behavior.

The van der Waals interaction is fairly long range and always negative (attractive), the electrostatic interaction is fairly long range and always positive (repulsive), while the steric interaction is short range and highly positive (strongly repulsive). The overall interaction potential has a complex dependence on separation because it is the sum of these three different interactions, and it may be attractive at some separations and repulsive at others (Figure 4.4). When the two droplets are separated by a large distance, there is no effective interaction between them. As they move closer together, the van der Waals attraction dominates initially and there is a shallow minimum in the profile, which is referred to as the *secondary minimum*. If this minimum is sufficiently deep compared to the thermal energy ($|w(h_{2^\circ})| > 5kT$), the droplets tend to be flocculated. However, if it is small compared to the thermal energy, the droplets tend to remain as non-aggregated individual entities. At closer separations the repulsive electrostatic interactions dominate, and there is an energy barrier $w(h_{EB})$ that must be overcome before the droplets can come any closer. If this energy barrier is sufficiently large compared to the thermal energy ($w(h_{EB}) > 20kT$), it will prevent the droplets from falling into the deep primary minimum at close separations. On the other hand, if it is not large compared to the thermal energy, the droplets will tend to fall into the primary minimum, leading to strong flocculation or coagulation of the droplets. In this situation the droplets would be prevented from coalescing because of the domination of the strong short-range repulsion at close separations. If the droplets could overcome the short-range repulsion they would merge together and become coalesced (Figure 4.4).

4.5.2 Mechanisms of Emulsion Instability

As mentioned earlier, emulsions are thermodynamically unstable systems that tend to revert back to the separate oil and water phases from which they were made (Friberg et al., 2004; McClements, 2005). The rate at which this process occurs, and the route that is taken, depend on the composition and physicochemical properties of the emulsion, as well as the prevailing environmental conditions. The most important mechanisms of physical instability are creaming, flocculation, coalescence, Ostwald ripening, and phase inversion (Figure 4.2). In practice, all these mechanisms act in concert and can influence one another. For example, droplet flocculation and coalescence often accelerate the rate of creaming (by increasing the particle size), whereas rapid creaming accelerates the rate of flocculation or coalescence (by bringing the droplets into close proximity in the cream layer). However, one mechanism often dominates the others, and identification of this mechanism will aid in establishing the most effective means of controlling the emulsion stability.

The length of time an emulsion must remain stable depends on the nature of the final product. Some food emulsions (e.g., cake batters, ice cream mix, and margarine premix) are formed during intermediate steps of a manufacturing process and only need to remain stable for a few seconds, minutes, or hours. Other emulsions (e.g., mayonnaise, dressings, and cream liqueurs) must persist in a stable state for days, months, or even years prior to sale and consumption. Some food processing operations (e.g., the production of butter, margarine, whipped cream, and ice cream) rely on a controlled destabilization of an emulsion. The origin of the major destabilization mechanisms, the factors that influence them and methods of controlling them are briefly reviewed in this section. This type of information is useful for food scientists because it facilitates the selection of the most appropriate ingredients and processing conditions required to produce a food emulsion with particular stability characteristics. The role of biopolymers in influencing these various mechanisms will also be stressed.

4.5.2.1 Gravitational Separation: Creaming and Sedimentation

The droplets in an emulsion usually have a different density than the liquid that surrounds them, and so a net gravitational force acts upon them (Friberg et al., 2004; McClements, 2005). If the droplets have lower density than the surrounding liquid, they tend to move upwards, which is referred to as 'creaming'. Conversely, if the droplets have a higher density than the surrounding liquid they tend to move downwards, which is referred to as 'sedimentation'. Most liquid oils have densities that are lower than that of water, and so there is a tendency for oil to accumulate at the top of an emulsion and water to accumulate at the bottom. Thus droplets in an oil-in-water emulsion tend to cream, whereas those in a water-in-oil emulsion tend to sediment. Nevertheless, there may be situations where the opposite is true, e.g., if the density of the oil phase is increased due to addition of weighting agent or due to fat crystallization. To a first approximation, the creaming velocity of a single isolated spherical droplet in a viscous liquid can be described by the Stokes equation:

$$\nu = -\frac{2gr^2(\rho_2 - \rho_1)}{9\eta_1} \tag{4.9}$$

where ν is the creaming velocity, g is the acceleration due to gravity, ρ is the density, η is the shear viscosity, and the subscripts 1 and 2 refer to the continuous phase and droplet, respectively. The sign of ν determines whether the droplet moves upwards (+) or downwards (−), i.e., cream or sediment. The Stokes equation assumes that the droplets are isolated spherical rigid particles that move through a Newtonian fluid. Stokes equation can be used to estimate the stability of an emulsion to creaming or sedimentation. For example, an oil droplet ($\rho_2 = 910$ kg/m^3) with a radius of 1 μm suspended in water ($\eta_1 = 1$ mPa·s, $\rho_1 = 1000$ kg/m^3) will cream at a rate of about 5 mm/day. Thus one would not expect an emulsion containing droplets of this size to have a particularly long shelf life. As a useful rule of thumb, an emulsion in which the creaming rate is less than about 1 mm/day can be considered to be stable toward creaming (Dickinson, 1992b).

Creaming of emulsion droplets is usually an undesirable process, which food manufacturers try to avoid. Stokes equation indicates that creaming can be retarded by minimizing the density difference ($\rho_2 - \rho_1$) between the droplets and the surrounding liquid, reducing the droplet size, or increasing the viscosity of the continuous phase. The Stokes equation is strictly applicable only to isolated rigid spheres suspended in an infinite viscous liquid. Since these assumptions are usually invalid for food emulsions, the equation must be modified to take into account hydrodynamic interactions, droplet fluidity, droplet aggregation, non-Newtonian aqueous phases, droplet crystallization, the adsorbed layer, and Brownian motion (Dickinson, 1992b; Walstra, 2003; McClements, 2005).

4.5.2.2 *Flocculation and Coalescence*

The droplets in emulsions are in continual motion because of their thermal energy, gravitational forces or applied mechanical forces, and as they move about they collide with their neighbors. After a collision, emulsion droplets may either move apart or remain aggregated, depending on the relative magnitude of the attractive and repulsive forces between them. Droplets tend to aggregate when the net force acting upon them is strongly attractive, and remain as individual entities when the net force is strongly repulsive. Two types of droplet aggregation are commonly observed in emulsions: flocculation and coalescence (Figure 4.2). In flocculation, two or more droplets come together to form an aggregate in which the emulsion droplets retain their individual integrity. In coalescence, two or more droplets merge together to form a single larger droplet. Development of effective strategies to improve the stability of emulsion-based food products often depend on understanding the factors that promote droplet aggregation. The rate at which droplet aggregation proceeds within a particular emulsion depends on two factors: *collision frequency* and *collision efficiency* (Friberg et al., 2004; McClements, 2005).

The collision frequency is the number of encounters between droplets per unit time per unit volume. Any factor that increases the collision frequency is likely to increase the aggregation rate. The collision frequency of collisions between droplets depends on whether the emulsion is subjected to mechanical agitation. For dilute emulsions containing identical spherical particles, the collision frequency N has been calculated for both quiescent and stirred systems (Friberg et al., 2004; McClements, 2005):

$$\text{Brownian motion: } N = \frac{4kTn_0^2}{3\eta} \tag{4.10}$$

$$\text{Simple shear: } N = \frac{16}{3}Gr^3n_0^2 \tag{4.11}$$

where n_0 is the initial number of particles per unit volume, r is the droplet radius, G is the shear rate, and kT is the thermal energy.

The collision efficiency, E, is the fraction of encounters between droplets that lead to aggregation. Its value ranges from 0 (no aggregation) to 1 (fast aggregation) and depends on the droplet interaction potential and the nature of the interface surrounding the droplets (Dickinson, 1992b; Walstra, 2003; McClements, 2005). The equations for the collision frequency must therefore be modified to take into account droplet–droplet interactions:

$$N = \frac{4kTn_0^2}{3\eta}E \tag{4.12}$$

where the collision efficiency E is a function of the colloidal interactions between the droplets (Friberg et al., 2004; McClements, 2005). The equations above are applicable only to the initial stages of aggregation in dilute emulsions containing identical spherical particles. In practice, most food emulsions are fairly concentrated systems and interactions between flocs as well as between individual droplets are important. The equations above must therefore be modified to take into account the interactions and properties of flocculated droplets.

The nature of the droplet interactions also determines the structure of the flocs formed, and the rheology and stability of the resulting emulsion (Friberg et al., 2004; McClements, 2005). When the attractive force between the droplets is relatively strong, two droplets tend to become 'locked' together as soon as they encounter each other. This leads to the formation of flocs that have quite open structures and can be characterized by a low fractal dimension. When the attractive forces are not particularly strong the droplets 'roll around' each other after a collision, which allows them to pack more efficiently and form denser flocs with a higher fractal dimension. The structure of the flocs formed in an emulsion has a pronounced influence on its bulk physicochemical properties. An emulsion containing flocculated droplets has a higher viscosity than one containing non-flocculated droplets, since the water trapped between the flocculated droplets increases the effective diameter (and therefore volume fraction) of the particles. Flocculated particles also exhibit strong shear thinning behavior: as the shear rate is increased, the viscosity of the emulsion decreases because the flocs are deformed and disrupted and so their effective volume fraction decreases. If flocculation is extensive, a three-dimensional network of aggregated particles extends throughout the system and the emulsion has a yield stress that must be overcome before the system will flow. The creaming rate of droplets is also strongly dependent on flocculation. At low droplet concentrations, flocculation increases the creaming rate because the effective size of the particles is increased, but at high droplet concentrations, it retards creaming because the droplets are trapped within the three-dimensional network of aggregated emulsion droplets.

In coalescence, two or more liquid droplets collide and merge into a single larger droplet (Figure 4.2). Extensive coalescence eventually leads to *oiling off*, i.e., the formation of free oil on the top of an emulsion. Coalescence occurs rapidly between droplets that are not protected by emulsifier molecules; for example, if one homogenizes oil and water in the absence of an emulsifier, the droplets rapidly coalesce. When droplets are stabilized by a layer of emulsifier molecules, the tendency for coalescence to occur is governed by the droplet interactions and the stability of the interfacial layer to rupture. If there is a strong repulsive force between the droplets at close separations, or if the interfacial layer is highly resistant to rupture, then the droplets will not tend to coalesce. Most food emulsions are stable to coalescence, but they become unstable when subjected to high shear forces, during aeration, when they are centrifuged, or when droplets remain in contact for extended periods (e.g., droplets in flocs, creamed layers, or highly concentrated emulsions).

4.5.2.3 *Partial Coalescence*

As described above, normal coalescence involves the merging together of two or more liquid droplets to form a single larger liquid droplet. On the other hand, *partial coalescence* occurs when two or more partly crystalline droplets aggregate to form a single irregularly shaped particle (Figure 4.5).

An irregular-shaped particle is formed because of the mechanical rigidity provided by the fat crystal network within the initial droplets. Partial coalescence occurs when partly crystalline droplets encounter each other and a crystal from one of them penetrates through the intervening layers (emulsifier-aqueous phase-emulsifier) and protrudes into the liquid region of another droplet (Walstra, 2003). Normally, the crystal would protrude into the aqueous phase and be surrounded by water; however, when it penetrates into another droplet, it is surrounded by oil, which is thermodynamically more favorable (better crystal wetting), thereby promoting partial coalescence. Over time the droplets may partly fuse together, thereby reducing the total surface area of oil exposed to aqueous phase and leading to changes in particle morphology. Partial coalescence tends to occur most readily when the initial droplets have an optimum ratio of solid fat to liquid oil. If the solid fat content of the droplets is either too low or too high, the droplets will tend not to undergo regular coalescence or coagulation, respectively (Walstra, 2003).

Partial coalescence is an important type of emulsion instability in dairy products because milk fat globules are partly crystalline at temperatures found in foods during processing, storage and utilization. Partial coalescence may be either undesirable (since it reduces shelf life or negatively impacts product properties) or desirable (since it is an essential processing step). The application of shear forces or temperature cycling to dairy cream containing partly crystalline milk fat globules can cause extensive partial coalescence, leading to an appreciable increase in viscosity ('thickening') and eventually phase separation. On the other hand, partial coalescence is an essential process in the production of ice cream, whipped toppings, butter, and margarine. Oil-in-water emulsions are cooled to temperatures where the droplets are partly crystalline, and a shear force is applied that promotes droplet aggregation via partial coalescence. In butter and margarine, extensive partial coalescence eventually results in phase inversion, whereas in ice cream and whipped cream the aggregated fat droplets form a network around the air cells and within the aqueous phase that provides the mechanical strength needed to produce good product stability and texture.

4.5.2.4 *Ostwald Ripening*

Ostwald ripening (OR) manifests itself as the growth of large droplets at the expense of smaller ones, which is due to diffusion of

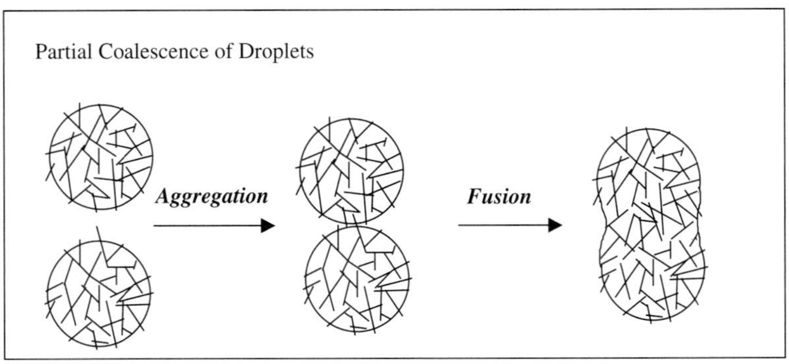

FIGURE 4.5 Schematic diagram of partial coalescence of partly crystalline droplets in oil-in-water emulsions. Initially, the droplets aggregate with each other when a crystal from one droplet penetrates into a liquid region in another droplet. The aggregated droplets may then partially fuse together over time.

disperse phase molecules through the continuous phase (Kabalnov et al., 1990; Kabalnov, 1998, 2001; McClements, 2005). OR occurs because the solubility of the material within a spherical droplet in the surrounding continuous phase increases as the radius of the droplet decreases:

$$S(r) = S(\infty)\exp\left(\frac{2\gamma V_m}{RT_r}\right) \quad (4.13)$$

Here V_m is the molar volume of the solute, γ is the interfacial tension, R is the gas constant, $S(\infty)$ is the solubility of the solute in the continuous phase for a droplet with infinite curvature (i.e., a planar interface), and $S(R)$ is the solubility of the solute when contained in a spherical droplet of radius r. In a polydisperse emulsion there is a greater solubility of dispersed phase molecules around smaller droplets than around large droplets. Consequently, disperse phase molecules move from small droplets to large droplets because of this concentration gradient, which causes the larger droplets to grow at the expense of the smaller ones. Once steady state conditions have been achieved, the rate of droplet growth due to OR is given by:

$$\frac{d\langle r\rangle^3}{dt} = \frac{8\gamma V_m S(\infty)D}{9RT} \quad (4.14)$$

where D is the diffusion coefficient of the disperse phase molecules through the continuous phase. This equation assumes that the emulsion is dilute and that the rate-limiting step is the diffusion of the disperse phase molecules through the continuous phase. Most food emulsions are fairly concentrated systems, and so the effects of the neighboring droplets on the droplet growth rate have to be considered. The OR rate may also be influenced by the nature of the interfacial layer surrounding the droplets (e.g., its thickness, permeability and rheology), and in such cases the above equation must be modified accordingly. OR is negligible in many foods because triacylglyercols have very low water solubilities, and therefore the mass transport rate is insignificant. Nevertheless, in emulsions that contain more water-soluble lipids, such as flavor or essential oils, Ostwald ripening is important.

4.5.2.5 Phase inversion

Phase inversion involves the conversion of an oil-in-water emulsion to a water-in-oil emulsion or vice versa (Figure 4.2). This process usually occurs as a result of some alteration in the system's composition or environmental conditions, such as dispersed phase volume fraction, emulsifier type, emulsifier concentration, temperature, ionic strength, pH, or application of mechanical forces. Phase inversion occurs through a series of complex physicochemical mechanisms involving droplet fragmentation and aggregation processes. At the point where phase inversion occurs, the system may briefly contain regions of oil-in-water emulsion, water-in-oil emulsion, multiple emulsions, and bicontinuous phases, before inverting to its final state. In foods, phase inversion can be divided into that promoted by fat crystallization and that promoted by surfactant transitions (McClements, 2005).

4.5.2.6 Chemical and Biochemical Stability

Various kinds of chemical and biochemical reactions can cause detrimental changes in the quality of food emulsions, e.g., lipid oxidation, color or flavor fading, biopolymer hydrolysis. Enzymes, transition metals, acids, bases, light, or various other components present in foods may catalyze these reactions. The reactions that are important in a given food emulsion depend on the concentration, type, and distribution of ingredients, and the thermal and shear history of the food. Chemical and biochemical reactions can alter the stability, texture, flavor, odor, color, and toxicity of food emulsions. Thus it is important to identify the most critical reactions that occur in each type of food so that they can be controlled in a systematic fashion. It should be stressed that chemical and biochemical reactions in structurally heterogeneous materials such as food emulsions may occur by a different pathway and at a different rate than in homogeneous solutions.

4.6 PHYSICOCHEMICAL PROPERTIES OF FOOD EMULSIONS

4.6.1 Rheology

Many of the desirable sensory attributes of emulsion-based food products are determined by their rheological properties, e.g., creaminess, thickness, smoothness, spreadability, pourability, flowability, brittleness, and hardness. Food manufacturers must therefore be able to design and create food products that have rheological properties expected by the consumer. The shelf life of many emulsion-based food products depends on the rheological characteristics of the component phases, e.g., the creaming of oil droplets in O/W emulsions depends on the viscosity of the aqueous phase. Rheological data are used by food engineers to design processing operations that depend on the way that a food emulsion behaves when it flows through a pipe, is stirred, or is packed into containers. Rheological measurements are used by food scientists as an analytical tool to provide fundamental information about the structural organization and interactions of the components within emulsions (Hunter, 1993; Tadros, 1994).

4.6.2 Modeling Emulsion Rheology

The category of food emulsions includes a variety of complex materials that exhibit different rheological properties, ranging from low-viscosity fluids (such as milk and fruit juice beverages) to hard solids (such as refrigerated margarine or butter). Our ability to control the rheological properties of food emulsions depends on a quantitative understanding of the relationship between their rheology, composition and microstructure. A variety of theories have been developed to relate the rheological properties of emulsions to their composition and microstructure. In general, the apparent viscosity of an emulsion can be described by the following equation:

$$\eta = f(\eta_1, \eta_2, \phi, r, w(h), \tau) \qquad (4.15)$$

where η_1 is the viscosity of the continuous phase, η_2 is the viscosity of the dispersed phase, ϕ is the dispersed phase volume fraction, r is the droplet radius, $w(h)$ is the droplet–droplet interaction potential, and τ is the applied shear stress. The most appropriate equation for describing the rheology of a particular emulsion depends on its characteristics, e.g. droplet concentration, droplet interactions, and continuous phase rheology. Fundamental expressions of the relationship between the rheology of colloidal suspensions and their composition/structure are only available for certain limiting cases, such as Einsteins' equation for a dilute suspension of rigid spherical particles:

$$\eta = \eta_1(1 + 2.5\phi) \qquad (4.16)$$

This equation shows that the rheology of a dilute emulsion is proportional to the rheology of the continuous phase and increases with increasing droplet concentration. In concentrated emulsions, the rheology is influenced by hydrodynamic interactions associated with the relative motion of neighboring particles (Larson, 1999). As the particle concentration increases the measured viscosity becomes larger than that predicted by the Einstein equation because these additional hydrodynamic interactions lead to a greater degree of energy dissipation. The Einstein equation can be extended to account for the effects of these interactions by including additional volume fraction terms (Pal et al., 1992):

$$\eta = \eta_0(1 + a\phi + b\phi^2 + c\phi^3 + \ldots) \qquad (4.17)$$

The value of the constants, *a, b, c*, etc., can either be determined experimentally or theoretically (Larson, 1999). This equation can be used up to particle concentrations of about 10% with $a = 2.5$ and $b = 6.2$ for colloidal dispersions of rigid spherical particles in the absence of long-range colloidal interactions (Larson, 1999). It is difficult to theoretically determine the value of higher-order terms in the equation because of the mathematical complexities involved in describing interactions between three or more particles. Instead, a semi-empirical approach is used to develop equations that describe the viscosity of more concentrated colloidal dispersions. One of the most widely used equations was derived by Dougherty and Krieger, which is applicable across the entire volume fraction range (Figure 4.6) (Hunter, 1986a; Mewis and Macosko, 1994).

$$\eta = \eta_0\left(1 - \frac{\phi}{\phi_c}\right)^{-[\eta]\phi_c} \qquad (4.18)$$

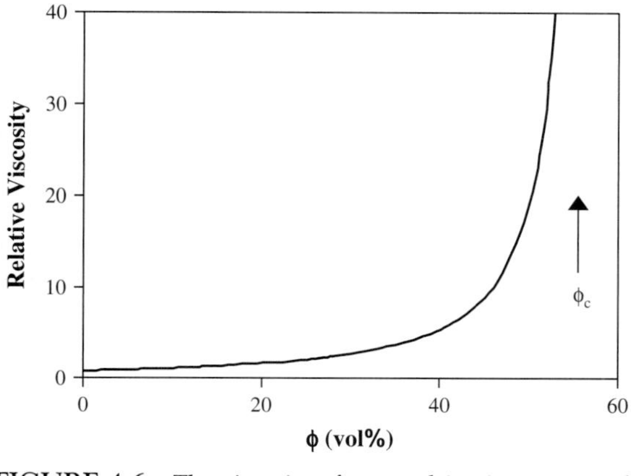

FIGURE 4.6 The viscosity of an emulsion increases with increasing droplet concentration, particularly when the droplets become close packed. At droplet concentrations above the critical packing value emulsions may have solid-like characteristics.

where $[\eta]$ is the intrinsic viscosity and ϕ_c is the *maximum packing* volume fraction, which is usually taken to be an adjustable parameter that is determined experimentally. Physically, ϕ_c is related to the particle volume fraction at which the spheres become close packed. The intrinsic viscosity is 2.5 for spherical particles, but may be appreciably larger for non-spherical or aggregated particles (Hiemenz and Rajagopalan, 1997). Typically, the value of ϕ_c is between about 0.6 and 0.7 for spheres which do not interact via long-range colloidal interactions (Hunter, 1986a), but it may be considerably lower for suspensions in which there are strong long-range attractive or repulsive interactions between the particles. This is because the effective volume fraction of the particles in the colloidal dispersion is greater than the actual volume fraction of the particles, so that the maximum packing volume fraction is reached at lower particle concentrations (Weiss and McClements, 2000).

4.6.3 Factors Influencing Emulsion Rheology

A variety of factors influence the rheology of emulsion-based food products. Some of the most important of these factors are briefly described below.

4.6.3.1 Continuous Phase Rheology

The viscosity of most food emulsions is dominated by the rheology of the continuous phase (Eq. 4.16). One of the most effective means of modifying the rheology of an emulsion is therefore to add a thickening or gelling agent to the continuous phase. The main exception to this rule is in systems that contain a network of closely associated or highly aggregated droplets. In these systems the rheological properties are strongly influenced by the number and strength of the forces between the droplets.

4.6.3.2 *Disperse Phase Volume Fraction*

The viscosity of food emulsions usually increases with increasing droplet concentration (Figure 4.6). The viscosity increases relatively slowly with ϕ at low droplet concentrations, but increases steeply when the droplets become closely packed together. At higher droplet concentrations the particle network formed has predominantly elastic characteristics.

4.6.3.3 *Droplet–Droplet Interactions*

The nature (sign, magnitude and range) of the colloidal interactions between the droplets is an important factor determining the rheological behavior of many food emulsions. When the interactions are long-range and repulsive, the effective volume fraction of the dispersed phase may be significantly greater than its actual volume fraction ($\phi_{\text{eff}} = \phi(1+\delta/r)^3$), and so the emulsion viscosity increases. When the interactions between the droplets are sufficiently attractive, the effective volume fraction of the dispersed phase is also increased due to droplet flocculation, which results in an increase in emulsion viscosity (Figure 4.7a) and shear thinning behavior (Figure 4.7b).

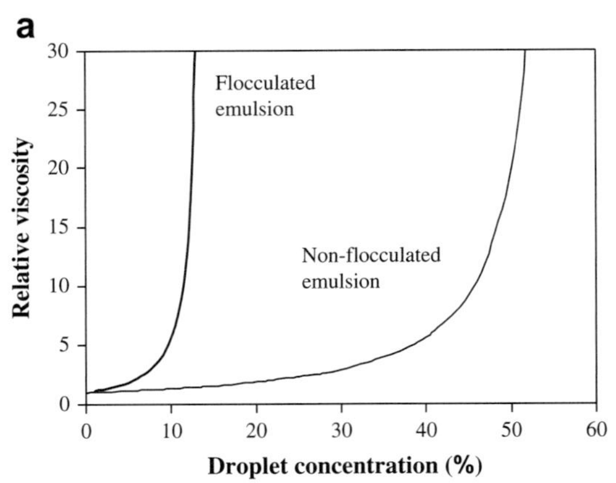

FIGURE 4.7a The viscosity of an emulsion increases dramatically when the droplet concentration exceeds a critical value due to particle packing and interactions. This critical concentration occurs at lower droplet concentrations in flocculated emulsions.

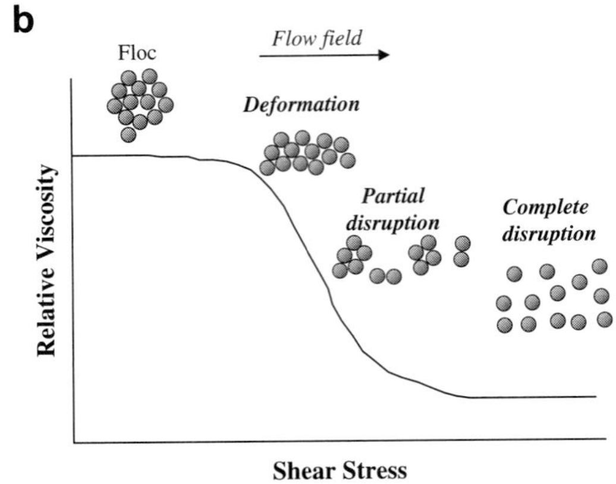

FIGURE 4.7b The viscosity of flocculated emulsions decreases appreciably with increasing shear stress due to progressive deformation and disruption of flocs. Eventually, a balance between floc formation and disruption may be reached.

The rheological properties of an emulsion therefore depend on the relative magnitude of the attractive (mainly van der Waals, hydrophobic and depletion interactions) and repulsive (mainly electrostatic, steric and thermal fluctuation interactions) colloidal interactions.

4.6.3.4 *Droplet Size*

Droplet size may influence the rheology of emulsions in a variety of ways. First, the viscosity of relatively concentrated suspensions (> 30%) tends to decrease with increasing droplet size due to Brownian motion effects (Pal et al., 1992; Mewis et al., 1994; Liu and Masliyah, 1996). This effect also causes emulsions to exhibit shear thinning behavior. At low shear stresses, the particles have a three-dimensional isotropic and random distribution because of their Brownian motion (Hunter, 1993). As the shear stress increases the particles become more ordered along the stream lines to form 'strings' or 'layers' of particles that offer less resistance to the fluid flow and therefore a decrease in viscosity. The viscosity decreases from a high constant value at low shear stresses (η_0) to a low constant value at high shear stresses (η_∞). The shear thinning behavior of an emulsion is characterized by a critical shear stress, which corresponds to the stress where the viscosity is half way between the low and high shear stress values. This critical shear stress increases with decreasing particle size, i.e. Brownian motion effects are more important for smaller droplets. Second, the droplet size influences the relative importance of the attractive and repulsive interactions between droplets, which may influence the rheology because it changes the effective volume fraction of the droplets (see above).

4.6.4 Appearance

The appearance of an emulsion is one of the most important factors influencing its perceived quality, as it is usually the first sensory impression that a consumer makes of a product (Hutchings, 1994; McClements, 1999). Knowledge of the factors that determine emulsion appearance therefore aids in the design of emulsion-based food products with improved quality or novel optical properties. When a light beam is incident upon the surface of an emulsion a portion of the incident light beam is transmitted through the emulsion while another portion is reflected. The relative proportions of light transmitted and reflected at different wavelengths depend on the scattering and adsorption of the light wave by the emulsion. Light scattering and absorption depend on size, concentration, refractive index and spatial distribution of droplets, as well as the presence of any chromophoric materials (e.g., dyes). Hence, the overall appearance of an emulsion is influenced by its structure and composition. Scattering is largely responsible for the 'turbidity', 'opacity' or 'lightness' of an emulsion, whereas absorption is largely responsible for its 'chromaticity' (blueness, greenness, redness, etc.). It should be stressed that the overall appearance of an emulsion also depends on the nature of the light source and detector used (Judd and Wyszecki, 1963; Billmayer and Saltzman, 1981; Wyszecki and Stiles, 1982).

4.6.5 Mathematical Modeling of Emulsion Color

Human beings have great difficulty objectively quantifying and expressing the color of objects. For this reason, color is normally scientifically quantified in terms of 'tristimulus coordinates', such as the $L^*a^*b^*$ system (Wyszecki et al., 1982). The color of an object can then be objectively described and reported in terms of just three parameters. For example, in the $L^*a^*b^*$ color space, L^* is lightness, and a^* and b^* are color coordinates: where $L^* = 0$ is black, $L^* = 100$ is white, $+a^*$ is the red direction, $-a^*$ is the green direction, $+b^*$ is the yellow direction, and $-b^*$ is the blue direction (Wyszecki et al., 1982). The overall color intensity of an object is characterized in terms of its chroma, $C = (a^2 + b^2)^{1/2}$. A theoretical approach has recently been developed to relate the tristimulus color coordinates of emulsions to their composition (dye and droplet concentration) and microstructure (particle size distribution) (McClements, 2002). This approach has led to the development of relationships of the form:

$$L^* = f(\alpha(\lambda), c, \phi, r, n) \tag{4.19}$$

$$a^* = f(\alpha(\lambda), c, \phi, r, n) \tag{4.20}$$

$$b^* = f(\alpha(\lambda), c, \phi, r, n) \tag{4.21}$$

where $\alpha(\lambda)$ is the absorption spectrum of a dye solution, c is the concentration of dye present, ϕ is the disperse phase volume fraction, r is the droplet radius, and n is the ratio of the refractive indices of the dispersed and continuous phases. These equations can be used to predict the influence of emulsion composition and microstructure on product appearance. Studies of the influence of composition and microstructure on the color of oil-in-water emulsions are in excellent qualitative agreement with predictions made using the above theory (Chantrapornchai, 1998, 1999, 2001; McClements, 1998; Chanamai, 2000; Chantrapornchai et al., 1999). However, quantitative agreement is still fairly poor, mainly because of problems associated with accounting for the optical measurement system, although a number of methods of overcoming this problem have been proposed (McClements, 2002).

4.6.6 Factors Influencing Emulsion Color

4.6.6.1 *Disperse Phase Volume Fraction*

The extent of light scattering by an emulsion usually increases as the droplet concentration is increased. As a result, emulsion lightness (*L*) increases and emulsion chromaticity (*C*) decreases with increasing disperse phase volume fraction (Figure 4.8).

L and *C* change steeply when the droplet concentration is increased from 0 to 5 wt%, but then remain relatively constant at higher droplet concentrations (Chantrapornchai, 1998, 1999; Chantrapornchai et al., 1999). This has important consequences for designing reduced-fat versions of full-fat emulsion-based products.

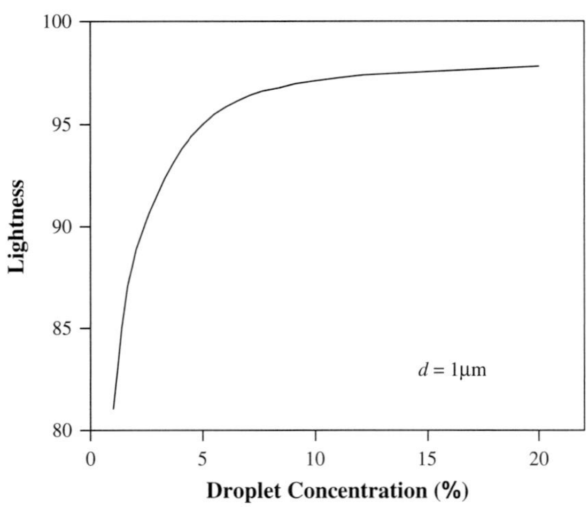

FIGURE 4.8 The lightness (opacity) of an emulsion increases with increasing droplet concentration due to increased light scattering. The increase in lightness is particularly strong between 0 and 5 wt% droplets.

4.6.6.2 *Droplet Size*

The intensity of light scattered at different angles by a droplet depends on the ratio of its radius to the wavelength of light. Theoretical predictions and experimental measurements show that emulsion lightness increases with droplet radius from 0 to 100 nm, has a maximum value around 100 nm, and then decreases as the droplet radius is increased further (Chantrapornchai, 1998).

4.6.6.3 *Relative Refractive Index*

The scattering efficiency of a droplet increases as the contrast in refractive index between it and the surrounding liquid increases (McClements, 2005). Emulsion lightness is therefore high when the refractive index of the droplets is either much smaller or much greater than the refractive index of the continuous phase, but decreases when the refractive index ratio tends towards unity. Indeed, it is possible to prepare optically transparent emulsions with high droplet concentrations by matching the refractive index of the continuous phase to that of the disperse phase by adding water-soluble solutes, such as glycerol or sucrose (Weiss et al., 2000; Chantrapornchai, 2001). Refractive index matching is particularly useful in some emulsion studies because it means that the emulsions can be analyzed using spectroscopy techniques that require optically transparent solutions, such as UV-visible spectrophotometry, fluorescence and CD.

4.6.7 Flavor

Flavor plays a crucial part in determining the perceived quality of food emulsions. The term 'flavor' refers to those volatile components in foods that are sensed by receptors in the nose (aroma) and those non-volatile components that are sensed by receptors on the tongue and the inside of the mouth (taste) (Thomson, 1986; Taylor and Linforth, 1996). Certain structural components within foods may also contribute to flavor because of their influence on the perceived texture (mouthfeel) within the mouth (Kokini, 1987). The flavor of a food is therefore a combination of aroma, taste and mouthfeel, with the former usually being the most important (Taylor et al., 1996). Flavor perception is an extremely complicated process, that depends on a combination of physicochemical, physiological and psychological phenomena (Bell, 1996). Before a food is placed in the mouth its flavor is perceived principally through those volatile components that are inhaled directly into the nasal cavity. After the food is placed in the mouth the flavor is determined by non-volatile molecules that leave the food and are sensed by receptors on the tongue and the inside of the mouth, as well as by those volatile molecules that are drawn into the nasal cavity through the back of the mouth (Taylor et al., 1996). The interactions between flavor molecules and human receptors that lead to the perceived flavor of a food are extremely complicated and are still poorly understood, although considerable progress has been made in this area recently (Thomson, 1986). In addition, expectations and eating habits vary from individual to individual, so that the same food may be perceived as tasting differently by two different individuals or by a single individual at different times. This section focuses on the physicochemical aspects of flavor partitioning and release in foods, because these are the most relevant topics to emulsion science.

4.6.7.1 Flavor Partitioning

The perception of a flavor depends on the precise location of the flavor molecules within an emulsion (McClements, 2005). The aroma is determined by the presence of volatile molecules in the vapor phase above an emulsion (Overbosch et al., 1991; Taylor et al., 1996), whereas the taste is mainly determined by the presence of certain non-volatile molecules in the aqueous phase of an emulsion (McNulty, 1987; Kinsella, 1989). Some flavor molecules prefer to be located within the interfacial layers surrounding emulsion droplets, which alters their concentration in the vapor and aqueous phases (Wedzicha, 1988). It is therefore important to establish the factors that determine the partitioning of flavor molecules between the different molecular environments within an emulsion. An emulsion system can be conveniently divided into four phases between which the flavor molecules distribute themselves: the interior of the droplets; the continuous phase; the oil–water interface; and, the vapor phase above the emulsion. The relative concentration of the flavor molecules in each of these regions depends on their molecular characteristics (e.g., molecular weight, polarity, charge) and the properties of each of the bulk phases (Baker, 1987; Bakker, 1995).

A number of the most important factors that influence the equilibrium distribution of flavors in food emulsions are listed below:

- *Flavor partition coefficients.* The equilibrium distribution of a particular flavor molecule between two phases (e.g., oil–water, air–water or air–oil) is characterized by an equilibrium partition function (K). These partition coefficients determine the distribution of the flavor molecules between the oil, water and headspace phases of an emulsion.
- *Surface activity.* Many flavor molecules are amphiphilic in character, having both non-polar and polar regions, and hence they can accumulate at oil–water interfaces.
- *Droplet concentration.* The concentration of flavor molecules in the headspace of an

emulsion depends on the disperse phase volume fraction, i.e. the ratio of oil to water. Studies have shown that there is a decrease in the fraction of a non-polar flavor in the vapor phase as the oil content increases, whereas the amount of a polar flavor is increased slightly. Thus the non-polar flavors in an emulsion become more odorous as the fat content is decreased, whereas the polar flavors remain relatively unchanged. This has important consequences when deciding the type and concentration of flavors to use in low-fat analogs of existing emulsion-based food products.

- *Flavor binding.* Many proteins and carbohydrates are capable of binding flavor molecules, and therefore altering their distribution within an emulsion (Franzen and Kinsella, 1975; Hansen and Booker, 1986; Bakker, 1995; Hau et al., 1996; O'Neill, 1996). Flavor binding can cause a significant alteration in the perceived flavor of a food. This alteration is often detrimental to food quality because it changes the characteristic flavor profile, but it can also be beneficial when the bound molecules are off-flavors. A flavor chemist must therefore take binding effects into account when formulating the flavor of a particular product.
- *Solubilization.* Surfactants are normally used to physically stabilize emulsion droplets against aggregation by providing a protective interfacial layer around the droplets. Nevertheless, there is often enough free surfactant present in an aqueous phase to form surfactant micelles. These surfactant micelles are capable of solubilizing non-polar molecules in their hydrophobic interior, which increases the affinity of non-polar flavors for the aqueous phase. By a similar argument, reverse micelles in an oil phase are capable of solubilizing polar flavor molecules.

4.6.7.2 Flavor Release

Flavor release is the process whereby flavor molecules move out of a particular molecular environment within a food and into the surrounding saliva or vapor phase (McNulty, 1987; Overbosch et al., 1991). The release of the flavors from a food material during mastication occurs under extremely complex and dynamic conditions (Land, 1996). A food usually spends a relatively short period (typically 1–30 seconds) in the mouth before being swallowed. During this period it is diluted with saliva, experiences temperature changes, and is subjected to a variety of complex mechanical forces. Mastication therefore usually causes dramatic changes in the structural characteristics of food emulsions.

During mastication non-volatile flavor molecules must move from within the food, through the saliva to the taste receptors on the tongue and the inside of the mouth, whereas volatile flavor molecules must move from the food, through the saliva and into the gas phase, where they are carried to the aroma receptors in the nasal cavity. The two major factors that determine the rate at which these processes occur are the equilibrium partition coefficient (because this determines the initial flavor concentration gradients at the various boundaries) and the mass transfer coefficient (because this determines the speed at which the molecules move through the various phases from one location to another). A variety of mathematical models have been developed to describe the release of flavor molecules from oil-in-water emulsions (McClements, 2005).

4.7 BIOPOLYMER EMULSIFIERS

The term 'emulsifier' is used to describe any surface-active substance that is capable of adsorbing to an oil–water interface and protecting emulsion droplets from aggregation (flocculation and/or coalescence) (McClements, 2004;

McClements, 2005). The most commonly used emulsifiers in the food industry are small-molecule surfactants, amphiphilic biopolymers and surface-active particulate matter. These emulsifiers vary widely in their ability to form and stabilize emulsions depending on their molecular and physicochemical characteristics. Ideally, an emulsifier should rapidly adsorb to the oil–water interface during homogenization, reduce the interfacial tension by an appreciable amount, and prevent droplet coalescence from occurring within the homogenizer. In addition, it is usually important that the emulsifier forms an interfacial layer around the droplets that prevents them from aggregating under the environmental conditions that the product experiences during its manufacture, transport, storage, and utilization. In this section, we will review the major types of biopolymer emulsifiers used in emulsion-based food products, and discuss some of the factors that should be considered in selecting an emulsifier for a particular application.

4.7.1 Interfacial Activity and Emulsion Stabilization

Usually, amphiphilic biopolymers must be fully dispersed and dissolved in an aqueous solution before they are capable of exhibiting their desirable emulsifying properties. Solvation of biopolymer ingredients prior to homogenization is therefore an important step in the formation of many food emulsions. This process usually involves a number of stages, including dispersion, wetting, swelling and dissolution. The effectiveness and rate of dissolution depends on many factors, including the nature of the ingredient (e.g., liquid, powder or granules), biopolymer molecular characteristics (e.g., molecular weight, polarity, hydrophobicity, charge), solution conditions (e.g., pH, ionic strength, composition, temperature), as well as the application of shearing forces. Generally, factors that favor biopolymer–biopolymer interactions tend to oppose good dissolution, whereas factors that favor biopolymer–solvent interactions tend to promote good dissolution. These factors are primarily governed by the nature of the molecular interactions that dominate in the particular system, which depends strongly on biopolymer type and solvent composition.

After a biopolymer ingredient has been adequately dissolved in the aqueous phase it is important to ensure that the solution and environmental conditions (e.g., pH, ionic strength, temperature, and solvent composition) will not promote droplet aggregation during homogenization or after the emulsion is formed. For example, it is difficult to produce protein-stabilized emulsions at pH values close to the isoelectric point of the proteins or at high salt concentrations, because the electrostatic repulsion between the droplets is insufficient to prevent droplet aggregation once the emulsions are formed.

The interfacial activity of many biopolymers is due to the fact that they have both hydrophilic and lipophilic regions distributed along their backbones. For example, most proteins have significant numbers of exposed non-polar amino acid side groups, whereas some polysaccharides have non-polar side chains attached to their polar backbones (Dickinson, 1992a, 2003). The major driving force for adsorption of these amphiphilic biopolymers to oil–water interfaces is therefore the hydrophobic effect. When the biopolymer is dispersed in an aqueous phase some of the non-polar groups are in contact with water, which is thermodynamically unfavorable because of hydrophobic interactions. When a biopolymer adsorbs to an interface it can adopt a conformation where the non-polar groups are located in the oil phase (away from the water) and the hydrophilic groups are located in the aqueous phase (in contact with the water). Adsorption also reduces the contact area between the oil and water molecules at the oil–water interface, which lowers the interfacial tension. Both of these factors favor the adsorption of amphiphilic biopolymers to oil–water interfaces. The conformation of a biopolymer at an interface, and the physicochemical and structural properties of the interfacial layer formed, depend on its molecular structure and interactions (Norde, 2003). Flexible random-coil biopolymers adopt an arrangement where the predominantly non-polar segments protrude into the oil phase, the predominantly polar segments protrude into the aqueous phase, and the neutral regions lie flat against the interface (Figure 4.9).

The interfacial layers formed by these types of molecules tend to be relatively open, thick and of low viscoelasticity. Globular biopolymers (usually proteins) adsorb to an interface so that the predominantly non-polar regions on the surface of the molecule face the oil phase, while the predominantly polar regions face the aqueous phase, and so they tend to have a particular orientation at an interface. Once they have adsorbed to an interface, biopolymers often undergo structural rearrangements so that they can maximize the number of favorable contacts between non-polar groups and oil (Norde, 2003). Random-coil biopolymers are relatively flexible molecules and can therefore rearrange their structures fairly rapidly, whereas globular biopolymers are more rigid molecules and therefore rearrange more slowly. The unfolding of a globular protein at an interface often exposes amino acids that were originally located in the hydrophobic interior of the molecule, which can lead to enhanced interactions with neighboring protein molecules through hydrophobic attraction or disulfide bond formation. Consequently, globular proteins tend to form relatively thin and compact layers that have high viscoelasticities. This may account for the fact that layers formed by globular proteins are more resistant to rupture than those formed by more random-coil proteins.

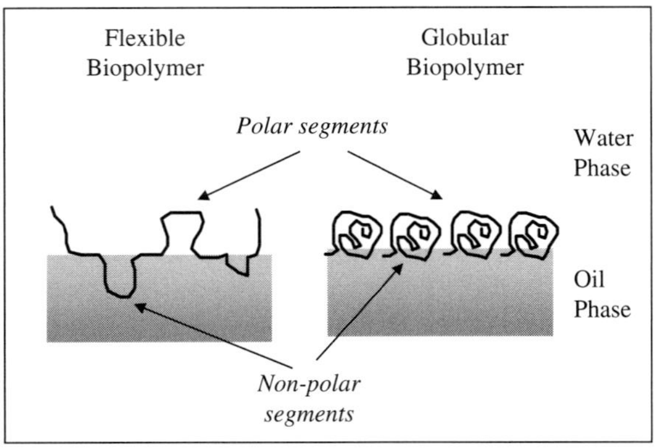

FIGURE 4.9 Schematic representation of conformations of flexible and globular biopolymer adsorbed molecules at an oil–water interface. Non-polar segments tend to protrude into the oil phase, while polar segments protrude into the water phase.

To be effective emulsifiers, biopolymers must rapidly adsorb to the surface of the emulsion droplets created during homogenization, and then form an interfacial layer that prevents the droplets from aggregating with one another. The interfacial layers formed around droplets by biopolymers can stabilize emulsion droplets against aggregation by a variety of different mechanisms. For example, they may increase the repulsive interactions (e.g., steric, electrostatic and hydration forces) or they may decrease the attractive interactions (e.g., van der Waals forces). The stabilizing mechanism that dominates within a particular emulsion is determined by the characteristics of the interfacial layer formed, e.g., its thickness, electrical charge, internal packing, and exposed reactive groups. The dominant stabilizing mechanism operating in an emulsion determines its sensitivity to droplet aggregation under different solution and environmental conditions, e.g., pH, ionic strength, temperature, solvent quality. For example, sterically stabilized emulsions have good stability to pH and ionic strength changes, whereas electrically stabilized emulsions do not. In the following sections, the interfacial properties and emulsion-stabilizing abilities of proteins and polysaccharides commonly used as food emulsifiers will be outlined.

4.7.2 Common Biopolymer Food Emulsifiers

Many food emulsions are stabilized by surface-active biopolymers that adsorb to droplet surfaces and form protective coatings. Some of these functional biopolymers are integral components of more complex food ingredients used in food manufacture (e.g., milk, eggs, meat, fish and flour), whereas others have been isolated from their natural environments and possibly modified before being sold as specialty ingredients (e.g., protein concentrates or isolates, hydrocolloid emulsifiers). In this section, we will focus primarily on those surface-active biopolymers that are sold as functional ingredients specifically designed for use as emulsifiers in foods. In addition, we will focus on the ability of biopolymers to create stable oil-in-water emulsions, rather than on their interfacial activity, since the former is more relevant to their application as emulsifiers in the food industry. This point can be clearly illustrated by considering the interfacial characteristics of globular proteins near their isoelectric point. Globular proteins are capable of rapidly adsorbing to oil–water interfaces and forming thick viscoelastic interfacial layers near their isoelectric points, but they will not form stable emulsions because the electrostatic repulsion between the droplets is insufficient to prevent droplet aggregation.

4.7.3 Proteins

The interfacial coatings formed by proteins are usually relatively thin and electrically charged, hence the major mechanism preventing droplet flocculation in protein-stabilized emulsions is electrostatic repulsion (Dickinson, 1992a, 2002). Consequently, protein-stabilized emulsions are particularly sensitive to pH and ionic strength effects (Figure 4.10), and will tend to flocculate at pH values close to the isoelectric point of the adsorbed proteins and when the ionic strength exceeds a certain level (McClements, 2005).

Emulsions stabilized by globular proteins are also particularly sensitive to thermal treatments, because these proteins unfold when the temperature exceeds a critical value exposing reactive non-polar and sulfhydryl groups (McClements, 2005). These reactive groups increase the attractive interactions between droplets through hydrophobic forces and disulfide bond formation, which may lead to droplet flocculation. It should be noted that a number of methods have been developed to attempt to improve the emulsifying properties of protein ingredients, including limited hydrolysis to form

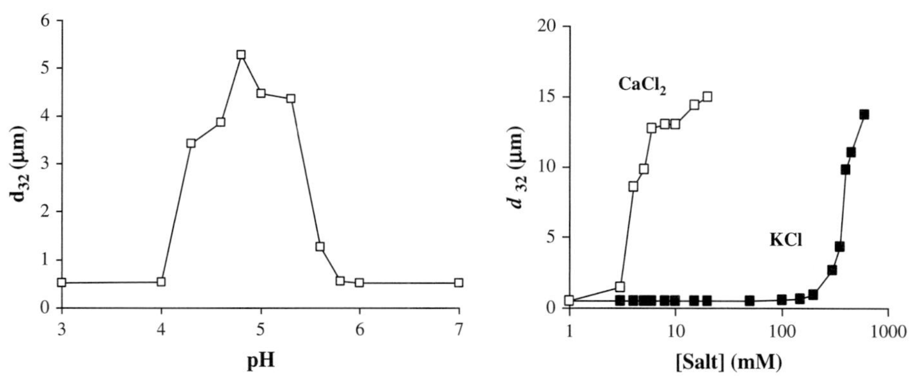

FIGURE 4.10 (a) pH-stability and (b) salt-stability of whey protein stabilized oil-in-water emulsions. These emulsions are mainly stabilized by electrostatic repulsion, and so they are highly sensitive to pH and ionic strength. They tend to aggregate near the protein isoelectric point and at high salt concentrations.

peptides, modification of protein structure by chemical, physical, enzymatic or genetic means, and blending of proteins with other ingredients, although not all of these processes are currently legally allowed.

4.7.3.1 *Milk Proteins*

Protein ingredients isolated from bovine milk are commonly used as emulsifiers in emulsion-based food products, including beverages, frozen desserts, ice creams, sports supplements, infant formula, and salad dressings. Milk proteins can be conveniently divided into two major categories: caseins ($\sim$ 80 wt%) and whey proteins ($\sim$ 20 wt%). A variety of milk protein ingredients are available for utilization as emulsifiers in foods, including whole milk, whey proteins, and caseins (Dickinson, 1999). These ingredients are usually sold in a powdered form that is light cream-to-white in appearance and has a bland flavor. These powders are normally available in the form of protein concentrates (25–80% protein) or protein isolates (>90% protein). It should be noted that there are a relatively large number of different kinds of proteins in both casein and whey, and that it is possible to fractionate these proteins into individual purified fractions (e.g., β-lactoglobulin, α-lactalbumin, bovine serum albumin, β-casein). Purified fractions are normally too expensive to be used as emulsifying ingredients in the food industry, but they are frequently used in research studies because they facilitate the development of a more fundamental understanding of protein functionality in emulsions.

4.7.3.2 *Meat and Fish Proteins*

Meat and fish contain a number of proteins that are surface-active and capable of stabilizing emulsions, e.g., gelatin, myosin, actomyosin, sarcoplasmic proteins and actin (Friberg et al., 2004; McClements, 2005). Many of these proteins play an important role in stabilizing meat emulsions, i.e., products formed by blending or homogenizing fat, meat and other ingredients. Emulsion stabilization is partly due to its ability to adsorb to the oil–water interface and partly due to its ability to increase the aqueous phase viscosity or to form a gel in the aqueous phase.

4.7.3.3 *Egg Proteins*

Both egg yolk and egg white contain a mixture of protein and non-protein components that are surface-active (Aluko and Mine, 1998; Mine, 1998, 2002; Anton and Gandemer, 1999; Castellani et al., 2006). Egg ingredients can be purchased in a variety of different forms for utilization in food emulsions, including fresh, frozen or dried forms of whole eggs, egg yolks, or egg whites. The processing treatments used to prepare these ingredients may influence their effectiveness at stabilizing emulsions, e.g., thermal processing, freezing or drying conditions. In the food industry, egg white is more commonly used for stabilizing foams, whereas egg yolk is more commonly used for stabilizing emulsions. Nevertheless, a number of studies have shown that egg white proteins can also be used to stabilize oil-in-water emulsions.

4.7.3.4 *Plant Proteins*

Surface active proteins can be extracted from a variety of plant sources, including legumes and cereals (Friberg et al., 2004; McClements, 2005). A considerable amount of research has been carried out to establish the ability of these proteins to stabilize emulsions, and whether they could be made into commercially viable value-added ingredients for utilization as emulsifiers in foods. One of the most widely studied proteins extracted from a plant source is soy protein, which is commercially available as a protein concentrate or isolate. Soy protein ingredients are a complex mixture of many individual protein fractions with different molecular and functional characteristics. In addition, each of these fractions contains a mixture of different protein sub-units that also have different molecular and functional characteristics.

4.7.4 Polysaccharides

4.7.4.1 Gum Arabic

Gum arabic is a biopolymer emulsifier that is extracted from the exudate of a shrub mainly found in Africa (*Acacia Senegal*). It is widely used in the beverage industry to stabilize cloud and flavor emulsions. Gum arabic is actually a complex mixture of different biopolymers, with at least three major high-molecular-weight fractions. It has been proposed that the surface-active fraction consists of branched arabinogalactan blocks attached to a polypeptide chain (Jayme et al., 1999; Dror et al., 2006; Yadav et al., 2007). The polypeptide chain has hydrophobic side groups that are believed to be responsible for gum arabic's ability to adsorb to lipid surfaces, while the hydrophilic arabinogalactan blocks extend into the aqueous phase. The interfacial layer formed by gum arabic is believed to provide stability against droplet aggregation mainly through steric repulsion, but with some contribution from electrostatic repulsion also (Jayme et al., 1999; Chanamai and McClements, 2002). It has been shown that gum arabic stabilized emulsions remain stable to droplet flocculation when exposed to a wide range of environmental stresses, e.g., pH (3–9), ionic strength (0–25 mM $CaCl_2$) and thermal treatment (30–90 °C) (Chanamai et al., 2002). Nevertheless, gum arabic has a relatively low affinity for oil–water interfaces compared to most other surface-active biopolymers, which means that it has to be used at relatively high concentrations to form stable emulsions. For example, gum arabic-to-oil ratios from about 1:1 to 2:1 are usually required to produce stable oil-in-water emulsion (Tse and Reineccius, 1995), whereas protein-to-oil ratios are typically only around 0.1:1 to 0.2:1 (Kim et al., 2004). For this reason, the utilization of gum arabic as an emulsifier in foods is usually restricted to products with relatively low droplet concentrations, e.g., beverage emulsions. In addition, there have been frequent problems associated with variations in the quality, cost and availability of gum arabic that have led many food scientists to investigate alternative sources of biopolymer emulsifiers for use in beverages (Garti, 1999; Tan, 2004).

4.7.4.2 Modified Starches

Natural starches usually have poor surface activity because they are predominantly hydrophilic molecules. Nevertheless, they can be chemically modified to produce highly effective emulsifiers by attaching hydrophobic side-groups along their backbones (Trubiano, 1995; Nilsson and Bergenstahl, 2006). These modified starches are widely used as emulsifiers in the beverage industry, often as an alternative or replacement to gum arabic. One of the most commonly used modified starches is an octenyl succinate derivative of waxy-maize (Trubiano, 1995; Prochaska et al., 2007). It consists primarily of amylopectin that has been chemically modified to contain side-groups that are anionic and non-polar. These non-polar side-groups attach to lipophilic droplet surfaces, while the polar starch chains protrude into the aqueous phase. These polymeric hydrophilic starch chains provide protection again droplet aggregation primarily through steric repulsion, although electrostatic repulsion may also play some role. Because the dominant stabilizing mechanism is steric repulsion, emulsions stabilized by modified starch are resistant to changes in pH (3–9), ionic strength (0–25 mM $CaCl_2$) and temperature (30–90 °C) (Chanamai et al., 2002). Like gum arabic, modified starch has a relatively low interfacial activity (compared to proteins or surfactants), and so a large excess must be added to ensure that all the droplet surfaces are adequately coated (Nilsson et al., 2006). For example, it is recommended that about a 1:1 ratio of modified starch-to-oil is required to prepare stable oil-in-water emulsions (Tse et al., 1995). Modified starches usually come in powdered or granular forms that are easily dispersible in cold water.

4.7.4.3 *Modified Celluloses*

Natural cellulose is not suitable as an emulsifier because it forms strong intermolecular hydrogen bonds, which make it insoluble in water. Nevertheless, cellulose can be isolated and modified in a variety of ways to produce food-grade ingredients that do have interfacial activity and so can be used as emulsifiers (Akiyama et al., 2005, 2007; Sun et al., 2007). The most commonly used surface-active cellulose derivatives are methyl cellulose (MC), hydroxypropyl cellulose (HPC) and methyl hydroxypropyl cellulose (MHPC). These ingredients are all non-ionic polymers that are soluble in cold water, but tend to become insoluble when the solution is heated above a critical temperature (around 50–90 °C). They have good stability to pH (2–11), salt and freeze–thaw cycling, which may be beneficial in a number of food emulsion applications.

4.7.4.4 *Other Polysaccharides*

Some other types of polysaccharides have been shown to be surface active, and capable of stabilizing oil-in-water emulsions, e.g., galactomannans, pectin, chitosan (Garti et al., 1993; Huang et al., 2001; Dickinson, 2003; Leroux et al., 2003; van Nieuwenhuyzen et al., 2006). Nevertheless, there is still some debate about the molecular origin of their surface activity (e.g., non-polar regions on the polysaccharide molecule itself, protein contaminants, or protein moieties bound to the polysaccharides), and about whether their ability to form stable emulsions is primarily due to their surface activity or the ability to thicken the aqueous phase (Dickinson, 2003).

4.7.5 Protein–Polysaccharide Complexes

Proteins tend to be better at producing small emulsion droplets when used at low concentrations than polysaccharides, whereas polysaccharides tend to be better at producing emulsions that are stable to a wider range of environmental conditions than proteins, e.g., pH, ionic strength, temperature, freeze–thaw cycling (McClements, 2005). It may therefore be advantageous to combine the beneficial attributes of these two kinds of biopolymer to produce small emulsion droplets with good environmental stability. Protein–polysaccharide complexes may have better emulsifying properties than either of the individual biopolymers. These complexes may be held together either by physical or covalent interactions, and may be formed either before or after homogenization (Guzey et al., 2006; McClements, 2006; Akhtar and Dickinson, 2007; Benichou et al., 2007; Kika et al., 2007; Turgeon et al., 2007; Jourdain et al., 2008). Ingredients based on protein–polysaccharide interactions will have to be legally acceptable, economically viable and show benefits over existing ingredients before they find widespread utilization in the food industry. It should be noted that gum arabic is a naturally occurring protein–polysaccharide complex that is already widely used in the food industry as an emulsifier.

4.7.6 Selection of an Appropriate Emulsifier

In this section, some schemes for classifying and comparing the effectiveness of different types of biopolymer emulsifier are briefly mentioned, as well as some of the factors that should be considered when selecting a biopolymer emulsifier for a particular application. As has been mentioned earlier, an effective emulsifier should have the following general characteristics:

- it should rapidly adsorb to the surface of freshly formed droplets during homogenization, so as to facilitate droplet disruption and prevent droplet coalescence;
- it should reduce the interfacial tension by a significant amount, so as to facilitate droplet disruption and prevent emulsifier desorption;
- it should form an interfacial layer around the droplets that protects them from aggregation, both within the homogenizer and during the subsequent processing, storage and utilization of the final product.

A number of food-grade biopolymers exhibit these characteristics and can be used as emulsifiers. Nevertheless, these biopolymers can vary considerably in their ability to form emulsions, as well as in their ability to stabilize emulsions. For example, many globular proteins rapidly adsorb to droplet surfaces during homogenization and form small droplets at low usage levels, but once formed the resulting emulsions are highly sensitive to alterations in pH and ionic strength. On the other hand, surface-active polysaccharides tend to adsorb more slowly during homogenization and thus form larger droplets, but the emulsions formed have greater resistance to alterations in pH and ionic strength.

It is useful for the food industry to have standardized means of assessing the relative efficiency of different types of emulsifiers for specific applications. Unfortunately, there has been little attempt to systematically compare the advantages and disadvantages of different biopolymer emulsifiers under standardized conditions, so that it is currently difficult for food manufacturers to rationally select the most suitable emulsifier for particular products. This section highlights some of the criteria that could be conveniently used to form the basis of such a comparison.

The functional properties of emulsifiers can be compared in terms of parameters that depend on the formulation and processing procedures of the food product that they will be used in, e.g.:

- *Minimum droplet size:* The minimum droplet size (d_{min}) that can be produced by a certain amount of emulsifier for a specified emulsion system using specified homogenization conditions.
- *Minimum emulsifier load:* The minimum amount of emulsifier (c_{min}) required to produce a desired droplet size for a specified emulsion system using specified homogenization conditions.
- *Stability index:* The long-term stability (e.g., to creaming, flocculation or coalescence) of a specified emulsion produced by an emulsifier using specified homogenization conditions.

The characteristics of the specified emulsion system (e.g., oil type, oil concentration, aqueous phase composition) used to establish the efficiency of an emulsifier depends on the food product being manufactured, and will vary considerably from product-to-product. In addition, the specified homogenization conditions will also vary according to the type of homogenizer used (e.g., high-speed blender, high-pressure valve homogenizer, microfluidizer or colloid mill) and the precise operating conditions (e.g., energy input, flow rate, temperature). The above approach is particularly suited for food manufacturers trying to determine the best emulsifier for utilization in their specific product, but it is not particularly suitable for development of a general classification scheme because of the wide variation in the composition and processing operations used to produce different foods. This approach could be adapted to develop a general classification scheme for comparing emulsifiers by stipulating standardized emulsion compositions and homogenization conditions. Analytical methods developed to measure *emulsifier capacity* and *emulsion stability index* are attempts at developing emulsifier classification schemes based on this principle (McClements, 2005).

At a more fundamental level, emulsifier properties can be characterized in terms of quantitative physical parameters that can be measured using analytical instruments under well-defined environmental conditions (McClements, 2005):

- *Surface load*, Γ_{sat}. The surface load at saturation is the mass of emulsifier adsorbed per unit surface area of interface when the interface is saturated with emulsifier, and is usually expressed as mg m^{-2}. The surface load provides a measure of the minimum amount of emulsifier required to produce an emulsion with a given surface area (or droplet size): the higher Γ, the greater the amount of emulsifier required to completely cover the same surface area.
- *Maximum surface pressure*, π_{max}. The maximum surface pressure is the interfacial tension of an oil–water interface in the absence of emulsifier minus the interfacial tension of the same interface when it is saturated with emulsifier. It provides a measure of the ability of an emulsifier to decrease the oil–water interfacial tension, and thereby facilitate droplet disruption: the higher π_{max}, the smaller the droplets that can be produced in a homogenizer at a fixed energy input, provided there is sufficient emulsifier present and that it adsorbs rapidly to the droplet surfaces.
- *Binding affinity*, $c_{1/2}$. The binding affinity is a measure of how strongly an emulsifier adsorbs to an oil–water interface. It can be expressed as the emulsifier concentration at which the surface pressure is half the maximum surface pressure. The stronger the binding affinity (the lower $c_{1/2}$), the lower the concentration of emulsifier required to reach interfacial saturation.
- *Adsorption kinetics*, τ_{ads}. Adsorption kinetics can be defined in terms of the average time required for an interface to become saturated with emulsifier. It is important that this time be measured under conditions that adequately represent the highly dynamic conditions that occur in most homogenizers. In practice, it is difficult to establish an accurate measure of the adsorption kinetics of different emulsifiers under realistically dynamic conditions.
- *Droplet aggregation stability*. The aggregation stability is a measure of the tendency for droplets to become aggregated (flocculated or coalesced) under a specified set of environmental conditions, e.g. pH, ionic strength, temperature, shearing rate. It can be expressed in a number of different ways, e.g., the percentage of droplets that are flocculated or coalesced, the percentage of droplets larger than a specified size, or the percentage increase in the mean size of the particles in an emulsion due to droplet aggregation.

One of the major challenges of food scientists is to relate these more fundamental parameters to the more practical parameters mentioned above that are of interest to food manufacturers.

A food manufacturer must decide which emulsifier is the most suitable for utilization in each particular product. In addition to the physicochemical characteristics considered above, a food manufacturer must also consider a number of economic, legal and marketing factors when selecting a suitable emulsifier.

4.8 BIOPOLYMER TEXTURE MODIFIERS

The ability of biopolymers to modify the texture of aqueous solutions by increasing the viscosity or forming a gel has been discussed in detail in other chapters in this book, so it will not be covered here. Instead, we will focus on some of the unique effects associated with texture modifiers when they are used in food emulsions:

- *Depletion flocculation.* When non-adsorbing biopolymers are added to the aqueous phase of an oil-in-water emulsion they increase the attraction between the oil droplets through an osmotic (depletion) effect. If the osmotic attraction generated is sufficiently high then the droplets will aggregate, which is known as depletion flocculation (Figure 4.11). This kind of droplet aggregation may alter the texture and stability of the overall system. For example, if the droplets are free to move, then depletion flocculation may promote creaming instability because of the increase in net particle size. On the other hand in a concentrated emulsion, the osmotic attraction may promote the formation of a three-dimensional network of aggregated droplets that is more stable to creaming and has gel-like rheological properties.
- *Bridging flocculation.* When biopolymers that are attracted to the droplet surfaces are added to an emulsion they may promote bridging flocculation (Figure 4.12). For example, if an anionic biopolymer is added to an emulsion containing cationic droplets, a single biopolymer may link two or more droplets together through an electrostatic attraction. In dilute emulsions bridging flocculation may lead to rapid creaming instability because the droplets are free to move upwards (Figure 4.12), whereas in concentrated emulsions bridging flocculation may lead to the formation of a three-dimensional network of aggregated droplets that prevents droplet movement and alters rheological properties.
- *Multilayer formation.* If sufficient charged biopolymer is added to an emulsion containing oppositely charged droplets it may completely saturate the droplet surfaces and form a stable system since the droplets are completely coated with polymer (Figure 4.12). These so-called multilayer emulsions may be highly stable to environmental stresses, such as pH, salt, heating, chilling, freezing and dehydration.
- *Filled gel effects.* The presence of oil droplets within a gelled aqueous biopolymer solution may alter the rheological properties of the system. If the oil droplets are actively incorporated into the gel network, then they will strengthen the gel, but if they are not then they may weaken the gel (depending on their particle size).
- *Droplet partitioning effects.* In a phase-separated mixed-biopolymer system oil droplets may preferentially partition into one of the phases. Thus multiphase systems may be formed, such as oil-in-water-in-water emulsions or (oil-in-water)-(water-in-water) systems.

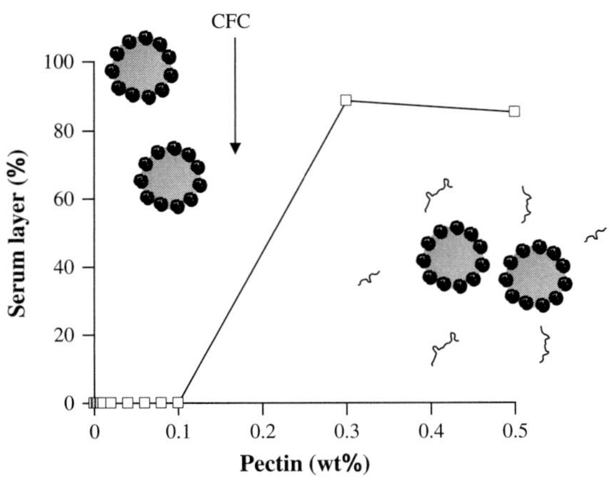

FIGURE 4.11 Thickening agents may induce emulsion instability when present at a sufficiently high concentration through a depletion flocculation mechanism. This figure shows the impact of pectin concentration on droplet flocculation in whey protein stabilized oil-in-water emulsions at pH 7, where both biopolymers are negative.

A manufacturer must therefore be aware that adding a biopolymer to an emulsion to thicken or gel the aqueous phase might also have other physicochemical effects on food emulsions,

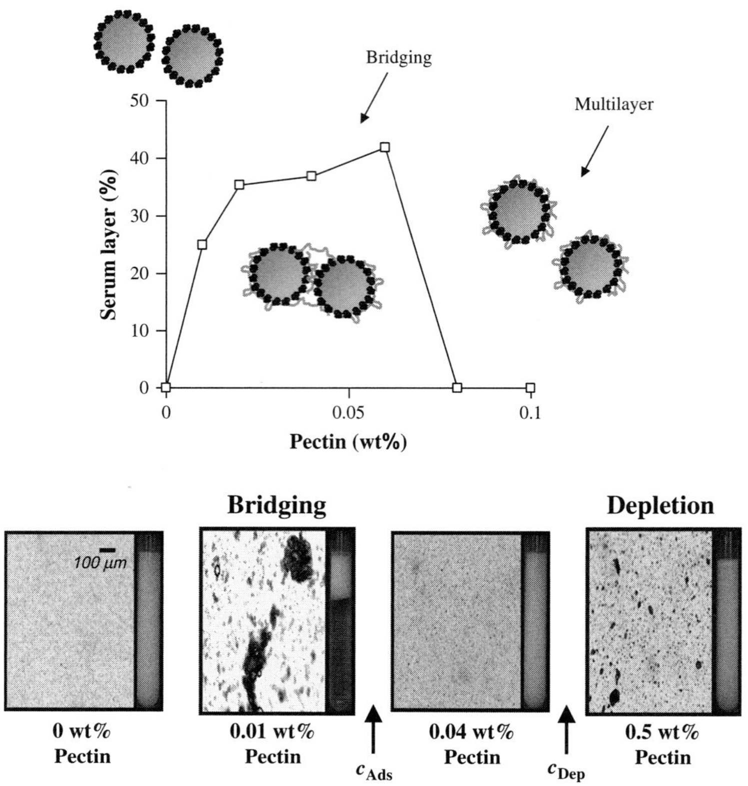

FIGURE 4.12 Thickening agents may induce emulsion instability through a bridging flocculation mechanism if they have an opposite charge to the adsorbed emulsifier. This figure shows the impact of pectin concentration on droplet flocculation in whey protein stabilized oil-in-water emulsions at pH 3.5, where the protein is positive and the pectin is negative.

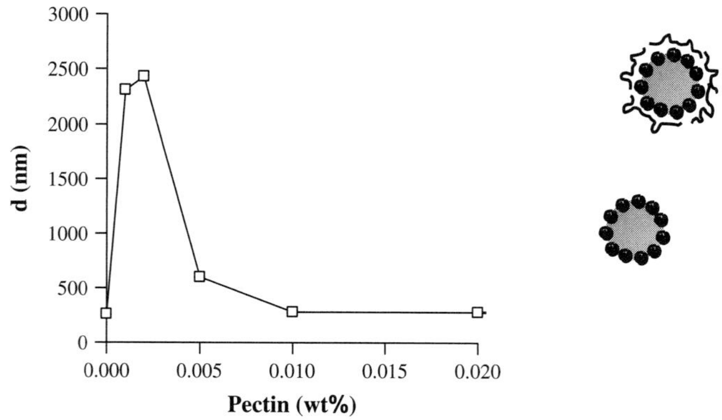

FIGURE 4.13 Microstructure of whey protein stabilized oil-in-water emulsions at pH 3.5, where the protein is positive and the pectin is negative. Highly complex aggregation behavior is observed when the pectin concentration is increased, e.g., stable (no pectin), unstable (bridging), stable (multilayer formation), unstable (bridging flocculation).

which may be either beneficial or detrimental to product quality. Extremely complex behavior can be observed when a charged biopolymer is added to an emulsion containing oppositely charged droplets (Figure 4.13). A better understanding of these phenomena may allow food manufacturers to rationally design food systems with improved or novel properties.

4.9 CONCLUSIONS

Biopolymers are widely used as emulsifiers in the food industry to improve the stability and physicochemical properties of food emulsions. A major advantage of biopolymers is that they are natural ingredients, which is often required for 'consumer-friendly' labels. There are many different kinds of biopolymers available in nature varying in their molecular characteristics and functional properties. In addition, the molecular and functional properties of biopolymers depend on the extraction, purification and processing treatments used to produce functional ingredients. This provides food scientists with a large palette of potential biopolymer-based functional ingredients to design improved or novel foods. On the other hand, this natural variation in molecular and functional properties can cause sample-to-sample variations that can cause problems with reproducibility during the manufacturing process. Improved knowledge of the molecular basis of biopolymer functionality in food emulsions will aid in the design of foods and food processing operations.

References

Akhtar, M. and Dickinson, E. (2007). Whey protein-maltodextrin conjugates as emulsifying agents: An alternative to gum arabic. *Food Hydrocolloids* **21** (4), 607–616.

Akiyama, E., Kashimoto, A., Fukuda, K., Hotta, H., Suzuki, T., and Kitsuki, T. (2005). Thickening properties and emulsification mechanisms of new derivatives of polysaccharides in aqueous solution. *Journal of Colloid and Interface Science* **282** (2), 448–457.

Akiyama, E., Yamamoto, T., Yago, Y., Hotta, H., Ihara, T., and Kitsuki, T. (2007). Thickening properties and emulsification mechanisms of new derivatives of polysaccharide in aqueous solution 2. The effect of the substitution ratio of hydrophobic/hydrophilic moieties. *Journal of Colloid and Interface Science* **311** (2), 438–446.

Aluko, R. E. and Mine, Y. (1998). Characterization of oil-in-water emulsions stabilized by hen's egg yolk granule. *Food Hydrocolloids* **12** (2), 203–210.

Anton, M. and Gandemer, G. (1999). Effect of pH on interface composition and on quality of oil-in-water emulsions made with hen egg yolk. *Colloids and Surfaces B-Biointerfaces* **12** (3-6), 351–358.

Baker, R. (1987). Controlled Releae of Biologically Active Agents. John Wiley and Sons, New York.

Bakker, J. (1995). Flavor interactions with the food matrix and their effects on perception. *In* Ingredient Interactions: Effects on Food Quality., Gaonkar, A. G. (ed.). Marcel Dekker, New York.

Bell, G. A. (1996). Molecular mechanisms of olfactory perception: Their potential for future technologies. *Trends in Food Science and Technology* **7**, 425.

Benichou, A., Aserin, A., Lutz, R., and Garti, N. (2007). Formation and characterization of amphiphilic conjugates of whey protein isolate (WPI)/xanthan to improve surface activity. *Food Hydrocolloids* **21** (3), 379–391.

Billmayer, F. W. and Saltzman, M. (1981). Principles of Color Technology. John Wiley and Sons, New York.

Castellani, O., Belhomme, C., David-Briand, E., Guerin-Dubiard, C., and Anton, M. (2006). Oil-in-water emulsion properties and interfacial characteristics of hen egg yolk phosvitin. *Food Hydrocolloids* **20** (1), 35–43.

Chanamai, R. and McClements, D. J. (2002). Comparison of gum arabic, modified starch, and whey protein isolate as emulsifiers: Influence of pH, CaCl2 and temperature. *Journal of Food Science* **67** (1), 120–125.

Chanamai, R. and McClements, D.J. (2000). Prediction of emulsion color from droplet characteristics: Dilute monodisperse oil-in-water emulsions. *Food Hydrocolloids* I.

Chantrapornchai, W., Clydesdale, F., and McClements, D. J. (1999). Theoretical and experimental study of spectral reflectance and color of concentrated oil-in-water emulsions. *Journal of Colloid and Interface Science* **218** (1), 324–330.

Chantrapornchai, W., Clydesdale, F., and McClements, D. J. (1999). Theoretical and experimental study of spectra reflectance and color of concentrated oil-in-water emulsions. *J. Colloids Int. Sci.* **218**, 324–330.

Chantrapornchai, W. and McClements, D.J. (2001). Influence of refractive index on the color of concentrated oil-in-water emulsions. *J. Food Sci.*

Chantrapornchai, W. C. F. and McClements, D. J. (1998). Influence of droplet size and concentration on the color of oil-in-water emulsions. *J. Agric. Food Chem.* **46**, 2914–2920.

Dickinson, E. (1992a). *Introduction to Food Colloids*. Cambridge, UK: Royal Society of Chemistry.

Dickinson, E. (1992b). *An Introduction to Food Colloids*. Oxford: Oxford Science Publishers.

Dickinson, E. (1999). Caseins in emulsions: interfacial properties and interactions. *International Dairy Journal* **9** (3-6), 305–312.

Dickinson, E. (2002). Food colloids. *Current Opinion in Colloid & Interface Science* **7** (5-6), 410–412.

Dickinson, E. (2003). Hydrocolloids at interfaces and the influence on the properties of dispersed systems. *Food Hydrocolloids* **17** (1), 25–39.

Dickinson, E. and Pawlowsky, K. (1997). Effect of l-carrageenan on flocculation, creaming, and rheology of a protein-stabilized emulsion. *Journal of Agricultural and Food Chemistry* **45** (10), 3799–3806.

Dickinson, E. and Pawlowsky, K. (1998). Influence of kappa-carrageenan on the properties of a protein-stabilized emulsion. *Food Hydrocolloids* **12** (4), 417–423.

Dror, Y., Cohen, Y., and Yerushalmi-Rozen, R. (2006). Structure of gum arabic in aqueous solution. *Journal of Polymer Science Part B-Polymer Physics* **44** (22), 3265–3271.

Franzen, K. L. and Kinsella, J. E. (1975). Physicochemical aspects of food flavouring. *Chemistry and industry* **21**, 505.

Friberg, S., Larsson, K., and Sjoblom, J. (2004). Food Emulsions. Marcel Dekker, New York.

Garti, N. (1997). Progress in stabilization and transport phenomena of double emulsions in food applications. *Food Science and Technology-Lebensmittel-Wissenschaft & Technologie* **30** (3), 222–235.

Garti, N. (1999). Hydrocolloids as emulsifying agents for oil-in-water emulsions. *Journal of Dispersion Science and Technology* **20** (1-2), 327–355.

Garti, N. and Bisperink, C. (1998). Double emulsions: progress and applications. *Current Opinion in Colloid & Interface Science* **3** (6), 657–667.

Garti, N., Reichman, D., Hendrickx, H., Dickinson, E., Jackson, L. K., and Bergenstahl, B. (1993). Hydrocolloids as Food Emulsifiers and Stabilizers. *Food Structure* **12** (4), 411–426.

Gu, Y. S., Decker, E. A., and McClements, D. J. (2004a). Influence of iota-carrageenan on droplet flocculation of beta-lactoglobulin-stabilized oil-in-water emulsions during thermal processing. Langmuir, **20** (22), 9565–9570.

Gu, Y. S., Decker, E. A., and McClements, D. J. (2004b). Influence of pH and iota-carrageenan concentration on physicochemical properties and stability of beta-lactoglobulin-stabilized oil-in-water emulsions. *Journal of Agricultural and Food Chemistry*, **52** (11), 3626–3632.

Guzey, D., Kim, H. J., and McClements, D. J. (2004). Factors influencing the production of O/W emulsions stabilized by beta-lactoglobulin-pectin membranes. *Food Hydrocolloids* **18** (6), 967–975.

Guzey, D. and McClements, D. J. (2006). Formation, stability and properties of multilayer emulsions for application in the food industry. *Advances in Colloid and Interface Science* **128**, 227–248.

Guzey, D. and McClements, D. J. (2007). Impact of electrostatic interactions on formation and stability of emulsions containing oil droplets coated by beta-lactoglobulin-pectin complexes. *Journal of Agricultural and Food Chemistry* **55** (2), 475–485.

Hansen, A. P. and Booker, D. C. (1986). Flavor interactions with casein and whey protein. *In* Flavor-Food Interactions, R. J. a. L. McGorrin, J. V. (ed.). American Chemical Society, Washington.

Hau, M. Y. M., Gray, D. A., and Taylor, A. J. (1996). Binding of volatiles to starch. *In* Flavor-Food Interactions, R. J. a. L. McGorrin, J. V. (ed.). American Chemical Society, Washington.

Hiemenz, P. C. and Rajagopalan, R. (1997). Principles of Colloid and Surface Chemistry. Marcel Dekker, New York.

Huang, X., Kakuda, Y., and Cui, W. (2001). Hydrocolloids in emulsions: particle size distribution and interfacial activity. *Food Hydrocolloids* **15** (4-6), 533–542.

Hunter, R. J. (1986a). *Foundations of Colloid Science*. Oxford University Press, Oxford.

Hunter, R. J. (1986b). *Foundations of Colloid Science: Volume 1*. Oxford University Press, Oxford.

Hunter, R. J. (1989). Foundations of Colloid Science, Volume 2. Oxford University Press, Oxford.

Hunter, R. J. (1993). Introduction to Modern Colloid Science. Oxford University Press, Oxford.

Hutchings, J. B. (1994). Food Colour and Appearence. Blackie Academic and Professional, London.

Israelachvili, J. N. (1992). Intermolecular and Surface Forces 2nd Edition, Academic Press, London, UK.

Jayme, M. L., Dunstan, D. E., and Gee, M. L. (1999). Zeta potentials of gum arabic stabilised oil in water emulsions. *Food Hydrocolloids* **13** (6), 459–465.

Jourdain, L., Leser, M. E., Schmitt, C., Michel, M., and Dickinson, E. (2008). Stability of emulsions containing sodium caseinate and dextran sulfate: Relationship to complexation in solution. *Food Hydrocolloids* **22** (4), 647–659.

Judd, D. B. and Wyszecki, G. (1963). Color in Business, Science, and Industry. Wiley, New York.

Kabalnov, A. (1998). Coalescence in Emulsions. *In* Modern Aspects of Emulsion Science, 205–260, Binks, B. P. (ed.). Royal Society of Chemistry, Cambridge, UK.

Kabalnov, A. (2001). Ostwald ripening and related phenomena. *Journal of Dispersion Science and Technology* **22** (1), 1–12.

Kabalnov, A. S., Gervits, L. D., and Makarov, K. N. (1990). Ostwald Ripening in Emulsions of Fluorocarbons. *Colloid Journal of the Ussr* **52** (6), 915–921.

Kika, K., Korlos, F., and Kiosseoglou, V. (2007). Improvement, by dry-heating, of the emulsion-stabilizing properties of a whey protein concentrate obtained through carboxymethylcellulose complexation. *Food Chemistry* **104** (3), 1153–1159.

Kim, H. J., Decker, E. A., and McClements, D.J. (2002a). Impact of protein surface denaturation on droplet flocculation in hexadecane oil-in-water emulsions stabilized by beta-lactoglobulin. *Journal of Agricultural and Food Chemistry,* **50** (24), 7131–7137.

Kim, H.J., Decker, E.A., and McClements, D.J. (2002b). Role of postadsorption conformation changes of beta-lactoglobulin on its ability to stabilize oil droplets against flocculation during heating at neutral pH. *Langmuir,* **18** (20), 7577–7583.

Kim, H. J., Decker, E. A., and McClements, D. J. (2004). Influence of free protein on flocculation stability of beta-lactoglobulin stabilized oil-in-water emulsions at neutral pH and ambient temperature. *Langmuir* **20** (24), 10394–10398.

Kinsella, J. E. (1989). Flavor Chemistry of Lipid Foods. Min, T. H. (ed.). American Oil Chemists Society, Champaign, IL. D.B. a. S.

Kokini, J. L. (1987). The physical basis of liquid food texture and texture-taste interactions. *Journal of Food Engineering* **6**, 51.

Land, D. G. (1996). Perspectives on the effects of interactions on flavour perception: An overview. *In* Flavor-Food Interactions, McGorrin, J. V. (ed.). American Chemical Society, Washington, DC. R.J. a. L.

Larson, R. G. (1999). The Structure and Rheology of Complex Fluids. Oxford University Press, Oxford.

Leroux, J., Langendorff, V., Schick, G., Vaishnav, V., and Mazoyer, J. (2003). Emulsion stabilizing properties of pectin. *Food Hydrocolloids* **17** (4), 455–462.

Liu, S. and Masliyah, J. H. (1996). Rheology of suspensions. *In* Suspensions: Fundamentals and Applications in the Petroleum Industry, Schramm, L. L. (ed.). American Chemical Society, Washington, D.C. (p. Chap 3).

McClements, D. J. (1999). Food Emulsions: Principles, Practice and Techniques. CRC Press, Boca Raton, Florida.

McClements, D. J. (2002). Colloidal basis of emulsions color. *Current Opinion in Colloid and Interface Science* **7** (5-6), 451–455.

McClements, D. J. (2004). Protein-stabilized emulsions. *Current Opinion in Colloid & Interface Science* **9** (5), 305–313.

McClements, D. J. (2005). Food Emulsions: Principles, Practice, and Techniques. CRC Press, Boca Raton.

McClements, D. J. (2006). Non-covalent interactions between proteins and polysaccharides. *Biotechnology Advances* **24** (6), 621–625.

McClements, D. J., Chantrapornchai, W., and Clydesdale, F. (1998). Prediction of food emulsion color using light scattering theory. *J. Food Sci.* **63**, 935–939.

McNulty, P. B. (1987). Flavor release-Elusive and dynamic. *In: Food Structure and Behaviour.*

Mewis, J. and Macosko, C.W. (1994). Suspension rheology. *In: Rheology: Principles, Measurements and Applications,* Chapter 10.

Mine, Y. (1998). Emulsifying characterization of hens egg yolk proteins in oil-in-water emulsions. *Food Hydrocolloids* **12** (4), 409–415.

Mine, Y. (2002). Recent advances in egg protein functionality in the food system. *Worlds Poultry Science Journal* **58** (1), 31–39.

Nilsson, L. and Bergenstahl, B. (2006). Adsorption of hydrophobically modified starch at oil/water interfaces during emulsification. *Langmuir* **22** (21), 8770–8776.

Norde, W. (2003). Colloids and Interfaces in Life Sciences. Marcel Dekker, New York, NY.

O'Neill, T.E. (1996). Flavor binding to food proteins: An overview. *In: Flavor-food Interactions*, Chapter 6.

Overbosch, P., Afterof, W. G. M., and Haring, P. G. M. (1991). Flavor release in the mouth. *Food Reviews International* **7**, 137.

Pal, R., Yan, Y., and Masliyah, J. H. (1992). Rheology of emulsions. *In: Emulsions: Fundamentals and Applications in the Petroleum Industry,* Chapter 4.

Prochaska, K., Kedziora, P., Le Thanh, J., and Lewandowicz, G. (2007). Surface activity of commercial food grade modified starches. *Colloids and Surfaces B-Biointerfaces* **60** (2), 187–194.

Schubert, H., Ax, K., and Behrend, O. (2003). Product engineering of dispersed systems. *Trends in Food Science & Technology* **14** (1), 9–16.

Schubert, H. and Engel, R. (2004). Product and formulation engineering of emulsions. *Chemical Engineering Research & Design* **82** (A9), 1137–1143.

Sun, W. B., Sun, D. J., Wei, Y. P., Liu, S. Y., and Zhang, S. Y. (2007). Oil-in-water emulsions stabilized by hydrophobically modified hydroxyethyl cellulose: Adsorption and thickening effect. *Journal of Colloid and Interface Science* **311** (1), 228–236.

Tadros, T. F. (1994). Fundamental principles of emulsion rheology and their applications. *Colloids and Surfaces* **91**, 30.

Tan, C. T. (2004). Beverage Emulsions. *In* Food Emulsions, 4th Edition, 485–524, Friberg, S., Larsson, K. and Sjoblom, J. (eds.) 4th Edition, Marcel Dekker, New York.

Taylor, A. J. and Linforth, R. S. T. (1996). Flavor release in the mouth. *Trends in Food Science and Technology* **7**.

Thomson, D.M.H. (1986). The meaning of flavor. *IN: Developments in Food Flavours*, pg. 1.

Trubiano, P. C. (1995). The role of specialty food starches in flavor emulsions. *Flavor Technology* **610**, 199–209.

Tse, K. Y. and Reineccius, G. A. (1995). Methods to predict the physical stability of flavor – Cloud emulsion. *Flavor Technology* **610**, 172–182.

Turgeon, S. L., Schmitt, C., and Sanchez, C. (2007). Protein-polysaccharide complexes and coacervates. *Current Opinion in Colloid & Interface Science* **12**, 166–178.

van Nieuwenhuyzen, W., Budnik, V., Meier, T., and Popper, L. (2006). Functional use of hydrocolloids in food systems. *Agro Food Industry Hi-Tech* **17** (5), 39–43.

Walstra, P. (1993). Principles of emulsion formation. *Chemical Engineering Science* **48**, 333.

Walstra, P. (2003). Physical Chemistry of Foods. Marcel Decker, New York, NY.

Wedzicha, B.L. (1988). Distribution of low-molecular weight food additives in dispersed systems. *In: Advances in Food Emulsions*, Chapter 10.

Weiss, J. and McClements, D. J. (2000). Influence of Ostwald ripening on rheology of oil-in-water emulsions containing electrostatically stabilized droplets. *Langmuir* **16**, 2145–2150.

Wyszecki, G. and Stiles, W. S. (1982). Color science: concepts and methods, quantitative data and formulae. Wiley, New York.

Yadav, M. P., Igartuburu, J. M., Yan, Y. C., and Nothnagel, E. A. (2007). Chemical investigation of the structural basis of the emulsifying activity of gum arabic. *Food Hydrocolloids* **21** (2), 297–308.

CHAPTER

5

Functional Interactions in Gelling Biopolymer Mixtures

Edwin R. Morris

Department of Food and Nutritional Sciences, University College Cork, Cork, Ireland

5.1 INTRODUCTION

This chapter is based heavily on an invited plenary lecture presented at the 9th International Hydrocolloids Conference, Singapore, 15–20 June 2008. Its aim is to set out some general principles governing the behavior of biopolymer mixtures, and to illustrate these with specific examples drawn mainly from the work of my own research group, focusing on gelling mixtures with two polymeric constituents.

5.1.1 Possible Outcomes of Mixing Two Different Biopolymers

When two biopolymers are mixed together, it is unusual for each to behave in exactly the same way as it would in the absence of the other polymer. In some systems, counterions to one charged polymer may promote gelation of another (e.g. gelation of alginate by Ca^{2+} ions from gum arabic, or of gellan by Na^{+} ions from sodium carboxymethylcellulose). Charged polymers may also affect one another by competing for ions from added salt. For example, in a recent investigation of mixtures of κ-carrageenan with whey protein isolate (Harrington et al., 2009) $CaCl_2$ was added at the concentration required to give optimum gelation of the carrageenan (Doyle et al., 2002), but it was found, surprisingly, that the Ca^{2+} cations were bound preferentially by undenatured whey protein molecules, inhibiting formation of the carrageenan network. Interactions of this type, between individual polyelectrolytes and simple ions, will not be discussed further.

The remainder of the chapter will concentrate on direct interactions between unlike polymer molecules (Braudo et al., 1986; Grinberg and Tolstoguvzov, 1972, 1977; Suchkov et al., 1981; Tolstoguzov and Braudo, 1983; Morris, 1986; Tolstoguzov, 1986, 1988a, 1988b, 1991, 1992a, 1992b, 2003; Picullel et al., 1995; Polyakov et al., 1997; Zasypkin et al., 1997). These can be classified (Piculell et al., 1994; Bergfeldt et al., 1996) as 'associative' or 'segregative', depending on whether they are enthalpically more favorable or less favorable than homotypic interactions between molecules of each type. When the heterotypic interactions are enthalpically unfavorable, there will be a tendency for the system to segregate into regions where the individual chains are surrounded by others of the same type, whereas enthalpically favorable heterotypic

Kasapis, Norton, and Ubbink: Modern Biopolymer Science
ISBN: 978-0-12-374195-0

interactions will promote association between the two polymers.

5.1.2 Associative Interactions

In a few systems, association occurs by formation of ordered heterotypic junctions analogous to the homotypic junction zones in single-component polysaccharide gels, and the resulting networks are often referred to as 'synergistic gels'. One such system, which was first investigated over 25 years ago (Thom et al., 1982; Toft, 1982) but does not appear to have attracted much interest since then, is alginate of high polyguluronate content in combination with high-methoxy pectin at low pH. The synergistic gels that have received greatest attention (as reviewed by, for example, Morris, 1995), however, are those formed by locust bean gum, konjac glucomannan and related (1 → 4)-diequatorially linked plant polysaccharides in combination with xanthan or with conformationally ordered (double helical) polysaccharides in the agar/carrageenan series.

A more general mechanism of association is by electrostatic attraction between polyanions, such as negatively charged polysaccharides, and polycations, such as proteins below their isoelectric point (Muchin et al., 1976; Tschumak et al., 1976; Imeson et al., 1977; Stainsby, 1980; Chilvers and Morris, 1987; Chilvers et al., 1988; Michon et al., 1995; Galazka et al., 1999; Gilsenan et al., 2003a; Ye, 2008). In some cases the association may lead to formation of a gel involving a 'coupled network' (Morris, 1986) of the constituent polymers. A more likely outcome, however, is co-precipitation of the two polymers as a 'complex coacervate'.

5.1.3 Segregative Interactions

Segregative interactions (Morris, 1998) are far more common than associative interactions. They occur in virtually all biopolymer mixtures where there is no over-riding drive to heterotypic binding. The tendency for individual molecules to be surrounded by others of the same type, which is often called 'thermodynamic incompatibility' (e.g. Syrbe et al., 1995) arises from differences in polarity (or polarizability).

The reason why such differences should introduce an enthalpic preference for clustering of like with like may be illustrated by considering the pairwise interactions that can occur between two polar species, A, with dipole strength I, and two species, B, with a higher dipole strength, I + d. The force of attraction between A and A will be proportional to I^2 and the attraction between B and B will be proportional to $(I + d)^2 = I^2 + 2Id + d^2$, giving a total of $2I^2 + 2Id + d^2$ for homotypic interaction. The corresponding value for heterotypic pairing (i.e. two A–B interactions) will be $2I\ (I + d) = 2I^2 + 2Id$. Thus interaction between species of the same type is more favorable, by an amount that depends on d^2 (i.e. with an increasing drive to segregation as the difference in the polarity or polarizability of the two species increases).

5.1.4 Phase Separation in Solutions and Gels

For small molecules, the enthalpic drive to segregation is normally outweighed by the entropic advantage of intimate mixing. For polymers, however, where (at equivalent concentrations) there are far fewer molecules free to move independently, entropy of mixing is much less significant, and mixed systems often separate into two co-existing phases, each enriched in one polymer and depleted in the other. Phase separation of biopolymers in the solution state can usually be detected by immediate development of turbidity on mixing, due to formation of a 'water-in-water emulsion' in which one phase exists as a continuous matrix with the other dispersed through it as small liquid droplets (e.g. Grinberg and Tolstoguzov, 1972, 1997; Polyakov et al., 1997).

After initial phase separation, mixtures of non-gelling polymers normally show gradual resolution into two clear layers (in response to differences in density between the phases), but for gelling biopolymers the water-in-water emulsion structure can be trapped by network formation, giving a biphasic co-gel with one phase continuous and the other dispersed. Phase-separated co-gels can also result from gelation of a single-phase mixture. Under these conditions, the first component to gel will give a continuous network, permeated by a solution of the second polymer. Subsequent gelation of the other component will then create a second gel phase. If the first network is weak, there may be an enthalpic advantage from the second, stronger, gel becoming the continuous phase, with the original network being broken down into dispersed particles (Mohammed et al., 1998b). A more likely outcome is for the first network to remain continuous, and for the second component to form the dispersed phase.

In a few mixed biopolymer systems (Piculell et al., 1992; Gilsenan et al., 2003b, 2003c), however, there is evidence that the first gel remains intact, and that the other component forms a second continuous phase running through the interstices of the original network, to give a bicontinuous co-gel (like a jelly set in the pores of a sponge).

5.1.5 Polymer Blending Laws

For phase-separated bicontinuous networks, the overall modulus of the composite gel (G_C) can be related to the moduli of the individual phases (G_X and G_Y) by a theoretical equation developed by Davies (1971):

Bicontinuous:

$$(G_C)^{1/5} = \phi_X(G_X)^{1/5} + \phi_Y(G_Y)^{1/5} \quad (5.1)$$

where ϕ_X and ϕ_Y denote the volume-fractions ('phase volumes') of the constituent phases, X and Y (with $\phi_X + \phi_Y = 1$). The moduli of composites with a dispersed phase in a continuous matrix are normally analyzed by the isostrain and isostress blending laws of Takayanagi et al. (1963).

In developing these relationships, Takayanagi and co-workers used sandwich structures with thin layers of two different condensed polymers (polyvinylchloride and rubber) arranged either in parallel or in series relative to the direction of deformation. In the parallel arrangement, the deformation of the weaker component is limited by the rigidity of the stronger material, so that both components are deformed to the same extent (isostrain conditions), and the overall modulus is the weighted average of the individual moduli:

$$\text{Isostrain: } G_C = G_X\phi_X + G_Y\phi_Y \quad (5.2)$$

In the series arrangement, the strength of the weaker component limits the force transmitted to the stronger material, so that both are subjected to the same stress (isostress conditions), and the overall deformation (compliance) of the composite ($J_C = 1/G_C$) approximates to the weighted-average of the individual compliances, giving:

$$\text{Isostress: } 1/G_C = \phi_X/G_X + \phi_Y/G_Y \quad (5.3)$$

It should be noted that all three blending laws (isostrain, isostress and bicontinuous) have the same general form:

$$(G_C)^n = \phi_X(G_X)^n + \phi_Y(G_Y)^n \quad (5.4)$$

with $n = 1$ for isostrain, $n = -1$ for isostress, and $n = 0.2$ for bicontinuous, and can be applied to Young's modulus (E) and to G′ from oscillatory measurements, as well as to shear modulus (G), which, for simplicity, is used in all the blending law relationships shown in this chapter.

Although the Takayanagi isostress and isostrain blending laws are strictly valid only when applied to the series and parallel arrangements for which they were derived, or when used to set upper (isostrain) or lower (isostress) bounds on the behavior of other morphologies (Clark et al., 1982, 1983), they are expected (Manson and Sperling, 1976) to apply reasonably well to composites with 'filler' particles (phase Y) dispersed through a continuous matrix (phase X). This expectation has been verified experimentally in a series of investigations of well-defined composites formed by trapping gelatinized starch granules in a biopolymer gel (Abdulmola et al., 1996a, 1996b; Mohammed et al., 1998a). One of these studies is summarized below, to illustrate some important general principles governing the rheology of biphasic networks with one phase continuous and the other dispersed.

5.2 APPLICABILITY OF POLYMER BLENDING LAWS TO BIPHASIC NETWORKS

The gelling biopolymer in the system used for illustration (Mohammed et al., 1998a) was agarose. The procedure adopted was as follows. Starch was gelatinized in an agarose solution by stirring at 80°C; the sample was then split; part was loaded onto oscillatory rheometer and cooled to gel the agarose; the other part was centrifuged to sediment the swollen granules, and the agarose supernatant was gelled in the same way as the agarose–starch mixture. Highly crosslinked waxy starch was used to avoid release of starch polysaccharides into the agarose phase. The modulus of the gelled supernatant was compared with a standard calibration curve of G' versus concentration for agarose alone, to give the concentration of agarose in the continuous phase (c_X); comparison with the original concentration in the mixture (c) then gives the phase volumes:

$$\phi_X = c/c_X \quad (5.5)$$

$$\phi_Y = 1 - \phi_X \quad (5.6)$$

The moduli of the composite (G_C) and the continuous phase (G_X) are known from experiment. The only unknown is G_Y (the modulus of the dispersed phase of gelatinized starch granules); this was determined by assuming that when both phases have the same modulus, that will also be the modulus of the composite ($G_X = G_Y = G_C$). Figure 5.1 shows the values of G_C measured for composites formed by 5% starch in mixtures with 0.10, 0.15, 0.20 and 0.25% agarose, in comparison with the corresponding values of G_X for the gelled supernatants from these mixtures. At low concentrations of agarose, the composites are strengthened by the dispersed starch phase ($G_C > G_X$); at high concentrations they are weakened ($G_C < G_X$). The two curves cross ($G_C = G_X$) at $G' \approx 850$ Pa, which was therefore taken as the modulus of the dispersed starch phase (G_Y).

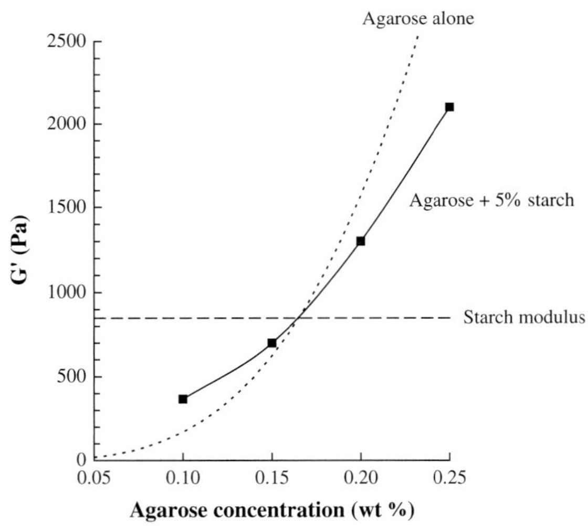

FIGURE 5.1 Determination of the modulus (G'_Y) of gelatinized starch granules in composite gels with agarose. The solid line is interpolated empirically through the observed moduli (G'_C) of composites containing 5% starch gelatinized in agarose solutions of different concentrations; the dotted line shows the modulus (G'_X) of the agarose phase; the modulus of the starch phase (– – –) is given by the point of intersection, where $G'_X = G'_Y = G'_C$ (Mohammed et al., 1998a).

The phase volumes derived (Eq. (5.6)) for the dispersed starch granules were unaffected by the concentration of agarose, and increased in direct proportion to the amount of starch present, reaching a value of $\phi_Y \approx 0.5$ (i.e. with the gelatinized granules occupying about half the total volume) at a starch concentration of 5 wt%. This was the maximum concentration of starch used, to avoid forcing the swollen granules into contact with one another. Mixtures were prepared at two concentrations of agarose: 0.10 and 0.25 wt%.

When the modulus (G_X) of the continuous agarose phase (derived experimentally by gelation of the supernatant from centrifugation) was higher than the (constant) modulus (G_Y) of the dispersed starch phase (derived by the procedure shown in Figure 5.1), experimental values of overall modulus (G_C) were compared with values calculated using Eq. (5.2), to test the expectation (Manson and Sperling, 1976) that deformation of the weaker 'filler' particles of gelatinized starch would be limited by the strength of the surrounding agarose matrix, with both therefore being subjected to the same strain (isostrain conditions). Conversely, when the modulus of the continuous phase was lower than that of the dispersed phase, calculated values of G_C were derived by Eq. (5.3), to test the expectation that the stress transmitted to the dispersed particles would be dictated by the strength of the weaker surrounding matrix (isostress conditions). The results obtained are shown in Figure 5.2.

At both concentrations of agarose studied (0.10 and 0.25 wt%) the overall moduli of the composite gels (G_C) increased by about an order of magnitude as the starch concentration was increased from 0 to 5 wt%, and were in good agreement with values calculated from the moduli of the constituent phases (G_X and G_Y) and their phase volumes (ϕ_X and ϕ_Y) by application of the Takayanagi blending laws. The modulus of the continuous agarose phase (gelled supernatant from centrifugation), however, also showed a large, progressive increase with

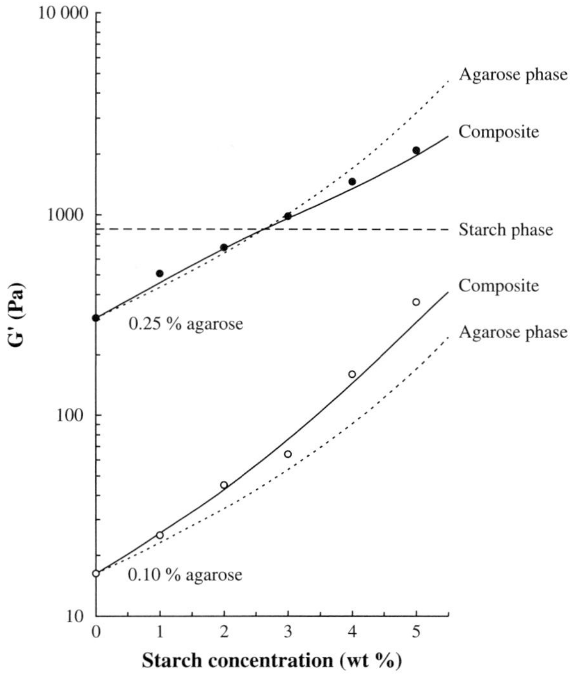

FIGURE 5.2 Observed values of composite modulus (G'_C) for gelatinized crosslinked starch in combination with 0.10 (O) and 0.25 (●) wt % agarose, in relation to the modulus (G'_Y) of the starch phase (– – –), the moduli (G'_X) of the agarose phase (- - - - -), and the calculated values of G'_C (—) from the isostrain model (for $G'_X > G'_Y$) or isostress model (for $G'_X < G'_Y$). (Mohammed et al., 1998a).

increasing concentration of starch, similar in magnitude to the increase in G_C.

At the lower concentration of agarose (0.10%), the modulus of the agarose matrix (G_X) remained lower than that of the dispersed phase of gelatinized starch granules (G_Y) up to the highest concentration of starch studied (5 wt%), and the composite gels were strengthened by the filler particles ($G_C > G_X$). For the higher agarose concentration (0.25 wt%), the modulus of the agarose matrix rose above that of the starch phase mid-way through the range of starch concentrations, and the filler particles then decreased the strength of the composite gels ($G_C < G_X$). In both cases, however, the overall moduli did not deviate by more than about a factor of

2 from those of the continuous (agarose) phase. This can be readily explained by the Takayanagi blending laws.

For composites where the filler particles (phase Y) are much softer than the surrounding matrix (phase X), the isostrain relationship (Eq. (5.2)) reduces to:

$$G_C \approx G_X \phi_X \quad \text{if } G_Y \ll G_X \tag{5.7}$$

Similarly, if they are much harder than the matrix the isostress relationship (Eq. (5.3)) reduces to:

$$1/G_C \approx \phi_X / G_X$$

$$\text{i.e.} \quad G_C \approx G_X/\phi_X \quad \text{if } G_Y \gg G_X \tag{5.8}$$

Thus at the maximum concentration of starch used (5 wt%), where the swollen granules occupied about half the total volume ($\phi_Y \approx \phi_X \approx 0.5$), the maximum effect of the dispersed phase would be to raise (Eq. (5.8)) or lower (Eq. (5.7)) the overall modulus by about a factor of 2, from $G_C \approx 2G_X$ (very hard filler) to $G_C \approx 0.5G_X$ (very soft filler), as observed experimentally (Figure 5.2). Even at the close-packing limit of $\phi_Y \approx 2/3$ ($\phi_X \approx 1/3$), beyond which the filler particles would be forced into contact with one another and could no longer, therefore, be regarded as forming a dispersed phase, their maximum effect would be to raise or lower the modulus of the composite by a factor of ~3.

The implications for understanding the rheology of biphasic co-gels can be summarized as follows:

- The strength of biopolymer composites is dominated by the continuous phase.
- The dispersed phase has little direct effect on overall modulus.
- However, it can have an enormous indirect effect, by occupying part of the total volume, and therefore increasing the polymer concentration in the dominant continuous phase.
- Overall moduli of composite gels can be matched with good precision by the Takayanagi blending laws, using the isostrain relationship (Eq. (5.2)) if the dispersed phase is weaker than the continuous matrix, and the isostress relationship (Eq. (5.3)) if it is stronger.
- Melting of a very strong dispersed gel phase is unlikely to reduce the overall modulus by more than about a factor of 4 (from $G_C \approx 2G_X$ to $G_C \approx G_X/2$).
- Any larger change would indicate a bicontinuous network structure.

5.3 PHASE COMPOSITION

5.3.1 Segregation in the Solution State

In biphasic solutions, the two phases can normally be separated (e.g. by centrifugation) and their composition can then be determined by standard analytical methods. The composition of the co-existing phases obtained from different starting compositions can be defined by a 'binodal' or 'cloud-point curve' which represents the boundary between monophasic and biphasic states of the system (see, for example, Grinberg and Tolstoguvzov, 1972; Tolstoguvzov, 1986, 1988a, 1991, 2003; Antonov et al., 1996; Musampa et al., 2007). In the monophasic region, at polymer concentrations below the binodal, the entropic advantage of random mixing outweighs the enthalpic advantage (Section 5.1.3) of polymer molecules being surrounded by others of the same type. In the biphasic region, at higher concentrations, enthalpy becomes dominant, promoting progressive segregation of the two polymers until the enthalpic advantage of further segregation is balanced by the entropic advantage of residual mixing. Exact balance (i.e. minimum free energy) is reached at the binodal, giving thermodynamically stable solutions with a small equilibrium concentration of each polymer in the phase where the other predominates.

Lowering temperature decreases the relative importance of entropy ($\Delta G = \Delta H - T\Delta S$), and therefore promotes more complete segregation of the two polymers, shifting the binodal towards the concentration axes (Figure 5.3). Increase in molecular weight, giving a smaller number of species free to move independently and therefore decreasing the entropy of mixing, also displaces the binodal to lower concentrations.

In mixtures where one polymer is neutral and the other is charged, preservation of electrical neutrality requires the charged polymer and its associated counterions to be present in the same relative proportions throughout the system, and since the number of counterions is, of course, very much greater than the number of charged chains, the entropic disadvantage of segregation is correspondingly greater than in mixtures of two neutral polymers. Phase separation, however, can be induced by addition of extraneous salt, since the imbalance in counterion concentration becomes progressively less significant as the overall concentration of ions is increased (Piculell et al., 1991, 1995; Antonov et al., 1996). The requirement for polyelectrolytes to be accompanied by their stoichiometric complement of counterions introduces an analogous (although less severe) barrier to segregation in mixtures of two charged biopolymers (Piculell et al., 1991), and can again be offset by addition of salt.

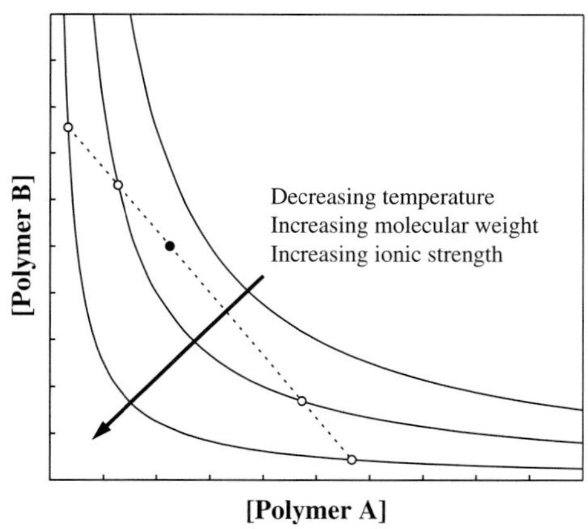

FIGURE 5.3 Schematic phase diagram, illustrating the effect of temperature, ionic strength (salt concentration) and effective molecular weight on the position of the binodal (—) and hence on the equilibrium composition (○) of co-existing phases formed from the same starting composition (●) along a common tie line (······).

Macroscopic resolution of biphasic solutions from the initial 'water-in-water emulsion' state normally results in formation of two liquid layers, with the denser phase as the lower layer. In a recent study of mixtures of konjac glucomannan (KGM) with disordered gelatin (i.e. at a temperature above the onset of gelation), however, Harrington and Morris (2009b) observed resolution into a spherical gel-like 'lump', consisting predominantly of KGM, suspended in a solution consisting predominantly of disordered gelatin. Rheological characterization showed that the 'lump' was, in reality, a highly concentrated solution, with the same density as the surrounding mobile solution, and adoption of spherical geometry was attributed to minimization of surface contact between the two phases.

5.3.2 Partition of Polymers and Solvent in Biphasic Co-Gels

Determination of phase composition in the agarose–starch composites described in Section 5.2 was straightforward: the agarose could not penetrate the gelatinized starch granules, so the composition of each phase was known, and the phase volumes (and hence the concentration of agarose in the continuous phase) could be determined experimentally by sedimenting the dispersed starch phase. The same approach can be applied to biopolymer co-gels where the continuous phase melts at lower temperature than the gelled particles of the dispersed phase: the two phases can be separated by heating to a temperature where the matrix has melted, and centrifuging to sediment the dispersed particles. The relative proportions of supernatant and sediment give the phase volumes, and the composition of both phases can be measured by standard analytical procedures.

Apart from these special cases, however, there is no satisfactory, general way of determining the composition and relative volumes of the co-existing phases in biphasic co-gels. It would be tempting to assume that when co-gels are formed from biphasic solutions the phases retain the same volumes and composition as they had prior to gelation. That, however, is unlikely. When gel formation is induced by heating (e.g. globular proteins; methylcellulose) or cooling (e.g. gelatin; carrageenan; gellan), the change in temperature will itself cause a shift in the binodal (Figure 5.3), with consequent redistribution of polymers and solvent between the phases. The changes in macromolecular organization required for formation of intermolecular junctions (conformational ordering of polysaccharides and gelatin; denaturation and aggregation of globular proteins) may also perturb phase equilibria by altering the enthalpic interactions between the constituent polymers.

An approach that was first used by Dr. Allan Clark and his colleagues in Unilever Research around 25 years ago (Clark et al., 1982, 1983) is to assume, as a working expedient, that the two constituent polymers in biphasic co-gels segregate completely into their respective phases. The parameters required for analysis of co-gel moduli by polymer blending laws (Eqs. (5.1)–(5.3)) are then determined solely by the way in which the solvent (water) is shared between the two phases. Solvent partition dictates the phase volumes (ϕ_X and ϕ_Y), which in turn dictate the concentration of polymer within the two phases, and hence their moduli (G_X and G_Y), determined using standard calibration curves of modulus versus concentration for the individual polymers. A second working assumption introduced by Dr. Clark was that solvent partition could be characterized by a single parameter, p, defined (Clark, 1987) as the ratio of water/polymer in one phase divided by the corresponding ratio for the other phase, irrespective of the amount of each polymer in the mixture. The same approach has been adopted in many subsequent investigations by myself and others.

The assumption that each polymer is confined entirely to its own phase is perhaps not as unrealistic as it may appear at first sight. It can be rationalized by two important factors:

- in the initial stages of gelation, after the onset of intermolecular association but before development of a continuous network, there will be a massive increase in effective molecular weight, which, as discussed above, will promote further segregation of biphasic solutions, or indeed may trigger phase separation in mixtures that were initially homogeneous (Loren et al., 2001; Fitzsimons et al., 2008c);
- the segregation process may be driven towards completion by progressive migration of non-gelling concentrations of the minor component in each phase to join the network being formed by the same material in the other phase, where it is dominant.

The real justification, however, is that analyses incorporating the assumption of complete segregation, with characterization of solvent partition by a p-factor, have given close agreement between observed moduli and values calculated by application of the appropriate blending law. Some illustrative examples are described in the two following sections.

5.4 BLENDING LAW ANALYSES OF GELATIN–CALCIUM PECTINATE CO-GELS

5.4.1 Experimental Procedure

The investigation summarized here was reported in full by Gilsenan et al. (2003b, 2003c). The materials used were type B gelatin (pI $\approx$ 4.9) and low-methoxy (LM) pectin (degree of esterification $\approx$ 31%). As described in Section 5.6.1, these form electrostatic complexes at low pH, by association between positively charged gelatin and negatively charged pectin, but association can be abolished by raising the pH. In the studies of co-gelation described in this section, the pH used (3.9) was high enough to eliminate evidence (Section 5.6.1) of complex formation.

Mixtures were prepared at 85°C, and incorporated $CaHPO_4$ at the concentration required for stoichiometric equivalence of Ca^{2+} cations to the unesterified carboxyl groups of the pectin. Some mixtures also included NaCl at concentrations up to 1 M. Immediately after mixing, the samples were loaded onto an oscillatory rheometer at 85°C, coated with light silicone oil to minimize loss of water by evaporation, cooled to 5°C at 1°C/min, held for 100 min (to promote gelation of the gelatin component), and heated back to 85°C (at 1°C/min). Formation and dissociation of gel networks by each polymer alone was characterized in the same way, under the same ionic conditions.

Figure 5.4a shows the changes in G′ observed for 1.0 wt% LM pectin during cooling and heating with stoichiometric Ca^{2+}. Formation of a calcium pectinate gel on cooling is evident in a large, sigmoidal increase in modulus as the

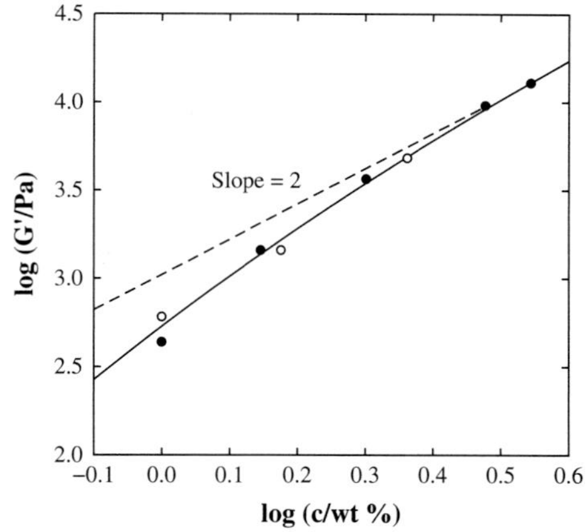

FIGURE 5.5 Concentration-dependence of G′ (10 rad s^{-1}; 2% strain) for LM pectin (pH 3.9; stoichiometric Ca^{2+}) after 100 min at 5°C, in water (●) and in 1 M NaCl (O) (Gilsenan et al., 2003b).

temperature drops below ~45°C, with little further increase during the holding period of 100 min at 5°C. Loss of network structure on heating occurs at higher temperature than gelation on cooling (by ~10°C).

As shown in Figure 5.5, the concentration-dependence of G′ for calcium pectinate alone has the form typical (Clark and Ross-Murphy, 1985) of gelling biopolymers, with log G′ versus log c

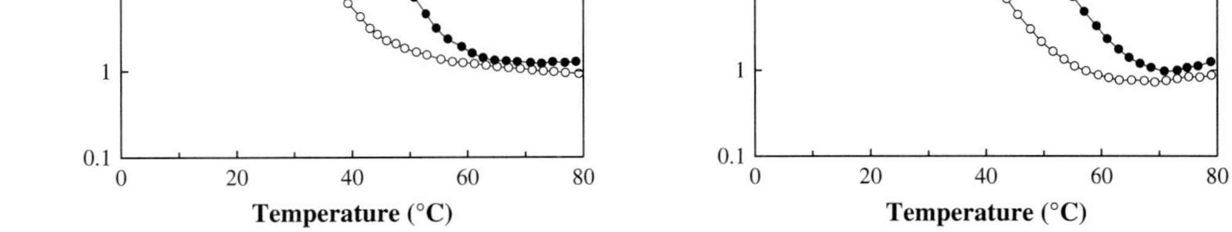

FIGURE 5.4 Variation of storage modulus, G′ (10 rad s^{-1}; 2% strain) on cooling (O) and heating (●) at 1°C/min for 1.0 wt% LM pectin (DE 31.1; pH 3.9) with stoichiometric Ca^{2+}, (a) in the absence of gelatin, and (b) in the presence of 3.0 wt% type B gelatin. Samples were held for 100 min at 5°C between the end of the cooling scan and the start of the heating scan (Gilsenan et al., 2003b).

approaching a limiting, minimum slope of 2 (c^2-dependence) at high concentrations. Incorporation of 1 M NaCl had no significant effect on G′. The modulus–concentration curve in Figure 5.5 was fitted to a third-order polynomial function, which was then used in the blending law analyses described in Section 5.4.3, to calculate G′ for any concentration of calcium pectinate within the range of the experimental calibration. Values of G′ for gelatin in the absence of pectin were expressed as a continuous function of concentration by the same procedure.

5.4.2 Phase Structure

As illustrated in Figure 5.4b, cooling and heating curves recorded for calcium pectinate (1.0 wt% with no added NaCl) in mixtures with gelatin (0.5–10.0 wt%) were similar to those observed (Figure 5.4a) for calcium pectinate alone at temperatures above ~25°C. At low temperatures, however, gelation of gelatin can be seen as a second wave of increase in G′ on cooling, with a corresponding reduction on heating, before dissociation of the calcium pectinate network.

Survival of gel structure after melting of the gelatin network demonstrates that calcium pectinate must be present as a continuous phase in the co-gels. However, as demonstrated by Eqs. (5.7) and (5.8), and the accompanying discussion in Section 5.2, the changes in modulus (Figure 5.4b) over the temperature-range of formation and dissociation of network structure by the gelatin component of the co-gels are much greater than could arise from a dispersed gelatin phase. The obvious conclusion is that the co-gels formed by gelatin and calcium pectinate in water are bicontinuous.

On progressive incorporation of NaCl (Figure 5.6) there was a progressive reduction in modulus of the network remaining after melting of gelatin, until, when the concentration of NaCl reached 1 M, decrease in G′ on heating was confined to the temperature-range of the gel–sol

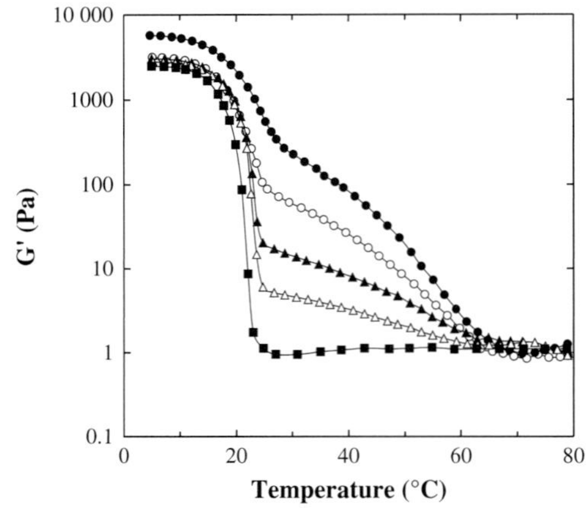

FIGURE 5.6 Variation of G′ (10 rad s^{-1}; 2% strain) during heating (1°C/min after 100 min at 5°C) for mixtures of 3.0 wt% gelatin and 1.0 wt% LM pectin with stoichiometric Ca^{2+} in the absence of NaCl (●), and at NaCl concentrations of 0.3 M (○), 0.5 M (▲), 0.75 M (Δ) and 1.0 M (■) (Gilsenan et al., 2003b).

transition of gelatin alone. This cannot be explained by loss of calcium pectinate structure at high concentrations of NaCl, since the moduli recorded for calcium pectinate alone (Figure 5.5) were unaffected by incorporation of 1 M NaCl.

The likely interpretation is again obvious. In the absence of NaCl, the mixtures of gelatin with pectin in the solution state at 85°C remain single phase, giving rise to bicontinuous network structure on cooling. However, as discussed in Section 5.3.1, and illustrated schematically in Figure 5.3, addition of salt promotes phase separation (by reducing the entropy penalty of counterions being confined to individual phases), with the extent of segregation increasing progressively as the concentration of added salt is increased. The salt-dependent changes shown in Figure 5.6 can then be explained by progressive resolution of the pre-gel solutions into a gelatin-rich continuous phase and pectin-rich dispersed phase, with the concentration of pectin in the continuous phase (which, as discussed in Section 5.2, dominates the overall rheology) decreasing progressively (Figure 5.3) with increasing concentration of NaCl.

In the blending-law analyses described in the following section, the co-gels formed in 1 M NaCl were therefore regarded as consisting of a dispersed phase of calcium pectinate in a continuous gelatin matrix, and those formed in water as being bicontinuous.

5.4.3 Analysis of Co-Gel Moduli by Polymer Blending Laws

The p-factor characterizing the distribution of solvent between the constituent phases of the gelatin–calcium pectinate co-gels was defined as the ratio of water/polymer in the gelatin phase divided by the corresponding ratio for the calcium pectinate phase:

$$p = (\text{water/gelatin})/(\text{water/pectin}) \quad (5.9)$$

The experimental values of G' (measured after 100 min at 5°C) for co-gels of 1.0 wt% calcium pectinate with 0.5–10 wt% gelatin in 1 M NaCl (gelatin continuous; calcium pectinate dispersed) were analyzed first.

The only adjustable parameter in the analysis was p, which was varied systematically, using the 'Solver' function in Microsoft Excel, to minimize the root-mean-square difference between observed values of log G' and values calculated by application of the Takayanagi blending laws (Eqs. (5.2) and (5.3)). For each trial value of p, the concentration of polymer in each phase was calculated from the amounts of gelatin and pectin used in preparing the mixtures and the distribution of water between the phases (Eq. (5.9)), which also gives the phase volumes. The moduli of the individual phases were then calculated using the polynomial fit to the calibration curve of G' versus concentration for calcium pectinate alone (Figure 5.5) and the corresponding polynomial for gelatin.

If the modulus derived in this way for the continuous gelatin phase was higher than that of the dispersed phase of calcium pectinate, the calculated value of co-gel modulus was derived

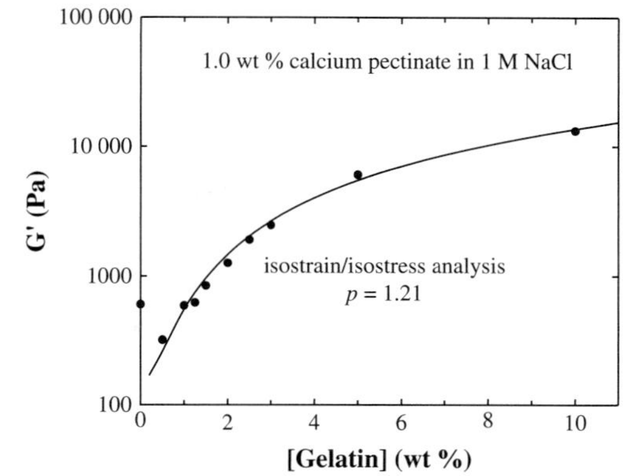

FIGURE 5.7 Observed (●) and calculated (—) moduli for co-gels of 1.0 wt % calcium pectinate (stoichiometric Ca^{2+}; 1 M NaCl) with varying concentrations of gelatin (Gilsenan et al., 2003c).

using the isostrain blending law (Eq. (5.2)), and the isostress blending law (Eq. (5.3)) was used if it was lower. Best agreement between observed and calculated values of log G' was obtained at $p = 1.21$; the quality of fit is shown in Figure 5.7.

For comparison of observed and calculated moduli of the co-gels formed in water, where the cooling and heating traces shown in Figure 5.4b indicate bicontinuous network structure, p was held constant at the value of 1.21 derived for those formed in 1 M NaCl, and co-gel moduli were calculated by the Davies blending law (Eq. (5.1)). As shown in Figure 5.8, there is again good agreement between the calculated values of log G' and those obtained experimentally. It should be emphasized that the comparison shown in Figure 5.8 came directly from application of the relevant blending law, with no adjustable parameter(s).

The quality of agreement achieved (Figures 5.7 and 5.8) using a single value of p to characterize co-gels comprising the same two polymers, but with very different topologies, suggests that the underlying assumptions of complete segregation and distribution of solvent according to a 'p-factor' relationship (Eq. (5.9)) are reasonably valid. As described in the

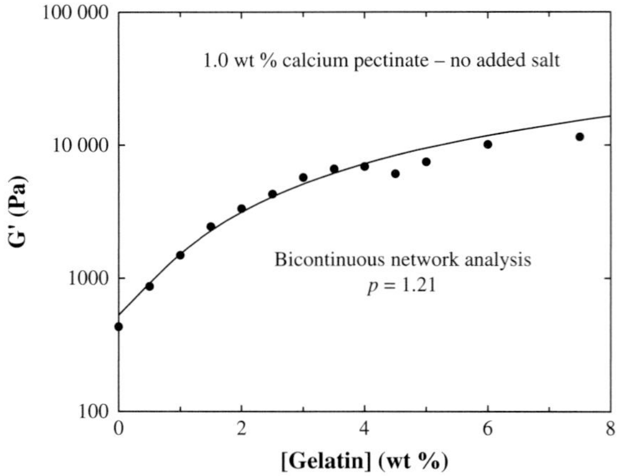

FIGURE 5.8 Experimental values of G′ (10 rad s^{-1}; 2% strain; 100 min at 5°C) for co-gels of gelatin with 1.0 wt% calcium pectinate in the absence of NaCl (●), in comparison with values calculated (Eq. (5.1)) for a bicontinuous network (—) (Gilsenan et al., 2003c).

following section, an even better standard of agreement between observed moduli and values calculated by an analysis incorporating the same assumptions has been achieved recently (Fitzsimons et al., 2008b) for co-gels formed by heating whey protein isolate (WPI) in mixtures with crosslinked starch.

5.5 CO-GELATION OF WHEY PROTEIN ISOLATE (WPI) WITH CROSSLINKED STARCH

5.5.1 Experimental Procedure

The WIP used was 'Bipro' from Davisco. The starch was of the same type as that used in the agarose–starch composites described in Section 5.2: phosphate-crosslinked hydroxypropylated waxy maize starch, which, for brevity, will be denoted as PCS (phosphate-crosslinked starch). Slurries of PCS and solutions of WPI (both adjusted to pH 7.0) were prepared at twice the concentration required in the final samples, and were then mixed in equal weights. Composite gels were prepared by heating to 80°C at 1°C/min, holding for 30 min, cooling (1°C/min) to 20°C and holding for 30 min. Thermogelation of globular proteins, including those in whey, is a two-stage process. The first stage is partial unfolding (denaturation) of the globule structure; the second is association (aggregation) of the denatured protein molecules to form a continuous network (Clark and Lee-Tuffnell, 1986; Mulvihill and Donovan, 1987). At neutral pH (as in the investigation summarized here) aggregation is inhibited by intermolecular electrostatic repulsion, but the repulsion can be screened by salt, allowing gelation to occur. 100 mM NaCl was found to be sufficient to induce gelation of WPI at concentrations down to 1.0 wt%, and was incorporated in all the samples studied. The swelling volume (Q) of PCS in 100 mM NaCl was determined by the polymer-exclusion method (Nayouf et al., 2003) using blue dextran, and the value obtained was Q ≈ 11.9 ml/g.

Studies by differential scanning calorimetry (DSC) showed no evidence of any association between WPI and PCS: as illustrated in Figure 5.9, the mixtures gave heating traces that

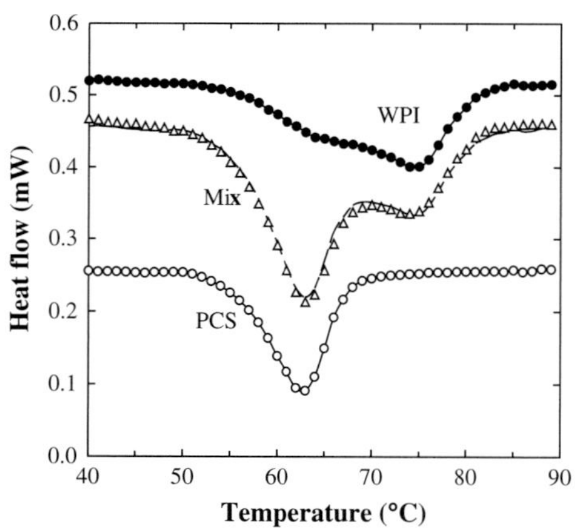

FIGURE 5.9 DSC heating curves (0.4°C/min) for 2 wt% PCS alone (O), 2 wt% WPI alone (●) and a mixture of both (Δ). The dashed line shows the curve obtained by addition of the individual traces for PCS and WPI. For clarity of presentation, the traces have been displaced vertically by arbitrary amounts (Fitzsimons et al., 2008b).

agreed well with a simple linear combination of the gelatinization endotherm for PCS and the denaturation endotherm for WPI. Gelatinization of PCS starts (Figure 5.9) at a temperature just above 50°C. To minimize sedimentation of ungelatinized starch granules during rheological characterization, samples were loaded onto an oscillatory rheometer at 50°C, and heating commenced immediately.

5.5.2 Starch Networks

Figure 5.10 shows illustrative plots of G′ versus time for 4 wt% WPI (in 100 mM NaCl), alone and in the presence of 3 wt% PCS, during heating (1°C/min) from 50 to 80°C and holding at 80°C. Thermogelation of WPI alone can be seen as a sharp increase in G′ which begins towards the end of heating and continues into the holding period at 80°C. In the mixture with PCS, however, this is preceded by an initial increase over the temperature-range of PCS gelatinization (Figure 5.9). Similar sharp increases in G′ during heating were observed for PCS alone, with little further change during holding at 80°C, cooling, and holding at 20°C.

Final values of G′ at 20°C for PCS alone are shown in Figure 5.11, plotted double-logarithmically against concentration (c). The corresponding plot for WPI alone is also shown, and has the form typical (Clark and Ross-Murphy, 1985) of gelling biopolymers, with the slope of log G′ versus log c approaching a limiting value of 2 at high concentration, and increasing progressively as the concentration is decreased towards the minimum value needed to give a continuous network. The concentration-dependence of modulus for PCS is broadly similar in form, and in particular shows no evidence of a discontinuity at the concentration where fully swollen granules would occupy the entire volume (cQ = 1), or at the close-packing limit (cQ ≈ 0.65).

Figure 5.12 shows mechanical spectra recorded (at 20°C) for PCS (in 100 mM NaCl) at a concentration (7 wt%) between close packing and complete space-occupancy, and at a much

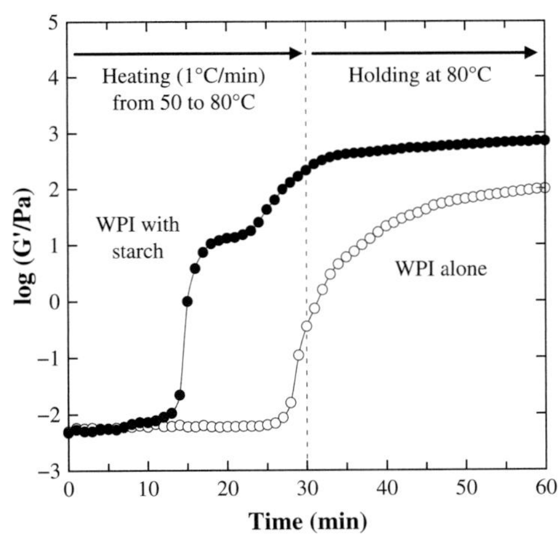

FIGURE 5.10 Changes in G′ (1 rad s^{-1}; 0.5% strain) observed for 4.0 wt% WPI in 100 mM NaCl on heating from 50°C to 80°C at 1°C/min and holding for 30 min at 80°C, alone (O) and in a mixture with 3.0 wt% PCS (●) (Fitzsimons et al., 2008b).

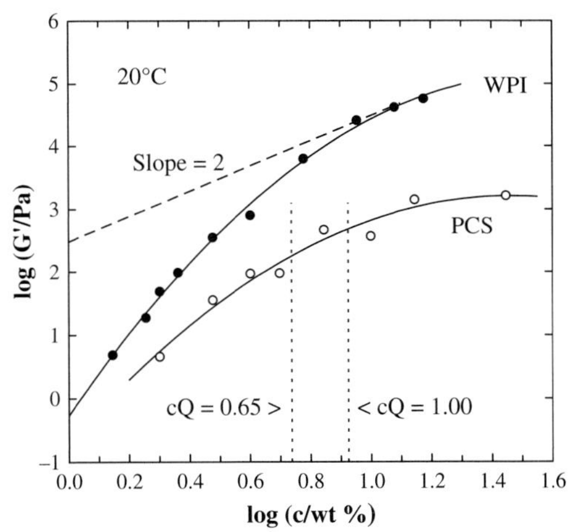

FIGURE 5.11 Variation of log G′ (1 rad s^{-1}; 0.5% strain; 20°C) with the logarithm of polymer concentration (c) for WPI (●) at concentrations of 1.4, 1.8, 2.0, 2.3, 3, 4, 6, 9, 12, and 15 wt% and for PCS (O) at concentrations of 2, 3, 4, 5, 7, 10, 14 and 28 wt%. The solid lines show second-order polynomial fits used in blending law analyses (Fitzsimons et al., 2008b).

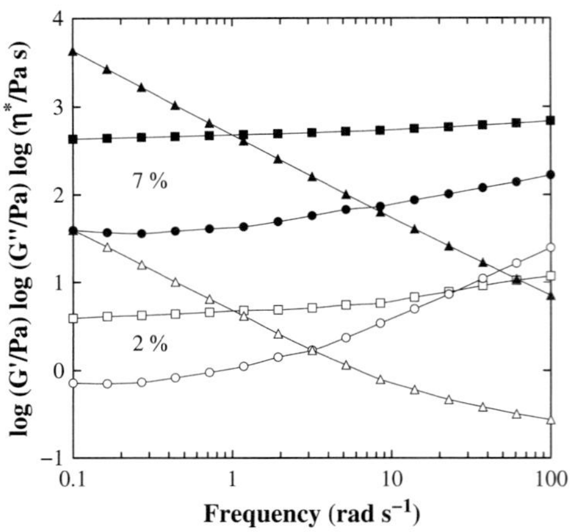

FIGURE 5.12 Mechanical spectra showing the frequency-dependence of G′ (squares), G″ (circles) and η* (triangles), measured at 0.5% strain and 20°C, for gelatinized PCS (in 100 mM NaCl) at concentrations of 2 wt% (open symbols) and 7 wt% (filled symbols) (Fitzsimons et al., 2008b).

lower concentration (2 wt%). The spectrum for 7 wt% PCS has the form typical (Ross-Murphy, 1984) of a strong gel (G′ > G″; little variation of either modulus with frequency, ω; linear dependence of log η* on log ω, with a slope close to −1). The spectrum for 2 wt% PCS also has obvious gel-like character, although at this concentration individual gelatinized granules would be widely separated from one another.

Gelation cannot be attributed to formation of an amylose network since PCS, like all waxy starches, is essentially devoid of amylose. The obvious interpretation (Hasan et al., 2008) is that continuous networks are formed by adhesive interactions (i.e. by gelatinized granules sticking to one another). Network formation at concentrations far below those required for close packing was also observed in a recent investigation of wheat starch (Hasan, 2006), but only when gelatinization was carried out under non-disruptive conditions (low-amplitude oscillation): gelatinization under shear (in making rotational measurements of viscosity) gave only the behavior expected for a dispersion of individual swollen granules. Oscillatory measurements after shear showed no indication of recovery of network structure, demonstrating that disruption of the associations that can form between starch granules on gelatinization under quiescent conditions is irreversible (analogous to the irreversible difference between stirred and set-style yogurts).

With hindsight, a critical factor in the success of the analysis of agarose–PCS composites described in Section 5.2 was that the starch was gelatinized by stirring in a hot solution of agarose, giving the dispersion of individual swollen granules that was assumed in the analysis. Preliminary attempts (Fitzsimons et al., 2008b) to apply the same analysis to the WPI–PCS composites (which were formed under non-destructive conditions of low-amplitude oscillation, to avoid disruption of WPI gel structure) were totally unsuccessful. However, as described below, good agreement between observed and calculated moduli was achieved by treating PCS as a second gelling biopolymer, rather than as a dispersion of separate gelatinized granules.

5.5.3 Phase Structure

As the concentration of WPI in the mixtures was increased above 2 wt% the increase in G′ accompanying gelatinization of PCS became progressively smaller (Figure 5.13), until at WPI concentrations of 8 wt% and above it was no longer detectable (Figure 5.13a). This can be explained by phase separation in the initial mixtures, with PCS being confined to a dispersed phase where, as discussed in Section 5.2, it would make little direct contribution to overall moduli. At the lowest concentrations of WPI studied (1.0 and 2.0 wt%), the increase in G′ on gelatinization of PCS was clearly evident (Figure 5.13b), but the subsequent further increase on gelation of WPI was too large to be explained by a dispersed phase of WPI in a continuous network of gelled PCS. As a working postulate it was proposed that the

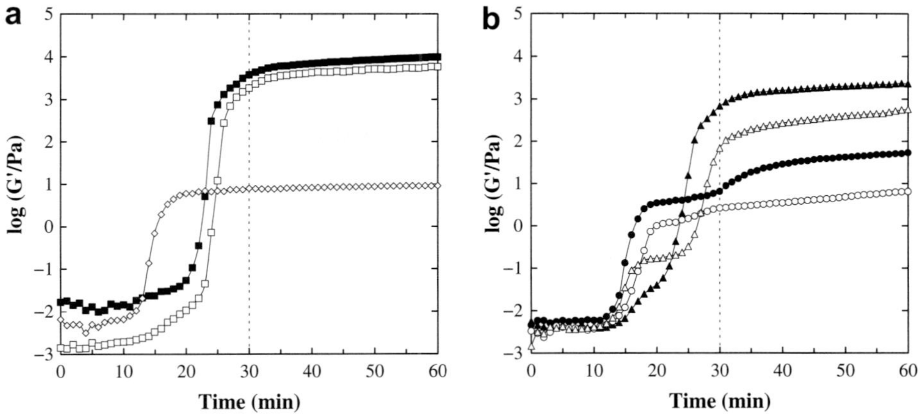

FIGURE 5.13 Changes in G′ (1 rad s^{-1}; 0.5% strain) during heating from 50 to 80°C at 1°C/min and holding for 30 min at 80°C for 2 wt% PCS (a) alone (◇) and in mixtures with WPI at concentrations of 8 wt% (□) and 10 wt% (■), and (b) with WPI at concentrations (wt%) of 1 (O), 2 (●), 4 (Δ) and 6 (▲) (Fitzsimons et al., 2008b).

PCS network formed initially was fragmented by subsequent gelation of WPI, and indeed fragments of gelled starch in a continuous protein matrix were visualized directly by light microscopy. In the blending law analysis described in the following section the composite gels were therefore treated as comprising a continuous phase of gelled WPI surrounding a dispersed phase of gelled PCS.

5.5.4 Blending Law Analysis

As in the analysis of gelatin–calcium pectinate co-gels described in Section 5.4.3, the polymers were assumed to segregate completely into their respective phases, and partition of solvent between the phases was characterized by a *p*-factor, which was defined as the ratio of water/polymer in the continuous protein phase divided by the corresponding ratio for the dispersed starch phase:

$$p = (\text{water/WPI})/(\text{water/PCS}) \qquad (5.10)$$

The calibration curves (Figure 5.11) of modulus versus concentration for WPI and PCS alone were fitted to (second order) polynomial functions to allow the polymer concentrations calculated for the individual phases at each trial value of p to be converted to the corresponding values of G′. Calculated values of G′ for the composite gels were derived using the isostrain blending law (Eq. (5.2)) if the modulus obtained for the continuous protein phase was higher than that of the dispersed starch phase, and the isostress relationship (Eq. (5.3)) if it was lower. Best agreement with experimental values of G′ for the composite gels was obtained at $p = 3.3$.

The concentration of PCS in the mixtures was varied across the range 1–5 wt% and the WPI concentrations from 1–10 wt%, giving experimental moduli spanning four orders of magnitude. Throughout this 10 000-fold range, the moduli calculated for the optimum value of p (3.3) remained close (Figure 5.14) to the experimental values, with maximum deviations comparable to the experimental error of the measurements. This almost perfect standard of agreement ($r^2 = 0.994$) between observed and calculated values of G′, obtained with a single adjustable parameter (p), strongly indicates that the assumptions on which the analysis was based (complete segregation; characterization of

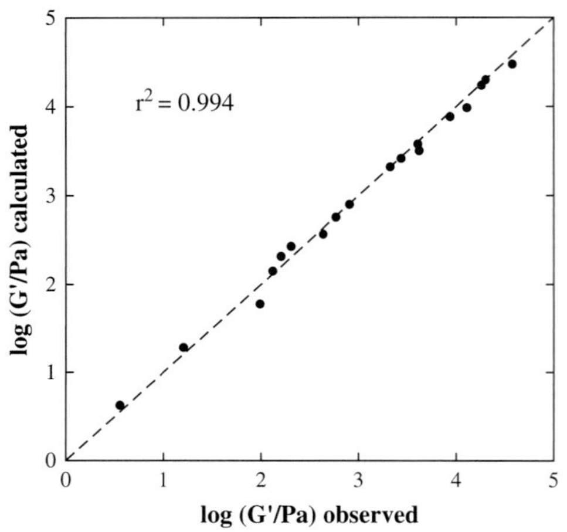

FIGURE 5.14 Comparison of observed values of log G′ (20°C) for gelled mixtures of WPI and PCS with the best fit ($p \approx 3.3$) obtained from the analysis (Section 5.5.4) based on composites of gelled PCS dispersed through a continuous matrix of gelled WPI. The dashed line corresponds to perfect agreement between observed and fitted moduli (Fitzsimons et al., 2008b).

solvent partition by a *p*-factor) cannot be seriously wrong.

5.6 ASSOCIATIVE INTERACTIONS

As outlined in Section 5.1.2, associative interactions between two different biopolymers can occur either by electrostatic attraction between polyanions and polycations or by formation of ordered junctions. An illustrative example of each is given below.

5.6.1 Electrostatic Association in Mixtures of Gelatin with Pectin

Section 5.4 described the formation of biphasic co-gels by Ca^{2+}-induced gelation of LM pectin (degree of esterification ≈ 31%) in the presence of type B gelatin (pI ≈ 4.9), and analysis of co-gel moduli by polymer blending laws (Gilsenan et al., 2003b, 2003c). The purpose of the investigation summarized here (Gilsenan et al., 2003a) was to explore the behavior of mixtures of the same two polymers in the absence of Ca^{2+}, one specific aim being to identify conditions under which the effect of segregative interactions in co-gelation could be studied without interference from electrostatic association.

The first step was to obtain titration curves characterizing the charge on each polymer as a function of pH. These are shown in Figure 5.15. The maximum charge on the gelatin at low pH is 0.88 mmol/g, passing through zero at pH 4.9 (the isoelectric point), and the maximum charge on the pectin is −3.13 mmol/g, dropping towards zero at low pH. From these curves, the ratio of charge on the two polymers could be determined for any combination of pectin and gelatin concentrations at any pH within the range of the titrations.

When mixtures were prepared at around charge balance (which for equal concentrations of gelatin and pectin occurs at ~pH 2.75) dense flocs were formed, as observed in earlier studies of mixtures of gelatin with pectin (Tshumak et al., 1976) or alginate (Muchin et al., 1976). These obviously arise from formation of an electrostatic complex by attraction between negatively

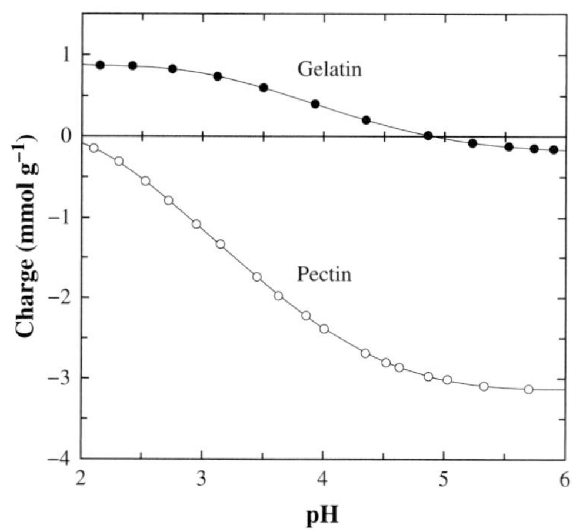

FIGURE 5.15 Titration curves showing the pH-dependence of charge for low-methoxy pectin (O) and type B gelatin (●) (Gilsenan et al., 2003a).

charged pectin and positively charged gelatin, with little net charge to promote solubility. For compositions well away from charge balance, however, the mixtures remained soluble, and could be characterized rheologically.

As an illustrative example, Figure 5.16 shows the changes in G′ observed for 2 wt% pectin, 2 wt% gelatin, and a mixture of both, on cooling (1°C/min) at pH 3.0. Gelation of gelatin alone can be seen as a sharp increase in modulus at temperatures below ~20°C. For pectin alone, there is a slight sigmoidal increase in G′ between ~45 and ~25°C. Larger increases were observed for higher concentrations of pectin at the same pH (3.0) and arise (Gilsenan et al., 2000) from formation of a pectinic acid gel. For the mixture of 2 wt% pectin with 2 wt% gelatin, the increase in modulus is much greater than for pectin alone, suggesting that the pectinic acid network is augmented by electrostatic coupling to gelatin, and is followed by a second increase which can be attributed to reinforcement of the coupled network by gelation of gelatin.

Formation of a hydrated gel rather than an insoluble floc can be explained by the difference in (absolute) charge between the two polymers. As shown in Figure 5.15, the negative charge density of pectin at pH 3.0 is substantially greater than the positive charge density of gelatin, so for a mixture where the two are present in equal concentrations (as in Figure 5.16) electrostatic association will give a complex solubilized by the surplus negative charge on the pectin. Hydrated networks were also obtained when the positive charge from the gelatin was substantially greater than the negative charge from the pectin, with solubilization then arising from the surplus positive charge.

As illustrated in Figure 5.17, dissociation of the electrostatic complex between gelatin and pectin at pH 3.0 could be seen in DSC heating scans as a second endotherm following the main endotherm observed over the temperature range of the gelatin gel–sol transition. On incorporation of salt (0.1 M NaCl) in the mixtures, however, the second endotherm was abolished, and there was a large decrease in the intensity of the main endotherm, with the residual peak being closely similar to that observed for the same concentration of gelatin alone. Loss of the second endotherm demonstrates that the electrostatic complex is dissociated by salt; reduction

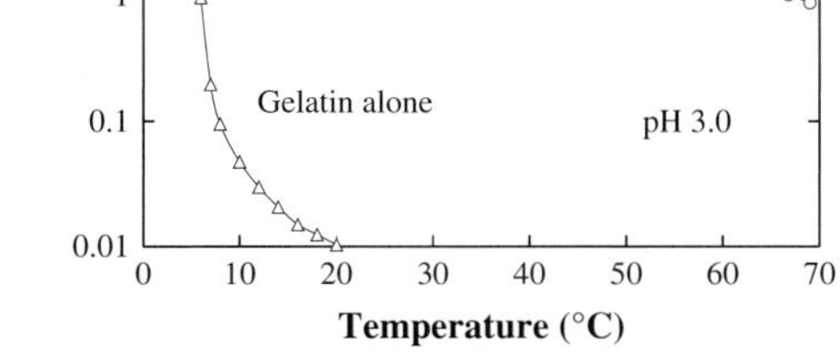

FIGURE 5.16 Changes in G′ (10 rad s^{-1}; 2% strain) during cooling at 1°C/min, for 2.0 wt% gelatin (Δ), 2.0 wt% pectin (O) and a mixture of both (●) at pH 3.0 (Gilsenan et al., 2003a).

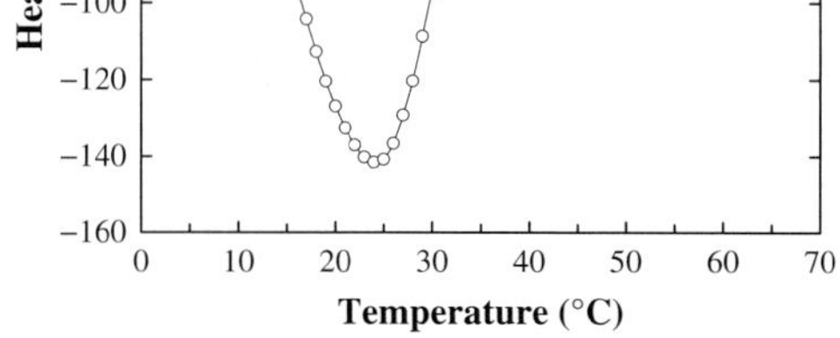

FIGURE 5.17 DSC heating scans (0.1°C/min, after 100 min at 5°C) for a mixture of 2.0 wt% gelatin with 1.0 wt% pectin at pH 3.0, in the presence (●) and absence (O) of 0.1 M NaCl (Gilsenan et al., 2003a).

in the main endotherm indicates that the ordered (triple helix) structure of gelatin is stabilized by electrostatic associations with pectin, and that these are also abolished by salt. An important general principle illustrated by these observations is that addition of salt inhibits associative interactions between polymers of opposite charge, by screening electrostatic attraction, but promotes segregation, by swamping the contribution of counterions to overall entropy (Section 5.3.1).

For gelatin–pectin mixtures prepared at pH 3.0 (but away from the composition range around charge balance where insoluble flocs were formed) the two endothermic process seen in DSC heating scans (Figure 5.17) were accompanied by two waves of reduction in turbidity (Figure 5.18). When the pH was raised to 3.9, however, the second process was abolished, leaving only the reduction in turbidity over the temperature range of the order–disorder transition of gelatin. The second endotherm in DSC also became undetectable at around the same pH. The conclusion drawn (Gilsenan et al., 2003a) was that gelatin does not develop sufficient positive charge for electrostatic association with pectin until it reaches a pH substantially below the isoelectric point of ~4.9, and that the range over which the two polymers remain unassociated extends (Figure 5.18) to at least pH 3.9, which was the value used in the studies of segregated co-gels described in Section 5.4.

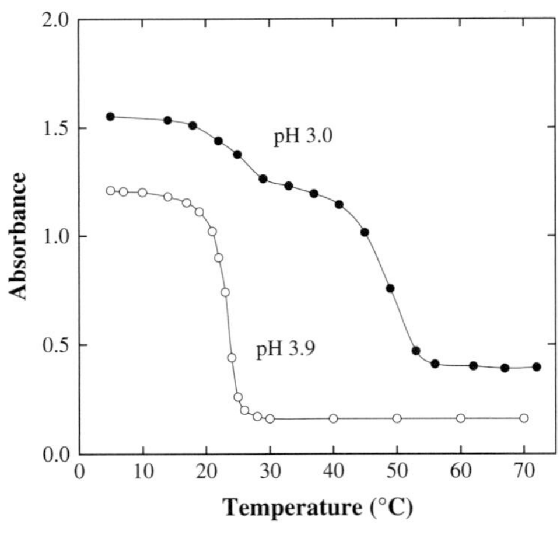

FIGURE 5.18 Temperature-dependence of turbidity, characterized by measurements of absorbance (400 nm; 1 cm pathlength) during heating for type B gelatin in combination with low-methoxy pectin at pH 3.0 (●) and at pH 3.9 (O) (Gilsenan et al., 2003a).

5.6.2 Ordered Association of Xanthan with Konjac Glucomannan (KGM)

Xanthan has a polymer backbone identical to cellulose, but is solubilized by charged trisaccharide sidechains attached at O(3) of alternate glucose residues. The sidechains have the structure: β-D-Man*p*-(1 → 4)-β-D-GlcA*p*-(1 → 2)-α-D-Man*p*-(1 →) with variable, non-stoichiometric substitution by *O*-acetate at C(6) of the inner mannose and 4,6-linked pyruvate ketal groups on terminal mannose residues. At high temperature and low ionic strength xanthan exists in solution as disordered coils, but undergoes a co-operative transition to a 5-fold helix structure on cooling and/or addition of salt. The transition is fully reversible, with no detectable thermal hysteresis. Solutions of ordered xanthan, although pourable, give gel-like mechanical spectra, and are termed 'weak gels' (Ross-Murphy, 1984).

However, xanthan forms true, cohesive gels when cooled in the presence of konjac glucomannan (KGM) or galactomannans with a low content of (single-sugar) galactose sidechains, such as locust bean gum (LBG). Like xanthan, these plant polysaccharides have linear (1 → 4)-diequatorially linked backbones. In the galactomannans the backbone consists entirely of D-mannose residues and is solubilized by the galactose sidechains; in KGM it consists of an irregular sequence of D-mannose and D-glucose residues and is solubilized by *O*-acetyl substituents and some limited branching.

There is compelling evidence that the 'synergistic' gelation (Morris, 1995) occurs by formation of ordered junctions between xanthan and the plant polysaccharide to give a coupled network, rather than by a mechanism based on phase separation.

1. Partially depolymerized LBG has been shown to co-elute with xanthan in gel permeation chromatography of mixed solutions (Cheetham et al., 1986).
2. Strong interactions can be detected by solution rheology at very low polymer concentrations, where phase separation is vanishingly unlikely (Goycoolea et al., 1995a).
3. Maximum interaction, as characterized by DSC, occurs (Goycoolea et al., 1995b) at a specific stoichiometric ratio of the two polymers, irrespective of their absolute concentrations, whereas phase separation should become increasingly apparent (Figure 5.3) as the concentration of either or both polysaccharides is raised.

An early proposal was that coupled networks are formed by attachment of mannan (Morris et al., 1977) or glucomannan (Dea et al., 1977) chains to the surface of the xanthan 5-fold helix. Later, however, it was concluded (Brownsey et al., 1988; Cairns et al., 1986, 1987; Morris, 1992) that attachment is to the cellulosic backbone of the xanthan molecule. This interpretation was based mainly on X-ray fiber diffraction, but the absence of a cohesive network in samples mixed at room temperature was adduced as a second line of evidence, with the conclusion that association can occur only when the xanthan is in its disordered form. The mechanism of synergistic gelation has been explored further in two recent investigations of mixtures of xanthan with KGM (Agoub et al., 2007; Fitzsimons et al., 2008a). Some of the main findings from these investigations are summarized below.

In the study by Fitzsimons et al. (2008a) solutions of xanthan and KGM were prepared separately and mixed at room temperature (20°C), and the rheology of the mixtures was characterized by low-amplitude oscillatory measurements and by creep–recovery experiments. Rheological changes on heating to a temperature where the xanthan component was converted to its disordered form, cooling, and re-heating were then monitored, along with the accompanying thermal transitions in DSC.

Figure 5.19 shows mechanical spectra recorded for a mixture of 0.5 wt% xanthan with 0.5

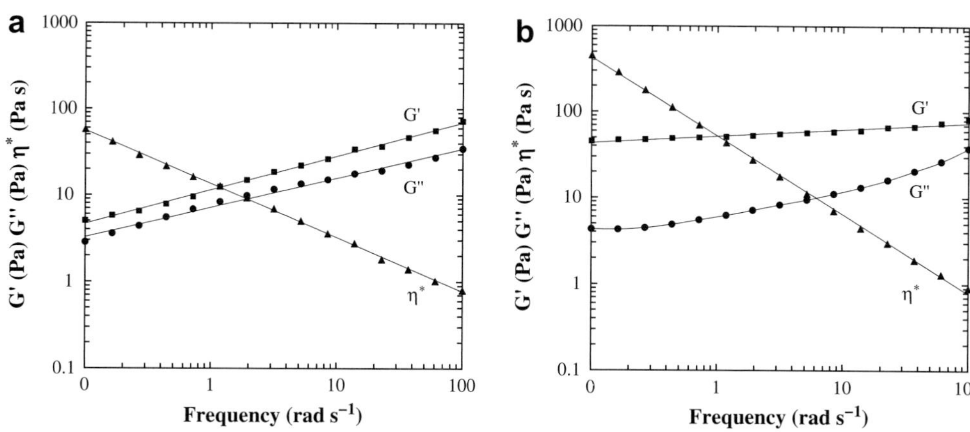

FIGURE 5.19 Mechanical spectra (20°C; 0.5% strain) showing the frequency-dependence of G′ (■), G″ (●) and η* (▲) for 0.5 wt% xanthan with 0.5 wt% KGM in 30 mM KCl, (a) after mixing at 20°C, and (b) after heating to 95°C and cooling again to 20°C (Fitzsimons et al., 2008a).

wt% KGM in 30 mM KCl (to enhance the stability of the xanthan helix) after mixing at 20°C and after heating to 95°C and cooling again to 20°C. As expected from numerous previous studies, the spectrum (Figure 5.19b) obtained on cooling from the solution state at high temperature is typical (Ross-Murphy, 1984) of a strong gel. The spectrum recorded immediately after mixing at 20°C (Figure 5.19a), however, also has obvious gel-like character. This cannot be attributed to the 'weak gel' properties of the xanthan component of the mixture, since the same concentration of xanthan in the absence of KGM gave a mechanical spectrum which, although also gel-like, had substantially lower moduli than those shown in Figure 5.19a. The creep–recovery curves obtained for the same (unheated) mixture (Figure 5.20) also showed obvious gel-like response, with sharp elastic recovery on removal of applied stress, and only slight irreversible deformation at the end of the recovery period.

The changes in G′ observed when the mixture was heated, cooled, and re-heated are shown in Figure 5.21 and the accompanying thermal transitions in DSC are shown in Figure 5.22, in direct comparison with those observed for the same concentration (0.5 wt%) of xanthan alone under the same ionic conditions (30 mM KCl). The DSC traces for xanthan alone show an exotherm at ~75°C on cooling and a corresponding endotherm on heating. These obviously arise from formation and melting of the 5-fold helix, with the slight difference in mid-point temperature between the two traces reflecting rate of heat transfer in the calorimeter ('thermal lag'), rather than genuine thermal hysteresis. The same conformational transitions are evident in the DSC traces for the mixture of xanthan with KGM, but are now accompanied by a second thermal process at lower temperature (exothermic on cooling; endothermic on heating). Comparison with Figure 5.21 shows that this arises from formation and melting of the synergistic gel.

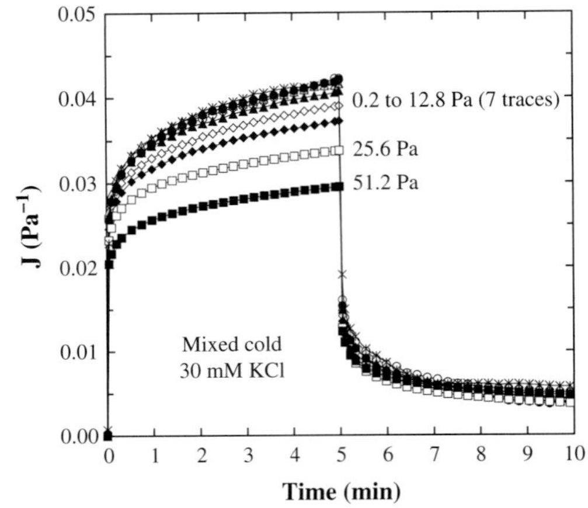

FIGURE 5.20 Creep–recovery curves recorded for 0.5 wt% xanthan with 0.5 wt% KGM in 30 mM KCl after mixing at 20°C. The curves show the variation of compliance (J) in response to applied stress values (Pa) of 0.2 (∗), 0.4 (O), 0.8 (●), 1.6 (Δ), 3.2 (▲), 6.4 (◇), 12.8 (◆), 25.6 (□) and 51.2 (■). Stress was held constant for 5 min, and recovery was monitored over a further period of 5 min after the stress was removed (Fitzsimons et al., 2008a).

Three features of the data in Figures 5.21 and 5.22 are of particular significance.

1. Conformational ordering of xanthan is complete well before the onset of synergistic

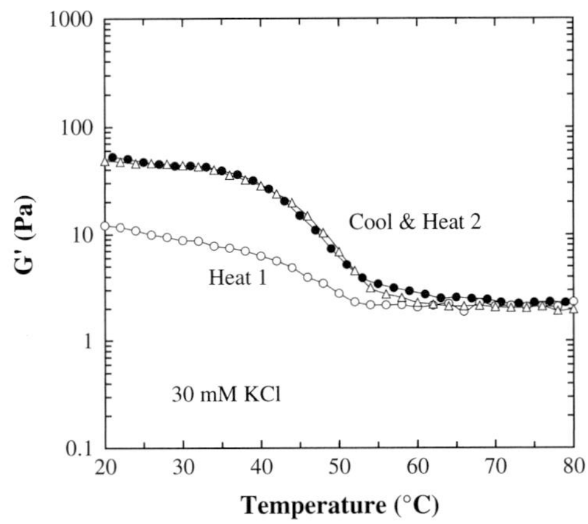

FIGURE 5.21 Changes in G′ (1 rad s^{1}; 0.5% strain) observed for 0.5 wt% xanthan with 0.5 wt% KGM in 30 mM KCl, on initial heating after mixing at 20°C (O), on cooling from 95°C (●) and on re-heating (Δ). The heating and cooling rate used was 1.0°C/min (Fitzsimons et al., 2008a).

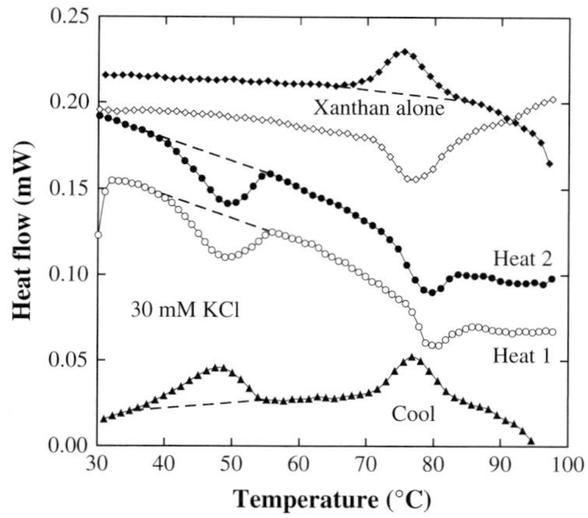

FIGURE 5.22 DSC traces recorded at 0.5°C/min for 0.5 wt% xanthan with 0.5 wt% KGM in 30 mM KCl on initial heating (O) after mixing at 20°C, on subsequent cooling (▲) back to 20°C, and on re-heating (●). The heating (◇) and cooling (◆) traces recorded for 0.5 wt% xanthan in 30 mM KCl in the absence of KGM are also shown. Dashed lines show baselines used in numerical integration of peak area and calculation of transition enthalpy. For clarity of presentation, the individual traces have been displaced vertically by arbitrary amounts (Fitzsimons et al., 2008a).

gelation, demonstrating that association of xanthan with KGM does not require the xanthan to be in its disordered form.

2. The DSC trace recorded on initial heating after mixing at 20°C (labeled 'Heat 1' in Figure 5.22) is virtually identical to the second heating scan ('Heat 2') for the cohesive gel formed on cooling.
3. There is a sigmoidal reduction in G′ when the initial mixture is heated, which follows the same temperature-course as the subsequent increase on cooling and decrease on heating, although the starting value is about four times lower than the modulus of the cohesive gel.

The proposed interpretation (Fitzsimons et al., 2008a) was that the same molecular interactions occur when xanthan and KGM are mixed at room temperature and when they are cooled together from the high-temperature solution state, but that the mixing process prevents formation of a continuous network, giving instead a dispersion of microgel particles with only 'weak gel' rheology, but with the same internal structure as the 'true' cohesive gels formed on quiescent cooling.

The investigation by Agoub et al. (2007) focused on the effect of acidic pH on the synergistic gelation of xanthan with KGM. A complication encountered in this research was loss of pyruvate ketal groups from xanthan on heating under acidic conditions. To overcome this problem, most of the experiments were carried out using pyruvate-free xanthan (PFX), obtained by genetic engineering. The findings summarized below are confined to PFX, alone and in combination with KGM. However, the loss of pyruvate from normal commercial xanthan on heating at low pH was sufficient to give essentially the same gel properties as PFX in cooled mixtures with KGM.

As shown in Figure 5.23, the order–disorder transition of PFX was displaced to progessively higher temperatures on reduction in pH, moving to above the experimentally accessible temperature range (i.e. to above 100°C) at values of pH below ~4.0. This can be explained, at least in part, by suppression of charge on the trisaccharide sidechains of the xanthan molecule (conversion of COO^- to COOH), with the consequent reduction in intramolecular electrostatic repulsion stabilizing the 5-fold helix. Synergistic gelation with KGM (Figure 5.24), however, was displaced in the opposite direction, moving to progressively lower temperature with decreasing pH.

As outlined above, two conflicting models have been proposed for interaction of KGM (or galactomannans) with xanthan: (i) attachment to the surface of the 5-fold helix, and (ii) binding to the cellulosic backbone of the xanthan molecule. These models came from research groups in, respectively, Unilever Research, Colworth Laboratory, Bedford, UK and the Institute of Food Research, Norwich, UK. In terms of the

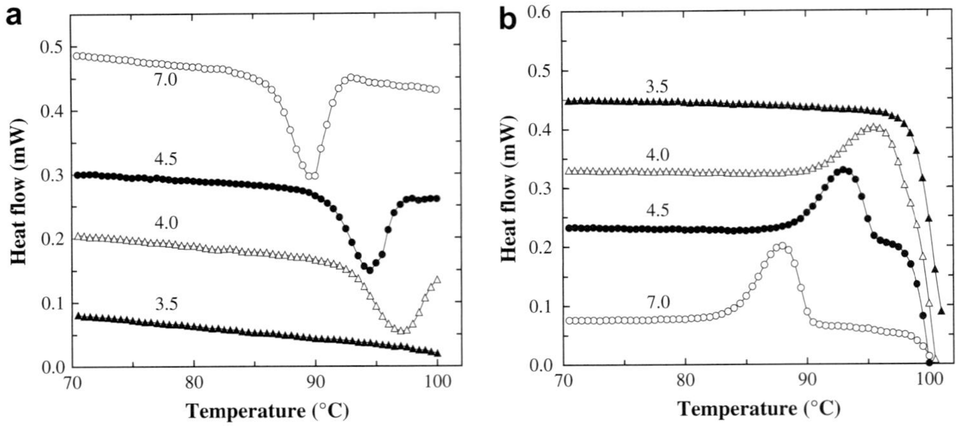

FIGURE 5.23 DSC curves obtained for 1.0 wt% PFX on (a) heating and (b) cooling (0.5°C/min) at pH values of 7.0 (O), 4.5 (●), 4.0 (Δ) and 3.5 (▲). For clarity of presentation, the individual traces have been displaced vertically by arbitrary amounts (Agoub et al., 2007).

'Unilever' model, it could be argued that interaction of xanthan with KGM occurs in competition with self-association of xanthan helices, and that self-association is promoted by reduction in electrostatic repulsion with decreasing pH. The 'Norwich' model, however, seems to provide a more direct and convincing explanation of the inverse correlation between thermal stability of the xanthan helix and the temperature at which gels are formed on cooling in the presence of KGM. As the stability of the helix increases, its resistance to unfolding to allow KGM chains to bind to the cellulosic backbone would also be expected to increase, which is entirely consistent with the changes observed (Figures 5.23 and 5.24) on varying pH. However, the huge separation between completion of conformational ordering of xanthan (Figure 5.23) and onset of synergistic gelation (Figure 5.24) on cooling at low pH reinforces the conclusion (Fitzsimons et al., 2008a) that the proposed requirement for xanthan to be disordered before association can occur is wrong, and is an unnecessary embellishment of the otherwise convincing 'Norwich' model.

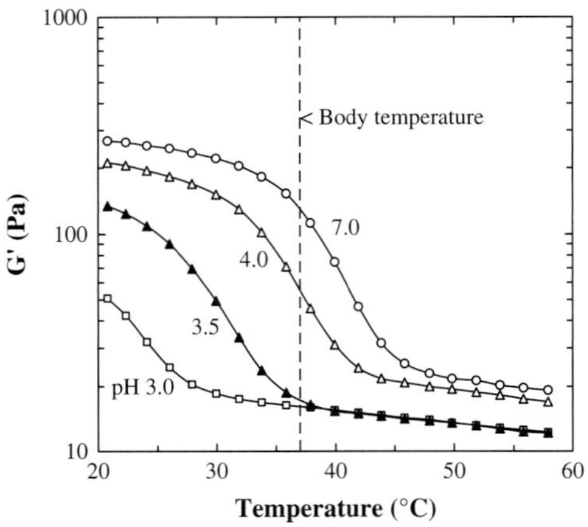

FIGURE 5.24 Temperature-dependent changes in G′ (1 rad s[1]; 1% strain) for 0.5 wt% PFX in combination with 0.5 wt% KGM at pH values of 7.0 (O), 4.0 (Δ), 3.5 (▲) and 3.0 (□) (Agoub et al., 2007).

Finally, it should be noted that softening or melting of xanthan–KGM gels at pH values in the range ~3.5 to ~4.0 occurs (Figure 5.24) at around body temperature, suggesting possible practical uses as a replacement for gelatin in applications where 'melt-in-the-mouth' properties are important for product quality, and where moderate acidity is acceptable or necessary (e.g. fruit jellies).

5.7 SEGREGATIVE INTERACTIONS IN SINGLE-PHASE MIXTURES

As illustrated in Sections 5.4 and 5.5, phase separation resulting from segregative interactions in mixtures of two different biopolymers can cause large enhancements in rheology, by confining each polymer to only part of the total volume and therefore increasing its effective local concentration. There is accumulating evidence, however, that segregative interactions in single-phase mixtures can also have huge effects.

In an early study by Kasapis et al. (1993), concentrated mixtures of gelatin with a gelling maltodextrin of comparatively low molecular weight (dextrose equivalent ≈ 6) were found to show classic phase separation when held at a temperature (45°C) above the onset of conformational ordering and gelation of either component. However, at concentrations below the binodal (Figure 5.3) massive precipitation of maltodextrin was observed, with the fraction precipitated increasing in direct proportion to the gelatin concentration, and reaching values as high as 60%. As outlined below, the ability of disordered polymers to drive self-association of a second polymeric material in single-phase mixtures has also been observed in several later investigations.

5.7.1 Effect of Soluble Biopolymers on Self-Association of Calcium Pectinate

Figure 5.25 shows changes in G′ (Figure 5.25a) and G" (Figure 5.25b) observed (Giannouli et al., 2004a) for 2 wt% LM pectin (degree of esterification = 31%) with stoichiometric Ca^{2+} on cooling (1°C/min) alone, and in the presence of 0.5–5.0 wt% guar gum of molecular weight ~540 kDa. The moduli recorded at high temperature (85°C) and on completion of cooling to 5°C are shown in Figure 5.26. It should be noted first that although the mixtures, prepared at 90°C, remained homogeneous and could be poured onto the (heated) plate of the rheometer, they already had gel-like character (G′ > G") before cooling, and that the strength of the initial 'weak gel' network increased with increasing concentration of guar gum. This increase, however, was

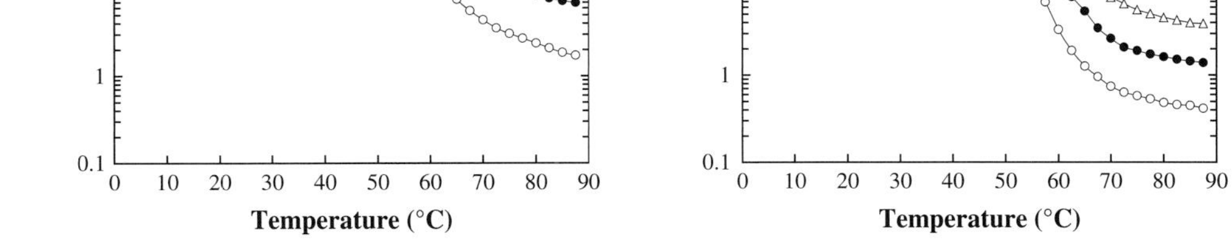

FIGURE 5.25 Changes in (a) G′ and (b) G" (10 rad s^{-1}; 0.5% strain) during cooling (1°C/min) from 90°C to 5°C for 2.0 wt% calcium pectinate alone (O) and in the presence of guar gum of molecular weight ≈ 540 kDa at concentrations (wt%) of 0.5 (●), 1.0 (Δ), 2.0 (▲), 3.0 (◇), 4.0 (◆) and 5.0 (□) (Giannouli et al., 2004a).

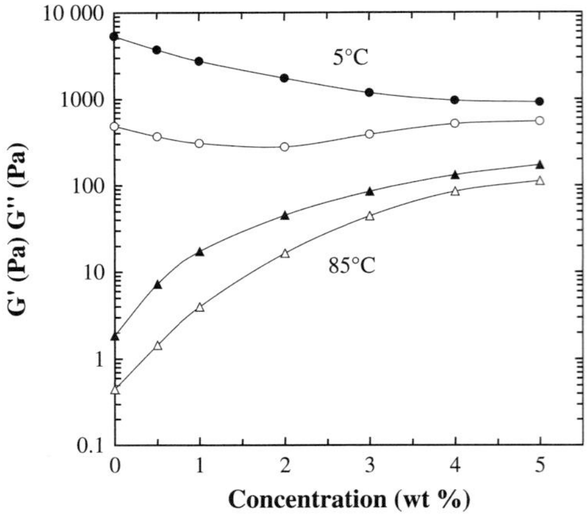

FIGURE 5.26 Measured values (10 rad s^{-1}; 0.5% strain) of G′ (filled symbols) and G″ (open symbols) at 85°C (triangles) and at 5°C (circles) for 2.0 wt% calcium pectinate in the presence of a range of concentrations of guar gum of molecular weight ≈ 540 kDa (Giannouli et al., 2004a).

accompanied by a progressive reduction in the final values of G′ at 5°C. Qualitatively similar behavior was observed (Giannouli et al., 2004b, 2004c) for single-phase mixtures of calcium pectinate with dextrans, inulin and gum arabic.

Similar behavior was also observed in earlier studies of calcium pectinate in mixtures with oxidized starch (Picout et al., 2000a) or potato maltodextrin (Picout et al., 2000b) and the following interpretation was proposed.

- The 'weak gel' structure formed at high temperature is crosslinked by thermally stable dimeric junctions (Morris et al., 1982) formed by 'egg-box' binding (Grant et al., 1973) of Ca^{2+} between runs of unesterified carboxyl groups along the inner face of each of the participating chains.
- The cohesive 'true' gel structure formed on cooling (Figure 5.25) involves weaker (thermally labile) association of dimers into larger assemblies that include some esterified residues (Powell et al., 1982).
- Self-association of calcium pectinate at high temperature is promoted by segregative interactions with the soluble polymeric constituent of the homogeneous (single-phase) mixture.
- Reduction in modulus at low temperature arises from excessive association of calcium pectinate in response to the same segregative interactions, giving large aggregated bundles that make little contribution to network crosslinking.

A particularly striking and unexpected feature of the investigation by Giannouli et al. (2004a) was that the effectiveness of guar gum in promoting self-association of calcium pectinate at high temperature was found to increase with decreasing molecular weight (as characterized by intrinsic viscosity, [η]). Indeed, as shown in Figure 5.27, a linear relationship was observed between 1/[η] and the values of log G′ recorded at 85°C for mixtures of 2 wt% calcium pectinate with 1 wt% guar gum at molecular weights between ~60 and ~1800 kDa. The reason the comparison was made at 1 wt% guar gum was that at higher concentrations the sample of lowest molecular weight (60 kDa) induced formation of cohesive gels which could not be loaded onto the rheometer. The effect of guar

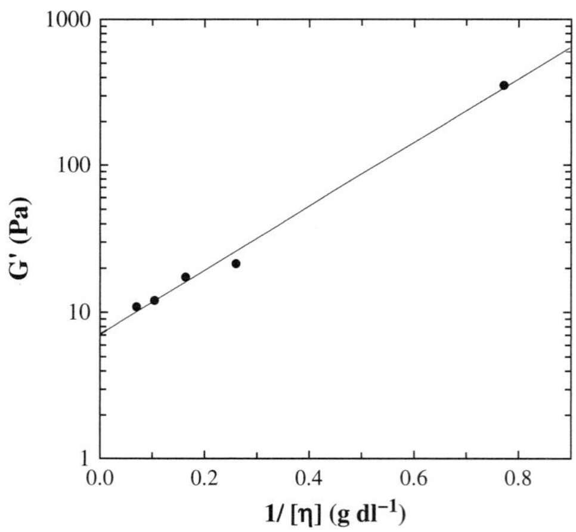

FIGURE 5.27 Variation of G′ at 85°C (●) with the reciprocal of intrinsic viscosity, 1/[η], for guar gums present at 1.0 wt% in single-phase mixtures with 2.0 wt% calcium pectinate (Giannouli et al., 2004a).

gum on an entirely different gelling system is described in the following section.

5.7.2 Effect of Guar Gum on Thermal Gelation of WPI

In a recent investigation by Fitzsimons et al. (2008d) incorporation of low concentrations of a standard commercial sample of guar gum in single-phase mixtures with WPI (3.0 wt% in 100 mM NaCl) was found to cause large changes in modulus of gels formed by heating to 80°C, holding for 30 min, and cooling to 20°C. Maximum enhancement was observed at 0.1 wt% guar gum, with progressive decrease in modulus at higher concentrations. The initial increase was attributed to enhanced self-association of thermally denatured WPI in response to segregative interactions with guar gum, and the subsequent decrease to excessive aggregation, analogous to the mechanism proposed (Section 5.7.1) for reduction in modulus of calcium pectinate. As shown in Figure 5.28, guar gum at the optimum concentration of 0.1 wt% increased the moduli of the WPI gel by more than a factor of 10.

In a continuation of this work, the effect of the molecular weight of the guar gum was explored (McCarthy et al., unpublished), using samples identical to those studied by Giannouli et al. (2004a). Optimum gelation of WPI (again at 3.0 wt% in 100 mM NaCl) from single-phase mixtures with these materials occurred at different concentrations of guar gum, but results obtained at a fixed concentration of 0.5 wt% will be used to provide a simple, direct comparison.

Figure 5.29 shows the changes in G′ observed during heating from 50 to 80°C at 1°C/min, and then holding for 30 min at 80°C, for 3.0 wt% WPI (in 100 mM NaCl), alone and in the presence of 0.5 wt% of an illustrative sample of guar gum (molecular weight ≈ 280 kDa). Enhanced self-association of WPI in the mixture with guar gum is clearly evident from the much earlier onset of increase in G′ and the much higher modulus reached by the end of the holding period at 80°C.

The final values of G′ at 80°C are shown in Figure 5.30, plotted against the reciprocal of the intrinsic viscosity of the guar gum used (as in Figure 5.27). There is again a linear relationship between log G′ and $1/[\eta]$, with the effectiveness

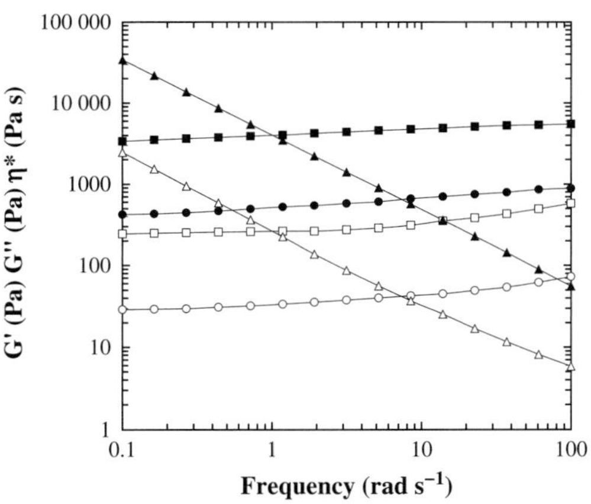

FIGURE 5.28 Mechanical spectra (20°C; 0.5% strain) showing the frequency-dependence of G′ (squares), G″ (circles) and η* (triangles) for gels formed by thermal denaturation of 3.0 wt% WPI (in 100 mM NaCl), alone (open symbols) and in the presence of 0.1 wt% guar gum (filled symbols) (Fitzsimons et al., 2008d).

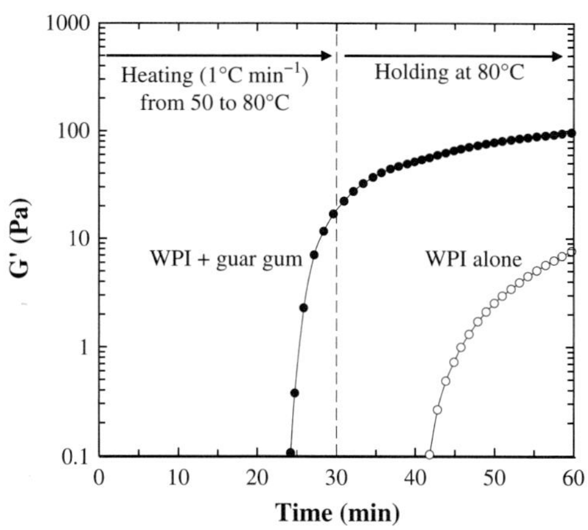

FIGURE 5.29 Changes in G′ (1 rad s^{-1}; 0.5% strain) observed for 3.0 wt% WPI in 100 mM NaCl on heating from 50°C to 80°C at 1°C/min and holding for 30 min at 80°C, alone (O) and in a mixture with 0.5 wt% guar gum of molecular weight ≈ 280 kDa (●) (McCarthy, Fitzsimons and Morris unpublished).

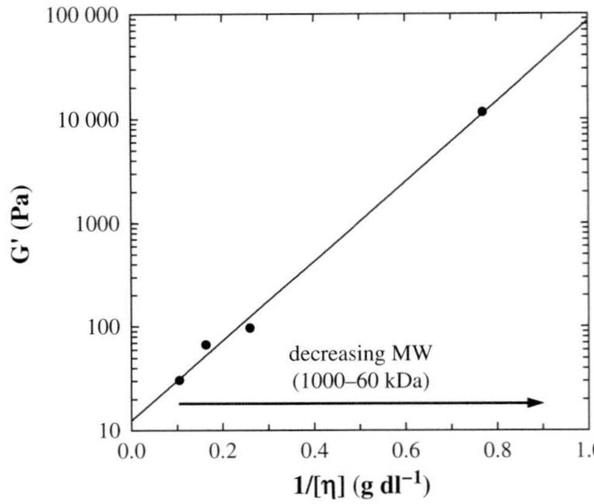

FIGURE 5.30 Variation of G′ at 80°C (●) with the reciprocal of intrinsic viscosity, 1/[η], for guar gums present at 0.5 wt% in mixtures with 3.0 wt% WPI in 100 mM NaCl (McCarthy, Fitzsimons and Morris unpublished).

of guar gum in promoting aggregation of denatured WPI increasing as the individual coils become smaller. The possible origin of this puzzling inverse correlation between molecular weight and segregative interactions in single-phase mixtures is discussed later (Section 5.8.1). The next section gives a brief summary of another system showing evidence of enhanced self-association in response to segregative interactions within a single phase.

5.7.3 Segregative Interactions between Deacetylated KGM and Disordered κ-Carrageenan

As described in Section 5.6.2, KGM has the same (1 → 4)-diequatorial backbone geometry as cellulose but, unlike cellulose, it is soluble in water. The main factor in conferring solubility is the presence of acetate groups on some of the sugar residues. It is well established (Nishinari et al., 1992) that when these are removed (by alkaline hydrolysis) the deacetylated chains can associate to form a gel network.

In a recent investigation by Penroj et al. (2005), single-phase mixtures of 0.3 wt% KGM with 0.3 wt% κ-carrageenan were treated with alkali at 90°C, which is well above the disorder–order transition temperature of the carrageenan. No gelation was observed for this low concentration of KGM on deacetylation under the same conditions in the absence of carrageenan, but the mixtures gave strong gels. As in the earlier studies of mixtures of soluble polymers with calcium pectinate summarized in Section 5.7.1, this was attributed to enhanced self-association of the gelling component (deacetylated KGM) to reduce thermodynamically unfavorable contacts (segregative interactions) between the two polymers.

The extent of deacetylation of KGM was varied by varying hydrolysis time. In most of the resulting mixtures the disorder–order transition of the carrageenan component (as monitored by DSC) was displaced to higher temperature by the presence of KGM, which was again attributed to enhanced self-association in response to segregative interactions within a single phase. At the highest degree of deacetylation attained, however, no displacement of the carrageenan transition was observed. The proposed interpretation was that the extensively deacetylated KGM chains were aggregated to such an extent that they could no longer promote formation of κ-carrageenan helices.

5.7.4 Effect of Soluble Biopolymers on Gelation of Gelatin

An early publication by Tolstogozov et al. (1974) reported a large increase in the rate of conformational ordering and associated gelation of gelatin on incorporation of small amounts of dextran. The possible generality of this behavior was explored in a recent investigation (Harrington and Morris, 2009a) of mixtures of gelatin with a wide range of soluble biopolymers: guar gum, KGM, dextran, inulin, gum arabic, carboxymethylcellulose (CMC) and hydroxypropylmethylcellulose (HPMC). Conformational ordering of gelatin after rapid quenching to 5°C from the disordered coil state at 45°C was monitored by optical rotation. Gel strength was characterized by compression testing after the samples had been held for 16 h at 5°C.

As mentioned in Section 5.3.1, mixtures of disordered gelatin with KGM showed the unusual phase-separation phenomenon of resolving into a single gel-like lump, consisting predominantly of highly entangled disordered KGM, surrounded by a mobile solution, consisting predominantly of gelatin. The results were otherwise entirely negative. Gel strength (Young's modulus) for the mixtures remained the same, to within experimental error, as the value obtained for the same concentration of gelatin alone, except at the highest concentration of inulin studied (20 wt%) where the modulus was increased by gelation of the inulin. The rate of change in optical rotation after quenching similarly remained close (to within ~10%) to the rate observed for gelatin alone.

A puzzling feature is the clash of experimental evidence with the study by Tolstoguzov et al. (1974), which I have cited in many of my research papers and reviews as an example of the ability of disordered polymers to promote self-association of a gelling component in single-phase mixtures. Both investigations used type B gelatin, and the molecular weight of the dextran was the same (65 kDa). There is no obvious explanation for this disagreement, but the results of Harrington and Morris (2009a) seem internally consistent in also showing little, if any, effect on conformational ordering and gelation of gelatin for any of the other soluble polymers studied.

5.8 CURRENT UNDERSTANDING AND FUTURE CHALLENGES

5.8.1 Effect of Soluble Biopolymers on Gelation From Single-Phase Mixtures

As described in Sections 5.7.1 and 5.7.2, low concentrations of guar gum (and other soluble biopolymers) can cause large increases in modulus when incorporated in single-phase mixtures with a gelling biopolymer. As shown in Figure 5.30, the increase can be as much as 1000-fold. This magnitude of rheological enhancement suggests that interactions between a gelling and a non-gelling (soluble) biopolymer, when present together in a single phase, may be of considerable potential value for future use in industry.

A common feature of the systems where these enhancements have been observed is that gel formation involves self-association of molecules with little flexibility: extended polysaccharide chains (Sections 5.7.1 and 5.7.3) or thermally denatured protein globules (Section 5.7.2). The reason no such enhancements were observed (Section 5.7.4) in the investigation of mixtures of gelatin with soluble biopolymers by Harrington and Morris (2009a) could be that the enthalpic advantage of segregation is sufficient to overcome the small loss of entropy from association of comparatively rigid species, but not sufficient to overcome the much greater reduction in entropy on conversion of flexible gelatin coils to conformationally immobile triple helices.

The inverse correlation (Figures 5.27 and 5.30) with hydrodynamic volume (intrinsic viscosity) of the disordered polymer argues against a mechanism based on phase separation, which, as discussed in Section 5.3.1 and illustrated schematically in Figure 5.3, is promoted by increasing molecular weight. It also argues against depletion flocculation (Dickinson, 1992), which arises from the inability of polymer coils to enter the gap between adjacent particles, and should again, therefore, become increasingly apparent as the size of the coils increases. It does, however, seem qualitatively consistent with a mechanism based on segregative interactions, as proposed in Section 5.7.

For disordered polymers of high molecular weight, a large proportion of the chain sequences will be buried in the interior of the polymer coil, where they cannot make segmental contacts with the other polymeric constituent in single-phase mixtures. As the molecular size decreases, however, the proportion of 'wasted' interior sequences will also decrease, which could explain why small coils are more effective (Figures 5.27 and 5.30) than larger ones. However, there is obvious scope for further research, and development of a more rigorous interpretation.

5.8.2 Rheology and Phase Composition of Biopolymer Co-Gels

As illustrated in Sections 5.2, 5.4 and 5.5, polymer blending laws (Eqs. (5.1)– (5.3)) seem to work well for both (comparatively rare) bicontinuous co-gels and the more common topology of a dispersed phase in a continuous matrix. The success of blending law analyses based on the assumption that each polymer is confined entirely to one phase (Figures 5.7, 5.8 and 5.14) indicates that segregation in biphasic gels is more complete than in the solution state, but it would be better to have more direct experimental evidence.

Some progress has been made in determining phase composition in biphasic gels by Fourier transform infrared spectroscopy (Durani and Donald, 1994, 1995), confocal laser Raman spectroscopy (Pudney et al., 2003), and microsampling with analysis by capillary electrophoresis (Zhong et al., 1997). However, these approaches are not yet suitable for routine use, and their further refinement, or development of new techniques of analysis, is an obvious target for future research.

5.8.3 Electrostatic Coupling

There is extensive literature on use of polyanion–polycation complexes for encapsulation (e.g. Chilvers and Morris, 1987; Chilvers et al., 1988). However, there seems to have been far less research on electrostatic coupling in biopolymer co-gels. There is obvious scope for further work on tailoring gel properties by informed manipulation of pH, ionic strength and ratio of polyanion/polycation (Section 5.6.1).

5.8.4 Failure Properties

With a few notable exceptions (e.g. McEvoy et al., 1985; Brink et al., 2007), most research on the rheology of biphasic co-gels has focused on small-deformation properties (low-amplitude oscillatory measurements). A future challenge is to develop improved understanding of the failure properties (fracture under stress) that are central to many practical applications.

5.8.5 Where have we got to?

Current understanding provides a useful framework for rationalizing the observed behavior of specific gelling mixtures, manipulating the rheology of these mixtures, and directing future research. However, we cannot yet predict with confidence what will happen in mixtures that have not been studied previously. My personal, perhaps uncharitable, impression is that we are still at the stage of saying 'You tell me what happens and I'll tell you why it happens'. There is obviously more to be done in extending the experimental data base and developing greater predictive understanding.

ACKNOWLEDGMENTS

The recent findings summarized in Sections 5.5, 5.6.2, 5.7.2 and 5.7.4 emanated from research conducted with the financial support of Science Foundation Ireland.

References

Abdulmola, N., Hember, M.W.N., Richardson, R.K., and Morris, E.R. (1996a). Composite rheology in mixed biopolymer systems. *In* "Gums and Stabilisers for the Food Industry 8" (G.O. Phillips, P.A. Williams, and D.J. Wedlock, Eds.), 275–285. IRL Press, Oxford, UK.

Abdulmola, N.A., Hember, M.W.N., Richardson, R.K., and Morris, E.R. (1996b). Application of polymer blending laws to starch–gelatin composites. *Carbohydr. Polym.* **31,** 53–63.

Agoub, A. A., Smith, A. M., Giannouli, P., Richardson, R. K., and Morris, E. R. (2007). "Melt in the mouth" gels from mixtures of xanthan and konjac glucomannan under acidic conditions: a rheological and calorimetric study of the mechanism of synergistic gelation. *Carbohydr. Polym.* **69**, 713–724.

Antonov, Y. A., Lashko, N. P., Glotova, Y. K., Malovikova, A., and Markovich, O. (1996). Effect of the structural features of pectins and alginates on their thermodynamic compatibility with gelatin in aqueous media. *Food Hydrocolloids* **10**, 1–9.

Bergfeldt, K., Piculell, L., and Linse, P. (1996). Segregation and association in mixed polymer solutions from Flory–Huggins model calculations. *J. Phys. Chem.* **100**, 3680–3687.

Braudo, E. E., Gotlieb, A. M., Plashina, I. G., and Tolstoguzov, V. B. (1986). Protein-containing multicomponent gels. *Die Nahrung* **30**, 355–364.

Brink, J., Langton, M., Stading, M., and Hermansson, A. M. (2007). Simultaneous analysis of the structural and mechanical changes during large deformation of whey protein isolate/gelatin gels at the macro and micro levels. *Food Hydrocolloids* **21**, 409–419.

Brownsey, G. J., Cairns, P., Miles, M. J., and Morris, V. J. (1988). Evidence for intermolecular binding between xanthan and the glucomannan konjac mannan. *Carbohydr. Res.* **176**, 329–334.

Cairns, P., Miles, M. J., and Morris, V. J. (1986). Intermolecular binding of xanthan gum and carob gum. *Nature (London)* **322**, 89–90.

Cairns, P., Miles, M. J., Morris, V. J., and Brownsey, G. J. (1987). X-ray fibre-diffraction studies of synergistic, binary polysaccharide gels. *Carbohydr. Res.* **160**, 411–423.

Cheetham, N. W. H., McCleary, B. V., Teng., G., Lum, F., and Maryanto (1986). Gel-permeation studies on xanthan–galactomannan interactions. *Carbohydr. Polym.* **6**, 257–268.

Chilvers, G. R. and Morris, V. J. (1987). Coacervation of gelatin–gellan gum mixtures and their use in microencapsulation. *Carbohydr. Polym.* **7**, 111–120.

Chilvers, G. R., Gunning, A. P., and Morris, V. J. (1988). Coacervation of gelatin–XM6 mixtures and their use in microencapsulation. *Carbohydr. Polym.* **8**, 55–61.

Clark, A. H. (1987). The application of network theory to food systems. *In* "Food Structure and Behaviour", 13–34, Blanshard, J. M. V. and Lillford, P. (eds.). Academic Press, London.

Clark, A. H. and Lee-Tuffnell, C. D. (1986). Gelation of globular proteins. *In* "Functional Properties of Food Macromolecules", 203–272, Mitchell, J. R. and Ledward, D. A. (eds.). Elsevier, London.

Clark, A. H. and Ross-Murphy, S. B. (1985). The concentration dependence of biopolymer gel modulus. *British Polymer Journal* **17**, 164–168.

Clark, A. H., Richardson, R. K., Robinson, G., Ross-Murphy, S. B., and Weaver, A. C. (1982). Structural and mechanical properties of agar/BSA co-gels. *Progress in Food and Nutrition Science* **6**, 149–160.

Clark, A. H., Richardson, R. K., Ross-Murphy, S. B., and Stubbs, J. M. (1983). Structural and mechanical properties of agar/gelatin co-gels. Small-deformation studies. *Macromolecules* **16**, 1367–1374.

Davies, W. E. A. (1971). The theory of elastic composite materials. *J. Phys. D: Appl. Phys.* **4**, 1325–1339.

Dea, I. C. M., Morris, E. R., Rees, D. A., Welsh, E. J., Barnes, H. A., and Price, J. (1977). Associations of like and unlike polysaccharides: mechanism and specificity in galactomannans, interacting bacterial polysaccharides, and related systems. *Carbohydr. Res.* **57**, 249–272.

Dickinson, E. (1992). An Introduction to Food Colloids. Oxford University Press, Oxford, UK.

Doyle, J., Giannouli, P., Philp, K., and Morris, E. R. (2002). Effect of K^+ and Ca^{2+} cations on gelation of κ-carrageenan. *In* "Gums and Stabilisers for the Food Industry 11", 158–164, Williams, P. A. and Phillips, G. O. (eds.). Royal Society of Chemistry, Cambridge, UK.

Durani, C. M. and Donald, A. M. (1994). Fourier transform infrared microspectroscopy of phase-separated mixed biopolymer gels. *Macromolecules* **27**, 110–119.

Durani, C. M. and Donald, A. M. (1995). Compositional mapping of mixed gels using FTIR microspectroscopy. *Carbohydr. Polym.* **28**, 297–303.

Fitzsimons, S. M., Tobin, J. T., and Morris, E.R. (2008a). Synergistic binding of konjac glucomannan to xanthan on mixing at room temperature. *Food Hydrocolloids* **22,** 36–46.

Fitzsimons, S. M., Mulvihill, D. M., and Morris, E.R. (2008b). Co-gels of whey protein isolate with crosslinked waxy maize starch: analysis of solvent partition and phase structure by polymer blending laws. *Food Hydrocolloids* **22,** 468–484.

Fitzsimons, S. M., Mulvihill, D. M., and Morris, E.R. (2008c). Segregative interactions between gelatin and polymerised whey protein. *Food Hydrocolloids* **22**, 485–491.

Fitzsimons, S. M., Mulvihill, D. M., and Morris, E.R. (2008d). Large enhancements in thermogelation of whey protein isolate by incorporation of very low concentrations of guar gum. *Food Hydrocolloids* **22**, 576–586.

Galazka, V. B., Smith, D., Ledward, D. A., and Dickinson, E. (1999). Complexes of bovine serum albumin with sulphated polysaccharides: effects of pH, ionic strength and high pressure treatment. *Food Chem.* **64**, 303–310.

Giannouli, P., Richardson, R. K., and Morris, E. R. (2004a). Effect of polymeric cosolutes on calcium pectinate gelation. Part 1. Galactomannans in comparison with partially depolymerised starches. *Carbohydr. Polym.* **55,** 343–355.

Giannouli, P., Richardson, R. K., and Morris, E.R. (2004b). Effect of polymeric cosolutes on calcium pectinate gelation. Part 2. Dextrans and inulin. *Carbohydr. Polym.* **55,** 357–365.

Giannouli, P., Richardson, R. K., and Morris, E.R. (2004c). Effect of polymeric cosolutes on calcium pectinate gelation. Part 3. Gum arabic and overview. *Carbohydr. Polym.* **55,** 367–377.

Gilsenan, P. M., Richardson, R. K., and Morris, E. R. (2000). Thermally-reversible acid-induced gelation of low-methoxy pectin. *Carbohydr. Polym.* **41**, 339–349.

Gilsenan, P. M., Richardson, R. K., and Morris, E. R. (2003a). Associative and segregative interactions between gelatin and low-methoxy pectin: Part 1 – Associative interactions in the absence of Ca^{2+}. *Food Hydrocolloids* **17**, 723–737.

Gilsenan, P. M., Richardson, R. K., and Morris, E. R. (2003b). Associative and segregative interactions between gelatin and low-methoxy pectin: Part 2 – Co-gelation in the presence of Ca^{2+}. *Food Hydrocolloids* **17**, 739–749.

Gilsenan, P. M., Richardson, R. K., and Morris, E. R. (2003c). Associative and segregative interactions between gelatin and low-methoxy pectin: Part 3 – Quantitative analysis of co-gel moduli. *Food Hydrocolloids* **17**, 751–761.

Goycoolea, F. M., Morris, E. R., and Gidley, M. J. (1995a). Screening for synergistic interactions in dilute polysaccharide solutions. *Carbohydr. Polym.* **28,** 351–358.

Goycoolea, F. M., Richardson, R. K, Morris, E. R., and Gidley, M. J. (1995b). Stoichiometry and conformation of xanthan in synergistic gelation with locust bean gum or konjac glucomannan. *Macromolecules* **28,** 8308–8320.

Grant, G. T., Morris, E. R., Rees, D. A., Smith, P. J. C., and Thom, D. (1973). Biological interactions between polysaccharides and divalent cations: the egg-box model. *FEBS Lett.* **32**, 195–198.

Grinberg, V. Y. and Tolstoguvzov, V. B. (1972). Thermodynamic incompatibility of gelatin with some D-glucans in aqueous media. *Carbohydr. Res.* **25**, 313–321.

Grinberg, V. Ya and Tolstoguzov, V. B. (1997). Thermodynamic incompatibility of proteins and polysaccharides in solutions. *Food Hydrocolloids* **11**, 145–158.

Harrington, J. C. and Morris, E. R. (2009a). Conformational ordering and gelation of gelatin in mixtures with soluble polysaccharides. *Food Hydrocolloids* **23**, 327–336.

Harrington, J. C. and Morris, E. R. (2009b). An unusual manifestation of phase separation in mixtures of disordered gelatin with konjac glucomannan. *Food Hydrocolloids* **23**, 460–467.

Harrington, J. C., Foegeding, E. A., Mulvihill, D. M., and Morris, E. R. (2009). Segregative interactions and competitive binding of Ca^{2+} in gelling mixtures of whey protein isolate with Na^+ κ-carrageenan. *Food Hydrocolloids* **23**, 468–489.

Hasan, S. (2006). Heat Treatment of Wheat Flour. PhD thesis, University College Cork, Ireland.

Hasan, S., Fitzsimons, S. M., O'Neill, E., and Morris, E. R. (2008). Adhesive interactions between gelatinised starch granules. *In* "Gums and Stabilisers for the Food Industry 14", 105–113, Williams, P. A. and Phillips, G. O. (eds.). Royal Society of Chemistry, Cambridge, UK.

Imeson, A. P., Ledward, D. A., and Mitchell, J. R. (1977). On the nature of the interaction between some anionic polysaccharides and proteins. *Journal of the Science of Food and Agriculture* **28**, 661–668.

Kasapis, S., Morris, E. R., Norton, I. T., and Gidley, M. J. (1993). Phase equilibria and gelation in gelatin/maltodextrin systems – Part II: polymer incompatibility in solution. *Carbohydr. Polym.* **21**, 249–259.

Loren, N., Hermansson, A.-M., Williams, M. A. K., Lundin, L., Foster, T. J., Hubbard, C. D., Clark, A. H., Norton, I. T., Bergstrom, E. T., and Goodall, D. M. (2001). Phase separation induced by conformational ordering of gelatin in gelatin/maltodextrin mixtures. *Macromolecules* **34,** 289–297.

Manson, J. A. and Sperling, L. H. (1976). Polymer Blends and Composites. Plenum Press, New York.

McEvoy, H., Ross-Murphy, S. B., and Clark, A. H. (1985). Large deformation and ultimate properties of biopolymer gels: 2. Mixed gel systems. *Polymer* **26**, 1493–1500.

Michon, C., Cuvelier, G., Launay, B., Parker, A., and Takerkart, G. (1995). Study of the compatibility/incompatibility of gelatin/iota-carrageenan/water mixtures. *Carbohydr. Polym.* **28**, 333–336.

Mohammed, Z. H., Hember, M. W. N., Richardson, R. K., and Morris, E. R. (1998a). Application of polymer blending laws to composite gels of agarose and crosslinked waxy maize starch. *Carbohydr. Polym.* **36**, 27–36.

Mohammed, Z. H., Hember, M. W. N., Richardson, R. K., and Morris, E. R. (1998b). Co-gelation of agarose and waxy maize starch. *Carbohydr. Polym.* **36**, 37–48.

Morris, E. R. (1995). Polysaccharide synergism: more questions than answers? *In* Biopolymer Mixtures, 247–288, Harding, S. E., Hill, S. E. and Mitchell, J. R. (eds.). Nottingham University Press, Nottingham, UK.

Morris, E. R. (1998). Segregative interactions in biopolymer co-gels. *In* "Phase/State Transitions in Foods: Chemical, Structural, and Rheological Changes", 159–186, Rao, M. A. and Hartel, R. W. (eds.). Marcel Dekker, New York.

Morris, E. R., Powell, D. A., Gidley, M. J., and Rees, D. A. (1982). Conformations and interactions of pectins. I. Polymorphism between gel and solid states of calcium polygalacturonate. *J. Mol. Biol.* **155**, 507–516.

Morris, E. R., Rees, D. A., Young, G., Walkinshaw, M. D., and Darke, A. (1977). Order-disorder transition for a bacterial polysaccharide in solution. A role for polysaccharide conformation in recognition between *Xanthomonas* pathogen and its plant host. *J. Mol. Biol.* **110**, 1–16.

Morris, V. J. (1986). Multicomponent gels. *In* "Gums and Stabilisers for the Food Industry 3", 87–99, Phillips, G. O., Wedlock, D. J. and Williams, P. A. (eds.). Elsevier, London.

Morris, V. J. (1992). Designing polysaccharides for synergistic interactions. *In* "Gums and Stabilisers for the Food Industry 6", 161–171, Phillips, G. O., Williams, P. A. and Wedlock, D. J. (eds.). IRL Press, Oxford, UK.

Muchin, M. A., Wajnermann, E. S., and Tolstogusow, W. B. (1976). Complex gels of proteins and acidic polysaccharides. *Die Nahrung* **20**, 313–319.

Mulvihill, D. M. and Donovan, M. (1987). Whey Proteins and their thermal denaturation – a review. *Irish Journal of Food Science and Technology* **11**, 43–75.

Musampa, R. M., Alves, M. M., and Maia, J. M. (2007). Phase separation, rheology and microstructure of pea protein–kappa-carrageenan mixtures. *Food Hydrocolloids* **21**, 92–99.

Nayouf, M., Loisel, C., and Doublier, J. L. (2003). Effect of thermomechanical treatment on the rheological properties of crosslinked waxy corn starch. *Journal of Food Engineering* **59**, 209–219.

Nishinari, K., Williams, P. A., and Phillips, G. O. (1992). Review of the physico-chemical characteristics and properties of konjac mannan. *Food Hydrocolloids* **6**, 199–222.

Penroj, P., Mitchell, J. R., Hill, S. E., and Ganjanagunchorn, W. (2005). Effect of konjac glucomannan deacetylation on the properties of gels formed from mixtures of kappa carrageenan and konjac glucomannan. *Carbohydr. Polym.* **59**, 367–376.

Picout, D. R., Richardson, R. K., Rolin, C., Abeysekera, R. M., and Morris, E. R. (2000a). Ca^{2+}-induced gelation of low methoxy pectin in the presence of oxidised starch: Part 1– Collapse of network structure. *Carbohydr. Polym.* **43**, 113–122.

Picout, D. R., Richardson, R. K., and Morris, E. R. (2000b). Co-gelation of calcium pectinate with potato maltodextrin. Part 1 – Network formation on cooling. *Carbohydr. Polym.* **43**, 133–141.

Piculell, L., Bergfeldt, K., and Nilsson, S. (1995). Factors determining phase behaviour of multi component biopolymer systems. *In* "Biopolymer Mixtures", 13–35, Harding, S. E., Hill, S. E. and Mitchell, J. R. (eds.). Nottingham University Press, Nottingham, UK.

Piculell, L., Iliopoulos, I., Linse, P., Nilsson, S., Turquois, T., Viebke, C., and Zhang, W. (1994). Association and segregation in ternary polymer solutions and gels. *In* "Gums and Stabilisers for the Food Industry 7", 309–322, Phillips, G. O., Williams, P. A. and Wedlock, D. J. (eds.). IRL Press, Oxford, UK.

Piculell, L., Nilsson, S., Falck, L., and Tjerneld, F. (1991). Phase separation in aqueous mixtures of similarly charged polyelectrolytes. *Polymer Communications* **32**, 158–160.

Piculell, L., Nilsson, S., and Muhrbeck, P. (1992). Effects of small amounts of kappa-carrageenan on the rheology of aqueous iota-carrageenan. *Carbohydr. Polym.* **18**, 199–208.

Polyakov, V. I., Grinberg, V. Ya., and Tolstoguzov, V. B. (1997). Thermodynamic incompatibility of proteins. *Food Hydrocolloids* **11**, 171–180.

Powell, D. A., Morris, E. R., Gidley, M. J., and Rees, D. A. (1982). Conformations and interactions of pectins. II. Influence of residue sequence on chain association in calcium pectate gels. *J. Mol. Biol.* **155**, 517–531.

Pudney, P. D. A., Hancewicz, T. M., Cunningham, D. G., and Gray, C. (2003). A novel method for measuring concentrations of phase separated biopolymers: the use of confocal Raman spectroscopy with self-modelling curve resolution. *Food Hydrocolloids* **17**, 345–353.

Ross-Murphy, S. B. (1984). Rheological methods. *In* Critical Reports on Applied Chemistry: Biophysical Methods in Food Research, 195–290, Chan, H. W. S. (ed.). SCI, London.

Stainsby, G. (1980). Proteinaceous gelling systems and their complexes with polysaccharides. *Food Chem.* **6**, 3–14.

Suchkov, V. V., Grinberg, V. Ya., and Tolstoguzov, V. B. (1981). Steady-state viscosity of the liquid two-phase disperse system water–casein–sodium alginate. *Carbohydr. Polym.* **1**, 39–53.

Syrbe, A., Fernandes, P. B., Dannenberg, F., Bauer, W., and Klostermeyer, H. (1995). Whey protein and polysaccharide mixtures: polymer incompatibility and its applications. *In* "Food Colloids and Macromolecules", 328–339, Dickinson, E. and Lorient, D. (eds.). The Royal Society of Chemistry, Cambridge, UK.

Thom, D., Dea, I. C. M., Morris, E. R., and Powell, D. A. (1982). Interchain associations of alginate and pectins. *Progress in Food and Nutrition Science* **6**, 97–108.

Toft, K. (1982). Interactions between pectins and alginates. *Progress in Food and Nutrition Science* **6**, 89–96.

Takayanagi, M., Harima, H., and Iwata, Y. (1963). Viscoelastic behaviour of polymer blends and its comparison with model experiments. *Memoirs – Faculty of Engineering Kyushu University* **23**, 1–13.

Tolstoguzov, V. B. (1986). Functional properties of protein–polysaccharide mixtures. *In* "Functional Properties of Food Macromolecules", 385–415, Mitchell, J. R. and Ledward, D. A. (eds.). Elsevier, London.

Tolstoguzov, V. B. (1988a). Concentration and purification of proteins by means of two-phase systems: membraneless osmosis process. *Food Hydrocolloids* **2**, 195–207.

Tolstoguzov, V. B. (1988b). Some physico-chemical aspects of protein processing into foodstuffs. *Food Hydrocolloids* **2**, 339–370.

Tolstoguzov, V. B. (1991). Functional properties of food proteins and role of protein–polysaccharide interaction. *Food Hydrocolloids* **4**, 429–468.

Tolstoguzov, V. B. (1992). The functional properties of food proteins. *In* "Gums and Stabilisers for the Food Industry 6", 241–266, Phillips, G. O., Williams, P. A. and Wedlock, D. J. (eds.). IRL Press, Oxford, UK.

Tolstoguzov, V. B. (1994). Some physico-chemical aspects of protein processing into foods. *In* "Gums and Stabilisers for the Food Industry 7", 115–124, Phillips, G. O., Williams, P. A. and Wedlock, D. J. (eds.). IRL Press, Oxford, UK.

Tolstoguzov, V. B. (2003). Some thermodynamic considerations in food formulation. *Food Hydrocolloids* **17**, 1–23.

Tolstoguzov, V. B. and Braudo, E. E. (1983). Fabricated foodstuffs as multicomponent gels. *Journal of Texture Studies* **14**, 183–212.

Tolstoguzov, V. B., Belkina, V. P., Gulov, V. Ja., Grinberg, V. Ja., Titova, E. F., and Belavzeva, E. M. (1974). State of phase, structure and mechanical properties of the gelatinous system water–gelatin–dextran. *Die Stärke* **26**, 130–137.

Tschumak, G. Ja., Wajnermann, E. S., and Tolstogusow, W. B. (1976). Structure and properties of complex gels of gelatin and pectin. *Die Nahrung* **20**, 321–328.

Ye, A. (2008). Complexation between milk proteins and polysaccharides via electrostatic interaction: principles and applications – a review. *Int. J Food Sci. Technol.* **43**, 406–415.

Zasypkin, D. V., Braudo, E. E., and Tolstoguzov, V. B. (1997). Multicomponent biopolymer gels. *Food Hydrocolloids* **11**, 159–170.

Zhong, H. J., Williams, M. A. K., Keenan, R. D., Goodall, D. M., and Rolin, C. (1997). Separation and quantification of pectins using capillary electrophoresis: a preliminary study. *Carbohydr. Polym.* **32**, 27–32.

CHAPTER

6

Effect of Processing on Biopolymer Interactions

Phil W. Cox, Fotios Spyropoulos, and Ian T. Norton
Department of Chemical Engineering, University of Birmingham, Edgbaston B15 2TT, UK

6.1 INTRODUCTION

Although it is well known that many biopolymers undergo conformational transitions to form ordered states, what is less known, and has only recently received attention by the academic community and industry, is that the material properties of hydrocolloids can be modified by applying shear during the conformational ordering process. The effect of applied forces depends upon the relative timescales and dimensions of the ordering process and the applied mechanical forces (e.g. shear fields, etc.).

6.1.1 Biopolymer Conformational Ordering (Timescales, Length Scales)

The studies carried out so far have shown that the hydrocolloids that can be influenced by processing to produce new structures and material properties all undergo aggregation processes to form bulk three-dimensional gels. Although other materials can be influenced by processing, this is often as a consequence of molecular damage. This type of effect will not be covered in this chapter.

So at what time and length scales do these conformational changes occur? The work carried out by Norton et al. (Norton et al., 1978, 1979, 1980, 1983a,b,c, 1984; Morris and Norton, 1983) showed conformational transitions in the millisecond or microsecond timescale. The dimensions of the ordered states are in the region of nm to 10s of nanometers. This means that most of the processes that have timescales in the order of seconds to minutes and applied forces acting at the micron (or above) length scales have little effect on these molecular conformation events.

However, hydrocolloids which form aggregated structures (e.g. agar, carrageenans, gellan, alginate, etc.) do so on time and length scales similar to those applied by the process. It is therefore highly likely that the effect of processing is to modify the extent and size of the aggregates. However, as will be discussed throughout this chapter, these effects, if at all occurring, are of minor consequence.

6.1.2 Gelation Processes (Time and Length Scales)

The biggest effect of applied mechanical forces is seen to be on gelation. This of course is what we

Kasapis, Norton, and Ubbink: Modern Biopolymer Science
ISBN: 978-0-12-374195-0

would intuitively expect as the timescale for gelation is between seconds and hours (Arnott et al., 1974) and gelation is a bulk phenomenon. This area will constitute the majority of this chapter and we will discuss how the gelation process must be largely complete within the process. By doing so we are capable of producing a diverse range of novel structures and rheological performance from what is a rather limited range of hydrocolloids. This is commercially very important, particularly in the fast-moving consumer goods industry, as the number of allowed additives is largely fixed and there is always consumer demand to reduce the number of additives in any given product.

6.1.3 Phase Separation

The use of two or more hydrocolloids is often common in products to impart the desired texture, appearance or in-use properties of the final product, such as cooking behavior in-mouth or in-body effects. This introduces an extra level of complexity that we need to consider. So what happens when two or more hydrocolloids are mixed (details given in Chapter 5) and then what happens if mechanical forces are applied to this mixture; during conformational ordering of one or more of the hydrocolloids and/or during synergistic interaction of the polymer chains or segregative phase separation?

In general terms, synergistic interactions can be treated in the same way as the aggregation and gelation of single polymer systems. However, for phase-separating systems we need to consider a number of different aspects.

6.1.4 General Considerations and Thermodynamic Aspects of Phase Separation

The reason that phase separation is such a common phenomenon in aqueous mixed biopolymer systems lies within the thermodynamics of these systems. Miscibility in a multicomponent system is governed by the free energy of mixing and therefore the entropy and enthalpy of mixing. In a mixture of small molecules (or low-molecular-weight components) the large entropy of mixing is the dominating factor resulting in complete miscibility. However, in a system containing relatively large components (such as a high-molecular-weight biopolymer aqueous mixture) the much lower entropy of mixing (per unit weight) no longer dominates and de-mixing can occur. Although this is entropically unfavorable, the enthalpy term is advantageous, as ultimately the molecules prefer to have neighbors of similar structure. This is particularly true when, as the temperature is lowered the hydrocolloid molecules start to interact and order thus driving up the effective molecular weight.

Within a process it is important to know the type of interaction (if any) taking place between the segments of the polymers in the mixture. If the interaction is repulsive (Zeman and Patterson, 1972), then phase separation occurs above certain polymer concentrations and the two polymers will collect separately in two opposite phases; what is known as 'segregative type' of phase separation (Piculell and Lindman, 1992). If, however, the interaction between the two polymers is attractive, as is the case between two oppositely charged polyelectrolytes, the two polymers will collect in the same phase and the result is a system with one phase enriched in both polymeric components existing in equilibrium with another phase poor in both; what is known as an associative type of phase separation (Piculell and Lindman, 1992). As will be discussed later this type of separation can be useful in processes to give mixed and sheared gels and beads for encapsulation. Since, in both types of phase separation, the main component of each of the resulting immiscible phases is water, these aqueous mixed biopolymer systems have been known as 'water-in-water' emulsions.

The presence of charged groups on one of the polymers has a profound influence on the phase behavior (Perrau et al., 1989). If no added salt is present in the system, the counterions resulting from the dissociating polyelectrolyte are forced to follow their parent macromolecules into the same phase in order to render the system electroneutral. The resulting entropic disadvantage of segregating the co-ions though is so large that phase separation is often suppressed (Piculell et al., 1995). Addition of salt to the system reduces the impact of this increased entropic term and phase separation increases as the salt concentration increases (Perrau et al., 1989; Piculell and Lindman, 1992; Piculell et al., 1995).

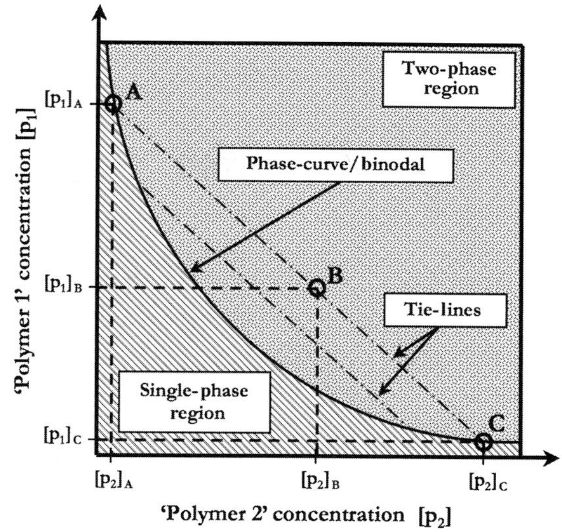

FIGURE 6.1 Phase diagram of a segregating aqueous 'Polymer 1'–'Polymer 2' mixture.

6.1.5 Phase-Diagrams

Knowledge of the polymer compositions where phase separation is taking place as well as the composition of polymers in the co-existing phases is essential if we are to use processes to impact on microstructure and kinetically trap structures. The composition of polymers needed to induce phase separation in an aqueous biopolymer mixture can be determined by the cloud point method (Aymard et al., 2000). Alternatively, the composition of the separated phases can be measured by direct chemical and/or physical analysis (e.g. Pudney et al., 2004).

The obtained polymer compositions can be then used to produce the phase diagram of the system; e.g. Figure 6.1 shows a phase diagram for a segregative aqueous polymer mixture. The solid curve is called the phase curve or binodal and it represents the borderline between systems that are completely miscible (single-phase region) and those that exhibit phase separation (two-phase region). For example, an aqueous mixture containing polymer concentrations depicted by point B will phase separate, giving two co-existing phases of concentrations depicted by point A and point C. Points A, B and C are connected by a straight line (dashed line in Figure 6.1) known as the tie-line of the phase diagram.

As with any phase diagram, moving the composition of the system along the same tie-line changes the relative volumes of the resulting co-existing phases, according to the relative lengths of the tie-line to the binodal, but the composition of the polymeric components in these individual phases remains the same. Therefore, at the middle point of the tie-line the co-existing phases each occupying 50% of the system's volume. It is in this region that the microstructure of the system becomes 'confused' as both phases are in volumetric balance and both exhibit the same tendency towards becoming the continuous phase. As a result bi-continuous systems are often observed. As with oil/water systems, either side of the midpoint will give phases of different volumes with the phase with the greatest phase volume forming the continuous phase. This region of bicontinuity can occur over phase volume ranges as high as 10% (Foster et al., 1996). This is consistent with the low interfacial tension in the mixed biopolymer systems, which is later discussed. It is around this 50:50 phase volume that processing parameters can have a significant impact and where shear-induced phase inversion is observed (Foster et al., 1996).

6.1.6 Mechanisms of Demixing

In order to understand and control the behavior of mixed biopolymers in processes we need to understand the mechanism(s) by which phase separation initially occurs and then also those by which the formed structures ripen. The initial stages of phase separation can occur via either a nucleation and growth process or via spinodal decomposition (Butler, 2002). A reduction in temperature reduces the entropy of the mixture, leading to an increased incompatibility in the system. In biopolymer systems, molecular ordering often occurs resulting in increased effective molecular weight. This reduces the entropy penalty for the formation of two phases and may also change the Flory-Huggins interaction (χ) parameters (Lorén et al., 2001). It has been shown for gelatin/maltodextrin (Aymard et al., 2000; Lorén et al., 2001; Norton and Frith, 2001; Williams et al., 2001), a system that has been widely studied, that, depending on the composition of the mixture and the temperature history applied to it, either of the above mechanisms can drive the separation process. The mechanism driving separation is dependent upon the depth of the quench. If a jump is carried out just inside the two-phase region of the phase diagram, then nucleation and growth occurs. Growth can be influenced by the applied process, as once the droplets have formed, they can start to ripen. Ultimately, in many systems, if this process is allowed to continue unchecked then bulk phase separation is expected. Light scattering evidence (Butler and Heppenstall-Butler, 2001) indicates that the increase in size goes as $t^{0.5}$ suggesting a diffusion controlled mechanism, as might be expected. As the system continues to ripen then microscopic evidence shows that a coalescence of droplets also occurs (Foster et al., 1996) either by controlling the size to which the droplet phase can grow or by inducing a phase inversion process.

If the depth of the quench is much further into the two-phase region then spinodal decomposition occurs. This is a much more 'spontaneous' process and thus is less affected by the applied process. However, as the structure ripens then applied shear forces influence the dimensions of the final structure.

As with synthetic polymers, it is possible to construct a spinodal on the phase diagram (Figure 6.2), which separates the metastable and unstable regions of the two-phase system. If, after the change in conditions, the mixture lies within the spinodal line on the phase diagram (i.e. in the unstable region, point A in Figure 6.2) then the mechanism of separation will be via spinodal decomposition. However, if we move to a position between the spinodal and the binodal (the metastable region, point B in Figure 6.2) then phase separation will follow via nucleation and growth. Although this diagram is not absolute it is very useful for process engineers who need to construct specific microstructures by modifying the process, e.g. temperature and shear.

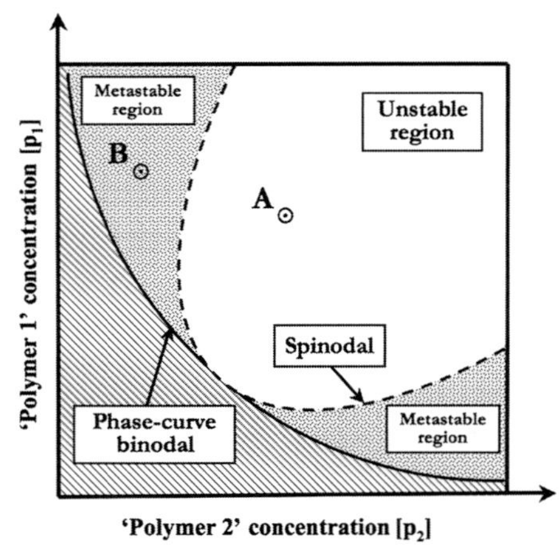

FIGURE 6.2 Phase diagram and spinodal line separating the metastable and unstable regions of the aqueous 'Polymer 1'–'Polymer 2' two-phase mixture.

6.1.7 Processing Aspects

If we are to modify the material properties of hydrocolloids both singularly and when mixed we need to consider the process that will be used. However, rather than considering each type of equipment, we should concentrate on the flow that the equipment induces and how the conformational ordering and gelation process can be initiated. Much of the equipment in use across all industry sectors is mixing devices. These can have impellers such as Rushton turbines, or be scraped surface devices or pin stirrers. All these types of process are designed for mixing or mixing in combination with cooling. The area that is interesting, from a hydrocolloid point of view, is when the process is intense enough to result in turbulent flow. The reason for this will be discussed in the following sections. A second type of flow that is interesting for the formulation of new and novel microstructures and textures is elongational flow; easily induced for example in extrusion processes. If aggregation and gelation are induced under these types of flow then asymmetrical structures can be produced. Again this will be discussed later. The processes of value for modification of hydrocolloid physical properties need to be capable of inducing conformational transitions, such as salt jumps or temperature ramps.

6.2 FLUID/SHEARED GELS

Modern food formulations are driven by a consumer demand for simplified and more natural products with a reduced list of components often with the additional demand for a controlled and balanced nutrition, e.g. calorie-controlled and reduced-energy-density products. To meet these demands, whilst maintaining the textural/physical properties of the original formulation, is challenging. One possible route is to form microparticles/spheres/fragments of gelled material so that the physical dimensions and strength of the fragments resembles those of oil droplets contained in typical emulsion-based products. From a hydrocolloid perspective the procedure can be simple; a heat (Brown et al., 1990) or ionically set (Kawachi et al., 1993) gel is sheared while undergoing conformational transition and subsequent aggregation/gelation. This causes the gel to cyclize and thus produces gel particles which can be used in a range of different food and personal product formulations (Norton et al., 2006a). The process and conformational ordering kinetics are controlled so that the final particle radius and particle physical properties are precisely designed to fine-tune the microstructure and material properties of the product. Current findings suggest that the particles produced via such processes are ideally spherical and that inter-particle interactions can be controlled to give viscoelastic fluids, i.e. 'fluid gels'.

It is worth noting that the term 'fluid gel' has been also used to describe thickened (by the addition of low levels of, possibly mixed, gelling materials) fluids (Martínez-Padilla et al., 2004). This type of system will not be described in this section as we consider these to be classical thickened liquids.

The use of gels in the food industry is commonplace, but has traditionally been restricted to rheological modifications of liquids, or to the stabilization of products for either processing or for long-term storage. However, with the introduction of fluid gels (systems based on gelled particles), hydrocolloids can now be used for different effects, such as microstructural replacement of fat in emulsion-based products. To do this effectively and in a controllable fashion requires knowledge not only of the gel properties but also an understanding of the molecular interactions and the subsequent colloidal interactions of these systems.

The basic process for producing fluid gels is conceptually simple, although this simplicity hides the competing physical and chemical

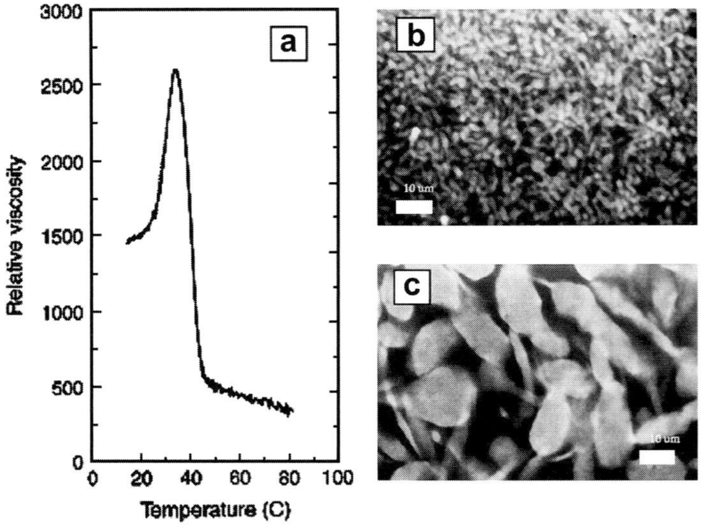

FIGURE 6.3 (a) Changes in viscosity during a cooling/shearing cycle (3% agar at 100 s^{-1} cooled at 1.5°C min^{-1}). 3% agar solutions cooled at 1°C min^{-1} and subjected to 40 s^{-1} (b) and 750 s^{-1} (c) shear rates.

processes which occur during their formation, i.e. the competition between structuring and the applied shear. The simplest way of considering this is of a gel which is sheared whilst it passes through its conformational transition and gelation temperatures (setting temperature). Here the two processes are combined to give the final microstructure and therefore the fluid gel performance. This net result is demonstrated in Figure 6.3a where the viscosity of a fluid gel during its production process was monitored (3% agar at 100 s^{-1}). As the process temperature approaches agar's setting temperature (35°C), the gel starts to order and below this onset point it is primarily the shear field that defines the size of the resultant particles. This is illustrated by Figures 6.3b and 6.3c that show a 3% agar fluid gel cooled at a rate of 1°C min^{-1} but exposed to either 40 s^{-1} or 750 s^{-1}. Thus the differences in microstructure obtained from modifying a single process variable become obvious. What is to follow is an exploration of how these greatly different structures can be designed and produced and also what benefits these structures can give.

The complex events that occur during fluid gel formation are shown schematically in Figure 6.4. This figure shows the formation process of fluid gels and indicates the ordering of the gels at a quasi-molecular level. In addition, the viscosity profile from Figure 6.3 is also indicated in relation to a decreasing temperature (from right to left). Initially, the free gel elements are in a molten state, are separated and remain relatively unaffected by the shear at this point (shown schematically as structure 'a' and indicated as point 'A' on the viscosity/temperature profile); as the temperature decreases, the molecules start to order. The conformational ordering process takes the form of individual molecules entangling together to form double helixes or ion bridging to give dimmers as the building blocks for egg box formation; this is indicated as process 'x' on the figure (position 'B' on the viscosity/cooling curve) and the resulting structures are shown as 'b'.

As cooling proceeds aggregation of the ordered conformations occurs to a point where extensive aggregation and gelation both take place (indicated as process 'y' to form the

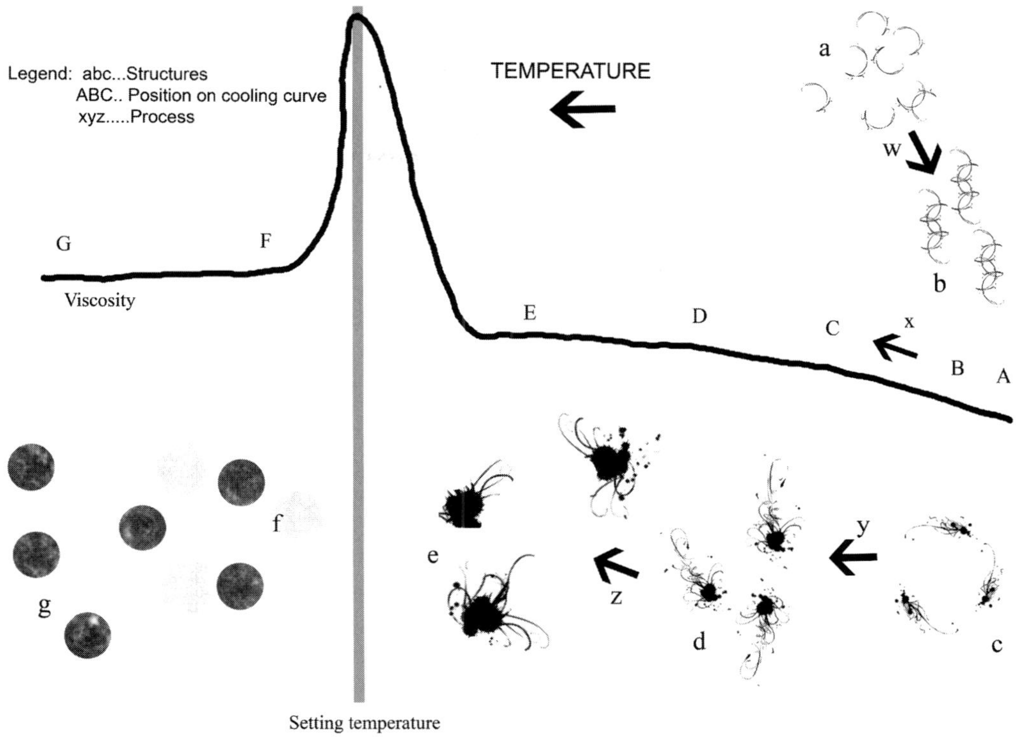

FIGURE 6.4 Schematic representation of the molecular and meso-scale events during fluid gel formation.

structures labeled 'c' and through point 'C' on the viscosity/cooling curve). This results in an increase in the viscosity, as indicated in Figures 6.3a and 6.4. From this point on, it seems that the shear field determines the size of the final gel particles. However, it also appears that after the particles are initially formed they continue to increase in size (process 'y'). This may be achieved through two sub-processes: either by the addition of helixes or free element to the growing particles (structures 'a' and 'b'), or by coalescence of particles (structures 'c'). As the particles are forming they may be considered as having a 'hairy' nature, due to none-ordered chain segments at the surface of the particle. These disordered regions are slowly lost as process 'z' proceeds and more conformational ordering and aggregation has occurred. At the same time the particles grow and as a consequence reach dimensions that are increasingly affected by the applied shear forces. Moreover, as they become very large (100s of microns) they are likely to be broken up by the shear and so their size is physically limited by the chemical engineering process (see Figures 6.3b and 6.3c). It seems most likely that the final particle size is defined by the eddy size of the turbulent flow in the shearing/mixing device. As the conformational ordering nears completion the structure of the gelled particles becomes fixed. It has been suggested by Norton et al. (1998) that the hydrocolloid chains that constitute the particles become completely ordered within single particles effectively smoothing off the particles as the temperature is lowered further. This results in the decreased viscosity after the peak so the final particle form is set at point F on the curve. Further cooling does not affect the particles ('G' and 'g').

By taking a closer look at what is happening during the process of fluid gel production, it seems that as the midpoint temperature for conformational ordering is approached several concomitant processes occur which can affect the morphology, size and subsequent 'reactivity' of the final gel particles. This means that if the cooling rate is very high then the gelation process dominates and large particles are produced. Conversely, if the cooling is more gentle then the applied shear dominates and small particles are produced with a narrower particle size distribution. Most notable are the combined effects of the shear field and the rate of aggregation of the free chains. This can lead to bridging between particles which has a dramatic effect on the fluid gel properties. By manipulating this type of bridging particulate, gels can be produced which have properties very similar to milk protein.

The next parameter which can be controlled through the applied process is particle size distribution which is affected by the rate of cooling as well as by shear. As mentioned earlier if cooling is carried out slowly then narrower size distributions are obtained. A problem that then emerges is that complete ordering of a gel may take in excess of hours or even days (Arnott et al., 1974) and so commercial applications may have to accept an incomplete particle formulation for realistic processes. This will result in particles that can interact via their hairy nature and can continue to grow as the aggregation and gelation processes complete in the final product. However, for most practical purposes, a timescale of seconds/minutes is usually sufficient for the majority of the gel to set. Therefore a controlled and designed process can be developed to deliver fluid gels of desired properties for a wide range of hydrocolloids allowing the material properties to be tailored to the product needs, e.g. a predefined size distribution of the particle, controlled melting temperatures, the behavior in physiological environments, physical, mechanical and textural properties. Norton et al. (1998) surveyed the initial selections of available gels for use in foods and clearly outline the basic principles of particle formation as seen above. Moreover, they also described the responses of sheared gelled colloidal systems, and found interesting novel properties of 'fluid gel emulsions', which again may be used in the design process; this is discussed in greater detail below.

If cooling is carried out too quickly in a process running at low Reynolds numbers then a confused gel (a fluid gel incorporated in a bulk three-dimensional gel) is formed. The formation process has been well characterized by the work of Gabrile et al. (in preparation) who showed at a fixed shear rate different properties are obtained by changing the cooling rate. An example of this can be seen in Figure 6.5 which shows shear stress as a function of temperature for fluid gels produced at a shear rate of 100 s^{-1} using 3% k-carrageenan. The results obtained

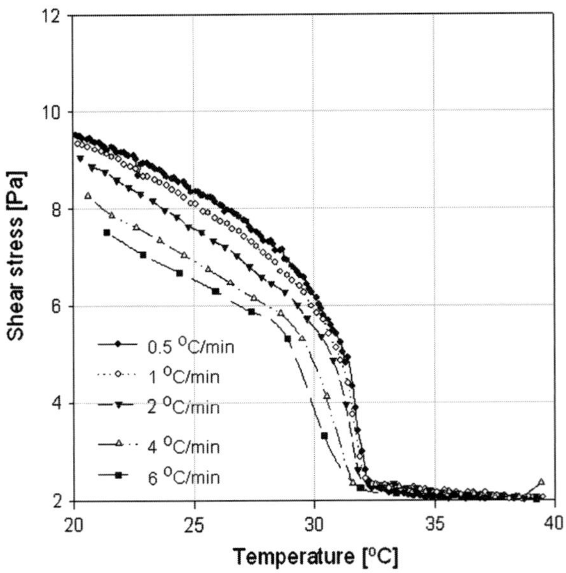

FIGURE 6.5 The effects of cooling rates upon the formation process and therefore final properties of an agar fluid gel produced at different cooling rates (data from Gabriele et al., in preparation).

show that at higher cooling rates a fluid gel is produced that has a higher storage modulus and so is more solid-like, indicating that either bigger and more irregularly shaped particles are formed at the higher cooling rates, or there is more inter-particle bridging. Care is still required, as the effects of the cooling rate can be 'over-ridden' if an inappropriate shear rate is used during the processing. Other work from our laboratory (not shown here) showed that combined use of appropriate shear and cooling rates could give a range of responses from only small changes in the processing conditions.

The results shown in Figure 6.5 were obtained after the fluid gels had been stored for a short period of time. This demonstrates, as previously reported, that once the setting temperature has been passed, and the gel within the particles is fully ordered, the structure is fixed, allowing the physical properties of a system to be clearly definable.

In order to control the physical properties of fluid gels other parameters must be considered, for example, the effects of gel concentration, the particle shape, the type of hydrocolloid or even mixtures of hydrocolloids (Norton et al., 1999; Gabrile et al., in preparation). Their experiments showed that under similar conditions for gel formation (cooling rate and shear) a lowering of the gel concentration initially gave a reduced viscosity for the transition from liquid to particulate (going from point 'E' to 'F' in Figure 6.4) with the final product viscosity (point 'G') being reduced in accordance with the reduction in concentration. Norton et al. (1999) and Sworn et al. (1995) showed that underlying these changes, for various gels, was a reduction in the particulate phase volume, but with a gross increase in the particulate size. The increased particle size at lower polymer concentrations was argued to be a result of a decreased viscosity in the process and therefore a lower power input in the process. This then gave larger eddies and less break-up of particles during their formation. Moreover, the rheological/material responses of the systems changed, not only was the viscosity lower, but there was, concomitantly, a change in the storage modulus; which in turn will affect the physical/mechanical/textural properties. This effect is shown here in several of the figures where processing variables, e.g. gel concentration (Figure 6.6) and shear rate applied (Figure 6.7), can be tailored to give the desired mechanical properties for a fluid gel.

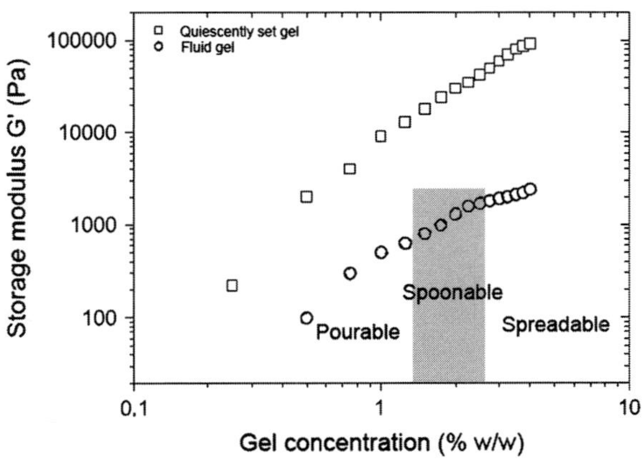

FIGURE 6.6 The response of the material properties of fluid gels with varying gel concentration. The figure compares quiescently set gels against fluid gels and indicates the relative properties of the fluid gel, bounded by the shaded region (redrawn from Norton et al., 2006a).

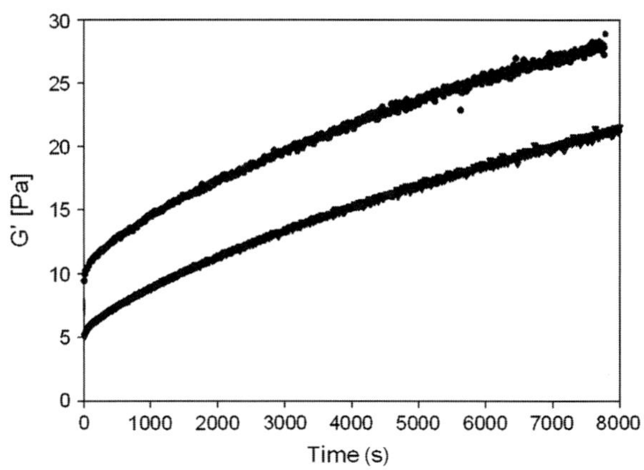

FIGURE 6.7 The evolution, with time, of the storage modulus of similar agar gels formed at 1 s^{-1} (upper curve) and 100 s^{-1} (lower curve). Data from Gabriele et al. (in preparation).

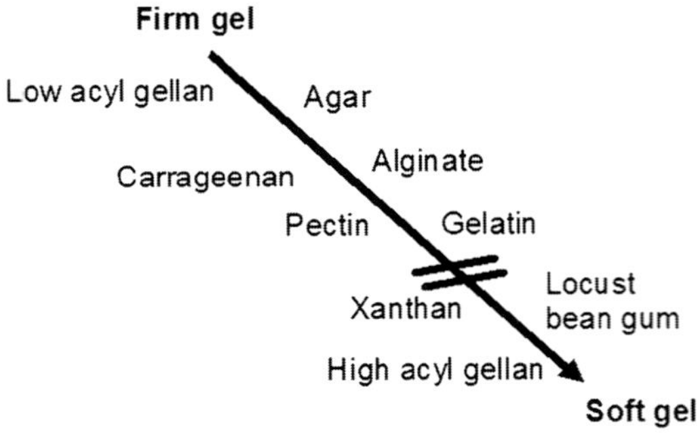

FIGURE 6.8 Schematic relationship between gel type and their physical properties (redrawn from Sworn, 2000). The double line indicates the physical limit for fluid gel formation.

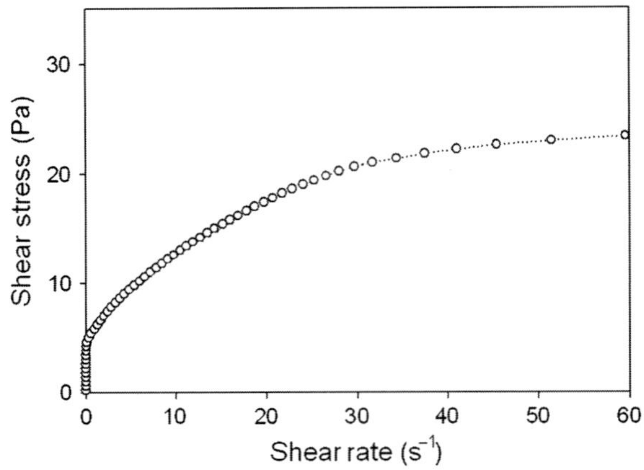

FIGURE 6.9 The persistence of an agar-based fluid gel's yield stress after three shear/rest cycles. The data are for a gel initially produced at 100 s^{-1} (data from Gabriele et al., in preparation).

Norton et al. (1998) also discuss the limits to which the concept of fluid gel formation is bound. This has been touched upon earlier in this chapter. That is, the hydrocolloid used to produce a fluid gel must under quiescent conditions produce a gel via aggregation. Figure 6.8 shows the relative strengths of different gelling hydrocolloids. Although this is from quiescent measurements it should follow that different gels also present their relative physical attributes when formed as a fluid gel. With regard to this, Norton et al. (1998) described the physical range of materials, seen in Figure 6.8, which have successfully produced gel particle and indeed showed that gel type, amongst the other variables does affect the final properties of the fluid gel. A practical limit is that only hydrocolloids above the break in the line form true gels (i.e. where the elastic modulus is at least an order of magnitude above the loss modulus, and the elastic modulus is independent of frequency over an order of magnitude change) and can be used to produce fluid gels. It has been seen (unpublished data) that gelatin, although capable of forming a true fluid gel, is difficult to use due to the highly dynamic nature of the gel junction zones at ambient temperature and thus the innate surface reactivity even when in suspension as separate particles. However, the inter-particle reactivity is not restricted to gelatin alone. Although, inter-particle interactions diminish with continued cooling and shearing during processing, as shown by the reduced final product viscosity shown in Figures 6.3a and 6.4, some interactions will persist. This continued level of interaction has been shown to exist as the storage modulus of these systems can change over a considerable storage time until an asymptote of the structure is reached (this is shown graphically in Figure 6.7), although some discussion remains as to whether this is an inter- or intra-particle interaction.

The inter-particle reactivity of the set gel can be best seen as a yield stress of the final particulate suspensions. The most likely cause for this is the partial retention of surface filaments post cooling. These filaments would then be free to interact until a sufficient stress is applied to either break or disentangle them. This is demonstrated by Figure 6.9 which shows the yield stress of an agar-based fluid gel. Notably, this yield stress has persisted through repeated shear and recovery experiments. Here the fluid gel has been formed, and then sheared, post shearing the gel is left to stand (for a time much greater than the shearing time) and then re-sheared. The data in

Figure 6.9 are from a gel after three cycles through this regimen and yet a noticeable (although reduced) yield stress persists between the particles (Gabriele et al., in preparation). Moreover, the experiments used to derive Figure 6.9 suggested that gel entanglements were mediated by ionic interactions between adjacent particle and their annular hairs (un-published data). Although ionic concentration effect are not to be covered in depth here, the use of multivalent counterions for stabilizing these systems into new materials may be of future interest.

It was also noted by Norton et al. (1999) that the size differences generated by changes in processing shear could produce a strange behavior in the final particulate systems, and this may or may not be implicated in the presence of annular hairs on the surface of the larger particles. The paper shows that the bulk viscosity of samples with smaller particles, produced from gels at a higher shear rate, had a lower viscosity than the larger particles made with reduced shear. This counterintuition for particulate systems may be, as suggested, due to closer packing of the forming filaments within each particle and so causing an effective reduction in phase volume (a reduction of the order of 4% was suggested). Or it might be the presence of persistence of the annular hairs on the larger particles which form a stabilizing network which was not present on the smaller particles. These smaller particles may have experienced different shear regimens during production and so are morphologically and chemically distinct from the larger ones. However, this is also slightly counterintuitive to the model formation system seen in Figure 6.4. What is clear is that more work in this field is required.

The majority of this section has centered on the process–gel interactions during the formation of fluid gels: shear rate, cooling rate, gel type and concentration. All these have been discussed, as have their influences upon the gel properties. Along with these discussions has been the imperative that the knowledge gained

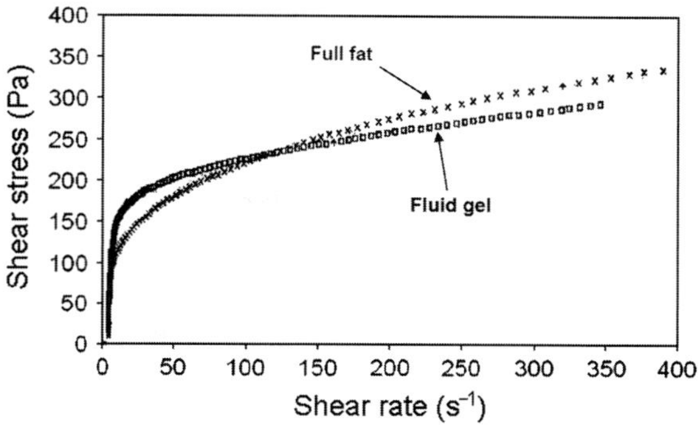

FIGURE 6.10 The comparative rheology of a full fat mayonnaise and a fluid gel, reduced fat alternative (redrawn from Norton et al., 2008).

should allow for novel products to be designed. The properties of the gels might then be used to achieve a fat level reduction or improved satiety in novel foods (Brown et al., 1990; Frith et al., 2002; Norton et al., 2006b). To this end Norton et al. (2006a) used the approaches described above to produce a range of materials with varying physical, mechanical and by extension oral properties, which could match existing products (Figure 6.10). Moreover, these variable materials were all made with similar ingredients. At the outset of this section it was stated that both these effects are currently important drivers for the food industry. What is important and has been alluded to and seen for a number of products is that matching of the oral properties of the new healthier product to those of the unhealthy version is a successful route for novel product entry into the market place (Norton et al., 2008a, 2008b). This remains a non-trivial task. However, Figure 6.10 shows the result of one such attempt to match the textural response, of a fluid gel and an original product, in this case of a mayonnaise. Although the match is not exact there is clear evidence that this approach is worth pursuing (Norton et al., 2008a, 2008b).

In conclusion, this section has shown the possible uses of fluid gels, i.e. for fat replacement, etc. It has also shown the design space within which a single-component system can be used to produce a variety of physical structures (Figures 6.3b and 6.3c) with variable physical properties (Figure 6.6). This design space is then bounded by the type, concentration and processing that is applied or used with a single-component system. Fluid gels, due to these fascinating properties, look set to find further use in the food industry and almost certainly beyond.

6.3 WATER-IN-WATER EMULSIONS

It has been long appreciated that biopolymers when mixed in aqueous solutions often phase separate to give two immiscible phases that mainly contain water (Albertsson, 1995). In many ways these systems are comparable to conventional oil and water emulsions, hence the term 'water-in-water' emulsions. Their microstructure therefore has been found to depend on the same physical principles as classical emulsions (Brown et al., 1995), such as droplet break-up and droplet coalescence, and as suggested before their behavior can be modeled using the same theories (Foster et al., 1996).

When phase separation occurs in the absence of an applied flow field, the included phase normally forms spherical domains, the size of which usually ranges between 2 and 20 μm (Norton and Frith, 2001). These structures are similar to conventional emulsions, however, some differences do exist and they usually arise from the fact that both immiscible phases in aqueous two-phase biopolymer systems are mainly water. This results in very small interfacial tensions between the two phases of water-in-water emulsions (Forciniti et al., 1990; Ding et al., 2002; Ding et al., 2005), which were shown to be in the range of a few μN/m.

Additionally, in the case of water-in-water emulsions, there is less need for stabilization of the droplets against creaming or sedimentation as the densities of the two (mainly water) phases are essentially the same. Furthermore, due to the absence of a hydrophobic phase, the microstructural properties of water-in-water emulsions can not be modified by surfactants; added emulsifiers show no preference to adsorb at the interface and tend to remain in the bulk water phases in self-aggregated form (micelles). This means that any contact between droplets in these systems will result in coalescence (Ding et al., 2002) in the same way as for classical oil and water emulsions. The best way to stabilize the structures is to gel one or both of the phases. This kinetically traps the emulsion structure and induces long-term stability (Loren et al., 2001).

The quality of water-in-water emulsion-based products often depends on the morphology and structure of these mixtures, which in turn strongly depends on the interfacial tension between the two aqueous phases. It is therefore very important to understand the phase behavior, rheological behavior and other factors affecting phase morphology and structure of such systems (e.g. effects of processing).

6.3.1 Interfacial Tension

The droplet size and extent to which the shape of an included phase can be modified via a process depends upon the interfacial tension (Ding et al., 2002). Surprisingly enough there is only a small number of studies on the interfacial tension in aqueous two-phase systems (Bamberger et al., 1984; Ding et al., 2002, 2005; Spyropoulos et al., 2008). This is probably associated to the fact that techniques used for the measurement of interfacial tensions in conventional o/w or w/o emulsions cannot be used in water-in-water emulsions because of the partial solubility of the phases and also since the interfacial tensions in the latter systems are very low.

A number of groups (Guido and Villone, 1999; Wolf et al., 2000; Ding et al., 2002) have adopted the approach of measuring droplet deformation in a controlled applied flow field to calculate interfacial tensions in a number of different systems. The technique involves observation of single drop deformation in an applied flow field, and the relaxation of shape after cessation of flow (Guido and Villone, 1999); thus problems associated with small density differences and partial solubilities are avoided.

The calculated interfacial tensions have been found to be in the range of 0.5–500 μN/m (Ding et al., 2002), i.e., two to three orders of magnitude lower than in typical oil/aqueous systems (Zaslavsky, 1995). This is mainly because interfaces in water-in-water emulsions are formed by two phases high in water content (~80–90%). As yet knowledge of the structure of the interface between these aqueous phases is limited, although it has been reported (Bamberger et al., 1984; Forciniti et al., 1990) that interfacial tension will depend on the molecular weight and total concentration of the polymers in the system. Interfacial tensions in water-in-water emulsions have been correlated either with the tie-line length (TLL) or the difference between the concentrations of the polymers in the co-existing phases (Bamberger et al., 1984; Forciniti et al., 1990; Ding et al., 2002). The interfacial tension in these systems has also been found to somewhat depend on the concentration of additives such as salts (Bamberger et al., 1984).

In conclusion what has become clear during our discussion is that the morphology of water-in-water emulsions is determined, almost exclusively, by the choice of the processing parameters. Other contributing factors such as rheological behavior and phase sense are discussed in the following sections.

6.3.2 Rheological Behavior

In order to facilitate the use of biopolymer mixtures in industrial applications knowledge of their flow properties and morphology during flow processing is of great importance (Lundin et al., 2000; Norton and Frith, 2001; Stokes et al., 2001; Wolf et al., 2001; Wolf and Frith, 2003; Ding et al., 2005). The apparent viscosity of these aqueous two-phase systems depends on the viscosity of the continuous phase, the viscosity ratio (λ), the chemical compositions of the individual phases (Wolf and Frith, 2003), the shape (aspect ratio) and the phase volume of the dispersed phase (Wolf and Frith, 2003). Hence there is a close link between flow behavior and composition with the shear viscosity/composition dependence patterns for aqueous biopolymer mixtures exhibiting great similarity to those of conventional o/w and w/o emulsions (Pal, 2001).

The blend viscosity also depends on the phase sense or in other words on which of the two phases in the system is dispersed and which continuous. As a result, at the phase inversion composition, the viscosity of water-in-water emulsions should 'change'. For this reason there have been reports in the literature where the phase inversion composition (under shear) is deduced from rheo-optical investigations (Wolf and Frith, 2003) or calculated by simple empirical models (Utracki, 1991).

6.3.3 Phase Inversion

The phase morphology and structure in many biopolymer mixtures can be kinetically trapped by the gelation of one or both of the phases. Monitoring the droplet size evolution through turbidity measurements has demonstrated that gelation of one or both phases will arrest the process of ripening of the phase structure (Williams et al., 2001). By variation of the quench depth, it then becomes possible to produce biopolymer gel 'composites' with a range of trapped microstructures and properties (Norton and Frith, 2001).

When biopolymer mixtures are subjected to industrial processes, the additional factor of a flow history is superimposed on the phase separation/gelation history. This can of course have quite dramatic effects on the final microstructures and textures produced. A considerable literature already exists on this subject for

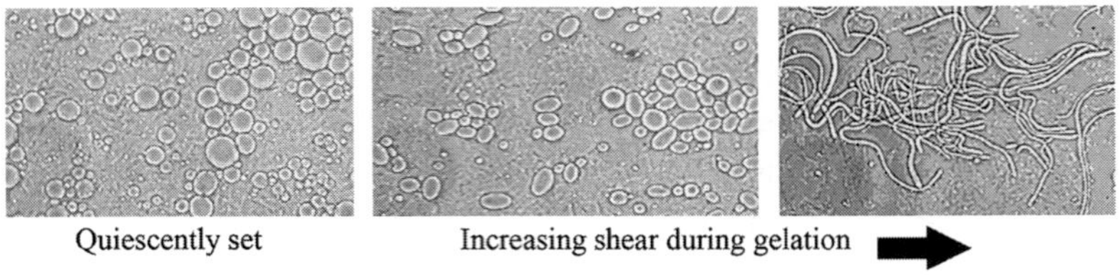

FIGURE 6.11 Different particle morphologies produced by increasing the shear regimen during gelation.

biopolymers (de Carvalho and Djabourov, 1997; Norton et al., 1999; Wolf et al., 2000). In mixed systems the simplest effect of shear while separation, molecular ordering and gelation are occurring is to reduce the particle size of the included phase and to stabilize the microstructure. This can be used to make a phase-separated biopolymer system that has droplet sizes in the range of 1–10 μm (Foster et al., 1996).

The very application of shear to biopolymer mixtures undergoing gelation can also influence the phase sense of the system and in some cases induce phase inversion. For instance, if gelation of both phases in a mixed system (7% gelatin/9.5% maltodextrin) is induced quiescently then the mixture will normally form a gelatin continuous composite (Foster et al., 1996). In the liquid state, the application of shear does not cause any phase inversion. However, if shear is applied as the mixture is cooled through the gelation temperature, then a phase inversion to a maltodextrin continuous system occurs. This is believed to occur because, as the mixture is cooled, the gelatin-rich phase orders first and its viscosity increases, becoming greater than that of the maltodextrin-rich phase. As with oil/water emulsions and polymer blends, the application of shear to such a mixture leads to the phase with the highest viscosity becoming the included phase (Han et al., 1998). By confirming the idea with other concentrations and different biopolymer mixtures it appears that this general rule for o/w emulsions and blends also applies to mixed biopolymers (Foster et al., 1996).

By controlling the phase sense of mixed biopolymers in the shear field and then by gelling one or both phases, either at the very end of the process or very quickly after the process, it is possible to kinetically trap the dispersed phase in an anisotropic morphology (Antonov et al., 1980; Wolf and Frith, 2003). These trapped structures can range from simple ellipsoidal droplets to even string phases (Figure 6.11).

Once these asymmetric particles have been formed they will exhibit different rheological properties to those possessing a quiescently cooled, spherical phase structure, e.g. higher viscosities at lower phase volumes, complex flow patterns, etc.

6.4 PROCESSING INSIDE PEOPLE

During the human digestive process, foods undergo major size reduction, by both mechanical forces and chemical reactions to help release embedded nutrients, which are eventually absorbed by the body. The kinetics of digestion depends on the chemical and physical characteristics of food and their interaction with the physiological events occurring within the gas-trointestinal (GI) tract. Mouth and stomach are the major compartments where foods are disintegrated into small size, while the small intestine is the most important site of nutrient absorption (Kong and Singh, 2008a).

It is well known that hydrocolloids can have a dramatic effect on the processing of foods in the oral cavity and the rest of the GI tract. What is starting to emerge is an area of research where hydrocolloids are added to foods in order to undergo 'changes' during a specific biological process(es) taking place within the human body. Understanding what 'processing inside people' involves could offer scientists the capability to design and formulate foods that have the potential to impact either texturally/sensorally or that break-down and release their micro-, and probably more importantly, macronutrients in a controlled fashion.

The present section considers the environment and processes that occur during the digestive process and how these influence or are influenced by hydrocolloids. The discussion that follows has been segmented in order to follow the specifics of the three main parts of the GI tract: processing under the mouth, stomach and small intestine conditions.

6.4.1 Processing in the Mouth

Once in the mouth, food is subjected to a complex series of manipulations by the tongue, teeth, lips and cheeks, during which it is converted into a form suitable for swallowing (Guinard and Mazzucchelli, 1996). Sensory assessments of food begin prior to ingestion when the food is seen, handled and smelled, and continue in the mouth to assess taste and texture during oral processing and swallowing. Some sensations persist after swallowing and are termed 'after-taste' (de Wijk et al., 2003).

The importance of oral processing has long been recognized and studied for solid foods. Mastication is the initial digestion stage and the main process by which solid food is broken down into smaller particles. It is a complex function that utilizes not only the muscles and teeth, but also the lips, cheeks, tongue, palate, and salivary glands. It involves well-controlled compressive, shearing and tensile forces brought about by the intermittent movements of the teeth (Malone et al., 2003). The forces applied to food during mastication depend on the texture properties of the food itself (food of brittle, elastic and/or plastic behavior) and usually range from 30–60 N. The bite duration (the length of time that the food is subjected to the applied force during each bite) is equally important and also relative to the properties of the food being consumed (Mioche and Peyron, 1995). Short bite durations ($\sim$0.8 s) are recorded for foods of elastic or brittle behavior, while the bite duration for plastic products is longer since, during the first bite, samples tend to flow under the constant level of applied force (Mioche and Peyron, 1995).

Due to the importance of mastication in the oral processing of solid foods there have been efforts to mathematically model the process (van der Bilt et al., 1987; Prinz and Lucas, 1997; Lucas et al., 2002). Even though some of these models are considered to adequately describe the process of mastication (for instance giving relatively successful predictions for the particle size resulting from mastication), they 'suffer' from the large number of assumptions which, due to the complexity of the process, are introduced and which sometimes over-simplify the problem.

Also important in the oral processing of solid foods is the effect of mixing with saliva. As soon as food is placed in the mouth, saliva is secreted in response to mechanical stimulation (chewing) and to the taste qualities of the food (e.g. pH). Saliva mixes with the food, acting as lubricant during the masticatory process, but also affecting the textural changes induced to the food during oral manipulation. Mixing with saliva also affects bolus formation, thus the decision for swallowing, and as a result influences the residence time of the food in the mouth (Guinard and Mazzucchelli, 1996).

The degree of mixing achieved during mastication has been experimentally assessed using food samples (e.g. chewing gums) constructed from two identical parts each containing different (contrasting) colors. These food samples are placed in the mouth, processed orally (e.g. chewing) and then spat out and analyzed using image-processing techniques. The same technique has been extended for use with semi-solid foods such as custards and mayonnaise (de Wijk et al., 2006).

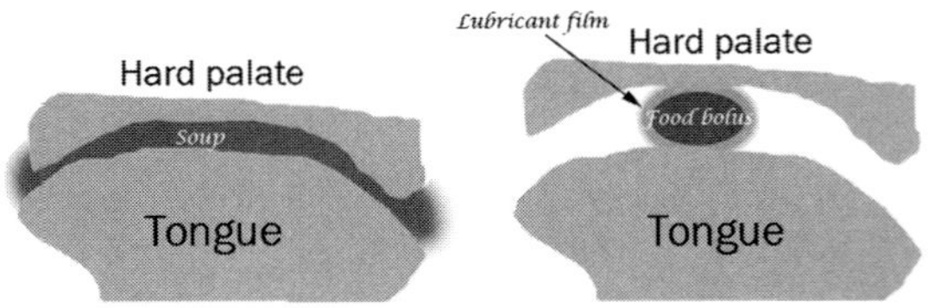

FIGURE 6.12 Squeezing/shearing between hard palate and tongue for a liquid and semi-solid food respectively. Note that in both cases food also acts as a lubricant between the two surfaces.

For soft food materials, which typically are those structured with hydrocolloids, including liquid, semi-liquid and semi-solid foods, mastication is not as important since it does not involve fragmentation by the teeth. Crucial in the oral processing of soft food materials are the structural changes resulting from heating (or cooling) to mouth temperature, mixing with saliva (dilution and chemical breakdown) and the flow involved in squeezing and shearing between palate and tongue (Figure 6.12). During oral processing, food is also subjected to the flow regimen in the mouth. These flow conditions are of specific importance to liquid and semi-liquid foods as they determine to a great extend how the food is perceived and broken down in the mouth. Determining the flow regimen(s) during oral processing though is far from an easy task since the flow conditions involved are both product and consumer specific but moreover because it is impossible to monitor (without any interference) the degree of mixing in vivo. The approach usually taken is to deduce the flow characteristics, involved during oral processing of the food, by correlation to certain sensory evaluated food attributes, such as its perceived viscosity (Shama and Sherman, 1973; Guinard and Mazzucchelli, 1996; Engelen and van der Bilt, 2008).

But these types of correlations are not trivial either. Sensory evaluation of viscosity for example does not occur at a constant shear rate since the effective shear rate strongly depends on the viscosity of the liquid and semi-liquid food products, which in their majority are of non-Newtonian (usually shear-thinning) nature. A study by Shama and Sherman (1973) concludes that low-viscosity liquids (viscosity roughly below 0.1 Pa s^{-1}) are evaluated at a constant shear stress of about 10 Pa in the mouth, while products with a viscosity above 10 Pa s^{-1} are evaluated at a constant shear rate of 10 s^{-1}. What the authors assumed was that the liquid is squeezed between the tongue and the palate and then sheared by the movements of the tongue; therefore that the prevailing flow type between the tongue and palate is shear flow. However other types of deformations, such as turbulent or elongational flow, also exist and may be equally or even more important than shear flow (van Vliet, 2002).

This was the basis behind the work of Malone et al. (2003) who introduced the concept of tribology for the study of mouth behavior. These researchers carried out a series of studies on a range of different hydrocolloid systems and showed that the concentration of the polymer can have significant effects on the thin film behavior (Figure 6.13). These studies also showed that the nature of the hydrocolloid itself can affect thin film behavior; e.g. xanthan, as a consequence of its more rigid structure, lowered the frictional coefficient more than a system containing the same concentration of guar, a more random polymer (Figure 6.14). In

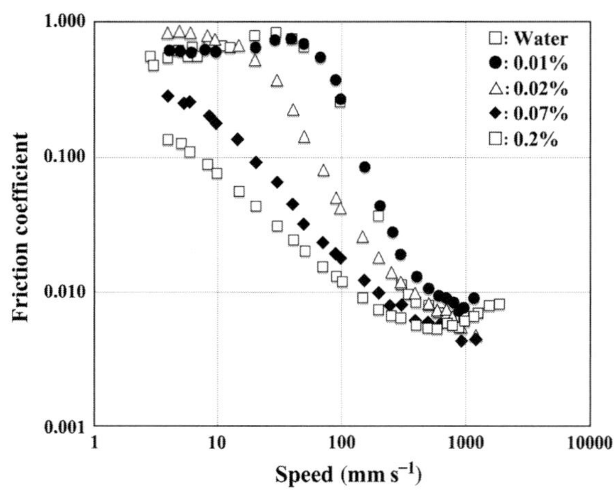

FIGURE 6.13 Stribeck curves obtained for water (reference) and for a range of xanthan concentrations.

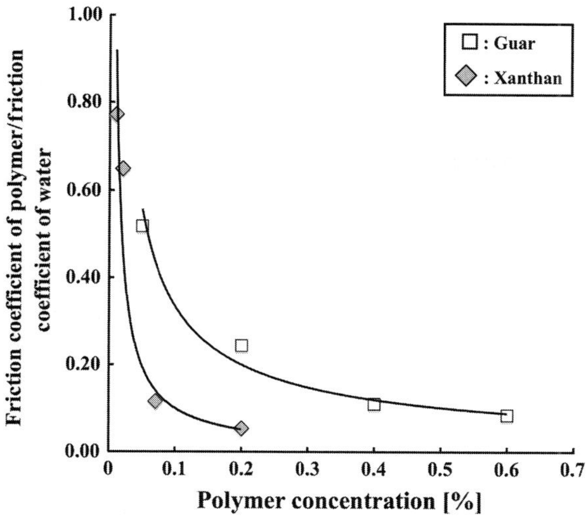

FIGURE 6.14 Friction coefficient for guar and xanthan (divided by the friction coefficient of water) as a function of the polymer concentration.

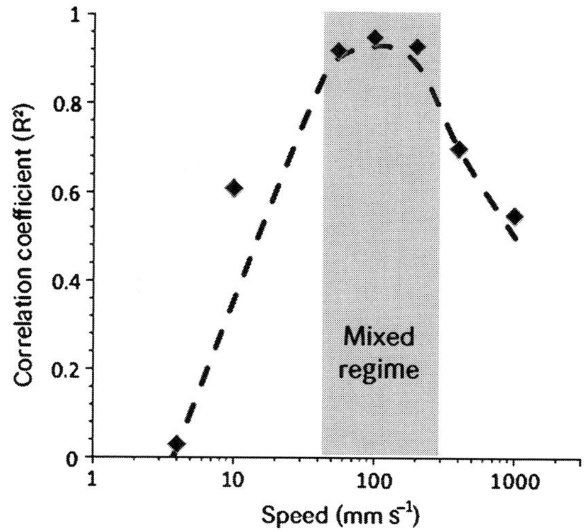

FIGURE 6.15 Correlation coefficient between tribology data of guar gum solutions and sensory data for their perceived slipperiness. Redrawn from Malone et al. (2003).

addition, this group of workers introduced the concept of 'soft tribology' in which soft (silicon rubber) materials are used as the surfaces in the tribometer so as to mimic the soft surfaces of the buccal cavity. These tribological measurements on different hydrocolloids were then correlated with a set of sensory measurements. From this correlation (Figure 6.15) it was concluded that the tongue and hard palate move against each other at speeds between 10 and 100 mm s^{-1} (Malone et al., 2003). Work within this research area is still at an early stage and future research seems set to reveal exciting results and insights into, amongst others, how food interacts with biosurfaces.

The ways by which foods break down in the mouth can also have significant effects on the perception of certain flavors in semisolids (Engelen and van der Bilt, 2008). In a study by Engelen et al. (2002) it is reported that increasing the product's temperature results in an increased intensity of perceived flavor and melting mouthfeel, while perceived thickness is decreased. These can be explained by a possible increased enzymatic action (leading to more structure breakdown and therefore flavor release) and/or a decrease in viscosity (and hence a more melting sensation).

Mixing the food with saliva, including breakdown and dilution, is considered to be of great importance for soft solids based on hydrocolloid structuring agents (Koliandris et al., 2008), as these products are masticated without chewing. Mixing with saliva increases flavor perception by assisting the breakdown of the food structure (due to the action of enzymes, e.g. by the action of amylase in starch-containing products or by inducing flocculation of emulsion-based products) and afterwards facilitating the transport of the tastants to the taste buds (van Ruth and Roozen, 2000; van Aken et al., 2007; Engelen and van der Bilt, 2008). The recent work on mixing and its dependence on the hydrocolloid structure is clearly important in this area. Fluid gels and mixed gels, as discussed earlier, are expected to have a major impact, although no systematic studies have been carried out to date.

With oral processing of hydrocolloids-based foods the composition and volume of saliva have also been found to influence flavor release (van Ruth and Roozen, 2000). In combination these effects result in a change in the viscosity of the food under oral processing. As a rule of thumb it is expected that mixing with saliva will increase the viscosity of liquids with a low viscosity and decrease the viscosity of those with a high viscosity, under mouth conditions (Prinz et al., 2007). However, when considering saliva effects on the viscosity of soft-solids, one should also take into account the residence time of such formulations in the mouth; the residence time for low-viscosity liquids is a few seconds at most.

A study by de Bruijne et al. (1993) on the break-up process in the oral cavity of small samples of oils of different viscosity showed that the effective flow, causing break-up, was purely elongational. This implies that foods of a high yield stress will not be broken up and dispersed by the saliva flow during oral processing, but more likely will initially be compressed between the tongue and the palate. The nature and extent of such a deformation (resembling squeezing flow between parallel plates of a narrow gap) will depend on the amount of saliva (amongst others providing lubrication) present, the surface roughness of the tongue and the palate, and the extent, type and rate of the tongue movements (van Vliet, 2002). Thus hydrocolloids that gel or are highly entangled in solution are expected to be more resilient to breakdown in the mouth.

6.4.2 Processing in the Stomach

Compared to oral mastication, gastric disintegration of foods has received less attention partly due to its complexity; it involves numerous influencing factors such as fed/fast state, gastric acid, bile salts, enzymatic reactions, and hydrodynamic and mechanical forces. On the other hand, stomach physiology has not been fully understood; the stomach wall movement, rheological properties of gastric content, the flow state of gastric fluid, and hydrodynamic/mechanical forces acting on foods require further clarification. Most of the studies have been connected with medical and nutritional research. Recently, the notion of healthy foods and bioavailability is gaining wide recognition (Norton et al., 2007).

Once in-mouth processing of the food has finished the formed bolus is transported through the esophagus to the stomach where gastric digestion occurs. The stomach performs three functions: storage, mixing, and emptying its contents into the small intestine (Kong and Singh, 2008a) and it is divided into four major regions: fundus, body, antrum, and pylorus (Figure 6.16). The proximal part of the stomach made of the fundus and body acts as a reservoir for undigested material, responsible for the emptying of liquids, whereas the distal stomach (antrum) is the grinder, mixer, and siever of solid food, and acts as a pump for gastric emptying of solids by propelling actions (Arora et al., 2005). The reservoir part of the stomach has a flexible volume with the ability to expand from as low as 25 mL (in the fasted state) to about 4 L (Kong and Singh, 2008b). The mixing and homogenizing function is achieved through the secretion of gastric juice and stomach contraction that produces grinding and crushing of the food.

Typical gastric juice in the stomach is a fluid of non-Newtonian (shear-thinning) behavior with viscosity roughly in the range 0.01 to 2 Pa s^{-1} (Marciani et al., 2000). Although ingestion of a high-viscosity meal will increase the

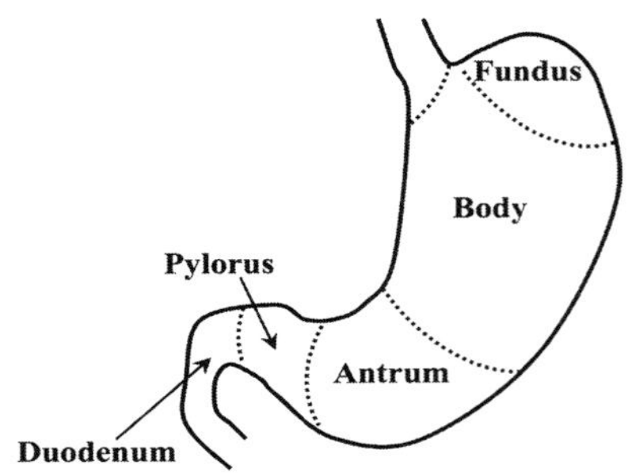

FIGURE 6.16 Schematic of the stomach and its different regions.

apparent viscosity of the contents of the stomach, the effect is minimized as the stomach responds to high-viscosity meal ingestion by rapid intragastric dilution causing a reduction of meal viscosity. Marciani et al. (2000) reported that the zero-shear viscosity of a meal containing locust bean gum fell from 11 to 2 Pa s^{-1} immediately after ingestion, and further decreased to 0.3 Pa s^{-1} after 30 min. In the fasted state, intragastric pH (in healthy subjects) is close to 2, while in the initial stages of the fed state it can increase to about 5. Within 1 h after eating though, the pH of the stomach decreases to less than 3 (Dressman, 1986). Food composition and quantity play a major role in deciding the time required to restore the fasting pH levels, although food pH value also has an influence (Kalantzi et al., 2006).

The stomach contraction imposes a considerable mechanical force on food particulates and thus plays a significant role on the disintegration of solids. Peristaltic waves originate from the stomach wall and spread toward the antrum, mixing and forcing the antral contents toward the pylorus. The frequency of the contractions is approximately 3 cycles per minute (Marciani et al., 2001a). Meanwhile, the pylorus contracts and the sphincter narrows, so that the pyloric opening is small on the arrival of the peristaltic wave. The contents are thus squirted back into the stomach, an action called 'retropulsion'. Repeated propulsion, grinding, and retropulsion, together with the function of acids and enzymes, reduce the size of food particles. In the gastric antrum, solids are ground to particles of a size less than 1–2 mm (Carneiro et al., 1999). Size reduction is a rate-limiting step in emptying digestible solids into the intestine (Meyer et al., 1981). Researchers have measured contraction forces present in the stomach. Marciani et al. (2001b) used magnetic resonance imaging (MRI) to directly visualize the breakdown of agar gelled beads in the antrum and reported forces close to 0.65 N exerted by the antral walls of the stomach.

The gastric wall motion itself, associated with antral contractile activity, pyloric opening, and fundic contractions, is also primarily responsible for the fluid motion within the stomach (Pal et al., 2004). The characteristic flow velocity is established by the propagation speed of the antral contraction waves (2–3 mm s^{-1}) (Kwiatek et al., 2006). Computational modeling has been used to construct flow paths for fluids and particles in the fluid motion, determine fluid forces in the fed stomach, and evaluate the stresses involved. Pal et al. (2003, 2004) developed a model of the human stomach and demonstrated that antrum contraction waves are central to gastric mixing. The strongest fluid motion is at the lower parts of the stomach, where the retropulsive jet is generated by contractions in the antrum, with velocities of up to 7.5 mm s^{-1}. Food particles receive the highest fluid surface shear stresses (10–30 Pa). On the contrary, the fluid motion was very low within the fundus with low shear forces on food particles (<1 Pa). Another pattern of fluid motion identified was flow vortices (or eddies) that circulated particles between successive contraction waves.

It is within these process parameters that we need to think about the way that hydrocolloid structures can both influence and be influenced by the biological, chemical and physical processes occurring in the stomach. One particular approach is to produce foods that structure the stomach contents and thus provide a feeling of satisfaction followed by prolonged satiety. If successful the meal then provides a pleasant sensation of 'fullness' for a prolonged period and so snacking on high-fat/calorie foods may be reduced. To fully achieve this requires design rules to manipulate new foods and give a reduced calorie intake within a day and as a function of time between meals.

Initial work has shown that a potential way of having soft or liquid foods that change the way that people feel and their energy intake is to use materials that respond to the environment they find themselves in; e.g. in this case the stomach environment. Consumption of liquid-like foods structured with a hydrocolloid that is sensitive to a shift in pH, so that it self structures in the stomach and gels (acid gelation) part of the stomach's content content, has been shown to increase satiety (Hoad et al., 2004; Norton et al., 2006a). The observed fullness effect was found to be directly related to the strength of the formed gel; i.e. weak gels only provoked a weak response similar to that of a viscous meal, while stronger gels, which were more resistant to breakdown in the stomach, were more successful in extending satiation. These results were obtained with calcium-sensitive alginate and with little or no control over gelation rates. In a recent work by Norton et al. (2006a) there was a conscious attempt to understand the gelation process and also control the gelation rates for a range of acid-sensitive hydrocolloids both singularly and in combination.

Instead of using liquid-based products to provide the above-mentioned satiety effect, a similar approach can be taken by using a solid food formulation (solid bar or cereal-like breakfast replacement); this could prove to be a more apt system to consider, especially if snacking between breakfast and lunch is to be reduced. But if the use of a solid, albeit palatable product, is to be explored then three kinetic processes must be controlled. The first one should be the process of dissolution of the product throughout the stomach. Although this is readily achievable with a liquid product in a stomach with little or no contents, the use of other food formats on a full stomach presents difficulties in ensuring their homogeneous distribution, which is set to prove itself as a non-trivial task. The second process that should be controlled is the rate of gelation of the food, so as to ensure that the whole of the structure is saturated with gastric juice prior to gelation. If the gelation rate is too rapid then the structuring material is likely to gel only on the outer surface. This will result in the formation in the stomach of an enlarged form of what is commonly referred to as a 'fish eye' and describes the exterior partial wetting of powders when dissolved. Presumably these structures would then slowly be broken (physically, chemically or enzymatically) to expose fresh surfaces, which in turn will gel and break (and so on), thus losing the controlled structuring process that is required. This might be addressed through the kinetics of gelation of single biopolymers that can be controlled by designing the required molecular structure. Also, new approaches might use mixed acid-sensitive biopolymers or an acid-sensitive biopolymer with a non-acid-sensitive biopolymer. Here the kinetics would not only be based on the molecular structure of both components but on the relative phase volumes of the polymers. Finally, the process by which the de-structuring of the stomach's contents occurs (to allow for what is central to any self-structuring methodology – the emptying and loss of stomach distension and satiety) should be controlled; this process will be affected by the other two discussed above. Here all of the factors affecting satiety and stomach emptying must be co-ordinated and controlled so that any structured materials are only meta-stable and will break appropriately to relieve stomach distension and allow the feeling of hunger to return in a timely fashion. Controlling the timing of the process is of primary importance. This is because, as stated at the outset, the intention of a self-structuring approach is to reinforce the concept of a normal/regular diet. So the re-establishment of a hunger pattern is paramount to this whole process, so that moderate/well-distributed intakes of food are consumed – i.e. regular meal times. Thus the diet is then controlled to incorporate satisfying meals which can be ingested unhampered by the structuring/de-structuring process and fulfills its task of limiting between-meal snacking on high-fat/high-sugar foods.

Here we have considered a single, albeit strong, approach to self-structuring in the stomach in order to provide satiety and thus control food intake. Recently, Foster and Norton (in press) have taken this argument further by introducing other important considerations for the design of self-structuring foods. Amongst those they suggest such approaches to self-structuring as, for example, the use of emulsions, the use of mixed hydrocolloids and physical effects such as dilution, viscosity and temperature effects.

6.4.3 Processing in the Small Intestine

The small intestine (or small bowel) plays a crucial role in human digestion, as it is responsible for the absorption of essentially all dietary organic molecules (including glucose, amino acids and fatty acids), electrolytes (sodium, chloride, potassium), water and other compounds such as drugs (Cunningham et al., 1998). It has an average length of approximately 2–6 m and it is loosely divided into three sections: the duodenum, the jejunum, and the ileum, with each comprising approximately 5%, 50%, and 45% of its length, respectively (Ganong, 1995). The average external diameter of the small bowel is approximately 2.5 cm, but it varies from 5 cm in the upper section of the duodenum to less than 2 cm in the distal ileum.

The pH environment in the (healthy) small intestine is highly individual and state (fasting or fed state) specific. pH values of the contents of the small intestine have also been found to be influenced by parameters such as the type of food (being digested) and the gastric emptying process. Moreover pH also depends on the section of the small intestine under consideration and it is generally lower in the duodenum, ~6.3 in the fasting period and ~5.3 after a meal (Rune and Viskum, 1969), but gradually increases towards neutral or even slightly alkaline pH values in the lower ileum (Karr et al., 1935).

From an engineering point of view, the small intestine is a complex system involving the flow and mixing of fluids having non-linear rheological properties under time-dependent conditions. Digestion and absorption in the intestine require constant mixing of the ingesta, propulsion of the intestinal contents and local microcirculation across the absorbing surface of the epithelium, where enzymatic digestion is completed and absorption occurs (Gregersen, 2002). The movements of the small intestine, responsible for the fluid dynamics, have been divided into 'mixing' and 'propulsive' contractions (Guyton and Hall, 2000). The mixing contractions, known as 'segmentation contractions', occur when sections of the small intestine become filled with digesta (the food and intestinal secretions in the small intestine) causing a localized concentric contraction; with the process being repeated at spaced intervals along the intestine (Guyton and Hall, 2000). On the other hand, propulsive contractions, also known as 'peristalsis', are responsible for propelling the digesta through the small intestine (Guyton and Hall, 2000). Propulsive contractions move intestinal contents at a velocity of 0.5–2.0 cm s^{-1}, and they are normally weak contractions that do not extend in length more than 3–5 cm from their point of origin (Bakalis et al., 2007).

The intestinal motility in the small intestine affects the fluid flow profiles of the digesta. The digesta is subjected to a wide range of shear rates and other deformation modes, such as extensional flow, at different sites of the gut and at different times (Ellis et al., 1996); this is further complicated by the extreme heterogeneity of the digesta system. Some information about the rheological properties of the digesta has been provided by animal studies; by using pigs that were fed meals with and without guar gum Roberts et al. (1990) measured the maximum zero shear viscosity of jejeunal digesta and found it to range from 18 to 1454 mPa s^{-1}.

Digestion and absorption in the intestine require the mixing of the intestinal content and also the microcirculation across the absorbing surface of the epithelium (Gregersen, 2002). To influence these processes it is important to understand and control the rheology of the fluid at the shear rates and flow profiles found in the GI tract. Although the shear rates and flow profiles for the human small intestine have not yet been determined, there is some information about flow profiles from numerical modeling of the small intestinal wall movements in animals (Jeffrey et al., 2003). The numerical findings by Jeffrey et al. (2003) showed that the redistribution of particles and mixed liquids in the intestine is caused by downstream and reverse flow, and vortical flow patterns. They also found that contractions generated pressures and shear stresses (maximum magnitude of 1.2 Pa) in particular along the moving section of the wall (Jeffrey et al., 2003). However, the fluid used and modeled in this study was Newtonian in nature and therefore does not exhibit the shear dependence expected for the intestinal content.

The physical form of the food itself being ingested as well as the physical properties of the produced intestinal contents can also influence the rate of intestinal digestion, absorption and transit. For instance certain soluble dietary fibers, giving viscous solutions, are known to reduce the rise in postprandial (after eating) glucose and insulin levels in man (Jenkins et al., 1978). Taylor et al. (1980) showed that guar gum, compared to other fibers or fiber analogs, had the greatest effect on glucose absorption. Blackburn et al. (1984) showed that the presence of guar gum in glucose solutions fed to humans reduced the peak increase in blood glucose level by 50%. The authors went on to conclude that guar improved glucose tolerance by reducing glucose absorption in the small intestine, which in turn was ascribed to inhibition of the intestinal motility effect on fluid convection.

There are many questions about how hydrocolloids influence the behavior of the intestine and how the processing parameters in the intestine impact on the hydrocolloid structure and properties. Clearly these are questions that need to be answered if foods are to be constructed for specific function in this part of the human GI tract. Recently researchers (Bakalis et al., 2007; Tharakan et al., 2007) have constructed a mixing device specifically targeted to simulate the flow, mixing, chemical reactivity and nutrient absorption of the intestine. Preliminary work has produced some quite interesting results (Figure 6.17), which leads us to believe that it is this type of experimental setup that should allow foods to be designed in the future that are well targeted for their behavior and nutrient delivery within the intestine.

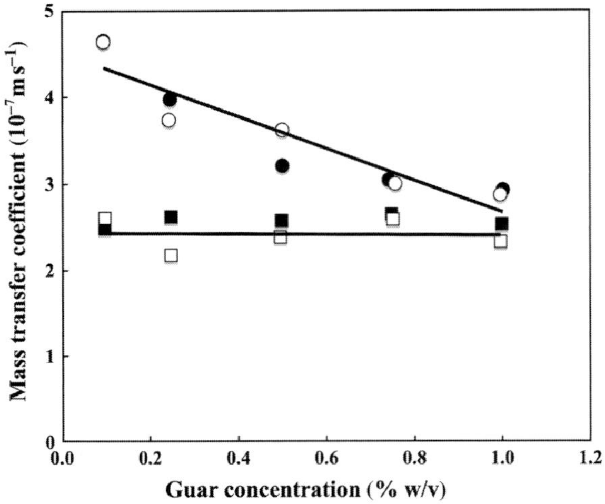

FIGURE 6.17 Mass transfer coefficient as a function of guar concentration for a range of processing conditions: no flow/no squeeze (■), flow/no squeeze (□), no flow/squeeze (●) and flow/squeeze (○). Redrawn from Tharakan et al. (2007).

6.5 THE FUTURE

As discussed in this chapter, flow can have a considerable effect on the properties of hydrocolloids. It is likely that in the future this will become more interesting to the academic and industrial communities. This will be driven on one hand by the industrial desire to achieve better control over the materials used and to develop the ability to induce new material properties from a limited number of acceptable ingredients. On the other hand, the area of consumer perception and what happens inside people will drive the scientific community into the areas of self-structuring systems and into consideration of the binding of hydrocolloids to biosurfaces, both at the molecular level and at the materials level (nm to 10s nm scale). Specifically we expect that future developments will be:

1. Construction of asymmetric structures – textured products.
2. Process control in mixed (emulsion) systems – better understanding of behavior.
3. Surfactants for water-in-water emulsions – block co-polymers molecular weight dependency.
4. Structures designed for GI tract processing.
5. Sheared/fluid gels as delivery systems.

ACKNOWLEDGMENTS

The authors wish to thank Mr A. Gabriele for his input and for carrying out part of the experimental work.

References

Albertsson, P-Å. (1995). Aqueous polymer phase systems: properties and applications in bioseparation. *In* Biopolymer Mixtures, 1–12, Harding, S. E., Hill, S. E. and Mitchell, J. R. (eds.). Nottingham University Press, Nottingham.

Antonov, Y. A., Grinberg, V. Y., Zhuravskaya, N. A., and Tolstoguzov, V. B. (1980). Liquid two-phase water-protein-polysaccharide systems and their processing into textured protein products. *Journal of Texture Studies* **11**, 199–215.

Arnott, S., Fulmer, A., Scott, W. E., Dea, I. C., Moorhouse, R., and Rees, D. A. (1974). The agarose double helix and its function in agarose gel structure. *Journal of Molecular Biology* **90**, 269–272.

Arora, S., Ali, J., Ahuja, A., Khar, R. K., and Baboota, S. (2005). Floating drug delivery systems: A review. *AAPS PharmSciTech* **6**, E372–E390.

Aymard, P., Williams, M. A. K., Clark, A. H., and Norton, I. T. (2000). A turbidimetric study of phase separating biopolymer mixtures during thermal ramping. *Langmuir* **16**, 7383–7391.

Bakalis, S., Thakaran, A., Norton, I. T., and Fryer, P. J. (2007). Modelling of physical processes in the small intestine. *FASEB Journal* **21**, 524.

Bamberger, S., Seaman, G. V. F., Sharp, K. A., and Brooks, D. E. (1984). The Effects of Salts on the Interfacial Tension of Aqueous Dextran-Poly(ethylene glycol) Phase Systems. *Journal of Colloid and Interface Science* **99**, 194–200.

Blackburn, N. A., Redfern, J. S., Jarjis, H., Holgate, A. M., Hanning, I., Scarpello, J. H. B., Johnson, I. T., and Read, N. W. (1984). The Mechanism of Action of Guar Gum in Improving Glucose-Tolerance in Man. *Clinical Science* **66**, 329–336.

Brown, C. R. T., Cutler, A. N., and Norton, I. T. (1990). Liquid based composition comprising gelling polysaccharide capable of forming a reversible gel and a method preparing such composition. *EP0355908*.

Brown, C. R. T., Foster, T. J., Norton, I. T., and Underdown, J. (1995). Influence of shear on the microstructure of mixed biopolymer systems. *In* Biopolymer mixtures, 65–85, Harding, S. E., Hill, S. A. and Mitchell, J. R. (eds.). Nottingham University Press, Nottingham.

Butler, M. F. (2002). Mechanism and Kinetics of Phase Separation in a Gelatin/Maltodextrin Mixture Studied by Small-Angle Light Scattering. *Biomacromolecules* **3**, 676–683.

Butler, M. F. and Heppenstall-Butler, M. (2001). Phase Separation in Gelatin/Maltodextrin and Gelatin/Maltodextrin/Gum Arabic Mixtures Studied Using Small-Angle Light Scattering, Turbidity, and Microscopy. *Biomacromolecules* **2**, 812–823.

Carneiro, A. A., Baffa, O., and Oliveira, R. B. (1999). Study of stomach motility using the relaxation of magnetic tracers. *Physics in Medicine and Biology* **44**, 1691–1697.

Cunningham, J. J., Organ, G. M., and Mooney, D. J. (1998). Engineering Small Intestine. *In* Frontiers in Tissue Engineering, 610–631, Patrick Jr., C. W., Mikos, A. G. and McIntire, L. V. (eds.). Elsevier Science Ltd, Oxford.

de Bruijne, D. W., Hendrickx, H. A. C. M., Anderliesten, L., and de Looff, J. (1993). Mouthfeel of foods. *In* Food colloids and polymers: stability and mechanical properties, 204–213, Dickinson, E. and Walstra, P. (eds.). Royal Society of Chemistry, Cambridge.

de Carvalho, W. and Djabourov, M. (1997). Physical gelation under shear for gelatin gels. *Rheologica Acta* **36**, 591–609.

de Wijk, R. A., Engelen, L., and Prinz, J. F. (2003). The role of intra-oral manipulation in the perception of sensory attributes. *Appetite* **40**, 1–7.

de Wijk, R. A., Prinz, J. F., and Janssen, A. M. (2006). Explaining perceived oral texture of starch-based custard desserts from standard and novel instrumental tests. *Food Hydrocolloids* **20**, 24–34.

Ding, P., Wolf, B., Frith, W. J., Clark, A. H., Norton, I. T., and Pacek, A. W. (2002). Interfacial tension in phase-separated gelatin/dextran aqueous mixtures. *Journal of Colloid and Interface Science* **253**, 367–376.

Ding, P., Pacek, A. W., Frith, W. J., Norton, I. T., and Wolf, B. (2005). The effect of temperature and composition on the interfacial tension and rheology of separated phases in gelatin/pullulan mixtures. *Food Hydrocolloids* **19**, 567–574.

Dressman, J. B. (1986). Comparison of canine and human gastrointestinal physiology. *Pharmaceutical Research* **3**, 123–131.

Ellis, P. R., Rayment, P., and Wang, Q. (1996). A physico-chemical perspective of plant polysaccharides in relation to glucose absorption, insulin secretion and the entero-insular axis. *Proceedings of the Nutrition Society* **55**, 881–898.

Engelen, L. and van der Bilt, A. (2008). Oral physiology and texture perception of semisolids. *Journal of Texture Studies* **39**, 83–113.

Engelen, L., de Wijk, R. A., Prinz, J. F., van der Bilt, A., Janssen, A. M., and Bosman, F. (2002). The effect of oral temperature on the temperature perception of liquids and semisolids in the mouth. *European Journal of Oral Sciences* **110**, 412–416.

Forciniti, D., Hall, C. K., and Kula, M. R. (1990). Interfacial-tension of polyethyleneglycol-dextran-water systems-influence of temperature and polymer molecular-weight. *Journal of Biotechnology* **16**, 279–296.

Foster, T. J., Brown, C. R. T., and Norton, I. T. (1996). Phase inversion of water-in-water emulsions. *In* Gums and stabilisers for the food industry 8, 297–306, Phillips, G. O., Williams, P. A. and Wedlock, D. J. (eds.). Oxford University Press, Oxford.

Foster, T. J. and Norton, I. T., *in press*. Self-assembling structures in the GI tract. In: McClements, D. J. and Decker, E. A. (eds.). *Designing Functional Foods*. Woodhead Publishing

Frith, W. J., Garijo, X., Foster, T. J., and Norton, I. T. (2002). Microstructural origins of the rheology of sheared gels. *In* Gums and Stabilisers in the Food Industry 11, 95–103, Wiliams, P. A. and Phillips, G. O. (eds.). Royal Society of Chemistry Special Publication No. 278, Cambridge.

Gabrielle, A., Norton, I. T., and Spyropoulos, F. 2009. Kinetic study of fluid gel formation and viscoelastic response. *Food Hydrocolloids*, doi:10.1016/jfoodhyd.2009.03.018.

Ganong, W. F. (1995). Review of medical physiology 17th ed., Appleton & Lange, Norwalk, Connecticut.

Gregersen, H. (2002). Biomechanics of the gastrointestinal tract. Springer, London.

Guido, S. and Villone, M. (1999). Measurement of interfacial tension by drop retraction analysis. *Journal of Colloid and Interface Science* **209**, 247–250.

Guinard, J. X. and Mazzucchelli, R. (1996). The sensory perception of texture and mouthfeel. *Trends in Food Science & Technology* **71**, 213–221.

Guyton, A. C. and Hall, J. E. (2000). Textbook of Medical Physiology. W.B. Saunders Company, Philadelphia.

Han, C. D., Sun, J. S., Chuang, H. K., and Lee, J. K. (1998). Prediction of equilibrium polymer blend morphology in dispersed two-phase flow and comparison with experiment. *Polymer Engineering and Science* **38**, 1154–1166.

Hoad, C. L., Rayment, P., Spiller, R. C., Marciani, L., Alonso, B. D., Traynor, C., Mela, D. J., Peters, H. P. F., and Gowland, P. A. (2004). *In* vivo imaging of intragastric gelation and its effect on satiety in humans. *Journal of Nutrition* **134**, 2293–2300.

Jeffrey, B., Udaykumar, H. S., and Schulze, K. S. (2003). Flow fields generated by peristaltic reflex in isolated guinea pig ileum: impact of contraction depth and shoulders. *The American Journal of Physiology – Gastrointestinal and Liver Physiology* **285**, G907–G918.

Jenkins, D. J. A., Wolever, T. M. S., Leeds, A. R., Gassull, M. A., Haisman, P., Dilawari, J., Goff, D. V., Metz, G. L., and Alberti, K. G. M. M. (1978). Dietary Fibers, Fiber Analogs, and Glucose-Tolerance – Importance of Viscosity. *British Medical Journal* **1**, 1392–1394.

Kalantzi, L., Goumas, K., Kalioras, V., Abrahamsson, B., Dressman, J. B., and Reppas, C. (2006). Characterization of the human upper gastrointestinal contents under conditions simulating bioavailability/bioequivalence studies. *Pharmaceutical Research* **23**, 165–176.

Karr, W. G., Osler-Abbott, W., and Sample, A. B. (1935). Intubation studies of the human small intestine. IV. Chemical characteristics of the intestinal contents in the fasting state and as influenced by the administration of acids, of alkalies and of water. *The Journal of Clinical Investigation* **14**, 893–900.

Kawachi, K., Takeuchi, M., and Nishiya, T. (1993). Solution containing whey protein, whey protein gel, whey protein powder and processed food product produced by using the same. US Pat. 5217741.

Koliandris, A., Lee, A., Ferry, A. L., Hill, S., and Mitchell, J. (2008). Relationship between structure of hydrocolloid gels and solutions and flavour release. *Food Hydrocolloids* **22**, 623–630.

Kong, F. and Singh, R. P. (2008a). A Model Stomach System to Investigate Disintegration Kinetics of Solid Foods during Gastric Digestion. *Journal of Food Science*, doi:10.1111/j.1750-3841.2008.00745.x.

Kong, F. and Singh, R. P. (2008b). Disintegration of Solid Foods in Human Stomach. *Journal of Food Science*, doi:10.1111/j.1750-3841.2008.00766.x.

Kwiatek, M. A., Steingoetter, A., Pal, A., Menne, D., Brasseur, J. G., Hebbard, G., Boesiger, P., Thumshirn, M., Fried, M., and Schwizer, W. (2006). Quantification of distal antral contractile motility in healthy human stomach with magnetic resonance imaging. *Journal of Magnetic Resonance Imaging* **24**, 1101–1109.

Lorén, N., Hermansson, A.-M., Williams, M. A. K., Lundin, L., Foster, T. J., Hubbard, C. D., Clark, A. H., Norton, I. T., Bergström, E. T., and Goodall, D. M. (2001). Phase separation induced by conformational ordering of gelatin in gelatin/maltodextrin mixtures. *Macromolecules* **34**, 289–297.

Lucas, P. W., Prinz, J. F., Agrawal, K. R., and Bruce, I. C. (2002). Food physics and oral physiology. *Food Quality and Preference* **13**, 203–213.

Lundin, L., Norton, I. T., Foster, T. J., Williams, M. A. C., Harmansson, A.-M., and Bergström, E. (2000). Phase separation in mixed biopolymer systems. *In* Gums and Stabilisers for the Food Industry *10*, 167–180, Williams, P. A. and Phillips, G. O. (eds.). Royal Society of Chemistry, Cambridge.

Malone, M. E., Appelqvist, I. A. M., and Norton, I. T. (2003). Oral behaviour of food hydrocolloids and emulsions. Part 1. Lubrication and deposition considerations. *Food Hydrocolloids* **17**, 763–773.

Marciani, L., Gowland, P. A., Spiller, R. C., Manoj, P., Moore, R. J., Young, P., Al-Sahab, S., Bush, D., Wright, J., and Fillery-Travis, A. J. (2000). Gastric response to increased meal viscosity assessed by echo-planar magnetic resonance imaging in humans. *Journal of Nutrition* **130**, 122–127.

Marciani, L., Young, P., Wright, J., Moore, R., Coleman, N., Gowland, P. A., and Spiller, R. C. (2001a). Antral motility measurements by magnetic resonance imaging. *Neurogastroenterology & Motility* **13**, 511–518.

Marciani, L., Gowland, P. A., Fillery-Travis, A., Manoj, P., Wright, J., Smith, A., Young, P., Moore, R., and Spiller, R. C. (2001b). Assessment of antral grinding of a model solid meal with echo-planar imaging. *The American Journal of Physiology – Gastrointestinal and Liver Physiology* **280**, G844–G849.

Martínez-Padilla, L. P., Lopez-Araiza, F., and Tecante, A. (2004). Steady and oscillatory shear behaviour of fluid gels formed by binary mixtures of xanthan and gellan. *Food Hydrocolloids* **18**, 471–481.

Meyer, J., Ohashi, H., Jehn, D., and Thomson, J. B. (1981). Size of liver particles emptied from the stomach. *Gastroenterology* **80**, 1489–1496.

Mioche, L. and Peyron, M. A. (1995). Bite force displayed during assessment of hardness in various texture contexts. *Archives of Oral Biology* **40**, 415–423.

Morris, E. R. and Norton, I. T. (1983). Polysaccharide aggregation in solutions and gels. *Studies in Physical and Theoretical Chemistry* **26**, 549–593.

Norton, I. T. and Frith, W. J. (2001). Microstructure design in mixed biopolymer composites. *Food Hydrocolloids* **15**, 543–553.

Norton, I. T., Goodall, D. M., Morris, E. R., and Rees, D. A. (1978). Dynamics of salt-induced random coil to helix transition in segmented iota-carrageenan. *Journal of the Chemical Society-Chemical Communications*, 515–516.

Norton, I. T., Goodall, D. M., Morris, E. R., and Rees, D. A. (1979). Kinetic evidence consistent with double helix formation in k-carrageenan solutions. *Journal of the Chemical Society-Chemical Communications,* DOI: 10.1039/C39790000988, 988–990.

Norton, I. T., Goodall, D. M., Morris, E. R., and Rees, D. A. (1980). Kinetic evidence for intramolecular conformational ordering of the extracellular polysaccharide (xanthan) from *Xanthomonas campestris*. *Journal of the Chemical Society-Chemical Communications,* DOI: 10.1039/C39800000545, 545–547.

Norton, I. T., Goodall, D. M., Morris, E. R., and Rees, D. A. (1983a). Dynamics of cation-induced conformational ordering in solutions of segmented iota-carrageenan. *Journal of the Chemical Society-Faraday Transactions I,* **79**, 2501–2515.

Norton, I. T., Goodall, D. M., Morris, E. R., and Rees, D. A. (1983b). Equilibrium and dynamic studies of the disorder order transition of kappa-carrageenan. *Journal of the Chemical Society-Faraday Transactions I* **79**, 2489–2500.

Norton, I. T., Goodall, D. M., Morris, E. R., and Rees, D. A. (1983c). Role of cations in the conformation of iota-carrageenan and kappa-carrageenan. *Journal of the Chemical Society-Faraday Transactions I* **79**, 2475–2488.

Norton, I. T., Goodall, D. M., Frangou, S. A., Morris, E. R., and Rees, D. A. (1984). Mechanism and dynamics of conformational ordering in xanthan polysaccharide. *Journal of Molecular Biology* **175**, 371–394.

Norton, I. T., Foster, T., and Brown, R. (1998). The Science and technology of fluid gels. *Royal Society of Chemistry Special Publication* **218**, 259–268.

Norton, I. T., Jarvis, D. A., and Foster, T. J. (1999). A molecular model for the formation and properties of fluid gels. *International Journal of Biological Macromolecules* **26**, 255–261.

Norton, I. T., Frith, W. J., and Ablett, S. (2006a). Fluid gels, mixed gels and satiety. *Food Hydrocolloids* **20**, 229–239.

Norton, I. T., Fryer, P. J., and Moore, S. (2006b). Product/process integration in food manufacture: engineering sustained health. *AIChE Journal* **52**, 1632–1640.

Norton, I. T., Moore, S., and Fryer, P. J. (2007). Understanding food structuring and breakdown: engineering approaches to obesity. *Obesity Reviews* **8**, 83–88.

Norton, I. T., Cox, P. W., and Spyropoulos, F. (2008). The role of hydrocolloids in the formation of health foods. *In* Gums and Stabilizers for the Food Industry *14*, 293–305, Wiliams, P. A. and Phillips, G. O. (eds.). Royal Society of Chemistry Special Publication No. 316, Cambridge.

Norton, I. T., Spyropoulos, F., and Cox, P. W. *in press.* Rheological control and understanding necessary to formulate healthy everyday foods. *Proceedings of the Norwegian Rheological Society.*

Pal, A., Abrahamsson, B., Schwizer, W., Hebbard, G. S., and Brasseur, J. G. (2003). Application of a virtual stomach to evaluate gastric mixing and breakdown of solid food. *Gastroenterology* **124**, A673–A674.

Pal, A., Indireshkumar, K., Schwizer, W., Abrahamsson, B., Fried, M., and Brasseur, J.G. (2004). Gastric flow and mixing studied using computer simulation. *Proceedings of the Royal Society of London Series B: Biological Sciences* **271**, 2587–2594.

Pal, R. (2001). Evaluation of theoretical viscosity models for concentrated emulsions at low capillary numbers. *Chemical Engineering Journal* **81**, 15–21.

Perrau, M. B., Iliopoulos, I., and Audebert, R. (1989). Phase separation of polyelectrolyte/nonionic polymer systems in aqueous solution: effects of salt and charge density. *Polymer* **30**, 2112–2117.

Piculell, L. and Lindman, B. (1992). Association and segregation in aqueous polymer/polymer, polymer/surfactant, and surfactant/surfactant mixtures: similarities and differences. *Advances in Colloid and Interface Science* **41**, 149–178.

Piculell, L., Bergfeldt, K., and Nilsson, S. (1995). Factors determining phase behaviour of multi component polymer systems. *In* Biopolymer Mixtures, 13–35, Harding, S. E., Hill, S. E. and Mitchell, J. R. (eds.). Nottingham University Press, Nottingham.

Prinz, J. F. and Lucas, P. W. (1997). An optimisation model for mastication and swallowing in mammals. *Proceedings of the Royal Society of London Series B: Biological Sciences* **264**, 1715–1721.

Prinz, J. F., Janssen, A. M., and de Wijk, R. A. (2007). In vitro simulation of the oral processing of semi-solid foods. *Food Hydrocolloids* **21**, 397–401.

Pudney, P. D. A., Hancewicz, T. M., Cunningham, D. G., and Brown, M. C. (2004). Quantifying the microstructures of soft solid materials by confocal Raman spectroscopy. *Vibrational spectroscopy* **34**, 123–135.

Roberts, F. G., Smith, H. A., Low, A. G., and Ellis, P. R. (1990). Influence of wheat breads containing guar flour supplements of high and low molecular weights on viscosity of jejunal digesta in the pig. *In* Dietary Fibre: Chemical and Biological Aspects, 164–168, Southgate, D. A. T., Waldron, K., Johnson, I. T. and Fenwick, G. R. (eds.). Royal Society of Chemistry, Cambridge.

Rune, S. J. and Viskum, K. (1969). Duodenal pH values in normal controls and inpatients with duodenal ulcer. *Gut* **10**, 569–571.

Shama, F. and Sherman, P. (1973). Identification of stimuli controlling the sensory evaluation of viscosity. *Journal of Texture Studies* **4**,111–118.

Spyropoulos, F., Ding, P., Frith, W. J., Norton, I. T., Wolf, B., and Pacek, A. W. (2008). Interfacial Tension in Aqueous Biopolymer-Surfactant Mixtures. *Journal of Colloid and Interface Science* **317**, 604–610.

Stokes, J. R., Wolf, B., and Frith, W. J. (2001). Phase-separated biopolymer mixture rheology: Prediction using a viscoelastic emulsion model. *Journal of Rheology* **45**, 1173–1191.

Sworn, G. (2000). Gellan gum. *In* Handbook of Hydrocolloids, 117–136, Phillips, G. O. and Williams, P. (eds.). Woodhead Publishing Ltd, Cambridge.

Sworn, G., Sanderson, G. R., and Gibson, W. (1995). Gellan gum fluid gels. *Food Hydrocolloids* **9**, 256–271.

Taylor, R. H., Wolever, T. M. S., Jenkins, D. J. A., Ghafari, H., and Jenkins, M. J. A. (1980). Viscosity and Glucose Diffusion: Potential for Modification of Absorption in the Small-Intestine. *Gut* **21**, A452.

Tharakan, A., Rayment, P., Fryer, P. J., and Norton, I. T. (2007). Modelling of physical and chemical processes in the small intestine. *Proceedings of European Congress of Chemical Engineering* (ECCE-6), Copenhagen, 16–20 September.

Utracki, L. A. (1991). On the viscosity-concentration dependence of immiscible polymer blends. *Journal of Rheology* **35**, 1615–1637.

van Aken, G. A., Vingerhoeds, M. H., and de Hoog, E. H. A. (2007). Food colloids under oral conditions. *Current Opinion in Colloid & Interface Science* **12**, 251–262.

van der Bilt, A., Olthoff, L. W., van der Glas, H. W., van der Weelen, K., and Bosman, F. (1987). A mathematical description of the comminution of food during mastication in man. *Archives of Oral Biology* **32**, 579–586.

van Ruth, S. M. and Roozen, J. P. (2000). Influence of mastication and saliva on aroma release in a model mouth system. *Food Chemistry* **71**, 339–345.

van Vliet, T. (2002). On the relation between texture perception and fundamental mechanical parameters for liquids and time dependent solids. *Food Quality and Preference* **13**, 227–236.

Williams, M. A. K., Fabri, D., Hubbard, C., Lundin, L., Foster, T. J., Clark, A. H., Norton, I. T., Lorén, N., and Hermansson, A. M. (2001). Kinetics of droplet growth in gelatin/maltodextrin mixtures following thermal quenching. *Langmuir* **17**, 3412–3418.

Wolf, B. and Frith, W. J. (2003). String phase formation in biopolymer aqueous solution blends. *Journal of Rheology* **47**, 1151–1170.

Wolf, B., Scirocco, R., Frith, W. J., and Norton, I. T. (2000). Shear-induced anisotropic microstructure in phase-separated biopolymer mixtures. *Food Hydrocolloids* **14**, 217–225.

Wolf, B., Frith, W. J., and Norton, I. T. (2001). Influence of gelation on particle shape in sheared biopolymer blends. *Journal of Rheology* **45**, 1141–1157.

Zaslavsky, B. Y. (1995). Aqueous two-phase partitioning; Physical Chemistry and Bioanalytical applications. Marcel Dekker, New York.

Zeman, L. and Patterson, D. (1972). Effect of the Solvent on Polymer Incompatibility in Solution. *Polymer Incompatibility* **5**, 513–516.

CHAPTER

7

Unified Application of the Materials-Science Approach to the Structural Properties of Biopolymer Co-Gels throughout the Industrially Relevant Level of Solids

Stefan Kasapis

School of Applied Sciences, RMIT University, Bundoora West Campus, Melbourne, Vic 3083, Australia

7.1 INTRODUCTION AND OVERVIEW OF PRODUCT DEVELOPMENT CONCERNS THAT NECESSITATED WORK IN PHASE-SEPARATED BIOPOLYMER GELS

7.1.1 Fat, Calories and Healthful Diet

Research in the area of biopolymer mixtures has grown rapidly in the last 25 years or so along with the continuous expansion of the food industry in the areas of ingredients, quality control and nutrition (Tolstoguzov, 2003). Thus the design of novel composite gels is a child of the industry that responded to consumer awareness of the need to reduce fat intake in the diet (Pszczola, 2006). In this respect, early use of polymeric 'additives' to improve a product's mouthfeel and nutritional profile focused on the commercial development of low-fat spreads and soft cheeses.

Levels of fat in cheese-like embodiments can reach up to 33% in the formulation, and this unacceptably high-fat content was challenged in early 1980s and the 1990s, among others, by the British Advisory Committee for Nutritional Education, which considered fat as the most likely component of the diet to damage health (Committee of Medical Aspects of Food Policy, 1994). Today, a plethora of medical reports have demonstrated that rich-in-fat diets lead to obesity, high plasma cholesterol and heart disease (Jonnalagadda and Jones, 2005). The second drive, of course, for a low-fat diet is the desire to be slim. Another point of interest to consumers in this junction is the protein levels that can be as high as 38% in traditional cheese

Kasapis, Norton, and Ubbink: Modern Biopolymer Science
ISBN: 978-0-12-374195-0

products. Thus partial replacing of fat and protein with complex carbohydrates that serve as dietary fiber (e.g., alginates, pectins, plant cellulose) or even the energy reserves (e.g., starch) can assist with the prevention of coronary heart disease and, in general, benefits health (Sloan, 2005).

A point should also be made about the efficient gelification of polysaccharides, which allows structure formation at relatively small concentrations. For example, gels comprising 99 parts of water and 1 part of deacylated gellan, with optimized additions of sodium or calcium counterions, are equivalent in strength to the semi-crystalline/liquid matrix of butter at 5°C (75–80% solids) or a 20% milk protein curd (Evageliou et al., 1997). Classification of polysaccharide or protein hydrogels implies the presence of a crosslinked network with soft but solid-like properties supporting the aqueous phase which is commonly found in substantial quantity. This operational definition of a hydrogel is not compatible with the partially crystalline fat matrix in butter or margarine. In the case of cheese, it is recognized that the first stage of its making reflects a typical biological gel which is formed by connecting milk proteins through a series of enzymatic reactions (chymosin-induced). The cheese itself, following a period of aging, is viewed as a 'closely packed system' that lacks the connectivity of a polymeric network structure. With or without the presence of reactive sites per primary molecule that form crosslinks, the above examples of structure formation fall within the umbrella of 'soft solids' and, as such, they will be considered collectively in the remainder of this treatise.

Clearly, substantial savings can be accomplished during product development by replacing part of fat and/or protein with an inexpensive polysaccharide formulation. It was inevitable, then, that in this climate the market for low-fat, healthy products would flourish, with the total European/US business for such products being worth tens of billions of dollars (Stewart-Knox and Mitchell, 2003). Dairy products account for about 90% of this market and consumers being partial to cheeses and spreads are prepared to look for alternatives and to a good extent pay extra for light foods across the board if they were available.

7.1.2 Textural and Mouthfeel Implications of Fat/Protein Reduction

In addition to the increase in the caloric content of a foodstuff, fat is a major contributor to organoleptic properties. These cover a wide range of attributes such as appearance, texture (small- and large-deformation properties), mouth-coating that relates to creaminess, and flavor release (Renard et al., 2006). It should be remembered that, in the main, consumers show a taste preference for a well-established 'full-fat' product and are prepared to consider a low-calorie alternative if it imitates adequately the mouthfeel of the standard variety. Removal of fat from the traditional embodiment will start gradually to affect the product characteristics and, clearly, there are limits to how much a consumer is prepared to compromise on taste. Therefore, the challenge facing a skilled food technologist is to come up with new ideas for fat replacers that match the eating quality of the full-fat formulation.

Traditionally, butter and margarine contain 70–80% 'solids', which form a continuous phase of liquid oil and triglyceride crystals. Fat crystals are generally not spherical but more like thin platelets with various ratios in length ($\approx$ 0.20 μm), width ($\approx$ 0.10 μm) and thickness ($\approx$ 0.05 μm). It is the shape and size of crystals that imparts creaminess in the mouth and successful replacement with biopolymer assemblies should reproduce this morphology (Cain et al., 1989). Water droplets ranging in size from 0.1 to 5 μm are evenly distributed throughout the material. Electron micrographs show that small crystals move to the droplets covering the water surface with a shield and preventing them from flocculating (Clarke, 2004). Upon mastication, the water-in-oil emulsion (w/o) in butter and margarine phase inverts into a continuous water phase that facilitates infusion of flavor and nutrients for fortified/enriched products (Aryana and Haque, 2001).

Reduction of the amount of fat to 40% (halvarines), 20–25% (quarterines) or even less than 10%, resulting in water-continuous spreads with the aqueous matrix supporting the inclusions of fat globules, relies increasingly on biopolymer technology that manipulates structure. In particular, protein and polysaccharide properties in a composite gel become key issues. As discussed in the preceding section, this is based on the ability of biopolymers to cross-link and/or entangle, thus creating a structural entity at high enough concentrations (Yan et al., 2006). However, too low of a viscosity and the matrix (product) will become unstable whereas overdosing results in unacceptable textures. A single biopolymer system behaves either as a viscoelastic liquid that can not support its shape against gravity (Sittikijyothin et al., 2005), or as a solid-like material whose brittle or elastic character is unsuitable for fat replacing (Barrangou et al., 2006; Badii and Howell, 2006).

Given the financial burden and legalistic obstacles in securing GRAS status for the use of new microbial or crustacean polysaccharides as food additives, industrialists must rely on creative manipulation of existing functional ingredients. Clearly, this requires basic understanding of the functional properties and molecular interactions of the available proteins and polysaccharides in mixtures that may balance out the elastic or brittle characteristics of the individual constituents into an overall profile of desirable texture and mouthfeel.

7.2 EXPERIMENTAL METHODS OF PINPOINTING PHASE-SEPARATION PHENOMENA IN MIXED GELS

Observations lying scattered in the literature indicate that the mechanism of phase separation by which polymeric stabilizers accomplish their function in the aqueous phase of reduced fat products is increasingly understood. Development of mimetic or novel textures would be further facilitated by a systematic consideration of the research tools available in monitoring the many changes which occur throughout the 'food cycle'. These include thermal processing, shearing/pumping through production lines and storage at ambient or refrigeration temperatures. In this respect, the present work will provide an outline of the attempts made to rationalize phase-separation phenomena in biopolymer gels on the basis of experimental evidence and physical theories from the realm of synthetic polymers leading to potential applications, rather than a comprehensive review of all the published data on these systems.

7.2.1 Kinetic Influences on the State of Phase Separation in Gels

Before we delve into the approaches utilized to obtain mechanistic understanding, it should be emphasized that classic phase-separation in solution leading to thermodynamic equilibrium between two polymer phases is not carried over to the gel state upon subsequent cooling. In other words, there is no time–temperature superposition principle able to sustain the state of thermodynamic equilibrium in a composite gel along the lines epitomized by the phase diagram for solutions. Formation of the former is accomplished by disorder-to-order transitions and possible aggregation of the polysaccharide helical segments. These alter dramatically the conditions of miscibility and polymer–polymer or polymer–solvent interactions described in solution by the Flory-Huggins theory.

Systematic studies into the extent of thermodynamic incompatibility between proteins and polysaccharides in solution was the natural follow up having developed in 1960s and 1970s a good understanding of the conformation–function relationships in single biopolymer systems (Anderson et al., 1969; Grant et al., 1973). Today, it is known that for a great number of binary mixtures phase separation into two liquid layers can occur when the total polymer concentration exceeds 4% depending, of course, on pH and ionic strength (Antonov and Goncalves, 1999; Alves et al., 2000; Zhang and Foegeding, 2003). Thus limited compatibility between proteins and ionic polysaccharides is observed when the pH value is higher than the isoionic point of the protein or, regardless of pH, the ionic strength is greater than 0.25 (Antonov et al., 1977). Following bulk-induced phase separation via centrifugation, each liquid layer (phase) is found to be rich in one polymer and depleted in the other, and vice versa, with both polymers having reached conditions of thermodynamic equilibrium in solution (Loret et al., 2005).

Phase equilibria facilitate the construction of a phase diagram. Points on the phase diagram derived, for example, from optical-rotation analysis define the composition of each phase and lie along the so-called 'bimodal' or 'cloud point' curve. The tie line joining a pair of these points, which reflect the polymer-A-rich with polymer-B-depleted phase and the polymer-A-depleted with polymer-B-rich phase, passes through another (experimental) point (Higgins et al., 2005). This corresponds to the initial composition of the solution, i.e., the overall concentrations of polymer A and B in the solution. As the overall concentration of the two polymers decreases, the family of tie lines degenerates to a single critical point defined by the intercept of the bimodal with the curve linking the mid-points of each tie line. The phase diagram can be highly asymmetric, with the asymmetric shape indicating the molecular weight differences between, for example, a globular protein and a plant polysaccharide.

As mentioned above, however, certain experimental procedures of which the cooling regimen constitutes a prime example, demonstrate that conditions of thermodynamic equilibrium in solution are readily disturbed in the co-gel. Figure 7.1 illustrates this by means of a starch hydrolysis product (maltodextrin) that gives a thermally reversible gel with a 'fat-like' texture, in combination with a gelatin gel, which allows lower concentrations of maltodextrin to be used, avoiding a 'starchy' mouthfeel (Singh et al., 2007). Following solubilization of the two materials at 90 and 70°C, respectively, mixed solutions at 45°C exhibit phase behavior of a single liquid phase, liquid phase plus precipitate in the form of aggregated amylose-like helices, or two co-existing liquid phases with

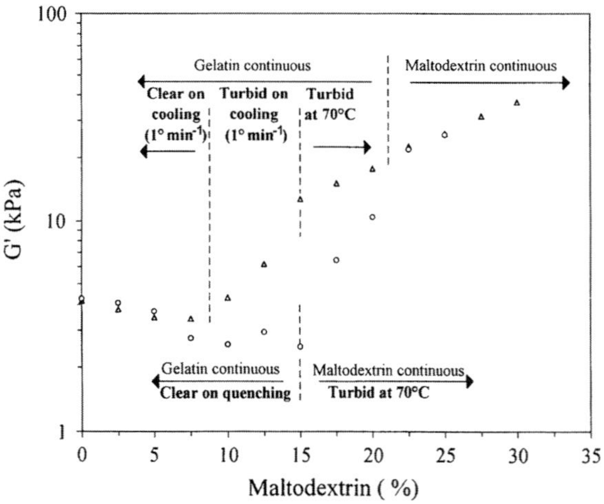

FIGURE 7.1 Storage modulus (G') data of 5% gelatin series with increasing amounts of maltodextrin (0 to 30%) taken from controlled cooling at 1 deg/min (Δ) and quenching (○) from 70 to 5°C. In both cases, samples were left at 5°C for 7 h (with permission from Alevisopoulos et al., 1996).

increasing level of solids. This led to a construction of a gelatine–maltodextrin phase diagram (Kasapis et al., 1993).

The main part of the experimental work focused on quench cooling from 70 to 5°C at about 33°C/min (Figure 7.1). This produced a sharp increase in the values of storage modulus on shear (G') at about 15% maltodextrin, which is coincident with the phase inversion from gelatin to maltodextrin continuous co-gels (Alevisopoulos et al., 1996). Below the phase inversion point, solutions remain clear and upon gelation produce a steady reduction in experimental moduli which was rationalized on the basis of gradually deswelling gelatin networks due to the slow ordering of the polysaccharide. At the maltodextrin continuous combinations, however, samples become cloudy upon mixing at 70°C, and 'blending law' analysis taking into account the compositional dependence of mixed-gel moduli indicates immediate phase separation in solution followed by separate gelation of the two biopolymers at low temperatures (Kasapis et al., 1993a).

In the event of slow controlled cooling at 1°/min, samples go turbid during the cooling cycle at concentrations of maltodextrin as low as 10% and there is an immediate reinforcing effect on the composite strength (Figure 7.1). However, there is no question that they are protein continuous since, upon heating, the expected gelatin-related melting profile at ≈ 30°C is obtained at carbohydrate levels up to 20% (Giraudier et al., 2004). At higher concentrations of the latter (22.5–30%), mixtures exhibit a prolonged melting behavior (> 80°C) characteristic of the maltodextrin network (Tharanathan, 2005). Supporting light microscopy evidence and theoretical modeling argue that the system has now phase inverted to a maltodextrin suspending matrix (Kasapis et al., 1993a).

It appears, therefore, that there is an antagonistic effect operating between ordering-aggregation leading to gelation and the extent of phase separation. In the case of gelatin-maltodextrin, early phase separation and gel reinforcement are recorded for slow-cooled samples in the way observed only beyond the phase inversion point for the quenched counterparts. The resulting composite gel is under kinetic control and the nature of phase topology is determined by the thermal treatment, which to a great extent negates the state of thermodynamic equilibrium of the mixture in solution. The above conclusions of kinetically trapped binary gels have been confirmed using confocal/electron microscopy supported by image analysis (Loren et al., 1999; Loren and Hermansson, 2000).

Cooling rate is not the only experimental parameter that can control the phase behavior of composite gels. Thus a combination of decreasing molecular-weight fractions of gelatin (from 2.1×10^5 to 0.8×10^5 M_W) and increasing molecular weight of 'polymeric' maltodextrin materials (dextrose equivalent from 4.7 to 3.9) can reduce to a third the amount of maltodextrin required in the presence of gelatin for phase inversion to a polysaccharide continuous phase (Alevisopoulos and Kasapis, 1999). Micromolecular studies employing small-angle light scattering (SALS) also demonstrated a relationship between molar weight (mass) and coarseness levels (i.e., the development of the dominant polymeric length scale) in systems where phase-separation phenomena were induced by gelation (Anderson and Jones, 2001; de Hoog and Tromp, 2003). Further texturizing of the discontinuous filler is achieved by applying a shear flow to a phase-separated biopolymer solution and then entrapping the anisotropic macrostructures by gelling either or both phases (Wolf et al., 2000; Norton and Frith, 2001). Depending on shear history, inclusions vary from spheroids to elongated, or irregularly shaped particles formed at high shear stresses, an outcome that may find application in particle coalescence and flavor release of low-fat processed foods (Caserta et al., 2005; Ding et al., 2005; Michon et al., 2005).

7.2.2 Tangible Evidence of Non-Interacting Phase-Separated Gels

The discussion in the preceding section indicated that food processing can lead to composite gels of disparate structural hence textural properties. It follows that background understanding is required for intelligent manipulation and refinement of the structural morphology of systems with industrial interest. In this respect, microscopy images can offer valuable qualitative evidence of changes in the distribution of the two components as a function of polymeric composition and other external stimuli (Butler et al., 2006; Braga et al., 2006; Mathew et al., 2006; Tischer et al., 2006; Lofgren and Hermansson, 2007). Figure 7.2 supports the discussion of gelatin or maltodextrin continuous phases with increasing concentrations of the latter presented in Figure 7.1 on the basis of evidence from mechanical measurements.

Contrasting refractive indices of the two component networks yield upon quenching a smooth continuous gelatin phase surrounding the maltodextrin aggregates at low levels of the polysaccharide, for example, 5 and 10% w/w in the top row of Figure 7.2 (Kasapis et al., 1996). At

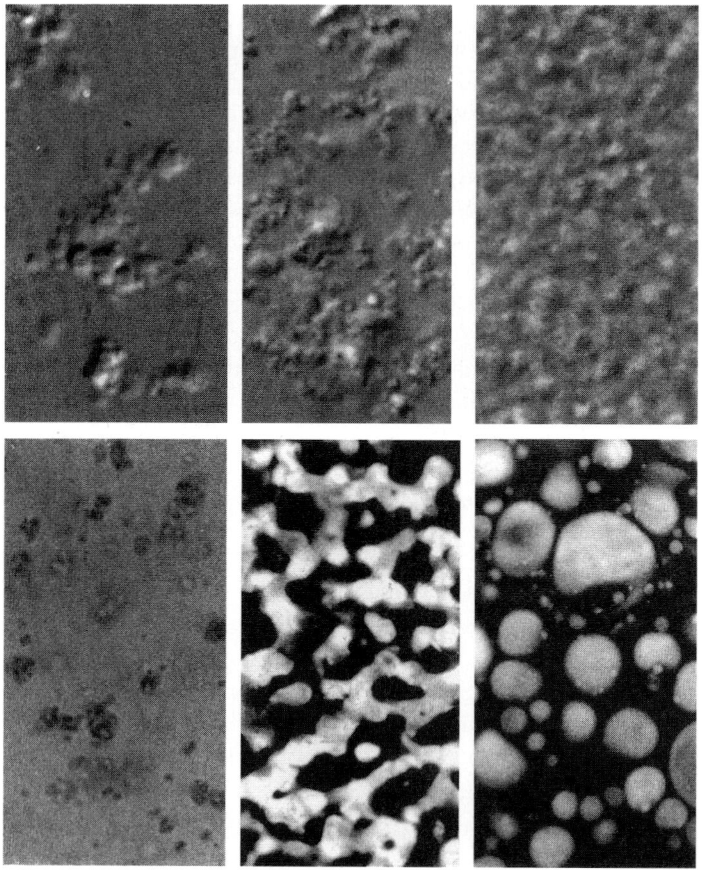

FIGURE 7.2 Micrographs of the 5% gelatin series shown in Figure 7.1. Quenched samples (top row; 250X magnification) were prepared at 5, 10 and 17.5% maltodextrin, and low-rate scanned gels (bottom row; 100X magnification) were made at 5, 20 and 22.5% maltodextrin (left to right) (with permission from Kasapis et al., 1996).

higher concentrations (e.g., 17.5% w/w) where a maltodextrin continuous matrix is indicated in Figure 7.1, the entirety of the micrograph is dominated by the optically dense maltodextrin network. Staining the control cooled mixtures with iodine vapor produces dark areas for the maltodextrin agglomerates whereas application of sirious red to the same preparation unveils a white gelatin phase. Bottom-row images in Figure 7.2 show a continuous, dark background formed first at about 22.5% maltodextrin. This suspends the white gelatin spheroids, an outcome which is congruent with the concept of delayed phase inversion in the slow scanned blends of Figure 7.1.

Experimental evidence from differential scanning calorimetry (DSC) can also provide a firm footing regarding the nature of the topology in polymeric composites (Icoz et al., 2005). Figure 7.3 illustrates typical exothermic peaks for single and mixed preparations of κ-carrageenan and gelatin, which were cooled from temperatures above the coil-to-helix transition to 0°C at 1°C/min. Samples were prepared at pH 6.7, hence rendering both polymers primarily negatively charged (gelatin pI = 4.5) (Mohanty and Bohidar, 2005).

A relatively sharp peak with a maximum heat flow temperature (T_{max}) of 38°C denotes the co-operative conformational transition of the polysaccharide segments (Figure 7.3a). By comparison, development of the gelatin network is relatively gradual culminating at lower temperatures (T_{max} is about 17.5°C in Figure 7.3b). Upon mixing, the temperature sequence and the overall peak form for each transition was maintained (Figure 7.3c), a result which argues (i) both components form gel networks in the mixture as in the individual preparations, and (ii) there are no specific interactions between them (i.e., no formation of heterotypic junctions) (Kasapis and Al-Marhoobi, 2005). In contrast, heterotypic associations usually manifest themselves by distorting the peaks of the individual gels and

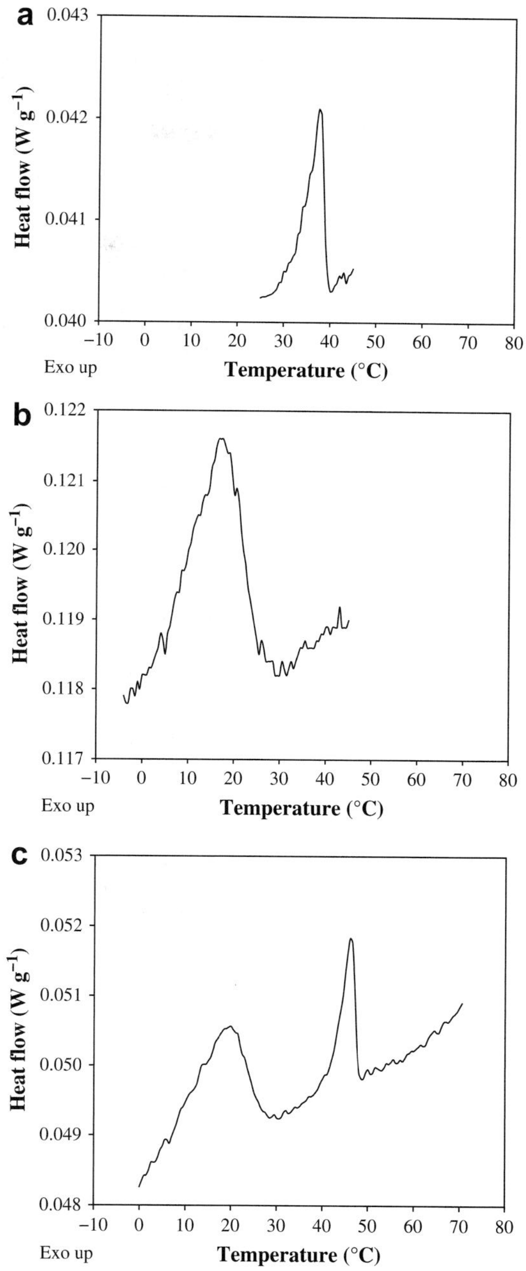

FIGURE 7.3 DSC exothermic peaks for (a) 1.5% κ-carrageenan (25 mM KCl added), (b) 10% gelatin and (c) a mixture of 1.5% κ-carrageenan (25 mM KCl) plus 10% gelatin (cooling rate: 1°C/min) (with permission from Kasapis and Al-Marhoobi, 2005).

generating a new thermal event in the DSC spectrum (Williams et al., 1992). Furthermore, the temperature band of the κ-carrageenan gel in the mixture is shifted to 46°C, a result which argues for phase separation in accordance with the increased thermal stability of the concentrated preparations of the polysaccharide (van de Velde et al., 2005).

7.2.3 Mechanical Measurements as an Empirical Indicator of Phase-Separation Patterns in Binary Gels

The validity of conclusions made on the basis of DSC and microscopy evidence can be tested using complementary rheological techniques thus building a database of experimental observations as a function of polymer composition and experimental stimuli (Baruch and Machluf, 2006). Figure 7.4 demonstrates how distinct temperature paths in the development and disintegration of deacylated gellan and gelatin structures answer the question of phase continuity in this material. In the instance of a gelatin continuous phase, the composite follows the gelling and melting profiles of the protein samples alone, with the onset of gelation and final collapse of structure occurring at about 22 and 29°C, respectively (Haug et al., 2004). In contrast, gellan continuous arrangements go through the polysaccharide gelling transition at ≈ 43°C, followed by a second 'wave' of structure formation as the gelatin gels at lower temperatures (Papageorgiou et al., 1994). In the heating direction, gels weaken significantly over the temperature-range of melting of the gelatin component, but remain intact until completion of melting of the gellan network (> 90°C) (Ikeda et al., 2004).

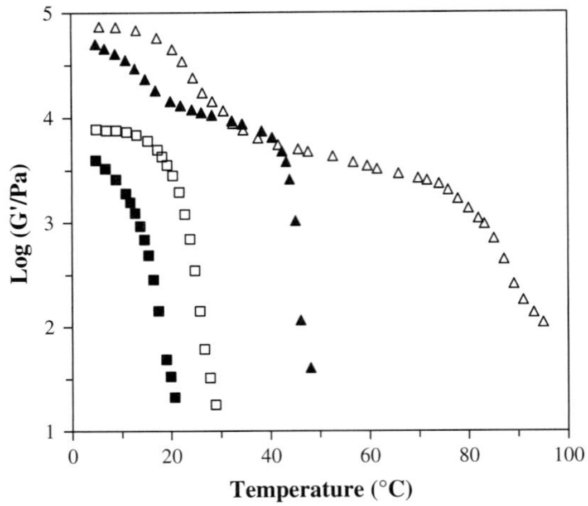

FIGURE 7.4 Gel forming and melting for mixed gels of 5% gelatin plus 0.5% deacylated gellan at 1 M NaCl added (G' on cooling [■] and heating [□]), and 5% gelatin plus 0.5% deacylated gellan at 0.1 M NaCl added (G' on cooling [▲] and heating [△]). Samples were monitored at a scan rate of 1°/min and set between runs at 5°C for 2 h. The stoichiometric equivalent of 0.1% deacylated gellan is about 1.5 mM of NaCl (with permission from Papageorgiou et al., 1994).

Phase separation in the gel state of non-interacting gelatine–gellan systems is due to the hindrance in conformational/spatial mobility of chain segments which results in excluded volumes for each polymeric constituent. In a nutshell, conditions in solution can be such so that two macromolecules are completely miscible, as determined for disordered coils by the entropy of mixing in the well-known Gibbs equation of free energy ($\Delta G = \Delta H - T\Delta S$). During cooling, a critical solution temperature will be reached at which chain segments of the same species would exhibit attractive interactions. Energy is released when these molecules associate, and this is represented by a negative enthalpy change which is required if the association is to be possible thermodynamically. The loss of conformational freedom (fall in entropy) and the tendency of molecules to cross-link with their own kind results in the formation of continuous network-like regions comprising ordered sequences. These restrict at a local level the presence of the corresponding regions of the second polymeric constituent and vice versa.

Proceeding along these lines and by employing a combination of distinct concentrations of gellan (0–0.5% w/w) and added NaCl

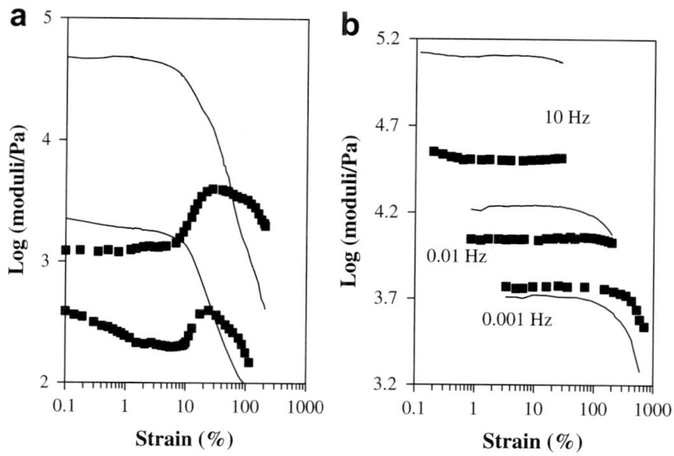

FIGURE 7.5 Strain sweeps of storage (—) and loss (■) modulus at 5°C for (a) a 30% SA-2 gel (top spectrum) and a maltodextrin continuous gel (15% SA-2 + 7.5% SC; bottom spectrum), and (b) a sodium caseinate continuous dispersion (15% SA-2 + 20% SC) at the frequencies shown on the graph (with permission from Manoj et al., 1996).

(≈ 1.0–1510 mM) the general framework of steric exclusion in the presence of 5% gelatin was unravelled. Thus gelatin continuous structures were formed when the ratio of gelatin to gellan concentration is higher than 50 or when the amount of added salt is more than 100 times the stoichiometric requirement of gellan, which results in precipitated polysaccharide agglomerates (filler particles) instead of a homogeneous, three-dimensional network. In the technological front, the small levels of the polysaccharide required to formulate a gellan continuous dispersion allow the production of thermally stable water desserts and confectioneries, during ambient-temperature storage (Giavasis et al., 2000), with the required long-range elasticity of the gelatin body (Helcke, 1997; Strauss and Gibson, 2004).

Besides the temperature profiles, biopolymers exhibit a range of rheological behavior depending on timescale and magnitude of the applied deformation (Tabilo-Munizaga and Barbosa-Canovas, 2005). Prediction of the viscoelastic properties over a wide range of shear regimens can identify the key macroscopic factors influencing a filled biopolymer system. Figure 7.5a, b shows how the investigation of maltodextrin (SA-2) with an approximate dextrose equivalent (DE) value of 2 (i.e., content of reducing end-groups relative to glucose as 100), and sodium caseinate (SC) mixtures has benefited from such a fundamental treatment (Manoj et al., 1996; Chronakis, 1998). Strain sweeps on maltodextrin and maltodextrin continuous gels are characteristic of brittle networks with a short linear viscoelastic region (up to about 5% strain), and a viscous flow as early as 40% deformation when the loss modulus on shear, G'', overtakes the storage modulus, G' (Lefebvre, 2006). This picture remains unaltered throughout a wide frequency range (from 0.001 to 10 Hz in Figure 7.5a) demonstrating that the polysaccharide interchain associations are stable over the experimental timescale of observation.

Contrary to that, the response to increasing amplitude of oscillation in a sodium caseinate continuous suspension is very much time-dependent. At the highest experimental frequency (10 Hz), the short-lived interactions of the protein yield a 'pseudo-gel' with a strong mechanical response; recording was terminated

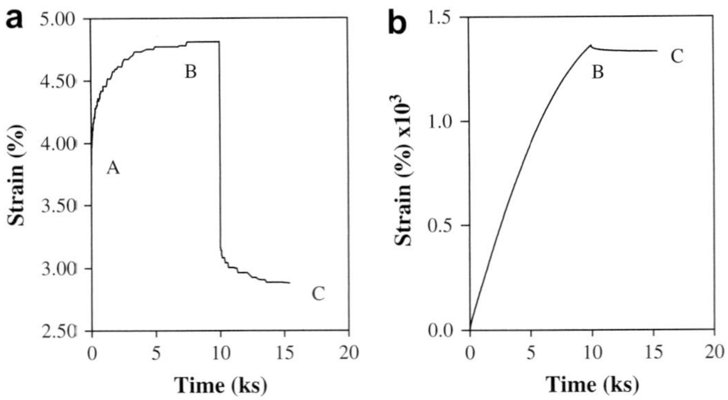

FIGURE 7.6 Instantaneous deformation (0A, 1 s), retardation period (AB, 10^4 s) and relaxation time (BC, 5.5×10^3 s) for (a) 17.5% SA-2 plus 7.5% SC gel and (b) a 7.5% SA-2 plus 17.5% SC dispersions upon creep compliance testing (5°C). In both cases the initial deformation (OA) was about 3.5% strain (with permission from Manoj et al., 1996).

at 30% strain due to exhaustion of stress available on the rheometer (Figure 7.5b). However, the pronounced frequency dependence of shear moduli in single caseinate dispersions is reflected in the mixed preparations. Thus protein continuous samples revert into viscoelastic liquids ($G'' > G'$) at relaxation times between 0.1 and 10^3 s (10 and 0.001 Hz, respectively) across the experimentally accessible amplitude of deformation, which reaches 800% strain (Braga and Cunha, 2004).

In parallel with strain-controlled experiments, stress-controlled (transient) measurements can assist in the elucidation of the state of phase separation in binary systems (Perez-Orozco et al., 2004). In Figure 7.6a, creep testing has been used to confirm conclusions reached in the maltodextrin/caseinate mixture discussed in the preceding paragraphs. Thus, application of a constant force to a 17.5% SA-2 plus 7.5% SC gel yields an instantaneous deformation of about 3.5%, which approaches asymptotically an equilibrium value of 4.8% strain at the end of a 10^4 s retardation period (Manoj et al., 1996). Upon removal of the stress, the sample recovers 41% of its structure in 5.5×10^3 s, a result which is compatible with the concept of active, semi-permanent interactions of the maltodextrin network surviving within the experimental constraints (Loret et al., 2004).

By comparison, the instantaneous deformation of 3.5% on the 7.5% SA-2 plus 17.5% SC sample transforms into an extensive flow, at almost a constant shear rate in the first half of the retardation curve. As a result an enormous deformation is recorded over 10^4 s ($\approx$ 1400%), which recovers very little during the subsequent relaxation time (2% of the original arrangement) due to the destabilization of associations with shorter relaxation functions than the experimental retardation time (Manski et al., 2007).

Overall the effect of amplitude of deformation and the time-dependent breakdown/recovery properties illustrated in Figures 7.5 and 7.6 argue convincingly for maltodextrin or caseinate continuous phases. Similar work was carried out in more than 40 combinations, which covered a concentration range in the mixture related to applications in low-fat dairy products (SA-2: 5–22.5%; SC: 2.5–35%). It was verified that phase inversion from a maltodextrin continuous network with discontinuous protein inclusions to a caseinate dispersion suspending the polysaccharide particles is determined by the relative polymer composition and occurs at a one-to-one weight ratio. Today the use of complementary mechanical techniques, and in combination with an appropriate protocol of thermal analysis and microscopy, has demonstrated its utility in a wide range of biopolymer mixtures including starch, milk proteins, galactomannans, pectin and carrageenans (Donato et al., 2005; Monteiro et al., 2005; Bhattacharya et al., 2006; Li et al., 2006; Lizarraga et al., 2006).

7.3 UTILIZATION OF REACTION KINETICS TO IDENTIFY PHASE-SEPARATION PHENOMENA IN BIOPOLYMER MIXTURES

It was emphasized that experimentation with a variety of techniques can be used to validate phase-separation phenomena in the mixture. These include cooling or heating traces from differential scanning calorimetry obtained at a low scan rate (typically 1°C/min), and dynamic oscillatory profiles obtained during gelling or melting of biopolymer networks at similarly low rates. It is advantageous to work on binary mixtures where polymeric components exhibit distinct molecular mechanisms thus facilitating resolution of experimental data. In the same vein, light microscopy images usually taken between 100 and 400x magnification can be informative in resolving the topology of biopolymer networks with individual refraction indices.

Nevertheless, replacing gelatin and/or the cold gelling polysaccharide with globular proteins creates a composite system where prior industrial processing of the ingredients may have caused extensive and irreversible destruction (denaturation) of the tertiary/quaternary structure of the molecules (Nunes et al., 2003; Batista et al., 2005). At high enough concentrations, such an event will result in aggregated spheroidal or fibrillar clusters that remain largely impervious to subsequent thermal treatments within the normal constraints of a laboratory experimentation (Clark et al., 2001; Veerman et al., 2003). Thus thermomechanical characterization of order–disorder transitions discussed earlier on

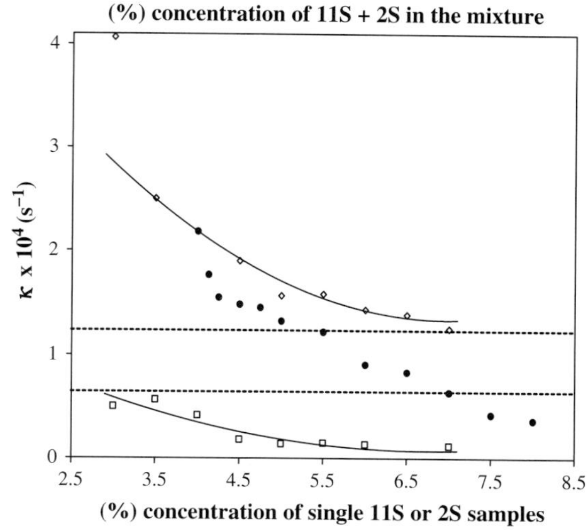

FIGURE 7.7 First-order reaction rate constants plotted as a function of protein concentration for single 11S (◇) and 2S (□) preparations, and their mixtures of 4% 11S with increasing concentrations of 2S (●) (unpublished results of Kasapis et al., with permission).

for polysaccharides and gelatin is no longer a viable proposition, an outcome that may impair identification of phase-separated microstructures in the binary gel state.

To circumvent this problem and offer a helpful device for the rationalization of the phase behavior in thermally denatured proteins, isothermal gelation rates of the polymeric constituents can be utilized (van Assche et al., 2001). The point is made in Figure 7.7, which illustrates data on two molecular fractions of soy protein, 11S (glycinin) and 2S a 'biologically active' protein (Lin et al., 2004; Mondor et al., 2004), where the researcher is unable to utilize the conventional experimental characterization of composite gels. The alternative working protocol involves small-deformation dynamic oscillatory work on soy systems in the presence of 0.4% glucono-δ-lactone (GDL), which were left to cure at 25°C for 1500 min. This was followed by similar work on mixtures of the two components at which the glycinin content was fixed at 4% (Tay et al., 2006). It was verified that the gelation curves fitted first-order reaction kinetics, a reasonable result considering that network development is based on 'collisions' of functional groups of the same reactant, i.e., polymeric segments of 11S or 2S. The integrated form of the first order rate law is as follows (Atkins and De Paula, 2006):

$$\kappa t = \ln \frac{a}{a - x} \qquad (7.1)$$

where, a and x represent the initial concentration of the reactant and the concentration of the product at time t, respectively. It is seen from equation (7.1) that the value of rate constant, κ, depends only on the ratio or the relative values of two concentrations thus obtaining the dimensions of reciprocal time, e.g., s^{-1}.

Since the original concentration of the reactant is not needed in order to find κ, equation (7.1) can be expressed as the ratio of two experimental parameters:

$$\frac{a}{a - x} = \frac{X_f}{X_f - X} \qquad (7.2)$$

where, X_f and X are these parameters at the end of the experiment and after time t, respectively. Combining equations (7.1) and (7.2) and implementing a certain rearrangement gives the operational form for modeling the experimental observations of dynamic oscillation, which is:

$$G'(t) = G'_f[1 - \exp[-\kappa(t - t_0)]] \qquad (7.3)$$

where, G'_f is the final storage modulus on shear, with t_o denoting the onset of gelation in the single and binary systems. This allows calculation of the rate constant, which is plotted in Figure 7.7 as a function of protein concentration.

A considerable difference of one-to-two orders of magnitude was observed in the values of κ at equal contents of the two molecular fractions of soy, with those of the high-molecular-weight material (11S) achieving the highest rates of gelation (Kuipers et al., 2005). Normally, gelation would proceed faster at higher concentrations of a given reactant but the opposite occurrence is evaluated graphically in Figure 7.7 (Tay and Perera, 2004). The drop of pH due to the GDL hydrolysis to gluconic acid is very much the requirement for gelation and the determinant of gelation kinetics in these systems, but the increasing buffering capacity of high protein concentrations diminishes acidification and hence coagulation of the macromolecules (Rickert et al., 2004).

Clearly, two distinct regions are noted that can be attributed to the gelation kinetics of glycinin or 2S at rate constant values above 1.25×10^{-4} s^{-1} and below 0.64×10^{-4} s^{-1}, respectively. Changes in the molecular organization of the mixture are now evident with increasing additions of 2S to 4% (w/w) 11S. In Figure 7.7, there is a transformation from a glycinin continuous matrix at 2S levels below 1%, to a system where the low-molecular-weight counterpart is capable of forming the dominant phase at concentrations higher than 3% thus determining kinetics of initial structure formation in the mixture.

The ability of the low-molecular-weight, hence slow-gelling, soy fraction to dictate the structural characteristics of the mixture at comparable concentrations of the two components is rather unseen in standard investigations of exclusion phenomena of biopolymers. Blending-law modeling of dynamic oscillatory data indicates that 2S possesses a high affinity for solvent as compared to 11S (Kasapis et al., unpublished results). The quantitative interpretation of the experimental evidence from rheology is congruent with the concept of a disproportionate aqueous 2S phase entrapping the polymeric segments of 11S, which are increasingly unable to participate in the formation of a continuous glycinin assembly. Further evidence from scanning electron microscopy demonstrates that the high water-holding capacity of increasing concentrations of 2S leads to a distinct three-dimensional topology in the mixture where noticeable remnants of the 11S network eventually are non-existent.

Conclusions of that nature, i.e., 2S dictates the structural characteristics of 11S at comparable concentrations of the two components in the mixture, are useful in processed foodstuffs, since tailor-made formulations of protein fractions can serve as value-added ingredients (LeBlanc et al., 2004; Thomas et al., 2004).

7.4 QUANTITATIVE ANALYSIS OF THE STRUCTURAL PROPERTIES OF BINARY COMPOSITE GELS

7.4.1 Blending Laws for Two-Phase, Synthetic Polymer Systems

Once it is established that polymeric constituents create a phase-separated organization in the mixture, and the nature of phase continuity is reasonably understood, the next step is to relate the physical properties of the components and their individual domains to the overall behavior of the composite gel. Such an achievement will facilitate intelligent manipulation, for example, of the textural attributes of firmness and plasticity thus contributing to a desirable mouthfeel in a variety of novel food preparations (Zoulias et al., 2002).

An approach that found considerable utility in the quantitative description of the structural properties of biopolymer co-gels relates the overall storage modulus on shear to those of the two phases by applying the so-called 'blending laws' proposed by Takayanagi and colleagues (Takayanagi et al., 1963; Fitzimons et al., 2007). These workers demonstrated that the viscoelastic behavior of polymer blends, formed by bonding together films of polybutadiene (PBD) and styrene-acrylonitrile copolymer (STAN), could be calculated according to the parallel or series model depending on the direction of the force passing through the specimen (Brown et al., 1995). The composite modulus (G'_c) for the parallel and series arrangements can be expressed as follows:

$$G'_c = \phi_X G'_x + \phi_y G'_y \quad (7.4)$$

and

$$G'_c = (\phi_X / G'_x + \phi_y / G'_y)^{-1} \quad (7.5)$$

where G'_x and G'_y are the storage moduli on shear, with ϕ_X and ϕ_Y ($\phi_X + \phi_Y = 1$) being the phase volumes of the two components X and Y, respectively.

Macroscopically homogeneous dispersions of PBD in STAN copolymer form a phase-separated mixture due to the non-molecular miscibility of the two components. The tendency to minimize the interfacial area results in spherical PBD inclusions, a topology which is critical for successful application of the blending laws to binary preparations (Gilsenan et al., 2003). Above the glass transition temperature (T_g) of PBD, but below that of STAN (where $G'_{PBD} << G'_{STAN}$), the modulus of the composite can be estimated from the known properties of the two component polymers on the basis of a parallel model. Therefore, by assuming extreme cases in the distribution of strain and stress within the mixed system equation (7.4) applies to isostrain conditions, where the continuous phase is more rigid than the disperse phase and the strain is uniform throughout the material (parallel model). On the other hand, equation (7.5) refers to isostress conditions, where the supporting phase is weaker than the discontinuous phase and both phases are now subjected to the same stress (series model) (Richardson and Kasapis, 1998).

Obviously, when the continuous phase is stronger than the disperse phase the composite modulus achieves values closer to the upper bound limit of the stronger component in isolation, whereas phase inversion in the mixture produces viscoelastic functions which are closer to the lower bound limit set by single preparations of the weaker component. In practical applications, the usefulness of the Takayanagi model lies in defining an area of modulus versus composition in which experimental results should be confined if the physical organization of the material is determined by the mechanism of phase separation (Manson and Sperling, 1976). As it will be discussed later on, however, the model is unable to predict modulus–composition relationships arising from different shapes and possibly size or orientation of the discontinuous filler particles for a given phase volume.

7.4.2 Addressing the Issue of Dynamic Solvent Partition Between Phases in Biological Mixtures

Direct application of this semi-theoretical procedure to elucidation of the phase topology of biopolymers is frustrated by the inclusion of solvent (water) in the composite, thus introducing one further complication that must be addressed first, namely: variable phase volumes (ϕ_X and ϕ_Y), depending on the partition (relative holding capacity) of solvent between the two gelling components (Ghiou et al., 2005). Obviously, effective concentrations following phase separation are higher than the initial, nominal amounts of the two constituents, and have to be estimated for use in the quantitative modulus–concentration relationship of single phases in relation to the overall modulus of the final mixture in the gel (Giannouli et al., 2004).

Clark and colleagues tackled the problem of solvent distribution in relatively dilute aqueous systems for food applications (up to 10% solids) by introducing the '*p*' parameter, i.e., the ratio of solvent to polymer in one phase divided by the corresponding ratio in the other phase (Clark et al., 1983):

$$p = (w_X/x)/(w_Y/y) \tag{7.6}$$

In samples where both the weights (x and y) of two polymers and water ($w = w_X + w_Y$) are known the *p* factor defines the phase volumes, hence the effective concentrations and real moduli of each biopolymer within its own phase. The analysis was extended to combinations within the range of intermediate level of solids (up to 45%) by calculating the concentration dependence of gel density thus incorporating this contribution to phase volume (Kasapis et al., 1993a; Manoj et al., 1997).

Utilization of this working protocol in aqueous three-component systems (X–Y–water) was based on the following two assumptions: (i) bulk phase separation to thermodynamic equilibrium takes place first in solution with gelation then occurring subsequently and independently in each phase, or (ii) solutions of biopolymer X set during cooling at higher temperatures than Y, effectively forming gels across the whole system at the nominal (original) concentration, and then taken to higher concentrations (process of deswelling) by partial removal of water as the second slower component also gels (Morris, 1998).

A further requirement in blending law modeling stipulates that the two macromolecules should be largely confined to their respective phases (Kasapis et al., 1999). It appears that this is the case, since there is evidence that polymer-entrapping gelation is minimized and extensive phase separation is allowed to develop by slowing down the rate of cooling of a binary solution, for example, from about 30 to 1°C/min. This has been demonstrated using light microscopy and observing at 15% maltodextrin in Figure 7.1 a tenfold increase in the mechanical strength of the slow scanned mixture at which the chain segments of the protein and polysaccharide networks link effectively (Alevisopoulos et al.,

1996). Reduction in the entropy of mixing, as outlined in the Gibbs equation of free energy, is one drive leading to structure formation and extensive segregation of the two polymers. More importantly, though, formation of minute helices serves as a nucleation center for formation of larger associations from the same kind of molecules, which, in solution, would have remained as cross contaminants of the second polymeric phase.

Based on the above, insights into the interplay between molecular conformation leading to gelation and solvent distribution in binary mixtures with relevance to the food industry may be outlined as follows: entropically driven species in solution like the disordered coils of maltodextrin and gelatin, or the globular structures of the largely denatured milk and soy protein isolates 'tolerate' each other at low concentrations in a monophasic solution. During controlled cold-setting, the rapid-gelling macromolecule develops its continuous network prior to its counterpart (Chronakis and Kasapis, 1993). The concept was successful in modeling the phase behavior of low-range concentrations in these systems, where gelatin and the commercially available milk protein were shown to employ relatively high rates of gelation.

Subsequent maltodextrin and soy-protein structure formation leads to deswelling of the already established networks but systems remain under kinetic control as judged by repeated weighings which showed a continuous, slow diffusion of water molecules from the rapid to the slow setting component. At high-solid combinations leading to phase inversion in the mixture, the kinetic effect is swamped by the enthalpic disadvantage in polymer segments being surrounded by others of a different type and phase separation at equilibrium occurs in solution (Abdulmola et al., 2000). Cooling of these solutions produces composite gels exhibiting a single pattern of water partition throughout the high concentration range, with the estimates of p-factor revealing the relative solvent affinity of the two components in their respective phases.

Prediction of the distribution of solvent between the two biopolymer phases was facilitated following a suggestion by Morris to calculate the values of storage modulus on shear for all possible distributions and finding which ones matched the experimental values of the mixture (Morris, 1992). A recent application of this approach is reproduced in Figure 7.8 which illustrates the use of blending laws in a sodium caseinate (SC)/barley β-glucan (LB) mixture. Clearly the amount of water held in each phase is tightly clustered around a single p value (p = 1.55, log p = 0.19) throughout the range of concentrations studied, with the exception of the 2% w/w SC – 2% w/w LB sample (Kontogiorgos et al., 2006). The latter makes the case of the level of uncertainty in modeling, since the effective concentration of the caseinate in this mixture will be at its farthest from the experimental range of

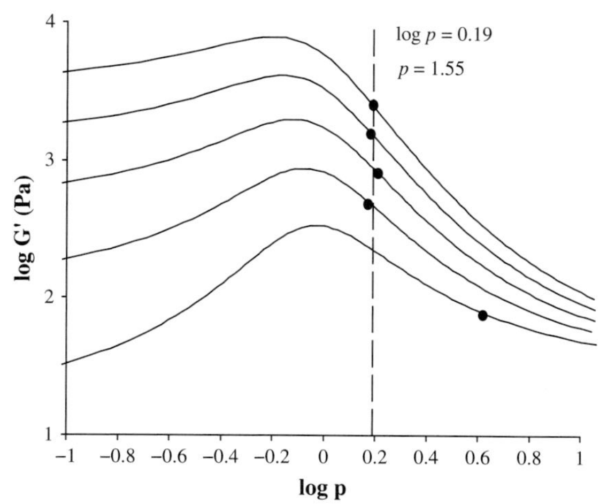

FIGURE 7.8 Analysis of the effect of solvent partition on the storage modulus of 2% sodium caseinate with increasing levels of β-glucan (2, 3, 4, 5 and 6% w/w leading to reinforced isostress-model bounds) in mixed gels as function of the p factor. Increasing concentrations of β-glucan in the mixture lead to higher composite moduli hence to an upward shift of the isostress bounds. The dashed line reflects the prediction of p value which remains constant at 1.55 (with permission from Kontogiorgos et al., 2006).

the concentration (5.5 to 13%) – storage modulus calibration curve utilized in that publication.

Besides the aforementioned examples, constant p values have been reported for the solvent partition in gelatin/low-methoxy pectin preparations (Gilsenan et al., 2003). In general, it was observed that true phase inversion in a given system due to increasing concentration of one component, resulted in two different clusters of p values, one for the X-continuous composite and the other for the Y-continuous composite following phase inversion (Chronakis et al., 1996). This school of thought appears to be promising for the rationalization of phase separation phenomena in co-gels especially when it is combined with competent experimental evidence from thermomechanical analysis, light/electron microscopy, etc. (de Bont et al., 2002; Petersson et al., 2005; Nickerson et al., 2006; Zhang et al., 2006).

As a final remark in this section, it should be mentioned that direct determination of phase composition is an obvious major target for future research on biopolymer composite gels. This is a notoriously difficult task to achieve owing to the requirements of performing reproducible scans into the z-direction (x–y is the horizontal observation plane) of a translucent or turbid, in all likelihood, material, and the subsequent segmentation of the recorded image based on the intensity value of the evaluated pixel. Nevertheless, in the regimen of dilute solutions, where only pair particle interactions are involved and with a colloid volume fraction of less than 0.2, certain progress has been made by utilizing confocal microscopy in combination with a powerful image analysis procedure (method of the 'variance measure'). This allowed determination of the phase boundary (i.e., phase volumes) of colloidal dispersions made of skim milk protein and amylopectin (de Bont et al., 2002). Solutions of casein–guar gum and casein–dextran were treated in a similar manner, thus unveiling clear differences in microstructures ranging from emulsion-like protein droplets in the former to protein aggregates in the latter. Clearly, successful extension of this work to the more difficult area of composite gels will allow interesting comparisons between tangible evidence and predictions of the aforementioned modeling. The outcome should facilitate fine-tuning of our ideas on the molecular interactions and topology of composite gels.

7.4.3 Refining the Blending-Law Analysis: Effect of the Shape of the Filler

The discussion has so far argued that modified blending laws can successfully predict the mechanical properties of composite gels provided that these are isotropic dispersions of spherical inclusions. This appears to be the case since the drive to reduce interfacial tension between two polymeric phases resolves the system into a continuous matrix dispersed with spheroidal fillers (Ding et al., 2002). As it was implied in the original Takayanagi paper, however, the basic blending laws cannot follow the change in structural properties once spheres aggregate into orientated particles or fibers are used as the filler phase. Today, microcrystalline cellulose (MCC) is widely available as a food ingredient and can serve as an 'organic' filler for modeling the structural properties of a blend (Watanabe et al., 2006; El-Sakhawy and Hassan, 2007). Another 'biological' example is the inclusion of mastic gum, a rather hard and brittle material at ambient temperature, in protein-based gel matrices in order to develop novel textures with a nutraceutical property (Triantafyllou et al., 2007). Combining thermal treatment with shear, and manipulation of surface-absorption properties can alter the degree of flocculation or shape of the gum particles (Mavrakis and Kiosseoglou, 2008).

Figure 7.9 reproduces the relationship between experimental storage modulus on shear and MCC concentration for 2% gelatin samples.

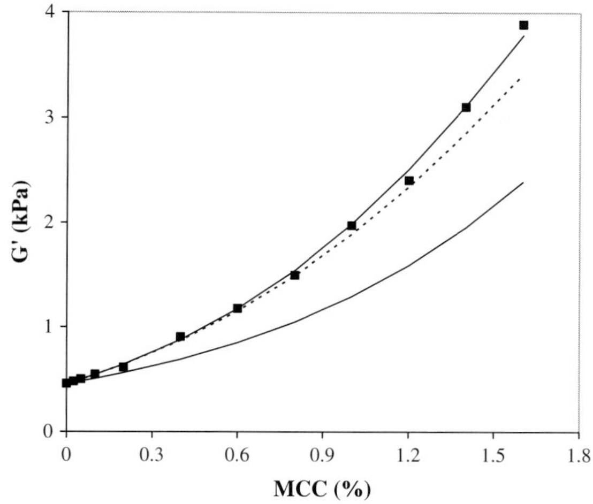

FIGURE 7.9 Experimental modulus of the 2% gelatin composite in the presence of increasing MCC concentrations (■) treated with the Lewis/Nielsen (top solid line), Kerner/Halpin (dashed line), and Takayanagi (bottom solid line) fits (with permission from Kasapis, 1999).

In the absence of MCC, gelatin forms a gel with G' values of about 460 Pa, whereas the network strength of the reinforced composite obtained at the maximum addition of the polysaccharide is $\approx$ 3900 Pa (Kasapis, 1999). Attempts to quantify the only unknown parameter (modulus of the filler phase, G'_2) from the isostress arrangement of equation (7.5) using the experimental phase volumes obtained by centrifugation and G'_1 values of the continuous phase via a gelatin concentration–modulus calibration curve, return negative values. This may be attributed to the inability of the blending laws to relate rigidity to the topology of a fiber-reinforced mixture.

In synthetic polymer research, a method that takes into account the shape of filler was first introduced by Kerner and then by Halpin, as follows (Wu et al., 2004):

$$G'/G'_1 = (1 + AB\phi_2)/(1 - B\phi_2) \qquad (7.7)$$

and

$$B = [(G'_2/G'_1) - 1]/[(G'_2/G'_1) + A] \qquad (7.8)$$

where G' is the modulus of the composite, ϕ_2 is the volume fraction of the filler phase, and the constant A is related to the generalized Einstein coefficient k_E ($A = k_E - 1$). The constant A takes into account such factors as geometry (shape) of the filler phase and Poisson's ratio of the matrix. Thus the value of A for any type of matrix in a composite with rigid spherical fillers is $(7 - 5\,v)/(8 - 10\,v)$, where v is the Poisson's ratio. In a viscoelastic liquid matrix (soft rubber), v is 0.5 hence k_E becomes 2.5. A point of interest is that the values of Einstein coefficient have not been determined experimentally for matrices with a Poisson's ratio less than 0.5 (e.g., a rigid epoxy resin or polystyrene). For soft rubbery matrices ($v = 0.5$) with inclusions in the geometry of aggregates of two spheres or fibers with rods–axial ratio up to 15, values of the Einstein coefficient vary from 2.58 to 9.4, respectively.

A further level of refinement was achieved by Lewis and Nielsen who considered the concept of maximum packing fraction of the filler phase, ϕ_m (Petersson and Oksman, 2006):

$$G'/G'_1 = (1 + AB\phi_2)/(1 - B\psi\phi_2) \qquad (7.9)$$

and

$$\psi = 1 + [(1 - \phi_m)/\phi_m^2]\phi_2 \qquad (7.10)$$

with the maximum packing fraction being related to the shape of the filler and its degree of polydispersity (Fornes and Paul, 2003). Thus ϕ_m for spheres in hexagonal close packing is 0.74, but in most practical situations related to food products is more like 0.64 (random close packing of spheres). The maximum packing fraction of fibers in random orientation is about 0.52. These values are generally obtained from sedimentation or vibratory motion measurements based on the ratio of true volume of the filler to the apparent volume occupied by the filler. Today, tabulated data of the maximum packing fraction and the generalized Einstein coefficient are available for inclusion in modeling algorithms as a function of the relevant experimental settings.

Reverting to the gelatin/MCC work, Figure 7.9 illustrates the experimental moduli of this gel and the standard of agreement obtained using composite theories. In the Lewis and Nielsen fit, the value utilized for ϕ_m is 0.52 which corresponds to the random orientation of fibers. It is rather difficult to estimate experimentally the maximum volumetric packing fraction in food-related mixtures with an elongated filler phase. As mentioned in the preceding paragraph, however, tabulated data from studies on block polymers and polyblends showed that varying the filler shape between fibers and spheres increases the ϕ_m value from 0.52 to 0.91 thus facilitating modeling (Luo and Daniel, 2003). Light microscopy work demonstrated that the average length of MCC fibers was 40 μm with the diameter being from 4–4.5 μm (overall axial ratio close to 10), which corresponds to an Einstein coefficient of 4.93 (Nielsen, 1974).

A computerized algorithm then minimized the value of G'_2 and good agreement with the experimental data was achieved for $G'_2 \approx 9{,}100$ Pa, in line with the series arrangement of the two components ($G'_2 > G'_1$). On the other hand, the Kerner/Halpin approach underestimated the rigidity of the composite at the upper range of the MCC concentration (>1% w/w in Figure 7.9). This estimate of G'_2 was also used in equation (7.5) to unveil the lower bound least able to follow the progress of structural properties in the system. Clearly, the topology of a co-gel (equations 7.7 to 7.8) is a more important consideration than the refinement of the packing arrangement of the filler (equations 7.9 to 7.10), as compared to the difficulty experienced by the original application of the blending laws (equations 7.5 to 7.6). It would be interesting to examine the strength of this approach in other biological fillers as they become available through processing of established food biopolymers or the structural characterization of new ones ultimately seeking food approval.

7.5 BRIDGING THE DIVIDE BETWEEN THE LOW- AND HIGH-SOLID ANALYSES IN BINARY CO-GELS

7.5.1 Experimental Observations Across the Range of Solids for Industrial Formulations

Besides research and product development in low and intermediate levels of solids described thus far, a considerable amount of work has been carried out over the years in several laboratories on the measurement of structural properties of biopolymers in the presence of high levels of co-solute (Doyle et al., 2006; Rogers et al., 2006; Seo et al., 2006). An example of the latter would be confectionery products having 5–15% moisture in the finished formulation and a high proportion of sugars. These products are almost exclusively manufactured by the process known as 'starch molding' in which a hot (e.g., 80°C) solution of all the ingredients at typically 25–30% moisture (liquor) is deposited into impressions made in the surface of 'dry' starch powder filling a shallow tray. The excess moisture is extracted by 'stoving' the sweets in the starch at a moderate temperature (about 50°C) for an extended period of typically several days leading to materials with rubbery or glassy consistency (Kritzer, 2006/2007).

High-solid systems are of increasing academic and industrial interest and several concepts (state diagram, glass transition temperature, etc.) have already been utilized in acquiring mechanistic understanding of molecular processes in dehydrated and partially frozen foods/biomaterials (Matveev et al., 1999; Burin et al., 2004; Kasapis, 2004; Rahman, 2006). Within this context of matching science to application, one feels compelled to note that a dividing line has emerged, which is quite rigorous, with researchers in the structure–function relationships of biopolymers opting to address issues largely in either high- or low-solid systems. This divide is becoming more and more pronounced, as scientists working in the high-solid regimen are increasingly inspired by the apparently 'universal' molecular physics of glassy materials, which may not consider much of the chemical detail at the vicinity of the glass transition temperature. By comparison, their colleagues working on low-solid systems are shifting their focus from the relatively universal structure–function relationships of biopolymers in solution to the much more specific ones involving multiscale assembly, complexation and molecular interactions.

An effort, however, should be made to better share the expertise of the two camps under the unified framework of the materials science approach. This is a prerequisite to ensure fully 'functional solutions' to contemporary needs, spanning the full range of relevant time, length and concentration scales. Otherwise, a gap may emerge between the voluminous literature on basic studies and a clear pathway for processing, preservation and innovation in high-solid preparations. It is our view that fundamental interpretation of mechanisms in low-solid (biopolymer/water) systems is well advanced and this understanding should be extended to resolving issues in the high-solid (biopolymer/sugar/water) mixtures. It is hoped that the following demonstrations can make a case of the direct link and safe passage in the structural properties from one type of material to another.

7.5.2 Structural Basis for the Mechanistic Understanding of the Transformation from Low- to High-Solid Biomaterials

To a certain extent this has been achieved in recent studies on the structural properties of gelling polysaccharides and gelatin in the presence of co-solute. Small deformation dynamic oscillation on shear was used to identify the nature of structural variation (major shift on properties or a gradual change) using the master (or composite) curve of viscoelasticity from the rubbery plateau through the transition region to the glassy state (Kasapis and Sablani, 2005). As summarized in Figure 7.10, the effect of low levels of sugar, i.e., addition of sucrose, fructose, glucose or glucose syrup up to 40% solids, on κ-carrageenan, agarose, deacylated gellan and gelatin is to create stronger (and more thermally stable) structures. Replacing sugar with urea or guanidine hydrochloride (hydrogen bond disruptors) reduces dramatically the network strength, thus arguing for the importance of polymer–polymer interactions supported via a proliferation of hydrogen bonding (Kasapis et al., 2003).

The nature of modulus development is entirely different at intermediate levels of co-solute (glucose syrup with or without sucrose). Thus there is an abrupt drop in the values of

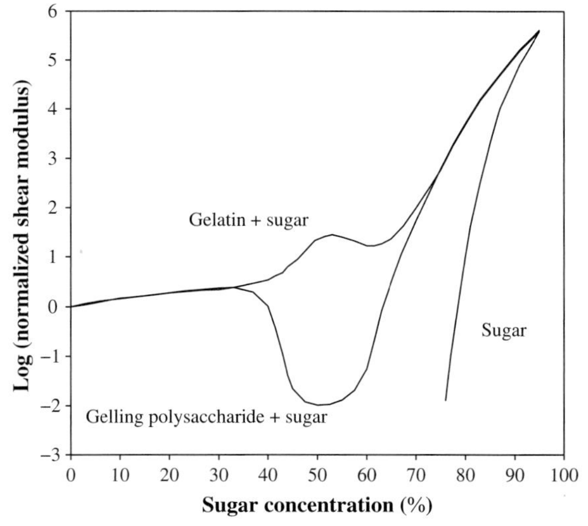

FIGURE 7.10 Variation of normalized shear modulus (frequency of 100 rad s^{-1}) as a function of sugar concentration for agarose/sugar, κ-carrageenan/sugar, deacylated gellan/sugar and gelatin/sugar mixtures, and single sugar preparations (with permission from Kasapis et al., 2003).

storage modulus between 40 and 60% sugar concentration in the polysaccharide mixture. Gelatin, on the other hand, shows an accelerated strengthening of the network which produces a wide loop in the structural profile as the shear modulus for the protein develops in the opposite direction to that of gelling polysaccharides. Supporting evidence from differential scanning calorimetry, scanning electron microscopy and turbidity-related visual observations argue for disaggregation in polysaccharides but of increased order in the gelatin network (Al-Amri et al., 2005).

The rationale behind the polysaccharide response is that the increasing shortage of water molecules and their relatively stable hydrogen bonding with sugar deprives gradually the polysaccharide associations of a hydration layer. This layer is required for the thermodynamic stability of intermolecular helices and the subsequent building of a network (Chandrasekaran and Radha, 1995). In the case of gelatin, addition of sugar raises the chemical potential of the protein to a thermodynamically unfavorable situation. Thus the drive for self-association is an attempt to stabilize the system by decreasing the surface of contact between gelatin and solvent (Oakenfull and Scott, 1986).

The dramatic drop in the mechanical strength of polysaccharide gels at intermediate levels of co-solute can be conceived as a transformation from a highly enthalpic aggregated structure to a network of reduced cross-linking (Whittaker et al., 1997). Increased flexibility of chain segments between intermolecular associations would render important the entropic contribution to a network's elasticity at the upper range of co-solute (> 70%). This is seen as an increasingly vitrified system with decreasing temperature or time of measurement. Figure 7.10 reproduces the mechanical manifestation of vitrification, which includes a progressive five-to-six-decade enhancement of shear modulus in normalized values. These correspond to a network strength of about 10^4 Pa in the rubbery plateau and from 10^9–10^{10} Pa in the glassy state of biological glasses (Kasapis, 2007).

In the same figure, acceleration of vitrification properties (at subzero temperatures) yields shear modulus values for the gelatin/sugar system which are coincident with those of polysaccharides. In the case of single sugar preparations, which are based on amorphous glucose syrup with or without sucrose, results converge belatedly with those of the polymeric mixtures at the upper range of solids (> 90% in the formulation). This transformation in the structural properties of biopolymers as a function of added co-solute has significant implications for the properties of foodstuffs and pharmaceuticals and will be dealt with in the following sections.

7.5.3 Changing Viscoelasticity as a Function of Increasing Levels of Co-Solute in Biopolymer Mixtures

An attempt was made to discuss what are high and what are low solids in biopolymer/sugar samples based on physicochemical considerations. It appears that the 'divide' in structural and hence textural properties can be bridged using scientific reasoning, and this school of thought will be extended to binary biopolymer composites with changing 'solvent quality'.

Figure 7.11 depicts the cooling and heating profiles of 1.5% κ-carrageenan (25 mM KCl added) with 10% gelatin ($pI = 4.5$) at various levels of co-solute obtained using dynamic oscillation at a scan rate of 1°C/min. Earlier work using calorimetry and light microscopy demonstrated the non-interactive nature of the two polymers in an aqueous system (pH of 6.7) without co-solute (Kasapis and Al-Marhoobi, 2005; Pranoto et al., 2007). Temperature sweeps between 60 and 0°C of the polymeric mixture unveil a thermally stable transition due to κ-carrageenan and a second wave of structure formation related to gelatin, with the cooling and heating bimodal profiles resulting in considerable thermal hysteresis (Funami et al.,

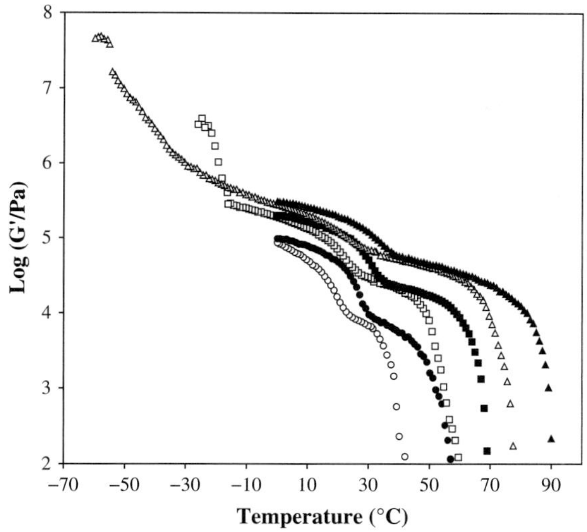

FIGURE 7.11 Temperature variation of G′ for 1.5% κ-carrageenan (25 mM KCl) plus 10% gelatin in the presence of 0% (cool (○); heat (●)), 28.5% (cool (□); heat (■)), and 56.5% (cool (△); heat (▲)) sugar, which is a 50:50 mixture of glucose syrup and sucrose (scan rate: 1°C/min; frequency: 1 rad/s; strain: 0.007%) (with permission from Kasapis and Al-Marhoobi, 2005).

2007). Gradual addition of co-solute at 28.5 and 56.5% solids indicates similar steps of gel formation and melting but with increasing thermal stability. Therefore, this type of experimentation can serve as a guide of changing viscoelasticity with increasing solids content.

In a second experiment, identical formulations were cooled down to subzero temperatures, thus looking for further molecular processes. Ice formation is observed in the sample with 28.5% co-solute at –16°C pinpointed by a mounting of the values of storage modulus. Soon after that (–26°C), increasing crystallinity makes the sample slippery, proper adhesion to the surface of the measuring geometry is lost and the experiment is abandoned (Walton, 2000). At 56.5% sugar content, the dependence of elasticity on temperature is quite spectacular and it develops unabated down to –54°C in Figure 7.11. At temperatures below –54°C, some crystallinity was again evidenced by the sudden jump in the values of *G′*. Meanwhile, a broad-temperature transition is unveiled between 0 and –50°C that bears resemblance to the glass-like consistency encountered in a high-solid regimen (Contreras Lopez et al., 2005).

Similar inroads into the structural properties of aqueous/high-solid materials have been reported for several biopolymer mixtures (Papageorgiou and Kasapis, 1995; Kasapis, 2002; Nickerson and Paulson, 2005). Table 7.1 summarizes progress made in co-gels of 1% w/w pectin and 3% w/w gelatin. Addition of sugar up to 58% generates a range of structure-formation ($t_{form} \leq 35°C$) and melting ($t_{melt} \leq 40°C$) indices that are covered by the temperature fingerprint of single gelatin gels (Al-Ruqaie et al., 1997; Chatterjee and Bohidar, 2005). It appears, therefore, that low and intermediate levels of co-solute mixtures are supported by a continuous gelatin phase. The step from the gelatin-dominated gelling and melting profiles to those in the range of a pectin gel is achieved when the co-solute content is increased to 66 and 74% (Willats et al., 2006). In the latter, there is an elevated onset of structuring ($t_{form} = 65°C$) but completion of melting is not observed within the experimentally accessible temperature range ($t_{melt} > 95°C$).

The dramatic transformation in the morphology of the composite is also seen in the values of tan δ at 5°C which increase from about 0.1 at low levels of co-solute to 0.82 at 74% sugar and maintain a value of 0.56 at 90°C. Therefore, despite the early melting profile of gelatin, gel mixtures retain overall cohesion at elevated temperatures, a result which argues that the polysaccharide has formed a continuous network at high levels of co-solute.

7.5.4 Quantitative Explorations of the Structural Properties in Biopolymer Mixtures at High Levels of Co-Solute

Experimental data of viscoelasticity, calorimetry and microscopy unveil phase-separation

TABLE 7.1 Small Deformation Viscoelastic Parameters for Mixtures of 1% w/w High Methoxy Pectin and 3% w/w Gelatin Preparations made with Increasing Levels of Co-solute. The Concentration of Glucose Syrup Refers to Dry Solids (with Permission from Al-Ruqaie et al., 1997)

Sucrose (%)	Glucose syrup (%)	Total solute (%)	Pseudoequilibrium modulus (G'; Pa) 85°C	5°C	Tan δ at 1 Hz 85°C	5°C	90°C	t_{form} (°C) 1°C/min	t_{melt} (°C) 1°C/min	Onset of glass transition (°C)
0	—	—	—	2079	>1	0.094	>1	13	24	—
20	—	20	—	2132	>1	0.097	>1	16	26	—
30	—	30	—	3081	>1	0.088	>1	24	30	—
40	—	40	—	3966	>1	0.109	>1	26	32	—
50	—	50	—	4387	>1	0.108	>1	28	34	—
50	8	58	—	7076	>1	0.160	>1	35	40	—
50	16	66	—	6939	>1	0.373	0.296	55	>95	—
50	24	74	—	9714	>1	0.817	0.557	65	>95	—
50	28	78	229	(158600)	0.851	1.831	0.813	>95	>95	52

phenomena in low-solid blends and these can form the basis of extending the work to intermediate and high-solid counterparts. Results in Table 7.1, for example, record the gradual increase in tan δ values, which argue for 'liquid-like' behavior of the mixture at 78% co-solute (tan $\delta \approx 1.8$ at 5°C). This is accompanied by a considerable rise in the values of storage modulus at the same temperature ($G' \approx 159$ kPa), an outcome that demarcates the advent of the glass transition region (Chanvrier et al., 2005). Such an event is also monitored in the subzero temperature runs of 1.5% κ-carrageenan/10% gelatin/56.5% co-solute in Figure 7.11.

Consideration of the phase topology allows utilization of the blending laws as discussed in Section 7.4, and this constitutes the first step in making headway from pictorial rheology to quantitative analysis in high-solid co-gels. A second requirement focuses on the determination of the glass transition temperature (T_g) of the material. In synthetic polymer research and more recently in biological 'rubbers and glasses' this is accomplished through the interlinking of the contributions of time and temperature to the overall mechanical response in the form of a basic function of time alone and a basic function of temperature alone. The concept used extensively to achieve a superposition between the time and temperature domains in the glass transition region is known as the free volume/Williams, Landel and Ferry theory (Ferry, 1991; Kilburn et al., 2005). The quantitative expression of this combined framework is given for the shear storage modulus by the so-called WLF equation, as follows:

$$\log a_T = \log[G'(T)/G'(To)] = -\frac{(\mathrm{B}/2.303\mathrm{f_o})(\mathrm{T}-\mathrm{T_o})}{(\mathrm{f_o}/\alpha_f)+\mathrm{T}-\mathrm{T_o}} \quad (7.11)$$

where, at any reference temperature, To, the fractional free volume, f_o, is the ratio of free to total volume of the molecule, αf is the thermal expansion coefficient, and B is usually set to one (Nickerson et al., 2004).

The usefulness of this protocol is demonstrated in Figure 7.12 that reproduces the experimental moduli of a mixed gel of 1.5% κ-carrageenan (25 mM KCl) plus 10% gelatin in the presence of 61.5% sugar covering part of the rubbery plateau, the transition zone and the

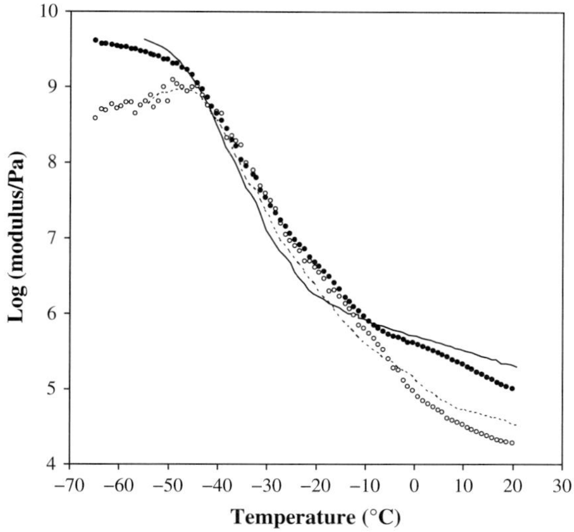

FIGURE 7.12 Modulus variation as a function of temperature for 1.5% κ-carrageenan (25 mM KCl) plus 10% gelatin in the presence of 61.5% sugar recorded experimentally (G′ (●); G″ (○)), and calculated at 73% total solids using the prediction of the isostress blending law (G′ (—); G″ (- - -)) (remaining experimental conditions as in Figure 7.11; with permission from Kasapis and Al-Marhoobi, 2005).

glassy state (Kasapis and Al-Marhoobi, 2005). In contrast to a crystallization element observed for the mixtures of reduced solids in Figure 7.11, the present composition follows a single pattern of vitrification.

At the high temperature end (>0°C), the observed pattern of response is qualitatively similar to that of the rubbery zone i.e., $G' > G''$, with relatively little frequency-dependence of either modulus (Tarrega et al., 2005). On further cooling at subzero temperatures, values of both moduli increase steeply leading to convergent lines and tan δ (G''/G') estimates of about 1. The spectacular development in viscoelasticity is thermally reversible and it is governed by configurational vibrations of segments of the molecule that are shorter than the distance between cross-links or points of entanglement (Kasapis et al., 2003a). Finally, the elastic component of the network dominates once more at temperature below −43°C in Figure 7.11. This corresponds to the glassy state at which molecular movement from one lattice site to another requires to overcome a considerable energy barrier (Kasapis, 2006). Completion of the sigmoidal modulus–temperature curve is accompanied by a large drop in the values of tan δ (≈ 0.08 at −65°C) further emphasizing the characteristic solid-like behavior of the 'foot' region of the viscoelastic spectrum.

The effect of changing temperature at a fixed frequency (1 rad/s) can be paralleled by implementing a series of frequency sweeps (0.1 to 100 rad/s) at constant temperature intervals (3–4 °C) and superposing the viscoelastic data at T_o thus creating a set of shift factors, a_T in equation 7.11 (Mano and Viana, 2006). The factor a_T is a fundamental descriptor of the temperature/time dependence of viscoelastic functions and its plotting as a function of temperature for the experimental data of gelatin/κ-carrageenan/co-solute yields a T_g value of about −42°C (Kasapis and Al-Marhoobi, 2005). This further confirms that the threshold observed between the glass transition region and the glassy state at ≈ −43°C in Figure 7.12 corresponds to completion of vitrification hence being accompanied by a true molecular transformation from free-volume dominated effects (WLF equation) to the predictions of the reaction rate theory. Modeling of the latter process is achieved by the modified Arrhenius equation utilizing a set of experimental temperatures and the reference temperature (Tanaka, 2005).

Acquiring such understanding facilitates application of the phase-topology protocol to high-solid materials in conjunction with supporting evidence from concentration–modulus calibration curves, patterns of solvent distribution between polymeric domains, and microscopy images. It was found that predictions of the isostress model of the blending laws were in good agreement with experimental observations spanning the entirety of the rubber-to-glass transformation (Figure 7.12). An unexpected result in this type of analysis was that the G'' trace of the predicted response within the glass transition region was more pronounced than the recorded modulus. This observation may prove to be of interest considering that current modeling of structural properties in low-solid systems is carried out on G' data since by comparison the G'' contribution to overall modulus can be considered to be insignificant (Chen and Evans, 2005). This, however, is no longer the case in high-solid preparations where the glass transition region includes a substantial viscous element (Chen et al., 2007; Kasapis et al., 2007).

7.6 MOLECULAR DYNAMICS OF BIOACTIVE COMPOUNDS IN A HIGH-SOLIDS CARBOHYDRATE MATRIX

This section attempts to provide insights into the wealth of information to be gleaned from a combination of mechanical and spectroscopic measurements in complex biomaterials with a nutraceutical function. Understanding the molecular dynamics of high-solid carbohydrates and proteins at the vicinity of T_g leads to the desire to improve the stability and quality of processed foods at which chemical reaction pathways and enzymatic processes are critical considerations. In this context, biopolymeric matrices act as inactive drug additives (excipients) and enzymatic activity relates to the diffusion-controlled substrate–enzyme interaction (Burin et al., 2004). Chemical considerations in these systems are mainly interested in the prevention of flavor/color degradation and oxidative reactions such as fat rancidity (Burin et al., 2000). Thus, studies on non-enzymatic browning (NEB) monitored the interactions of reducing sugars (fructose, glucose and xylose) with lysine in several matrices made of trehalose, maltodextrin or poly(vinylpyrrolidone) (Lievonen et al., 2002). Other examples include the degradation of aspartame, the acid-catalyzed hydrolysis of sucrose, the deterioration of saffron quality, the methylesterase catalyzed de-esterification of pectin, and the alkaline phosphatase catalyzed hydrolysis of disodium p-nitrophenyl phosphatase (Terefe et al., 2004).

Interpretation of results paid special attention to the extent of reaction in relation to the amplitude of the difference between glass and experimental temperature ($T - T_g$). Attempts were made to follow the rates of reactant consumption with the predictions of the reaction rate theory (Arrhenius equation) or the Williams-Landel-Ferry (WLF) equation. There have been reports, however, where neither the Arrhenius nor the WLF framework are able to follow convincingly the kinetics of the molecular processes in the glass transition region. It is now clear that the so-called 'universal' C_1 and C_2 values of the WLF equation should not be employed in fundamental research that probes molecular properties. It is also doubtful that the aforementioned constants obtained by fitting kinetic rates above the T_g to the WLF equation relate directly to the free volume concept. Problems encountered include negative estimates, physically unrealistic high values, etc.

The following constitutes a demonstration of utilizing mechanical spectroscopy and the principle of time–temperature superposition (TTS), as advocated by the synthetic polymer research, to derive fundamentally valid C_1 and C_2 values for a carbohydrate matrix. These were then used to assess whether the concept of free volume is the overriding consideration in the kinetics of diffusional mobility of a bioactive compound in the softening dispersion of the said matrix. The example of a bioactive compound is that of caffeine, since many food materials like tea, coffee, sodas and chocolate contain this mild stimulant in amounts that vary widely from 100 to over 1000 μg/ml (Belay et al., 2008).

The diffusion patterns of caffeine can be monitored using UV spectroscopy, which provides a sensitive method for quantification purposes over the various temperatures of experimental interest. The UV absorption spectrum of caffeine in dichloromethane exhibits a pair of absorption bands peaking at about 230 nm and 275 nm. The intensity of the latter peak (λ_{max}) is considerably higher and it is chosen for subsequent experimentation. In doing so, it is first verified that instrumental readings at λ_{max} = 275 nm remain linear as a function of caffeine concentration (i.e. follow the Beer-Lambert law) within the conventional experimental range of 0–3.0 AU.

Figure 7.13 shows the variation in absorbance of 0.4% caffeine diffused from 79.6% glucose syrup to dichloromethane over 8 h of observation. The experimental temperature range of –40°C to 30°C was chosen to coincide with the glass transition region of the sample matrix (< –43°C for 80.0% glucose syrup) as identified by DSC and theoretical modeling of mechanical data (Kasapis, 2005). Absorbance behavior is very evident with recording time and spectra approach asymptotically a fixed value at the upper range of readings. Furthermore, spanning a temperature range of 70°C results in a 60-fold increase in caffeine absorbance at the end of the experimental routine.

To further demonstrate the effect of matrix vitrification on the diffusional mobility of caffeine, data in Figure 7.13 were recast as a function of temperature for selected recording times. This parallels the approach of plotting thermodynamic properties such as volume, enthalpy and free energy of glass-forming liquids thus looking for a reduction in the magnitude of these primary properties and a change in their derivatives with respect to temperature (Allen, 1993). Clearly, the higher the amplitude of difference between T_g and experimental temperatures the more pronounced the diffusion of caffeine in the carbohydrate matrix (Figure 7.14). This becomes minimal at –40°C, a temperature that lies close to the DSC or mechanical T_g data mentioned in the preceding paragraph.

The main issue dealt with in this section is that of assessing the control of a glassy matrix over the mobility of a bioactive compound, and the outcome of such an effort is illustrated in Figure 7.15. In doing so, the extent of glucose syrup/caffeine interplay was determined by considering the first part (20 min) of spectra in

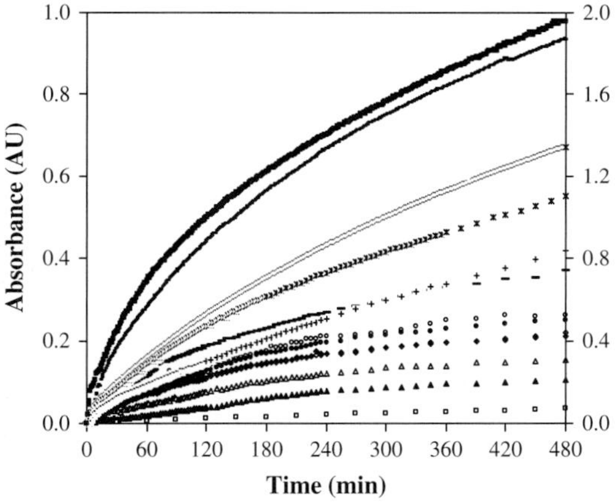

FIGURE 7.13 Absorbance of 0.4% caffeine diffused from 79.6% glucose syrup to dichloromethane as a function of time of observation (8 h) at –40 (□), –35 (▲), –30 (△), –25 (◆), –20 (◇), –15 (●), –10 (○), 0 (+), 5 (—), 10 (*), 15 (X), 20 (–) (left y axis), and 30°C (■; right y axis) obtained at 275 nm.

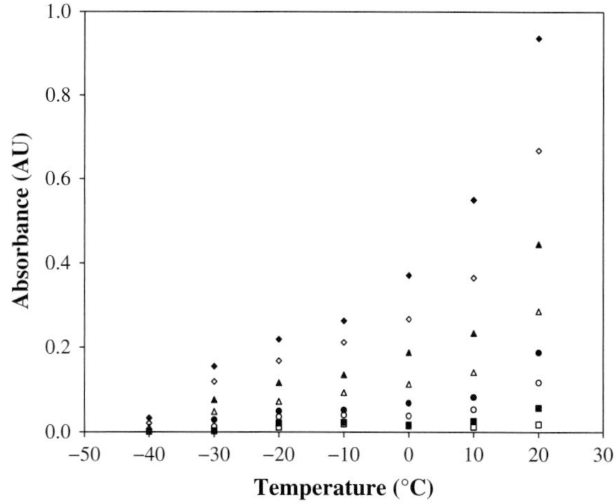

FIGURE 7.14 Absorbance of 0.4% caffeine diffused from 79.6% glucose syrup to dichloromethane as a function of temperature for the time periods of 3 (□), 7 (■), 15 (○), 30 (●), 60 (△), 120 (▲), 240 (◇) and 480 (◆) min obtained at 275 nm.

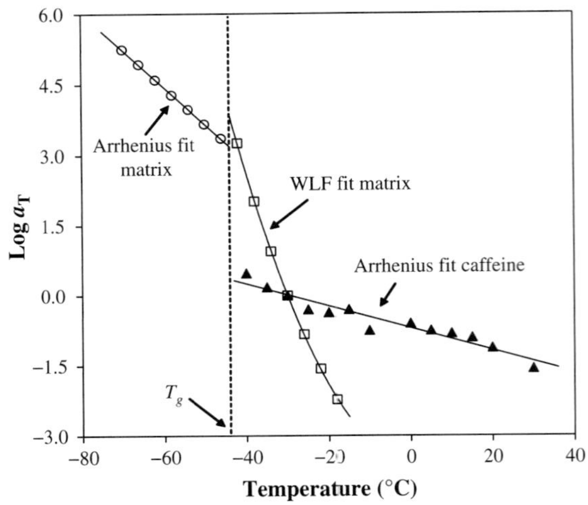

FIGURE 7.15 Temperature variation of the factor a_T within the glass transition region (□) and the glassy state (○) for 80.0% glucose syrup, and for kinetic data of 0.4% caffeine (▲) diffused from the carbohydrate matrix to dichloromethane (reference temperature for both systems is −30°C).

Figure 7.13, which exhibited a very acceptable linearity between absorbance and recording time; a high R^2 value is required and, in this case, it ranged between 0.974 and 0.993. This is equivalent to an appropriate rate law of a zero-order reaction for this part of the spectrum where the gradient (i.e. rate) is constant at $dx/dt = k$. Gradient or rate constant values were found to develop from 2.0×10^{-4} and through 24×10^{-4} to 223×10^{-4} at −40°C, 0°C and 30°C, respectively. Finally, the spectroscopic shift factor (log $\frac{k_o}{k}$) was calculated for each experimental temperature, with k_o being the rate constant at the reference temperature of −30°C. This allowed direct comparison with the molecular dynamics of glucose syrup in the glass transition region being available in the literature using mechanical spectroscopy.

It was felt that such a comparison was justified since the WLF parameters ($C_1^o = B/2.303f_o = 14.13$; $C_2^o = f_o/\alpha_f = 64.0$ deg) for 80% glucose syrup at the reference temperature of −30°C were properly derived using equation (7.11) and the principle of time–temperature superposition on mechanical spectra (Kasapis, 2005). Thus the use of inappropriate 'universal constants' of 17.44 and 51.6° was avoided, since according to Ferry for amorphous synthetic polymers and available data for biomaterials 'it is evident that the actual variation from one polymer (or biomaterial) to another is too great to permit use of these universal values' (Ferry, 1980). Furthermore, WLF parameters of this approach yield valid predictions of fractional free volume at T_o and T_g (quoted in the ensuing discussion), as compared to some kinetic data in the literature that correspond to physically unrealistic values of f_o or f_g within the glass transition region.

Figure 7.15 summarizes the molecular dynamics of sample matrix and bioactive compound, with data being normalized at the reference temperature of −30°C (log $a_T = 0$). The rates of molecular diffusion of caffeine were assessed against the predictions of the modified Arrhenius equation, which produces a good-quality fit ($R^2 = 0.968$) and a constant value for the energy of activation, $E_a = 0.47$ kJ/mole. This is a relatively low estimate and reflects the freedom of mobility of the compound within the carbohydrate matrix. In contrast, the activation energy of the latter ranges from 98–140 kJ/mol, thus indicating a strong temperature effect on the rubber-to-glass transition of polysaccharides or the melt-to-glass transformation of glucose syrup (Kasapis, 2001).

Finally, it has been suggested in the literature that an operational Arrhenius function is a sign of a comparatively high level of free volume, but it has been proved difficult to assign values to that both for the matrix and the suspended material. As discussed presently, the fractional free volume of 80.0% glucose syrup was found to vary from 0.031 at T_o to 0.024 at T_g. In addition, a WLF fit on the kinetic data of caffeine was performed ($R^2 = 0.847$), which yielded a fractional free volume of 0.209 at the reference temperature of −30°C. Conditions of molecular relaxation and diffusional mobility correspond to a free volume towards 20 and 30% of the total volume of the molecule (Cangialosi et al., 2003), and the value of 20.9% indicates that this is indeed the case for caffeine within the glassy carbohydrate matrix. The latter, on the other hand, is characterized by a collapsing free volume (e.g. 3.1% at T_0) that makes volume related or WLF considerations the overriding molecular mechanism of vitrification.

7.7 STRUCTURAL PROPERTIES OF NON-AQUEOUS SYSTEMS USED IN CONTROLLED TOPICAL DELIVERY

Besides considerations of structure formation in aqueous/sugar systems in relation to bioactive compound release described in the preceding section, non-aqueous solvents are of increasing interest since they can be used as vehicles of topical drug delivery (Qiu and Bae, 2006). Water poses a risk of hydrolytic reactions and the alternative formulations of hydro-alcoholic compounds (e.g., ethanol) can neither eliminate the possibility of drug hydrolysis nor prevent undesirable variation in formulation owing to high evaporative potential (Agrawal et al., 2003). An example of improved chemical stability is the non-aqueous ethylcellulose matrix that is able to structure the hydrophobic solvent, and the last section of this chapter aims to discuss this in brief. Ethylcellulose (EC) has found application as a polymeric coating, in controlled release of oral dosage forms, in transdermal films and patches, and as a particulate emulsion stabilizer (Wu et al., 2003). For example, it has been demonstrated that a formulation containing EC, the non-volatile solvent of propylene glycol dicaprylate (PGD), and low levels of minocycline hydrochloride (0.05% w/w MH) provides satisfactory antibacterial activity against *Propionibacterium acnes*, as compared to a blank gel matrix and a standard MH solution.

Very often, industrial operations identify inconsistencies in the structural properties of ethyl cellulose in a propylene glycol dicaprylate/dicaprate solution with storage time, an outcome that affects adversely the ability of the system to be used as vehicle for topical drug delivery. This is demonstrated here by reproducing the outcome of changes in the particle size of ethylcellulose over a prolonged period of observation (27 days in Figure 7.16) for untreated and thermally treated samples. The immediate observation is that the values of the arithmetic mean roughness of polymer particles (R_a) are substantially lower than for the corresponding periods of observation in unheated materials indicating that the structural characteristics of the two systems are distinct. There is an initial increase in the roughness values of freshly made samples (R_a = 267 nm), which culminates at 3 days of observation (R_a = 379 nm). This is followed by a sharp increase in light transmission (i.e., decrease in particle roughness) that reduces appreciably the size of measured entities after 27 days of observation (R_a = 159 nm). Overall, the relative magnitude of these changes argues for a time–temperature equivalence, since it appears that under a longer period of observation (approximately 45 days), both time and

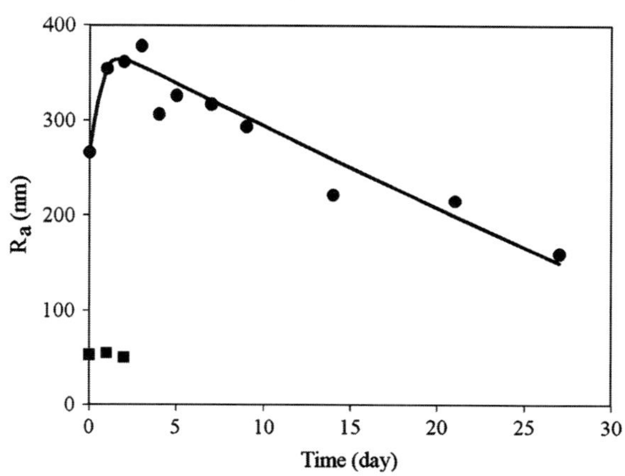

FIGURE 7.16 Arithmetic mean roughness (R_a) of the fresh (●) and heated (■) gel of 12% EC in PGD with respect to time of observation.

temperature treatments should yield comparable morphologies for the ethylcellulose gel.

The partial disintegration of structure demonstrated with increasing time of observation in Figure 7.16 is, of course, interesting but this trend is not apparent in similar isothermal routines that follow the mechanical properties of preparations. As shown in Figure 7.17, there is considerable increase in the values of storage and loss modulus in the initial stages of the time sweep, which is mirrored by the corresponding decrease in the damping factor, tan δ. At longer times, experimental traces level out thus achieving G' and tan δ values of 3.1 kPa and 0.28, respectively, after 6 days of continuous monitoring, or even at longer periods utilizing frequency sweeps between 0.1 and 100 rad/s at regular time intervals of 24 h.

The nature of the molecular processes leading to structural disintegration could be in the form of topological disentanglements, or helix-to-coil transitions of a low energetic content (van de Velde et al., 2005). Micro DSC is ideally poised to ascertain this and in the case of κ-carrageenan, for example, thermodynamic transitions recorded on cooling are likely to produce well-resolved peaks, which are sharp at the beginning and quite broad towards the end. The former development is due to a co-operative process of structure formation with the latter being attributed to polydispersity and the formation of a 'heterogeneous' network encompassing a range of temperature-induced relaxation processes (Kasapis and Al-Marhoobi, 2005). By contrast, thermograms of EC/PGD mixtures produce flat baselines on heating and cooling, which argue for molecular interactions broadly similar to topological entanglements. Extending the temperature range of observation to subzero temperatures records monotonic traces, which are devoid of sigmoidal patterns of heat flow that are associated with glass transition phenomena. At the lower range of experimental temperatures, a cooling exotherm denotes the freezing point of propylene glycol dicaprylate at –59°C.

A mechanistic model that may rationalize experimental observations is illustrated in Figure 7.18. This should consider that ethylcellulose/propylene glycol dicaprylate mixtures shows a continuous increase in rigidity as a function of time of observation at ambient

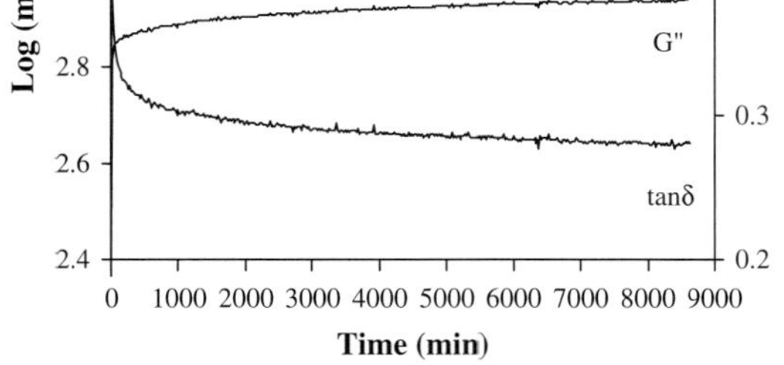

FIGURE 7.17 Storage modulus, loss modulus and tan δ variation as a function of time at 25°C for the 12% EC/PGD gel (frequency: 1 rad/s; strain: 0.5%).

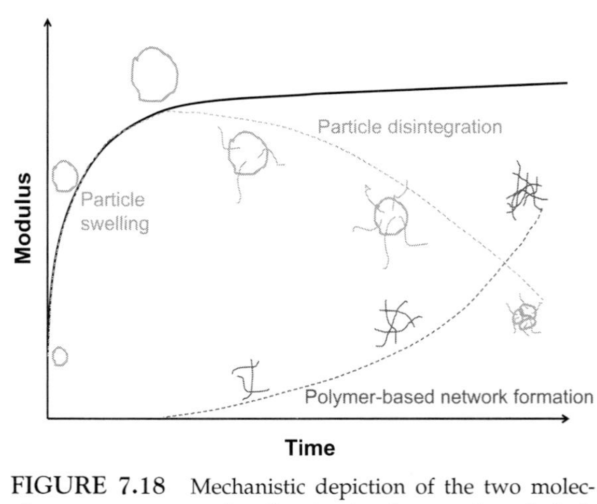

FIGURE 7.18 Mechanistic depiction of the two molecular processes governing the structural properties of ethylcellulose/propylene glycol dicaprylate gel as a function of time at ambient temperature.

temperature (Figure 7.17), with the obvious interpretation of this behavior being that there are dynamic structures present in the system developing continuously. This, in combination with the increase in particle roughness at the initial stages of 7.16, which is overtaken by structural disintegration at longer periods of observation, is suggestive of contributions from two distinct molecular processes. The passage from one process to the other should correspond to the maximum in the detectable values of particle roughness. Prior to that point, solvent diffusion and consequent particle swelling is the driving force of recorded viscoelasticity but a critical stage is reached at which partial dissociation of solvated regions will commence.

A direct analogy to this would be the thermal gelatinization of aqueous starch dispersions leading to granule swelling and subsequent disintegration followed by leaching of amylose-like sequences in the surrounding environment (Tan et al., 2008). According to the micro DSC data, increasing particle erosion results in topological entanglements, rather than any enthalpic interactions, among the released polymeric strands. Incorporation of the polymeric strands at concentrations higher than the point of coil overlap and entanglement further underpins the increase in rigidity of the solvated system. This evolving composite structure of topological interactions between individual species supporting particle remnants should be of critical importance in elucidating the functional properties of preparations related to topical drug delivery.

7.8 CONCLUDING REMARKS

Phase separation in protein and polysaccharide gels remains one of the basic tools of achieving the required structural properties and textural profile in food product formulations. As ever, the industrialist is faced with the challenge of innovation in an increasingly competitive market in terms of ingredient cost, product added-value, and expectations of a healthy lifestyle to mention but a few. It appears, however, that a gap persists between fundamental knowledge and direct application to food-related concepts with a growing need for scientific input. Furthermore, within the context of materials science, there is a tendency to examine research findings in either low- or high-solid systems without considering synergistic insights/benefits to contemporary needs, spanning the full range of relevant time, length and concentration scales.

Recent understanding of the structural properties of single and composite biopolymer gels in a high-sugar environment indicated the suitability of a unified school of thought to link molecular dynamics and phase topology across the range of solids with industrial interest. Fundamental research in high-viscosity, low-water-content materials has recently seen rapid growth mainly due to work on biopolymer composites in relation to novel formulation of soft confectionery products. Proper documentation of the transformation from rubber-like to glassy consistency led to a fundamental definition of the mechanical glass transition temperature. This concept assisted in addressing the issues of the physics and chemistry of biopolymers in industrial preparations. Earlier on, blending-law modeling advanced knowledge, prediction and control of steric exclusion phenomena thus allowing use of biopolymer composites in the manufacture of low- and intermediate-solid mimetics. It is hoped that this partnership between bio- and synthetic polymer science will provide further opportunities to glean and utilize information for work in food materials.

Thus, one of the main objectives of this treatise has been to raise awareness that phenomenological and theoretical aspects of polymeric structure and functionality can be correlated and extended to binary systems in a high-solid environment. It is further anticipated that research will increasingly incorporate bioactive compounds in the polymeric matrix in order to achieve fundamental understanding of chemical, biological and physical processes occurring in real foodstuffs and topical drug delivery systems. Finally, the empirical and fundamental tools described here should be able to bridge the divide in the analysis of the two 'types' of composite materials (from low to high solids) thus dealing effectively with the specific and often intricate problems of their science and applications.

ACKNOWLEDGMENTS

The author is indebted to Dr. Johan B. Ubbink/Nestle Research Center and Professor Ian T. Norton/University of Birmingham for stimulating discourses during drafting this manuscript.

References

Abdulmola, N. A., Richardson, R. K., and Morris, E. R. (2000). Effect of oxidised starch on calcium pectinate gels. *Food Hydrocolloids* **14**, 569–577.

Agrawal, A. M., Manek, R. V., Kolling, W. M., and Neau, S. H. (2003). *AAPS PharmSciTech* **4**, 1.

Al-Amri, I. S., Al-Adawi, K. M., Al-Marhoobi, I. M., and Kasapis, S. (2005). Direct imaging of the changing polysaccharide network at high levels of co-solute. *Carbohydrate Polymers* **61**, 379–382.

Alevisopoulos, S. and Kasapis, S. (1999). Molecular weight effects on the gelatin/maltodextrin gel. *Carbohydrate Polymers* **40**, 83–87.

Alevisopoulos, S., Kasapis, S., and Abeysekera, R. (1996). Formation of kinetically trapped gels in the maltodextrin-gelatin system. *Carbohydrate Research* **293**, 79–99.

Allen, G. (1993). A history of the glassy state. *In* The Glassy State in Foods, 1–12, Blanshard, J. M. V. and Lillford, P. J. (eds.). Nottingham University Press, Nottingham, UK.

Al-Ruqaie, I. M., Kasapis, S., and Abeysekera, R. (1997). Structural properties of pectin-gelatin gels. Part II: Effect of sucrose/glucose syrup. *Carbohydrate Polymers* **34**, 309–321.

Alves, M. M., Antonov, Yu. A., and Goncalves, M. P. (2000). Phase equilibria and mechanical properties of gel-like water-gelatin-locust bean gum systems. *International Journal of Biological Macromolecules* **27**, 41–47.

Anderson, N. S., Gampbell, J. W., Harding, M. M., Rees, D. A., and Samuel, J. W. B. (1969). X-ray diffraction studies of polysaccharide sulphates: Double helix models for κ- and ι-carrageenans. *J. Mol. Biol.* **45**, 85–99.

Anderson, V. J. and Jones, R. A. L. (2001). The influence of gelation on the mechanism of phase separation of a biopolymer mixture. *Polymer* **42**, 9601–9610.

Antonov, Yu. A. and Goncalves, M. P. (1999). Phase separation in aqueous gelatin-κ-carrageenan systems. *Food Hydrocolloids* **13**, 517–524.

Antonov, Yu. A., Grinberg, V. Ya., and Tolstoguzov, V. B. (1977). Phase equilibria in water-protein-polysaccharide systems II. Water-casein-neutral polysaccharide systems. *Colloid and Polymer Science* **255**, 937–947.

Aryana, K. J. and Haque, Z. U. (2001). Effect of commercial fat replacers on the microstructure of low-fat cheddar cheese. *International Journal of Food Science and Technology* **36**, 169–177.

Atkins, P. and De Paula, J. (2006). *Physical Chemistry*, 8th Edition, Oxford University Press, Oxford, UK.

Badii, F. and Howell, N. K. (2006). Fish gelatin: Structure, gelling properties and interaction with egg albumen proteins. *Food Hydrocolloids* **20**, 630–640.

Barrangou, L. M., Daubert, C. R., and Foegeding, E. A. (2006). Textural properties of agarose gels. I. Rheological and fracture properties. *Food Hydrocolloids* **20**, 184–195.

Baruch, L. and Machluf, M. (2006). Alginate-chitosan complex coacervation for cell encapsulation: Effect on mechanical properties and on long-term viability. *Biopolymers* **82**, 570–579.

Batista, A. P., Portugal, C. A. M., Sousa, I., Crespo, J. G., and Raymundo, A. (2005). Accessing gelling ability of vegetable proteins using rheological and fluorescence techniques. *International Journal of Biological Macromolecules* **36**, 135–143.

Belay, A., Ture, K., Redi, M., and Asfaw, A. (2008). Measurement of caffeine in coffee beans with UV/vis spectrometer. *Food Chemistry* **108**, 310–315.

Bhattacharya, S., Narasimha, H. V., and Bhattacharya, S. (2006). Rheology of corn dough with gum arabic: Stress relaxation and two-cycle compression testing and their relationship with sensory attributes. *Journal of Food Engineering* **74**, 89–95.

Braga, A. L. M. and Cunha, R. L. (2004). The effects of xanthan conformation and sucrose concentration on the rheological properties of acidified sodium caseinate-xanthan gels. *Food Hydrocolloids* **18**, 977–986.

Braga, A. L. M., Azevedo, A., Marques, M. J., Menossi, M., and Cunha, R. L. (2006). Interactions between soy protein isolate and xanthan in heat-induced gels: The effect of salt addition. *Food Hydrocolloids* **20**, 1178–1189.

Brown, C. R. T., Foster, T. J., Norton, I. T., and Underdown, J. (1995). Influence of shear on the microstructure of mixed biopolymer systems. *In* Biopolymer Mixtures, 65–83, Harding, S. E., Hill, S. E. and Mitchell, J. R. (eds.). Nottingham University Press, Nottingham.

Burin, L., Jouppila, K., Roos, Y., Kansikas, J., and del Pilar Buera, M. P. (2000). Color formation in dehydrated modified whey powder systems as affected by compression and T_g. *J. Agric. Food Chem.* **48**, 5263–5268.

Burin, L., Jouppila, K., Roos, Y. H., Kansikas, J., and Buera, M. P. (2004). Retention of β-galactosidase activity as related to Maillard reaction, lactose crystallization, collapse and glass transition in low moisture whey systems. *International Dairy Journal* **14**, 517–525.

Butler, M. F., Clark, A. H., and Adams, S. (2006). Swelling and mechanical properties of biopolymer hydrogels containing chitosan and bovine serum albumin. *Biomacromolecules* **7**, 2961–2970.

Cain, F. W., Clark, A. H., Dunphy, P. J., Jones, M. G., Norton, I. T., and Ross-Murphy, S. B. (1989). Edible plastic dispersion. *European Patent* 0,298,561 B1.

Cangialosi, D., Schut, H., van Veen, A., and Picken, S. J. (2003). Positron annihilation lifetime spectroscopy for measuring free volume during physical aging of polycarbonate. *Macromolecules* **36**, 142–147.

Caserta, S., Sabetta, L., Simeone, M., and Guido, S. (2005). Shear-induced coalescence in aqueous biopolymer mixtures. *Chemical Engineering Science* **60**, 1019–1027.

Chandrasekaran, R. and Radha, A. (1995). Molecular architectures and functional properties of gellan gum and related polysaccharides. *Trends in Food Science & Technology* **6**, 143–148.

Chanvrier, H., Colonna, P., Valle, G. D., and Lourdin, D. (2005). Structure and mechanical behaviour of corn flour and starch-zein based materials in the glassy state. *Carbohydrate Polymers* **59**, 109–119.

Chatterjee, S. and Bohidar, H. B. (2005). Effect of cationic size on gelation temperature and properties of gelatin hydrogels. *International Journal of Biological Macromolecules* **35**, 81–88.

Chen, B. and Evans, J. R. G. (2005). Thermoplastic starch-clay nanocomposites and their characteristics. *Carbohydrate Polymers* **61**, 455–463.

Chen, D. Z., Tang, C. Y., Chan, K. C., Tsui, C. P., Yu, P. H. F., Leung, M. C. P., and Uskokovic, P. S. (2007). Dynamic mechanical properties and in vitro bioactivity of PHBHV/HA nanocomposite. *Composites Science and Technology* **67**, 1617–1626.

Chiou, B.-S., Yee, E., Glenn, G. M., and Orts, W. J. (2005). Rheology of starch-clay nanocomposites. *Carbohydrate Polymers* **59**, 467–475.

Chronakis, I. S. (1998). On the molecular characteristics, compositional properties, and structural-functional mechanisms of maltodextrins: A review. *Critical Reviews in Food Science* **38**, 599–637.

Chronakis, I. S. and Kasapis, S. (1993). Structural properties of single and mixed milk/soya protein systems. *Food Hydrocolloids* **7**, 459–478.

Chronakis, I. S., Kasapis, S., and Richardson, R. K. (1996). Small deformation rheological properties of maltodextrin-milk protein systems. *Carbohydrate Polymers* **29**, 137–148.

Clark, A. H., Kavanagh, G. M., and Ross-Murphy, S. B. (2001). Globular protein gelation – theory and experiment. *Food Hydrocolloids* **15**, 383–400.

Clark, A. H., Richardson, R. K., Ross-Murphy, S. B., and Stubbs, J. M. (1983). Structural and mechanical properties of agar/gelatin co-gels – small deformation studies. *Macromolecules* **16**, 1367–1374.

Clarke, C. (2004). The Science of Ice Cream. The Royal Society of Chemistry, Cambridge, UK.

Committee of Medical Aspects of Food Policy (1994). Nutritional Aspects of Cardiovascular Disease, DHSS Report on Health and Social Subjects no 46. HMSO, London.

Contreras Lopez, E., Champion, D., Blond, G., and Le Meste, M. (2005). Influence of dextran, pullulan and gum arabic on the physical properties of frozen sucrose solutions. *Carbohydrate Polymers* **59**, 83–91.

de Bont, P. W., van Kempen, G. M. P., and Vreeker, R. (2002). Phase separation in milk protein and amylopectin mixtures. *Food Hydrocolloids* **16**, 127–138.

de Hoog, E. H. A. and Tromp, R. H. (2003). On the phase separation kinetics of an aqueous biopolymer mixture in the presence of gelation: The effect of the quench depth and the effect of the molar mass. *Colloids and Surfaces A: Physicochem. Eng. Aspects* **213**, 221–234.

Ding, P., Pacek, A. W., Frith, W. J., Norton, I. T., and Wolf, B. (2005). The effect of temperature and composition on the interfacial tension and rheology of separated phases in gelatin/pullulan mixtures. *Food Hydrocolloids* **19**, 567–574.

Ding, P., Wolf, B., Frith, W. J., Clark, A. H., Norton, I. T., and Pacek, A. W. (2002). Interfacial tension in phase-separated gelatin/dextran aqueous mixtures. *Journal of Colloid and Interface Science* **253**, 367–376.

Donato, L., Garnier, C., Novales, B., and Doublier, J. L. (2005). Gelation of globular protein in presence of low methoxy pectin: effect of Na^+ and/or Ca^{2+} ions on rheology and microstructure of the systems. *Food Hydrocolloids* **19**, 549–556.

Doyle, J. P., Giannouli, P., Martin, E. J., Brooks, M., and Morris, E. R. (2006). Effect of sugars, galactose content and chainlength on freeze-thaw gelation of galactomannans. *Carbohydrate Polymers* **64**, 391–401.

El-Sakhawy, M. and Hassan, M. L. (2007). Physical and mechanical properties of microcrystalline cellulose prepared from agricultural residues. *Carbohydrate Polymers* **67**, 1–10.

Evageliou, V., Alevisopoulos, S., and Kasapis, S. (1997). Application of stress-controlled analysis to the development of low-fat spreads. *Journal of Texture Studies* **28**, 319–335.

Ferry, J. D. (1980). Dependence of viscoelastic behavior on temperature and pressure. *In* Viscoelastic Properties of Polymers, 264–320. John Wiley, New York, USA.

Ferry, J. D. (1991). Some reflections on the early development of polymer dynamics: Viscoelasticity, dielectric dispersion, and self-diffusion. *Macromolecules* **24**, 5237–5245.

Fitzsimons, S. M., Mulvihill, D. M., and Morris, E. R. (2007). Co-gels of whey protein isolate with crosslinked waxy maize starch: analysis of solvent partition and phase structure by polymer blending laws. *Food Hydrocolloids* **21**, 638–644.

Fornes, T. D. and Paul, D. R. (2003). Modeling properties of nylon 6 / clay nanocomposites using composite theories. *Polymer* **44**, 4493–5013.

Funami, T., Hiroe, M., Noda, S., Asai, I., Ikeda, S., and Nishinari, K. (2007). Influence of molecular structure imaged with atomic force microscopy on the rheological behavior of carrageenan aqueous systems in the presence or absence of cations. *Food Hydrocolloids* **21**, 617–629.

Giannouli, P., Richardson, R. K., and Morris, E. R. (2004). Effect of polymeric cosolutes on calcium pectinate gelation. Part 2. Dextrans and inulin. *Carbohydrate Polymers* **55**, 357–365.

Giavasis, I., Harvey, L. M., and McNeil, B. (2000). Gellan gum. *Critical Reviews in Biotechnology* **20**, 177–211.

Gilsenan, P. M., Richardson, R. K., and Morris, E. R. (2003). Associative and segregative interactions between gelatin and low-methoxy pectin. Part 3. Quantitative analysis of co-gel moduli. *Food Hydrocolloids* **17**, 751–761.

Giraudier, S., Hellio, D., Djabourov, M., and Larreta-Garde, V. (2004). Influence of weak and covalent bonds on formation and hydrolysis of gelatin networks. *Biomacromolecules* **5**, 1662–1666.

Grant, G. T., Morris, E. R., Rees, D. A., Smith, P. J. C., and Thom, D. (1973). Biological interactions between polysaccharides and divalent cations: The egg-box model. *FEBS Letters* **32**, 195–198.

Haug, I. J., Draget, K. I., and Smidsrod, O. (2004). Physical and rheological properties of fish gelatin compared to mammalian gelatin. *Food Hydrocolloids* **18**, 203–213.

Helcke, T. Summer. (1997). Ingenious gelatin – a multitalented ingredient. *The European Food & Drink Review*, 57–60.

Higgins, J. S., Tambasco, M., and Lipson, J. E. G. (2005). Polymer blends; stretching what we can learn through the combination of experiment and theory. *Prog. Polym. Sci.* **30**, 832–843.

Icoz, D. Z., Moraru, C. I., and Kokini, J. L. (2005). Polymer-polymer interactions in dextran systems using thermal analysis. *Carbohydrate Polymers* **62**, 120–129.

Ikeda, S., Nitta, Y., Temsiripong, T., Pongsawatmanit, R., and Nishinari, K. (2004). Atomic force microscopy studies on cation-induced network formation of gellan. *Food Hydrocolloids* **18**, 727–735.

Jonnalagadda, S. S. and Jones, J. M. (2005). Position of the American Dietetic Association: Fat Replacers, ADA Reports. *Journal of the American Dietetic Association*, 266–275. February.

Kasapis, S. (1999). The elastic moduli of the microcrystalline cellulose – gelatin blends. *Food Hydrocolloids* **13**, 543–546.

Kasapis, S. (2001). The use of Arrhenius and WLF kinetics to rationalise the rubber-to-glass transition in high sugar/κ-carrageenan systems. *Food Hydrocolloids* **15**, 239–245.

Kasapis, S. (2002). Viscoelasticity of oxidised starch/low methoxy pectin mixtures in the presence of glucose syrup. *International Journal of Food Science and Technology* **37**, 403–413.

Kasapis, S. (2004). Definition of a mechanical glass transition temperature for dehydrated foods. *Journal of Agricultural and Food Chemistry* **52**, 2262–2268.

Kasapis, S. (2005). Glass transition phenomena in dehydrated model systems and foods: A review. *Drying Technology* **23**, 731–757.

Kasapis, S. (2006). Building on the WLF/free volume framework: Utilisation of the coupling model in the relaxation dynamics of the gelatin/co-solute system. *Biomacromolecules* **7**, 1671–1678.

Kasapis, S. (2007). The effect of pressure on the glass transition of biopolymer/co-solute. Part I: The example of gelatin. *International Journal of Biological Macromolecules* **40**, 491–497.

Kasapis, S. and Al-Marhoobi, I. M. (2005). Bridging the divide between the high- and low-solid analyses in the gelatin/κ-carrageenan mixture. *Biomacromolecules* **6**, 14–23.

Kasapis, S. and Sablani, S. S. (2005). A fundamental approach for the estimation of the mechanical glass transition temperature in gelatin. *International Journal of Biological Macromolecules* **36**, 71–78.

Kasapis, S., Al-Marhoobi, I.M., and Mitchell, J.R. (2003a). Molecular weight effects on the glass transition of gelatin/co-solute mixtures. *Biopolymers* **70**, 169–185.

Kasapis, S., Morris, E. R., Norton, I. T., and Gidley, M. J. (1993). Phase equilibria and gelation in gelatin/maltodextrin systems. Part II: polymer incompatibility in solution. *Carbohydrate Polymers* **21**, 249–259.

Kasapis, S., Morris, E.R., Norton, I.T. and Clark, A.H. (1993a). Phase equilibria and gelation in gelatin/maltodextrin systems. Part IV: composition-dependence of mixed-gel moduli. *Carbohydrate Polymers* **21**, 269–276.

Kasapis, S., Al-Marhoobi, I. M., Deszczynski, M., Mitchell, J. R., and Abeysekera, R. (2003). Gelatin vs. polysaccharide in mixture with sugar. *Biomacromolecules* **4**, 1142–1149.

Kasapis, S., Sablani, S. S., Rahman, M. S., Al-Marhoobi, I. M., and Al-Amri, I. S. (2007). Porosity and the effect of structural changes on the mechanical glass transition temperature. *Journal of Agricultural and Food Chemistry* **55**, 2459–2466.

Kasapis, S., Alevisopoulos, S., Abeysekera, R., Manoj, P., Chronakis, I. S., and Papageorgiou, M. (1996). Scientific and technological aspects of polymer incompatibility in mixed biopolymer systems. *In* Gums and Stabilisers for the Food Industry 8, 195–206, Phillips, G. O., Williams, P. A. and Wedlock, D. J. (eds.). IRL Press, Oxford.

Kasapis, S., Giannouli, P., Hember, M. W. N., Evageliou, V., Poulard, C., Tort-Bourgeois, B., and Sworn, G. (1999). Structural aspects and phase behaviour in deacylated and high acyl gellan systems. *Carbohydrate Polymers* **38**, 145–154.

Kilburn, D., Claude, J., Schweizer, T., Alam, A., and Ubbink, J. (2005). Carbohydrate polymers in amorphous states: An integrated thermodynamic and nanostructural investigation. *Biomacromolecules* **6**, 864–879.

Kontogiorgos, V., Ritzoulis, C., Biliaderis, C. G., and Kasapis, S. (2006). Effect of barley β-glucan concentration on the microstructural and mechanical behaviour of sodium caseinate gels. *Food Hydrocolloids* **20**, 749–756.

Kritzer, J. (2007). The science behind the success: Sugar replacement with isomalt. *Food & Beverage Asia*, 24–30. December 2006/January.

Kuipers, B. J., Van Koningsveld, G. A., Alting, A. C., Driehuis, F., Gruppen, H., and Voragen, A. G. J. (2005). Enzymatic hydrolysis as a means of expanding the cold gelation conditions of soy proteins. *Journal of Agricultural and Food Chemistry* **53**, 1031–1038.

LeBlanc, J. G., Garro, M. S., Savoy de Giori, G., and Font de Valdez, G. (2004). A novel functional soy-based food fermented by lactic acid bacteria: Effect of heat treatment. *Journal of Food Science* **69**, 246–250.

Lefebvre, J. (2006). An outline of the non-linear viscoelastic behaviour of wheat flour dough in shear. *Rheol. Acta* **45**, 525–538.

Li, J., Eleya, M. M. O., and Gunasekaran, S. (2006). Gelation of whey protein and xanthan mixture: Effect of heating rate on rheological properties. *Food Hydrocolloids* **20**, 678–686.

Lievonen, S. M., Laaksonen, T. J., and Roos, Y. H. (2002). Nonenzymatic browning in food models in the vicinity of the glass transition: Effects of fructose, glucose, and xylose as reducing sugars. *J. Agric. Food Chem.* **50**, 7034–7041.

Lin, J., Fido, R., Shewry, P., Archer, D. B., and Alcocer, M. J. C. (2004). The expression and processing of two recombinant 2S albumins from soybean (*Glycine max*) in the yeast *Pichia* Pastoris. *Biochimica et Biophysica Acta* **1698**, 203–212.

Lizarraga, M. S., Piante Vicin, D. De, Gonzalez, R., Rubiolo, A., and Santiago, L. G. (2006). Rheological behaviour of whey protein concentrate and λ-carrageenan aqueous mixtures. *Food Hydrocolloids* **20**, 740–748.

Lofgren, C. and Hermansson, A.-M. (2007). Synergistic rheological behaviour of mixed HM/LM pectin gels. *Food Hydrocolloids* **21**, 480–486.

Loren, N. and Hermansson, A.-M. (2000). Phase separation and gel formation in kinetically trapped gelatin/maltodextrin gels. *International Journal of Biological Macromolecules* **27**, 249–262.

Loren, N., Langton, M., and Hermansson, A.-M. (1999). Confocal laser scanning microscopy and image analysis of kinetically trapped phase-separated gelatin/maltodextrin gels. *Food Hydrocolloids* **13**, 185–198.

Loret, C., Meunier, V., Frith, W. J., and Fryer, P. J. (2004). Rheological characterisation of the gelation behaviour of maltodextrin aqueous solutions. *Carbohydrate Polymers* **57**, 153–163.

Loret, C., Schumm, S., Pudney, P. D. A., Frith, W. J., and Fryer, P. J. (2005). Phase separation and molecular weight fractionation behaviour of maltodextrin/agarose mixtures. *Food Hydrocolloids* **19**, 557–565.

Luo, J.-J. and Daniel, I. M. (2003). Characterization and modeling of mechanical behavior of polymer/clay nanocomposites. *Compos. Sci. Technol.* **63**, 1607–1616.

Mano, J. F. and Viana, J. C. (2006). Stress-strain experiments as a mechanical spectroscopic technique to characterise the glass transition dynamics in poly(ethylene terephthalate). *Polymer Testing* **25**, 953–960.

Manoj, P., Kasapis, S., and Chronakis, I. S. (1996). Gelation and phase separation in maltodextrin-caseinate systems. *Food Hydrocolloids* **10**, 407–420.

Manoj, P., Kasapis, S., and Hember, M. W. N. (1997). Sequence-dependent kinetic trapping of biphasic structure in maltodextrin-whey protein gels. *Carbohydrate Polymers* **32**, 141–153.

Manski, J. M., van der Goot, A. J., and Boom, R. M. (2007). Influence of shear during enzymatic gelation of caseinate-water and caseinate-water-fat systems. *Journal of Food Engineering* **79**, 706–717.

Manson, J. A. and Sperling, L. H. (1976). *Polymer Blends and Composites*. Plenum Press, New York.

Mathew, S., Brahmakumar, M., and Abraham, T. E. (2006). Microstructural imaging and characterisation of the mechanical, chemical, thermal, and swelling properties of starch-chitosan blend films. *Biopolymers* **82**, 176–187.

Matveev, Y. I., Slade, L., and Levine, H. (1999). Determination of the main technological parameters of food substances by means of the additive contribution method. *Food Hydrocolloids* **13**, 381–388.

Mavrakis, C. and Kiosseoglou, V. (2008). The structural characteristics and mechanical properties of biopolymer/mastic gum microsized particles composites. *Food Hydrocolloids* **22**, 854–861.

Michon, C., Chapuis, C., Langendorff, V., Boulenguer, P., and Cuvelier, G. (2005). Structure evolution of carrageenan/milk gels: Effect of shearing, carrageenan concentration and nu fraction on rheological behavior. *Food Hydrocolloids* **19**, 541–547.

Mohanty, B. and Bohidar, H. B. (2005). Microscopic structure of gelatin coacervates. *International Journal of Biological Macromolecules* **36**, 39–46.

Mondor, M., Ippersiel, D., Lamarche, F., and Boye, J. I. (2004). Effect of electro-acidification treatment and ionic environment on soy protein extract particle size distribution and ultrafiltration permeate flux. *Journal of Membrane Science* **231**, 169–179.

Monteiro, S. R., Tavares, C., Evtuguin, D. V., Moreno, N., and Lopes da Silva, J. A. (2005). Influence of galactomannans with different molecular weights on the gelation of whey proteins at neutral pH. *Biomacromolecules* **6**, 3291–3299.

Morris, E. R. (1992). The effect of solvent partition on the mechanical properties of biphasic biopolymer gels: an approximate theoretical treatment. *Carbohydrate Polymers* **17**, 65–70.

Morris, E. R. (1998). Segregative interactions in biopolymer co-gels. *In* Phase/State Transitions in Foods: Chemical, Structural, and Rheological Changes, 159–186, Rao, M. A. and Hartel, R. W. (eds.). Marcel Dekker, New York.

Nickerson, M. T. and Paulson, A. T. (2005). Time-temperature studies of κ-carrageenan gelation in the presence of high levels of co-solutes. *Carbohydrate Polymers* **61**, 231–237.

Nickerson, M. T., Paulson, A. T., and Speers, R. A. (2004). A time-temperature rheological approach for examining food polymer gelation. *Trends in Food Science & Technology* **15**, 569–574.

Nickerson, M. T., Farnworth, R., Wagar, E., Hodge, S. M., Rousseau, D., and Paulson, A. T. (2006). Some physical and microstructural properties of genipin-crosslinked gelatin-maltodextrin hydrogels. *International Journal of Biological Macromolecules* **38**, 40–44.

Nielsen, L. E. (1974). Morphology and the elastic modulus of block polymers and polyblends. *Rheol. Acta* **13**, 86–92.

Norton, I. T. and Frith, W. J. (2001). Microstructure design in mixed biopolymer composites. *Food Hydrocolloids* **15**, 543–553.

Nunes, M. C., Batista, P., Raymundo, A., Alves, M. M., and Sousa, I. (2003). Vegetable proteins and milk puddings. *Colloids and Surfaces B: Biointerfaces* **1**, 1–9.

Oakenfull, D. and Scott, A. (1986). Stabilization of gelatin gels by sugars and polyols. *Food Hydrocolloids* **1**, 163–175.

Papageorgiou, M. and Kasapis, S. (1995). The effect of added sucrose and corn syrup on the physical properties of gellan-gelatin gels. *Food Hydrocolloids* **9**, 211–220.

Papageorgiou, M., Kasapis, S., and Richardson, R. K. (1994). Steric exclusion phenomena in gellan/gelatin systems. Part I: Physical properties of single and binary gels. *Food Hydrocolloids* **8**, 97–112.

Perez-Orozco, J. P., Beristain, C. I., Espinosa-Paredes, G., Lobato-Calleros, C., and Vernon-Carter, E. J. (2004). Interfacial shear rheology of interacting carbohydrate polyelectrolytes at the water-oil interface using an adapted conventional rheometer. *Carbohydrate Polymers* **57**, 45–54.

Petersson, L. and Oksman, K. (2006). Biopolymer based nanocomposites: Comparing layered silicates and microcrystalline cellulose as nanoreinforcement. *Composites Science and Technology* **66**, 2187–2196.

Petersson, M., Loren, N., and Stading, M. (2005). Characterisation of phase separation in film forming biopolymer mixtures. *Biomacromolecules* **6**, 932–941.

Pranoto, Y., Lee, C. M., and Park, H. J. (2007). Characterisations of fish gelatin films added with gellan and κ-carrageenan. *LWT* **40**, 766–774.

Pszczola, D. E. (2006). Future strategies for fat replacement. *Food Technology* **6**, 61–84.

Qiu, L. Y. and Bae, Y. H. (2006). *Pharmaceutical Research* **23**, 1.

Rahman, M. S. (2006). State diagram of foods: Its potential use in food processing and products stability. *Trends in Food Science & Technology* **17**, 129–141.

Renard, D., van de Velde, F., and Visschers, R. W. (2006). The gap between food gel structure, texture and perception. *Food Hydrocolloids* **20**, 423–431.

Richardson, R. K. and Kasapis, S. (1998). Rheological methods in the characterisation of food biopolymers. *In* Instrumental Methods in Food and Beverage Analysis, 1–48, Wetzel, D. L. B. and Charalambous, G. (eds.). Elsevier, Amsterdam.

Rickert, D. A., Johnson, L. A., and Murphy, P. A. (2004). Functional properties of improved glycinin and β-conglycinin fractions. *Journal of Food Science* **69**, 303–311.

Rogers, M. A., Roos, Y. H., and Goff, H. D. (2006). Structural heterogeneity and its effect on the glass transition in sucrose solutions containing protein and polysaccharide. *Food Hydrocolloids* **20**, 774–779.

Seo, J.-A., Kim, S. J., Kwon, H.-J., Yang, Y. S., Kim, H. K., and Hwang, Y. H. (2006). The glass transition temperatures of sugar mixtures. *Carbohydrate Research* **341**, 2516–2520.

Singh, J., Kaur, L., and McCarthy, O. J. (2007). Factors influencing the physico-chemical, morphological, thermal and rheological properties of some chemically modified starches for food applications – A review. *Food Hydrocolloids* **21**, 1–22.

Sittikijyothin, W., Torres, D., and Goncalves, M. P. (2005). Modelling of the rheological behaviour of galactomannan aqueous solutions. *Carbohydrate Polymers* **59**, 339–350.

Sloan, A. E. (April 2005). Top 10 global food trends. *Food Technology* **59**, 20–32.

Stewart-Knox, B. and Mitchell, P. (2003). What separates the winners from the losers in new food product development? *Trends in Food Science & Technology* **14**, 58–64.

Strauss, G. and Gibson, S. M. (2004). Plant phenolics as cross-linkers of gelatin gels and gelatin-based coacervates for use as food ingredients. *Food Hydrocolloids* **18**, 81–89.

Tabilo-Munizaga, G. and Barbosa-Canovas, G. V. (2005). Rheology for the food industry. *Journal of Food Engineering* **67**, 147–156.

Takayanagi, M., Harima, H., and Iwata, Y. (1963). Viscoelastic behaviour of polymer and its comparison with model experiments. *Mem. Fac. Eng. Kyushu Univ.* **23**, 1–13.

Tan, I., Torley, P. J., and Halley, P. J. (2008). *Carbohydrate Polymers* **72**, 272.

Tanaka, H. (2005). Two-order-parameter model of the liquid-glass transition. II. Structural relaxation and dynamic heterogeneity. *Journal of Non-Crystalline Solids* **351**, 3385–3395.

Tarrega, A., Duran, L., and Costell, E. (2005). Rheological characterization of semisolid dairy desserts. Effect of temperature. *Food Hydrocolloids* **19**, 133–139.

Tay, S. L. and Perera, C. O. (2004). Effects of glucono-δ-lactone on 7S, 11S proteins and their protein mixture. *Journal of Food Science* **69**, 139–143.

Tay, S. L., Kasapis, S., Perera, C. O., and Barlow, P. J. (2006). Functional and structural properties of 2S soy protein in relation to other molecular protein fractions. *Journal of Agricultural and Food Chemistry* **54**, 6046–6053.

Terefe, N. S., Nhan, M. T., Vallejo, D., Loey, A. V., and Hendrickx, M. (2004). Modeling the kinetics of the pectin methylesterase catalyzed de-esterification of pectin in frozen systems. *Biotechnol. Prog.* **20**, 480–490.

Tharanathan, R. N. (2005). Starch – Value addition by modification. *Critical Reviews in Food Science and Nutrition* **45**, 371–384.

Thomas, M. E. C., Scher, J., Desobry-Banon, S., and Desobry, S. (2004). Milk powders ageing: Effect on physical and functional properties. *Critical Reviews in Food Science and Nutrition* **44**, 297–322.

Tischer, P. C. S. F., Noseda, M. D., de Freitas, R. A., Sierakowski, M. R., and Duarte, M. E. R. (2006). Effects of iota-carrageenan on the rheological properties of starches. *Carbohydrate Polymers* **65**, 49–57.

Tolstoguzov, V. (2003). Some thermodynamic considerations in food formulation. *Food Hydrocolloids* **17**, 1–23.

Triantafyllou, A., Chaviaras, N., Sergentanis, T. N., Protopapa, E., and Tsaknis, J. (2007). Chios mastic gum modulates serum biochemical parameters in a human population. *Journal of Ethnopharmacology* **111**, 43–49.

van Assche, G., Verdonck, E., and van Mele, B. (2001). Interrelations between mechanism, kinetics, and rheology in an isothermal cross-linking chain-growth copolymerisation. *Polymer* **42**, 2959–2968.

van de Velde, F., Antipova, A. S., Rollema, H. S., Burova, T. V., Grinberg, N. V., Pereira, L., Gilsenan, P. M., Tromp, R. H., Rodolph, B., and Grinberg, V. Ya (2005). The structure of κ/ι-hybrid carrageenans II. Coil-helix transition as a function of chain composition. *Carbohydrate Research* **340**, 1113–1129.

Veerman, C., Sagis, L. M. C., Heck, J., and van der Linden, E. (2003). Mesostructure of fibrillar bovine serum albumin gels. *International Journal of Biological Macromolecules* **31**, 139–146.

Walton, A. (2000). Modern rheometry in characterising the behaviour of foods. *Food Science and Technology Today* **14**, 144–146.

Watanabe, A., Morita, S., Kokot, S., Matsubara, M., Fukai, K., and Ozaki, Y. (2006). Drying process of microcrystalline cellulose studied by attenuated total reflection IR spectroscopy with two-dimensional correlation spectroscopy and principal component analysis. *Journal of Molecular Structure* **799**, 102–110.

Whittaker, L. E., Al-Ruqaie, I. M., Kasapis, S., and Richardson, R. K. (1997). Development of composite structures in the gellan polysaccharide/sucrose system. *Carbohydrate Polymers* **33**, 39–46.

Willats, W. G. T., Knox, J. P., and Mikkelsen, J. D. (2006). Pectin: New insights into an old polymer are starting to gel. *Trends in Food Science & Technology* **17**, 97–104.

Williams, P. A., Clegg, S. M., Langdon, M. J., Nishinari, K., and Phillips, G. O. (1992). Studies on the synergistic interaction of konjac mannan and locust bean gum with kappa carrageenan. *In* Gums and Stabilisers for the Food Industry 6, 209–216, Phillips, G. O., Wedlock, D. J., and Williams, P. A. (eds.). IRL Press, Oxford.

Wolf, B., Scirocco, R., Frith, W. J., and Norton, I. T. (2000). Shear-induced anisotropic microstructure in phase-separated biopolymer mixtures. *Food Hydrocolloids* **14**, 217–225.

Wu, Y.-P., Jia, Q.-X., Yu, D.-S., and Zhang, L.-Q. (2004). Modeling young's modulus of rubber-clay nanocomposites using composite theories. *Polym. Test.* **23**, 903–909.

Wu, P. C., Huang, Y. B., Chang, J. I., Tsai, M. J., and Tsai, Y. H. (2003). *International Journal of Pharmaceutics* **260**, 115.

Yan, H., Saiani, A., Gough, J. E., and Miller, A. F. (2006). Thermoreversible protein hydrogel as cell scaffold. *Biomacromolecules* **7**, 2776–2782.

Zhang, G. and Foegeding, E. A. (2003). Heat-induced phase behavior of β-lactoglobulin/polysaccharide mixtures. *Food Hydrocolloids* **17**, 785–792.

Zhang, J., Jiang, L., and Zhu, L. (2006). Morphology and properties of soy protein and polylactide blends. *Biomacromolecules* **7**, 1551–1561.

Zoulias, E. I., Oreopoulou, V., and Tzia, C. (2002). Textural properties of low-fat cookies containing carbohydrate- or protein-based fat replacers. *Journal of Food Engineering* **55**, 337–342.

CHAPTER

8

Mapping the Different States of Food Components Using State Diagrams

Yrjö H. Roos

Department of Food and Nutritional Sciences, University College Cork, Cork, Ireland

8.1 INTRODUCTION

The physical state and physicochemical properties of biological materials, as reviewed by Roos (1995, 2008) and Slade and Levine (1995), affect their behavior in various stages of processing and storage of particularly food and pharmaceutical systems (Yu, 2001). Water interacts primarily with hydrophilic compounds, i.e., carbohydrates and proteins, and to a lesser extent with the more hydrophobic lipids. The composition of the carbohydrate fraction in biological materials often varies from low-molecular-weight sugars, which are often completely dissolved in water, to cell wall polymers which may be plasticized or softened by water, but are not fully water soluble. Proteins also have varying hydrophilic and hydrophobic characteristics and a complex polymeric structure. These materials can often be considered as miscible or partially miscible biological materials (Roos, 1995; Slade and Levine, 1995).

As biological materials are complex mixtures of molecules and molecular structures, their components are likely to exist in disordered, non-equilibrium, amorphous states (Levine and Slade, 1986; Roos, 2008). The importance of the amorphous structures to processing and stability has been shown to apply particularly to systems with long-term shelf-life requirements, such as frozen and low-water foods (Slade and Levine, 1995; Roos et al., 1996). The present chapter describes the importance of state diagrams in mapping processing requirements of biological materials and their shelf-life control. This includes effects of glass transition, i.e., transition between the solid glassy and supercooled liquid states of amorphous biological materials, and their water plasticization (softening caused by water), on physicochemical properties. As the amorphous states are non-equilibrium states, their time dependence with some important temperature- and water-content-dependent phenomena are reviewed.

8.2 GLASS TRANSITION

Glass transition is a universal property of disordered, supercooled liquids indicating a reversible change of solid-like and liquid-like states. The glass transition occurs over a temperature range, although it is often referred to with a single, glass transition temperature, T_g. The glass transition involves a significant change in

Kasapis, Norton, and Ubbink: Modern Biopolymer Science
ISBN: 978-0-12-374195-0

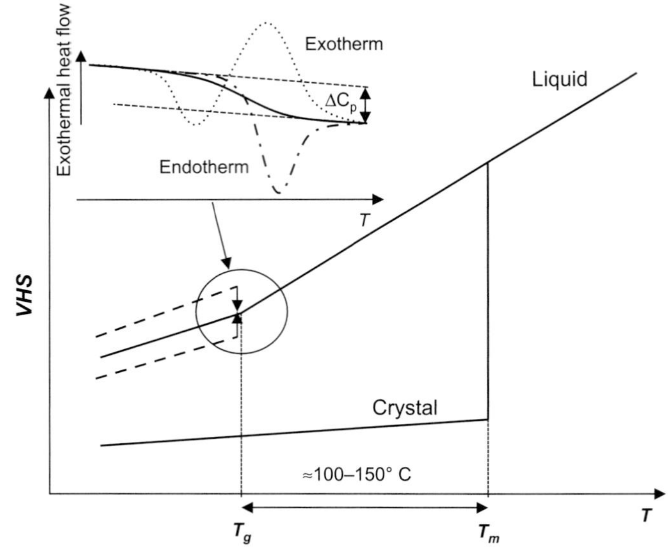

FIGURE 8.1 A schematic representation of enthalpy, H, entropy, S, volume, V and heat capacity Cp in various states of materials. Glass transition shows a change in the temperature dependence of H, S and V and a step change in Cp. However, glass transition is a property of a non-equilibrium state and various relaxations in thermal properties may occur around the glass transition as illustrated in the inset figure.

the translational mobility of molecules (Sperling, 1992), and it has thermodynamic characteristics of a second-order phase transition (White and Cakebread, 1966; Wunderlich, 1981; Roos, 1995). A second-order phase transition exhibits no latent heat but a change in heat capacity and other second derivatives of thermodynamic functions, including thermal expansion coefficient (Roos, 1995). Hence, glass transition can be observed from enthalpy, entropy and volume changes and corresponding changes in heat capacity and thermal expansion coefficient (Figure 8.1). The importance of glass transitions to food systems was emphasized by White and Cakebread (1966) but only the development of modern thermal analytical techniques has led to numerous studies of glass transitions and a wider acceptability of the glass transition concept to the food product development, process optimization and shelf-life control (Levine and Slade, 1986; Roos and Karel, 1991; Roos, 1995, 2008; Slade and Levine, 1995; Roos et al., 1996).

Food solids form amorphous or partially amorphous structures in the removal of solvent water (Figure 8.2). This takes place in dehydration processes and freezing as well as in many processes involving high-solids contents, e.g., baking, concentration, manufacturing of confectionary and extrusion (White and Cakebread, 1966; Slade and Levine, 1991; Roos, 1995). The removal of solvent water by evaporation in candy manufacturing or separation as ice during freezing produces supersaturated states of dissolved food components. The supercooled, amorphous components exist in a non-equilibrium state and they exhibit time-dependent changes as they exhibit viscous flow or approach the equilibrium state. These changes include mechanical changes resulting from the viscous flow, i.e., various collapse phenomena observed from changes in structure, stickiness, caking, and loss of porosity (e.g., White and Cakebread, 1966; Tsourouflis et al., 1976; To and Flink, 1978a, 1978b; Levine and Slade, 1988; Slade and Levine,

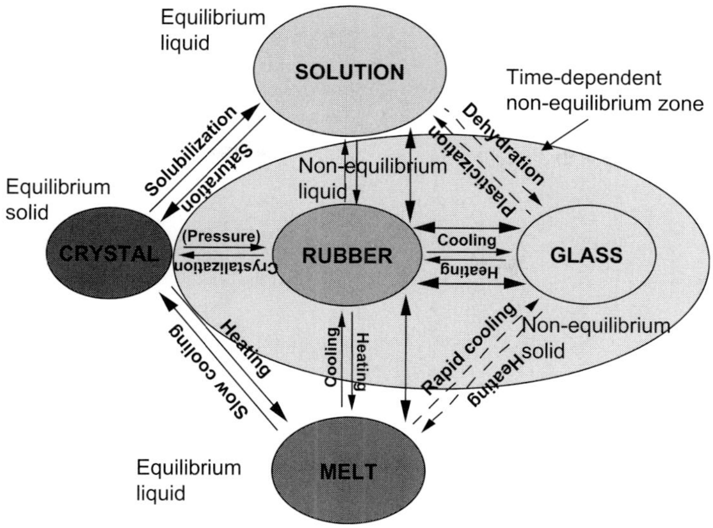

FIGURE 8.2 Relationships of equilibrium and non-equilibrium states of food systems.

1991; Roos, 2008). These changes as well as the glass transition also produce changes in molecular mobility, diffusion and molecular organization, i.e., crystallization of amorphous sugars, flavor retention and release, and, possibly, reaction kinetics (Roos and Karel, 1990, 1992; Slade and Levine, 1991; Roos, 1995, 2008; Roos et al., 1996). Reactions affected directly by diffusion of reactants and glass transition include non-enzymatic browning and enzymatic changes while some reaction rates may be reduced as a result of reactant entrapment in glassy matrices, e.g., lipid oxidation (Karel et al., 1994). Crystallization processes often compete with diffusion and glass formation and it may take place even during an extremely rapid cooling to temperatures well below the equilibrium melting temperature, T_m. Therefore, kinetics is critical in glass formation (vitrification), i.e., solidi-fication of the non-crystalline, non-equilibrium state (Figure 8.2).

Mechanical properties of supercooled materials change significantly below the T_m (Figures 8.3 and 8.4). The equilibrium melting temperature of several glass formers, including most sugars, is located at about $T_g + 100°C$ (Slade and Levine, 1991; Roos, 1993a, 1995, 2008). Rapid cooling of a melt to more than 100°C below the T_m increases viscosity solidifying the disordered supercooled liquid (vitrification) when the T_g is approached. At temperatures below the T_g viscosities of glassy materials are extremely high and approximately constant, 10^{12} Pa s^{-1} (White and Cakebread, 1966; Sperling, 1992).

Understanding the glass transition and introduction of state diagrams has improved understanding of the physicochemical principles which affect relaxation times of mechanical changes in foods (Levine and Slade, 1986; Roos and Karel, 1990, 1992; Slade and Levine, 1991; Roos, 1995). Relaxation times refer to rates of changes caused by existing driving forces in materials, i.e., relaxation times of structural changes responding to an external disturbance or towards an equilibrium state can be extremely long in the glassy state, but they may dramatically decrease within and above the glass transition temperature range as a result of increasing molecular mobility (Roos, 2008). The practical consequence of the extended relaxation times below glass transition is the solid-like appearance of glassy materials as well as their resistance to flow and ability to support their own

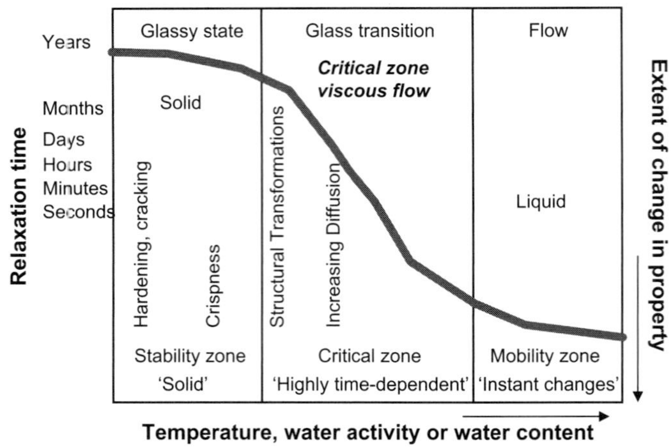

FIGURE 8.3 Relaxation times in food systems and their dependence on temperature, water activity and water content. Significant changes in molecular mobility and material state occur as a result of plasticization and glass transition.

weight. For example, freeze-dried foods can withstand flow below the glass transition while rapid collapse of structure occurs as a result of viscous flow when thermal or water plasticization reaches a sufficient level above the glass transition to allow liquid flow of the plasticized system (Figure 8.4). Combined effects of glass transition, water content, and temperature often control the kinetics of various chemical reactions and deterioration (e.g., White and Cakebread, 1966; Simatos and Karel, 1988; Slade and Levine, 1991; Karel et al., 1994). A major assumption related to shelf life and quality prediction of non-crystalline foods is that stability is maintained in the glassy state and various changes occur above the T_g with rates determined by the temperature

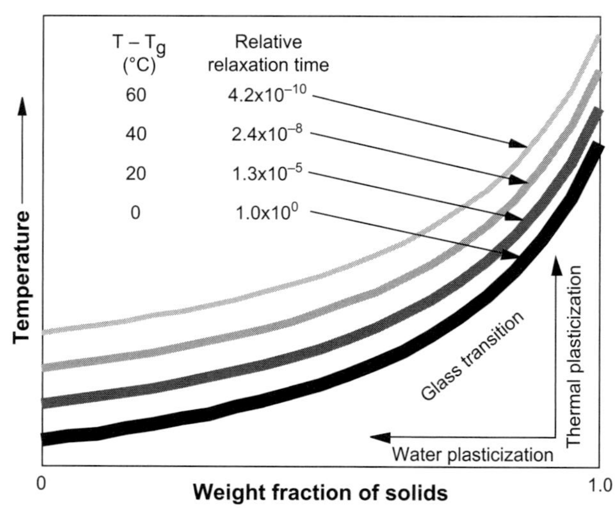

FIGURE 8.4 A state diagram showing glass transition and effects of thermal and water plasticization on relaxation times and time-dependent material properties at various levels of plasticization.

difference T – T_g (Levine and Slade, 1986; Slade and Levine, 1991; Roos, 1995, 2008).

8.3 GLASS FORMATION

Amorphous solids can be obtained by several techniques, which all provide a rapid change of the material from a liquid state to the non-equilibrium glassy state. This can be achieved by any method allowing the freezing of molecules of the material at a rate which is higher than needed by the molecules to arrange to form crystals. The common methods for producing amorphous materials include rapid cooling from melt to temperatures well below T_m or rapid removal of solvent. Cooling from a melt requires that the process has a rate which is faster than the rate of nucleation and crystal growth. This allows the molecules to remain in the supercooled, liquid, amorphous state that has a higher enthalpy and volume than the equilibrium, crystalline material at the same conditions. In solutions, molecules exist in a random order. The solute molecules may become 'frozen' in the solid, glassy state as a result of a sufficiently rapid removal of the solvent. Both cooling from melt and rapid removal of solvent are typical of food manufacturing processes. Food processes that result in amorphous or partially amorphous structures of food components include baking, evaporation, extrusion, dehydration, and freezing. A typical feature of all these processes is that they produce a melt of mixtures of food solids and water at a high temperature followed by rapid cooling or removal of water, producing amorphous solids with low water content. Glass formation in these processes can be clearly explained with the aid of state diagrams (Figure 8.5).

Various glassy states of the same material may be formed depending on the rate of glass formation, i.e., the state of the material at the time of freezing of the molecules to enter the glassy state. The physical state of the glass produced may be affected by the time-dependent characteristics of the 'freezing' of molecules to the solid, glassy state. The rate of glass formation affects the free volume of the molecules within the amorphous matrix, as may be observed from various relaxations when rapidly formed glasses are heated to above their T_g (Figure 8.1). The time-dependent nature of the glass formation may also be

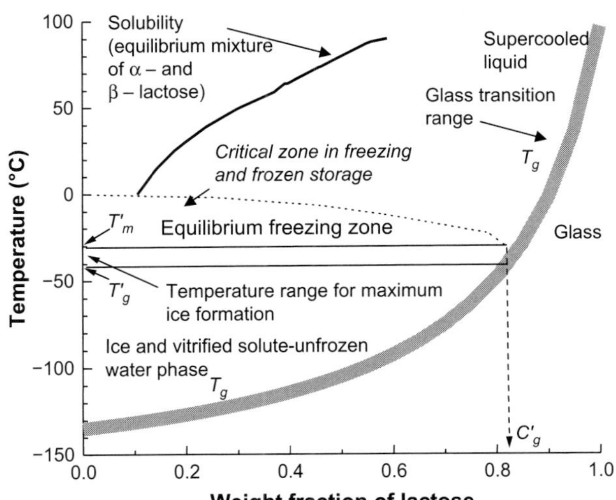

FIGURE 8.5 State diagram for lactose showing water content dependence of glass transition and state transitions of freeze-concentrated states of lactose.

observed from changes in the physical state that result from annealing (thermal treatment) in the vicinity of the T_g. Food stability is often related to the location of the T_g in relation to storage temperature, but it should also be noted that the glass transition is a time-dependent property of the amorphous, non-equilibrium state and that the relative rate of glass formation may contribute to the textural properties of rapidly cooled amorphous foods.

Glass formation by the removal of the solvent is probably the most common and natural process of food preservation. The removal of water concentrates the food solids, which, at low water contents, may enter the glassy state. Dehydration of most foods results in the formation of at least partially amorphous materials. The most rapid removal of solvent occurs in drum drying and spray drying. These processes probably produce quantitatively the highest amounts of amorphous food solids, e.g., most spray-dried food powders. The proper control of the amorphous state of lactose has been one of the key factors for maintaining stability of dehydrated dairy foods.

Ice formation in food materials results in freeze-concentration of solutes. The extent of freeze-concentration is dependent on temperature according to the melting temperature depression of water caused by the solute phase. Eutectic crystallization of solutes in most foods is unlikely during normal freezing processes which may decrease the temperature to sufficiently low temperatures for glass formation by the freeze-concentrated solids. The glass transition temperature of freeze-concentrated solutes is dependent on the extent of freeze-concentration (e.g., Levine and Slade, 1986, 1989; Roos and Karel, 1991c; Roos, 2008). Maximum freeze-concentration may occur at temperatures slightly below the onset temperature of ice melting, T_m', within the maximally freeze-concentrated material, while the glass transition of the maximally freeze-concentrated systems is located at T_g' (Roos and Karel, 1991b). Foods which contain high amounts of low-molecular-weight sugars have low T_g' values. Generally, the T_g' increases with increasing molecular weight of the solute fraction which agrees with the hydrogen bonding and ice melting temperature depression properties of carbohydrates (Levine and Slade, 1986, 1989; Roos and Karel, 1991b; Roos, 1993a, 1995).

The formation of amorphous materials in proper freeze-drying processes is based on the separation of water from solution by sufficiently rapid freeze-concentration to the maximally freeze-concentrated state. Thus, frozen materials to be freeze-dried should contain pure ice crystals within a maximally freeze-concentrated solute matrix which has glass transition at T_g' (Roos, 2002). The solute fraction at temperatures below T_g' remains in the solid, glassy state and the freeze-concentrated solute phase can support its own weight, although ice crystals are removed by sublimation. A successful freeze-drying process produces materials which have the T_g close to the T_g of the anhydrous solutes due to the removal of ice and most of the plasticizing water. It should be noted that any exposure of a freeze-drying material to temperatures above T_m' allows ice melting and plasticization of the freeze-concentrated solutes. The freeze-concentrated solute fraction at temperatures above T_m' cannot support its own weight, which results in flow, collapse, reduced water removal, and poor product quality, as suggested by the state diagram shown in Figure 8.5.

8.4 DETERMINATION OF GLASS TRANSITION

Glass transition is typically detected by observing changes in dielectric (dielectric constant), mechanical (modulus, viscosity), and thermodynamic (enthalpy, free volume, heat capacity, thermal expansion coefficient) properties of amorphous materials (White and

Cakebread, 1966; Wunderlich, 1981; Sperling, 1992; Roos, 2008). One individual amorphous material can have an indefinite number of solid, glassy states and corresponding relaxations can be observed at the solid–liquid change in state around the glass transition. Changes in material properties around the glass transition are time-dependent and may differ depending on frequency, temperature, and time-scale of experiments (Roos, 1995). Hence, glass transition is a time-dependent change of non-equilibrium states which can be characterized by the dramatic change in relaxation times around the transition (Slade and Levine, 1995).

The most common method for observing a glass transition is differential scanning calorimetry (DSC). DSC techniques can be used to detect the change in heat capacity around the glass transition (e.g., Wunderlich, 1981; Roos, 1995). The glass transition temperature, T_g, is a temperature value characteristic to the temperature range of the changes in thermodynamic material properties. As shown in Figure 8.1, glass transition is often associated with enthalpy changes resulting from the non-equilibrium characteristics of the glassy state and freezing of glass-forming molecules at various thermodynamic states in the vitrification process (Roos, 2008). DSC scans are commonly repeated to observe and confirm heat capacity changes apart from endothermal and exothermal changes around the glass transition (Roos, 1995, 2008). Glass-transition-associated changes in relaxation times can be derived from material responses to dielectric and mechanical perturbations. Relaxations can be observed by dynamic mechanical thermal analysis (DMA/DMTA) and mechanical spectroscopy which measure the effects of a sinusoidally varying stress on dynamic moduli (Roos, 2008). Changes in dielectric properties can be detected with dielectric thermal analysis (Laaksonen et al., 2002; Roos, 2008). Other important techniques in observing glass transition from changes in molecular mobility and diffusion are electron spin resonance spectroscopy (ESR) and nuclear magnetic resonance (NMR) spectroscopic methods. Fourier transfer infra-red (FTIR) and Raman spectroscopies may also be used to observe changes in molecular bonding occurring in amorphous systems over the glass transition (Söderholm et al., 1998).

Although novel thermal analytical and spectroscopic techniques allow detection of the glass transition, relaxations and molecular mobility associated with the transition, numerous empirical methods are commonly used to observe changes in flow resulting from the change in molecular mobility and the state of the material around the glass transition. Examples of empirical measurements of solid to liquid changes in foods are the observation of sticky points of powders (Lazar et al., 1956) and measurements of collapse in frozen (Bellows and King, 1971) and dehydrated systems (To and Flink, 1978a, 1978b, 1978c). The state diagrams with glass transitions and relaxations data have advanced understanding of the relationships of the various methods. The state diagrams also explain differences in various time-dependent characteristics of materials associated with empirical observations of material properties.

8.5 WATER PLASTICIZATION AND PLASTICIZERS

Water is the main softener (plasticizer) of amorphous, hydrophilic food components. Many glass-forming food components are water-soluble. Hence, water as a plasticizer must have its own glass transition at a lower temperature than any of the dissolved substances. The onset temperature of glass transition, T_g, of supercooled water has been found at about –135°C with a ΔCp of 1.94 J/g°C (Sugisaki et al., 1968). Water as a small molecular glass former exhibits a very sharp glass transition over a narrow temperature range with a substantial change in heat capacity. Water as a glass former shows highly 'fragile' properties (Ito et al., 1999). At around −135°C, the non-crystalline water softens to the supercooled, viscous liquid which as a result of the extensive supercooling ($T \ll 0°C$), enhanced molecular mobility and diffusion, crystallizes almost instantly as translational mobility of the small water molecules appears. Polysaccharides and proteins are large, polymeric food components and they exhibit glass transitions well above the boiling temperature of water. For example, anhydrous starch is expected to have glass transition around 200–250°C (Orford et al., 1989; Roos and Karel, 1991b). The T_g values of sugars (Slade and Levine, 1991; Roos, 1993a), oligosaccharides (Orford et al., 1989; Roos, 1995), and proteins (Kokini et al., 1994) are relatively well known and located between those of water and polymeric food components (Roos, 1995). Amorphous carbohydrates and sugars are highly sensitive to water, and water plasticization can often be observed from the dramatic depression of the T_g at extremely low water contents.

The glass transition and water plasticization properties of food components have significant practical importance. They often indicate appropriate temperatures and water contents for food processing. They may also be used as parameters defining storage stability and to predict shelf life. Prediction of T_g depression as a result of water plasticization is useful in the evaluation of effects of food composition on the T_g. Numerous empirical relationships for the modeling of polymer plasticization by diluents are available (Sperling, 1992). The Gordon-Taylor equation (Eq. 8.1) (Gordon and Taylor, 1952) has proved to be particularly useful in modeling experimental T_g data. It has fitted to water plasticization data of amorphous sugars (Roos and Karel, 1991a, 1991b; Roos, 1993a, 1993b), maltodextrins (Roos and Karel, 1991c), cereal proteins (Kokini et al., 1994), and complex food solids (Roos, 1993b; Jouppila and Roos, 1994). The glass transition temperature of anhydrous solids, T_{g1}, may be obtained experimentally and $T_{g2} = -135°C$ is often used for amorphous water (Slade and Levine, 1991; Roos and Karel, 1991a, 1991b, 1991c).

$$T_g = (w_1 T_{g1} + k w_2 T_{g2})/(w_1 + k w_2) \quad (8.1)$$

where T_{g1}, T_{g2} and w_1 and w_2 refer to glass transitions and weight fractions of components 1 and 2, respectively. The parameter k is a constant.

Compositional effects on the glass transition temperature are useful in the design of food materials that are subject to well-defined processing and storage conditions. The T_g and water plasticization may affect such properties as stickiness or collapse, e.g., one of the main applications of maltodextrins is their use as drying aids. Maltodextrins are able to decrease stickiness and improve storage stability of food powders, which is due to their ability to increase the T_g of the solids. The glass transition temperature of mixtures of miscible food solids is dependent on the component compounds. Unfortunately, the number of various compounds may be large and the combined effect on the T_g can be unpredictable. However, the effect of composition on the T_g of binary or tertiary mixtures of food solids is often sufficient for the evaluation of compositional effects on physical properties. The glass transition temperature of such mixtures, e.g. sugars and starch, can be successfully predicted (Roos and Karel, 1991c). Roos and Karel (1991c) studied glass transition temperatures of sucrose and maltodextrin mixtures. Maltodextrins of various molecular weights were significantly plasticized by sucrose. It was also found that addition of maltodextrins to sucrose caused a fairly small and almost molecular-weight-independent increase of the T_g up to a concentration of 50% (w/w). Thus, in practical applications such as dehydration of fruit juices fairly high concentrations of maltodextrins are needed. The increase of the T_g also increases the critical water content and water activity for stability during storage (Roos and Karel, 1991c; Roos, 1993b).

It should also be noted that a number of hydrophilic food components other than water may exist as solvents and plasticizers of food solids (Roos, 1995; Bhandari and Roos, 2003). These may occur in dehydrated and extruded foods and many plasticizers, particularly glycerol and sorbitol are used as plasticizers to improve film-forming properties of starch and other food polymers in edible films (Talja et al., 2007).

8.6 GLASS TRANSITION AND WATER ACTIVITY

Water activity is thermodynamically defined as the activity coefficient of water in a food system. It can be shown that water activity equals the ratio of water vapor pressure of water in a system and that of pure water at the same temperature. Water activity is a temperature-dependent equilibrium property of water in a food and it can be derived from the measured vapor pressure of water above the food at equilibrium. A sorption isotherm shows the water content against a_w and, therefore, the relationship of a_w and water content at a constant temperature.

Modeling studies of sorption properties are particularly important in predicting shelf life of low- and intermediate-water foods (e.g., Labuza et al., 1970; Labuza, 1980). The Brunauer-Emmett-Teller (BET) (Brunauer et al., 1938) and Guggenheim-Anderson-deBoer (GAB) (van den Berg and Bruin, 1981) relationships are well-known sorption models which provide the monolayer value, m_m, often considered as the optimal water content for stability of low-water foods (Labuza et al., 1970; Labuza, 1980). The monomolecular water content correlates with the onset of increased molecular mobility in low-water foods. The applicability of the BET model is limited to the a_w range of 0.1–0.5 (Labuza, 1968), but the GAB model fits to experimental sorption data over a wider a_w range (van den Berg, 1981).

Both the physical state of food solids and stability are extremely sensitive to water at low water contents. Roos (1987) showed that the glass transition temperature plotted against a_w showed linearity over a wide a_w range. The linearity for high-sugar systems often applies over the a_w range of 0.1–0.8, but the relationship over the whole a_w range appears to be sigmoid (Roos and Karel, 1991c; Roos, 1993b). Roos (1993b) suggested that a combined use of sorption models and the Gordon-Taylor equation was applicable to model water sorption and water plasticization. The models could be fitted to experimental data to locate critical values for a_w and water content, defined as those decreasing the T_g to the observation temperature (Roos, 1993b, 1995; Jouppila and Roos, 1994). Some food components, however, have anhydrous T_g at temperatures below room temperature which suggests that they cannot form glassy structures at normal storage conditions.

The use of the glass transition and water activity data to establish stability maps based on sorption isotherms and state diagrams principles provides a further tool to describe food properties and their stability characteristics in diagram (Figure 8.6). These diagrams also define critical values for water activity and water content at normal storage temperatures (Roos, 1993b, 2008). Their use and development are also benefiting from development of new modeling options and dynamic determination of water sorption isotherms, which may reflect and include information on structural transformations.

8.7 MECHANICAL PROPERTIES AND RELAXATIONS

Mechanical properties of amorphous food systems change dramatically around the glass

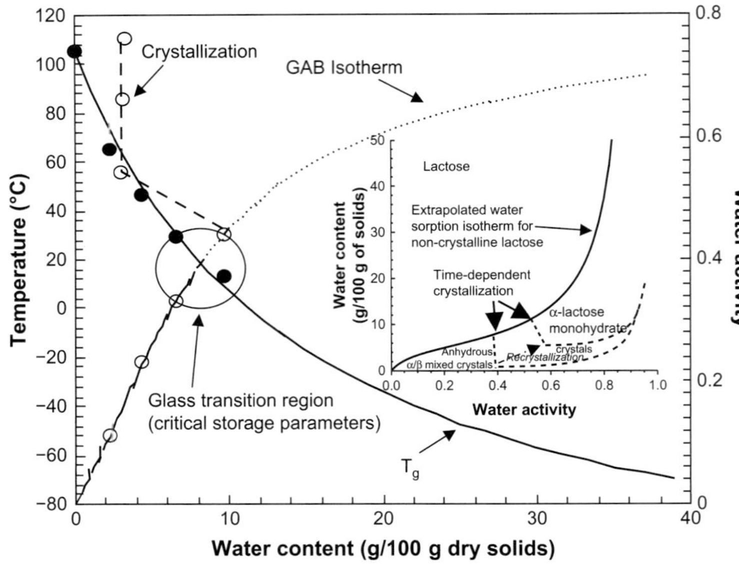

FIGURE 8.6 Water sorption and glass transition data for lactose which indicate critical water activity and water content values for stability. Excessive plasticization above the critical values results in time-dependent flow and crystallization.

transition. Williams et al. (1955) suggested an empirical relationship to describe the temperature dependence of mechanical properties of amorphous materials above their T_g. This relationship is known as the Williams-Landel-Ferry (WLF) equation. It was reported to be applicable over the temperature range from T_g to T_g + 100°C (Levine and Slade, 1986). The applicability of T_g as a reference temperature with 'universal constants' for food materials has been emphasized by Slade and Levine (1991, 1995). The rapid drop in viscosity assumed above T_g seems not to be valid (Peleg, 1992) as the T_g is not a well-defined exact temperature and no instant decrease in viscosity at the T_g may occur. Other models have also become important, such as the Vogel-Tammann-Fulcher (VTF) and power-law equations (Ollett and Parker, 1990). Peleg (1992) suggested the use of the Fermi relationship for the modeling of changes around and above the T_g. Similarities in the physical properties of various amorphous materials, however, suggest that the use of the WLF or other models for predicting temperature dependence of relaxation times (e.g., viscosity and flow) allows the establishment of diagrams showing time-dependent material characteristics above the T_g. Such diagrams can be used, for example, to evaluate effects of composition on relaxation times, i.e., rates of changes in molecular mobility-related material properties, at a constant temperature or to establish critical temperatures for food processing (e.g., agglomeration, extrusion, and dehydration) and storage (e.g., storage of low-water foods and powders at high relative humidity/temperature environments) (Roos and Karel, 1991a; Slade and Levine, 1991; Roos, 1995). The understanding of food systems and particularly mixtures of food components as fragile or strong glass-forming systems is useful in predicting stability of food systems at conditions in the vicinity of the glass transition as well as their crystallization characteristics. The fragile/strong classification of glass-forming materials was based on the Arrhenius–non-Arrhenius behavior of amorphous systems and the VTF modeling of relaxation times as introduced by Angell (e.g. Angell, 1995).

8.8 STIFFNESS

Peleg (1993) used stiffness as a general term referring to the response of food materials to an external stress. Solid, glassy systems were expected to exhibit solid-like properties and high stiffness values. Above the T_g, the stiffness values decreased rapidly as a result of plasticization and softening. The effects of glass transition on viscosity, flow characteristics, stiffness and collapse are obvious (Figure 8.6). Changes in stiffness often follow the typical drop in modulus around glass transition. Collapse phenomena can be described by the use of stiffness parameters which may be plotted against temperature, water content or a_w, and time (Slade and Levine, 1995; Peleg, 1993; Roos, 1995, 2008). The general change in a stiffness parameter is a result of glass transition-dependent changes in relaxation times around the glass transition.

8.9 COLLAPSE PHENOMENA

Collapse phenomena in food systems refer to their time-dependent changes and viscous flow in processing and storage. According to Levine and Slade (1988), collapse phenomena cover various time-dependent structural transformations that may occur in amorphous food and other biological materials at temperatures above T_g. Typical collapse phenomena included or were considered to affect stickiness and caking of food powders, plating of particles on amorphous granules, crystallization of food components, viscous flow and collapse of dehydrated structures, release and oxidation of encapsulated lipids and flavors, enzymatic activity, non-enzymatic browning, graining of boiled sweets, sugar bloom in chocolate, ice recrystallization, and solute crystallization during frozen storage (Levine and Slade, 1988; Roos, 1995).

8.10 STICKINESS AND CAKING

Stickiness and caking are phenomena that may occur when amorphous food products are heated or exposed to humid conditions. Stickiness and caking problems are typical of amorphous food powders and they occur when plasticization of particle surfaces results in a sufficient decrease of surface viscosity for adhesion. Stickiness and caking may occur both during production of dehydrated foods and during food storage but they are useful in the manufacturing of agglomerated and instant food powders. The difference between stickiness and caking is not always clearly defined, but it may be assumed that stickiness refers to a surface property and the tendency of a material to adhere on a surface, including interparticle adhesion. Adhesion may also be temporary and it does not necessarily involve subsequent caking. Caking may be considered as a collapse phenomenon which occurs as a result of stickiness where particles form permanent aggregates and harden into a solid structure. Stickiness often results in loss of free-flowing properties of powders. Stickiness of food materials may occur and deteriorate material behavior and quality during both production and storage. Stickiness is also an important factor in the control of agglomeration and production of 'instant' powders.

Caking of sticky powders occurs when sufficient time is allowed for the surface contact of sticky particles. According to Peleg (1977) liquid bridging is one of the main interparticle phenomena that are responsible for caking in food powders. Factors which may cause liquid bridging are: (1) water sorption; (2) melting of component compounds (e.g., lipids); (3) chemical reactions which produce liquids (e.g., non-enzymatic browning); (4) excessive liquid ingredient; (5) water released as a result of crystallization of amorphous sugars; and (6) wetting of the powder or equipment. The most common caking mechanism in food powders is plasticization as a result of water sorption and subsequent interparticle fusion (Peleg, 1983).

The sticky point of amorphous food solids is located at an isoviscosity state above T_g (Downton et al., 1982; Roos and Karel, 1991a) and the measurement of the sticky point with the method of Lazar et al. (1956) can be considered to be a method which, in fact, locates the glass transition of the food solids (Chuy and Labuza, 1994). Thus, the sticky point of food solids with various compositions can be predicted with the same principles which are used in evaluating effects of composition on glass transition temperatures of food solids and polymers. Roos and Karel (1991a) found that the sticky point of food solids correlated with the end point temperature of the glass transition and surface viscosity of 10^7 Pa s^{-1} of amorphous components on particle surfaces.

Collapse of macroscopic structure during dehydration is one of the most detrimental changes, which (1) reduces drying rate; (2) causes slow and difficult rehydration; and (3) decreases the quality of dehydrated foods. Karanthanos et al. (1993) found that the T_g of celery was very low and significantly lower than the drying temperature during most of the drying process due to the high initial water content of the material. They observed that the extent of shrinkage increased proportionally with the decrease in volume that was caused by the loss of water. However, air-drying at 5°C resulted in a product with higher porosity than drying at 60°C although drying occurred well above T_g even at the lower temperature. Collapse during dehydration is due to structural mobility, which is a function of viscosity governed by the temperature difference $T - T_g$. These results suggest the importance of state diagrams in mapping appropriate dehydration parameters for particular food solids.

8.11 GLASS TRANSITIONS IN FROZEN FOODS

Ice formation in high-water foods occurs at temperatures below 0°C with gradually increasing amount of ice with decreasing temperature. The maximum amount of ice formed depends on composition, solutes and temperature. Freeze-concentration in most foods results in the formation of a highly freeze-concentrated amorphous phase. This continuous, hydrophilic phase contains unfrozen water, dissolved, freeze-concentrated solutes and dispersed ice crystals (Levine and Slade, 1986; Roos, 1995). At a sufficiently low temperature, the freeze-concentrated liquid may transfer into the glassy state. Around the glass transition, ice formation ceases as translational mobility of molecules and diffusion become limited causing kinetic limitations for crystallization processes. Roos and Karel (1991d) suggested that ice formation in freeze-concentrated sucrose solutions was time-dependent.

Viscosity of the freeze-concentrated solute phase in frozen foods is an important factor that may affect time-dependent recrystallization phenomena, ice formation, and material properties in such processes as freeze-concentration and freeze-drying. The WLF-type temperature dependence of mechanical properties of freeze-concentrated materials has been emphasized by Levine and Slade (1986). Levine and Slade (1988) pointed out that the retarding effect of added maltodextrins in ice cream on ice recrystallization was based on the elevation of T_m'. Levine and Slade (1988) suggested that mechanical properties of frozen materials followed the WLF-type temperature dependence above T_m'. The WLF-type temperature dependence above T_g' predicts that the viscosity at the onset of ice melting is 10^8–10^9 Pa s^{-1}, which agrees fairly well with the viscosity found by Bellows and King (1973) to allow collapse in freeze-drying. The onset of ice melting is also observed with other methods such as dielectric analysis (DEA; DETA) and dynamic mechanical analysis (DMA; DMTA) as reviewed by Roos (2008). The onset of the T_g' can be found at lower temperatures than the T_m' as the viscosity and other kinetic limits for crystallization at T_g' (approximately 10^{12} Pa s^{-1}) are higher than at the T_m' (approximately 10^7 Pa s^{-1}). Both the T_g' and T_m' may be used to explain factors affecting the shelf life and quality of frozen foods, although the most dramatic and practically important changes occur at temperatures above T_m' (Roos, 1995). The T_m' has also been defined as a 'mobility transition' referring to increased 'mobility' resulting from increased amounts of unfrozen water above T_m'.

Several methods have been used to determine solute and unfrozen water contents in maximally freeze-concentrated systems. The maximum solute concentration in a maximally freeze-concentrated phase, C_g', may be obtained from DSC measurements (Slade and Levine, 1991; Roos, 1995). However, a more accurate value can be derived from the state diagrams which show changes in physical state at various temperatures and water contents for hydrophilic food systems (Figure 8.5). Most studies on sugars and other hydrophilic food components have reported C_g' values around 80% w/w (Roos, 1993a, 1995; Kokini et al., 1994).

8.12 CRYSTALLIZATION AND RECRYSTALLIZATION

Crystallization and recrystallization in amorphous foods are time-dependent phase transitions, which are governed by molecular mobility, diffusion and therefore by the glass transition and physical state. These processes include: (1) crystallization of amorphous components such as sugars; (2) retrogradation phenomena in gelatinized starch; (3) solute crystallization in frozen foods; (4) time-dependent ice formation at low temperatures; and (5) recrystallization of ice during frozen storage. Both crystallization and recrystallization processes belong to phase transitions which may dramatically decrease food quality during storage.

Water plasticization and depression of T_g to below ambient temperature are responsible for crystallization of amorphous sugars in foods (Roos and Karel, 1990, 1992; Jouppila and Roos, 1994; Roos, 1995; Haque and Roos, 2005). Crystallization seems to initiate at T_g and it proceeds with a rate determined by $T - T_g$. Crystallization above T_g also released entrapped compounds and, consequently, caused loss of flavors or exposed lipids to atmospheric oxygen (Shimada et al., 1991). Oxidation of non-encapsulated lipids may significantly decrease the shelf life of low-water foods.

8.13 STATE DIAGRAMS AND STABILITY

Deteriorative changes typical of low-water and frozen foods include enzyme-catalyzed changes, non-enzymatic browning, and oxidation. The rate of deteriorative changes in low-water foods is likely to depend on molecular mobility. Water as a plasticizer may have a significant effect on molecular mobility above a critical, temperature-dependent value. Slade and Levine (1991) suggested that reaction rates may become diffusion-controlled in amorphous foods. At temperatures below the T_g, the rates of diffusion-controlled reactions were expected to be extremely low. A significant increase in reaction rates above T_g could be observed as a result of enhanced diffusion. Therefore, the temperature dependence of diffusion-controlled reactions was suggested to differ from Arrhenius-type kinetics and in some cases follow the WLF-type temperature dependence above the T_g (Slade and Levine, 1991; Roos, 1995; Roos et al., 1996).

Karmas et al. (1992) analyzed temperature dependence of non-enzymatic browning rates of foods with various water activities. Browning rates below T_g were low, but the rates increased with increasing $T - T_g$. The rate of the reaction was also dependent on temperature, water content, crystallization, and other structural changes which occurred concomitantly with plasticization. Karmas et al. (1992) observed changes in activation energies in the vicinity of T_g. This suggested that the reaction rate was affected by the T_g. Roos and Himberg (1994) found that browning in food models occurred also at temperatures below T_g. The rate of browning increased both with increasing temperature and increasing $T - T_g$. Therefore, reactions in amorphous foods are controlled by several factors, including the T_g. Roos et al. (1996) found that the amount of water produced in the non-enzymatic browning reaction in milk powders was significant and enhanced the non-enzymatic reaction as a result of additional water plasticization.

Rates of deteriorative changes in frozen foods are reduced as a result of the decreased temperature, although freeze-concentration of solutes and phase separation of component compounds, e.g., lipids, may increase reaction rates. Levine and Slade (Levine and Slade, 1986, 1988; Slade and Levine, 1991) stated that frozen foods are stable below T_g', but at temperatures above T_g' rates of deteriorative changes increase according to the WLF temperature dependence. However, only few data on reaction rates in frozen foods are available. Lim and Reid (1991) reported reduced rates for enzymatic activity and oxidation of ascorbic acid and stabilization of actomyosin at temperatures below T_g' or rather T_m' in freeze-concentrated maltodextrin matrices. In sucrose matrices, protein insolubilization was prevented below the mobility transition, but oxidation of ascorbic acid occurred below the transition temperature. Reaction rates were not affected by T_g' in freeze-concentrated carboxymethylcellulose matrices. Ice melting occurs in frozen foods above T_m' (Roos and Karel, 1991a, 1991b, 1991d; Roos, 1993a, 1995) which causes dilution of the freeze-concentrated solutes, depression of the rate controlling T_g, and a significant decrease in viscosity (Roos, 1995). Therefore, temperatures above T_m' may increase and the concomitant dilution decrease rates of diffusion-controlled changes in frozen foods (Roos and Karel, 1991a, 1991b, 1991c; Kerr and Reid, 1994; Roos, 1995).

Characterization of the physical state of food materials and application of the polymer science theories to the description of food properties and various kinetic phenomena have significantly contributed to the present understanding of food stability. Knowledge of material properties is extremely useful in the production of encapsulated flavors, extruded products, confectionery; development of totally new products, such as dehydrated enzymes or starters, and avoiding quality changes that may result from mechanical changes, e.g., loss of crispness and recrystallization phenomena. The temperature, water content-, and time-dependent changes which have been problems in manufacturing and storage of food powders and other low-moisture foods can be reduced by avoiding exceeding their critical values based on the T_g determination or by compositional adjustments that provide sufficiently high values for critical a_w and T_g. The most important applications of producing high-quality dehydrated foods include reduced collapse and improved flavor retention in dehydration processes. The kinetics of enzyme activity is important to food quality and applying the knowledge in food industry may allow the design of improved products with extended shelf lives or even improved retention of activity, e.g., in enzyme preparations.

State diagrams are effective tools in establishing relationships between the physical state of food materials, temperature, and water content. State diagrams show the glass transition temperature as a function of water content and the effect of ice formation on T_g and on the equilibrium ice melting temperature, T_m. State diagrams may also show solubility as a function of temperature and information on various changes that may occur due to the metastable state of amorphous food solids and their approach towards the equilibrium state. In food formulation and design state diagrams are invaluable tools for understanding of the effects of food composition and water content on the physical state and physicochemical properties during processing and storage.

References

Angell, C. A. (1995). Formation of glasses from liquids and biopolymers. *Science* **267**, 1924–1935.

Bellows, R. J. and King, C. J. (1973). Product collapse during freeze drying of liquid foods. *AIChE Symp Ser* **69** (132), 33–41.

Bhandari, B. R. and Roos, Y. H. (2003). Dissolution of sucrose crystals in the anhydrous sorbitol melt. *Carbohydr. Res.* **338**, 361–367.

Brunauer, S., Emmett, P. H., and Teller, E. (1938). Adsorption of gases in multimolecular layers. *J. Am. Chem. Soc.* **60**, 309–319.

Chuy, L. E. and Labuza, T. P. (1994). Caking and stickiness of dairy-based food powders as related to glass transition. *J. Food Sci.* **59**, 43–46.

Downton, D. P., Flores-Luna, J. L., and King, C. J. (1982). Mechanism of stickiness in hygroscopic, amorphous powders. *Ind. Eng. Chem. Fundam* **21**, 447–451.

Gordon, M. and Taylor, J. S. (1952). Ideal copolymers and the second-order transitions of synthetic rubbers. I. Non-crystalline copolymers. *J. Appl. Chem.* **2**, 493–500.

HaqueMdKamrul and Roos, Y. H. (2005). Crystallization and X-ray diffraction of spray-dried and freeze-dried amorphous lactose. *Carbohydrate Research* **340**, 293–301.

Ito, K., Moynihan, C. T., and Angell, C. A. (1999). Thermodynamic determination of fragility in liquids and a fragile-to-strong liquid transition in water. *Nature* **398**, 492–495.

Jouppila, K. and Roos, Y. (1994). Glass transitions and crystallization in milk powders. *J. Dairy Sci.* **77**, 2907–2915.

Karathanos, V., Anglea, S., and Karel, M. (1993). Collapse of structure during drying of celery. *Drying Technol* **11**, 1005–1023.

Karel, M., Anglea, S., Buera, P., Karmas, R., Levi, G., and Roos, Y. (1994). Stability-related transitions of amorphous foods. *Thermochimica Acta* **246**, 249–269.

Karmas, R., Buera, M. P., and Karel, M. (1992). Effect of glass transition on rates of nonenzymatic browning in food systems. *J. Agric. Food Chem.* **40**, 873–879.

Kerr, W. L. and Reid, D. S. (1994). Temperature dependence of the viscosity of sugar and maltodextrin solutions in coexistence with ice. Lebensm.Wiss. u. *-Technol* **27**, 225–231.

Kokini, J. L., Cocero, A. M., Madeka, H., and de Graaf, E. (1994). The development of state diagrams for cereal proteins. *Trends Food Sci. Technol* **5**, 281–288.

Laaksonen, T. J., Kuuva, T., Jouppila, K., and Roos, Y. H. (2002). Effects of arabinoxylans on thermal behavior of frozen wheat doughs as measured by DSC, DMA, and DEA. *Journal of Food Science* **67** (1), 223–230.

Labuza, T. P. (1968). Sorption phenomena in foods. *Food Technol* **22**, 263–265, 268, 270, 272.

Labuza, T. P. (1980). The effect of water activity on reaction kinetics of food deterioration. *Food Technol* **34** (4), 36–59.

Labuza, T. P., Tannenbaum, S. R., and Karel, M. (1970). Water content and stability of low-moisture and intermediate-moisture foods. *Food Technol* **24**, 543–544, 546–548, 550.

Lazar, M., Brown, A. H., Smith, G. S., Wong, F. F., and Lindquist, F. E. (1956). Experimental production of tomato powder by spray drying. *Food Technol* **10**, 129–134.

Levine, H. and Slade, L. (1986). A polymer physico-chemical approach to the study of commercial starch hydrolysis products (SHPs). *Carbohydr. Polym* **6**, 213–244.

Levine, H. and Slade, L. (1988). Principles of "cryostabilization" technology from struc-ture/property relationships of carbohydrate/water systems – A review. *Cryo-Lett.* **9**, 21–63.

Levine, H. and Slade, L. (1989). A food polymer science approach to the practice of cryostabilization technology. *Comments Agric. Food Chem.* **1**, 315–396.

Lim, M. H. and Reid, D. S. (1991). Studies of reaction kinetics in relation to the T′g of polymers in frozen model systems. *In* "Water Relationships in Foods," 103–122, Levine, H. and Slade, L. (eds.). Plenum Press, New York.

Ollett, A. L. and Parker, R. (1990). The viscosity of supercooled fructose and its glass transition temperature. *J. Texture Stud* **21**, 355–362.

Orford, P. D., Parker, R., Ring, S. G., and Smith, A. C. (1989). Effect of water as a diluent on the glass transition behavior of malto-oligosaccharides, amylose and amylopectin. *Int. J. Biol. Macromol* **11**, 91–96.

Peleg, M. (1977). Flowability of food powders and methods for its evaluation. *J. Food Process Eng* **1**, 303–328.

Peleg, M. (1992). On the use of the WLF model in polymers and foods. *Crit. Rev. Food Sci. Nutr* **32**, 59–66.

Peleg, M. (1993). Mapping the stiffness-temperature-moisture relationship of solid biomaterials at and around their glass transition. *Rheol. Acta* **32**, 575–580.

Roos, Y. H. (1987). Effect of moisture on the thermal behavior of strawberries studied using differential scanning calorimetry. *J. Food Sci.* **52**, 146–149.

Roos, Y. (1993a). Melting and glass transitions of low molecular weight carbohydrates. *Carbohydr. Res.* **238**, 39–48.

Roos, Y. (1993b). Water activity and physical state effects on amorphous food stability. *J. Food Process. Preserv.* **16**, 433–447.

Roos, Y. H. (1995). Phase Transitions in Foods. Academic Press, San Diego, CA.

Roos, Y. H. (2002). Importance of glass transition and water activity to spray drying and stability of dairy powders. *Le Lait* **82**, 475–484.

Roos, Y. H. (2008). The glassy state. *In* Food Materials Science. Principles and Practice, 67–94, Aguilera, J. M. and Lillford, P. J. (eds.). Springer Science + Business Media, New York.

Roos, Y. H. and Himberg, M. J. (1994). Nonenzymatic browning behavior, as related to glass transition, of a food model at chilling temperatures. *J. Agric. Food Chem.* **42**, 893–898.

Roos, Y. and Karel, M. (1990). Differential scanning calorimetry study of phase transitions affecting the quality of dehydrated materials. *Biotechnol. Prog* **6**, 159–163.

Roos, Y. and Karel, M. (1991a). Applying state diagrams to food processing and development. *Food Technol.* **45,** 66, 68–71, 107.

Roos, Y. and Karel, M. (1991a). Plasticizing effect of water on thermal behavior and crystallization of amorphous food models. *J. Food Sci.* **56**, 38–43.

Roos, Y. and Karel, M. (1991b). Amorphous state and delayed ice formation in sucrose solutions. *Int. J. Food Sci. Technol.* **26**, 553–566.

Roos, Y. and Karel, M. (1991c). Water and molecular weight effects on glass transitions in amorphous carbohydrates and carbohydrate solutions. *J. Food Sci.* **56**, 1676–1681.

Roos, Y. and Karel, M. (1992). Crystallization of amorphous lactose. *J. Food Sci.* **57**, 775–777.

Roos, Y. H., Karel, M., and Kokini, J. L. (1996). Glass transitions in low moisture and frozen foods: Effects on shelf life and quality. *Food Technology* **50** (11), 95–108.

Shimada, Y., Roos, Y., and Karel, M. (1991). Oxidation of methyl linoleate encapsulated in amorphous lactose-based food model. *J. Agric. Food Chem.* **39**, 637–641.

Simatos, D. and Karel, M. (1988). Characterization of the condition of water in foods – physico-chemical aspects. *In* "Food Preservation by Water Activity," 1–41, Seow, C. C. (ed.). Elsevier, Amsterdam.

Slade, L. and Levine, H. (1991). Beyond water activity: recent advances based on an alternative approach to the assessment of food quality and safety. *Crit. Rev. Food Sci. Nutr* **30**, 115–360.

Slade, L. and Levine, H. (1995). Glass transitions and water-food structure interactions. *Adv. Food Nutr. Res.* **38**, 103–269.

Sperling, L. H. (1992). Introduction to Physical Polymer Science. John Wiley & Sons, Inc., New York.

Sugisaki, M., Suga, H., and Seki, S. (1968). Calorimetric study of the glassy state. IV. Heat capacities of glassy water and cubic ice. *Bull. Chem. Soc. Jpn* **41**, 2591–2599.

Talja, R. A., Helén, H., Roos, Y. H., and Jouppila, K. (2007). Effect of various polyols and polyol contents on physical and mechanical properties of potato starch-based films. *Carbohydr. Polym* **67**, 288–295.

To, E.T. and Flink, J.M. (1978a). 'Collapse,' a structural transition in freeze dried carbohydrates. I. Evaluation of analytical methods. *J. Food Technol.* **13**, 551–565.

To, E.T. and Flink, J.M. (1978b). 'Collapse,' a structural transition in freeze dried carbohydrates. II. Effect of solute composition. *J. Food Technol.* **13**, 567–581.

To, E.T. and Flink, J.M. (1978c). 'Collapse,' a structural transition in freeze dried carbohydrates. III. Prerequisite of recrystallization. *J. Food Technol.* **13**, 583–594.

Tsourouflis, S., Flink, J. M., and Karel, M. (1976). Loss of structure in freeze-dried carbohydrates solutions: Effect of temperature, moisture content and composition. *J. Sci. Food Agric* **27**, 509–519.

van den Berg, C. and Bruin, S. (1981). Water activity and its estimation in food systems: Theoretical aspects. *In* "Water Activity: Influences on Food Quality," 1–61, Rockland, L. B. and Stewart, G. F. (eds.). Academic Press, Inc., New York.

White, G. W. and Cakebread, S. H. (1966). The glassy state in certain sugar-containing food products. *J. Food Technol* **1**, 73–82.

Williams, M. L., Landel, R. F., and Ferry, J. D. (1955). The temperature dependence of relaxation mechanisms in amorphous polymers and other glass-forming liquids. *J. Am. Chem. Soc.* **77**, 3701–3707.

Wunderlich, B. (1981). The basis of thermal analysis. Chpt. 2 in "Thermal Characterization of Polymeric Materials," Turi E. A. (ed.). Academic Press, Inc., New York.

Yu, L. (2001). Amorphous pharmaceutical solids: preparation, characterization and stabilization. *Adv. Drug Deliv. Rev.* **48**, 27–42.

CHAPTER

9

Structural Advances in the Understanding of Carbohydrate Glasses

Johan B. Ubbink

Nestlé Research Center, Vers-chez-les-Blanc, CH-1000 Lausanne 26, Switzerland

9.1 CARBOHYDRATE PHASE BEHAVIOR IN THE PREDICTION OF FOOD AND PHARMACEUTICAL STABILITY

Over the last couple of decades, important progress has been made in the understanding of the physical properties of carbohydrate–water systems in terms of their temperature and water-dependent phase transitions (Figure 9.1). This has resulted in an extensive use of phase and state diagrams to predict both behavior during processing and the storage stability of carbohydrate-based foods (Roos, 1995) and pharmaceutics (Roberts and Debenedetti, 2002). A property which has attracted particular attention is the glass transition temperature (T_g) of amorphous carbohydrates. The glass transition temperature is of particular importance for the stability of amorphous foods and pharmaceutics as it is strongly depressed by water (Slade and Levine, 1995).

The concept of the glass transition of a food or pharmaceutical product has had such an impact that currently 'stability' is assumed to be guaranteed when an amorphous food or pharmaceutical product is in the glassy state (Table 9.1). This is obviously correct when viscosity-dependent processes are considered (Angell, 1995). For example, a powder consisting of a matrix of amorphous carbohydrates, such as milk powder (Vuataz, 2002), or a glass-encapsulation system (Ubbink and Schoonman, 2007), is physically stable below its glass transition temperature, as below T_g, large-scale molecular rearrangements and macroscopic flow of the matrix are inhibited and the powder will keep its physical state for years or even decades. In addition, many of the systems will also have an increased chemical stability below T_g, as the rate of chemical reactions will be very low in the glassy state. This is in particular the case for bimolecular reactions involving high-molecular-weight species, such as carbohydrates and proteins in the Maillard reaction (Eichner and Karel, 1972).

In the glassy state, several dynamic processes are still proceeding with significant rates. These processes include the diffusion of water (Tromp et al., 1997) and other small molecules such as gases (Schoonman et al., 2002), and local rearrangements of the carbohydrate molecules resulting in glassy-state aging of the matrix

Kasapis, Norton, and Ubbink: Modern Biopolymer Science
ISBN: 978-0-12-374195-0

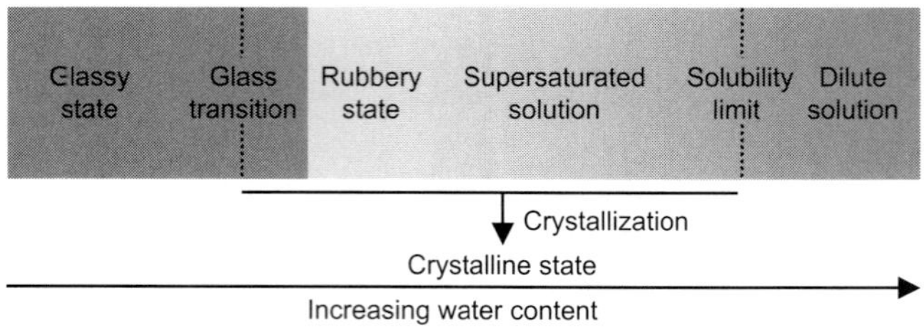

FIGURE 9.1 Outline of phase transitions and physical states of carbohydrate–water systems in the non-frozen state at constant (e.g. ambient) temperature. The glass transition, here indicated to occur at a distinct water content of the carbohydrate–water system, will occur at higher temperatures for lower water contents and vice-versa. Crystallization of the carbohydrates which have propensity to do so will occur between the glass transition and the solubility limit of the carbohydrate, but with kinetics which depends strongly on the viscosity of the matrix and which thus may vary by many orders of magnitude. Complex mixtures of carbohydrates often do not crystallize, or only with significant delay.

(Noel et al., 2005). In addition, in the glassy state, variations are observed in the protective effects of various carbohydrates for complex and sensitive biological systems such as membrane and lipid systems (Crowe et al., 1997) and proteins (Cicerone and Soles, 2004). It is often observed that the stability of such systems in the glassy state does not correlate with the matrix T_g.

Several mechanisms have been proposed to explain the stabilizing effects of carbohydrates below T_g. These include the so-called 'water-replacement' hypothesis (Crowe et al., 1992), which states that those carbohydrates that can replace the hydrogen bonding between water and lipid membranes. Carbohydrates are therefore particularly effective in stabilizing encapsulated lipid complexes (Table 9.1). Although partially corroborated by various experimental studies using Fourier-transform infrared (FTIR) spectroscopy, which show a positive correlation between increases in the interaction strength between carbohydrates and the lipid complexes and the stability of the encapsulated complexes (Crowe et al., 1997), it does not provide detailed physical insight in the structure and properties of the carbohydrate matrix itself.

TABLE 9.1 Dual Role of Glassy Carbohydrates in the Stabilization and Delivery of Active Ingredients

	Encapsulation	Biostabilization
Principle	Macroscopic: Entrapment by formation of vitreous barrier	Molecular: Stabilization by hydrogen bonding between vitreous scaffold and active ingredient 'Water-replacement' hypothesis
Bioactive ingredient	Flavors Drugs PUFAs Various nutrients	Drugs Vesicles, membranes Proteins, peptides Nucleic acids
Technologies	Spray drying Melt extrusion Vacuum drying (Freeze drying)	Mainly freeze drying In nature exploited by microorganisms (yeasts) and plants

Apart from FTIR spectroscopy, several other techniques have successfully been applied to determine the properties of carbohydrates in the glassy state. These include dielectric spectroscopy (Noel et al., 2000), dilatometry (Benczédi et al., 1998), NMR (Derbyshire, 2004), ESR (van den Dries et al., 2000) and neutron scattering (Cicerone and Soles, 2004). These techniques cover α- and β-relaxations (dielectric spectroscopy), matrix and solute dynamics (NMR, ESR, neutron scattering) and several structural aspects (NMR, neutron scattering).

One important aspect of glassy carbohydrates which has not yet been covered in significant detail in the scientific literature, is the local structure at the molecular level. In line with the extensive literature which exists on e.g. synthetic polymers, this local structure is expected to significantly influence the properties in the glassy state. Apart from dilatometry (Benczédi et al., 1998), which provides quantitative information on the specific volume of the matrix, and from which some structural conclusions can be drawn provided a statistical-mechanical model is assumed, the techniques mentioned above do not provide sufficient insight in the structure of the carbohydrate matrices, in particular at the molecular level in the glassy state. For this reason, we have recently introduced positron annihilation lifetime spectroscopy (PALS) as technique to explore the molecular structure of amorphous carbohydrates in the glassy state (Kilburn et al., 2004).

In this paper, I review our recent investigations of the glassy state structure using PALS, and show that useful insight in the physics of glassy carbohydrate matrices is obtained from this technique. Our PALS studies are then related with the experimental results of other groups, and with molecular dynamics (MD) simulations performed both by us (Limbach and Ubbink, 2008) and by others (Roberts and Debenedetti, 1999; Molinero and Goddard, 2005), in order to arrive at an interpretation of the structure, interactions and dynamics of carbohydrate–water systems in the glassy state and just above the glass transition temperature. Recent experiments on the aging of glassy carbohydrate matrices will also be discussed.

In addition, an extended framework which will facilitate a rational use of glassy carbohydrates in food and pharmaceutical applications will be presented. In this framework, a distinction is made between processes which influence the product stability which are dependent on the viscosity of the system and those which are not. For the latter class of processes, the recently introduced concept of 'molecular packing' (Kilburn et al., 2005; Townrow et al., 2007) is demonstrated to be useful to interpret and predict stability-related properties, such as the modulation of water sorption in the glassy state (Ubbink et al., 2007), the stability of encapsulated oxidation-sensitive compounds (Anandaraman and Reineccius, 1986) and the stabilization of biological systems and surfactants complexes in glassy carbohydrate matrices (Crowe et al, 1997).

9.2 EFFECTS OF WATER ON THE STRUCTURE OF CARBOHYDRATE GLASSES

In Figure 9.2a, the hole size of thermally annealed maltodextrin DE 12 matrices prepared by solvent casting is shown as a function of the temperature for various water contents. These water contents are prepared by equilibrating the matrices at various water activities at $T = 25°C$. The hole size as probed by PALS is shown to linearly expand with increasing temperature. Two regimens are to be discerned, a glassy state at low temperatures and a rubbery state at high temperatures. The temperature at which the cross-over between the two linear branches of the hole volume as a function of temperature occurs is identified as the glass transition temperature (Figure 9.2b). It is of interest to observe the very good agreement between the T_g determination from the thermal expansion of the molecular holes and by DSC, as PALS is essentially a static technique and DSC provides a dynamic measure of the T_g.

An important additional observation from Figure 9.2a is that, at constant temperature, the hole size increases with increasing water content. This is the case in both the glassy and the rubbery states. This expansion of the average hole size with increasing water content is most likely a general feature of amorphous carbohydrates, as it is observed for both maltodextrins (Figure 9.2a), which are both highly polydispersed and branched, and for simple disaccharides such as trehalose (Figure 9.3). The effect

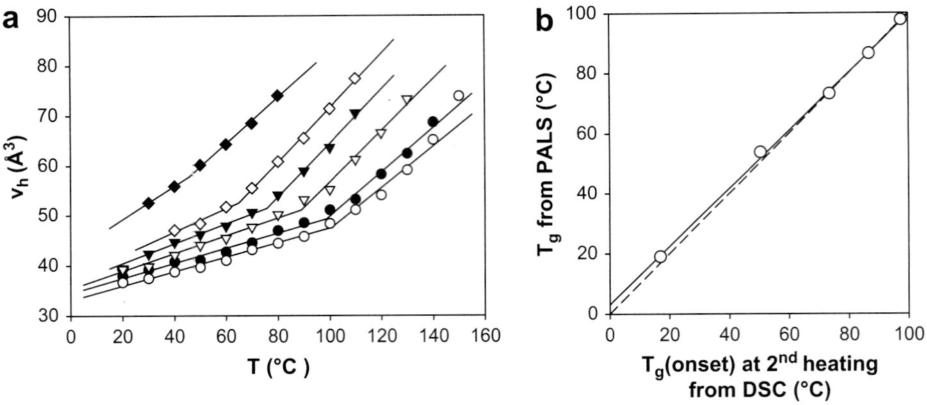

FIGURE 9.2 (a) Hole volume of annealed maltodextrin matrices (DE 12) as a function of the temperature for various water activities equilibrated at 25°C. ○: $a_w = 0.11$; ●: $a_w = 0.22$; Δ: $a_w = 0.33$; ▲: $a_w = 0.54$; ◇: $a_w = 0.54$; ◆: $a_w = 0.75$. (b) Comparison of the glass transition temperatures from PALS and DSC for malrodextrin DE 12 matrices at various water concentrations. – – –: slope = 1. ——: linear regression. (a) is reproduced from Kilburn et al. (2005) and (b) from Kilburn et al. (2005), both with permission from the American Chemical Society.

of water on the expansion of the hole size in glassy carbohydrate matrices was interpreted as resulting from the interference of water with the hydrogen bonding between the carbohydrate molecules (Figure 9.4a) (Kilburn et al., 2005; Townrow et al., 2007). This explanation is tentative, however, as no direct experimental evidence on water-content-dependent changes in the hydrogen bonding pattern in glassy carbohydrates is available. In a recent paper (Townrow et al., 2007), we have addressed this issue in some more depth, by considering not only the average hole size as determined by PALS, but also the width of the hole size distribution. In combination with a number of distinct models for the effect of water on the average hole size and the width of the hole size distribution in glassy carbohydrate matrices (Figure 9.4), we have strengthened our initial conclusion that the principal mechanism by which water acts on the structure and dynamics of glassy carbohydrate matrices is by influencing the hydrogen bonding between the carbohydrate molecules. Further proof using additional techniques, such as FTIR and NMR spectroscopy, will be needed to completely resolve this issue.

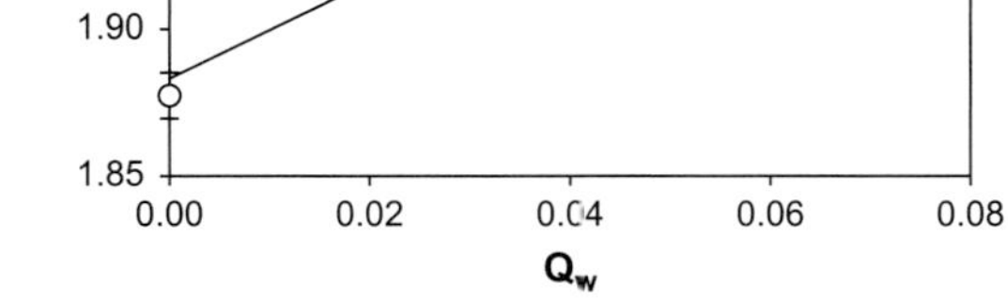

FIGURE 9.3 Mean radius of holes in amorphous trehalose as a function of the water content at 25°C. The solid line is the linear regression of the experimental data; $R^2 = 0.87$. The figure is reproduced from Kilburn et al. (2006).

When considering the structural information provided by PALS, it should be realized that this technique probes only the holes between the carbohydrate and water molecules making up the matrix. It thus provides only an impression of the unoccupied part of the system, and it does

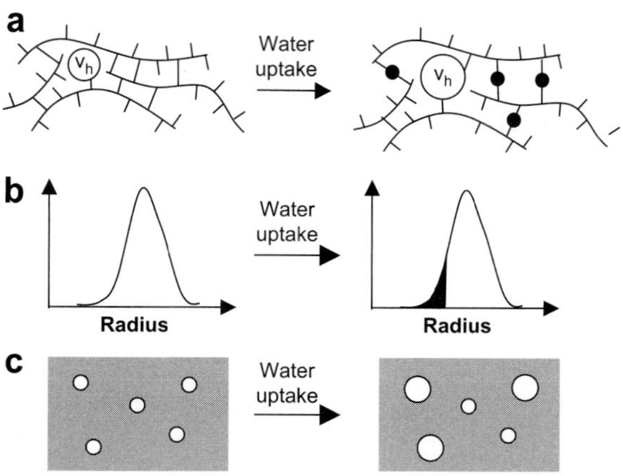

FIGURE 9.4 Suggested mechanisms leading to hole carbohydrate sizes increasing with water content in the glassy state. (a) Hole expansion because of interference of water with the hydrogen bonding between the carbohydrate molecules in the glassy state. (b) Preferential filling of small holes in a glassy matrix with a polydisperse hole size distribution. (c) Increase in water content leads to an effective two-phase system with distinct hole size distributions, with the molecular holes in the water phase being larger than the holes in the carbohydrate system.

not directly probe the carbohydrate molecules themselves. Full insight in the structure of a carbohydrate–water system can consequently only be obtained by combining PALS with techniques which directly probe properties of the carbohydrate molecules, such as FTIR, NMR and EPR spectroscopy, and scattering techniques such as neutron and X-ray scattering.

The complexity of the structure of an amorphous carbohydrate–water system may be gauged from snapshot images plotted from the positions of atoms in MD simulations (Figure 9.5a). From such MD simulations, the actual holes in a carbohydrate–water system can be derived by filling the spaces between the atoms (whose sizes are defined by their Van der Waals radii) with virtual spheres which are smaller than the smallest detail of the molecular holes one aims to determine (Limbach and Ubbink, 2008) (Figure 9.5b). Such representations of the holes between the molecules allow one to assess the complexity of the structures for which only simplified information is obtained from PALS. In particular, from a first inspection of Figure 9.5b, it is obvious that the shape of the molecular holes is far from spherical, as is conventionally assumed in PALS, and furthermore, the hole size distribution is far from monodisperse, as is also often assumed.

9.3 MOLECULAR PACKING IN GLASSY CARBOHYDRATES

Apart from water content and temperature, we may anticipate that the carbohydrate composition influences the structure of the carbohydrate matrix. In Figure 9.6a, the average hole size from PALS is plotted as a function of temperature for maltodextrins of various degrees of hydrolysis (Kilburn et al., 2005). An important observation is that in the glassy state significant differences in hole size occur at the same temperature, whereas in the rubbery state, no significant differences can be detected. In the glassy state, the hole size decreases with increasing degree of maltodextrin hydrolysis, and decreases thus with decreasing molecular weight. In effect, by lowering the (average) molecular weight, the local packing of the carbohydrates becomes denser. It is of interest to note that the dependence of the specific volume on temperature and carbohydrate composition is very similar to the behavior of the hole volume (Figure 9.6b), with the specific volume decreasing with decreasing molecular weight in the glassy state, whereas it is independent of the carbohydrate molecular weight in the rubbery state. In fact, we find a direct correspondence between the hole volume (a molecular property) and the specific volume (a macroscopic property) (Kilburn et al., 2005; Townrow et al., 2007) (Figure 9.6c).

MD simulations may be used to verify the observations from the PALS experiments. For instance, when a series of carbohydrate systems of varying degree of polymerization is cooled

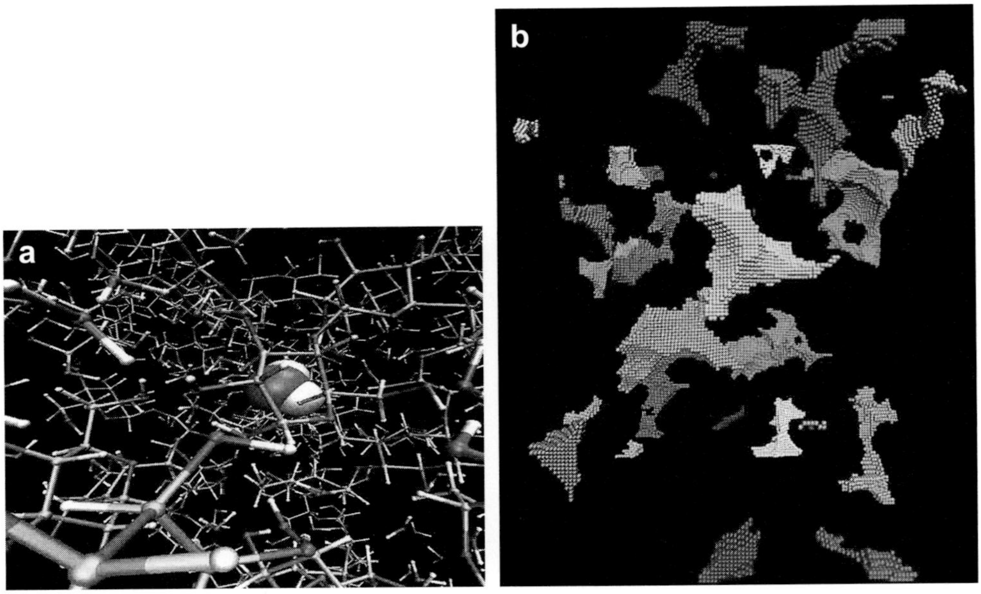

FIGURE 9.5 (a) Snap-shot from a molecular dynamics (MD) simulation of a carbohydrate–water system (80 wt% maltotriose in water, $T = 25°C$). All molecules (carbohydrates and water) are shown in a non-space-filling stick-and-ball representation; one water molecule is shown with its actual size as given by the Van der Waals radii of the atoms. (b) Free volume holes in an anhydrous carbohydrate matrix as determined from MD simulations. The matrix consists of maltopentose and the temperature is 310 K. (b) is reproduced from Limbach and Ubbink (2008) with permission from the Royal Society of Chemistry.

down in computer simulations, the free volume of the matrix is seen to be independent of the degree of polymerization in the 'rubbery' state, whereas the free volume decreases with decreasing degree of polymerization in the 'glassy' state, in line with the observations from PALS (Figure 9.7). It should nevertheless be noted that MD simulations unfortunately cannot be performed on timescales which allow a precise determination of a glass transition temperature (Limbach and Ubbink, 2008).

In a recent PALS study, the effects of variations in carbohydrate composition and water content have been systematically investigated for blends of a fairly monodisperse maltopolymer ($M_w/M_n = 2.2$, where M_w and M_n are the weight and number average molecular weight, respectively) and the disaccharide maltose (Townrow et al., 2007). When the molecular hole size is plotted as a function of the water content for a series of such maltopolymer-maltose blends, a rather complex picture of the structural behavior of the carbohydrate matrices is obtained (Figure 9.8a). Two of the previous observations on maltodextrins are confirmed: (1) in the rubbery state, the hole size tends to be independent of the carbohydrate composition, and (2) the hole size in the glassy state decreases with decreases carbohydrate molecular weight (see also Figure 9.8b). Several additional observations can also be made. First, at very low water content, hole filling is observed not only for low maltose but, interestingly, also for water (Townrow et al., 2007) (Figure 9.8a). This hole-filling mechanism can possibly be related to the mechanism of 'anti-plasticization' observed at very low water contents (Benczédi et al., 1998). Secondly, in the

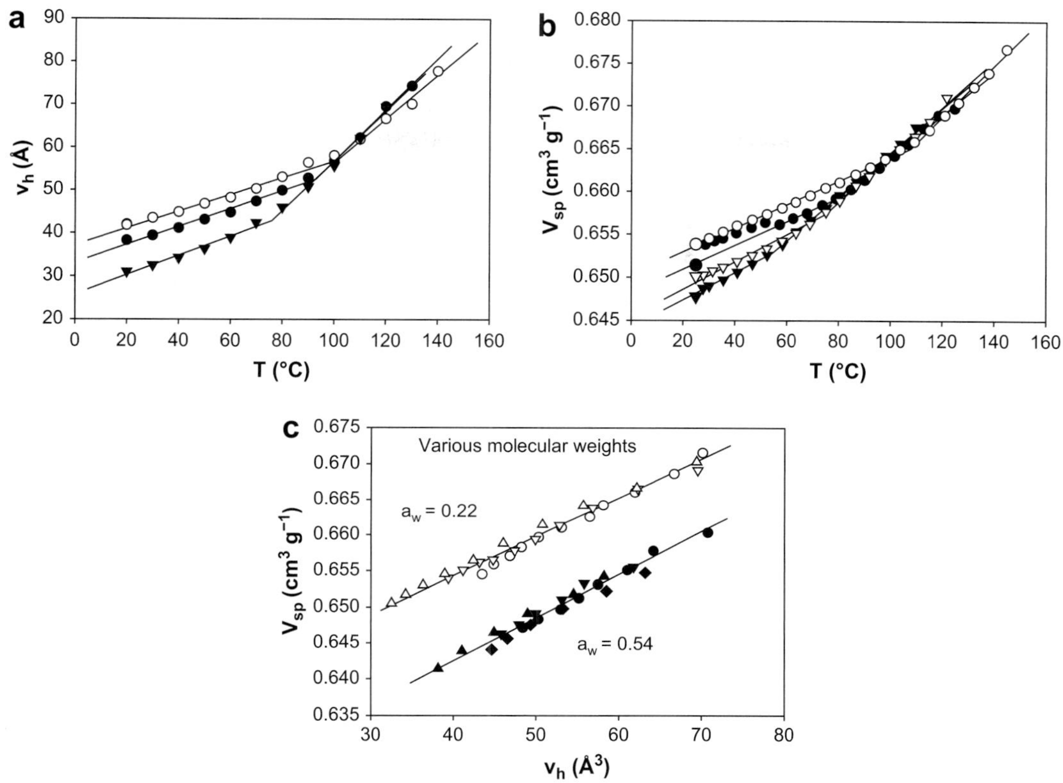

FIGURE 9.6 Hole volume (a) and specific volume (b) of annealed maltodextrin matrices equilibrated at $a_w = 0.22$ (at $T = 25°C$) as a function of the temperature for various molecular weight distributions. ○: DE 6; ●: DE 12; ▲: DE 33. Specific volume versus hole volume (c) of annealed maltodextrin matrices as a function of the temperature for various molecular weight distributions. Open symbols: aw = 0.22 (at T = 25°C); filled symbols: aw = 0.54. ●: DE 6; ▼: DE 12; ◆: DE 21; ▲: DE 33. Reproduced from Kilburn et al. (2005) with permission from the American Chemical Society.

glassy state at non-zero water content, the hole size of a matrix consisting of carbohydrate polymers decreases rapidly with the first ~30% of low-molecular-weight carbohydrates which are added (Figure 9.8b).

The disparate effect of water and low-molecular-weight sugars as plasticizers for carbohydrate polymers strongly suggests that water acts as a plasticizer mainly by interacting with the carbohydrate polymers and interfering with their hydrogen bonding. Low-molecular-weight sugars, conversely, do not modify the interactions between the carbohydrate polymers so much, but owing to their small size, they reduce the average number of entanglements experienced by a polymer chain and thus enable molecular reorganizations under conditions where the polymer chains themselves would already be frozen into a glassy state.

The differences in glassy state properties observed for the various molecular weight distributions in the temperature dependence of both hole volume and specific volume are thus essentially determined by value of the glass transition temperature for the specific combination of molecular weight distribution and water content. As discussed below, this has important technological implications.

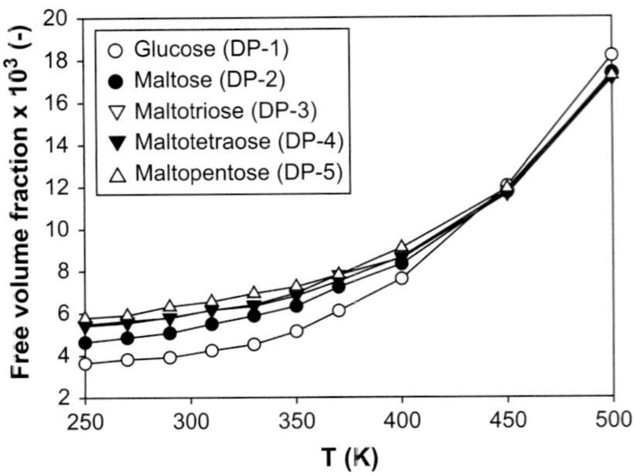

FIGURE 9.7 Temperature dependence of the free volume fraction f of maltooligomer–water mixtures at a carbohydrate concentration of 90 wt%. Each data point is the average of 20 independent quench runs. Symbols: □ – glucose; O – maltose; Δ – maltotriose; ∇ – maltotetraose; ◇ – maltopentose. Reproduced from Limbach and Ubbink (2008) with permission from the Royal Society of Chemistry.

9.4 STRUCTURAL ASPECTS OF THE AGING OF CARBOHYDRATE GLASSES

Amorphous materials below the glass transition exhibit a structural evolution which is known as aging (Struik, 1977; Hodge, 1995; Hutchinson, 1995). Even though in the glassy state large-scale reorganizations are inhibited, small local motions are still possible. As an amorphous matrix becomes constrained in a non-equilibrium state when passing though the glass transition, the residual motions below T_g will favor a slow evolution of the glassy state towards a hypothetical equilibrium state. This approach towards equilibrium is generally accompanied by a small, but measurable decrease of the specific volume of the matrix (Hutchinson, 1995).

Volumetric relaxations in carbohydrate–water systems below T_g are only small: the typical volume decrease due to aging is usually less than 0.1% (Lourdin et al., 2002) (Figure 9.9). As these changes in specific volume are essentially due to molecular rearrangements, it is of course of interest to assess if PALS could be used to quantify the changes in free volume during aging. Here we have to be cautious, though. In previous work on the temperature dependence of the specific volume and the hole volume, we have observed that in glassy carbohydrates, an increase in specific volume of about 1.5% is accompanied by a change in hole size of about

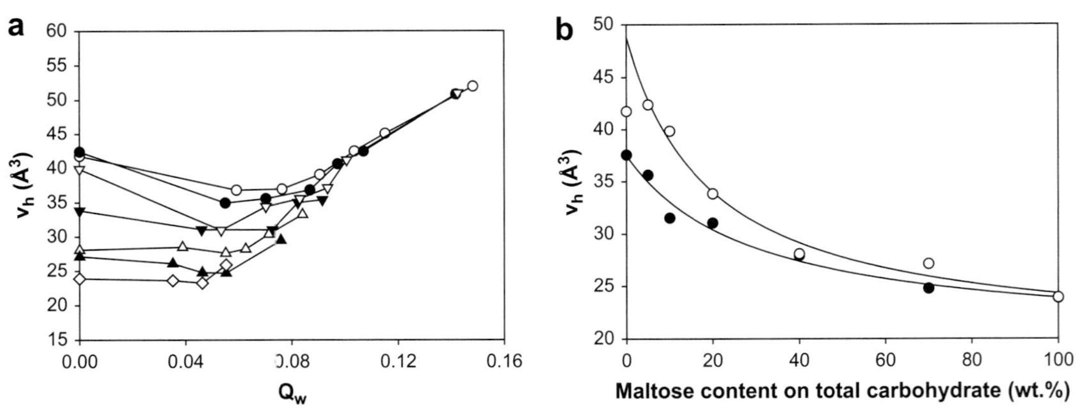

FIGURE 9.8 (a) Hole volume as a function of the weight fraction of water for a range of maltopolymer–maltose matrices equilibrated at water activities between 0 and 0.75 (T = 25 °C). (b) Hole size at 25°C as a function of the weight fraction of maltose on total carbohydrate basis for the anhydrous matrices (open symbols) and for the matrices at a water content of 5 wt. % (filled symbols) (T = 25°C). The line is a fit of a simple 'overlap' model as proposed by Townrow et al. (2007). The maltopolymer used for the experiments reported in the figures is a chromatographically fractioned maltodextrin ($M_w/M_m = 2.2$). Reproduced from Townrow et al. (2007) with permission from the American Chemical Society.

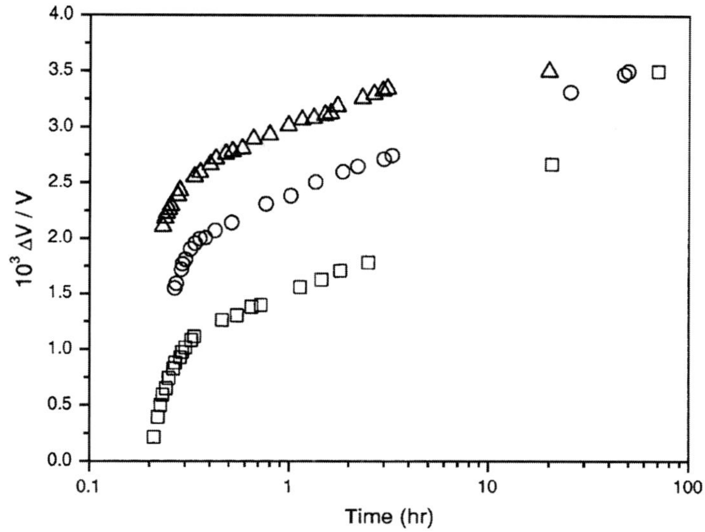

FIGURE 9.9 Structural effects during aging in glassy carbohydrates. Relative volume change as a function of time for maltose–5% water at 297 K (□), 300 K (○) and 303 K (——). The data at 300 and 303 K are displaced upwards by 0.5 and 1.0 ordinate units, respectively. Reproduced from Lourdin et al. (2002) with permission from Elsevier.

20–25% (Kilburn et al., 2005). If we translate these findings to the aging phenomenon, we would expect a maximum change in hole size of about 1–2%, which at a typical hole size in a glassy carbohydrate matrix of about 40 $Å^3$, will give an absolute change in hole size of about 1 $Å^3$. Because of various issues (see e.g. Townrow et al., 2007), this is clearly at the limit of detectable changes in a typical PALS experiment.

Even though the structural changes during aging are small, several physical properties of the system show significant changes in the evolution to the equilibrium state. Because of the closer and more optimal packing after aging, the interactions between the molecules in the matrix tend to increase, giving rise to the so-called phenomenon of enthalpy relaxation (Noel et al., 2005; Liu et al., 2006). At the same time, the mechanical properties also evolve (Noel et al., 2005). Generally, it is observed that the elastic moduli and the strength at break of glassy materials increase with aging time, whereas the deformation at break decreases significantly (Lourdin et al., 2002; Noel et al., 2005).

9.5 DYNAMIC PROPERTIES CLOSE TO THE GLASS TRANSITION

It is of importance to investigate whether the structural variations as discussed in the sections above have an impact on the dynamics of the carbohydrate–water systems. Several techniques, such as NMR and EPR spectroscopy, and scattering techniques such as neutron and X-ray scattering can in principle be used to obtain quantitative information on such structure–property relations. Unfortunately, whereas NMR spectroscopy is perfectly feasible in aqueous solutions of carbohydrates, and even in the rubbery state, it becomes very difficult to obtain structural and dynamic information from NMR in the glassy state. Difficulties are also encountered in the application of neutron and X-ray scattering. A major issue with neutron scattering on carbohydrate matrices is the large incoherent scattering length of hydrogen, and from X-ray scattering only limited information may be obtained on the amorphous state of carbohydrates as generally only a broad amorphous peak is observed. ESR spectroscopy has delivered some important information on the dynamics of glassy carbohydrates (van den Dries et al., 2000).

MD simulations have recently been used with increasing success to predict the structure and dynamics of concentrated carbohydrate–water systems (Roberts and Debenedetti, 1999; Molinero and Goddard, 2005; Limbach and Ubbink, 2008). MD simulations are also very useful to investigate structure and dynamics of complex food systems at the molecular level, as it allows incorporating molecular structure, physical properties and interactions in a highly detailed manner unattainable in analytical statistical mechanics (Limbach and Kremer, 2006). Such MD simulations can be either based on all-atom models or use coarse-graining procedures to reduce the level of molecular detail and thus the required computer power (Limbach and Kremer, 2006). Recent advances in computing performance and in force fields and simulation methods now enable the simulation of systems of a complexity which is sufficiently high to allow direct application for quite complex carbohydrate–water systems. Detailed mechanisms for the diffusion of water in such systems and for the hydrogen-bonding between the carbohydrate and water molecules have been derived from such simulations (Roberts and Debenedetti, 1999; Molinero and Goddard, 2005; Limbach and Ubbink, 2008).

In Figure 9.10, the Arrhenius plot of the diffusion coefficient of water in maltooligomers is shown for three carbohydrate concentrations. It is of interest to note that, in solution and in the rubbery state, the diffusion coefficient of water is independent of the degree of polymerization of the carbohydrate and solely determined by the carbohydrate concentration.

In Figure 9.11, the ratio τ_{gluc}/τ_{H_2O} of the rotational correlation times of water τ_{H_2O} and glucose τ_{gluc} are shown for different carbohydrate concentrations. For low concentrations ratio τ_{gluc}/τ_{H_2O} is relatively constant upon changing

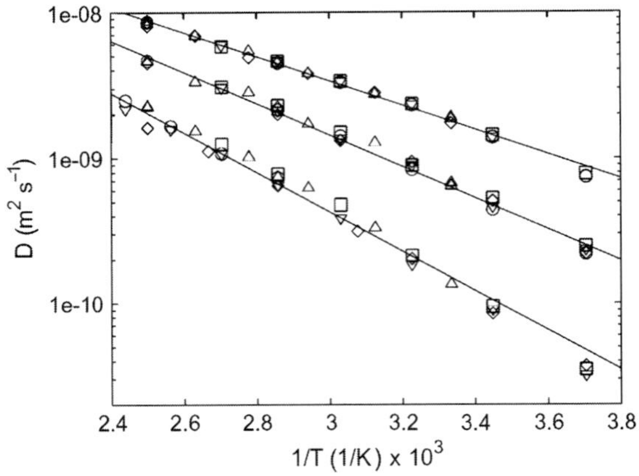

FIGURE 9.10 Arrhenius plot for the temperature dependence of the water diffusion coefficient for different carbohydrate concentrations and different molecular weights. From top to bottom: 20, 50 and 70 wt% malto-oligomer content. □ – glucose; ○ – maltose; △ – maltotriose; ▽ – maltotetraose; ◆ – maltopentose. Reproduced from Limbach and Ubbink (2008) with permission from the Royal Society of Chemistry.

the temperature. The situation changes, when the glucose concentration is increased. Above about 50 wt% glucose, the ratio τ_{gluc}/τ_{H_2O} rather suddenly increases by a factor of 3–4 in

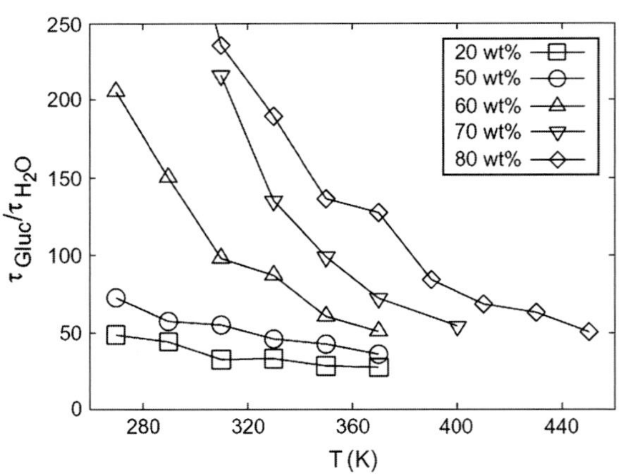

FIGURE 9.11 Temperature dependence of the ratio of the rotational time of glucose τ_{gluc} and the rotational time for water τ_{H_2O} for various glucose concentrations in glucose–water mixtures. The decoupling of the mobilities at higher glucose concentrations with decreasing temperature can be seen. Reproduced from Limbach and Ubbink (2008) with permission from the Royal Society of Chemistry.

a temperature window of about 100 K. With increasing concentration and lower temperatures the errors in the ratio τ_{gluc}/τ_{H_2O} become larger since the error in determining τ_{gluc} is increasing. Qualitatively similar behavior is observed for the ratio of the diffusion constants of the two molecules (Limbach and Ubbink, 2008), but a quantitative analysis is impossible because of the rapidly increasing uncertainty in the value of the diffusion coefficients when the glass transition is approached.

This decoupling of mobilities in the approach to the glass transition, which has also been observed in experiment (van den Dries et al., 2000), explains that the diffusion of water shows an Arrhenius-type behavior although the mobility of the maltooligomer molecules is almost frozen in. In a recent paper (Limbach and Ubbink, 2008), we have also investigated if differences in carbohydrate molecular weight result in differences in the structural packing at the molecular level and, hence, in differences in the water mobility. Unfortunately, significant molecular weight differences in the molecular packing of maltooligomers occur only below the glass transition temperature, and in this regimen, MD simulations do not yet provide sufficiently quantitative results.

As discussed in the section below, the decoupling of the mobility of water, and most likely also other small molecules such as gases, from the mobility of the carbohydrate molecules in the approach to the glass transition has important implications in the food and pharmaceutical industry, since the mobility of these compounds is, e.g., related to the shelf-life stability of such products or to the barrier properties of membranes.

9.6 TECHNOLOGICAL IMPLICATIONS

The current findings on the structure of carbohydrates in the glassy state and the dependence of the structure on, in particular, composition, may help to define rational strategies to optimize the stability of products in the food and pharmaceutical fields. In these applications, the first distinction which needs to be made is between viscosity-controlled stability issues, and stability issues which are related to the structure and dynamics of the carbohydrate glassy state (Table 9.2). The first category comprises stability issues such as the collapse of powder structure, the undesired crystallization of the carbohydrates constituting the product matrix (such as lactose in milk powder), and to some extent also chemical reactions (namely those between larger-molecular-weight species whose mobility is controlled by the mobility of the carbohydrate matrix). For this category of stability issues, the highest stability is obtained by increasing the glass transition temperature to the highest practical value, by either increasing the carbohydrate molecular weight or by lowering the water content of the matrix.

The second category encompasses rather diverse phenomena such as those in which the mobility of low-molecular-weight compounds or the local structure and dynamics of the

TABLE 9.2 Criteria for Physical Product Stability

Compromise between density and T_g	Only T_g requirement*
Encapsulation/biostabilization	Physical stability of powders
Modulation of moisture level	Caking
Matrix expansion below T_g	Agglomeration/compaction
Protection against oxygen	(Avoidance of) crystallization
Mechanical properties	Flowability

*To quantitatively predict collapse, caking and agglomeration, the analysis of the water-content-dependent glass transition is usually combined with the rheology of the system in the rubbery state.

carbohydrate matrix (in particular hydrogen bonding) play the central role, and includes situations in which the moisture uptake, the diffusion of oxygen, and irreversible transitions in biomolecular complexes are limiting the stability of the system. In these situations, a subtle compromise between two important factors need to made in order to arrive at the optimal stability of the product (Figure 9.12). On the one hand, it is required that the product matrix remains in the glassy state during the shelf life of the product. As for the viscosity-related stability issues, this is best achieved by increasing the carbohydrate molecular weight and by lowering the water content. On the other hand, the molecular packing of the carbohydrate matrix, as reflected, e.g., by its density or the molecular hole size, is controlling the product stability. If for example, the barrier properties against oxygen are limiting the stability of the product (as it for instance contains an encapsulated oxidation-sensitive compound such as a polyunsaturated fatty acid (PUFA)), the uptake of oxygen may be minimized by maximizing the matrix density (at constant water content) and therefore by decreasing the molecular weight of the carbohydrates constituting the encapsulation matrix (Figure 9.12).

This consequently leads to a situation where a compromise needs to be made between the T_g-requirement and the matrix density.

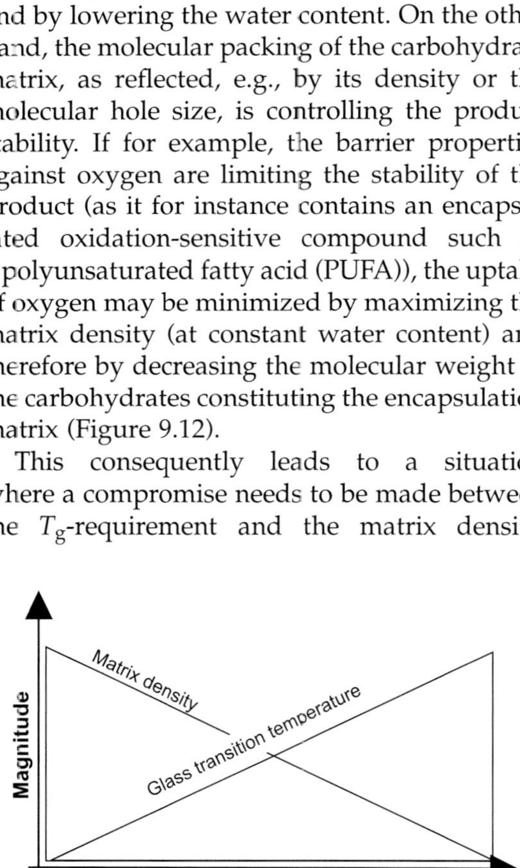

FIGURE 9.12 Diagram schematically showing the effect of carbohydrate molecular weight on the matrix density and on the glass transition temperature.

Consequently, the best strategy to maximize the stability in cases where it is limited by processes occurring in the glassy state is to strive for the lowest average molecular weight of the carbohydrate matrix which still satisfies the requirement that the product remains in the glassy state during the shelf life, as at this molecular weight, the barrier properties for oxygen are the highest.

Several examples illustrate our approach. First, the sorption of water by amorphous carbohydrates is considered. The sorption of water by amorphous food and pharmaceutical products is generally fitted using the BET model or one of its variants, in particular the GAB model (Roos, 1995). These sorption models were originally derived for the adsorption of low-molecular-weight compounds on porous surfaces, and in their physical description, they combine a Langmuir-type monolayer formation followed by a multilayer condensation on top of this monolayer. Although fitting the water sorption by carbohydrate-based system relatively well for water activities in the range $0 < a_W < 0.8$, these models fundamentally do not apply for the absorption of water by amorphous carbohydrate systems as they lack a correct physical description of the sorption process. In particular, these models do not relate to the glass transition which is encountered when at constant temperature amorphous carbohydrate matrices absorb sufficient amounts of water (Ubbink et al., 2007).

We recently observed that the sorption of water in the glassy state is directly related to the specific volume of the matrix (Figure 9.13a). This relation is lost in the rubbery state. This means that, in the glassy state, the sorption of water is dependent on the molecular packing of the carbohydrates. Consequently, by adapting the specific volume of the matrix by changing its composition, the amount of water absorbed at a given water activity and temperature may be modulated. Specifically, in the glassy state, we can reduce the amount of water absorbed by increasing the content of low-molecular-weight carbohydrates (in this case maltose).

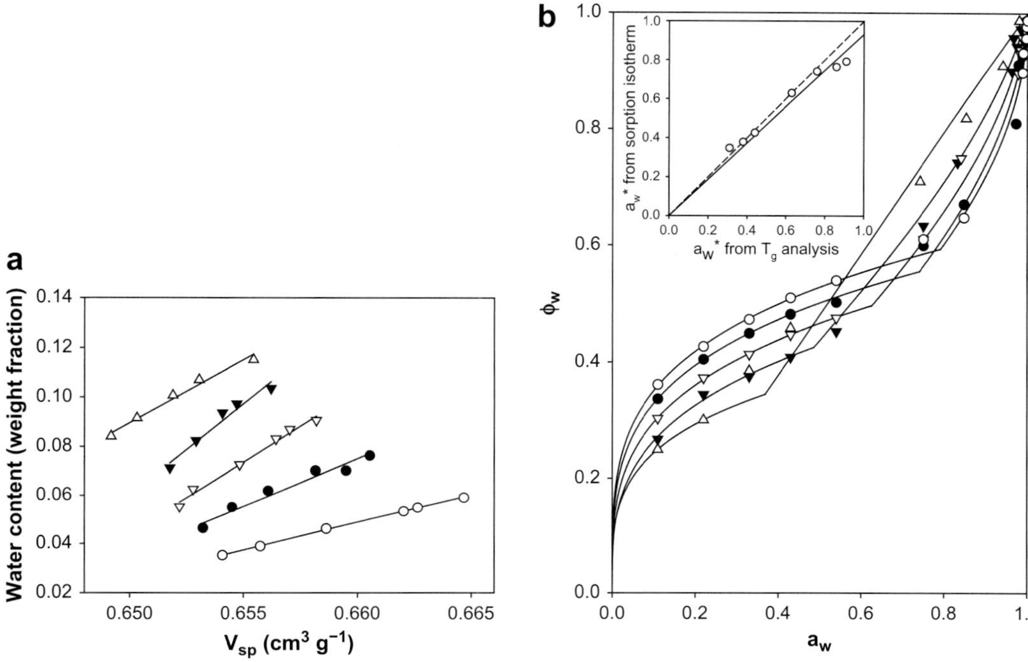

FIGURE 9.13 Water sorption in bidisperse glassy carbohydrates consisting of blends of varying composition of a fractioned maltopolymer and the disaccharide maltose. (a) Relation between the specific volume of the maltopolymer–maltose matrices and the water content for various water activities ($T = 25$ °C). At each water activity, the specific volume varies with blend composition (varying from 100% maltopolymer–0% maltose to 0% maltopolymer–100% maltose). ○: $a_w = 0.11$; ●: $a_w = 0.22$; ▽: $a_w = 0.33$; ▼: $a_w = 0.43$; △: $a_w = 0.54$. (b) Water sorption isotherms of the bidisperse carbohydrate matrices. The sorption data are fitted using a two-phase sorption model. In the glassy state, the water sorption data are fitted using a Freundlich isotherm model, and in the rubbery state, the water sorption data are fitted using a Flory-Huggins isotherm. The blends vary in composition: open circles: 100% maltopolymer–0% maltose; filled circles: 90–10; open triangles (down): 80–20; filled triangles (down): 60–40; open triangles (up): 0% maltopolymer–100% maltose. The inset shows the correlation between the critical water activity at which the carbohydrate blends pass into the rubbery state (horizontal axis) and the water activity at which the Freundlich and Flory-Huggins fits of the sorption data intersect.

On the other hand, increasing the content of low-molecular-weight carbohydrates in the blend will reduce the water activity at which the carbohydrate matrix will pass into the rubbery state (Figure 9.13b). In order to modulate the water content of a glassy carbohydrate matrix to the desired value, the dependence of both the molecular packing and the glass transition on the carbohydrate molecular weight need to be taken into account, in line with the scheme proposed in Figure 9.12.

Given the explicit dependence of the water sorption by amorphous carbohydrate matrices on the physical state of the matrix, we have recently proposed a novel approach in which sorption models are explicitly developed for the sorption of water in the rubbery and glassy states (Ubbink et al., 2007). In the rubbery state, an adaptation of the Flory-Huggins sorption isotherm for a ternary system (maltopolymer, maltose and water) was shown to correctly fit the water sorption data. In the glassy state, the sorption was modeled using a Freundlich model which was originally derived for the sorption on systems with an energetically heterogeneous distribution of sorption sites (Ubbink et al.,

2007). The overall fit of the sorption data using the two models shows good agreement with the sorption data (Figure 9.13b) and, in particular, with the critical moisture content at which the glass transition temperature is 25°C (inset of Figure 9.13b).

A second example is on the oxygen uptake by carbohydrate-encapsulated citrus oils. Here I base myself on the data published by Anandaraman and Reineccius (1986). By spray drying they prepared a series of maltodextrin-based capsules with a variation in the degree of

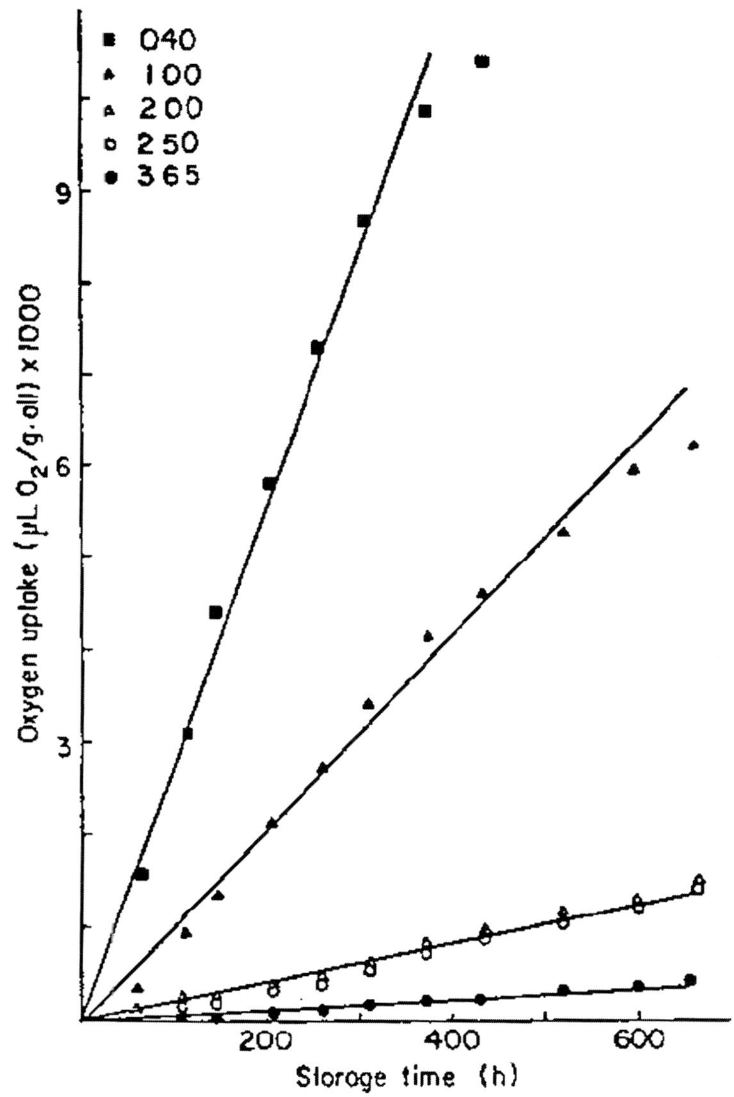

FIGURE 9.14 Oxidation of citrus oil encapsulated in maltodextrins of varying molecular weight distribution at $T = 45°C$. ■: DE 4; ▲: DE 10; △: DE 20; ○: DE 25; ●: DE 37. Reproduced from Anandaraman and Reineccius (1986) with permission of the Institute of Food Technologists.

hydrolysis of the maltodextrins. Interestingly, they observed that during storage at $T = 45\ ^\circ C$, the rate of oxygen uptake varied by more than one order of magnitude (Figure 9.14). The rate of oxygen uptake decreased with increasing degree of maltodextrin hydrolysis, and decreased thus with decreasing carbohydrate molecular weight, in line with our results on the carbohydrate molecular packing in the glassy state. As data on the water content of the capsules are lacking in this study, and as no density measurements were performed, further, better controlled studies are warranted.

A last example relates to the stabilization of lipid vesicles and related biological lipid membrane systems such as red blood cells by glassy carbohydrates. Crowe et al. (1997) have observed that by varying the molecular weight of the carbohydrate matrix, the membrane phase transition temperature may be influenced. Specifically, they observed that by lowering the average molecular weight of the carbohydrate matrix by increasing the fraction of glucose in hydroxyethyl starch (HES) – glucose mixtures, T_m is significantly depressed, resulting in an increased membrane fluidity, and consequently an improved liposome integrity during storage, and thus to an increased retention of active ingredients encapsulated within the liposomes (in this case a fluorescent dye) (Figure 9.15). At the same time, they observe that the interactions between the carbohydrate matrix and the phospholipids are increasingly modified with increasing glucose levels (as observed from FTIR analysis of the phosphate asymmetric stretch vibrations). Again, this is in line with our results on the molecular packing of carbohydrates in the glassy state. By reducing the average molecular weight of the carbohydrate matrix, one increases the local density of packing of the carbohydrate molecules, thereby allowing for a higher density of hydroxyl groups in the vicinity of the phospholipid head group and thus to increased levels of hydrogen bonding between the phospholipids and the carbohydrate encapsulation matrix. This in turn influences the phase behavior of the

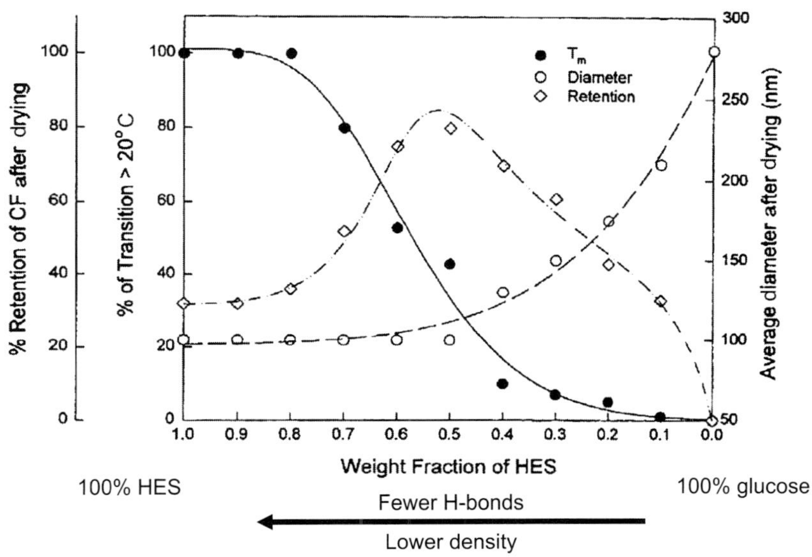

FIGURE 9.15 Effect of the weight fraction of hydroxethyl starch (HES) on total carbohydrate (HES and glucose) on the retention of carboxyfluorescein in liposomes, membrane phase transition temperature and average diameter following rehydration. Reproduced from Crowe et al. (1997) with permission from Academic Press. The annotations below the figure have been added for the present manuscript.

liposomes and increases the stability during storage under anhydrobiotic conditions.

9.7 CONCLUSIONS AND PERSPECTIVES

In this paper, recent developments in the understanding of the properties of amorphous carbohydrates have been discussed with emphasis on their properties in the glassy state. The impact of the glass transition concept on food and pharmaceutical technology is briefly discussed and it is argued that those aspects of the performance of amorphous carbohydrates which are exclusively related to the glassy state are only indirectly related to the glass transition temperature of the matrix. Our investigations using positron annihilation lifetime spectroscopy reveal a molecular-weight-dependent structural signature of carbohydrates in the glassy state, and a relation with structural and dynamic properties as determined from molecular dynamics (MD) simulations is established. The molecular packing of carbohydrates in the glassy state is discussed as a new concept to rationalize several hitherto unexplained phenomena related to the performance of glassy carbohydrates in food and pharmaceutical applications. It turns out that in particular the molecular weight distribution of the carbohydrates influences the molecular packing and consequently the properties in the glassy state. The molecular packing of carbohydrates in the glassy state has an impact for important applications, such as the modulation of moisture sorption, the encapsulation of oxidation-sensitive compounds and in the stabilization of biological systems and surfactants complexes.

ACKNOWLEDGMENTS

I would like to take the opportunity to thank Prof. Ashraf Alam, Dr. Duncan Kilburn, Sam Townrow and Mina Roussenova from the University of Bristol (UK) for stimulating collaboration. I also thank my colleagues at the Nestlé Research Center who have contributed to the experimental and theoretical program on the molecular physics of carbohydrate systems at low water content. Thanks are in particular due to Dr. Hans-Jörg Limbach, Maria-Isabelle Giardiello, Dr. Vincent Meunier, Jean-Pierre Marquet and Johanna Claude.

References

Anandaraman, S. and Reineccius, G. A. (1986). Stability of encapsulated orange peel oil. *Food Technol.* **40**, 88–93.

Angell, C. A. (1995). Formation of Glasses from Liquids and Biopolymers. *Science* **267**, 1924–1935.

Benczédi, D., Tomka, I., and Escher, F. (1998). Thermodynamics of Amorphous Starch-Water Systems. *Macromolecules* **31**, 3055–3061. 1. Volume Fluctuations.

Cicerone, M. T. and Soles, C. L. (2004). Fast Dynamics and Stabilization of Proteins: Binary Glasses of Trehalose and Glycerol. *Biophys. J.* **86**, 3836–3845.

Crowe, J. H., Hoekstra, F. A., and Crowe, L. M. (1992). Anhydrobiosis. *Annu. Rev. Physiol.* **54**, 579–599.

Crowe, J. H., Oliver, A. E., Hoekstra, F. A., and Crowe, L. M. (1997). Stabilization of dry membranes by mixtures of hydroxyethyl starch and glucose: the role of vitrification. *Cryobiol.* **35**, 20–30.

Derbyshire, W., Van den Bosch, M., Van Dusschoten, D., MacNaughtan, W., Farhat, I. A., Hemminga, M. A., and Mitchell, J. R. (2004). Fitting of the beat pattern observed in NMR free-induction decay signals of concentrated carbohydrate–water solutions. *J. Magn. Res.* **168**, 278–283.

van den Dries, I. J., van Dusschoten, D., Hemminga, M. A., and van der Linden, E. (2000). Effects of Water Content and Molecular Weight on Spin Probe and Water Mobility in Malto-oligomer Glasses. *J. Phys. Chem. B* **104**, 10126–10132.

Eichner, K. and Karel, M. (1972). The influence of water content and water activity on the sugar-amino browning reaction in model systems under various conditions. *J. Agric. Food Chem.* **20**, 218–223.

Hodge, I. M. (1995). Physical Aging in Polymer Glasses. *Science* **267**, 1945–1947.

Hutchinson, J. M. (1995). Physical aging of polymers. *Prog. Polym. Sci.* **20**, 703–760.

Kilburn, D., Claude, J., Mezzenga, R., Dlubek, G., Alam, A., and Ubbink, J. (2004). Water in Glassy Carbohydrates: Opening It Up at the Nanolevel. *J. Physi. Chem. B* **108**, 12436–12441.

Kilburn, D., Claude, J., Schweizer, T., Alam, A., and Ubbink, J. (2005). Carbohydrate Polymers in Amorphous States: An integrated Thermodynamic and Nanostructural Investigation. *Biomacromolecules* **6**, 864–879.

Kilburn, D., Townrow, S., Meunier, V., Richardson, R., Alam, A., and Ubbink, J. (2006). Organization and Mobility of Water in Amorphous and Crystalline Trehalose. *Nature Materials* **5**, 632–635.

Limbach, H. J. and Kremer, K. (2006). Multi-scale modelling of polymers: Perspectives for food materials. *Trends Food Sci. Technol.* **17**, 215–219.

Limbach, H. J. and Ubbink, J. (2008). Structure and Dynamics of Maltooligomer-Water Solutions and Glasses. *Soft Matter* (in press).

Liu, L., Bhandari, B., and Zhou, W. (2006). Glass Transition and Enthalpy Relaxation of Amorphous Food Saccharides: A Review,. *J. Agric. Food Chem.* **54**, 5701–5717.

Lourdin, D., Colonna, P., Brownsey, G., Noel, T. R., and Ring, S. G. (2002). Structural relaxation and physical aging of starchy materials. *Carbohydr. Res.* **337**, 827–833.

Molinero, V. and Goddard, W. A. (2005). Microscopic Mechanism of Water Diffusion in Glucose Glasses. *Phys. Rev. Lett.* **95**, 045701-1–045701-4.

Noel, T. R., Parker, R., and Ring, S. G. (2000). Effect of molecular structure and water content on the dielectric relaxation behaviour of amorphous low molecular weight carbohydrates above and below their glass transition. *Carbohydr. Res.* **329**, 839–845.

Noel, T. R., Parker, R., Brownsey, G. J., Farhat, I. A., MacNaughtan, W., and Ring, S. G. (2005). Physical Aging of Starch, Maltodextrin, and Maltose. *J. Agric. Food Chem.* **53**, 8580–8585.

Roberts, C. J. and Debenedetti, P. G. (1999). Structure and Dynamics in Concentrated, Amorphous Carbohydrate-Water Systems by Molecular Dynamics Simulation. *J. Phys. Chem. B* **103**, 7308–7318.

Roberts, C. J. and Debenedetti, P. G. (2002). Engineering Pharmaceutical Stability with Amorphous Solids. *AICHE J.* **48**, 1140–1144.

Roos, Y. H. (1995). *Phase Transitions in Foods*. Academic Press, New York.

Schoonman, A., Ubbink, J., Bisperink, C., Le Meste, M., and Karel, M. (2002). Solubility and Diffusion of Nitrogen in Maltodextrin/Protein Tablets. *Biotechnol. Prog.* **18**, 139–154.

Slade, L. and Levine, H. (1995). Glass transition and water–food structure interactions. *Adv. Food Nutr. Res.* **38**, 103–269.

Struik, L. C. E. (1977). Physical Aging in Plastics and Other Glassy Materials. *Polym. Engng. Sci.* **17**, 165–173.

Townrow, S., Kilburn, D., Alam, A., and Ubbink, J. (2007). Molecular packing in bidisperse carbohydrate matrices. *J. Phys. Chem. B* **111**, 12643–12648.

Tromp, R. H., Parker, R., and Ring, S. G. (1997). Water diffusion in glasses of carbohydrates. *Carbohydr. Res.* **303**, 199–205.

Ubbink, J., Giardiello, M. I., and Limbach, H. J. (2007). Sorption of Water by Bidisperse Mixtures of Carbohydrates in Glassy and Rubbery States. *Biomacromolecules* **8**, 2862–2873.

Ubbink, J. and Schoonman, A. (2007). Flavor Delivery Systems. *In* "Kirk-Othmer Concise Encyclopedia of Chemical technology," 5th Edition, 940–944, Wiley, New York.

Vuataz, G. (2002). The phase diagram of milk: a new tool for optimizing the drying process. *Le Lait* **82**, 485–500.

CHAPTER

10

Biopolymer Films and Composite Coatings

Amos Nussinovitch

Institute of Biochemistry, Food Science and Nutrition, Faculty of Agricultural, Food and Environmental Quality Sciences, The Hebrew University of Jerusalem, P.O. Box 12, Rehovot 76100, Israel

10.1 INTRODUCTION

Water-soluble gums are valuable in many fields, including adhesives, agriculture, biotechnology, ceramics, cosmetics, explosives, food, paper, textiles and texturization, among many others (Nussinovitch, 2003). The most recent developments and progress in the utilization of gums can be found in the field of edible coatings. Edible and biodegradable films have the potential to reduce packaging and limit moisture, aroma and lipid migration between food components (Krochta and De Mulder-Johnston, 1997). Such films can contain antioxidants, preservatives or other additives to improve foods' mechanical integrity, handling and quality, and to change surface gloss (Nussinovitch, 1998). Coatings are not limited to the food industry and thus gums are also used in coatings for fiberglass, fluorescent lamps, glass, metals, optical products, paper products, latex and textiles (Davidson, 1980). In the past 5 years, there has been an enormous expansion in both the development and marketing of edible films. In 1999, sales were just $1 million, mostly from niche products such as edible underwear – a popular gag gift at bachelor parties. In 2005, market revenues of more than $100 million were recorded. Retail sales of edible films are expected to hit at least $350 million in 2008 (http://www.ceepackaging.com/2006/08/23/film-going-down). The demand for edible packaging films is also on the rise and could reach $25 million in sales revenues over the next 3 or 4 years (IMR, 2007).

10.2 MECHANISMS OF FILM FORMATION

Edible films can be produced from hydrocolloids, lipids, resins, and composites. There are many methods for forming films directly on food surfaces. For film-forming materials dispersed in aqueous solutions, solvent removal is required to achieve solid film formation and control of its properties (Hernandez-Izquierdo and Krochta, 2008). For example, the temperature and rate of drying influence the mechanical properties and crystallinity of cellulosic films (Reading and

Kasapis, Norton, and Ubbink: Modern Biopolymer Science
ISBN: 978-0-12-374195-0

Spring, 1984; Greener, 1992). Proteins are heteropolymers that usually contain most of the 20 amino acids, allowing for an enormous number of sequential arrangements with a wide range of interactions and chemical reactions (Stevens, 1999; Pommet et al., 2003). In contrast, polysaccharides contain only a few monomers, e.g., cellulose and starch contain only one monomer, glucose (Hernandez-Izquierdo and Krochta, 2008). In polysaccharides, the hydroxyl is the only reactive group, while proteins present a large variety of possible interactions and chemical reactions (Hernandez-Izquierdo and Krochta, 2008): they may participate in chemical reactions through covalent (peptide and disulfide) linkages and non-covalent interactions (ionic, hydrogen, and van der Waals bonding). In addition, hydrophobic interactions occur between non-polar groups of amino acid chains (Kokini et al., 1994). Interlinkages between proteins participating in the formation of a film can lead to improved film properties (Okamoto, 1978). Edible films can be formed via two main processes: a 'wet process' in which biopolymers are dispersed or solubilized in a film-forming solution (solution casting) followed by evaporation of the solvent, and a 'dry process' which relies on the thermoplastic behavior exhibited by some proteins and polysaccharides at low moisture levels in compression molding and extrusion (Cuq et al., 1997a; Pommet et al., 2003; Liu et al., 2006).

A number of proteins have received particular attention for the production of edible coatings. The prolamin fraction of corn is known as zein (Esen, 1987). Zein can form films when cast from appropriate solvent systems. The mechanisms that come into play upon solvent evaporation include hydrophobic and hydrogen bond development in the formed film matrix (Reiners et al., 1973). A limited number of disulfide bonds, due to low cystine content, may also be present. The produced films are tough, glossy, and scuff- and grease-resistant, and plasticizers should be included in their formulation to decrease brittleness (Reiners et al., 1973). Cross-linking agents may be involved in improving water resistance and tensile properties (Szyperski and Gibbons, 1963).

Wheat gluten films can be manufactured by the deposition and subsequent drying of wheat gluten dispersions. The solvent in many cases is aqueous ethanol under alkaline or acidic conditions. Homogeneous solutions are produced by heating and mixing (Gontard et al., 1992; Gennadios et al., 1993a). The following mechanisms are involved in the film's formation: upon dispersing the gluten in alkaline environments, disulfide bonds in the gluten solution are reduced by the reducing agents (Okamoto, 1978). Upon casting of the film-forming solution, disulfide bonds re-form, linking together polypeptide chains and producing a film structure. Mechanisms contributing to the re-formation of disulfide bonds include re-oxidation in air and sulfhydryl–disulfide interchange reactions (Wall et al., 1968; McDermott et al., 1969). Hydrogen and hydrophobic bonds also add to the film structure. Without plasticizer, such films are brittle; however, plasticizer addition decreases the rigidity of such films by mediating between polypeptide chains, disrupting some of the extensive intermolecular associations (Wall and Beckwith, 1969).

Edible films based on soybeans have long been produced in the Orient. Protein is the major component of these films, but lipids and carbohydrates are also incorporated. Consequently, soymilk films are in fact multicomponent films. The protein content in soymilk films was found to be higher than that in the initial soymilk (Wu and Bates, 1972). Therefore, the film-formation mechanism was proposed to consist of the isolation and partial concentration of the proteins, probably via the mechanism of endothermic polymerization of heat-denatured protein together with surface dehydration (Wu and Bates, 1972). Another description of this mechanism proposed formation of a protein matrix by heat-catalyzed protein–protein interactions, with hydrogen disulfide and hydrophobic bonds being the major associative forces in the film network (Farnum et al., 1976). Yet another suggested mechanism of polymerization involved intermolecular disulfide and hydrophobic bonds, whereby heating is required to alter the three-dimensional structure, thereby exposing sufhydryl groups and hydrophobic side chains. Once dried, the unfolded protein macromolecules move toward each other and are linked through the hydrophobic and disulfide bonds (Fukushima and Van Buren, 1970).

Films can be produced from soy protein isolates by heating aqueous dispersions of same to form surface films or by the deposition and drying of soy protein solutions. Film production continues with polymerization and solvent evaporation at the film–air interface (Okamoto, 1978). The mechanism of film formation is explained thus: when the protein is in solution, hydrophobic groups are oriented towards the interior of the protein molecule, away from water. At the air–water interface, these hydrophobic groups extend out of the water into the air where they interact with each other, while hydrophilic groups remain submerged (Cheesman and Davies, 1954). Another proposed mechanism involves interfacial coagulation which may occur when the protein concentration in the interface monolayer exceeds a certain limit and the protein coagulates, forming a three-dimensional coagulum at the interface from the two-dimensional monolayer (Hernandez-Izquierdo and Krochta, 2008). Proof that polymerization takes place through such bonds was provided by demonstrating its inhibition by cleavage of disulfide bonds (Circle et al., 1964). Blockage of sulfhydryl with appropriate reagents also inhibited polymerization (Wolf and Tamura, 1969). pH may also influence film formation. Near the isoelectric point of soy protein (~4.6), protein coagulation and, consequently, solution casting are impossible (Gennadios et al., 1993b). A method similar to film formation on soymilk surfaces was used to develop peanut protein–lipid films on the surface of heated peanut milk (Aboagye and Stanley, 1985).

Cross-linking is responsible for the production of collagen films with desirable properties. In the presence of formaldehyde, this mechanism involves the combining of free amino groups of basic amino acids. Treatment with glyceraldehyde (non-toxic, in contrast to glutaraldehyde), which promotes cross-linking, increased the mechanical properties of the produced films and demonstrated, in part, the approach by which such films are produced (Jones and Whitmore, 1972). Addition of lower alkyl diols with 4–8 carbon atoms also improved the mechanical properties of the collagen films (casings) by reducing internal hydrogen bonding while increasing intermolecular spacing (Boni, 1988). Collagen films extruded from acid dispersions were subjected to irradiation in the presence of a photosensitive dye which catalyzed protein cross-linking and as a result, improved film properties (Kuntz, 1964). The protein gelatin forms through partial hydrolysis of collagen. Gelatin forms a clear, flexible, strong and oxygen-impermeable film, whose properties are influenced by drying temperature. Above 35°C, gelatin exists as a single molecule in a configuration that cannot form interchain hydrogen bonds, whereas at lower temperatures, gelatin has a 'collagen fold' configuration capable of forming interchain hydrogen bonds (Robinson, 1953).

Milk film is formed from milk proteins covering food surfaces, possibly via the effects of high relative humidity and elevated temperatures which decrease lactose solubility and increase its granulation (Mabesa et al., 1980). Milk film can also form at the air–water interface from small amounts of insoluble milk powder, the underlying mechanism thought to be the formation of sugar–protein complexes (Leach et al., 1966). Casein is the major protein in milk, including αs1-, αs2-, β- and κ-caseins and a small amount of γ-casein (Brunner, 1977). α-casein may form flexible films via increased intermolecular hydrogen bonding of its low hydrophobic and flexible random coil structure (Dalgleish, 1989). Electrostatic interactions can also play an important role. In general, edible coatings based on proteins may serve in many potential applications, such as for meat products, nuts, seafood, confectionary products, fruits and vegetables, and as ingredients for microencapsulation and controlled-release purposes (Nussinovitch, 2003).

10.3 OBTAINING A WELL-MATCHED COATING

The coating procedure involves wetting a surface with a coating gum solution, followed by the solution's possible penetration (Hershko et al., 1998), and potential adhesion between the two commodities. The wetting stage is the shortest and most significant; if the solution used for spreading is suited to the food that is being coated, spreadability is spontaneous (Mittal, 1997). Nevertheless, it is nearly impossible to find gum solutions (or their combinations) that are perfectly suited to the surface properties of a particular object, i.e., in terms of surface tension and polarity. As a result, the closest likely combination should be sought to obtain compatibility. Creating a successful coating that adheres to a food surface requires an estimation of the interfacial tension between the coating solution and the surface. This depends on the surface tension of the surface, the surface tension of the coating solution and the contact angle between the two (Adamson, 1976). To estimate a solid's surface tension, the critical surface tension needs to be calculated. Prior to this, the critical surface tension of the surface to be coated should be derived from Zisman plots followed by extrapolation. These plots are obtained by calculating the cosine of each measured contact angle of the pre-chosen liquid on this surface and plotting it against the already-known surface-tension values of the solvents being used (Dann, 1970; Barton, 1983).

Finding an appropriate coating solution can be quite a challenge. Coating solutions are water-based, i.e. they include mostly water with a surface tension of 72.8 dyne cm^{-1}. For many solid surfaces, lower surface-tension values must be taken into account. Hydrocolloids, as a general rule, have the potential to lower the surface tension of solutions designated for use as coating agents (Gaonkar, 1991). The lower the surface or interfacial tensions of a gum solution, the higher its surface or interfacial activity (Gaonkar, 1991). Bearing in mind that competent coating involves compatibility between liquid and solid surface tensions (Wu and Salunkhe, 1972), a logical step is to reduce the surface tension of the coating solution to conform to the lower surface tension of the food surface, thereby lowering the interfacial tension and improving adhesion (Oliver and Mason, 1977). In previous studies (Nussinovitch et al., 1994), it was demonstrated that the addition of sterols effectively yields better adhesion between fruit and vegetable surfaces and coating films due to better compatibility with respect to the hydrophobicity of the two adhering surfaces (Nussinovitch and Hershko, 1996). Spreadability is an additional feature to be considered. It is enhanced by a food's surface roughness for coating solutions having contact angles smaller than 90° (Hershko and Nussinovitch, 1998), and inhibited by rough surfaces for solutions with contact angles greater than 90°. Surface roughness has been defined as the ratio of the true area of the solid to the apparent area. Wenzel was the first to propose a relationship between the contact angle of a liquid on a rough surface and its contact angle on an ideally smooth surface (Wenzel, 1936). Surface roughness can be evaluated with a roughness tester (Ward and Nussinovitch, 1996) or by image-processing of atomic force microscopy micrographs (Hershko et al., 1998). Roughness decreases the interfacial tension as a result of improved spreadability (Oliver and Mason, 1977). The importance of the magnitude of the interfacial tension is well recognized by the polymer coating industry. Better compatibility between the coated object and the coating film can be achieved by incorporating surface-active agents within the coating gum solution. It can be concluded from the compatibility requirements that tailor-made hydrocolloid coatings for different food materials can be achieved only by further determination of the chemical and physical properties of the coating solutions and the objects to be coated.

10.4 FILM-APPLICATION STAGES AND METHODS FOR TESTING FILMS

Films can be applied by dipping or spraying (Krochta et al., 1994). Brushes, falling-film enrobing technique, panning or rollers can also be used to apply films to the surfaces of the coated objects (Guilbert, 1986). Immersion of fresh produce in a gum solution takes between 15 and 120 s for a complete coating, the duration depending on wettability, concentration and viscosity of the hydrocolloid solution, surface roughness of the biological specimen, and possible penetration of the coating solution into the specimen (Hershko et al., 1996). In general, foods to be coated are dipped in a hydrocolloid solution followed by draining and drying. A coating solution can be dried to obtain a dry film adhering to the food surface; thus a film that never passes through a gel state is formed. An additional alternative is to use gum solutions which should be cross-linked before drying. In this case, a second immersion of the hydrocolloid-coated food in a cross-linking bath to induce gel formation (i.e., in alginate, LMP, κ- or λ-carrageenan or gellan) takes between 30 s and 2 min, depending on the concentration and temperature of the cross-linking agent, the thickness of the coating gum solution and the geometric complexity of the coated object (Nussinovitch, 2003). The third stage of manufacturing a cross-linked coating is the continuous strengthening of the gel coating layer at high relative humidity. The fourth stage, namely drying, can result in different dry-film textures and structures, depending on the length of drying. The texture and structure of the dried film will vary according to the time required for the gelled film layer to dry. Properties of the dried films also depend on properties of the drying equipment and on thickness and composition of the coating film (Nussinovitch, 2003).

There are many methods to evaluate coatings' properties. Sometimes films need to be produced by casting to obtain appropriate specimens for testing. Gas permeability of packaging films can be monitored in several ways (Stern et al., 1964; Karel, 1975; Aydt et al., 1991). Many devices for measuring film permeability to oxygen are commercially available (Landroac and Proctor, 1952; Quast and Karel, 1972). Water-vapor transmission rates through dried coatings can be determined by ASTM E96-93 (standard testing methods for water-vapor transmission of materials used for gums destined for coatings and adhesives).

Peel testing, i.e. the force necessary to peel off the coating, is used to estimate the film's degree of adhesion to a surface (Figure 10.1). The coating is peeled at 90° from the substrate, and the adhesion strength is estimated by the force per unit width necessary to peel the coating. It is important to study surface wetting and adhesion properties of coated commodities to obtain 'good' coatings. The roughness of film surfaces can be studied using a roughness tester, and electron and atomic force microscopy (the latter being used for finer mapping of surface roughness). An important parameter is Ra, the arithmetic mean of the absolute values of the roughness profile's deviation from the center line within the length being evaluated. The surface tension of gelling and inducing solutions, and their contact angles on food and other objects' surfaces, can be studied with surface-tension instruments (maximum adhesion requires a contact angle of 0°). It is important to note that 'if the coating does not

FIGURE 10.1 A coating is peeled at 90° from the substrate, and the adhesion strength is estimated by the force per unit width necessary to peel the coating.

spread spontaneously over the substrate surface, so that there is intermolecular contact between the substrate surface and the coating, there cannot be interactions and hence no contribution to adhesion' (Wicks et al., 1994).

10.5 SELECTING BIOPOLYMERS FOR SPECIFIC APPLICATIONS

Edible films and coatings should be chosen based on their suitability to the task at hand. As a general rule, if the aim is to retard moisture migration, lipid-based or composite films are chosen, i.e. films consisting of a combination of lipid and hydrocolloid components present in a bilayer or conglomerate (Greener and Fennema, 1994). To retard oil and fat migration, hydrocolloid-type films are chosen. For all other applications, i.e. to retard gas or solute migration, improve structural integrity or handling properties, retain volatile flavor components or convey food additives, hydrocolloid, lipid or composite combinations are chosen in accordance with their suitability (Greener and Fennema, 1994). Hydrocolloid films can be used when control of water-vapor migration is not the objective. Such films are good barriers to oxygen, carbon dioxide and lipids. Lipids, in addition to being a barrier against water vapor, might be added to increase the gloss of coated products (Greener and Fennema, 1994). Composite films combine the advantages of lipids and hydrocolloids while each compensates for the other's disadvantages (Greener and Fennema, 1989).

In the storage and marketing of fruits and vegetables, one must bear in mind that after harvest, these products are not 'static materials' – they consist of living tissue that needs to 'breathe', or it will undergo certain anaerobic reactions and, as a consequence, 'suffocate' (Nisperos-Carriedo et al., 1992). Fruits and vegetables use up oxygen and release carbon dioxide as they respire, and lose water (transpiration), the amount depending on temperature, gaseous makeup and humidity of the surrounding environment. To extend the shelf life of fruits and vegetables, hydrocolloids can be chosen to produce new types of coatings. Fruit and vegetable respiration is reduced by these films' selective permeabilities to oxygen and carbon dioxide, and thus the films serve as atmosphere modifiers (Nisperos-Carriedo et al., 1992). The formulation of such films can include a 'waxy' material to mimic the natural waxy coating of produce, and to give them a shiny appearance. Nevertheless, the emphasis should be on controlling gas exchange and creating a modified atmosphere inside the fruit that delays ripening and senescence, similar to the more costly practice of maintaining a controlled atmosphere (CA) (Nisperos-Carriedo et al., 1992). Wax coatings were developed to mimic the natural coating of fruits and vegetables. However, these coatings inhibited respiratory gas exchange to such an extent that fermentation was induced. As a result, ethanol build-up was detected along with that of other volatiles, coinciding with fermented and bitter taste. At high levels, these volatiles are considered off-flavors and they reduce fruit quality (Nussinovitch, 2003). To solve this problem, disturbances need to be created in the ordered, regular structure of the traditional wax coating, thereby improving fruit respiration (Chen and Nussinovitch, 2000, 2001).

After processing, i.e., washing, sorting, trimming, peeling, slicing, coring, etc., the freshness quality of the produce changes (Shewfelt, 1987), because it remains biologically and physiologically active. Cutting a fruit or vegetable leads to tissue breakdown caused by enzymatic action, the formation of secondary metabolites, increased production of ethylene, increased respiration and changes in microbial flora. Approaches for the preservation of minimally processed products include storage at low temperatures, special preparation procedures, the use of additives, and atmospheric modification or control (Krochta et al., 1994). To minimize the undesirable changes in the processed product, coatings should be selected that are capable of forming an efficient barrier to moisture loss, exhibit selective permeability to gases, control migration of water-soluble solutes, and enable the incorporation of additives such as flavor, preservatives or coloring (Krochta et al., 1994). In practice, achieving such an ideal coating is not a simple matter, and one needs to define which factors are undesirable or problematic. In general, dry films made up of layers may swell, dissolve or disintegrate upon contact with fluids and these are therefore not appropriate (Guilbert, 1986). An emulsion coating might be appropriate, depending upon the stability of the preparation. Another factor to consider with cut surfaces is that because they are covered with fluid, the binding of lipid materials becomes problematic. In this case, biopolymers with functional groups for ionic cross-linking that include acetylated monoglycerides might be helpful. Combinations such as a caseinate/acetylated monoglyceride/alginate emulsion, or replacing the alginate with low methoxy pectin provide alternative options (Guilbert, 1986; Kester and Fennema, 1988).

Meat and meat products may suffer from shrinkage, microbial contamination and surface discoloration. For simply delaying moisture transport, a thin coating film (gel) produced from any of a variety of polysaccharides (i.e. alginate, carrageenan, pectin, starch, etc.) can be used successfully due to evaporation of water within the gel (Baker et al., 1994). If extended periods of storage are required, the hydrophobicity of the coating needs to be increased (Kester and Fennema, 1988). The same or different hydrocolloids can be used for such applications, but the formulation should include some lipids. These can be oils, mono-, di- and triglycerides, waxes or water-in-oil emulsions. If an antimicrobial agent needs to be incorporated into the coating, the water activity in the coating should not be high (i.e. $0.8 > a_W$), so as to avoid instability, but it should also not be lower than ~0.65, so that good permeability of the preservative can be achieved (Rico-Pena and Torres, 1991). Many food products may contain high oil content, such as nuts or fried products. To eliminate oxidative off-flavors, the coating should have low oxygen permeability (Swenson et al., 1953) and incorporation of an antioxidant in the coating is also recommended (Cosler, 1957, and see previous examples in this chapter). One general conclusion that can be drawn is that although edible films and coatings find uses in a variety of applications, the technical information on them is far from adequate, leaving the food scientist with the formidable task of developing a film for each food application (Greener and Fennema, 1994).

10.6 EDIBLE PROTECTIVE FILMS

10.6.1 Packaging Materials Fit for Human Consumption

Solutions, highly viscous suspensions or gels can be produced from gums (hydrocolloids), which are high-molecular-weight molecules (Nussinovitch, 1997a). Due to the hydrophilic nature of many gums, the coatings that are produced from them have limited moisture-barrier abilities. However, in gel form, when no drying is applied, the films can retard moisture loss by serving as sacrificing agents, as found, for example, in soft white-brined cheeses and meats (Kampf and Nussinovitch, 2000). Reviews on food-packaging materials based on natural polymers can be found elsewhere (McHugh, 1996; Cuq et al., 1997a; Nussinovitch, 1997b; Debeaufort et al., 1998; Lin and Zaho, 2007). The macromolecules found in edible films are mainly polysaccharides and proteins. Polysaccharide and protein films are fine gas barriers, but poor moisture barriers. Conversely, pure lipid films are good moisture barriers, but poor gas barriers. This is the main reason for the interest in developing composite edible coatings that include the virtues of each class of ingredients (McHugh, 1996).

Manufacturers of novel edible films need to take many issues into consideration. These include barrier stability, mechanical properties, simplicity of application, biodegradability (environmentally friendly vs. plastic), nontoxicity, safety and cost for the manufacturer and consumer, as well as organoleptic properties. Usually, coatings are tasteless; however they can be designed with a unique taste that then influences the taste of the final product (Debeaufort et al., 1998; Ghanbarzadeh et al., 2007). Biopolymers can be used as an alternative source in the production of edible coatings. However, these polymers are sensitive to temperature and relative humidity: solutions blending gelatin and polyvinyl alcohol (PVA) provide a way around this problem, at least with respect to humidity (Bergo et al., 2006). Biodegradable films obtained from chitosan and methylcellulose (MC) can also reduce environmental problems associated with synthetic packaging. However, the cost of biodegradable films is still prohibitive; the use of chitosan, a waste by-product of the fishing industry, may provide a good alternative (Pinotti et al., 2007). Biodegradable films which could serve as an environmentally friendly alternative to synthetic plastic packaging films can be prepared from kafirin, the prolamin protein of sorghum. The resultant inferior functional properties can be improved by adding tannic acid or sorghum-condensed tannins as modifying agents during casting (Emmambux et al., 2004). Vegetable starches, and fish-muscle protein prepared from blue marlin meat and even low-quality fish meat can also serve as a good source for biodegradable films (Bae et al., 2008). In the latter case, the bacterial population in the films can be dramatically reduced by the addition of polylysine (Hamaguchi et al., 2007). Natural biopolymers are suitable for the construction of different types of wrappings and films. The produced biodegradable packaging can help in controlling physiological, microbiological, and physicochemical changes in food products. This is accomplished by controlling mass transfer between the food product and the ambient atmosphere or between components in heterogeneous food products, and by modifying and controlling the food's surface conditions (pH, level of specific functional agents, and slow release of flavor compounds) (Guilbert et al., 1997). The material's characteristics (polysaccharide, protein, or lipid, plasticized or not, chemically modified or not, used alone or in combination) and manufacturing procedures (casting of a film-forming solution, thermoforming) must be tailored to each specific food product and its surroundings (relative humidity, temperature). A few possible uses of these materials might be the wrapping of fabricated foods, protection of fruits and vegetables by controlling maturation, protection of meat and fish, control of internal moisture transfer in pizzas, all of which hinge on the organoleptic, mechanical, and gas- and solute-barrier properties of the films (Guilbert et al., 1997).

10.6.2 Inclusion of Food Additives within Edible Coatings

Edible films can be used as carriers for a range of food additives, including antibrowning and antimicrobial agents, colorants, flavors, nutrients and spices (Pena and Torres, 1991; Wong et al., 1996; Li and Barth, 1998; Pranoto et al., 2005). Such additives could serve as a possible treatment for reducing the deleterious effects of minimal processing of fresh-cut fruit (Baldwin et al., 1996; Wong et al., 1996; Alzamora and Guerrero, 2003; Burt and Reinders, 2003; Pranoto et al., 2005). Different processing operations, including peeling, cutting and shredding, induce enzymatic browning, which influences quality but can be successfully controlled by sulfites. However, ever since the ban on the use of sulfites for fresh fruits and vegetables (FDA, 1987), a replacement has been urgently sought. Several alternative chemical compounds have been suggested as enzymatic browning inhibitors (Nicolas et al., 1994). Other common antibrowning agents include citric acid and oxyresveratrol. Ascorbic acid is extensively used to inhibit enzymatic browning of fruit: it reduces the o-quinones generated by action of the polyphenol oxidase enzymes back to their phenolic substrates (McEvily et al., 1992). Several thiol-containing compounds, such as cysteine, N-acetylcysteine, and reduced glutathione, have also been investigated as inhibitors of enzymatic browning (Oms-Oliu et al., 2006). These compounds react with quinones formed during the initial phase of the enzymatic browning reactions to yield colorless products or reduce o-quinones to o-diphenols (Richard et al., 1991). Antibrowning agents incorporated into edible coatings are aimed at minimizing the browning of vegetative tissue on cut and exposed surfaces (Havenainen, 1996). Consequently, the incorporation of antibrowning agents into edible films in minimally processed fruits was studied by several groups (Wong et al., 1994; Lee et al., 2003). Most antibrowning agents are hydrophilic compounds that can enhance water-vapor permeability and water loss when incorporated into edible coatings (Ayranci and Tunc, 2004). In a study on fresh-cut Fuji apples, edible films based on alginate proved to be good carriers for antibrowning agents (N-acetylcysteine and glutathione) (Rojas-Grau et al., 2007a). In this particular case, the edible coating was generally applied before the antibrowning agents such that the gel coating adhered to the fruit and the antibrowning agents were then incorporated in the dipping solution, which contained calcium for cross-linking and instant gelling of the coating (Rojas-Grau et al., 2007a).

Peeling, cutting or slicing of minimally processed fruits increases the product's functionality but induces wounding. Microorganisms present on the food surface may be involved in spoilage (Del Rosario and Beuchat, 1995; Thunberg et al., 2002). Therefore, incorporation of antimicrobial compounds into edible films provides a novel way of improving the safety and shelf life of ready-to-eat foods (Cagri et al., 2004). Frequently used antimicrobials include conventional preservatives such as benzoic and sorbic acids, bacteriocins (nisin and pediocin), and plant-derived secondary metabolites, such as essential oils and phytoalexins (Burt, 2004; Fenaroli, 1995). Essential oils have been evaluated for their ability to protect food against pathogenic bacteria in, for example, contaminated apple juice (Burt, 2004; Friedman et al., 2004; Raybaudi-Massilia et al., 2006). Essential oils can be added to edible films and coatings to modify flavor, aroma and/or odor, as well as to introduce antimicrobial properties (Cagri et al., 2004). The effects of malic acid and essential oils of cinnamon, palmarosa and lemon grass and their main active compounds as natural antimicrobial substances incorporated into an alginate-based edible coating on the shelf life and safety of fresh-cut 'Piel de Sapo' melon were investigated. Palmarosa oil incorporated at 0.3% into the coating appeared to be a promising preservation alternative for fresh-cut melon, since it had good acceptance by panelists, maintained the fruit's quality parameters, inhibited the growth of native flora and reduced the population of *Salmonella enteritidis* (Raybaudi-Massilia et al., 2008).

Twelve years ago, edible films made from fruit purees were developed and shown to be a promising tool for improving the quality and extending the shelf life of minimally processed fruit (McHugh et al., 1996; McHugh and Senesi, 2000; Salcini and Massantini, 2005). The inclusion of plant essential oils within coatings improved the shelf life of edible apple-puree films (Rojas-Grau et al., 2006). The effect of lemon grass, oregano oil and vanillin incorporated into the edible apple puree-alginate coatings on the shelf life of fresh-cut 'Fuji' apples was also investigated (Rupasinghe et al., 2006; Rojas-Grau et al., 2007a). In general, all antimicrobial coatings significantly inhibited the growth of psychrophilic aerobes, yeasts and molds. The antimicrobial effect of essential oils against *Listeria innocua* inoculated into apple pieces before coating was also proven (Rojas-Grau et al., 2007b). When antimicrobial agents such as benzoic acid, sorbic acid, propionic acid, lactic acid and nisin were incorporated into edible films, the coatings retarded surface growth of bacteria, yeasts, and molds on a wide range of products, including meats and cheeses (Cagri et al., 2004). Benzoic acid ($C_7H_6O_2$) included in edible coatings inhibits the growth of molds, yeasts and some bacteria. It was either added directly or created from reactions with its sodium, potassium or calcium salt. Skinless tilapia fillet shelf life was prolonged under refrigeration after being coated with a gelatin coating containing benzoic acid as an antimicrobial agent (Ou et al., 2002). Lipid materials such as wax, fatty acids, and neutral lipid resins and waxes are used to improve the water-barrier properties of biopolymer-based edible films. The water-vapor barrier properties of lipid-based biopolymer films are affected by the nature of the utilized lipid materials, the film structure, and factors such as temperature, vapor pressure or physical state of the water contacting the films (Rhim, 2004). In laminated MC/corn zein-fatty acid films, water-vapor permeability was reported to decrease with increasing chain length and concentration of the fatty acids (Park et al., 1994).

Another manuscript on composite films composed of fatty acids and soy protein isolate demonstrated that their physical properties are, to a great extent, subject to the type of fatty acid added and its concentration (Rhim et al., 1999). Many other studies reached the conclusion that lipid-based edible films and coatings can play an important role in the food industry by controlling the moisture-barrier properties of biopolymer-based edible films and coatings. Emulsifiers are extensively used in the food industry to improve texture and stability, among other features. Sucrose esters are manufactered by esterifying sucrose with edible fatty acids from palm oil. They are neutral in taste, odor and color, stable at high temperatures for short times, soluble in cold water, kosher, non-GMO (genetically modified organisms) and vegetarian. When sucrose esters served as part of the composition of gelatin films, they decreased the films' water-vapor permeability and tensile strength (Jongjareonrak et al., 2006). Films containing fatty acids and their sucrose esters exhibited superior water-vapor permeability relative to those containing only the fatty acids (palmitic and stearic). Films that included fatty acids and their sucrose esters were generally more transparent than those with only fatty acids. The chain lengths of the fatty acids and their sucrose esters affected the properties of the films differently, depending upon the gelatin source (Jongjareonrak et al., 2006). Another manuscript demonstrated that sucrose esters have a large influence on the stabilization of emulsified edible-film structures containing arabinoxylans and hydrogenated palm kernel oil (The et al., 2002a). Sucrose esters also improved the moisture-barrier properties of these coatings. Both lipophilic (90% di- and triester) and hydrophilic (70% monoester) sucrose esters can ensure the stability of the emulsion used to form the film, especially during preparation and drying (The et al., 2002b).

The practice of adding probiotics to obtain functional edible films and coatings is in its infancy. Apparent health benefits and biological functions of bifidobacteria in humans include, among others, the intestinal manufacture of lactic and acetic acids, reduction of colon cancer risk, inhibition of pathogens, lessening of serum cholesterol, enhancement of calcium absorption, and activation of the immune system (Mitsuoka, 1990; Gibson and Roberfroid, 1995; Kim et al., 2002). A viable bifidobacterial population of 10^5 cfu g^{-1} in the final product has been specified as the therapeutic minimum needed to gain the aforementioned health advantages (Naidu et al., 1999). A recent study described the formulation of alginate- and gellan-based edible films containing viable bifidobacteria for coating fresh-cut fruits. Values above 10^6 cfu g^{-1} *Bifidobacterium lactis* Bb-12 were maintained for 10 days during refrigerated storage of these fresh-cut fruits, demonstrating the ability of alginate- and gellan-based edible coatings to carry and support viable probiotics on these items (Tapia et al., 2007).

10.6.3 Meat, Seafood and Fish Coatings

Meat is an animal tissue that is used as food. Most often it refers to the skeletal muscle and associated fat, but it may also include non-muscle organs, such as the lungs, liver, skin, brain, bone marrow and kidneys. The components of meat are muscle fiber, connective tissue, fat tissue, bone and pigment. The proportions of these ingredients affect the eating quality of the meat. Commonly eaten animal meats include beef, veal, lamb, pork, fowl, and less often, game animals (Gates, 1987). Over several hundred different species of seafood are consumed in the United States. Seafood may be divided into two general groups, fish and shellfish. Fish, as a food, include the edible parts of water-dwelling, cold-blooded vertebrates with gills. Other edible water-dwelling animals, such as mollusks and crustaceans, are often collectively referred to as shellfish. Crustaceans include crab, lobster and shrimp. Mollusks include abalone, clam, oyster and scallops. Crabs and lobsters are purchased fresh or cooked. Meat from cooked crustaceans is available chilled, frozen and canned (Gates, 1987).

Alginate-based coatings have been used to coat meats (Allen et al., 1963; Williams et al., 1978; Wanstedt et al., 1981; Glicksman, 1983; King, 1983) and are good oxygen barriers (Conca and Yang, 1993). Calcium-alginate coating of lamb carcasses helped reduce surface microbial growth and achieve a faster chill rate (Lazarus et al., 1976). Alginates, carrageenans, cellulose ethers, pectin and starch derivatives have been used to improve stored meat quality. Such coatings serve as sacrificing agents, i.e. moisture loss is delayed until it evaporates from the film (Kester and Fennema, 1988). Meat, poultry and seafood coated with calcium alginate exhibit reduced shrinkage, and decreased oxidative rancidity, moisture migration and oil absorption during processing (Baker et al., 1994). Alginate coatings have been used to retard the development of oxidative off-flavors in precooked meat patties (Kester and Fennema, 1988). In a study of the physicochemical, microbiological and sensorial qualities of cooked pork patty coated with pectin-based material containing green tea leaf extract powder (Kang et al., 2007), the coated patties had higher moisture contents than the controls in both air- and vacuum-packaging. The numbers of total aerobic bacteria were significantly reduced by the coating treatments as well as by irradiation. No difference was detected in sensory characteristics due to gamma irradiation or coating treatments (Kang et al., 2007). Combined treatments (irradation and coating) for extending the shelf life of meat products are an option. For example, moist beef biltong strips were inoculated with *Staphylococcus aureus* or sprayed with distilled water (non-inoculated controls). Both the non-inoculated and inoculated biltong strips were coated with an edible casein-whey protein coating followed by irradiation to a target dose of 4 kGy. This level of irradiation effectively ensured the safety of moist beef biltong and, providing the initial fungal counts are not excessive, may extend the food's shelf life. The edible coating had no significant effect on microbial counts, possibly because the high moisture content of the biltong diminished the coating's oxygen-barrier properties (Nortje et al., 2006). Another combined treatment includes using high pressure plus an edible coating to inhibit microbial growth and prevent oxidation. Coating muscle with gelatin-based films enriched with oregano or rosemary extracts increased the phenol content and the antioxidant power of muscle, particularly when used in association with high pressure, due to migration of antioxidant substances from the film (Gomez-Estaca et al., 2007). The edible films with the included plant extracts lowered lipid-oxidation levels (Gomez-Estaca et al., 2007). The extracts served as antioxidants, i.e., substances that terminate chain reactions by removing free radical intermediates and inhibit other oxidation reactions by being oxidized themselves (German, 1999), since they contained reducing agents (i.e. thiols, polyphenols, etc.). This finding was not surprising since plants maintain complex systems of multiple types of antioxidants, such as glutathione, vitamin C, and vitamin E, as well as enzymes, for instance various peroxidases (Beecher, 2003; Ortega, 2006). In comparison, the gelatin-chitosan-based edible films reduced microbial growth (total count) more successfully, due to the antimicrobial activity of chitosan. This activity varies considerably with the type of chitosan, the target organism and the environment in which it is applied (Allan and Hadwiger, 1979). The combination of high pressure and edible films yielded the best results in terms of both preventing oxidation and inhibiting microbial growth (Gomez-Estaca et al., 2007).

The shelf lives of frozen shrimp, fish and sausages are extended by the use of alginate coatings (Earle and Snyder, 1966; Daniels, 1973). In Japan, an edible polysaccharide film (Soafil) has found wide use in the meat industry as a casing for processed smoked meats. Carrageenan coatings have also been used to reduce off-odor development in chicken carcasses (Nussinovitch, 2003). Carrageenan coatings with soluble antibiotics have been found to be effective spoilage-retarding agents (Pearce and Lavers, 1949; Nussinovitch, 1997a). Agar coatings with water-soluble antibiotics were found to effectively extend the shelf life of poultry parts (Meyer et al., 1959). Agar coatings that included nisin (a bacteriocin) reduced contamination of *Salmonella typhimurium* in fresh poultry stored at 4°C (Natrajan and Sheldon, 1995). The effect of incorporating the additives trisodium phosphate into pea starch and acidified sodium chlorite into calcium alginate on their antimicrobial activity was studied against a three-strain cocktail of *Salmonella* inoculated on chicken skin. Coatings with 0.5% pea starch were absorbed quickly by the chicken skin and showed high skin adhesion, whereas those with >3.5% pea starch exhibited low skin adhesion and slow absorption (Mehyar et al., 2007). Alginate coatings with or without acidified sodium chlorite were stable, and about 50% of the coating weight was retained at 120 h. The latter coatings appeared to have low absorptiveness because the skin gained approximately 1.0% of its weight within 60 min of application. These findings indicate that effects of the agents in coatings on skin pH, the extent of coating adhesion, and absorption may contribute to overall antimicrobial behaviors (Mehyar et al., 2007).

Antimicrobial packaging materials could offer a potential alternative solution to preventing the spread of spoilage and pathogenic microorganisms via meat foodstuffs. Instead of mixing antimicrobial compounds directly with the food, incorporating them in films allows them to have their functional effect at the food surface – where most microbial growth is found (Lee et al., 2003; Coma, 2008). Antimicrobial packaging includes systems such as adding a sachet to the package, dispersing bioactive agents within the packaging itself, coating bioactive agents on the surface of the packaging material, or utilizing antimicrobial macromolecules with film-forming properties or edible matrices (Coma, 2008). Chitosan has film-forming abilities as well as a certain amount of antimicrobial activity, and therefore can serve as a potential and promising natural food-coating material. It was applied to improve the quality of various foods of agricultural, poultry and seafood origin (No et al., 2007). Chitosan films are clear, tough, flexible and good oxygen barriers (Sanford, 1989; Kaplan et al., 1993). Chitosan has been used for the preparation of safely and stably coated mutton kebabs and streaky bacon (Rao et al., 2005). The use of chitosan as an edible film was also evaluated for its antimicrobial activity against *Listeria monocytogenes* on the surface of ready-to-eat roast beef. An acetic acid–chitosan coating was more effective at reducing *L. monocytogenes* counts than a lactic acid–chitosan coating (Beverly et al., 2008). This is not surprising since on an equimolar basis, acetic acid generally exhibits higher antimicrobial activity than other organic acids (Sorrells et al., 1989; Conner et al., 1990; Ray and Sandine, 1992). Coating of pink salmon fillets with chitosan was effective at reducing relative moisture loss compared to control non-coated fillets. Chitosan postponed lipid oxidation, and there were no significant effects of coating on the color parameters or whiteness values of the cooked fillets after 3 months of frozen storage (Sathivel, 2005). Chitosan film was also used for glazing skinless pink salmon fillets. Chitosan glazing delayed lipid oxidation in the salmon fillets after 8 months of frozen storage (Sathivel et al., 2007). Chitosan is also beneficial in the development of coatings for the shelf-life extension of fresh Atlantic cod and herring fillets. A significant reduction in relative moisture loss was also observed with cod samples coated with chitosan after up to 12 days of storage. Chitosan coating also significantly reduced chemical spoilage due to lipid oxidation and growth of microorganisms (Jeon et al., 2002). In general, these observations can be explained by the unique properties of chitosan in comparison to other hydrocolloids. Chitosan has a positive charge in acidic solutions, due to the presence of primary amines on the molecule that bind protons (Kubota and Kikuchi, 1998). Being cationic, chitosan has the potential to bind to many different food components, such as pectins, proteins and inorganic polyelectrolytes (e.g. polyphosphate), allowing it to serve as a surface coating on meat products and fruits, or as an additive in acidic foods (Kubota and Kikuchi, 1998). Another unique feature of chitosan is its natural antimicrobial activity against a wide range of target organisms. Yeasts and molds are the most sensitive group, followed by Gram-positive and then Gram-negative bacteria (Ralston et al., 1964; Kendra and Hadwiger, 1984). Therefore, it is not surprising that at the experimental level, there are numerous reports describing the use of chitosan as a food preservative (El Ghaouth et al., 1991; Zhang and Quantick, 1998; Sagoo et al., 2002). Ferulic acid-incorporated films made from a starch–chitosan blend can find possible application as edible films or coatings. The barrier properties of the film blend improved considerably upon incorporation of oxidized ferulic acid. The amorphous nature of the ferulic acid-containing film blend supported good miscibility of the components, and Fourier transform infrared spectroscopy studies indicated intermolecular interactions between the different components. The ferulic acid-incorporated films were also found to reduce lipid-peroxide formation (Mathew and Abraham, 2008). An edible coating composed of whey proteins and acetylated monoglycerides protected smoked salmon fillets against dehydration and lipid oxidation. A multicomponent coating comprised of collagen, casein and cellulose derivatives reduced oil absorption during frying. Microbial proliferation, in particular *Listeria monocytogenes* growth, was reduced by coating fish products with a formulation of hydrocolloids, organic acids (lactic and acetic acids) and antimicrobial compounds (Sensidoni and Peressini, 1997). In conclusion, edible films can have many potential applications in meat and fish processing, with their functional characteristics depending on their constituents.

10.6.4 Edible Coatings for Fruits and Vegetables

A seed is the product of the ripened ovule of gymnosperm and angiosperm plants which develops after fertilization and some growth in the mother plant. A major use of hydrocolloids in agriculture is as a seed coating. The seed surface is coated with a hydrophilic polymer, which can absorb water after planting, thereby increasing the rates, as well as the probabilities of germination. Since coating composition and manufacture can be controlled, coatings can be designed to delay germination, inhibit rot, control pests, fertilize or bind the seed to the soil. Most of the reports on seed coatings deal with alginate coatings; reports on agar, various water-soluble cellulose ethers and hydrolyzed starch–polyacrylonitrile copolymers can also be found (Ferraris, 1989). Soybean, cotton, corn, sorghum, sugar beet and different vegetable seeds have been coated. As a result of coating, cotton and soybean yields increased by 20–30%. The poorer the growing conditions, the more pronounced the advantage of the coated seeds over controls (Mark et al., 1985; Nussinovitch, 1998).

A fruit is the ripened ovary – together with the seeds – of a flowering plant. The definition of a vegetable is traditional rather than scientific, and is somewhat arbitrary and subjective. Fresh fruits include 75–90% water. With the exception of a few fruits, their fat content is generally low. Fruits also include carbohydrates, minerals, acids, enzymes, pigments, aromatic compounds and vitamins (Gates, 1987). About 25% of freshly harvested fruits and vegetables are lost through spoilage (Wills et al., 1981). Post-harvest shelf-life extension can be achieved by applying edible coatings, which are semi-permeable to water vapor and gases. Such coatings can enhance or replace other techniques used for the same purpose, such as modified-atmosphere or controlled-atmosphere storage (Smith and Stow, 1984; Barkai-Golan, 1990). Other achievements of coating applications include improvement of mechanical handling properties and retention of volatile flavor compounds (Mellenthin et al., 1982; Nisperos-Carriedo and Baldwin, 1988). As a result of the coating, the permeability to oxygen and carbon dioxide changes and the coated commodity becomes an individual package with a modified atmosphere (Nussinovitch, 2003). Respiration of the fruit or vegetable causes a reduction in oxygen and an increase in carbon dioxide. Therefore, care must be taken in designing the coating: if oxygen levels become too low, anaerobic reactions will occur, resulting in off-flavors and abnormal ripening (Kedar, 1986). Ethanol and acetaldehyde concentrations in the tissue can be used as monitors for final and next-to-final products of anaerobic respiration (Kader, 1986). Therefore, if the coating can create moderate modified-atmosphere conditions, a climacteric fruit will exhibit decreased respiration, lower ethylene production, slower ripening and extended shelf life (Nussinovitch, 2003).

The scientific literature is replete with examples of fruit and vegetable coatings. Therefore, the rest of this section is devoted to a few examples involving different gums and other ingredients, as well as a few commodities. Sodium alginate is the gum of choice for many experimentalists due to its easy dissolution and simple cross-linking reaction. Sodium-alginate coatings postponed the drying of mushroom tissue, thereby preventing changes in its texture during short periods of storage. As a result of the coating, fresh mushrooms had a better appearance and gloss, as did garlic (Nussinovitch and Kampf, 1992, 1993; Nussinovitch and Hershko, 1996). Alginate or zein were used as edible coatings to delay post-harvest ripening and to maintain tomato (*Solanum lycopersicon* Mill.) quality (Zapata et al., 2008). Coated tomatoes showed lower respiration rate and ethylene production than controls, with a twofold lower concentration of ethylene precursor. In addition, the evolution of parameters related to losses in tomato quality, such as softening, color evolution and weight loss, was significantly delayed in coated tomatoes as compared to controls (Zapata et al., 2008). In peaches, sodium-alginate and MC coatings contributed to reductions in respiration rate and moisture loss, as well as to other changes relative to non-coated peaches. The sodium-alginate- and MC-coated samples maintained their acceptability up to 21 and 24 days, respectively, compared to 15 days for controls. These observations are the result of successful attempts to create modified atmospheres inside the coated fruit. The polysaccharide coatings retard gas migration, and delay ripening and senescence as effectively as the more costly controlled-atmosphere environments (in which specific levels of gases are maintained and regulated by external sources) (Kader, 1986). In general, environmental levels of oxygen below 8% decrease ethylene production, and carbon dioxide levels above 5% prevent or delay many fruit-tissue responses to ethylene, including ripening (Kader, 1986). Further discussion on post-harvest physiology and effects of atmosphere and temperature can be found elsewhere (Hulme, 1970), and the reader is also referred to the section 'Selecting biopolymers for specific applications'.

The reduced respiration and transpiration rates as a result of the coatings were considered responsible for maintaining peach quality and increasing their shelf life (Maftoonazad et al., 2008). Incorporation of ingredients found naturally in garlic skin (Figure 10.2) to the hydrocolloid (alginate- or gellan-based) gum solution before coating improved film adhesion to the surface of the coated commodity (Nussinovitch et al., 1994). This was due to a reduction in the surface tension of the coating solution, and its improved wettability (Wu, 1973). The effect of alginate- and gellan-based edible coatings on the shelf life of fresh-cut Fuji apples packed in trays with a plastic film of known permeability to oxygen was investigated by measuring changes in headspace atmosphere, color, firmness and microbial growth during storage at 4°C. Ethylene concentration in coated apples appeared to be delayed, while production of this gas was detected in uncoated apples from the very first day of storage. The coating included N-acetylcysteine (as an antibrowning agent), which helped maintain the firmness and color of the apple wedges throughout the storage period. Application of the edible coatings also retarded the microbiological deterioration of the fresh-cut apples, and effectively prolonged their shelf life by 2 weeks in storage (Rojas-Grau et al., 2008). In parallel to the use of alginate and gellan or other familiar gums, the coating industry is trying to locate additional gums and/or raw materials, and less frequently, wastes or by-products from the food industry for use in future coatings. The effect of a whey

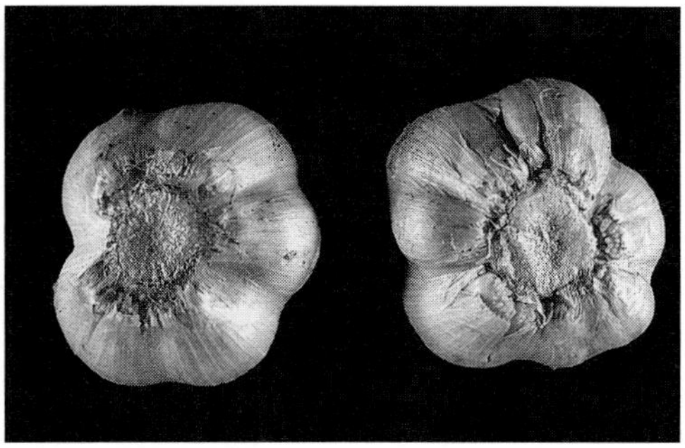

FIGURE 10.2 Alginate-coated garlic (left) versus non-coated garlic (right). The thin coating (tens of microns) glued onto the garlic epidermis is strong and transparent. It eliminates the initial opening of the cloves seen in the non-coated commodity (courtesy of V. Hershko).

protein isolate–pullulan coating on fresh-roasted chestnut and roasted freeze-dried chestnut quality and shelf life was studied. Coatings were formed directly on the fruit surface, giving good adherence and coverage. The coating had a low, albeit significant effect on reducing moisture loss and decay incidence. The quality of the harvested chestnuts was improved and their shelf life increased (Gounga et al., 2008). The effects of cassava starch films and polyvinyl chloride (PVC) on the maintenance of post-harvest quality of bell peppers stored at room temperature were evaluated. Fruit firmness and pH decreased while titratable acidity and soluble solids increased towards the end of storage at room temperature. The treatments did not promote significant changes in total pectin content during the storage period, although lower soluble pectin content was observed in the fruits covered with PVC (Hojo et al., 2007). The application of gelatin–starch coatings delayed the ripening process of avocados, as indicated by better pulp firmness and retention of skin color, and lower weight loss of coated fruits in comparison with control avocados. The coatings also produced a delayed respiratory climacteric pattern (Aguilar-Mendez et al., 2008).

Fruits and vegetables can be bought whole or sliced. The use of alginate or gellan-based coating formulations on fresh-cut papaya pieces was studied to determine their ability to improve resistance to water vapor, to affect gas exchange and to carry agents to help maintain the overall quality of the minimally processed fruit. Formulations containing glycerol plus ascorbic acid exhibited slightly improved water-barrier properties relative to the uncoated samples (Tapia et al., 2008). The incorporation of sunflower oil into the formulations resulted in increased water-vapor resistance of the coated samples. In general, coatings improved firmness of the fresh-cut product during the period studied (Tapia et al., 2008). Water loss and browning of cut apple slices were also inhibited by coatings of chitosan and lauric acid (Pennisi, 1992). Treatments to inhibit browning and decay and to extend the shelf life of fresh-cut 'Keitt', 'Kent' and 'Ataulfo' mango cultivars were also investigated. Combinations of calcium chloride, the antioxidants ascorbic acid and citric acid, and two commercial film coatings resulted in reduced browning and deterioration of fresh-cut mangoes stored at 5°C, especially for cv. Ataulfo (Gonzalez-Aguilar et al., 2008). A chitosan derivative used as a post-harvest edible coating for fresh foods selectively forms permeable non-toxic films (Meheriuk and Lau, 1988; Meheriuk, 1990; El Ghaouth et al., 1991). Synthetic polymers can modify the permeation response of a chitosan membrane to O_2 and CO_2 (Bai et al., 1988). Extracellular microbial polysaccharides such as pullulan can be used to produce edible and biodegradable films which are clear, odorless and tasteless and act as efficient oxygen barriers (Kaplan et al., 1993). Transparent water-soluble films, with low permeability to oxygen, can be produced from hydroxypropylated amylomaize starch. The produced coatings exhibit increased bursting strength and elongation, and reduced tensile strength (Roth and Mehltretter, 1967). Other components have served to develop coatings of this nature for prunes, raisins, dates, figs, nuts and beans. Other starch hydrolysate formulations were developed by the National Starch and Chemical Co (West Bridgewater, NJ) to coat dried apricots, almonds and apple slices (Murray and Luft, 1973). A novel approach to developing coatings composed of fruit pulp is currently being developed (McHugh et al., 1996). One example of this is a mango film that provides a good oxygen barrier with sufficient mechanical properties to wrap whole and minimally processed mangoes. This film was found to reduce weight loss and extend the ripening period and shelf life of whole fresh mangoes (Sothornvit and Rodsamran, 2008).

Composite coatings of carboxymethylcellulose (CMC) with fatty-acid ester emulsifiers have been used for pears and bananas (Ukai et al., 1975; Banks, 1984a, 1984b; Smith and Stow, 1984; Drake et al., 1987; Mitsubishi-Kasei Ryoto, 1989). Delayed ripening and changes in internal gas levels were detected (Banks, 1983, 1984a, 1984b). This type of marketable coating was first called Tal Pro-long (Courtaulds Group, London) and later, Pro-long. It increases resistance to some types of fungal rot in apples, pears and plums, but was not effective at decreasing respiration rate or water loss in tomato or sweet pepper (Lowings and Cutts, 1982; Nisperos-Carriedo and Baldwin, 1988). Valencia oranges coated with Tal Pro-long had improved essence and lower ethanol levels than controls (Nisperos-Carriedo et al., 1990). Another coating with a similar composition, Semperfresh (United Agriproducts, Greely, CO), contains a higher proportion of short-chain unsaturated fatty-acid esters in its formulation (Drake et al., 1987). These coatings retarded color advance and retained acids and firmness in apples relative to controls (Smith and Stow, 1984). Semperfresh also extended the storage life of citrus (Curtis, 1988). The addition of waxes to Semperfresh produces more glossy fruits with higher turgidity, less decay and good flavor. On the other hand, Semperfresh is not effective at retarding water loss in melons (Edwards and Blennerhassett, 1990). A novel approach to extending the shelf life of fruits is the use of special ingredients within the coating (see later). Aloe vera gel coating has been suggested as a means of preserving the quality and safety of cv. Crimson Seedless table grapes during cold storage and subsequent shelf life (Valverde et al., 2005). A similar approach for post-harvest treatment of sweet cherry has been applied. Sweet cherry treated with aloe vera gel exhibited significant delays in the following: increases in respiration rate, rapid weight loss and color changes, accelerated softening and ripening, stem browning and increased microbial populations, without any detrimental effect on taste, aroma or flavor (Martinez-Romero et al., 2006).

10.6.5 Coatings for Fried Products

Frying is used to alter the edible quality of a food, to thermally destroy microorganisms and enzymes, and to reduce water activity on the food's surface (Fellows, 1990). Fats and oils are the cooking mediums. As an outcome of the high temperatures involved, fried foods cook rapidly, producing a unique flavor and texture. Oil uptake depends, among many other things, on frying temperature. It can increase up to 20% as the frying temperature is decreased (Pedreschi et al., 2007), as also found in several other studies that looked at specific temperature ranges (Zeddelmann and Olendorf, 1979; Fan and Arce, 1986). However, increasing the frying temperature is not always beneficial, as frying time is independent of oil temperatures in the range of 155–200°C (Pravisani and Calvelo, 1986). Thus many other factors affect oil uptake and these need to be considered on a case-by-case basis (Selman and Hopkins, 1989). Since fried foods are high in calories relative to the same foods cooked in water or by other methods, the consistent demand for lower-calorie content of fried products can be satisfied, at least to a certain extent, by using coatings that can retard oil and fat migration (Gates, 1987). In addition to their properties as good film-formers, non-ionic cellulose ethers are capable of yielding tough and flexible transparent films owing to the linear structure of the polymer backbone. Such coatings are soluble in water and resistant to fats and oils (Krumel and Lindsay, 1976). Hydrocolloids that serve in the construction of such films are gellan and LMP (low-methoxy pectin), among many others. Gellan-based coatings have been used for several years in Japan and other Asian countries with tempura-type fried foods (Kelco, 1990). An edible coating system called Fry Shield, developed and patented by Kerry

Ingredients (Beloit, WI) and Hercules (Wilmington, DE), is based on calcium-reactive pectin and reduces fat intake during frying. French fries treated with pectin take up half the usual amount of oil (Nussinovitch, 2003).

MC and hydroxypropylmethylcellulose (HPMC) films are effective at reducing oil absorption by French fries, onion rings, and other fried, processed products (Sanderson, 1981) as a consequence of their resistance to fat and oil migration. The effect of an edible MC coating on reduced oil uptake during frying was analyzed on a dough system. The reduction in oil uptake was 30% for coated dough discs compared to non-coated ones; the coating did not modify either the water content of the samples or the quality attributes of the fried dough, such as color and texture. The coating reduced oil uptake, modifying the wetting properties relative to the interfacial tension and also becoming a mechanical barrier to lipids (Suarez et al., 2008). In addition to decreasing oil penetration and absorption by dry foods, the cellulose films can reduce weight loss and improve the adhesion of batter to products (Nussinovitch, 2003). The effect of type, molecular weight, and concentration of cellulose ethers changed the microstructure of fried batter-coated potatoes (Naruenartwongsakul et al., 2008). For controlled-viscosity batters, the structure of fried batter containing MC of higher molecular weight and concentration with simultaneously higher moisture content showed greater hole sizes, which allowed a larger amount of oil penetration through the batter into the food substrate (Naruenartwongsakul et al., 2008). In contrast, the structure of batters with controlled initial moisture content with a higher molecular weight and concentration of MC was relatively more continuous; therefore, the batter helps in preventing oil penetration into the food substrate (Naruenartwongsakul et al., 2008).

In addition to decreasing the oil content of a fried product, MC can be utilized to change its color. As an example, Akara, which is a popular West African fried food, is prepared from a cowpea flour and soy flour blend. Application of an edible coating caused the product to absorb 26% less oil during frying than akara made from 100% cowpea flour. MC-coated akara was also found to be significantly different in total color from the control (Huse et al., 2006): the 6% soy/MC-coated akara was significantly darker than the control. Akara formulated with 6% soy and coated with corn zein or MC was significantly more yellow and less red, respectively, than the control. These changes were due to the different oil absorption and reflectance of the coated surface of the fried product (Huse et al., 2006). Frying in a gaseous environment influences the crispness of the product. When MC or whey protein isolate were incorporated into either the pre-dust or batter before frying, fried chicken nuggets, using nitrogen gas, were crispier than those fried under steam (Ballard and Mallikarjunan, 2006). Combinations of HPMC/MC with corn flour, seasoning, and salt were used to coat chicken pieces before frying at 180°C for 10 min. The enrobing and frying time had a significant effect on the moisture and fat contents of the chicken pieces. The coating containing a 0.5% gum mixture was most acceptable from a sensory point of view. Enrobing increased the moisture content up to 42.6%, while fat uptake was reduced to 32% (Sudhakar et al., 2006).

10.6.6 Miscellaneous Coatings

Studies of edible-film properties have progressed appreciably in the last century; the results are expected to be applied to a wide variety of foods, as well as to other applications (Krochta and De Mulder-Johnston, 1997; Hershko and Nussinovitch, 1998). Edible films can be utilized to coat nuts and peanuts, confectionery products and lightly processed agricultural products, to control the migration of preservatives and dough additives, in composite bilayers, blends and biopolymer films, and for many biotechnological uses (Nussinovitch, 2003). More than 100 years ago (Figure 10.3), a very interesting type of edible coating was developed in Japan. 'Oblate' is the registered tradename for this edible paper which is prepared from starch and agar. It is 10–15 μm thick, and it is usually sold in drug stores in Japan for dosing medicines in powdered form. Oblate can also be employed as a convenient vehicle for wrapping condensed, sticky, sweet agar jelly: the non-sticky edible film coats the sticky confection and permits its consumption without having it stick to the fingers (Davidson, 1980). Gum arabic is used extensively in the coating industry due to the ease of preparation of solutions with over 50% gum, its compatibility with other plant hydrocolloids and its ability to produce stable emulsions with most oils over a wide range of pHs (Nussinovitch, 1997a). Gum arabic with or without gelatin has been used to produce protective films for chocolates, nuts, cheeses and pharmaceutical tablets (Colloides Naturels Inc., 1988; Mazza and Qi, 1991) and has also been reported to inhibit darkening of cooked potatoes (Mazza and Qi, 1991). A composite film of gum acacia and glycerol monostearate was reported to have good water-vapor barrier properties (Martin-Polo and Voilley, 1990). A carrageenan-based coating, Soageena (edible polysaccharide), was developed by Mitsubishi International Corp. (Tokyo, Japan) for fresh produce (IFT, 1991; Baldwin, 1994). Other carrageenan coatings have been used to retard moisture loss from coated foods (Torres et al., 1985). A carrageenan-based coating applied to cut grapefruit halves successfully decreased shrinkage and taste deterioration (Bryn, 1972). Carrageenan, gellan and alginate have been used to coat fresh, soft-brined cheeses (Figure 10.4) (Kampf and Nussinovitch, 2000). Similar to coated meats, these hydrocolloid coatings offer only a weak barrier against moisture transport. However, the water contained within the different coatings serves as a sacrificing agent, and loss of product moisture is delayed until this liquid has been evaporated (Kampf and Nussinovitch, 2000).

Chemically modified celluloses can be used to manufacture edible and biodegradable films. MC, HPMC, hydroxypropyl cellulose (HPC) and CMC are examples of raw materials for the production of coatings with moderate strength, resistance to oils and fats, elasticity and transparency, without odor or taste, and the ability to serve as a moderate barrier to moisture and oxygen (Hagenmaier and Shaw, 1990; Nisperos-Carriedo, 1994). Information on MC and HPMC can be found elsewhere (Vojdani and Torres, 1990). MC films make fine barriers to oil and fat migration (Nelson and

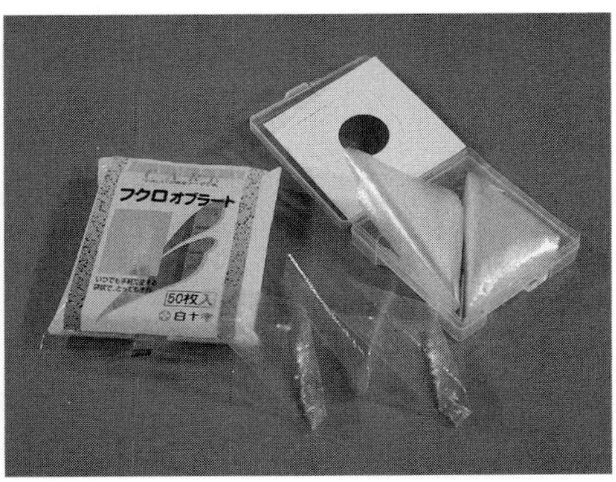

FIGURE 10.3 'Oblate' is a registered tradename for edible paper prepared from starch and agar. The product is generally sold in drug stores in Japan for dosing powdered medicines (courtesy of Madoka Hirashima (Mie University, Japan)).

FIGURE 10.4 Coated white-brined cheese (courtesy of Nir Kampf).

Fennema, 1991). MC and HPMC can be used to manufacture composite films with solid lipids (Koelsch and Labuza, 1991; Martin-Polo et al., 1992; Debeaufort et al., 1993; Park et al., 1994). Bilayer films consisting of a solid lipid and a layer of MC or HPMC have been used to decrease the migration of water in model foods (Rico-Pena and Torres, 1990). Non-ionic cellulose ethers are capable of producing tough, transparent and flexible films that are both water-soluble and fat- and oil-resistant (Krumel and Lindsay, 1976). MC films prevent lipid migration (Nussinovitch, 1994). HPMC and a bilayer film consisting of stearic-palmitic acid slowed moisture transfer from tomato paste to crackers (Kamper and Fennema, 1985). Formulations involving MC, HPMC and HPC delayed browning and increased volatile flavor components (Nisperos-Carriedo et al., 1992). HPC films retarded oxidative rancidity and moisture absorption in nut meats, coated nuts and candies (Ganz, 1969; Krumel and Lindsay, 1976).

Whey is the watery, fluid part of the milk that remains after the curd, or casein precipitate, has been removed. α-Lactalbumin and β-lactoglobulin (whey protein fractions) and whey protein isolate are used to manufacture edible films due to their beneficial functional properties and industrial surplus (Chen, 1995). Whey proteins produce transparent, flexible, colorless and odorless edible films which can serve as moderately good moisture barriers and excellent oxygen barriers (Fairley et al., 1996; Mate and Krochta, 1996). Incorporation of plasticizers (e.g. polyols and mono-, di- and oligosaccharides) reduces their brittleness. Whey–protein–isolate films have limited barrier properties which can be enhanced by the inclusion of other film-forming polysaccharides (Ciesla et al., 2006). Protein, sorbitol, beeswax and potassium sorbate were found to influence the water-vapor permeability and water solubility of whey protein films. Beeswax was the most important factor influencing the stickiness and appearance of the films. The amount of protein had no effect on stickiness or appearance, while the amount of sorbitol (35–50%, w/w) in the films had no influence on appearance (Ozdemir and Floros, 2008). Scanning electron microscopy demonstrated these films' favorable structure and the addition of pululan at low concentrations was sufficient to significantly modify their barrier properties and improve their potential characteristics for food applications (Gounga et al., 2007).

10.7 NOVEL PRODUCTS

Novel edible-coating products are ubiquitous in the marketplace. An example of these is Listerine PocketPaks®: small patches of edible film that melt instantly on the tongue, releasing breath freshener. PocketPaks contains at least four hydrocolloids: carrageenan, locust bean gum, pullulan and xanthan gum. Listerine PocketPaks pushed the dental-care category into new territory by being the first item of this kind available to active consumers. Its success has paved the way for many other such products. Strips could prove appealing as a carrier for drug delivery. In 2004, the Swiss pharmaceutical giant Novartis introduced Triaminic® and Theraflu® Thin Strips™. Thin Strips were the first multi-symptom cough and cold medicines to deliver an accurate dose in thin-film format and were selected from more than 100 applicants for their taste and quality ingredients, packaging, consumer acceptance, innovation, differentiation, and convenience (http://www.webpackaging.com; http://www.iqdurableink.com). Clarifoil, the producers of films for print and packaging uses, launched Clarisol, an innovative new range of water-soluble films: these are cast PVA/PVOH water-soluble films that are offered as consistent-quality products suitable for a wide range of functional uses with the following requirements: convenience, delivery of measured doses and easy handling for consumers and industrial operatives. They could potentially also have uses in specialty products. Pets may be the next target for these 'drugs-on-film', as edible films could serve as a convenient alternative to stuffing pills down pets' throats. Potentially, drugs could also be given at lower doses on films because they are better absorbed through the tongue.

The food industry has provided many new uses for edible-film technology. Meat producers are using films to cure and glaze hams. Athletes consume electrolyte strips in lieu of sports drinks to fight dehydration. Films may someday be used for separating the tomato sauce from the crust on a frozen pizza so that the crust stays crisp. Origami Foods, in cooperation with the USDA's Agricultural Research Service, has developed a new, unique food product in which almost any fruit, vegetable or their combination can be used to create edible films. Such products are low-fat, low-calorie, tasty and healthy, and were developed for people who have an aversion to seaweed, or are otherwise interested in an alternative (http://www.origami-foods.com; http://www.ceepackaging.com). Another development in the area of edible films relates to decorations. Today, it is possible to decorate cakes with computer-designed images. These images (sometimes clip-art) are printed with new high-quality food-grade inks on edible paper, using a standard inkjet printer. The attractive print-outs can be placed on any cake or other baked good. The same process can be used to produce designed candy wrappers, rolls of paper for pies or quiches, or decorative protective covers for condiments and tortes, and in hundreds of other applications where a cosmetic or personal touch is desired on a dessert product.

Novel products, developments and research directions that could be important in the field of coatings or that use coatings as part of their achievement have been developed in recent years. These developments include (but are not limited to) a novel composite film for potential food-packaging applications that was prepared by plasticization of protein coatings on polypropylene film (Lee et al., 2008). The optical and tensile properties of the films depend on the types of proteins and plasticizers used. High glossy surfaces were observed on films coated with whey protein isolate and corn zein, with the sucrose-plasticized whey protein isolate coating giving the highest gloss. Whey protein isolate-coated films also showed greater transparency and tensile strength than the other coated films. The additive nisin has an E number of 234 and is a polycyclic peptide antibacterial agent with 34 amino acid residues, used as a food preservative. It is a 'broad-spectrum' bacteriocin used in processed cheese, meat, beverages, etc., at levels ranging from ~1–25 ppm, depending on the food type and regulatory approval during production, in order to extend shelf life by suppressing Gram-positive spoilage and pathogenic bacteria (http://en.wikipedia.org/wiki/Nisin). Nisin-incorporated whey protein isolate coatings on polypropylene film exhibited significant growth inhibition of the bacterium *Lactobacillus plantarum* (Lee et al., 2008). Another product in the packaging realm is a novel, quickly dissolvable, edible and heat-sealable film consisting of a blend of konjac glucomannan and gelatin. This product was successfully prepared by using a solvent-casting technique with different blending ratios of the two polymers. Taking the degradability into account, this film blend might be a perfect material for edible inner packaging (Li et al., 2006).

A novel research approach makes use of alginate coating to monitor the uptake of solids during osmotic dehydration of a model food system (Lazarides et al., 2007). A combination of product coating with alternative settings of product/solution contact was investigated to monitor solids uptake during osmotic dehydration. Potato was used as a model plant material for short-term (i.e. 3 h) osmotic treatment in a series of solutions with decreasing or increasing concentrations of sucrose to simulate co-current or counter-current product/solution contact (flow) (Lazarides et al., 2007). The alginate coating yielded significantly decreased solids uptake, without negatively affecting water removal. Moreover, the overall, 'dehydration efficiency' was drastically improved by combined coating and counter-current contact. The importance of this approach is that a mass-exchange prediction can be applied under alternative treatment scenarios regarding initial product solids (Lazarides et al., 2007).

10.8 NON-FOOD GUM COATINGS

Paper is a thin material used mainly for writing on, printing on or packaging. It is produced by pressing together moist fibers, typically cellulose pulp derived from wood or grasses, and drying them into flexible sheets. Many hydrocolloids can be utilized to obtain successful paper coatings. Agar has been found suitable for use in photographic papers when esterified with succinic or phthalic anhydride and after enzymic hydrolysis. Agar can also be used as an adhesive in the gloss-finishing of paper products. HPC is used for coating, due to its thermoplasticity and solvent solubility. It serves as an oil and fat barrier and is responsible for thermoplastic coating. Polyethylene oxide is also used for paper coating and sizing. As a processing additive, it is used as a fiber-formation aid (Nussinovitch, 1998). Another manuscript discusses the associative behavior of CMCs, hydroxy ethyl celluloses and hydrophobically modified cellulosic thickeners in clay-based coatings and their effects on coating rheology and coated-paper properties (Young and Fu, 1991). PVA is used in paper sizings and coatings. Polyvinylpyrrolidone is used in all types of paper manufacture, mostly as an economical fluidizer and antiblocking agent in paper coating. The starches used in coating colors can be enzymatically converted, thermochemically converted or oxidized, to dextrins, hydroxyethyl starch ethers and starch acetates (Davidson, 1980).

Polyacrylic acid can be used as a film former. One important application is the use of the sodium salt of polyacrylic acid as the major component of non-glare coatings for headlights. Antifogging coatings for glass and transparent plastics for optical use have also been created by cross-linking polyacrylic acid with aminoplast resins to produce scratch- and water-resistant coatings (Davidson, 1980). In the manufacture of fluorescent lights, phosphor bonding to glass tubes is enhanced by using polyethylene oxide resin as a temporary binder in combination with barium nitrate as a bonding agent (Nussinovitch, 1997a). For the manufacture of coatings, paints, foams or adhesives, xanthan gum (a polysaccharide produced by a process involving fermentation of glucose or sucrose by the *Xanthomonas campestris* bacterium) is compatible with the common types of latex emulsions, making it effective as a stabilizer, thickener and modifier of rheological properties (Nussinovitch, 1997a).

10.9 NEXT GENERATION OF EDIBLE FILMS

A nanometer equals one billionth of a meter, and can be written as 1×10^{-9} m (Uldrich, 2003). Inclusion of nanoparticles in edible and nanocomposite films is expected to improve the mechanical and oxidative stability, barrier properties, and biodegradability of conventional polymeric matrices (Sorrentino et al., 2007). Four different types of chitosan-based nanocomposite films were prepared using a solvent-casting method by incorporation with four types of nanoparticles. A certain degree of intercalation occurred in the nanocomposite films, with the highest intercalation found in the unmodified montmorillonite-incorporated films, followed by films with organically modified montmorillonite and silver-zeolite (Rhim et al., 2006). In all nanocomposite films except those incorporating silver, nanoparticles were dispersed homogeneously throughout the chitosan polymer matrix. Consequently, mechanical and barrier properties of chitosan films were affected through intercalation of nanoparticles. In addition, chitosan-based nanocomposite films, especially silver-containing ones, showed a promising range of antimicrobial activities (Rhim et al., 2006).

An example from a different product area is double-core tennis balls with a nanocomposite coating that keeps them bouncing twice as long as the old-style balls and also substantially extends the ball's shelf life (Uldrich, 2003). To date, not many studies have been performed on the possibility of incorporating nanoparticles to improve the physical properties of biodegradable nanocomposites (Mangiacapra et al., 2005). Adding clay montmorillonite to pectins could lower the diffusion of oxygen (Mangiacapra et al., 2005). Similarly, a considerable improvement in physical properties was recorded for nanocomposites prepared with gelatin and montmorillonite (Lee and Fu, 2003; Zheng et al., 2002). Edible coatings and films constitute a viable means of incorporating food additives and other substances in order to enhance product color, flavor, and texture and to control microbial growth (Siragusa and Dickson, 1992). To this aim, nanoparticles can be used as carriers of antimicrobials and additives. Nanotechnology provides a number of ways to create novel laminate films to be used in the food industry. A nanolaminate consists of two or more layers of material with nanometer dimensions that are physically or chemically bonded to each other. A deposition technique can be utilized to coat a charged surface with interfacial films consisting of multiple nanolayers of different materials (Decher and Schlenoff, 2003). This technology may have a number of important applications in the food industry. Nanolaminates could be used as coatings that are attached to food surfaces, rather than as self-standing films (Kotov, 2003; Weiss et al., 2006). To produce these layers, a variety of different adsorbing substances can be used. The functionality of the final film will depend on many factors, including type of materials, number of layers, their sequences, preparation conditions, etc. Thus research and development of bio-nanocomposite materials for food applications such as packaging and other food contact surfaces is expected to grow in the next decade with the advent of new polymeric materials and composites with inorganic nanoparticles (Sorrentino et al., 2007).

An appreciable increase in stability in chitosan-layered nanocomposites has also been reported (Darder et al., 2003). Another publication described the inclusion of different-sized fillers, all less than 1.0 μm, in HPMC-based films, to improve the film's mechanical properties (Dogan and McHugh, 2007). Such results will enable food scientists to envision a new generation of composite edible films and barriers; they will no longer be restricted to emulsified and bilayer films for current and novel applications (Dogan and McHugh, 2007). The application of nanocomposites promises to expand the use of edible and biodegradable films (Lagaron et al., 2005; Sinharay and Bousmina, 2005) and to help reduce the packaging waste associated with processed foods (Labuza and Breene, 1988; Vermeiren et al., 1999). In addition, inorganic particles may be used to add multiple functionalities, such as color and odor, but also to act as reservoirs for the controlled-release functions of drugs or fungicides (Lee et al., 2003; Li et al., 2004). In conclusion, future research directions will make use of nanotechnological tools to improve tensile, barrier and delivery properties of edible films, to develop hydrophobic nanoparticles for improving water-barrier properties and to use nanoparticles to deliver health-promoting compounds, antimicrobials, sweetness, flavors and other active agents. It is expected that future edible films will include antimicrobial edible films, added-value fruit and vegetable films and invisible edible films, and that the tools of nanotechnology will be used to achieve developments in packaging and film-production technologies.

References

Aboagye, Y. and Stanley, D. W. (1985). Texturization of peanut proteins by surface film formation. Influence of process parameters on film forming properties. *Can. Inst. Food Sci. Technol. J.* **18**, 12–20.

Adamson, A. W. (1976). *In* "Physical Chemistry of Surfaces," 3rd edn., 1–43, Adamson, A. W. and Gast, A. P. (eds.) 3rd edn., John Wiley and Sons, New York.

Aguilar-Mendez, M. A., Martin-Martinez, E. S., Tomas, S. A., Cruz-Orea, A., and Jaime-Fonseca, M. R. (2008). Gelatine-starch films: physicochemical properties and their application in extending the post-harvest shelf life of avocado (*Persea americana*). *J. Sci. Food Agric* **88**, 185–193.

Allan, C. R. and Hadwiger, L. A. (1979). The fungicidal effect of chitosan on fungi of varying cell wall composition. *Exp. Mycol.* **3**, 285–287.

Allen, L., Nelson, A. I., Steinberg, M. P., and McGill, J. N. (1963). Edible corn-carbohydrate food coatings. Evaluation on fresh meat products. *Food Technol* **17**, 442–443.

Alzamora, S. M. and Guerrero, S. (2003). Plant antimicrobials combined with conventional preservatives for fruit products. *In* "Natural Antimicrobials for the Minimal Processing Foods," 235–249, Roller, S. (ed.). CRC Press, Boca Raton, FL.

Aydt, T. P., Weller, C. L., and Testin, R. F. (1991). Mechanical and barrier properties of edible corn and wheat protein films. *Am. Soc. Agric. Eng* **34**, 207–211.

Ayranci, E. and Tunc, S. (2004). The effect of edible coatings on water and vitamin C loss of apricots (*Armeniaca vulgaris* Lam.) and green peppers (*Capsicum annuum* L.). *Food Chem.* **87**, 339–342.

Bae, H. J., Cha, D. S., Whiteside, W. S., and Park, H. J. (2008). Film and pharmaceutical hard capsule formation properties of mungbean, water chestnut, and sweet potato starches. *Food Chem.* **106**, 96–105.

Bai, R., Huang, M., and Jiang, Y. (1988). Selective permeabilities of chitosan-acetic acid complex membrane and chitosan-polymer complex membranes for oxygen and carbon dioxide. *Polymer Bulletin* **20**, 83–88.

Baker, R. A., Baldwin, E. A., and Nisperos-Carriedo, M. O. (1994). Edible coatings and films for processed foods. *In* "Edible Coatings and Films to Improve Food Quality," 89–105, Krochta, J. M., Baldwin, E. A. and Nisperos-Carriedo, M. O. (eds.). Technomic Publishing Co., Basel, Switzerland.

Baldwin, E. A. (1994). Edible coatings for fresh fruits and vegetables: past, present and future. *In* "Edible Coatings and Films to Improve Food Quality," 25–65, Krochta, J., Baldwin, E. and Nisperos-Carriedo, M. (eds.). Technomic Publishing Co, Basel, Switzerland.

Baldwin, E. A., Nisperos, M. O., Chen, X., and Hagenmaier, R. D. (1996). Improving storage life of cut apple and potato with edible coating. *Postharvest Biol. Technol* **9**, 151–163.

Ballard, T. S. and Mallikarjunan, P. (2006). The effect of edible coatings and pressure frying using nitrogen gas on the quality of breaded fried chicken nuggets. *J. Food Sci.* **71**, 259–264.

Banks, N. H. (1983). Evaluation of methods for determining internal gases in banana fruit *Musa acuminata* x. *Musa ballisiana*. *J. Exp. Bot* **34**, 871–879.

Banks, N. H. (1984a). Some effects of Tal Pro-long coating on ripening bananas. *J. Exp. Bot.* **35**, 127–137.

Banks, N. H. (1984b). Studies of the banana fruit surface in relation to the effects of Tal Pro-long coating on gaseous exchange. *Scientific Hort.* **24**, 279–286.

Barkai-Golan, R. (1990). Postharvest disease suppression by atmospheric modifications. *In* "Food Preservation by Modified Atmospheres," 238–265, Calderon, M. and Barkai-Golan, R. (eds.). CRC Press, Boca Raton, FL.

Barton, A. F. M. (1983). "CRC Handbook of Solubility Parameters and Other Cohesion Parameters". CRC Press, Boca Raton, FL. 427, 441.

Beecher, G. (2003). Overview of dietary flavonoids: nomenclature, occurrence and intake. *J. Nutr* **133**, 3248–3254.

Bergo, P., Carvalho, R. A., Sorbal, P. J. A., and Bevilacqua, F. R. S. (2006). Microwave transmittance in gelatin-based films. *Measurement Sci. Technol* **17**, 3261–3264.

Beverly, R. L., Janes, M. E., Prinyawlwatkula, W., and No, H. K. (2008). Edible chitosan films on ready-to-eat roast beef for the control of *Listeria monocytogenes*. *Food Microbiol.* **5**, 534–537.

Boni, K.A. (1988). Strengthened edible collagen casing and method of preparing same. US Patent 4,794,006.

Brunner, J. R. (1977). *In* "Milk Proteins in Food Proteins," 175–208, Whitaker, J. R. and Tannenbaum, S. R. (eds.). Avi Publishers Inc., Westport, CT.

Bryn, D. S. (1972). US Patent 3,707,383.

Burt, S. (2004). Essential oils: their antibacterial properties and potential applications in foods—a review. *Int. J Food Microbiol.* **94**, 223–253.

Burt, S. A. and Reinders, R. D. (2003). Antibacterial activity of selected plant essential oils against *Escherichia coli* O157:H7. *Lett. Appl. Microbiol.* **36**, 162–167.

Cagri, A., Uspunol, Z., and Ryser, E. (2004). Antimicrobial edible films and coating. *J. Food Prot* **67**, 833–848.

Cheesman, D. F. and Davies, J. T. (1954). Physiochemical and biological aspects of proteins at interfaces. *Adv. Protein Chem.* **9**, 439–501.

Chen, H. (1995). Functional properties and applications of edible films made of milk proteins. *J. Dairy Sci.* **78**, 2563–2583.

Chen, S. and Nussinovitch, A. (2000). Galactomannans in disturbances of structured wax-hydrocolloid-based-coatings of citrus fruit (easy-peelers). *Food Hydrocolloids* **14**, 561–568.

Chen, S. and Nussinovitch, A. (2001). Permeability and roughness of wax-hydrocolloid coatings and their limitations in determining citrus fruit overall quality. *Food Hydrocolloids* **15**, 127–137.

Ciesla, K., Salmieri, S., and Lacroix, M. (2006). Modification of the properties of milk protein films by gamma radiation and polysaccharide addition. *J Sci. Food Agric* **86**, 908–914.

Circle, S. J., Meyer, E. W., and Whitney, R. W. (1964). Rheology of soy protein dispersions. Effect of heat and other factors on gelation. *Cereal Chem.* **41**, 157–172.

Colloides Naturels Inc. (1988). Sealgum: Something New in Films. Product Bulletin. Colloides Naturels Inc., Bridgewater, NJ.

Coma, V. (2008). Bioactive packaging technologies for extended shelf life of meat-based products. *Meat Science* **78**, 90–103.

Conca, K. R. and Yang, T. C. S. (1993). Edible food barrier coatings. *In* "Biodegradable Polymers and Packaging," 357–369, Ching, C., Kaplan, D. and Thomas, E. (eds.). Technomic Publishing Co., Inc., Lancaster, Pennsylvania.

Conner, D. E., Scott, V. N., and Bernard, D. T. (1990). Growth, inhibition, and survival of *Listena rnonocytogenes* as affected by acidic conditions. *J. Food Prot* **53**, 652–655.

Cosler, H.B. (1957). Methods of producing zein-coated confectionery. US Patent 2,791,509.

Cuq, B., Gontard, N., and Guilbert, S. (1997a). Packaging materials based on natural polymers. *Industries Alimentaires et Agricoles* **114**, 110–116.

Cuq, B., Gontard, N., and Guilbert, S. (1997b). Thermoplastic properties of fish myofibrillar proteins: application to biopackaging fabrication. *Polymer* **38**, 4071–4078.

Curtis, G. J. (1988). Some experiments with edible coatings on the long-term storage of citrus fruits. *In* proc. 6th Int. Citrus Congress 3, 1514–1520, Goren, R. and Mendel, K. (eds.). Balaban Publishers, Margraf Scientific Book, Tel Aviv, Israel.

Dalgleish, D. G. (1989). Milk proteins—chemistry and physics. *In* "Food Proteins," 155–178, Kinsella, J. E. and Soucie, W. G. (eds.). American Oil Chemists Society, Champagne, IL.

Daniels, R. (1973). "Edible Coatings and Soluble Packaging" Noyes Data Corp., Park Ridge, NJ. 360.

Dann, J. R. (1970). Forces involved in the adhesive process. Critical surface tensions of polymeric solids as determined with polar liquids. *J Colloid Interface Sci.* **32**, 302–319.

Darder, M., Colilla, M., and Ruiz-Hitzky, E. (2003). Biopolymer-clay nanocomposites based on chitosan intercalated in montmorillonite. *Chem. Mat* **15**, 3774–3780.

Davidson, R. L. (1980). Handbook of Water-Soluble Gums and Resins. McGraw-Hill Book Company, New York.

Debeaufort, F., Martin-Polo, M., and Voilley, A. (1993). Polarity homogeneity and structure affect water vapor permeability of model edible films. *J. Food Sci.* **58**, 426–434.

Debeaufort, F., Queada-Gallo, J. A., and Voilley, A. (1998). Edible films and coatings: tomorrow's packaging: a review. *Crit. Rev. Food Sci.* **38**, 299–313.

Decher, G. and Schlenoff, J. B. (2003). "Multilayer Thin Films: Sequential Assembly of Nano Composite Materials" Wiley-VCH, Weinheim, Germany, p. 543.

Del Rosario, B. A. and Beuchat, L. R. (1995). Survival and growth of enterohemorrhagic *Escherichia coli* O157:H7 in cantaloupe and watermelon. *J. Food Prot* **58**, 105–107.

Dogan, N. and McHugh, T. H. (2007). Effects of microcrystalline cellulose on functional properties of hydroxy propyl methyl cellulose microcomposite films. *J. Food Sci.* **72**, 16–22.

Drake, S. R., Fellman, J. K., and Nelson, J. W. (1987). Postharvest use of sucrose polyesters for extending the shelf life of stored Golden Delicious apples. *J. Food Sci.* **52**, 1283–1285.

Earle, R. D. and Snyder, C. E. (1996). US Patent 3,255,021.

Edwards, M. E. and Blennerhassett, R. W. (1990). The use of postharvest treatments to extend storage life and to control postharvest wastage of Honey Dew melons (*Cucumis melo L.* var. inodorus Naud.) in cool storage. *Aust. J. Exp. Agric* **30**, 693–697.

El Ghaouth, A., Arul, J., Ponnampalan, R., and Boulet, M. (1991). Chitosan coating: effect on storability and quality of fresh strawberries. *J. Food Sci.* **56**, 1618–1624.

Emmambux, M. N., Stading, M., and Taylor, J. R. N. (2004). Sorghum kafirin film property modification with hydrolysable and condensed tannins. *J. Cereal Sci.* **40**, 127–135.

Esen, A. (1987). A proposed nomenclature for the alcohol-soluble proteins (zeins) of maize (*Zea mays* L.). *J. Cereal Sci.* **5**, 117–128.

Fairley, P., Monahan, F. J., German, J. B., and Krochta, J. M. (1996). Mechanical properties and water vapor permeability of edible films from whey protein isolate and sodium dodecyl sulfate. *J Agric. Food Chem.* **44**, 438–443.

Fan, L. L., and Arce, J. A. (1986). Preparation of fried food products with oil containing emulsifiers. US Patent 4,608,264.

Farnum, C., Stanley, D. W., and Gray, J. L. (1976). Protein-lipid interactions in soy films. *Can. Inst. Food Sci. Technol. J.* **9**, 201–206.

FDA (1987). Chemical preservatives. *In* "Code of Federal Regulation". US Government Printing Office, Washington, DC title 21, part 182.

Fellows, P. J. (1990). Food Processing Technology. Ellis Harwood, NY.

Fenaroli, G. (1995). Fenaroli's Handbook of Flavor Ingredients. CRC Press, Boca Raton, FL.

Ferraris, R. (1989). Gel seeding of sorghum into Mywybilla clay. *In* "Proc. Aust. Sorghum Workshop, Toowoomba, Queensland," 66–73, Foale, M. A., Hare, B. W. and Henzell, R. G. (eds.). Australia, Brisbane. Australian Institute of Agricultural Science.

Friedman, M., Henika, P. R., Levin, C. E., and Mandrel, R. E. (2004). Antibacterial activities of plant essential oils and their components against *Escherichia coli* O157:H7 and *Salmonella enterica* in apple juice. *J. Agric. Food Chem.* **52**, 6042–6048.

Fukushima, D. and Van Buren, J. (1970). Mechanisms of protein insolubilization during the drying of soy milk. Role of disulfide and hydrophobic bonds. *Cereal Chem.* **47**, 687–696.

Ganz, A. J. (1969). CMC and hydroxypropylcellulose versatile gums for food use. *Food Prod. Dev.* **3**, 65–71.

Gaonkar, A. G. (1991). Surface and interfacial activities and emulsion of some food hydrocolloids. *Food Hydrocolloids* **5**, 329–337.

Gates, J. C. (1987). Basic Foods 3rd edn. Holt, Rinehart and Winston, NY.

Gennadios, A., Brandenburg, A. H., Weller, C. L., and Testin, R. F. (1993a). Effect of pH on properties of wheat gluten and soy protein isolate films. *J. Agric. Food Chem.* **41**, 1835–1839.

Gennadios, A., Weller, C. L., and Testin, R. F. (1993b). Temperature effect on oxygen permeability of edible protein-based films. *J. Food Sci.* **58**, 212–214.

German, J. (1999). Food processing and lipid oxidation. *Adv. Exp. Med. Biol.* **459**, 23–50.

Ghanbarzadeh, B., Oromiehie, A. R., Musavi, M., Falcone, P. M., D-Jomeh, Z. E., and Rad, E. R. (2007). Study of mechanical properties, oxygen permeability and AFM topography of zein films plasticized by polyols. *Packaging Technol. Sci.* **20**, 155–163.

Gibson, G. R. and Roberfroid, M. B. (1995). Dietary modulation of the human colonic microbiota: introducing the concept of probiotics. *J. Nutr* **125**, 1401–1412.

Glicksman, M. (1983). Red seaweed extracts. *In* Food Hydrocolloids, Glicksman, M. (ed.). CRC Press Inc, Boca Raton, FL **2**, 73.

Gomez-Estaca, J., Montero, P., Gimenez, B., and Gomez-Guillen, M. C. (2007). Effect of functional edible films and high pressure processing on microbial and oxidative spoilage in cold-smoked sardine (*Sardina pilchardus*). *Food Chem.* **105**, 511–520.

Gontard, N., Guilbert, S., and Cuq, J. L. (1992). Edible wheat gluten films: influence of the main process variables on film properties using response surface methodology. *J. Food Sci.* **57**, 190–195, 199.

Gonzalez-Aguilar, G. A., Celis, J., Sotelo-Mundo, R. R., de la Rosa, L. A., Rodrigo-Garcia, J., and Alvarez-Parrilla, E. (2008). Physiological and biochemical changes of different fresh-cut mango cultivars stored at 5 degrees C. *Int. J Food Sci. Technol.* **43**, 91–101.

Gounga, M. E., Xu, S. Y., and Wang, Z. (2007). Whey protein isolate-based edible films as affected by protein concentration, glycerol ratio and pullulan addition in film formation. *J. Food Eng* **83**, 521–530.

Gounga, M. E., Xu, S. Y., Wang, Z., and Yang, W. G. (2008). Effect of whey protein isolate-pullulan edible coatings on the quality and shelf life of freshly roasted and freeze-dried Chinese chestnut. *J. Food Sci.* **73**, 155–161.

Greener, I. (1992). *Physical Properties of Edible Films and Their Components*. Ph.D. dissertation. University of Wisconsin.

Greener, I. and Fennema, O. (1989). Barrier properties and surface characteristics of edible bilayer films. *J. Food Sci.* **54**, 1393–1399.

Greener, D. and Fennema, O. (1994). Edible films and coatings: characteristics, formation, definitions and testing methods. *In* "Edible Coatings and Films to Improve Food Quality," 1–24, Krochta, J. M., Baldwin, E. A. and Nisperos-Carriedo, M. O. (eds.). Technomic Publishing Co. Inc., Lancaster, Basel.

Guilbert, S. (1986). Technology and application of edible protective films. *In* "Food Packaging and Preservation Theory and Practice," 371, Mathlouthi, M. (ed.). Elsevier Applied Science Publishing Co., London.

Guilbert, S., Cuq, B., and Gontard, N. (1997). Recent innovations in edible and/or biodegradable packaging materials. *Food Additives and Contaminants* **14**, 741–751.

Hagenmaier, R. D. and Shaw, P. E. (1990). Moisture permeability of edible films made with fatty acid and (hydroxypropyl) methylcellulose. *J. Agric. Food Chem.* **38**, 1799–1803.

Hamaguchi, P. Y., Weng, W., Kobayashi, T., Runglertkreingkrai, J., and Tanaka, M. (2007). Effect of fish meat quality on the properties of biodegradable protein films. *Food Sci. Technol. Res.* **13**, 200–204.

Havenainen, R. (1996). New approaches in improving the shelf life of minimally processed fruit and vegetables. *Trends Food Sci. Technol* **7**, 179–187.

Hernandez-Izquierdo, V. M. and Krochta, J. M. (2008). Thermoplastic processing of proteins for film formation—a review. *J. Food Sci.* **73**, 30–39.

Hershko, V., Klein, E., and Nussinovitch, A. (1996). Relationships between edible coatings and garlic skin. *J. Food Sci.* **61**, 769–777.

Hershko, V. and Nussinovitch, A. (1998). Relationships between hydrocolloid coating and mushroom structure. *J. Agric. Food Chem.* **46**, 2988–2997.

Hershko, V., Weisman, D., and Nussinovitch, A. (1998). Methods for studying surface topography and roughness of onion and garlic skins for coating purposes. *J. Food Sci.* **63**, 317–321.

Hojo, E. T. D., Cardoso, A. D., Hojo, R. H., Boas, E. V. D. V., and Alvarenga, M. A. R. (2007). Use cassava starch films and PVC on post-harvest conservation of bell pepper. *Ciencia E Agrotechnologia* **31**, 184–190.

Hulme, A. C. (1970). Biochemistry of Fruits and their Products. Academic Press, London. Volume 1, 620.

Huse, H. L., Hung, Y. C., and Mcwatters, K. H. (2006). Physical and sensory characteristics of fried cowpea (*Vigna unguiculata* L. Walp) paste formulated with soy flour and edible coatings. *J. Food Qual* **29**, 419–430.

IFT (1991). New from Mitsubishi. *In* "Annual Meeting and Food Expl. Program and Exhibit Directory". Dallas Convention Center, Chicago, IL.

IMR (2007). Food Hydrocolloids Conference Proceedings. Lisbon, Portugal. April 22–24.

Jeon, Y. J., Kamil, J. Y. V. A., and Shahidi, F. (2002). Chitosan as an edible invisible film for quality preservation of herring and Atlantic cod. *J. Agric. Food Chem.* **50**, 5167–5178.

Jones, H. W. and Whitmore, R. A. (1972). US Patent 3,694,234.

Jongjareonrak, A., Benjakul, S., Svisessanguan, W., and Tanaka, M. (2006). Fatty acids and their sucrose esters affect the properties of fish skin gelatin-based film. *Eur. Food Res. Technol* **222**, 650–657.

Kader, A. A. (1986). Biochemical and physiological basis for effects of controlled and modified atmospheres on fruits and vegetables. *Food Technol* **40**, 99–104.

Kamper, S. L. and Fennema, O. R. (1985). Use of edible films to maintain water vapor gradients in foods. *J. Food Sci.* **50**, 382–384.

Kampf, N. and Nussinovitch, A. (2000). Hydrocolloid coatings of cheeses. *Food Hydrocolloids* **14**, 531–537.

Kang, H. J., Jo, C., Kwon, J. H., Kim, J. H., Chung, H. J., and Byun, M. W. (2007). Effect of a pectin-based edible coating containing green tea powder on the quality of irradiated pork patty. *Food Control* **18**, 430–435.

Kaplan, D. L., Mayer, J. M., Ball, D., McCassie, J., Allen, A. L., and Stenhouse, P. (1993). Fundamentals of biodegradable polymers. *In* "Biodegradable Polymers and Packaging," 1–42, Ching, C., Kaplan, D. and Thomas, E. (eds.). Technomic Publishing Co., Inc., Lancaster, PA.

Karel, M. (1975). Protective packaging of foods. *In* Principles of Food Science, 399–464, Fennema, O. (ed.). Marcel Dekker, NY.

Kelco. (1990). *Gellan Gum Multifunctional Gelling Agent*. Technical Bulletin, Rahway, NJ. Merck & Co.

Kendra, D. F. and Hadwiger, L. A. (1984). Characterisation of the smallest chitosan oligomer that is maximally antifungal to *Fusarium solani* and elicits pisatin formation in *Pisium satvium*. *Exp. Mycol.* **8**, 276–281.

Kester, J. J. and Fennema, O. R. (1988). Edible films and coatings: a review. *Food Technol.* **42**, 47–59.

Kim, W. S., Tanaka, T., Kumura, H., and Shimazaki, K. (2002). Lactoferrin-binding proteins in *Bifidobacterium bifidum*. *Biochem. Cell Biol.* **80**, 91–94.

King, A. H. (1983). *In* "Food Hydrocolloids" *Vol. 2* Brown seaweed extracts 115–, Glicksman, M. (ed.). CRC Press Inc., Boca Raton, FL.

Koelsch, C. M. and Labuza, T. P. (1991). Packaging, waste disposal and food safety. II: Incineration or degradation of plastics and a possible integrated approach. *Cereal Foods World* **36**, 284–298.

Kokini, J. L., Cocero, A. M., Madeka, H., and de Graaf, E. (1994). The development of state diagrams for cereal proteins. *Trends Food Sci. Technol* **5**, 281–288.

Kotov, N. A. (2003). Layer-by-layer assembly of nanoparticles and nanocolloids: intermolecular interactions, structure and materials perspective. *In* "Multilayer Thin Films: Sequential Assembly of Nanocomposite Materials," 207–243, Decher, G. and Schlenoff, J. B. (eds.). Wiley-VCH, Weinheim, Germany.

Krochta, J. M., Baldwin, E. A., and Nisperos-Carriedo, M. (1994). Edible Coatings and Films to Improve Food Quality. Technomic Publishing Co. Switzerland, Basel.

Krochta, J. M. and De Mulder-Johnston, C. (1997). Edible and biodegradable polymer films. *Food Technol* **51**, 61–73.

Krumel, K. L. and Lindsay, T. A. (1976). Nonionic cellulose ethers. *Food Technol* **30**, 36–43.

Kubota, N. and Kikuchi, Y. (1998). Macromolecular complexes of chitosan. *In* "Polysaccharides: Structural Diversity and Functional Versatility," 595–628, Dumitriu, S. (ed.). Marcel Dekker Inc, New York. Polysaccharides.

Kuntz, E. (1964). Preparation of collagenous materials. US Patent 3,152,976.

Labuza, T. P. and Breene, W. M. (1988). Application of active packaging for improvement of shelf-life and nutritional quality of fresh and extended shelf-life foods. *J. Food Proc. Preservation* **13**, 1–69.

Lagaron, J. M., Cabedo, L., Cava, D., Feijoo, J. L., Gavara, R., and Gimenez, E. (2005). Improving packaged food quality and safety. *Part 2: Nanocomposites. Food Additives and Contaminants* **22**, 994–998.

Landroac, A. H. and Proctor, B. E. (1952). Gas permeability of films. *Mod. Packaging* **25** (10), 131–135, 199–201.

Lazarides, H. N., Mitrakas, G. E., and Matsos, K. I. (2007). Edible coating and counter-current product/solution contacting: a novel approach to monitoring solids uptake during osmotic dehydration of a model food system. *J. Food Eng.* **82**, 171–177.

Lazarus, C. R., West, R. L., Oblinger, J. L., and Palmer, A. Z. (1976). Evaluation of a calcium alginate coating and a protective plastic wrapping for the control of lamb carcass shrinkage. *J Food Sci.* **41**, 639–641.

Leach, B. E., Rainey, N. H., and Pallansch, M. J. (1966). Physical and chemical analysis of film formed at air-water interface on reconstitution of whole milk powders. *J. Dairy Sci.* **49**, 1465–1468.

Lee, C. H., An, D. S., Park, H. J., and Lee, D. S. (2003). Wide spectrum antimicrobial packaging materials incorporating nisin and chitosan in the coating. *Packaging Technol. Sci.* **16**, 99–106.

Lee, J. W., Son, S. M., and Hong, S. I. (2008). Characterization of protein-coated polypropylene films as a novel composite structure for active food packaging application. *J. Food Eng.* **86**, 484–493.

Lee, J. Y., Park, H. J., Lee, C. Y., and Choi, W. Y. (2003). Extending shelf life of minimally processed apples with edible coatings and antibrowning agents. *Lebensmittel-Wissenschaft und-Technologie* **36**, 323–329.

Lee, W. F. and Fu, Y. T. (2003). Effect of montmorillonite on the swelling behavior and drug-release behavior of nanocomposite hydrogels. *J. Appl. Polymer Sci.* **89**, 3652–3660.

Li, B., He, J., Evans, D. G., and Duan, X. (2004). Inorganic layered double hydroxides as a drug delivery system intercalation and in vitro release of fenbufen. *Appl. Clay Sci.* **27**, 199–207.

Li, B., Kennedy, J. F., Jiang, Q. G., and Xie, B. J. (2006). Quick dissolvable, edible and heatsealable blend films based on konjac glucomannan-gelatin. *Food Res. Int* **39**, 544–549.

Li, P. and Barth, M. M. (1998). Impact of edible coatings on nutritional and physiological changes in lightly processed carrots. *Postharvest Biol. Technol* **14**, 51–60.

Lin, D. and Zaho, Y. Y. (2007). Innovations in the development and application of edible coatings for fresh and minimally processed fruits and vegetables. *Comprehensive Rev. Food Sci. Food Safety* **6**, 60–75.

Liu, L., Kerry, J. F., and Kerry, J. P. (2006). Effect of food ingredients and selected lipids on the physical properties of extruded edible films/casings. *Int. J. Food Sci. Technol* **41**, 295–302.

Lowings, P. H. and Cutts, D. G. (1982). The preservation of fresh fruits and vegetables. "Proc. Inst. Food Sci. Tech. Ann." Symp. UK, Nottingham, p. 52.

Mabesa, R. C., Marshall, R. T., and Anderson, M. E. (1980). Milk films exposed to high humidity: studies with electron microscopy and electrophoresis. *J. Food Prot* **43**, 29–35.

Maftoonazad, N., Ramaswamy, H. S., and Marcotte, M. (2008). Shelf-life extension of peaches through sodium alginate and methyl cellulose edible coatings. *Int. J Food Sci. Technol.* **43**, 951–957.

Mangiacapra, P., Gorrasi, G., Sorrentino, A., and Vittoria, V. (2005). Biodegradable nanocomposites obtained by ball milling of pectin and montmorillonites. *Carbohydrate Polymers* **64**, 516–523.

Mark, H. F., Othmer, D. F., Overberger, C. G., and Seaborg, G. T. (1985). Kirk-Othmer Encyclopedia of Chemical Technology 3rd edn., Wiley-Interscience, NY. **20**, 219–220.

Martinez-Romero, D., Alburquerque, N., Valverde, J. M., Guillen, F., Castillo, S., Valero, D., and Serrano, M. (2006). Postharvest sweet cherry quality and safety maintenance by Aloe vera treatment: a new edible coating. *Posharvest Biol. Technol* **39**, 93–100.

Martin-Polo, M., Mauguin, C., and Voilley, A. (1992). Hydrophobic films and their efficiency against moisture transfer. Influence of the film preparation technique. *J. Agric. Food Chem.* **40**, 407–412.

Martin-Polo, M. and Voilley, A. (1990). Comparative study of the water permeability of edible films composed of gum arabic and glycerolmonostearate. *Sci. Aliments* **10**, 473–483.

Mate, J. I. and Krochta, J. M. (1996). Comparison of oxygen and water vapor permeabilities of whey protein isolate and β-lactoglobulin edible films. *J. Agric. Food Chem.* **44**, 3001–3004.

Mathew, S. and Abraham, T. E. (2008). Characterisation of ferulic acid incorporated starch-chitosan blend films. *Food Hydrocolloids* **22**, 826–835.

Mazza, G. and Qi, H. (1991). Control of after-cooking darkening in potatoes with edible film-forming products and calcium chloride. *J. Agric. Food Chem.* **39**, 2163–2166.

McDermott, E. E., Stevens, D. J., and Pace, J. (1969). Modification of flour proteins by disulfude interchange reactions. *J. Sci. Food Agric* **20**, 213–217.

McEvily, A. J., Iyengar, R., and Otwell, W. S. (1992). Inhibition of enzymatic browning in foods and beverages. *Crit. Rev. Food Sci. Nutr* **32**, 253–273.

McHugh, T. H. (1996). Effects of macromolecular interactions on the permeability of composite edible films. *In* "Macromolecular Interactions in Food Technology," 134–144, Parris, N., Kato, A., Creamer, L. and Pearce, J. (eds.). American Chemical Society, Washington, DC.

McHugh, T. H., Huxsoll, C. C., and Krochta, J. M. (1996). Permeability properties of fruit puree edible films. *J. Food Sci.* **61**, 88–91.

McHugh, T. H. and Senesi, E. (2000). Apple wraps: a novel method to improve the quality and extend the shelf life of fresh-cut apples. *J. Food Sci.* **65**, 480–485.

Meheriuk, M. (1990). Skin color in Newton apples treated with calcium nitrate, urea, diphenylamine and a film coating. *Hortscience* **25**, 775–776.

Meheriuk, M. and Lau, O. L. (1988). Effect of two polymeric coatings on fruit quality of Bartlett and d'Anjou pears. *J. Amer. Soc. Hort. Sci.* **113**, 222–226.

Mehyar, G. F., Han, J. H., Holley, R. A., Blank, G., and Hydamaka, A. (2007). Suitability of pea starch and calcium alginate as antimicrobial coatings on chicken skin. *Poultry Sci.* **86**, 386–393.

Mellenthin, W. M., Chen, P. M., and Borgic, D. M. (1982). In line application of porous wax coating materials to reduce friction discoloration of Bartlett and d'Anjou pears. *Hort. Sci.* **17**, 215–217.

Meyer, R. C., Winter, A. R., and Wiser, H. H. (1959). Edible protective coatings for extending the shelf life of poultry. *Food Technol* **13**, 146–148.

Mitsubishi-Kasei Ryoto (1989). Sugar ester. In "Mitsubishi-Kasei Technical Information", 1–20.

Mitsuoka, T. (1990). Bifidobacteria and their role in human health. *J. Ind. Microbiol.* **6**, 263–267.

Mittal, K. L. (1997). The role of the interface in adhesion phenomena. *Polymer Eng. Sci.* **17**, 467–473.

Murray, D. G. and Luft, L. R. (1973). Low D.E. corn starch hydrolysates. *Food Technol* **27**, 32–40.

Naidu, A. S., Bidlack, W. R., and Clemens, L. R. A. (1999). Probiotic spectra of lactic acid bacteria (LAB). *Crit. Rev. Food Sci. Nutr.* **38**, 13–126.

Naruenartwongsakul, S., Chinnan, M. S., Bhumiratana, S., and Yoovidhya, T. (2008). Effect of cellulose ethers on the microstructure of fried wheat flour-based batters. *LWT-Food Sci.Technol.* **41**, 109–118.

Natrajan, N. and Sheldon, B. W. (1995). Evaluation of bacteriocin-based packaging and edible film delivery systems to reduce Salmonella in fresh poultry. *Poultry Sci.* **74** (1), 31.

Nelson, K. L. and Fennema, O. R. (1991). Methylcellulose films to prevent lipid migration in confectionery products. *J. Food Sci.* **56**, 504–509.

Nicolas, J. J., Richard-Forget, F. C., Goupy, P. M., Amiot, M. J., and Aubert, S. Y. (1994). Enzymatic browning reactions in apple and apple products. *Crit. Rev. Food Sci. Nutr* **34**, 109–157.

Nisperos-Carriedo, M. O. (1994). Edible coatings and films based on polysaccharides. *In* "Edible Coatings and Films to Improve Food Quality," 305–336, Krochta, J. M., Baldwin, E. A. and Nisperos-Carriedo, M. O. (eds.). Technomic Publishing Co., Basel, Switzerland.

Nisperos-Carriedo, M. O. and Baldwin, E.A. (1988). Effect of two types of edible films on tomato fruit ripening. Proc. Florida State Horticultural Society, **101**, 217–220.

Nisperos-Carriedo, M. O., Baldwin, E. A., and Shaw, P. E. (1992). Development of edible coating for extending postharvest life of selected fruits and vegetables. *Proc. Florida State Horticultural Society* **104**, 122–125.

Nisperos-Carriedo, M. O., Shaw, P. E., and Baldwin, E. A. (1990). Changes in volatile flavor components of pineapple orange juice as influenced by the application of lipid and composite film. *J. Agric. Food Chem.* **38**, 1382–1387.

No, H. K., Meyers, S. P., Prinyawiwatkul, W., and Xu, Z. (2007). Applications of chitosan for improvement of quality and shelf life of foods: a review. *J. Food Sci.* **72**, 87–100.

Nortje, K., Buys, E. M., and Minnaar, A. (2006). Use of gamma-irradiation to reduce high levels of *Staphylococcus aureus* on casein-whey protein coated moist beef biltong. *Food Microbiol.* **23**, 729–737.

Nussinovitch, A. (1994). Extending the shelf life of mushrooms by hydrocolloid coating. *Hassadeh* **74** (10), 1131–1132.

Nussinovitch, A. (1997a). Hydrocolloid Applications. Gum Technology in the Food and Other Industries. Blackie Academic & Professional, London.

Nussinovitch, A. (1997b). Agricultural uses of hydrocolloids. *In* Hydrocolloid Applications. Gum Technology in the Food and Other Industries (A. Nussinovitch, (ed.), 169–184. Blackie Academic & Professional, London.

Nussinovitch, A. (1998). Hydrocolloid coating of foods: a review. *The Leatherhead Food Industry J.* **1**, 174–188.

Nussinovitch, A. (2003). Water-Soluble Polymer Applications in Foods. Blackwell Science Ltd., London.

Nussinovitch, A. and Hershko, V. (1996). Gellan and alginate vegetable coatings. *Carbohydrate Polymers* **30**, 185–192.

Nussinovitch, A., Hershko, V., and Rabinowitch, H.D. (1994). Israel Patent Application no. 111,495 PCT/US95/14252.

Nussinovitch, A. and Kampf, N. (1992). Shelf-life extension and texture of alginate coated mushrooms. *In* IFTEC, p. 118. The Hague, The Netherlands.

Nussinovitch, A. and Kampf, N. (1993). Shelf-life extension and conserved texture of alginate coated mushrooms (*Agaricus bisporus*). *J. Food Sci. Technol* **26**, 469–475.

Okamoto, S. (1978). Factors affecting protein film formation. *Cereal Foods World* **23**, 256–262.

Oliver, J. F. and Mason, S. G. (1977). Microspreading studies of rough surface by scanning electron microscopy. *J. Colloid Interface Sci.* **60**, 480–487.

Oms-Oliu, G., Aguiló-Aguayo, I., and Martín-Belloso, O. (2006). Inhibition of browning on fresh-cut pear wedges by natural compounds. *J. Food Sci.* **71**, 216–224.

Ortega, R. M. (2006). Importance of functional foods in the Mediterranean diet. *Public Health Nutr* **9**, 1136–1140.

Ou, C.-Y., Shwu-Fen, T., Chieh-Hsien, L., and Yih-Ming, W. (2002). Using gelatin based antimicrobial edible coating to prolong shelf life of tilapia fillet. *J. Food Qual* **25**, 213–222.

Ozdemir, M. and Floros, J. D. (2008). Optimization of edible whey protein films containing preservatives for water vapor permeability, water solubility and sensory characteristics. *J. Food Eng* **86**, 215–224.

Park, J. W., Testin, R. F., Park, H. J., Vergano, P. J., and Weller, C. L. (1994). Fatty acid concentration effect on tensile strength, elongation and water vapor permeability of laminated edible films. *J. Food Sci.* **59**, 916–919.

Pearce, J. A. and Lavers, C. G. (1949). Frozen storage of poultry. V. Effects of some processing factors on quality. *Canadian J. Res.* **27**, 253–265.

Pedreschi, F., Moyano, P., Santis, N., and Pedreschi, R. (2007). Physical properties of pre-treated potato chips. *J. Food Eng.* **79**, 1474–1482.

Pena, D. C. R. and Torres, J. A. (1991). Sorbic acid and potassium sorbate permeability of an edible methylcellulose-palmitic acid films: water activity and pH effects. *J. Food Sci.* **56**, 497–499.

Pennisi, E. (1992). Sealed in edible film. *Science News* **141**, 12.

Pinotti, A., Garcia, M. A., Martino, M. N., and Zaritzky, N. E. (2007). Study on microstructure and physical properties of composite films based on chitosan and methylcellulose. *Food Hydrocolloids* **21**, 66–72.

Pommet, M., Redl, A., Morel, M. H., Domenek, S., and Guilbert, S. (2003). Thermoplastic processing of protein-based bioplastics: chemical engineering aspects of mixing, extrusion and hot molding. *Macromol. Symp.* **197**, 207–217.

Pranoto, Y., Salokhe, V., and Rakshit, K. S. (2005). Physical and antibacterial properties of alginate-based edible film incorporated with garlic oil. *Food Res. Int* **38**, 267–272.

Pravisani, C. I. and Calvelo, A. (1986). Minimum cooking time for potato strip frying. *J. Food Sci.* **51**, 614–617.

Quast, D. G. and Karel, M. (1972). Technique for determining oxygen concentration within packages. *J. Food Sci.* **37**, 490–491.

Ralston, G., Tracey, M., and Wrench, P. (1964). The inhibition of fermentation in baker's yeast by chitosan. *Biochim. Biophys. Acta* **93**, 652–655.

Rao, M. S., Chander, R., and Sharma, A. (2005). Development of shelf-stable intermediate-moisture meat products using active edible chitosan coating and irradiation. *J Food Sci.* **70**, 325–331.

Ray, B. R. and Sandine, W. E. (1992). Acetic, propionic, and lactic acids of starter culture bacteria as biopreservatives. *In* "Food Biopreservatives of Microbial Origin," 103–136, Ray, B. R. and Daeschel, M. (eds.). CRC Press, Boca Raton, FL.

Raybaudi-Massilia, R., Mosqueda-Melgar, J., and Martin-Belloso, O. (2006). Antimicrobial activity of essential oils on *Salmonella enteritidis, Escherichia coli,* and *Listeria innocua* in fruit juices. *J. Food Prot* **69**, 1579–1586.

Raybaudi-Massilia, R. M., Mosqueda-Melgar, J., and Martin-Belloso, O. (2008). Edible alginate-based coating as carrier of antimicrobials to improve shelf-life and safety of fresh-cut melon. *Int. J. Food Microbiol.* **121**, 313–327.

Reading, S. and Spring, M. (1984). The effects of binder film characteristics on granule and tablet properties. *J. Pharm. Pharmacol* **36**, 421–426.

Reiners, R. A., Wall, J. S., and Inglett, G. E. (1973). Corn proteins, potential for their industrial use. *In* "Industrial Uses of Cereals," 285–302, Pomerantz, Y. (ed.). American Association of Cereal Chemists, Inc., St. Paul, MN.

Rhim, J. W. (2004). Increase in water vapor barrier property of biopolymer-based edible films and coatings by compositing with lipid materials. *Food Sci. Biotechnol* **13**, 528–535.

Rhim, J. W., Hong, S. I., Park, H. M., and Ng, P. K. W. (2006). Preparation and characterization of chitosan-based nanocomposite films with antimicrobial activity. *J. Agric. Food Chem.* **54**, 5814–5822.

Rhim, J. W., Wu, Y., Weller, C. L., and Schnepf, M. (1999). Physical characteristics of emulsified soy protein fatty acid composite films. *Science des Aliments* **19**, 57–71.

Richard, F. C., Goupy, P. M., Nicolas, J. J., Lacombe, J. M., and Pavia, A. A. (1991). Cysteine as an inhibitor of enzymatic browning. 1. Isolation and characterization of addition compounds formed during oxidation of phenolics by apple polyphenol oxidase. *J. Agri. Food Chem.* **39**, 841–847.

Rico-Pena, D. C. and Torres, J. A. (1990). Edible methylcellulose-based films as moisture impermeable barriers in sundae ice cream cones. *J. Food Sci.* **55**, 1468–1469.

Rico-Pena, D. C. and Torres, J. A. (1991). Sorbic acid and potassium sorbate permeability of an edible methylcellulose-palmitic acid film: water activity and pH effects. *J. Food. Sci.* **56**, 497–499.

Robinson, C. (1953). The hot and cold forms of gelatin. *In* "Nature and Structure of Collagen," 96–105, Randel, J. T. (ed.). Academic Press, Inc., New York.

Rojas-Grau, M. A., Avena-Bustillos, R. J., Friedman, M., Henika, P. R., Martin-Belloso, O., and McHugh, T. H. (2006). Mechanical, barrier and antimicrobial properties of apple puree edible films containing plant essential oils. *J. Agric. Food Chem.* **54**, 9262–9267.

Rojas-Grau, M. A., Raybaudi-Massilia, R. M., Soliva-Fortuny, R. C., Avena-Bustillos, R. J., McHugh, T. H., and Martin-Belloso, O. (2007). Apple puree-alginate edible coating as carrier of antimicrobial agents to prolong shelf-life of fresh-cut apples. *Postharvest Biol. Technol* **45**, 254–264.

Rojas-Grau, M. A., Tapia, M. S., and Martin-Belloso, O. (2008). Using polysaccharide-based edible coatings to maintain quality of fresh-cut Fuji apples. *LWT-Food Sci. Technol* **41**, 139–147.

Rojas-Grau, M. A., Tapia, M. S., Rodriguez, F. J., Carmona, A. J., and Martin-Belloso, O. (2007). Alginate and gellan based edible coatings as support of anti-browning agents applied on fresh-cut Fuji apple. *Food Hydrocolloids* **21**, 118–127.

Roth, W. B. and Mehltretter, C. L. (1967). Some properties of hydroxypropylated amylomaize starch films. *Food Technol.* **21**, 72–74.

Rupasinghe, H. P., Boulter-Bitzer, J., Ahn, T., and Odumeru, J. A. (2006). Vanillin inhibits pathogenic and spoilage microorganisms in vitro and aerobic microbial growth in fresh-cut apples. *Food Res. Int* **39**, 575–580.

Sagoo, S., Board, R., and Roller, S. (2002). Chitosan inhibits growth of spoilage micro-organisms in chilled pork products. *Food Microbiol.* **19**, 175–182.

Salcini, M. C. and Massantini, R. (2005). Minimally processed fruits: an update on browning control. *Stewart Postharvest Rev.* **3**, 1–7.

Sanderson, G. R. (1981). Polysaccharides in foods. *Food Technol.* **35**, 50–57.

Sandford, P. A. (1989). Commercial uses and potential applications. *In* Chitin and Chitosan: Sources Chemistry, Biochemistry, Physical Properties and Applications, 51–69, Skjak Braek, T., Anthonsen, T. and Sandford, P. (eds.). Elsevier Applied Science, New York. Chitosan.

Sathivel, S. (2005). Chitosan and protein coatings affect yield, moisture loss, and lipid oxidation of pink salmon (*Oncorhynchus gorbuscha*) fillets during frozen storage. *J. Food Sci.* **70**, 455–459.

Sathivel, S., Liu, Q., Huang, J., and Prinyawiwatkul, W. (2007). The influence of chitosan glazing on the quality of skinless pink salmon (*Oncorhynchus gorbuscha*) fillets during frozen storage. *J. Food Eng* **83**, 366–373.

Selman, J. D. and Hopkins, M. (1989). Factors affecting oil uptake during the production of fried potato products. Tech. Memorandum 475. Campden Food & Drink Res. Assoc. Chipping Campden, Gloucestershire, UK.

Sensidoni, A. and Peressini, D. (1997). Edible films: potential innovation for fish products. *Industrie Alimentari* **36**, 129–133.

Shewfelt, R. L. (1987). Quality of minimally processed fruits and vegetables. *J. Food Qual.* **10**, 143–156.

Sinharay, S. and Bousmina, M. (2005). Biodegradable polymers and their layered silicate nanocomposites: in greening the 21st century materials world. *Progr. Mat. Sci.* **50**, 962–1079.

Siragusa, G. R. and Dickson, J. S. (1992). Inhibition of *Listeria monocytogenes* on beef tissue by application of organic acids immobilized in a calcium alginate gel. *J. Food Sci.* **57**, 293–298.

Smith, S. M. and Stow, J. R. (1984). The potential of a sucrose ester coating material for improving the storage and shelf life qualities of Cox's Orange Pippin apples. *Ann. Appl. Biol.* **104**, 383–391.

Sorrells, K. M., Enigel, D. C., and Hatfield, J. R. (1989). Effect of pH, acidulant, time, and temperature on the growth and survival of *Listeria monocytogenes*. *J. Food Prot* **52**, 571–573.

Sorrentino, A., Gorrasi, G., and Vittoria, V. (2007). Potential perspectives of bio-nanocomposites for food packaging applications. *Trends Food Sci. Technol* **18**, 84–95.

Sothornvit, R. and Rodsamran, P. (2008). Effect of a mango film on quality of whole and minimally processed mangoes. *Postharvest Biol. Technol.* **47**, 407–415.

Stern, S. A., Sinclair, T. P., and Gaeis, T. P. (1964). An improved permeability apparatus of the variable-volume type. *Mod. Plastics* **10**, 50–53.

Stevens, M. P. (1999). Polymer Chemistry. An Introduction. Oxford Univ. Press, New York.

Suarez, R. B., Campanone, L. A., Garcia, M. A., and Zaritzky, N. E. (2008). Comparison of the deep frying process in coated and uncoated dough systems. *J. Food Eng.* **84**, 383–393.

Sudhakar, S. G., Pandey, M. C., Manral, M., Radhakrishna, K., and Bawa, A. S. (2006). Effect of enrobing with carboxy methyl cellulose or hydroxy propyl methyl cellulose in corn or gram flour on moisture and fat content of chicken during frying. *J. Food Sci. Technol.* **43**, 377–381.

Swenson, H. A., Miers, J. C., Schultz, T. H., and Owens, H. S. (1953). Pectinate and pectate coatings. Applications to nuts and fruit products. *Food Technol* **7**, 232–235.

Szyperski, R. T. and Gibbons, J. P. (1963). Zein systems developed for heat cured coatings. *Paint Varnish Prod* **53**, 65–73.

Tapia, M. S., Rojas-Graue, M. A., Carmona, A., Rodriguez, F. J., Soliva-Fortuny, R., and Martin-Belloso, O. (2008). Use of alginate- and gellan-based coatings for improving barrier, texture and nutritional properties of fresh-cut papaya. *Food Hydrocolloids* **22**, 1493–1503.

Tapia, M. S., Rojas-Grau, M. A., Rodriguez, F. J., Ramirez, J., Carmona, A., and Martin-Belloso, O. (2007). Alginate- and gellan-based edible films for probiotic coatings on fresh-cut fruits. *J. Food Sci.* **72**, 190–196.

The, D. P., Debeaufort, F., Peroval, C., Despre, D., Courthaudon, J. L., and Voilley, A. (2002a). Arabinoxylan-lipid-based edible films and coatings. 3. Influence of drying temperature on film structure and functional properties. *J. Agric. Food Chem.* **50**, 2423–2428.

The, D. P., Peroval, C., Debeaufort, F., Despre, D., Courthaudon, J. L., and Voilley, A. (2002b). Arabinoxylan-lipids-based edible films and coatings. 2. Influence of sucroester nature on the emulsion structure and film properties. *J. Agric. Food Chem.* **50**, 266–272.

Thunberg, R. L., Tran, T. T., Bennett, R. W., Matthews, R. N., and Delay, N. (2002). Microbial evaluation of selected fresh produce obtained at retail markets. *J. Food Prot* **65**, 677–682.

Torres, J. A., Motoki, M., and Karel, M. (1985). Microbial stabilization of intermediate moisture food surfaces. Control of surface preservative concentration. *J. Food Proc. Pres* **9**, 75–92.

Ukai, N.T., Tsutsumi, T., and Marakami, K. (1975). US Patent 3,997,674.

Uldrich, J. (2003). The Next Big Thing Is Really Small: How Nanotechnology Will Change the Future of Your Business. Crown Publishing Group, New York.

Valverde, J. M., Valero, D., Martinez-Romero, D., Guillen, F., Castillo, S., and Serrano, M. (2005). Novel edible coating based on Aloe vera gel to maintain table grape quality and safety. *J. Agric. Food Chem.* **53**, 7807–7813.

Vermeiren, L., Devlieghere, F., Van Beest, M., de Kruijf, N., and Debevere, J. (1999). Developments in the active packaging of foods. *Trends Food Sci. Technol* **10**, 77–86.

Vojdani, F. and Torres, J. A. (1990). Potassium sorbate permeability of methylcellulose and hydroxypropyl methylcellulose coatings: effect of fatty acids. *J. Food Sci.* **55**, 841.

Wall, J. S. and Beckwith, A. C. (1969). Relationship between structure and rheological properties of gluten proteins. *Cereal Sci. Today* **14**, 16–18.

Wall, J. S., Friedman, M., Krull, L. H., Cavins, J. F., and Beckwith, A. C. (1968). Chemical modification of wheat gluten proteins and related model systems. *J. Polymer Sci. Part C* **24**, 147–161.

Wanstedt, K. G., Seideman, S. C., Donnelly, L. S., and Quenzer, N. M. (1981). Sensory attributes of precooked, calcium alginate-coated pork patties. *J. Food Protection* **44**, 732.

Ward, G. and Nussinovitch, A. (1996). Gloss properties and surface morphology relationships of fruits. *J. Food Sci.* **61**, 973–977.

Weiss, J., Takhistov, P., and McClements, D. J. (2006). Functional materials in food nanotechnology. *J Food Sci.* **71**, 107–116.

Wenzel, R. N. (1936). Surface roughness and contact angle. *Industry Eng. Chem.* **28**, 988–993.

Wicks, Z.W., Jones, F.N., and Peter-Pappas, S. (1994). Organic Coatings, Science and Technology. Wiley and Sons, Inc., New York.

Williams, S. K., Oblinger, J. L., and West, R. L. (1978). Evaluation of calcium alginate film for use on beef cuts. *J. Food Sci.* **43**, 292.

Wills, R., Lee, T., Graham, D., McGlasson, B., and Hall, E. (1981). "Postharvest, an Introduction to the Physiology and Handling of Fruit and Vegetables". New South Wales University Press, Kensington, New South Wales, Australia. 1–2.

Wolf, W. J. and Tamura, T. (1969). Heat denaturation of soybean 11S protein. *Cereal Chem.* **46**, 331–344.

Wong, D. W. S., Camirand, W. P., and Pavlath, A. E. (1994). Development of edible coatings for minimally processed fruit and vegetables. *In* "Edible Coatings and Films to Improve Food Quality," 65–88, Krochta, J. M., Baldwin, E. A. and Nisperos-Carriedo, M. O. (eds.). Technomic Publishing Co, Switzerland.

Wong, D. W., Gregorski, K. S., Hudson, J. S., and Pavlath, A. E. (1996). Calcium alginate films: thermal properties and permeability to sorbate and ascorbate. *J. Food Sci.* **61**, 337–341.

Wu, L. C. and Bates, R. P. (1972). Soy protein lipid films. 1. Studies on the film formation phenomenon. *J. Food Sci.* **37**, 36–39.

Wu, S. (1973). Polar and nonpolar interactions in adhesion. *J. Adhesion* **5**, 39–55.

Wu, M. T. and Salunkhe, D. K. (1972). Control of chlorophyll and solanine synthesis and sprouting of potato tubers by hot paraffin wax. *J. Food Sci.* **37**, 629–630.

Young, T. S. and Fu, E. (1991). Associative behavior of cellulosic thickeners and its implications on coating structure and rheology. *TAPPI J.* **74**, 197–207.

Zapata, P. J., Guillen, F., Martinez-Romero, D., Castillo, S., Valero, D., and Serrano, M. (2008). Use of alginate or zein as edible coatings to delay postharvest ripening process and to maintain tomato (*Solanum lycopersicon* Mill) quality. *J. Sci. Food Agric.* **88**, 1287–1293.

Zeddelmann, H. V. and Olendorf, P. (1979). Beeinflusung des fettvebrauches der herstellung von fettgeback. *Getreide, Mehl und Brot* **33**, 24–26.

Zhang, D. and Quantick, P. C. (1998). Antifungal effects of chitosan coating on fresh strawberries and raspberries during storage. *J. Hort. Sci. Biotechnol.* **73**, 763–767.

Zheng, J. P., Li, P., Ma, Y. L., and Yao, K. D. (2002). Gelatine/montmorillonite hybrid nanocomposite. I. Preparation and properties. *J. Appl. Polymer Sci.* **86**, 1189–1194.

CHAPTER

11

Protein + Polysaccharide Coacervates and Complexes: From Scientific Background to their Application as Functional Ingredients in Food Products

Sylvie L. Turgeon and Sandra I. Laneuville

Dairy Research Centre STELA and Institute of Nutraceutical and Functional Foods INAF, Laval University, Quebec, Canada

11.1 INTRODUCTION

Proteins and polysaccharides are widely utilized in the food, pharmaceutical and biomedical industry due to their many functional properties. In food products proteins and polysaccharides can determine the stability, structure, viscoelastic properties, texture and in-mouth perception of numerous food systems (Tolstoguzov, 2003). Moreover, they can be processed into functional ingredients, e.g. for interfacial stabilization (Girard et al., 2002b; Laplante et al., 2006; Gu et al., 2007; Dickinson, 2008a), to form edible films (Blaise, 1997), for vitamin and flavor microencapsulation (Burova et al., 1999; Lamprecht et al., 2000; McClements, 2006) to obtain texturizers such as fat replacers (Chen et al., 1989; Laneuville et al., 2005a), meat or caviar analogs (Chen et al., 1985; Soucie et al., 1988; Tolstoguzov, 2003) and to form novelty gels, e.g. fish protein gels or electrostatic gels (Haug et al., 2004; Laneuville et al., 2006a).

When proteins and polysaccharides are mixed together in an aqueous environment, two different types of interaction can arise: thermodynamic incompatibility or thermodynamic compatibility mainly depending on the electrical charges on both biopolymers, and therefore on the factors affecting them, such as the pH and the ionic strength. Whether the mixture is under one or the other condition, it will present completely different functional properties. In general, thermodynamic incompatibility, also known as segregative phase separation, occurs at high concentrations and high ionic strengths, when both molecules carry similar charges, thus resulting in an electrostatic repulsion between molecules and leading to the formation of two different phases, one rich in protein and the other one rich in polysaccharide. For the interested

Kasapis, Norton, and Ubbink: Modern Biopolymer Science
ISBN: 978-0-12-374195-0

reader there are several reviews on protein + polysaccharide thermodynamic incompatibility (Doublier et al., 2000; Tolstoguzov, 2003; Kasapis, 2008). On the other hand, thermodynamic compatibility, also known as associative phase separation or complex coacervation, usually occurs at relatively lower concentrations (< 3 to 4 wt% total solids), low ionic strengths (< 0.4 M), and when both molecules carry net opposite electric charges (Tolstoguzov, 2007). This occurs at a pH between the proteins' isoelectric point (Ip) and the pK_a of the polysaccharide (Tolstoguzov, 1997). Under such conditions, molecules spontaneously attract each other and separate into two phases, one rich in both protein and polysaccharide and the other one depleted in biopolymers but rich in solvent. The protein and polysaccharide in the biopolymer rich phase are held together by electrostatic forces and can take the form of a coacervate or a precipitate. There are several recent reviews on the subject (Schmitt et al., 1998; Turgeon et al., 2003, 2007; de Kruif et al., 2004; Cooper et al., 2005; Dickinson, 2008a).

In its first part, this chapter will present the basic notions on protein + polysaccharide attractive electrostatic interaction, including recent advancements that have shed light on the mechanism of protein + polysaccharide associative phase separation and the structure of the separated phases. Subsequently, it will deal with the current or potential applications in the food industry to produce either functional ingredients or for microencapsulation. The aim of this chapter is to illustrate how fundamental research can provide the tools to manipulate and tailor the functionality of protein + polysaccharide complexes and coacervates.

11.2 HISTORICAL BACKGROUND

Protein + polysaccharide complex coacervation was first reported at the beginning of the century by Tiebackx (Tiebackx, 1911) who worked with the gelatin + acacia gum system. However it was Bungenberg de Jong (1936, 1949) who studied complex coacervation in great detail. He coined the term 'coacervate' to identify the phase rich in colloids and the term 'equilibrium liquid' to identify the phase depleted in colloids. His complete and consistent study on complex coacervation by mixing gelatin or serum albumin with acacia gum had, already from the beginning of the century, showed the effect of the pH, the ionic strength (and salt type), molecular ratio (*r*) and temperature on the yield, composition and morphology of coacervates. He also found the existence of an optimal pH and *r* for coacervation, which were related to charge neutralization. Since then, numerous studies suggest that these properties are common to a large number of protein + polysaccharide systems. It should be noted that complex coacervation is different from simple coacervation, the latter occurs when a single polymer or colloid is placed under desolvation conditions (Bungenberg de Jong, 1936). This induces a molecular collapse through sequential intermolecular charge neutralization events or through hydrophobic interactions to finally separate into a liquid coacervated phase, e.g. aqueous gelatin in the presence of alcohol or acetone (Mohanty et al., 2005) or a surfactant in the presence of a specific type of salt (Menger et al., 1998).

The works by Bungenberg de Jong were followed by several theoretical models on complex coacervation by Overbeek and Voorn (Overbeek et al., 1957), Veis and Aranyi (Veis et al., 1960), Tainaka (Tainaka, 1979) and others. A full discussion of the theoretical aspects of protein polysaccharide coacervation is beyond the scope of this chapter, the interested reader is referred to the reviews by Burgess (1990), de Kruif et al. (2004) and Schmitt et al. (1998) for further enlightenment.

11.3 STRUCTURES FORMED DURING PROTEIN + POLYSACCHARIDE ASSOCIATIVE PHASE SEPARATION

Depending on several factors, but mainly on the protein–polysaccharide binding affinity, molecular charge density and molecular conformation (contour length, chain flexibility and molecular weight), different types of structures can be formed upon protein + polysaccharide electrostatic associative interaction:

1. Coacervates (liquid in nature);
2. Complexes (either soluble or insoluble, the latter are also known as interpolymeric complexes or co-precipitates); and
3. Gels (network stabilized by electrostatic interactions).

In this section these structures will be defined and differentiated to avoid some confusion that has been brought about in literature where sometimes complexes are referred to as coacervates and vice versa. Even though the three structures are formed through electrostatic interactions between proteins and polysaccharides, they bear enormous structural differences and functional properties.

Figure 11.1 presents phase contrast micrographs of these three structures. Coacervates in Figure 11.1a were obtained by mixing β-lactoglobulin (βlg) + an exopolysaccharide (EPS) produced by *Lactobacillus rhamnosus* RW-9595M (Van Calsteren et al., 2002) at a mixing $r = 2$, and at a total solids concentration of 1 wt%. Interpolymeric complexes (Figure 11.1b, precipitate) were obtained by mixing βlg + Fucoidan, an algal anionic polysaccharide (Rioux et al., 2007), at $r = 5$ and 1 wt%. Finally, the electrostatic gel (Figure 11.1c) was obtained by mixing BSA + xanthan gum at $r = 2$ and 0.2 wt%. All systems were gradually acidified using glucono-delta-lactone acid (GDL) under quiescent conditions and have a final pH ~ 4.0 ± 0.2. The structural differences are clearly visible with coacervates separating as droplets, interpolymeric complexes separating as fractal aggregates and the gel consisting of an interconnected network of complexes.

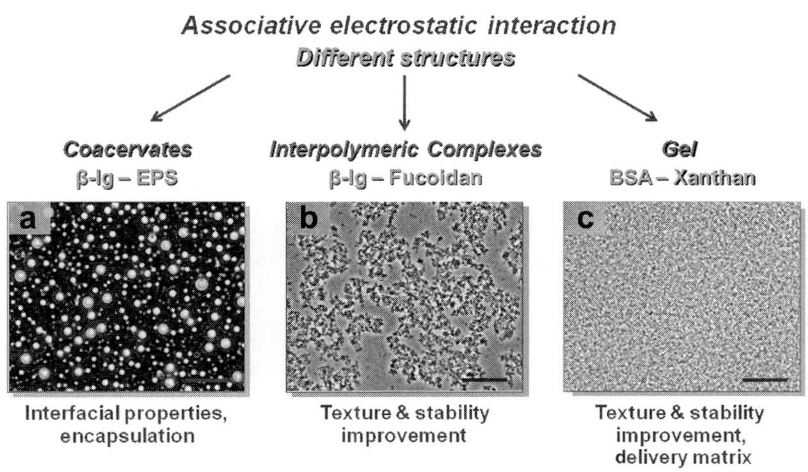

FIGURE 11.1 Phase contrast micrographs of the different structures that can be obtained when proteins and polysaccharides interact under associative phase separation conditions. (a) Coacervates; (b) interpolymeric complexes (fractal aggregates); and (c) electrostatic gel. Micrographs were taken at 40X, bars = 40 μm. See text for details.

Coacervates are dense spherical aggregates (vesicles), which are liquid in nature and remain in a liquid state (Figure 11.1a). Coacervates can issue from the associative phase separation in biopolymeric, colloidal or polyelectrolyte systems under suitable conditions. Systems that form coacervates have sometimes been referred to as liquid/liquid emulsions. Interaction between coacervates can lead to coalescence and the formation of transient multivesicular structures (Figure 11.3a) that will coalesce further and eventually completely separate into a dense coacervated phase, illustrating its liquid-like character (Sanchez et al., 2002; Schmitt et al., 2001b). Protein + polysaccharide systems that form coacervates have great interfacial properties; otherwise coacervate droplets can be stabilized and utilized for ingredient encapsulation as will be discussed in Section 11.7.2.

Complexes are protein + polysaccharide aggregates of fractal nature that phase separate as a solid precipitate (Figure 11.1b). These aggregates may remain soluble or not, depending on the overall charge they bear. *Soluble complexes* are formed at low protein to polysaccharide ratios, and at moderate ionic strengths, i.e. when the biopolymers have low charge densities or when the pH of the system is relatively far from the proteins' Ip. Under such conditions, few charged moieties are available on each molecule, consequently when interactions do occur, the formed complexes consist of single polysaccharide chains bearing only a few protein molecules. Such complexes still carry enough overall negative charge to remain soluble. If environmental conditions are favorable, more protein can be bound to the polysaccharide until charge neutralization is approached. Subsequently, soluble complexes can interact with each other to form *interpolymeric complexes*. Light-scattering studies have shown that soluble complexes present different internal structure and fractal degree (d_f) (they can be denser or looser) compared to their respective interpolymeric complexes (Girard et al., 2004; Laneuville et al., 2005b). Soluble and interpolymeric complexes have excellent texturing and stabilizing properties (see Section 11.7.1), additionally, soluble complexes can also have good interfacial properties (Section 11.7.2).

More recently, it has been found that an *electrostatic gel* can be obtained from the associative interaction between proteins and polysaccharides. The three-dimensional structure of this type of gel is stabilized primarily by attractive electrostatic interactions (Laneuville et al., 2006). The gelation process occurs under specific conditions (slow acidification, quiescent conditions) and follows, at least initially, a similar path as that of complexation, i.e. with the formation of primary soluble complexes followed by interpolymeric electrostatic complexation. However, interpolymeric complexes are able to form junction zones that results in the freeze-in of the whole structure and at the point of gelation, coarsening and phase separation (complexes precipitation) is completely arrested. Furthermore, it has been reported (Laneuville et al., 2007) that the internal structure and firmness of electrostatic gels can be modulated by choosing proteins and polysaccharides of different origin or by adjusting the mixing parameters (e.g. biopolymeric *r*, ionic strength, concentration). The principal distinction of this gel with other protein-containing gels is that it can be formed at extremely low concentrations, <0.5 wt%, and that the protein in the system is found in a 'native' state, since it has not been subjected to any denaturing process (thermal, enzymatic, or the like). This type of gel has potential application as a texturizing agent or as delivery matrix.

11.4 PROTEIN + POLYSACCHARIDE ASSOCIATIVE PHASE SEPARATION KINETICS

Protein + anionic polysaccharide electrostatic attractive interactions will occur at a pH below the proteins' Ip, under such conditions the binding sites on the polysaccharide include the negatively charged (ionized) carboxylate, sulfate, pyruvate, and/or acetate groups, which interact with the (protonated) positively charged ϵ-amino, α-amino, guanidinium and imidazole groups on the proteins (Samant et al., 1993). Electrostatic interaction is the main force in protein + polysaccharide associative phase separation. However, entropic factors (e.g. molecular conformation and molecular weight of the interacting molecules) and enthalpic factors (e.g. molecular charge density and macromolecular ratio) influence phase separation (Girard et al., 2003b; Ou et al., 2006). More specifically, isothermal titration calorimetric (ITC) studies have shown that when weakly charged biopolymers interact, e.g., during the formation of soluble complexes, the interaction is enthalpically (charge) driven, weakly charged molecules have lesser counterion condensation, the entropy gain due to counterions release is low and the interaction is mainly driven by the opposite charges on the molecules (Girard et al., 2003b; Allen and Warren, 2004). At high charge densities, e.g., during the formation of interpolymeric complexes, enthalpic and entropic factors are involved (Girard et al., 2003b; Dickinson, 2008a). Complex coacervation is driven by the entropy gain that arises from the molecular rearrangements that occur during electrostatic interaction in order to form a random aggregated phase (Burgess, 1990). Any entropy loss, caused by a reduction in the randomness of molecular conformations owing to a reduced molecular mobility and to conformational flexibility loss upon protein + polysaccharide complexation is compensated to some extent by the important release of counterions and occluded water molecules from both the protein and polyelectrolyte and also by favorable enthalpy contributions due to solvent reorganization (Burgess, 1994b; Li et al., 1994; Tolstoguzov, 1997; Turgeon et al., 2007). Counterion release has been studied by computer simulations (Allen and Warren, 2004; Ou et al., 2006). Recently, Gummel et al. (2007) were the first to demonstrate experimentally the release of counter ions following protein + polyelectrolyte complexation using neutron scattering and a specific labeling technique. The enthalpic contribution is related to the interaction of proteins with residual charges on the soluble complexes and to the energy gain resulting from the increased ion density in overlapping soluble complexes (Burgess, 1990; Girard et al., 2003b). These entropic and enthalpic contributions were predicted by theory (Tainaka, 1979) and have also been found in computer simulation studies (Ou et al., 2006).

Two procedures can be utilized to produce protein + anionic polysaccharide complexes or coacervates. The first one, designated the *titration method*, consists in dispersing and mixing both molecules at conditions far from the optimal interaction pH, followed by subsequent acidification, generally by acid or GDL addition, to a pH below the proteins' Ip. The second procedure, termed the *mixing method*, consists in preparing the biopolymer dispersions separately, adjusting their pH to the optimal interaction pH and then mixing them together; in this case the interaction occurs very rapidly. Both procedures result in the formation of complexes, or coacervates, with different sizes, composition and yield (Tolstoguzov, 1997; Plashchina et al., 2007; Bedie et al., 2008). In general, complexes formed by the titration method tend to be easily re-dispersed whereas complexes obtained by the mixing method are larger, and tend to be non-stoichiometric due to the abruptness of the interaction, although complexes and coacervates are known to rearrange in time (Weinbreck et al., 2003b; Schmitt et al., 2005b; Tolstoguzov, 2007). It has also been proposed that these differences may be more important at high polyelectrolyte concentrations, where segregative phase-separation phenomena occurring in the system (at pHs far from the Ip) may induce polysaccharide self-association before interaction and as a result change the composition of the final complex (Plashchina et al., 2007).

To study the kinetics of associative phase separation, the titration method has usually been applied since it allows for a more gradual interaction, thus facilitating the determination of the different steps taking place during the interaction. An important advancement in understanding protein + polysaccharide (polyelectrolyte) associative interaction kinetics occurred in the 1980s where experimental results on pH-induced interaction confirmed the existence of clearly differentiated transitions at different critical pHs (Figure 11.2) (Burgess et al., 1984; Dubin et al., 1988; Park et al., 1992; Xia et al., 1993; Mattison et al., 1995) as predicted by the theory of Tainaka (1979), which is based on earlier studies by Veis and Aranyi (1960). Turbidity and light-scattering measurements allowed to determine that protein + polysaccharide interaction begins at a critical pH_c, characterized by a slow increase in turbidity and related to the formation of soluble complexes. This transition occurs at the molecular level, hence also called primary or intrapolymeric complexation (Mattison et al., 1995). Intrapolymeric complexes still carry an overall negative charge, and may remain charged until a pH well below the proteins' Ip, allowing them to remain soluble (Xia et al., 1993; Kaibara et al., 2000). As the pH continues to decrease and charges on the protein molecules increase, more and more protein attaches to the polysaccharide up until the second critical pH_Φ is achieved, at this point soluble complexes approach neutrality and have the tendency to self-associate yielding neutral interpolymeric complexes that tend to precipitate. This step appears as an intermediate process before the system undergoes extensive higher-order aggregation and bulk phase separation (Dubin and Murrell, 1988; Park et al., 1992; Laneuville et al., 2005b). Complexes can achieve neutrality at the point of charge equivalence or stoichiometry, usually at this point the complex yield is highest. In a βlg + xanthan gum system it was found that complexes could bind excess protein only to a certain extent, after which a decline in repulsive forces caused by protein over-aggregation onto the complex induced extensive secondary

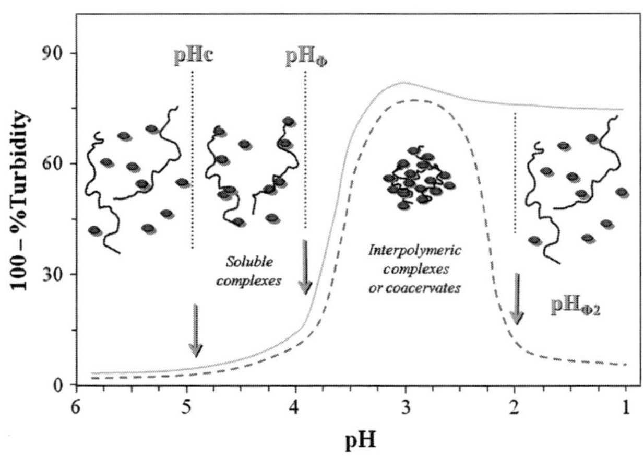

FIGURE 11.2 Schematic representation of the turbidity evolution over the course of acidification for a hypothetical system containing a protein interacting with a strong polyelectrolyte (solid curve) or with a polyelectrolyte with pK_a fixed at pH = 2 (dotted curve). pHc denotes the formation of soluble complexes, pH_Φ denotes the formation of interpolymeric complexes and $pH_{\Phi 2}$ denotes the attainment of the polysaccharides' pK_a and the concomitant complex dissociation. Adapted from Mattison et al. (1999) and Weinbreck et al. (2003a).

aggregation. Accordingly, the final size of the complexes can be controlled by adjusting the mixing *r* (Laneuville et al., 2005b). If the pH is decreased even further the critical $pH_{\Phi 2}$ is reached, which is related to the attainment of the pK_a of the polysaccharides' charged groups; below this pH, complexes begin to breakdown since the polysaccharide will be positively charged (Figure 11.2, dotted curve) (Weinbreck et al., 2003a). For strong polyelectrolytes such as sulfated polysaccharides and other polysaccharides such as xanthan gum, $pH_{\Phi 2}$ is absent (Figure 11.2, solid curve) since these polysaccharides' dissociation is not suppressed at low pH (i.e. they remain charged until pH 1 or lower). Other authors have found other critical pHs, notably for protein conformational changes following binding or for morphological changes in the coacervates droplets (Kaibara et al., 2000; Mekhloufi et al., 2005).

It is currently established that the insensitivity of pH_c to the mixing *r* or to the Mw, as found in several systems (Park et al., 1992; Xia et al., 1993; Mattison et al., 1998, 1999; Girard et al., 2002a; Weinbreck et al., 2003a; Laneuville et al., 2005b), supports the notion that complexation begins with a single protein molecule binding to a single segment on the polysaccharide. As a consequence, the process is not influenced by mass-action effects and is only dependent on the local charge densities and the flexibility of the interacting molecules, which are in turn influenced only by the pH and the ionic strength (Burgess and Carless, 1984; Mattison et al., 1999). On the other hand, pH_{Φ} is affected by the mixing *r* and the Mw; at high protein content more molecules are available to neutralize the complex, consequently pH_{Φ} is attained sooner (Mattison et al., 1995; Weinbreck et al., 2003a). High-molecular-weight polyelectrolytes are able to form interpolymeric complexes at lower concentrations; whereas if the polyelectrolyte is too small, pH_{Φ} may not be attained, i.e. no interpolymeric complexes can be formed (Li et al., 1994). From apparent ellipticity and light-scattering measurements it could be deduced that coacervation was largely controlled by the neutralization of aggregated complexes following the reduction of the negative charges on the polysaccharide as the pH drew closer to its pK_a (Mekhloufi et al., 2005).

To study the kinetics of interpolymeric complexation, light or neutron scattering and turbidity measurements have been the techniques of choice, they allow the evolution of the interaction and the internal structure of the developing complexes to be followed by measuring the apparent fractal degree (d_f) from the slope of the scattering profile at large scattering angles (Girard et al., 2004; Laneuville et al., 2005b). At the beginning of complexation the measured d_f values are not a real fractal dimension, as given in the strict definition of a fractal structure (Kim and Berg, 2000), since the complexes are still evolving and have not yet attained equilibrium. Nevertheless, over small periods of time the formed structures are relatively stable allowing one to measure an 'apparent' d_f (Girard et al., 2004; Laneuville et al., 2005b). In βlg + xanthan mixtures, loose and highly tenuous aggregates (primary or soluble complexes) with $d_f \sim 1.80$ formed at pHc, indicating a diffusion-limited cluster aggregation mechanism (DLCA) (Laneuville et al., 2005b). This type of mechanism occurs when repulsive forces are negligible, and the colloidal aggregation model assumes that every collision results in particles sticking irreversibly upon contact (Tang et al., 2000). Then phase separation ensued following a nucleation and growth mechanism (Laneuville et al., 2005b). Other studies on a βlg + pectin system presented a clear induction time, where the size of the emerging structures decreased, indicating the formation of a large quantity of small primary complexes while the turbidity increased drastically, demonstrating that complexes also phase separated following a nucleation and growth mechanism (Girard et al., 2004). Interpolymeric complexes structuration will be discussed in more detail in Section 11.5.

An interesting property common to most protein + polyanion systems, is that the interaction begins above the proteins' Ip; basically this would be possible due to the presence of charged patches on the proteins' surface allowing for ion–dipole interactions to overcome repulsive ion–ion interactions (Park et al., 1992; Xia et al., 1993; Grymonpre et al., 2001). Studies by Joanny (1994) showed that absorption of a polyampholyte chain onto an oppositely charged surface can occur if the charge density of the surface is large enough. Moreover, several of the molecular configurations available upon interaction were thermodynamically favorable. Literature reports on several proteins with charged patches, for example Grymonpre et al. (2001) and Seyrek et al. (2003) provided representations of the electrostatic potential or surface charge distribution (as generated by protein electrostatic modeling by DelPhi – Molecular Simulations Inc.) clearly showing charged patches in several proteins including βlg, BSA, and lysozyme. For βlg, it has been specifically determined that it possesses several charged patches (basic peptides 1–14, 41–60, 76–83 and 132–148, the latter being part of the α-helix), which are susceptible to interact with polyanions above the proteins' Ip (Girard et al., 2003a). In agreement, it has been reported that βlg displays structural changes in its α-helix during complexation with acacia gum as determined by circular dichroism and apparent ellipticity measurements (Schmitt et al., 2001a; Mekhloufi et al., 2005).

It has also been proposed that due to the polyampholyte character of proteins, they can become polarized by the electric field of a polyelectrolyte when they bind or come in close proximity to it. This would induce proton migration from the NH_3^+ to the CO_2 groups, thus providing an additional polarization-induced mechanism of attraction (Bowman et al., 1997; Mattison et al., 1998). Such an effect has been reported on biomacromolecules like gelatine and evidenced by a molecular conformational change into an extended rod-like shape in the vicinity of agar molecules (Singh et al., 2007b); or on polyelectrolytes such as polyacrylic acid, hyaluronic acid and pectin, which were able to adjust their charge distribution near an oppositely charged micelle, resulting in the effective increase of their charge density and enabling them to bind like a highly charged polyion (Yoshida et al., 1998; Cooper et al., 2006). Titration studies at low ionic strength further allowed measuring this effect as a slight pK shift resulting in an effective change in the proteins' Ip, favoring complex formation (Wen et al., 1997; Girard et al., 2002a). More recently, computer simulation studies have shown that the proteins' ability to regulate the electrical charges on its surface (capacitance) may be stronger around its Ip, and that charge-induced interactions can in fact be stronger than ion–dipole interactions (daSilva et al., 2006). Simulation studies have also shown an equivalent effect in polyelectrolytes in which proton mobility (charge migration) was favored at a pH relatively close to the pK_a (i.e. low charge densities) resulting in an optimization of the binding affinity (Cooper et al., 2006). However, if the polyelectrolyte is fully ionized (at pH values at least two units far from the pK_a), the charges become quenched and do not move, the resulting interaction would then be weaker compared to that of a polyelectrolyte of the same charge density but having motile protons (Cooper et al., 2006). Finally, polarizability of intrapolymeric complexes has also been proposed as a factor influencing interpolymeric complexation leading to higher-order aggregation (Takahashi et al., 2000).

11.5 INTERNAL STRUCTURE OF COACERVATES AND INTERPOLYMERIC COMPLEXES

In recent years, research has focused on understanding the internal structure of the coacervated or the complexed phases. A summary of the main results will be presented in this section.

11.5.1 Coacervates

When coacervates are found as independent droplets and at low protein content, a layer of polyelectrolyte can form around them, providing steric and electrostatic stabilization; as a result, coacervates retain their small size and are stable against coalescence. At higher protein content, the stabilization provided by the polysaccharide is no longer efficient and coarsening follows swiftly (Sanchez et al., 2002). Confocal microscopy has shown that both the protein and the polysaccharide are found at the interface of the coacervates (Lamprecht et al., 2000; Schmitt et al., 2001b). Coacervates that are under vesicular conformation can be stabilized and used to encapsulate a variety of ingredients (see Section 11.7.2).

Weinbreck et al. (2004c) were among the first authors to study the internal structure of the separated coacervate phase, in a βlg + acacia gum system, using fluorescence recovery after photobleaching, diffusing wave spectroscopy and nuclear magnetic resonance. They found that proteins and polysaccharides can diffuse in the liquid coacervated phase albeit with more difficulty than in the dilute biopolymer mixture, because of the higher concentration and also due to the electrostatic interactions. More interestingly, they found that protein molecules diffused ten times faster than the polysaccharide, indicating that both diffused independently and that the structure was in continual movement (Weinbreck et al., 2004c). This result suggests that the proteins change their binding location on the polymer chains very rapidly. However, the diffusivity of the protein greatly decreases around the optimal pH of interaction (Bohidar et al., 2005; Kayitmazer et al., 2007a). The coacervated phase can be considered as a network of interconnected intrapolymeric complexes with contacts that are transiently stabilized by the association with protein molecules (Bohidar et al., 2005) or as a heterogeneous network containing entangled and crosslinked protein and polysaccharide rich phases (Singh et al., 2007a). The works by Bohidar et al. (2005) and Kayitmazer et al. (2007a, 2007b) further showed that the coacervated phase is heterogeneous and consists of extensive dilute domains in which partially interconnected dense domains are embedded. Figure 11.3a presents a schematic representation of the internal structure of a coacervated phase as described by Kayitmazer et al. The dense domains can disintegrate and reform in time and under strong interacting conditions they become larger and more stable (dissolve more slowly) compared to the domains formed under pH or ionic strength conditions that are less favorable for interaction (Menjoge et al., 2008). Moreover, Kayitmazer et al. (2007b) were able to measure differences in the mesophase structures formed with different polyelectrolytes. In another study, the presence of protein aggregates in the structure of βlg + pectin coacervates has been shown to influence the structure (Wang et al., 2007c). Finally, it was found that the coacervated phase composition can change with time (Weinbreck et al., 2004b). These molecular rearrangements can have a profound effect on coacervate's functional properties, e.g. it has been reported that freshly prepared coacervates have better interfacial properties than aged coacervates (Schmitt et al., 2005b).

11.5.2 Complexes

The internal structure of interpolymeric complexes is much denser and the molecules in them are less motile compared to a coacervated phase. Although molecular rearrangements have been found to occur during the course of interpolymeric aggregation (Laneuville et al., 2004) the final equilibrium structure is achieved faster (in the order of hours instead of days). However, depending on the mixing conditions and the

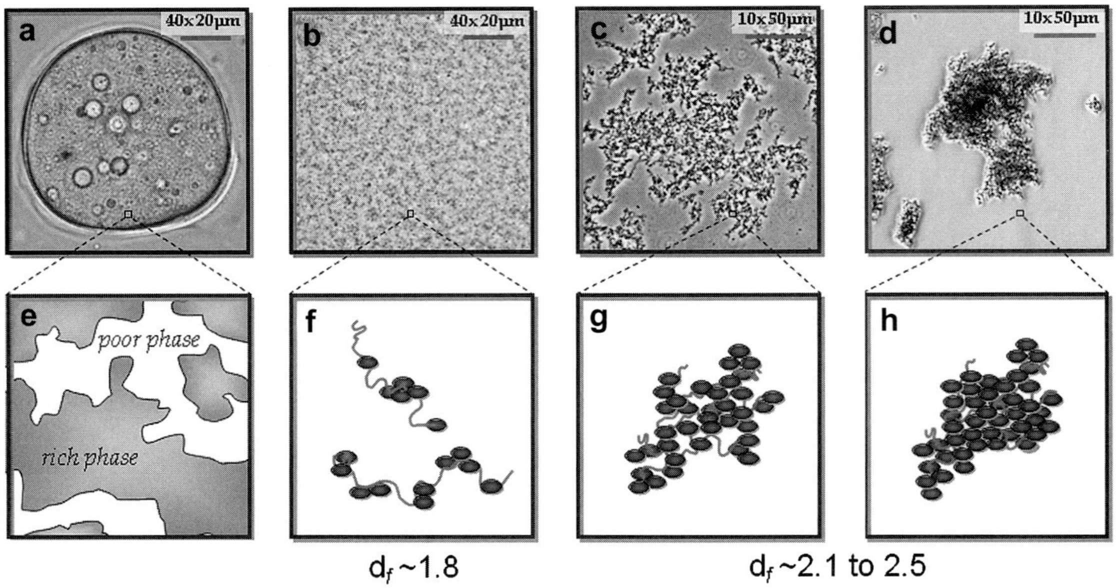

FIGURE 11.3 Phase contrast micrographs of different protein + polysaccharide systems. (a) βlg + EPS coacervate at 1 wt% showing a multivesicular structure. (b) βlg + propylene glycol alginate at 0.5 wt% form soluble complexes. (c) Sodium caseinate + xanthan gum at 0.04 wt% form interpolymeric complexes presenting looser structure compared to (d) sodium caseinate + gellan gum at 0.5 wt% which forms interpolymeric complexes with very dense structure. (e–h) A highly schematic representation of the internal structure of coacervates compared to that of complexes (not at the same scale). Micrographs (a) and (b) were taken at 40X, bars = 20 μm, micrographs (c) and (d) were taken at 10X, bars = 50 μm.

protein content of the system, equilibrium attainment may take longer, e.g., Weinbreck et al. (2003b) report that at low protein content, complexes took up to fours days to rearrange to finally form a few neutralized complexes. Apparent fractal degree (d_f) measurements by light-scattering techniques allowed to follow the evolution of the internal structure of the emerging complexes (Girard et al., 2004; Laneuville et al., 2005b). The d_f value indicates the degree of compactness of a structure: if particles aggregate in linear arrays the d_f value tends to 1; whereas if they form compact spherical aggregates, the d_f tends to 3 (Bushell et al., 2002). At values higher than 3, the surface of the structure is probed (Sanchez et al., 2002). For βlg in mixtures with pectin or with xanthan gum it was found that complexes underwent a restructuring process over the course of interaction (Girard et al., 2004; Laneuville et al., 2005b). In βlg + xanthan mixtures, tenuous soluble complexes formed at pHc, with a $d_f \sim 1.8$ (1.65–1.90, depending on r) then at pH_Φ, a rapid compaction of the structures took place, resulting in denser interpolymeric complexes with a final d_f ranging from 2.2 to 2.4. Figures 11.3c–h presents micrographs and schematic representations of these concepts. The restructuring process was driven by the increasing charges in the molecules, as the pH decreased, and may have been amplified by shear. Another factor favoring interpolymeric complexes' compact structure has its roots in thermodynamics. If the polyelectrolyte has a stiff conformation, the entropy loss upon complexation will be greater than if it was a random coil, the latter being more flexible has more entropically favorable conformations, many of which may facilitate interaction. On the other hand, stiff polyelectrolytes will try to retain their original conformation to minimize entropy loss. As a result the protein will tend to occupy the interstitial spaces of the polyelectrolyte chain resulting in denser complexes (Li et al., 1994). However, the evolution of the internal structure of βlg + pectin complexes showed that, contrary to βlg + xanthan gum complexes, interpolymeric complexes had a looser structure (final $d_f \sim 1.5$–2.2) compared to the initial intrapolymeric complexes. For comparison, light-scattering studies on a system undergoing coacervation showed that the aggregation and precipitation of coacervates can lead to the formation of a heterogeneous structure with fractal interfaces. In a βlg + acacia gum system, the d_f increased from 2.7 to 3.7, indicating very sharpened surfaces, and then decreased to 2.3 (Sanchez et al., 2002).

It has also been found that, as in a coacervated phase, proteins bound in an interpolymeric complex can experience re-positioning. Proteins with lower surface charge density (with longer-range interactions) will form complexes with labile ionic bonds, whereas highly charged proteins will form stronger bonds. Therefore, protein charge density plays an important role in determining the size and density of interpolymeric complexes (Tsuboi et al., 1996).

11.5.3 Molecular Conformational Changes Upon Protein + Polysaccharide Associative Interaction

Protein and polysaccharide molecules can experience conformational changes upon associative interaction. The preferred methods to measure these changes are usually spectroscopic and/or high-sensitivity differential scanning calorimetric techniques. Relative ellipticity measurements during βlg + acacia gum interaction showed that in soluble complexes, βlg presented a loss in its α-helix structure, then as bulk phase separation ensued, it regained secondary structure suggesting a molecular reorganization probably induced by protein concentration in the coacervate (Mekhloufi et al., 2005). Tryptophan fluorescence and Raman spectroscopy studies on BSA + anionic polysaccharides or pea globulin and α-gliadin + acacia gum complexes have shown an important conformational change in the proteins' molecular structure upon interaction (Teramoto et al., 1999; Chourpa et al., 2006). Additionally, it has been reported that complexation can cause the loss of micellization-ability in proteins such as β-casein (Burova et al., 2007). Finally, polysaccharides also experience conformational changes, e.g. κ- or ι-carrageenan showed a reduction of their helical structure upon complexation with casein (Burova et al., 2007).

Often complexation induces increased protein stability against thermal denaturation (Zhang et al., 2004; Capitani et al., 2007). Several mechanism have been proposed for this effect, including the formation of a complex with a partially unfolded protein, which would suppress further aggregation (Chung et al., 2007); or that the presence of a specific binding site on the protein relative to a polysaccharide may be responsible. In the absence of such a site, complexation would have the opposite effect, increasing the thermal sensitivity of the protein, probably due to a higher bonding affinity when the protein is in the unfolded state (van de Weert et al., 2004). Conversely, other studies have found an opposite effect; the authors proposed that the stretching of the protein, due to a lack in the complementary spacing between charges in the protein and polyelectrolyte, causes the protein to become more sensitive to high temperatures (Ivinova et al., 2003). Interestingly, it has been reported that the stabilization/destabilization effect of complexation may depend on the pH and ionic strength (Mounsey et al., 2008).

11.5.4 Coacervation vs. Complexation

Generally speaking, coacervate and complex development follows the same initial path, i.e.

with the formation of soluble intrapolymeric complexes at a particular pHc (specific to each protein and polysaccharide pair and at a given ionic strength), then as the pH continues to approach the proteins' Ip, soluble complexes begin to interact with each other to form interpolymeric complexes followed by bulk phase separation. However, it is still not clearly understood why some protein + polysaccharide systems phase separate as interpolymeric complexes whereas others result in coacervates. Nonetheless, a general trend seems to be emerging with coacervates being formed when polysaccharides or proteins with a low charge density and/or very flexible backbone are utilized, e.g. gelatin, acacia gum, dextran sulfate, sodium hyaluronate and some specific varieties of pectin. Whereas interpolymeric complexes (co-precipitates) form when the polysaccharide and/or the protein in the system is highly charged and/or have a very stiff structure, e.g. κ- and ι-carrageenan, sodium alginate, gellan, or xanthan gum.

The formation of complexes may also be related to a large binding affinity between molecules, which results in a strong interaction. When the binding affinity is high, proteins bond tightly to the polysaccharide inducing extensive counterion expulsion. This phenomenon, which is entropically driven, favors desolvation to the point that precipitation is induced instead of coacervation (Wang et al., 2000; Kayitmazer et al., 2007b; Kumar et al., 2007). A special case is noted in polyelectrolyte poly(dimethyldiallylammonium chloride) (PDADMAC) + micelle interaction where the boundaries for coacervate and precipitate formation seems to be easily differentiated and controlled by varying the ionic strength and charge density of the micelles, with coacervates forming preferentially at intermediate charge densities and precipitates forming at high charge densities (Wang et al., 2000). Furthermore, it appears that in some systems, hydrogen bonding or hydrophobic interactions may also have an influence on the formation of coacervates versus interpolymeric complexes (Biesheuvel et al., 2006; Kumar et al., 2007; Nigen et al., 2007). It has been reported that the molecular weight (Mw) of the polyelectrolyte can play an important role in a polycation + mixed micelle system, with larger Mw being necessary for the formation of coacervates and lower Mw favoring complexes, the latter was attributed to the greater impact of chain configurational entropy loss when complexes were formed (Kumar et al., 2007). On the other hand, for a βlg + xanthan gum system, interpolymeric complexes were always obtained when using xanthan degraded to different extents. The main effect observed when degraded xanthan was utilized was the formation of smaller interpolymeric complexes, although no tests were carried out with samples having very low Mw (Laneuville et al., 2005b).

It now appears that the factors influencing interpolymeric complexes vs. coacervates formation are unique to each protein + polyelectrolyte system and depend on the specific characteristics of the macromolecules in the mixture; e.g. low-methoxyl-pectin (LM-pectin) will form coacervates with gelatin (a low-charge-density polyampholyte), whereas it will form interpolymeric complexes with βlg. Moreover, the critical parameters change for each protein + polysaccharide pair and may involve a balance between several factors. Thus some systems are more sensitive to molecular charge densities or charge distribution complementarity, whereas others depend more on the macromolecular conformation or Mw. As an example, it has been found that human serum albumin possesses a flat positive domain as the main binding site, therefore when this molecule is allowed to interact with oppositely charged polyelectrolytes differences in the polyelectrolytes' stiffness may not have a great impact on the binding affinity (Cooper et al., 2006).

With the advent of new theoretical approaches and computer simulations, the factors controlling complex or coacervate formation are beginning to be understood for each protein and polysaccharide pair. A great challenge still concerns the crucial transition region between soluble complexes and coacervate or interpolymeric complex formation. A shrewd choice to study this subject would be to apply a multimethodological approach. Field-theoretic simulation methods have considerable promise for the investigation of polyelectrolyte complexation phenomena; the interested reader is referred to the recent review by de Vries et al. (2006).

11.6 PARAMETERS AFFECTING PROTEIN + POLYSACCHARIDE ATTRACTIVE ELECTROSTATIC INTERACTION

The factors affecting the course of protein + polysaccharide associative phase separation and the characteristics of the formed structures can be classified as extrinsic or intrinsic factors. Extrinsic factors include the macromolecular mixing *r*, pH, ionic strength, total solids content, temperature, rate of acidification and rate of shear during acidification. Intrinsic factors are related to the nature and the characteristics of the interacting molecules such as the molecular weight (Mw), net charge and flexibility of chains (Samant et al., 1993; Tolstoguzov, 1997) as will be discussed hereafter.

11.6.1 Mixing Ratio (r) and Molecular Concentration

Several studies have shown that the protein + polysaccharide mixing *r* has a major effect on the characteristics of the resulting complexes, namely, the complexes size, composition, and viscosity in solution (Weinbreck et al., 2004d; Laneuville et al., 2005b; Ganzevles et al., 2007a; Turgeon et al., 2007; Wang et al., 2007c). At low *r*, only intrapolymeric complexes are formed (Wang et al., 2000), whereas at higher *r* the coacervate or interpolymeric complexes volume fraction is increased. The structure of complexes or coacervates was found to be greatly influenced by the mixing *r*, at low *r*, smaller coacervates form and coalesce rapidly into very large coacervates, whereas at higher *r* the coacervates did not coalesce as readily (Schmitt et al., 2001b). For complexes, mixing at low *r* also results in the formation of smaller complexes that remain charged over a wider pH range. On the other hand, at high protein content, very large complexes containing an excess amount of protein form (Laneuville et al., 2005b). Wang et al. (2007b) reports that increasing the proportion of proteins in a βlg + pectin system results in the formation of coacervates with a higher elastic character.

Electrostatic complexation can occur at a wide biopolymer concentration, starting at extremely low concentrations (10^{-2} mg ml^{-1}) if the ionic strength is low enough (< 0.2M), compared to the concentration needed for a segregative phase separation (Tolstoguzov, 1986). However, at too high concentrations, an auto-suppression of the interaction occurs, since complex coacervation is favored by entropic factors, i.e. when complexes rearrange into a randomly distributed coacervated phase. Once the mixture composition reaches the concentration of the complexed or coacervated phase, the entropy gain is completely lost and there is no force to drive the phase separation (Burgess, 1994b; Li et al., 1994).

11.6.2 pH, Ionic Strength and Macromolecular Charge Density

Since coacervation/complexation is electrostatic in nature, it is greatly affected by factors that modify the electrical charge density on the interacting molecules such as the pH and the ionic strength (Burgess and Carless, 1984; Mattison et al., 1995; Wang et al., 1999; Weinbreck et al.,

2003b). Counterions shield charged groups on both proteins and polysaccharides, effectively reducing their charge density and therefore reducing the strength of the interaction. Studies on the effect of molecular charge density on complexation have shown that the strength of the interaction is proportional to the number of charged moieties, i.e. more charged molecules form stronger complexes (Girard et al., 2002a), however the charge distribution can also play an important role (Samant et al., 1993). Capillary electrophoresis results (Girard et al., 2002a) showed that LM-pectin bonded more strongly to βlg than the high-methoxyl-pectin (HM-pectin), the latter having less charged carboxylic groups than LM-pectin. Furthermore, HM-pectin became saturated more quickly with proteins limiting the electrostatic interaction (Girard et al., 2002a). Similarly, Burova et al. (2007) and Galazka et al. (1999) report that the strength of protein + carrageenan complexation is related to the density of sulfate groups on the polysaccharide with ι-carrageenan binding more strongly than κ-carrageenan. Nevertheless, it should be noted that increasing the ionic strength by addition of small amounts of salts can in some cases have beneficial effects. Wang et al. (2007b) reports that the addition of $<$ 0.21M NaCl has resulted in a coacervate with higher βlg and pectin content. Laneuville et al. (2007) also report that the addition of small amounts of NaCl in a BSA + xanthan gum electrostatic gel allows a stable gel with no syneresis to be obtained.

Charge density can also play a role in the type of structure to be formed. Studies on the interaction between PDADMAC, a strong cationic polyelectrolyte, and negatively charged triton micelles with sodium dodecyl sulfate, showed that at too low charge densities, coacervation did not occur, on the other hand at very high charge densities precipitation of interpolymeric complexes ensued (Wang et al., 1999, 2000).

Other studies have found that the nature of the charged group, carbonyl vs. sulfated groups, also influences the interactions strength. From tryptophan fluorescence studies, it was determined that BSA interacted more strongly with polysaccharides bearing sulfonic groups (over a wide pH range), with a fairly strong interaction being detected even above the proteins' Ip, compared to polysaccharides bearing only carboxylic moieties, which were able to strongly interact with the protein only below its Ip (Teramoto et al., 1999).

11.6.3 Molecular Conformation, Molecular Weight and Charge Distribution

The number of proteins a polyelectrolyte can bind depends largely on the polymer contour length, chain flexibility, and protein dimensions (Dubin and Murrell, 1988). It has been suggested that a more compact conformation of the polyelectrolyte results in a higher charge density, i.e. larger number of charges per unit length, and therefore greater polarizing effect on the protein, which promotes a stronger attraction (Bowman et al., 1997). However, molecular simulations showed that increased chain flexibility and charge mobility resulted in stronger binding, this effect apparently being more important for micelles than for single proteins (Grymonpre et al., 2001; Kayitmazer et al., 2003; Cooper et al., 2006). Mattison et al. (1998) suggested that the length of the binding segment was controlled by the intrinsic stiffness of the polymer chain and that this parameter may be stronger than the effect of the polyelectrolyte linear charge density. Finally, there is the effect on the formation of complexes vs. coacervation, as Burgess (1990) reported, to obtain coacervation, molecules should be in a random coil configuration.

It has also been determined that charge distribution can affect the interaction affinities (Kayitmazer et al., 2003). Comparative studies on the binding of polyelectrolytes with two proteins with similar Mw and the same number of basic groups: RNase (ribonuclease) and lysozyme showed that these two proteins interacted differently, owing to the different distribution of their basic residues. RNase has its basic residues located relatively close to one another whereas lysozyme has its residues randomly distributed. This affected the polarizability of the resulting primary soluble complex, i.e. the soluble complexes formed with lysozyme were more prone to aggregation due to a superior polarizability mainly attributable to the lack of complementarities in the charges spacing (Takahashi et al., 2000). Field-theoretic simulation results have shown remarkable differences between the phases formed with uniformly or unevenly charged chains bearing the same charge density. Uniformly charged chains formed large homogeneous phases (a dilute phase and a coacervate phase); whereas the unevenly charged polyelectrolyte (block copolymer) formed a mesophase with micellar structure (Popov et al., 2007).

Studies on the effect of the Mw of the complex characteristics found that polymers with higher Mw resulted in the formation of larger primary complexes that aggregated more readily into interpolymer complexes (Li et al., 1994; Laneuville et al., 2005b; Wang et al., 2000), presumably since the polyelectrolyte acts as the backbone for the formation of primary complexes (Park et al., 1992; Laneuville et al., 2005b) and possibly also due to a higher entropy gain (Tainaka, 1979). Li et al. (1994) reported that higher Mw favors interpolymer vs. soluble complexes formation and that at a given ionic strength, increasing the polyelectrolyte Mw resulted in an increase in the coacervation region and in the coacervate volume fraction (increase in protein recovery) (Li et al., 1994; Wang et al., 2000). Studies on PDADMAC, a strong cationic polymer, interacting with charged micelles, showed that an increase in the Mw of the micelle reduced the required charge density for coacervation and increased the coacervation yield (Wang et al., 2000). Furthermore, a critical Mw for coacervate formation was required, below this boundary, coacervation was not possible (Wang et al., 2000). Therefore, by adjusting the size of the polyelectrolyte or the micelles, the final size of the complexes/coacervates can be controlled. Moreover, the internal structure of complexes can be controlled by varying the polyelectrolytes' Mw, lower Mw results in denser complexes, thus allowing additional control of their properties (Laneuville et al., 2005b).

The presence of protein aggregates also influences coacervation (Schmitt et al., 2001b), protein aggregates are not able to form coacervates, instead they precipitate and form structures composed of a protein core surrounded by a polysaccharide layer. Thus, samples containing aggregates will present different functional properties.

11.6.4 Other Interactions: Hydrogen and Hydrophobic Bonding

Even though the main force at play during proteins + polysaccharide associative phase separation is electrostatic, in some cases other types of interactions, such as hydrogen or hydrophobic interactions can be involved to stabilize and/or influence the forming structures (Dickinson, 1998). However, few studies have been carried out on the subject and this area should be very interesting to explore. In general, the presence of hydrogen or hydrophobic bonding seems to be dependent on the choice of macromolecules, i.e. some systems show strong hydrogen or hydrophobic interactions, in addition to electrostatic interactions, whereas others completely lack them.

Some interesting research was made on a βlg + alginate system, where complexation was allowed to occur in the presence of urea (to shield hydrogen interaction) or SDS (to shield hydrophobic interaction). It was found that the gelified structures formed at pH 6 were strongly influenced by hydrogen bonding and featured a balance between electrostatic attractive interactions and hydrogen bonding, the latter being principally controlled by the amount of alginate in the system (El-Haddad, 2002).

In other systems such as in βlg + LM– or HM–pectin complexes, the ester groups of pectin could form hydrogen bonds with the hydroxyl, amide, phenyl and carboxylic groups of the βlg (Girard et al., 2002a). Potentiometric studies showed that urea and high temperature decreased the amount of bound βlg onto pectin molecules, indicating the importance of hydrogen bonding, particularly for HM–pectin, which has less carboxyl groups than LM–pectin (Girard et al., 2002a). In a gelatin + low ester-pectin system, urea did not prevent protein + polysaccharide complex formation, conversely it decreased the degree of aggregation in the concentrated complex phase (Plashchina et al., 2007). In mixtures of synthetic polyelectrolytes + BSA, the formation of soluble complexes seemed to be greatly influenced by hydrogen bonding between the carboxyl and phenolic OH groups on the protein and the ether groups on the polyelectrolyte (Azegami et al., 1999). It has been suggested that hydrogen bonding would be involved, particularly at pHs above the proteins' Ip (Girard et al., 2002a) and that H-bonding is only favored when the charge densities are low (Yoshida et al., 1998).

Kaibara et al. (2000) reported that in a BSA + PDADMAC system, heating to temperatures above 50°C, where thermal denaturation induces protein unfolding and hydrophobic interactions are favored, complexes were formed instead of coacervates, the latter were able to form at lower temperatures. It has also been shown that the addition of small amounts of salt (5–10 mM) can strengthen hydrophobic interactions by screening electrostatic repulsions (Seyrek et al., 2003). Other systems, such as gelatin + κ-carrageenan or βlg + xanthan gum were insensitive to urea addition and high temperatures (Antonov et al., 1999; Laneuville et al., 2007).

11.6.5 Processing Factors

11.6.5.1 Homogenization and Shear (Pumping)

Shear forces can have an impact on the properties of complexes and coacervates and are an important parameter to control for industrial-scale productions; however few studies have dealt with these issues in detail. Particularly, it was found that when shear forces were applied during complexation, restructuring processes of interpolymeric complexes took place set by a competition between attractive electrostatic forces and rupture forces caused by shear (Laneuville et al., 2005b). In systems forming coacervates, finding the right conditions of temperature and shear allowed stabilization of the system against flocculation (Sanchez et al., 2001). Also, by adjusting the polysaccharide aggregation degree or by changing its Mw, e.g. by mechanical degradation prior to the complexation process, can allow the formation of particulated complexes instead of fibrous ones (Laneuville et al., 2005b).

Galazka et al. (1999) found that by applying a pressure treatment to the complexes led, upon pressure release, to the strengthening of the interaction, due in part to a partial denaturation of the protein, which exposed more charged groups. Similarly, BSA + anionic polysaccharide interactions were stronger following heat denaturation due to the increased molecular flexibility in the denatured state, which permitted configurational adjustments that allowed interactions to be maximized, yielding more stable complexes than those formed with the native proteins (Samant et al., 1993).

11.6.5.2 Acidification Rate

Some studies have found that the acidification method can have an impact on the formed structures, e.g. using HCl even if diluted (0.1 M) and added slowly with enough time between drops results in coarser complexes than those obtained with GDL, which allows for a gradual acidification. In an extreme case, as in a βlg + xanthan gum system, when HCl is used, fibrous complexes are obtained, whereas when GDL is used, particulated complexes can be obtained (Figure 11.4).

The advances in understanding the effect of the intrinsic and extrinsic factors on the internal structure of coacervates or complexes and the kinetics of interaction has endowed the food scientist with the necessary tools to be able to choose the adequate biomacromolecules, which coupled with the control of the extrinsic factors can, in general, be employed to obtain mixed systems with tailored functionalities for specific applications, as such they could be utilized either as a texture agent, as an emulsifier or foaming stabilizing agent, and even for the encapsulation of ingredients, as will be discussed in the following section.

Knowledge of the effect that intrinsic and extrinsic factors have on protein + polysaccharide associative interactions endows the food scientist with the necessary tools to be able to choose the adequate macromolecules and experimental conditions to control the reaction kinetics and the structuration process to form the desired structures and obtain mixed systems with tailored functionalities for specific applications. Real applications in food and pharmaceutics are beginning to emerge, as will be discussed in the following section.

11.7 FUNCTIONAL PROPERTIES AND POTENTIAL APPLICATIONS OF PROTEIN + POLYSACCHARIDE COMPLEXES AND COACERVATES

Associative conditions between proteins and polysaccharides have been identified as a way to produce novel ingredients with properties differing from those of the initial biopolymers (Sanchez et al., 1997). The current state of knowledge for the application of protein + polysaccharide complexes and coacervates, based on their functional properties, is presented in this section. Rheological and textural properties as well as interfacial properties in emulsions and foams are the prerequisite for the successful use of protein + polysaccharide mixed systems. They represent an important tool for food manufacturers to enhance food quality and even nutritional attributes. Novel microstructures can be generated using the wide spectrum of food biopolymers available, mixing conditions (environmental conditions, biopolymer r) and by varying processing parameters such as shearing rate or heat treatment.

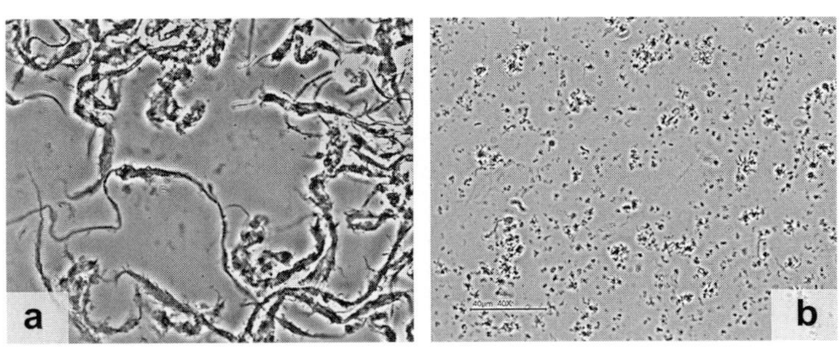

FIGURE 11.4 Phase contrast micrographs showing (a) fibrous complexes obtained by slowly adding HCl 1N to a whey protein + xanthan gum system at r = 2 and 1 wt%. (b) Particulated complexes obtained using the same biopolymers but by gradual acidification with GDL. Final pH ~ 4.0 in both cases. Micrographs were taken at 40X, bars = 40 μm.

11.7.1 Rheological and Textural Properties: Water Binding in Relation to Food Viscosity, Texture and Heat Stability Control

Rheological properties can be defined as the behavior of a liquid or a solid under shear. The viscosity characterizes fluid resistance to flow and is defined as the ratio between the shear stress applied to the fluid and the resulting shear rate. A constant response to an applied shear stress is typical of a Newtonian behavior, while apparent viscosity values of non-Newtonian fluids depend on the shear rate. The viscosity of a macromolecule depends on its concentration, its interactions with water, and its molecular characteristics (size, shape, flexibility). Polysaccharides commonly generate high viscosity at low concentrations in water, explained by their high molecular weight (Lapasin et al., 1995). Proteins have lower viscosity and must be present at higher concentrations to reach high viscosity values. Associative interactions between proteins and polysaccharides result in complexes or coacervates with different viscosity characteristics (rheological behavior) as compared to each individual component. The behavior depends on the nature of the associative system, coacervate vs. interpolymeric complexes, and the food environment.

11.7.1.1 Rheology of Coacervates

Protein + polysaccharide coacervates form a dense and structured phase enriched in biopolymers. Its rheological behavior ranges from viscous to viscoelastic and can even result in a gelified structure. The rheological properties of these systems can be modulated by polysaccharide characteristics, protein to polysaccharide r and pH (Weinbreck et al., 2004c; Turgeon et al., 2007). For whey protein + acacia gum coacervates, the highest viscosity was obtained at pH 4.0, the pH at which electrostatic interactions were the strongest (Weinbreck et al., 2004d). The small deformation behavior of this system points to a highly viscous character with a viscous modulus (G″) three to seven times higher than the elastic modulus (G′) (Weinbreck et al., 2004e). However, BSA + κ-carrageenan coacervates (r of 10:1, pH 4.5 and 0.1M NaCl) had an elastic behavior, G′ being more than two times greater than G″ (Wang et al., 2007a). Similarly, Wang et al. (2007b) reported a strong elastic behavior of low esterified pectin complexes with βlg. At the extreme, xanthan gum + βlg mixtures formed an elastic gel at low concentrations by in situ acidification (Laneuville et al., 2006). It can therefore be hypothesized that the structural characteristics of the polysaccharide influence the rheological behavior of the formed complexes or coacervates. Protein + globular polysaccharides (e.g. acacia gum) are more viscous in nature while linear polysaccharides (e.g. carrageenan) interact with proteins to give viscoelastic solutions and even elastic gels (Turgeon et al., 2007).

11.7.1.2 Rheology of Soluble Complexes

In some instances, instead of coacervates, proteins and polysaccharides may associate to form soluble or interpolymeric insoluble complexes. Soluble complexes are highly charged and are the result of the early interaction between proteins and polysaccharides at macromolecular ratios far from the equimolar charge ratio. Electrostatic repulsions can then serve to stabilize the system against protein aggregation (from heat or isoelectric conditions). This property has been exploited in several well-known food and dairy applications. The interaction of milk proteins with carrageenan is a classical example of the beneficial effect of protein + polysaccharide associative interactions. Carrageenans as sulfated polysaccharides are able to form complexes with casein micelles at pHs higher than the proteins' Ip (Teramoto et al., 1999). Carrageenans interact with positively charged patches on κ-casein and by doing so, are able to stabilize and improve the rheological properties of milk systems. Furthermore, carrageenan molecules adsorbed at the surface of casein micelles or cocoa particles can still interact with other carrageenan molecules that are free in the bulk to form a stable connected network (Syrbe et al., 1998). This has been illustrated by Langerdorff et al. (1997) who reported a decrease from 1 to 0.5 wt% in the minimum ι-carrageenan concentration necessary for gelation in the presence of caseins in milk as compared to gelation in the same ionic environment (milk permeate). In systems containing carrageenans, temperature has an impact on the interaction and the rheological behavior because of the coil–helix transition of these polysaccharides (Langendorff et al., 1997). The presence of cations is also important as they modulate carrageenan–carrageenan interactions.

In fermented dairy products, the controls of protein + polysaccharide interactions represent a powerful tool to modulate viscosity and heat stability. HM pectin is used in acidic dairy products, such as sour milk and yogurt drinks where it stabilizes positively charged casein aggregates by associative interactions (Tromp et al., 2004). In these systems, stabilization is achieved since negatively charged pectins link to positively charged caseins, creating a charged layer around casein aggregates, thus hindering an over-aggregation of the system by electrostatic repulsion. The choice of a pectin sample with the appropriate degree of methoxylation allows control of the interaction strength and optimizes the stabilization effect (Pereyra et al., 1997). Similarly, lactic acid bacteria producing exopolysaccharides (EPS) are now widely used in the dairy industry to avoid the addition of commercial polysaccharides in yoghurt (Lucey, 2004; Ruas-Madiedo et al., 2002). EPS production in fermented dairy products allows improving the viscosity, texture and the overall sensory and water-retention properties of the product. EPS with different monosaccharide composition, structure (type of linkage, ramification), charge and molecular weight can be produced by numerous strains depending on the culture conditions (De Vuyst et al., 2001; Ruas-Madiedo et al., 2002). The differences in backbone stiffness, molecular weight and size, nature of the side groups, monomer composition and type of linkages between monomers have been related to their functional properties (Faber et al., 1998; Tuinier et al., 2001). It was hypothesized that the different structural features among EPS and their different interaction with the protein matrix explain their behavior in milk (Marshall et al., 1999). Recently, it was shown that sulfated EPS having a high negative charge and a high molecular weight were the most efficient to decrease the gelation time and increase the firmness of milk gels (Ding et al., 2002; Girard et al., 2008). Other studies showed that the ropy texture of fermented milks could be associated to the adsorption of charged EPS onto the casein network (Dyrby et al., 2004).

11.7.1.3 Rheology of Interpolymeric Insoluble Complexes

The rheological behavior of interpolymeric insoluble complexes is less documented. Sanchez et al. (1995) found that the viscosity of a soy protein mixed with xanthan gum under conditions favoring complexation was 2.4 times higher than the viscosity of the unmixed biopolymer solutions. Other authors reported either a decrease in pseudoplasticity and apparent viscosity upon mixing ovalbumin + glutamate glucan or an increase in viscosity for a β-casein + glutamate glucan at acidic pH (Dickinson, 1998). Tolstoguzov (1996) proposed that when proteins and polysaccharides interact to form a complex, the molecules are found in a more compact conformation and the junction zones are less hydrophilic, reducing the hydration capacity, and thus resulting in a viscosity decrease. Such effect has been clearly detected by rheological measurements and previously reported in literature (Delben et al., 1997; Laneuville et al., 2000). It is also known that the protein content in the mixture affects the complexes' size and therefore the final viscosity of the system. For a βlg + xanthan gum system, at low protein content, the viscosity decreased after complexation whereas at high protein content ($r = 20$), the viscosity increased. This increase was caused by an important protein aggregation and/or the over-aggregation of protein onto the complexes surface at acidic pH and was further intensified by the heat treatment (applied to stabilize the complexes against environmental changes) (Laneuville et al., 2000). It is expected that the viscosity of complexes relates to the strength of the interaction and to the protein:polysaccharide r (Mann et al., 1996), as polysaccharide dispersions display generally higher viscosity than the protein dispersions. Stronger interactions could result in more compact, less hydrated complexes having lower viscosity and presenting lower pseudoplastic character, while weaker interactions could be more prone to a shear thinning behavior.

11.7.1.4 Gelation of Proteins + Polysaccharides Under Associative Conditions

In mixed protein + polysaccharide systems, if one or both biopolymers are under gelling conditions, a mixed gel can be produced. Mixed gels have been classified into three types: interpenetrating, coupled, and phase-separated networks (Morris, 1986; Zasypkin et al., 1997). Interpenetrating networks are formed when the two components gel separately and form independent networks. Coupled networks are formed in the presence of favorable intermolecular interactions between the different types of biopolymers. In contrast, phase-separated gels are produced in conditions where interactions between the different polymers are repulsive in nature (Piculell et al., 1992; Zasypkin et al., 1997). This latter behavior is very common and has been reported for numerous protein + polysaccharide systems (Zasypkin et al., 1997; Turgeon et al., 2001, 2003; Kasapis, 2008). Coupled gels can be differentiated using small deformation rheology during gel formation (Ould Eleya and Turgeon, 2000; Kasapis, 2008). By applying temperature fluctuations to a βlg + κ-carrageenan gel, it was possible to identify the specific contribution of each biopolymer in a phase-separated mixed gel. On cooling, an additional contribution to G' by the polysaccharide was clearly identified on the rheological profile and the gelling temperature, this was not seen in the case of coupled gels. Furthermore, for a gel with liquid inclusions, a centrifugation step allows the measurement of the separation of one biopolymer in the liquid phase. If one biopolymer is able to diffuse freely in the gel, there are no associative interactions between the two types of biopolymer (Ould Eleya and Turgeon, 2000).

Few examples of coupled gels could be found in the literature. Stronger gels were formed when proteins were mixed with carrageenan at acidic pH due to electrostatic interaction between the polymers (Ould Eleya and Turgeon, 2000; Baeza et al., 2002). Strong translucent gels stabilized by electrostatic interactions were obtained at very low concentrations and a wide range of r, for mixtures of βlg, BSA or ovalbumin with xanthan or gellan gum, such gels have potential applications as structuring ingredients or as delivery matrices (Laneuville et al., 2007). Negative effects can also be obtained. The presence of pectin in a casein acid gel hindered casein aggregation into a coherent network and completely inhibited gelation at a concentration of 0.8 wt% (Matia-Merino et al., 2004), while gel rigidity was reduced at lower pectin concentrations. Similarly, if protein + polysaccharide complexes are added to milk, the size of the complexes should be controlled since larger complexes can disturb mixed gels and result in weaker gels with a gritty texture (Bertrand, 2008).

Coupled gels have not received a lot of attention until now. However several food systems could benefit from protein + polysaccharide associative favorable effects. In acidic dairy products EPS-producing strains are now widely used in Europe as substitutes of commercial polysaccharides due to their possible interactions in milk systems. In North America, their use arises in combination with commercial stabilizers and the effect of such protein + polysaccharide + exopolysaccharide mixed system still needs to be better understood. Very diversified EPS structures can be produced by different strains widening the possibilities to control functionality. The use of polysaccharide in cheese-making is prohibited in several countries, therefore, EPS-producing strains represent a way to control the texture and water retention in some cheese products (Duboc et al., 2001; Hassan, 2008).

11.7.1.5 Protein + Polysaccharide Complexes As Texturing Agents

The development of protein + polysaccharide complexes as food ingredients arose in the 1980s to produce meat analogs and fat substitutes (Chen and Soucie, 1985; Chen et al., 1989). Fibrous protein + xanthan gum complexes were used to produce meat mimetics that retain their firmness and fibrous quality upon prolonged storage in salad dressings at pH 4.1 (Soucie et al., 1988). Particulated complexes were also obtained to produce fat replacers; the complexes' size was controlled by mechanical fragmentation before drying (Chen et al., 1989). Another patent described the production of protein + polysaccharide complexes and their stabilization by a drying step or a heat treatment (Chen and Soucie, 1985; Chen et al., 1989). As the form of protein + polysaccharides complexes can be fibrous and difficult to disrupt by high pressures, an alternative means to control the size has been previously proposed (Laneuville et al., 2000). A high-pressure pre-treatment was applied to the polysaccharide before the complexation step to prevent formation of fibrous complexes. These complexes have been used as fat replacers in cake frostings (Laneuville et al., 2005a). Casein + carrageenan complexes obtained by the addition of small quantities of carrageenan (< 0.1 wt%) to milk under shear were recently proposed as potential texturing ingredients. The shear treatment strength and the κ-carrageenan concentration affected the particles' size distribution (Ji et al., 2008). Sanchez and Paquin (1997) reviewed the process and presented the main applications and limitations of protein and polysaccharide microparticles.

Other applications as texturing agents include protein + polysaccharide core-shell microparticles utilized as a cream or fat substitute (Perrau et al., 1989; Bishay et al., 1996). The microparticles were spherical and composed of a gelified polysaccharide core (e.g. pectin, gellan, and alginate) and a protein shell. The polysaccharide was gelified by heat treatment or the addition of ions such as calcium, subsequently the protein layer was added by interaction at a pH below the proteins' Ip. Protein + polysaccharide complexes have also been utilized as a structuring agent with the capacity to bind and/or structure water in low-fat cocoa confectionery products (Rey et al., 2007). Finally, water-soluble complexes were produced to take advantage of the properties of insoluble proteins such as corn-derived prolamines, gluten or zein (Blaise, 1997). Such complexes allowed the formation of films, which could then be applied to preserve foodstuffs.

Ingredients based on protein + polysaccharide complexes represent a powerful tool to modulate rheological properties of several products. The main drawback is the sensitivity of their functional properties to pH. However, considering the large spectrum of protein + polysaccharide couples available there are numerous possibilities to obtain the desired effect at a determined ratio. Moreover, some interesting approaches have been studied to stabilize complexes for micro- or nano-encapsulation applications, which will be treated in the next section, these could offer new possibilities.

11.7.2 Interfacial Properties of Protein + Polysaccharide Mixed Systems

Proteins and polysaccharides are known to play an important role in the formation and stabilization of foams and emulsions. Several studies have been aimed at understanding how these biopolymers adsorb and interact at the air–water and oil–water interfaces in order to control the product morphology and stability. The use of protein + polysaccharide mixed systems is a complex process in which several factors must be controlled to obtain the desired effect (as the order of protein and polysaccharide addition, the formation of complexes at the interface or in the bulk, the time of complexes' addition, etc.). Some studies have begun to clarify the puzzle but additional studies with other protein + polysaccharide systems will help the food technologists to employ complexes' attributes more easily.

11.7.2.1 Foaming Properties

To generate a foam, gas, water, a surfactant and energy are needed. The energy input contributes to enlarge the interfacial area between the two phases whereas the surfactant will adsorb at the interface and contribute to its stabilization (Walstra, 2003). The study of the interfacial properties at air–water interfaces constitutes the fundamental approach to understand how biopolymers adsorb and interact in foams. Several food proteins, such as egg-white proteins and gelatin, are widely used foaming agents. Readers are referred to recent reviews on protein foams for more details (Foegeding et al., 2006; Murray, 2007). To enlarge the functionalities and applications of proteins and also to use novel protein sources, the addition of polysaccharides has been studied. Polysaccharides could act in two manners, by controlling the rheology of the aqueous phase or by increasing the thickness of the adsorbed layer (Dickinson et al., 1996a). This reduces the thinning rate of the aqueous films between bubbles and increases the stability of the foam. Ganzevles et al. (2006a, 2006b, 2007c, 2008) have used a comprehensive approach to study protein + polysaccharide mixed systems at the A/W interface using pectin, pullulan and βlg as a model. The effect of protein + polysaccharide mixing r, polysaccharide charge density, Mw and sequence of adsorption on adsorption kinetics, surface rheology and structural characteristics of the film were systematically studied at pH 4.5 and at low ionic strength (< 10 mM). A model for mixed protein + polysaccharide adsorption was proposed, as influenced by the complexes charge (Figure 11.5). Protein molecules diffuse towards the interface, adsorb and reorganize, if proteins are complexed with a polysaccharide, slower diffusion to the interface is expected (larger hydrodynamic volume). Proteins located on the outer regions of complexes can adsorb to the interface forming a first dense layer (layer 1, Figure 11.5a). More complexes are present behind layer 1 but with lower density. For highly charged complexes (Figure 11.5b), electrostatic repulsions result in an interface with lower protein density compared to an interface formed with neutral complexes (Figure 11.5c). Consequently, for charged complexes, surface rheology is affected and the dilatational elasticity (variation of surface pressure upon change in surface area) is lower, indicating a less stable interface. A similar behavior is observed when complexes are added after a protein interface has already been created (Figures 11.5d and 11.5e). In this case, neutral complexes also form a denser layer.

Polysaccharides have an impact on the interface structure. Kudryashova et al. (2007) have recently demonstrated that proteins do not organize the same way at the interface when complexed with a polysaccharide. Complexation allowed the protein to preserve its initial conformation, even at the interface, by maintaining a favorable aqueous microenvironment. βlg + acacia gum complexes obtained at charge

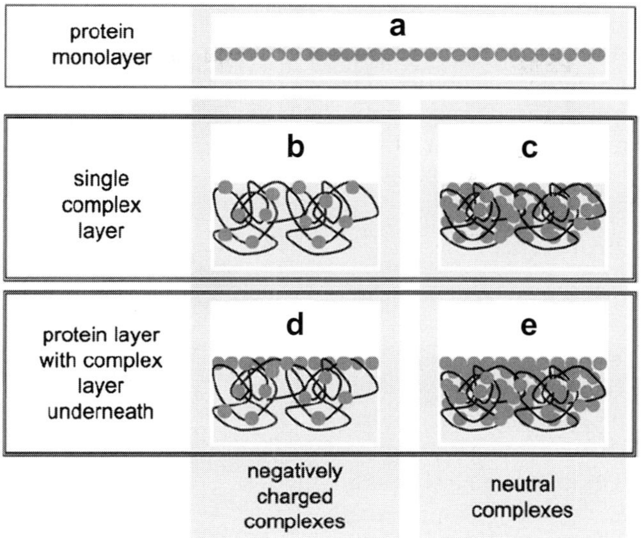

FIGURE 11.5 Schematic representation of different adsorbed layers at the air–water interface: (a) protein monolayer, (b) and (c) mixed layers from simultaneous adsorption where (b) is from negatively charged complexes and (c) from net neutral complexes, (d) and (e) complexes adsorbed at previously formed protein layer where (d) are negatively charged complexes and (e) net neutral complexes. From Ganzevles et al. (2006) with permission.

neutralization ratio were more efficient than βlg alone to stabilize A/W interfaces through higher surface dilatational elasticity and viscosity (Schmitt et al., 2005a). The same authors revealed an important effect of coacervate reorganization over time on their interfacial properties (Schmitt et al., 2005b). Just after interaction, soluble electrostatic complexes were formed, while over time they associated, and reorganized to form insoluble dense liquid coacervates. This reorganization affected foam stability importantly: soluble electrostatic complexes resulted in stable foams with more homogeneous, more hydrated and less rigid interfaces than the interfaces obtained with the coacervated phase. The latter were less stable, experiencing fast liquid drainage, film rupture and bubble coalescence (Schmitt et al., 2005b).

Protein + polysaccharide coacervates have been utilized to stabilize interfaces in products such as mayonnaises and foamed emulsions (Kolodziejczyk et al., 2004). In these cases care was taken to mix both biopolymers at the same time as the interface they had to stabilize was being formed. Under such conditions, soluble complexes had smaller sizes and higher diffusion coefficients, allowing them to efficiently stabilize the forming interface.

Protein + polysaccharide interactions can be used to control protein adsorption at the air–water interface. Parameters affecting protein + polysaccharide interaction offer a means to modulate adsorption kinetics, surface rheology and the net charge of the surface layer. Process parameters as simultaneous protein + polysaccharide adsorption or sequential adsorption represent another control parameter in the functionality of mixed adsorbed layers. Studies in complex food systems are still needed to assess food applications (Foegeding et al., 2006) however, the fundamental results offer promising behavior in several products, such as sherbet and aerated foams obtained from acidic dairy products. For applications at neutral pHs, such as in ice creams and meringues, complexes need to be stabilized against neutral pH. Otherwise, complexes could be obtained at higher pHs using protein and chitosan, a cationic polysaccharide, which forms complexes at a pH above than the proteins' Ip.

11.7.2.2 Emulsifying Properties

The addition of polysaccharides to proteins to modify their emulsifying properties has been more widely studied than foaming properties. Several reviews have been devoted to this subject (Dickinson, 1998, 2003, 2006, 2008a; Dickinson et al., 1996b). Protein + polysaccharide complexes seem to organize similarly at O/W interfaces as at A/W interfaces (Ganzevles et al., 2007a). The addition of a polysaccharide to emulsions to increase the continuous-phase viscosity is of current usage and contributes to emulsion stability by slowing the creaming rate. Otherwise, as with A/W interfaces, when proteins are used to emulsify O/W interfaces, polysaccharides can be added to interact with proteins in the bulk before adsorption at the interface or can be added after the interface has been formed to interact with the adsorbed protein layer. These two processes have been recently named mixed emulsions and bilayer emulsions respectively (Dickinson, 2008b). The case of proteins and polysaccharides bearing opposite charges will be presented here, as it is related to associative interactions. For other situations, with neutral proteins or polysaccharides bearing the same charges, readers are referred to reviews where segregative conditions prevail (Dickinson and McClements, 1996b; Dickinson, 1998).

11.7.2.3 Mixed Emulsions

When a charged polysaccharide is added to a protein-stabilized emulsion, several scenarios are possible depending on the concentration (Figure 11.6). At low polysaccharide concentrations, when insufficient polysaccharide is present to cover all emulsified droplets, bridging flocculation occurs because more than one droplet is attached to a single polysaccharide molecule (Figure 11.6a). At intermediate concentrations, a steric stabilization is possible. The surface is completely saturated and there is no free polysaccharide in the bulk (Figure 11.6b). At higher concentrations, additional polysaccharide remains free in the bulk and depletion-flocculation could be induced, destabilizing the whole system (Figure 11.6c). The amount of polysaccharide to be used has to be determined for each protein + polysaccharide couple based on both biopolymers' charge density and the amount of interfacial surface to be covered.

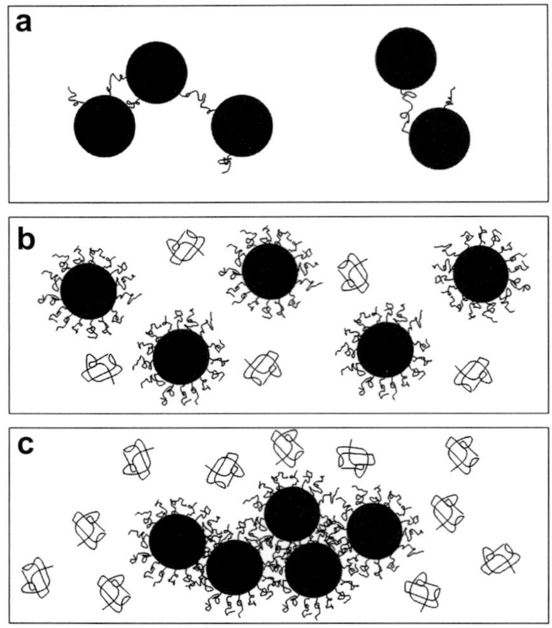

FIGURE 11.6 Adsorbing, non-gelling polysaccharide in colloidal dispersions. With increasing polysaccharide concentration, the systems evolves from (a) bridging flocculation; (b) stabilization to (c) instability caused by depletion flocculation (adapted from Syrbe et al., 1998).

An interface stabilized with protein + polysaccharide complexes has larger dilatational elasticity, as shown with βlg + HM-pectin

(Ganzevles et al., 2006a) and pea protein + acacia gum (Ducel et al., 2004). Charged interfaces favor emulsion stability by creating repulsive conditions, as shown by studies on whey proteins + chitosan (a cationic polysaccharide) mixtures at a pH above the proteins' Ip (Laplante et al., 2005a, 2005b). Whey protein isolate (WPI) + pectin complexes were also reported to be more stable when bearing charges (Neirynck et al., 2007).

Numerous foods are emulsion-based and an enumeration would be difficult to achieve. Usually, food processors adjust formulations and process parameters to obtain stable and palatable food products. Emulsions can also be used as delivery systems for bioactive or sensible compounds (McClements et al., 2007). Complex coacervation represents a method of choice for encapsulation given that during a liquid/liquid phase separation, the coacervated phase can spontaneously form a coating layer around the suspended or emulsified active ingredient in the solution media (Burgess and Carless, 1984; Burgess, 1994b; Daniels et al., 1995; Gouin, 2004). Agitation allows limiting coacervate sedimentation and assures a homogeneous coating. An interesting fact is the very high load (amount of oil in the capsule), up to 99%, that can be obtained (Gouin, 2004).

11.7.2.4 Mixed Emulsions for Encapsulation

The gelatin + acacia gum system was the first to be reported for encapsulation purposes and certainly the most studied (Thies, 2007). This system possesses several key characteristics profitable for numerous applications: it forms viscous coacervates during microcapsule formation, usually prepared at 50–60°C, a temperature at which gelatin melts. On cooling, the rigidity increases resulting in a stable gelified shell around the compound to protect. If the delivery must proceed in the human body, gelatin is readily melted in the mouth. The applications of coacervated emulsions can then range from flavor release (Yeo et al., 2005; Prata et al., 2008) and nutrient protection to masking compounds with unpleasant taste or odor (Junyaprasert et al., 2001; Lamprecht et al., 2001; Pierucci et al., 2007). Yeo et al. (2005) tested gelatin + arabic gum to encapsulate and vehiculate a flavor to be released during cooking in baked goods. The oil remained stable for 4 weeks at room temperature and was released at temperature $\geq$ 100°C. Interestingly, oil droplets coated with coacervates could be formed again upon cooling.

Since the 1990s, other protein + polysaccharide systems have been introduced for similar encapsulation applications (Schmitt et al., 1998). Weinbreck et al. (2004a) used whey protein + arabic gum coacervates to encapsulate lemon and orange flavors. The pH window allowing encapsulation was similar to the one needed to obtain coacervation. The optimal pH was 4.0, where the viscosity of the complex was at its highest and most of the biopolymers were in a complexed form (Weinbreck and Wientjes, 2004e). Burgess (1994b) also reported that higher viscosity confers better stability. Furthermore, Weinbreck et al. (2004a) have shown that droplet size could be used to modulate flavor release since the larger the coacervates' sizes (> 200 μm), the easier the flavor release during chewing. Ducel et al. (2004) have confirmed the potential of using plant proteins (α-gliadin and pea globulin) to form coacervates with arabic gum and carboxymethylcellulose. Another system forming coacervates, gelatin + a modified guar gum (sodium carboxymethyl guar) was utilized to encapsulate clove oil (Thimma et al., 2003). The addition of surfactants was suggested to improve encapsulation yield (Mayya et al., 2003), decrease droplet diameter and accelerate coacervation in a gelatin + arabic gum system (Tan et al., 2008).

Recently, Bedie et al. (2008) have used WPI + LM-pectin interpolymeric complexes to evaluate their functionality as an entrapment matrix for thiamine. Two processes to form complexes were compared, pre-blending acidification (mixing method) and post-blending acidification (titration method). Complexes characteristics (particle size, charge, yield) and microscopic appearance were related to pH, protein + polysaccharide *r* and whether acidification was performed before or after polymer mixing. Entrapment efficiency (percent of thiamine entrapped) was higher for complexes obtained by the mixing method. This was related to the formation of larger complexes during the sudden and rapid mixing of WPI and LM-pectin solutions. Post-blending acidification resulted in smaller and more homogeneous complexes that should be preferred for applications in liquids where sedimentation is a concern (Bedie et al., 2008). Finally, the use of protein + polysaccharide complexes and coacervates could be utilized to protect a compound in the upper digestive tract (in acidic pH conditions) to then be released in the intestine, where the prevailing higher pH may be more suitable for the bioavailability and bio-stability of some molecules (Lamprecht et al., 2006).

From these applications, possibly the most successful in food industry has been in flavor delivery. Coacervate capsules for flavor delivery usually consist of a vegetable oil droplet surrounded by a thin coacervate layer. This layer swells to different extents, depending on the water content of the medium, and becomes permeable for the flavor compounds, as in the swollen state, the mesh size of the coacervate layer is much larger than the typical size of a flavor molecule (Mw generally below 200 Da). In the swollen state, the coacervate layer is thus not much of a physical barrier, however the release of the flavor compounds into the surrounding (food) matrix will still be rather slow due to the presence of the oil droplet. Conversely, in the dry state, the coacervate layer will be very densely packed, and will form a physical barrier for the flavor compounds, thus providing a good shelf life (Ubbink J., 2008, personal communication).

However, one of the principal challenges for applying coacervates for flavor encapsulation is that flavor compounds partition to the oil phase to varying degrees depending on their hydrophobicity; in turn, partitioning dictates the release rate (Ubbink J., 2008, personal communication; Taylor, 2002). It is possible then to optimize a compounds' release rate by adjusting the coacervate layer properties (swelling rate, porosity, density, thickness, etc.) to compensate for the constraints induced by the partitioning of the chosen flavor compound. This can be achieved by controlling the environmental conditions during coacervation (pH, ionic strength, macromolecular ratio), by applying diverse mechanical treatments (spray-drying, homogenization, etc.) and by the appropriate choice of the coacervating macromolecules (Taylor, 2002; Malone and Appelqvist, 2003; Madene et al., 2006; Kaushik and Ross, 2007).

11.7.2.5 Bilayer Emulsions

In the case of bilayer emulsions, the oil phase is emulsified using an aqueous phase containing one biopolymer (usually protein constitutes the first adsorbed layer), the emulsion is then rinsed to eliminate non-adsorbed biopolymer, afterwards the second biopolymer, bearing opposite charges, is added (Guzey et al., 2006). These steps can be repeated using the same biopolymers or alternatively with different biopolymers to obtain a multilayered–stabilized interface (McClements, 2005). As compared with the mixed emulsion, the interface is made of superimposed biopolymer layers instead of a mixed protein + polysaccharide layer. Among the advantages of the bilayer approach, is a better control of the interfacial layer properties by modulating the interface structure, charge, thickness and permeability (Guzey et al., 2006). The control of the interface composition allows to improve the stability against pH variations (McClements et al., 2007), salt (Guzey et al., 2004, 2007), thermal processing (Ogawa et al.,

2003; Pajonk et al., 2003; Gu et al., 2005), chilling (Ogawa et al., 2003), freezing (Mun et al., 2008), dehydration (Mun et al., 2008) and mechanical agitation (McClements et al., 2007). Limitations include the additional processing steps required to rinse the emulsion and the limited range of useful concentrations, which must remain low, since such systems present a tendency to flocculate (McClements et al., 2007).

11.7.2.6 Bilayer Emulsions for Encapsulation

The main application of bilayer emulsions resides in the conception of tailor-made delivery systems. For example, the possibility to improve the chemical stability of the emulsion offers a solution to protect oxidation-sensitive compounds (McClements et al., 2007). Recently, Sagis et al. (2008) investigated a method of producing fiber-reinforced capsules based on layer-by-layer adsorption on emulsion droplets. Whey protein fibrils (produced by a pre-heat treatment) and HM-pectin were added in alternate layers over oil droplets to obtain up to seven layers. As the number of layers increased the stability to freeze-drying of the microcapsules improved. The authors presented this process as a flexible method to produce microcapsules for which the size can be controlled by the emulsification method and the strength by the number of layers. Furthermore, the production involved usual equipment and food-grade ingredients, which is appealing for the food industry. However, a process with multiple steps has an impact on the price of the resulting product; therefore this approach is principally oriented towards higher value products as delivery systems.

Coacervation has been widely utilized in the pharmaceutical industry for drug encapsulation and controlled delivery, for the interested reader, several good reviews on the subject are available (Agnihotri et al., 2004; Dumitriu and Chornet, 1998; Liu et al., 2008).

11.8 MAIN LIMITATIONS FOR THE USE OF COACERVATES AND COMPLEXES IN FOOD APPLICATIONS AND ENCAPSULATION

Although promising functional properties and a wide range of possibilities to modulate protein + polysaccharide electrostatic interactions (biopolymer nature, *r*, environmental conditions etc.) several potential applications are restrained by different limiting factors. The main weakness of complexes and coacervates for food applications is the narrow pH and ionic strength range that ensures complexes stability (Turgeon et al., 2007; Bedie et al., 2008). Other concerns include the tendency of droplets stabilized with protein + polysaccharide complexes to coalesce, limiting product shelf-life stability. These constraints are also encountered in encapsulation applications.

Several approaches have been evaluated to stabilize complexes and coacervates. Production of stable protein + polysaccharide complexes has been studied through the formation of covalent conjugates via the Maillard reaction (non-enzymatic browning) (Dickinson, 2003). This technique allows the terminal or side-chain amine groups on a protein molecule to be linked to the reducing end of a glucosidic chain (monosaccharide, oligosaccharide or polysaccharide). Discussing the properties of these conjugates' properties is beyond the scope of this chapter, readers can refer to Dickinson (2008) for more details.

In non-food applications, reticulating agents as glutaraldehyde or formaldehyde are currently used, however most are not permitted in food because of toxicity concerns. Some alternatives have been developed for food applications, e.g. glyceraldehyde has been proposed as a biocompatible cross-linker reagent (Bulgarelli et al., 1999), it reacts with the amino groups in proteins to form covalent bonds with polysaccharides through the Maillard reaction (Tessier et al., 2003). Glyceraldehyde has also been used to crosslink protein matrices (Caillard et al., 2008) and could potentially be used to stabilize coacervates or interpolymeric complexes. However, the reaction is long and the number of formed links is inferior compared to glutaraldehyde reticulation. The ability to form bridges between proteins has also been exploited to reinforce casein + gelatin microspheres (Bulgarelli et al., 1999). Transglutaminase is an enzyme that catalyzes the crosslinking of proteins, promoting covalent bond formation between the γ-carbonyl group of a glutamine residue and the ϵ-amino group of a lysine residue. This method has been utilized to crosslink the proteins in the microcapsules' shell material (Soper et al., 1997). Another interesting approach for food applications is the use of plant phenolics to crosslink gelatin + pectin coacervates (Strauss et al., 2004). The resulting microparticles had superior mechanical strength, reduced swelling, fewer free amino groups and were stable to temperatures up to 200°C. Genipin, a component of traditional Chinese medicine, is a natural crosslinking agent obtained from a fruit extract. Genipin itself is colorless but it reacts spontaneously with amino acids to form edible blue pigments and has been proposed as a hardener for gelatin coacervates (Kyogoku et al., 1994). Another reticulating agent is dehydroascorbic acid, which can form covalent bonds with primary amines through its carbonyl groups (Larisch et al., 1996; Aiedeh et al., 1997). Finally, a possibility arises from water extraction by solvent, e.g. ethanol has been used to harden the wall of protein + polysaccharide coacervates as a non-solvent for the polymers (Lamprecht et al., 2001) compared hardening methods (by ethanol or spray-drying), and reticulation (by dehydroascorbic acid or glutaraldehyde) to stabilize gelatine + acacia gum coacervates. Microparticles fabricated with hardened walls using the ethanol treatment were more efficient to hinder lipid oxidation than microparticles hardened with dehydroascorbic acid or spray-drying.

One patented application includes the production of βlg + sodium caseinate or βlg + arabic gum coacervates to encapsulate a lipophilic core to transport and protect sensitive compounds such as vitamins or oils and also to reduce the bad taste of some ingredients (Mellema, 2005). Coacervates were stabilized by crosslinking with glutardialdehyde or transglutaminase, which allowed for a better retention of the encapsulated material. Subsequently, coacervates were incorporated into a spread formulation to enrich it with β-carotene.

Heat treatment, a common food process, has also been explored to stabilize protein + polysaccharide complexes. It has been postulated that an appropriate heat treatment could induce protein denaturation and allow protein–protein interactions, freezing the complexed structure. However, conditions have to be determined as the change in conformation could favor protein over-aggregation at the expense of protein + polysaccharide complex stabilization. Heat treatment of WPI + xanthan gum complexes provoked over-aggregation of proteins (Laneuville et al., 2000). It is however difficult to differentiate protein aggregates from protein + polysaccharide complexes and methods based on light microscopy observations cannot discriminate between the two (Laneuville et al., 2000). Gentes (2007) studied the applicability of a heat treatment to stabilize protein + polysaccharide complexes. The integrity of complexes was deduced using a centrifugation method coupled with the analytical determination of protein and polysaccharide content of each phase. It was found that the strength of the electrostatic interaction was important as WPI + LM-pectin complexes were disrupted by heat treatment (90°C/2 min) at pH 5.0 but remained stable at pH 4.5. The stabilized complexes were added to milk for yogurt production. The structure of acidic gels was modified and syneresis increased due to the large size of the formed aggregates. Optimization of production parameters could permit to control particle size and their functionality. Similarly, chitosan + ovalbumin mixtures were heat-treated (80°C/20 min) at different pHs ranging from 4.3 to 5.9 (Yu et al., 2006). This treatment promoted the formation of nanogels with a stable size distribution (in the order of nanometers) even after long storage periods and from a pH ranging from 2.0 to 10.5. The authors proposed that chitosan chains were crosslinked by ovalbumin, which gelled upon heat treatment forming the core of the nanogels (size ranging from 120 to 500 nm). The same authors further tested this concept by applying a heat treatment to lysozyme + dextran conjugates (covalently linked) to obtain nanogels of an average size of 200 nm. The use of these nanogels to load drugs (e.g. ibuprofen) was presented.

11.9 PERSPECTIVES

Formulated foods become more and more sophisticated due to the diversity of available ingredients, processes and the emphasis placed on their nutritional value. Numerous nutritive ingredients are now added to enrich products and processes; ensuring their quality as well as their stability is required. Furthermore, the level of difficulty rises when the delivery has to be realized in a specific part of the digestive tract to ensure the beneficial effects. Functionality of individual food ingredients is advanced and at present, research is needed for multi-ingredient systems optimization. Mixture knowledge wisely applied in combination with appropriate process technology could be the key to produce tailor-made food microstructure. Protein + polysaccharide complexes and coacervates are powerful and versatile tools to address this complexity; they can contribute to food functionality and procure various structural features, form stable capsules, gels, etc. From the diverse protein + polysaccharide systems studied, the next step will be to explore protein interaction with a mixture of polysaccharides. Combination of low- and high-molecular-weight polyelectrolytes was suggested (Prokop et al., 1998) to adjust the viscosity and reactivity of mixtures. The same approach could be attempted by reacting a polysaccharide with a mixture of proteins; otherwise modifications as protein hydrolysis prior to complexation are avenues to explore. Interesting results have recently shown that in-mouth behavior of emulsions could be triggered by the charge of the interfacial layer (Silletti et al., 2007). Protein + polysaccharide at the interface could then be involved in these phenomena.

ACKNOWLEDGMENTS

The authors acknowledge Dr. Ubbink for the insightful comments about flavor release from coacervated emulsions.

References

Agnihotri, S. A., Mallikarjuna, N. N., and Aminabhavi, T. M. (2004). Recent advances on chitosan-based micro- and nanoparticles in drug delivery. *Journal of Controlled Release* **100**, 5–28.

Aiedeh, K., Gianasi, E., Orienti, I., and Zecchi, V. (1997). Chitosan microcapsules as controlled release systems for insulin. *Journal of Microencapsulation* **14**, 567–576.

Antonov, Y. A. and Gonçalves, M. P. (1999). Phase separation in aqueous gelatin-κ-carrageenan systems. *Food Hydrocolloids* **13**, 517–524.

Azegami, S., Tsuboi, A., Izumi, T., Hirata, M., Dubin, P. L., Wang, B., and Kokufuta, E. (1999). Formation of an Intrapolymer Complex from Human Serum Albumin and Poly(ethylene glycol). *Langmuir* **15**, 940–947.

Baeza, R. I., Carp, D. J., Perez, O. E., and Pilosof, A. M. R. (2002). κ-carrageenan – Protein interactions: Effect of proteins on polysaccharide gelling and textural properties. *Lebensmittel-Wissenschaft Und-Technologie-Food Science and Technology* **35**, 741–747.

Bedie, G. K., Turgeon, S. L., and Makhlouf, J. (2008). Formation of native whey protein isolate-low methoxyl pectin complexes as a matrix for hydro-soluble food ingredient entrapment in acidic foods. *Food Hydrocolloids* **22**, 836–844.

Bertrand, M-E. (2008). Étude des propriétés gélifiantes et viscosifiantes de systèmes mixtes isolat de protéines de lactosérum-polysaccharides en conditions associatives. Ph.D. thesis. Université Laval, Québec, Canada. 139p.

Biesheuvel, P. M., Lindhoud, S., deVries, R., and CohenStuart, M. A. (2006). Phase Behavior of Mixtures of Oppositely Charged Nanoparticles: Heterogeneous Poisson-Boltzmann Cell Model Applied to Lysozyme and Succinylated Lysozyme. *Langmuir* **22**, 1291–1300.

Bishay, I. E., Clark, D. R. (1996). Carbohydrate/Protein Cream Substitutes. The Nutrasweet Co. ed. US Patent 5,536,514.

Blaise, M. (1997). Food Preserving Composition and method of use. WO/1997/023137

Bohidar, H., Dubin, P. L., Majhi, P. R., Tribet, C., and Jaeger, W. (2005). Effects of Protein-Polyelectrolyte Affinity and Polyelectrolyte Molecular Weight on Dynamic Properties of Bovine Serum Albumin-Poly(diallyldimethylammonium chloride) Coacervates. *Biomacromolecules* **6**, 1573–1585.

Bowman, W. A., Rubinstein, M., and Tan, J. S. (1997). Polyelectrolyte-Gelatin Complexation: Light-Scattering Study. *Macromolecules* **30**, 3262–3270.

Bulgarelli, E., Forni, F., and Bernabei, M. T. (1999). Casein/gelatin beads: I. Cross-linker solution composition effect on cross-linking degree. *International Journal of Pharmaceutics* **190**, 175–182.

Bungenberg de Jong, H. G. (1936). La coacervation complexe et son importance en biologie. *In* "Tome I: Généralités et coacervats complexes. Actualité Scientifiques et industrielles VI. Exposés de Biologie", Fauré-Fremiet, E. (ed.). Hermann et Cie. Paris.

Bungenberg de Jong, H. G. (1949). Crystallisation-coacervation-flocculation. *In* Colloid Science, Vol. 2, 232–258, Kruyt, H. G. (ed.). Elsevier, Amsterdam.

Burgess, D. J. (1990). Practical analysis of complex coacervate systems. *Journal of Colloid and Interface Science* **140**, 227–238.

Burgess, D.J. (1994b). Complex Coacervation: Micro-Capsule Formation. *In* Macromolecular Complexes in Chemistry and Biology, 285–300, Dubin, P. L., Bock, J., Davis, R. M., Schulz, D., Thies, C. (eds.). Springer-Verlag, Berlin.

Burgess, D. J. and Carless, J. E. (1984). Microelectrophoretic studies of gelatin and acacia for the prediction of complex coacervation. *Journal of Colloid and Interface Science* **98**, 1–8.

Burova, T. V., Grinberg, N. V., Golubeva, I. A., Mashkevich, A. Y., Grinberg, V. Y., and Tolstoguzov, V. B. (1999). Flavour release in model bovine serum albumin/pectin/2-octanone systems. *Food Hydrocolloids* **13**, 7–14.

Burova, T. V., Grinberg, N. V., Grinberg, V. Y., Usov, A. I., Tolstoguzov, V. B., and deKruif, C. G. (2007). Conformational Changes in ι- and κ-Carrageenans Induced by Complex Formation with Bovine β-Casein. *Biomacromolecules* **8**, 368–375.

Bushell, G. C., Yan, Y. D., Woodfield, D., Raper, J., and Amal, R. (2002). On techniques for the measurement of the mass fractal dimension of aggregates. *Advances in Colloid and Interface Science* **95**, 1–50.

Caillard, R., Remondetto, G. E., Mateescu, M. A., and Subirade, M. (2008). Characterization of Amino Cross-Linked Soy Protein Hydrogels. *Journal of Food Science* **73**, C283–C91.

Capitani, C., Pérez, O. E., Pacheco, B., Teresa, M., and Pilosof, A. M. R. (2007). Influence of complexing carboxymethylcellulose on the thermostability and gelation of α-lactalbumin and β-lactoglobulin. *Food Hydrocolloids* **21**, 1344–1354.

Chen, W. S., Henry, G. A., Gaud, S. M., Miller, M. S., Kaiser, J. M., Balmaceda, R. G., Morgan, R. G., Baer, C. C., Borwanker, R. P., Helgeth, L. C., Strandholm, J. J., Hasenheuttl, G. L., Kerwin, P. J., Chen, C. C., Kratochvíl, J. F., and Lloyd, W. L. (1989). Microfragmented Ionic Polysaccharide/Protein Complexes Dispersions. KRAFT Inc. European Patent Application EP0340035.

Chen, W. S. and Soucie, W. G. (1985). Edible Fibrous Serum Milk Protein – Xanthan Gum Complexes. USA. KRAFT Inc, Glenview, IL. US Patent 4,559,233.

Chourpa, I., Ducel, V., Richard, J., Dubois, P., and Boury, F. (2006). Conformational Modifications of α-Gliadin and Globulin Proteins upon Complex Coacervates Formation with Gum Arabic as Studied by Raman Microspectroscopy. *Biomacromolecules* **7**, 2616–2623.

Chung, K., Kim, J., Cho, B.-K., Ko, B.-J., Hwang, B.-Y., and Kim, B.-G. (2007). How does dextran sulfate prevent heat induced aggregation of protein? The mechanism and its limitation as aggregation inhibitor. *Biochimica et Biophysica Acta (BBA) – Proteins & Proteomics* **1774**, 249–257.

Cooper, C. L., Dubin, P. L., Kayitmazer, A. B., and Turksen, S. (2005). Polyelectrolyte-protein complexes. *Current Opinion in Colloid & Interface Science* **10**, 52–78.

Cooper, C. L., Goulding, A., Ulrich, Kayitmazer A. B., Stoll, S., Turksen, S., Yusa, S.-I., Kumar, A., and Dubin, P. L. (2006). Effects of Polyelectrolyte Chain Stiffness, Charge Mobility and Charge Sequences on Binding to Proteins and Micelles. *Biomacromolecules* **7**, 1025–1035.

Daniels, R. and Mittermaier, E. M. (1995). Influence of pH adjustment on microcapsules obtained from complex coacervation of gelatin and acacia. *Journal of Microencapsulation* **12**, 591–599.

daSilva, F. L. B., Lund, M., Jonsson, B., and Akesson, T. (2006). On the Complexation of Proteins and Polyelectrolytes. *Journal of Physical Chemistry B* **110**, 4459–4464.

de Kruif, C. G., Weinbreck, F., and de Vries, R. (2004). Complex coacervation of proteins and anionic polysaccharides. *Current Opinion in Colloid & Interface Science* **9**, 340–349.

de Vries, R. and Cohen Stuart, M. (2006). Theory and simulations of macroion complexation. *Current Opinion in Colloid & Interface Science* **11**, 295–301.

De Vuyst, L., De Vin, F., Vaningelgem, F., and Degeest, B. (2001). Recent developments in the biosynthesis and applications of heteropolysaccharides from lactic acid bacteria. *International Dairy Journal* **11**, 687–707.

Delben, F. and Stefancich, S. (1997). Interaction of food proteins with polysaccharides. I. Properties upon mixing. *Journal of Food Engineering* **31**, 325–346.

Dickinson, E. (1998). Stability and rheological implications of electrostatic milk protein-polysaccharide interactions. *Trends in Food Science & Technology* **9**, 347–354.

Dickinson, E. (2003). Hydrocolloids at interfaces and the influence on the properties of dispersed systems. *Food Hydrocolloids* **17**, 25–39.

Dickinson, E. (2006). Colloid science of mixed ingredients. *Soft Matter* **2**, 642–652.

Dickinson, E. (2008). Interfacial structure and stability of food emulsions as affected by protein-polysaccharide interactions. *Soft Matter* **4**, 932–942.

Dickinson, E., Izgi, E. (1996a). Foam stabilization by protein-polysaccharide complexes. *Colloids and Surfaces A: Physicochemical and Engineering Aspects* **113**, 191–201.

Dickinson, E. and McClements, D. (1996b). Protein-polysaccharide interactions *in Advances in food colloids*, 81–101. Blackie academics and professional, London.

Ding, P., Wolf, B., Frith, W. J., Clark, A. H., Norton, I. T., and Pacek, A. W. (2002). Interfacial tension in phase-separated gelatin/dextran aqueous mixtures. *Journal of Colloid and Interface Science* **253**, 367–376.

Doublier, J.-L., Garnier, C., Renard, D., and Sanchez, C. (2000). Protein-polysaccharide interactions. *Current Opinion in Colloid & Interface Science* **5**, 202–214.

Dubin, P. L. and Murrell, J. M. (1988). Size distribution of complexes formed between poly(dimethyldiallylammonium chloride) and bovine serum albumin. *Macromolecules* **21**, 2291–2293.

Duboc, P. and Mollet, B. (2001). Applications of exopolysaccharides in the dairy industry. *International Dairy Journal* **11**, 759–768.

Ducel, V., Richard, J., Saulnier, P., Popineau, Y., and Boury, F. (2004). Evidence and characterization of complex coacervates containing plant proteins: application to the microencapsulation of oil droplets. *Colloids and Surfaces A. Physicochemical and Engineering Aspects* **232**, 239–247.

Dumitriu, S. and Chornet, E. (1998). Inclusion and release of proteins from polysaccharide-based polyion complexes. *Advanced Drug Delivery Reviews* **31**, 223–246.

Dyrby, M., Petersen, R. V., Larsen, J., Rudolf, B., Norgaard, L., and Engelsen, S. B. (2004). Towards on-line monitoring of the composition of commercial carrageenan powders. *Carbohydrate Polymers* **57**, 337–348.

El-Haddad, N. (2002). Étude des forces impliqués dans les gels mixtes d'alginate et de protéines de lactosérum Thèse M.Sc. Laval University, Québec. 73p.

Faber, E. J., Zoon, P., Kamerling, J. P., and Vliegenthart, J. F. G. (1998). The exopolysaccharides produced by Streptococcus thermophilus Rs and Sts have the same repeating unit but differ in viscosity of their milk cultures. *Carbohydrate Research* **310**, 269–276.

Foegeding, E. A., Luck, P. J., and Davis, J. P. (2006). Factors determining the physical properties of protein foams. *Food Hydrocolloids* **20**, 284–292.

Galazka, V. B., Smith, D., Ledward, D. A., and Dickinson, E. (1999). Complexes of bovine serum albumin with sulphated polysaccharides: effects of pH, ionic strength and high pressure treatment. *Food Chemistry* **64**, 303–310.

Ganzevles, R. A., Cohen Stuart, M. A., van Boekel, M. A. J. S., and De Jongh, H. H. J. (2007a). Manipulation of adsorption behaviour at liquid interfaces by changing protein-polysaccharide electrostatic interactions. *In* Food Colloids: Self-Assembly and Material Science, 195–208. Leser, M. E., and Dickinson, E. (ed.). Royal Society of Chemistry, Cambridge.

Ganzevles, R. A., Cohen Stuart, M. A., Vliet, T., de Jongh, H. H. J. (2006a). Use of polysaccharides to control protein adsorption to the air-water interface. *Food Hydrocolloids* **20**, 872–878.

Ganzevles, R. A., Fokkink, R., van Vliet, T., Cohen Stuart, M. A., and de Jongh, H. H. J. (2008). Structure of mixed β-lactoglobulin/pectin adsorbed layers at air/water interfaces; a spectroscopy study. *Journal of Colloid and Interface Science* **317**, 137–147.

Ganzevles, R. A., Kosters, H., van Vliet, T., Stuart, M. A. C., and de Jongh, H. H. J. (2007c). Polysaccharide charge density regulating protein adsorption to air/water interfaces by protein/polysaccharide complex formation. *Journal of Physical Chemistry B* **111**, 12969–12976.

Ganzevles, R. A., Zinoviadou, K., van Vliet, T., Stuart, M. A. C., and de Jongh, H. H. J. (2006b). Modulating surface rheology by electrostatic protein/polysaccharide interactions. *Langmuir* **22**, 10089–10096.

Gentes, M. C. (2007). Utilisation de complexes formés de pectine et d'isolat de protéines sériques dans la formulation de yogourts brasseés. Thèse MSc. Université Laval, Quebec. 160p.

Girard, M., Sanchez, C., Laneuville, S. I., Turgeon, S. L., and Gauthier, S. F. (2004). Associative phase separation of β-lactoglobulin/pectin solutions: a kinetic study by small angle static light scattering. *Colloids and Surfaces B: Biointerfaces* **35**, 15–22.

Girard, M. and Schaffer-Lequart, C. (2008). Attractive interactions between selected anionic exopolysaccharides and milk proteins. *Food Hydrocolloids* **22**, 1425–1434.

Girard, M., Turgeon, S.L., and Gauthier, S.F. (2002a). Interbiopolymer complexing between β-lactoglobulin and low- and high-methylated pectin measured by potentiometric titration and ultrafiltration. *Food Hydrocolloids* **16**, 585–591.

Girard, M., Turgeon, S. L., and Gauthier, S. F. (2003a). Quantification of the Interactions between β-Lactoglobulin and Pectin through Capillary Electrophoresis Analysis. *Journal of Agricultural & Food Chemistry* **51**, 6043–6049.

Girard, M., Turgeon, S. L., and Gauthier, S. F. (2003b). Thermodynamic Parameters of β-Lactoglobulin-Pectin Complexes Assessed by Isothermal Titration Calorimetry. *Journal of Agricultural and Food Chemistry* **51**, 4450–4455.

Girard, M., Turgeon, S. L., and Paquin, P. (2002b). Emulsifying Properties of Whey Protein-Carboxymethylcellulose Complexes. *Journal of Food Science* **67**, 113–119.

Gouin, S. (2004). Microencapsulation: industrial appraisal of existing technologies and trends: *Trends in food science and technology* **15** (7–8), 330–347.

Grymonpre, K. R., Staggemeier, B. A., Dubin, P. L., and Mattison, K. W. (2001). Identification by Integrated Computer Modeling and Light Scattering Studies of an Electrostatic Serum Albumin-Hyaluronic Acid Binding Site. *Biomacromolecules* **2**, 422–429.

Gu, Y. S., Decker, E. A., and Julian McClements, D. (2007). Application of multi-component biopolymer layers to improve the freeze-thaw stability of oil-in-water emulsions: β-Lactoglobulin-ι-carrageenan-gelatin. *Journal of Food Engineering* **80**, 1246–1254.

Gu, Y. S., Regnier, L., and McClements, D. J. (2005). Influence of environmental stresses on stability of oil-in-water emulsions containing droplets stabilized by β-lactoglobulin-ι-carrageenan membranes. *Journal of Colloid and Interface Science* **286**, 551–558.

Gummel, J., Cousin, F., and Boue, F. (2007). Counterions Release from Electrostatic Complexes of Polyelectrolytes and Proteins of Opposite Charge: A Direct Measurement. *Journal of the American Chemical Society* **129**, 5806–5807.

Guzey, D., Kim, H. J., and McClements, D. J. (2004). Factors influencing the production of o/w emulsions stabilized by β-lactoglobulin-pectin membranes. *Food Hydrocolloids* **18**, 967–975.

Guzey, D. and McClements, D. J. (2006). Formation, stability and properties of multilayer emulsions for application in the food industry. *Advances in Colloid and Interface Science* **128**, 227–248.

Guzey, D. and McClements, D. J. (2007). Impact of electrostatic interactions on formation and stability of emulsions containing oil droplets coated by β-lactoglobulin-pectin complexes. *Journal of Agricultural and Food Chemistry* **55**, 475–485.

Hassan, A. N. (2008). ADSA Foundation Scholar Award: Possibilities and challenges of exopolysaccharide-producing lactic cultures in dairy foods. *Journal of Dairy Science* **91**, 1282–1298.

Haug, I. J., Draget, K. I., and Smidsrød, O. (2004). Physical behaviour of fish gelatin-k-carrageenan mixtures. *Carbohydrate Polymers* **56**, 11–19.

Ivinova, O. N., Izumrudov, V. A., Muronetz, V. I., Galaev, I. Y., and Mattiasson, B. (2003). Influence of Complexing Polyanions on the Thermostability of Basic Proteins. *Macromolecular Bioscience* **3**, 210–215.

Ji, S., Corredig, M., and Goff, H. D. (2008). Aggregation of casein micelles and κ-carrageenan in reconstituted skim milk. *Food Hydrocolloids* **22**, 56–64.

Joanny, J.-F. (1994). Adsorption of a polyampholyte chain. *Journal de Physique II*, 1281–1288.

Junyaprasert, V. B., Mitrevej, A., Sinchaipanid, N., Boonme, P., and Wurster, D. E. (2001). Effect of process variables on the microencapsulation of vitamin a palmitate by gelatin-acacia coacervation. *Drug Development and Industrial Pharmacy* **27**, 561–566.

Kaibara, K., Okazaki, T., Bohidar, H. B., and Dubin, P. L. (2000). pH-Induced Coacervation in Complexes of Bovine Serum Albumin and Cationic Polyelectrolytes. *Biomacromolecules* **1**, 100–107.

Kasapis, S. (2008). Phase separation in biopolymer gels: A low- to high-solid exploration of structural morphology and functionality. *Critical Reviews in Food Science and Nutrition* **48**, 341–359.

Kaushik, V. and Roos, Y. H. (2007). Limonene encapsulation in freeze-drying of gum Arabic-sucrose-gelatin systems. *LWT – Food Science and Technology* **40**, 1381–1391.

Kayitmazer, A. B., Bohidar, H. B., Mattison, K. W., et al. (2007a). Mesophase separation and probe dynamics in protein-polyelectrolyte coacervates. *Soft Matter* **3**, 1064–1076.

Kayitmazer, A. B., Seyrek, E., Dubin, P. L., and Staggemeier, B. A. (2003). Influence of Chain Stiffness on the Interaction of Polyelectrolytes with Oppositely Charged Micelles and Proteins. *Journal of Physical Chemistry B* **107**, 8158–8165.

Kayitmazer, A. B., Strand, S. P., Tribet, C., Jaeger, W., and Dubin, P. L. (2007b). Effect of Polyelectrolyte Structure on Protein-Polyelectrolyte Coacervates: Coacervates of Bovine Serum Albumin with Poly(diallyldimethylammonium chloride) versus Chitosan. *Biomacromolecules* **8**, 3568–3577.

Kolodziejczyk, E., Schmitt, C. (2004). Interface stabilisation of a product with 2 or more phases with a protein-polysaccharide complex. NESTEC S.A. European Patent EP1402790.

Kruif, CGd, Weinbreck, F., and Vries, Rd (2004). Complex coacervation of proteins and anionic polysaccharides. *Current Opinion in Colloid and Interface Science* **9** (5), 340–349.

Kudryashova, E. V., Visser, A. J. W. G., vanHoek, A., and deJongh, H. H. J. (2007). Molecular Details of Ovalbumin-Pectin Complexes at the Air/Water Interface: A Spectroscopic Study. *Langmuir* **23** (15), 7942–7950.

Kumar, A., Dubin, P. L., Hernon, M. J., Li, Y., and Jaeger, W. (2007). Temperature-Dependent Phase Behavior of Polyelectrolyte-Mixed Micelle Systems. *Journal of Physical Chemistry B* **111**, 8468–8476.

Kyogoku, N., Saeki, T., and Fujikawa, S. (1994). Genipin as hardener for gelatin coacervates in producing microcapsules. Can. *Pat.* 1322133 A1.

Lamprecht, A. and Kawashima, Y. (2006). pH-sensitive microparticles for oral drug delivery. *Journal of Drug Delivery Science and Technology* **16**, 25–34.

Lamprecht, A., Schafer, U., and Lehr, C. M. (2001). Influences of process parameters on preparation of microparticle used as a carrier system for Omega-3 unsaturated fatty acid ethyl esters used in supplementary nutrition. *Journal of Microencapsulation* **18**, 347–357.

Lamprecht, A., Schäfer, U. F., and Lehr, C.-M. (2000). Characterization of microcapsules by confocal laser scanning microscopy: structure, capsule wall composition and encapsulation rate European. *Journal of Pharmaceutics and Biopharmaceutics* **49**, 1–9.

Laneuville, S. I., Paquin, P., and Turgeon, S. L. (2000). Effect of preparation conditions on the characteristics of whey protein-xanthan gum complexes. *Food Hydrocolloids* **14** (4), 305–314.

Laneuville, S. I., Paquin, P., Turgeon, S. L. (2005a). Formula optimization of a low-fat food system containing whey protein isolate-xanthan gum complexes as fat replacer *Journal of Food Science* **70**, S513–S9.

Laneuville, S. I., Sanchez, C., Turgeon, S. L., Hardy, J., and Paquin, P. (2005b). Small-angle Static Light-Scattering Study of Associative Phase Separation Kinetics of β-Lactoglobulin+Xanthan Gum Mixtures under Shear. *In* Food Colloids: Interaction, microstructure and processing, 443–465, Dickinson, E. (ed.). The Royal Society of Chemistry, Cambridge. Chapter 35.

Laneuville, S. I., Turgeon, S. L., Sanchez, C., and Paquin, P. (2006). Gelation of Native β-Lactoglobulin Induced by Electrostatic Attractive Interaction with Xanthan Gum. *Langmuir* **22**, 7351–7357.

Laneuville, S. I., Turgeon, S. L., Sanchez, C., Paquin, P. (2007). Gelation of undenatured proteins with polysaccharides. WO/2007/014447. Laval University.

Langendorff, V., Cuvelier, G., Launay, B., and Parker, A. (1997). Gelation and flocculation of casein micelle/carrageenan mixtures. *Food Hydrocolloids* **11**, 35–40.

Lapasin, R. and Pricl, S. (1995). *Rheology of industrial polysaccharides: theory and applications*. Blackie Academic & Professional, Glasgow. 620p.

Laplante, S., Turgeon, S.L., and Paquin, P. (2005a). Effect of pH, ionic strength, and composition on emulsion stabilising properties of chitosan in a model system containing whey protein isolate. *Food Hydrocolloids* **19**, 721–729.

Laplante, S., Turgeon, S. L., and Paquin, P. (2005b). Emulsion stabilizing properties of various chitosans in the presence of whey protein isolate. *Carbohydrate Polymers* **59**, 425–434.

Laplante, S., Turgeon, S. L., and Paquin, P. (2006). Emulsion-stabilizing properties of chitosan in the presence of whey protein isolate: Effect of the mixture ratio, ionic strength and pH. *Carbohydrate Polymers* **65**, 479–487.

Larisch, B., Pischetsrieder, M., and Severin, T. (1996). Reactions of dehydroascorbic acid with primary aliphatic amines including N-α-acetyllysine. *Journal of Agricultural and Food Chemistry* **44**, 1630–1634.

Li, Y., Xia, J., and Dubin, P. L. (1994). Complex Formation between Polyelectrolyte and Oppositely Charged Mixed Micelles: Static and Dynamic Light Scattering Study of the Effect of Polyelectrolyte Molecular Weight and Concentration. *Macromolecules* **27**, 7049–7055.

Liu, Z., Jiao, Y., Wang, Y., Zhou, C., and Zhang, Z. (2008). Polysaccharides-based nanoparticles as drug delivery systems. *Advanced Drug Delivery Reviews* **60**, 1650–1662.

Lucey, J. A. (2004). Cultured dairy products: an overview of their gelation and texture properties. *International Journal of Dairy Technology* **57** (2–3), 77–84.

Madene, A., Jacquot, M., Scher, J., and Desobry, S. (2006). Flavour encapsulation and controlled release International. *Journal of Food Science and Technology* **41**, 1–21.

Malone, M. E. and Appelqvist, I. A. M. (2003). Gelled emulsion particles for the controlled release of lipophilic volatiles during eating. *Journal of Controlled Release* **90**, 227–241.

Mann, B. and Malik, R. C. (1996). Studies on some functional characteristics of whey protein-polysaccharide complex. *Journal of Food Science and Technology (India)* **33**, 202–206.

Marshall, V. M. and Rawson, H. L. (1999). Effects of exopolysaccharide-producing strains of thermophilic lactic acid bacteria on the texture of stirred yoghurt. *International Journal of Food Science and Technology* **34**, 137–143.

Matia-Merino, L., Lau, K., and Dickinson, E. (2004). Effects of low-methoxyl amidated pectin and ionic calcium on rheology and microstructure of acid-induced sodium caseinate gels. *Food Hydrocolloids* **18**, 271–281.

Mattison, K. W., Brittain, I. J., and Dubin, P. L. (1995). Protein-Polyelectrolyte Phase Boundaries. *Biotechnology Progress* **11**, 632–637.

Mattison, K. W., Dubin, P. L., and Brittain, I. J. (1998). Complex Formation between Bovine Serum Albumin and Strong Polyelectrolytes: Effect of Polymer Charge Density. *Journal of Physical Chemistry B* **102**, 3830–3836.

Mattison, K. W., Wang, Y., Grymonpré, K., and Dubin, P. L. (1999). Micro- and Macro-phase behavior in protein-polyelectrolyte complexes. *Macromolecular Symposia* **140**, 53–76.

Mayya, K. S., Bhattacharyya, A., and Argillier, J. F. (2003). Micro-encapsulation by complex coacervation: influence of surfactant. *Polymer International* **52**, 644–647.

McClements, D. J. (2005). Theoretical analysis of factors affecting the formation and stability of multilayered colloidal dispersions. *Langmuir* **21**, 9777–9785.

McClements, D. J. (2006). Non-covalent interactions between proteins and polysaccharides. *Biotechnology Advances* **24**, 621–625.

McClements, D. J., Decker, E. A., and Weiss, J. (2007). Emulsion-Based Delivery Systems for Lipophilic Bioactive Components. *Journal of Food Science* **72**, R109–R24.

Mekhloufi, G., Sanchez, C., Renard, D., Guillemin, S., and Hardy, J. (2005). pH-Induced Structural Transitions during Complexation and Coacervation of β-Lactoglobulin and Acacia Gum. *Langmuir* **21**, 386–394.

Mellema, M. (2005). Complex coacervate encapsulate comprising lipophilic core. Univeler European Patent EP1585592.

Menger, F. M. and Sykes, B. M. (1998). Anatomy of a Coacervate. *Langmuir* **14**, 4131–4137.

Menjoge, A. R., Kayitmazer, A. B., Dubin, P. L., Jaeger, W., and Vasenkov, S. (2008). Heterogeneity of Polyelectrolyte Diffusion in Polyelectrolyte-Protein Coacervates: A 1H Pulsed Field Gradient NMR Study. *Journal of Physical Chemistry B* **112**, 4961–4966.

Mohanty, B. and Bohidar, H. B. (2005). Microscopic structure of gelatin coacervates. *International Journal of Biological Macromolecules* **36**, 39–46.

Morris, V. J. (1986). Multicomponent gels. *In* Gums and stabilizers for the food industry-9, 87–99, Philips, G. O., Wedlock, D. J. and Williams, P. A. (eds.). Elsevier, London.

Mounsey, J. S., O'Kennedy, B. T., Fenelon, M. A., and Brodkorb, A. (2008). The effect of heating on β-lactoglobulin-chitosan mixtures as influenced by pH and ionic strength. *Food Hydrocolloids* **22**, 65–73.

Mun, S., Cho, Y., Decker, E. A., and McClements, D. J. (2008). Utilization of polysaccharide coatings to improve freeze-thaw and freeze-dry stability of protein-coated lipid droplets. *Journal of Food Engineering* **86**, 508–518.

Murray, B. S. (2007). Stabilization of bubbles and foams. *Current Opinion in Colloid & Interface Science* **12**, 232–241.

Neirynck, N., Van lent, K., Dewettinck, K., and Van der Meeren, P. (2007). Influence of pH and biopolymer ratio on sodium caseinate-guar gum interactions in aqueous solutions and in O/W emulsions. *Food Hydrocolloids* **21**, 862–869.

Nigen, M., Croguennec, T., Renard, D., and Bouhallab, S. (2007). Temperature Affects the Supramolecular Structures Resulting from α-Lactalbumin-Lysozyme Interaction. *Biochemistry* **46**, 1248–1255.

Ogawa, S., Decker, E. A., and McClements, D. J. (2003). Influence of environmental conditions on the stability of oil in water emulsions containing droplets stabilized by lecithin-chitosan membranes. *Journal of Agricultural and Food Chemistry* **51**, 5522–5527.

Ou, Z. and Muthukumar, M. (2006). Entropy and enthalpy of polyelectrolyte complexation: Langevin dynamics simulations. *The Journal of Chemical Physics* **124**, 154902–154911.

Ould Eleya, M. and Turgeon, S. L. (2000). The effects of pH on the rheology of β-lactoglobulin/k-carrageenan mixed gels. *Food Hydrocolloids* **14** (3), 245–251.

Overbeek, J. T. G. and Voorn, M. J. (1957). Phase separation in polyelectrolyte solutions: Theory of complex coacervation. *Journal of Cellular and Comparative Physiology* **49**, 7–26.

Pajonk, A. S., Saurel, R., Andrieu, J., Laurent, P., and Blanc, D. (2003). Heat transfer study and modeling during Emmental ripening. *Journal of Food Engineering* **57**, 249–255.

Park, J. M., Muhoberac, B. B., Dubin, P. L., and Xia, J. (1992). Effects of protein charge heterogeneity in protein-polyelectrolyte complexation. *Macromolecules* **25**, 290–295.

Pereyra, R., Schmidt, K. A., and Wicker, L. (1997). Interaction and stabilization of acidified casein dispersions with low and high methoxyl pectins. *Journal of Agricultural and Food Chemistry* **45**, 3448–3451.

Perrau, M. B., Iliopoulos, I., and Audebert, R. (1989). Phase separation of polyelectrolyte/nonionic polymer systems in aqueous solution: effects of salt and charge density. *Polymer* **30**, 2112–2117.

Piculell, L. and Lindman, B. (1992). Association and Segregation in Aqueous Polymer/Polymer, Polymer Surfactant, and Surfactant Surfactant Mixtures – Similarities and Differences. *Advances in Colloid and Interface Science* **41**, 149–178.

Pierucci, A., Andrade, L. R., Farina, M., Pedrosa, C., and Rocha-Leao, M. H. M. (2007). Comparison of α-tocopherol microparticles produced with different wall materials: pea protein a new interesting alternative. *Journal of Microencapsulation* **24**, 201–213.

Plashchina, I. G., Zhuravleva, I. L., and Antonov, Y. A. (2007). Phase behavior of gelatin in the presence of pectin in water-acid medium. *Polymer Bulletin* **58**, 587–596.

Popov, Y. O., Lee, J., and Fredrickson, G. H. (2007). Field-theoretic simulations of polyelectrolyte complexation.

Journal of Polymer Science Part B: Polymer Physics **45**, 3223–3230.

Prata, A. S., Menut, C., Leydet, A., Trigo, J. R., and Grosso, C. R. F. (2008). Encapsulation and release of a fluorescent probe, khusimyl dansylate, obtained from vetiver oil by complex coacervation. *Flavour and Fragrance Journal* **23**, 7–15.

Prokop, A., Hunkeler, D., Powers, A., Whitesell, R., and Wang, T. (1998). Water Soluble Polymers for Immunoisolation II: Evaluation of Multicomponent Microencapsulation In Systems Microencapsulation Microgels Iniferters. *Advances in Polymer Science* **136**, 53–73.

Rey, B., Rossi-vauthey, P., Rousset, P., Sandoz, l., Schafer, O., and Schmitt, C.J. (2007). Low-fat confectionery product. European Patent EP1759591 NESTEC S.A.

Rioux, L.-E., Turgeon, S. L., and Beaulieu, M. (2007). Characterization of polysaccharides extracted from brown seaweeds. *Carbohydrate Polymers* **69**, 530–537.

Ruas-Madiedo, P., Hugenholtz, J., and Zoon, P. (2002). An overview of the functionality of exopolysaccharides produced by lactic acid bacteria. *International Dairy Journal* **12**, 163–171.

Sagis, L. M. C., de Ruiter, R., Miranda, F. J. R., et al. (2008). Polymer microcapsules with a fiber-reinforced nanocomposite shell. *Langmuir* **24**, 1608–1612.

Samant, S. K., Singhal, R. S., Kulkarni, P. R., and Rege, D. V. (1993). Protein-polysaccharide interactions a new approach in food formulations. *International Journal of Food Science & Technology* **28**, 547–562.

Sanchez, C., Despond, S., Schmitt, C., and Hardy, J. (2001). Effect of Heat and Shear on β-Lactoglobulin-Acacia Gum Complex Coacervation. *In* Food Colloids: Fundamentals of Formulation, 332–343, Dickinson, E. and Miller, R. (eds.). Royal Society of Chemistry, Cambridge, UK.

Sanchez, C., Mekhloufi, G., Schmitt, C., Renard, D., Robert, P., Lehr, C.-M., Lamprecht, A., and Hardy, J. (2002). Self-Assembly of β-Lactoglobulin and Acacia Gum in Aqueous Solvent: Structure and Phase-Ordering Kinetics. *Langmuir* **18**, 10323–10333.

Sanchez, C. and Paquin, P. (1997). Protein and protein-polysaccharide microparticles. *In* Food proteins and their applications; 503–528, Damodaran, S. and Paraf, A. (eds.). Marcel Dekker, New York.

Sanchez, V. E., Bartholomai, G. B., and Pilosof, A. M. R. (1995). Rheological Properties of Food Gums As Related To Their Water Binding-Capacity and To Soy Protein-Interaction. *Food Science and Technology-Lebensmittel-Wissenschaft & Technologie* **28**, 380–385.

Schmitt, C., Kolodziejczyk, E., and Leser, M.E. (2005a). Interfacial and foam stabilization properties of β-lactoglobulin-Acacia gum electrostatic complexes. *In* Food Colloids: Interactions, microstructure and processing, 289–300, Dickinson, E. (ed.). Royal Society of Chemistry, Cambridge.

Schmitt, C., PalmadaSilva, T., Bovay, C., Rami-Shojaei, S., Frossard, P., Kolodziejczyk, E., and Leser, M. E. (2005b). Effect of Time on the Interfacial and Foaming Properties of β-Lactoglobulin/Acacia Gum Electrostatic Complexes and Coacervates at pH 4.2. *Langmuir* **21**, 7786–7795.

Schmitt, C., Sanchez, C., Desobry-Banon, S., and Hardy, J. (1998). Structure and Technofunctional Properties of Protein-Polysaccharide Complexes. *A Review Critical Reviews in Food Science and Nutrition* **38**, 689–753.

Schmitt, C., Sanchez, C., Despond, S., Renard, D., Robert, P., and Hardy, J. (2001a). Structural modification of β-lactoglobulin as induced by complex coacervation with acacia gum. *In* Food Colloids 2000-Fundamentals of Formulation, 323–331, Dickinson, E. and Miller, R. (eds.). Royal Society of Chemistry, Cambridge.

Schmitt, C., Sanchez, C., Lamprecht, A., Renard, D., Lehr, C.-M., de Kruif, C. G., and Hardy, J. (2001b). Study of β-lactoglobulin/acacia gum complex coacervation by diffusing-wave spectroscopy and confocal scanning laser microscopy. *Colloids and Surfaces B: Biointerfaces* **20**, 267–280.

Seyrek, E., Dubin, P. L., Tribet, C., and Gamble, E. A. (2003). Ionic Strength Dependence of Protein-Polyelectrolyte Interactions. *Biomacromolecules* **4**, 273–282.

Silletti, E., Vingerhoeds, M. H., Norde, W., and Van Aken, G. A. (2007). The role of electrostatics in saliva-induced emulsion flocculation. *Food Hydrocolloids* **21**, 596–606.

Singh, S. S., Aswal, V.K., and Bohidar, H. B. (2007a). Structural studies of agar-gelatin complex coacervates by small angle neutron scattering, rheology and differential scanning calorimetry. *International Journal of Biological Macromolecules* **41**, 301–307.

Singh, S. S., Bohidar, H. B., and Bandyopadhyay, S. (2007b). Study of gelatin-agar intermolecular aggregates in the supernatant of its coacervate. *Colloids and Surfaces B: Biointerfaces* **57**, 29–36.

Soper, J. C. and Thomas, M. T. (1997). Enzymatically protein encapsulating oil particles by complex coacervation. US Pat. 6-039-901.

Soucie, W. G., Chen, W. S., Witte, V. C., Henry, G. A., and Drehkoff, W. D. (1988). Shelf Stable Acid Food Dressings Containing Fibrous Protein Complexes. KRAFT Inc., Glenview, IL, USA. U.S. Patent 4,762,726.

Strauss, G. and Gibson, S. A. (2004). Plant phenolics as cross-linkers of gelatin gels and gelatin-based coacervates for use as food ingredients. *Food Hydrocolloids* **18**, 81–89.

Syrbe, A., Bauer, W. J., and K., H. (1998). Polymer science concepts in dairy systems – an overview of milk protein and food hydrocolloid interaction. *International Dairy Journal* **8**, 179–193.

Tainaka, K.-I. (1979). Study of Complex Coacervation in Low Concentration by Virial Expansion Method. I. Salt Free Systems. *Journal of Physical Society Japan* **46**, 1899–1906.

Takahashi, D., Kubota, Y., Kokai, K., Izumi, T., Hirata, M., and Kokufuta, E. (2000). Effects of Surface Charge Distribution of Proteins in Their Complexation with Polyelectrolytes in an Aqueous Salt-Free System. *Langmuir* **16**, 3133–3140.

Tan, B. H. and Tam, K. C. (2008). Review on the dynamics and micro-structure of pH-responsive nano-colloidal systems. *Advances in Colloid and Interface Science* **136**, 25–44.

Teramoto, A., Takagi, Y., Hachimori, A., and Abe, K. (1999). Interaction of albumin with polysaccharides containing ionic groups. *Polymers for Advanced Technologies* **10**, 681–686.

Taylor, A. J. (2002). Release and Transport of Flavors *In Vivo*: Physicochemical, Physiological, and Perceptual Considerations. *Comprehensive Reviews in Food Science and Food Safety* **1**, 45–57.

Tessier, F. J., Monnier, V. M., Sayre, L. M., and Kornfield, J. A. (2003). Triosidines: novel Maillard reaction products and cross-links from the reaction of triose sugars with lysine and arginine residues. *Biochemical Journal* **369**, 705–719.

Thies, C. (2007). Microencapsulation of Flavors by Complex Coacervation. *In* Encapsulation and Controlled Release Technologies in Food Systems, Lakkis, J. M. (ed.). Blackwell Publishing 149–70.

Thimma, R. T. and Tammishetti, S. (2003). Study of complex coacervation of gelatin with sodium carboxymethyl guar gum: microencapsulation of clove oil and sulphamethoxazole. *Journal of Microencapsulation* **20**, 203–210.

Tiebackx, F. W. Z. (1911). Gleichzeitige Ausflockung zweier. *Zeitschrift fur Chemie und Industrie der Kolloide* **8**, 198–201.

Tolstoguzov, V. (1996). Structure-property relationships in foods. *Macromolecular Interactions in Food Technology* **650**, 2–14.

Tolstoguzov, V. B. (1986). Functional properties of protein-polysaccharide mixtures. *In* Functional properties of food macromolecules, 385–415, Mitchell, J. R. and Ledward, D. A. (eds.). Elsevier Applied Science, London.

Tolstoguzov, V. B. (1997). Protein-Polysaccharide Interactions. *In* Food Proteins and Their Applications in Foods, 171–198, Damodaran, S. and Paraf, A. (eds.). Marcel Dekker Inc., New York.

Tolstoguzov, V. B. (2003). Some thermodynamic considerations in food formulation. *Food Hydrocolloids* **17**, 1–23.

Tolstoguzov, V. B. (2007). Ingredient interactions in complex foods: aggregation and phase separation. *In* Understanding and controlling the microstructure of complex foods, McClements, D. J. (ed.). Woodhead Publishing, Cambridge.

Tromp, R. H., de Kruif, C. G., van Eijk, M., and Rolin, C. (2004). On the mechanism of stabilisation of acidified milk drinks by pectin. *Food Hydrocolloids* **18**, 565–572.

Tsuboi, A., Izumi, T., Hirata, M., Xia, J., Dubin, P. L., and Kokufuta, E. (1996). Complexation of Proteins with a Strong Polyanion in an Aqueous Salt-free System. *Langmuir* **12**, 6295–6303.

Tuinier, R., van, C. W., Looijesteijn, P., Schols, H., Voragen, A., and Zoon, P. (2001). Effects of structural modifications on some physical characteristics of exopolysaccharides from *Lactococcus lactis Biopolymers* **59**, 160–166.

Turgeon, S. L. and Beaulieu, M. (2001). Improvement and modification of whey protein gel texture using polysaccharides. *Food Hydrocolloids* **15** (4-6), 583–591.

Turgeon, S. L., Beaulieu, M., Schmitt, C., and Sanchez, C. (2003). Protein-polysaccharide interactions: phase-ordering kinetics, thermodynamic and structural aspects. *Current Opinion in Colloid & Interface Science* **8**, 401–414.

Turgeon, S. L., Schmitt, C., and Sanchez, C. (2007). Protein-polysaccharide complexes and coacervates. *Current Opinion in Colloid & Interface Science* **12**, 166–178.

Van Calsteren, M.-R., Pau-Roblot, C., Bégin, A., and Roy, D. (2002). Structure determination of the exopolysaccharide produced by *Lactobacillus rhamnosus* strains RW-9595M and R. *Biochemistry Journal* **363**, 7–17.

van de Weert, M., Andersen, M. B., and Frokjaer, S. (2004). Complex Coacervation of Lysozyme and Heparin: Complex Characterization and Protein Stability. *Pharmaceutical Research* **21**, 2354–2359.

Veis, A. and Aranyi, C. (1960). Phase Separation in Polyeletrolyte Systems I. Complex Coacervates of gelatin. *Journal of Physical Chemistry* **64**, 1203–1210.

Walstra, P. (2003). *Physical chemistry of foods*. Marcel Dekker Inc., New York.

Wang, X., Jiang, Y. M., and Huang, Q. (2007a). Encapsulation technologies for preserving and controlling the release of enzymes and phytochemicals. *In* Encapsulation and controlled release technologies in food systems, 135–147. Lakkis, J. M. (ed.). Blackwell Publisher.

Wang, X., Lee, J., Wang, Y.-W., and Huang, Q. (2007b). Composition and Rheological Properties of β-Lactoglobulin/Pectin Coacervates: Effects of Salt Concentration and Initial Protein/Polysaccharide Ratio. *Biomacromolecules* **8**, 992–997.

Wang, X., Li, Y., Wang, Y.-W., Lal, J., and Huang, Q. (2007c). Microstructure of β-Lactoglobulin/Pectin Coacervates Studied by Small-Angle Neutron Scattering. *Journal of Physical Chemistry B* **111**, 515–520.

Wang, Y., Kimura, K., Dubin, P. L., and Jaeger, W. (2000). Polyelectrolyte-Micelle Coacervation: Effects of Micelle Surface Charge Density, Polymer Molecular Weight: and Polymer/Surfactant Ratio. *Macromolecules* **33**, 3324–3331.

Wang, Y., Kimura, K., Huang, Q., Dubin, P. L., and Jaeger, W. (1999). Effects of Salt on Polyelectrolyte-Micelle Coacervation. *Macromolecules* **32**, 7128–7134.

Weinbreck, F., de Vries, R., Schrooyen, P., and de Kruif, C. G. (2003a). Complex Coacervation of Whey Proteins and Gum Arabic. *Biomacromolecules* **4**, 293–303.

Weinbreck, F., Minor, M., and de Kruif, C.G. (2004a). Microencapsulation of oils using whey protein/gum arabic coacervates. *Journal of Microencapsulation* **21**, 667–679.

Weinbreck, F., Nieuwenhuijse, H., Robijn, G. W., and deKruif, C. G. (2003b). Complex Formation of Whey Proteins: Exocellular Polysaccharide EPS B40. *Langmuir* **19**, 9404–9410.

Weinbreck, F., Nieuwenhuijse, H., Robijn, G.W., and deKruif, C. G. (2004b). Complexation of Whey Proteins with Carrageenan. *Journal of Agricultural and Food Chemistry* **52**, 3550–3555.

Weinbreck, F., Rollema, H.S., Tromp, R.H., and deKruif, C. G. (2004c). Diffusivity of Whey Protein and Gum Arabic in Their Coacervates. *Langmuir* **20**, 6389–6395.

Weinbreck, F., Tromp, R.H., and deKruif, C.G. (2004d). Composition and Structure of Whey Protein/Gum Arabic Coacervates. *Biomacromolecules* **5**, 1437–1445.

Weinbreck, F. and Wientjes, R. H. W. (2004e). Rheological properties of whey protein/gum arabic coacervates. *Journal of Rheology* **48**, 1215–1228.

Wen, Y.-P. and Dubin, P. L. (1997). Potentiometric Studies of the Interaction of Bovine Serum Albumin and Poly(dimethyldiallylammonium chloride). *Macromolecules* **30**, 7856–7861.

Xia, J., Dubin, P. L., Kim, Y., Muhoberac, B. B., and Klimkowski, V. J. (1993). Electrophoretic and quasi-elastic light scattering of soluble protein-polyelectrolyte complexes. *Journal of Physical Chemistry* **97**, 4528–4534.

Yeo, Y., Bellas, E., Firestone, W., Langer, R., and Kohane, D. S. (2005). Complex coacervates for thermally sensitive controlled release of flavor compounds. *Journal of Agricultural and Food Chemistry* **53**, 7518–7525.

Yoshida, K., Sokhakian, S., and Dubin, P. L. (1998). Binding of Polycarboxylic Acids to Cationic Mixed Micelles: Effects of Polymer Counterion Binding and Polyion Charge Distribution. *Journal of Colloid and Interface Science* **205**, 257–264.

Yu, S. Y., Hu, J. H., Pan, X. Y., Yao, P., and Jiang, M. (2006). Stable and pH-sensitive nanogels prepared by self-assembly of chitosan and ovalbumin. *Langmuir* **22**, 2754–2759.

Zasypkin, D. V., EE, B., and VB, T. (1997). Multicomponent biopolymer gels. *Food Hydrocolloids* **11**, 159–170.

Zhang, G., Foegeding, E. A., and Hardin, C. C. (2004). Effect of Sulfated Polysaccharides on Heat-Induced Structural Changes in β-Lactoglobulin. *Journal of Agricultural and Food Chemistry* **52**, 3975–3981.

CHAPTER

12

Single Molecule Techniques: Atomic Force Microscopy and Optical Tweezers

Victor J. Morris

Institute of Food Research, Norwich Research Park, Colney, Norwich NR4 7UA, UK

12.1 ATOMIC FORCE MICROSCOPY

The atomic force microscope (AFM) is the most useful and versatile type of probe microscope for studying biological systems (Morris et al., 1999). In food science the use of AFM has allowed new systems to be observed in new ways that have enabled the solution of previously intractable problems.

Like all probe microscopes the AFM generates images by 'feeling' a surface with a probe (Binnig et al., 1986). By monitoring the interaction between the probe and the sample surface it is possible to generate a variety of images of the surface structure. Conventional microscopes amplify the way we see specimens. The AFM amplifies the way a blind person would create an image by touching a specimen.

Because the AFM generates images by feeling the surface structure and texture with a sharp probe or tip it is possible to visualize samples in either a liquid or a gaseous environment. This reduces the level of sample preparation and results in images of samples in a 'near-native' state.

12.1.1 Image Formation

In order to generate an image of a specimen it is necessary to move the probe relative to the sample surface. This can be done in one of two ways: either the sample is scanned beneath a probe or the probe is scanned over the sample. The highest resolution is generally achieved in an apparatus that scans the sample beneath the probe.

The essential elements of a typical AFM are shown in Figure 12.1. In the type of instrument shown the specimen is positioned on top of a piezoelectric device, usually a tube scanner (Binnig and Smith, 1986), which can be used to move the sample in three dimensions (3D). The specimen is moved by causing the expansion or contraction of the tube scanner. The practical resolution of the instrument will be determined by the noise level associated with the power supply used to control the tube scanners. The probe or tip used to feel the specimen is microfabricated, usually from silicon nitride (Albrecht and Quate, 1988). The tips are pyramidal in shape and typically ~ 3 μm in height. The radius of curvature at the end of the tip is typically ~ 30–50 nm. The tips are attached at one end of

Kasapis, Norton, and Ubbink: Modern Biopolymer Science
ISBN: 978-0-12-374195-0

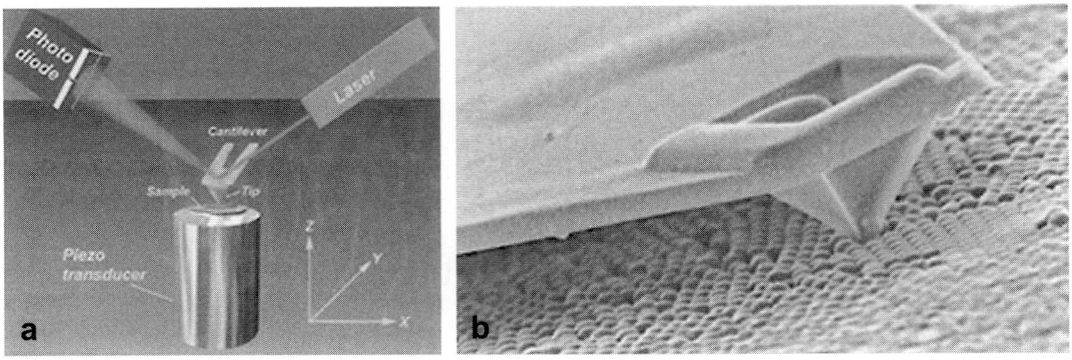

FIGURE 12.1 (a) The essential features of an atomic force microscope. (b) A scanning electron micrograph of the tip-cantilever assembly.

a cantilever (Figure 12.1b), which bends or twists as the probe experiences changes in force between the tip and the sample surface during scanning. Any movement of the cantilever can be detected through a variety of methods (Meyer, 1992) but the normal method of detection in modern instruments is the use of an optical lever. A low-power laser beam is reflected from the end of the cantilever onto a four-quadrant photodiode (Figure 12.1a). As the cantilever bends, or twists, the change in position of the reflected laser beam on the photodiode maps the motion of the cantilever. The output from the photodiode, in the form of a potential difference, is used to generate images of the sample.

In order to map the structure of the specimen the sample is raster-scanned underneath the probe. Typically a single scan will consist of an array of 256 × 256 image points. The sample is moved by the tube scanner to an image point and then dwells at that point whilst data are collected. The sample is then moved to the next image point and scanned line by line beneath the probe.

At each image point there will be a change in the 'up and down' deflection of the cantilever, due to a change in the force of interaction between the tip and the sample. If the sample is homogeneous in structure, then the change in force will arise purely due to changes in the separation of the tip and sample: the changes in force then reflect changes in the topography of the sample surface. Thus, by observing the changes in deflection of the cantilever, it becomes possible to produce various 3D images of the sample.

12.1.2 Imaging Modes

A number of different types of images can be generated (Meyer, 1992; Morris et al., 1999; Garcia and Perez, 2002; Holscher and Schirmeisen, 2005). The most common method of generating an image is called the dc imaging mode. At the first image point the applied force (cantilever deflection) is pre-set. At each subsequent image point, the potential difference, generated by the photodiode, and corresponding to the up and down deflection of the cantilever (the error signal), is fed into a feedback circuit which is then used to control the vertical (z) motion of the tube scanner. By moving the tip and sample together, or apart, it is possible to reduce the error signal to zero. This movement of the tip relative to the sample surface corrects for the change in tip-sample separation (surface topography) observed on moving between the image points. Thus, if the dwell time at each image point is sufficiently long, then the resultant motions of the tube scanner in x, y and z, generated by the feedback loop, can be amplified to produce a 3D image of the topography of the sample surface. If the sample is homogeneous in structure then the gray levels in the image will be directly related to the variations in height of the surface. In practice the sample is often heterogeneous and other surface features will affect the measured changes in force between the tip and the sample: these factors can include variations in the charge distribution or the relative hardness of the sample. Because these material properties affect the measured force they contribute to the apparent height or contrast in the image. The imaging modes can be modified to emphasize these features: a technique called force mapping can be used to map distributions of elasticity or charge on biological surfaces (Hoh et al., 1997; Heinz and Hoh, 1999).

Another type of image can be generated by switching off the feedback circuit. In this case, at each image point, the error signal can be measured and used to generate an image of the surface. The gray levels in the image now correspond to changes in force and are only indirectly related to the topography of the sample. In this 'deflection mode' the images can be acquired more quickly, because the dwell time at each image point is zero. Monitoring the deflection has allowed the development of high-speed video rate imaging (Humphris et al., 2005).

The AFM can be operated in an intermediate mode often called the 'error signal mode'. The feedback circuit is switched on, but, under the normal operating conditions, the dwell times are insufficient to allow the error signal to be completely reduced to zero at each image point. The residual error signal is then used to generate an image. This mode of operation is particularly useful for imaging rough samples. Generally images are normalized in order to maximize the use of gray levels to accommodate the variation in height across the sample surface. For rough samples the fine detail in the image is obscured, due to the need to use most of the range of gray levels to accommodate the relatively large, slowly varying background in the image. The slowly varying background can be corrected by the feedback circuit, and the feedback circuit effectively acts as a filter, removing the background, and enhancing fine detail in the image. This type of image provides a quick pictorial representation of the structure, which facilitates processing of true topographical images to reveal the fine detail in the image.

In addition to the 'up and down' motion it is also possible to monitor any twisting of the cantilever. Twisting of the cantilever arises from friction or localized adhesion between the tip and the sample surface. For heterogeneous samples the changes in adhesion can be monitored and used to produce frictional or adhesive maps of the surface. If the adhesive forces between the tip and the sample become too large then the act of scanning the sample beneath the tip can damage the sample. Such effects can be serious when the AFM is used to image biological structures or single molecules deposited onto flat substrates: imaging in air can distort, destroy or displace deposited structures. In air the most common cause of adhesion arises from capillary forces. When biological samples are deposited from solution onto the substrate most of the aqueous solvent evaporates leaving the biological material on the substrate. A thin layer of water remains, coating both the substrate and the surface of the tip. When the tip is brought close to the substrate coalescence of these surface layers occurs, effectively sticking the tip to the substrate. When the sample is scanned beneath the tip it becomes difficult to lift the tip over the deposited sample: the tip rips through the sample or displaces it. This effect can be eliminated if images are obtained under liquids, or by using techniques known as ac imaging modes (Meyer, 1992; Morris et al., 1999; Garcia and Perez, 2002; Holscher and Schirmeisen, 2005).

In order to image under liquids the sample and substrate need to be contained within a liquid cell, the nature of which will vary for different makes of AFM. Different types of liquids can be used. When high-resolution images of biopolymers are required it is often useful to image the deposited molecules under a precipitant. The liquid prevents any desorption from the surface and can often be used to induce the molecules to adopt an ordered conformation which, in turn, restricts molecular motion and blurring of the image. When the liquid used is a solvent for the molecules it is necessary to develop methods for pinning the molecules to the substrate surface, in order to prevent desorption or motion on the substrate surface.

Even in air it is possible to eliminate the effects of capillary forces through the use of what are termed ac imaging modes (Meyer, 1992; Morris et al., 1999; Garcia and Perez, 2002; Holscher and Schirmeisen, 2005). AC imaging modes involve the use of external stimulus to oscillate the cantilever close to its resonant frequency. Any interactions between the tip and the sample will change the amplitude and the phase of the oscillation. Both factors can be used to generate images of the sample. The details of the acquisition and display of these images varies considerably between different makes of commercial AFM. Changes in phase result from differences in energy dissipation in different regions of the sample. Phase images are often very sensitive to differences in material properties and hence contrast in the images. However, the interpretation of such images is still not straightforward.

The most familiar form of ac imaging is Tapping mode™ (Garcia and Perez, 2002). The amplitude of the oscillation is large and the tip only momentarily taps onto the sample surface. The use of stiff cantilevers prevents adhesion and limits specimen damage. Whilst 'Tapping mode' largely eliminates frictional or adhesive forces, it is more difficult to control the applied normal force, which can also lead to sample damage. However, Tapping mode is probably the best method for imaging in air. Tapping and ac images can also be obtained under liquids.

By modifying the nature of standard tips it is possible to select different forms of interaction and to introduce new types of contrast into images. Standard tips can be altered by chemical modification or by the attachment of molecular species to the tip. Modified tips can be used to locate or map specific structures on a surface. The procedures are known as chemical or affinity mapping (Dufrene and Hinterdorfer, 2008). For example, an antibody-labeled tip can be used to search for a particular molecular species on a sample surface or chemically modified tips can be used to map hydrophilic or hydrophobic areas on surfaces.

The AFM is a microscope which is used primarily in food science to image biological samples or to map surface roughness or composition. However, at a given image point, variation of the tip-sample separation can be used to generate a force–distance curve. For modified tips this can provide information on specific chemical or molecular interactions. It is also possible to adapt the AFM to probe interactions between colloidal particles. These types of studies have been critically reviewed by Ralston and co-workers (Ralston et al., 2005). Although such studies have led to improved understanding of colloidal forces and the molecular origins of friction and adhesion they are predominantly focused on studies between hard spheres and coated hard spheres as models for biological systems. In soft matter systems such as foods there is considerable interest in the interactions between deformable particles and this type of study will be addressed later in this article.

A penalty that has to be paid for high resolution is that it restricts the maximum scan range of the tube scanner. This is not a problem when imaging individual isolated molecules. However, high-resolution imaging of more complex biological structures is complicated by the difficulty in locating the region of interest within the biological specimen. This type of problem can be overcome by building combined AFM–optical microscopes. In the early first-generation models the AFM was simply added onto an inverted optical microscope. This design compromised the mechanical stability of the AFM and introduced noise which restricted the achievable resolution. In the design of the latest second generation of hybrid microscopes the AFM is now an integral part of the microscope, resulting in resolution as good as that of a standalone AFM. For these instruments the light microscope can be used to examine a biological sample and identify areas for further investigation. It is then possible to zoom in on a chosen feature and image it with the AFM.

Although the AFM achieves a resolution equivalent to that of the electron microscope the nature of the images obtained are very different. The AFM delivers a surface profile whereas, in the electron microscope, the electron beam interacts with all of the material as it passes through the object. In the electron microscope the information that can be obtained allows 3D reconstruction of the object. This is not possible with AFM images and, in addition, the image obtained depends on the shape and size of the probe tip, and the size of the object being scanned. Because of the finite size and shape of the tip, different regions of the tip interact with the sample as it is scanned. This leads to a broadening of the image termed 'probe broadening' (Figure 12.2). For larger objects or rough surfaces interactions will occur with the side-walls of the tip giving an angular appearance to the images. With smaller objects, such as molecules deposited onto atomically flat substrates such as mica, the interaction will be largely with the hemispherical tip of the probe leading to an apparent increase in the width of the molecules. This effect can be corrected through the use of calibration standards or simple geometric models (Vesenka et al., 1996) allowing true sample widths to be measured. Advantages of the AFM are the ability to image under more natural conditions and to produce 3D images from which heights can be measured. However, imaging involves pressing on the sample and, if the object is deformable, then it will be compressed to some degree during scanning and the measured height will be smaller than the true height. This effect can be reduced by optimizing the normal force applied during scanning.

12.1.3 Applications

Microscopic methods provide information on structure. The use of new microscopic methods in food science has always led to new scientific insights, which have spawned new technological applications. The use of AFM has allowed new information to be obtained on food systems and has led to the solution of previously intractable problems. The following account of applications is intended to illustrate the sort of information that can be obtained using the technique and to indicate areas where use of AFM has generated significant new findings.

12.1.4 Macromolecules

AFM is ideally suited to imaging food macromolecules and methods have been developed for

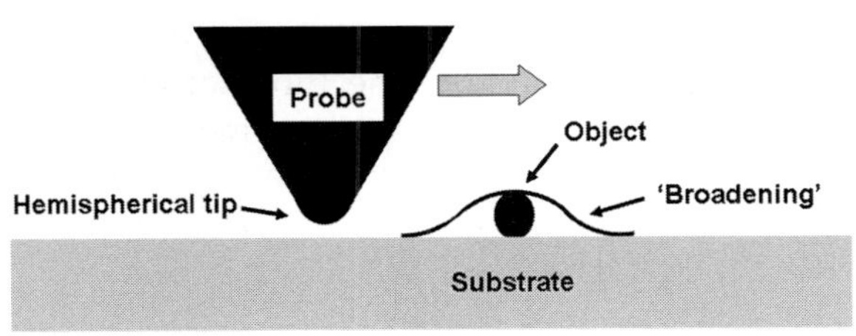

FIGURE 12.2 Schematic diagram illustrating probe broadening effects.

imaging food polysaccharides and proteins: in general the molecules are deposited onto atomically flat substrates such as freshly cleaved mica (Morris et al., 1999).

12.1.5 Polysaccharides

The information that can be obtained for individual polysaccharides will depend on the primary structure of the polysaccharide. For random-coil polysaccharides, even when imaged in air, the thin layer of solvent retained on the substrate will allow the molecules to access all configurations over periods of time rapid compared to the scan time. In this case the time-averaged structure is seen: the molecules will appear as blurred spherical objects.

Certain polysaccharides such as amylose, xyloglucans, arabinoxylans, alginates or pectin exist as semi-flexible coils in solution, but can be induced to form ordered conformations in the solid state. If these molecules are imaged under buffers then the molecules will generally desorb from the substrate (Gunning et al., 2000). It is possible to use a range of strategies to pin the molecules onto the substrate and allow imaging under buffer. A common approach for negatively charged polysaccharides is the use of divalent ions to bind the polymer to the substrate. An alternative approach is to image the molecules under a precipitant such as butanol. The contour length distributions can be measured and used to estimate molecular weight distributions. Various procedures can be used to examine the shape of the molecules and to estimate their stiffness, through the calculation of parameters such as the Kuhn statistical segment length (L_K). It is important to establish whether the method of deposition and imaging has influenced the shape of the molecules. The use of precipitants such as butanol may induce the formation of ordered conformations: this can be assessed by comparing images obtained under precipitants and under buffer. Even when imaging under buffer it is necessary to establish that the procedure used to bind the molecules to the substrate does not alter their shape.

New information has been obtained (Round et al., 2001; Gunning et al., 2003; Ikeda et al., 2005; Kirby et al., 2007) on the branching of polysaccharides (Figure 12.3). It is difficult to directly visualize short sidechains, or sidechains

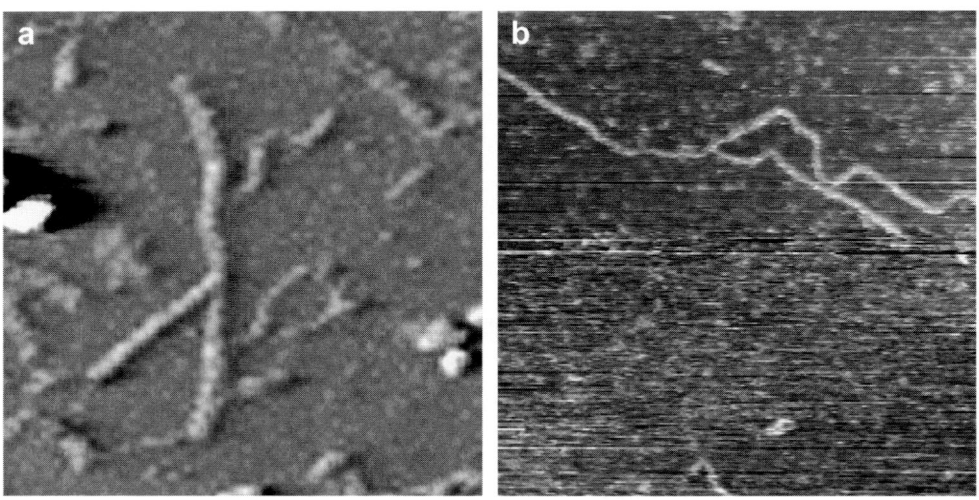

FIGURE 12.3 AFM images of (a) a branched pectin molecule, error signal mode, scan size 230 × 230 nm and (b) an amylose molecule, topography, scan size 500 × 500 nm.

attached to the backbone via flexible linkages. However, the presence of a heterogeneous or block-like distribution of sidechains will lead to local variations in height or stiffness of the polysaccharides. In the case of mucins these variations have been used to analyze the glycosylation patterns (Round et al., 2004) and such methods may be adaptable to analyze branching distributions on polysaccharides. An alternative approach, used to analyze heterogeneity of mucins and arabinoxylans, is to image and analyze the binding of antibodies (Round et al., 2007), or inactivated enzymes (Adams et al., 2004) to selected regions of polymer chains, in order to infer the distribution of sidechains.

Infrequent branching of the polysaccharide backbone has been reported for pectin (Round et al., 2001; Kirby et al., 2007) (Figure 12.3a), arabinoxylans (Adams et al., 2003) and also amylose molecules (Gunning et al., 2003) (Figure 12.3b). Quite complex multi-branched structures have been observed for the surface-active soyabean polysaccharide (Ikeda et al., 2005). For these structures new information can be obtained on the number of branches per molecule, the branch length distribution and the relative numbers of linear, branched or multi-branched molecules within a population. Knowledge of the shape and size of molecules, and the extent and distribution of branches, is important in understanding the viscosity and viscoelastic behavior of these molecules in solution.

Some of the most useful functional food polysaccharides are those that can be induced to adopt ordered conformations in solution. Examples include the helix-forming polysaccharides xanthan, gellan, agar and the carrageenans. These ordered conformations are retained on deposition and these types of molecules are easier to image and characterize. In general the major problem is to inhibit self-association, in order to observe individual molecules, and obtain information on their shape (stiffness) and contour length (molecular weight). Occasionally, under favorable conditions it is possible to visualize and measure the pitch of the helix (Kirby et al., 1995). However, most of these polysaccharides have been very well characterized by standard physical–chemical techniques. In this case it is much more useful to use AFM to probe the self-association of the polysaccharides, and to use this information to understand functional behavior such as thixotropy and gelation.

12.1.6 Proteins

In general proteins adsorb fairly easily to substrates such as mica and are relatively easy to image under a variety of liquids. The majority of food proteins are globular and have dimensions similar in size to the radius of curvature of the AFM tip. Because of their near-spherical shape and the effects of probe broadening, the AFM is unlikely to yield new information on molecular size and shape. For certain unusual-shaped meat and plant proteins, such as collagen, actin, myosin, gelatin, or soya proteins the AFM provides similar information to that obtainable from TEM studies, although under milder preparative conditions. For these systems the AFM has proved to be more useful in investigating self-association of these molecules, in order to understand the nature of bulk protein gels, or the structures formed at air–water or oil–water interfaces and responsible for the stability of food foams and emulsions.

12.1.7 Protein–Polysaccharide Complexes

Proteins tend to be surface-active and are used as emulsifiers and foam stabilizers. In general polysaccharides are not surface-active. There are a few polysaccharide extracts that show useful surface activity. These include gum arabic, sugar beet pectin and water-soluble wheat pentosans (arabinoxylans). All of these materials are considered to contain protein–polysaccharide complexes, which are believed to contribute, at least in part, to the surface activity of the extracted polysaccharide. AFM has the potential to visualize and characterize such protein–polysaccharide complexes. In practice such surface-active polysaccharide extracts are often difficult to image.

Gum arabic extracts consist of three fractions: arabinogalactan which is the major fraction, ~ 10% of an arabinogalactan–protein complex and ~ 1% glycoprotein. Drop deposition of gum arabic onto fresh mica results in the formation of aggregated structures, possibly because the complex becomes concentrated at the air–water interface leading to self-association of the protein component as the aqueous solvent evaporates. If drop deposition is carried out in the presence of Tween 20, which should disrupt protein structures formed at the air–water interface, then small individual molecules can be seen. Since the surfactant should disrupt any protein aggregates and displace the complex from the mica, it is likely that the molecules seen are, as suggested by the authors (Ikeda et al., 2005), predominately the arabinogalactan, and not the protein–polysaccharide complex. Water-soluble arabinoxylans and sugar beet pectin extracts also contain protein which is difficult to remove by conventional separation methods. Both materials are also difficult to image after drop deposition. Where images have been obtained of the entire extract there is evidence for the existence of protein–polysaccharide complexes. For arabinoxylans treatment of the samples to remove protein eliminated the complexes and yielded images of the polysaccharide component alone (Adams et al., 2003). For sugar beet pectin ~ 67% of the pectin molecules in the extract were found to contain what has been attributed (Kirby et al., 2006a) to be a protein molecule attached to the end of the polysaccharide chains (Figure 12.4). The direct observation of these complexes provides a basis for understanding the types of structures they could form at air–water or oil–water interfaces and thus their role in stabilizing food foams and emulsions.

As discussed earlier the binding of inactivated enzymes to polysaccharides can be imaged and used to assess the structural heterogeneity of the polysaccharide molecules. The formation of complexes between proteins and polysaccharides can sometimes yield new information on the mechanism of action of enzymes. For example the starch-binding domain (SBD) in glucoamylase facilitates the breakdown of crystalline starch structures. AFM images of the complexes formed on the binding of SBDs to the starch polysaccharide amylose revealed novel cyclic structures (Figure 12.5a). These images provided new information on how SBDs bound amylose (Figure 12.5b) and hence suggested new molecular models (Figure 12.5c–f) by which glucoamylases degraded crystalline starch (Morris et al., 2005). The SBDs acted as a template for the

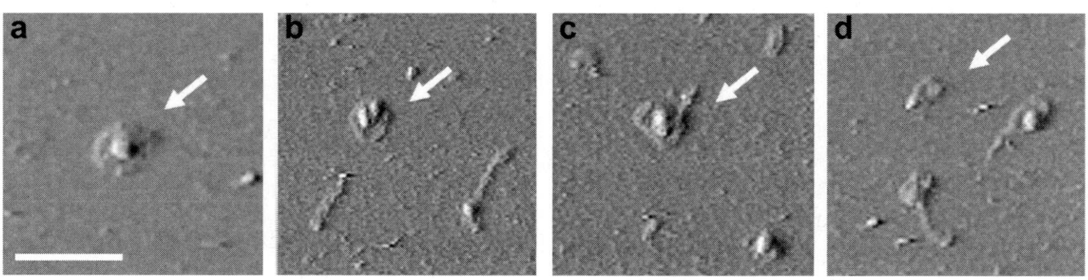

FIGURE 12.4 AFM error signal mode images of sugar beet pectin–protein complexes. The bar marker is 100 nm. The arrowed structures have been attributed to complexes in which a protein is attached to one end of the pectin chain which is coiled to different extents about the protein.

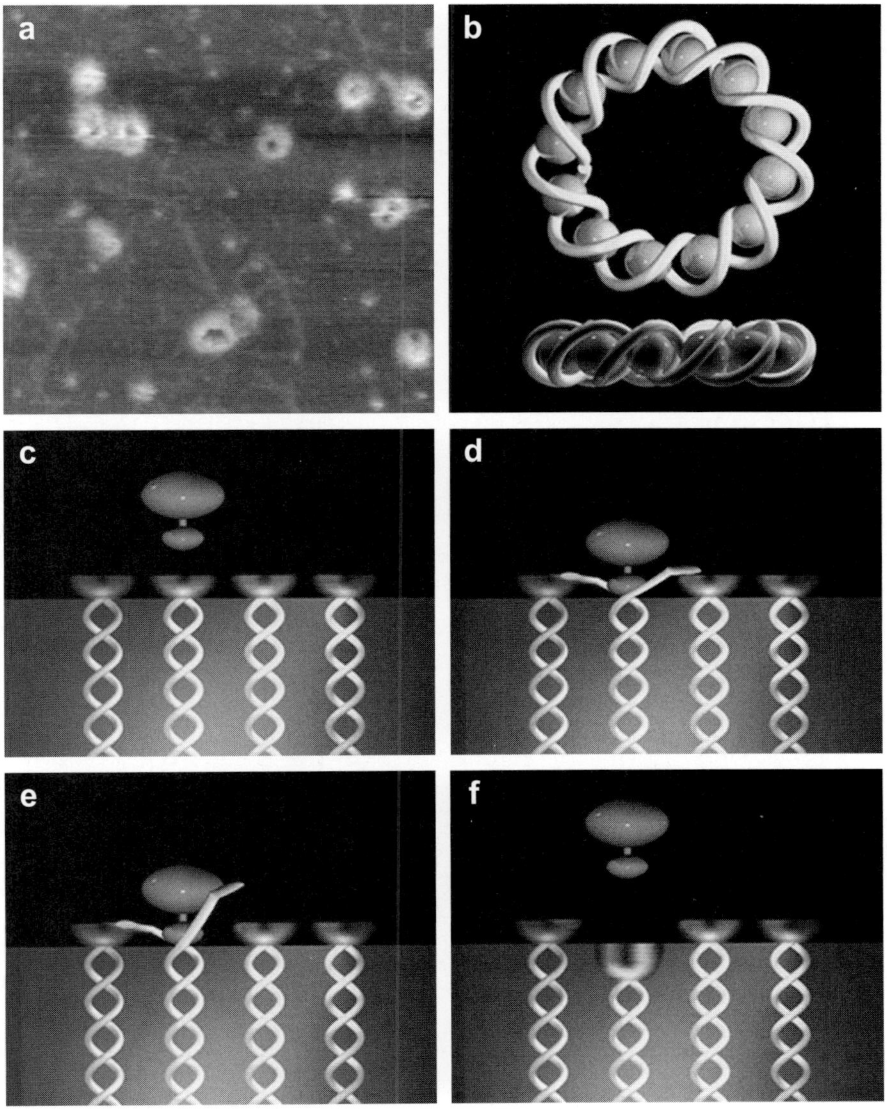

FIGURE 12.5 Glucoamylase contains a catalytic domain linked to the starch-binding domain (SBD) through a flexible linker region. (a) AFM topography images of circular amylose–SBD complexes, scan size 800 × 800 nm. (b) A model of the amylose–SBD complex. (c) Approach of the SBD on the glucoamylase to the end of an amylosic double helix at the surface of a starch crystal, the free ends of the amylosic chains are shown blurred to indicate molecular motion. (d) Docking of the SBD locates the ends of the amylosic chains near the catalytic domain of glucoamylase. (e) Cleavage of the amylosic chain by the catalytic domain. (f) Release of the SBD.

formation of an expanded amylosic double helix, suggesting that the SBD could recognize and dock onto the ends of amylosic double helices at the surface of starch crystals (Figure 12.5b–f).

12.1.8 Bulk Gels and Networks

The methods developed for imaging individual proteins and polysaccharides can be adapted to use AFM to probe their self-association. Such studies provide new information on biopolymer gelation (Morris, 2007a).

12.1.9 Polysaccharide Gels

The charged helix-forming polysaccharides such as the carrageenans and gellan form thermo-reversible gels. Gels form on cooling from the sol state and the gels melt on heating (Morris, 2007a). The molecular basis of gelation is similar and can be illustrated through studies on gellan gum. Substantial physical–chemical studies have established that gelation involves two separate stages: namely helix formation on cooling and helix aggregation. Because the polysaccharides are charged, helix formation is enhanced with increasing ionic strength. The presence of certain 'gel-promoting cations' induces helix aggregation and gelation. Some degree of gelation can occur in the absence of 'gel-promoting cations', but the networks are weak, often not self-supporting, and they break easily at low deformation. AFM can be used to investigate the mechanism of gelation (Gunning et al., 1996a; Morris, 2007a) through studies on gel precursors, films or, in certain cases, bulk gels (Figure 12.6). In the absence of 'gel-promoting cations' the gel precursors are thin-branched fibrils (Figure 12.6a). The formation of these structures is believed to be due purely to double helix formation, usually between gellan chains of unequal length. The addition of 'gel-promoting cations' is observed to result in the formation of thicker-branched fibers (Figure 12.6b). The variable thickness of the fibers is attributed to a cation-induced side-by-side association of the fibrils.

The studies on gel precursors at sub-gelling concentrations suggest a basis for understanding the form of the association of the polysaccharides in gels formed at higher polysaccharide concentrations. It is difficult to image the network structure in hydrated bulk gels: this is because the gels are likely to deform on scanning, blurring the image. At higher polysaccharide concentrations it is possible to induce association and form a thin hydrated film on the mica substrate. In this case it is possible to image the network structure, because the imaging process will compress the polymer network down against the hard mica substrate. The polymeric structure observed within the film is seen to be a continuous branched fibrous network composed of aggregated gellan helices (Figure 12.6c). In the case of gellan it has also been possible to observe similar networks directly in a bulk aqueous gel. The high-resolution structure seen in Figure 12.6d was obtained because, under acidic conditions, gellan forms high-modulus gels, which show negligible distortion on scanning. The AFM image reveals the network structure present on the upper surface of the gel, which is seen to be equivalent to the branched fibrous structure observed in hydrated films. The nature of the junction zones formed by cation-induced association of gellan helices has been extremely well characterized by physical chemical methods and through the modeling of X-ray fiber diffraction data. The AFM images provide new information on the long-range structure within the gels. The gels are found to be fibrous networks in which the junction zones act as glue, binding the helices together. In previous models little was known about the long-range structure and the gels were generally pictured as rubber-like structures with extended junction zones linked by disordered polysaccharide chains. Similar fibrous structures have also been observed by AFM in other gelling polysaccharides such as pectin (Fishman et al., 2007).

Xanthan gum also forms double helices on cooling from the sol state. Rather than forming strong gels it tends to form weak gel-like structures similar to those formed by gellan gum in the absence of gel-promoting cations. Thus xanthan gum is used as a thickening and stabilizing agent rather than as a gelling agent.

In general it is quite difficult to dissolve commercial samples of xanthan gum. If the

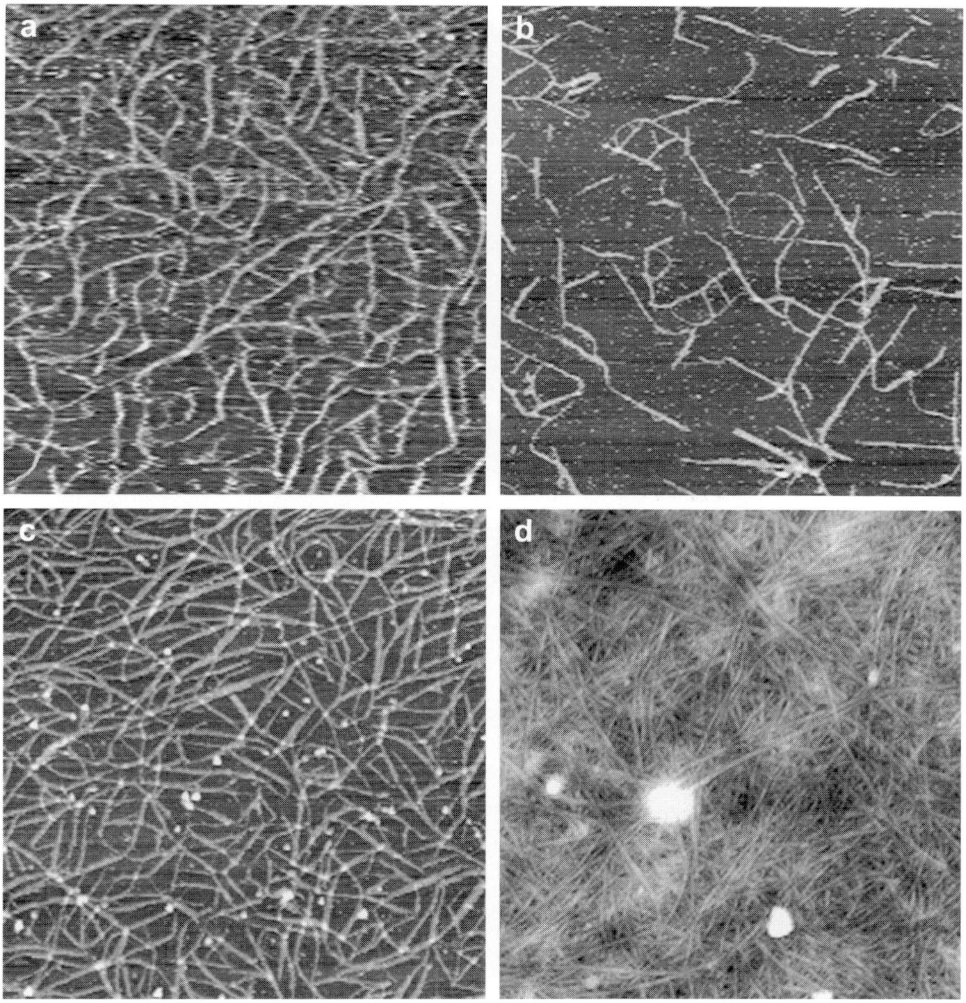

FIGURE 12.6 AFM images of gellan samples. (a) Topgraphy image of TMA gellan aggregates, image size 1 × 1 μm. (b) Topography image of gellan aggregates in the presence of potassium cations, image size 800 × 800 nm. (c) Topography image of a hydrated gellan film formed on mica, image size 800 × 800 nm. (d) Topography image of the network structure at the surface of a hydrated bulk acid-set gellan gel, image size 2 × 2 μm.

material is dispersed at high temperature, followed by centrifugation or filtration, then it is possible to obtain true solutions of xanthan gum. AFM images of xanthan molecules, drop deposited onto mica, show stiff extended rod-like structures (Morris, 2007b). In these images the ends of individual molecules are clearly visible (Figure 12.7a). The centrifugation or filtration steps remove material, suggesting that commercial xanthan preparations, used as thickeners and stabilizers, contain xanthan aggregates or microgels. AFM images of such structures (Morris, 2007b) show continuous networks for which it is difficult to detect the ends of individual molecules (Figure 12.7b). These 'microgels' are believed to be formed during the precipitation of the polysaccharide from the fermentation broth and on subsequent

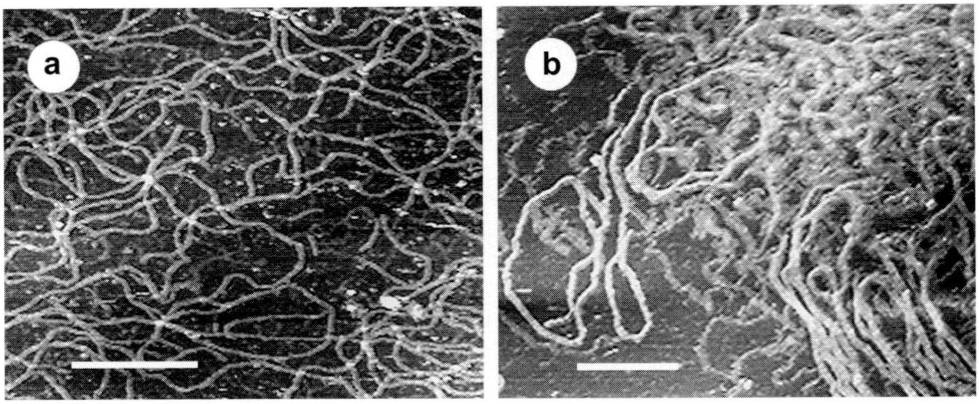

FIGURE 12.7 AFM topography images of xanthan samples. (a) An entangled network of xanthan molecules – note that the ends of individual xanthan molecules are visible. (b) A portion of a xanthan microgel – note that the xanthan forms a continuous network with few chain ends visible. Bar marker is 300 nm.

drying, solely due to double helix formation. On this basis, samples of xanthan are considered to be dispersions of swollen microgels, which can associate into a weak gel network. Shearing would break the inter-microgel links, allowing the microgels to deform and flow past each other, rather like deformable red blood cells. Upon removal of the applied shear the particulate network would be able to reform, explaining the reversible shear-thinning behavior observed for xanthan gum samples. In this context it is instructive to note that xanthan-like behavior can be induced in preparations of gelling polysaccharides if the gelation process is disrupted by stirring during cooling from the sol state (Norton et al., 1999). Such a process would break up the gel structure into a dispersion of microgels.

12.1.10 Protein Gels

Proteins form heat-set gels (Morris, 2007a). The proteins start to aggregate when they are heated above a characteristic temperature and this aggregation continues on cooling the sample. The gels are thermo-irreversible and usually opaque, although transparent gels are formed under certain conditions. The common opaque gels are formed when the repulsive interactions between the proteins are reduced or screened. TEM and light microscopy suggest that these gels consist of largely colloidal networks formed from aggregated protein particles. It is difficult to image such gels or even gel precursors by AFM because of the large size and deformability of the aggregates (Ikeda, 2003). If the repulsions between proteins are not screened then the proteins can form transparent gels. Under these conditions, in the early stages of association, linear protein aggregates are formed (Ikeda and Morris, 2002; Ikeda, 2003a, 2003b; Gosal et al., 2004). These types of structure are fairly easy to image and new information can be obtained on the extent, and nature of the aggregation process, as a function of preparative conditions. In the simplest case the aggregates are long chains of protein but more sophisticated structures, such as hollow tubes, have been reported and characterized for certain milk proteins. If the linear gel precursors are to form gels then they must aggregate further to form fairly open branched networks. At present there is no clear AFM data suggesting how this occurs, and there are very few AFM studies on the molecular microstructure in bulk films and gels. There is clearly scope for further research in this area.

The fibrous protein gelatin is widely used as a gelling agent, but it is unusual in that it forms thermo-reversible gels, similar to those formed by polysaccharides such as gellan or the carrageenans. Like the polysaccharides, gelatin undergoes a reversible helix-coil-helix transition on heating and cooling. AFM studies have been used to study network formation both in the bulk and at air–water interfaces (Mackie et al., 1998). The aggregation and assembly process involves helix formation and association. The networks are fibrous structures similar to those reported for thermo-reversible polysaccharides.

12.1.11 Interfacial Protein Networks

The formation of protein structures at air–water and oil–water interfaces is important in stabilizing food foams and emulsions. This is an area where AFM has made an impact, mainly due to the ability to image, for the first time, interfacial structures at the molecular level (Morris and Gunning, 2008).

In order to visualize the types of structures formed at interfaces it is necessary to use a model system. Interfacial structures can be created on a Langmuir trough and Langmuir-Blodgett methods used to sample and visualize the interfacial structures by depositing them onto substrates. Freshly cleaved mica or graphite can be used as hydrophilic or hydrophobic substrates. Either air–water or oil–water interfaces can be studied. The interfacial structures can be created by spreading the molecules at the interface or by adsorption from the bulk phases. In the latter case the interfacial structures can become obscured due to passive adsorption of protein from the bulk phase during dipping, but methods have been devised to eliminate such effects (Morris et al., 1999).

An unexpected result of using AFM to image proteins at air–water interfaces was the observation (Gunning et al., 1996b) that the networks contained defects (holes). This heterogeneity at the molecular level proved to be crucial in understanding competitive displacement of proteins by surfactants; a process of importance in determining the stability and lifetime of food foams and emulsions. In food systems the interfacial structures can be very complex with the interfaces populated with various species including proteins, surfactant-like molecules, and even surface-active polysaccharides. By using AFM to probe interactions between different types of molecules at interfaces it is possible to explain the behavior observed for the types of structure which occur in food foams and emulsions.

12.1.12 Protein–Surfactant Interactions

A very important type of interaction that leads to the destabilization of foams and emulsions is competitive displacement of proteins from interfaces by surfactants. Given sufficient time, and sufficiently high surfactant concentrations, surfactants will eventually displace proteins from an interface. Understanding how this process occurs allows the development of strategies for stabilizing interfaces and hence the resultant foams or emulsions.

Figure 12.8 shows AFM images of the competitive displacement of the protein β-lactoglobulin from an air–water interface by the non-ionic surfactant Tween 20. The images show the progressive colonization of the interface by the surfactant. In the early stages of displacement small surfactant domains nucleate within the protein structure. After nucleation the surfactant domains grow in size. An important result of the AFM images is the observation that the colonization of the interface by the surfactant is heterogeneous, rather than homogeneous, and the AFM data explain why this is so (Mackie et al., 1999).

The AFM images are 3D representations of the interfacial structure. The images shown were obtained under butanol, which removes the surfactant and freezes the displacement process. Analysis of the images can be used to measure

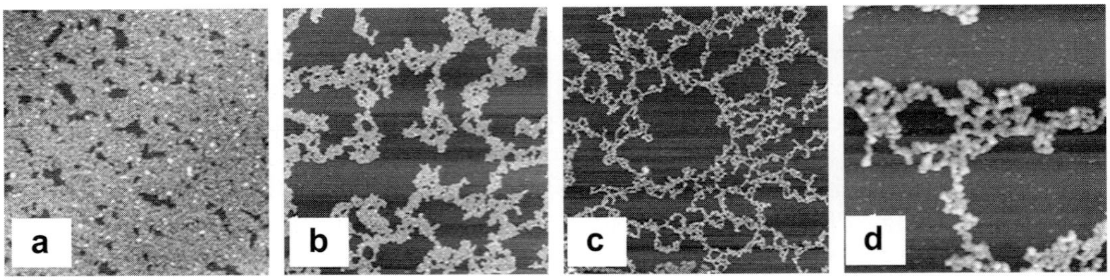

FIGURE 12.8 AFM topography images showing the competitive displacement of β-lactoglobulin protein film (white) from an air–water interface by the non-ionic surfactant Tween 20 (black) as a function of increasing surface pressure. Image sizes (a) 1 × 1 μm, (b) 3.6 × 3.6 μm, (c) 5.2 × 5.2 μm and a high-resolution image (600 × 600 nm) showing individual proteins in the network structure.

the area occupied by the protein and also the height of the network. Thus it is possible to monitor changes in the protein volume and hence the surface concentration of protein during the displacement process. Given that the surfactants are more surface active than individual proteins one might expect a progressive drop in protein concentration as the size of the surfactant domains increases: one would expect that the surfactant would displace individual exposed proteins, particularly at the boundaries between the protein and surfactant domains. However, what the AFM clearly shows is that, as the surfactant domains grow, the protein concentration remains unchanged, and then suddenly, at a certain surface pressure (concentration of adsorbed surfactant concentration), the protein is expelled from the interface (Figure 12.9).

At the higher surfactant concentrations there is clear evidence that the proteins form a network structure and it is this network structure that dictates the mode of displacement. By linking together into a network the proteins resist displacement. The expanding surfactant domains have to stretch the network until it breaks, and only then can they displace protein from the interface into the bulk.

If the proteins are linked together, and hence are protected from displacement, then how does the surfactant initially gain access to the interface? The answer is that they can occupy the defects (holes) in the protein network. The holes will arise because the proteins partially unfold and interact when they adsorb onto the interface. As they adsorb and start to form networks the space which is available for further adsorption will begin to decrease. Eventually there will be insufficient room for further protein adsorption, leaving holes in the network.

Once the surfactant has gained a foothold on the interface more surfactant can enter the nucleated domains forcing them to expand. As the surfactant domains expand they compress the protein network. Energy is stored by refolding of individual proteins and then folding and buckling of the network structure. Eventually, the protein network breaks, releasing individual proteins, or protein aggregates, that can be displaced by the surfactant.

It was the ability of the AFM to demonstrate heterogeneity of the protein networks and

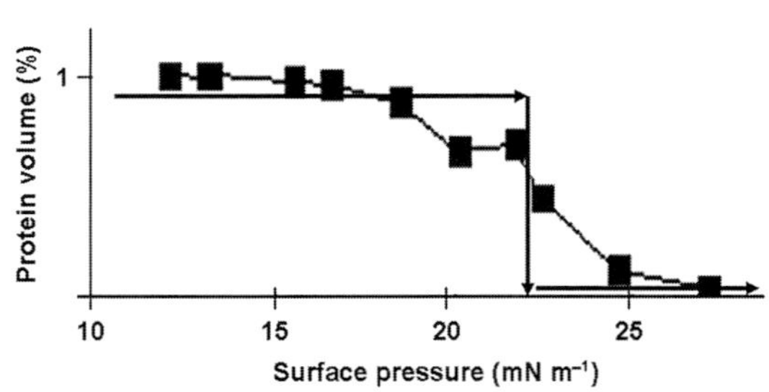

FIGURE 12.9 Graph showing changes in surface volume during the displacement of β-lactoglobulin by Tween 20 from an air–water interface.

heterogeneity in the displacement process at the molecular level that allowed this novel displacement mechanism to be discovered (Mackie et al., 1999). The process which involves the folding, buckling and eventual failure of the protein network has thus been termed 'orogenic displacement' (Figure 12.10).

The displacement mechanism has been found to be generic for all proteins studied and for different types (neutral, charged, water-soluble and oil-soluble) of surfactants, at both air–water and oil–water interfaces: the proteins always form networks and these networks need to be broken to release and displace protein (Morris and Gunning, 2008). For charged surfactants domain nucleation is favored over domain growth and little expansion of the nucleated domains is seen to occur. However, the growing occupancy of the interface with surfactant still leads to compression of the protein network

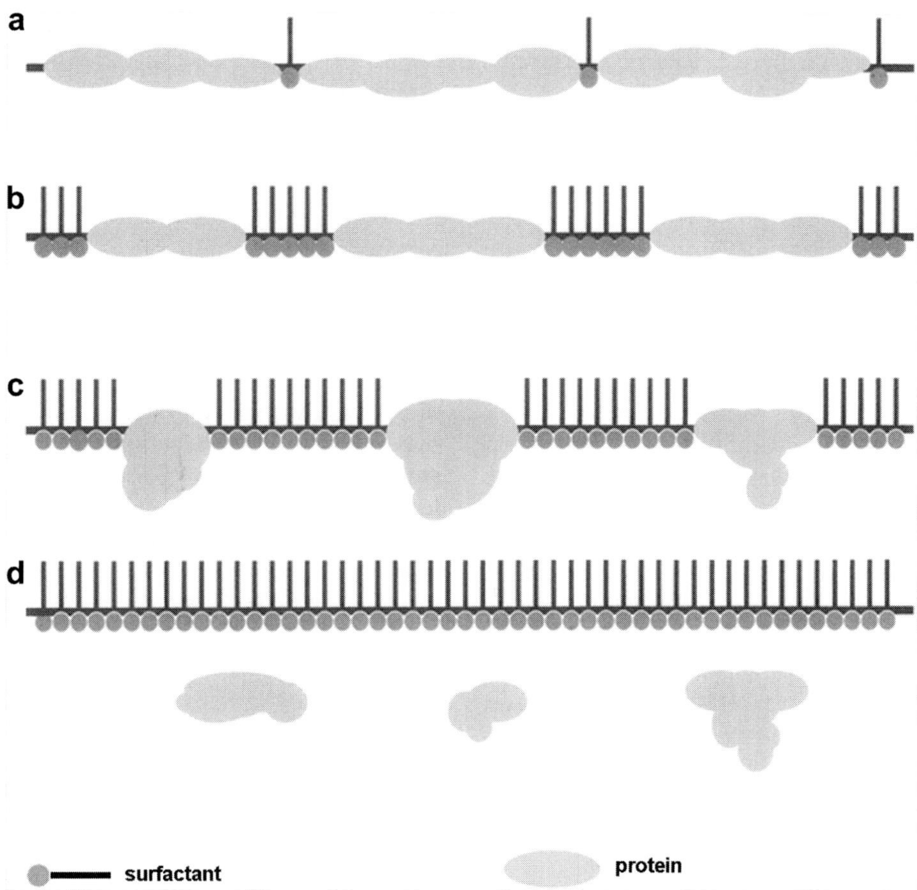

FIGURE 12.10 A schematic diagram illustrating the various stages during orogenic displacement of proteins by surfactant from an interface. (a) Surfactants colonize defects in the protein network. (b) As the surfactant domains expand they compress the proteins causing them to refold. (c) Further expansion of the surfactant domains causes the network to buckle and fold, forcing protein to extend out into the bulk whilst remaining inter-connected at the interface. (d) Failure of the protein network allows protein to be completely displaced into the bulk media.

which eventually fails (Mackie et al., 2000; Gunning et al., 2004a). The generic nature of the displacement process means that interfaces can be stabilized either by strengthening the protein network or by preventing surfactants reaching the interface.

In commercial food products the interfaces will contain a mixture of proteins and surfactants and even possibly some surface-active polysaccharides. Most protein samples used as commercial foam stabilizers or emulsifiers will be isolates, such as whey protein or sodium caseinate, rather than pure proteins. The question is whether the simple models of protein displacement can be extended to accommodate this level of structural complexity?

Using a combination of AFM and fluorescence microscopy it has been possible to examine the structures formed by binary mixtures of proteins and the displacement of such structures from interfaces by surfactants. These studies suggest that, when mixtures of proteins adsorb at interfaces, they rapidly interact and they become kinetically trapped in an immobile network, with little evidence for phase separation. Displacement of the mixtures with surfactant occurs through an orogenic mechanism. Within the mixed networks different proteins appear to interact with each other to different extents and the expanding surfactant domains appear to break the weakest links, preferentially displacing those proteins which, on their own, would form the weakest networks. An important observation is that the final failure of the mixed network is dominated by the protein components which, on their own, would form the strongest network (Morris and Gunning, 2008; Mackie et al., 2001). Surprisingly this appears to be true even if this protein is only a minor component of the mixture. Thus it is possible to predict the general behavior of networks formed by protein isolates, provided the composition is known (Woodward et al., 2004). In practice the preparation and isolation of pure proteins and isolates will be different and this can modify the quality of the network formed.

12.1.13 Starch and Plant Cell Walls

As well as imaging individual biopolymers, and the aggregated structures that they form, it is also possible to image self-assembled biopolymer structures formed in plant tissue. Examples of such structures are starch and plant cell walls.

12.1.14 Starch

Plants store energy in the form of starch. Starch is produced as water-insoluble spheroidal granules, the shape, size and size distribution of which varies depending on the plant source (Morris, 2007a). The granules are ordered and partially crystalline. Two polysaccharides can be extracted from starch granules: an essentially linear polysaccharide (amylose), believed to be largely located in the amorphous regions of the granule, and a highly branched polysaccharide (amylopectin), which is considered to give rise to the crystalline component of the granule. The chain length of the branches of the amylopectin is regulated during biosynthesis, resulting in crystalline lamellae which are ~ 9 nm in size. These crystallites are within annular bands (growth rings), with the molecular axes oriented in a radial direction within the granule. In the presently accepted model of the starch granule structure (Waigh et al., 2000; Daniels and Donald, 2003) the growth rings are considered to alternate between amorphous and partially crystalline bands. In normal starches, within the partially crystalline growth rings, the crystallites are considered to be embedded in a matrix of largely amorphous amylose (Figure 12.11).

An alternative model for the granule structure, based largely on electron microscopy studies, suggests that the crystallites are organized into further higher-order structures called blocklets (Gallant et al., 1997). AFM provides a new method for visualizing the internal structure of starch granules and testing models for granular

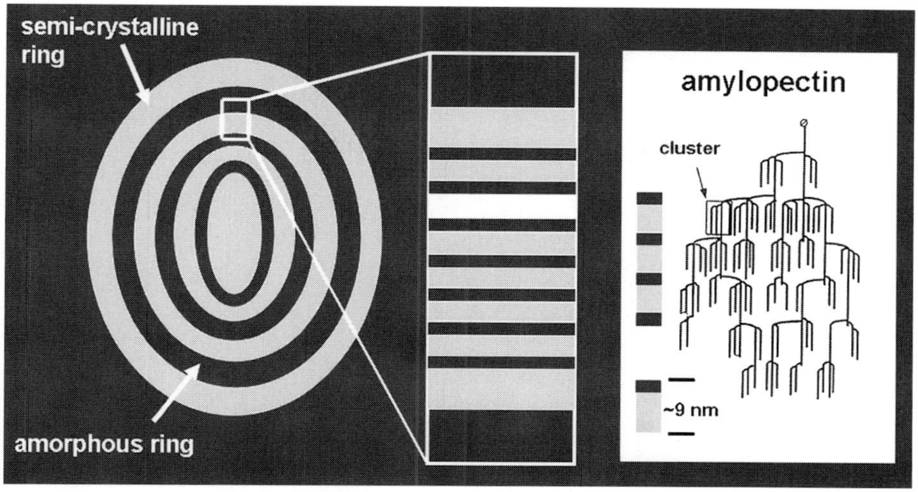

FIGURE 12.11 The conventional model for the structure of starch granules. The crystal lamellae are considered to be formed by the branches of the amylopectin molecules and contained within semi-crystalline growth rings. The semi-crystalline growth rings are considered to be interspersed by amorphous growth rings.

structure. In developing and testing methodology for imaging complex structures such as starch it is important to identify and explain the origin of contrast in the AFM images. To observe the internal structure of the starch granule it is necessary to open up the granule. For isolated starches this can be done by encasing the granules in a rapidly setting resin, which does not penetrate into the granule (Ridout et al., 2002). The starch is encased in small cylindrical blocks which are set at room temperature. The top of the block can be honed down with a freshly prepared glass knife to produce a flat, shiny surface, consisting of starch granules with cut internal surfaces exposed. For plants which produce dry seeds it is possible to make use of the natural encasement of the starch within the seeds in order to image the starches in situ. The seeds can be cut open and honed down to a flat surface using a glass knife (Parker et al., 2008).

Preliminary experiments (Ridout et al., 2004; Parker et al., 2008) have shown that the contrast in the AFM images arises due to the preferential absorption of water into certain regions of the exposed face of the cut granules: dry samples show little evidence for internal structure (Figure 12.12a–c). Minimal wetting results in the appearance of a banded structure (Figure 12.12d–f) and, at higher resolution, globular structures attributed to blocklets or the packets of crystalline lamellae formed from the branches of amylopectin molecules (Figure 12.13).

Thus AFM studies provide independent evidence in support of the blocklet model of starch granule, a structure first proposed from electron microscopy studies (Gallant et al., 1997). The AFM images provide new information that suggests that the blocklets are distributed uniformly throughout the granule and that all the bands (growth rings) are semi-crystalline (Parker et al., 2008). Contrast in the images is attributed to the preferential absorption of water in the amorphous (amylose) regions. Thus the banding or growth ring structure is considered to result from the fact that, in alternate bands, the swelling of the amorphous (amylose) regions is slightly different: alternate bands are considered to contain different levels of amylose and hence different levels of crystallinity. The higher-resolution topography and

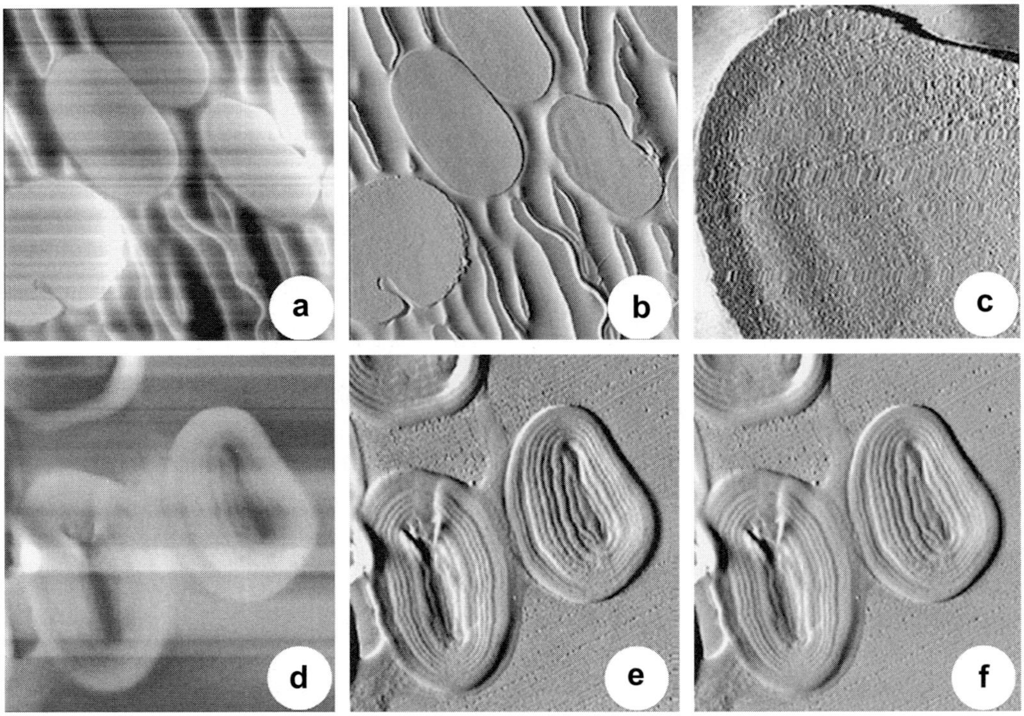

FIGURE 12.12 AFM images of encased pea starch. Freshly cut dry samples show no internal structure: (a) topography image 80 × 80 μm, (b) error signal mode image 80 × 80 μm and (c) error signal mode image 8.7 × 8.7 μm. Wetting of the sample surface induces contrast revealing the internal structure of the granules: (d) topography, (e) error signal and (f) left-shaded topography images, scan sizes 35.5 × 35.5 μm.

phase images (Figure 12.13) reveal that the most crystalline bands are heterogeneous containing poorly swelling patches, believed to be deficient in amylose (Parker et al., 2008). The detailed explanation for the heterogeneous distribution of amorphous regions throughout the granule is at present unclear.

The picture that emerges of the granular structure (Figure 12.14) is in contrast to the conventional model (Figure 12.11), which envisages alternate crystalline and amorphous bands. The idea for amorphous and crystalline bands arose from microscopy data on granules treated with acid or α-amylase: etching appeared to remove alternate growth rings. Etching should preferentially cleave amorphous regions, suggesting the presence of alternate amorphous and crystalline bands. The AFM data suggests that this explanation may not be clear-cut. Those bands that contain the most amorphous regions would swell most, and absorb preferentially the acid or enzyme solution, and hence would be etched more rapidly. Breakdown of the swollen amorphous regions would release blocklets from the granule. Thus preferential etching of the more swollen bands could account for the selective removal of material from alternate bands.

As well as providing new insights into the structure of granules produced by normal starches, the same methods can be used to explore structural changes resulting from mutations in starch biosynthesis (Ridout et al., 2003). In particular the ability to image starch within

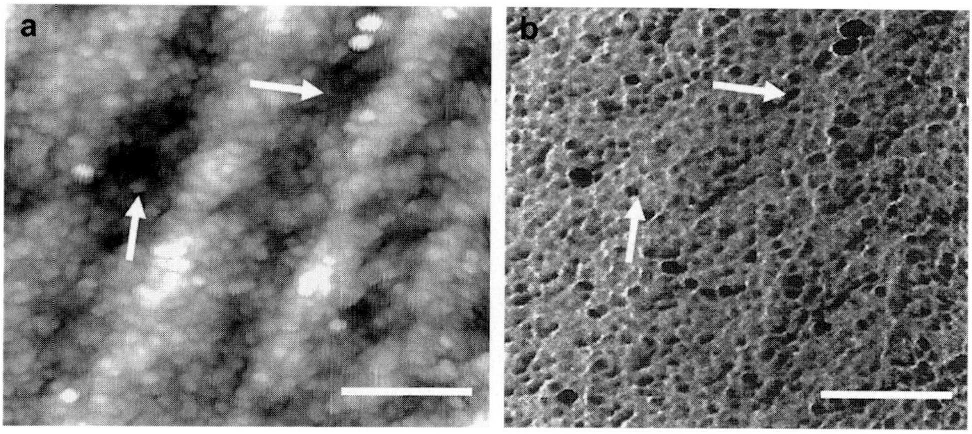

FIGURE 12.13 High-resolution images of the internal structure of pea starch granules naturally encased within the dry seed. (a) Topography image showing bright and dark bands and spheroidal blocklet structures. Note the dark bands contain bright swollen regions and dark, unswollen patches (arrowed). (b) Phase image of the same region within the starch granule. Within the phase image the blocklets are seen to be uniformly distributed throughout the granules within both the dark and bright bands. Within the arrowed 'dark patches' the contrast between the blocklets and surrounding media is much reduced. Bar marker 400 nm.

single seeds offers a method for screening for interesting changes in starch structure. The approach has been tested by investigating isogenic pea mutants, where the mutations in individual biosynthetic enzymes are expressed in a constant genetic background. Mutations at the *r* loci (*r* mutants) affect a branching enzyme involved in the synthesis of the amylopectin

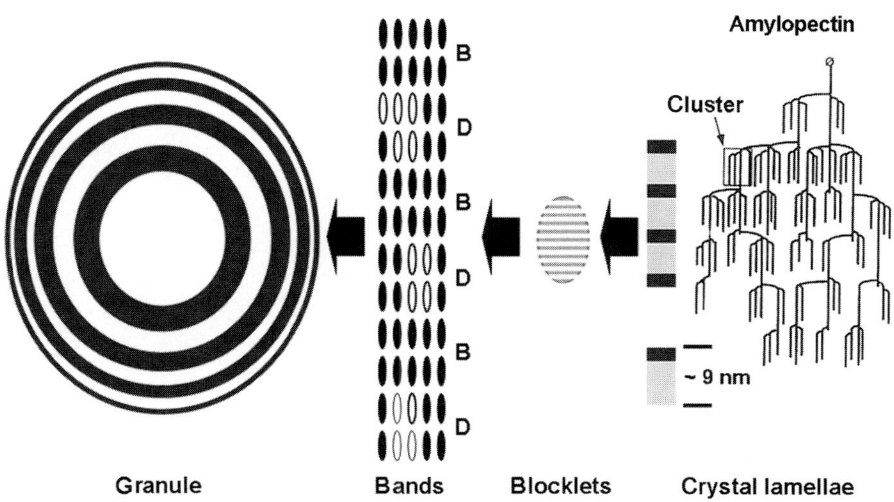

FIGURE 12.14 A new model of starch granule structure. The crystal lamellae are contained within blocklet structures which are distributed uniformly throughout the granule. The banding pattern arises due to preferential swelling of selected regions within the granules. All the bands are semi-crystalline. The dark bands result from restricted swelling of localized patches within the bands. The banding pattern is considered to arise from the distribution of amorphous amylose within the granule.

structure (Bogracheva et al., 1999; Ridout et al., 2003). The mutation reduces the level of amylopectin in the granule, producing a high-amylose starch. This mutation also results in dramatic changes in the functional properties of the starch. Despite causing an ~50% reduction in amylopectin content, the overall level of crystallinity within the granule remains largely unaltered. An acid treatment (lintnerization) can be used to isolate and characterize the molecular structure within the crystalline regions of the granule. Such studies suggest that the crystalline regions contain additional chains, larger in length than the branches of the amylopectin molecules. The length of the chains in the crystals determines the melting point of the crystals (Moates et al., 1997). When starch is heated in water, the melting of the crystals leads to disruption of the granule structure, swelling and the release of amylose. This process called gelatinization occurs for wild-type starches at about 65°C. For the *r* mutant the gelatinization of the starch is broadened and complete gelatinization and swelling of the granules only occurs for temperatures above 100°C: presumably because the longer-length chains in the crystals give higher melting point crystals.

AFM images (Ridout et al., 2003) of wetted sections cut from the blocks of encased *r* mutant pea starch show that the granules are fractured and lack visible growth rings (Figure 12.15a–d). Very little swelling occurs on wetting the samples which are flat. This makes it difficult to see detail in the image: the contrast is poor. For cut sections deposited onto mica it is possible, but difficult, to map the local hardness of the surface, in order to try to localize hard crystalline regions and softer amorphous regions. This can be achieved by oscillating the tip-cantilever assembly and generating a force–distance curve at each image point. From the initial slope of these curves it is possible to extract a parameter that depends on the local modulus. Thus it is possible to map the hardness of the surface. These types of images are difficult to obtain and can be subject to artefacts, but the result

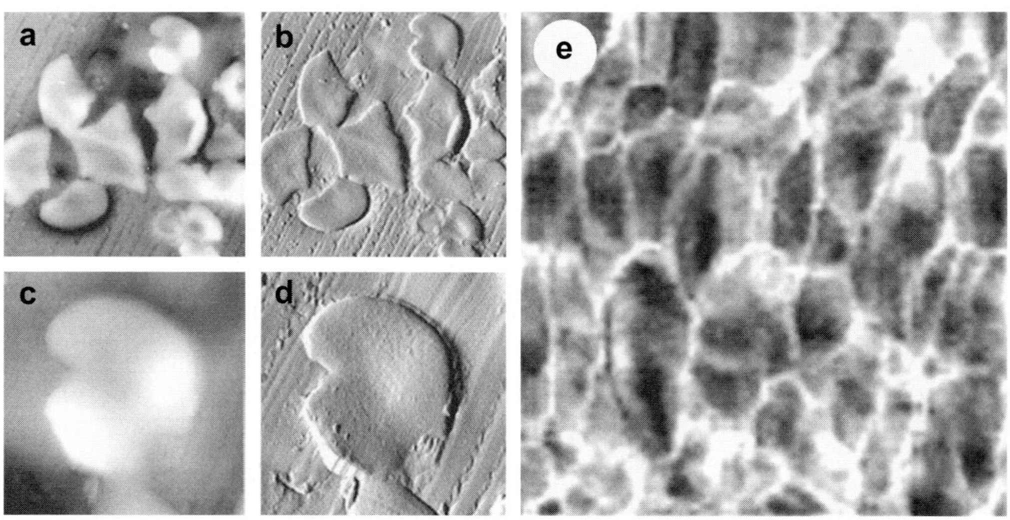

FIGURE 12.15 AFM images of the interior of sectioned encased *r* mutant pea starch granules. (a) Topography image and (b) error signal mode images, 30 × 30 μm, showing cracked and fissured granules. (c) Topography and (c) error signal mode images, 10 × 10 μm, showing the absence of banding even after wetting of the surface. (e) High-resolution force modulation image of the interior of the starch granule. The bright (hard) regions suggest the presence of a novel network structure. Image size 778 × 778 nm.

(Figure 12.15e) appears to show an unexpected fine network structure permeating throughout the granule.

The existence of such novel networks within the granule would explain the changes in properties of the starches. Partially crystalline networks would make the granules fragile and likely to fracture. The small pore size would restrict ingress of water and reduce swelling of the wetted sections. Since the melting temperature of the crystals depends on chain length, the presence of a broad range of chain lengths within the network would account for the broadened gelatinization behavior of these starches.

The presence of such high melting point networks within granules would mean that the granular structure would be retained on processing and thus these starches would be expected to show interesting nutritional properties. Thus the ability to screen granular structures in isolated starches or seeds offers a method for screening for starches with novel interesting properties.

12.1.15 Plant Cell Walls

Cutting open seeds reveals other structures besides starch (Figure 12.16). Even in the dry state it is possible to distinguish between starch granules, protein bodies and cell wall structures (Figure 12.16a–c). The different hardness of the different regions, and the way they respond to cutting by the glass knife, generates contrast in

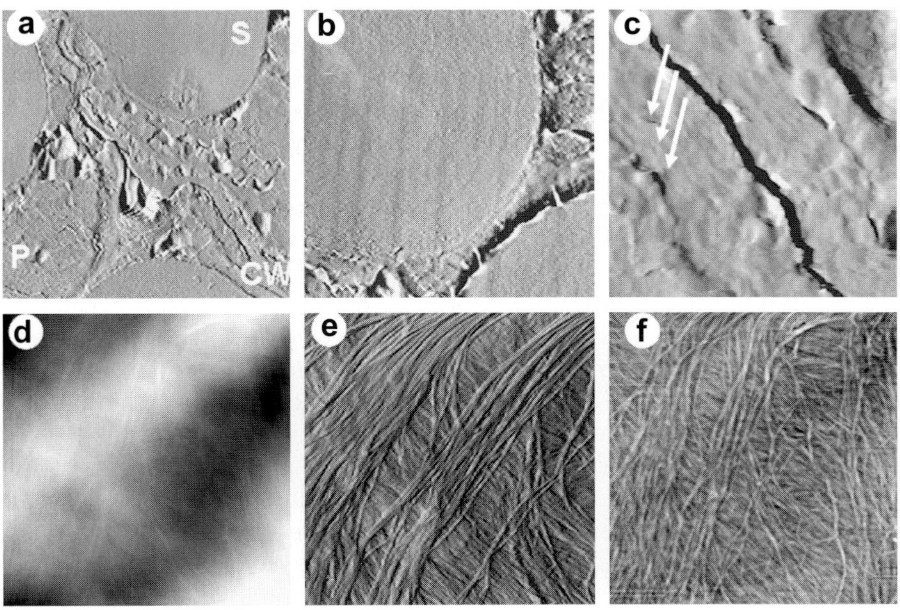

FIGURE 12.16 AFM images of plant tissue. Error signal mode images of a cut surface of a dry pea seed. (a) Scan size 18.5 × 18.5 μm showing the presence of starch granules (S), cell walls (CW) and protein bodies (P). (b) Higher-resolution image 10 × 10 μm showing starch granules. (c) A higher-resolution image of the cell wall region, image size 3 × 3 μm. The image shows two cell walls pushed together. The arrows indicate layers of cellulose fibers. Isolated fragments of cell walls can be imaged whilst wet: images of Chinese water chestnut cell walls clearly show the layers of cellulose fibers. (d) Topography image, scan size 2 × 2 μm dominated by the roughness of the sample. (e) Error signal mode image of the same region revealing the layered cellulose structures, scan size 2 × 2 μm. (f) Background subtracted image of a similar area of the cell wall showing the layered cellulose structure. Scan size 2 × 2 μm. This reconstructed image is a true height image.

the images. In Figure 12.16c it is just possible to see where two cell walls are pushed together and to see individual cellulose layers in the cell wall. If the cell walls are extracted they can be deposited onto substrates such as glass and imaged whilst still wet (Kirby et al., 1996; Round et al., 1996). In this case images can be obtained perpendicular to planes containing the cellulose fibers (Figure 12.16d–f). The raw images show little molecular detail because the surface of the cell wall fragment is too rough (Figure 12.16d). The molecular detail can be seen in error signal mode images (Figure 12.16e) or if the topography images are shadowed. It is possible to extract true height information from the raw images (Round et al., 1996) by constructing and subtracting a background, representing the general roughness of the cell wall fragment, from the raw images (Figure 12.16f).

For non-graminaceous plants the AFM images reveal the cellulose fibers within the cell wall, but it is not possible to image the other components of the cell wall, such as the xyloglucan tethers between the fibers, and the interpenetrating pectin network. This is presumably because these structures undergo rapid thermal motion and are deformed on scanning by the tip. However, it is possible to selectively and sequentially remove other cell wall components and observe the effects on the cellulose networks. Because the cellulose fibers are fairly close-packed the AFM can be used to monitor the inter-fiber spacing in the hydrated cell wall fragment. Removal of the pectin network leads to shrinkage, consistent with a role for the polyelectrolyte pectin network in swelling or de-swelling the cell wall (Kirby et al., 2006b).

12.2 SURFACE FORCES

The AFM was originally designed to image flat surfaces at high resolution. The instruments have been used to image molecular structures deposited onto flat surfaces. Relatively large objects such as bacteria can also be imaged on a range of substrates, in gaseous or liquid environments, and as a function of time. It should be possible to obtain information on structures such as biofilms. Because images can be obtained under realistic hydrated conditions it is easier to assess the roles of bacterial cell wall structures or extracellular slimes on adhesion to surfaces. As has been described earlier it is becoming possible to prepare and image sectioned material and the contrast seen for such samples can be quite good. With the availability of hybrid AFM–optical microscopes it is becoming possible to adapt methods, developed for light microscopy and electron microscopy to prepare complex biological samples for higher-resolution imaging by AFM. As with starch the nature of the contrast mechanisms need to be better understood. However, AFM offers the prospect of new information based on novel contrast mechanisms and the possibility of obtaining images of the samples, in the 'near-native' state seen in the light microscope, but at the higher resolution presently normally only achievable in the electron microscope. At present there are few published studies on processed food materials. Certainly one might expect harder structures such as exudates or glassy materials to provide interesting materials to image. Through the use of phase imaging, or the adaption of various imaging modes to map charge, hardness, adhesion or friction, it may be possible to characterize such materials. It may be possible to map localized crystalline regions through selective melting of the crystalline regions or the swelling of amorphous regions.

Surface roughness determines the reflection and scattering of light and as such the appearance of food samples. The AFM can be used qualitatively or quantitatively to measure surface roughness for quality control, or as a scientific tool to understand the effects of processing on surface structure. A good example of this type of study is the use of AFM to monitor changes in surface structure for products such as chocolates and confectionery. In the case of chocolate the AFM can be used to characterize surface pores, the presence of embedded cocoa butter or sugar crystals, or to follow phase transitions or crystallization and growth, during processes such as tempering or storage (Rousseau, 2007). These types of studies are seldom published in the scientific literature but, although not yet a routine tool for quality control, they can be used in industrial laboratories for investigating problems involved in quality control and for underpinning internal product development.

There is also potential for product development in the food industry based on the controlled preparation of surface structure by sequential deposition of different molecular layers (Guzey, 2006). This is a form of nanotechnology which can be applied to the design of interfacial structures, coatings or barriers, and possibly the development of novel environmentally responsive release devices. Although this process involves the deposition of molecular layers the actual preparation of such structures is relatively straightforward, and uses conventional methods and materials. AFM is one of a variety of surface techniques which can be used to characterize these structures. If the layers are prepared on hard, atomically flat substrates such as mica then, whilst the number of layers is small, it is possible to obtain images of the structures at high resolution. The imaging process and the quality of images obtained is similar to those shown earlier for hydrated gellan films. However, rather than drop depositing the material onto mica the material is allowed to adsorb by incubating the mica in appropriate solutions. As the number of layers becomes larger the multilayers become diffuse and tend to deform on scanning. There is also a tendency for the tips to become contaminated, making imaging difficult, and causing artefacts. With very diffuse layers the AFM will eventually only provide an indication of surface thickness and roughness, or possibly new information on the softness of the multilayer.

12.2.1 Force Spectroscopy and Force Mapping

The AFM images by measuring changes in force between the tip and the sample. At each image point it is possible to generate a force–distance curve by bringing the tip down towards the surface and then retracting the tip away from the surface. This creates a stack of images generated at different distances from the surface. This type of approach can be used to map features such as surface charge distribution or local hardness of the surface (Hoh et al., 1997; Heinz and Hoh, 1999). As has been discussed for starch, hardness maps may, for example, be used to visualize crystalline and amorphous regions on a surface. If the tip sticks to the surface then, on retraction, a finite force needs to be applied to release the tip from the surface. The adhesive force can be determined and used to generate an adhesion map of the surface. In addition to probing the up and down motion of the cantilever it is also possible to monitor the twisting of the cantilever. This allows frictional maps of the surface to be produced. These types of measurements, which can also be made with derivatized tips, are a form of micro-tribology. This type of measurement has not really been exploited much in the food area.

12.2.2 Molecular Force Measurements

Although the AFM is generally used as a microscope it is also a useful tool for measuring forces even at the molecular level. If the spring constant of the cantilever is known, then the forces applied to a surface, or required to remove the tip from a surface, can be calculated by measuring the deflection of the cantilever. Standard methods are available for calibrating the cantilevers and the tube scanners. By attaching molecules to both the tip and the sample it becomes possible to study the deformation of single molecules (Liu et al., 2005; Wei, 2008). These types of studies have been reported for a range of polysaccharides and proteins including food biopolymers: the approach can be used to investigate the unfolding of protein structures (Borgia et al., 2008) or the elasticity or conformational changes of individual polysaccharide chains (Zhang and Marszalek, 2006; Giannotti and Vancso, 2007). The studies on protein unfolding are extremely elegant but have as yet not delivered new information relevant to the applications of proteins in soft systems such as food materials.

Force spectroscopy has been used to investigate elasticity and conformational changes of individual polysaccharide chains (Rief et al., 1997; Brant, 1999; Fisher et al., 2000; Abu-Lail and Camesano, 2003; Sietmoen et al., 2003; Zhang and Zhang, 2003; Yuasa, 2006; Zhang and Marszalek, 2006; Giannotti and Vancso, 2007). These studies provide complementary information to that obtained by imaging of polysaccharide structures. The simplest type of experiment involves drop depositing molecules onto a substrate and then using the tip, rather like a fishing rod, to pick up and stretch the molecules. This is easy to do. However, it is not certain whether single molecules, or groups of molecules, are attached to the tip, or at what point on the molecule any attachment occurs. A more thorough approach is to derivatize the ends of the molecules so that they can be specifically attached to a substrate and then also collected by the tip. In this case the whole molecule is stretched. The simplest form of analysis uses theories for worm-like coils or freely jointed chain models (Bouchiat et al., 1999; Abu-Lail and Caesano, 2003) which allow calculation of the persistence length (L_P). Currently there is a need to reconcile differences in the values found for L_P, determined from force spectroscopy, with values found from imaging or solution scattering studies.

When the contour length of the molecules is large compared to the persistence length, then the stretching will be largely entropic in nature, with enthalpic effects appearing at high extensions. The transition region will depend on the molecular weight and the detailed secondary structure of the polysaccharide. Data have been reported for polysaccharides such as native and derivatized dextrans (Rief et al., 1997; Frank and Belfort, 1997; Li et al., 1998), cellulose derivatives (Li et al., 1998; Kuhner et al., 2006), extracellular polysaccharide from *Pseudomonas atlantica* (Frank and Belfort, 1997), heparin (Marszalek et al., 2003), amylose and modified amyloses (Lu et al., 2004; Zhang and Marszalek, 2006; Zhang et al., 2006), alginate (Williams et al., 2008), curdlan (Zhang et al., 2003) and xanthan (Li et al., 1998). An analysis of the data on dextrans at low extensions suggested Kuhn lengths ($L_K = 2L_P$) of 0.6 nm (Rief et al., 1997). Force spectroscopy can be used to follow the unwinding of helical structures and to distinguish between different types of helical structure (Zhang et al., 2003). Measurements have been made on both native and denatured xanthan: denatured xanthan behaved similarly to the structurally related carboxymethylcellulose, whereas native xanthan showed quite different behavior, which was attributed to helix formation (Li et al., 1998). Investigations of the force–distance curves for a number of polysaccharides (amylose, pullulan, dextran, pectin and methylcellulose) have considered the role of the pyranose ring in determining the elasticity of the polymer chain (Marszalek et al., 1998, 2002). For dextran, pullulan and amylose it was shown that the enthalpic component of the elasticity could be eliminated by periodate cleavage of the glucose rings, suggesting that the force-induced distortion of the sugar ring, and a chair-boat conformational transition, determine this contribution. In the case of polysaccharides (e.g. pectin or amylose) which can adopt ordered helical secondary structures, a periodate oxidation will inhibit helix formation, and this will contribute to the loss of the enthalpic contribution. Whereas polysaccharides such as dextran, pullulan, amylose, heparin and pectin all show enthalpic contributions to their elasticity, at least at large extensions, carboxymethylcellulose and methylcellulose behave entropically (Li et al., 1998; Marszalek et al., 1998): in the cellulose derivatives the pyranose rings are already fully extended and cannot contribute further to extension of the chain. Comparative studies of λ-, ι- and κ-carragennans showed how the presence of an anhydride bridge can alter the conformation and conformational transitions of the polysaccharide on extension (Xu et al., 2002). It has been proposed that the characteristic transitions associated with sugar rings can be used as a fingerprint for particular polysaccharides (Li et al., 1999; Marszalek et al., 2001; Zhang and Zhang, 2003; Lee et al., 2004). Alginate is an example of a naturally occurring block copolymer. Force spectroscopy has been explored as a tool for characterizing the heterogeneity of alginates by monitoring the signatures of the different conformational changes from the different 'blocks' within the polysaccharide (Williams et al., 2008).

Rather than probing the forced conformational changes of the molecules by direct extension it is also possible to induce oscillations of molecules through the use of magnetically oscillated cantilevers (Humphris et al., 2000; Kawakami et al., 2005) or through analysis of thermally driven oscillations (Kawakami et al., 2004) of the AFM cantilevers.

In the area of soft solids it seems that this type of study is more likely to provide new understanding of the molecular origins of friction and adhesion than to contribute directly to further understanding of the functionality of individual polysaccharides.

Rather than studying the deformation of individual molecules it is also possible to measure the binding between different macromolecules. An example of this type of interaction would be the specific binding between carbohydrates and lectins (Touhami et al., 2003). In this case the carbohydrates or proteins are covalently attached to either the substrate or the tip. Tips can be difficult to derivatize and an alternative is to covalently attach the molecule to a silica sphere and then glue this to the tip. The measurements can be performed on model systems at the molecular level (Gunning et al., 2009), to probe interactions on cellular systems under physiological conditions (Gunning et al., 2008) or to map molecules such as adhesins on bacterial surfaces (Dupres et al., 2005). The molecules are brought together and then pulled apart. The adhesive force (or energy) required to separate them can be used to characterize the specific interaction (Figure 12.17). A number of plant cell wall carbohydrates are claimed to be bioactive and there are suggestions that this bioactivity may be due to specific interactions between fragments of the carbohydrates and mammalian lectins expressed on the surface of cells (Nangia-Makker et al., 2002). Force spectroscopy provides one of a number of methods for testing such molecular mechanisms (Gunning et al., 2009).

12.2.3 Colloidal Forces

The AFM can be adapted to measure colloidal interactions (Leite and Herrmann, 2005; Ralston et al., 2005; Liang et al., 2007). Rather than coupling molecules to the tip it is possible to glue colloidal particles, usually spheres, to the end of a tip-less cantilever. To investigate colloidal interactions it is usually convenient to measure the force between the sphere and a flat substrate. The force is monitored through the cantilever deflection as the sphere approaches the substrate surface. At the point at which the sphere comes into physical contact with the surface the movement of the probe complies with the movement of the tube scanner: this is called the constant compliance region, which can be used to define the zero separation point in the force–distance curve. Adhesive forces, if present, can be determined from the finite force required to detach the colloidal probe from the surface (Leite and Herrmann, 2005). The types of particles used, or the nature of the surface of the

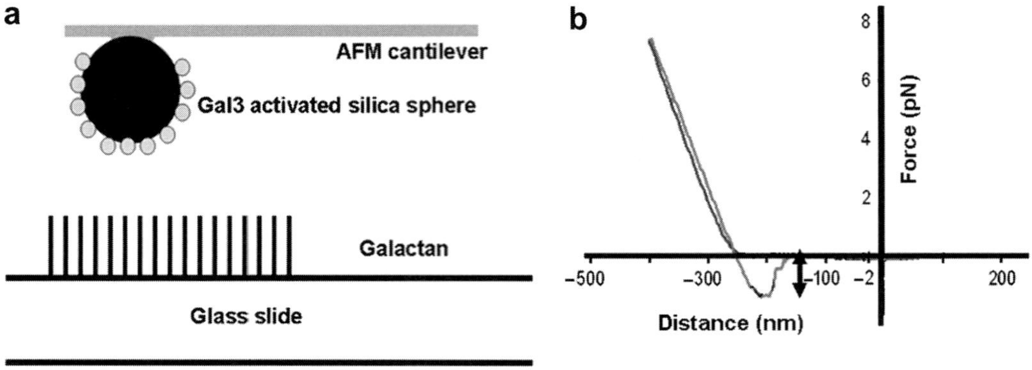

FIGURE 12.17 A diagram illustrating the use of force spectroscopy to measure polysaccharide–lectin interactions. (a) In preliminary experiments the mammalian lectin Gal3 has been covalently bound to a sphere glued to the cantilever. A pectin-derived galactan has been covalently bound to the substrate. (b) Force–distance curves showing a finite adhesive force attributed to specific binding between the lectin and the polysaccharide fragment. The data support a model whereby binding and inhibition of Gal3 by pectin fragments has been proposed as a mechanism for the anti-cancer action of pectin (Nangia-Makker et al., 2002).

sphere and the substrate can be modified in a variety of ways, and the interactions can be studied in liquid media. Experiments can be performed to model the effects of biological surfaces on inter-particle interactions. Spores or bacteria have been glued onto tips to investigate their attachment to surfaces (Ong et al., 1999; Bowen et al., 2002). Simple model colloidal systems, created by coating spheres and surfaces with protein or surfactant layers, can be used to investigate the role of such layers in flocculation or, in the case of oil droplets, coalescence.

In the case of foams and emulsions a better model system can be achieved by studies on deformable particles. Spheres attached to tip-less cantilevers can be used to probe the deformation of air bubbles and oil droplets. Recently it has become possible to attach oil droplets to tip-less cantilevers and also to glass surfaces and then to probe the interaction between them in a liquid medium (Gunning et al., 2004b). In this type of experiment it is also possible to control and modify the composition and structure of the interfacial layers at the oil–water interface. This type of measurement between two coated droplets provides a realistic model of the effect of interfacial structure on droplet interactions that might be expected to occur in an emulsion. The use of droplets allows the unfolding and interaction between proteins at the interface, the diffusion of surfactant-like molecules on close approach of the droplets and the effects of the distortion of the droplets themselves to be monitored. This type of experiment, which has allowed the surfactant-induced displacement of protein from oil droplets to be monitored, showed that the orogenic displacement mechanism, discussed earlier in the section on interfacial protein networks, and developed on models using finite-sized flat interfaces, actually applies for the closed, curved surfaces of droplets. A difficulty with measurements between droplets is in calculating the mass or volume of the droplets and determining the 'zero' point of contact between deformable particles. Despite these difficulties the studies have already yielded new insights into the behavior of deformable droplets. It is generally accepted that the deformability of droplets is determined by the Laplace pressure within the droplet which depends on the interfacial tension. Studies on the interaction

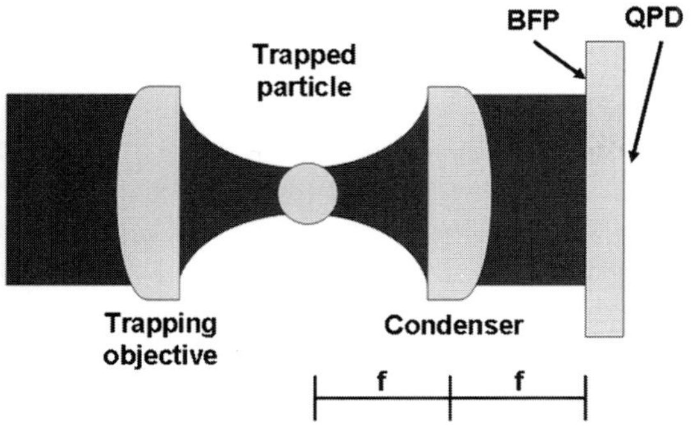

FIGURE 12.18 Schematic diagram illustrating the essential elements of an optical tweezers apparatus. BFP is the back focal plane, QPD is the four-quadrant photodiode and f is the focal length of the condenser lens.

between droplets coated with elastic protein films have clearly shown that, contrary to perceived wisdom, it is the mechanical properties of the films which dominate and dictate the deformability of the droplets on close approach: as the surface film forms and consolidates the interfacial tension is reduced but the droplets become harder.

The relatively high stiffness of the cantilevers means that as a force transducer the AFM is best suited to measuring forces in the range 10^2–10^4 pN. Certain molecular interactions, or weak colloidal forces such as depletion forces, may thus be difficult to measure by AFM. An alternative approach is the use of a technique called optical tweezers. Optical tweezers can be used to measure forces between 0.1–100 pN and have been used in elegant studies of molecular processes such as studies on molecular motors (Berry and Armitage, 1999), transcription of DNA (Herbert et al., 2008) and muscle action (Thomas and Thornhill, 1998).

12.2.4 Optical Tweezers

Optical tweezers (Neuman and Nagy, 2008) are a device attached to a microscope for observing and manipulating particulates trapped in a focused laser beam (Figure 12.18). Key features are a suitable optical microscope and, for a single optical trap, a trapping laser and a high numerical aperture (NA) microscope objective. The optical trap is created by focusing a laser beam to a diffraction-limited spot through the objective lens. Interaction between the light and a dielectric particle in the vicinity of the focused spot creates a three-dimensional force, dependent on the laser power and the polarizability of the particle, which acts to move the particle towards the focal point of the laser beam. The laser used to trap the particles should have a Gaussian intensity profile in order to achieve the smallest size focused spot and hence the largest optical gradient. In order to achieve a stable trap the microscope objective lens should have an NA of at least 1.2, requiring the use of water or oil immersion objectives. Water objectives do not suffer from the spherical aberration effects experienced with oil immersion objectives and will permit trapping of particles deep within the solution. Near infrared lasers are usually used to trap particles in order to minimize optical-induced damage to biological samples. Additional factors that will influence the choice of the laser are variations in the laser power, or any movement of the focused spot, both of which will generate noise in the experimental measurements. For biological studies the preferred type of laser is a diode-pumped neodymium yttrium aluminium garnet (Nd:YAG) laser with a wavelength of 1.064 μm. The wavelength employed will determine the choice of microscope objective.

12.2.5 Modes of Operation

The optical assembly between the laser and the microscope can be used to steer the laser beam and to position the optical trap in three dimensions. The objective lens is used to focus the laser beam and to view the trapped sphere. The simplest application is to use the tweezers as a nano-manipulator to non-invasively collect and position particles.

By projecting the image of the particles held in the trap onto a four-quadrant photodiode it is possible to monitor the motion of a trapped particle: the back focal plane of the microscope condenser lens is imaged on the photodiode. Changes in the interference pattern between the reflected and transmitted light can be used to track changes in position of the particles in three dimensions. For small displacements of the particles from the equilibrium position within the trap (~ 150 nm) the restoring force is linearly proportional to the displacement. The restoring force will be asymmetric. Along the beam direction the restoring force will be determined by the degree of focus of the laser beam, and hence the laser power and the NA of the objective lens. Within the image plane, perpendicular to the beam direction, the restoring force depends on the Gaussian profile of the laser and will be axially symmetric. By determining the 'spring constant' or stiffness of the system it is possible to use displacement data to measure the forces experienced by a particle as it senses the local environment. The instrument is capable of measuring forces in the range 0.1–100 pN at a temporal resolution of 10^{-4} s. The achievable resolution will be dependent on the characteristics of the trap and the dielectric properties of the particles.

The use of the instrument can be enhanced by active control of the position of the trap in the specimen plane and the position of the trapping chamber. Through use of a feedback circuit it is possible to maintain constant force acting on the particle or to compensate for problems such as thermal drift. There are two distinct modes of operation. The position of the trap can be moved relative to the sample chamber, through the use of piezoelectric controlled mirrors, or acousto-optic deflectors: the motion of the trap can be very fast (~ 10 μs) but the extent of displacement within a single axial plane is limited to a few micrometers. The alternative is to use piezoelectric devices to move the trapping chamber relative to the trap. This allows three-dimensional control of the displacement over a larger spatial range (~ 100 μm) but the temporal resolution is reduced (~ 10 ms).

Use of the instrument as a force transducer requires accurate determination of the position of the particles and determination of the stiffness of the optical trap. The position detector can be calibrated by directly observing the motion of a bead over a known distance. There are a range of methods for determining the stiffness of the trap. Common approaches involve modeling the thermal motion of the particle within the trap or, by use of Stokes law, to compare the trapping force with the viscous drag experienced due to the relative motion of the particles with respect to their liquid environment.

12.2.6 Applications

In the 'passive' mode of operation the system can be used to monitor the motion of particles as they sense their local environment. The trap determines the local volume within which the particles can diffuse and by probing the thermally induced 3D motion of the particles it is possible to measure the local viscosity. If the medium within which the particle is placed is viscoelastic then information on the viscous and elastic components can be determined (Gisler and Weitz, 1998; Furst, 2005). The probe could be used to follow the development of an elastic structure such as occurs during gelation. Here the information obtained will depend on the particle size relative to the pore size of the network. In a polymer network the probe particles can diffuse through the pores of the network but are excluded from the regions occupied by the polymer network. Hence the motion of the particles can be used to map the network structure within the gel (Tischer et al., 2001). Large particles will be trapped in the gel network and can be used to monitor the viscoelastic deformations of the gel.

The optical tweezers can be used to bring a trapped particle close to another surface and to monitor the interaction between the surface and the particle. In food science there is a growing interest in particulate interactions in the gut. These include the means by which pathogens invade or 'friendly' bacteria colonize the gut. It also includes interactions involved in drug delivery or the uptake of nutrients from particulate food samples. Optical tweezers can be used to probe bacterial adhesion to surfaces (Aspholm et al., 2006; Camesano et al., 2007) and, through the use of optical tweezers, both specific and non-specific particle interactions with surfaces could be monitored, and quantitative information obtained on the nature of the interactions. There are reported examples of studies of the interaction of virus-coated particles with erythrocytes (Mammen et al., 1996). By using beads derivatized with particular molecular species it would be possible to monitor or map specific interactions with cellular surfaces. Other cellular processes could also be studied and quantified such as phagocytosis. The optical tweezers can also be used to observe the local motion of organelles such as granules or the biomechanics within living cells (Addae-Mensah and Wikswo, 2008).

By attaching molecules to the surfaces of beads it is possible to follow molecular motion and interactions. Thus a bead on the surface of a membrane can be used to probe molecular motion within the membrane or observe interactions. Similarly tethered molecules could be used to explore specific molecular interactions involved in recognition processes, or to test molecular models of the bioactivity of certain food molecules. Beads derivatized with tethered molecules can be used to 'fish' for interactions with other molecules on natural or model surfaces. Alternatively through the use of two optical traps it is possible to measure molecular interactions between beads derivatized with interacting pairs of molecules.

At present most researchers use home-built optical tweezers designed for particular applications. With the availability of commercial instruments such as the JPK NanoTracker™ the use of this technique should become more widespread.

12.3 CONCLUSIONS

Single-molecule techniques such as AFM and optical tweezers are becoming increasingly important for studying biological systems. In soft solid systems such as food materials AFM has been used primarily as an imaging tool. The ability to probe complex systems at the molecular level under near-native conditions has opened new areas of research and generated new molecular understanding. With the advent of second-generation hybrid AFM–optical microscopes it is becoming possible to extend imaging from studies of simple molecules to more complicated food and plant samples. The AFM can also be used as a force transducer and force spectroscopy is emerging as a useful tool for probing molecular, colloidal and cellular interactions. The AFM can probe forces in the range 10^2–10^4 pN. Optical tweezers can measure forces in the range 0.1–100 pN and can be used to probe structure in 3D under physiological conditions. The use of optical tweezers in the soft solids area is likely to expand as new commercial instruments become available.

References

Abu-Lail, N. I. and Camesano, T. A. (2003). Polysaccharide properties probed with atomic force microscopy. *J. Microscopy, Oxford* **212**, 217–238.

A-Hassan, E., Heinz, W. F., Antonio, M. D., D'Costa, N. P., Nageswaraan, S., Schoenberger, C.-A., and Hoh, J. H. (1998). Relative microelastic mapping of living cells by atomic force microscopy. *Biophys. J.* **74**, 1564–1578.

Adams, L. L., Kroon, P., Williamson, G., and Morris, V. J. (2003). Characterisation of heterogeneous arabinoxylans by direct imaging of individual molecules by atomic force microscopy. *Carbohydrate Research* **338**, 771–780.

Adams, E. L., Kroon, P. A., Williamson, G., Gilbert, H. J., and Morris, V. J. (2004). Inactivated enzymes as probes of the structure of arabinoxylans as observed by atomic force microscopy. *Carbohydrate Research* **339**, 579–590.

Addae-Mensah, K. A. and Wikswo, J. P. (2008). Measurement techniques for cellular biomechanics *in vitro*. *Experimental Biology & Medicine* **233**, 782–808.

Albrecht, T. R. and Quate, C. F. (1988). Atomic resolution with the atomic force microscope on conductors and nonconductors. *J. Vac. Sci. & Technol* **6**, 271–274.

Aspholm, M., Kalia, A., Ruhl, S., Schedin, S., Arnqvist, A., Linden, S., Sjostrom, R., Gerhard, M., Semino-Mora, C., Dubois, A., Unemo, M., Danielsson, D., Teneberg, S., Lee, W.-K., Berg, D., and Boren, T. (2006). *Helicobacter pylori* adhesion to carbohydrates. *Methods in Enzymology* **417**, 293–339.

Berry, R. M. and Armitage, J. P. (1999). The bacterial flagella motor. *Advances Microbial Physiology* **41**, 291–337.

Binnig, G., Quate, C. F., and Gerber, C. (1986). Atomic force microscope. *Phys. Rev. Letts* **56**, 930–933.

Binnig, G. and Smith, D. (1986). Single tube three-dimensional scanner for scanning tunneling microscopy. *Rev. Sci. Instrum* **57**, 1688–1689.

Bogracheva, T. Y., Cairns, P., Noel, T., Hulleman, S., Wang, T. L., Morris, V. J., Ring, S. G., and Hedley, C. L. (1999). The effect of mutant genes at the r, rb, rug3, rug4, rug5 and lam loci on the granular structure and physicochemical properties of pea seed starch. *Carbohydrate Polymers* **39**, 303–314.

Borgia, A., Williams, P. M., and Clarke, J. (2008). Single-molecule studies of protein folding. *Annual Reviews of Biochemistry* **77**, 101–125.

Bouchiat, C., Wang, M. D., Allemand, J.-F., Strick, T., Block, S. M., and Croquette, V. (1999). Estimating the persistence length of a worm-like chain molecule from force-extension data. *Biophys. J* **76**, 409–413.

Bowen, W. R., Fenton, A. S., Lovitt, R. W., and Wright, C. J. (2002). The measurement of Bacillus mycoides spore adhesion using atomic force microscopy, simple counting methods, and a spinning disk technique. *Biotechnology & Bioengineering* **79**, 170–179.

Brant, D. A. (1999). Novel approaches to the analysis of polysaccharide structures. *Current Opinion Structural Biol.* **9**, 556–562.

Camesano, T. A., Liu, Y. T., and Datta, M. (2007). Measuring bacterial adhesion at environmental interfaces with single-cell and single-molecule techniques. *Advances Water Resources* **30**, 1470–1491.

Daniels, D. R. and Donald, A. M. (2003). An improved model for analyzing the small angle x-ray scattering of starch granules. *Biopolymers* **69**, 165–175.

Defrene, Y. F. and Hinterdorfer, P. (2008). Recent progress in AFM molecular recognition studies. *Pflugers Archiv-European J. Physiol* **456**, 237–245.

Dupres, V., Menozzi, F. D., Locht, C., Clare, B. H., Abbott, N. L., Cuenot, S., Bompard, S., Raze, D., and Dufrene, Y. F. (2005). Nanoscale mapping and functional analysis of individual adhesins on living bacteria. *Nature Methods* **2**, 515–520.

Fisher, T. E., Marszalek, P. E., and Fernadez, J. M. (2000). Stretching single molecules into novel conformations using the atomic force microscope. *Nature Structural Biology* **7**, 719–724.

Fishman, M., Cooke, P. H., Chau, H. K., Coffin, D. R., and Hotchkiss, A. T. (2007). Global structures of high methoxyl pectin from solution and in gels. *Biomacromolecules* **8**, 573–578.

Frank, B. P. and Belfort, G. (1997). Intermolecular forces between extracellular polysaccharides measured using the atomic force microscope. *Langmuir* **13**, 6234–6240.

Furst, E. M. (2005). Applications of laser tweezers in complex fluid rheology. *Current Opinion Colloid & Interface Science* **10**, 79–86.

Gallant, D. J., Bouchet, B., and Baldwin, P. M. (1997). Microscopy of starch: Evidence of a new level of granule organization. *Carbohydrate Polymers* **32**, 177–191.

Garcia, R. and Perez, R. (2002). Dynamic atomic force microscopic methods. *Surface Sci. Reports* **47**, 197–301.

Giannotti, M. I. and Vancso, G. J. (2007). Interrogation of single synthetic polymer chains and polysaccharides by AFM-based force spectroscopy. *CHEMPHYSCHEM* **8**, 2290–2307.

Gisler, T. and Weitz, D. A. (1998). Tracer microrheology in complex fluids. *Current Opinion Colloid & Interface Science* **3**, 586–592.

Gosal, W. S., Clark, A. H., and Ross-Murphy, S. B. (2004). Fibrillar beta-lactoglobulin gels: Part 1. *Fibril formation and structure*. *Biomacromolecules* **5**, 2408–2419.

Gunning, A. P., Kirby, A. R., Ridout, M. J., Brownsey, G. J., and Morris, V. J. (1996a). Investigation of gellan networks and gels by atomic force microscopy. *Macromolecules* **29**, 6791–6796.

Gunning, A. P., Wilde, P. J., Clark, D. C., Morris, V. J., Parker, M. L., and Gunning, P. A. (1996b). Atomic force microscopy of interfacial protein films. *J. Colloid & Interface Science* **183**, 600–602.

Gunning, A. P., Mackie, A. R., Kirby, A. R., Kroon, P., Williamson, G., and Morris, V. J. (2000). Motion of a cell wall polysaccharide observed by atomic force microscopy. *Macromolecules* **33**, 5680–5685.

Gunning, A. P., Giardina, T. P., Faulds, C. B., Juge, N., Ring, S. G., Williamson, G., and Morris, V. J. (2003). Surfactant mediated solubilisation of amylose and visualisation by atomic force microscopy. *Carbohydrate Polymers* **51**, 177–182.

Gunning, P.A., Mackie, A.R., Gunning, A.P., Woodward, N.C., Wilde, P.J., and Morris, V.J. (2004a). The Effect of Surfactant Type on Surfactant-Protein Interactions at the Air-Water Interface. *Biomacromolecules* **5**, 984–991.

Gunning, A. P., Mackie, A. R., Wilde, P. J., and Morris, V. J. (2004b). Atomic force microscopy of emulsion droplets: probing droplet-droplet interactions. *Langmuir* **20**, 116–122.

Gunning, A. P., Chambers, S., Pin, C., Man, A. L., Morris, V. J., and Nicoletti, C. (2008). Mapping specific adhesive interactions on living intestinal epithelial cells with atomic force microscopy. *FASEB J* **22**, 2331–2339.

Gunning, A. P., Bongaerts, R. J. M., and Morris, V. J. (2009). Recognition of galactan components of pectin by galectin-3. *FASEB J* **23**, 415–424.

Guzey, D. (2006). Formation, stability and properties of multilayer emulsions for application in the food industry. *Advances Colloid & Interface Science* **128**, 227–248.

Heinz, W. F. and Hoh, J. H. (1999). Spatially resolved force spectroscopy of biological surfaces using the atomic force microscope. *Trends Biotechnol.* **17**, 143–150.

Herbert, K. M., Greenleaf, W. L., and Block, S. M. (2008). Single-molecule studies of RNA polymerase: Motoring along. *Annual Reviews Biochemistry* **77**, 149–176.

Hoh, J. H., Heinz, W. F., and A-Hassan, E. (1997). Force Volume. *In* Digital Instruments Support Note N. 240, Digital Instruments, 112 Robin Hill Roas, Santa Barbara, CA 93117, California, USA.

Holscher, H. and Schirmeisen, A. (2005). Dynamic force microscopy and spectroscopy. *Advances Imaging & Electron Physics* **135**, 41–101.

Humphris, A. D. L., Tamayo, J., and Miles, M. J. (2000). Active quality factor control in liquids for force spectroscopy. *Langmuir* **16**, 7891–7894.

Humphris, A.D.L., Miles, M.J., and Hobbs, J.K. (2005). A mechanical microscope: high-speed atomic force microscopy. *Appl. Phys. Letts.* 86 Issue 3 article number 034106.

Ikeda, S. (2003a). Torsional fracture properties of heat-induced whey protein gels and microstructures of gel precursors visualized by atomic force microscopy. *Nippon Shokuhin Kagaku Kogaku Kaishi* **50**, 237–242.

Ikeda, S. (2003b). Heat-induced gelation of whey proteins observed by rheology, atomic force microscopy, and Raman scattering spectroscopy. *Food Hydrocolloids* **17**, 399–406.

Ikeda, S., Funami, T., and Zhang, G. Y. (2005). Visualizing surface active hydrocolloids by atomic force microscopy. *Carbohydrate Polymers* **62**, 192–196.

Ikeda, S. and Morris, V. J. (2002). Fine-stranded and particulate aggregates of heat-denatured whey proteins visualized by atomic force microscopy. *Biomacromolecules* **3**, 382–389.

Kawakami, M., Byrne, K., Khatri, B., McLeish, T. C., Radford, S. E., and Smith, D. A. (2004). Viscoelastic properties of single polysaccharide molecules determined by analysis of thermally driven oscillations of an atomic force microscope cantilever. *Langmuir* **20**, 9299–9303.

Kawakami, M., Byrene, K., Khatri, B. S., McLeish, T. C. B., Radford, S. E., and Smith, D. A. (2005). Viscoelastic measurements of single molecules on a millisecond time scale by magnetically driven oscillation of an atomic force microscope cantilever. *Langmuir* **21**, 4765–4772.

Kirby, A. R., Gunning, A. P., Morris, V. J., and Ridout, M. J. (1995). Observation of the helical structure of the bacterial polysaccharide acetan by atomic force microscopy. *Biophysical J.* **68**, 359–362.

Kirby, A. R., Gunning, A. P., Waldron, K. W., Morris, V. J., and Ng, A. (1996). Visualisation of plant cell walls by atomic force microscopy. *Biophysical J.* **70**, 1138–1143.

Kirby, A. R., MacDougall, A. J., and Morris, V. J. (2006a). Sugar Beet Pectin – Protein Complexes. *Food Biophysics* **1**, 51–56.

Kirby, A. R., Ng, A., Waldron, K. W., and Morris, V. J. (2006b). AFM investigations of cellulose fibres in Bintje potato (*Solanum tuberosum L*) cell wall fragments. *Food Biophysics* **1**, 163–167.

Kirby, A. R., McDougall, A. J., and Morris, V. J. (2007). Atomic Force Microscopy of Tomato and Sugar Beet pectin molecules. *Carbohydrate Polymers* **71**, 640–647.

Kuhner, F., Erdmann, M., and Gaub, H.E. (2006). Scaling exponent and Kuhn length of pinned polymers by single molecule force spectroscopy. *Phys. Rev. Letts.* 97 issue **21**, article number 218301.

Lee, G., Novak, W., Jaroniec, J., Zhang, G. M., and Marszalek, P. E. (2004). Molecular dynamics simulations of forced conformational transitions in 1,6-linked polysaccharides. *Biophys. J* **87**, 1456–1465.

Li, H., Rief, M., Oesterhelt, F., and Gaub, H. E. (1998). Single-molecule force spectroscopy on xanthan by AFM. *Adv. Mater* **3**, 316–319.

Li, H. B., Rief, M., Oesterhelt, F., Gaub, H. E., Zhang, X., and Shen, J. C. (1999). Single-molecule force spectroscopy on polysaccharides by AFM – nanomechanical fingerprint of α-(1,4)-linked polysaccharides. *Chem. Phys. Letts* **305**, 197–201.

Liang, Y., Hilal, N., Langston, P., and Starov, V. (2007). Interaction forces between colloidal particles in liquid: Theory and experiment. *Advances Colloid & Interface Science* 134–135, 151–166.

Liete, F. L. and Herrmann, P. S. P. (2005). Application of atomic force spectroscopy (AFS) to studies of adhesion phenomena: a review. *J. Adhesion Science & Technology* **19**, 365–405.

Liu, C. J., Shi, W. Q., Cui, S. X., Wang, Z. Q., and Zhang, X. (2005). Force spectroscopy of polymers: Beyond single chain mechanics. *Current Opinion Solid State Materials* **9**, 140–148.

Lu, Z. Y., Nowak, W., Lee, G. R., Marszalek, P. E., and Yang, W. T. (2004). Elastic properties of single amylose chains in water: A quantum mechanical and AFM study. *J. Amer. Chem. Soc.* **126**, 9033–9041.

Mackie, A. R., Gunning, A. P., Ridout, M. J., and Morris, V. J. (1998). Gelation of gelatin – Observations at the air/water interface and in the bulk. *Biopolymers* **46**, 245–252.

Mackie, A. R., Gunning, A. P., Wilde, P. J., and Morris, V. J. (1999). The orogenic displacement of protein from the air/water interface by competitive adsorption. *J Colloid & Interface Science* **210**, 157–166.

Mackie, A. R., Gunning, A. P., Wilde, P. J., and Morris, V. J. (2000). The competitive displacement of β-lactoglobulin from the air-water interface by SDS. *Langmuir* **16**, 8176–8181.

Mackie, A. R., Gunning, A. P., Ridout, M. J., Wilde, P. J., and Morris, V. J. (2001). Orogenic displacement in mixed β-lactoglobulin /β-casein films at the air/water interface. *Langmuir* **17**, 6593–6598.

Mammen, M., Helmerson, K., Kishore, R., Chai, S. K., Phillips, W. D., and Whitesides, G. M. (1996). Optically controlled collisions of biological objects to evaluate potent polyvalent inhibitors of virus-cell adhesion. *Chemistry & Biology* **3**, 757–763.

Marszalek, P. E., Oberhauser, A. F., Pang, Y.-P., and Fernandez, J. M. (1998). Polysaccharide elasticity governed by chair-boat transitions of the glucopyranose ring. *Nature* **396**, 661–664.

Marszalek, P. E., Li, H., and Fernandez, J. M. (2001). Fingerprinting polysaccharides with single-molecule atomic force microscopy. *Nature Biotechnology* **19**, 258–262.

Marszalek, P. E., Li, H. B., Oberhauser, A. F., and Fernandez, J. M. (2002). Chair-boat transitions in single polysaccharide molecules observed with force-ramp AFM. *Proc. Natl. Acad. Sci* **99**, 4278–4283.

Marszalek, P. E., Oberhauser, A. F., Li, H. B., and Fernandez, J. M. (2003). The force-driven conformations of heparin studied with single molecule force microscopy. *Biophys. J* **85**, 2696–2704.

Meyer, E. (1992). Atomic force microscopy. *Progress Surface Sci* **41**, 3–49.

Moates, G. K., Noel, T. R., Parker, R., and Ring, S. G. (1997). The effect of chain length and solvent interactions on the dissolution of the B-type crystalline polymorph of amylose in water. *Carbohydrate Research* **298**, 327–333.

Morris, V. J. (2007a). Gels. *In* "The Chemical Physics of Food" (P. S. Belton, Ed.). Chapter 6, pp. 151–198, Blackwell Pub., Oxford.

Morris, V. J. (2007b). Atomic force microscopy (AFM) techniques for characterising food structure. *In* "Understanding and Controlling the Microstructure of Complex Foods" pp. 209–231, (McClements, D. J. (ed.). chapter 8, CRC Press, New York.

Morris, V. J. and Gunning, A. P. (2008). Microscopy, microstructure and displacement of proteins from interfaces: implications for food quality and digestion. *Soft Matter* **4**, 943–951.

Morris, V. J., Kirby, A. R., and Gunning, A. P. (1999). Atomic Force Microscopy for Biologists. Imperial College Press, London.

Morris, V. J., Gunning, A. P., Faulds, C. B., Williamson, G., and Svensson, B. (2005). AFM images of complexes between amylose and *Aspergillus niger* glucoamylase mutants, native and mutant starch binding domains: a model for the action of glucoamylase. *Starke* **57**, 1–7.

Nangia-Makker, P., Conklin, J., Hogan, V., and Raz, A. (2002). Carbohydrate-binding proteins in cancer and their ligands as therapeutic agents. *Trends Molecular Medicine* **8**, 187–192.

Neuman, K. C. and Nagy, A. (2008). Single-molecule force spectroscopy: optical tweezers, magnetic tweezers and atomic force microscopy. *Nature Methods* **5**, 491–505.

Norton, I. T., Jarvis, D. A., and Foster, T. J. (1999). A molecular model for the formation and properties of fluid gels. *International J. Biological Macromolecules* **26**, 255–261.

Ong, Y. L., Razatos, A., Georgiou, G., and Sharma, M. M. (1999). Adhesion forces between *E-coli* bacteria and biomaterial surfaces. *Langmuir* **15**, 2719–2725.

Parker, M. L., Kirby, A. R., and Morris, V. J. (2008). *In situ* imaging of pea starch in seeds. *Food Biophysics* **3**, 56–76.

Ralston, J., Larson, I., Rutland, M. V. V., Feiller, A. A., and Kleijn, M. (2005). Atomic force microscopy and direct surface force measurements – (IUPAC technical report). *Pure & Applied Chemistry* **77**, 2149–2170.

Ridout, M. J., Gunning, A. P., Wilson, R. H., Parker, M. L., and Morris, V. J. (2002). Using AFM to image the internal structure of starch granules. *Carbohydrate Polymers* **50**, 123–132.

Ridout, M. J., Parker, M. L., Hedley, C. L., Bogracheva, T. Y., and Morris, V. J. (2003). Atomic force microscopy of pea starch granules: Granule architecture of wild-type parent,*r*, and *rb* single mutants, and the *rrb* double mutant. *Carbohydrate Research* **338**, 2135–2147.

Ridout, M. J., Parker, M. L., Hedley, C. L., Bogracheva, T. Y., and Morris, V. J. (2004). Atomic Force Microscopy of Pea Starch: Origins of Image Contrast. *Biomacromolecules* **5**, 1519–1527.

Rief, M., Oesterhelt, F., Heymann, B., and Gaub, H. E. (1997). Single molecule force spectroscopy on polysaccharides by atomic force microscopy. *Science* **275**, 1295–1297.

Round, A. N., Kirby, A. R., and Morris, V. J. (1996). Collection and processing of AFM images of plant cell walls. *Microscopy & Analysis* **55**, 33–35.

Round, A. N., Rigby, N. M., Ring, S. G., and Morris, V. J. (2001). Investigating the nature of branching in pectins by atomic force microscopy and carbohydrate analysis. *Carbohydrate Research* **331**, 337–342.

Round, A. N., Berry, M., McMaster, T. J., Corfield, A. P., and Miles, M. J. (2004). Glycopolymer charge density determines conformation in human ocular mucin gene products: an atomic force microscope study. *J. Structural Biology* **145**, 246–253.

Round, A. N., McMaster, T. J., Miles, M. J., Corfield, A. P., and Berry, M. (2007). The isolated MUC5AC gene product from human ocular mucin displays intramolecular conformational heterogeneity. *Glycobiology* **17**, 578–585.

Rousseau, D. (2007). The microstructure of chocolate. *In* Understanding and Controlling the Microstructure of Complex Foods, 590–648, McClements, D. J. (ed.). CRC Press, New York. chapter 24.

Sietmoen, M., Maurstad, G., Sikorski, P., Paulson, B. S., and Stokke, B. T. (2003). Characterisation of bacterial polysaccharides: steps towards single-molecular studies. *Carbohydrate Research* **336**, 2459–2475.

Thomas, N. and Thornhill, R. A. (1998). The physics of biological motors. *J. Physics D Applied Physics* **31**, 253–256.

Tischer, C., Altmann, S., Fisinger, S., Horber, J. K. H., Stelzer, E. H. K., and Florin, E. L. (2001). Three-dimensional thermal noise imaging. *Applied Physics Letters* **79**, 3878–3880.

Touhami, A., Hoffman, B., Vasella, A., Denis, F. A., and Dufrene, Y. F. (2003). Probing specific lectin-carbohydrate interactions using atomic force microscopy imaging and force measurements. *Langmuir* **10**, 1745–1751.

Vesenka, J., Marsh, T., Miller, R., and Henderson, E. (1996). Atomic force microscopy of G-wire DNA. *J. Vac. Sci. & Technol* **14**, 1413–1417.

Waigh, T. A., Kato, K. L., Donald, A. M., Gidley, M. J., Clarke, C. J., and Riekel, C. (2000). Side-chain liquid-crystalline model for starch. *Starke* **52**, 450–460.

Wei, H. and van de Ven, T. G. M. (2008). AFM-based single molecule force spectroscopy of polymer chains: Theoretical models and applications. *Applied Spectroscopy Reviews* **43**, 111–133.

Williams, M. A. K., Marshall, A., Haverkamp, R. G., and Graget, K. I. (2008). Stretching single polysaccharide molecules using AFM: a potential method for the investigation of the intermolecular urinate distribution of alginate? *Food Hydrocolloids* **22**, 18–23.

Woodward, N. C., Wilde, P. J., Mackie, A. R., Gunning, A. P., Gunning, P. A., and Morris, V. J. (2004). The effect of processing on the displacement of whey protein; applying the orogenic model to a real system. *J. Agric. Food Chemistry* **52**, 1287–1292.

Xu, Q. B., Zhang, W., and Zhang, X. (2002). Oxygen bridge inhibits conformational transition of 1,4-linked α-D-galactose detected by single-molecule atomic force microscopy. *Macromolecules* **35**, 871–876.

Yuasa, H. (2006). Ring flip of carbohydrates: Function and applications. *Trends Glycosci. & Glycotechnol* **18**, 353–370.

Zhang, L., Wang, C., Cui, S. X., Wang, Z. Q., and Zhang, X. (2003). Single-molecule force spectroscopy on curdlan: unwinding helical structures and random coils. *Nano Letts* **3**, 1119–1124.

Zhang, Q. M. and Marszalek, P. E. (2006). Solvent effects on the elasticity of polysaccharide molecules in disordered and ordered states by single-molecule force spectroscopy. *Polymer* **47**, 2526–2532.

Zhang, Q. M., Lu, Z. Y., Hu, H., Yang, W. T., and Marszalek, P. E. (2006). Direct detection of the formation of V-amylose helix by single molecule force spectroscopy. *J. Amer. Chem. Soc.* **128**, 9387–9393.

Zhang, W. and Zhang, X. (2003). Single molecule mechanochemistry of macromolecules. *Progr. Polym. Sci.* **28**, 1271–1295.

CHAPTER

13

Dietary Fiber: Fulfilling the Promise of Added-Value Formulations

Steve W. Cui and Keisha T. Roberts

Guelph Food Research Centre, Agriculture and Agri-Food Canada, 93 Stone Road West, Guelph, Ontario, Canada N1G 5C9

13.1 RECENT DEVELOPMENTS IN DIETARY FIBER RESEARCH

13.1.1 Definitions and Analytical Methods

The functions of dietary fiber or roughage have been described in ancient herbs and medicinal literature around the world for hundreds of years, however, it was not formally acknowledged as 'dietary fiber' until the middle of the 20th century: it is generally accepted that Hipsley introduced the term 'dietary fiber' for the first time in 1953 in describing the non-digestible constituents that make up the plant cell wall in foods (Hipsley, 1953). After nearly 20 years a so-called 'dietary fiber hypothesis' emerged which linked dietary fiber with a number of health conditions, such as diverticular disease and colon cancer; meanwhile, a new definition was given to dietary fiber: 'the skeletal remains of plant cells that are resistant to digestion (hydrolysis) by enzymes of man' (Trowell, 1972). Later, Jenkins and co-workers demonstrated that some non-digestible storage polysaccharides, e.g. guar gum, reduced serum cholesterol concentration and flattened the postprandial glycemia in clinical trials (Jenkins et al., 1975; Gassull et al., 1976). Trowell and colleagues realized that the earlier definition excluded many other plant materials such as mucilages, storage polysaccharides, and algal polysaccharides, which were also not hydrolyzed by the alimentary enzymes. They broadened their definition to include all digestion-resistant polysaccharides (mostly plant storage saccharides), such as gums, modified celluloses, hemicellulose, mucilages, oligosaccharides, pectins, and associated minor substances, such as waxes, cutin, and suberin (Trowell et al., 1976; Devries et al., 2001). This definition became the base for many versions of definitions which emerged in the next 30 years by health and regulatory organizations around the world (Health and Welfare Canada, 1985; US Food and Drug Administration (US-FDA), 1987; Health Canada, 1988; AACC, 2001; AFSSA, 2002; Health Council of The Netherlands, 2006). However, there is no universally acceptable definition of dietary fiber, regardless

Kasapis, Norton, and Ubbink: Modern Biopolymer Science
ISBN: 978-0-12-374195-0

numerous efforts were made. Perhaps the most accepted definition would be the one proposed by AACC in 2001:

> 'Dietary fibre is the remnants of the edible part of plants and analogous carbohydrates that are resistant to digestion and absorption in the human small intestine with complete or partial fermentation in the human large intestine. It includes polysaccharides, oligosaccharides, lignin and associated plant substances. Dietary fibre exhibits one or more of either laxation (faecal bulking and softening; increased frequency; and/or regularity), blood cholesterol attenuation, and/or blood glucose attenuation.'

In addition to the AACC's definition, several other countries also specified oligosaccharides in their definitions, such as Japan, Australia and New Zealand (ANZFA, 2000, proposed).

In the most recent effort of providing a universally acceptable definition of dietary fiber, the latest version of the dietary fiber definition was proposed by the CODEX ALIMENTARIUS COMMISSION during the 32nd Session at FAO Headquarters, Rome, 29 June – 4 July 2009 (ALINORM 09/32/A):

> Dietary fibre means carbohydrate polymers[1] with ten or more monomeric units[2], which are not hydrolysed by the endogenous enzymes in the small intestine of humans and belong to the following categories:
>
> Edible carbohydrate polymers naturally occurring in the food as consumed;
>
> Carbohydrate polymers, which have been obtained from food raw material by physical, enzymatic or chemical means and which have been shown to have a physiological effect of benefit to health as demonstrated by generally accepted scientific evidence to competent authorities;
>
> Synthetic carbohydrate polymers which have been shown to have a physiological effect of benefit to health as demonstrated by generally accepted scientific evidence to competent authorities.

However, there are still different opinions regarding what should be included as dietary fibre. For example, arguments have been presented to include resistant starch as part of dietary fiber because it shows all the characteristics of dietary fibre and it is not digested in the stomach, but passes on to the colon where it is fermented, promoting healthy digestive bacteria (Topping, 2007). Other groups of experts insisted that the definition of dietary fiber should be more clearly linked to fruits, vegetables and wholegrain cereals and proposed the definition 'Dietary fibre consists of intrinsic plant cell wall polysaccharides' because they believed that the epidemiological support for the health benefits of dietary fiber was based on diets containing fruits, vegetables and wholegrain cereal foods which contained plant cell walls (Geneva Meeting, 2006). Regardless of the many versions of definitions of dietary fiber that have been proposed by scientific and regulatory agencies worldwide, it has become clear that an internationally acceptable definition of dietary fiber must link specific physiological effects of dietary fiber with workable analytical methods which are accurate, reproducible and easy to use for quantitative measurement of all dietary fibers.

[1] When derived from a plant origin, dietary fibre may include fractions of lignin and/or other compounds when associated with polysaccharides in the plant cell walls and if these compounds are quantified by the AOAC gravimetric analytical method for dietary fibre analysis: Fractions of lignin and the other compounds (proteic fractions, phenolic compounds, waxes, saponins, phytates, cutin, phytosterols, etc.) intimately "associated" with plant polysaccharides are often extracted with the polysaccharides in the AOAC 991.43 method. These substances are included in the definition of fibre insofar as they are actually associated with the poly- or oligosaccharidic fraction of fibre. However, when extracted or even re-introduced into a food containing non digestible polysaccharides, they cannot be defined as dietary fibre. When combined with polysacchrides, these associated substances may provide additional beneficial effects (pending adoption of Section on Methods of Analysis and Sampling).

[2] Decision on whether to include carbohydrates with monomeric units *LOWER THAN 10* should be left to national authorities.

Dietary fiber is a complex group of food components as it does not exist as a single entity. Precise evolution of the fiber from food is determined by the analytical method employed. Parallel to the slow perfection process of the definition of dietary fiber, the methods for determining soluble, insoluble and total dietary fiber have experienced a long history of debate. Many methodologies for measuring dietary fiber have been proposed over the years, with specific methods being adopted as AOAC International Official Methods. These include (1) 985.29 – Total dietary fibre in foods, enzymatic-gravimetric method; (2) 991.42 – Insoluble dietary fibre in foods and food products, enzymatic-gravimetric method; phosphate buffer (3) 993.19 – Soluble dietary fibre in foods and food products, enzymatic-gravimetric method; phosphate buffer, (4) 991.43 – Total, soluble and insoluble dietary fibre in foods, enzymatic-gravimetric method; MES-tris buffer, (5) 992.16 – Total dietary fibre, enzymatic-gravimetric method; (6) 993.21 – Total dietary fibre in foods and food products with $\leq$ 2% starch, non-enzymatic-gravimetric method and (7) 994.13 – Total dietary fibre (determined as neutral sugar residues, uronic acid residues, and klason lignin) Gas chromatographic Colorimetric-Gravimetric Method. Since there is no universally accepted method available yet, the following summary of several of the most popular methods provide readers with alternative choices.

13.1.1.1 *Enzymatic Gravimetric Method*

The main steps of the enzymatic gravimetric method involve (1) enzymatic treatments for starch and protein removal, (2) collection of soluble fiber, through precipitation with aqueous ethanol, (3) isolating and weighting the dietary residue and (4) correction for protein and ash. This method was originally proposed in the 1930s by McCance et al. and has evolved over the years. The way to success for this method came when high purity and specific enzymes become commercially available, especially with respect to ensuring that there were no enzymes present to digest dietary fiber. However, one must also precisely follow the digestion steps as shown in Figure 13.1 (Prosky, 2001).

Total fiber determination
Gelatinization and Termamyl incubation,
pH 6.0, 30 min, 100° C
↓
Protease incubation, pH 7.5, 30 min, 60°C
↓
Amyloglucosidase incubation, pH 4.5, 30 min, 60°C
↓
Precipitation with 4 vol. ethanol
↓
Filtration
↓
Washing with alcohol and acetone
↓
Drying
Correction for undigestible protein and ash

FIGURE 13.1 Flow diagram of the AOAC method for total fiber.

13.1.1.2 *Detergent Methods*

Initially, the neutral detergent (ND) system was used for ruminant forages but has undergone alterations to ensure its applicability to human foods. There were limitations with this method as starch was incompletely solubilized. This was solved by amylase incorporation and eventually a method was approved by the American Association of Cereal Chemists (AACC method 32-20).

The acid detergent fiber (ADF) method has not had much alteration for human analysis. This method has usually been executed for lignin determination but is also employed to predict cell wall digestibility. Depending on the fiber source being tested initial extraction with ND may be required before refluxing with acid detergent (Asp, 2001; Robertson and Horvath, 2001).

13.1.1.3 *The Southgate Method*

This method was developed in the 1950s to facilitate the calculation of the energy value by measuring the 'unavailable carbohydrate' of the human diet (Southgate, 1969). Before this method the energy value of the diet was

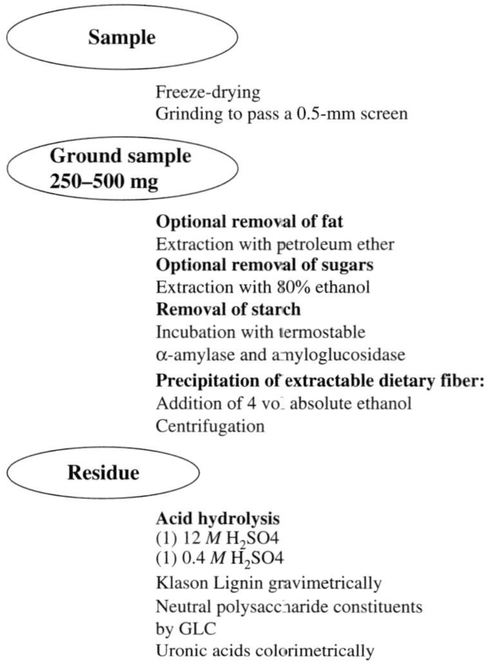

FIGURE 13.2 Flow diagram of the analysis of dietary fiber by the Uppsala method.

calculated using carbohydrate by difference in the USA and by available carbohydrate in the UK. This caused major problems with energy values of cereals on different sides of the Atlantic (Asp, 2001; Southgate, 2001).

13.1.1.4 The Uppsala Method

This method has been approved for dietary fiber analysis by the AOAC and AACC, being methods 994.13 and 32-25 respectively. This method was proven to be acceptable through a collaborative study, on various samples like foods, feeds digest and feces. It was concluded this method was not more time consuming than the gravimetric enzymatic method (Theander et al., 2001). Figure 13.2 shows the analysis of dietary fiber by the Uppsala method.

13.1.1.5 The Crude Fiber Method

This method is used to determine fiber in grains, cereals, flours, feeds and other foods containing fiber, from which fat can be removed, while leaving a workable residue (Furda, 2001). Figure 13.3 is a flow chart for determining crude fiber.

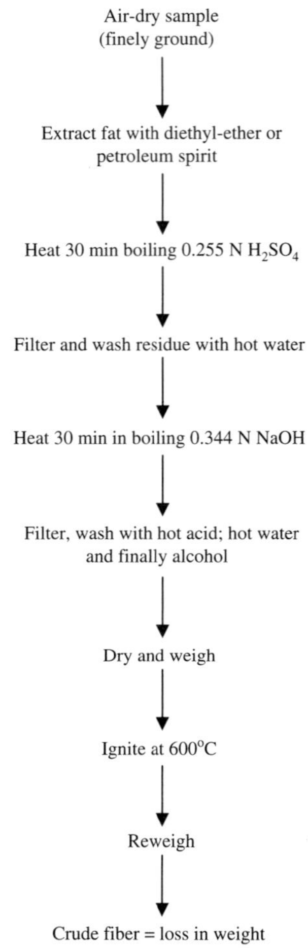

FIGURE 13.3 Flow diagram for determining crude fiber.

In summary, thoughtfulness is essential to ensure the method includes all the dietary fiber but excludes all non-dietary fiber components. A method has been described for the measurement of total, soluble, and insoluble dietary fiber, including resistant starch (RS) and low-molecular-weight non-digestible oligosaccharides (LMWNDO) of DP >3. This method was recently proposed by McCleary and is currently under cross laboratory testing. This method combines the key attributes of AOAC Official

Methods of Analysis 991.43, 2001.03, and 2002.02. The advantage of the current method is to capture the LMWNDO which is not precipitable in ethanol. The ethanol soluble filtrate is recovered by desalting through ion exchange resins, concentrating and quantitating the oligosaccharides by HPLC methods (McCleary, 2008, personal communication).

13.1.2 Physiological Effects of Dietary Fiber

Dietary fiber has been shown to have numerous benefits which are not easily disputed like improved intestinal function, cholesterol reduction and increased microbial biomass. Other benefits of fiber include weight reduction, improved satiety, regulating glucose and insulin responses as well as improved microbiota composition, inflammatory markers such as C-reactive protein, interleukin 6 and tumor necrosis factor alpha in obesity and diabetes. The mechanisms proposed for the benefits of dietary fiber in humans are improved insulin sensitivity through its effects on glycemic and insulinemic responses, increased viscosity of the food bolus, its gel-forming capacity and fermentation causing short-chain fatty acid (SCFA) production. However, some studies have also shown negative effects of dietary fiber consumption such as hyperglycemia and gastrointestinal disturbances, while others showed no benefits from fiber consumption. Lack of and/or negative effects may be associated with the food form consumed and the bioavailability of the fiber in the food products: e.g., solid versus liquid, lack of information on molecular weight (MW), solubility and viscosity of the soluble fiber. In other cases researchers may not account for confounders, such as varying dietary fiber concentrations among studies.

Mechanisms have been proposed for the effect of dietary fiber in type 2 diabetes risk reduction, namely better insulin sensitivity, tempering of inflammatory markers and influences on gut microbiota. Mechanisms for the effects of soluble fiber have been related to its viscosity. On mixing with water, the viscosity of these fiber solutions in the gastrointestinal tract increases, trapping nutrients and slowing gastric emptying. The viscosity of fiber is dependent on its molecular weight, solubility and its concentration in a food product. These three factors are therefore important for elucidating the source of conflicting results in studies employing fiber. For insoluble fiber its effects on satiety and weight control are believed to be associated with increased hormonal secretion, for example cholecystokinin and gastric inhibitory polypeptide (GIP) (Dikeman and Fahey, 2006; Weickert and Pfeiffer, 2008).

The value of the role of dietary fiber in the prevention of chronic non-communicable diseases is not only beneficial for improved lifestyles and longevity but has deep economic roots as well. These diseases are economically significant due to the burden on healthcare systems internationally. According to Brennan (2005), the cost of diabetes, obesity and other associated factors to the United States was approximately $128 billion (Can) in 2004. In the UK direct costs were $710 million (Can), while in New Zealand costs for diabetics in 2001/2 was $176 million (Can) (2008 values). In Canada the estimated cost of diabetes in 1997 was $423.2 million (Can), while it was estimated that for 1997 the total direct cost of obesity was $1.8 billion; the costs of hypertension and coronary artery disease were $656.6 million and $346.0 million respectively (Birmingham et al., 1999).

Diabetics have been found to have an increased risk of heart disease, while the trend in the increase in diabetes seems to pattern the increase in obesity in developed and some developing countries, including China and India, where recent economic developments resulted in dramatic changes in lifestyle and dietary habits. Therefore, when looking at the financial burden of diabetes on healthcare systems, it may be prudent to also evaluate the costs of associated conditions like cardiovascular disease (CVD) and obesity. According to the World Health Organization (WHO) in 1985 there were approximately 30 million people living with diabetes worldwide, this estimate increased to 171 million in 2000 and is projected to be around 366 million persons in 2030 living with this disease. Current and projected data worldwide for persons with CVD and obesity who are also diabetic would give a better picture of the impact of this cluster of diseases from an economic and quality-of-life perspective.

More recently the term metabolic syndrome (MS) has been used to describe the clustering of some diseases in individuals. Some features of the metabolic syndrome are hyperglycemia such as insulin resistance and type 2 diabetes mellitus, as well as high blood pressure, central obesity, decreased high-density lipoproteins and elevated triglycerides. Exhibiting three of these five components constitutes MS. This clustering increases one's chance for cardiovascular disease and diabetes mellitus (Maki et al., 2007; Galisteo et al., 2008). Currently, soluble dietary fiber's efficacy in the prevention and treatment of high cholesterol and by extension CVD has been legitimized by the United States Food and Drug Administration for psyllium and β-glucan from oats and barley. The effect of dietary fiber from different sources on features of the MS should be thoroughly investigated through both prospective and retrospective analyses. Fortunately, there is growing consensus for the benefits of dietary fiber especially due to its inferred role in influencing most of the metabolic disturbances in metabolic syndrome (MS) (Galisteo et al., 2008).

However, the pressing issue that must be addressed is the need for enhanced collaboration between food scientists and nutritionists. Complementarity between these disciplines would ensure dietary fiber is added to food at physiologically relevant quantities without adversely affecting the organoleptic qualities of food. This would further facilitate the transfer of science to industrial applications.

13.1.2.1 Dietary Fiber in Food: Influence in Diabetes Mellitus

Dietary fiber has been shown to reduce the postprandial glycemic and insulinemic responses. This is attributable to its viscosity, based on research highlighting more consistent reduction with viscous versus non-viscous fibers. To perform these functions viscous fibers must be mixed with the carbohydrate in a hydrated form. When dry guar or psyllium were dispersed over foods or consumed before the main carbohydrate part of the meal, there was no influence on blood glucose response (Wolever and Jenkins, 2001).

The mechanism involved is slowed gastric emptying and carbohydrate absorption within the gastrointestinal tract, associated with viscosity. Soluble fiber results in increased viscosity of the food bolus, decreasing transit speed and therefore influencing the rate of intestinal absorption. Additionally this is believed to cause improved cholecystokinine (CCK) release. Persons with non-insulin-dependent diabetes mellitus display rapid gastric emptying of meals due to reduced CCK levels (McCleary and Prosky, 2001; Schneeman, 2001; Spiller, 2001; Wolever and Jenkins, 2001; Borderias et al., 2005). Fiber may also physically trap nutrients limiting access of digestive enzymes, thus preventing the breakdown of carbohydrates, lipids and proteins.

Viscous properties of fiber are highly correlated with its molecular weight (MW). This was highlighted in a study where 5 g β-glucan of MW 70 000 considerably reduced postprandial glucose and insulin, while β-glucan at MW 40 000 did not produce this effect. Therefore any activity that affects the viscosity, molecular weight and solubility of fiber must be determined and accounted for to ensure reliable results (Wood, 2007). The lack of consistent results that has been seen in studies employing fiber could be due to a lack of knowledge of the physicochemical characteristics of the fiber.

Structure–function relationships must be emphasized in future work as well as determinations of changes in the fiber's matrix in vivo, in order to facilitate the advancement in fiber research (Guillon and Champ, 2000).

Physicochemical attributes of fiber may be altered on addition to food and during storage, for example oat bran muffins were found to have reduced solubility of the β-glucan fraction due to frozen storage (Lan-Pidhainy et al., 2007). Furthermore, the nature of the food matrix may also affect the efficiency of fiber.

The influence of food matrix on properties attributed to dietary fiber has been confirmed in some studies but not in others. Benini et al. (1995) showed fiber naturally present in food slowed gastric emptying. These authors believed the mechanisms could be associated with the food's resistance to grinding. This resistance is due to the food's structure consistency, as well as its size and not necessarily solely the fiber, with more cohesive food forms resulting in delayed glucose responses. Other proposed mechanisms highlighted were increased viscosity and delayed fat absorption. Studies have shown fat and proteins reduce postprandial glycemia. Tuomasjukka et al. (2007) found the addition of fat to rolled oat did not affect glycemic response. However, the fat content of this fiber-supplemented meal was low at 6 grams of fat for the corresponding 50 g of available carbohydrate, while studies showing an effect of fat on glucose and insulin used about 15 g of fat per 50 g available carbohydrate. Therefore, the higher ratio of fat to available carbohydrate may have supplemented fiber's glycemic effect.

Additionally, Moghaddam et al. (2006) showed there was an effect of fat and protein on glycemic responses. They found that fiber intake up-regulated protein's glycemic effect, while this association was not seen with fat. These authors proposed the mechanisms for the glucose-lowering effect of fat and protein must be different. Therefore the effect of fat and protein on glycemia and insulinemia may have confounded previous work on the efficacy of dietary fiber. This means when evaluating the efficiency of dietary fiber the ratio of fat and protein to available carbohydrate must be low to prevent the additional lowered glycemic effect of the fat and protein.

When Panlasigni and Thompson (2006) evaluated the effects of brown versus milled rice in healthy and diabetic subjects, food structure was related to glycemic response. Milled rice had a higher volume expansion than brown rice on cooking. This increased the hydration and gelatinization of the milled rice, which subsequently makes starch more easily accessible for digestion. Furthermore, the bran coat surrounding the brown rice grain may act as a barrier to water and digestive enzymes. Consequently, brown rice gave a lower glycemic response both in healthy and diabetic subjects. It must be realized that food structure and fiber may work symbiotically to attenuate the rise in glycemic response.

Conversely, Behall et al. (1999) did not find a significant difference in glucose or insulin responses when breads made from whole wheat flour and fine whole wheat flour were used, indicating food particle size was not associated with the mechanism for glycemic and insulinemic responses in this case.

Some experiments have focused on the glycemic index (GI) of fiber-supplemented foods. According to Jenkins et al. (2002) the GI is the measurement of an individual's glucose response to a known quantity of available carbohydrate from a test food, in comparison to that person's glycemic response from a standard food, usually glucose or white bread. It is believed that a low glycemic index results in a corresponding low glucose and insulin response. According to Makelainen et al. (2007) this is not always true. Low GI does not always correspond to low insulin index (II) in some studies (Makelainen et al., 2007). Therefore, the use of the glycemic index is not always beneficial or practical, as foods are complex and fat and protein also affect glucose and insulin responses.

TABLE 13.1 Showing Glycemic Indexes of Four Oat Products

	iAUC* Mean	s.e.m**
Glucose		
Oat product 2 g***	83.7	6.1
Oat product 4 g	58.3	6.0
Oat product 4 g (frozen)	64.3	8.4
Oat product 6 g	63.6	8.5
Insulin		
Oat product 2 g	75.0	9.8
Oat product 4 g	57.6	9.7
Oat product 4 g (frozen)	41.8	8.3
Oat product 6 g	51.1	9.1

Source: Makelainen et al. (2007).
*iAUC – incremental Area Under the Curve.
**s.e.m – Standard error of the mean.
***Amount of β-glucan.

The effect of oat products with increasing β-glucan content, i.e. 2, 4 and 6 grams, on the glycemic and insulin indexes was evaluated, on comparison to a glucose load (control) (Makelainen et al., 2007). It was found that oat products when compared to the control reduced the rate of carbohydrate absorption. There was a reduced glycemic and insulinemic response compared to the control. However, the glycemic value of the products was not correlated with the insulin response. For example, the product containing 4 g of β-glucan had the lowest GI but caused a higher insulin response than anticipated. See Table 13.1 showing glycemic and insulin indexes of the four oat products.

However, the use of the GI provides a tool to evaluate the possible effects of a single food item on health. From a mixed meal perspective, when conducting research it may not be practical, but in a clinical health setting it may be a valuable tool. The GI may aid consumers in choosing food items for glycemic complementarity, e.g. low-GI foods could be paired with high-GI foods in a meal to temper the effect of the high-GI food. Additionally, the concept of glycemic load has been proposed to deal with different glycemic effects from mixed meals. These may be especially relevant for persons with diabetes mellitus. It is noteworthy that based on experiments it is believed that habitual, long-term consumption of high-GI foods is correlated with an enhanced risk of chronic diseases like obesity, cardiovascular diseases and diabetes (Makelainen et al., 2007).

Low GI foods may facilitate blood glucose control in diabetics and also could attenuate serum lipids (Jenkins et al., 2002). This dual effect is important in diabetics as according to Narender et al. (2006), diabetes mellitus increases one's risk of developing coronary artery disease, myocardial infarction, hypertension and dyslipidemia. Consequently, diabetics are also known to develop dyslipidemia with the seriousness of this secondary condition being dependent on the severity of the diabetes.

Therefore, the consistent consumption of low-GI foods may be especially beneficial for diabetic subjects. Furthermore, according to Frost et al. (2003), lately studies have insinuated reducing carbohydrate absorption rate can lower the risk of developing diabetes in men and women by 2.5 times. A corresponding lowered insulin demand might be important as even short periods of hyperinsulinemia cause insulin resistance in healthy subjects.

Prospective studies have found a correlation between dietary fiber intake from whole grains and a reduced risk of type 2 diabetes mellitus (T2DM). This risk could be reduced by as much as 30% in men after adjusting for confounders like body mass index (BMI). This reduced risk was also seen in women. However, this positive correlation was not found for fruits and vegetables which are also high in dietary fiber. Additionally, for women there was a 59% higher chance of having DM with consistent high glycemic food consumption (Fung et al., 2002; Liu, 2003; Schulze et al., 2004).

The efficacy of fiber from grains has been shown in a number of studies involving non-insulin-dependent diabetics, obese persons with elevated fasting blood glucose and diabetics treated with insulin and oral agents. All subjects had better glycemic responses after consumption of whole grain dietary fiber (Jenkins et al., 1988; Kamble and Shinde, 2004; Rave et al., 2007). In addition to the beneficial effect on glycemia, Rave et al. (2007) and Ostman et al. (2006) also found there was better insulin economy. Ostman et al. (2006) further highlighted that impaired glucose tolerant (IGT) women with previous gestational diabetes had reduced insulin responses by an average of 35%.

Qi et al. (2006) evaluated the effect of dietary fiber on plasma adiponectin concentrations in lean, overweight and obese type 2 diabetic women. Adiponectin has been shown to improve insulin sensitivity but may be down-regulated in obese individuals. Persons with higher levels of adiponectin are shown to have a lower risk of cardiovascular diseases and diabetes mellitus. Furthermore, diabetics with higher levels of this cytokine exhibit an improved lipid profile, glycemic control and less inflammation. Qi et al. (2006) found lean women had higher adiponectin levels and adiponectin was positively correlated with cereal fiber consumption. Also, BMI did not affect the association between adiponectin and cereal fiber consumption, while high glycemic index was associated with a lower level of adiponectin. Although there has been variability in results with respect to the effect of fiber on adiponectin levels, it was realized that there is a positive correlation between long-term fiber intake and adiponectin levels, while fiber has no effect on adiponectin in short-term studies (Galisteo et al., 2008).

Dietary fiber consumption has also been inversely associated with the inflammatory markers C-reactive protein and tumor necrosis factor alpha (TNF-alpha) levels. The metabolic syndrome has been associated with increased levels of the inflammatory markers C-reactive protein (CRP), tumor necrosis factor alpha and interleukin (IL) 6.

Studies on persons with type 2 diabetes have shown a decrease in the levels of CRP and TNF alpha receptor 2 due to whole grain and bran consumption. The suggested mechanisms for dietary fiber's effects on inflammatory cytokines were:

1. Its effect on weight loss which has been associated with lower levels of CRP and TNF alpha in obese individuals, while increasing adiponectin in women.
2. Its ability to reduce lipid oxidation, which is believed to also reduce inflammation.
3. The production of SCFA, especially butyrate which has shown anti-inflammatory properties. Butyrate has shown these properties in inflammatory bowel disease and is believed to inhibit nuclear factor-*k*B activation through PPAR alpha (Galisteo et al., 2008).

13.1.2.2 Dietary Fiber in Food: Influence in CVD

Foods with high quantities of dietary fiber are believed to reduce the risk of cardiovascular disease, type 2 diabetes and various cancers. However, the most promising effects are in fiber's ability to control glucose and insulin sensitivity. This improved insulin sensitivity and the corresponding reduced insulin production is responsible for attenuated serum cholesterol levels over time (Wood, 2002).

Soluble fibers have been shown to have hypocholesterolemic properties. Furthermore, dietary fractions from cereals have proven to temper or prevent hypercholesterolemia. Cereals with this quality include fiber from oat bran, rice bran and barley, despite many compositional differences. Consequently, the United States Food and Drug Administration (FDA) has endorsed the soluble fibers β-glucan and psyllium at 0.75 g per serving and 1.78 g per serving respectively, as products that might reduce the risk of cardiovascular disease (CVD). This endorsement is based on consumption of 4 servings per day, as part of a low-saturated-fat, low-cholesterol diet (Kahlon, 2001; Jenkins et al., 2002; Boderias et al., 2005; Wood, 2007).

Psyllium fiber, which is soluble, has been shown to reduce low-density lipoprotein (LDL). This effect was related to the quantity of psyllium ingested, with 9.1, 9.4 and 10.2 grams per day causing a reduction of LDL by 6.0%, 7.4% and 7.2% respectively. In a study conducted from 1985 to 1995 on the Coronary Artery Risk Development in Young Adults (CARDIA), dose–response associations were seen with dietary fiber consumption, for blood pressure, LDL, high-density lipoprotein (HDL), cholesterol and triglycerides. A similar association was seen for cereal fibers. Men in the highest tertile consuming cereal fiber had an 82% reduced chance of coronary heart disease (CHD). In a nurses' health study conducted on women there was a reduced risk by 34% for coronary heart disease, for persons in the highest quintile of fiber consumption. It was also realized with every 5 gram increase in cereal fiber consumption there was a 37% reduced chance of heart disease. Additionally, cereal fiber was also associated with a reduced risk of stroke and ischemic stroke in elderly persons (Pereira et al, 2001; Flight and Clifton, 2006). Other studies attribute the CHD risk reduction to whole grains and not cereal grains generally; while other studies find no risk-reducing effects (Flight and Clifton, 2006). However, the classification and corresponding quantities for high versus low fiber intake varies among experiments. It is worth noting that in retrospective studies there were no available data such as physicochemical characteristics of the fiber-rich foods.

There are proposed mechanisms for soluble fiber's ability to reduce blood lipids, despite no general consensus among scientists. The most popular of the proposed mechanisms is the binding of bile acids, which results in higher quantities of cholesterol in feces. The binding of bile acids in the small intestine is believed to affect enterohepatic bile acid recycling, thus reducing the bile acid pool, causing the liver to increase LDL uptake from circulation (Jenkins et al., 1998; Schneeman, 2001; Wood, 2002; Borderias et al., 2005).

Another mechanism involves the fermentation of fibers, oligosaccharides and undigested starches. This fermentation in the large intestine causes the production of the metabolite, short-chain fatty acids (SCFA), by indigenous bacteria (Woods and Gorbach, 2001). The SCFA propionate has been associated with reduced cholesterol levels, while acetate has been shown to increase cholesterol levels (Jenkins et al., 1998). Therefore a high production of propionate compared with acetate from fermentation of fiber in the large intestine may attenuate cholesterol formation. Over time, with the elucidation of which dietary fibers cause the production of specific SCFA at known ratios, specific fibers could be consumed for the attenuation or prevention of specific chronic non-communicable diseases.

In the final analysis soluble dietary fiber is associated with reduced serum cholesterol and may reduce heart disease. One must consider the lack of support for the efficacy of soluble dietary fiber in CVD risk reduction could be associated with a lack of physicochemical data on the fibers used in many studies. Most studies employing soluble fiber did not take into account the viscosity, concentration, solubility and MW of the soluble fiber. On the other hand there seems to be a strong association between insoluble dietary fiber consumption and reduced risk of heart disease. However, the mechanisms for this feature of insoluble fiber are unknown. Furthermore extensive research needs to be conducted on the composition of SCFA from the fermentation of various fibers and the properties of dietary fiber that are responsible for these SCFA compositional differences needs to be elucidated.

13.1.2.3 Dietary Fiber as Prebiotics: Fermentation and SCFA

By definition fiber is not broken down in the small intestine, however, almost all fiber fractions are broken down to some degree by microorganisms in the large intestine. Soluble fibers such as pectin, gum arabic and guar gum ferment rapidly and disappear almost completely in transit. Insoluble fibers like dietary cellulose ferment slowly and in humans approximately 40% of cellulose is broken down. Also fermentability seems to be affected by physical structure, as fruit and vegetable fibers seem to be more fermentable than those from cereals (Oakenfull, 2001).

Consumption of dietary fiber leads to an increase in microbial mass. Wheat bran fed to young men increased stool weight due to fiber residue, while oat bran increased stool weight due to microbial mass (Schneeman, 2001). This fermentative capacity in turn affects the production of metabolites, such as short-chain fatty acids (SCFA) which are believed to be important to the health of the large intestine (McCleary and Prosky, 2001; Spiller, 2001).

The term, 'prebiotic' was defined as: 'a non-digestible food ingredient that beneficially affects the host by selectively stimulating the growth, activity and the limited number of bacteria in the colon, and thus improves host health' (Gibson and Roberfroid, 1995). A prebiotic effect has been attributed to too many food components, sometimes without consideration of the criteria required. In order to claim an ingredient to be a prebiotic, the following physiological effects must de demonstrated (Roberfroid, 2005):

1. the ingredient resists hydrolysis by mammalian enzymes and gastric acidity, hence, no gastrointestinal absorptions occurs;
2. the ingredient can be fermented by the intestinal microflora;
3. the ingredient can selectively stimulate the growth and activity of intestinal bacteria associated with the health and well-being of the host.

According to the above criteria, inulins clearly belong to prebiotics.

13.1.3 Regulatory Issues and New Challenges

13.1.3.1 Regulatory Issues

Over the years studies into the health benefits of dietary fiber consumption have shown conflicting results. Brennan and Samyue (2004) found that although inulin, β-glucan enriched fraction, potato fiber and resistant starch, added to biscuits, reduced sugar release in vitro, there was a small increase in sugar release as more fiber was included in the biscuit formula. This opposes food research that showed increasing fiber reduces sugar evolution. Frost et al. (2003) also found results of their study contradicted previous research. They found employing 1.7 g of psyllium fiber in a meal, despite being the US FDA recommended level for a health claim, did not reduce carbohydrate absorption. However, the previous study showed positive results on blood glucose and insulin, when 5–10 g of psyllium fiber were employed in a meal.

This has led to regulatory agencies like the United States Department of Agriculture (USDA) and the Health Canada, enforcing strict regulations on claims associated with these products. Studies must be verified before health claims are acknowledged by these agencies in an effort to protect consumers. Health claims on foods have significant implications from an economic perspective for companies, as sales may soar due to the competitive advantage.

According to DeVries (2001) oat bran epitomizes the influence of health claims on foods. The problem though was the array of forms of oat bran due to varying production methods. Consequently, a definition for oat bran and its products that is directly related to its ability to reduce the risk of heart disease was imposed. Food companies therefore have to ensure their products actually contribute to a reduction in cholesterol when a specified quantity is consumed.

Labeling regulations, policies and practices are based on an array of work and define minimal quantities of dietary fiber for specific classifications. Countries like Australia, the European Union and Mexico do not seem to have general quantities for these claims but proposals have been made. In Canada a food product is considered a source of dietary fiber when it contains 2 grams per serving, high fiber is 4 grams per serving and very high fiber is 6 grams per serving. In Japan and Korea a food source of dietary fiber must contain 3 g per 100 g of food or 1.5 g per 100 kcal. A food source rich in fiber must contain 6 g dietary fiber per 100 g of food or 3 g dietary fiber per 100 kcal (DeVries, 2001).

Due to other factors such as phytates, saponins, low-fat diets, cholesterol levels, etc. that may play a role in fiber's physiological effects, claims can be made within a specific context. The following are examples of claims that can be made:

1. Fruits, vegetables and grain products with dietary fiber, especially soluble fiber which have not been fortified, but are low in saturated fat, cholesterol and generally low in fat, may impair the chance of coronary heart disease.
2. Fruits and vegetables low in fat with a substantial source of vitamin A, C or innate dietary fiber could hinder the risk of some modes of cancer; however the soluble fiber content must be stated.
3. Soluble fiber from oat bran, rolled oats (oatmeal) and whole oat flour, in the form of soluble β-glucan, as well as psyllium husks may limit the risk of heart disease. However, these products must be low in fat, saturated fat and cholesterol. Additionally there must be 0.75 g of whole oat soluble fiber or 1.7 g of psyllium husk soluble fiber per reference amount. The food must also state the soluble fiber content.
4. A diet abundant in whole grain foods and various plant foods and low in total and saturated fat and cholesterol may protect against heart disease and certain cancers. Whole grain ingredients must make up, for the least, 51% of the food, by weight per serving. The dietary fiber content must be equal to or more than 3.0 g per reference amount (RA) of 55 g, 2.8 g per RA of 50 g, 2.5 g per RA of 45 g and 1.7 g per RA of 35 g (McCleary and Prosky, 2001).

The ability for these claims to be made accurately worldwide rests with the analytical methodologies (DeVries, 2001). This ensures accuracy in labeling by food manufacturers and facilitates enforcement authorities ensuring labeling regulations are followed. Thus approved method(s) ideally should be easy to follow, cost-efficient and robust. However, this may not always be possible, as in the case of dietary fiber, there are various components associated with it, hence the analysis is complex.

13.1.3.2 New Challenges

In addition to the old, well-established challenges such as the incorporation of physiologically relevant quantities of dietary fiber into food without adverse effects on organoleptic qualities, recent challenges are due to more rigorous studies on the basic chemical structures of dietary fiber and its physiological/clinical effects, in order to understand the structure–functionality relationships. These new challenges hinge on a marriage between the disciplines of food science and nutrition. Food scientists need to determine the physicochemical properties of fibers before and after processing and during storage and understand the molecular basis of the action mechanisms. Data on molecular weight, solubility, shape and conformation as well as particle/pore size and viscosity will affect the properties of dietary fiber in food as well as in human physiology. Nutritionists must then determine the physiological relevance of these factors if any. The recent advancement in mapping the genome of humans makes it possible to foresee or predict some of the diseases that are related to dietary fiber intake: a tailored diet and lifestyle could be made for those high-risk groups to minimize the risk of diseases, such as diabetes, CVD and some cancers. However, in reality, consumer preferences and individual requirements dictate the success of a functional food product: the product must be palatable and affordable. This requires the development of a dietary fiber industry which can provide a stable supply of a variety of well-characterized (chemically, physically and physiologically) dietary fibers that can be used by the food industries.

In summary the following are the challenges for both food scientists and nutritionists:

1. A stable supply of well-characterized dietary fiber – development of a dietary fiber industry.
2. Establishment of the structure–function relationships for all dietary fibers.
3. Processing and formulation: products palatability and for targeted populations.
4. Regulatory issues (definition, measurements and nutrition labeling).

13.2 TECHNOLOGICAL PROPERTIES OF DIETARY FIBER

The technological functions of dietary fiber include water-holding capacity (WHC), fat-binding capacity (FBC), viscosity, gel-forming capacity (GFC), chelating capacity and texturizing capacity. These functional properties are highly influenced by the physicochemical characteristics of fiber like MW, porosity, particle size, ionic form as well as dietary fiber constituent polysaccharides. This is highlighted where the quantity of lignin in grape antioxidant dietary fiber was found to impact swelling capacity, while varying the MW of β-glucan and applying to beverages and soups altered the viscosity of these foods and by extension the texture, through alterations in thickness of these products (Cho et al., 1998; Lazaridou and Biliaderis, 2007; Sanchez-Alonso and Borderias, 2008).

13.2.1 Water-Holding Capacity (WHC)

When fiber is mixed with water, the ability to hold this water within its matrix after being subjected to stress, for example centrifugation, is a measure of WHC. Polysaccharides are hydrophilic and their abundant free hydroxyl groups form hydrogen bonds with water. Therefore the pH of a medium affects WHC. Both soluble and insoluble polysaccharides exhibit WHC; however soluble fibers, for example pectins and gums, have a higher WHC than cellulose fibers (Schneeman, 2001; Borderias et al., 2005; Ahmad et al., 2008). WHC is therefore related to the porous matrix structure of polysaccharide chains, which through hydrogen bonding trap and retain water. Therefore factors such as grinding, or anything that causes alteration of fiber's matrix structure, would affect WHC and swelling capacity (Figuerola et al., 2005).

Neutral, low-water-holding fibers like those in oat or pea bran improve crispiness, reduce breakage and improve workability, in extruded products like chips, crisps, crackers and snacks.

13.2.2 Viscosity, Gel-Forming Capacity and Texturizing

This is the ability of particular polysaccharides to thicken on mixing with liquid. The extent of thickening is affected by the chemical composition of the polysaccharides. Gums, pectins, β-glucans and polysaccharides from algae form viscous solutions and are usually used as thickeners (Phillips et al., 2000).

Gels are colloidal systems which entangle or immobilize water and other solutes. This phenomenon is very significant in the food industry as it gives rigidity to many products. Numerous soluble fibers form gels, for example carrageenans (Iota and Kappa), pectins, konjac, etc. A classic example of gel formation in the food industry is the use of pectin gels in jam. Gel formation depends on polysaccharide (soluble dietary fiber) concentration, temperature, presence of particular ions and pH (Fox and Cameron, 1991; Borderias et al., 2005).

The viscous and gel-forming properties of soluble fibers can aid the restructuring of foods whose main ingredients are meat or fish muscle. Soy protein concentrate with approximately 25% fiber was added to restructured products made from minced muscle in order to simulate muscle myotome (Borderias et al., 2005). Fiber adds texture, especially in low-fat comminuted meat products. Fruit fibers, when added to low-fat dry fermented sausage, gave products with sensory properties comparable to conventional sausage (Borderias et al., 2005; Turhan et al., 2005).

Fibers have also been shown to prevent the distortion and shrinkage of restructured products. Turhan et al. (2005) stated the addition of hazelnut pellicle to low-fat beef burgers improved cooking yield and there was less reduction in diameter and thickness when compared to control beef burgers without fiber.

Gums have been used extensively for improving food texture and structure, especially in bakery products. In the snack food industry these properties, along with dietary fiber's fat mimetic properties, have transformed the image of some snack foods to be associated with healthful lifestyles (Gimeno et al., 2004).

Dietary fiber exhibits the preceding technological functions which are fortunately being exploited through industrial production of dietary fiber as food ingredients. These functional characteristics are spurring the development of high-dietary-fiber formulated foods. However, certain general criteria must be followed, these being no nutritionally objectionable components, the dietary fiber ingredient must be concentrated, bland in taste, color and odor, have a good shelf life, be compatible with the methods of food processing employed, while still having positive physiological effects (Figuerola et al., 2005).

13.2.3 Fat-Binding Capacity (FBC)

In the food industry, this quality is more dependent on the porosity of the fiber, than molecular affinity. FBC is reliant on surface properties, overall charge density, thickness and the hydrophobic nature of the fiber. Fibers which are components of batters should initially be placed in water to prevent the entry of excessive fat during frying (Borderias et al., 2005; Figuerola et al., 2005). Some viscous polysaccharides due to their bulk and WHC have a cholesterol-lowering effect, by increasing bile acid excretion and modifying lipid absorption (Schneeman, 2001).

13.2.4 Chelating Capacity

A number of fibers are capable of binding other polar molecules and ions, e.g. minerals through cationic exchange in vitro. This causes reduced mineral availability and electrolyte absorption in vivo. This capacity to bind ions appears to be associated with the number of free carboxyl groups and especially the uronic acid content.

For example Pectins interact with bivalent ions like iron, calcium, copper and zinc in vitro. It has been reported that absorption of zinc and iron is improved by sugar beet fiber, while absorption inhibition seen with wheat bran is due to phytate (Oakenfull, 2001; Borderias et al., 2005).

13.2.5 Soluble and Insoluble Dietary Fibre

It is generally accepted that the technical functionality of dietary fiber in food is reliant on the ratio of soluble to insoluble dietary fiber. When dietary fiber is used as a food ingredient it is suggested the ratio of soluble to insoluble dietary fiber be 1:2. This ratio is more reflected in fiber from fruits and vegetables, while fibers from cereals typically contain higher levels of insoluble dietary fiber as cellulose and hemicellulose (Figuerola et al., 2005). When Sanchez-Alonso and Borderias (2008) evaluated the technological effects of red grape antioxidant dietary fiber (GADF) added to minced fish muscle, they found this fiber, composed of 20.78% soluble fiber and 51.04% insoluble dietary fiber had increased WHC and reduced thawing drip, increased cooking yield and reduced the chewiness of the product. Products with 2% GADF were preferred to the control on sensory evaluation. In the study by Figuerola et al. (2005) on fiber concentrates from apple pomace and citrus peel the ratio of soluble to insoluble dietary fiber ranged from 4.5:1 to 12.9:1, respectively. Despite this ratio fiber concentrates from both apple and citrus fruit exhibited technical characteristics which make them suitable for food applications such as thickening, texturizing and volume replacement. It was found that individual concentrates had unique features which would determine suitable food applications (Figuerola et al., 2005). Therefore broad generalizations should be used as guidelines as tests must be conducted to determine the functional properties of individual dietary fibers from the same and varying sources in order to accurately delineate individual characteristics.

13.3 DIETARY FIBER PRODUCTS: CHEMISTRY, FUNCTIONAL PROPERTIES AND APPLICATIONS IN FOODS

Over the years extensive research has been conducted on the effects of applying dietary fiber to food. This research has seen mixed results with fiber improving organoleptic properties in some cases, while causing deleterious effects on organoleptic characteristics of fiber-fortified foods in others. This even occurs when the same fiber is added to the same food matrix, for example, there have been mixed results on the application of β-glucan to bread. Research has shown that many factors such as processing conditions of food and fiber, methods of fiber extraction, plant genetics and environment during harvesting and the degree of purity of the fiber would affect not only physicochemical characteristics such as viscosity and molecular weight of dietary fiber, but by extension the physiological effects of dietary-fiber-supplemented foods, on consumption (Brennan and Cleary, 2005). In the current section, applications of high molecular weight soluble fiber (HMw), low molecular weight soluble fiber (LMw) and insoluble fiber in foods and their physiological as well as organoleptical properties will be covered.

13.3.1 Soluble Dietary Fiber (HMw)

13.3.1.1 Cereal β-Glucans

13.3.1.1.1 Structures Cereal β-glucan is a polysaccharide that occurs in the subaleurone and endosperm cell walls of the seeds of cereals, including oats, barley, rye and wheat. A typical distribution of the β-glucans in grain seeds is illustrated in Figure 13.4. β-glucan in oats is more concentrated in the subaleurone layers than the endosperms; in contrast, a more even distribution of β-glucan is observed for barley and rye grains. Although there is a lack of information on where the β-glucan resides in the seed of wheat, it is likely concentrated in the walls of the subaleurone cells as evidenced by the enrichment of β-glucans in the bran fractions using debranning or conditional milling processes (Dexter and Wood, 1996). The level of β-glucans in cereals varied from as low as 0.5–1% in whole wheat to as high as 3–11% in barley, as shown in Table 13.2.

Cereal β-glucans are typical linear homopolysaccharides which only have a single type of sugar unit, i.e. β-D-glucopyranose (β-D-Glcp). Over 90% of the β-D-Glcp residues in cereal

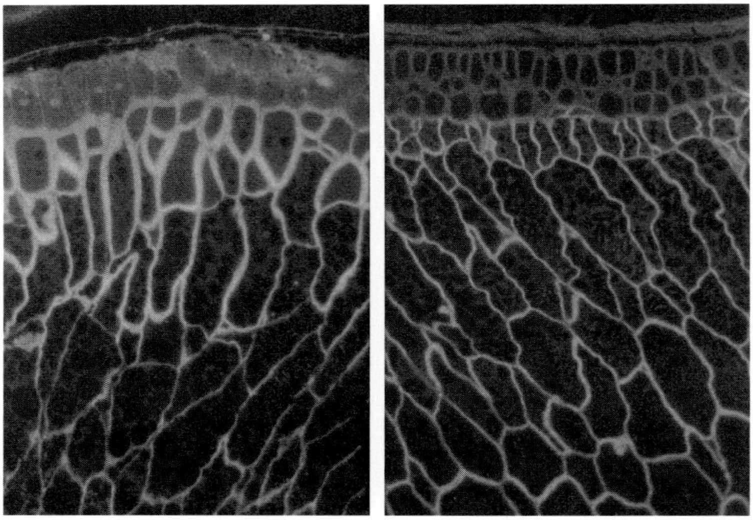

FIGURE 13.4 Typical distribution of β-glucans in grain seeds adapted from Wood 1991.

β-glucans are arranged as blocks of two or three consecutive (1→4)-linked units separated by a single (1→3)-linkage, which forms the two building blocks of cereal β-glucans: a cellotriosyl unit and a cellotetraosyl unit, as shown in Figure 13.5. The remaining less than 10% of the polymer chain is mainly composed of longer cellulosic sequences ranging from 5 up to 14 β-D-Glc*p* residues. The fingerprint of structural features of cereal β-glucans is represented by the oligosaccharide profiles released by a specific enzymatic hydrolysis (lichenase: EC 3.2.1.73., endo-1,3(4)-β-D-Glucanase). Of the oligomers released, the trisaccharide and tetrasaccharide units are particularly important and their ratio constitutes the fingerprint of a specific grain. For example, the tri/tetrasaccharide ratios for β-glucan from wheat, barley and oats are 4.5, 3.0 and 2.3, respectively (Table 13.3).

TABLE 13.2 Amount (w/w) of β-Glucans in Four Common Cereals

	β-glucan content
Barley	3–11%
Oat	3.2–6.8%
Rye	1–2%
Wheat	0.5–1%

13.3.1.1.2 Molecular Weight and Conformational Properties Cereal β-glucans in aqueous solution adopt a disordered random coil conformation. Recent studies suggest that there are no significant differences in molecular conformation between cereal β-glucans when measured in extreme dilute solution in a good solvent. The Mark-Houwink-Sakurada exponents obtained for oat, barley and wheat fall in the range of 0.70 ± 0.05, which implies an expanded, semi-flexible chain conformation (Li, 2007). With such a conformation, the estimated persistence length, which is a measure of the chain stiffness, should not be much bigger than ~ 4 nm (Gomez et al., 1997). However, the measured values of persistence length were sometimes 10–20 times higher than 4 nm (Grimm et al., 1995). This is the result of molecular aggregation among β-glucan molecules through hydrogen bonding, which imparts great

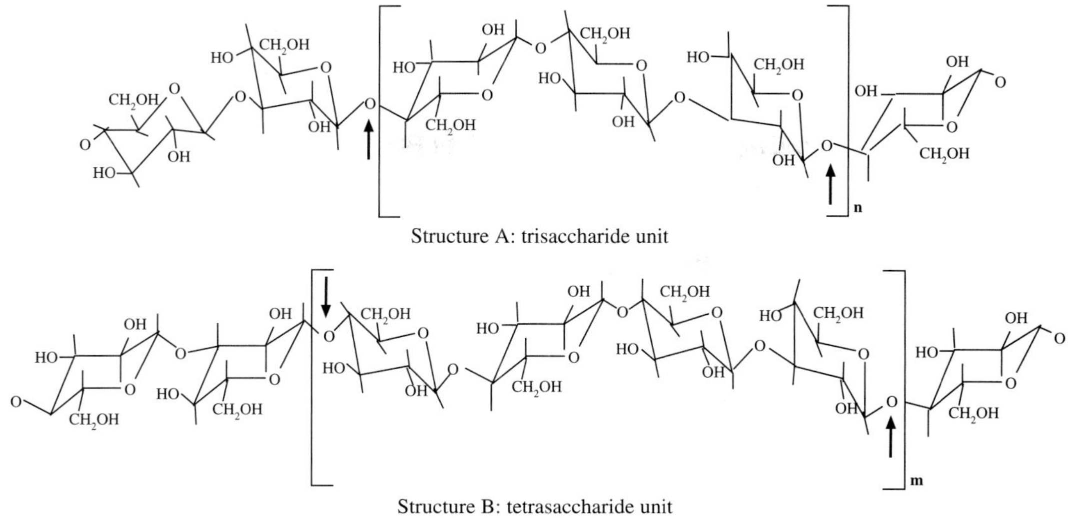

FIGURE 13.5 Building blocks of cereal β-glucans.

stiffness (thus high solution viscosity) to the polymer. Recently, the existence of molecular aggregates in water solution was clearly demonstrated by dynamic light-scattering measurements (Figure 13.6). (Li et al., 2006). These aggregates could be effectively eliminated in 0.5 M sodium hydroxide, but not with other treatments, including the use of 6 M urea as a solvent and a repeated filtration process or ultrasonic treatment. The success of obtaining an aggregate-free solution allowed the measurement of true molecular weight and conformational properties of single molecules; it confirmed an extended random coil conformation for all cereal β-glucans and the persistence length of wheat β-glucan was determined to be in the range of ~ 5 nm (Li et al., 2006).

The molecular weight of β-glucans in the cell wall matrix has so far not been measured. The apparent molecular weight obtained for isolated β-glucans scattered in the range of 10^4–10^6 g mol^{-1}, depending on sources of raw material, methods of isolation, degree of aggregation and determination techniques. Generally, the molecular weights of carefully isolated β-glucans follow the trend of oat > barley > rye > wheat, coinciding with their ease of extractability. One of the popular methods for determination of molecular weight and its distribution is size exclusion chromatography (SEC). In the absence

TABLE 13.3 Oligosaccharide Compositions of β-Glucan from Different Sources

	Peak Area%		Total%		Molar Ratio
β-D-Glucan Source	tri	tetra	Tri+tetra	Penta~nona	tri/tetra
Wheat	72.3	21.0	93.3	6.7	4.2–4.5
Barley	63.7	28.5	92.2	7.8	3.0
Oat	58.3	33.5	91.9	8.1	2.3
Rye	73.7	21.4	95.1	4.9	3.4
Lichenan	86.3	2.7	89.1	10.9	31.9

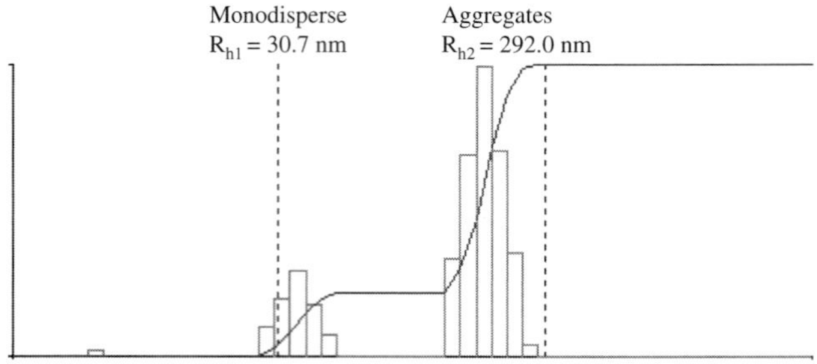

FIGURE 13.6 Molecular aggregates in water solution demonstrated by dynamic light scattering.

of a molecular weight detector, molecular weight standards are required to calibrate the columns. Isolated β-glucans are highly polydisperse in molecular weight, giving a broad molecular weight distribution. Wang and co-workers developed a method which allowed separation of β-glucans into many fractions with low polydispersity, which are particularly suitable for the use as molecular weight standards (Figure 13.7) (Wang et al., 2003).

13.3.1.1.3 *Functional Properties of β-Glucan and Arabinoxylans: Flow Behavior, Viscoelastic and Gelation Properties* Semi-dilute and concentrated solutions of cereal β-glucans are appreciably different from one another as a consequence of varied chemical structures and molecular weight. For example, high-molecular-weight oat β-glucan solutions exhibit typical viscoelastic flow behavior, and do not form a gel within a reasonable time period. Freshly prepared barley and wheat β-glucan solutions are also viscoelastic fluids as demonstrated in Figure 13.8. It is interesting to observe that a thermo-reversible gel could be formed if the β-glucan solutions are allowed to stand for a period of time under lower temperature. (Cui, 2001). Partially hydrolyzed oat β-glucans can also form a gel, but the gel development time is much longer (Lazaridou and Biliaderis, 2004; Tosh et al., 2004). Figure 13.9 shows the mechanical spectrum of a gel formed by 5% wheat β-glucan fraction 1 (Mw = 340 kDa).

Accumulating evidence suggests that the arrangement of cellotriosyl and cellotetraosyl units and their ratio in the polymer chain are important factors controlling the solution properties of these polysaccharides (Bohm and Kulicke, 1999; Cui et al., 2000; Lazaridou and Biliaderis, 2004; Tosh et al., 2004) although

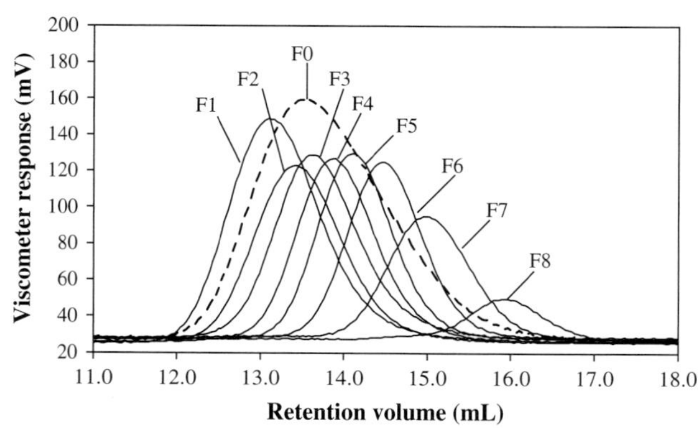

FIGURE 13.7 Separation of β-glucans into many fractions with low polydispersity.

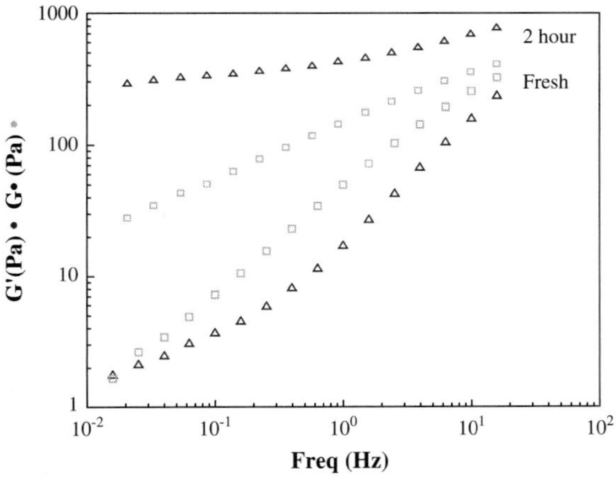

FIGURE 13.8 Viscoelastic properties of wheat β-glucan: freshly prepared versus standing for 2 hours at room temperature (25°C) (M_W=3.4×10^4, measured at 25°C).

earlier studies ascribed this mainly to the long runs of cellulosic segments (Doublier and Wood, 1995). Recent evidence from our laboratory suggested that sections of consecutive cellotriose units were mainly responsible for forming stable junction zones that lead to aggregation and gelation (Cui and Wang, 2006; Li, 2007). According to this mechanism, the gelation capacity of cereal β-glucans is in the order of wheat > barley (rye) > oat, which has been repeatedly demonstrated experimentally (Cui et al., 2000; Lazaridou and Biliaderis, 2004; Tosh et al., 2004). Molecular weight is another important factor that exerts significant influence on the solution properties of β-glucans. The gelation rate generally increases with decrease in molecular weight as shown in Figure 13.10 (must be above the critical gelation chain length); this phenomenon is in disagreement with regular gelling polysaccharides, in which longer polymer chains favor the formation of gel networks. The hypothesis for this phenomenon is that small β-glucan molecules have higher mobility (less restriction of diffusion) than their large counterparts, thus more readily interact with each other to form a stable junction zone, hence, the formation of aggregates. The further aggregation of the aggregates forms the three-dimensional gel networks at concentrate solution regime.

13.3.1.4 *Physiological and Organoleptic Properties*

13.3.1.4.1 *Physiological properties* Oat and barley β-D-glucans have demonstrated health benefits, including lowering cholesterol levels and attenuating postprandial glycemic response. The Food and Drug Administration of the USA

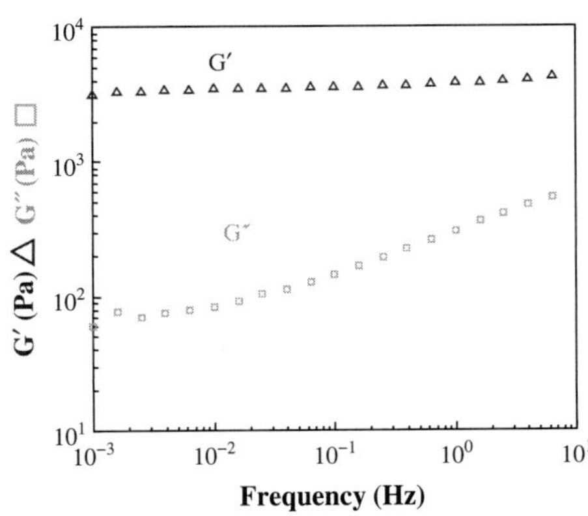

FIGURE 13.9 Mechanical spectrum of a gel formed by 5% wheat β-glucan.

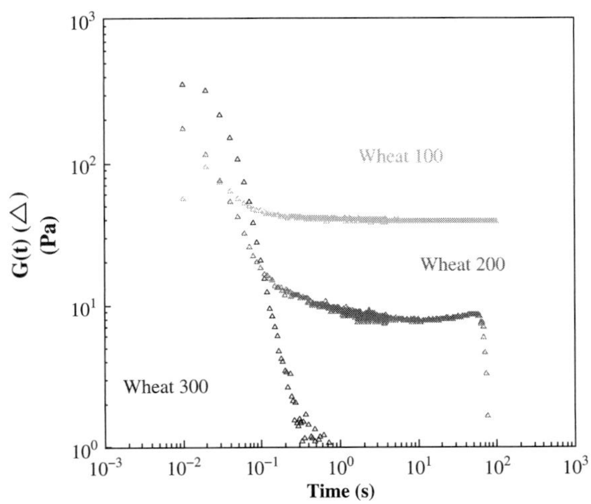

FIGURE 13.10 Increase in gelation rate corresponding to decrease in molecular weight.

allowed a health claim for both oat and barley β-D-glucans for lowering the risk of coronary heart disease (FDA, 1997, 2006). These claims were based on substantial scientific evidence from both animal models and human clinical trials. For example, Kalra and Jood (2000) showed that barley β-glucan lowered the levels of total cholesterol, low-density lipoprotein (LDL)-cholesterol and triglycerides in rats, while Kahlon et al. (1993) demonstrated that barley as well as oat β-glucan significantly lowered the cholesterol levels of hamsters.

Bourdon et al. (1999) reported that barley β-glucan lowered the cholesterol concentration and attenuated the insulin response in humans. Cavallero et al. (2002) studied the effect of barley β-glucan on human glycemic response and found a linear decrease in glycemic index with the increase of β-glucan content. The presence of 5 g of oat β-glucan in extruded breakfast cereals caused a 50% decrease in glycemic response (Tappy et al., 1996). In a drinking model, oat gum significantly decreased the postprandial glucose rise and reduced the total and LDL-cholesterol levels in human subjects (Wood et al., 1990; Braaten et al., 1994). Wood later demonstrated that the observed physiological effect of oat β-glucan was positively correlated to the viscosity of the drink, which is determined by molecular weight and concentration of β-glucan used (Wood, 2004). Cereal β-glucans also demonstrated other health benefits. For example, oat fiber prolongs satiety after meals and alleviates constipation (Malkki and Virtanen, 2001).

The solubility and viscosity of β-glucans are critical for exerting the reported positive physiological effect. By the same principle, the structural features, molecular weight, or any characteristics that could affect the viscosity-producing properties of β-glucans would have impact on their physiological effect. One of the structural features affecting the gelation property of β-glucan is the ratio of cellotriosyl to cellotetraosyl units in the polymer. It has been shown that the ratio of cellotriosyl to cellotetraosyl units affects chain association and, hence, the viscosity. A higher ratio of cellotriosyl to cellotetraosyl causes reductions in solubility, promoting the gel-forming properties (Cui and Wood, 1998; Tosh et al., 2004; Brennan and Cleary, 2005; Lazaridou and Biliaderis, 2007).

Therefore differences in MW cannot be taken in isolation, as the actual viscosity and solubility must also be accounted for. Brennan and Cleary (2005) stated any factors that may impact on β-glucan's rheological parameters would also affect viscosity, which could affect organoleptic and nutritional properties of the final products. However, there is lack of information on whether β-glucan gels have any significant influence on physiological effects.

Food matrix may also influence the physiological effects of food but this is still debated. Liquid food fiber preparations have been shown to be more effective physiologically than solid food fiber preparations. In mildly hypercholesterolemic subjects orange juice fortified with β-glucan from oats gave lower total and low-density lipoprotein cholesterol (LDL) and a better ratio of total to high-density lipoprotein cholesterol (HDL), than the same preparation of β-glucan, when added to bread and cookies. However, these results may also be attributed to the effects of food processing. One study has shown that the baking of barley- and oat-fortified bread has been found to partially degrade β-glucan from 1200–2300 $\times 10^3$ to 240–1920 $\times 10^3$ (Lazaridou and Biliaderis, 2007), while another study found β-glucan degradation in bread occurred with increased mixing and fermentation time (Brennan and Cleary, 2005).

The efficacy of β-glucans in food systems is believed to be dependent on MW and concentration and these affect viscosity, which is related to insulinemia and glycemia. Viscosities in the range of 20–8000 m Pa s^{-1} for aqueous oat gum solutions were found to improve glycemia and insulinemia. However, in the gastrointestinal tract there are a number of factors which may affect viscosity, solubility and molecular weight.

An in vitro assay mimicking the human digestion process may predict the efficacy of β-glucan-enriched solid foods on reducing glycemia. Additionally, the GI effect of bread products containing β-glucans may be correlated with the fluidity index (FI) of this product based on in vitro enzymatic digestion, using a Bostwick consistometer (Lazaridou and Biliaderis, 2007).

13.3.1.4.2 Organoleptic Properties Food formulation and food matrix would influence the rheology and sensory properties of fiber-fortified foods. Gelation in food systems has been found to be influenced by the type and concentration of sugar in the formula. Glucose, fructose, sucrose, xylose and ribose (30% w/v) when added to aqueous dispersions of barley β-glucan (6% w/v) increased gelation time. Another study has shown inhibition in gelation was more potent with xylose and fructose than glucose and sucrose, while sorbitol facilitated gelation. These effects on gelation are associated with the MW of β-glucans, with decreased effects being observed with reduced MW (Lazaridou and Biliaderis, 2007).

Consequently, differences in formula in the same fiber-enriched food system may result in organoleptic and nutritional differences. Commercially, this may give some companies a competitive advantage if they can highlight the physiological benefits of their product over a competitor's within the same product line. Breads and pasta were shown to have different physiological effects despite similar ingredients (Wolever and Jenkins, 2001).

The organoleptic effects of β-glucan have been extensively researched but oftentimes these effects are not delineated to show the relationship between organoleptic properties and β-glucan botanical origin, processing conditions, pore size and physicochemical properties. In yeast-leavened breads β-glucan caused high water retention, poor crumb texture and loaf volume. These effects are dependent on the concentration of β-glucan in the bread as well as the molecular weight, with HMW β-glucan producing breads of superior quality to its LMW counterpart.

Brennan and Cleary (2007) enriched bread with β-glucan in the form of glucagel. Three breads were tested, a control (no glucagel), and breads with 2.5% and 5% glucagel. However, there were negative alterations to dough quality and baking performance, in the fiber-supplemented bread. Breads had a reduced volume and height and a firmer crumb texture compared to the control. However, in vitro studies showed a significant reduction in the liberation of reducing sugar. Therefore, there can be positive health benefits from the inclusion of glucagel into bread. However, if studies like these are to be transferred to commercial applications the organoleptic characteristics of the fiber-supplemented breads should be comparable with the control, in order to facilitate acceptance by consumers.

Izydorczyk et al. (2005) incorporated fiber-rich fractions of barley into Asian noodles at 250 g kg^{-1}. This addition produced 4.25–5 g β-glucans and around 7.2–8.5 g dietary fiber per serving, at a serving size of 85–100 g. Asian noodles enriched with fiber-rich barley fractions were more brown and less uniform in color than the control (no fiber). The luminosity was only slightly lower at 90.4 and 89.8 for fiber-enriched noodles and 92.7 for the control. The viscosity of noodles with fiber-rich fractions was increased. This is attributable to the high swelling and water-holding capacity of the fiber, as well as its large particle size and partial solubilization of the non-starch polysaccharides. The maximum cutting stress or firmness of the cooked noodles and the resistance to compression or chewiness of the noodles increased with fiber addition. (Izydorczyk et al., 2005).

For Asian noodles containing fiber-rich fractions of barley there was a significant reduction in glucose liberation. This could have been due to the partial solubilization of non-starch polysaccharides. This theory was supported by micrographs showing a more compact matrix surrounding the fiber, compared with a relatively porous control matrix (Izydorczyk et al., 2005).

13.3.1.4.3 *Commercial Applications of β-*Glucans Oat and barley flours and minimally processed bran products are available as food ingredients in North America, Europe and other parts of the world. Advanced processing and fermented oat products have emerged during the last decade, including Oatrim from oats, Glucagel™ from barley, and more recently Viscofiber® from both oats and barley. Oatrim is a product prepared by treating oat bran or oat flour with a thermo-stable β-amylase at high temperatures. The application of Oatrim in food includes bakery products, frozen desserts, processed meats, sauces and beverages. Oatrim is also used as a fat replacer (Lee et al., 2005). Glucagel™ is based on barley β-D-glucans prepared by extraction and partial hydrolysis. The low-molecular-weight β-D-glucans (15 000 to 150 000 Daltons) form gels at 2% polymer concentration and above. The major functionality of Glucagel™ is its gelling and fat mimetic properties, which is used as fat substitutes in bakeries, dairy products, dressings, and edible films (Morgan et al., 1999). β-D-glucan isolated from oats can also be used in the personal care industry, as a moisturizer in lotions and hand creams. Viscofiber® is a product recently developed by a Canadian company. It is a highly viscous product produced by a new technology using a combination of water–alcohol treatment of barley and oat flours with or without enzymes. The product is basically the intact cell walls of the endosperms. The current primary market for this product is as dietary supplements for the functional food and nutraceutical industry (Vasanthan, 2005). In Scandinavia countries, oat-based dairy-substitute products can be found on the market, such as beverages (Oatly), fermented oat products (e.g. oat yogurt) and oat ice creams. With more evidence on the health benefits of cereal β-glucans and the development of novel processing technologies, it is predictable that the worldwide application of β-glucans in the food, nutraceutical and cosmetic industries will surge in the near future.

13.3.1.5 Psyllium

Psyllium is commercially available as particles or powder form prepared from the husk of ispaghula or psyllium seeds of the *Plantago* genus, which are widely distributed throughout the temperate regions of the world, including the Mediterranean regions of southern Europe and the Indian subcontinent. It is conventionally used as a laxative agent for promoting the healthy function of the colon due to its superior capacity to interact with water (Marteau et al., 1994; Alabaster et al., 1996; Satchithanandam et al., 1996; Al-Assaf et al., 2003; Nakamura et al., 2004) and the products produced by the fermentation in the large intestine (Velázquez et al., 2000). Recently, it has been utilized as a dietary fiber supplement in the food and nutraceutical industries. There is sufficient documented evidence supporting the claim that psyllium can reduce low-density lipoprotein (LDL) cholesterol levels and ameliorate the serum lipid profile (Kritchevsky and Tepper, 1995; Satchithanandam et al., 1996; Olson et al., 1997; Pérez-Olleros et al., 1999; Fang, 2000; Gaw, 2002; Romero et al., 2002). It was also found that psyllium can effectively reduce amoebic growth and the severity of diarrhea (Belknap et al., 1997; Hayden et al., 1998; Zaman et al., 2002). In 1998, the US FDA approved the health claim that consumption of psyllium can reduce the risk of heart disease by reducing LDL cholesterol levels (US Food and Drug Administration, 1998). This health claim allowed by the FDA led to a significant increase in the use of psyllium in the functional food and nutraceutical industries, including psyllium-enriched breakfast cereals, bakery and snack foods.

13.3.1.5.1 Hydration and Extraction of Psyllium Polysaccharides Commercial psyllium contains about 75–85% carbohydrates, depending on the quality of supply. Psyllium does not dissolve in water but can absorb large amounts of water to form a paste or gel. Many methods have been reported to extract psyllium polysaccharides. The polysaccharides can be extracted from the husk using hot water and mild or strong alkaline solutions (Hefti, 1954). The polysaccharides obtained from various extraction conditions showed differences in physicochemical properties. The molecular weight of the water-extracted polysaccharide was much smaller than that of alkaline-extracted polysaccharide (Edwards et al., 2003). Kennedy et al. (1979) extracted polysaccharides from psyllium by 1.2 M NaOH solution and the extract was subjected to further fractionation. A 2.5 M NaOH solution was used by Haque and coworkers to extract the polysaccharides from the seed husk at room temperature for 30 min, and the extract was neutralized, dialyzed, and freeze-dried to obtain psyllium polysaccharide (Haque et al., 1993). However, the above extraction procedures did not afford a complete picture of all psyllium polysaccharides: some polysaccharides remained in the residue (e.g., water extraction) where others were mixtures of soluble and insoluble fractions (alkaline extraction). This led to a sequential extraction by water, 0.5 M NaOH, 1.2 M NaOH and 2.0 M NaOH solutions, which yielded three main fractions: water extractable fraction (WE) and 0.5 NaOH solution extractable but not soluble in water (gel-forming fraction, AEG), and 0.5 NaOH solution extractable and water-soluble fraction (AES). The yields of the three fractions were 18%, 61% and 9%, respectively, with a recovery rate of over 90% (Guo et al., 2008).

13.3.1.5.2 Structural Features and Molecular Characteristics of Psyllium Polysaccharides Psyllium polysaccharides are composed mainly of arabinose (22%) and xylose (57%), 10–15% uronic acid and smaller and/or trace amounts of galactose, rhamnose, glucose and mannose. Although there was discrepancy on the fine structure of psyllium polysaccharides, concerning whether the (1→3)-linked xylosyl residue was in the backbone chain, previous and recent structural analysis confirmed that psyllium polysaccharides are heterogeneous highly branched arabinoxylans: some fractions contained uronic acids while other fractions were neutral polysaccharides (Guo et al., 2008). The xylan backbone is substituted by branches including arabinose, xylose, and an aldobiouronic acid at O-2 and/or O-3 positions (Kennedy et al., 1979; Tomoda et al., 1981; Samuelsen et al., 1999; Edwards et al., 2003; Fischer et al., 2004). Other studies suggested that psyllium polysaccharides have several kinds of side chains, including α-D-glucuronopyranose-(1→2)-α-L-Rha*p*-(1→4)-β-D-Xyl*p*, α-D-GalA*p*-(1→3)-α-L-Ara*f*-(1→4)-β-D-Xyl*p*, and α-L-Ara*f*-(1→3)-β-D-Xyl*p*-(1→4)-β-D-Xyl*p* (Edwards et al., 2003). Fisher et al. (2004) suggested psyllium polysaccharide was a neutral highly branched arabinoxylan, and the polysaccharide carried single xylopyranosyl at either position 2 or position 3 as side chains, as well as trisaccharide branches with a sequence of Ara*f*-α-(1→3)-Xyl*p*-β-(1→3)-Ara*f*. These apparent discrepancies in the literature on the fine structure of psyllium polysaccharides were partially resolved by a recent study by Guo et al. (2008). These authors fractionated psyllium polysaccharides into three fractions using sequential extraction methods of water and 0.5 M NaOH solutions. Chemical analysis of the three fractions indicated that both WE and AES contained about 15% uronic acid while AEG was found as a neutral polysaccharide. Combined methylation analysis and NMR spectroscopy were used to elucidate the structure feature of AEG: it is a highly branched arabinoxylan with (1→4) linkage in the backbone chain. The main side chains were xylose, arabinose, and a disaccharide branch, (1→3)-*β*-D-Xyl*p*-Ara*f*. The majority of the side chains were connected to the backbone through O-2 or O-3 positions (Guo et al., 2008). This study confirmed that psyllium is a heterogeneous material which contains both neutral and acidic polysaccharides and the discrepancies in the literature may reflect the structures of different fractions of psyllium polysaccharides extracted. It is likely that the source of the material could also cause differences in chemical structure. The presence of soluble and gel fractions observed by Guo et al. (2008) is in agreement with the findings of Al Assaf and co-workers (2003) who fractioned psyllium polysaccharides into three fractions: a water-soluble fraction (Mw 1–3 million Dalton), a gel fraction (Mw 9–20 million Dalton), and a matrix fraction (insoluble). Edward et al. (2003) reported molecular weights of 2.2×10^6 for water extracted and 1.6×10^6 for 0.1 M NaOH extracted psyllium polysaccharides using size-exclusion chromatography coupled with multi-angle laser light scattering. Comparing the data of Al-Assaf's group and Edward's group, it was found that the water-extractable polysaccharides were in the same order and the difference existed between the polysaccharides extracted from alkaline solutions. The flow-field-flow fractionation method may provide more accurate measurement on the supermacromolecules because this method does not have Mw limitation as size-exclusion chromatography does, which can exclude big molecules in the void volume (Al Assaf et al., 2003) (Figure 13.11).

13.3.1.5.3 Hydration, Viscoelastic and Gelling Properties The polysaccharide from psyllium husk forms a gel-like paste or dispersion when hydrated. These dispersions showed 'weak-gel' properties as demonstrated by Haque and co-workers (Figure 13.12, Haque et al., 1993; Guo et al., 2008). Psyllium gels exhibited unique thermal stability: both G', and G'' decreased as the temperature increased, but no melting of the gel was observed until the temperature reached above 80°C, where a deeper decrease of G' was observed; however, a complete melting process was not observed within the testing temperature range. The cooling curve of G', G'' and tan δ of psyllium

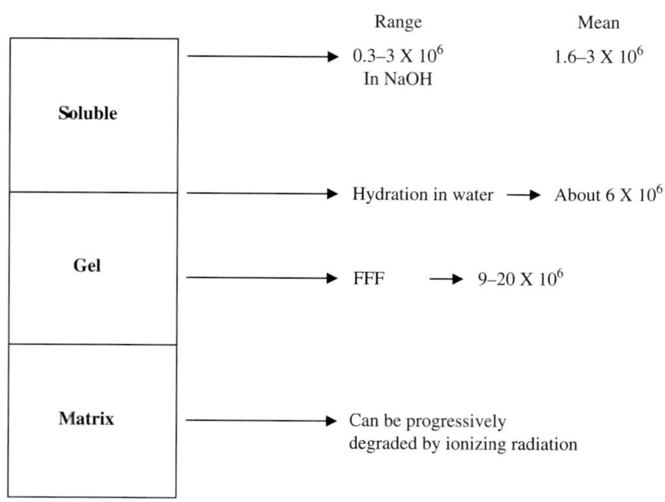

FIGURE 13.11 Molecular weights (Da) of the ispaghula husk solid matrix, the gel and soluble fraction formed by interaction with water. FFF, field-flow fractionation.

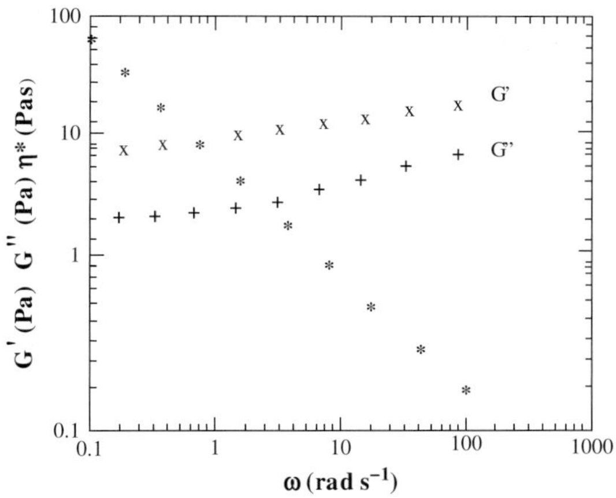

FIGURE 13.12 Frequency dependence of G′, G″ and η^* and shear rate dependence of steady shear viscosity η for 2% (w/v) psyllium at 25°C.

gel superimposed the heating curve indicating that psyllium gel did not show significant thermal hysteresis in the temperature course of the rheological test (Figure 13.13) (Haque et al., 1993).

The behavior of psyllium polysaccharides in combination with hydroxypropylmethylcellulose (HPMC) was more resistant to temperature change. The 1% (w/v) psyllium dispersion in the presence of 1% (w/v) HPMC also exhibited weak gel properties (G′> G″) at low frequency (Figure 13.14). The rheological behavior of the mixed solution on heating was fully reversible on cooling (Figure 13.15), but with thermal hysteresis, which was attributed to the present of HPMC (Haque et al., 1993). The rheological behavior of the mixed system with temperature resembled those of wheat gluten during the baking process; as a result, the mixed system was recommended to replace wheat gluten in gluten-free breads.

13.3.1.5.4 *Calcium-Ion-Binding Capacity of Psyllium Polysaccharides*

Psyllium binds with calcium, reducing the bioavailability and absorption of calcium (Weber et al., 1993). The bioavailability of calcium was reduced to less than 90% by diets containing 10% fiber from psyllium, possibly via the ionic interactions between Ca^{2+} and the carboxyl groups of uronic acids (Luccia and Kunkel, 2002a). It seems that there is a maximum amount of Ca^{2+} which could be reduced or bonded by a certain amount of psyllium polysaccharides (Weber et al., 1993; Luccia and Kunkel, 2002b) and the capacity of psyllium to reduce the bioavailability and adsorption of calcium is greatly affected by the

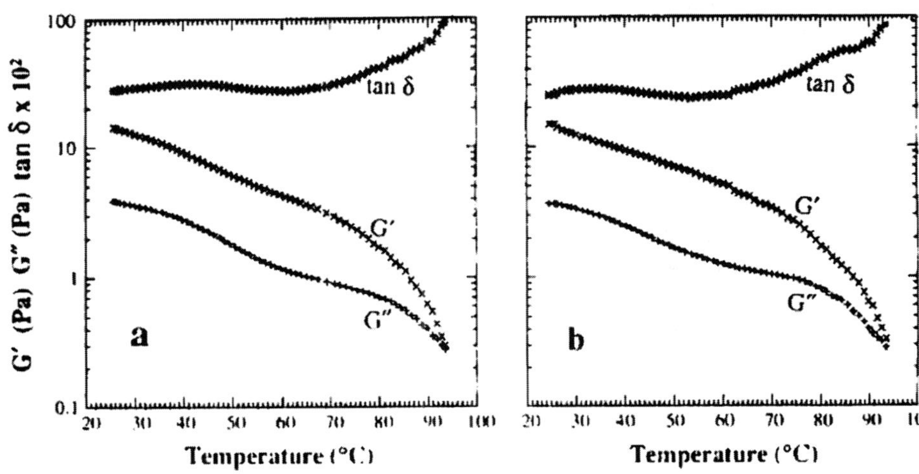

FIGURE 13.13 The temperature dependent changes in G′, G″ and tan δ observed for psyllium dispersions of 1% (w/v).

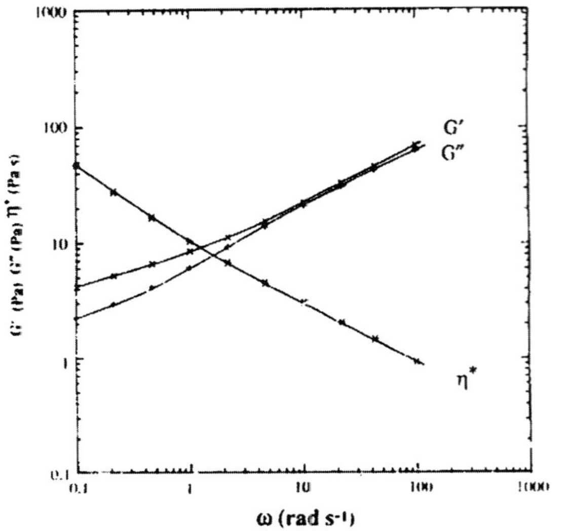

FIGURE 13.14 The mechanical spectrum (25°C) of a 1% (w/v) dispersion of isabgol in combination with 1% (w/v) HPMC (adapted from Haque et al., 1993).

pH value of the system (Weber et al., 1993). However, a recent study by Guo and co-workers (unpublished data) indicated that a uronic acid-free neutral polysaccharide fraction from psyllium also exhibited strong interactions with calcium ions. This psyllium polysaccharide (AEG) formed a weak gel in aqueous solution which resembled the gelling properties of psyllium husk. Similar to psyllium polysaccharide gels, AEG gel did not have a sharp melting point and no thermal hysteresis was observed in the heating and cooling cycle. When Ca^{2+} was added to the gel, it changed the gelling properties and the microstructure of the gel: AEG gel became more resistant to temperature changing, and the microstructure of the gel changed from fibril strand gel to aggregated gel (Guo et al., 2008).

13.3.1.5.5 Physiological and Organoleptic Properties of Psyllium

From animal experiments to human clinical trails, many studies have proven that psyllium can reduce the risk of heart disease by improving the serum lipid profile. In 1998, the Food and Drug Administration (FDA) approved the health claim stating the effective role of psyllium in reducing the risk

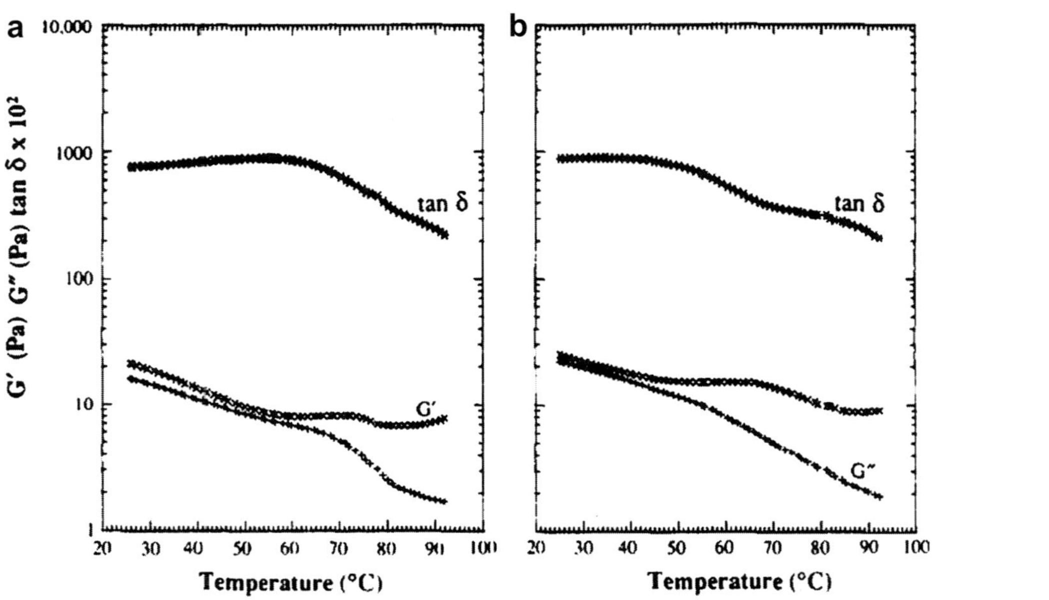

FIGURE 13.15 Temperature dependence of G′, G″ and tan δ for 1% (w/v) psyllium in combination with 1% (w/v) HPMC on (a) heating and (b) cooling at 1 degree/min (adapted from Haque et al., 1993).

of heart diseases (US Food and Drug Administration, 1998). Psyllium could significantly lower the total cholesterol by reducing fat adsorption, therefore, could reduce the severity of atherosclerosis (Kritchevsky and Tepper, 1995; Satchithanandam et al., 1996; Fang, 2000). It was also found that the same amount of psyllium was much more effective than cellulose for decreasing the severity of atherosclerosis (Kritchevsky and Tepper, 1995). The reduction of total blood cholesterol (TC) by psyllium ranged from 5% to 7.1% according to the different studies (Jensen et al., 1997; Olson et al., 1997; Taylor et al., 1999; Kris-Etherton et al., 2002; Oetjen and Haseley, 2004). The ability of psyllium to improve the serum lipid profile, however, may be influenced by gender and hormonal status (Roy et al., 2002; Vega-Lopez et al., 2002). Two principal mechanisms have been postulated for the lipid-lowering action of psyllium. The first suggested that psyllium could bind, absorb, or otherwise sequester bile salts during passage through the intestinal lumen. This would prevent the normal reabsorption of bile salts and increase fecal bile acid content (Turley and Dietschy, 1995; Vergara-Jimenez et al., 1999; Romero et al., 2002; Roy et al., 2002). The second mechanism involves psyllium physically disrupting the intraluminal formation of micelles which may decrease cholesterol absorption and bile acid reabsorption (Roy et al., 2002). In either mechanism, bound bile acids are moved to the terminal ileum and colon. This interruption of enterohepatic circulation leads to an increase in hepatic conversion of cholesterol into bile acid. The reduction in cholesterol leads to an up-regulation of the LDL receptor and, in turn, to an enhanced uptake of LDL cholesterol from the plasma. The net result is a decrease in serum LDL cholesterol, hence, in total cholesterol (Turley and Dietschy, 1995; Vergara-Jimenez et al., 1999; Romero et al., 2002; Roy et al., 2002) (Figure 13.16).

Psyllium is highly beneficial to large bowel functionality. Recent studies supported the fact that psyllium can eliminate large bowel disorders and prevent diarrhea (Belknap et al., 1997; Hayden et al., 1998; Zaman et al., 2002). Research also showed that psyllium could reduce the severity of diarrhea induced by enterotoxigenic *E. coli* (Hayden et al., 1998). Zaman et al., 2002 suggested that psyllium could effectively decrease the amoebic growth, which is the major reason for amoebic dysentery.

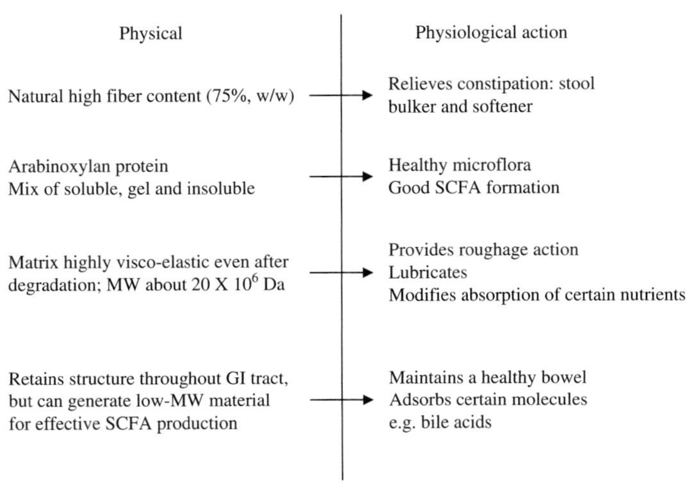

FIGURE 13.16 Relationship between physical properties of ispaghula husk and physiological action in the colon. MW, molecular weight; GI, gastrointestinal; SCFA, short-chain fatty acids.

Psyllium can improve colon health by increasing gastric and colonic mucin levels and producing short-chain fatty acids by microflora fermentation (Hara et al., 1996; Velázquez et al., 2000). Psyllium has been linked to the prevention of colon cancer by suppression of tumor formation (Alabaster et al., 1996; Nakamura et al., 2004). It is believed that tumor promotion is possible due to the clonal expansion of an initial cell. Such expansion is caused by removing growth suppression of the initial cell through inhibition of gap junctional intercellular communication (GJIC). Psyllium could restore normal GJIC in the cells, thus maintaining the homeostatic set point of growth suppression in a tissue in order to prevent colon cancer (Nakamura et al., 2004).

Since the FDA's health claim on psyllium, the food industry became more interested in using psyllium as a new source of dietary fiber. However, incorporating psyllium into food, especially at the level required to allow a health claim, is challenging due to its strong water-absorbing and gelling properties. Additionally, an unpleasant slimy mouth feel and undesirable flavor characteristics were reported for food containing psyllium. To promote the application of psyllium in foods or other consumer products, it is necessary to improve its functional and sensory properties. Some researchers developed an enzyme treatment method to produce psyllium with improved sensory properties and reduced water-adsorbing capacity (Yu and Perret, 2003a, 2003b; Yu et al., 2003) without altering its hypolipidemic capacity (Allen et al., 2004).

13.3.2 Low Mw/Viscosity Soluble Dietary Fiber – Inulin

Inulins refer to a group of naturally occurring fructose-containing oligosaccharides which belong to a class of carbohydrates known as fructans. Inulin was first discovered by a German scientist, Rose, who in 1804 found 'a peculiar substance' from plant origin in a boiling water extract from the roots of *Inula helenium*, a genus of perennial herbs of the Compositae family, natives of the temperate regions of Europe, Asia and Africa (Flamm et al., 2001). Inulins are found naturally in more than 36 000 types of plants worldwide, including 1200 native grasses belonging to 10 families. It is estimated that approximately one-third of the earth's vegetation contains inulins (Boeckner et al., 2001). Most of the inulins are found as storage foods in taproots and tubers in chicory (*Cichorium intybus*), elecampane (*inula hellenum*), dandelion (*Taraxacum officinale*), Jerusalem artichoke (*Helianthus tuberosus*), murnong (*Microseris lanceolata*), salsify (*Tragopogon porrifolius*), and yacon (*Polymnia sonchifolia*) (Roberfroid, 2005). Inulins have received increased attention during the last few decades as a source of dietary fiber and functional food ingredients due to their important health attributes, such as resistance to digestion and being totally fermented in the large bowel. Hence inulins have many beneficial physiological effects, such as improving the balance of friendly bacteria, improving mineral absorption and bone mineralization, inhibiting of colonic carcinogenesis, reduction of blood cholesterol levels, immune stimulation and enhancing vitamin synthesis (Boeckner et al., 2001; Flamm et al., 2001; Cherbut, 2002; Roberfroid, 2002; Raschka, 2005; Bosscher et al., 2006).

13.3.2.1 *Process and Preparation of Inulins*

Until the late 1990s, inulin was not available as a pure compound on a large industrial scale due to high processing costs. With the improvements in separation and purification technology, inulin was prepared into a relatively pure compound by ultrafiltration to remove low-molecular-weight nitrogenous species and minerals from the aqueous extract. So far the only plants that have been used industrially for the extraction of fructans belong to the Compositae family, i.e. chicory, jerusalem artichoke, and dahlia. However, commercial inulin production is essentially from chicory (*Cichorium intybus* L.). Chicory is a biennial plant requiring soil and climatic conditions resembling those for sugar beets. The plants chicory and jerusalem artichoke share similarities in agronomic practices and inulin production technologies. The inulin production process from these plants involves three general steps: (1) extraction with hot water from the roots; (2) purification of the raw inulin using ultrafiltration and ion-exchange resin; (3) evaporation and spray drying of the purified juice to give an inulin powder. An outline of the process is given in Figure 13.17.

The best conditions for high yields of inulin are mild maritime-like climates, rich sandy-clay soils and long periods of daily illumination, similar to that in western European countries such as France, Holland and Belgium (Roberfroid, 2005). The inulin content (15–20%) is high and fairly constant from year to year for a given region and the yields for the crop are around 45 ton roots per hectare. Jerusalem artichoke has a higher inulin content (17–20.5%), but only 20% of the inulin has a degree of polymerization (DP) longer than 10, which appeared to be affected by harvesting time and storage temperature (Boeckner et al., 2001). The tubers are difficult to store outside the soil because of rapid onset of rotting (Roberfroid, 2002). Early-harvested fresh tubers contain a greater amount of highly polymerized fructan fractions, which has more industrial value than late-harvested tubers or those after storage. Therefore, it is recommended that the crop should be harvested according to the daily capacity of processing facilities or develop new preservation techniques to prevent the post-harvest changes in quality (Saengthongpinit and Sajjaanantakul, 2005). Many dahlia cultivars are available, but they have all been selected for their flowers rather than for inulin production. Inulins are also present in significant amounts in many fruits and vegetables (Flamm et al., 2001). Table 13.4 summarizes the content and compositions of inulins from miscellaneous plants, most of which are edible.

13.3.2.2 *Chemical, Functional Properties and Application of Inulins*

13.3.2.2.1 Chemical properties Inulins are mainly comprised of fructose units, typically with a terminal glucose. Chemically, inulin is known as alpha-D-glucopyranosyl-[β-D-fructofuranosyl] (n-1)-D-fructofuranosides, and is abbreviated as GpyFn. In the inulin molecules D-fructofuranosyl

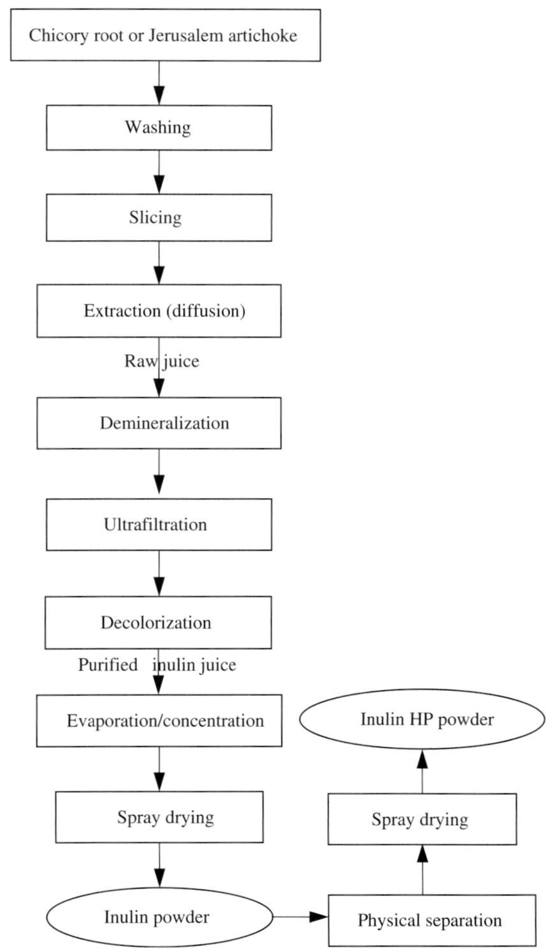

FIGURE 13.17 Outline of inulin production.

TABLE 13.4 Inulin Content and Chain Length of Miscellaneous Plants

Plant	Inulin (g/100 g)	Chain length degree of polymerization (DP)
Globe artichoke	2–7	DP ≥ 5, content up to 95%
		DP ≥ 40, content up to 87%
Banana	1	DP < 5, content up to 100%
Barley	0.5–1.0	
Very young kernels	22	
Chicory	15–20	DP < 40, content up to 83%
	mean 16.2	DP 2–65
		DP ≥ 40, content up to 17%
Dandelion (leaves)	12–15	
Garlic	16	DP ≤ 5, content up to 75%
	mean 13	
Jerusalem artichoke	17–20.5	DP < 40, content up to 94%
		DP 2–50
		DP ≥ 40, content up to 6%
Leek	3–10	DP 12 is most frequent
Onion	1–7.5	DP 2–12
	mean 3.6	
Salsify	20	DP ≤ 5, content up to 75%
Wheat	1–4	DP ≤ 5, content up to 50%

Adapted from Flamm et al. (2001).

β-(1→2) linkages are the main glycosidic bonds (Roberfroid, 2000). Lower case n refers to the number of fructose residues in the inulin chain; py is the abbreviation for pyranosyl. The chain length of plant inulins could range from 2 to 170 fructose units. Inulin is naturally synthesized from sucrose by adding a fructosyl unit. The smallest inulin is 1-kestose, which is composed of two residues of fructose and one glucose. The basic structural formula of inulin is given in Figure 13.18.

GpyFn are both a mixture of oligomers and polymers that are best characterized by the average and the maximum degree of polymerization (DPav and DPmax, respectively). The DP of inulin and the presence of branches are important properties that influence its functional properties. The DPmax of plant inulin is much lower (maximal DP < 200) than inulins produced by bacteria. The DPmax and DPav vary according to plant species, weather conditions, harvest time, and the physiological age of the plant. Plant inulin is considered to be a linear molecule; however, recent structural analysis demonstrated that even native chicory inulin (DPav = 12) has a very small degree of branching (~1–2%) through β-(2→6) linkages (VanLaere and Van Den Ende, 2002; Stephane et al., 2004).

13.3.2.3 *Functional Properties*

Although inulin is often spray dried, it crystallizes easily due to the flexibility of the molecules. The long-chain inulin (DP 10 to 60; $DP_{[av]}$ = 25), also known as inulin of high performance (inulin HP), is produced by

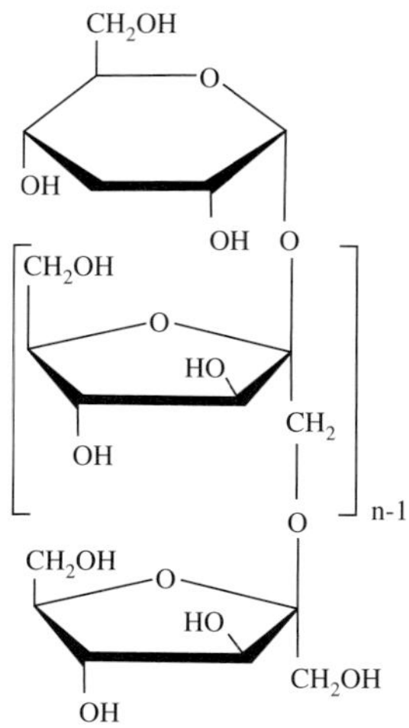

FIGURE 13.18 Basic structural formula of inulin.

physical and chromatographic separation techniques (ion-exchangers and ultrafiltration) to eliminate oligomers with a DP < 10. Inulin has a bland, neutral taste, without any off-flavor or aftertaste (Franck, 2002). Because it also contains fructose, glucose and sucrose, native inulin is slightly sweet (10% sweetness in comparison with sucrose). In contrast, inulin HP is not sweet. Inulin mixes easily with other ingredients without modifying their flavors. The solubility of inulin in water is moderate (maximum 12% at room temperature) and it produces low viscosity (less than 2 mPa for a 5% w/w solution in water) (Franck, 2002; Zimeri and Kokini, 2002).

When thoroughly mixed with proper amounts of water or other aqueous liquid, the submicron crystalline inulin (especially inulin HP) particles can form a 3-dimensional gel network and result in a white creamy structure with a short spreadable texture (Andre et al., 1996; Kim et al., 2001); this unique characteristic makes inulin an excellent fat replacer and it has been incorporated into foods to replace up to 100% fat (Roberfroid, 2005). The gel network is physically stable because large amounts of water are immobilized in the network. Instant inulin has been developed using a specific spray-drying process – a stable homogeneous gel can be produced with minimum shearing (Roberfroid, 2005). Inulin works in synergy with most gelling agents, including gelatin, alginate, carrageenan, gellan gum, and maltodextrin (Roberfroid, 2005). It also improves the stability of foams and emulsions, such as aerated desserts, ice cream, table spreads and sauces. In summary, inulin gels can provide a short spreadable texture, a smooth fatty mouth feel, as well as a glossy appearance and a well-balanced flavor release. It is interesting to note that as a fat replacer, long-chain inulin exhibited about twice the functionality as the short chains, thus allowing for lower dosage levels (Roberfroid, 2005).

The gel-formation ability of inulin is due to precipitation of inulin molecules from solution into submicron crystalline particles, probably through crowding effect (Kim et al., 2001). The gel strength and texture obtained depend on the concentration of inulin and total dry matter content. Other factors, such as shearing parameters (i.e. temperature, time, speed, or pressure) and types of shearing device used also affect the organoleptic properties of the final products. The best conditions for preparing inulin gels are 20–30% (w/v) inulin, heated at 80–90°C for 3–5 min at pH 6–8, then cooled down to room temperature (Kim et al., 2001). In severe conditions such as high temperature or low pH (100°C, pH 1 and 2), inulin could be hydrolyzed into shorter chains which are non-gel-forming components. In a low concentration (≤5%, w/v), inulin–water mixture does not form a gel because the system does not have enough particle and/or molecular density of inulin chains to reach the critical crowding effect (Kim et al., 2001; Zimeri and Kokini, 2002). A sketch demonstration of a slice of an inulin gel is shown in Figure 13.19 (Bot et al., 2004). Inulin gels exhibited visco-elastic behavior and showed shear-thinning and thixotropic properties.

FIGURE 13.19 Sketch of a slice of an inulin gel.

Moisture content in inulin has a significant effect on its crystallinity and some physicochemical properties, such as glass transition temperature (Tg). Native inulin was found to be a semi-crystalline material; however, after being stored at 25°C and water activity values (a_W) < 0.75, inulin is in a glassy state. When presolubilized, dried and stored at low moisture contents, inulin's relative crystallinity is low (<13%); however, when it is stored at conditions above Tg (aw < 0.75), the crystallinity level will increase up to native status (~ 42%) (Zimeri and Kokini, 2002). The semi-crystalline property and glass transition temperature of inulin are believed responsible for inulin's fat-mimetic properties. The physicochemical parameters of some commercial inulins are summarized in Table 13.5 (Franck, 2002; Zimeri and Kokini, 2002; Lopez-Molina, 2005).

13.3.2.4 *Applications of Inulins*

Inulins can be used for either their nutritional advantages or fat mimic properties, however, their applications often offer a dual benefit: an improved organoleptic property and a better-balanced nutritional composition of the food. A summary of applications of inulins in foods and beverages is given in Table 13.6 (Franck, 2002).

TABLE 13.5 Physicochemical and Technological Properties of Chicory Inulin, Inulin HP and Globe Artichoke Inulin in Powder Form

Chemistry	Chicory inulin	Chicory inulin HP	Globe artichoke inulin
DPav	12	25	46
Content (% dry matter)	92	99.5	99.5
Dry matter (%)	95	95	95
Sugars (% dry matter)	8	< 0.5	< 0.5
pH (10% in H_2O)	5–7	5–7	5–7
Ash (% dry matter)	< 0.2	< 0.2	< 0.2
Heavy metals (% dry matter)	< 0.2	< 0.2	< 0.2
Color	White	White	White
Taste	Neutral	Neutral	Neutral
Sweetness versus sucrose (%)	10	None	None
Water solubility (% at 25 °C)	12	2.5	5
Water viscosity (5% at 10 °C)	1.6 MPa	2.4 MPa	2.4 MPa
Food application (specific)	Fat replacers	Fat replacers	Fat replacers
Food application (synergism)	Gelling agent ↑	Gelling agent ↑	Gelling agent ↑

Adapted from Franck (2002).

TABLE 13.6 Typical Examples of Food Technology Applications of Chicory Inulins

Food products	Applications
Dairy products	Body and mouth feel
	Foam stability
	Sugar and fat replacement
	Synergy with sweeteners
Frozen desserts	Sugar and fat replacement
	Synergy with sweeteners
	Texture and melting
Table spreads	Fat replacement
	Texture and spreadability
	Emulsion stability
Baked foods and breads	Sugar replacement
	Moisture retention
Breakfast cereals	Crispness and expansion
Fruit preparations	Body and mouth feel
	Sugar replacement
	Synergy with sweeteners
Meat products	Fat replacement
	Texture and stability
Chocolate	Sugar replacement
	Heat resistance

Adapted from Franck (2002).

The use of inulin as a dietary fiber ingredient in bakery products and breakfast cereals often leads to an improved taste and texture; for example, it gives more crispiness to the products. These ingredients also help to extend the shelf life and freshness of breads and cakes. Their high solubility allows fiber incorporation in aqueous systems such as in beverages, dairy products, and table spreads. Inulin also has found applications in chocolate as a low-calorie bulk agent without added sugar (Roberfroid, 1999).

Because of its specific gelling characteristics, inulin is used for developing low-fat foods without compromising texture and taste. For example, it can be applied in fat-reduced spreads containing dairy products, as well as in butter-like recipes and other dairy-based spreadable products. In low-fat dairy products, e.g. fresh cheese, cream cheese, or processed cheese, the addition of a few percents of inulin gives a creamier mouth feel and imparts a better-balanced rounded flavor. Inulin can also be used as a fat replacer in frozen desserts, providing a fatty mouth feel and excellent melting properties, as well as freeze–thaw stability (Niness, 1999). Other applications as fat replacement include in meat products, sauces, soup, and yogurts – the resulting products retain their typical structure longer and exert a fat-like mouth feeling (Guven et al., 2005; Kipa et al., 2006). Due to the above-mentioned advantages and easy processing characteristics, inulin has become an important ingredient in the food industry: it offers new opportunities for developing nutritionally balanced, yet better-tasting products.

13.3.2.5 Digestion of Inulins

Inulins are known as 'colonic foods' or foods that feed the microflora in the large bowel. Inulins are resistant to digestion in the upper part of the gastrointestinal system due to its almost exclusively β-(2$\rightarrow$1) linkages which cannot be hydrolyzed by enzymes in the oral cavity and small intestine. There is no single bacterium capable of degrading inulins in the large bowel; in fact, most bacteria contribute to the complex metabolic interactions. This makes the colonic microflora a true interactive ecosystem in which many different microorganisms play distinct and complementary roles. Inulins are first hydrolyzed, most likely inside the bacterial cells – primarily inside the bifidobacteria. The hydrolysates are then fermented to produce short-chain fatty acids (SCFAs), lactates, and gases. The main final products of the fermentation process include SCFAs (mainly acetate, propionate, and butyrate), ethanol and gases (CO_2, H_2, CH_4, etc.) (Boeckner et al., 2001). Some intermediate metabolites, including formate, lactate, and succinate, usually do not normally accumulate (Boeckner et al., 2001; Roberfroid, 2005).

Because of the rapid fermentation in the large bowel, inulins can significantly influence colonic functions. For example, inulins can increase stool weight with a bulking index of 1–2 g of additional stool per gram of ingested inulin, equivalent to the bulking index of other fermentable dietary fibers (Roberfroid, 1999). Inulins can also stimulate bowel movements and normalize stool frequency, especially for slightly constipated individuals. Additional physiological activities exhibited by inulins include colonic mucosa protection and repairing. Inulin can modulate the composition of gut microflora, which plays a major role in gastrointestinal physiology (Roberfroid et al., 1998; Gibson, 1999; Rao, 1999; Saavedra and Tschemia, 2002). Inulins modulate other key physiological functions such as mineral absorption, bone mineralization (Roberfroid et al., 2002; Griffin et al., 2003; Courdray et al., 2006) and lipid metabolism (Trautwein et al., 1998; Davidson and Maki, 1999; Williams, 1999; Delzenne et al., 2002). These functions may contribute to reducing the risk of intestinal diseases and to maintaining the health and well-being of the host, and might also play a role in reducing the risk of colon cancer (Taper and Roberfroid, 1999; Buddington et al., 2002; Taper and Roberfroid, 2002); these functions will be discussed separately in the following sections.

13.3.2.6 Physiological Functions of Inulins

13.3.2.6.1 The Prebiotic Effect Inulins enter the colon almost as intact molecules; it acts as a fermentable substrate for colonic microflora, and elicits systemic physiological functions. In fact, inulins are amongst the most studied and well-established prebiotics (Gibson and Roberfroid, 1995). Various in vitro and in vivo studies have shown that a diet supplemented with inulins provides an effective means to promote the growth of bifidobacteria and lactobacilli; meanwhile, it selectively reduces the growth of pathogenic microorganisms, hence it can be potentially used for treating intestinal dysfunctions (Levrat et al., 1993; Roland et al., 1995; Elmer et al., 1996; Pedersen et al., 1997). However, the challenge is to understand how to evaluate the overall function of inulin because the prebiotic effects are not only dependent on the numbers of beneficial bacteria increased, but, more importantly, the activities associated with these bacteria. The health benefits for the host rely on what these bacteria do and how they interact with other bacteria, and essentially, how they modulate the intestinal functions.

13.3.2.6.2 Production of SCFA and Effect on Host Health During the fermentation process, energy is provided for bacterial proliferation and increasing cell mass. The major fermentation products are short-chain fatty acids (SCFAs), along with lactate. In animal trials (rat cecum) inulin has demonstrated significantly higher efficiency in producing SCFAs compared with other dietary fibers ($p < 0.05$), including wheat bran, pea hull, oat husk, cocoa seed and carrot fiber (Roland et al., 1995). SCFAs act as sources of anions in the colonic lumen, affecting both colonocyte morphology and function. For example, SCFAs can stimulate sodium and water absorption and minimize the effects caused by diarrhea. SCFAs may also enhance ileal motility and increase intestinal cell proliferation by local action and increasing mucosal blood flow (Boeckner et al., 2001). In addition to their effects on gut morphology and function, the SCFAs are also absorbed through the colonic epithelial cells into the portal blood and become a source for host energy and regulators of several metabolic processes (Tannock, 2002).

13.3.2.6.3 Effect on Lipid Metabolism Another significant physiological effect of inulins is to modulate either the digestion/absorption or the metabolism of lipids related to triglyceridemia and cholesterolemia, as well as the distribution of lipids between different lipoproteins (Delzenne et al., 2002).

A series of animal studies demonstrated that inulins affect the metabolism of lipids primarily by decreasing triglyceridemia both in the fasted and the postprandial state. In animal trials, diets supplemented with 10% inulin reduced triglyceridemia by 36% (mean value of eight different studies); comparatively, the effect on cholesterolemia was minor – the same diets only resulted in a slight decrease in cholesterol in both normal and (slightly) hypertriglyceridemic conditions (Flamm et al., 2001). Studies on human subjects largely confirm the animal experiments (Wolever et al., 1995; Letexier et al., 2003). The postulated mechanism was that inulins in the diet possibly reduced liver lipogenesis by reducing the expression of the coding of genes for the lipogenic enzymes; this was evidenced in a study where subjects receiving 10 g inulin daily demonstrated a reduced hepatic lipogenesis but not cholesterol synthesis (Letexier et al., 2003).

The possible mechanisms for the lipid-lowering effects may result from the propriónate absorption in the colon which could suppress hepatic synthesis (Wolever et al., 1995). Another route by which inulin may decrease serum lipid levels is via lowered serum insulin and glucose, both known to regulate lipogenesis. For example, lower levels of glucose and insulin were found after feeding a diet containing inulin at 10 g/100 g to rats, which contributed to the reduced hepatic fatty acid and triglyceride synthesis, and are part of the mechanism of the hypolipidemic effect of inulin (Kok et al., 1998).

In addition to the mechanisms involved, the question still remains open on the links between the gastrointestinal site of the fermentation of inulins and their effect on lipid homeostasis inside the body (the so-called systemic effect). A number of hypotheses have been proposed to answer this question:

1. Modifications of glucose and insulin levels; however, the effects of inulins on glycemia and insulinemia are not yet fully understood and available data are still conflicting, indicating that they may depend on physiological (fasted vs. postprandial state) or disease (diabetes) conditions (Daubioul et al., 2000).
2. Modifications of the absorption of macronutrients, especially carbohydrates, which delay gastric emptying and increase small intestinal transit time. It must be emphasized, however, that inulins do not have the high viscosity of other non-starch polysaccharides – a physical property that is usually correlated with their effect on absorption of macronutrients.
3. Increased production of fermentation end products, especially propionate, which is absorbed via the portal vein (Boeckner et al., 2001); it is needed to increase its concentration by more than two-fold in rats in order to reach a concentration which could inhibit the carrier-mediated acetate uptake and decrease the concentration of fatty acid synthase mRNA in cultured isolated hepatocytes (Delzenne et al., 2002).
4. Increased concentration of polyamines, especially putrescine, as observed in the cecum of rats (Delzenne et al., 2000) or
5. Changes in the production of gut peptides (Delzenne et al., 2005).

The decrease in triglyceride content in the liver tissue, which appeared in most of the studies performed, requires further investigation. Non-alcoholic liver steatosis is a frequent pathology, occurring mainly in overweight and obese people, which has recently been proposed as a new feature of the metabolic syndrome. These studies suggest that the systemic effects of inulins are rather important for the health and well-being of humans.

13.3.2.6.4 Immunomodulation and Anticancer Effects of Inulins Non-digestible inulins may have an impact on the immune system, especially in the area of the gut-associated lymphoid tissue (GALT) (Watzl et al., 2005). A major outcome of many studies is that inulin supplementation increases cell number and cell composition in Peyer's patches (PP); however, the underlying mechanism of prebiotic-induced alterations of the cellular structures of PP is not clear. Substantial experimental data suggest that there are at least two different types of mechanism of prebiotics that mediate immunological effects.

1. Inulins are known to increase the amount of lactic-acid-producing bacteria, especially bifidobacteria (Gibson and Roberfroid, 1995). The changes in the number of bifidobacteria induced by inulin supplementation are a prerequisite for changes of immunological functions, such as IgA production. In addition to the increase in bacterial cell number, prebiotics promote the growth in bacterial cell wall components as well as DNA derived from luminal bacteria, which in turn, may stimulate mucosal immune cells.
2. Another possible mechanism is based on the enhanced production of SCFAs in inulin-supplemented animals. The total concentration of SCFAs produced in the colon ranges from 70 to 140 mM, and SCFAs are rapidly transferred to the bloodstream. Usual SCFA concentrations in the bloodstream are 100–150 mM for acetate, 4–5 mM for propionate and 1–3 mM for butyrate (Wolever et al., 1997). Two receptors for SCFAs have been identified on leucocytes, opening up new perspectives for understanding how SCFAs may activate leucocytes and induce signal transduction. For the G-protein-coupled receptor GPR43, acetate and propionate have been found to be the most potent ligands (Brown et al., 2003). However, butyrate and isobutyrate were more active on the receptor GPR41 (Le Poul et al., 2003). While GPR41 is expressed in a wide range of tissues including neutrophils, GPR43 is highly expressed in immune cells. The average concentrations of propionate and butyrate in blood are too low to activate GPR41 or GPR43. However, the blood concentrations reached by acetate are well within the active range for GPR43 (Roller et al., 2004). Enhanced SCFA production in the gut after inulin supplementation may increase SCFA supply to immune cells located along the GALT and activate these cells via the SCFA receptors.

The investigations (Taper and Roberfroid, 1999, 2002) indicated that dietary treatment with inulin incorporated in the basal diet for experimental animals: (i) reduced the incidence of mammary tumors induced in Sprague-Dawley rats by methylnitrosourea; (ii) inhibited the growth of transplantable malignant tumors in mice; and (iii) decreased the incidence of lung metastases of a malignant tumor implanted intramuscularly in mice. Inulin and SYN (combination of probiotic (PRO) and inulin) on the immune system of rats were investigated in azoxymethane (AOM)-induced colon cancer model (Roller et al., 2004). Proliferative responsiveness of lymphocytes (PP) from AOM-treated rats was suppressed in SYN-supplemented rats ($p < 0.01$). Overall, SYN supplementation in carcinogen-treated rats primarily modulated immune functions in the PP, coinciding with a reduced number of colon tumors. Inulin and PRO provided in combination as SYN may contribute to the suppression of colon carcinogenesis by modulating the gut-associated lymphoid tissue.

Although data from human studies are still scarce, the results from recent animal studies clearly suggest that inulins have a strong impact on the immune system, especially the intestinal immune functions, in which immune cells of the PP are primarily activated by inulins. Data from tumor models further demonstrated that a reduced number of colonic tumors in inulin-supplemented animals coincided with enhanced NK cell cytotoxicity. However, whether humans with a daily intake of prebiotics also benefit in improved host resistance and reduced colon cancer risk remains to be answered (Watzl et al., 2005).

13.3.2.6.5 *Side Effects of Fermentation of Inulins*

Consumption of high dosages of inulins will produce a significant amount of gases (mainly CO_2, H_2, and CH_4), which could create some unwanted symptoms in the gut, such as flatulence, bloating, borborygmi, and cramps in the gut (Pedersen et al., 1997). Data collected so far suggest that, on average, these symptoms are mild to moderate; however, a small percentage (1–4%) of the adult population may experience more severe reactions (Boeckner et al., 2001).

13.3.2.7 *Summary*

Inulin has a long history of human consumption and has been recently classified as dietary fiber due to its non-digestible characteristics and fermentability in the intestine. The most pronounced health contributions of inulin are its ability to selectively stimulate the growth of bacterial genera and species known to be beneficial for health in vivo and in humans, such as *Bifidobacterium* and *Lactobacillus*, at the expense of potential pathogenic microorganisms. The ability to selectively promote the growth of *Bifidobacterium* and *Lactobacillus* in human makes inulin a prebiotic. In addition to the effects on gut microflora, inulin also influences mineral bioavailability, lipid homeostasis, and exerted immunomodulation and anticancer activities. A summary of the physiological properties of inulins (similar to other fermentable dietary fibers) is given in Table 13.4 (Jenkins et al., 1999). Considering all the favorable physiological effects combined with the good physicochemical and processing properties described earlier, inulin has been known as a food with potential health-promoting effects or a 'physiologically functional food'.

TABLE 13.7 Potential Effects of Inulin and Nonabsorbable Carbohydrates

Local physiological effects	Systemic physiological effects
↑ Fecal bulk	↓ (↑) Cholesterol
↑ Bacteria	↓ TG (↓ insulin; ↓ glucose)
Selective ↑ bacteria	↓ NH_3
↑ SCFA production	↓ Urea
Selective ↑ In SCFA	↑ B vitamins
↑ Mineral absorption	↑ Immune function
↑ B vitamin synthesis	(↑ Glutamine?)

Abbreviations: SCFA, short-chain fatty acids; TG, triglyceride.
Adapted from Jenkins, et al. 1999.

13.3.3 Insoluble Dietary Fiber

Insoluble dietary fiber has traditionally been associated with fecal bulking and increases in microbial biomass, but studies are increasingly associating physiological benefits with its consumption. This means the addition of insoluble dietary fiber to food may have a plethora of benefits which may be advantageous to the food industry. Insoluble dietary fiber has been associated with improved satiety and weight reduction, but this is debated (Weickert and Pfeiffer, 2008). Some studies have found whole grain consumption results in improved body mass index and insulin sensitivity (Flight and Clifton, 2006). However, prospective cohort studies have consistently shown a significant reduction in diabetes risk with whole grain consumption. Additionally, these benefits were also associated with the outer bran fraction of the kernel and not the germ fraction. Theoretically, this effect should be associated with soluble fiber. Furthermore, whole grain consumption has been found to have protective effects against coronary heart disease (CHD) and cardiovascular disease (CVD), with a 20% to 40% risk reduction conferred for habitual wholegrain consumers versus persons who seldom consume whole grains (Flight and Clifton, 2006; Weickert and Pfeiffer, 2008).

Many studies have been conducted in academia and food industries, sometimes even through collaborative efforts on the incorporation of dietary fiber into foods. These efforts have borne both positive and negative results. With a surge in interest in healthful lifestyles coupled with the projected rise in degenerative diseases there is a need to try to elucidate the factors that affect both positive and negative results on applying dietary fiber to food. Fibers with potential commercial viability which are abundant but considered waste, such as lemon albedo and white grape dietary fiber, as well as industrially produced dietary fibers like wheat bran and resistant starch are of great interest to food scientists. Although there may be differences in extraction and/or processing of dietary fiber and the fortified food products, this section will attempt to relate the physicochemical data, when given, with the organoleptic properties of the foods.

13.3.3.1 *Wheat Bran*

Wheat bran is found in the outer layer of the wheat grain and is produced during wheat milling. Wheat bran was either treated with alkaline hydrogen peroxide (AHP) or untreated before wheat flour substitution at 5%, 10%, 15% and 20% to produce chocolate and plain cakes. Substitution caused a reduction in moisture and nitrogen-free extracts (mostly carbohydrate), while increasing protein, fat, fiber and ash contents. The fiber content was better in AHP-treated than -untreated composite flours. The calorific value of cakes decreased with increasing wheat bran addition. This calorific reduction was higher in AHP-treated cakes which had more dietary fiber due to the treatment (Anjum et al., 2006).

Cakes containing treated fiber scored higher among sensory panelists, than untreated fiber cakes. Cakes with up to 20% supplementation were acceptable by panelists but there was a preference for chocolate cakes. Cakes produced from treated bran had higher sensory scores as AHP removed lignin and hemicelluloses producing a softer structure. These two fiber components are associated with hard structures as they absorb more moisture. Addition of wheat bran also affected cake cell size and uniformity, with untreated bran having a more negative effect on cakes than treated bran. The tenderness and softness of cakes increased with increasing bran addition, with treated wheat bran cakes being softer and more tender than cakes from untreated bran. Wheat bran caused darkening of the cake crumb, which was masked in chocolate cakes. However, the flavor of cakes was also negatively impacted by increased bran addition (Anjum et al., 2006).

Highly purified insoluble wheat fiber, commercially called Vitacel and composed mainly of cellulose and hemicellulose was used in restructured fish products. This dietary fiber is white, odorless and tasteless and inert towards other ingredients. It is composed of 74% cellulose, 26% hemicellulose and less than 0.5% lignin. The particle sizes used were 250 micrometers long and 25 micrometers wide (WF 200) and 80 micrometers long and 20 micrometers wide (WF 600). These fibers have a high water-binding capacity (Sánchez-Alonso et al., 2007).

The shorter fiber retained moisture more efficiently than the longer fiber when added to hake and horse mackerel. However, when extra water was added to the formula fiber did not effectively bind this extra water causing obvious differences between the control and the fiber-supplemented products. However, in these samples only products containing long fibers absorbed thaw drip. The ability of fiber to reduce water loss is important, especially for products using minced beef as the use of insoluble fibers can improve yield and preserve the integrity of a product's texture. This is manifested when fiber prevents the disintegration of breadcrumb coatings during frying.

Wheat fiber was used in minced hake and horse mackerel muscle at enrichment levels of

6% and 3%. At 6% wheat fiber bound water more effectively than 3% enrichment and the control. Sensory panelists gave good ratings to products containing 3% fiber and poor ratings to products with 6% fiber. However, muscles for control samples shrank when water was expelled during heating. Chewiness of the control sample was higher than those with dietary fiber. As the level of wheat fiber increased the products became whiter. This is beneficial as some European consumers relate whiteness to good quality (Sánchez-Alonso et al., 2007).

Vitacel WF 200 and WF 600 were also used in this experiment. Bulk density was 85 g l^{-1} for WF 200 and 210 g l^{-1} for WF 600, swelling (SW) was 11.5 ml g^{-1} and 6.2 ml g^{-1} for WF 200 and WF 600 respectively. Fat absorption capacity (FAC) was 6.3 g g^{-1} and 3.1 g g^{-1} for WF 200 and WF 600 respectively, while water retention (WRC) was 7.7 g g^{-1} and 3.5 g g^{-1} for WF 200 and WF 600 respectively. The long dietary fiber sample had higher SW, WRC and FAC values, probably due to the differences in the bulk density between the two fibers; therefore the shorter fiber absorbed less water than the open structure long fiber. It was also found that the long fiber samples were more hydrophilic than lipophilic which may translate to the long dietary fiber having better functional properties (Sánchez-Alonso et al., 2007).

Three and six percent wheat dietary fiber at different particles sizes were used in surimi gels received from giant squid. Although the final moisture in the samples was not significant, it was realized that long wheat dietary fiber had higher water retention than oil retention. However, when samples were frozen, surimi gels containing short-particle dietary fiber bound significantly more water. There was no significant difference detected by panelists for appearance, however, there were differences for taste and texture as dietary-fiber-enriched samples were softer and more deformable (Sánchez-Alonso et al., 2007).

When wheat dietary fiber was added to surimi gel obtained from giant squid, there was a significant decrease in breaking force, indicating weaker gel strength than the control. There was a loss of hardiness due to the presence of fiber, indicated by texture profile analysis. This compromised gel strength and hardiness is related to the reduced protein content, in addition to a disruption in the protein matrix. Chewiness decreased proportionally to the quantity of fiber present and the cohesiveness of control and gels with 3% fiber were similar, while products with 6% fiber had reduced cohesiveness. Wheat dietary fiber reduced the whiteness of the gels but the whiteness values were higher than premium quality surimi (Sánchez-Alonso et al., 2007).

Beeswing fiber is the outer layer of the pericarp of the wheat grain, i.e. the epidermis, hypodermis and remnants of the thin-walled cells. When wheat is milled beeswings can be either mixed with fine bran to form red beeswings or mixed with fine endosperm to form white beeswings.

Bran and white and red beeswings fibers were added to low-fat beef burgers, at 5%, 10% and 15% fortification. There was an increase in moisture, protein and carbohydrate, while fat and cholesterol decreased with fiber addition. Consequently, there was a reduction in caloric values for beef burgers with fiber. In this trial various levels of wheat fiber addition did not differ to the control for tenderness and firmness. However, there was a significant difference among fiber types for tenderness, juiciness and overall acceptability. Beef burgers containing white and red beeswings were more tender and juicy than that containing bran; therefore, low-fat beef burgers containing beeswings had better overall acceptability than those with bran (Mansour and Khalil, 1997).

When Mansour and Khalil (1997) fortified low-fat beef burgers with bran and beeswings the color was similar to that of the control. Also the difference in tenderness seen in beef burgers with bran, opposed to those with beeswings was supported by shear force data. Shear force for bran beef burgers was significantly higher than the control and other treatment. Shear force for red and white beeswings was lower than the control, except for 5% white beeswing fortification. Beeswings were remarkably soft in texture and the shear force for bran burgers should improve if the hydration ratio of water: fiber is increased by more than 2:1 respectively.

13.3.3.2 *Rice Bran*

Rice bran is located on the outer layers of the rice grain and is a by-product from the milling of rice (Pacheco De Deláhaye et al., 2005). Rao et al. (2007) studied the functional properties of non--starch polysaccharides (NSP) from rice (*Oryza sativa* var. *Jaya*) and ragi (*Eleusine coracana* var. *Indaf-15*). These authors looked at the effects of these soluble fiber components in their native state and after malting. Malting increased the yields of NSP by 3- and 5-fold for rice and ragi, respectively. These water-soluble NSP had over 98% sugars and less than 1% protein. Malting also increased the arabinose to xylose ratio, xylanase activity and decreased the molecular weight of polysaccharides, which have been found to increase the solubility of NSP. NSP in this experiment with the exception of ragi malt were found to have low viscosity, but generally the viscosity was increased by malting and increased concentration. NSP showed low gelling ability but good foam stabilization which was comparable with that of highly viscous gums.

Relative viscosity, oxidative gelation, farinograph and extensograph characteristics as well as bread quality were determined for bread with rice bran. Dough containing 0.25 and 0.50% NSP had increased water absorption, while decreasing dough stability and dough development time based on farinograph measurements. Extensograph tests showed there was improved resistance to extension and extensibility, comparable with the effects of adding dough improvers.

There was a significant increase in loaf volume and softness. It was determined that due to improvements in bread quality, especially from malted samples at 0.50% concentration, these dietary fiber sources were appropriate for supplementation. These authors also noted that the low viscosity observed makes rice and ragi NSP suitable for fiber-deprived health drinks. However, when Abdul-Hamid and Luan (2000) evaluated the functional properties of dietary fiber from defatted rice bran, they found substitution of wheat flour at 5% and 10% significantly reduced loaf volume and increased bread firmness.

The dietary fiber preparation as well as the concentration used to bake bread by Abdul-Hamid and Luan (2000) was different to that of Rao et al. (2007). Composition of rice bran fiber was total dietary fiber 65% of which 9% was soluble dietary fiber, protein 17% and ash 18%. Water-holding capacity was 4.89 ml g^{-1}, fat-binding capacity was 4.54 ml g^{-1} and emulsifying capacity was 14.43%. Rice bran dietary fiber was found to have low viscosity compared to gum solutions, attributable to its low soluble fiber content. Despite negative changes in loaf volume and bread firmness, sensory panellists found the breads were comparable with commercially available high-fiber bread.

Pacheco De Deláhaye et al. (2005) enriched pizza dough with high levels of dietary fiber, stabilized rice bran flour. Supplementation levels were 5% and 10% rice bran flour, and the proximate composition of rice bran flour in grams/100 grams was: moisture 8.82%, crude protein 16.00%, crude fat 18.00%, ash 8.52%, starch 17.00% and total dietary fiber 27.00%. In this study viscosity decreased with increased concentration of rice bran flour in pizza dough. The fiber had a softening effect on the dough due to moisture retention. Based on farinograph measurements there was a reduction in stability and water absorption. However, pizza dough with 5% supplementation gave good acceptability on sensory evaluation.

13.3.3.3 *Other Insoluble Dietary Fiber*

13.3.3.3.1 *Lemon Albedo* Lemon albedo is the white, spongy, cellulosic tissue of the citrus peel that is high in fiber. Commercially, the peel of citrus fruits is the most popular source for the extraction of pectin (Fernández-Gines et al., 2004; Schroder et al., 2004). The characteristics of albedo include particle size 1–3 mm, raw albedo moisture 65.7%, ash 2.96%, fiber 30%, pH 4.53. CIELAB determinations by Minolta CM-2600D were L^*80.48, a^*-6.31, b^* 221.28. Cooked albedo values were moisture 72.03%, ash 2.80%, fiber 22%, pH 5.02, L^*78.53, a^*-6.06 and b^*16.82.

Lemon albedo (raw or cooked) was used in bologna sausages at 2.5%, 5%, 7.5% and 10%. Moisture increased, probably due to water retention in the meat matrix by albedo, due to its soluble constituents, chiefly pectin and correspondingly high water-holding capacity. Protein, fiber and ash of sausages also increased. In sausages lower concentrations of albedo (2.5%, 5%) resulted in higher moisture contents than higher concentrations of albedo (7.5%, 10%). Albedo addition resulted in a reduced fat content. There was also a significant reduction in residual nitrites probably due to its reaction with biocompounds in the albedo. This is a further advantage to fiber addition as there would be a possible reduction in nitrosamine formation. Albedo caused bologna sausages to be lighter in color expressed by changes in L^* and a^* values for the bologna sausages.

Albedo addition resulted in a reduction in juiciness perception and increased hardness perception. This increased hardness and reduced juiciness was perceived more in raw than cooked albedo. Concentrations of raw albedo up to 5% had acceptable scores, while for cooked albedo concentrations up to 7.5% had acceptable scores (Fernández-Gines et al., 2004).

13.3.3.3.2 *White Grape Dietary Fiber (WGDF)* White grape pomace (WGDF) is comprised of the peels and seeds of grapes, usually the by-products of the wine, juice, liquor, jam, canned grape and raisin industry (Bravo and Saura-Calixto, 1998; Sánchez-Alonso et al., 2007). Micrographs indicated WGDF was a compressed matrix so the components were not discernible, but had small grooves and holes in the structure. WGDF (*Vitis vinifera* var. *Airen*) was milled to a particle size smaller than 0.25 mm. Other physicochemical characteristics were protein 7.3%, ash 3.5%, oil 5.5%, insoluble dietary fiber 53.25%, soluble dietary fiber 23.01% and water retention capacity 5.9 grams of water per gram of dry matter.

When WGDF was added to minced fish muscle, at 2% and 4% the protein content was reduced and water retention increased with increased enrichment. There was a difference in texture, but this was insignificant based on sensory evaluation. Panelists also found that at 2% enrichment, flavor of the fish products was better than the control (no fiber) and fiber enrichment at 4%. For overall acceptability, products with 2% fiber enrichment received a score of 7.0 ± 0.8, controls received a score of 6.5 ± 0.9, while the lowest score was 6.1 ± 0.8 for fish with 4% dietary fiber enrichment (Sánchez-Alonso et al., 2007).

The addition of white grape dietary fiber (WGDF) to minced fish muscle changed the texture characteristics, compared to the control. Samples containing fiber were softer and less springy and cohesive. There was also an increase in cooking yield. This change in texture received positive feedback on sensory evaluation. There were changes in color, highlighted by L^* a^* b^* evaluation. The most pertinent change was an increase in yellowness due to the yellow-brown color of the WGDF (Sánchez-Alonso et al., 2007).

13.3.3.3.3 *Resistant Starch (RS)* There are different types of resistant starch (RS) classified as RS1, RS2, RS3 and RS4. Resistant starch 3 commercially available as Novelose 330 from National Starch Food Innovation was used in the following experiment due to its high thermal stability. It contained 8% moisture, 33% dietary fiber and had a pH of 5.5. Sanz et al. (2007) substituted resistant starch (RS3) for wheat flour at 10% and 20%, in fried battered squid rings. This caused a reduction in the fat content of the product and an increase in the moisture content. This could have been the result of RS3 strengthening the methylcellulose barrier against fat update and moisture loss. The fiber content of the product increased significantly from 5% for controls to 13.2% for squid batter with 20% supplementation. There was no significant difference for the attributes appearance, color, crispness, oiliness, flavor and overall acceptability between the three products.

The addition of resistant starch (RS3) to fried battered squid rings resulted in a progressive increase in hardness and fragility of the crust. Thus products with fiber were crispier than the control. There was a significant increase in redness (a^* value) and a significant decrease in luminosity or clarity (L^*). These instrumental changes translated into a more intense golden-brown color in the RS3 products. This color change may be due to larger volumes of reducing sugar for maillard and caramelizing reactions (Sanz et al., 2007).

Resistant starch commercially available as HI-maize 1043 from National Starch Food Innovation with a composition of 12% moisture and 63.9% dietary fiber was incorporated into muffin batter. When Baixauli, Sanz, Slavador and Fiszman (2007) evaluated the effect of resistant starch (RS) addition on muffin quality, they found a reduction in height with increasing RS concentration. Resistant starch also caused a reduction in apparent viscosity with increasing RS concentration. This affected the consistency of the muffin batter associated with an increase in buoyancy of the air bubbles from the batter; therefore the batter became more similar to water which would cause the texture of the muffins to be denser. Resistant starch was substituted for wheat flour at 0%, 5%, 10%, 15% and 20% in this experiment. It was determined that products with less than 10% RS had negligible effects on muffin height or the quantity of gas cells in the crumb.

13.4 CONCLUDING REMARKS

The current chapter provides the most recent updates of the field of dietary fiber and covers progressive development of the definition of dietary fiber in different countries and regions and the improvements of analytical methods, validations of nutritional status and the technological properties that are important for preparation and production of food products. However, the biggest challenge lying ahead of food scientists and nutritionists working in the field of dietary fiber are: (1) to produce a stable supply of varieties of dietary fibers which are well-characterized (insoluble, soluble-HMw and soluble-LMw); (2) to provide complete chemical, technological/functional and physiological information of these three types of dietary fiber and this information could be used by product developers to incorporate these dietary fibers in a broad range of food products to exert the physiological benefits as well as maintaining the organoleptic characteristics acceptable by consumers; (3) to provide physiological properties for all dietary-fiber-enriched products which will allow the manufactures to make a health claim where regulations permit from their respective regulatory bodies. Health claims will not only benefit the consumers but also the functional food industry. With the current trends in the demographic changes of the populations around the world, people with obesity, diabetics and old age become the major social problem in countries across the world. As a result, specific needs for targeted food products arise rapidly, such as low-calorie foods for obesity, high-fiber and low-glycemic foods for diabetics and high-fiber healthy foods for the elderly. This will bring challenges as well as opportunities, and will take the research and utilization of dietary fiber into a new area.

References

Abdul-Hamid, A. and Luan, Y. S. (2000). Functional properties of dietary fibre prepared from defatted rice bran. *Food Chemistry* **68**, 15–19.

Ahmad, A., Anjum, F. M., Zahoor, T., Nawaz, H., and Din, A. (2008). Physicochemical and functional properties of barley β-glucan as affected by different extraction procedures. *International Journal of Food Science and Technology*, 1–7.

Al Assaf, S., Phillips, G. O., Williams, P. A., Takigami, S., Dettmar, P., and Havler, M. (2003). Molecular weight, tertiary structure, water binding and colon behaviour of ispaghula husk fibre. *Proceedings of the Nutrition Society* **62** (1), 211–216.

Alabaster, O., Zhaocheng, T., and Shivapurkar, N. (1996). Dietary fiber and the chemopreventive modelation of colon carcinogenesis. *Mutation Research* **350** (1), 185–197.

Allen, K. G. D., Bristow, S. J., and Yu, L. (2004). Hypolipidemic effects of modified psyllium preparations. *Journal of Agricultural and Food Chemistry* **52** (16), 4998–5003.

Andre, I., Putaux, J. L., Chanzy, H., Taravel, F. R., Timmermans, J. W., and deWit, D. (1996). Single crystals of inulin. *J. Bio. Macr.* **18**, 195–204.

Anjum, F. M., Rauf, M., Khan, M. I., and Hussain, S. (2006). Preparation of low calorific fiber rich cakes by wheat bran supplementation. *Nutrition & Food Science* **36**, 438–444. No. 6.

ANZFA (Australia New Zealand Food Authority). (2000). Notice of a Proposed Change to Food Regulation and Further Invitation for Submissions. Application 227. *Inulin and Fructooligosaccharides as Dietary Fibre* Canberry: ANZFA.

Asp, N. G. (2001). Enzymatic gravimetric method. *In* "CRC Handbook of Dietary Fibre", 3rd Edition, 51–62, Spiller, G. (ed.). CRC Press, Florida.

Baixauli, R., Sanz, T., Salvador, A., and Fiszman, S. M. (2007). Muffins with resistant starch: Baking performance in relation to the rheological properties of the batter. *Journal of Cereal Science* **47**, 502–509.

Behall, K. M., Scholfield, D. J., and Hallfrisch, J. (1999). The effect of particle size of whole-grain flour on plasma glucose, insulin, glucagon and thyroid-stimulating hormone in humans. *Journal of the American College of Nutrition* **18** (No. 6), 591–597.

Belknap, D., Davidson, L. J., and Smith, C. R. (1997). The effects of psyllium hydrophilic mucilloid on diarrhea in enterally fed patients. *Heart & Lung: The Journal of Acute and Critical Care* **26** (3), 229–237.

Benini, L., Castellani, G., Brighenti, F., Heaton, K. W., Brentegani, M. T., Casiraghi, M. C., Sembenini, C., Pellegrini, N., Fioretta, A., Minniti, G., Porrini, M., Testolin, G., and Vantini, I. (1995). Gastric emptying of a solid meal is accelerated by the removal of dietary fibre naturally present in food. *Gut* **36**, 825–830.

Birmingham, C. L., Muller, J. L., Palepu, A., Spinelli, J. J., and Anis, A. H. (1999). The cost of obesity in Canada. *Canadian Medical Association Journal* **160** (4), 483–488.

Boeckner, L. S., Schnepf, M. I., and Tungland, B. C. (2001). Inulin: A review of nutritional and health implications. *Adv. Food Nutr. Res.* **43**, 1–68.

Boeckner, L. S., Schnepf, M. I., and Tungland, B. C. (2001). Inulin: A review of nutritional and health implications. *Adv. Food Nutr. Res.* **43**, 1–68.

Bohm, N. and Kulicke, W. M. (1999). Rheological studies of barley (1-3)(1-4)-α-glucan in concentrated solution: investigation of the viscoelastic flow behaviour in the sol-state. *Carbohydr. Res.* **315**, 293–301.

Borderias, A. J., Sanchez-Alonso, I., and Pérez-Mateos, M. (2005). New applications of fibres in foods: Addition to fishery products. *Trends in Food Science & Technology* **16**, 458–465.

Bosscher, D., Loo, J. V., and Franck, A. (2006). Inulin and oligofructose as functional ingredients to improve bone mineralization. *International Dairy Journal* **16**, 1091–1097.

Bot, A., Erle, U., and Vreeker, R. (2004). Agterof WGM. Influence of crystallisation conditions on the large deformation rheology of inulin gels. *Good hydrocolloids* **18** (4), 547–556.

Bourdon, P., Yokoyama, W., Davis, P., Hudson, C., Backus, R., and Richter, D. (1999). Postprandial lipid, glucose, insulin, and cholecystokinin responses in men fed barley pasta enriched with β-glucan. *American Journal of Clinical Nutrition* **69**, 55–63.

Braaten, J. T., Wood, P. J., Scott, F. W., Wolynetz, M. S., Lowe, M. K., and Bradleywhite, P. (1994). Oat β-glucan reduces blood cholesterol concentration in hypercholesterolemic subjects. *European Journal of Clinical Nutrition* **48**, 465–474.

Brennan, C. and Cleary, L. (2005). The potential use of cereal 1-3, 1-4) β D glucans as functional food ingredients. *Journal of Cereal Science* **42**, 1–13.

Brennan, C. S. and Cleary, L. J. (2007). Utilisation Glucagel in the β-glucan enrichment of breads: A physicochemical and nutritional evaluation. *Food Research International* **40**, 291–296.

Brennan, C. S. (2005). Dietary fibre, glycaemic response and diabetes. *Mol. Nutr. Food Res.* **49**, 560–570.

Brown, A. J., Goldsworthy, S. M., and Barnes, A. (2003). The orphan G protein-coupled receptors GPR41 and GPR43 are activated by propionate and other short chain carboxylic acids. *J. Biol. Chem.* **278**, 11312–11319.

Buddington, K. K., Donahoo, J. B., and Buddington, R. K. (2002). Dietary oligofructose and inulin protect mice from enteric and systemic pathogens and tumor inducers. *J. Nutr.* **132**, 472–477.

Cavallero, A., Empilli, S., Brighenti, F., and Stanca, M. (2002). High (1-3, 1-4)-β-glucan barley fractions in bread making and their effects on human glycemic response. *Journal of Cereal Science* **36**, 59–66.

Cherbut, C. (2002). Inulin and oligofructose in the dietary fibre concept. *Bri. J. Nutr.* **87**, S159–S162.

Cho, Y. I., No, H. K., and Meyers, S. (1998). Physicochemical characteristics and functional properties of various commercial chitin and chitosan products. *Journal of Agricultural Food Chemistry* **46**, 3839–3843.

Courdray, C., Feillet-Coudray, C., Gueux, E., Mazur, A., and Rayssiguier, Y. (2006). Dietary inulin intake and age can affect intestinal absorption of zinc and copper in rats. *J. Nutr.* **136**, 117–122.

Cui, S. (2001). Polysaccharide gums from agricultural products: processing, structures and applications, 103–166. Boca Raton: CRC Press.

Cui, W., and Wang, Q. (2006). Structure-Function Relationships of Cereal β-Glucans, Oral presentation at the 8th International Hydrocolloids Conference, held in Trondheim, Norway.

Cui, W., Wood, P. J., Blackwell, B., and Nikiforuk, J. (2000). Physicochemical properties and structural characterization by 2 dimensional NMR spectroscopy of wheat β-D-glucan-comparison with other cereal β-D-glucans. *Carbohydr. Polym.* **41**, 249–258.

Daubioul, C., De Wispelaere, L., Taper, H., and Delzenne, N. (2000). Dietary oligofructose lessens hepatic steatosis but does not prevent hypertriglyceridemia in obese Zucker rats. *J. Nutr.* **130**, 1314–1319.

Davidson, M. H. and Maki, K. C. (1999). Effects of dietary inulin on serum lipids. *J. Nutr.* **129**, 1474S–1477S.

Delzenne, N., Kok, N., Deloyer, P., and Dandrifosse, G. (2000). Polyamines as mediators of physiological effects of dietary oligofructose in rats. *J. Nutr.* **130**, 2456–2460.

Delzenne, N. M., Cani, P. D., Daubioul, C., and Neyrinck, A. M. (2005). Impact of inulin and oligofructose on gastrointestinal peptides. *Bri. J. Nutr.* **93**, S157–S161.

Delzenne, N. M., Daubioul, C., Neyrinck, A., Lasa, M., and Taper, H. S. (2002). Inulin and oligofructose modulate lipid metabolism in animals: review of biochemical events and future prospects. *Bri. J. Nutr.* **87**, S255–S259.

DeVries, J. (2001). Analytical issues regarding the regulatory aspects of dietary fibre nutrition labeling. *In* "Advanced Dietary Fibre Technology", 123–138, McCleary, B. and Prosky, L. (eds.). Blackwell Science, UK.

DeVries, J. W. (2001). Analytical issues regarding the regulatory aspects of dietary fibre nutrition labelling. *In* "Advanced Dietary Fibre Technology", 123–138, McCleary, B. and Prosky, L. (eds.). Blackwell Science, UK, Oxford.

Dexter, J. E. and Wood, P. J. (1996). Recent applications of debranning of wheat before milling. *Trends Food Sci. & Techn.* **7**, 35–41.

Doublier, J. L. and Wood, P. J. (1995). Rheological properties of aqueous-solutions of (1-3)(1-4)-β-D-glucan from oats (*Avena-Sativa L*). *Cereal Chemistry* **72**, 335–340.

Edwards, S., Chaplin, M. F., Blackwood, A. D., and Dettmar, P. W. (2003). Primary structure of arabinoxylans of ispaghula husk and wheat bran. *Proceedings of the Nutrition Society* **62** (1), 217–222.

Elmer, G. W., Surawicz, C. M., and McFarland, L. V. (1996). Biotherapeutic Agents: Review, a neglected modality of the treatment and prevention of selected intestinal and vaginal infections. *J. Am. Med. Assoc.* **275**, 870–876.

Elmer, G. W., Surawicz, C. M., and McFarland, L. V. (1996). Biotherapeutic Agents: Review, a neglected modality of the treatment and prevention of selected intestinal and vaginal infections. *J. Am. Med. Assoc.* **275**, 870–876.

Fang, C. (2000). Dietary Psyllium Reverses Hypercholesterolemic Effect of *Trans* Fatty Acids in Rats. *Nutrition Research* **20** (5), 695–705.

FDA News. (2006). FDA Finalizes health claim associating consumption of barley products with reduction of risk of coronary heart disease. P06–70.

FDA Talk Paper. (1997). FDA allows whole oat foods to make health claim on reducing the risk of heart disease. T97–5.

Fernández-Ginés, J. M., Fernández-Lopez, J., Sayas-Barberá, E., Sendra, B., and Perez-Alvarez, J. A. (2004). Lemon albedo as a new source of dietary fiber: Application to bologna sausages. *Meat Science* **67**, 7–13.

Figuerola, F., Hurtado, M. L., Estevez, A. M., Chiffelle, I., and Asenjo, F. (2005). Fibre concentrates from apple pomace and citrus peel as potential fibre sources for food enrichment. *Food Chemistry* **91** (3), 395–401.

Fischer, M. H., Yu, N., Gray, G. R., Ralph, J., Anderson, L., and Marlett, J. A. (2004). The gel-forming polysaccharide of psyllium husk (*Plantago ovata* Forsk). *Carbohydrate Research* **339** (11), 2009–2017.

Flamm, G., Glinsmann, W., Kritchevsky, D., Prosky, L., and Roberfroid, M. (2001). Inulin and oligofructose as Dietary fiber: A review of the evidence. *Crit. Rev. Food Sci. Nutr.* **41**, 353–362.

Flight, I. and Clifton, P. (2006). Cereal grains and legumes in the prevention of coronary heart disease and stroke: a review of the literature. *European Journal of Clinical Nutrition* **60**, 1145–1159.

Fox, N. A. and Cameron, A. G. (1991). Food Science, Nutrition and Health. Edward Arnold, London.

Franck, A. (2002). Technological functionality of inulin and oligofructose. *Bri. J. Nutr.* **87**, S287–S291.

Frost, G. S., Brynes, A. E., Dhillo, W. S., Bloom, S. R., and McBurney, M. I. (2003). The effects of fiber enrichment of pasta and fat content on gastric emptying, GLP-1, glucose, and insulin responses to a meal. *European Journal of Clinical Nutrition* **57**, 293–298.

Fung, T. T., Hu, F. B., Pereira, M. A., Liu, S., Stampfer, M. J., Colditz, G. A., and Willet, W. C. (2002). Whole-grain intake and the risk of type 2 diabetes: a prospective study in men. *The American Journal of Clinical Nutrition* **76**, 535–540.

Furda, I. (2001). The crude fibre method. *In* CRC Handbook of Dietary Fibre, 3rd Edition, 111–112, Spiller, G. (ed.). CRC Press, Florida.

Galisteo, M., Duarte, J., and Zarzuelo, A. (2008). Effects of dietary fibers on disturbances clustered in the metabolic syndrome. *Journal of Nutritional Biochemistry* **19**, 71–84.

Gassull, M. A., Goff, D. V., Haisman, P., Hockaday, T. D. R., Jenkins, D. J. A., Jones, K., Leeds, A. R., and Wolever, T. M. S. (1976). The effect of unavailable carbohydrate gelling agents in reducing the post-prandial glycaemia in normal volunteers and diabetics. *J Physiol* **259**, 52P–53P.

Gaw, A. (2002). A new reality: achieving cholesterol-lowering goals in clinical practice. Atheroscler Suppl;. *Apr* **2** (4), 5–8.

Gibson, G. R. (1999). Dietary modulation of the human gut microflora using the prebiotics oligofructose and inulin. *J. Nutr.* **129**, 1438S–1441S.

Gibson, G. R. and Roberfroid, M. R. (1995). Dietary modulation of the human colonic microbiota: introducing the concept of prebiotics. *J. Nutr.* **125**, 1401–1412.

Gibson, G. R. and Roberfroid, M. R. (1995). Dietary modulation of the human colonic microbiota: introducing the concept of prebiotics. *J. Nutr.* **125**, 1401–1412.

Gimeno, E., Moraru, C. I., and Kokini, J. L. (2004). Effect of xanthan gum and CMC on the structure and texture of corn flour pellets expanded by microwave heating. *Cereal Chemistry* **81** (1), 100–107.

Gomez, C., Navarro, A., Garnier, C., Horta, A., and Carbonell, J. V. (1997). Physical and structural-properties of barley (1-3),(1-4)-β-D-glucan III – formation of aggregates analyzed through its viscoelastic and flow behavior. *Carbohydr. Polym* **34** (3), 141–148.

Griffin, I. J., Hicks, P. M. D., Heaney, R. P., and Abrams, S. A. (2003). Enriched chicory inulin increases calcium absorption mainly in girls with lower calcium absorption. *Nutr. Res.* **23**, 901–909.

Guillon, F. and Champ, M. (2000). Structural and physical properties of dietary fibres, and consequences of processing on human physiology. *Food Research International* **33**, 233–245.

Guo, Q., Cui, S. W., Wang, Q., and Young, C. J. (2008). Fractionation and physicochemical characterization of psyllium gum. *Carbohydrate Polymers* **73** (1), 35–43.

Guo, Q., Cui, S. W., Wang, Q., Goff, H. D., and Smith, A. (2008). Microstructure and Rheological Properties of Psyllium Polysaccharide Gel. *Food Hydrocolloid*, (in press).

Guo, Q., Cui, S. W., Wang, Q., Goff, H. D., and Smith, A. (2008). Microstructure and Rheological Properties of Psyllium Polysaccharide Gel. *Food Hydrocolloid*, (in press).

Guven, M., Yasar, K., Karaca, O. B., and Hayaloglu, A. A. (2005). The effect of inulin as a fat replacer on the quality of set-type low-fat yogurt mawnufacture. *Int. J. Dairy Tech.* **58**, 180–184.

Haque, A., Richardson, R. K., Morris, E. R., and Dea, I. C. M. (1993). Xanthan-like 'weak gel' rheology from dispersions of ispaghula seed husk. *Carbohydrate Polymers* **22** (4), 223–232.

Hara, H., Suzuki, K., Kobayashi, S., and Kasai, T. (1996). Fermentable property of dietary fiber may not determine cecal and colonic mucosal growth in fiber-fed rats. *Journal of Nutritional Biochemistry* **7** (10), 549–554.

Hayden, U. L., Mcguirk, S. M., West, S. E. H., and Carey, H. V. (1998). Psyllium Improves Fecal Consistency and Prevents Enhanced Secretory Responses in Jejunal Tissues of Piglets Infected with ETEC. *Digestive Diseases and Sciences* **43** (11), 2536–2541.

Health and Welfare Canada (1985). Report of the Expert Advisory Committee on Dietary Fibre. Supply and Services Canada, Ottawa.

Health, Canada. (1988). Guideline Concerning the Safety and Physiological Effects of Novel Fibre Sources and Food Products Containing Them. Food Directorate, Health Protection Branch, Health Canada, Ottawa.

Hefti, M. C. (1954). Separation of nonsugar polysaccharide from plant seeds of fruits. *Chemistry* **51**, 8459–8469.

Hipsley, E. H. (1953). Dietary "fibre" and pregnancy toxaemia. *Br Med J* **2**, 420–422.

Izawa, M., Kano, Y., and Koshino, S. (1993). Relationship between structure and solubility of (1-3)(1-4)- β-D-glucan from barley. *Journal of the American Society for Brewing Chemistry* **51**, 123–127.

Izydorczyk, M. S., Lagassé, S. L., Hatcher, D. W., Dexter, J. E., and Rossnagel, B. G. (2005). The enrichment of Asian noodles with fiber-rich fractions derived from roller milling of hull-less barley. *Journal of the Science of Food and Agriculture* **85**, 2094–2104.

Jenkins, D. J. A., Kendall, C. W. C., and Ransom, T. P. P. (1998). Dietary fibre, the evolution of the human diet and coronary heart disease. *Nutrition Research* **18** (4), 633–652.

Jenkins, D. J. A., Kendall, C. W. C., and Vuksan, V. (1999). Inulin, oligofructose and intestinal function. *The Journal of Nutrition* **129**, 1431S–1433S.

Jenkins, D. J. A., Kendall, C. W. C., Augustin, L. S. A., Franceschi, S., Hamidi, M., Marchie, A., Jenkins, A. L., and Axelsen, M. (2002). Glycemic index: overview of implications in health and disease. *American Journal of Clinical Nutrition* **76**, 266s–273s.

Jenkins, D. J. A., Kendall, C. W. C., Vuksan, V., Vidgen, E., Parker, T., Faulkner, D., Mehling, C. C., Garsetti, M., Testolin, G., Cunnane, S. C., Ryan, M. A., and Corey, P. (2002). Soluble fibre intake at a dose approved by the US Food and Drug Administration for a claim of health benefits: serum lipid risk factors for cardiovascular disease assessed in a randomized controlled crossover trial. *American Journal of Clinical Nutrition* **75**, 834–839.

Jenkins, D. J. A., Newton, C., Leeds, A. R., and Cummings, J. H. (1975). Effect of pectin, guar gum, and wheat fibre on serum cholesterol. *Lancet* **1**, 1116–1117.

Jenkins, D. J. A., Wesson, V., Wolever, T. M. S., Jenkins, A. L., Kalmusky, J., Guidici, S., Csima, A., Josse, R. G., and Wong, G. S. (1988). Wholemeal versus wholegrain breads: proportion of whole or cracked grain and glycaemia response. *BMJ* **297**, 958–960.

Jensen, C. D., Haskell, W., and Whittam, J. H. (1997). Long-Term Effects of Water-Soluble Dietary Fiber in the Management of Hypercholesterolemia in Healthy Men and Women. *The American Journal of Cardiology* **79** (1), 34–37.

Kahlon, T. (2001). Cholesterol lowering properties of cereal fibres and fractions. *In* Advanced Dietary Fibre Technology, 206–220, McCleary, B. and Prosky, L. (eds.). Blackwell Science, UK.

Kahlon, T. S., Chow, F. I., Knuckles, B. E., and Chiu, M. M. (1993). Cholesterol-lowering effects in hamsters of β-glucan-enriched barley fraction, dehulled whole barley, rice bran, and oat bran and their combinations. *Cereal Chemistry* **70**, 435–440.

Kalra, S. and Jood, S. (2000). Effect of dietary barley β-glucan on cholesterol and lipoprotein fractions in rat. *Journal of Cereal Science* **31**, 141–145.

Kamble, R. M. and Shinde, U. V. (2004). Utility of Bran Products in Non Insulin Dependant Diabetes Mellitus (NIDDM) Patients. *Journal of Human Ecology* **16** (3), 219–222.

Kennedy, J. F., Sandhu, J. S., and Southgate, D. A. T. (1979). Structural data for the carbohydrate of ispaghula husk *ex Plantago ovata* forsk. *Carbohydrate Research* **75**, 265–274.

Kim, Y., Faqih, M. N., and Wang, S. S. (2001). Factors affecting gel formation of inulin. *Carbohydrate Polymers* **46**, 135–145.

Kipa, P., Meyerb, D., and Jellema, R. H. (2006). Inulins improve sensoric and textural properties of low-fat yoghurts. *Int. Dairy J.* **16**, 1098–1103.

Kok, N. N., Taper, H. S., and Delzenne, N. M. (1998). Oligofructose modulates lipid metabolism alterations induced by a fat-rich diet in rats. *J. Appl. Tox.* **18**, 47–53.

Kris-Etherton, P. M., Taylor, D. S., Smiciklas-Wright, H., Mitchell, D. C., Bekhuis, T. C., Olson, B. H., and Slonim, A. B. (2002). High-Soluble-Fiber Foods in Conjunction With a Telephone-Based, Personalized Behavior Change Support Service Result in Favorable Changes in Lipids and Lifestyles After 7 Weeks. *Journal of the American Dietetic Association* **102** (4), 503–510.

Kritchevsky, D. and Tepper, S. A. (1995). Influence of dietary fiber on establishment and progression of atherosclerosis in rabbits. *The Journal of nutritional biochemistry* **6** (9), 509–512.

Lazaridou, A. and Biliaderis, C. G. (2007). Molecular aspects of cereal β-glucan functionality: Physical properties, technological applications and physiological effects. *Journal of Cereal Science* **46**, 101–118.

Lazaridou, A. and Biliaderis, C. G. (2004). Cryogelation of cereal β-glucans: structure and molecular size effects. *Food Hydrocolloids* **18**, 933–947.

Le Poul, E., Loison, C., and Struyf, S. (2003). Functional characterization of human receptors for short chain gatty acids and their role in polymorphonuclear cell activation. *J. Biol. Chem.* **278**, 25481–25489.

Lee, S., Kim, S., and Inglett, G. E. (2005). Effect of shortening replacement with oatrim on the physical and rheological properties of cakes. *Cereal Chem.* **82**, 120–124.

Leeds, A. R. (1995). Psyllium- a super source of soluble dietary fibre. *Food technology in Australia* **47** (2), S2–S4.

Letexier, D., Diraison, F., and Beylot, M. (2003). Addition of inulin to a moderately high-carbohydrates diet reduces hepatic lipogenesis and plasma triacylglycerol concentrations in humans. *Am. J. Clin. Nutr.* **77**, 559–564.

Levrat, M. A., Remesy, C., and Demigne, C. (1993). Influence of inulin on urea and ammonia nitrogen fluxes in the rat cecum: consequences on nitrogen excretion. *J. Nutr. Biochem.* **4**, 351–356.

Levrat, M. A., Remesy, C., and Demigne, C. (1993). Influence of inulin on urea and ammonia nitrogen fluxes in the rat cecum: consequences on nitrogen excretion. *J. Nutr. Biochem.* **4**, 351–356.

Li, W. (2007). PhD Thesis. University of Guelph, Guelph, Ontario, Canada.

Li, W., Cui, S., and Wang, Q. (2006). Solution and conformational porperties of wheat β -D-glucan studied by light scattering and viscometry. *Biomacromolecules* **7**, 446–452.

Lin, K. and Huang, H. (2003). Konjac/gellan gum mixed gels improve the quality of reduced-fat frankfurters. *Meat Science* **65**, 749–755.

Liu, S. (2003). Whole-grain foods, dietary fiber, and type 2 diabetes: searching for a kernel of truth. *The American Society of Clinical Nutrition* **77**, 527–529.

Lopez-Molina, D., Navarro-Martinez, M. D., Melgarejo, F. R., Hiner, A. N. P., Chazarra, S., and Rodriguez-Lopez, J. N. (2005). Molecular properties and prebiotic effect of inulin obtained fromartichoke (*Cynara scolymus L.*). *Phytochem.* **66**, 1476–1484.

Luccia, B.H.D. and Kunkel, M.E. (2002a). Psyllium reduces relative calcium bioavailability and induces negative changes in bone composition in weanling Wistar rats. *Nutrition Research* **22**(9), 1027–1040.

Luccia, B.H.D. and Kunkel, M.E. (2002b). In vitro availability of calcium from sources of cellulose, methylcellulose, and psyllium. *Food Chemistry,* **77**(2), 139–146.

Mäkeläinen, H., Anttila, H., Sihvonen, J., Hietanen, R. M., Tahvonen, R., Salminen, E., Mikola, M., and Sontag-Strohm, T. (2007). The effect of β-glucan on the glycemic and insulin index. *European Journal of Clinical Nutrition* **61**, 779–785.

Maki, K. C., Galant, R., Samuel, P., Tesser, J., Witchger, M. S., Ribaya-Mercado, J. D., Blumberg, J. B., and Geohas, J. (2007). Effects of consuming foods containing oat β-glucan on blood pressure, carbohydrate metabolism and biomarkers of oxidative stress in men and women with elevated blood pressure. *European Journal of Clinical Nutrition* **61**, 786–795.

Mälkki, Y. and Virtanen, E. (2001). Gastrointestinal effects of oat bran and oat gum: a review. *Lebensmittel-Wissehschaft Und-Technologie* **34**, 337–347.

Mansour, E. H. and Khalil, A. H. (1997). Characteristics of low-fat beefburger as influenced by various types of wheat fibers. *Food Research International* **30**, 199–205. No. 3/4.

Marteau, P., Flourié, B., Cherbut, C., Corrèze, J.-L., Pellier, P., Seylaz, J., and Rambaud, J.-C. (1994). Digestibility and bulking effect of ispaghula husks in healthy humans. *Gut* **35**, 1747–1752.

Moghaddam, E., Vogt, J. A., and Wolever, T. M. S. (2006). The Effects of Fat and Protein on Glycemic Responses in Nondiabetic Humans Vary with Waist Circumference, Fasting Plasma Insulin and Dietary Fiber Intake. *The Journal of Nutrition* **136**, 2500–2511.

Morgan, K. R., Roberts, C. J., Tendler, S. J. B., Davies, M. C., and Williams, P. M. (1999). A ^{13}C CP/MAS NMR spectroscopy and AFM study of the structure of Glucagel TM, a gelling β-glucan from barley. *Carbohydr. Res.* **315**, 169–179.

Nakamura, Y., Trosko, J. E., Chang, C., and Upham, B. L. (2004). Psyllium extracts decreased neoplastic phenotypes induced by the *Ha-Ras* oncogene transfected into a rat liver oval cell line. *Cancer Letters* **203** (1), 13–24.

Narender, T., Puri, A., Shwet, T., Khaliq, T., Saxena, R., Bhatia, G., and Chandra, R. (2006). 4-Hydroxyisoleucine an unusual amino acid as antidyslipidemic and antihyperglycemic agent. *Bioorganic and Medicinal Chemistry Letters* **16** (2), 293–296.

Niness, K. R. (1999). Inulin and oligofructose: What are they? *J. Nutr.* **129**, 1402S–1406S.

Oakenfull, D. (2001). Physical chemistry of dietary fibre. *In* "CRC Handbook of Dietary Fibre", 3rd Edition, 33–44, Spiller, G. (ed.). CRC Press, Florida.

Oetjen, G. and Haseley, P. (2004). Freeze-drying. Wiley-VCH, Weinheim.

Olson, B. H., Anderson, S. M., Becker, M. P., and Anderson, J. W. (1997). Psyllium-enriched cereals lower blood total cholesterol and LDL cholesterol, but not HDL cholesterol, in hypercholesterolemic adults: Results of a meta-analysis. *The Journal of Nutrition* **127** (10), 1937–1945.

Osburn, W. N. and Keeton, J. T. (2004). Evaluation of low-fat sausage containing desinewed lamb and konjac gel. *Meat Science* **68**, 221–233.

Östman, E. M., Frid, A. H., Groop, L. C., and Björck, I. M. E. (2006). A dietary exchange of common bread for tailored bread of low glycaemic index and rich in dietary fibre improved insulin economy in young women with impaired glucose tolerance. *European Journal of Clinical Nutrition* **60**, 334–341.

Pacheco de Delahaye, E., Jiménez, P., and Perez, E. (2005). Effect of enrichment with high content dietary fiber stabilised rice bran flour on chemical and functional properties of storage frozen pizzas. *Journal of Food Engineering* **68**, 1–7.

Panlasigui, L. N. and Thompson, L. U. (2006). Blood glucose lowering effects of brown rice in normal and diabetic subjects. *International Journal of Food Sciences and Nutrition* **57** (3/4), 151–158.

Pedersen, A., Sandstrom, B., and Van Amelsvoort, J. M. M. (1997). The effect of ingestion of inulin on blood lipids and gastrointestinal symptoms in healthy females. *Bri. J. Nutr.* **78**, 215–222.

Pedersen, A., Sandstrom, B., and Van Amelsvoort, J. M. M. (1997). The effect of ingestion of inulin on blood lipids and gastrointestinal symptoms in healthy females. *Bri. J. Nutr.* **78**, 215–222.

Pereira, M. A., Pins, J. J., Jacobs, D. R., Jr., Marquart, L., and Keenan, J. (2001). Whole grains, cereal fibre, and chronic diseases: Epidemiological evidence. *In* CRC Handbook of Dietary Fibre, 3rd Edition, 461–479, Spiller, G. (ed.). CRC Press, Florida.

Peres-Olleros, L. (1999). Comparative study of natural carob fibre and psyllium husks in rats. Influence on some aspects of nutritional utilization and lipidemia. *Journal of the Science of Food and Agriculture* **79** (2), 173–178.

Poutanen, K. (2001). Effect of processing on the properties of dietary fibre. *In* Advanced Dietary Fibre Technology, 277–297, McCleary, B. and Prosky, L. (eds.). Blackwell Science, UK.

Prasad-Rao, R. S., Manohar, R. S., and Muralikrisha, G. (2007). Functional properties of water-soluble non-starch polysaccharides from rice and ragi: Effect on dough characteristics and baking quality. *LWT – Food Science and Technology* **12**, 1–9.

Prosky, L. (2001). What is dietary fibre? A new look at the definition. *In* Advanced Dietary Fibre Technology, 63–76, McCleary, B. and Prosky, L. (eds.). Blackwell Science, UK.

Qi, L., Meigs, J. B., Liu, S., Manson, J. E., Mantzoroz, C., and Hu, F. B. (2006). Dietary fibers and Glycemic load, obesity and plasma adiponectin levels in women with type 2 diabetes. *Diabetes Care* **29** (No. 7), 1501–1505.

Rao, A. V. (1999). Dose-response effects of inulin and oligofructose on intestinal bifidogenesis effects. *J. Nutr.* **129**, 1442S–1445S.

Rao, A. V. (1999). Dose-response effects of inulin and oligofructose on intestinal bifidogenesis effects. *J. Nutr.* **129**, 1442S–1445S.

Raschka, L. and Daniel, H. (2005). Mechanisms underlying the effects of inulin-type fructans on calcium absorption in the large intestine of rate. *Bone* **37** (5), 728–735.

Rave, K., Rogen, K., Dellweg, S., Heise, T., and Dieck, H. (2007). Improvement of insulin resistance after diet with a whole-grain based dietary product: results of a randomized, controlled cross-over study in obese subjects with elevated fasting blood glucose. *British Journal of Nutrition* **98**, 929–936.

Roberfroid, M. B. (1999). Caloric value of inulin and oligofructose. *J. Nutr.* **129**, 1436S–1437S.

Roberfroid, M. B. (1999). Concepts in functional foods: The case of inulin and oligofructose. *J. Nutr.* **129**, 1398S–1401S.

Roberfroid, M. B. (1999). Concepts in functional foods: The case of inulin and oligofructose. *J. Nutr.* **129**, 1398S–1401S.

Roberfroid, M. B. (2000). Nondigestible oligosaccharides. *Critic. Rev. Food Sci. Nutr.* **87**, 461–480.

Roberfroid, M. B. (2002). Functional foods: concepts and application to inulin and oligofructose. *Bri. J. Nutr.* **87**, S139–S143.

Roberfroid, M. B. (2005). Inulin-type fructans: Functional Food Ingredients. *CRC Series, Boca Raton, FL CRC Press.*

Roberfroid, M. B. (2005). Inulin-type fructans: Functional Food Ingredients. *CRC Series, Boca Raton, FL CRC Press.*

Roberfroid, M. B., Cupms, J., and Devogelaer, J. P. (2002). Dietary chicoty inulin increases whole-body bone mineral density in growing male rats. *J. Nutr.* **132**, 3599–3602.

Roberfroid, M. B., van Loo, J. A. E., and Gibson, G. R. (1998). The bifidogenic nature of chicory inulin and its hydrolysis products. *J. Nutr.* **128**, 11–19.

Roberfroid, M. B., van Loo, J. A. E., and Gibson, G. R. (1998). The bifidogenic nature of chicory inulin and its hydrolysis products. *J. Nutr.* **128**, 11–19.

Robertson, J. B. and Horvath, P. J. (2001). Detergent analysis of foods. *In* "CRC Handbook of Dietary Fibre", 3rd Edition, 63–66, Spiller, G. (ed.). CRC Press, Florida.

Roland, N., Nugon-Baudon, L., Andrieux, C., and Szylit, O. (1995). Comparative study of the fermentative characteristics of inulin and different types of fibre in rats inoculated with a human while faecal flora. *Br. J. Nutr.* **74**, 239–249.

Roland, N., Nugon-Baudon, L., Andrieux, C., and Szylit, O. (1995). Comparative study of the fermentative characteristics of inulin and different types of fibre in rats inoculated with a human while faecal flora. *Br. J. Nutr.* **74**, 239–249.

Roller, M., Femia, A. P., Caderni, G., Rechkemmer, G., and Watzl, B. (2004). Intestinal immunity of rats with solon cancer is modulated by oligofructose-enriched inulin combined with Lactobacillus rhamnosus and Bifidobacterium lactis. *Bri. J. Nutr.* **92**, 931–938.

Romero, A. L., West, K. L., Zern, T., and Fernandez, M. L. (2002). The seeds from Plantago ovata lower plasma lipids by altering hepatic and bile acid metabolism in guinea pigs. *Journal of Nutrition* **132** (6), 1194–1198.

Roy, S., Freake, H. C., and Fernandez, M. L. (2002). Gender and hormonal status affect the regulation of hepatic cholesterol 7α-hydroxylase activity and mRNA abundance by dietary soluble fiber in the guinea pig. *Atherosclerosis* **163** (1), 29–37.

Saavedra, J. M. and Tschernia, A. (2002). Human studies with probiotics and prebiotics: clinical implications. *Bri. J. Nutr.* **87**, S241–S246.

Saavedra, J. M. and Tschernia, A. (2002). Human studies with probiotics and prebiotics: clinical implications. *Bri. J. Nutr.* **87**, S241–S246.

Saengthongpinit, W. and Sajjaanantakul, T. (2005). Infuence of harvest time and storage temperature on characteristics of inulin from Jerusalem artichoke (*Helianthus tuberosus* L.) tubers. *Posthavest Biol. Tech.* **37**, 93–100.

Samuelsen, A. B., Hanne, C. E., Smestad, P. B., Brull, L. P., and Thomas-Oates, J. E. (1999). Structural studies of a heteroxylan from Plantago major L. seeds by partial hydrolysis, HPAEC-PAD, methylation and GC-MS, ESMS and ESMS/MS. *Carbohydrate research* **315** (3-4), 312–318.

Sánchez-Alonso, I., Haji-Maleki, R., and Borderias, A. J. (2007). Wheat fiber as a functional ingredient in restructured fish products. *Food Chemistry* **100**, 1037–1043.

Sánchez-Alonso, I., Jimenez-Escrig, A., Saura-Calixto, F., and Borderias, A. J. (2008). Antioxidant protection of white grape pomace on restructured fish products during frozen storage. *LWT- Food Science and Technology* **41**, 42–50.

Sánchez-Alonso, I., Jiménez-Escrig, A., Saura-Calixto, F., and Borderlas, A. J. (2007). Effect of grape antioxidant dietary fibre on the prevention of lipid oxidation in minced fish: Evaluation by different methodologies. *Food Chemistry* **101**, 372–378.

Sánchez-Alonso, I., Solas, M. T., and Borderias, A. J. (2007). Physical Study of Minced Fish Muscle with a White-Grape By-Product Added as an Ingredient. *Journal of Food Science* **72,** *Nr* **2**, 94–101.

Sánchez-Alonso, I., Solas, M. T., and Borderias, A. J. (2007). Technological implications of addition of wheat dietary fibre to giant squid (*Dosidicus gigas*) surimi gels. *Journal of Food Engineering* **81**, 404–411.

Sánz, T., Salvador, A., and Fiszman, S. M. (2007). Resistant starch (RS) in battered fried products: Functionality and high-fibre benefit. *Food Hydrocolloids*, 1–7.

Satchithanandam, S., Klurfeld, D. M., Calvert, R. J., and Cassidy, M. M. (1996). Effects of dietary fibers on gastrointestinal mucin in rats. *Nutrition Research* **16** (7), 1163–1177.

Schneeman, B. (2001). Dietary fibre and gastrointestinal function. *In* Advanced Dietary Fibre Technology, 168–176, McCleary, B. and Prosky, L. (eds.). Blackwell Science, UK.

Schulze, M. B., Liu, S., Rimm, E. B., Manson, J. E., Willet, W. C., and Hu, F. B. (2004). Glycemic index, glycemic load and dietary fiber intake and incidence of type 2 diabetes in younger and middle-aged women. *The American Journal of Clinical Nutrition* **80**, 348–356.

Southgate, D. A. T. (2001). Food components that behave as dietary fibre. *In* CRC Handbook of Dietary Fibre, 3rd Edition, 15–18, Spiller, G. (ed.). CRC Press, Florida.

Stephane, C., Kervarec, N., Pichon, R., Magne, C., Bessieresa, M. A., and Deslandesa, E. (2004). NMR charaterisation of inulin-type fructooligosaccharides as the major water-soluble carbohydrates from *matricaria maritime (L.). Carbohydr. Res.* **339**, 2445–2449.

Tannock, G. W. (2002). Molecular methods for exploring the intestinal ecosystem. *Bri. J. Nutr.* **87**, S199–S201.

Taper, H. S. and Roberfroid, M. B. (1999). Influence of inulin and oligofructose on breast cancer and tumor growth. *J. Nutr.* **129**, 1488S–1491S.

Taper, H. S. and Roberfroid, M. B. (2002). Inulin/oligofructose and anticancer therapy. *Bri. J. Nutr.* **87**, S283–S286.

Taylor, D. S., Kris-Etherton, P. M., Smiciklas-Wright, H., Mitchell, D. C., Bekhuis, T. C., Slonim, A. B., Olson, B. H., and Bryson, W. L. (1999). A Behavior Change Support Service and Food Sources of Soluble Fiber Favorably Affect Cholesterol Management Lifestyle Changes. *Journal of the American Dietetic Association* **99** (9), A115.

Theander, O., Westerlund, E., Andersson, R., and Aman, P. (2001). Determination of total dietary fibre and its individual components by the uppsala method. *In* CRC Handbook of Dietary Fibre, 3rd Edition, 87–109, Spiller, G. (ed.). CRC Press, Florida.

Tomoda, M., Shimizu, N., Shimada, K., Gonda, R., and Sakabe, H. (1984). Plant mucilage. XXXIV. The location of o-acetyl groups and the structure features of plantago-mucilage A, the mucous polysaccharide from the seeds of *plantago magor* var. *asiatica* **32** (6), 2182–2186.

Tosh, S. M., Brummer, Y., Wood, P. J., Wang, Q., and Weisz, J. (2004). Evaluation of structure in the formation of gels by structurally diverse (1-3)(1-4)- β -D-glucans from four cereal and one lichen species. *Carbohydr. Polym.* **57**, 249–259.

Tosh, S. M., Wood, P. J., Wang, Q., and Weisz, J. (2004). Structural characteristics and rheological properties of partially hydrolyzed oat β-glucan: the effects of molecular weight and hydrolysis method. *Carbohydr. Polym.* **55**, 425–436.

Trautwein, E. A., Rieckhoff, D., and Erbersdobler, H. F. (1998). Dietary inulin lowers plasma cholesterol and triacylglycerol and alters biliary bile acid profile in hamsters. *J. Nutr.* **128**, 1937–1943.

Trowell, H. (1972). Crude fibre, dietary fibre and atherosclerosis. *Atherosclerosis* **16**, 138–140.

Tuomasjukka, S., Viitanen, M., and Kallio, H. (2007). The glycaemia response to rolled oat is not influenced by the fat content. *British Journal of Nutrition* **97**, 744–748.

Turhan, S., Sagir, I., and Ustun, N. S. (2005). Utilization of hazelnut pellicle in low-fat beef burgers. *Meat Science* **71**, 312–316.

Turley, S. D. and Dietschy, J. M. (1995). Mechanisms of LDL-cholesterol lowering action of psyllium hydrophillic mucilloid in the hamster. *Biochimica et Biophysica Acta (BBA)/Lipids and Lipid Metabolism* **1255** (2), 177–184.

US Food and Drug Administration. (1998). *Food labeling: health claims; soluble fiber from certain foods and coronary heart disease.*

USFDA (U.S. Food and Drug Administration). (1987). Nutrition labeling of food; calorie content. Federal Register **52**, 28590–28691.

Van Laere, A. and Van Den Ende, W. (2002). Inulin metabolism in dicots: chicory as a model system. *Plant, Cell Environ.* **25**, 803–813.

Vasanthan, T. (2005). Grain fiber compositions and methods of use. US Patent Application No. 20050208145.

Vega-Lopez, S., Conde-Knape, K., Vidal-Quintanar, R., Shachter, N. S., and Fernandez, M. L. (2002). Sex and hormonal status influence the effects of psyllium on lipoprotein remodeling and composition. *Metabolism* **51** (4), 500–507.

Velázquez, M., Davies, C., Marett, R., Slavin, J., and Feirtag, J. M. (2000). Effect of oligosaccharides and fibre substitutes on short chain fatty acis production by human faecal microflora. *Anaerobe* **6**, 87–92.

Velázquez, M., Davies, C., Marett, R., Slavin, J. L., and Feirtag, J. M. (2000). Effect of Oligosaccharides and Fibre Substitutes on Short-chain Fatty Acid Production by Human Faecal Microflora. *Anaerobe* **6** (2), 87–92.

Vergara-Jimenez, M., Furr, H., and Fernandez, M. L. (1999). Pectin and psyllium decrease the susceptibility of LDL to oxidation in guinea pigs. *Journal of Nutritional Biochemistry* **10** (2), 118–124.

Wang, Q., Wood, P. J., Haun, X., and Cui, S. (2003). Preparation and characterization of molecular wight standards of low polydispersity from oat and barley (1-3)(1-4)- β -D-glucan. *Food Hydrocolloids* **17**, 845–853.

Watzl, B., Girrbach, S., and Roller, M. (2005). Inulin, oligofructose and immunomodulation. *Bri. J. Nutr.* **93**, S49–S55.

Weber, C. W., Kohlhepp, E. A., Idouraine, A., and Ochoa, L. J. (1993). Binding capacity of 18 fiber sources for calcium. *Journal of Agricultural and Food Chemistry* **41** (3), 1931–1935.

Weickett, M. O. and Pfeiffer, A. F. H. (2008). Metabolic effects of dietary fiber consumption and prevention of diabetes. *The Journal of Nutrition* **138**, 439–442.

WHO http://www.who.int/diabetes/facts/en/.

Williams, C. M. (1999). Effects of inulin on lipid parameters in humans. *J. Nutr.* **129**, 1471S–1473S.

Williams, P. A. and Phillips, G. O. (2001). Introduction. *In* Handbook of Hydrocolloids, 1–3, Phillips, G. O. and Williams, P. A. (eds.). CRC Press, Bota Raton, FL.

Wolever, T., Spadafora, P., Cunnane, S., and Pencharz, P. (1995). Propionate inhibits incorporation of colonic [1, 2-^{13}C] acetate into plasma lipids in humans. *Am. J. Clin. Nutr.* **61**, 1241–1247.

Wolever, T. M., Josse, R. G., Leiter, R. A., and Chiasson, J. L. (1997). Time of day and glucose tolerance status affect serum short-chain fatty acid concentrations in humans. *Metabolism* **46**, 805–811.

Wolever, T. M. S. and Jenkins, D. J. A. (2001). Effect of dietary fibre and foods on carbohydrate metabolism. *In* CRC Handbook of Dietary Fibre, 3rd Edition, 321–360, Spiller, G. (ed.), CRC Press, Florida.

Wood, P. J. (1991). Oat β-glucan–physicochemical properties and physiological effects. *Trends in Food Science and Technology*, 311–314.

Wood, P. J. (2004). Relationships between solution properties of cereal β-glucans and physiological effects – a review. *Food Sci. & Techn.* **13**, 313–320.

Wood, P. J. (2007). Cereal β-glucans in diet and health. *Journal of Cereal Science* **46**, 230–238.

Wood, P. J., Braaten, J. T., Scott, F. W., Riedel, D., and Poste, L. M. (1990). Comparisons of viscous properties of oat and guar gum and the effects of these and oat bran on glycemic Index. *Journal of Agricultural and Food Chemistry* **38**, 753–757.

Woods, M. N. and Gorbach, S. L. (2001). Influences of fibre on the ecology of the intestinal flora. *In* CRC Handbook of Dietary Fibre, 3rd Edition, 257–269, Spiller, G. (ed.). CRC Press, Florida.

Yu, L. and Perret, J. (2003a). Effects of Xylanase Treatments on Gelling and Water-Uptaking Properties of Psyllium. *Journal of Agricultural and Food Chemistry* **51**(2), 492–495.

Yu, L. and Perret, J. (2003b). Effects of solid-state enzyme treatments on the water-absorbing and gelling properties of psyllium. *Lebensmittel-Wissenschaft und -Technologie* **36**(2), 203–208.

Yu, L., Perret, J., Parker, T., and Allen, K. G. D. (2003). Enzymatic modification to improve the water-absorbing and gelling properties of psyllium. *Food Chemistry* **82** (2), 243–248.

Zaman, V., Manzoor, S. M., Zaki, M., Aziz, N., and Gilani, A. (2002). The presence of antiamoebic constituents in Psyllium husk. *Phytotherapy Research* **16** (1), 78–79.

Zimeri, J. E. and Kokini, J. L. (2002). The effect of moisture content on the crystallinity and glass trasition temperature of inulin. *Carbohydrate Polymers* **48**, 299–304.

Zimeri, J. E. and Kokini, J. L. (2003). Rheological properties of inulin-waxy maize starch systems. *Carbohydrate polymers* **52**, 67–85.

CHAPTER

14

Resistant Starch in Vitro and in Vivo: Factors Determining Yield, Structure, and Physiological Relevance

Anthony R. Bird [1], *Amparo Lopez-Rubio* [2], *Ashok K. Shrestha* [3], *and Michael J. Gidley* [3]

[1]Commonwealth Scientific and Industrial Research Organisation, Food Futures National Research Flagship, and CSIRO Human Nutrition, Kintore Ave, Adelaide, South Australia 5000, Australia
[2]Bragg Institute, Australian Nuclear Science and Technology Organisation, PMB 1, Menai, NSW 2234, Australia
[3]Centre for Nutrition and Food Sciences, The University of Queensland, St Lucia, 4072, Australia

14.1 INTRODUCTION

The term resistant starch (RS) is a relatively recent one that reflects the fact that its presence and importance have only been appreciated during the last 30 years. RS is normally defined as that portion of dietary starch that escapes digestion in the small intestine of a healthy human. Within the healthy small intestine there is typically an abundant supply of α-amylase in the soluble phase which serves to break many starches down to low-molecular-weight fragments. These fragments are then broken down at the gut wall into glucose, which is then taken up into the bloodstream. When some of the starch escapes this digestive process, there are several important consequences. One is that there is less starch that is converted to glucose and taken up from the small intestine with potential impact on diabetes and energy intake levels. A second is that RS is transported into the colon where it acts in a similar manner to dietary fiber, providing a source of fermentable carbohydrate for colonic bacteria. A third consequence is that the end products of fermentation (acetate, propionate, and especially butyrate) have direct benefits on the health of the colon and potentially affect systemic health after uptake into the bloodstream.

Four categories of starches are generally identified as having the potential to contribute to RS:

- RSI: Physically inaccessible starch within, e.g. intact plant tissues or dense food matrices.
- RSII: Intact granular starches that have slow digestion rates.
- RSIII: Starch that has been processed and partially recrystallized (retrograded).
- RSIV: Chemically modified starch that slows α-amylase attack.

Kasapis, Norton, and Ubbink: Modern Biopolymer Science
ISBN: 978-0-12-374195-0

This review will focus on RSII and RSIII as these two forms are those which are most influenced by native starch molecular and macroscopic structure. There have only been a limited number of studies on RS as defined above, due to the difficulty of obtaining samples. Nevertheless, it is only this definition of RS that has physiological meaning. The vast majority of studies has been carried out in vitro (enzyme-resistant starch), and may not accurately represent the structures formed in vivo. However, practical considerations mean that this is the only possible route in many cases.

In the following sections, measurement of resistant starch (Section 14.2), potential health benefits (Section 14.3), effects of a range of processing techniques on grains, foods (Section 14.4) and isolated starches (Section 14.5), and proposed mechanisms for the molecular and microstructural basis of RS (Section 14.6) are discussed.

14.2 MEASUREMENT OF RESISTANT STARCH

Starchy foods are a dietary cornerstone and the principal energy source for most humans. Until about the last 30 years, it was a widely held belief that starch was digested to completion in the small intestine, despite evidence to the contrary (Borchers, 1961; Sandstedt et al., 1962; Shainkin and Birk, 1970; Palmer, 1972; Mason and Just, 1976; Holloway et al., 1978; Snow and O'Dea, 1981). However, it was becoming increasingly clear from clinical and laboratory studies conducted in the early 1980s that starch comprises both indigestible and digestible components. Anderson and colleagues (1981) demonstrated, albeit indirectly using the breath hydrogen technique, that significant amounts of starch eaten by humans could not only escape digestion and absorption in the small intestine, but also serve subsequently as a substrate for the colonic microflora. Also about that time analysts in Englyst's laboratory in Cambridge were finding that their efforts to develop an assay for determination of non-starch polysaccharides (NSP, the principal components of dietary fiber) in processed foods were hampered by the presence of residual (retrograded) starch which proved highly resistant to removal by α-amylase and pullulanse (Englyst et al., 1982, 1983). It was evident from these investigations that enzyme hydrolysis alone was insufficient to remove starch completely and that an aggressive chemical treatment step was necessary. Englyst and colleagues (1982) coined the term 'resistant starch' to denote starch which resists enzymatic amylolysis in vitro. Research a short while later using human ileostomates, intubated subjects and various animal models provided irrefutable evidence that the complete assimilation of starch in the small intestine is not a forgone conclusion.

It is now well established that the amount of RS in a given food is not a fixed entity but is instead determined by many different factors. Food form and structure, botanical source of the starch and its amylose content, other dietary constituents, notably protein and fat, the presence of α-amylase inhibitors and other antinutrients, the methods used to manufacture and store foods as well as culinary practices all directly influence starch digestibility in the gut (Champ et al., 2003; Champ, 2004). The workings of the gastrointestinal tract of the consumer also play a major role in dietary starch disposal. In healthy individuals there is an excess hydrolytic and absorptive capacity of the small intestine relative to starch intake (Diamond and Hammond, 1992; Bird et al., 1996; Hammond and Diamond, 1997). Accordingly, the residence time of starch in the lumen of the upper gut is a primary determinant of the amount of that material ultimately reaching the terminal ileum. Increases or decreases in the passage rate of starch along the small bowel will shorten or extend, respectively, the exposure time of ingested starch to salivary and pancreatic amylases, and brush border hydrolases, which in turn may alter the yield of resistant starch delivered to the colonic microflora.

RS is clearly a physiological concept and it is primarily of interest, as is dietary fiber, because of its physiological properties and potential for improving human health and well-being. Although originally an analytical construct, RS was redefined in 1992 as that fraction of starch, together with any products of starch digestion, including glucose, maltose and malto-oligosaccharides, escaping the ileum and entering the colon of healthy individuals (Asp, 1992). Unlike other dietary plant polysaccharides, such as NSP and oligosaccharide, RS is not intrinsically indigestible and includes material which is potentially digestible. Starches escape digestion for a variety of reasons and RS has been categorized on this basis into four different types (see Bird and Topping, 2008). Furthermore, the proportion of ingested starch not absorbed in the upper gut is specific for any given individual. Exact measurement of the RS content of foods is undoubtedly a formidable technical challenge.

14.2.1 In Vivo Methods

By definition, determination of the RS content of human foods is predicated on the use of physiological techniques in healthy humans. However, food databases on RS are scant because such analyses are difficult to perform, owing to the limited accessibility of the human gastrointestinal tract. Perhaps it is worth noting at this point that of the different in vivo techniques currently available, none are ideal and each has inherent drawbacks. However they are all exceedingly slow, labor intensive and require a considerable investment in specialized resources and expertise.

The human ileostomy model is the most frequently used in vivo method for RS determination (Andersson, 1992; Champ, 2004). It is widely accepted as a reliable and direct means for investigating human small intestinal digestion of foods and meals. In essence it is a basic nutrient balance method and uses adults with conventional, well-established and well-functioning permanent ileostomies. Ideally, ileostomates who have had minimal (<10 cm) small bowel resection are recruited. Also, only those individuals whose ileostomy was performed to remedy ulcerative colitis should be recruited. Recruitment of suitable volunteers is difficult normally as the number of ileostomates in any given population is small and nowadays the surgical operation is less common owing to advances in medical science.

The usual protocol is to collect and analyze ileal digesta for starch or other nutrients of interest and, by comparison with the total amount of starch ingested during the study period, calculate its disappearance (digestibility) or conversely the amount which escapes (resists) digestion in the upper gastrointestinal tract. Most studies have used between four and eight volunteers who are maintained on a low- or no-starch diet during the experimental period. A specified amount of the starchy test food is eaten, usually at breakfast, and then stomal effluent is collected over the next 24 hours or thereabouts. It is essential that the contents of the ileostomy bag are collected at frequent intervals, usually every 2 hours or less, and frozen immediately to minimize possible starch degradation by human amylases or bacteria normally present in intestinal contents.

A period of 24 hours is considered ample time to ensure quantitative collection of starch escaping the small bowel. Although it takes about 3–4 hours for ingested food to first appear in the colon (Read et al., 1986; Wutzke and Glasenapp, 2004), the last remnants of a meal do not reach the colon for at least another 5 hours (Read et al., 1986). Almost all of the undigested test food will have been discharged from the terminal ileum 8 hours after ingestion. However, sample collection periods of at least 14 h and ideally 24 h duration are preferable as variation between and within individuals in the rate of passage of food through the upper gut, which is not inconsiderable (Silvester et al., 1995), must be accommodated. Also at least one study (Control) day is required in which test foods are not eaten in order to establish baseline starch excretion.

Because much of the starch exiting the end of the small intestine has proved rather impervious to amylolysis, retrograded and intact granular starches in particular, it is critical that the chosen quantification method is sufficiently aggressive to ensure complete solubilization and hydrolysis of all starch present in ileal digesta collections. Not only must the method be quantitative for starch per se, it must be also capable of detecting products of partial or complete starch digestion, including glucose. However, glucose in ileal digesta may not necessarily be of starch origin. This problem is dealt with indirectly by subtraction of baseline 'starch' excretion data from those values obtained on the days when test foods are eaten.

The human ileostomy model has been extensively validated and utilized by various research groups for studying human starch digestion (e.g. Muir et al., 1992, 1993; Englyst et al., 1996; Champ, 2004). The technique yields physiologically accurate quantitative information. The functional capacity of the gut of ileostomates appears to be reasonably similar to that of individuals with an intact gastrointestinal tract. Gastric emptying rate and small bowel transit time are comparable to those of healthy individuals (Read et al., 1986), although a recent study suggests that while the emptying rate of liquids may be similar that of solids is less (Robertson and Mathers, 2000). The terminal ileal bacterial population of ilesotomates has been reported to be 10-fold larger than that of normal individuals (about 10^7 organisms versus 10^6 per gram of contents, respectively; Finegold et al., 1970); however, the taxonomic composition of the microflora appears to be similar, as are concentrations of SCFA which are the major end products of bacterial fermentation of carbohydrates. SCFA levels in ileal effluent are normally low (e.g. ≤ 2 mM; Chapman et al., 1985) suggesting that any loss of carbohydrate due to bacterial activity is small (< 2 g day^{-1}; Brouns et al., 2007). We have estimated that starch degradation in stomal collections is insignificant when the 'gold standard' regimen for sample collection and processing is adhered to (Bird, Topping, Vuaran and Bushell, unpublished results).

The time it takes for starch to pass from the mouth to the terminal ileum (orocecal transit time) is a major determinant of RS yield in the upper gut (Chapman et al., 1985; Silvester et al., 1995) as is the amount of starch ingested. Indeed, there is a positive linear relationship between starch intake and ileal starch excretion (Chapman et al., 1985; Silvester et al., 1995) and transit time is independent of starch intake (Chapman et al., 1985). Even at high intakes (>100 g), the proportion of starch resisting digestion remained reasonably constant (Chapman et al., 1985). There is a substantial excess functional capacity to assimilate dietary starch, luminal hydrolysis apparently being the rate-limiting step in this process (Vonk et al., 2000).

Although the ileostomy method potentially may underestimate RS, the measurement error is small and, given the considerable within- and between-individual variation in ileal starch digestibility, undoubtedly inconsequential. An obvious advantage is that subjects are ambulatory and can go about their normal daily activities with few restrictions during testing periods. This is not the case for the other existing in vivo techniques. Indeed, intubated volunteers must restrict movement for several hours, a requirement that may modulate normal intestinal transit and, accordingly, alter the amount of starch passing through the end of the small bowel. Between-study comparisons suggest that the ileostomy technique is reasonably reliable, at least for determining the RS content of foods containing high-amylose maize starch. Despite its imperfections the ileostomy model is superior to the other options currently available.

The intestinal intubation method is also used but much less extensively than the ileostomy model (Stephen et al., 1983). Basically the tip of a multi-lumen nasogastric tube is positioned in the terminal ileum and the luminal contents aspirated at specified time intervals after food consumption by the volunteer. The technique requires co-administration of a non-absorbable marker. There are clear ethical considerations with this invasive approach as well as significant technical shortcomings, notably the potential for modifying intestinal transit. For food particles to be aspirated they must be smaller than the internal diameter of the lumenal catheter, hence homogenized (liquidized) foods are necessary. This requirement will facilitate digestion of the test food by luminal α-amylases, although this effect may be offset by a reduction in gut residence time due to the smaller particle size. The overall impact of these factors on digesta transit presumably is small in comparison with that caused by the presence in the lumen of a foreign body (i.e. catheter) which may encourage peristaltic action and thereby shorten transit time in the small bowel. It must be mentioned that for overly coarse foods the ileostomy model is also not appropriate because ileostomates are prone to intestinal blockage. Nevertheless this RS method is still applicable to most commonly consumed starchy foods.

Various intact and surgically modified animal models also have been used to determine the resistant starch content of human foods. The former include antibiotic-treated (Gudmand-Hoyer, 1991) and germ-free rat models (see Champ, 2004), whereas the ileal-cannulated pig and the colectomized rat models (Hildebramdt and Marlett, 1991; Marlett and Longacre, 1996) are examples of the latter. Although useful research tools, they too have various technical limitations. For instance rodents do not readily consume human foods in the form that they are normally eaten. Usually it is necessary to feed rats diets that are powdered and/or pelleted. Eating behavior also differs between humans and experimental species. Rats are mainly nocturnal feeders and nibble food, while pigs bolt their meals; these characteristics are likely to have an impact on starch digestibility. Furthermore, whether the results of studies using these models are transferable to humans has yet to be determined. Also, being feeding trials, they too are time-consuming, laborious and consequently too expensive to be used for routine analyses.

Various indirect techniques based on the measurement of hydrogen in expirated air have been developed (McBurney, 1991; Strocchi and Levitt, 1991; Rumessen, 1992). Breath hydrogen concentration rises in response to fermentation of unabsorbed carbohydrates by large-bowel bacteria. In contrast to other in vivo techniques, this one has the distinct advantage of being non-invasive and also can be applied to specific populations such as younger subjects (e.g. children, infants) or those with particular illnesses, for instance cystic fibrosis (Dewit et al., 1992; Weaver et al., 1995). However the breath hydrogen technique yields only semi-quantitative information (Rumessen, 1992). It is subject to considerable within- and between-subject variation, and quantitative estimates differ substantially from those produced by the ileostomy or intubation methods (Champ et al., 2003). Accordingly it has limited application. Recent advances in the use of stable isotopic (^{13}C) techniques and mass spectrometry have improved the quantitative accuracy of breath measurement methodology (Hiele et al., 1990; Normand et al., 1992; Vonk et al., 2000; Symonds et al., 2004). However, the analytical instrumentation is specialized and expensive, there are limitations on the types of foods that may be measured (Symonds et al., 2004), and the accuracy of these advanced breath testing methods is essentially unknown because comparative methodology and validation studies have yet to be performed.

Currently available in vivo methods are cumbersome research instruments applicable to specialized physiological, clinical or nutritional studies. They do not satisfy the specific needs of analysts in food research and development, regulatory and carbohydrate research laboratories.

Apart from the intrinsic ethical and technical constraints in conducting human (or animal) studies, these bioassays require particular adherence to stringent testing protocols and conditions in order to effectively manage the large intra- and inter-individual variation inherent in biological systems in general, and gastrointestinal starch assimilation in particular. Starch digestion and absorption involves a complex, multifaceted orchestration of numerous physicochemical processes. The functional efficiency and effectiveness of the gut with respect to starch digestion is further influenced by eating behavior, in addition to the general health of the consumer along with their reproductive and nutritional status.

14.2.2 In Vitro Methods

To overcome the limitations of the in vivo approach, numerous laboratory-based methods for routine determination of (enzymatic) resistant starch have been devised (see Champ et al., 2003). Each has its merits and shortcomings which will not be discussed in detail here. For a thorough description and assessment of the most commonly used methods the reader is referred to detailed reviews by Champ et al. (2003), Champ (2004) and Nugent (2005). The majority of the laboratory techniques use a direct approach to quantify RS and essentially attempt to simulate starch digestion in humans. Some of the earlier methods used assay conditions which did not reflect conditions in the gut lumen. For instance, some employed incubation temperatures of 100°C. Harsh sample-processing methods, for instance grinding or milling, were also often employed to produce a uniform particle size for RS analysis. Although this may have the desired effect of reducing sampling error, it invariably results in an underestimation of RS owing to the extensive disruption of plant cellular structures and consequent overexposure of entrapped starch to α-amylases.

Most in vitro assays for RS that are in use today are variations on Berry's (1986) modification of Englyst's original method (Englyst et al., 1982) and ultimately date back to Williams and Olmsted (1935) who were amongst the first to use enzymatic techniques for food analysis. Contemporary laboratory methods for RS typically incorporate the following steps: disruption of the food matrix, usually by mechanical means, such as grinding or mincing, prior to mixing the macerated sample with aqueous buffer and sometimes artificial saliva. The resultant bolus is subjected to a series of controlled incubations at physiological temperature and at a pH that reliably simulates the gastric (acid pH in conjunction with pepsin) and pancreatic (pancreatin or α-amylase; amyloglucosidase) phases of digestion. Protein, fat and starch are hydrolyzed using a combination of enzyme preparations of bacterial, fungal and/or mammalian origin. Residual starch is recovered by either centrifugation or washing with ethanol, dispersed in DMSO or KOH and hydrolyzed to completion using thermostable α-amylase and amyloglucosidase. The resultant glucose is quantified mostly by spectrophotometry using a glucose oxidase/peroxidase-chromogen buffered system, and the resultant yield of resistant starch then calculated. RS content is most frequently expressed as g per 100 g food on either an 'as eaten' or dry matter basis. It is also common for it be normalized to total starch content. Englyst and colleagues (1992) developed an indirect method of analysis in which RS is determined as the difference between total (TS) and digestible starch (DS), the latter fraction comprising rapidly and slowly digestible starches (RDS and SDS, respectively). Others have adopted a similar approach for quantifying RS (Tover et al., 1990).

Standardized assays for total dietary fiber, such as the enzymatic–gravimetric method of Prosky et al. (1985) (AOAC 985.29), are not quantitative for RS although they, or slight modifications thereof (Gelroth and Ranhotra, 2000), are often used for this purpose (see

McCleary, 2007). RS is underestimated to varying extents because the assay does not fully account for all of the RS that may be present in the food sample. Generally, these methods use aggressive sample preparation procedures, such as milling, so they detect only the more resilient forms of RS, notably a variable proportion of retrograded starches (Delcour and Eerlingen, 1996). As a consequence of underestimating RS, dietary fiber (as defined by numerous authorities, e.g. AACC, 2000; FSANZ, 2000) is also underestimated. Integrated analytical methods for total dietary fiber are being developed which quantify various fiber components (McCleary, 2007), but whether such methods accurately measure RS has yet to be confirmed.

As RS assays have evolved there has been increasingly greater emphasis on procedures and assay conditions which are more physiological, consistent with the concept of RS, and deliver results in tune with in vivo data. The methods of Englyst et al. (1992), McCleary and Monaghan (2002), Champ et al. (2001b) and Muir and O'Dea (1992) are germane in this sense. The defining analytical merits and comparative performance of these methods have been examined in detail (Champ et al., 2003). Importantly, there has been an attempt by the investigators to relate assay performance to reference values obtained from human studies and this sets these particular methods apart. Indeed, for the majority of RS assays, accuracy has not been determined and as such their scientific value is compromised.

Preparation of the food sample such that only the 'digestible' starch fraction is removed and 'resistant starch' retained is critical for assay accuracy. Muir and O'Dea (1992) improved upon traditional sample preparation procedures, which were especially harsh, by using volunteers to chew and expectorate test foods so as to better simulate the oral phase of digestion. Granfeldt et al. (1992) had earlier adopted this approach in their in vitro assay for estimating the glycemic properties of food carbohydrates. Using volunteers to provide masticated food samples for laboratory analyses has obvious shortcomings. Aside from its slowness and inherent variability, this approach is not feasible or practical for many laboratories and there are the attendant difficulties when experimental products are involved. Disruption of the food matrix has been accomplished with various mechanical mincers (e.g. Englyst et al., 1992). This approach is considered to mimic chewing and produces food particles of a size and consistency similar to that in masticated samples (Hoebler et al., 2000). However, sample heterogeneity can be a problem for certain types of starchy foods, especially some legumes (Bird, Topping and Usher, unpublished observations).

In contrast to other methods, that of McCleary and Monaghan (2002) has undergone extensive inter-laboratory testing, acquired international accreditation (AOAC Method 2002.02 and AACC Method 32-40) and can be performed using a commercially available assay kit (Cat. No. K-RSTAR, Megazyme International, Ireland). As a method for RS determination it is widely accepted and has proven reliable, robust and precise but, although suitable for routine determinations, it is time consuming, as is the case with other in vitro assays for predicting RS. Englyst and colleagues (1992) introduced (shorter) incubation periods, which is not only more in accordance with small intestinal transit times in humans (Read et al., 1986), but also offers the obvious advantage of higher sample throughput. However, the Englyst method (1992) is widely regarded as difficult to perform.

In comparison with in vivo methods, in vitro assays are rapid and, from a technical perspective, most are easy to perform. They achieve a higher level of precision than bioassays, most yield reproducible results, and generally assay precision is positively related to the mean (McCleary et al., 2002; Bird, Usher and Topping, unpublished results). It is important that acceptable analytical precision is achieved at low RS concentrations given that the majority of everyday starchy foods contain RS concentrations below 1% (Bird, Topping and Usher, unpublished results; Murphy et al., 2008). The exceptions are pulses and legumes (Muir and O'Dea, 1992; Englyst et al., 1992), and processed foods containing RS concentrates, such as high-amylose maize starch.

In vivo values, preferably obtained using the human ileostomy model, provide the only yardstick by which laboratory methods for RS can be assessed. Ideally, simultaneous in vitro and in vivo testing on identical food samples measured 'as consumed' and encompassing the broadest possible spectrum of food products must form the basis of assay validation studies given the multitude of potential factors that may impact RS content (e.g. food storage conditions, duration and preparation). Unfortunately, the accuracy of in vitro methods published to date has not been adequately verified. Although the more advanced in vitro methods produce data that correspond well with in vivo values, the validation programs for all four of the aforementioned methods are limited. Each involved narrow selections of foods, making it difficult to gauge overall accuracy and hence suitability of these methods for mainstream use. Indeed, most of the validation studies reported in the literature included obscure types of foods or foods of an experimental nature.

It would appear that in vitro procedures are inconsistent with the physiological definition of resistant starch. Specifically they neglect starch derivatives, i.e. smaller starch-derived oligosaccharides, disaccharides and glucose that, in the in vivo situation, escape digestion and absorption in the upper gut of healthy individuals (Birkett et al., 1997; Champ et al., 1999; Danjo et al., 2003). Free glucose and malto-oligosaccharides represent between about 10–20% of starch in ileal digesta (Silvester et al., 1995; Champ et al., 1998).

Although comparative studies are scarce, it is nevertheless clear that there are large discrepancies in the results obtained with different in vitro methods (Champ, 1992; Champ et al., 1998; Madrid and Arcot, 2000; Dysseler and Hoffem, 1994a; Marlett and Longacre, 1996; Walter et al., 2005). Indeed, the values generated are essentially method dependent. Although the emphasis has been on developing quantitative in vitro methods that are as close to physiological as possible, the litmus test is simply whether they yield values which agree well with those produced in vivo for all foods. This criterion is absolute and paramount. Qualitative accuracy of the analytical methodology is also becoming an increasingly important consideration given that the nature of RS delivered to the colon, as well as the amount, determine the metabolic response of the microflora and potential health benefits for the host. Although the data are very limited, there is possibly a disparity between the molecular properties of RS produced in vitro (enzymatic) and in vivo physiological resistant starch (Faisant et al., 1993). Recently it was shown that human recombinant maltase-glucoamylase differs markedly in its hydrolytic properties compared to fungal amyloglucosidase (Champ et al., 2003; Ao et al., 2007). The latter is typically used in in vitro assays for RS determination.

14.3 HEALTH BENEFITS OF RS

The nutritional importance of the discovery several decades ago of a starch fraction that proved resistant to digestion was not immediately evident, which is perhaps understandable given the prevailing view that dietary starch was digested to completion in the human small bowel. Early in vivo studies in humans showed that as much as 20% of the starch in conventional meals comprising legumes, potato and cereal foods escaped digestion and absorption in the small intestine (Anderson et al., 1981; Stephen et al., 1983). In vitro fermentation, dietary intervention and mechanistic studies in subsequent years have significantly expanded our knowledge of the physiological properties of RS and sparked interest in the potential of RS for promoting health and well-being. Recent reviews by Baghurst et al. (1996), Topping and Bird (1999), Bird et al. (2000), Bird and Topping (2001), Champ (2004), Kendall et al. (2004), Nugent et al. (2005), Sajilata et al. (2006) and Sharma et al. (2008) cover this area in detail. There is now a substantial body of evidence demonstrating that reducing the rate and extent of starch digestion in the upper GI tract is conducive to good health (Bird and Topping, 2001; Topping et al., 2006; Rahman et al., 2007; Regina et al., 2007). However, for RS, comparison of different studies is especially difficult owing to the fact that for most dietary interventions, and epidemiological investigations, intakes are not known with any certainty. Furthermore, preparation and storage of study foods and other factors that impact on their RS content are often not reported.

Also, from a physiological and human health perspective, all RS should not be considered identical. Although the vast majority of research on RS has involved high-amylose maize starch (HAMS), principally in the form of intact granules (known as RS2) and to a lesser extent as a retrograded HAMS (known as RS3), it is clear that different RS sources vary in their physiological properties. This is understandable given that RS comprise distinct molecular, structural and physical variants, so when those entities reach the colon their interaction with the microflora and, consequently, their metabolic fate will likely also differ. In vitro fermentation studies using pooled fecal inocula from healthy humans have shown that fermentation patterns differ among RS sources (Topping and Clifton, 2001). Duration of dietary interventions, in addition to the type and amount of starch ingested, most likely explain much of the variation observed.

Although most of the attention that RS has attracted from health professionals relates to its positive effects on bowel health, there is theoretical and mounting experimental evidence that RS has beneficial physiological properties which extend beyond the gastrointestinal tract. Indeed, RS appears to have some merit as a dietary adjunct in the prevention and management of diabetes and obesity, which are major health concerns in the developing as well as the developed world. Evidence from in vitro and animal studies that RS has favorable effects on mineral, lipid and glucose metabolism is accumulating. However corresponding evidence from studies in humans is either non-existent or contradictory, or where there are improvements in energy balance or indices of cardiovascular and metabolic health, the magnitude of the response is often not large. The physiological properties and major health effects of RS are summarized below. For more detailed information see the reviews referred to above.

14.3.1 Energy Intake, Weight Management and Obesity

Shifting the site of starch assimilation from the small intestine to the colon is potentially a strategy for weight management and the fight against obesity because RS yields about half as much energy as digestible starch (8 versus 15 kJ g^{-1}, respectively; Livesey, 1995; Elia and Cummings, 2007). Indeed, some animal studies suggest that the energy value of RS is close to negligible (Ranhotra et al., 1996). Starch efflux from the small bowel was increased about sevenfold when high-amylose maize starch (HAMS) replaced conventional cornstarch in cooked food products (Clarke et al., 2007). Accordingly, augmenting the RS content of foods is seen as a potential avenue for reducing their energy density. This would be expected to translate into a substantial reduction in digestible energy intake, especially for individuals on high-starch diets. However, Behall and Howe (1996) found that the metabolizable energy values of HAMS and a conventional cornstarch when consumed by humans were remarkably similar. Well-designed, long-term studies on the effects of RS consumption on body weight in humans are lacking. A practical difficulty is incorporating into foods or diets sufficient RS to have meaningful effect on energy intake of consumers. Most conventional starchy foods, including those that are minimally processed, contain low levels of RS. Although commercially available sources, such as HAMS, contain appreciable levels of RS, their level of inclusion in manufactured food products is often not particularly high.

In addition to diluting dietary energy, RS consumption has been reported to increase satiety, promote postprandial fat catabolism and reduce storage of lipid (Raben et al., 1994; Tapsell, 2004; Keenan et al., 2006). These actions may relate to the RS-induced increase in supply of SCFA from the gut (and possibly reduced bioavailability of potentially toxic bacterial metabolites) and consequent modulation of metabolism of extra-intestinal tissues. It has been shown in rats that RS stimulates secretion and regulates gene transcription of various gut hormones, including GLP-1 and PYY, which are involved in nutrient assimilation and body energy balance (Keenan et al., 2006; Zhou et al., 2006). However, evidence from feeding studies of an appropriate duration in humans on the long-term effects of RS on energy intake, whole-body energy expenditure or lipid oxidation is either inconclusive or lacking. The situation on whether RS actually promotes satiety also lacks clarity. Many of the investigations on RS and weight loss, satiety and related variables are confounded by dietary treatments which are not adequately balanced for macronutrient content and/or fiber intake or their duration is simply not long enough.

14.3.2 Glycemic Control and Diabetes

The benefits which stem from slowing the rate of carbohydrate digestion, including attenuation of risk of type II diabetes, are well known and slowly digested carbohydrates are commonly used to control postprandial hyperglycemia. RS per se does not contribute to postprandial hyperglycemia. It is important to remember that a proportion of the starch in commercially available RS sources, such as HAMS, is digestible and it is this fraction which has a direct effect on postconsumption blood glucose levels.

Reducing the glycemic load of the diet is important for metabolic health (Howlett and Ashwell, 2008) and RS has been shown to dampen postprandial blood glucose and insulin concentrations and increase peripheral insulin sensitivity (Raben et al., 1994; Behall and Hallfrisch, 2002; Robertson et al., 2005; Yamada et al., 2005; Behall et al., 2006). For RS to produce a meaningful change in glycemia or circulating insulin, it purportedly must account for at least 14% of total starch intake (see Nugent, 2005). In addition to having an acute effect, RS is capable of inducing the 'second meal effect' and so via this mechanism it may facilitate improved glycemia in the long term (Robertson et al., 2003). Although the effects of RS on metabolic health endpoints appear promising, the evidence is based on relatively few properly designed dietary interventions and more information is needed on responses to different types of RS and in cohorts at increased risk of developing type II diabetes.

Glycemic index (GI; a comparative measure of the glycemic properties of only those carbohydrates assimilated in the upper gut) does not directly apply because RS is not digested in the small bowel. Accordingly any RS-induced improvement in postprandial glycemia is presumably a consequence of a lower glucose yield (glycemic load) in the portal vein per serving of test food. It is especially difficult in practice to differentiate GI and glycemic load effects owing to the lack of accurate and convenient analytical techniques for RS as discussed earlier.

14.3.3 Blood Lipids and Heart Health

In addition to its effects on glycemia, RS has been shown to have a range of other systemic benefits, such as improving blood lipid profile.

RS provided as retrograded potato starch (Hashimoto et al., 2006) or as retrograded or granular high-amylose maize starches lowers plasma cholesterol and/or triglycerides in rats perhaps through modulation of bile acid and neutral sterol metabolism (Verbeek et al., 1995) or hepatic cholesterol uptake (Udagawa et al., 2008). But rats appear to respond differently in this respect from other animal species of interest, such as pigs and humans (Bird and Topping, 2001). Indeed, evidence for RS having an effect on blood cholesterol and triglyceride levels in humans is inconsistent (Champ et al., 2003; Nugent et al., 2005).

14.3.4 Bowel Health

Most interest in RS stems from its potential to improve large bowel health and reduce risk of serious bowel diseases such as colorectal cancer and, more recently, as a possible preventative and therapeutic for diarrheal diseases (Topping and Bird, 2001; Binder and Ramakrishna, 2004; Subramanya et al., 2006). Epidemiological information suggests that the high incidence rates of chronic bowel diseases may be linked to low intakes of starch, and by inference, RS (Cassidy et al., 1994; Topping and Bird, 1999; Bird et al., 2000a; Bird and Topping, 2001; Young and LeLeu, 2004). However, it must be acknowledged that estimating dietary intakes of RS at the population level reliably is next to impossible. Food composition tables and other databases for RS are limited as well as inaccurate, as is the method of using a fixed proportion of total starch consumption (usually 5%; Baghurst et al., 1996) as a measure of RS intake.

Modern food-processing methods, and the concomitant reduction in the amount of starchy foods eaten with urbanization and increasing disposable income, has effectively eliminated RS from contemporary Western diets (Bird and Topping, 2008). The dearth in starch reaching the colon, as distinct from the comparatively higher ileal yield of non-starch polysaccharides (the major component of dietary fiber), may be a possible contributing factor in the high prevalence of lifestyle-related diseases in industrialized societies (Segal, 2002; O'Keefe, 2008). Dietary intervention studies in humans and experimental feeding trials using small and large animal models demonstrate that RS has positive effects on a range of indices of bowel health. These are largely a consequence of the changes in the intracolonic environment brought about by the increase in proliferation and metabolic activity of the colonic microflora consistent with the provision of extra substrate (RS). The concomitant increase in the number of bacterial cells (biomass) may improve bowel habit through the production of softer and bulkier stools (Phillips et al., 1995; Topping and Bird, 2001). RS may also facilitate laxation by sparing NSP from degradation by colonic bacteria (Phillips et al., 1995). These mechanical effects are expected to lessen the likelihood of simple constipation, diverticular disease and related problems. Stool mass is inversely related to risk of colorectal cancer (Cummings et al., 1992), however fecal bulking actions of RS are modest compared to other insoluble NSP (Cummings et al., 1996; Birkett et al., 1997; Cummings, 1997).

The bowel health benefits of RS are believed to be mediated primarily through the products of its fermentation by gut bacteria. RS functions as a prebiotic in that ingestion may trigger a disproportionate increase in colonic bacterial populations considered important for host health, notably bifidobacteria and lactobacilli (Bird and Topping, 2008). Consumption of RS also stimulates colonic fermentation as evidenced by a lowering of luminal pH, and an increase in fecal concentrations and excretion of short-chain fatty acids (SCFA; principally acetate, propionate and butyrate), the major end products of carbohydrate fermentation in the large bowel (Topping and Clifton, 2001). One of the main reasons RS has sparked interest is because its fermentation favors the production of butyrate (Weaver et al., 1992; Brouns et al., 2002). Although all SCFA play an important role in the maintenance of normal colonic physiology and mucosal health, butyrate is particularly potent in the prevention and progression of carcinogenesis. A direct causal link between butyrate and colon cancer in humans has not yet been established, and indeed butyrate appears actually to have contradictory actions ('butyrate paradox'), nevertheless there is a substantial body of experimental evidence linking increased bioavailability of butyrate with prevention and progression of tumorigenesis in the large bowel (Topping and Clifton, 2001; Brouns et al., 2002). Butyrate regulates expression of a range of genes involved in cell growth, differentiation and apoptosis (Sengupta et al., 2006) and of the major SCFA, it is certainly the most effective at suppressing the growth of colon cancer cell lines (Topping and Clifton, 2001) and promoting the removal of genetically damaged cells through apoptosis (Hague et al., 1995).

There is good evidence from animal studies that RS suppresses development of colon cancer, including both the incidence and multiplicity of adenocarcinomas, and that butyrate plays a pivotal protective role in these effects (LeLeu et al., 2007). Although such direct evidence is lacking in humans, numerous dietary interventions have shown that increased consumption of RS from various sources is associated with improvement in a range of markers of bowel health, including, among others, acidification of feces, increased concentrations and excretion of SCFA, including butyrate, lowered production of protein fermentation products such as ammonia and phenols, and reduced fecal levels of secondary bile acids (van Munster et al., 1994; Birkett et al., 1996; Noakes et al., 1996; Muir et al., 2004; Young and LeLeu, 2004; Bird et al., 2008).

RS consumption appears to result in an intracolonic environment in which the production of metabolites that are beneficial for mucosal health is promoted and agents which are cytotoxic, mutagenic, or may compromise the integrity of the colonic mucus barrier are reduced. In a series of studies, Toden and colleagues (2005, 2006, 2007a, 2007b) have shown that damage to DNA in colonocytes was elevated, and colonic mucus barrier thinned in rats on high-protein diets whereas inclusion in the diet of high-amylose maize starch (a rich source of RS), mitigated these potentially harmful effects. Furthermore, attenuation of DNA damage was dependent on the level of RS intake (Toden et al., 2007c). It is worth noting that the diets which induced high rates of DNA damage, although low in RS, contained reasonable levels of NSP (as wheat bran). The protective mechanism is as yet unknown but butyrate appears to be closely involved given that low rates of genetic damage were associated with high butyrate concentrations in the large bowel (Toden et al., 2007c). In addition to playing an important role in maintaining chromosomal integrity, RS may actually prevent DNA damage from occurring in the first place. RS suppresses production and/or bioavailability of potentially toxic derivatives from fermentation of luminal protein, peptides and amino acids (Birkett et al., 1996; Bird and Topping, 2001). Instead, these substrates are directed to support the RS-induced increase in bacterial proliferation. RS promotes bacterial de novo protein synthesis by the large bowel microflora, thereby reducing exposure of the colonic epithelium to ammonia as well as the absorption of this potentially toxic metabolite into the bloodstream (Mosenthin et al., 1992). The net portal-drained flux of ammonia (as well as butyrate and other SCFA) is increased markedly in pigs fed RS as raw potato starch (Van der Meulen et al., 1997). By providing an alternative substrate for sulfomucin-degrading bacteria, RS may also act by ensuring that the functional integrity of the surface mucus barrier, and exposure of the epithelium to potentially harmful luminal agents, is not compromised (Ridlon and Hylemon, 2006).

There is limited evidence, mostly from animal studies, that RS has positive effects on other variables of relevance to nutrition and health, including promotion of mineral absorption from the large bowel, modulation of immune function and prevention of inflammatory bowel disease (Topping and Clifton, 2001; Nugent, 2005). In rats and pigs increased consumption of RS facilitates gastrointestinal assimilation of a range of minerals including, among others, calcium, magnesium, iron and zinc (Bird et al., 2000; Champ, 2004; Nugent, 2005). Acidification of the luminal contents of the large bowel in response to RS consumption is thought to increase the bioavailability of mineral salts which would otherwise be lost in the fecal stream. The limited information available from human studies on RS and mineral balance relates mainly to calcium and HAMS as the RS source, and suggests that in general RS produces a small improvement in mineral balance (Coudray et al., 1997, 2003; Behall et al., 2002).

There have been no reports of allergic reactions to RS ingestion (Goldring, 2004) and the few adverse reports relate to gastrointestinal upsets. As is the case with other fermentable fiber sources, high intakes of RS can cause abdominal bloating, distension and flatulence. Such negative responses have been reported in some (Muir et al., 1994; Phillips et al., 1995) but not all studies in which subjects consumed considerable quantities of foods containing RS (Yamada et al., 2005). Indeed, very high intakes of HAMS (up to 60 g day^{-1}) were not associated with adverse effects (Goldring, 2004). Inconsistencies between studies may relate to differences in the type and amount of RS eaten as well as the length of the dietary adaptation period, and also differences between individuals in their ability to digest starch. In pigs the site of fermentation in the large bowel has been shown to be dependent on the source of RS consumed (Bird et al., 2007). These and other findings dispel the notion that RS is readily and extensively fermented in the proximal large bowel.

14.3.5 RS Intakes

Gaining a better understanding of the link between dietary consumption of RS (let alone intake of particular types of RS) and its physiological actions and health outcomes is made all the more difficult because, as discussed, an accurate and practical quantitative method for RS is not presently available. Consequently, information on the quantity of RS that humans consume is limited and unreliable. RS consumption estimates for adult populations in various European countries (Tomlin and Reid, 1990; Dyesseler and Hoffem, 1994b; Brighenti et al., 1998), the United States (Murphy et al., 2008), Australia (Baghurst et al., 1996; Roberts et al., 2004), New Zealand (Baghurst et al., 1996) and Sweden (Liljeberg Elmståhl, 2002) are in the order of 3–9 g day^{-1}. Estimates based exclusively on reliable in vivo methodologies suggest that most Western diets are at the lower end of this range and <5 g RS day^{-1} is probably closer to reality (Birkett et al., 1997; Bird and Topping, unpublished results). In contrast, many rural-based populations consuming minimally processed traditional foods ingest about 2–10 times that amount of RS (Platen and Shurpelaker, 1994; Muir et al., 1998; Ahmed et al., 2000; Baghurst et al., 2001; Segal, 2002).

14.4 EFFECT OF PROCESSING ON RESISTANT STARCH FORMATION IN FOODS

Considering the significance of RS for human health, food scientists have attempted to enhance the level of enzyme-resistant retrograded starch in our diet. Although many plant foods are very rich in enzyme-resistant starch, most of them are physically inaccessible (known as RSI, such as in legumes, pulses and some cereals) or indigestible (RSII, such as unripe banana, raw potatoes) (Englyst et al., 1992). Processing of these raw starchy foods in most cases, however, destroys

RSI and RSII, leading to the formation of enzyme-resistant retrograded starch or RSIII. It is now well understood that RSIII is formed during processing due to gelatinization and subsequent retrogradation of starch (Berry, 1986; Sievert and Pomeranz, 1989). Knowledge of the food processing methods and conditions as well as the underlying mechanism that leads to the formation of RS in foods is of great importance for nutritionists as well as food industries, since it offers the possibility of increasing the RS content in processed foods.

14.4.1 Heat Treatment

Raw granular starch exists in a semi-crystalline state which is disrupted when heated in excess water. This characteristic temperature, the *gelatinization temperature*, for most starches is about 60°C and above. Gelatinization is an irreversible endothermic phenomenon during which water penetrates the starch granule, resulting in swelling (as much as 100 times) and solubilization of starch, particularly the amylose component, leading to loss of granule organization and crystallinity (birefringence and X-ray diffraction pattern), loss of double helical order, and increase in viscosity (Ring et al., 1988; Cooke and Gidley, 1992; Jenkins and Donald, 1998; Hancock and Tarbet, 2000). When cooled, the dispersed starch molecules undergo slow re-association forming a tightly packed structure, the process commonly being called retrogradation (Miles et al., 1985). Retrograded starch is thermally stable, and can only be re-solubilized at 80–150°C depending on the contribution of amylopectin (low dissolution temperature) and amylose (high dissolution temperature) (Haralampu, 2000). Concentrated amylose gels can be formed during retrogradation, and slowly crystallize to the double helical B-form with a melting temperature as high as 130–170°C (Miles et al., 1985; Szczodrak and Pomeranz, 1991).

Starches that differ in botanical origin respond differently to heating and cooling cycles and the water level of the cooked starch. Starch type markedly affects the swelling of granules, the loss of double helical order within starch molecules, and the solubilization of amylose and amylopectin chains. The manner in which re-association of these starch polymers occurs during cooling/storage conditions to form a gel network largely determines the subsequent resistance of the starch to enzyme digestion. Table 14.1 documents reported effects of various processing methods on RS formation as well as changes in physicochemical properties of retrograded starch from various plant sources.

14.4.2 Extrusion Cooking

Extrusion cooking is a common processing method for foods such as breakfast cereals, noodles and pasta products. Some of the advantages that have been attributed to this technique include low cost, high productivity, versatility and unique product shapes. Depending on the intended final product, various parameters such as temperature, screw speed, moisture, torque, shear rate can be set (Faraj et al., 2004). Although a significant number of studies have been carried out on the effect of extrusion on formation of RS, the results are mixed. Some studies showed that extrusion decreases RS content in, e.g., barley flour (Ostergard et al., 1989; Vasanthan et al., 2002; Faraj et al., 2004), rice and amaranth starch (Parchure and Kulkarni, 1997; Sagum and Arcot, 2000), potato products (Escarpa et al., 1997; Garcial-Alonso et al., 1999; Garcia-Alonso and Goni, 2000), wheat flour (Siljerstrom et al., 1986; Eerlingen 1993b), high-amylose starches (Chanvrier et al., 2007; Htoon et al., 2009). On the other hand, several other investigators have reported a higher yield of RS, such as in wheat pastry flour (Kim et al., 2006), banana and mango (Agustiniano et al., 2005; Bello-Perez, 2005; Gonzalez-Soto et al., 2006); waxy and

TABLE 14.1 Resistant Starch Formation and Structural Changes in Starches Processed by Various Techniques

Starch sources	Processing conditions	Structural change	Resistant starch formation
	Extrusion cooking		
Rice and Amaranth (extracted and purified)[1]	Single screw extruder (moisture 14%, 15 bar and 60 rpm screw speed, screw compression 4:1)		RS yield reduced from 6.2 to 3.8% in rice and 6 to 3.7% in Amaranth.
Banana[2]	Twin screw extruder (moisture 36%, heating up to 120°C, screw speed 300 rpm), extrudate immediately quenched in liquid N_2, stored from 0 to 72 h, ground and stored	FTIR and XRD showed starch extrudate progressively changed from amorphous to B-type and finally to A-type crystalline form. Melting enthalpy increase with storage time, plateau reached at 8 h	RS yield of extrudate progressively increased with time, e.g., 1.2% at 0 h to 5.3% at 24 h.
Mango (extracted and spray dried)[3]	Single screw extruder (moisture 15–40%, temperature 90–138°C, screw speed 58–90 rpm, barrel 19 mm and length/diameter ratio 24), extrudate equilibrated for 24 h, dried at 50°C and ground	Water absorption index (WAI) increased as moisture content and temperature decreased	Extrusion increased RS yield from 1.1% in native starch up to 9.7%, depending on screw speed, temperature and feed moisture.
Mango, banana (extracted and spray dried) and corn starch[4]	Single screw extruder (moisture 35%, heating up to 130°C, low shear, screw speed 30–65 rpm) dried at 40°C and ground	Solubility and swelling decrease with increase screw speed. No change in gelatinization temperature but big decrease in melting enthalpy for extruded products	Increasing screw speed decreased RS in corn and mango but had no effect on banana starch. Extrusion decreased RS in native banana (19 to 3.9%), did not change RS in corn starch, and increased RS in mango starch (2.7 to 4.1%).
Waxy and regular barley flours[5]	Twin screw extruder (moisture 20–40%, temperature 90–160°C, screw speed 60–100 rpm, barrel 35 cm and diameter 31.8/20 mm, compression ratio 3:1), cooled, dried overnight at 40°C and milled		RS1 and RS2 not totally destroyed on extrusion. RS content remain same or decreased after extrusion. Extrusion at 100°C favors RS formation. Refrigeration increased RS content of extrudates.

(*Continued*)

TABLE 14.1 Resistant Starch Formation and Structural Changes in Starches Processed by Various Techniques—*Cont'd*

Starch sources	Processing conditions	Structural change	Resistant starch formation
Pastry wheat flour[6]	Twin screw extruder (moisture 20–60%, temperature up to 120°C, screw speed 150–250 rpm), stored at 0, 7, and 14 at 4°C, dried at 50°C for 16 h, ground and stored at –20°C	Decreased thermal stability with increased feed moisture and screw speed. Set-back viscosity increased with moisture	Extrusion increased RS yield. RS content increased with feed moisture e.g. 0.5% at 20% moisture to 2.5% RS at 60% moisture; and storage e.g., 0.5% RS at 0 day to 1.2% RS after 14 days. No effect of screw speed noticed.
Waxy and regular barley[7]	Twin screw extruder (feed moisture 20–50%, temperature 90–140°C, screw speed 50 rpm, L/D ratio 20:1) extrudate dried at 75°C and ground	Extrusion increased soluble as well as total dietary fiber. Insoluble fiber increased in regular barley but almost no change in waxy barley	No RS formed in waxy barley extrudate. In regular barley, RS increased with temperature till 120°C (from 0.8 to 2.9%) then decreased. RS increased when moisture increased from 20 to 50%.
Waxy, normal and high amylose corn starch (HAMS) and wheat starch[8]	Capillary rheometer (moisure 28%, temperature 140°C, Shear rate 722 s^{-1}), stored at 4, 30 and 60°C for 72 h, dried at mild condition and ground	B-type in HAMS and A-type crystallinity in lower amylose starches. V-type crystallinity appears after high temperature storage. Crystallinity decreased with extrusion but increased with storage temperature. Retrograded amylose endothermic peak was about 145°C.	Extrusion increased RS content in normal maize (1.5 to 2.1%), wheat starch (0.8 to 2.8) but decrease in waxy (0.7 to 0.1%) and HAMS (69 to 20%). Higher temperature storage yielded more RS (~62% increase) in normal maize.
Heat treatment (moist heat/baking/frying)			
Rice with varying amylose (10 to 32%) contents[9]	Rice to water ratio 1:2 (w/v). Boiled for 20 min and pressure cooking in domestic pressure cooker for 10 min		Higher RS in high amylose rice. Similar decrease in RS content during pressure cooking and boiling (12.9 to 2% in high; 9.2 to 1.6% in medium; and 8.6 to 0.8% in low amylose rice).
Pea, potato, wheat and maize starches (waxy, regular and amylomaize V and VII)[10]	Starch-water at 1:2 1:3.5 and 1:10 ratios autoclaved at 121, 134 and 148°C for 1 h, cool, store overnight at 4°C (autoclave-cooling cycle repeated 20 times), freeze-dried and ground. RS isolated by treating with heat-stable amylase and amyloglucosidase followed by filtration and vacuum drying	Endothermic transition peak for amylomaize VII ~150°C for its residues ~153°C. Increase in RS yield increases melting enthalpy in amylomaize VII (2.7 to 8.8 J g^{-1}) and its isolate residues (8.2 to 19.7 J g^{-1})	RS yield proportional to amylose content. In amylomaize VII, slight decrease in RS observed with increase with starch: water ratio, whereas RS increased in repeatedly autoclave-cool samples (> 40% RS after 20 cycles). Very high temperature decreased RS yield.

Barley (43% amylose). Starch fractioned based on granule size[11]	Starch-water at 1:5 ratios autoclaved at 121°C for 1 h, cool, store overnight at 4°C, vacuum-dried and ground. RS isolated by treating with amylase, amyloglucosidase and protease followed by centrifugation and vacuum drying of residues	Isolated RS showed transition at 116–177°C, mean at 158°C. Increasing autoclave–cooling cycle increased temp and enthalpy in starch fractions e.g., 138 to 159°C and 9.6 to 38 J g^{-1}, transition temperature and enthalpy, respectively from 0 to 20 cycles	Increasing autoclaving-cooling cycle 1 to 20 cycle increased RS yield from 6 to 26%.
Wheat starch[12]	Wheat starch suspension (10%) autoclaved for 1 h, cooled to 100°C and incubated at 0, 68 and 100°C. Starch sample withdrawn at different times, treated with enzyme and RS residues removed (treat with heat-stable amylase, amyloglucosidase and protease followed by filtration and drying)	Crystallinity develops slowly at 68 and 100°C compared to 0°C. Incubation at 100°C resulted in A-type crystals while B-type crystals at 0 and 68°C	Initial 150 min incubation yield 4, 2.5 and 2% RS at temperature of 0, 68 and 100°C, respectively. RS yield increased to ~8% for sample incubated for 190 min at 100°C, but no RS change occurs at low temperatures.
Waxy rice[13]	Starch partially gelatinized by heating at 60, 65 and 70°C for 5 min and retrograded starch by cooling fully gelatinized (20 min boiling) paste at 4°C for 2, 4 and 7 days	Relative melting enthalpy decreased (69 to 9 J g^{-1}) with increase in gelatinization temperature and increased (37 to 80 J g^{-1}) with storage time of retrograded starch	RS content of native and gelatinized waxy rice starch 9.3 and 2.9%, respectively. RS yield decreased (8.6 to 7.7%) with increase in gelatinization temperature and increased with increasing storage time of retrograded starch (3 to 3.8).
Native corn, rice, wheat and potato starches and their flours[14]	Three different processing conditions used: 1. Starch suspension (5 g in 40 mL) autoclaved at 120°C for 20 min at 2 bar pressure with continuous stirring 2. Boiled in a water bath with stirring for 45 min 3. Gelatinized starch retrograded by cooling to room temperature, freezing for 16 h, drying in oven and milled	Stirring during cooking yielded more homogenous products than unstirred samples	Starches contained higher RS than flours except rice. High pressure autoclaving, boiling, or stirring did not affect the RS values of starches. pH (3.5–10.5) of the water/starch suspension during gelatinization did not affect the formation of resistant starch. Storing as well as drying conditions affected the RS content of the retrograded starch.

(*Continued*)

TABLE 14.1 Resistant Starch Formation and Structural Changes in Starches Processed by Various Techniques—*Cont'd*

Starch sources	Processing conditions	Structural change	Resistant starch formation
Common bean (*Phaseolus vulgaris*)[15]	Bean cooked for 2 h and also autoclaved for 20 min, cooled and stored from 24 to 96 h at 4°C. The processed beans analyzed for resistant starch, retrograded resistant starch and in vitro amylolysis		Retrograded resistant starch content was higher in beans cooked by the traditional process (2.7–2.8%) compared to autoclaved beans (1.6–1.9%). But RS content of beans from these two cooking methods were not significantly different. The initial in vitro α-amylolysis rate in freshly cooked beans was higher in the autoclaved preparation than in the beans cooked by the traditional process, but final hydrolysis indices (96 min) were similar for both samples.
Rice, corn, polenta, spaghetti, bread, potato, peas, beans, lentils and chickpeas[16]	Soaked corn, beans, lentils, chickpea and dry polenta, spaghetti, and potatoes were cooked from 28–70 min, dried at 60°C for 18 h, and ground. Rest of the samples were cooked and stored at −20°C for 30 days		Storage of products cooked and stored −20°C for 30 days favored RS formation over freshly-cooked products. The % increase in RS content were: whole rice (108), polished rice (43), corn (14), polenta 15), spaghetti (61), bread (8), potato (117), peas (24), beans (47), lentils (35) and chickpeas (8).
Potato products: Potato flours, boiled, boiled and cooled, raw flakes, mashed, baked, fried, crisp and retrograded flour[17]	Boiled potato: boiled, cool and refrigerate for 24 h Instant mashed potato: per package information Baked potato: bake at 200°C for 50 min, skin removed Fried potato: fried 180°C for 8 min, oil removed by adsorbent paper Potato crisp: product purchased from market Retrograded product: 5 g flour in 40 mL water, autoclaved at 120°C for 30 min at 200 Pa with stirring, cooled, frozen 16 h at −20°C, thawed 8 h at room temperature, drying 60°C/16 h and grinding		RS content of potato products differ with processing conditions, such as, Raw – 69.1%; boiled – 1.2%; boiled and cooled – 4.6%; raw flake – 2.8%; instant mashed – 2.1%; oven bake – 3.7; French fry – 6.6%; potato crisp – 3.3%; retrograded – 10.4%.

Barleys: High-amylose (HAB, 44%), Waxy (WB, 3%), Ordinary (OB, 23%), and Ordinary Glacier (OGB, 33%)[18]	Dough prepared with 70% whole meal barley and 30% wheat flour. Baked by conventional method (200°C/45 min) or low temp and long time (120°C/20 h) or 'pumpernickel condition'. Breads cut into slices, crumb removed and stored in the freezer	The water absorption index values decreased when the storage time increased, pattern that agrees with the higher RS content	High RS in high amylose containing breads (White bread RS 0.2%). Compared to conventional baking, 'pumpernickel condition' increased RS yield, e.g., in WB (0.6 to 2.5%); OB (2.4 to 6.2%); OGB (1.6 to 3.5%) and HAB (3.5 to 10.3%).
Heating in presence of adjuncts (including lipids)			
Potato starch with adjuncts e.g., lignin, cellulose, pectin, gum, phytic acid, albumen, sucrose, potassium, calcium and phytic acid[19]	Starch solution (5%) and adjuncts stirred, autoclaved and gelatinized at 123°C/20 min, cooled, frozen, vacuum dried and ground		Adjuncts mostly decreased RS content, depending on amount and type. No influence (lignin and cellulose), low influence (pectin, gum, phytic acid, albumen, sucrose), high influence (potassium, calcium, phytic acid).
Native and defatted high amylose corn starch, HACS (70% amylose, NHA and DHA)[20]	Autoclaved at 125°C for 10 min, cooled to 90°C, frozen at −20°C for 24 h, freeze thawed (FT) at 25°C for 4 h, ground, freeze dried (control sample not thawed) and cycle repeated up to 3 times	B-type crystal observed for NHA, DHA and their repeated FT samples. XRD peaks were less intense peak for both FT samples. Higher peak intensity and melting enthalpy for DHA FT samples as compared to NHA counterparts indicating more amylose–amylose interaction. Greater peak viscosity noticed for DHA. FT cycles increase overall viscosity	Autoclaving-cooling increased the RS content of native HACS from 15.8% to ~18% in NHA and 20.5% in DHA starches. Repeated autoclaving-cooling further increased RS content of starches.
Wheat starch and amylomaize VII[21]	Starch defatted with various solvents (methanol, petroleum ether, water saturated butanol; alone or in combination). Autoclaved for 1 h and stored overnight at 4°C (repeated). Also autoclaved with sodium dodecyl sulfate (SDS). RS residues isolated	De-fatting increased gelatinization temperature and also affected granule order. Native and defatted starch had A-type pattern. SDS-defatted amylose had V-type but SDS-RS had B-type crystal	De-fatting and autoclaving increased RS yield: from 5.9% in native to 7 to 9% in defatted starched. Adding SDS decreased RS significantly (3%) SDS decreased RS of autoclaved starch to almost 0% level).
Wheat starch (WS, defatted), amylomaize VII, glucose, sucrose, ribose, maltose, sorbitol[22]	Starch solution (20%) autoclaved for 1 h. Sugar solution at starch: water: sugar ratio (1:50:5 and 1:10:1.8) added and stored for 20–320 min. RS residue was isolated	Sugars did not affect the X-ray diffraction and DSC characteristics of isolated RS residues (except maltose)	High concentration of glucose or sucrose decreased RS in WS gel from 3.4 to 2.8%, which further decreased to 2.5% in presence of maltose or ribose (no effect at low sugar concentration). Sugars increase RS in amylomaize.

(Continued)

TABLE 14.1 Resistant Starch Formation and Structural Changes in Starches Processed by Various Techniques—*Cont'd*

Starch sources	Processing conditions	Structural change	Resistant starch formation
Amylomaize VII, lysophosphatidyl choline (LPC), sodium stereoyl lactylate (SSL) and hydroxylated lecithin (OHL)[23]	Starch suspension (10%) autoclaved 1 h, cooled to 100°C, lipid added at 10%, heated to 90°C, cool to 4°C overnight, freeze-dried, ground. RS analysis done using three separate amylolytic enzymes	DSC peaks at 95–110°C indicating amylose-lipid complex and ~155°C for RS observed. Amylose-lipid complex gave V-type patterns but enzyme extract gave B-type pattern	RS yield decrease in amylose-lipid complex (from 37% in autoclaved-cooled amylomaize VII to 6.3, 11.7 and 25.5% in LPC, SSL and OHL complexed starches, respectively, using pancreatic α-amylase).
Corn meal (wheat starch, potato starch, citric acid (CA), high amylose corn starch (HACS) as adjuncts)[24]	Twin screw extruder (moisture sufficient for extrusion, temperature maximum at 140°C, screw speed 200–300 rpm). Corn starch extruded with various adjuncts at various concentrations		RS+dietary fiber or DF (2.1%) not affected by addition of other starches. CA added at the level of 7.5% increase RS+DF to 5.2%. Adding further 30% HACS increased RS+DF up to 14.4%. RS content decreased with screw speed.
	Acid/enzyme hydrolysis		
Cassava starch[25]	Starch suspended in lactic acid solution (1–100 mmol L^{-1}), autoclaved at 121°C for 1 h and incubated at –20 to 100°C for 6 to 48 h	Melt transition of treated starch ranged between 158 to 175°C and enthalpies between 0.2–1.6 J g^{-1}. Melt transition indistinct for acid treated samples	Longer incubation time, high temp (except 100°C) and moderate LA (10mmol/L) yield more RSIII than others. Very long autoclaving times yield less RSIII.
Banana flour[26]	Starch debranched with pullulanase, autoclave for 30 min, cooled, stored at 24, 36 and 48 h at 4, 32 and 60°C and freeze dried	Increasing storage time decreased the water absorption index. The water solubility index (WSI) was affected by the storage temperature but not by the storage time. XRD peaks at 5 and 23° in native banana starch (C type crystal) diminished in autoclaved/stored starch. SEM shows more compact structure in starch at low temperature	Autoclave and storage increased RS yield. RS content was higher at lower storage temperature e.g. 4 and 32°C (~34% RS) than higher temperature e.g. 60°C (~24% RS). Native starch more hydrolysable than autoclaved (75 vs 35% at 90 min hydrolysis time).
Native and lintnerized banana starch[27]	Autoclaved 121°C for 1 h, cooled and store at 4°C for 24 h (repeat 3 times), freeze dry and ground	Less soluble lintnerized starches had lower pasting behavior than native-gelatinized starch. RS-rich powder more thermally stable. Thermal transition and enthalpy for lintnerized-autoclaved starch (146°C and 18.8 J g^{-1}) compared to lintnerized starch (76°C and 19 J g^{-1})	RS yield of native-gelatinized starch increase from 1.5% to 2.6% in lintnerized, 16% in autoclaved and 19.3% in lintnerized-autoclaved starches.

Amylose from potato[28]	Amylose (0.83% w/v) hydrolyzed by β amylase at different incubation periods, heat inactivated, cooled, residue separated and dried. Extracts solubilized, boiled and cooled at 4°C (RS formed). RS isolated by treating with hydrolyzing enzyme	Amylose average chain length (DPn) ranges from 40–610. RS residues showed DPn from 19 to 26 (independent of amylose chain length) and B-type crystallinity

[1]Parchure and Kulkarni (1997);
[2]Bello-Perez et al. (2005)
[3]Agustiniano-Osornio et al. (2005)
[4]Gonzalez-Soto et al. (2006a)
[5]Faraj et al. (2004)
[6]Kim et al. (2006)
[7]Vasanthan et al. (2002)
[8]Chanvrier et al. (2007)
[9]Sagum and Arcot (2000)
[10]Sievert and Pomeranz (1989)
[11]Szczodrak, J. and Pomeranz, Y. (1991)
[12]Eerlingen et al. (1993a)
[13]Chung et al. (2006)
[14]Garcia-Alonso et al. (1999)
[15]Landa-Habana et al. (2003)
[16]Rosin et al. (2002)
[17]Garcia-Alonso and Goni et al. (2000)
[18]Akerberg et al. (1998)
[19]Escarpa et al. (1997)
[20]Jeong and Lim (2003)
[21]Eerlingen et al. (1994a)
[22]Eerlingen et al. (1994b)
[23]Czuchajowska et al. (1991)
[24]Unlu and Faller (1998)
[25]Onyango et al. (2006)
[26]Gonzalez-Soto et al. (2006b)
[27]Aparicio-Saguilan et al. (2005)
[28]Eerlingen et al. (1993)

regular barley (Vasanthan et al., 2002; Faraj et al., 2004); corn meal (Ulnu and Faller, 1998); and barley (Vasanthan et al., 2002).

A number of extrusion parameters such as feed moisture, temperature, screw speed and storage conditions are known to affect the RS content of extrudates. Increasing feed moisture is reported to increase the yield of RS up to 500% in pastry wheat flour (Kim et al., 2006). This is consistent with moisture facilitating retrogradation of starch, particularly at high moisture. They also showed about a 2.5 times increase in RS content of pastry wheat flour following 7 day storage at 4°C. Huth et al. (2000) reported up to 6% RS formation during extrusion of barley flour followed by storage at 4°C or −18°C for 3–7 days. However, no further increase of RS was observed when storage was continued for 14 days. In this case the gelatinized and fragmented low-amylose starches underwent molecular re-association (retrogradation) for the first few days of storage, however no such phenomenon was found for high-amylose maize starch (Chanvrier et al., 2007). In this and many other cases, these comparisons are difficult because different methods for determining RS were used.

Agustiniano et al. (2005) suggested that the use of mango starch is a good alternative for RS preparation as extrusion of mango starch can increase RS from 1.1% up to 9.7%. In their study, lower screw speed and higher temperature resulted in higher RS content of the starch, possibly due to more opportunity for amylose chain association with a longer residence time. The change in RS content during extrusion is also affected by the source of starch. Extrusion at 30 rpm was found to increase RS content of mango starch, but there was no change for corn starch and a decrease for banana starch (RS decreased from 19% to 3.9%). Extruded starch samples obtained at higher screw rates showed further decrease in RS content. In another study, extrusion was reported to increase RS content in normal maize (up to 300%) and wheat starch (126%) but decrease in waxy maize and HAMS (up to 200%) (Chanvrier et al., 2007). They found that higher storage temperature yielded more RS, consistent with promotion of retrogradation. Htoon et al. (2009) also reported extrusion of high-amylose starches such as Gelose 80 and Hylon VII significantly reduces the RS value by 3–5 times from a very high level (60–70%) in uncooked starch to a still useful 15–20% after extrusion.

An increase of RS content of waxy as well as regular barley starch was reported by Vasanthan et al. (2002). They showed the initial increase of RS with temperature up to 120°C but decrease after further increase in temperature. RS increased with increased feed moisture, irrespective of temperature.

Extrusion of starch in the presence of sufficient water triggers a number of physicochemical and functionality changes in starch granules, such as the loss of granular structure associated with melting of crystallites and underlying helices, and generating an amorphous structure (Faraj et al., 2004). On its own, this would be expected to increase the vulnerability of starch to amylase digestion. Upon cooling, hydrated amylose (and amylopectin) chains, under ideal conditions, would undergo a process of molecular re-association into double helices, stabilized by hydrogen bonds, and may consequently acquire resistance to enzymic digestion. Therefore, extruded products may also lead to a higher RS content. It could be true that shearing action of the extruder screw in the presence of high temperature may have caused degradation of longer amylose chains into small molecular fragments that could not be incorporated into a crystalline structure (Eerlingen et al., 1993b; Gidley et al., 1995). The fragmented starches may have remained more or less in an amorphous condition. Another explanation could be that there was no new compensating or additional crystallites formed between amylose chains in high-amylose starches as extrusion temperatures (~120°C) may be below the threshold of starch fragmentation. Besides, the rapid drying that occurs in extrusion might not have resulted in the opportunity for extensive retrogradation to occur. Any of these factors could be responsible for lower RS values in extruded starch.

Extrusion processing affects the crystalline structure of the starches as shown by X-ray diffraction (XRD), small angle X-ray scattering (SAXS) and NMR and FTIR. Bello-Perez et al. (2005) studied the change in long-range order (XRD), short-range order (FTIR) as well as enzyme digestibility of freshly extruded banana starch. They reported a typical amorphous structure in the freshly extruded banana due to complete gelatinization which slowly acquires a small diffraction peak at about 7 h and when storage time increased the peak intensity increased giving a typical A-type crystalline pattern (Figure 14.1). The fast retrogradation in banana starch could have been related to the amylopectin chain length as longer chain length, such as in some waxy maize starches, have higher retrogradation rate compared to those with short chain lengths (Yuan et al., 1993). Chanvrier et al. (2007) also reported after 72 h storage, the levels and degree of crystallinity of processed starches varied as a function of storage temperature and starch type.

Extrusion processing markedly decreases crystallinity of high-amylose starches (Chanvrier et al., 2007; Lopez-Rubio, 2007; Htoon et al., 2009). Htoon et al. (2009) showed % crystallinity of raw Gelose 80 and Hylon VII (both high-amylose maize starches) reduced from 17.6% to 2.1% and 17.4% to 2.4% for extruded Gelose 80 and Hylon VII, respectively. They also reported a significant drop in the double helix content of the extruded high-amylose starches. Lopez-Rubio et al. (2007) reported that the gelatinization and subsequent retrogradation of starch is a function of feed moisture, shear rate and temperature during extrusion. The XRD and SAXS analysis showed that the B-type Gelose 80 crystal under 'mild' processing (50% moisture, 100°C and 150 s^{-1}) condition almost completely gelatinized, giving a nearly amorphous structure whereas 'extreme' processing (35% moisture, 140°C and 750 s^{-1}) resulted in some crystallinity of the single helical V-type (Chanvrier et al., 2007; Lopez-Rubio et al., 2007).

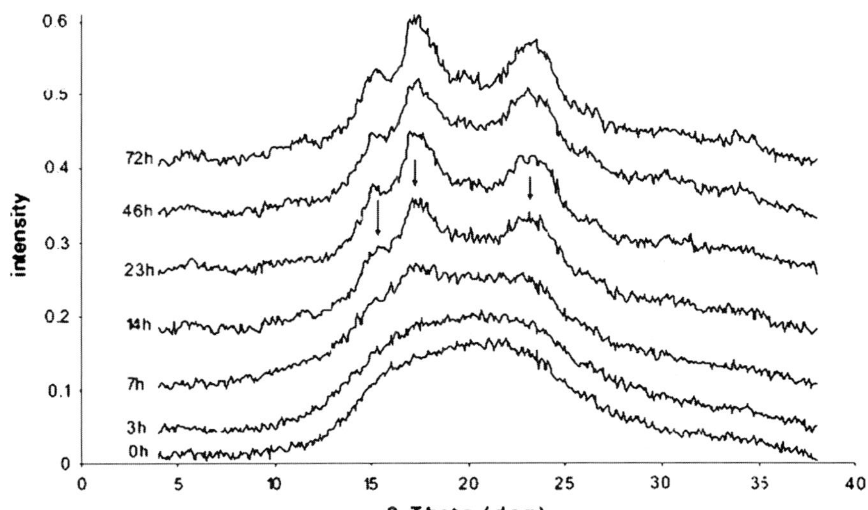

FIGURE 14.1 Development of crystallinity in banana starch extrudate stored at 25°C for different times (Bello-Perez et al., 2005).

Thermal analysis (DSC) results of the melting of retrograded samples in excess water showed an increase in the melting enthalpy on storage, believed to be related to the order formed in the sample during retrogradation (Bello-Perez et al., 2005; Chanvrier et al., 2007). Bello-Perez et al. (2005) reported an increase of melting enthalpy of freshly extruded starch from 0.5 J g^{-1} to 5 J g^{-1} after 10 h storage at 25°C with a plateau after that (24 h). The final retrograded sample after 24 h storage had a lower level of order than the native banana starch as shown by a much lower enthalpy compared to the native starch (ΔH = 23 J g^{-1}). In contrast, there was no change in melting enthalpies of extruded normal and high-amylose starch and wheat starch stored (72 h) from 4 to 60°C, which remained between 1 to 2.5 J g^{-1} (Chanvrier et al., 2007). Their endothermic peak was located between 145–160°C, independent of storage conditions. The peak disappeared on a second scan, consistent with the melting of retrograded amylose. Waxy maize, due to a higher amylopectin content, showed low melting peaks of 30–60°C with enthalpy of 0.4 J g^{-1}, irrespective of storage condition. Kim et al. (2005) reported two endotherm peaks for extrudates stored for 14 days. The first endotherm peak was ~52°C (ΔH ~0.9 Jg^{-1}) and the second one was 123–150°C (ΔH 17–57 Jg^{-1}). The melting peak, in general, decreased with feed moisture and screw speed. Enthalpy decreased with increasing screw speed and decreasing feed moisture.

Extrusion cooking of starch under high energy conditions can cause extensive degradation of starch macromolecules (Colonna et al., 1984; Chinnaswamy and Hanna, 1990). Chinnaswamy and Hanna (1990) showed that in both native and extruded starches differing in amylose contents, the GPC fraction I (amylopectin) content decreased as the fraction II content increased proportionately with increasing amylose content. Fractionation of pure amylose and amylopectin, before and after extrusion cooking, showed that the branched component, amylopectin, degraded to a greater extent than its linear counterpart, amylose.

14.4.3 Boiling and High-Pressure Cooking

Moisture and heat treatments such as steaming/cooking and autoclaving or high-pressure cooking are common methods of food preparation. Heat and moisture are two of the most important factors that affect the gelatinization and retrogradation of starch (Sievert and Pomeranz, 1989, 1990; Czuchajowska et al., 1991; Sievert et al., 1991; Szczodrak and Pomeranz, 1991; Eerlingen et al., 1993a, 1993b, 1994a, 1994b; Garcia-Alonso et al., 1999; Rosin et al., 2002; Landa-Habana et al., 2003; Aaparcio-Saguilan et al., 2005).

Parchure and Kulkarni (1997) reported that a 10 min pressure cooking of native rice and amaranth caused a significant increase in RS content; from 6.2 to 14% in rice and 6 to 10.2% in amaranth, by dry basis. These RS increases were almost 1.7 times higher that those achieved by boiling for 15 min. Sagum and Arcot (2000) reported that RS in rice is positively correlated with amylose content. Boiling as well as pressure cooking of rice markedly decreases the RS content of rice. The decrease was more prominent in low-amylose rice than in higher-amylose rice. In wheat starch, autoclaving reportedly increases RS content from less than 1% to 9% (Siljestrom and Asp, 1985). Siljestrom and Bjorck (1990) also investigated the RS formation in commercially available autoclaved corn, potato, and leguminous products and in autoclaved purees intended for consumption by infants. RS levels found (% dry basis) were as follows: 0.8–2.4 in purees, 0.2–3.2 in canned legumes, 1.9 in canned potatoes, 0.5–0.9% in reconstituted dried potatoes, <0.1% in canned corns, and 1.1% in corn flakes. In all these cases, it is probably not so much the heat/moisture treatment itself that is responsible for differences in RS content as much as the subsequent storage conditions (time, temperature, moisture) that promote retrogradation to different extents.

Considerable RS yield was reported following autoclaving of starches at 120°C (Berry, 1986). This study showed that autoclaving of a starch with high-amylose content (amylomaize, Hylon V and VII or purified potato amylose) led to significant RS formation (20–34% RS, dry weight basis), whereas no such effect was observed in autoclaved high-amylopectin starch (waxy maize or purified potato amylopectin) (0.2–4% RS, dry weight basis). When, however, starch with high amylopectin content was de-branched, the amount of RS increased noticeably, suggesting that linear molecules derived from amylopectin may undergo retrogradation.

Landa-Habana et al. (2003) showed that retrograded resistant starch content was higher in beans cooked with the traditional process (2.65–2.79%) than in autoclaved beans (1.62–1.94%). The initial in vitro α-amylolysis rate in freshly cooked beans was higher in the autoclaved preparation than in the beans cooked by the traditional method.

Sievert and Pomeranz (1989) autoclaved suspensions of wheat, maize, potato, waxy maize and amylomaize starches at different concentrations and different temperatures and allowed them to cool overnight at 4°C. The autoclave–cooling cycle was repeated up to 20 times and samples were freeze dried. The yield of RS was positively correlated with amylose content of the test starch, with amylomaize VII (70% amylose) giving 21.3% RS. Formation of RS in amylomaize was affected by autoclaving temperature, autoclaving–cooling cycle, and starch:water ratio. The number of autoclave–cooling cycles exerted the most significant effect as 20 cycles raised the RS content by more than 40%. Sievert and Pomeranz (1990) reported that the endothermic transition for autoclaved, cooled and freeze-dried regular maize, pea, wheat and potato starches lies between 42–72°C which is assigned to the melting of retrograded amylopectin. Amylomaize VII treated in the same way did not give any transition at that temperature but did at ~155°C. An increase in amylose content of starch is generally associated with an increase in melting temperatures and enthalpies. However, the RS residues from the above starches showed a prominent peak at 155°C and enthalpies ranging from 12.3 to 21.2 J g^{-1} that apparently was due to melting of recrystallized amylose. A small endothermic peak between 41 and 67°C and enthalpies of 3.3 to 6.3 J g^{-1} observed in RS residues from amylomaize, maize and wheat starch may originate from lipid components, presumably adhering to resistant starch residues, but are also consistent with retrograded amylopectin.

In a similar study, Szczodrak and Pomeranz (1991) found that the formation of RS in barley starch was strongly affected by the number of autoclaving–cooling cycles: increasing the number of cycles from one to 20 raised the RS yield from 6 to 26%. Bjorck et al. (1987) reported 6.2% (dry basis) RS in autoclaved wheat starch. The RS increased up to 7.8% when the autoclave–cooling cycle was carried out three more times. Increasing water content and increasing autoclave temperature from 121°C to 148°C resulted in slight declines in RS formation. However, Skrabanja et al. (1998) reported that a higher number of autoclaving–cooling cycles did not affect the proportion of slowly digested starch (SDS) or resistant starch (RS) fraction found in buckwheat.

Eerlingen et al. (1993a) studied the effect of incubation time and temperature on RS formation in autoclaved (1 h) wheat starch–water suspension incubated at 0, 68 and 100°C. The amounts and kinetics of RS formation in starch differed with incubation temperature. The starch incubated at 0°C showed rapid formation of RS (4% in 15 min) and a leveling off afterwards. At 100°C, initially there was some RS formed which rose very quickly after 150 min, to a maximum of 10% (Figure 14.2). This means at low temperature, the nucleation rate was high and the propagation rate was slow, whereas, at higher

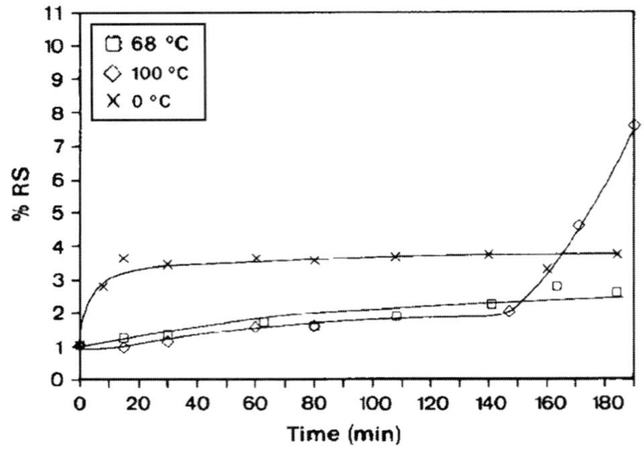

FIGURE 14.2 Enzyme-resistant starch formation in wheat starch–water mixture autoclaved for 1 h at 121°C and stored at different temperatures (0, 68 and 100°C) as a function of time (Eerlingen et al., 1993a).

temperature nucleation was slow but propagation was high. This phenomenon was interpreted in terms of crystallization in partially crystalline polymer systems.

Garcia-Alonso et al. (1999) assessed some parameters affecting RS formation. They reported that starches (from corn, rice, wheat and potato) contain higher RS levels than their flours. There was no difference on RS formation during high-pressure autoclaving (HPA) or boiling of starches. Stirring during gelatinization in HPA did not improve the RS content either. Apparently, gelatinization was unaffected by pH values between 3.5 and 10.5. They suggested that to obtain optimum RS yields during retrogradation, it is necessary to cool down starch gels prior to freezing, followed by thawing at room temperature and drying at 60°C. Garcia-Alonso (2000) also showed that boiling of native potato starch reduced RS content from 69% to 1.2%. However, boiling followed by cooling increased RS up to 4.6%.

Chung et al. (2006) prepared partially gelatinized starch by heating waxy maize at 60, 65 and 70°C for 5 min and retrograded starch by storing fully gelatinized paste at 4°C for 2, 4 and 7 days. The result showed that RS content decreased with increase in gelatinization temperatures and increased with retrogradation time. It also showed that the slowly digestible starch (SDS) and RS is correlated positively with the relative melting enthalpy of the partially gelatinized or retrograded starch sample. In the fully gelatinized starch samples, most of the swollen starch granules were completely disrupted by excess heat and mechanical agitation, resulting in transformation to a continuous amorphous structure, whereas partially gelatinized starch contained some starch granules that were not disrupted. Therefore, the hydrolyzing enzymes have more access to starch molecules in fully gelatinized starch, resulting in low RS values compared to partially gelatinized starch. They also related the enzyme resistance and ordered structure of cooked starches. The higher RS content as well as higher melting enthalpies in partially gelatinized starch compared to fully gelatinized starch seems to be linked to some inherent structure of the native granules.

Jeong and Lim (2003) showed that autoclaved, frozen and thawed (repeated three times) native high-amylose starch, HAS (70% amylose) increased the RS content from 15.8% to ~18% in native and 20.5% in defatted HA starches. The freeze–thawed native and defatted HAS had less intense B- and V-type peak compared to the original samples. Similar crystalline patterns and intensities in the freeze-dried starches indicate that the crystal structure was mainly formed during the freezing process and thawing did not largely affect crystal formation. In the DSC thermogram, the freeze–thawed starches exhibited two endothermic transitions in the temperature ranges of 90–110°C and 130–160°C, representing amylose–lipid complexes and amylose–amylose double helix crystals, respectively. By defatting, the melting enthalpy for the amylose double helices was increased, indicating that the residual lipids inhibited the amylose crystal formation. Ice crystals in the starch gel matrix became smaller and the ice cell membrane became thinner as freeze–thawing was repeated.

Ferrero et al. (1993) stored maize starch pastes frozen at different temperatures (–5°C, –10°C, and –20°C) and found that the sample stored at –5°C exhibited the highest levels of retrogradation and sponge matrix formation. At lower temperatures (–10°C and –20°C), starch mobility, which is required for chain association, was significantly reduced because the storage temperature was far below the glass transition temperature. Upon extended storage, however, transformation from small to large ice crystals may occur, which could help formation of concentrated starch regions in which retrogradation is favored. In comparison to cooked foods stored at ambient condition, cooked food kept under frozen condition (–20°C) (for 30 days) considerably increased the RS content in various cereals and cereal products and legumes (Rosin et al., 2002).

Tovar and Melito (1996) measured the RS content of black, red and lima beans. RS measured directly in conventionally and high-pressure steamed beans was 3–5 times higher than in the raw legumes suggesting retrogradation as the major mechanism behind the reduction in digestibility. Bravo et al. (1998) studied the effect of different processing treatments on the in vitro digestibility and the resistant starch formation in moth bean, horse gram and black gram. The RS content of the raw legumes decreased from 1.2 and 2.6% (dry basis) to 0.3 to 0.5% in freshly cooked and 0.4 to 0.8% cooked and stored samples.

Vasanthan and Bhatty (1998) have used annealing as a means of enhancing the resistant starch content of legume starches. Annealing retrograded gels at 100–140°C and 30% moisture for 1 h increased the resistant starch content of field pea and lentil starches from 8.4 to 14.1% and 6.5 to 9.5%, respectively.

14.4.4 Baking

Many studies have reported on RS formation during baking. Westerlund et al. (1989), to evaluate the effect of baking on RS formation, baked white bread and divided it into three fractions (crumb, inner crust, and outer crust). It was found after baking for 35 minutes that starch level was highest in dough and lowest in outer crust but conversely, RS level was lowest in dough and highest in crumb. Akerberg et al. (1998) studied the effects of baking conditions on barleys (waxy barley, WB and high-amylose barley, HAB) with varying amylose:amylopectin ratio on resistant starch formation. The RS content of bread increased with increasing amylose:amylopectin ratio. The amount of RS (total starch basis) varied from <1% (WB) to approximately 4% (HAB) in conventionally baked bread, and from about 2% to 10% in the corresponding long-time/low-temperature baked products. Liljeberg et al. (1996) also reported that bread baked under low-temperature, long-time conditions contained significantly higher amounts of RS (5.4%) than under conventional conditions (3.0%). Addition of lactic acid increased RS recovery further (6.6% starch basis), whereas malt had no impact on RS yield. The highest level of RS was found in the bread containing high-amylose barley flour (7.7%, starch basis). The high content of RS formed during long-time baking is in agreement with the amount (8.1%, starch basis) previously found in commercial sourdough fermented pumpernickel bread baked with an extended baking process (20 h) (Liljeberg and Bjorck, 1994). The authors suggested that prolonged exposure of high-amylose barley to higher temperature favors the propagation of amylose crystals resulting in greater enzyme resistance.

Attempts to increase the RS content in bread products have been made by adding defined starch fractions to white bread. Eerlingen et al. (1994d) formulated bread with added corn starch, high-amylose starch or retrograded corn starch containing 44.1, 83.2 and 29.2% RS, respectively. One day after baking, the RS content of breads containing wheat flour, corn starch, high-amylose starch and enzyme-resistant retrograded high amylose starch was 0.0, 0.4, 7.7, and 8.4%, respectively. After seven days of storage, the RS levels had increased to 4.0, 4.4, 10.2, and 11.0%, respectively. This study showed that RS levels of bread were lower than could be predicted from the analytical data of the starting materials and is evidence that some RS destruction occurs in the bread-making process. Differential scanning calorimetry measurements confirmed that the increase can probably be ascribed at least in part, to increases in the levels of retrograded amylopectin.

14.4.5 Effects of Non-Starch Materials on Resistant Starch Formation

Lipid is known to be one of the major factors that affect the crystallization of amylose and subsequent RS formation. Sarko and Wu (1978) suggested a competitive mechanism between amylose retrogradation and formation of amylose lipid complex (ALC). It is known that the formation of ALC decreases the susceptibility of amylose to amylolysis (Holm et al., 1983). Czuchajowska et al. (1991) reported the formation of ALC when amylomaize was autoclaved and reacted with some lipid complexing agents such as lysophosphatidylcholine (LPC), sodium stearoyl lactylate (SSL) or hydroxylated lecithin (OHL). The RS content of amylose complexed with lipids was lower than that of retrograded amylase, showing ALC are prone to amylolysis. In a similar study, the presence of LPC was also found to affect association of amylose chains due to formation of amylose–lysophosphatidyl complex (Sievert and Pomeranz, 1993). At an LPC concentration of 10% (w/w), the complex formation was predominant and no association of amylose was observed.

Defatting of wheat starch was reported to increase the RS yield in wheat starch (Eerlingen et al., 1994a). It is estimated that lipids can bind ~6 times their own weight of starch (Karkalas and Raphaelis, 1986; Eerlingen, 1994a). Addition of lipid to an amylose gel (up to 1.2%) lowered RS level substantially. X-ray diffraction and DSC studies have shown that ALC forms in the presence of endogenous as well as added lipids, which means that less amylose is available to form enzyme-resistant double helices (Czuchujaowska et al., 1991; Eerlingen et al., 1994a). It is likely that the complexing segments of the amylose chain and the resulting steric hindrance for formation of double helices in the uncomplexed parts of the chain are responsible for the observations (Eerlingen and Delcour, 1995). DSC studies showed a peak at around 95–110°C indicating the formation of amylose–lipid complexes (ALC) and another peak at 155°C indicating the presence of enzyme-resistant retrograded starch (Czuchajowska et al., 1991). The melting enthalpies of ALC samples were negatively correlated with the yield of RS. The addition of lipid to starch was reported to induce V-type crystal formation (Czuchujaowska et al., 1991; Eerlingen et al., 1994a).

From the studies of isolated barley starch autoclaved with lipid components, it was postulated that amylose crystallization (as measured by the enthalpy of a 158°C endotherm) that is involved in the formation of RS is competitively affected by its complexation with lipids (Szczodrak and Pomeranz, 1992).

Extrusion of corn starch with additives such as sodium stearoyl lactylate (SSL) and diaceytyl tartaric acid easters of mono-glycerides (DATEM) reduced the level of RS in the amylose lipid complexes (Adamu, 2001). However, extrudates containing citric acid gave higher RS values. In comparison, when guar gum was added to corn starch, the RS yield increased significantly. For example, extruded starch and SSL had 1.25% RS whereas inclusion of guar gum resulted in 10.1% RS. XRD of the extruded starches gave a V-diffraction pattern indicating the formation of amylose–lipid complex. Purification of the isolated RS by HPSEC indicated the additives to have a substantial effect on molecular weight: Datem and SSL increased molecular weight of RS, while citric acid decreased it.

The addition of soluble sugars such as glucose, maltose, sucrose, and ribose has been found to reduce the level of crystallization and subsequently reduce the RS yield (I'Anson et al., 1990; Kohyama and Nishi, 1991; Eerlingen et al., 1994b). Addition of sugars such as glucose, sucrose, maltose, or ribose at higher concentrations on autoclaved wheat followed by storage reduced the RS yield (Eerlingen et al., 1994b). In the presence of sucrose or glucose, RS yield of wheat starch gels decreased from ~3.4 to 2.8%, which further decreased to 2.5% in the presence of maltose or ribose. However, addition of sugars results in increasing RS content of amylomaize VII. The authors commented that the differences in response of the RS yield of sugars in high-amylose corn starch and wheat starch gels could not be rationalized, although it was clear that the difference in gelatinization temperature of the two starches were not the cause of the different behavior of the starches in the presence of sugars. No influence of the sugars on the XRD and DSC characteristics of the isolated starch was detected, except for a higher melting enthalpy of isolated RS when it was formed in the presence of ribose. Levine and Slade (1990) also found reduction of amylopectin retrogradation in wheat starch gel when sugars are added. Eerlingen and Delcour (1995) suggested that since crystallization of amylose takes place above the glass transition temperature of sugars, it might have an impact on the glass transition temperature affecting the retrogradation of starch.

Unlu and Faller (1998) investigated the effect of citric acid, high-amylose corn starch, wheat starch, and potato starch on RS formation in yellow corn starch during extrusion cooking. Addition of 30% high-amylose corn starch significantly increased the RS content of corn starch from 1.75 to 5.6%. Citric acid also increased the RS (7.5%) content but the product was unpalatable. No effect of wheat and potato starch on RS content of corn starch was observed. Escarpa et al. (1997) carried out high-pressure cooking of potato starch with a number of nutrients and other food constituents to study the RS formation. Except for insoluble dietary fiber constituents (cellulose and lignin), all the tested food ingredients reduced the formation of RS. Calcium ions, potassium ions and catechin showed the highest reduction of RS formation, while the nutrients studied (albumin, olive oil and sucrose) as well as phytic acid affected it to a lesser extent. These results were not significantly changed by varying amounts of the studied dietary components. Mercier (1980) and Holm et al. (1983) reported that processing of amylose with lipid or components such as lecithin, palmitic acid, oleic acid and soybean oil affected the retrogradation to a lesser extent than monoglycerides. Nevertheless, these authors found that pure potato amylose and oleic acid formed complexes highly resistant to amylolysis.

Mangala et al. (1999) studied the effect of amylopectin as well as protein and lipid on the RS content of processed rice and ragi. Autoclaving significantly increased the RS content of all the defatted starches, but not for deproteinized cereal starches. Native starches appeared to have an A-type pattern whereas defatted sample showed characteristics of V-type patterns. The endothermic temperature and enthalpies of defatted starches considerably decreased in comparison to native starches. The evidence that starch which was autoclaved and stored for longer time (more than 20 days) was relatively resistant to amylolysis, indicates that amylopectin may undergo higher degrees of entanglement, leading to greater crystallization.

14.4.6 Effects of Acid or Enzyme Hydrolysis

It has been shown that, after debranching, linear starch chains can contribute to a high RS content (Berry, 1986). Various authors have shown that degree of polymerization (DP) of the linear chains influences the retrogradation phenomenon (Eerlingen et al., 1993b; Gidley et al., 1995). Schmiedl et al. (2000) have suggested that a DP of 20 is optimal for a high RS3 output.

Gonzalez-Soto et al. (2006) debranched starch from previously gelatinized and autoclaved banana starch using pullulanase and autoclaved the product at 121°C for 30 min. Compared to the control sample (~9% RS), the debranched-autoclaved had a higher RS content (23–35% RS). RS content was higher at lower storage temperature e.g. 4 and 32°C (~34% RS) than higher temperature e.g. 60°C (~24% RS). Storage time did not affect the RS formation. It was suggested that the lower glass transition temperature of debranched starch in comparison to storage temperature may have hindered or slowed down the retrogradation of starch. The C-type crystallinity of native banana starch is altered in treated samples showing other peaks which are indicators of higher crystallinity. Debranching time may also affect the RS formation. Liu et al. (2003) observed that increasing debranching (pullulanase) time from 4 to 48 h increased the RS yield of the native corn starch from 18% to 36.4%.

Vasanthan and Bhatty (1998) lintnerized retrograded field pea and lentil starch gels with 2.2 N HCl for 30 min prior to annealing and found that this increased their RS content from 14.1 to 19.6% and 9.5 to 16.5%, respectively. This increase was explained as follows: acid hydrolysis of retrograded gel amylopectins at or closer to the branch points could cause the double helices that form the starch crystallites to unravel. This would then increase the amount of free linear chains, which could then associate together during annealing forming more perfect crystallites. The increase in RS content was attributed to the increase in the number of crystallites of greater perfection and stability.

Onyango et al. (2006) showed that RS formation in lintnerized cassava starch is affected by the strength of acid used, incubation time and temperature. Higher concentrations of lactic acid in an autoclaving starch solution (100 mmol L^{-1}) resulted in lower quantities of RS formation, in comparison to 1 and 10 mmol L^{-1}. The RS yields declined when the autoclaving time exceeded 45 min indicating the effects of continuing acid and thermal hydrolysis of cassava starch. The authors suggested that the RS yield was low because the polymers were not long enough to form enzyme-resistant crystallites (confirming Eerlingen et al., 1993b, that lower rate of crystallization when the degree of polymerization is less than 100). The melt transitions of acid-treated starches ranged between 158–175°C with a peak at about 164°C which is characteristic of dissociation of recrystallized amylose. However, the melting enthalpies were 0.2–1.6 J g^{-1}, values much lower than those previously reported for RS from retrograded cereals (Sievert and Wursch, 1993; Shamai et al., 2003). This study suggested that formation of RS takes place in two stages: starch hydrolysis during autoclaving in acidic condition and recrystallization during incubation. Formation of RS was low at low temperature. The result showed that incubation of autoclaved cassava starch at 4°C favored a high nucleation rate and therefore high initial yields of RS. However, the yield did not improve after 48 h incubation as at this temperature propagation (crystal growth from the nuclei formed) was limited by the temperature-induced high viscosity of the medium. The opposite mechanism, slow nucleation but faster propagation and crystal formation at higher temperature resulted in higher final RS contents from autoclaved starch incubated at higher temperatures.

Aparcio-Saguilan et al. (2005) prepared resistant starch from native and lintnerized banana starches by repetitive autoclave and cooling treatments. Lintnerization and autoclaving increased amylose content of native starch from 37 to 45% indicating partial debranching of amylopectin. The RS content of the native banana starch dramatically decreased from 38 to 1.5% after gelatinization (20 min boiling). Lintnerization, itself, caused only a slight increase in RS value (2.6%), but repetitive autoclaving–cooling significantly increased the RS value to 19.3%. Murphy (2000) suggested that lintnerization selectively removes amorphous regions of starch granules, thus yielding materials rich in amylopectin crystallites. Therefore, the lower RS content in the lintnerized crystallized starch fractions is largely susceptible to enzyme hydrolysis, a finding also reported by Fredriksson et al. (2000). The increased RS content in lintnerized–autoclaved starch is contributed to by enrichment of the linear glucan chains leading to retrogradation of starch.

14.4.7 Microwave Radiation

Microwave treatment causes a rise in the temperature within a penetrated medium as a result of rapid changes of the electromagnetic field at high frequency. Lewandowicz et al. (2000) studied the effect of microwave radiation on physicochemical properties of tuber and cereal starches (at 30% moisture). XRD analyses showed although A-type crystallinity remained unchanged in all cereal starches during microwave treatment, the degree of crystallinity decreased in wheat and corn starch and remained unchanged in waxy corn. Previous studies have reported that microwave radiation changed the crystallinity of the potato starch from B-type to A-type (Lewandowicz et al., 1997). The gelatinization transition of microwave-treated cereal starches was shifted to higher temperatures, compared to untreated ones. The total enthalpy change for the transition decreased substantially for irradiated wheat and corn starches, but remained the same for waxy corn. The authors concluded that a relatively higher susceptibility of amylose-rich starches to crystallinity, swelling and solubility of cereal and potato starches. The association or changes in the amorphous amylose part of the starch granule rather than crystallization is more likely to be responsible for changes in starch structure during microwave treatment.

Microwave cooking of legumes produced a redistribution of the insoluble non-starch polysaccharides to soluble fraction. Microwave radiation decreased the RS level of legumes from 32.5% (of total starch) to about 10% in the cooked samples. In contrast, the levels of rapidly digestible starch increased from 35.6% in chickpeas and 27.5% in common beans to about 80% (Marconi et al., 2000). In a similar study, the rate of hydrolysis of kudzu starch and corn starch increased following heat treatment, especially after microwave treatment (Geng et al., 2003).

14.4.8 Fermentation

Unlike enzyme and acid hydrolysis, fermentation of starch is found to lower the RS content of starchy foods. Abd-Elmoneim et al. (2004) mixed sorghum flour with water and previously fermented dough starter, and fermented at 37°C for a maximum of 36 h. The result showed a decrease in RS content, which is accompanied by the increase in the in vitro digestibility. Study of Indian fermented products such as Idlis and Dhokhalas, also showed that fermentation decreases the RS content (Kavita et al., 1998).

14.5 MODEL STUDIES OF ISOLATED STARCHES

The previous section has described a large number of studies that have shown that processing methods can lead to either an increase or decrease in RS values. Sometimes these variations are a logical consequence of processing conditions, but in other cases the logic is less obvious. There are two main reasons for this variation in results. One is that the RS analytical methods used are often different and hence hard to compare between laboratories. The second is that starch is an inherently complex material. In order to address the latter, studies of model starches have often been carried out and linked to molecular and microstructural probes of polymer structure.

One approach has been to prepare retrograded starch from various sources such as cereals, legumes, pure starch, etc., using various techniques such as heating, e.g., repeated autoclave–cooling/freezing; acid-hydrolysis, enzyme-hydrolysis, etc. The digestible fraction from the retrograded starch is then removed by amylolysis with amylase and the enzyme-resistant fraction is isolated by centrifugation and oven- or freeze-dried and further characterized for their physicochemical properties.

14.5.1 X-ray Diffraction Studies

Eerlingen et al. (1993a) isolated RS from wheat starch by autoclave–cooling; incubating at different temperature; starch hydrolysis with amylase, amyloglucosidase and protease; filtration and oven drying. X-ray diffraction showed the crystal patterns of RS residues differ with the incubating temperatures. The RS formed at 0°C was of B-type, whereas at 100°C it was A-type. The formation of B-type crystalline structure at low incubation temperature in the isolated RS residues has also been reported by several other investigators (Berry et al., 1988; Russel et al., 1989; Siljestrom et al., 1989; Sievert et al., 1991; Eerlingen et al., 1993b; Gidley et al., 1995). Eerlingen et al. (1993a) reported that increasing the storage temperature of the gelatinized wheat starch affects the degree of crystallinity in the isolated residues. There was no change in crystallinity of RS residue stored at 0°C, little change at 68°C but a large change at 100°C. The increased maturation rate of crystals at higher temperature was responsible for higher crystallinity at 100°C. X-ray diffraction data suggested that chain fragments packed in B-type crystallites with a large crystal lattice contribute to the formation of RS in autoclaved amylomaize residues (Sievert et al., 1991). The structure development in RS residues after repeated cycles of autoclaving and cooling may be due to increasing inter-chain amylose associations and to decreasing proportions of less-ordered domains (Sievert et al., 1991).

Quantitative analysis of enzyme-resistant retrograded starches from wheat and amylomaize V and amylomaize VII established that these starches contained 60–70% double helices (by ^{13}C CP/MAS NMR) and about 25–30% crystallinity (XRD) (Gidley et al., 1995). They suggested that more than half of the double helices in the samples are not present in domains sufficiently large and regularly arranged to diffract X-rays. This supports the view of Sievert et al. (1991) that RS residues comprise small and/or imperfect crystallites and interspersed amorphous material.

The original V-type crystallites disappeared when the RS residues from amylose–lipid complex were isolated and a weak B-type pattern appeared. The alteration in the pattern from V to B is due to enzymatic hydrolysis of amylose–lipid complex and concentration of RS structures (Czuchajowska et al., 1991).

Shamai et al. (2003) analyzed the polymorphic structure of RS residues from various starches after autoclaving and incubating at two different temperatures. They reported that retrogradation at 40°C leads to the formation of B-type, whereas incubation at higher temperature (95°C) resulted in A- and V-type polymorphs. There were differences in polymorph density and degree of hydration of the crystalline structure, DSC analysis showed no significant differences in the thermal behavior of the two polymorphs.

14.5.2 Differential Scanning Calorimetry Studies

DSC is used to quantitatively measure endothermic transitions as well as enthalpies associated with changes in the molecular order of starch in aqueous solution. Retrograded starch is known to be thermally very stable. Gidley and Bulpin (1989) showed heating up to 160°C is needed to dissolve retrograded amylose and the melting peak is assigned to the unwinding of double helical structures. Gidley et al. (1995) reported a wider range of endothermic transition for retrograded wheat and amylomaizes which started from 90°C and peaked at 150–160°C with enthalpies ranging from 17.5 mJ mg^{-1} for amylomaize V to 27.5 mJ mg^{-1} for amylomaize VII. The breadth of the endotherm together with the wide range of chain lengths present in enzyme-resistant starch was consistent with temperatures required for helix melting to be increasing with chain length. From the gelatinization of granular starches, Cooke and Gidley (1992) suggested that melting of double helices, or disruption of crystalline packing, is the primary determinant of the endotherm enthalpy. They found melting enthalpy as high as 35 mJ mg^{-1} is needed to melt the highly ordered (crystalline) structure. Gidley et al. (1995) emphasized that if the helix melting enthalpy is considered to be the sum of stabilizing structural features which are present in each turn of the helix irrespective of its length, then a constant enthalpy per gram of helix would be expected, so the major factor which may lead to helix length variation would be end effects.

Various studies have shown that an isolated RS fraction has a melting temperature of about 150°C and melting enthalpies ranging from 8 to 30 mJ mg^{-1} (Sievert and Pomeranz, 1989, 1990; Eerlingen et al., 1994).

DSC data for isolated RS residues obtained from different fractions of high-amylose barley showed a similar endothermic transition range (116–177°C), with a mean peak temperature of 158°C (Szczodrak and Pomeranz, 1991). The maximum melting enthalpy for RS from barley, 37 J g^{-1}, was obtained by 12 repeated autoclave–cooling cycles.

14.5.3 Chain Length Distribution

The formation of resistant starch in an aqueous system and its physicochemical properties is largely influenced by the chain (branch) length. There are a significant number of studies that relate processing with the alteration of chain lengths with the enzyme resistance of starch. Eerlingen et al. (1993a) hydrolyzed potato starch amylose to varying degrees by incubation with β-amylase for different periods to obtain amylose with various chain lengths (or degree of polymerization, DPn). RS material from these fractions was isolated and analyzed by gel permeation chromatography (GPC). They suggested that although the amylose fraction in potato contained a wide range of chain length (DPn 40–610), the fraction with low DPn (< 100) contained a relatively high concentration of chains that can be incorporated directly in the crystal structure. On the other hand, in the amylose fraction with high DPn, the long chains are likely to interact simultaneously over different zones of the amylose molecules which hinder the crystallization (Gidley and Bulpin, 1989). The yield of RS increased with DPn to plateau value of 23–28% within a range of DPn of 100–610. They concluded the chain lengths of the amylose molecules had no influence on the chain lengths of the enzyme-resistant starch and on the crystalline structure (B-type).

Cairns et al. (1996) characterized the RS produced in vitro by hydrolysis of retrograded pea starch gels and pea amylose gels by porcine pancreatic α-amylase. They showed that RS residue from in vitro digestion consisted of semi-crystalline, mostly linear material that was present in two main molecular size sub-fractions: (1) DP ca 100 (arising from semi-crystalline material present in the retrograded part of the gel) and (2) DP 20–30 (composed of recrystallized amylose fragments released during degradation by amylase). A third minor sub-fraction consisting of oligosaccharides (DP ~5) was also present. This model of RSIII showed that the resistant residue from hydrolyzed amylose gels closely resembled the resistant residue recovered from hydrolyzed starch gels, indicating the primary importance of retrograded amylose in determining the RS content of starch. Analysis of in vivo RS recovered during an ileostomy study produced results that were similar to those obtained from RSIII in vitro (Cairns et al., 1996).

Gidley et al. (1995) also reported a chain length ranging from minimum degree of polymerization (DP) of less than 10 to a maximum of c. 100 for enzyme-resistant starch residues from amylomaizes (V and VII) and wheat starch using high-performance anion exchange chromatography (HPAEC). The local maximum for the chain length varied with the substrate but was mainly at DP of 20–30. HPAEC analysis showed a periodicity in chain length for DP multiples of six above DP18 for enzyme-resistant retrograded starches.

Botham et al. (1995) assessed the digestibility of retrograded chick pea starch gel using an ileostomy model. The resistant starch contained molecules with a wide range of molecular weights. A major fraction consisted of fragments of amylose with an average degree of polymerization of 70–80. The resistant product was partially crystalline, giving a B-type pattern. The resistant product obtained in vivo was found to be comparable with the product obtained from the amylolysis of starch and amylose gels in vitro (Ring et al., 1988; Cairns et al., 1990).

RS residues isolated from wheat-based foods such as chapatti and phulka were structurally characterized as linear 1,4-linked α-D-glucan essentially derived from a retrograded amylose fraction, which was dependent on the severity of the processing treatments as well as the levels of gluten and damaged starch in the wheat flour (Tharanathan and Tharanathan, 2001).

Jane and Robyt (1984) studied the structure of the retrograded amylose obtained from dilute solutions ($\leq$0.35%, w/w for one to two weeks at 5°C) and their fragmented fraction obtained from α-amylase hydrolysis. The amorphous, folding areas on the surface of the lamellae of packed helices are hydrolyzed by α-amylase with the formation of resistant amylodextrin fraction. The degree of polymerization (DPn) of these fractions corresponds to the diameter of the helices and the folding length of the chain. They proposed a structure of retrograded amylose residues having double helical crystalline regions that are 10 nm long and separated from each other by amorphous regions (Figure 14.3). The average length of these enzyme-resistant fractions varies from 32 to 50. Similar chain lengths can arise from acid digestion of retrograded amylose as shown in Figure 14.3. This serves as a model for the formation of RSIII from retrogradation of amylose-containing starches. Although the chain lengths measured by various workers do not always coincide, the partially crystalline structure illustrated in Figure 14.3 provides a straightforward hypothesis for the finding of linear low-molecular-weight chains in RSIII materials.

Leloup et al. (1992) investigated the structural features of amylose gels from pea starch (2–8%, w/v) at different scales of organization. The amylose gel exhibited a macroporous structure with mesh size 100–1000 nm and containing filaments 20 ± 10 nm wide. The amount of amylose that resisted hydrolysis, assigned as the ordered fraction within the gel increased from 67 to 82% as the concentration of the amylose in the gel increased from 2.7 to 7.9%. This partially

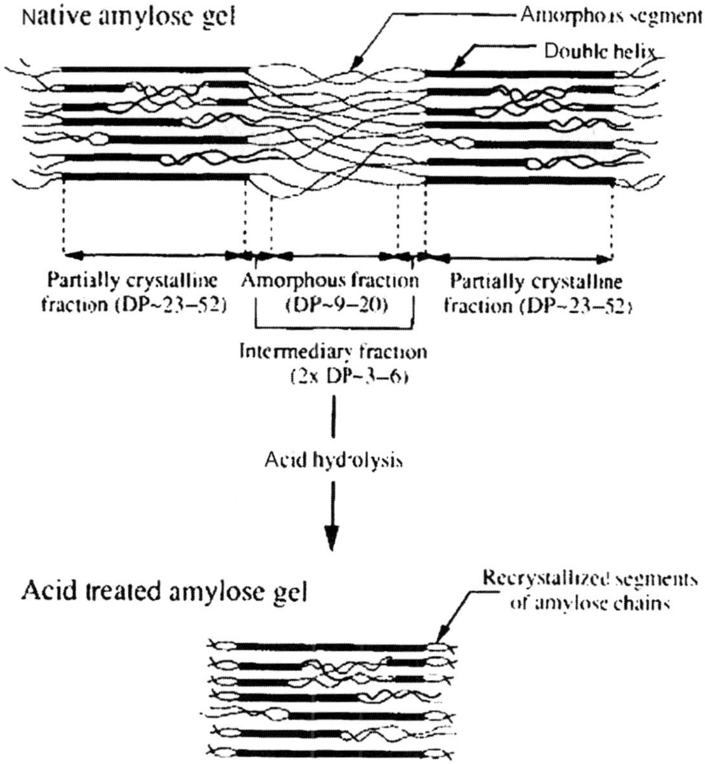

FIGURE 14.3 A model for amylose gels (Jane and Robyt, 1984).

organized B-type filamentous structure contained amylose chains of average degree of polymerization of 26–31 (DPn) and 56–73 (DPw). There was an easily digestible fraction having amorphous structure (6 < DP < 30). The amylose gels hydrolyzed by acid resulted in sharper endothermic peaks and higher melting temperatures (about 150°C).

Mild acid treatment has been used to dissolve and remove the amorphous fraction leading to an acid-resistant (or lintnerized starch, LS) crystalline structure. The structures of the acid-resistant residues provide information regarding the nature of the crystalline parts in the granule which corresponds to the cluster model of amylopectin. Srichuwong et al. (2005a) studied the molecular composition of acid-resistant residues prepared from starches showing A-, B- and C-type XRD patterns by fluorophore-assisted capillary electrophoresis (FACE). They showed that lintnerized starch not only retained the X-ray diffraction patterns of their native starches but also improved the diffraction intensity. FACE results show a bimodal distribution of chains, more in A- and C- than B-type. The first portion suggested that double helices originated from linear chains with an approximate average DP 14, 16 and 15 would span the length of the crystalline lamella of A-, B- and C-type starches, respectively. The single-chain distribution in A-type (10–17%) was slightly higher than C-type (8%) but almost double compared to B-type (6%). The reduced proportion of shorter chains of DP6 in lintnerized starches compared to native starches showed that short chains are readily hydrolyzed together with amorphous materials by acid, since they are not long enough to form a stable double helix (Gidley and Bulpin, 1987; Bertoft, 2001). The authors concluded that the major factor determining the degree of acid hydrolysis between A- and B-type starches would be the characteristics of amorphous lamellae, the proportion of B-type crystallites and granule size. Digestibility data showed that A- and C-type LS were almost completely (85–100%) hydrolyzed by amylase within 24 h, whereas B-type LS were hydrolyzed more slowly (53–58%). A high amount of single branch chain in A-type LS renders them more prone to enzyme hydrolysis. B-type crystallites are more enzyme resistant due to their thicker lamellar with rare branching points. Srichuwong et al. (2005b) reported that the large granule size of potato and edible canna starches partly contributed to their lower digestibility.

Faisant et al. (1993) characterized the structural features of the in vivo resistant starch fraction of beans (B), potato (P), retrograded (R) and complexed high amylose starches (C). SEC analysis of the RS residues showed three distinct populations of α-glucan molecules: oligosaccharides (DPn $<$ 5); mainly enzyme-resistant fraction of DPn 35 (5 $<$ DPn $<$ 100); and high-molecular-weight fraction (DPn $>$ 100). The proportion of glucans in these major fractions of starch, however, varied with resistant starch fraction, e.g., chain length of RS residues B, P, R and C were 53, 17, 79 and 83, respectively, in the major resistant starch fraction. The RS content in raw B, P, R and C were 9.5, 1.2, 49 and 2%, respectively. The lintnerized and ileal content of all the samples showed B-type pattern and melting enthalpy of about 150°C.

Themeier et al. (2005) analyzed the amylose as well as RS content of several cereal starches further suggesting that RS and TDF correlated with the amylose content. However, pea starches with medium level of amylose (30–34) showed remarkably high level of RS, 9–11.5%. Previously, Bogracheva et al. (1998) and Cairns et al. (1997) also reported peas with medium-level amylose level and C-type crystallinity to have high RS content. In contrast, high-amylose pea starch (~ 65% amylose) contained about 19% RS, which is far below about 50% RS present in high-amylose corn starches (~70% amylose). This study revealed that the size of starch granule, its morphology, extent of crystallinity or damage, and possibly other yet to be identified features may all have an impact on resistance to amylolysis.

14.6 MOLECULAR AND MICROSTRUCTURAL ORGANIZATION OF RESISTANT STARCHES

Numerous studies have been carried out to characterize the molecular and microstructural organization of resistant starch (RS) fractions, aiming at unravelling the mechanisms of amylolysis resistance. These have focused on RSII (granular starch) and RSIII (processed or retrograded starch) as the mechanisms behind RSI (physically inaccessible starch) are understood to be due to plant or food microstructure and not starch properties per se, and RSIV (chemically modified starch) is not often encountered. Some common features of RS from granular and processed starches together with the main findings explaining their mechanisms of resistance are summarized below.

14.6.1 Chain Length and Branching Characteristics of Resistant Fractions

Digestion of granular cereal starches does not affect their chain length profile, as observed for different starch varieties (Evans and Thompson, 2004; Zhang et al., 2006b). The chromatograms obtained for debranched granular starches were considered to consist of four different regions, and α-amylase treatment only caused relatively small changes in the proportions of the different regions (Evans and Thompson, 2004). Although enzymatic hydrolysis initially affects the less-ordered amorphous regions, several studies have shown that the crystalline and amorphous parts of the starch granule are evenly digested (Gerard et al., 2001; Zhang et al., 2006a), which would explain the small variation in chain length profile observed after hydrolysis. It is not yet clear whether these observations for maize starches extend to all granular starches.

In the case of processed starches, the chromatograms of the fractions left after digestion, both in vivo and in vitro, show three populations of α-glucan molecules: oligosaccharides (glucose, maltose, maltotriose and maltotetraose), a high-molecular-weight fraction (attributed to potentially digestible material) and an intermediate fraction, which contains the highest proportion of the isolated material and is attributed to retrograded amylose crystallites (Faisant et al., 1993a; Cairns et al., 1995, 1996). The latter fraction, which is considered to be RSIII, has been observed to consist of low-molecular-weight (MW) polydisperse linear chain segments with average degrees of polymerization (DP) varying between 19 for resistant amylose crystals (Eerlingen et al., 1993b), 20–30 for retrograded wheat and high-amylose maize starches (Faisant et al., 1993a; Gidley et al., 1995) and ~60 for amylose gels and autoclaved maize and wheat starches (Russell et al., 1989; Siljestrom et al., 1989; Cairns et al., 1995). The average chain lengths of the resistant fractions were observed to be affected by the hydrolytic reagent used (Jane and Robyt, 1984) and, thus, the differences between the different average DP of RS obtained by different authors can be rationalized by invoking differences in experimental conditions used for RS formation and in enzymes employed for the isolation of the resistant fractions. It is interesting to note that although the average DP of the resistant fractions is independent of the initial chain length of the amylose it originated from (Eerlingen et al., 1993b), the latter one has been observed to affect RS yield, which rises up to DP ~100 and remains rather constant above that (Eerlingen et al., 1993b). These results are entirely consistent with the earlier finding that the onset of network branching in retrograded amylose and the maximum rate of retrogradation as a function of amylose chain length (Gidley and Bulpin, 1989) both occur at around DP 100. Thus the enzyme-resistant fraction can be rationalized as being due to the linearized segments of retrograded amylose networks as illustrated in Figure 14.3. From the limited studies carried out, the average degree of polymerization of the retrograded amylose fraction obtained in vivo may be higher than the one in vitro, most likely due to more extended hydrolysis due to the specific conditions used in vitro (Faisant et al., 1993a).

Regarding the branching properties of the resistant fractions, differences between granular and processed starches are also noticeable. For granular starches, where the resistance is mainly due to the granular organization, amylopectin molecules may be present after amylolysis as observed in stained samples under polarized light (Evans and Thompson, 2004). However, with the exception of high-amylose variants, all cereal starch granules have very low (<3%) RS values showing that the rate of amylolysis of both amylose and amylopectin fractions within granules is fast enough to be complete after (simulated) small intestinal digestion.

For processed starches, where the resistance is thought to be due to amylose retrogradation, the situation is different and hardly any amylopectin is present after digestion (Rabe and Sievert, 1992; Fredriksson et al., 2000; Leeman et al., 2006). The branched component of starch interferes with amylose retrogradation (Berry, 1986) and is unfavorable to the formation of RS, as apart from the branches being hindered in movement, the typical lengths in them, 20–40 glucose units are far from the optimum of 100 (Eerlingen et al., 1993b). In autoclaved wheat, for instance, no signs of amylopectin crystals were observed after digestion and the degree of branching in the

RS fractions was comparable to that of the amylose reference (Siljestrom et al., 1989).

14.6.2 Characterization Techniques

A number of techniques have been used to characterize the physicochemical characteristics and the molecular and microstructural organization of the RS fractions from different starch varieties.

The thermal characteristics of the RS fractions have been mainly analyzed using differential scanning calorimetry (DSC). The DSC method, widely used to study gelatinization and retrogradation of starch, showed an endothermic transition over a similar temperature range (120–165°C) for the RS fractions obtained from processed starches (Sievert and Pomeranz, 1989; Sievert and Pomeranz, 1990; Leloup et al., 1992; Faisant et al., 1993b; Gidley et al., 1995), which has been ascribed to the dissociation of crystalline amylose with a melting temperature estimated at ~160°C (Biliaderis et al., 1985; Gudmundsson, 1994). In comparison with the endotherms of retrograded amylose, the RS residues exhibited endotherms over a broader temperature range with higher melting enthalpies (Sievert and Pomeranz, 1989). However, this endotherm is not observed in the RSII residues from native, ungelatinized starches (Sievert and Pomeranz, 1989), confirming the different mechanism of resistance to hydrolysis from starches in granular form and after processing. The enthalpy of the endothermic transition is related to the amount of molecular (double helical) order (Cooke and Gidley, 1992) and, generally, higher enthalpies are accompanied by greater resistance to enzymic digestion (Themeier et al., 2005).

Electron microscopy, both in scanning (SEM) and in transmission mode (TEM), has been extensively used to visualize RS fractions, which differ depending on the starch variety and the processing conditions experienced by the starches. Two main modes of enzymatic attack have been described in granular starches: exo-corrosion (i.e. the enzymes cause surface alterations eroding sections of the granule surface) and endo-corrosion (penetration of the enzymes through small pores observed in the surface of the granules, eroding the internal parts) (Valetudie et al., 1993). The enzymatic hydrolysis of starch granules from cereals is characterized by the formation of holes and a preferential disruption of the core of the granule (Planchot et al., 1995). In the case of potato or banana starch granules, the degradation is slower and holes are reported only after mild acid hydrolysis of starch (Helbert et al., 1996). Through immuno-gold labeling of the enzymes and cross-sectioning of hydrated starch granules, it was possible to visualize the enzymes at the sites of hydrolysis and get a better understanding of the mode of enzymatic attack (Helbert et al., 1996).

Upon processing and gelatinization, the granular structure of starch is no longer dominant and, usually, irregularly shaped particles are visible. After digestion, SEM images suggested the presence of denser structures (Sievert and Pomeranz, 1989); small particles with well-defined shapes were observed in the RS fractions from extruded high-amylose maize starch (Lopez-Rubio et al., 2007).

X-ray diffraction (XRD) has been used to describe the crystalline structure of the RS fractions. Upon digestion of granular high-amylose maize starches, the original crystal polymorph is maintained but the intensity of the diffraction peaks is reduced (Lopez-Rubio et al., 2007), indicating lower crystallinity of the resistant fractions after the enzymatic treatment. In the case of regular maize starch, an even digestion of the amorphous and crystalline regions was observed and, thus, the XRD patterns remained rather similar during the digestion process (Zhang et al., 2006). In processed starches, early studies on RSIII residues showed weak B-type crystalline patterns (Berry et al., 1988; Russell et al., 1989; Siljestrom et al., 1989; Sievert et al., 1991; Eerlingen et al., 1993a; Gidley et al., 1995),

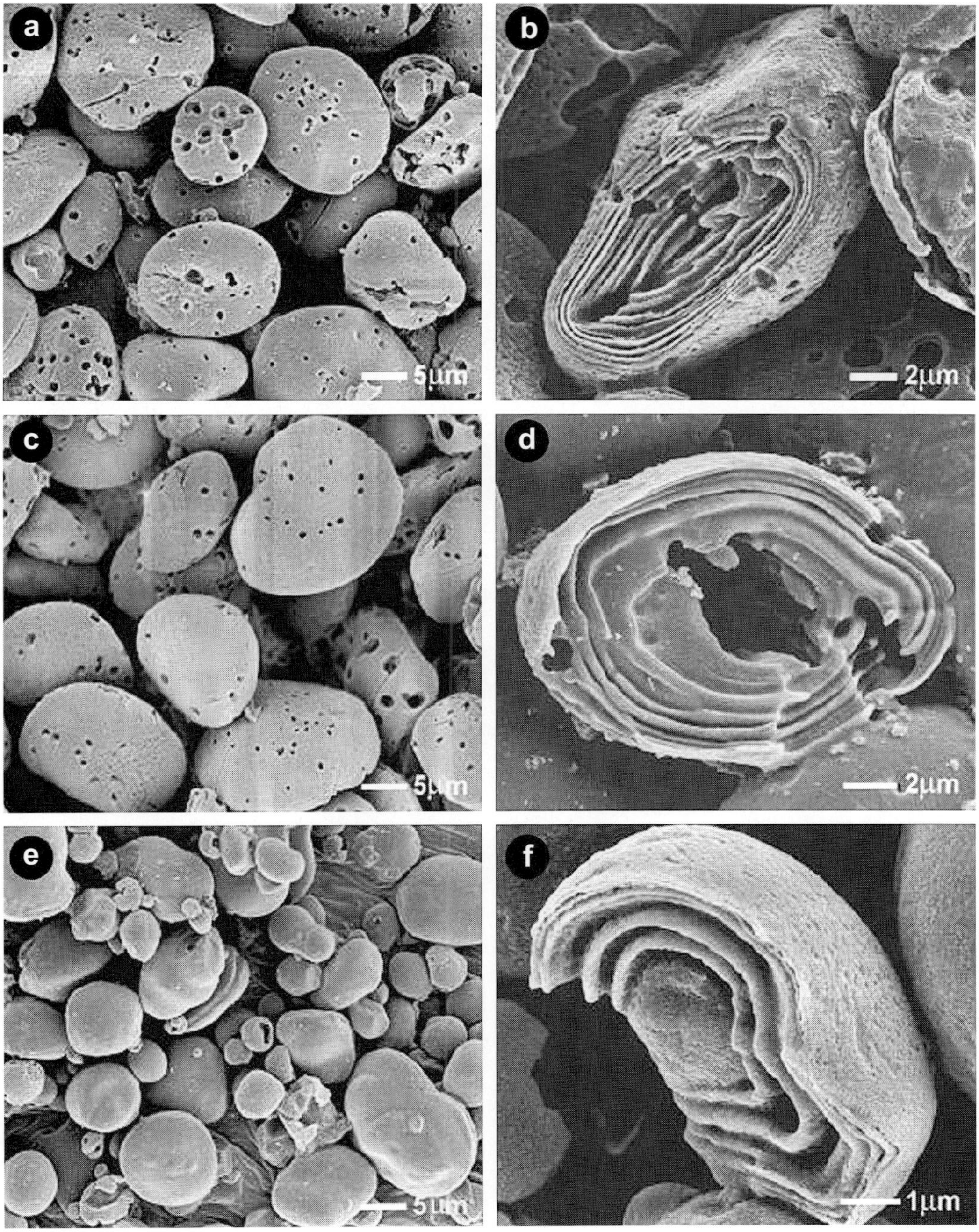

FIGURE 14.4 Scanning electron micrographs of (a,b) waxy, (c,d) normal, and (e,f) high-amylose barley starch granules following hydrolysis by porcine pancreatic amylase at 37°C for 1 hour (Li et al., 2004).

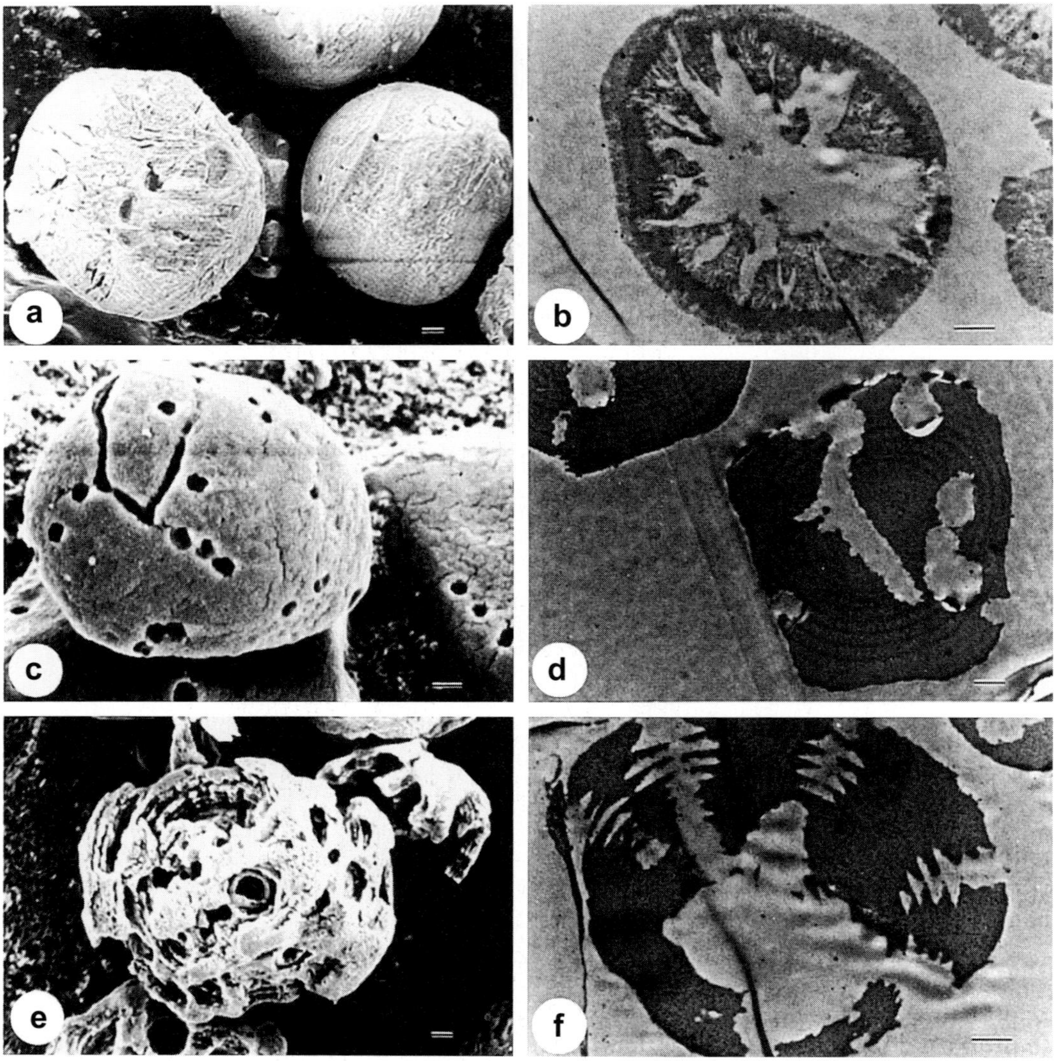

FIGURE 14.5 Electron micrographs of (a,b) high-amylose, (c,d) normal, and (e,f) waxy maize starch granules after hydrolysis by *Aspergillus fumigatus* [K-27] alpha-amylase visualized by scanning (a,c,e) and transmission (b,d,f) methods. Bars = 1μm (Planchot et al., 1995).

which have also been observed in the resistant fractions obtained in vivo from healthy human volunteers by the intubation technique (Faisant et al., 1993a; Molis et al., 1992). However, many factors influence the polymorphic form of α-(1 → 4)-glucans; the A-type crystals being favored, for instance, at high temperatures and shorter average chain lengths of amylose (Gidley and Bulpin, 1987; Eerlingen et al., 1993a). Recently, it has been reported that both the retrogradation conditions (Shamai et al., 2003) and the processing method of the starch (Lopez-Rubio et al., 2007) may give rise to different polymorphic crystals after the enzymatic digestion. Generally,

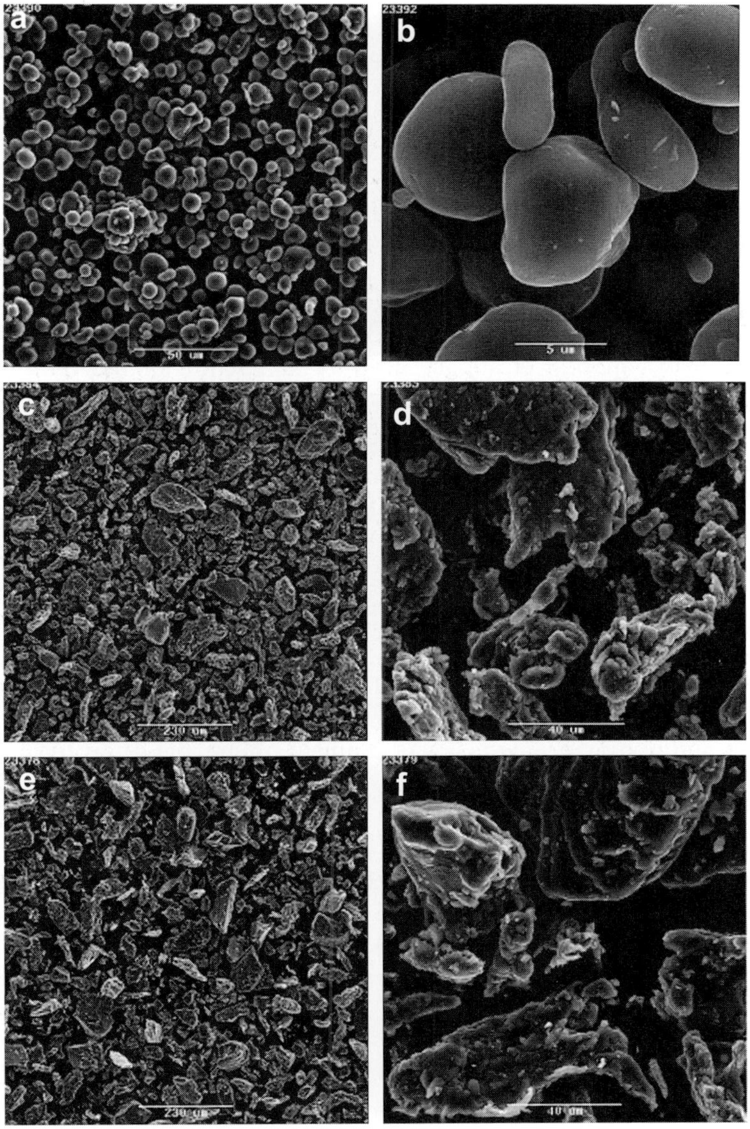

FIGURE 14.6 SEM images at low (a,c,e) and high (b,c,f) magnification of the enzyme-resistant fraction of high-amylose maize starch (a,b) non-extruded, (c,d) extruded under 'mild' conditions, and (e,f) extruded under 'extreme' conditions (Lopez-Rubio et al., 2007).

the diffraction patterns from RS display broader diffraction lines, suggesting that smaller and/or less perfect crystallites are present in the resistant fractions in comparison with the ones in native starches.

NMR spectra of starches can be interpreted as a composite of intensity features from ordered (double helical) and non-ordered (amorphous single chain) material (Gidley and Bociek, 1985). Results on RS samples from retrograded starch indicated a higher double helical content than in granular starch (Gidley et al., 1995). The technique has also proved useful to detect the presence of complexed lipid in the resistant fractions (Gidley et al., 1995).

FT-IR studies on starch have suggested that changes in intensity of bands present in the 'fingerprint' region of these polymers (i.e. the spectral region from 800–1200 cm^{-1}) can be used to quantify changes in polymer conformation and hydration of processed starches (Goodfellow and Wilson, 1990; vanSoest et al., 1995; Bello-Perez et al., 2005). Specifically, the bands at 1045 (or 995) and 1022 cm^{-1} have been linked with order/crystallinity and amorphous regions in starch, respectively (vanSoest et al., 1995; Bello-Perez et al., 2005). The intensity ratios of 1045/1022 and 995/1022 cm^{-1} have been observed to be higher for the resistant residues of processed starches, indicating a more ordered structure than before the enzyme treatment (Htoon et al., 2009). Another change in the spectra of RSIII is the increase in intensity of the bands around 1555 and 1653 cm^{-1}, which has been correlated with an increase in the number of reducing groups (Mahadevamma et al., 2003), coherent with the idea of a greater number of shorter chains obtained after hydrolysis.

Small-angle X-ray scattering (SAXS) has been shown to be useful for studying the lamellar arrangement in semi-crystalline starch granules (Waigh et al., 1996). SAXS patterns from hydrated native starches show a broad scattering peak, from which the average thickness of the lamellar repeat unit (that is the thickness of the crystalline plus amorphous region) can be calculated. In contrast with the well-defined long period present in granular starches which is kept (with smaller intensity) in RSII (Lopez-Rubio et al., 2007), after processing and digestion, the original lamellar organization is lost and the nanostructure of RSIII can be better described as crystals interspersed in an amorphous matrix (Shamai et al., 2004; Lopez-Rubio et al., 2007). This technique has also been used to follow the evolution of the nanostructure during the enzymatic process in extruded high-amylose maize starches, showing an increase in the molecular order as a function of digestion (Lopez-Rubio et al., 2008).

14.6.3 Mechanisms of Resistance to Enzymic Digestion of Starch Granules

The fraction of native (unprocessed) starch granules not digested in the small intestine is usually classified as RSII. In comparison with gelatinized starch, the hydrolysis of native granules is usually much slower (Ring et al., 1988). This can be explained by chains within the granules having restricted mobility and, thus, their constituent glucose units being more firmly locked into a specific configuration (Oates, 1997). In addition, the dense packing of polymers within granules is likely to provide some barrier to the access of hydrolytic enzymes. The mechanism of enzyme-catalyzed hydrolysis of $\alpha(1 \rightarrow 4)$-glycosidic bonds involves the enzyme-induced ring distortion of one D-glycosyl residue from the $^4C_1$-chair conformation to a 'half chair' conformation (Thoma, 1968). Potentially, amylases are able to unwind the double-helix structure in starches through the influence of ring distortion on neighboring units. This is likely to be more feasible at the ends of helices rather than in the middle, consistent with the more rapid hydrolysis of double helices based on amylopectin branches (probable helix DP of 10–20) than those based on amylose segments (probable average helix DP of 40–60). In gelatinized starch, where the chains are in random conformation, the enthalpy of activation is lower and, as a result, the glycosyl residues are more prone to nucleophilic attack by functional groups of the enzyme.

Enzymatic reaction involves several steps: diffusion to the solid surface, adsorption and catalysis. Therefore, hydrolysis of granular starch can be controlled by influencing each of these steps. The first step, i.e. the diffusion, is mainly determined by the food matrix, in a way that a high matrix viscosity or dense microstructure may prevent the enzymes from diffusing (quickly) to the granule surface. The adsorption and catalysis steps, on the other hand, are influenced both by enzyme-related factors and by the intrinsic properties of the granules.

Several studies have shown that the digestion of native starch is strongly influenced by the enzyme strain (Williamson et al., 1992; Valetudie et al., 1993; Planchot et al., 1995; Sarikaya et al., 2000). Enzyme concentration and reaction time are other factors influencing both rate and extent of hydrolysis. However, although the percentage of starch degradation and the mode of action of different enzymes might vary considerably, the trends in terms of rate and extent of hydrolysis when comparing different starch varieties remain the same (Valetudie et al., 1993; Planchot et al., 1995) and, therefore, for a given enzyme and digestion conditions, the degree of hydrolysis of a specific starch will only depend on its structural and molecular organization.

There are many factors which contribute to the relative resistance to digestion of granules from different starch sources. The interconnections between all these factors further complicate the story and there are still knowledge gaps to fill in order to fully understand the process.

The granular architecture and, more specifically, the surface organization is one of the main factors determining kinetics and degree of hydrolysis. Digestion of native starch proceeds in two stages: initial rapid hydrolysis which continues at a slower and more constant rate in the second stage (Bertoft and Manelius, 1992). Initial hydrolysis in most starches is superficial and will thus depend on the surface characteristics. The structural organization at the surface, for instance, conditions the adsorption of the enzymes, a step which is assumed to be a prerequisite for subsequent catalytic activity (Leloup et al., 1991). It is well known that native starch granules from potato or banana are highly resistant to digestion (Englyst et al., 1992), as well as high-amylose starches, like the commercially available high-amylose maizes or Novelose240 obtained from selective breeding and/or physicochemical processing techniques (Haralampu, 2000). Even though the composition and physical characteristics of them are different, the patterns of hydrolysis show some similarities. The main interesting feature common to highly resistant native starches is the heterogeneity in degradation, i.e. while some granules remain intact during the amylase treatment, others appear highly degraded (Valetudie et al., 1993; Faisant et al., 1995a, 1995b; Planchot et al., 1995). The organization of the outer shell in these granular varieties seems to pose difficulties for the adsorption of enzymes and/or subsequent catalysis which delays the digestion process. Once the outer shell barrier is overcome and catalysis initiated, hydrolysis occurs more rapidly but only affecting, as mentioned before, a few starch granules. Nevertheless, these highly resistant granules show different behavior during hydrolysis. In the case of potato starch, the surface of the granules is first roughened, some micropores are formed which evolve into larger pores and degradation occurs mainly due to exo-corrosion (Gallant et al., 1972, 1973, 1992). On the other hand, high-amylose starches show minor exo-corrosion and extensive endo-corrosion. Initially, rough surfaces are formed with superficial scratches and pores, through which the enzymes penetrate forming radially oriented channel patterns internally (Gallant et al., 1992). The outer shell of high-amylose maize granules is highly resistant, remaining practically intact (Evans and Thompson, 2004). In the case of banana starch granules, similar observations as for high-amylose starch were made, hydrolysis occurring mainly via endo-corrosion, with the granule surface showing less damage (Faisant et al., 1995b). A non-uniform enzyme attack of the granule and endo-corrosion was also observed in the most resistant tropical tuber starches (Valetudie et al., 1993). Surface porosity also affects the extent of hydrolysis as demonstrated for amylose and amylopectin gels (Leloup et al., 1990).

Another factor that is considered to influence digestion of granular starches is the starch composition (i.e. amylase:amylopectin ratio). Generally, the extent of hydrolysis is inversely correlated with the amount of amylose (Cone and Wolters, 1990; Vasanthan and Bhatty, 1996; Rendleman, 2000; Evans and Thompson, 2004; Li et al., 2004; Riley et al., 2004). Cereal starches with low-amylose contents are generally more susceptible to enzymatic hydrolysis than starches from roots, tubers and fruits (Valetudie et al., 1993; Themeier et al., 2005), while pea starches with medium amylose contents show markedly higher RS contents than cereal starches. However, the strong linear correlation between amylose and RS content valid for maize starches (Morita et al., 2007) cannot be found generally for pea starches from different varieties (Themeier et al., 2005). In contrast with the observations for the highly resistant starches, advanced and homogeneous degradation is observed for all granule sizes/shapes in the case of normal maize starch. In the case of waxy maize, extensive degradation, starting with large depressions on the surface and enlargement of channels by exo-corrosion followed by endo-corrosion occurs (Figures 14.4 and 14.5). It is therefore characteristic of poorly resistant starches, the absence of preferential hydrolysis zones and a uniform susceptibility to amylolysis (Planchot et al., 1995).

Crystal organization and, specially, the type of crystal polymorph are other factors that come into play when considering amylolysis. Starch resistance to α-amylase has been shown to be proportional to increasing amounts of B-type crystallites in starch (Hoover and Zhou, 2003). The B-polymorph is more resistant to digestion than the A- and C-polymorphs. Legume starches with intermediate resistance to hydrolysis display C-type diffraction patterns (Cairns et al., 1997; Bogracheva et al., 1998), while the A-type polymorph, even though it is characterized by a denser and better-packed monoclinic unit cell (Imberty et al., 1988), is considered to be the less resistant to digestion (medium- and low-amylose starches from cereals display A-type patterns, while potato and high-amylose starches are characterized by B-type crystallites). However, the polymorph alone cannot explain resistance to hydrolysis, as wrinkled pea starch, with 90% B-type polymorph (Cairns et al., 1997) only shows 18.4% RS, while the RS yield of a commercial high-amylose maize starch is 54.4% (Themeier et al., 2005).

Granular size also influences the digestion characteristics, but some contradictory results have been obtained when considering this parameter. Related to the size of the granules is the available specific surface: small granules, with more specific surface for the enzymes to adsorb, have been generally considered to be hydrolyzed more rapidly (Ring et al., 1988; Vasanthan and Bhatty, 1996). Apart from the rate, the extent of hydrolysis has also been negatively correlated with size for waxy barley starch (Tester and Qi, 2004). However, high-amylose maize starch is characterized by small granules but high resistance to hydrolysis (Planchot et al., 1995), and following the degradation of pea starch under a light microscope it was seen that, after 24 hours of digestion, the larger granules vanished and many smaller granules remained undigested (Themeier et al., 2005). Therefore, granule size has to be taken into consideration with other parameters to understand its influence in the digestion process. In a recent study, based on fractionation of several wheat starch varieties with similar amylose contents into large and small granule size populations, it was observed that, although the kinetics of digestion were initially faster for the small granules, after 4 hours of digestion, the rate of hydrolysis was faster for the large granules which also showed a greater extent of degradation (Salman et al., 2009). This seems to indicate that size will probably influence the kinetics during the first stages of digestion, which are mainly determined by the surface characteristics while, later on, the granular organization at the molecular level plays a major role.

The first stages of digestion are also apparently affected by the shape of the granules. For instance, in a study comparing different tropical tuber starches, it was observed that plane faces in the granules are more susceptible to attack, as hydrolysis often begins in the edges which delimit each plane of separation (Valetudie et al., 1993). When granules are small and rounded, hydrolysis is more difficult than when they are polyhedral. In the case of polyhedral granules the enzyme attack begins in the junction zones (Gallant et al., 1972, 1973; Fuwa et al., 1977). Moreover, the surface area per volume ratio is higher for polyhedral than for spherical granules (Tester et al., 2004).

Other factors that may influence digestion are, for instance, the presence of compound granules, which reduce the ability of amylases to bind, potentially restricting hydrolysis (Tester et al., 2004) and the presence of complexed lipids, which make amylose chains less readily accessible to the active site of the enzymes, although it does not confer complete resistance to hydrolysis, just a prolonged hydrolysis profile (Cheetham and Tao, 1998; Tester et al., 2004). Indeed, it is likely that all 'resistant' starches are capable of extensive further digestion with extended hydrolysis times. For instance, Htoon et al. (2009) showed that the RS recovered from both granular and processed high-amylose maize was extensively digested by a second round of in vitro digestion.

Many common cereal starches (regular and waxy maize, wheat, barley, sorghum) all have very low RS values in the granular form. This has been proposed to be due to a combination of numerous channels connecting the surface to the interior (Figures 14.4 and 14.5) and a side-by-side hydrolysis of both amorphous and crystalline regions as indicated by very similar XRD patterns after various stages of digestion (Zhang et al., 2006a, 2006b). This led Zhang et al. (2006a) to propose the model shown in Figure 14.7 to explain the slow but essentially complete digestion of most cereal starch granules (high-amylose variants are the obvious exception).

14.6.4 Mechanisms of Resistance to Enzymic Digestion of Processed Starches

During the 1980s, physicochemical studies were carried out to understand the generation of RS in processed starches (Siljestrom and Asp, 1985; Berry, 1986; Englyst and Macfarlane, 1986; Ring et al., 1988) and the accepted theory of RS formation was considered to be based on starch retrogradation or starch recrystallization (Figure 14.3). Upon cooling of gelatinized starch, the dispersed starch molecules re-associate spontaneously and the portion of the starch crystallites formed that resists enzymatic attack constituted the RS fraction. As with granular starches, the most important factor determining the extent of digestion is the starch composition, particularly amylase:amylopectin ratio, with the amylose content positively correlated with the RS yield (Sievert and Pomeranz, 1989). The crystal polymorph also plays an important role, with the B-polymorph the most resistant one, with a slower rate and lower extent of digestion than the A-polymorph as demonstrated using amylose spherulitic crystallites with the same DP and lintner starches (Williamson et al., 1992; Planchot et al., 1997).

The digestion conditions (enzyme strain used, time of digestion, temperature) also have an influence on the rate and extent of RSIII production (Jane and Robyt, 1984; Williamson et al., 1992; Eerlingen et al., 1993a), as well as food matrix density, which can influence the kinetics of hydrolysis by restricting the diffusion of enzymes. In food products, the presence of

TABLE 14.2 Chronological Compilation of Results from Selected Studies of Enzyme Action on Starch Granules (RSII)

Starch	Techniques used	Key finding	Literature ref
Potato, manioc, wheat, maize, waxy maize and high amylose maize starches	SEM and TEM	The type of attack is particular for each starch variety (some showing exocorrosion alone, others show endocorrosion at preferential sites as well)	Gallant et al., Starch (1973)
Tropical tuber starches	Light microscopy, scanning electron microscopy (SEM), transmission electron microscopy (TEM) and X-ray diffraction (XRD)	Degree of hydrolysis dependent on enzyme used and crystal polymorph. Polyhedric shaped granules more damaged than spherical ones. Hydrolysis by endocorrosion and pore size related to enzyme source	Valetudie et al., Starch (1993)
Wheat starch, maize starch, potato starch, wrinkled pea starch and smooth pea starch	XRD, SEM, TEM	Different extent of hydrolysis depending on enzyme type. Granular organization plays major role in enzymatic resistance	Planchot et al., Journal of Cereal Science (1995)
Raw banana starch flour	Ileostomy studies, TEM and SEM	Same hydrolysis patterns as potato and high amylose starches, i.e., heterogeneous hydrolysis and mainly endocorrosion	Faisant et al., European Journal of Clinical Nutrition (1995)
Maize starch	SEM and TEM	Visualization of enzyme attack by immuno-gold labelling of amylases	Helbert et al., International Journal of Biological Macromolecules (1996)
Potato and high amylose maize starch	HPSEC, DSC, RVA	Although both varieties show similar hydrolysis profiles, the RS fractions showed different structural features indicating a different mechanism of resistance	Jiang and Liu, Starch (2002)
4 commercial high amylose maize starches, normal maize starch and waxy maize starch	TEM, polarized light microscopy and SEC	Size and chain-length profile of RS fractions similar to those in native starches. Granular organization key to explain levels of RS	Evans and Thompson, Cereal Chemistry (2004)
7 cereal starches and 11 pea starches	DSC, SEM, granule size distribution, light microscopy	Besides amylose content, granule morphology, size, degree of crystallinity and extent of damage have a strong impact on amylase resistance	Themeier et al., Carbohydrate Polymers (2005)

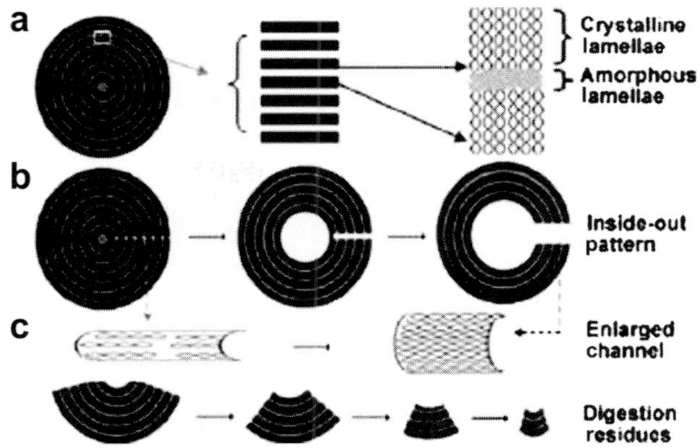

FIGURE 14.7 Schematic diagram to show the structure of cereal starch granule and the process of enzymatic digestion. (a) Crystalline and amorphous layer structure, (b) the inside-out layer-by-layer digestion starts at the channel and always ends at the crystalline layer (black), (c) side-by-side digestion to enlarge the internal channel (top) and the resulting pyramidal shape residuals with less and less layers after fragmentation of starch granules (bottom) (Zhang et al., 2006a).

other food components may retard the digestion of starch by slowing the approach of enzymes to starch or impeding adsorption of the enzymes through shielding of the carbohydrate chains. For instance, the kinetics of digestion have been observed to be slower in condensed products such as spaghetti or lasagne (Fardet et al., 1998). Furthermore, the RS values obtained in in vivo trials with healthy human volunteers were higher than the in vitro values obtained using three different in vitro validated digestion methods (Faisant et al., 1993a).

Several processing methods, aimed at increasing retrogradation and/or generating specific matrix structures that slow down digestion, have been implemented to increase the RS yields, as discussed in Section 14.4.

Nevertheless, recently it was observed that inducing retrogradation does not always lead to an increase in RS content and that, on the other hand, almost amorphous high-amylose starches provided high levels of the resistant fraction (Leeman et al., 2006; Chanvrier et al., 2007; Htoon et al., 2009). Based on structural studies of the starch samples before and after digestion (Lopez-Rubio et al., 2007), and following the evolution of the structure at the atomic, nano- and microscale, a new mechanism of RSIII formation from processed starches was proposed (Lopez-Rubio et al., 2008). It was observed that, during digestion, molecular order at the different length scales considered took place, suggesting that the shorter amylose chains, with intrinsic increased mobility, generated during the hydrolysis process are able to reorganize in the digestion medium itself, constituting the RSIII fraction. This enzymatic-induced molecular order was previously proposed by Cairns and co-workers based on studies with amylose gels, as the increased crystallinity observed after the α-amylase treatment, could not be totally ascribed to the elimination of amorphous fractions (Cairns et al., 1995). Further studies are required to examine the molecular structures associated with RSIII from near-amorphous high-amylose starches, but the results obtained to date are consistent with the model shown in Figure 14.8.

The presence of a significant amount of amorphous material in the RSIII fractions was usually explained on the basis of double helical resistance (Gidley et al., 1995) or due to protection of the amorphous portions by entrapment within the imperfect crystals (Jane

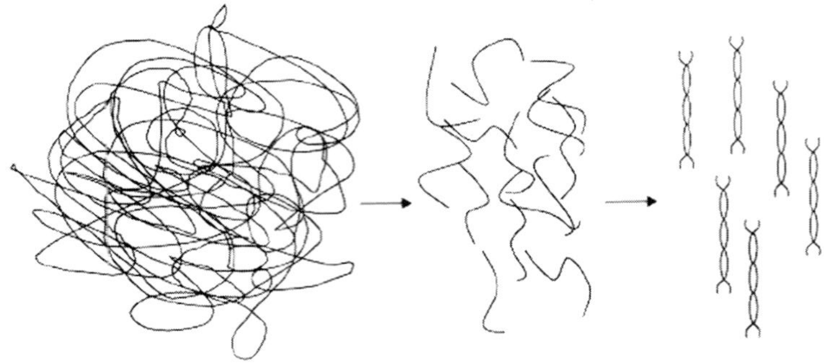

FIGURE 14.8 Schematic diagram of recrystallization during enzyme digestion of high-amylose starches. As the enzymes hydrolyze amylose molecules, shorter chains are formed, which will have increased mobility and reorganize/retrograde more rapidly in the digestion medium.

and Robyt, 1984; Leloup et al., 1992; Cairns et al., 1995), while the effect of storage in increasing the RS content was explained as a slower formation of more perfect crystals containing fewer internal defects (Cairns et al., 1995). Although it is clear that resistance is explained by the limited time the enzymes have to fully digest the starches, i.e. transit time, differences in resistant fractions between starch varieties are probably due to matrix density and/or organization, which control the extent to which the enzymes can access starch molecules delaying amylolytic degradation. Consequently, the RSIII yield of a specific starch variety will depend on the competition between (a) the kinetics of amylose retrogradation as these influence local molecular density and (b) the kinetics of amylase digestion.

14.7 CONCLUDING REMARKS

Whilst considerable progress has been made in understanding the biopolymer science underlying the variable resistance of starch-based materials to digestion by amylolytic enzymes, there are also several important unresolved questions. Some of the more important principles of RS formation are:

- Resistance of starch-based materials to amylolysis is rarely absolute; RS values correspond to a specific time point during digestion both in vitro and in vivo.
- Many cereal starch granules have low levels of RS (II) due to the accessibility of granule interiors to enzymes through pores or channels within the granule.
- Amylopectin-based crystallites are typically sufficiently susceptible to amylase to give very low RS yields whether present within granules (RSII) or after processing (RSIII). This is most likely due to the relatively short glucan chain lengths (10–20) involved in individual double helices, resulting in low stability.
- Amylose-based crystallites typically give significant RS (III) yields, due to the relatively long glucan chains (40–60?) involved in individual double helices, resulting in high stability.
- Recrystallization (retrogradation) of non-crystalline starch polymer segments can occur during the enzyme digestion process, notably for high-amylose starches. This is most likely due to the amylolytic production of intermediate-length amylose molecules which have sufficient mobility and rapid crystallization kinetics such that new resistant structures (RSIII) are formed.

TABLE 14.3 Chronological Compilation of Results from Selected Studies of Enzyme Action on Processed Starches (RSIII)

Starch	Process	Techniques used	Key finding	Literature ref
Amylose complexes and amylose from potato starch	Retrogradation of amylose (DP~1000) in fridge for 1–2 weeks. Formation of V-complexes with different organic molecules	Thin layer and gel permeation chromatography (GPC), X-ray diffraction (XRD)	Differences in sizes of resistant amylodextrins depend on differences in specificities of hydrolysing agents. In-depth description of retrograded amylose based on results	Jane and Robyt, Carbohydrate Research (1984)
Wheat starch, high-amylose maize starch and waxy maize starch	Debranching of waxy starch, autoclaving and drying (or freeze-drying)	Gel permeation chromatography (GPC), iodine binding	The RS fraction from debranched waxy maize had lower DP (~40) than the fractions obtained from wheat and high amylose maize (~60). Very low RS levels for branched waxy maize, even after lintnerization and autoclaving	Russell et al., Journal of Cereal Science (1989)
Wheat starch, maize starch	Autoclaving at 1.4 bar for 1 hour and cooled at room temperature	XRD, differential scanning calorimetry (DSC), iodine binding, GPC, gas liquid chromatography (GLC)	RS consist of crystallized linear α-glucans of relatively low molecular weight (DP~65)	Siljeström et al., Starch (1989)
Wheat starch, pea starch, potato starch, waxy maize starch and two high-amylose maize starches	Autoclaving (at 121°C, 134°C and148°C)/ cooling (4°C) cycles repeated up to 20 times. Different starch/water ratios were used	DSC and scanning electron microscopy (SEM)	RS yield increased with decreasing levels of water and increasing number of autoclaving-cooling cycles. Linear relationship between melting enthalpy and RS content	Sievert and Pomeranz, Cereal Chemistry (1989)
Wheat starch, pea starch, potato starch, waxy maize starch and two high-amylose maize starches	Autoclaving (at 121°C, 134°C and 148°C)/ cooling (4°C) cycles	DSC	The endothermic transition in RS at ~155°C (due to recrystallized amylose) could not be detected directly in processed starches by DSC	Sievert and Pomeranz, Cereal Chemistry (1990)
High-amylose maize starch	Autoclaving (at 121°C, 134°C and148°C)/ cooling (4°C) cycles	DSC and XRD	Changes in crystallinity as a consequence of repeated autoclaving/cooling cycles. Poor B-type XRD patterns of RS fractions	Sievert et al., Cereal Chemistry (1991)

(Continued)

TABLE 14.3 Chronological Compilation of Results from Selected Studies of Enzyme Action on Processed Starches (RSIII)—*Cont'd*

Starch	Process	Techniques used	Key finding	Literature ref
Oligosaccharides (DP~20) from potato starch	Lintnerization of potato starch and spherulitic crystallization	DSC, SEM and XRD	B-type structure hydrolysed more slowly than A-type. Different patterns of enzyme attack depending on the crystal polymorph	Williamson et al., Carbohydrate Polymers (1992)
Wheat starch	Autoclaving and incubation at different temperatures (0°C, 68°C and 100°C)	XRD	Different crystal polymorphs depending on the incubation temperature (A-type being favoured at high temperatures)	Eerlingen et al., Cereal Chemistry (1993a)
Potato starch (amylose)	Amylose with different chain lengths were heated to 100°C in water bath and allowed to cool overnight	GPC and XRD	DP of RS fractions is independent of the chain length of the amylose it was formed from but RS yield increases with amylose DP up to 100 and it levels above that	Eerlingen et al., Cereal Chemistry (1993b)
High-amylose maize starch	Extrusion and cooling. Autoclaving with monoglycerides ('complexed' starch)	Size exclusion chromatography (SEC), HPLC, DSC and XRD	Lower in vitro RS values, but similar thermal and crystalline organization after in-vivo and in-vitro digestion	Faisant et al., Carbohydrate Polymers (1993)
Amylose from pea starch	Formation of gels and dry films	XRD, SEC	Enzymic hydrolysis promoted crystallization. Three main molecular size sub-fractions present in RS	Cairns et al., Journal of Cereal Science (1995, 1996)
Wheat, 2 high-amylose maize starches, potato starch	Autoclaving and drying (or freeze-drying)	DSC, XRD, ^{13}C CP/MAS NMR spectroscopy, HPAEC	Increased double helical content and crystallinity in RS fractions in comparison with granular starches	Gidley et al., Carbohydrate Polymers (1995)
Commercial spaghetti and lasagne	Cooked for 10 minutes	SEC, light microscopy, SEM	Lower digestion rates in food products (pasta) explained by increased tortuosity of the enzyme provided by protein network	Fardet et al., Journal of Cereal Science (1998)
Rice starches and ragi starches	4 autoclaving/cooling cycles followed by retrogradation at different conditions	XRD and DSC	Defatting of starches significantly increases RS yield	Mangala et al., Food Chemistry (1999)

High-amylose maize starch, wheat starch and maize flour	Autoclaving and storing in oven at 40°C or 90°C for 24 hours	DSC and XRD	The polymorph structure of the RS fraction is determined by the crystallization temperature	Shamai et al., Carbohydrate Polymers (2003)
Potato and sweet potato starches	Partial acid hydrolysation, autoclaving/cooling cycles and heat-moisture treatment	DSC and XRD	Significant increase in RS contents after hydrolysis followed by autoclaving/cooling and annealing. New and sharper peaks in XRD from RS fractions	Shin et al., Cereal Chemistry (2004)
High-amylose maize starch	Autoclaving and storing in oven at 40°C or 90°C for 24 hours	Small angle X-ray scattering (SAXS)	Description of the nanostructure of RS fractions as a dilute lamellar system, i.e. crystallites dispersed in an amorphous matrix	Shamai et al., Biomacromolecules (2004)
Waxy rice starch	Partial gelatinization + retrogradation	DSC	At similar levels of melting enthalpy, partially gelatinized samples are more resistant than retrograded starch	Chung et al., Journal of Cereal Science (2006)
High-amylose maize starch	Extrusion at two different processing conditions	SEM, XRD, Ultra small angle X-ray scattering (USAXS) and SAXS	Increased molecular order and crystallinity of the RS fractions which seems to indicate that reorganization takes place during digestion	Lopez-Rubio et al., Biomacromolecules (2007)
Regular maize starch, high-amylose maize starch and wheat starch	Autoclaving and retrogradation at 40°C and 90°C	XRD) and SAXS	RS yield higher for the samples retrograded at 40°C, but hydrolysis of the RS fractions was lower for the starches retrograded at 90°C	Zabar et al., Macromolecular Bioscience (2008)
Two high-amylose starches	Extrusion in capillary rheometer at two different processing conditions	Fourier transformed infrared (FT-IR) spectroscopy, ^{13}C NMR, XRD	Measured levels of enzyme resistance are due to the interruption of a slow digestion process, rather than the presence of completely indigestible material	Htoon et al., Carbohydrate Polymers (2008)
High-amylose starch	Extrusion in capillary rheometer at two different processing conditions	SAXS, XRD, ^{13}C NMR, SEM and FT-IR	Development of crystallinity and molecular order as a function of digestion suggesting that the mechanism of enzyme resistance is based on the competition between the kinetics of digestion and the kinetics of amylose retrogradation	Lopez-Rubio et al., Biomacromolecules (2008)

Some of the important unresolved questions are:

- What are the most appropriate in vitro assay conditions that best mirror in vivo digestive conditions, and which can therefore form the basis for an accurate in vitro assay?
- How does the microstructure and molecular composition of in vivo RS differ from that of in vitro RS (II and III) from a range of sources?
- Why do some granules (potato, banana, high-amylose maize) have so much slower kinetics of amylolysis than most cereal starch granules?
- Why do isolated RS (II and III) samples have such high levels (>50%) of amorphous material that might be expected to be rapidly digested by amylase?

References

Abd-Eloneim-Elkhalifa, Schiffler, B. and Bernhard, R. (2004). Effect of fermentation on the starch digestibility, resistant starch, and some physicochemical properties of sorghum flour. *Narhung/Food* **48**, 91–104.

Adamu, B. O. A. (2001). Resistant starch derived from extruded corn starch and guar gum as affected by acid and surfactants: structural characterization. *Starch/Starke* 3, 582–591.

Agustiniano-Osornio, J. C., Gonzalez-Soto, R. A., Flores-Huicochea, E., Manrique-Quevedo, N., Sanchez-Hernandez, L., and Bello-Perez, L. A. (2005). Resistant starch production from mango starch using a single-screw extruder. *Journal of the Science of Food and Agriculture* **85**, 2105–2110.

Ahmed, R., Segal, I., and Hassan, H. (2000). Fermentation of dietary starch in humans. *American Journal of Gastroenterology* **95**, 1017–1020.

Akerberg, A., Liljeberg, H., and Bjorck, I. (1998). Effects of amylose/amylopectin ratio and baking conditions on resistant starch formation and glycemic index. *Journal of Cereal Science* **28**, 71–80.

Anderson, I. H., Levine, A. S., and Levitt, M. D. (1981). Incomplete absorption of the carbohydrate in all-purpose wheat flour. *New England Journal of Medicine* **304**, 891–892.

Andersson, H. (1992). The ileostomy model for the study of carbohydrate digestion and carbohydrate effects on sterol excretion in man. *European Journal of Clinical Nutrition* **46** (Suppl. 2), S69–S76.

Ao, Z., Quezada-Calvillo, R., Sim, L., Nichols, B. L., Rose, D. R., Sterchi, E. E., and Hamaker, B. R. (2007). Evidence of native starch degradation with human small intestinal maltase- glucoamylase (recombinant). *FEBS Letters* **581**, 2381–2388.

Aparicio-Saguilan, A., Flores_Huicochea, E., Juscelino, T., Garcia-Suarez, F., Gutierrez-Meraz, F., and Bello-Perez, L. A. (2005). Resistant starch-rich powders prepared by autoclaving of native and lintnerized banana starch: partial characterization. *Starch/Starke* **57**, 405–412.

Asp, N. G. (1992). Resistant starch. *European Journal of Clinical Nutrition* **46** (Suppl 2), S1.

Baghurst, P. A., Baghurst, K. I., and Record, S. J. (1996). Dietary fibre, non-starch polysaccharides and resistant starch – a review. *Supplement to Food Australia* **48**, S3–S35.

Behall, K. M. and Howe, J. C. (1996). Resistant starch as energy. *Journal of the American College of Nutrition* **15**, 248–254.

Behall, K. M. and Hallfrisch, J. (2002). Plasma glucose and insulin reduction after consumption of breads varying in amylose content. *European Journal of Clinical Nutrition* **56**, 913–920.

Behall, K. M., Howe, J. C., and Anderson, R. A. (2002). Apparent mineral retention is similar in control and hyperinsulinemic men after consumption of high amylose cornstarch. *Journal of Nutrition* **132**, 1886–1891.

Behall, K. M., Scholfield, D. J., Hallfrisch, J. G., and Liljeberg-Elmstahl, H. G. M. (2006). Consumption of both resistant starch and β-glucan improves postprandial plasma glucose and insulin in women. *Diabetes Care* **29**, 976–981.

Bello-Perez, L. A., Ottenhof, M. A., Agama-Acevedo, E., and Farhat, I. A. (2005). Effect of storage time on the retrogradation of banana starch extrudate. *Journal of Agriculture and Food Chemistry* **53**, 1081–1086.

Berry, C. S., I'Anson, K., Miles, M. J., and Morris, V. J. (1988). Physical chemical characterization of resistant starch from wheat. *Journal of Cereal Science* **8**, 203–206.

Berry, C. S. (1986). Resistant starch: formation and measurement of starch that survives exhaustive digestion with amylolytic enzymes during the determination of dietary fibre. *Journal of Cereal Science* **4**, 301–314.

Bertoft, E. (2001). Investigation of the fine structure of alpha-dextrins derived from amylopectin and their relation to the structure of waxy-maize starch. *Carbohydrate Research* **212**, 229–241.

Bertoft, E. and Manelius, R. (1992). A method for the study of the enzymatic-hydrolysis of starch granules. *Carbohydrate Research* **227**, 269–283.

Biliaderis, C. G., Page, C. M., Slade, L., and Sirett, R. R. (1985). Thermal-behavior of amylose-lipid complexes. *Carbohydrate Polymers* **5**, 367–389.

Biliarderis, C. G. (1992). Structures and phase transitions of starch in food systems. *Food Technology* **46**, 98–109.

Binder, H. J. and Ramakrishna, B. S. (2004). Resistant starch – an adjunct to oral rehydration solution: not yet ready for prime time. *Journal of Pediatric Gastroenterology and Nutrition* **39**, 325–327.

Bird, A. R. and Topping, D. L. (2001). Resistant starch, fermentation, and large bowel health. *In* Handbook of Dietary Fiber, 147–158, Cho & ML Dreher, S. S. (ed.). Marcel Dekker, NY.

Bird, A. R. and Topping, D. L. (2008). Resistant starch as a prebiotic. *In* Therapeutic Microbiology: Probiotics and Related Strategies, 159–173, Versalovic, J. and Wilson, M. (eds.). ASM Press, Washington.

Bird, A. R., Croom, W. J., Jr., Fan, Y. K., Black, B. L., McBride, B. W., and Taylor, I. L. (1996). Peptide regulation of intestinal glucose absorption. *Journal of Animal Science* **74**, 2523–2540.

Bird, A. R., Brown, I. L., and Topping, D. L. (2000). Starches, resistant starches, the gut microflora and human health. *Current Issues in Intestinal Microbiology* **1**, 25–37.

Bird, A. R., Vuaran, M. S., King, R. A., Noakes, M., Keogh, J., Morell, M. K., and Topping, D. L. (2007). A novel high-amylose barley cultivar (Himalaya 292) improves indices of bowel health in humans. *British Journal of Nutrition* **92**, 607–615.

Birkett, A., Muir, J., Phillips, J., Jones, G., and O'Dea, K. (1996). Resistant starch lowers fecal concentrations of ammonia and phenols in humans. *American Journal of Clinical Nutrition* **63**, 766–772.

Birkett, A. M., Jones, G. P., de Silva, A. M., Young, G. P., and Muir, J. G. (1997). Dietary intake and faecal excretion of carbohydrate by Australians: importance of achieving stool weights greater than 150g to improve faecal markers relevant to colon cancer risk. *European Journal of Clinical Nutrition* **51**, 625–632.

Bjorck, I., Nyman, M., Pedersen, B., Siljestrom, M., Asp, N. G., and Eggum, B. O. (1987). Formation of enzyme resistant starch during autoclaving of wheat-starch-studies in vitro and in vivo. *Journal of Cereal Science* **6**, 159–172.

Bogracheva, T. Y., Morris, V. J., Ring, S. G., and Hedley, C. L. (1998). The granular structure of C-type pea starch and its role in gelatinization. *Biopolymers* **45**, 323–332.

Borchers, R. (1961). A note on the digestibility of the starch of high amylose corn by rats. *Cereal Chemistry* **38**, 145–146.

Botham, R. L., Cairns, P., Morris, V. J., Ring, S. G., Englyst, H. N., and Cummings, J. H. (1995). A physicochemical characterization of chick pea starch resistant to digestion in the human small intestine. *Carbohydrate Polymers* **26**, 85–90.

Bravo, L., Siddhuraju, P., and Saura-Calixto, F. (1998). Effect of various processing methods on the in vitro starch digestibility and resistant starch content of Indian pulses. *Journal of Agricultural and Food Chemistry* **46**, 4667–4674.

Brighenti, F., Casiraghi, M. C., and Baggio, C. (1998). Resistant starch in the Italian diet. *British Journal of Nutrition* **80**, 333–341.

Brouns, F., Kettlitz, B., and Arrigoni, E. (2002). Resistant starch and "the butyrate revolution" *Trends in Food Science and Technology* **13**, 251–261.

Brouns, F., Arrigoni, E., Langkilde, A. M., Verkooijen, I., Fassler, C., Andersson, H., Kettlitz, B., van Nieuwenhoven, M., Philipsson, H., and Amadò, R. (2007). Physiological and metabolic properties of a digestion-resistant maltodextrin, classified as type 3 retrograded resistant starch. *Journal of Agriculture and Food Chemistry* **55**, 1574–1581.

Cairns, P., Sun, L., Morris, V. J., and Ring, S. G. (1995). Physicochemical studies using amylose as an in-vitro model for resistant starch. *Journal of Cereal Science* **21**, 37–47.

Cairns, P., Botham, R. L., Morris, V. J., and Ring, S. G. (1996). Physiochemical studies on resistant starch in vitro and in vivo. *Journal of Cereal Science* **23**, 265–275.

Cairns, P., Bogracheva, T. Y., Ring, S. G., Hedley, C. L., and Morris, V. J. (1997). Determination of the polymorphic composition of smooth pea starch. *Carbohydrate Polymers* **32**, 275–282.

Cairns, P., Leloup, V., and Miles, M. J. (1990). Resistant starch – an X-ray diffraction study into the effect of enzymatic-hydrolysis on amylose gels in vitro. *Journal of Cereal Science* **12**, 203–206.

Cassidy, A., Bingham, S. A., and Cummings, J. H. (1994). Starch intake and colorectal-cancer risk – an international comparison. *British Journal of Cancer* **69**, 937–942.

Champ, M. (1992). Determination of resistant starch in foods and food products: interlaboratory study. *European Journal of Clinical Nutrition* **46** (Suppl. 2), S51–S62.

Champ, M. M. J. (1996). The analysis of complex carbohydrates: relevance of values obtained *in vitro*. *Proceedings of the Nutrition Society* **55**, 863–880.

Champ, M. M.-J. (2004). Physiological aspects of resistant starch and in vivo measurements. *Journal of AOAC International* **87**, 749–755.

Champ, M. M.-J., Molis, C., Flourié, B., Bornet, F., Pellier, P., Colonna, P., Galmiche, J.-P., and Rambaud, J.-C. (1998). Small-intestinal digestion of partially resistant cornstarch in healthy subjects. *American Journal of Clinical Nutrition* **68**, 705–710.

Champ, M., Martin, L., Noah, L., and Gratas, M. (2001a). In vivo techniques to quantify resistant starch. In: "Complex Carbohydrates in Foods". S. Sungsoo Cho, L. Prosky and M. Dreher (eds.). Marcel Dekker, 157–168.

Champ, M., Martin, L., Noah, L., and Gratas, M. (2001b). Analytical methods for resistant starch. In: "Complex

Carbohydrates in Foods". S. Sungsoo Cho, L. Prosky and M. Dreher (eds.). Marcel Dekker, 169–187.

Champ, M., Langkilde, A. M., Brouns, F., Kettlitz, B., and Le Bail-Collet, Y. (2003). Advances in dietary fibre characterisation. 2. Consumption, chemistry, physiology and measurement of resistant starch; implications for health and food labelling. *Nutrition Research Reviews* **16**, 143–161.

Chanvrier, H., Uthayakumaran, S., Appelqvist, I. A. M., Gidley, M. J., Gilbert, E. P., and López-Rubio, A. (2007). Influence of storage conditions on the structure, thermal behavior, and formation of enzyme-resistant starch in extruded starches. *Journal of Agriculture and Food Chemistry* **55**, 9883–9890.

Chapman, R. W., Sillery, J. K., Graham, M. M., and Saunders, D. R. (1985). Absorption of starch by healthy ileostomates: effect of transit time and of carbohydrate load. *The American Journal of Clinical Nutrition* **41**, 1244–1248.

Cheetham, N. W. H. and Tao, L. P. (1998). Variation in crystalline type with amylose content in maize starch granules: an X-ray powder diffraction study. *Carbohydrate Polymers* **36**, 277–284.

Chinnaswamy, R. and Hanna, M. A. (1990). Macromolecular and functional-properties of native and extrusion-cooked corn starch. *Cereal Chemistry* **67**, 490–499.

Chung, H. J., Lim, H. S., and Lim, S. T. (2006). Effect of partial gelatinization and retrogradation on the enzymatic digestion of waxy rice starch. *Journal of Cereal Science* **43**, 353–359.

Clarke, J. M., Bird, A. R., Topping, D. L., and Cobiac, L. (2007). Excretion of starch and esterified short chain fatty acids by ileostomy subjects after ingestion of acylated starches. *American Journal of Clinical Nutrition* **86**, 1146–1151.

Colonna, P., Doublier, J. L., Melcion, J. P., Monredon, F., and Mercier, C. (1984). Extrusion-cooking and drum drying of wheat starch. *I. physical and macromolecular modifications Cereal Chemistry* **61**, 538.

Cone, J. W. and Wolters, M. G. E. (1990). Some properties and degradability of isolated starch granules. *Starch-Starke* **42**, 298–301.

Cooke, D. and Gildley, M. J. (1992). Loss of crystalline and molecular order during starch gelatinization: origin of the enthalpic transition. *Carbohydrate Research* **227**, 103–112.

Coudray, C., Demigné, C., and Rayssiguier, Y. (2003). Effects of dietary fibers on magnesium absorption in animals and humans. *Journal of Nutrition* **133**, 1–4.

Cummings, J. H., Bingham, S. A., Heaton, K. W., and Eastwood, M. A. (1992). Fecal weight, colon cancer risk, and dietary intake of nonstarch polysaccharides (dietary fiber). *Gastroenterology* **103**, 1783–1789.

Cummings, J. H. (1997). The large intestine in nutrition and disease. Institut Danone, Brussels, Belgium.

Czuchajowska, Z., Sievert, D., and Pomeranz, Y. (1991). Enzyme-resistant starch. IV. Effects of complexing starch. *Cereal Chemistry* **68**, 537–542.

Danjo, K., Nakaji, S., Fukuda, S., Shimoyama, T., Sakamoto, J., and Sugawara, K. (2003). The resistant starch level of heat moisture-treated high amylose cornstarch is much lower when measured in the human terminal ileum than when estimated in vitro. *Journal of Nutrition* **133**, 2218–2221.

Delcour, J. A. and Eerlingen, R. C. (1996). Analytical implications of the classification of resistant starch as dietary fiber. *Cereal Foods World* **41**, 85–86.

Dewit, O., Prentice, A., Coward, W. A., and Weaver, L. T. (1992). Starch digestion in young children with cystic fibrosis measured using a 13C breath test. *Pediatric Research* **32**, 45–49.

Diamond, J. and Hammond, K. (1992). The matches, achieved by natural selection, between biological capacities and their natural loads. *Experientia* **48**, 551–557.

Dysseler, P. and Hoffem, D. (1994a). Comparison between Englyst's method and Berry's modified method on 20 different starch foods. Proceedings of the Concluding Plenary Meeting of EURESTA. 95–98.

Dysseler, P. and Hoffem, D. (1994b). Estimation of resistant starch intake in Europe. Proceedings of the Concluding Plenary Meeting of EURESTA. 84–86.

Eerlingen, R. C. and Delcour, J. A. (1995). Formation, analysis, structure and properties of type III enzyme resistant starch. *Journal of Cereal Science* **22**, 129–138.

Eerlingen, R. C., Cillen, G., and Delcour, J. A. (1994). Enzyme-resistant starch. IV. Effect of endogenous lipid and added sodium dedecyl sulfate on formation of resistant starch. *Cereal Chemistry* **71**, 170–177.

Eerlingen, R. C., Bjorck, I. V. D., Delcour, J. A., Slade, L., and Levine, H. (1994). Enzyme-resistant starch. VI. Influence of sugars on resistant starch formation. *Cereal Chemistry* **71**, 472–476.

Eerlingen, R. C., Jacobs, H., and Delcour, J. A. (1994). Enzyme resistant starch. V. Impact of amylopectin retrogradation on enzyme susceptibility. *Cereal Chemistry* **71**, 351–355.

Eerlingen, R. C., Vanhesendonck, I. P., Depaepe, G., and Delcour, J. A. (1994). Enzyme-resistant starch. III. The quality of straight-dough bread containing varying levels of enzyme-resistant starch. *Cereal Chemistry* **71**, 165–170.

Eerlingen, R. C., Crombez, M., and Delcour, J. A. (1993a). Enzyme resistant starch. I. Quantitative and qualitative influence of incubation time and temperature of autoclaved starch on resistant starch formation. *Cereal Chemistry* **70**, 339–344.

Eerlingen, R. C., Deceuninck, M., and Delcour, J. A. (1993b). Enzyme resistant starch. II. Influence of amylose chain length on resistant starch formation. *Cereal Chemistry* **70**, 345–350.

Elia, M. and Cummings, J. H. (2007). Physiological aspects of energy metabolism and gastrointestinal effects of carbohydrates. *European Journal of Clinical Nutrition* **61** (Suppl. 1), S40–S74.

Englyst, H., Wiggins, H. S., and Cummings, J. H. (1982). Determination of the non-starch polysaccharides in plant foods by gas-liquid chromatography of constituent sugars as alditol acetates. *Analyst* **107**, 307–318.

Englyst, H. N., Anderson, V., and Cummings, J. H. (1983). Starch and non-starch polysaccharides in some cereal foods. *Journal of the Science of Food and Agriculture* **34**, 1434–1440.

Englyst, H. N., Kingman, S. M., Hudson, G. J., and Cummings, J. H. (1996). Measurement of resistant starch in vitro and in vivo. *British Journal of Nutrition* **75**, 749–755.

Englyst, H. N., Kingman, S. M., and Cummings, J. H. (1992). Classification and measurement of nutritionally important starch fractions. *European Journal of Clinical Nutrition* **46**, S33–S50.

Englyst, H. N. and Macfarlane, G. T. (1986). Breakdown of resistant and readily digestible starch by human gut bacteria. *Journal of the Science of Food and Agriculture* **37**, 699–706.

Escarpa, A., Gonzalez, M. C., Morales, M. D., and Saura-Calixto, F. (1997). An approach to the influence of nutrients and other food constituents on resistant starch formation. *Food Chemistry* **60**, 527–532.

Evans, A. and Thompson, D. B. (2004). Resistance to alpha-amylase digestion in four native high-amylose maize starches. *Cereal Chemistry* **81**, 31–37.

Faisant, N., Buléon, A., Colonna, P., Molis, C., Lartigue, S., Galmich, J. P., and Champ, M. (1995a). Digestion of raw banana starch in the small-intestine of healthy humans – structural features of resistant starch. *British Journal of Nutrition* **73**, 111–123.

Faisant, N., Champ, M., Colonna, P., and Buléon, A. (1993a). Structural discrepancies in resistant starch obtained in-vivo in humans and in-vitro. *Carbohydrate Polymers* **21**, 205–209.

Faisant, N., Gallant, D. J., Bouchet, B., and Champ, M. (1995b). Banana starch breakdown in the human small-intestine studied by electron-microscopy. *European Journal of Clinical Nutrition* **49**, 98–104.

Faisant, N., Champ, M., Colonna, P., Buleon, A., Molis, C., Langkilde, A. M., Schweizer, T., Schwizer, B. Flourie and Galmiche, J. P. (1993b). Structural features of resistant starch at the end of human small intestine. *European Journal of Clinical Nutrition* **47**, 285–296.

Faraj, A., Vasanthan, T., and Hoover, R. (2004). The effect of extrusion cooking on resistant starch formation in waxy and regular barley flours. *Food Research International* **37**, 517–525.

Fardet, A., Hoebler, C., Baldwin, P. M., Bouchet, B., Gallant, D. J., and Barry, D. J. (1998). Involvement of the protein network in the in vitro degradation of starch from spaghetti and lasagne: a microscopic and enzymic study. *Journal of Cereal Science* **27**, 133–145.

Ferrero, C., Martino, M. N., and Zaritzky, N. E. (1993). Stability of frozen starch pastes: Effect of freezing, storage and xanthan gum addition. *Journal of Food Processing and Preservation* **17**, 191–211.

Finegold, S. M., Sutter, V. L., Boyle, J. D., and Shimada, K. (1970). The normal flora of ileostomy and transverse colostomy effluents. *Journal of Infectious Disease* **122**, 376–381.

Fredriksson, H., Bjorck, R., Andersson, H., Liljeberg, H., Silverio, J., Eliasson, A. C., and Aman, P. (2000). Studies on α-amylase degradation of retrograded starch gels from waxy maize and high amylopectin potato. *Carbohydrate Polymers* **43**, 81–87.

Fried, M., Abramson, S., and Meyer, J. H. (1987). Passage of salivary amylase through the stomach in humans. *Digestive Diseases and Sciences* **32**, 1097–1103.

Fuwa, H., Glover, D. V., Sugimoto, Y., Nishimura, R., and Tanaka, M. (1977). Comparative susceptibility to amylases of starches from different plant species and several single endosperm mutants and their double-mutant combinations with opaque-2 inbred oh43 maize. *Cereal Chemistry* **54**, 230–237.

Gallant, D., Derrien, A., Aumaitre, A., and Guilbot, A. (1973). Dégradation in vitro de l'amidon par le suc pancréatique. Etude par microscopie électronique à transmission et à balayage. *Starch - Stärke* **25**, 56–64.

Gallant, D., Mercier, C., and Guilbot, A. (1972). Electron Microscopy of Starch Granules Modified by Bacterial Alpha-Amylase. *Cereal Chemistry* **49**, 354–365.

Gallant, D. J., Bouchet, B., Buléon, A., and Perez, S. (1992). Physical characteristics of starch granules and susceptibility to enzymatic degradation. *European Journal of Clinical Nutrition* **46**, S3–S16.

Garcia-Alonso, A. and Goni, I. (2000). Effect of processing on potato starch: In vitro availability. *Starch/Starke* **2–3**, 81–84.

Garcia-Alonso, A., Jimenez-Escrig, A., Martin-Carron, N., Bravo, L., and Saura-Calixto, F. (1999). Assessment of some parameters involved in the gelatinization and retrogradation of starch. *Food Chemistry* **66**, 181–187.

Gelroth, J. A. and Ranhotra, G. S. (2000). Determination of resistant starch in selected grain-based foods. *Journal of AOAC International* **83**, 988–991.

Geng, Z., Zongdao, C., and Toledo, R. (2003). Effects of different processing methods on the glycemic index of kudzu starch. *Journal of Chinese Cereals Oils Association* **18**, 5–10.

Gerard, C., Colonna, P., Buléon, A., and Planchot, V. (2001). Amylolysis of maize mutant starches. *Journal of the Science of Food and Agriculture* **81**, 1281–1287.

Gidley, M. J., Cooke, D., Darke, A. H., Hoffman, R. A., Russell, A. L., and Greenwell, P. (1995). Molecular order and structure in enzyme-resistant retrograded starch. *Carbohydrate Polymers* **28**, 23–31.

Gidley, M. J. and Bociek, S. M. (1985). Molecular-organization in starches – a c-13 cp mas nmr-study. *Journal of the American Chemical Society* **107**, 7040–7044.

Gidley, M. J. and Bulpin, P. V. (1987). Crystallization of maltooligosaccharides as models of the crystalline forms of starch: minimum chain length requirement for the formation of double helices. *Carbohydrate Research* **6**, 291–300.

Gidley, M. J. and Bulpin, P. V. (1989). Aggregation of amylose in aqueous systems-effect of chain-length on phase-behavior and aggregation kinetics. *Macromolecules* **22**, 341–346.

Goldring, J. M. (2004). Resistant starch: safe intakes and legal status. *Journal of AOAC International* **87**, 733–739.

Goñi, I., García-Diz, L., Mañas, E., and Saura-Calixto, F. (1996). Analysis of resistant starch: a method for foods and food products. *Food Chemistry* **56**, 445–449.

Gonzalez-Soto, R. A., Escobedo, R., Hernandez-Sanchez, H., Sanchez-Rivera, M., and Bello-Perez, L. A. (2006). The influence of time and storage temperature on resistant starch formation from autoclaved debranched banana starch. *Food Research International* **40**, 304–310.

Goodfellow, B. J. and Wilson, R. H. (1990). A fourier-transform ir study of the gelation of amylose and amylopectin. *Biopolymers* **30**, 1183–1189.

Granfeldt, Y., Björck, I., Drews, A., and Tovar, J. (1992). An in vitro procedure based on chewing to predict metabolic response to starch in cereal and legume products. *European Journal of Clinical Nutrition* **46**, 649–660.

Gudmand-Hoyer, E. (1991). Methodological aspects of in vivo methods for measurement of starch digestibility. Report of a European Flair Concerted Action Workshop. Vedback, Copenhagen, Denmark. EURESTA. 10–20.

Gudmundsson, M. (1994). Retrogradation of starch and the role of its components. *Thermochimica Acta* **246**, 329–341.

Hammond, K. A. and Diamond, J. (1997). Maximal sustained energy budgets in humans and animals. *Nature* **386**, 457–462.

Haralampu, S. G. (2000). Resistant starch – a review of the physical properties and biological impact of RS3. *Carbohydrate Polymers* **41**, 285–292.

Hashimoto, N., Ito, Y., Han, K. H., Shimada, K. I., Sekikawa, M., Topping, D. L., Bird, A. R., Noda, T., Chiji, H., and Fukushima, M. (2006). Potato pulps lowered the serum cholesterol and triglyceride levels in rats. *Journal of Nutrition Science Vitaminol (Tokyo)* **52**, 445–450.

Helbert, W., Schülein, M., and Henrissat, B. (1996). Electron microscopic investigation of the diffusion of Bacillus licheniformis alpha-amylase into corn starch granules. *International Journal of Biological Macromolecules* **19**, 165–169.

Hiele, M., Ghoos, Y., Rutgeerts, P., Vantrappen, G., and de Buyser, K. (1990). 13CO2 breath test to measure the hydrolysis of various starch formulations in healthy subjects. *Gut* **31**, 175–178.

Higgins, J. A., Higbee, D. R., Donahoo, W. T., Brown, I. L., Bell, M. L., and Bessesen, D. H. (2004). Resistant starch consumption promotes lipid oxidation. *Nutrition and Metabolism* **1**, 8.

Hilderbrandt, L. A. and Marlett, J. A. (1991). Starch bioavailability in the upper gastrointestinal tract of colectomized rats. *Journal of Nutrition* **121**, 679–686.

Hoebler, C., Karinthi, A., Devaux, M. F., Guillon, F., Gallant, D. J. G., Bouchet, B., Melegari, C., and Barry, J. L. (1998). Physical and chemical transformations of cereal food during oral digestion in human subjects. *British Journal of Nutrition* **80**, 429–436.

Hoebler, C., Devaux, M. F., Karinthi, A., Belleville, C., and Barry, J. L. (2000). Particle size of solid food after human mastication and in vitro simulation of oral breakdown. *International Journal of Food Sciences and Nutrition* **51**, 353–366.

Holm, J., Bjorck, I., Ostrouska, S., Eliasson, A. C., Asp, N., Larsson, K., and Lundquist, I. (1983). Digestibility of amylose lipid complexes in vitro and in vivo. *Starch/Starke* **35**, 294.

Holm, J., Björck, I., Asp, N., Sjöberg, L., and Lundquist, I. (1985). Starch availability in vitro and in vivo after flaking, steam-cooking and popping of wheat. *Journal of Cereal Science* **3**, 193–206.

Holm, J., Björck, I., Drews, A., and Asp, N. -G. (1986). A rapid method for the analysis of starch. *Starch/Stärke* **38**, 224–226.

Hoover, R. and Zhou, Y. (2003). In vitro and in vivo hydrolysis of legume starches by alpha-amylase and resistant starch formation in legumes – a review. *Carbohydrate Polymers* **54**, 401–417.

Howlett, J. and Ashwell, M. (2008). Glycemic response and health: summary of a workshop. *American Journal of Clinical Nutrition* **87** (suppl), 212S–216S.

Htoon, A., Shrestha, A. K., Flanagan, B. M., Lopez-Rubio, A., Bird, A. R., Gilbert, E. P., and Gidley, M. J. (2009). Effects of processing high amylose maize starches under controlled conditions on structural organization and amylase digestibility. *Carbohydrate Polymers* **75**, 236–245.

Huth, M., Dongowski, G., Gebhardt, E., and Flamme, W. (2000). Functional properties of dietary fibre enriched extrudates from barley. *Journal of Cereal Science* **32**, 115–128.

I'Anson, K. J., Miles, J. J. T., Morris, V. J., Besford, J. S., and Jarvis, D. A. (1990). The effects of added sugars on the retrogradation of starch gels. *Journal of Cereal Science* **11**, 243–248.

Imberty, A., Chanzy, H., Perez, S., Buléon, A., and Tran, V. (1988). The double-helical nature of the crystalline part of a-starch. *Journal of Molecular Biology* **201**, 365–378.

Jane, J. L. and Robyt, J. F. (1984). Structure studies of amylose-V complexes and retro-graded amylose by action of alpha amylases, and a new method for preparing amylodextrins. *Carbohydrate Research* **132**, 105–118.

Jenkins, P. J. and Donald, A. M. (1998). Gelatinization of starch: a combined SAX/WAXS/DSC and SANS study. *Carbohydrate Research* **308**, 133–147.

Jenkins, P. J., Cameron, R. E., and Donald, A. M. (1993). A universal feature in the structure of starch granules from different botanical sources. *Starch/Starke* **45**, 417–420.

Jeong, H. Y. and Lim, S. T. (2003). Crystallinity and pasting properties of freeze-thawed high amylose maize starch. *Starch/Starke* **55**, 511–517.

Karkalas, J. and Raphaelides, S. (1986). Quantitative aspects of amylose-lipid interactions. *Carbohydrate Research* **157**, 215–234.

Kavita, V., Varghese, S., Chitra, G. R., and Jamuna, P. (1998). Effects of processing, storage time and temperature on the resistant starch of foods. *Journal of Food Science and Technology* **35**, 299–304.

Keenan, M. J., Zhou, J., McCutcheon, K. L., Raggio, A. M., Bateman, H. G., Todd, E., Jones, C. K., Tulley, R. T., Melton, S., Martin, R. J., and Hegsted, M. (2006). Effects of resistant starch, a non-digestible fermentable fiber, on reducing body fat. *Obesity* **14**, 1523–1534.

Kendall, C. W. C., Emam, A., Augustin, L. S. A., and Jenkins, D. J. A. (2004). Resistant starches and health. *Journal of AOAC International* **87**, 769–774.

Kim, J. H., Tanhehco, E. J., and Ng, P. K. W. (2006). Effect of extrusion conditions on resistant starch formation from pastry wheat flour. *Food Chemistry* **99**, 718–723.

Kohoyama, K. and Nishinari, K. (1991). Effect of soluble sugars on gelatinization and retrogradation of sweet potato starch. *Journal of Agriculture and Food Chemistry* **39**, 1406–1410.

Landa-Habana, L., Pina-Hernandez, A., Agama-Acevedo, E., Juscelino, T., and Bello-Perez, L. A. (2004). Effect of cooking procedures and storage on starch bioavailability in common beans **59**, 133–136.

Leeman, A. M., Karlsson, M. E., Eliasson, A. C., and Björck, I. M. E. (2006). Resistant starch formation in temperature treated potato starches varying in amylose/amylopectin ratio. *Carbohydrate Polymers* **65**, 306–313.

Le Leu, R. K. and Young, G. P. (2007). Fermentation of starch and protein in the colon: Implications for genomic instability. *Cancer Biology and Therapy* **6**, 259–260.

Le Leu, R. K., Brown, I. L., Hu, Y., Esterman, A., and Young, G. P. (2007). Suppression of azoxymethane-induced colon cancer development in rats by dietary resistant starch. *Cancer Biology and Therapy* **6**, 1621–1626.

Leloup, V. M., Colonna, P., and Ring, S. G. (1991). Alpha amylase adsorption on crystalline starch. *Food Hydrocolloids* **5**, 225–228.

Leloup, V. M., Colonna, P., and Ring, S. G. (1990). Studies on probe diffusion and accessibility in amylose gels. *Macromolecules* **23**, 862–866.

Leloup, V. M., Colonna, P., Ring, S. G., Roberts, K., and Wells, B. (1992). Microstructure of amylose gels. *Carbohydrate polymers* **18**, 189–197.

Levine, H. and Slade, L. (1990). Influences of he glassy and rubbery states on thermal, mechanical and structural properties of doughs and baked products. *In* Dough Rheology and Baked Product Texture, 157–330, Faridi, H. and Faubion, J. M. (eds.). Van Nostrand Reinhold, NY.

Lewandowicz, G., Formal, J., and Walkowski, A. (1997). Effect of microwave radiation on potato and tapioca starches. *Carbohydrate Polymers* **34**, 213–217.

Lewandowicz, G., Jankoski, T., and Formal, J. (2000). Effect of microwave radiation on physico-chemical properties and structure of cereal starches. *Carbohydrate Polymers* **42**, 193–199.

Li, J. H., Vasanthan, T., Hoover, R., and Rossnagel, B. G. (2004). Starch from hull-less barley: V. In-vitro susceptibility of waxy, normal, and high-amylose starches towards hydrolysis by alpha-amylases and amyloglucosidase. *Food Chemistry* **84**, 621–632.

Liljeberg, H., Akerberg, A., and Bjorck, I. (1996). Resistant starch formation in bread as influenced by choice of ingredients or baking conditions. *Food Chemistry* **56**, 389–394.

Liljeberg Elmståhl, H. (2002). Resistant starch content in a selection of starchy foods on the Swedish market. *European Journal of Clinical Nutrition* **56**, 500–505.

Liu, Q., Weber, E., Fan, M., Wood, P., Cui, S., and Wei, D. (2003). A study of the formation, molecular structure and thermal properties of resistant starch type III from normal corn starch. *In* AACC Annual Meeting (P 139). Oregan, Portland.

Livesey, G. (1995). The impact of complex carbohydrates on energy balance. *European Journal of Clinical Nutrition* **49** (Suppl 3), S89–S96.

Livesey, G., Wilkinson, J. A., Roe, M., Faulks, R., Clark, S., Brown, J. C., Kennedy, H., and Marinos, E. (1995). Influence of the physical form of barley grain on the digestion of its starch in the human small intestine and implications for health 1-3. *American Journal of Clinical Nutrition* **61**, 75–81.

Lopez-Rubio, A., Htoon, A., and Gilbert, E. P. (2007). Influence of extrusion and digestion on the nanostructure of high-amylose maize starch. *Biomacromolecules* **8**, 1564–1572.

Lopez-Rubio, A., Flanagan, B. M., Shrestha, A. K., Gidley, M. J., and Gilbert, E. P. (2008). Molecular rearrangement of starch during in-vitro digestion: towards a better understanding of enzyme resistant starch formation. *Biomacromolecules* In press.

Macgregor, E. A. (1988). Alpha-amylase structure and activity. *Journal of Protein Chemistry* **7**, 399–415.

Madrid, J. and Arcot, J. (2000). Comparison of two in vitro analysis of resistant starch of some carbohydrate containing foods. *Proceedings of the Nutrition Society of Australia* **24**, 208.

Mahadevamma, S., Harish Prashanth, K. V., and Tharanathan, R. N. (2003). Resistant starch derived from processed legumes – purification and structural characterization. *Carbohydrate Polymers* **54**, 215–219.

Mangala, S. L., Udayasnakar, K., and Tharanathan, R. N. (1999). Resistant starch from processed cereals: the influence of amylopectin and non-carbohydrate constituents in its formation. *Food Chemistry* **64**, 391–396.

Marconi, E., Ruggeri, S., Cappelloni, M., Leonardi, D., and Carnovale, E. (2000). Related physiochemical, nutritional, and microstructural characteristics of chickpeas (*Cicer arietinum, L.*) and common beans (*Phaseolus vulgaris, L.*). *Journal of Agricultural and Food Chemistry* **48**, 5986–5994.

Marlett, J. A. and Longacre, M. J. (1996). Comparison of in vitro and in vivo measures of resistant starch in selected grain products. *Cereal Chemistry* **73**, 63–68.

Mason, V. C. and Just, A. (1976). Bacterial activity in the hind-gut of pigs. 1. Its influence on the apparent digestibility of dietary energy and fat. Z. Tierphysiol. *Tierernahrg Futtermittelkde* **36**, 301–310.

McBurney, M. I. (1991). Passage of starch into the colon of humans: quantitation and implications. *Canadian Journal of Physiology Pharmacology* **69**, 130–136.

McCleary, B. V. (2007). An integrated procedure for the measurement of total dietary fibre (including resistant starch), non-digestible oligosaccharides and available carbohydrates. *Anal. Bioanal. Chem.* **389**, 291–308.

McCleary, B. V. and Monaghan, D. A. (2002). Measurement of resistant starch. *Journal of AOAC International* **85**, 665–675.

McCleary, B. V., McNally, M., and Rossiter, P. (2002). Measurement of resistant starch by enzymatic digestion in starch and selected plant materials: collaborative study. *Journal of AOAC International* **85**, 1103–1111.

Mercier, C. (1980). *In* Food Process Engineering *1*, 795–807, Linko, P., Malkki, Y., Olkku, J. and Larinkari, J. (eds.). Applied Science Publishers Ltd, London.

Miles, M. J., Morris, V. J., and Ring, S. G. (1985). The roles of amylose and amylopectin in the gelatinization and retrogradation of starch. *Carbohydrate Research* **135**, 257–269.

Molis, C., Champ, M., Flourié, B., Pellier, P., Bornet, F., Colonna, P., Kozlowski, F., Rambaud, J. C., and Galmiche, J. P. (1992). Small intestinal digestibility of processed corn starches in healthy-human subjects. *European Journal of Clinical Nutrition* **46**, S131–S132.

Morita, T., Ito, Y., Brown, I. L., Ando, R., and Kiriyama, S. (2007). In vitro and in vivo digestibility of native maize starch granules varying in amylose contents. *Journal of Aoac International* **90**, 1628–1634.

Mosenthin, R., Sauer, W. C., Henkel, H., Ahrens, F., and de Lange, C. F. M. (1992). Tracer studies of urea kinetics in growing pigs: II. The effect of starch infusion at the distal ileum on urea recycling and bacterial nitrogen excretion. *Journal of Animal Science* **70**, 3467–3472.

Muir, J. G. and O'Dea, K. (1992). Measurement of resistant starch: factors affecting the amount of starch escaping digestion in vitro. *American Journal of Clinical Nutrition* **56**, 123–127.

Muir, J. G. and O'Dea, K. (1993). Validation of an in vitro assay for predicting the amount of starch that escapes digestion in the small intestine of humans. *American Journal of Clinical Nutrition* **57**, 540–546.

Muir, J. G., Young, G. P., and O'Dea, K. (1994). Resistant starch – implications for health. *Proceedings of the Nutrition Society of Australia* **18**, 23–32.

Muir, J. G., Lu, Z. X., Young, G. P., Cameron-Smith, D., Collier, G. R., and O'Dea, K. (1995). Resistant starch in the diet increases breath hydrogen and serum in human subjects. *American Journal of Clinical Nutrition* **61**, 792–799.

Muir, J. G., Walker, K. Z., Kaimakamis, M. A., Cameron, M. A., Govers, M. J., Lu, Z. X., Young, G. P., and O'Dea, K. (1998). Modulation of fecal markers relevant to colon cancer risk: a high-starch Chinese diet did not generate expected beneficial changes relative to a Western-type diet. *American Journal of Clinical Nutrition* **68**, 372–379.

Muir, J. G., Yeow, E. G. W., Keogh, J., Pizzey, C., Bird, A. R., Sharpe, K., O'Dea, K., and Macrae, F. A. (2004). Combining wheat bran with resistant starch has more beneficial effects on fecal indexes than does wheat bran alone. *American Journal of Clinical Nutrition* **79**, 1020–1028.

Murphy, M. M., Douglass, J. S., and Birkett, A. (2008). Resistant starch intakes in the United States. *Journal of the American Dietetic Association* **108**, 67–78.

Murphy, P. (2000). Starch. *In* Handbook of Hydrocolloids, 41–65, Phillips, G. O. and Williams, P. A. (eds.). CRC Press, Boca Raton.

Noakes, M., Clifton, P. M., Nestel, P. J., Le Leu, R., and McIntosh, G. (1996). Effect of high-amylose starch and oat bran on metabolic variables and bowel function in subjects with hypertriglyceridemia. *American Journal of Clinical Nutrition* **64**, 944–951.

Normand, S., Pachiaudi, C., Khalfallah, Y., Guilluy, R., Mornex, R., and Riou, J. P. (1992). 13C appearance in plasma glucose and breath CO2 during feeding with naturally 13C-enriched starchy food in normal humans. *American Journal of Clinical Nutrition* **55**, 430–435.

Nugent, A. P. (2005). Health properties of resistant starch. *Nutrition Bulletin* **30**, 27–54.

Oates, C. G. (1997). Towards an understanding of starch granule structure and hydrolysis. *Trends in Food Science & Technology* **8**, 375–382.

O'Keefe, S. J. D. (2008). Nutrition and colonic health: the critical role of the microbiota. *Current Opinion in Gastroenterology* **24**, 51–58.

Onyango, C., Bley, T., Jacob, A., Henle, T., and Rohm, H. (2006). Influence of incubation temperature and time on resistant starch type III formation from autoclaved and acid-hydrolyzed cassava starch. *Carbohydrate Polymer* **66**, 494–499.

Ostergard, K., Bjorck, I., and Vainiopaa, J. (1989). Effect of extrusion cooking on starch and dietary fibre in barley. *Food Chemistry* **34**, 215–227.

Parchure, A. A. and Kulkarni, P. R. (1997). Effect of processing treatments on generation of resistance starch. *International Journal of Food Science and Nutrition* **48**, 257–260.

Phillips, J., Muir, J. G., Birkett, A., Lu, Z. X., Jones, G. P., O''Dea, K., and Young, G. P. (1995). Effect of resistant starch on fecal bulk and fermentation-dependent events in humans. *American Journal of Clinical Nutrition* **62**, 121–130.

Planchot, V., Colonna, P., and Buléon, A. (1997). Enzymatic hydrolysis of alpha-glucan crystallites. *Carbohydrate Research* **298**, 319–326.

Planchot, V., Colonna, P., Gallant, D. J., and Bouchet, B. (1995). Extensive degradation of native starch granules by alpha-amylase from aspergillus-fumigatus. *Journal of Cereal Science* **21**, 163–171.

Platel, K. and Shurpalekar, K. S. (1994). Resistant starch content of Indian foods. *Plant Foods Human Nutrition* **45**, 91–95.

Prosky, L., Asp, N. G., Furda, I., DeVries, J. W., Shweizer, T. F., and Harland, B. F. (1985). Determination of total dietary fibre in foods and food products: collaborative study. *Journal of the Association of Official Analytical Chemists* **68**, 677–679.

Rabe, E. and Sievert, D. (1992). Effects of baking, pasta production, and extrusion cooking on formation of resistant starch. *European Journal of Clinical Nutrition* **46**, S105–S107.

Raben, A., Tagliabue, A., Christensen, N. J., Madsen, J., Holst, J. J., and Astrup, A. (1994). Resistant starch: the effect on postprandial glycemia, hormonal response, and satiety. *American Journal of Clinical Nutrition* **60**, 544–551.

Rahman, S. M., Morell, M., Topping, D., Bird, A., Li, Z., Dennis, E., and Peacock, J. (2007). Low glycaemic response cereals for enhanced human health. *International Diabetes Monitor* **19**, 21–25.

Ranhotra, G. S., Gelroth, J. A., and Glaser, B. K. (1996). Energy value of resistant starch. *Journal of Food Science* **61**, 453–455.

Read, N. W., Al-Janabi, M. N., Holgate, A. M., Barber, D. C., and Edwards, C. A. (1986). Simultaneous measurement of gastric emptying, small bowel residence and colonic filling of a solid meal by the use of the gamma camera. *Gut* **27**, 300–308.

Rendleman, J. A. (2000). Hydrolytic action of alpha-amylase on high-amylose starch of low molecular mass. *Biotechnology and Applied Biochemistry* **31**, 171–178.

Regina, A., Bird, A. R., Li, Z., Rahman, S., Mann, G., Chanliaud, E., Berbezy, P., Topping, D., and Morell, M. K. (2007). Bioengineering cereal carbohydrates to improve human health. *Cereal Foods World* **52**, 182–188.

Ridlon, J. M. and Hylemon, P. B. (2006). A potential role for resistant starch fermentation in modulating colonic bacterial metabolism and colon cancer risk. *Cancer Biology and Therapy* **5**, 273–274.

Riley, C. K., Wheatley, A. O., Hassan, I., Ahmad, M. H., Morrison, E. S., and Asemota, H. N. (2004). In vitro digestibility of raw starches extracted from five yam (Dioscorea spp.) species grown in Jamaica. *Starch-Starke* **56**, 69–73.

Ring, S. G., Gee, J. M., Whittam, M., Orford, P., and Johnson, I. T. (1988). Resistant Starch: its chemical form in foodstuffs and effect on digestibility in vitro. *Food Chemistry* **28**, 97–109.

Roberts, J., Jones, G. P., Rutishauser, I. H. E., Birkett, A., and Gibbons, C. (2004). Resistant starch in the Australian diet. *Nutrition and Dietetics* **61**, 98–104.

Robertson, M. D. and Mathers, J. C. (2000). Gastric emptying rate of solids is reduced in a group of ileostomy patients. *Digestive Diseases and Sciences* **45**, 1285–1292.

Robertson, M. D., Currie, J. M., Morgan, L. M., Jewell, D. P., and Frayn, K. N. (2003). Prior short-term consumption of resistant starch enhances postprandial insulin sensitivity in healthy subjects. *Diabetologia* **46**, 659–665.

Robertson, M. D., Bickerton, A. S., Dennis, A. L., Vidal, H., and Frayn, K. N. (2005). Insulin-sensitizing effects of dietary resistant starch and effects on skeletal muscle and adipose tissue metabolism. *American Journal of Clinical Nutrition* **82**, 559–567.

Rosin, P. M., Lajolo, F. M., and Menezes, E. W. (2002). Measurement and characterization of dietary starches. *Journal of Food Composition and Analysis* **15**, 367–377.

Rumessen, J. J. (1992). Hydrogen and methane breath tests for evaluation of resistant carbohydrates. *European Journal of Clinical Nutrition* **46** (Suppl.2), S77–S90.

Russell, P. L., Berry, C. S., and Greenwell, P. (1989). Characterization of resistant starch from wheat and maize. *Journal of Cereal Science* **9**, 1–5.

Sagum, R. and Arcot, J. (2000). Effect of domestic processing methods on the starch, non-starch polysaccharides and in vitro starch and protein digestibility of three varieties of rice with varying levels of amylose. *Food Chemistry* **70**, 107–111.

Sajilata, M. G., Singhai, R. S., and Kulkarni, P. R. (2006). Resistant starch – a review. *Comprehensive Reviews in Food Science and Food Safety* **5**, 1–17.

Salman, H., Blazek, J., Lopez-Rubio, A., Gilbert, E. P., Hanley, T., and Copeland, L. (2009). Structure-function relationships in A and B granules from wheat starches of similar amylose content. *Carbohydrate Polymers* **75**, 420–427.

Sarikaya, E., Higasa, T., Adachi, M., and Mikami, B. (2000). Comparison of degradation abilities of alpha- and beta-amylases on raw starch granules. *Process Biochemistry* **35**, 711–715.

Sarko, A. and Wu–, H.–C. H. (1978). The crystal structures of A-, B-, and C- polymorphs of amylose and starch. *Starch/Starke* **30**, 73.

Saura-Calixto, F., Goñi, I., Bravo, L., and Mañas, E. (1993). Resistant starch in foods: modified method for dietary fiber residues. *Journal of Food Science* **58**, 642–643.

Schmiedl, D., Bauerlein, M., Bengs, H., and Jacobasch, G. (2000). Production of heat stable, butyrogenic resistant starch. *Carbohydrate Polymer* **43**, 183–193.

Segal, I. (2002). Physiological small bowel malabsorption of carbohydrates protects against large bowel diseases in Africans. *Journal of Gastroenterology and Hepatology* **17**, 249–252.

Sengupta, S., Muir, J. G., and Gibson, P. R. (2006). Does butyrate protect from colorectal cancer? *Journal of Gastroenterology and Hepatology* **21**, 209–218.

Shamai, K., Bianco-Peled, H., and Shimoni, E. (2003). Polymorphism of resistant starch type III. *Carbohydrate Polymers* **54**, 363–369.

Shamai, K., Shimoni, E., and Bianco-Peled, H. (2004). Small angle X-ray scattering of resistant starch type III. *Biomacromolecules* **5**, 219–223.

Sharma, A. and Yadav, B. S. (2008). Resistant starch: physiological roles and food applications. *Food Reviews International* **24**, 193–234. and Ritika.

Sievert, D., Czuchajowska, Z., and Pomeranz, Y. (1991). Enzyme-resistant starch.III. x-ray-diffraction of autoclaved amylomaize vii starch and enzyme-resistant starch residues. *Cereal Chemistry* **68**, 86–91.

Sievert, D. and Pomeranz, Y. (1989). Enzyme-resistant resistant. I. Characterization and evaluation of enzymatic, thermoanalytical and microscopic methods. *Cereal Chemistry* **66**, 342–347.

Sievert, D. and Wursch, P. (1993). Thermal behavior of potato amylose and enzyme resistant starch from maize. *Cereal Chemistry* **70**, 333–350.

Sievert, D. and Pomeranz, Y. (1991). Enzyme-resistant resistant. 3. X-ray diffraction of autoclaved amylomaize VII starch and enzyme-resistant starch residues. *Cereal Chemistry* **68**, 86–91.

Sievert, D. and Pomeranz, Y. (1990). Enzyme-resistant starch .2. Differential scanning calorimetry studies on heat-treated starches and enzyme-resistant starch residues. *Cereal Chemistry* **67**, 217–221.

Siljestrom, M. and Asp, N. G. (1985). Resistant starch formation during baking – effect of baking time and temperature and variations in the recipe. *Zeitschrift Fur Lebensmittel-Untersuchung Und-Forschung* **181**, 4–8.

Siljerstrom, M., Eliasson, A. C., and Bjorck, I. (1989). Characterization of resistant starch from autoclaved wheat starch. *Starch/Starke* **41**, 147–151.

Siljerstrom, M., Westerlund, E., Bjorck, I., Holm, J., Asp, J. G., and Theander, O. (1986). The effects of thermal processes on dietary fibre and starch content of whole grain wheat and flour. *Journal of Cereal Science* **4**, 315–323.

Siljestrom, M. and Bjorck, I. (1990). Digestible and undigestible carbohydrates in autoclaved legumes, potatoes and corn. *Food Chemistry* **38**, 145–152.

Silvester, K. R., Englyst, H. N., and Cummings, J. H. (1995). Ileal recovery of starch from whole diets containing resistant starch measured in vitro and fermentation of ileal effluent. *American Journal of Clinical Nutrition* **62**, 403–411.

Skrabanja, V. and Kreft, I. (1998). Resistant starch formation following autoclaving of buckwheat (*Fagopyrum esculentum Moench*) groats. An in vitro study. *Journal of Agricultural and Food Chemistry* **46**, 2020–2023.

Snow, P. and O'Dea, K. (1981). Factors affecting the rate of hydrolysis of starch in food. *American Journal of Clinical Nutrition* **34**, 2721–2727.

Soper, N. J., Chapman, N. J., Kelly, K. A., Brown, M. L., Phillips, S. F., and Go, V. L. W. (1990). The íleal brake'after ileal pouch-anal anastomosis. *Gastroenterology* **98**, 111–116.

Srichuwong S, Isono, N., Mishima, T. and Hisamatsu, M. (2005a). Structure of lintnerized starch is related to X-ray diffraction pattern and susceptibility to acid and enzyme hydrolysis of starch granules. *International Journal of Biological Macromolecules* **37**, 115–121.

Srichuwong S, Sunarti T. C., Mishima, T., Isono, N. and Hisamatsu, M. (2005b). Starches from different botanical sources: Contribution of amylopectin fine structure to thermal properties and enzyme digestibility. *Carbohydrate Polymer* **60**, 529–538.

Stephen, A. M., Haddad, A. C., and Philips, S. F. (1983). Passage of carbohydrate into the colon. Direct measurement in humans. *Gastroenterology* **85**, 589–595.

Strocchi, A. and Levitt, M. D. (1991). Measurement of starch fermentation in the human large intestine. *Canadian Journal of Physiology and Pharmacology* **69**, 108–110.

Subramanya, S., Ramakrishna, B. S., Binder, H. J., Farthing, M. J., and Young, G. P. (2006). Evaluation of oral rehydration solution by whole-gut perfusion in rats: effect of osmolarity, sodium concentration and resistant starch. *Journal of Pediatric Gastroenterology and Nutrition* **43**, 568–575.

Symonds, E. L., Kritas, S., Omari, T. I., and Butler, R. N. (2004). A combined 13CO2/H2 breath test can be used to assess starch digestion and fermentation in humans. *Journal of Nutrition* **134**, 1193–1196.

Szczodrak, J. and Pomeranz, Y. (1991). Starch and enzyme-resistant starch from high-amylose starch. *Cereal Chemistry* **68**, 589–596.

Szczodrak, J. and Pomeranz, Y. (1992). Starch-lipid interactions and formation of resistant starch in high-amylose barley. *Cereal Chemistry* **69**, 626–632.

Tapsell, L. C. (2004). Diet and metabolic syndrome: where does resistant starch fit in? *Journal of AOAC International* **87**, 756–760.

Tester, R. F., Karkalas, J., and Qi, X. (2004). Starch-composition, fine structure and architecture. *Journal of Cereal Science* **399**, 151–165.

Tester, R. F., Karkalas, J., and Qi, X. (2004). Starch structure and digestibility enzyme-substrate relationship. *Worlds Poultry Science Journal* **60**, 186–195.

Tester, R. F. and Qi, X. (2004). Molecular basis of the gelatinisation and swelling characteristics of waxy barley starches grown in the same location during the same season. Part I. Composition and alpha-glucan fine structure. *Journal of Cereal Science* **39**, 47–56.

Tharanathan, M. and Tharanathan, R. N. (2001). Resistant starch in wheat-based products: Isolation and characterization. *Journal of Cereal Science* **34**, 73–84.

Themeier, H., Hollman, J., Neese, U., and Lindhauer, M. G. (2005). Structural and morphological factors influencing the quantification of resistant starch II in starches of different botanical origin. *Carbohydrate Polymers* **61**, 72–79.

Thoma, J. A. (1968). A possible mechanism for amylase catalysis. *Journal of Theoretical Biology* **19**, 297–310.

Toden, S., Bird, A. R., Topping, D. L., and Conlon, M. A. (2005). Resistant starch attenuates colonic DNA damage induced by higher dietary protein in rats. *Nutrition and Cancer* **51**, 45–51.

Toden, S., Bird, A. R., Topping, D. L., and Conlon, M. A. (2006). Resistant starch prevents colonic DNA damage induced by high dietary cooked red meat or casein in rats. *Cancer Biology and Therapy* **5**, 267–272.

Toden, S., Bird, A. R., Topping, D. L. and Conlon, M. A. (2007a). Differential effects of dietary whey, casein and soya on colonic DNA damage and large bowel short chain fatty acids in rats fed diets low and high in resistant starch. *British Journal of Nutrition* **97**, 535–543.

Toden, S., Bird, A. R., Topping, D. L., and Conlon, M. A. (2007b). High red meat diets induce greater numbers of colonic DNA double-strand breaks than white meat: attenuation by high amylose maize starch. *Carcinogenesis* **28**, 2355–2362.

Toden S., Bird, A. R., Topping, D. L. and Conlon, M. A. (2007c). Dose-dependent reduction of dietary protein-induced colonocyte DNA damage by resistant starch in rats is more highly correlated with levels of caecal butyrate than other short chain fatty acids. *Cancer Biology and Therapy* **6**, 253–258.

Tomlin, J. and Read, N. W. (1990). The effect of resistant starch on colon function in humans. *British Journal of Nutrition* **64**, 589–595.

Topping, D. L. and Bird, A. R. (1999). Foods, nutrients and digestive health. *Australian Journal of Nutrition and Dietetics* **56** (Suppl), S22–S34.

Topping, D. L. and Clifton, P. (2001). Short chain fatty acids and human colonic function – roles of resistant starch and non starch polysaccharides. *Physiological Reviews* **81**, 1031–1064.

Topping, D. L., Bird, T., Clarke, J., Keogh, J., Li, Z., Morell, M., Noakes, M., Regina, A., Rahman, S., and White, R. (2006). Controlling starch digestibility for human health. *Asia. Pacific Journal of Clinical Nutrition* **15** (Suppl), S43.

Tovar, J., Bjorck, I., and Asp, N. G. (1990). Starch content and alpha-amylolysis rate in precooked legume flours. *Journal of Agricultural Chemistry* **38**, 1818–1823.

Tovar, J. and Melito, C. (1996). Steam cooking and dry heating produce resistant starch in legumes. *Journal of Agricultural and Food Chemistry* **441**, 2642–2645.

Udagawa, H., Kitaoka, C., Sakamoto, T., Kobayashi-Hattori, K., Oishi, Y., Arai, S., and Takita, T. (2008). Serum cholesterol-decreasing effect of heat-moisture-treated high-amylose cornstarch in cholesterol-loaded rats. *Bioscience Biotechnology Biochemistry* **72**, 880–884.

Unlu, E. and Faller, J. F. (1998). Formation of resistant starch by a twin-screw extruder. *Cereal Chemistry* **37**, 346–350.

Valetudie, J. C., Colonna, P., Bouchet, B., and Gallant, D. J. (1993). Hydrolysis of tropical tuber starches by bacterial and pancreatic alpha-amylases. *Starch-Starke* **45**, 270–276.

van der Meulen, J., Bakker, G. C. M., Bakker, J. G. M., de Visser, H., Jongbloed, A. W., and Everts, H. (1997). *Journal of Animal Science* **75**, *2697–2704.*

van Munster, I. P., de Boer, H. M., Jansen, M. C., de Haan, A. F., Katan, M. B., van Amelsvoort, J. M., and Nagengast, F. M. (1994). Effect of resistant starch on breath-hydrogen and methane excretion in healthy volunteers. *American Journal of Clinical Nutrition* **59**, 626–630.

vanSoest, J. J. G., Tournois, H., de Wit, D., and Vliegenthart, J. F. G. (1995). Short-range structure in (partially) crystalline potato starch determined with attenuated total reflectance Fourier-transform IR spectroscopy. *Carbohydrate Research* **279**, 201–214.

Vasanthan, T. and Bhatty, R. S. (1996). Physicochemical properties of small- and large-granule starches of waxy, regular, and high-amylose barleys. *Cereal Chemistry* **73**, 199–207.

Vasanthan, T. and Bhatty, R. S. (1998). Enhancement of resistant starch (RS3) in amylomaize, barley, field pea and lentil starches. *Starke/Starch* **50**, 286–289.

Vasanthan, T., Gaosong, J., Yeung, J., and Li, J. (2002). Dietary fiber profile of barley flour as affected by extrusion cooking. *Food Chemistry* **77**, 35–40.

Verbeek, M. J. F., De Deckere, E. A. M., Tijburg, L. B. M., van Amelsvoort, J. M. M., and Beynen, A. C. (1995). Influence of dietary retrograded starch on the metabolism of neutral steroids and bile acids in rats. *British Journal of Nutrition* **74**, 807–820.

Vonk, R. J., Hagedoorn, R. E., de Graaff, R., Elzinga, H., Tabak, S., Yang, Y. X., and Stellaard, F. (2000). Digestion of so-called resistant starch sources in the human small intestine. *American Journal of Clinical Nutrition* **72**, 432–438.

Walter, M., Picollo da Silva, L. and Casagrande Denardin, C. (2005). Rice and resistant starch: different content depending on chosen methodology. *Journal of Food Composition and Analysis* **18**, 279–285.

Weaver, G. A., Krause, J. A., Miller, T. L., and Wolin, M. J. (1992). Cornstarch fermentation by the colonic microbial community yields more butyrate than does cabbage fiber fermentation; cornstarch fermentation rates correlate negatively with methanogenesis. *American Journal of Clinical Nutrition* **55**, 70–77.

Weaver, L. T., Dibba, B., Sonko, B., Bohane, T. D., and Hoare, S. (1995). Measurement of starch digestion of naturally 13C-enriched weaning foods, before and after partial digestion with amylase-rich flour, using a 13C breath test. *British Journal of Nutrition* **74**, 531–537.

Westerlund, E., Theander, O., Andersson, R., and Aman, P. (1989). Effects of baking on polysaccharides in white bread fractions. *Journal of Cereal Science* **10**, 149–156.

Williams, R. D. and Olmestedt, W. (1935). A biochemical method for determining indigestible residue (crude fibre) in faeces: lignin, cellulose, and non-water soluble hemicelluloses. *Journal of Biological Chemistry* **108**, 653–666.

Williamson, G., Belshaw, N. J., Self, D. J., Noel, T. R., Ring, S. G., Cairns, P., Morris, V. J., Clark, S. A., and Parker, M. L. (1992). Hydrolysis of a-type and b-type crystalline polymorphs of starch by alpha-amylase, beta-amylase and glucoamylase-1. *Carbohydrate Polymers* **18**, 179–187.

Wolever, T. M. S., Cohen, Z., Thompson, L. U., Thorne, M. J., Jenkins, M. J., Prokipchuk, E. J., and Jenkins, D. J. (1986). Ileal loss of available carbohydrate in man: comparison of a breath hydrogen method with direct measurement using a human ileostomy model. *American Journal of Gastroenterology* **81**, 115–122.

Wutzke, K. D. and Glasenapp, B. (2004). The use of 13C-labelled glycosyl ureides for evaluation of orocaecal transit time. *European Journal of Clinical Nutrition* **58**, 568–572.

Yamada, Y., Hosoya, S., Nishimura, S., Tanaka, T., Kajimoto, Y., Nishimura, A., and Kajimoto, O. (2005). Effect of bread containing resistant starch on postprandial blood glucose levels in humans. Bioscience Biotechnology. *Biochemistry* **69**, 556–566.

Young, G. P. and Le Leu, R. K. (2004). Resistant starch and colorectal neoplasia. *Journal of AOAC International* **87**, 775–786.

Yuan, R. C., Thompson, D. B., and Boyer, C. D. (1993). Fine structure of amylopectin in relation to gelatinization and retrogradation behavior of maize starches from the wx-containing genotypes in two inbred lines. *Cereal Chemistry* **70**, 81–89.

Zhang, G. Y., Ao, Z., and Hamaker, B.R. (2006a). Slow digestion property of native cereal starches. *Biomacromolecules* **7**, 3252–3258.

Zhang, G. Y., Venkatachalam, M., and Hamaker, B.R. (2006b). Structural basis for the slow digestion property of native cereal starches. *Biomacromolecules* **7**, 3259–3266.

Zhou, J., Hegsted, M., McCutcheon, K. L., Keenan, M. J., Xi, X. C., Raggio, A. M., and Martin, R. J. (2006). Peptide YY and proglucagon mRNA expression patterns and regulation in the gut. *Obesity* **14**, 683–689.

CHAPTER

15

Glycemic Response Reduction in Processed Food Products

Charles Stephen Brennan [1,2] *and Margaret Anne Brennan* [2]

[1]Hollings Faculty, Manchester Metropolitan University, Old Hall Lane, Manchester, UK

[2]Institute of Food, Nutrition and Human Health, Massey University, Palmerston North, New Zealand

15.1 INTRODUCTION

The concept of the glycemic index (GI) is well known amongst researchers and consumers (Vermeulen and Turnbull, 2000). It is based on the postprandial increase in the plasma glucose concentration (i.e., the glycemic response) from a fixed amount of available carbohydrate (starch) in a test food with the glycemic response elicited from the same amount of carbohydrate in a standardized reference food, i.e. glucose or white bread (Brand Miller et al., 2002). Bread (one of the oldest processed foods) has been researched extensively with regards to understanding the relationship between the starch content of the raw materials, food processing and formulation, and glycemic interactions (Brennan et al., 1996; Tudorica et al., 2002; Cleary et al., 2007). Much of this research has been conducted to attempt to develop a reliable and rapid method for the determination of carbohydrate digestibility, and hence glycemic impact determination. As such these experiments have focused on the use of in vivo and in vitro analysis in order to determine the glycemic response of carbohydrate rich foods (Bjorck et al., 1984; Gourgue et al., 1994; Brennan et al., 1996; Woolnough et al., 2007, 2008).

15.2 PROCESSING AND CARBOHYDRATE DIGESTIBILITY

It is generally regarded that carbohydrate-rich food products (for example those derived from cereal flour material) are high-GI food products due to the ease of starch digestibility from cereal food products. In turn, diets of high-GI foods have been linked to increased weight gain, lack of both insulin and blood glucose control, and increased levels of obesity (Brennan, 2005).

Consumer opinion suggests that the greater the number of processing steps involved in the manufacture of food, the higher the potential glycemic response of an individual to that food product. Table 15.1 illustrates the glycemic index values of some commonly consumed food items and appears to confer some verification of this assumption. This may be due to the increased degree of starch gelatinization and starch disorganization, following thermal processing, as well

Kasapis, Norton, and Ubbink: Modern Biopolymer Science
ISBN: 978-0-12-374195-0

TABLE 15.1 Glycemic Index of Commercially Available Processed Foods

Food product	GI (glucose = 100)	GI (bread = 100)
Bread (white)	95 ± 15	100
Bread (barley grain)	43	62 ± 4
Nachochips	74	103
Pretzels	83	119
Twisties (corn snack)	74	106
Chickpea	41	58
All bran (Kellogg's)	42 ± 5	60 ± 7
Bran flakes (Kellogg's)	74	106
Cheerios (General Mills, Canada)	74	106 ± 9
Coco Pops (Kellogg's)	77	110
Cornflakes (Kellogg's)	81 ± 3	116 ± 5
Rice Bubbles (Kellogg's)	87 ± 4	124 ± 6
Weetbix (Sanitarium)	69	99

Adapted from Foster-Powell et al. (2002).

as a reduction in potential particle size of carbohydrate fraction. Two of the most popular forms of processed foods are ready-to-eat snacks and breakfast cereal products. Both of these products are easy to use by the consumer, and deliver textural properties which fit specific consumer needs. The technology behind these products revolves around the extrusion process (Mercier and Feillet, 1975; Case et al., 1992; Bouvier, 2001).

15.3 THE EFFECT OF EXTRUSION PARAMETERS AND PROCESSING ON FOOD QUALITY

Extrusion can be regarded, simplistically, as a continuous cooking process. There are three major screw-types which are used in extrusion processing: single-screw, and intermeshing twin-screw in either counter-rotating or co-rotating styles. The screw within the extruder barrel constantly rotates during extrusion, mixing and molding the raw materials fed into the extruder. The food material is propelled forward under both pressure and shear, creating an opportunity to manipulate the structure of the components whilst in the extrusion barrel. At the end of the extruder sits a die assembly through which the product is forced, this die assembly can be shaped to enable the manufacture of products with the particular shape or size required for that particular product.

In the manufacture of extruded snacks and breakfast cereal products, melting (cooking) of the raw material occurs in the final stages of the extruder due to high pressure (above the vapor pressure of water) and high temperature (above 100°C). At the die face there is a considerable differential in pressure. It is this reduction of pressure at the interface between the die face (high pressure) and the atmosphere (relatively low pressure), together with the evaporation of the superheated water present within the exudates, which enables the molten product behind the die face to expand considerably at the die face, and hence creates a porous food structure (Alverez-Martinez et al., 1998).

The rate of expansion of such products depends on the rheological and thermal properties of the molten material and on the geometry of the die aperture. Thus, the rheology of the pastes within the extruder can have a significant influence on the product characteristics (degree of expansion and overall porosity of the product). Differences in flow rate and hence product characteristics can be created by changing the process parameters of the extruder such as die pressure, screw flow dynamics, or screw energy consumption (Lund, 1984). Figure 15.1 shows different factors that have an influence on the extrusion cooking process.

Ungelatinized starches are not easily digested by humans and are nutritionally regarded as potentially resistant starches. Although the extrusion process can cook and melt the raw material within the extruder, water levels are

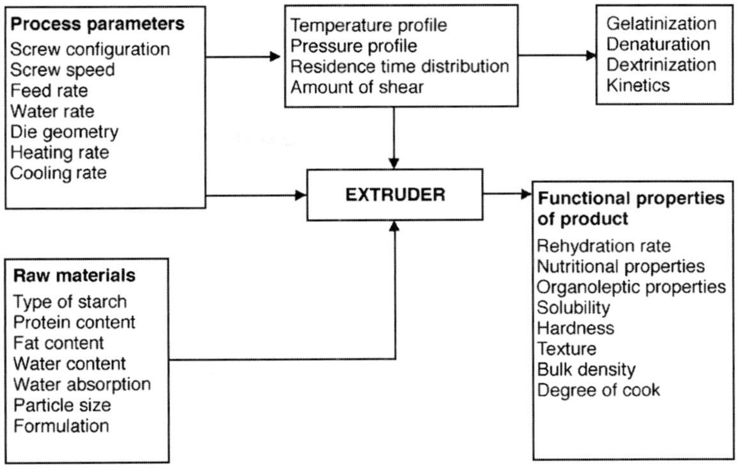

FIGURE 15.1 Interaction of raw material properties, process variables and product (adapted from Wang et al., 2001).

normally well regulated and therefore during the extrusion cooking complete gelatinization may not occur. However, the digestibility of starch can be improved due to partial gelatinization and fragmentation of starch (Wang et al., 1993). It is therefore possible to consider the extrusion process as a method to initiate the predigestion of starch.

Within the starch granule, the branches of amylopectin are susceptible to shear forces and can be cleaved from the main chain; this will cause a change in the functionality of the ingredient. Amylose and amylopectin have both been found to be reduced in molecular weight by extrusion (Politz et al., 1994). These shorter branches could cross-link forming novel, indigestible linkages and therefore lower the glycemic index (Theander and Westerlund, 1987). For instance, high-amylase rice has been extruded into noodles which exhibit a reduced glycemic index (Panlasigui et al., 1992).

The effect of screw configuration on the rate of sugar release can be seen in Figure 15.2 where a straight screw configuration, and a screw configuration incorporating a reverse screw paddle, were used to compare the effect of extrusion on the starch digestibility of control samples (wheat flour based) and samples with 5% wheat bran inclusion. The results are consistent with the previous research of Politz et al. (1994), and illustrate that the reverse screw configuration can reduce the rate of sugar release (hence potential glycemic impact) compared to the straight screw configuration. Potentially due to the complexation of starch during the controlled gelatinization process.

15.4 MANIPULATING THE GLYCEMIC IMPACT OF EXTRUDED SNACK PRODUCTS

As has been shown in Table 15.1, extruded snack products tend to be regarded as high-GI food products. Primarily, this is due to the low non-starch polysaccharide (NSP) or dietary fiber (DF) composition of such snack products combined with the high carbohydrate content of the cereals. It may also be related to the extrusion process which alters the chemical composition of the food product, and the digestibility of the starch within the carbohydrate food products. Cellulose is often regarded as the most widely used NSP in household food products including breakfast cereals, wholemeal bread and brown rice. Since cellulose is insoluble these foods have

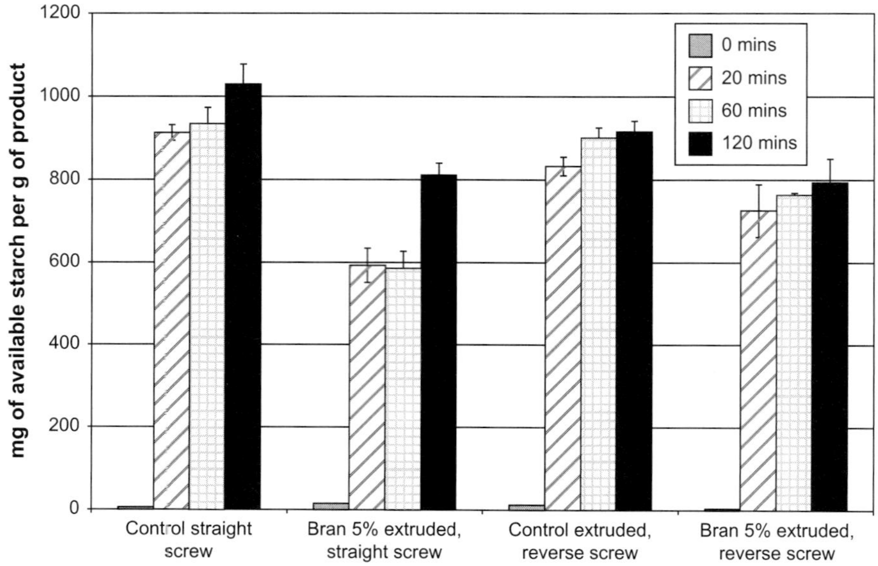

FIGURE 15.2 Carbohydrate digestibility of extruded-bran-enriched breakfast cereal products from straight and reverse screw configurations.

the same glycemic index whether or not their DF content is increased (Jenkins et al., 1983). Kelloggs All-Bran, an extruded wheat bran product, is an exception to this. Despite its high insoluble fiber content it has a low glycemic index (Foster-Powell and Miller, 1995). However it is possible to affect the digestibility of starch in breakfast cereal products by altering the soluble DF composition of extruded breakfast cereals (Brennan et al., 2008a, 2008b).

This effect is even more important when considering the ability of DF-enriched food products to impact on the glycemic response of the next meal, something that has been shown by Jenkins and his associates (1986). As such, the glycemic response to a lunch time meal can be reduced significantly when it is preceded by a low-GI breakfast meal compared to a breakfast containing high-GI foods. Björck and Elmståhl (2003) have also reported a glycemic impact of dietary fiber from one meal to another, where an evening meal rich in indigestible carbohydrates, with a low glycemic index will result in an improved glucose tolerance the following morning. It is therefore possible that the provision of a low-GI breakfast cereal could have advantages in regulating the glycemic impact of meals throughout the day.

15.5 THE LINK BETWEEN SLOWLY DIGESTIBLE AND RAPIDLY DIGESTIBLE CARBOHYDRATES AND THE GLYCEMIC IMPACT OF PROCESSED FOODS

When discussing the relevance of food processing to the glycemic impact of foods, one of the most important concepts to think about is the difference between rapidly digestible, slowly digestible and resistant starch fractions. The rapidly digestible fraction can lead to a rapid rise in blood sugar levels (hence a rapid glycemic response) once a food item is ingested. This 'spike' in blood sugar levels has several negative effects for the health of diabetics and also non-diabetic individuals. Most of the early research conducted by Jenkins et al. was aimed at developing strategies to attenuate this glycemic response through the manipulation of the rate of starch digestibility (Jenkins et al., 1978, 1983, 1987, 1997; Fairchild et al., 1996).

Subsequent work from our laboratory (Brennan et al., 2005, 2008a) has illustrated that by manipulation of product composition it is possible to increase the amount of slowly digestible starch components in processed foods, hence reduce the rate and extent of predicted glycemic response. For instance, the utilization of wholemeal flour in extruded snack products, instead of refined flour components, has been shown to reduce the amount of rapidly digestible starch components and increase the amount of slowly digestible starch in extruded breakfast cereal products (Figure 15.3).

Recently, Mishra et al. (2008) have also illustrated that differing the processing techniques employed in the cooking of potato can increase the amount of slowly digestible starch in a potato meal, hence reduce the potential glycemic impact of such a product. Particle size of the carbohydrate source can also impact on the digestibility of the starch fraction. This has been given as one of the reasons why chick pea and pasta grade flours generally produce lower glycemic responses compared to finely milled bread wheat flour (Hardacre et al., 2006).

Whilst true that whole grain, unrefined, unprocessed, carbohydrate dense material generally produces a lower glycemic response in an individual compared to refined and processed carbohydrate-rich foods (Anderson and Conley, 2007), the generalization that processing itself increases the glycemic impact of foods can be very misleading. Food processing usually involves the application of moisture followed by a thermal treatment, the essential conditions for starch gelatinization. Starch digestibility has been linked to an increase in starch gelatinization. Nevertheless, as has been shown in the previous examples, it is relatively easy to use processing parameters to modify the degree of gelatinization of the starch and hence modulate the glycemic response of individuals. By altering the degree of starch gelatinization, one can also look at regulating the rate and extent of starch retrogradation. In this way it is possible to inhibit the apparent digestibility of starch. The recent research of Aguliera and Parada clearly illustrates that by regulating the degree of starch gelatinization in a potato (or general starch)

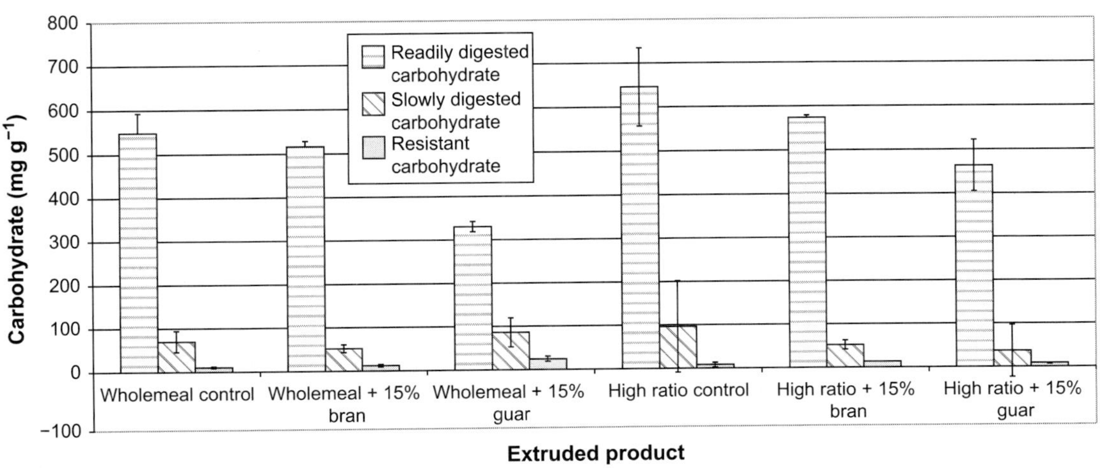

FIGURE 15.3 Carbohydrate digestibility of high ratio and wholemeal flour bases with guar and bran replacement before extrusion (Brennan et al., 2008a).

based system one can manipulate the glycemic index of such processed foods (Parada and Aguilera, 2007, 2008).

15.6 USE OF DIETARY FIBER IN MANIPULATING STARCH DIGESTIBILITY

Non-starch polysaccharides (NSP), and dietary fibers (DF), have been used for many years to modulate the extent of starch swelling, gelatinization and hence carbohydrate digestibility. For instance, clinical studies have shown that diets rich in soluble fibers (NSPs) such as guar gum, pectin and sugar beet fibers, result in lower post-prandial blood glucose and insulin levels (Jenkins et al., 1987). Soluble NSP from pulses, vegetables, whole fruits, oats and barley, form gelatinous gels within the stomach which appear to delay gastric emptying and enzymatic digestion (Jenkins et al., 1978), whereas insoluble NSP have little effect on gastric emptying and no effect on glucose absorption. This may in part explain why high-fiber/high-NSP diets alone are not necessarily synonymous with low-glycemic-index foods (Jenkins et al., 1983). As this research indicates, the use of NSP and DF to alter the glycemic impact of foods may be due to the viscosity-altering behavior of the polysaccharide.

The hydration dynamics of NSPs has been linked to the functionality of starches (Parvathy et al., 2007) with much attention focused on galactomannan conformations and the structure of entangled networks these components form (Ellis et al., 1996; Wang et al., 2003). However, the

TABLE 15.2 Viscosity of the Guar Gum (2%, 25°C)

	Minimum viscosity	Maximum viscosity
Guar 1	800	1200
Guar 2	2200	5500
Guar 3	4000	7000

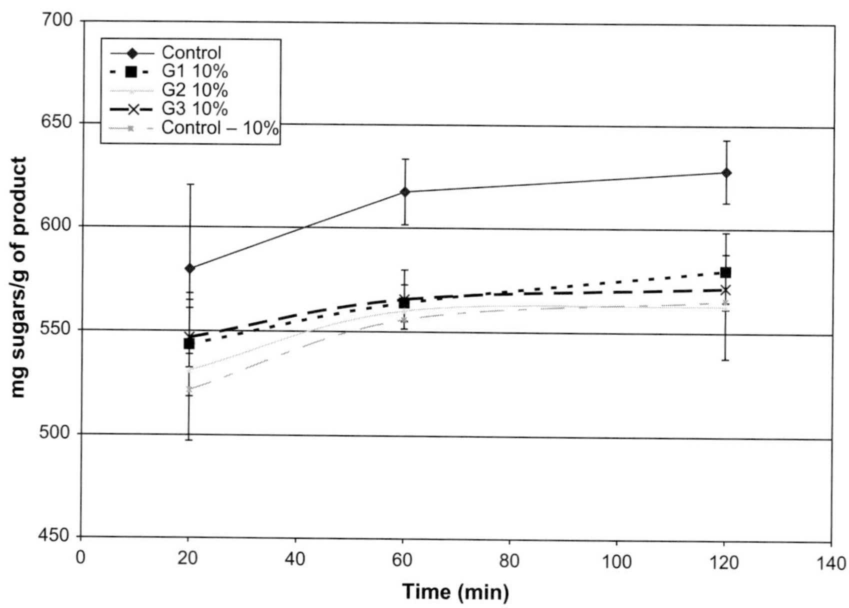

FIGURE 15.4 Reducing sugars release after 20, 60 and 120 min in the breakfast cereals.

effect NSPs have on the viscosity of products may not be the sole mode by which polysaccharides work in controlling the glycemic impact of carbohydrate-rich food products (Brennan et al., 1996). For instance, the viscosity-altering properties of guar gum have been investigated in relation to the potential glycemic effect of extruded snack products. In this study, a range of guar gum samples based on varying viscosity parameters (Table 15.2) were used at a 10% replacement level.

Figure 15.4 compares the rate of in vitro starch degradation between the samples. All the guar samples showed a reduction in the rate of sugar release (hence starch degradation) when compared to the control samples. However guar gum samples with higher viscosity profiles yielded similar sugar release profiles as those samples with low-viscosity guar gums (Figure 15.4). This leads us on to trying to determine the factors which contribute to the potential glycemic impact of processed foods such as extruded snack products.

If viscosity is not the only way we can control the glycemic impact of processed foods, then we need to investigate the role of ingredients (NSPs) in manipulating the degradability of starch granules during processing and post-ingestion. The link between food structure and gastrointestinal function is one which will deliver exciting results in the years to come.

15.7 CONCLUSION

As illustrated in this brief review of the area of processed foods, the nutritional impact of food processing can be manipulated so that the potential glycemic impact of foods can be regulated. In order to achieve this the practitioner needs to be aware of the role of selected food ingredients, and the role of processing conditions, on the structure, and hence function, of food products. The examples of the manipulation of starch digestibility of breakfast cereal products demonstrate the reality of obtaining functional foods for the health-conscious consumer market. Further work is continuing in order to establish the relationships between food composition, water mobility, food structure and starch digestion. Once this has been modeled, it will be possible to artfully create processed food products with enhanced nutritional values, the snack foods for tomorrow.

References

Alverez-Martinez, L., Kondury, K. P., and Harper, J. M. (1998). A general model for expansion of extruded products. *Journal of Food Science* **53**, 609–615.

Anderson, J.W. and Conley, S.B. (2007). Whole grains and diabetes. *In* Whole Grains and Health, 29–26. Marquart, Jacobs, McIntosh, Poutanen and Reicks (eds.). Blackwell Publishing Oxford, UK.

Björck, I. and Elmståhl, H. L. (2003). The glycaemic index: importance of dietary fibre and other food properties. *Proceedings of the Nutrition Society* **62**, 201–206.

Björck, I., Asp, N.-G., Birkhed, D., and Lundquist, I. (1984). Effects of processing on availability of starch for digestion in vitro and in vivo; I extrusion cooking of wheat flours and starch. *Journal of Cereal Science* **2**, 91–103.

Bouvier, J. M. (2001). Breakfast Cereals. *In* Extrusion Cooking Technologies and applications, 133–160, Guy, Robin (ed.). Woodhead Publishing Ltd, Oxford, UK.

Brand-Miller, J. C., Holt, S. H. A., Pawlak, D. B., and McMillan, J. (2002). Glycaemic index and obesity. *American Journal of Clinical Nutrition* **76**, 281S–285S.

Brennan, C. S., Blake, D. E., Ellis, P. R., and Schofield, J. D. (1996). Effects of guar galactomannan on wheat bread microstructure and on the *in vitro* and *in vivo* digestibility of starch in bread. *Journal of Cereal Science* **24**, 151–160.

Brennan, C. S. (2005). Dietary Fibre, Glycaemic response, and Diabetes. *Molecular Nutrition and Food Research* **49**, 560–570.

Brennan, M. A., Merts, I., Monro, J., Woolnough, J., and Brennan, C. S. (2008a). Impact of guar and wheat bran on the physical, and nutritional, quality of extruded breakfast cereals. *Starch* **60**: 248–256.

Brennan, M. A., Monro, J. A., and Brennan, C. S. (2008). Effect of inclusion of soluble and insoluble fibres into extruded breakfast cereal products made with reverse screw configuration. *International Journal of Food Science and Technology* **43**, 2278–2288.

Case, S. E., Hamann, D. D., and Schwartz, S. J. (1992). Effect of starch gelatinisation on physical properties of extruded wheat- and corn-based products. *Cereal Chemistry* **69**, 401–404.

Cleary, L. J., Andersson, R., and Brennan, C. S. (2007). A comparison of the behaviour and susceptibility to degradation of two barley β-glucan fibre fractions, Glucagel™ and an aqueous-ethanol extract, in a white wheat bread, during fermentation, baking and in vitro digestion. *Food Chemistry* **102**, 889–897.

Ellis, P. R., Rayment, P., and Wang, Q. (1996). A physicochemical perspective of plant polysaccharides in relation to glucose absorbtion, insulin secretion and the entero-insular axis. *Proceedings of Nutrition Society* **55**, 881–883.

Fairchild, R. M., Ellis, P. R., Byrne, A. J., Luzio, S. D., and Mir, M. A. (1996). A new breakfast cereal containing guar gum reduces the postprandial plasma glucose and insulin concentrations in normal-weight human subjects. *British Journal of Nutrition* **46**, 63–73.

Foster-Powell, K. and Miller, J. B. (1995). International tables of glycaemic index. *American Journal of Clinical Nutrition* **62**, 871S–893S.

Gourgue, C., Champ, M., Guillon, F., and Delort-Laval, J. (1994). Effect of extrusion cooking on the hypoglycaemic properties of citrus fibre: an *in vitro* study. *Journal of Science Food and Agriculture* **64**, 493–499.

Hardacre, A. K., Clark, S. M., Riviere, S., Monro, J. A., and Hawkins, A. J. (2006). Some textural, sensory and nutritional properties of expanded snack food wafers made from corn, lentil and other ingredients. *Journal of Texture Studies* **37**, 94–111.

Jenkins, D. J., Jenkins, A. L., Wolever, T. M., Collier, G. R., Rao, A. V., and Thompson, L. U. (1987). Starchy foods and fibre: reduced rate of digestion and improved carbohydrate metabolism. *Scand Journal of Gastroentrol Suppl* **129**, 132–141.

Jenkins, D. J., Popovich, D. G., Kendall, C. W., Rao, A. V., Wolever, T. M., and Tariq, N. (1997). Metabolic effects of non-absorbable carbohydrates. *Scand Journal of Gastroentrol Suppl* **222**, 10–13.

Jenkins, D. J., Wolever, T. M., Jenkins, A. L., Lee, R., Wong, G. S., and Josse, R. (1983). Glycaemic response to wheat products: reduced response to pasta but no effect of fibre. *Diabetes Care* **6**, 155–159.

Jenkins, D. J., Wolever, T. M., Leeds, A. R., Gassull, M. A., Haisman, P., and Dilawari, J. (1978). Dietary fibres, fibre analogues, and glucose tolerance: importance of viscosity. *British Medical Journal* **1**, 1392–1394.

Lund, D. (1984). Influence of time, temperature, moisture, ingredients, and processing conditions on starch gelatinisation. *Critical Reviews in Foods Science and Nutrition* **20**, 249–273.

Mercier, C. and Feillet, P. (1975). Modification of carbohydrate components by extrusion-cooking of cereal products. *Cereal Chemistry* **52**, 283–297.

Mishra, S., Monro, J., and Hedderley, D. (2008). Effect of processing on slowly digestible starch and resistant starch in potato. *Starch* **60**, 500–507.

Panlasigui, L. N., Thompson, L. U., Juliano, B. O., Perez, C. M., Jenkins, D. J. A., and Yiu, S. H. (1992). Extruded rice noodles: starch digestibility and glycaemic response of healthy and diabetic subjects with different habitual diets. *Nutritional Research* **12**, 1195–1204.

Parada, J. and Aguilera, J. M. (2007). Food microstructure affects the bioavailability of several nutrients. *Journal of Food Science* **72**, R21–R32.

Parada, J. and Aguilera, J. M. (2008). In vitro digestibility and glycemic response of potato starch is related to granule size and degree of gelatinization. *Journal of Food Science*. In Press.

Parvathy, K. S., Susheelamma, N. S., and Tharanathan, R. N. (2007). Hydration characteristics of guar gum samples and their fractions. *Food Hydrocolloids* **21**, 630–637.

Politz, M. L., Timpa, J. D., and Wasserman, B. P. (1994). Quantitative measurement of extrusion-induced starch fragmentation products in maize flour using nonaqueous automated gel-permeation chromatography. *Cereal Chemistry* **71**, 532–536.

Theander, O. and Westerlund, E. (1987). Studies on chemical modifications in heat processed starch and wheat flour. *Starch* **39**, 88–93.

Tudorica, C. M., Kuri, V., and Brennan, C. S. (2002). Nutritional and physicochemical characteristics of dietary fiber enriched pasta. *Journal of Agricultural and Food Chemistry* **50**, 347–356.

Vermeulen, A. and Turnbull, W. H. (2000). Feasibility of the gi guide to increase knowledge about the glycaemic index in practice. *Journal of Human Nutrition and Dieteics* **13**, 397–405.

Wang, S., Casulli, J., and Bouvier, J. M. (1993). Effect of dough ingredients on apparent viscosity and properties of extrudates in twin extrusion cooking. *International Journal of Food Science and Technology* **28**, 465–479.

Wang, L. P., Chessari, C., and Karpiel, E. (2001). Inferential control of product quality attributes: application to food cooking extrusion process. *Journal of Process Control* **11**, 621–636.

Wang, Q., Ellis, P. R., and Ross Murphy, S. B. (2003). Dissolution kinetics of guar gum powders- II. Effects of concentration and molecular weight. *Carbohydrate Polymers* **53**, 75–83.

Woolnough, J. W., Monro, J. A., Bird, T., and Brennan, C. S. (2007). Glycaemic carbohydrates: standardisation of *in vitro* methods. *Asia Pacific Journal of Clinical Nutrition* **16**, S47.

Woolnough, J., Monro, J., Bird, A., and Brennan, C.S. (2008). Simulating human carbohydrate digestion in vitro: a review of methods and the need for standardisation. *International Journal of Food Science and Technology* **43**, 2245–2256.

CHAPTER

16

Biopolymers in Controlled-Release Delivery Systems

Kunal Pal [1], *Allan T. Paulson* [2], *and Dérick Rousseau* [1]

[1]Department of Chemistry and Biology, Ryerson University, Toronto, Ontario, Canada
[2]Dept. of Process Engineering and Applied Science – Food Science Program, Dalhousie University, Halifax, NS, Canada

16.1 INTRODUCTION

In its ever-growing scope, polymer science includes various aspects of biology, chemistry, medicine and materials science. The flexibility in polymer processing plays an important role in the development of controlled delivery devices for a range of applications related to drugs, food-related bioactive ingredients and genes. The release of bioactive agents from a delivery system may be either diffusion-controlled (diffusion of drug through a rate-controlling barrier/matrix), degradation-controlled (chemical or physical breakdown of the matrix leads to bioactive agent release) or via an environmental trigger (change in pH, ionic strength or pressure tailors the release of the bioactive agent).

Biopolymers are generally synthesized by an organism or plant and have a more complex chemical structure than that of synthetic polymers. They are usually biocompatible when compared to synthetic polymers (Enderle et al., 2005) and can be easily tailored for the development of controlled-release matrices. A key disadvantage of biopolymers is their susceptibility to microbial contamination. Of late, the term biopolymer has been extended to polymers synthesized using naturally occurring monomers (e.g., lactic acid and glycolic acid) such as PLA [poly(lactic acid)], PGA [poly(glycolic acid)] and PLGA [poly(lactic-co-glycolic acid)] (Enderle et al., 2005). In general, these polymers are biocompatible, non-toxic and biodegradable. Some examples of biopolymers used for biomedical or pharmaceutical applications are shown in Table 16.1.

Rather than develop new drugs or bioactive compounds (which can be costly), many pharmaceutical companies now use controlled-/sustained-delivery technologies to improve existing therapeutic agents simply by controlling the rate at which they enter the bloodstream, thus reducing and ideally avoiding under- or overdosing (Brannon-Peppas, 1997).

Hydrogels represent a key means of controlled or sustained delivery. They consist of three-dimensional, insoluble (bio)polymeric networks capable of imbibing large amounts of

Kasapis, Norton, and Ubbink: Modern Biopolymer Science
ISBN: 978-0-12-374195-0

TABLE 16.1 Examples of Biopolymers Commonly Used for Controlled Delivery Systems

Material	Uses	Refs.
Collagen and gelatin	Wound dressings, tissue engineering, drug delivery	1–6
Chitin and chitosan	Wound dressings, tissue engineering, drug delivery	7–9
Alginate	Drug delivery, cell encapsulation	10,11
Cellulose derivatives	Tissue engineering, drug delivery	12,13
Starch	Tissue engineering, drug delivery	14,15
Carrageenan	Tissue engineering, drug delivery	16,17
Hyaluronic acid	Wound dressings, tissue engineering, drug delivery	18–20
Pectin	Wound dressings, tissue engineering, drug delivery	21–23
Dextran	Wound dressings, tissue engineering, cell encapsulation, drug delivery	24–27

1 = Bertone et al., 2008; 2 = Bolliet et al., 2008; 3 = Liu et al., 2008; 4 = Pawde and Deshmukh, 2008; 5 = Jeong et al., 2008; 6 = Bin Choy et al., 2008; 7 = Krucinska et al., 2007; 8 = Park and Kim, 2007; 9 = Jayakumar et al., 2008; 10 = Chan et al., 2008; 11 = Siti-Ismail et al., 2008; 12 = Rambo et al., 2008; 13 = Mayo-Pedrosa et al., 2008; 14 = Chung et al., 2008; 15 = Sadeghi et al., 2008a; 16 = Sadeghi et al., 2008b; 17 = Ferreira et al., 2008; 18 = Boateng et al., 2008; 19 = Nesti et al., 2008; 20 = Mayo et al., 2008; 21 = Ghosh and Jassal, 2002; 22 = Liu et al., 2004; 23 = Souto-Maior et al., 2008; 24 = Zhang et al., 2007; 25 = Lee et al., 2008; 26 = Wu et al., 2008; 27 = Bachelder et al., 2008.

water or biological fluids, and may be designed for stimuli-response release. Compared to other synthetic biomaterials, hydrogels closely mimic living tissues because of their similar chemical building blocks, thereby exhibiting a reduced incidence of toxicity and inflammation (Gupta et al., 2002; Enderle et al., 2005). The hydrophilic nature of biopolymers imparts water-binding properties whereas the presence of either physical or chemical cross-links results in the formation of a three-dimensional network that helps in retaining structural integrity when placed in an aqueous environment (Gupta et al., 2002; Enderle et al., 2005).

Parameters used to characterize the suitability of a hydrogel for a particular application include: (i) the polymer volume fraction in the swollen state, (ii) the average molecular weight of the polymer chain between two neighboring cross-linking points, and (iii) the mesh/pore size between polymer chains. The polymer volume fraction in the swollen state is a measure of the amount of fluid imbibed and retained by the hydrogel (Peppas et al., 2000; Enderle et al., 2005). The molecular weight between two consecutive cross-links can give a measure of the degree of cross-linking of the polymer. Finally, the mesh size, or the average distance between adjacent cross-links, can be described by a correlation length that provides a measure of the space between biopolymer chains available for diffusion (Peppas et al., 2000; Hoffmann, 2002).

The preparation of hydrogels typically involves cross-linking functional groups (e.g., hydroxyl, amine, amide, ether, carboxylate and sulfonate) along the polymer chains to increase network rigidity (Gupta et al., 2002; Hoffmann, 2002). The design of a hydrogel's microstructure in part depends on the target route of administration as well as the properties of the incorporated compound. Hydrogels are often formed using specific molds to obtain the appropriate size and conformation to maximize delivery effectiveness, depending on the administration routes, e.g., orally (spherical beads, cylinders and discs), in implants (drum-shaped, disc-shaped and cylindrical preparations), and through the skin (films and gel slabs).

16.2 DRUG LOADING AND RELEASE

Release mechanisms can be chemical or physical in nature, but always involve some form of diffusion. Diffusion is defined as the movement of individual molecules of a substance through a semi-permeable barrier from an area of higher concentration to an area of lower concentration (Subramanyam, 2006). It is fundamentally reliant on the properties of the polymeric network and the solvent–polymer interaction, and varies in magnitude depending on the phase; it is fastest in gases (~10 cm min^{-1}), slower in liquids (~0.05 cm min^{-1}) and slowest in solids (~0.00001 cm min^{-1}). Diffusion in hydrogels is more complex and the diffusivities of incorporated compounds will lie somewhere between those in liquids and solids.

For degradable matrices and systems loaded with the compound of interest, release will be controlled by the cleavage of the polymer bonds within the network, even though the diffusion of the liberated therapeutic compound may be rate-limiting (Gupta et al., 2002; Lih et al., 2008). For non-biodegradable systems, release will be diffusion-controlled and will depend on the concentration gradient. However, release may also depend on osmotic pressure and/or matrix swelling (Leong and Langer, 1988; Markland et al., 1999).

Physically controlled release is classified into two types, depending on its mode of release: (a) reservoir-type diffusion or (b) matrix-type diffusion (He et al., 2004; Juntanon et al., 2008).

Reservoir-type: In these hydrogels, therapeutic compounds (solid or liquid) are entrapped in a reservoir within a microporous or non-porous polymeric network. If the therapeutic agent is saturated, its transport will be constant (or follow zero-order release kinetics) since its chemical potential will remain unaffected (Leong and Langer, 1988; Dorski et al., 1997; Gupta et al., 2002; Isiklan et al., 2008). To maintain zero-order release, the therapeutic compound must remain in a solid or suspension state. If the compound has high water solubility, saturation will be difficult to maintain and release will deviate from zero-order kinetics. Even under ideal conditions, release is generally not constant during the initial and final release phases (Leong and Langer, 1988; Gupta et al., 2002). When immersed in a releasing medium, the system will take time to attain a steady state. If there is little or no therapeutic compound present near the surface of the hydrogel, then an induction period will be required for saturation of the surface. In contrast, if the therapeutic compound is concentrated near the surface, the initial release rate will be higher than the equilibrium-state value, caused by a 'bursting' effect. As release nears completion, the therapeutic compound concentration in the core will drop and the release rate will decrease. On the other hand, if boundary sink conditions are imperfect, i.e., if there is an appreciable concentration of the therapeutic compound in the medium at the interface, there will be a drop in the therapeutic compound chemical potential that will slow down release (Leong and Langer, 1988; Gupta et al., 2002).

Matrix-type: These hydrogels, where the drug is dissolved or dispersed within a polymer network, tend to be the most common. A decreased release rate over time due to the increased diffusion distance is a characteristic of these systems (Iskilan et al., 2008). The solubility of the therapeutic compound, its diffusivity in the polymer phase and polymer–compound interactions play an important role in the governing release. Along with these parameters, the shape and the path-length for diffusion also are critical (Leong and Langer, 1988; Gupta et al., 2002; Hoffman, 2002). In certain solvent-penetrating systems, release depends on polymer relaxation, i.e., the stress required to maintain the strain on the polymer decreases, which is a result of aqueous solution absorption (Korsmever and Peppas, 1983; Satish et al., 2006).

When a hydrogel is immersed in an aqueous solution, it will relax into a state that has little resistance to the diffusion of the incorporated compound. The key factor dictating release kinetics becomes swelling, which is a result of polymer relaxation. Important parameters associated with release from a hydrogel include its bioactive compound hydrophilicity and network cross-linking density. Matrix-type hydrogels are commonly used given their ease in development and cost-effectiveness. They tend to follow Higuchi's model, where bioactive compound release is proportional to the square root of time (Higuchi, 1963).

16.3 MODELING DIFFUSION

Diffusion (D) can be determined in a double diffusion cell, where a concentrated solute solution (kept in the donor compartment) is separated from a pure solvent (kept in the receptor compartment) by a semi-permeable membrane (Sloan, 1989; Nagy et al., 1989; Foldvari et al., 1998; Tenjarla et al., 1999; Liang et al., 2007; Pal et al., 2007a, 2007b). Two examples of such cells are the horizontal transport cell (Figure 16.1) and the vertical transport cell (Figure 16.2). During diffusion, the solute passes from the donor compartment (which is a region of high concentration) to the receptor compartment

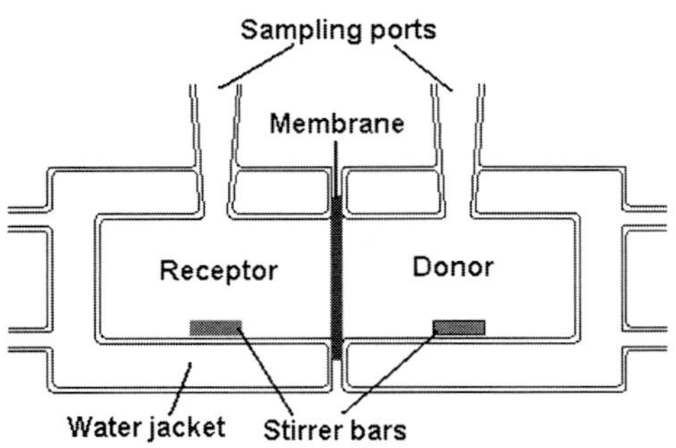

FIGURE 16.1 Schematic diagram of a horizontal diffusion cell.

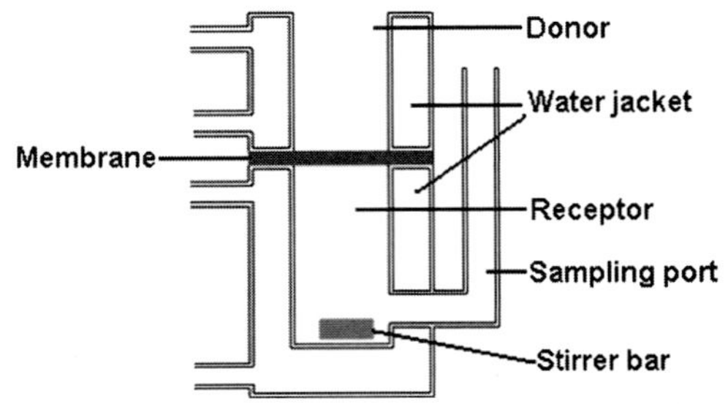

FIGURE 16.2 Schematic diagram of a vertical diffusion cell.

(which is a region of low concentration) through a semi-permeable barrier (membrane). This leads to the decrease of the solute concentration in the donor compartment and corresponding increase in the concentration in the acceptor compartment. This process continues until equilibrium is reached, i.e., the concentration of the solute in both compartments is the same. Thus, a steady state is reached. In general, measurements of D are carried out using a kinetic method, by sampling from the donor and acceptor compartments at set time intervals. The measured solute concentration in the two chambers is then used to calculate the solute D value through the semi-permeable matrix (Falk et al., 2004; Pal et al., 2006):

$$D = \frac{1}{\beta t} x \ln \frac{C_D(t) - C_R(t)}{C_D(0) - C_R(0)} \tag{16.1}$$

with

$$\beta = \frac{A_H}{W_H} x \left[\frac{1}{V_D} + \frac{1}{V_R} \right] \tag{16.2}$$

where $C_D(0)$ = initial concentration of drug in donor compartment; $C_R(0)$ = initial concentration of drug in receptor compartment; $C_D(t)$ = concentration of drug in donor compartment after time t; $C_R(t)$ = concentration of drug in receptor compartment after time t; A_H = effective cross-sectional area of diffusion in the hydrogel; W_H = width of the hydrogel sample;

V_D = volume of drug solution in donor compartment; V_R = volume of receptor compartment fluid.

The slope of a plot of $-\ln\frac{C_D(t)-C_R(t)}{C_D(0)-C_R(0)}$ vs. time is used for calculating D.

Diffusion can be studied either under steady-state or under sink conditions (Saylor et al., 2007; Gazzaniga et al., 2008). In steady-state diffusion, the diffusion coefficient does not vary with time, i.e., dc/dt or dM/dt is constant and so diffusion is not allowed to attain equilibrium. During such experiments, the concentration of the solute in the donor and the acceptor compartments (i.e., hydrogel and surrounding medium, respectively) has to be maintained. This can be achieved by connecting the donor and acceptor to large reservoirs of solutions and recirculating these during the experimental period (Subrahmanyam, 2006).

For experimental procedures under sink conditions, the concentration of the solute in the acceptor compartment is maintained at a lower level than that of the donor compartment (Christensen et al., 1982; Tongwen and Binglin, 2000). The donor acts as a source of solute and the receptor acts as a sink, which is accomplished by connecting the receptor to a large reservoir from which the solution is recirculated to maintain a constant concentration gradient. For such experimental studies, sink conditions are preferred as maintaining constant a donor concentration can be difficult, and experiments with recirculation in one compartment are easier than recirculation in both compartments (Subrahmanyam, 2006).

Diffusion coefficients can be described using Fick's 1st and 2nd laws of diffusion (Foldvari et al., 1998). These laws are applicable when a semipermeable barrier acts as a rate-controlling step for bioactive compound release.

16.3.1 Fick's First Law

The diffusion of molecules is denoted as dM/dt (rate of mass transfer) and is generally expressed as flux (J). J is defined as the rate of the mass transfer across a unit surface area of a barrier and is expressed as (Martin et al., 1991; Gennaro, 1990; Fournier, 1998; Dill and Drmberg, 2003; Hench and Jones, 2005; Subrahmanyam, 2006; Gazzaniga et al., 2008; see also http://cnx.org/content/m1035/latest/):

$$J = \frac{1}{2} \times \frac{dM}{dt} \tag{16.3}$$

where dM = change in mass, g; S = barrier surface area, cm^2; dt = change in time, s.

The unit for flux is:

$$J = \frac{1}{cm^2} \times \frac{g}{s} \tag{16.4}$$

which is often expressed as kg m^{-2} s^{-1}.

Fick's first law directly correlates flux with the concentration gradient and can be expressed as:

$$J = -D \times \frac{dC}{dx} \tag{16.5}$$

where dC = change in concentration of the solute, g/cm^3; D = diffusion coefficient of the solute, cm^2/s; dx = change in distance, cm.

The negative sign in Equation (16.5) indicates a decrease in the concentration in the donor compartment. Since flux increases continuously during the process, it is always a positive entity. Combining Equations (16.3) and (16.5) yields the rate of mass transfer:

$$\frac{dM}{dt} = -D\,S\frac{dC}{dx} \tag{16.6}$$

16.3.2 Fick's Second Law

Fick's second law deals with the change in the concentration with time at a definite location with respect to the x, y and z axes. This law states that the change in concentration with time in a particular region is proportional to the change in the concentration gradient at that point of time. In general, this equation is not used in the pharmaceutical and nutraceutical industries. Briefly, it is expressed mathematically as (Gennaro, 1990; Martin et al., 1991; Fournier, 1998; Dill and Drmberg, 2003; Hench and Jones, 2005; Subrahmanyam, 2006; Gazzaniga et al., 2008 see also http://cnx.org/content/m1036/latest/)

$$\frac{\partial C}{\partial t} = D\left[\frac{\partial^2 C}{\partial x^2} + \frac{\partial^2 C}{\partial y^2} + \frac{\partial^2 C}{\partial z^2}\right] \qquad (16.7)$$

where x, y and z are the coordinates.

D is affected by temperature, pressure, solvent properties and the chemical properties of the solute. Hence, it is regarded as a coefficient and not a constant. The effect of temperature on *D* can be expressed by the Arrhenius expression (see http://en.wikipedia.org/wiki/Fick's_law_of_diffusion):

$$D = D_0 \cdot e^{-\frac{E_a}{R.T}} \qquad (16.8)$$

where D = diffusion coefficient of the solute; D_0 = maximum diffusion coefficient (at infinite temperature); E_a = activation energy for diffusion; T = Temperature, K; R = Gas constant.

According to Hoffmann (2002), the structure of the polymer network, mesh/pore size and polymer composition play an important role in the determination of *D*. Parameters such as average pore size, size distribution and connectivity, which are often difficult to quantify, are usually included into a factor termed tortuosity. These factors are usually dependent on the properties of the polymer chains and the cross-link density of the polymer network. Tortuosity helps in the determination of the effective diffusion path length, which is defined as the product of film thickness and the ratio of porous fraction and tortuosity (Hoffman, 2002).

Release from a delivery system may either follow Fickian or non-Fickian diffusion. The three main release models used in the pharmaceutical industry include Higuchian kinetics, zero-order kinetics and first-order kinetics, with the former taking Fickian diffusion into account.

16.4 HIGUCHIAN MODEL

This model is suitable when the active agent has been dispersed in an insoluble matrix that can swell. When a delivery system is exposed to an external medium, the exposed active agent layer dissolves and diffuses out of the matrix. This process continues as the boundary between the dissolution medium and the dispersed bioactive compound evolves with time (Subrahmanyam, 2006). The region within the polymer matrix devoid of any drug is referred to as the polymer ghost (Figure 16.3).

Higuchi's model assumes that the rate of dissolution of the active agent is faster than the rate of diffusion of the active agent, which ensures the continuous release of the active agent to the surrounding medium. The rate equation for this kind of system is represented by (Subrahmanyam, 2006):

$$\frac{dM}{dx} = C_0.dx - \frac{C_s}{2} \qquad (16.9)$$

where dM = change in the amount of drug released per unit area; dx = change in polymer ghost thickness; C_0= total amount of drug in unit

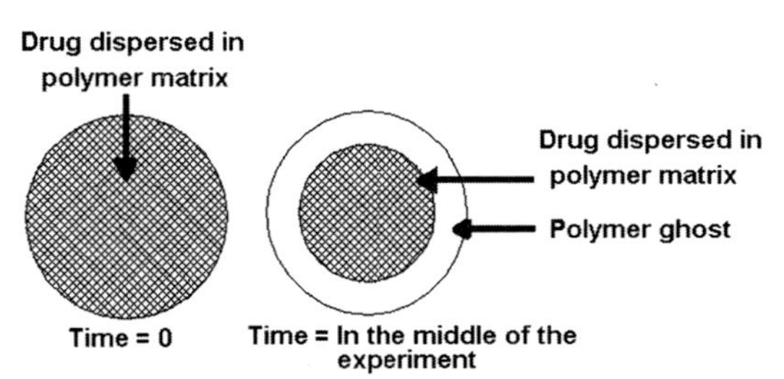

FIGURE 16.3 Change in the boundary between the dissolution medium and drug for diffusion-controlled release systems (modified from Hoffman, 2002).

volume of the matrix; C_s= saturated concentration of the drug in the matrix.

The diffusion theory expresses the rate of mass transfer as follows (Subrahmanyam, 2006):

$$k = 2C_s D_m C_0$$

$$\frac{dM}{dt} = \frac{D_m C_s}{x} \quad (16.10)$$

where D_m = diffusion coefficient in the matrix.

Equating, integrating and solving Equations (16.9) and (16.10) yields:

$$M = [C_s D_m (2C_0 - C_s)t]^{1/2} \quad (16.11)$$

When the amount of drug is in excess of the saturation concentration, i.e., $C_0 >> C_s$:

$$M = [C_s D_m 2C_0 t]^{1/2} \quad (16.12)$$

or

$$M = kt^{1/2} \quad (16.13)$$

where $k = 2C_s D_m C_0$ is a constant as Cs, D_m and C_0 are constant for a given delivery system (Subrahmanyam, 2006). If a plot of the amount of drug released vs. the square root of time is linear, then drug release from the matrix is said to be diffusion-controlled and Higuchian (Ali et al., 2007; Oladiran and Batchelor, 2007; Tebbe et al., 2007; Sriamornsak et al., 2008). For example, the release of ibuprofen when dispersed in polystyrene microspheres can result in a delivery system following Higuchian kinetics. Depending on the composition of the ibuprofen–polystyrene microspheres, 30–80% of the drug is released into a dissolution medium in 24 h (Saravanan et al., 2004). Similarly, the release of ketorolac tromethamine from albumin microspheres indicates diffusion-controlled release (Mathew et al., 2007). Pea protein has been used for preserving the biological activity of ascorbic acid and the prepared microparticles released the ascorbic acid following Higuchian kinetics (Pierucci et al., 2006) (Table 16.2).

TABLE 16.2 Summary of Equations for Release Kinetics

Model	Equation
Zero order	F=k.t
First order	ln F=k.t
Higuchian	F=k.√t

* F is the fraction of drug released from the system, k is the release constant and t is the time.

Zero-order release systems are structurally similar to the matrices following Higuchian kinetics, with the only exception being that release is constant over a period of time because the releasing surface remains constant (Mockkel and Lippold, 1993). The modeling of zero-order release systems is based on certain assumptions, namely a uniform distribution of the drug in the delivery system, and the average pore size in the delivery system being significantly smaller than the size of the bioactive compound. In practicality, however, zero-order release systems may often convert to first-order release systems (Ehtezazi et al., 2000).

16.5 SWELLING

It is possible to design a delivery system that is incapable of releasing its therapeutic agent until it is placed in an appropriate environment that promotes the diffusion of the solvent into the delivery system. These swelling-controlled delivery systems may be initially dry and swell after absorbing dissolution media. The swelling increases the aqueous solvent content within the formulation with an increase in the drug dissolution thereby enabling the drug to diffuse out of the swollen network into the external environment (Kashyap et al., 2005). As described later, there are numerous examples of food-related biopolymers used in the swelling-controlled-release of therapeutic compounds (Montisci et al., 2001; Liang et al., 2003; Gunasekaran et al., 2007; Hasegawa et al., 2007).

16.5.1 Stimuli-responsive Delivery

An ideal delivery system should respond to physiological requirements, sense the changes and, accordingly, alter its release profile. The symptoms of most diseases follow a cyclic pattern and require drug delivery to mirror these cycles. If a drug possesses side effects, release, when not required, poses an extra burden on the body's metabolic system. Thus, delivery patterns need optimization for self-regulated mechanisms. Hydrogels can display a dramatic effect in their swelling behavior, network structure, permeability or mechanical strength in response to different stimuli, such as temperature and pH (Peppas et al., 2000; Kavanagh et al., 2004; Hasegawa et al., 2007; Shang et al., 2007; Deng et al., 2008).

16.6 TEMPERATURE-SENSITIVE HYDROGELS

Temperature-sensitive hydrogels are probably the most commonly studied class of sensitive polymer systems in drug delivery research (Li et al., 2008a, 2008b; Vertommen et al., 2008; Wang et al., 2008). They have gained considerable attention in the pharmaceuticals field due to their ability to swell or shrink as a result of changes in temperature. Thermosensitive hydrogels can be classified as positive or negative temperature-sensitive systems. A positive temperature-sensitive hydrogel has an upper critical solution temperature (UCST). Such hydrogels contract upon cooling below the UCST. Negative temperature-sensitive hydrogels have a lower critical solution temperature (LCST) and contract upon heating above the LCST. Temperature can thus be utilized to modulate the volume of these hydrogels and consequently, their release profile can be altered (Peppas et al., 2000).

16.6.1 pH-sensitive Hydrogels

Hydrogels that exhibit pH-dependent swelling consist of networks comprised of ionic polymers. The polymers with a large number of ionizable groups are known as polyelectrolytes, and include food-grade materials (e.g., alginic acid and chitosan) (Dai et al., 2008a, 2008b). In an aqueous medium of appropriate pH and ionic strength, these groups can ionize, resulting in changes to polymer conformation and behavior. Electrostatic repulsion between ionized polymer chains may increase the uptake of solvent in the network (Katchalsky and Michaeli, 1955; Brannon-Peppas and Peppas, 1991; Tomita et al., 2008). Anionic hydrogels are composed of polymer chains containing negatively charged functional groups. Ionization occurs when the environmental pH is above the pK_a of the ionizable moiety. As ionization increases, there is a resultant increase in electrostatic repulsions between the chains, which results in greater swelling. Conversely, cationic materials contain groups such as amines. These groups ionize in media that are at a pH below the pK_b of the ionizable species. Thus, in a low-pH environment, ionization increases. There are many advantages to using ionic over neutral networks in drug delivery. Their pH sensitivities can be easily exploited in a wide variety of biomedical applications, including corneal implants, controlled-release devices and biocompatible materials (Peppas et al., 2006; Farooqui et al., 2007; Zhou et al., 2008). Drug/bioactive ingredient release is most often observed during the swelling of a hydrogel. However, a few instances have reported drug release during syneresis (Gutowska et al., 1997a, 1997b). Another interesting characteristic of many responsive hydrogels is the reversibility of the induced changes in network microstructure. Such elastic deformability allows hydrogels to return to their original shape at the end of a triggering stimulus (Gupta et al., 2002; Lang et al., 2006).

The majority of research efforts to date have focused on single stimulus response hydrogels for controlled delivery. The next few years should see developments in double or multiple stimuli response hydrogels. Already, progress is being made in this direction. An interpenetrating network of gelatin and dextran has been proposed as a dual stimuli-responsive biodegradable hydrogel, where lipid microspheres have been incorporated as reservoirs for bioactive compounds (Kurisawa and Yui, 1998a, 1998b, 1998c, 1998d). Hydrogels prepared below the sol–gel transition temperature were found to release the lipid microspheres in the presence of both α-chymotrypsin and dextranase; no release in the presence of either enzyme alone occurred.

16.7 EQUILIBRIUM SWELLING AND THE FLORY-REHNER THEORY

Swelling is characteristic of cross-linked polymer gels once they are immersed in a compatible fluid. If the network structure only swells and is not broken down in the presence of a solvent, then a state of equilibrium swelling may be attained, which may be explained by the Flory-Rehner theory (Flory and Rehner, 1943a, 1943b). This theory states that when a polymeric network swelling in a solvent is allowed to reach equilibrium, there are only two opposing forces at work – the force of thermodynamic mixing and the retractile force of the polymer. As more solvent penetrates the polymer network, the volume begins to increase and the network junction zones are forced to elongate and expand. The swelling effectively reduces the chain configurational entropy. Opposed to this is the increase in entropy due to the mixing of polymer and solvent during swelling. Once these opposing entropies or forces become equal in magnitude, equilibrium is reached. This situation is described in terms of the Gibbs free energy:

$$\Delta G_{total} = \Delta G_{elastic} + \Delta G_{mixing} \qquad (16.14)$$

where $\Delta G_{elastic}$ is the change in energy due to the deformation of network chains between network junction zones and ΔG_{mixing} is the result of the solvent–polymer mixing. In the case of polyelectrolytes, ionic groups may complicate swelling behavior, where the ionic nature of the polymers (ΔG_{ion}) also contributes to the total change in Gibbs free energy:

$$\Delta G_{total} = \Delta G_{elastic} + \Delta G_{mixing} + \Delta G_{ion} \qquad (16.15)$$

By taking the derivative of Equation (16.15) with respect to the number of moles of solvent, an expression for the chemical potential can be derived:

$$\mu_1 - \mu_{1,0} = \Delta\mu_{elastic} + \Delta\mu_{mixing} + \Delta\mu_{ion} \qquad (16.16)$$

where μ_1 is the chemical potential of the solvent in the polymer gel and $\mu_{1,0}$ is the chemical potential of the pure solvent. $\Delta\mu_{elastic}$, $\Delta\mu_{mixing}$ and $\Delta\mu_{ion}$ are the changes in chemical potential due to deformation of the polymer chain, solvent polymer mixing and ionic character of the hydrogel, respectively. At equilibrium, the difference between the chemical potentials outside and inside the gel must be zero. A modified version of the original Flory-Rehner theory was proposed by Peppas and Merrill for hydrogels prepared in the presence of water (Peppas and Merrill, 1977). Water effectively alters the change in chemical potential due to elastic forces and must be accounted for in the expression. In addition, the chemical potential has a strong dependency on the nature of the ions and the ionic strength of the surrounding medium. The following equation is derived for the swelling of an anionic hydrogel prepared in the presence of a solvent (Brannon-Peppas and Peppas, 1991):

$$\frac{V_1}{4I}\left(\frac{v_{2,s}}{\overline{v}}\right)^2\left(\frac{K_a}{10^{-pH}-K_a}\right)^2 = \left[\ln(1-v_{2,s})+v_{2,s}+\chi_1 v_{2,s}^2\right]$$

$$+v_{2,r}\left(\frac{V_1}{\overline{v}\overline{M}_c}\right)\left(1-\frac{2\overline{M}_c}{\overline{M}_n}\right)\left[\left(\frac{v_{2,s}}{v_{2,r}}\right)^{1/3} - \left(\frac{v_{2,s}}{2v_{2,r}}\right)\right] \quad (16.17)$$

and for the swelling of a cationic hydrogel:

$$\frac{V_1}{4I}\left(\frac{v_{2,s}}{\overline{v}}\right)^2\left(\frac{K_b}{10^{pOH}-K_a}\right)^2 = \left[\ln\left(1-v_{2,s}\right)+v_{2,s}+\chi_1 v_{2,s}^2\right]$$

$$+v_{2,r}\left(\frac{V_1}{\overline{v}\overline{M}_c}\right)\left(1-\frac{2\overline{M}_c}{\overline{M}_n}\right)\left[\left(\frac{v_{2,s}}{v_{2,r}}\right)^{1/3} - \left(\frac{v_{2,s}}{2v_{2,r}}\right)\right] \quad (16.18)$$

where $\overline{M}_n$ is the molecular weight of the polymer without cross-linking, $\overline{M}_c$ is the number average polymer molecular weight between two adjacent cross-links determined from Equation (16.17) or (16.18), $\overline{v}$ is the specific volume of the hydrogel prior to swelling, V_1 is the molar volume of the solvent, water (18 mL mol^{-1}), $v_{2,s}$ is the polymer volume fraction in the swollen state determined as approximately the inverse of the equilibrium swelling ratio, $v_{2,r}$ is the polymer volume fraction in the relaxed state (the state of the polymer immediately after cross-linking but before swelling), I is the ionic strength, K_a and K_b are the dissociation constants for the acidic and basic moieties on the polymer, and χ_1 is the Flory-Huggins parameter which describes the polymer-solvent interaction (Ofner and Bubnis, 1996).

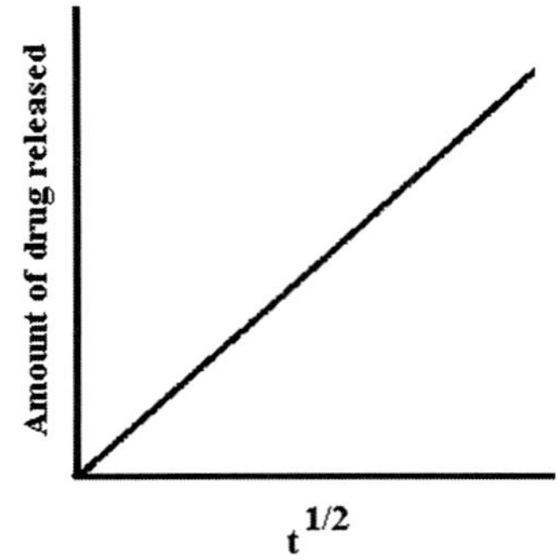

FIGURE 16.4 Schematic diagram of drug release.

Using the number average molecular weight between cross-links, $\overline{M}_c$, the cross-link density, q, can be determined from (Rajvaidya et al., 2006):

$$q = \frac{\overline{M}_n}{\overline{M}_c} \quad (16.19)$$

The parameter $v_{2,s}$ is determined from the volume-swelling ratio, q_v(Caykara and Kantoglu, 2004):

$$v_{2,s} = \frac{1}{q_v} \quad (16.20)$$

The volume-swelling ratio is calculated as:

$$q_v = 1+\frac{(q_w-1)\times\rho_2}{\rho_1} \quad (16.21)$$

where ρ_2 and ρ_1 are the densities of the polymer network and solvent, respectively. The weight-swelling ratio, q_w, is determined from:

$$q_w = \frac{m_s}{m_o} \quad (16.22)$$

where m_o and m_s are the mass of the unswollen gel and the mass of the swollen gel at equilibrium, respectively (Figure 16.4 near here).

16.8 APPROACHES TO CROSS-LINKING

Cross-linking improves the thermal and mechanical stability of the matrix, and can be tailored to modify the release rate of the incorporated active agents (Campoccia et al., 1998; Prestwich et al., 1998; Peppas et al., 2000; Hoffman, 2002; Trimukhe and Varma, 2008). Cross-linking methods can be broadly classified into two categories: physical (e.g., irradiation with UV and gamma radiation and dehydrothermal treatment) and chemical (use of cross-linkers like genipin, glutaraldehyde, etc.) (Moffat and Marra, 2004; Taleb, 2008). Desired release properties in a matrix modified with physical cross-linking methods are seldom achieved, as there is difficulty in controlling the cross-linking density. Chemical cross-linkers can be categorized into two types: non-zero-length cross-linkers and zero-length cross-linkers. Examples of the former include bi-/multi-functional molecules, which bridge free carboxylic acid groups, amino groups and hydroxyl groups between adjacent polymer molecules (e.g., glutaraldehyde, polyepoxides and isocyanates). With zero-length cross-linkers, reactive groups such as carboxylic acid and amine groups present in polymer network chains react with each other leading to formation of a covalent bond. Cross-linking with acyl azides and carbodiimides are classical examples of zero-length cross-linkers (Petite et al., 1994; Kao et al., 2004). The main advantage of chemical cross-linking includes the easy control of the cross-linking density by tailoring the reaction conditions either by changing the cross-linking reaction period or by changing cross-linker concentration (Peppas, 1987; Kashyap et al., 2005). The pore size of the gelled matrix is governed by the cross-linking density, which in turn affects the diffusion of the solute particles through the gelled matrix. The pore size of a matrix is decreased with the increase in the cross-linking density, which will reduce the swelling ability of the (bio)polymeric network.

Though far from extensive, the following section describes the mechanisms and properties of some well-known and more recently examined chemical cross-linking agents.

16.9 GLUTARALDEHYDE

Glutaraldehyde (GA) has seen extensive use as a cross-linking agent for biomedical applications such as enzyme and cell immobilization as well as hydrogel synthesis (Long et al., 2007; Rokhade et al., 2007; Sacks et al., 2007; Silva et al., 2007; Trimukhe and Varma, 2008; Wu and Brazel, 2008). GA is a colorless liquid with a pungent odor at room temperature. It is a dialdehyde whose aldehydic groups are highly reactive and can form covalent bonds with functional groups such as amines, thiols, phenols, hydroxyl and imidazoles. The schematic representation of the reaction of GA with hydroxyl and amino groups (which are very common in hydrogels) is shown in Figure 16.5 (Kim et al., 1994; Walt and Agayn, 1994).

The cross-linking of hydroxyl groups with GA must be carried out in acidic conditions. The presence of acid catalyzes acetalization amongst the hydroxyl and aldehydic groups (Tomihata and Ikada, 1997). Conversely, cross-linking of amines with GA in the presence of acidic conditions is slower compared to neutral and basic conditions. This phenomenon, i.e., slower reaction kinetics in acidic conditions, is used in developing hydrogels or scaffolds of various shapes and sizes (Chajra et al., 2008; Ju et al., 2008; Yu et al., 2008).

GA is the most commonly used cross-linking agent due to its effectiveness in the stabilization of biomaterials, and it is easily accessible, economical and its aqueous solutions can effectively cross-link collagenous tissues (Khor, 1997; Bigi et al., 2002). However, if released into the host due to degradation, GA is toxic. There is

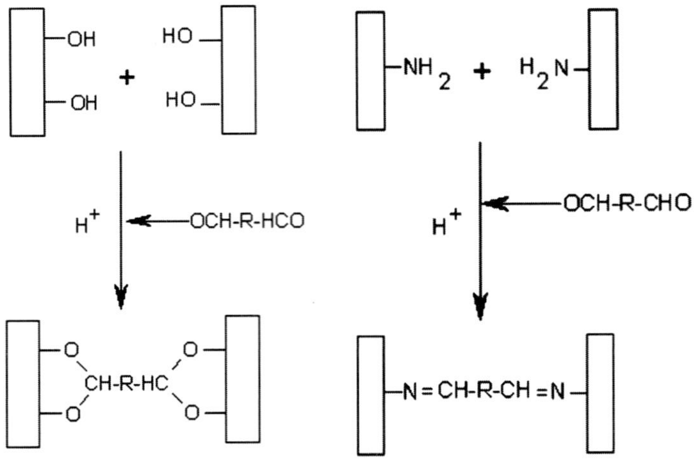

FIGURE 16.5 Schematic representation of the cross-linking reaction of glutaraldehyde with hydroxyl and amino groups.

also some local cytotoxicity, if not neutralized properly, and calcification of long-term implants has been reported (Pettenazzo et al., 2001; Sacks et al., 2007). For this and other reasons, there is increasing demand for cross-linkers that form stable, biocompatible cross-linked products and cytotoxicity.

16.10 GENIPIN

Genipin (Figure 16.6) is an aglycone derived from an iridoid glycoside called geniposide present in the fruit of *Gardenia jasminoides*, a common flower in many parts of Asia. It is widely used in herbal medicine, and the dark blue pigments obtained by its spontaneous

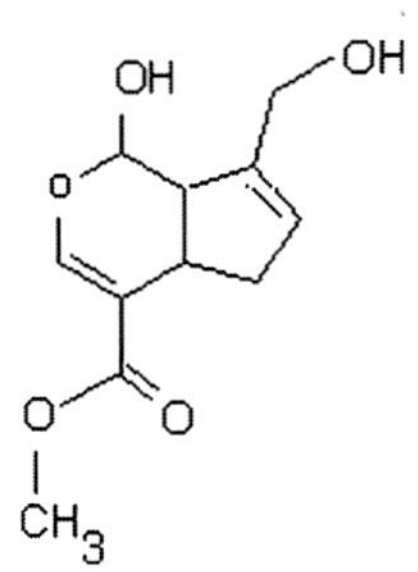

FIGURE 16.6 Molecular structure of genipin.

reaction with amino acids or proteins have been used in the fabrication of food dyes. Other applications involve the preparation of gelatin capsules and the immobilization of enzymes (Bigi et al., 2002).

It is a biodegradable molecule with low cytotoxicity (Linnes et al., 2007; see also http://www.wou.edu/las/physci/ch350/Projects_2006/Aaandering/Genipin.htm). The use of genipin as a biological glue has been evaluated extensively. It has been found that it results in lessened cytotoxicity and inflammatory responses as compared to aldehyde glues (e.g., formaldehyde and glutaraldehyde). In addition, wound healing has been shown to occur more rapidly when treated with genipin (Sung et al., 1999; Liang et al., 2003).

Genipin's cross-linking mechanism is complex and is not fully understood. It is known that it cross-links materials containing primary amine groups via two mechanisms (Butler et al., 2003; Mi et al., 2000, 2005). The mechanism proposed by Zhu and Park (Fournier, 1998) (Figure 16.7) is based on the ring-opening reaction of genipin, which can be initiated by an amino group via a nucleophilic attack on the olefinic carbon atom of genipin. This is subsequently followed by a covalent grafting of the

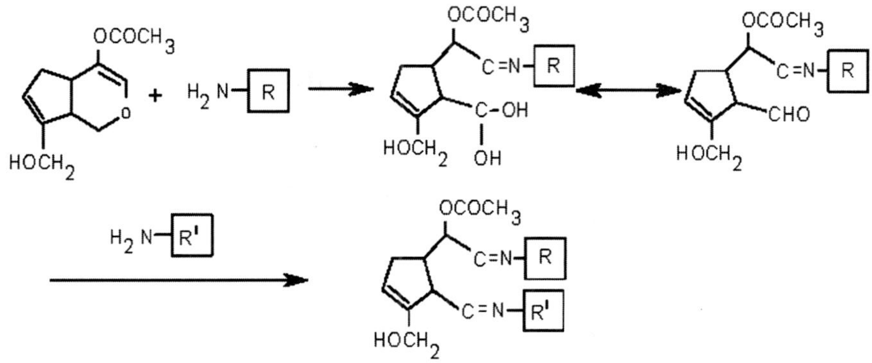

FIGURE 16.7 Schematic representation of mechanism of genipin cross-linking proposed by Zhu and Park (modified from Zhu and Chan-Park, 2007).

genipin onto the polymer with the amino group by a two-step reaction. An unstable intermediate formed during the reaction collapses to form a tautomeric aldehyde. The aldehyde group thus formed is again attacked by another amine group from another polymer forming another covalent bond resulting in the formation of the cross-link (Zhu and Chan-Park, 2007).

16.11 QUINONES AND PHENOLS

The tanning of biological materials with quinones (Figure 16.8) alters their mechanical and solubility properties, making them insoluble in water, detergents, organic solvents and in strong acids and alkalies. Furthermore, there are changes to texture, with the cross-linked materials evolving from soft and pliable to hard and tough. This phenomenon has been exploited to develop protein-based biomaterials with desirable textural attributes (Suyama and Tsunooka, 1998). Polymeric phenolics (e.g., catechol) are abundant in the plant kingdom. These polyphenols have the capacity to form networks with biopolymers (Figure 16.9) in the presence of suitable enzymes, such as catechol oxidase. The enzymes result in the formation of reactive intermediates like quinones and free radicals (Waite, 1990). Lignification in plants is a classic example of this reaction (Andersen et al., 1996; Chase et al., 2000; Kramer et al., 2001; Wu et al., 2006). Another example of this includes the

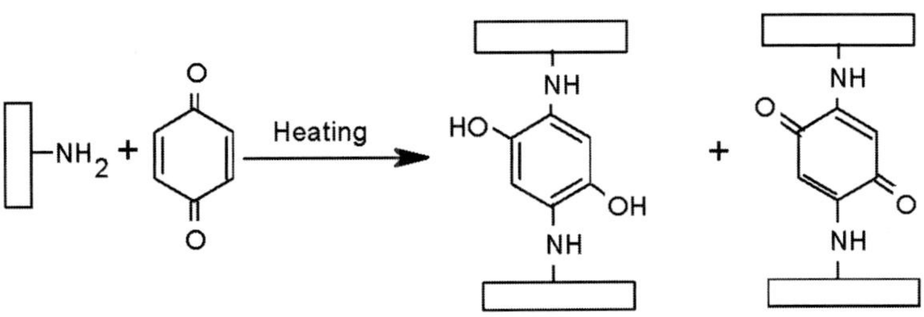

FIGURE 16.8 Schematic representation of mechanism of cross-linking of amino groups with quinones (modified from Suyama and Tsunooka, 1998).

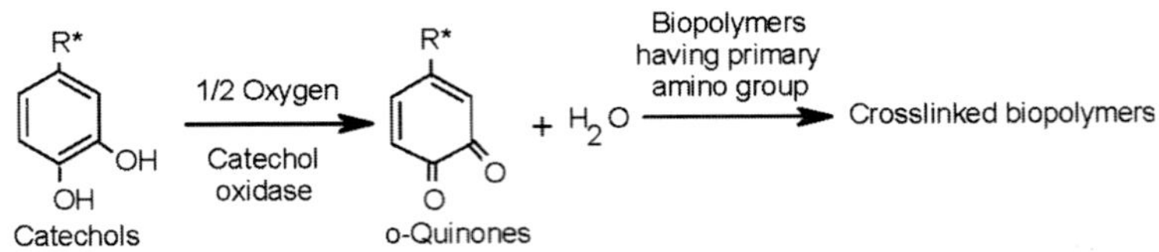

FIGURE 16.9 Schematic representation of the conversion of catechol into quinones, which subsequently takes part in the cross-linking of the biopolymers (modified from Waite, 1990).

curing of the mussel's adhesive proteins which leads to the formation of a cross-linked network (Waite, 1990; McDowell et al., 1999; Burzio and Waite, 2000; Wo et al., 2006). This phenomenon has been explored in the fabrication of hydrogels for biomedical and food applications (Strauss and Gibson, 2004). These cross-linkers are natural, non-toxic (in general) and easily available.

16.12 POLYELECTROLYTE CROSS-LINKING AND COMPLEXES

The use of ionic cross-linking has recently gained importance in medical and pharmaceutical applications. The main advantages of this approach are that it is simple and does not require the use of catalysts. In order to prepare an ionically cross-linked polymeric network, an ionic polymer (e.g., sodium alginate or chitosan) and a charged counterion (also known as ionic cross-linker) (e.g., calcium chloride) or ionic polymer having an opposite charge are dispersed in a solvent, after which ionic interactions occur (Figure 16.10) (Berger et al., 2004; Subrahmanyam, 2006).

Anionic polymers such as alginates can be cross-linked and gelled using multivalent cationic cross-linkers in the presence of divalent cations such as Ca^{2+}, Sr^{2+} or Ba^{2+}. Divalent cations bind with the guluronic acid residues of the alginates and form junction zones with other chains resulting in the formation of a cross-linked networked structure, which has the ability to retain its structure. Depending upon the concentration of the divalent cation used, alginates can give rise to either highly viscous thixotropic solutions (at low levels of Ca^{2+} ions) or a permanent gelled network (at high levels of Ca^{2+} ions) (Alexakis et al., 1995; Gombotz and Wee, 1998; Tonnesen and Karlsen, 2002).

Metallic anions [e.g., Mo(VI), Pt(II) and phosphate-bearing groups (e.g., tripolyphosphates)] may be used for inducing ionic cross-links in cationic polymers. The pH of the

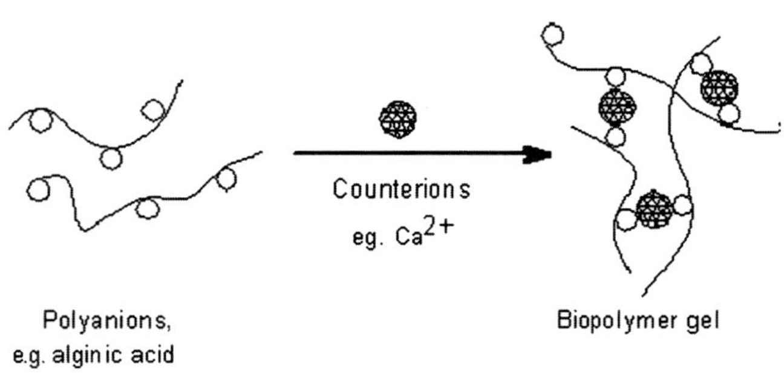

FIGURE 16.10 Schematic representation of hydrogel formation using a polyelectrolyte and an ionic cross-linker.

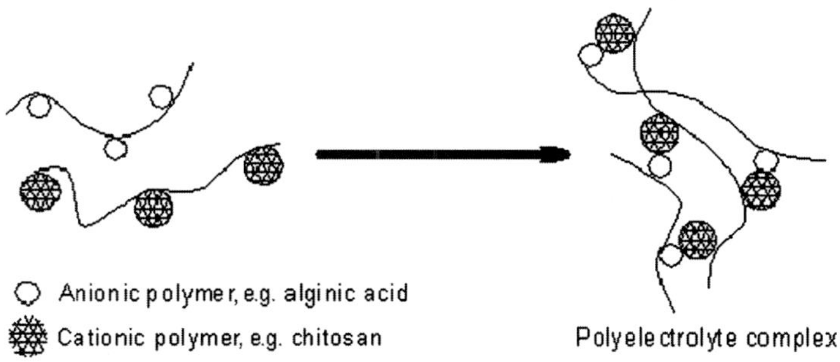

FIGURE 16.11 Schematic representation of hydrogel formation using polyelectrolytes with opposite charges.

polymer solution also plays an important role in cross-linking. For example, with a chitosan solution, if the pH is higher than the pK_a of the amino groups, the polymer solution will undergo coacervation-phase inversion without any cross-link formation, which is attributed to the precipitation of the cationic polymers (Berger et al., 2004).

The cross-links formed between two oppositely charged polymers leads to the formation of polyelectrolyte complexes (Figure 16.11) (Whateley, 1992; Thu et al., 1996; Dumitriu and Chornet, 1998; Girod et al., 2004). Gels stabilized with this approach have been successfully used for the development of biomedical products, such as matrices for drug/gene delivery systems and tissue engineering. For example, calcium alginate cores encapsulated with a polyelectrolyte complex of alginate and poly (L-lysine) as well as alginate-chitosan polyelectrolyte systems have been used in various biomedical applications (Thu et al., 1996; Sankalia et al., 2007; Sarmento et al., 2007). Ionic cross-linking can be easily controlled and hence is gaining importance in the development of hydrogels for biomedical and food-related applications.

Overall, the rate of cross-linking can be influenced by the size of the cross-linker. Smaller cross-linkers have the ability to diffuse more readily into the polymer solution, resulting in a faster rate of reaction. The overall charge on the polymer may affect the properties of the cross-linked gels. Larger polymers will also affect the final rheological properties of the resulting polymer network. Lastly, if the polymer solution is not fully cross-linked, the hydrogel may be pH-sensitive, given the presence of free charged groups in the polymer network (Berger et al., 2004). This can be achieved either by shortening the reaction time or by using a stoichiometrically lower quantity/amount of the cross-linker.

16.13 POLYMER–DRUG INTERACTIONS

The interactions between drugs and polymers can be tailored to modify the release profile of a bioactive agent. Such interactions may be either chemical or physical in nature. When an ionic polymer comes in contact with an ionic drug having an opposite charge, the polymer interacts with the drug, resulting in complexation. Drug release from this type of matrix is dependent on the exchange of drug molecules with counter-ions in the dissolution medium, if present. As the drug is released, the polymer undergoes dissolution, exposing the inner drug–polymer complex and ensuring complete drug release. This approach has been used to develop various delivery systems (Bonferoni et al., 2007a, 2007b). Bonferoni et al. (2007a, 2007b) prepared microparticles of chondroitin-6-sulfate (an anionic polymer) loaded with ciprofloxacin HCl (a cationic drug) via ionic interactions. These particles were subsequently entrapped in a Carbopol® matrix (a cross-linked acrylic acid matrix). The resulting complex was used in ocular delivery systems and showed a prolonged precorneal residence time and increased drug bioavailability in the anterior segment of the eye (Bonferoni et al., 2007a, 2007b). In another study, a complex of chitosan (which is cationic) and salicylic acid (an anionic drug) was tested as a transdermal delivery system (Puttipipatkhachorn et al., 2001).

In the following sections of this chapter, the physicochemical properties and applications of collagen, gelatin, chitosan and alginates in pharmaceutical and biomedical applications are discussed.

16.14 COLLAGEN

Collagen is connective tissue found in the hides of mammals. It is proteinaceous, occurs as fibers and constitutes ~25% of total protein in mammals. It is mainly present in the extracellular matrix, which helps in supporting most tissues and cells (Park and Lake, 1990; Young and Lovell, 1991; Hench and Jones, 2005; see also http://en.wikipedia.org/wiki/Collagen). The length of a typical collagen fiber is ~300 nm with a diameter of ~1.5 nm. Structurally, it consists of a right-handed coiled triple helix, which is made up of three left-handed polypeptide helix strands stabilized by hydrogen bonds (see http://en.wikipedia.org/wiki/Collagen). Collagen molecules form bundles which in turn are organized into fibrils that constitute the basic unit of collagen fibers. The component amino acids are arranged in a regular pattern within the chains, with the three main residues being glycine (every third amino acid), proline and hydroxyproline. The carboxyl and secondary amino groups of proline and hydroxyproline are responsible for the spontaneous formation of left-handed helices (see http://en.wikipedia.org/wiki/Collagen). Glycine residues are present along the interior axis of the helix whereas proline and hydroxyproline residues point outwards. These latter residues are responsible for the thermal stability of the triple helix. Collagen has high tensile strength (50–100 MPa) and Young's modulus (1–2 GPa).

Collagen-based matrices have been widely used in delivery systems, wound healing and various other biomedical applications given the biocompatibility and availability of the collagen. In the following section, applications of collagen in various delivery systems are discussed.

16.14.1 Cross-linked Collagen in Controlled Drug Delivery

Koob and Hernandez (2002) reported a novel method of stabilizing collagen-based materials with catechol-containing monomers. The authors oxidized *o*-catechols with *o*-quinone and polymerized bovine collagen with di-catechol nordihydroguaiaretic acid (NDGA), and reported an increase in the tensile strength and stiffness of the collagen fibers once treated with NDGA. The mechanical properties increased further with a second NDGA treatment. The group further reported that NDGA cross-linked fibers had better mechanical properties than the collagen fibers treated with other cross-linking agents, namely GA or carbodiimide (Koob and Hernandez, 2002). The group recently reported the potential use of the NDGA cross-linked fibers as a matrix for drug delivery. In the study, the authors loaded the cross-linked fibers with dexamethasone and dexamethasone-21-phosphate, and determined the diffusion coefficients of these drugs into the fibers from the respective aqueous solutions. The diffusion coefficients of dexamethasone and dexamethasone-21-phosphate into the fibers were 1.86×10^{-14} m^2 s^{-1} and 2.36×10^{-13} m^2 s^{-1}, respectively. The use of these kinds of fibers has also been proposed for the delivery of other active agents for the treatment of the human diseases in lieu of GA-treated fibers (Moussy et al., 2007).

pH-responsive collagen gels have been prepared using alkali-treated collagen reacted with naturally derived cross-linkers based on citric acid (CA) or malic acid (MA). In this study, the authors reported that as the concentration of the CA and MA was increased, there was a corresponding decrease in the amino groups. But with increased CA in the cross-linked gels, there was an increase in the residual carboxylic groups, which accounted for the pH-sensitive nature of the treated gels. The authors concluded that the CA-treated samples have potential in sustained oral drug delivery and tissue engineering applications (Saito et al., 2007). Collagen and algal sulfated polysaccharide films cross-linked with GA have been used in applications such as in the coating of cardiovascular prostheses, support for cellular growth and in systems for controlled drug delivery (Figueiro et al., 2006). Duan and Sheardown (2005) developed highly cross-linked collagen, reacted with polypropyleneimine octa-amine dendrimers. The cross-linking of collagen with multifunctional dendrimers was carried out by activating the carboxylic acid groups of glutamic and aspartic acid residues in collagen. The group subsequently reported the use of dendrimer-cross-linked collagen matrices for applications in cartilage tissue repair and in the development of corneal tissue engineering (Duan et al., 2007; Vinas et al., 2007).

16.14.2 Collagen in Gene and Hormone/Growth Factor Delivery Systems

Biopolymers, because of their ability to serve as gene carriers and tissue engineering scaffolds, are poised to play an important role in the field of regenerative medicine (Duan et al., 2007; Kleinmann et al., 2007; Vinas et al., 2007). Sarojini et al. (2007) developed silica colloidal particles for localized recombinant DNA release that were coated with a collagen-containing viral vector, and transferred to human lung fibroblast cultures. The authors reported that only cells in contact with the particles were infected due to the gradual release of the viral particles from the collagen matrix, resulting in apoptotic cell death. This result indicated that the particles could be used as a delivery system to deliver genes of choice to localized subgroups of specific cells of interest (Sarojini et al., 2007). In the same year, Takezawa et al. (2007) reported the use of collagen vitrigel membranes (which are rigid glassy membranes obtained by drying collagen gel membranes) for the subcutaneous delivery of vascular endothelial growth factor (VGF). In vivo release studies of VGF in rats indicated angiogenesis.

16.14.3 Collagen in Ophthalmic Drug Delivery

Ophthalmic drug delivery systems are normally based on aqueous drops of drugs, water-insoluble drug suspensions in ointments or oil drops containing drugs. With such approaches, most of the drug is lost due to reflex blinking and lacrimation. Improvements in the ocular bioavailability of drugs have been achieved with the use of collagen shields (Bu et al., 2007). Such shields consist of a collagen matrix loaded with a pharmacologically active agent (Taravella et al., 1999; Kleinmann et al., 2007). Studies on delivery of fourth-generation fluoroquinolones (broad-spectrum antibiotics such as gatifloxacin, ofloxacin and moxifloxacin) indicated that the shields could help in maintaining the desired drug concentration for a prolonged period of time in the aqueous and vitreous humor (Taravella et al., 1999; Hariprasad et al., 2005; Kleinmann et al., 2007).

16.14.4 Collagen-based Formulations in Oral Delivery Systems

The delivery of active agents to the stomach can be ineffective given the short residence time of the ingested delivery systems. Collagen sponges have been used successfully to increase gastric residence time. Tablets containing collagen sponges expand quickly after coming in contact with gastric fluids. The increase in the gastric residence time of the delivery system is attributed to an increase in the physical size of the matrix. For example, studies on riboflavin delivery from the collagen tablets have indicated sustained release for over 12 h (Groening et al., 2007). Collagen treated with citric acid has also shown potential as an oral delivery matrix, though little work has been reported (Saito et al., 2007).

16.14.5 Collagen as a Matrix/Scaffold for Drug Delivery

Collagen matrices have been used to deliver antimicrobials to wound surfaces to inhibit microbes capable of hindering wound healing. For example, sustained delivery of ciprofloxacin (a fourth-generation fluoroquinolone) from succinylated type-I collagen matrix has been shown to improve healing by eliminating bacteria present in the wound (Sripriya et al., 2007).

Collagen may also be used in association with other polymers, e.g., poly(caprolactone) (PCL) (Prabu et al., 2006). PCL-modified collagen matrices, which are transparent, can help in the easy monitoring of wound surfaces during healing. These have been used for the sustained release of amikacin and gentamycin (classical examples of antibiotics) for 48 h.

The network properties of collagen blends can be altered by sterilization, which may affect the release profile of incorporated bioactive agents. For example, the glass transition temperature of collagen/PLGA blends is reduced by ethylene oxide sterilization while there is no change in properties when these are subjected to beta- and gamma-irradiation (Friess and Schlapp, 2006).

16.15 GELATIN

Gelatin is a translucent, colorless and brittle powder that is nearly tasteless. It has long been used as a gelling agent in the food, pharmaceutical and cosmetic industries due to its ease of use and availability. The major source of gelatin is animal skin and bones and fish scales. It is prepared by the thermal degeneration of collagen present in these sources (see www.lsbu.ac.uk/water/hygel.html). The method of extraction of gelatin from collagen can be modified to yield either acidic or basic gelatin. Gelatin can undergo polyion complexation with either positively or negatively charged therapeutic agents, depending on the type of the gelatin. Acidic gelatin is generally used to deliver basic bioactive agents whereas basic gelatin is used to deliver acidic bioactive agents (Young et al., 2005). In general, gelatin is extracted from type I collagen (with a triple helix structure), which contains two $\alpha1$ (I) and one $\alpha2$ (I) chains. Each of the α-strands has a molecular weight of ~95 kDa and is present in the gelatin along with several polypeptides (see www.lsbu.ac.uk/water/hygel.html; Ikada, 2006). Like collagen, gelatin is also composed of mainly three amino acids, namely glycine, proline and 4-hydroxyproline (see www.lsbu.ac.uk/water/hygel.html). Gelatin having higher levels of pyrrolidines forms stronger gels due to lower water absorption. This is usually related to the presence of higher triple helix content (Ikada, 2006). The gels formed by gelatin are thermo-reversible in nature. Its gel-to-sol transition takes place at ~35°C, i.e., gelatin forms a gel at temperatures below 35°C and has a sol-like consistency at temperatures above 35°C (see www.lsbu.ac.uk/water/hygel.html; Bohidar and Jena, 1993).

As described earlier, gelatin's gelling properties can be altered with chemical cross-links, an approach that has been used by various researchers for the development of controlled-release drug delivery systems (Mukherjee and Banthia, 2006; Hayashi et al., 2007; Sutter et al., 2007; Yamamoto, 2007). A typical structural unit of gelatin is given in Figure 16.12.

16.15.1 Gelatin in Peptide Delivery

Recombinant gelatin (e.g., HU4 gelatin), modified with acrylates has been used as

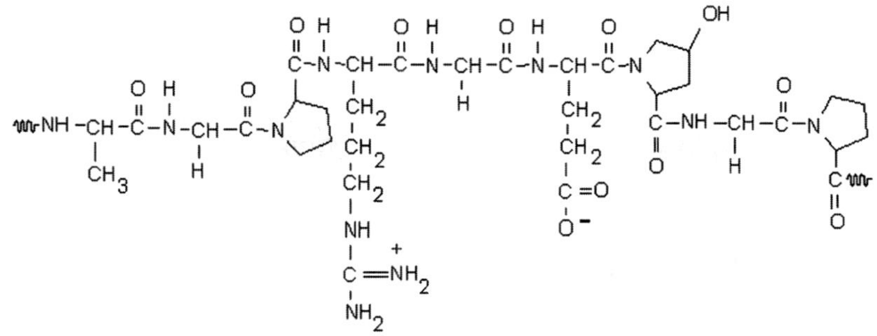

FIGURE 16.12 A typical structural unit of gelatin (reproduced with permission from www.lsbu.ac.uk/water/hygel.html).

a matrix for the delivery of proteins (e.g., lysozyme and trypsin inhibitor). The release of proteins loaded in gelatin matrices prepared in this manner has been found to be diffusion-controlled, with the complete release of the incorporated active agents taking place over 120 h. Furthermore, matrices prepared with this method have been found to be degradable in the presence of metalloproteinase 1, indicating biodegradability under in vivo conditions (Hayashi et al., 2007). Gelatin has also been used for the development of a release system for basic fibroblast growth factor (bFGF). The matrix system, so obtained, was used for the topical administration of bFGF in rabbit models to study the effect of bFGF on angiogenesis and tissue blood perfusion of hind limb ischemia. The results indicated sustained release of bFGF from the hydrogel, which augmented angiogenesis and improved tissue blood flow (Doi et al., 2007). Hayashi et al. (2007) used the above-mentioned concept of bFGF delivery and developed a bFGF-gelatin hydrogel complex to coat implants for bone augmentation. In vivo results indicated new bone growth around the implant. The group found that an optimal amount of bFGF was needed for the development of new bone. Seki et al. (2007) studied the effect of sperminated gelatin (SG), prepared by the addition of spermine to gelatin, on the nasal absorption of insulin in rats. They found a 5.3-fold increase in insulin absorption in the presence of 0.2% SG. The plasma glucose levels of the rats fell in a manner dependent on insulin levels, indicating the probable use of the SG for the intra-nasal delivery of insulin. These findings are promising examples for insulin-dependent diabetes treatment as insulin intake is mainly by subcutaneous injections and hence is one of the main reasons for non-compliance of patients.

Yamamoto (2007) compared the efficacy of gelatin and chitosan capsules as colon-specific drug delivery systems. He incorporated prednisolone into capsules and studied the bioavailability of the steroid in rats by measuring its plasma concentration. He reported that, though the intestinal concentration of the steroid was greater in the case of chitosan capsules, the plasma concentration of the drug was higher for the gelatin capsules.

The use of aminated gelatin microspheres (AGMS) was investigated as a nasal drug delivery system for peptide drugs by Wang et al. (2006). Fluorescein-labeled insulin and FITC-dextran (MW = 4.4 kDa) were used as a model drug for in vitro release studies. The investigators found that the release of labeled insulin from AGMS was significantly slower than from native gelatin microspheres (GMS). However, there was no significant difference in the release profile of the dextran from the AGMS and GMS microspheres. The absorption-enhancing effect of insulin was determined by measuring the plasma glucose concentrations of healthy rats following intranasal administration of insulin-incorporated microspheres. AGMS microspheres indicated a higher hypoglycemic effect when delivered as an insufflation powder formulation rather than as a suspension. The group indicated that AGMS might be a new candidate carrier for the nasal delivery of peptide drugs.

16.15.2 Gelatin in Wound Healing and Implantable Delivery Systems

Mukherjee and Banthia (2006) developed a PVA-gelatin hydrogel containing adrenochrome (a blood coagulant) for wound-healing applications. The hydrogel was prepared by mixing gelatin and PVA solutions in water followed by heating at 40°C and subsequent addition of the drug. They suggested that the gel would help in instant coagulation of blood at the wound surface and thus help in wound healing.

The presence of proteases in wound surfaces results in the degradation of granulation tissue and endogenous biologically active proteins thereby hindering the wound-healing process. The inhibition of these enzymes facilitates wound healing. Apart from its antimicrobial property, doxycycline has been found to inhibit proteases on wounds, when applied topically, and promote wound healing. Adhirajan et al. (2007) encapsulated doxycyline in microspheres of gelatin cross-linked with 1-ethyl-3-(3-dimethylaminopropyl) carbodiimide. It was found that the drug-loaded microspheres released the drug in a controlled manner. DeFail et al. (2006) encapsulated doxorubicin into PLGA microspheres (using a double emulsion/solvent extraction method) and incorporated these into gelatin scaffolds. The release profile of doxorubicin was determined in phosphate-buffered solution for 30 days. The microspheres were added to aqueous gelatin during cross-linking/gelation with GA. The microspheres and the scaffolds were suspended in murine mammary mouse tumor cell line 4T1 for 48 h. Results indicated that the release of the drug was controlled by the incorporation of PLGA microspheres into gelatin scaffolds. Liu et al. (2007) developed PLGA-placitaxel microspheres incorporated into cross-linked gelatin sponges, and then cross-linked with carbodiimide. The release of placitaxel (an anti-cancer drug) was governed by the degradation of PLGA. In vivo lymphatic delivery was performed in rats with orthotopic lung cancer. The implantation of the sponge intraperitoneally and intrapleurally resulted in spontaneous absorption of the particles in the lymphatic system, which indicated the potential use of these sponges for targeted delivery of drugs to the lymphatic system.

16.15.3 Gelatin in Stimuli-responsive Delivery System

Hu et al. (2007) developed ferro-scaffolds of different compositions using biodegradable gelatin and iron oxide nanoparticles, which were produced in situ. The pore sizes of the scaffolds ranged from 50 to 200 mm depending on gelatin concentration. The yield of the iron oxide nanoparticles decreased with increasing gelatin content. The release rate of incorporated vitamin B_{12} was reduced when in the presence of a magnetic field. The authors suggested that the ferro-scaffolds could be used as stimuli-responsive drug carriers.

16.15.4 Gelatin Micro- and Nanoparticles as Delivery Systems

Gelatin-based micro- and nanoparticles have been extensively studied for the controlled delivery of drugs and genetic material (Sivadas et al., 2008; Tsent and Lin, 2008). Such particles may be used to encapsulate and subsequently release genetic material when degraded or digested by enzymes. In addition to targeting tissues or cells in the body through surface functionalization, encapsulation helps in the protection of nucleic acids from harsh in vivo conditions, namely enzymatic degradation and phagocytosis (Chavanpatil et al., 2006). Sha and

Kaiming (2007) encapsulated plasmid DNA in poly-ethylene-glycolated gelatin nanoparticles through a mild water–ethanol solvent displacement method under controlled pH and temperature. The developed particles delivered plasmid DNA into NIH 3T3 murine fibroblast cells and targeted solid tumors. Gelatin nanoparticles have been also used successfully to target lung cancer cells. Tseng et al. (2007) grafted NeutrAvidin (FITC) (deglycosylated avidin having an affinity for biotin) on the surface of gelatin nanoparticles followed by a conjugation of biotinylated epithelial growth factor (bEGF) with NeutrAvidin (FITC), forming a core-shell structure (gelatin-avidin-bEGF). The conjugation with bEGF resulted in improved targeting efficiency for the detection of lung adenocarcinoma.

Pica et al. (2006) suggested that intratumor administration of gelatin-methotrexate microspheres may minimize the systemic toxicity of methotrexate and may also help in overcoming drug resistance. A gelatin-methotrexate conjugate was prepared by blocking amino groups in gelatin with citraconic anhydride followed by a conjugation of amino group of methotrexate with the carboxylate group of gelatin. This was subsequently followed by cross-linking with GA to form microspheres. The in vitro evaluation of the drug-loaded microspheres was performed by simulating tumor conditions at pH 6.5 and 37°C. The release of gelatin fragments in the presence of cathepsin B (a protease) and the drug was evaluated with HPLC. Results indicated that the release of fragments with molecular weights <10 kDa was low whereas release of free methotrexate was negligible. The authors suggested that this method may be used for the development of other proteinaceous microspheres.

Rokhade et al. (2006) successfully encapsulated ketorolac tromethamine, a non-steroidal anti-inflammatory drug, in semi-interpenetrating polymer network (IPN) microspheres of gelatin and sodium carboxymethylcellulose (CMC) using GA as a cross-linker, resulting in particles 247–535 μm in diameter. In vitro release studies indicated non-Fickian behavior, with release depending on the extent of cross-linking and the amount of sodium CMC used to produce the microspheres. Liang et al. (2003) made gelatin microspheres cross-linked with genipin, with GA-cross-linked gelatin microspheres used as a control. The biocompatibility of the microspheres was determined by injecting them intramuscularly in rats. Results indicated that the inflammatory reaction for genipin-cross-linked microspheres was significantly less than with GA-cross-linked microspheres. In addition, the degradation of the genipin-cross-linked microspheres was slower than their GA counterparts.

Gelatin alone or in conjunction with mucoadhesive polymers such as mucin may be used to encapsulate various acid-labile therapeutic substances in microspheres for rectal delivery. In vitro release studies of ceftrioxone (third-generation cephalosporin) encapsulated in gelatin and gelatin-mucin microspheres indicated diffusion-controlled non-Fickian release and sustained delivery (Ofokansi et al., 2007). Lu et al. (2007) developed a sustained-release drug delivery system for indometacin, where the drug was encapsulated in a polymer shell consisting of gelatin and cellulose acetate phthalate. In vitro release studies of the drug prepared by complex coacervation indicated Higuchian kinetics.

16.15.5 Gelatin as a Matrix for Biologically Active Agents

Hiroyuki et al. (2006) examined the improved efficacy of erythropoietin-gelatin hydrogel sheets in the treatment of myocardial infarctions. Using rabbits, they conducted in vivo studies where the control group was infused with saline in a first experimental group, with erythropoietin subcutaneously injected in a second test group, and with erythropoietin-gelatin hydrogel sheets applied to the affected area on the heart in a third group. The results indicated that the hydrogel sheet decreased myocardial infarctus size and improved the function of the left ventricle compared to erythropoietin systemic injection (Kobayashi et al., 2006).

Films of alginate and gelatin cross-linked with Ca^{2+} were developed by Dong et al. (2006) using a solvent casting method. They incorporated ciprofloxacin hydrochloride as a model drug and found that its release from the gel films was dependent on the pH and ionic strength of the release solution. The drug release rate was faster when the pH of the dissolution medium was pH 7.4 compared to pH 3.6 and was accelerated by higher ionic strengths. The other factors playing an important role in drug release were the component ratio of alginate and gelatin, the amount of ciprofloxacin hydrochloride loaded in the gel films, the thickness of the drug-loaded films and the cross-linking time with Ca^{2+}.

16.16 CHITIN AND CHITOSAN

Chitin is a polysaccharide found in the outer skeleton of insects, crabs, shrimps and lobsters, whereas chitosan is deacetylated chitin. Chitin resembles cellulose in structure, but it has an acetamido group instead of a hydroxyl group at the C-2 position of the backbone polymer chain. It is composed of 2-acetamido-2-deoxy-β-D-glucose attached with β (1→4) linkages (Figure 16.13) and is degraded by chitinase (Muzzarelli, 1973; Zikakis, 1984; Ravi Kumar, 2000). To improve solubility, it is necessary to deacetylate chitin by 80–85% or higher, thereby yielding chitosan. With further deacetylation, enhanced solubility is achieved. Both chitin and chitosan are biocompatible, biodegradable, non-toxic and have good adsorption properties, which make them suitable for various drug delivery applications. Chitosan is one of the few cationic polyelectrolytes found in nature (Takayanagi and Motomizu, 2006).

Chitin and chitosan vary in composition depending on the origin and manufacturing process. Their use has been limited due to the relatively laborious isolation process which can increase production costs. With a better understanding of their inherent biological and physicochemical characteristics, however, they have seen much use in various biomedical applications (e.g., wound-healing agent and a delivery vehicle for pharmaceuticals and genes) (Cravotto et al., 2005).

16.16.1 Chitin/Chitosan as a Matrix for Biologically Active Agents

Saito et al. (2006) developed adriamycin-containing chitosan sheets by mixing adriamycin with a chitosan suspension followed by freeze-drying. They inserted the chitosan sheet into the peritoneal cavity of mice and found that adriamycin was stable even after 2 months, indicating that the sheet could be used to improve therapeutic efficacy in the treatment of topical lesions.

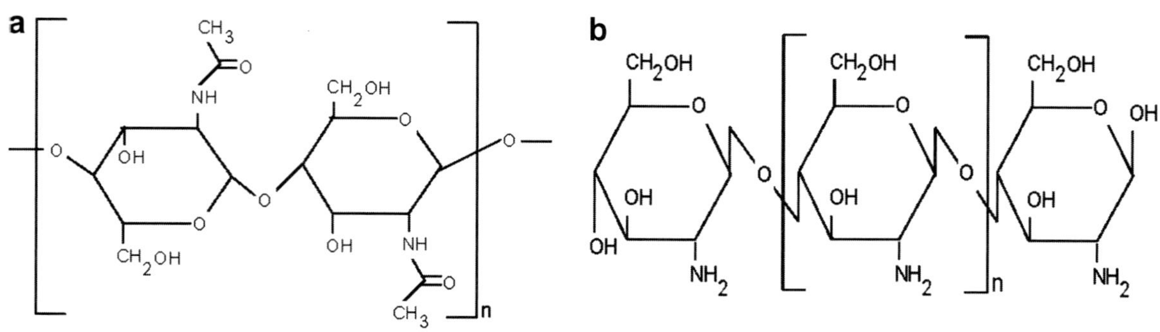

FIGURE 16.13 Chemical structure of (a) chitin and (b) chitosan (modified from Khor and Lim, 2003).

16.16.2 Chitin/Chitosan in Stimuli-Responsive Delivery Systems

Sun et al. (2007) developed chitosan-based hydrogel films having both thermal and pH sensitivity. They blended chitosan with poly-(N-isopropyl-acrylamide) (PNIPAAm), a temperature-sensitive polymer, and polyethylene glycol. X-ray diffraction studies revealed higher crystallinity in the blended films as compared to chitosan or PNIPAAm films alone. Scanning electron microscopy showed that the films could be made more porous upon heating, indicating their temperature sensitivity. The films were found to be pH sensitive because of the pendant amino groups present in the chitosan. Guo et al. (2007) also developed thermo- and pH-responsive hydrogels. They prepared a semi-IPN polyampholyte of carboxymethyl chitosan and poly (2-(dimethylamino) ethyl methacrylate) using N-N′-methylenebisacrylamide as cross-linker. Swelling and de-swelling of the hydrogel was reversible and consistent. The release profile of incorporated coenzyme A increased with temperature, but decreased as the proportion of carboxymethyl chitosan within the hydrogel was increased.

16.16.3 Chitin/Chitosan Particles as Delivery Systems

Reverchon and Antonacci (2006) successfully micronized chitosan in a 1% acetic acid aqueous solution using supercritical fluid-assisted atomization. Characterization of the microparticles revealed a size range of 0.1–1.5 μm when the chitosan was precipitated from solutions having concentrations ranging between 1 and 10 mg mL^{-1}. Higher precipitation temperatures resulted in a decrease in particle crystallinity. The authors considered these microparticles as promising vehicles for drug delivery. Yuan et al. (2007) studied the effects of genipin cross-linking on drug/protein release from chitosan microspheres. The release of incorporated albumin revealed that the amount of cross-linker and cross-linking duration played a significant role in release rate. Jayakumar et al. (2007) synthesized water-soluble thiol-containing chitosan using a graft copolymerization technique and made beads of the modified chitosan, loading them with indometacin. In vitro release studies indicated pH dependency, with slower release at pH 1.4 than 7.4. This was attributed to the ionization of the thiol groups and the high solubility of indometacin in an alkaline medium. Peng et al. (2006) prepared microspheres of N-methylated chitosan (NMC) cross-linked with GA to encapsulate ofloxacin, a fourth-generation fluoroquinolone. The ofloxacin was electrostatically cross-linked with the polymer, and its release depended on the molecular weight of the N-methylated chitosan, with higher NMC molecular weights slowing release. Release profiles indicated non-Fickan diffusion through the swollen microspheres. Shi et al. (2006) prepared beads via the polyelectrolyte complexation of chitosan and alginate solutions. Infrared spectroscopy revealed electrostatic interactions between the amino groups of chitosan and carboxyl groups of alginate. The beads were loaded with bovine serum albumin (BSA) and the in vitro release profile was studied at different pHs. Bead composition influenced the encapsulation and release of BSA. The release was also dependent on pH, with higher pH values leading to increased release.

16.16.4 Chemically Modified Chitin/Chitosan for Drug Delivery

Jayakumar et al. (2007) discussed the generation of new bifunctional materials by chemical modification of chitin and chitosan with sulfates, which have been used in a variety of biomedical applications (e.g., as anti-coagulants, in drug-delivery matrices and as anti-microbial polymers). N-alkyl-O-sulfated chitosan developed by Zhang et al. (2003) was used to form micelles

(100–400 nm diameter) in water. The group entrapped Taxol®, an anti-cancer drug, into the polymeric micelles by physical entrapment and suggested that the modified polymer could be used as a potential drug carrier. Another variant of sulfated chitosan, N-carboxymethylchitosan N,O-sulfate, has been found to inhibit HIV-1 replication and viral binding with the CD4 cells, thus inhibiting the progression of the virus (Sosa et al., 1991). Lee et al. (2006) developed nanoparticles of thiolated chitosan for intranasal delivery of theophylline to alleviate allergic asthma. They studied the efficacy of the delivery system to suppress inflammatory allergic response on ovalbumin (OVA)-sensitized BALB/c mice. The results revealed that the beneficial effects of theophylline were increased when it was complexed with the nanoparticles. Satoh et al. (2007) successfully modified chitosan in a highly regioselective manner. They synthesized water-soluble (at neutral pH) 6-amino-6-deoxy-chitosan using N-phthaloyl-chitosan as a starting material. It was evaluated for its capability as a gene carrier. The transfection results for COS-1 cells revealed that the synthesized product was superior to standard chitosan. Jayakumar et al. (2006) synthesized phosphorous-containing chitosan by graft copolymerization and prepared beads encapsulating indometacin by using tripolyphosphate at pH 4.0. In vitro drug release patterns were pH-dependent, with the release rate increasing as the pH of the dissolution media was increased, indicating the potential use of the polymer in developing a delayed drug delivery system for oral administration.

16.16.5 Chitin/Chitosan in Cardiovascular Delivery Systems

A novel biodegradable and rapidly expanding stent of chitosan cross-linked with an epoxy compound having a shape-memory property was developed by Chen et al. (2007). The stent prepared using cross-linked chitosan was able to withstand >30% deformation as compared to 10% in commercial metallic stents before loss of elasticity. Glycerol and poly(ethylene oxide) were used to reduce the crystallinity of the chitosan films. Preliminary in vivo studies in animals showed no thrombus formation and adverse reactions. The authors suggested the use of the degradable stent as a vehicle for local drug delivery.

16.16.6 Chitin/Chitosan Derivatives as in-situ Gelling Agent

Yu et al. (2007) discussed the development of a novel composite hydrogel using dialdehyde konjac glucomannan as a cross-linking agent for chitosan. Cross-linking with this compound led to rapid gelling (in minutes). Drug release studies using ofloxacin indicated sustained release of the active agent, which may be due to the association of the drug with the polymer.

16.16.7 Chitin/Chitosan in Ocular Delivery Systems

Verestiuc et al. (2006) prepared a series of hybrid polymeric hydrogels by reacting acrylic acid-functionalized chitosan with either N-isopropylacrylamide or 2-hydroxyethyl methacrylate monomers. The hybrid polymers prepared with this approach were pressed into minitablets and evaluated as a drug delivery system for ocular delivery. The matrices were loaded with chloramphenicol, atropine, norfloxacin or pilocarpine. In vitro release patterns revealed drug-specific carrier compositions for the controlled delivery of these compounds.

16.17 CELLULOSES

Cellulose is a structural polysaccharide with the chemical formula $(C_6H_{10}O_5)_n$ (Figure 16.14). β-Glucose molecules condense through β(1→4)-glycosidic bonds to give rise to cellulose. Cellulose is a linear polymer and is highly crystalline,

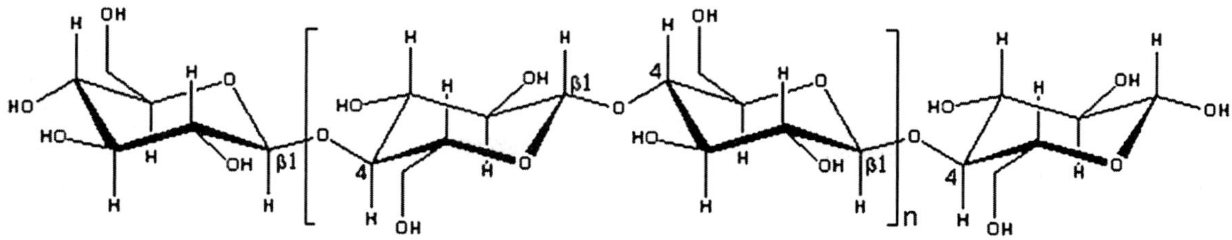

FIGURE 16.14 Chemical structure of cellulose (reproduced with permission from http://www.lsbu.ac.uk/water/hycel.html).

which is attributed to extensive intramolecular hydrogen bonding (see http://www.lsbu.ac.uk/water/hycel.html). Acetic acid bacteria, some forms of algae and oomycetes also have the capability to synthesize cellulose.

16.17.1 Cellulose as a Thermo-sensitive Polymer

Aqueous solutions of most natural polymers (e.g., gelatin and carrageenan) form gels below a critical temperature usually known as the sol-to-gel transition temperature. Some cellulose derivates (e.g., methylcellulose (MC) and hydroxypropyl methyl cellulose (HPMC)) form a gel upon heating, with gelation temperatures of 40–50°C and 75–90°C for MC and HPMC, respectively. The transition temperature of HPMC can be lowered to ~40°C by reducing the hydroxylpropyl molar substitution. The polymer chains are hydrated at lower temperatures while they start dehydrating as the temperature increases. Partial dehydration results in polymer–polymer association thereby resulting in a network structure (Ruel-Gariepy and Leroux, 2004). Tate et al. (2001) developed methylcellulose-based scaffolds for the repair of brain defects. Their gels were biocompatible in both in vitro and in vivo conditions. Carlsson et al. (1988, 1990) reported a change in the thermal behavior of aqueous solutions (1-4 wt%) of ethyl(hydroxyethyl)cellulose (EHEC) with the addition of ionic surfactants such as sodium dodecyl sulfate (SDS) or cetyl triammonium bromide (CTAB). The systems underwent sol-to-gel phase transitions at 30–40°C, resulting in the formation of gels, with micelle-like surfactant clusters influencing gel properties. Scherlund et al. (2000) used this system for the sustained delivery of lidocaine and prilocaine in periodontal pockets.

16.17.2 Cellulose Esters

Cellulose esters have been used in drug delivery, given their negligible cytotoxicity, stability, high water permeability, film-forming capability and compatibility with most bioactive agents (Edgar, 2007). Cellulose acetate phthalate (CAP) is widely used and was first described by Malm and Fordycee in 1940. In 1951, Malm's group reported that CAP could be used for development of enteric coated delivery systems (Malm et al., 1951). Other cellulose esters of importance in this regard are cellulose acetate trimellitate (CAT) and cellulose acetate succinate (CAS). Levine et al. (1987) coated capsules containing beclomethasone dipropionate with CAP and administered it to patients who underwent colostomies. The authors were able to recover more than triple the amount of the steroid and its metabolites from the patients administered with the CAP-coated capsules when compared to patients treated with capsules without coating. Liu et al. (2002) developed microcapsules of CAP and cellulose acetate with encapsulated nitrofurantoin. The microcapsules slowed the release of the drug, with lower pHs further slowing release. In vivo results indicated a decrease in the occurrence of stomach ulcers, which was attributed to the enteric nature of the formulation.

There are numerous examples of cellulose derivatives converted into microparticles for drug delivery. Pongpaibul and Whitworth (1986) made microparticles of cellulose acetate butyrate (CAB) with propranolol entrapped in the matrix. In vitro studies indicated a decrease in the release rate of the drug with an increase in polymer content. Sprockel et al. (1989) developed microparticles of CAB, which were used to entrap water-soluble drugs complexed with a sulfonated anion exchange resin. The release of the drug from the microparticles, when suspended in water, indicated sustained release. Palmieri et al. (2002) developed microparticles of CAP, CAT and hydroxypropyl methylcellulose phthalate (HPMCP) entrapping ketoprofen. The microparticles showed good protection at gastric pH and rapid release at small intestinal pH, demonstrating a potential method for the development of pH-dependent release systems without the need to coat them with enteric polymers.

Cellulose diacetate (CA) is highly permeable to water, but is impermeable to salts and many organic compounds. This property has been used to develop osmotic delivery systems. For example, Theeuwes (1975) reported that the release rate from such systems was not affected by agitation of the dissolution medium and the pH of the surrounding environment. Makhija et al. (2003) reported an osmotic drug delivery system for pseudoephidrine. The group used polyethylene glycol (PEG) and diethylphthalate as a dopant to create pores in the CA semipermeable membranes. The drug's release rate depended on CA film thickness and PEG content.

16.18 ALGINATES

Alginates are isolated from brown seaweed using dilute alkaline extraction. The resulting solutions are treated with mineral acids and are subsequently converted to sodium alginate. Alginic acid is a linear polymer consisting of D-mannuronic acid and L-guluronic acid residues (Draget et al., 1996; see also http://www.lsbu.ac.uk/water/hyalg.html) (Figure 16.15). Alginic acid forms a high-viscosity acid gel in the presence of water, which is attributed to the hydration of the polymer chain and intermolecular hydrogen bonding. Alginate polymers form gels in the presence of divalent and multivalent cations (except Mg^{2+}) by cross-linking of the carboxylate groups on the polymer backbone (Tonnesen and Karlsen, 2002).

Particles of alginates, cross-linked with calcium ions, have been studied extensively. In general, there are three different methods for the

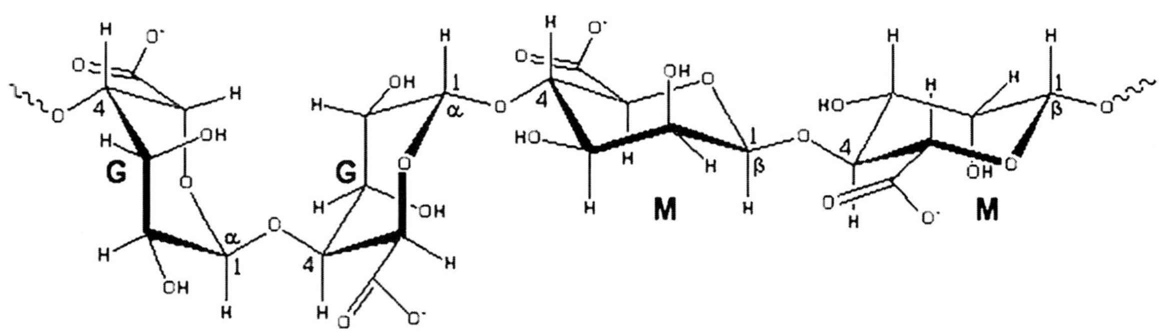

FIGURE 16.15 Chemical structure of alginic acid (reproduced with permission from http://www.lsbu.ac.uk/water/hycel.html).

preparation of the particles. In the first method, the aqueous alginate solution is added dropwise into an aqueous calcium chloride solution. The second method involves the in situ release of cross-linking calcium ions from calcium carbonate by an acidification process, whereas the third method involves the addition of a dispersed calcium chloride solution to an alginate solution.

Under acidic conditions, the swelling of calcium alginate particles is negligible, as alginic acid is an anionic polymer and the carboxyl groups are not ionized at pHs lower than the pKa of the polymer. Hence, any drug release will be mainly associated with the diffusion of the active agent through the insoluble matrix. Under neutral and basic conditions, however, the anionic polymer becomes ionized and swells. Drug release under these conditions will depend on the swelling of the beads and/or the erosion of the polymer matrix as well as the inherent properties of the drug. Because of the above properties, alginates have been used for the development of a multiple-unit, controlled-release drug delivery system. Higher concentrations of alginate in the beads and alginates rich in guluronic acid have been found to decrease the release rate of active agents. Low-molecular-weight alginates have been studied for their capacity to enhance the dissolution rate of acidic, basic and neutral drugs (Tonnesen and Karlsen, 2002).

16.18.1 Alginates in Diffusion-Controlled Delivery Systems

Alginates have been used to encapsulate drugs in microcapsules or for matrix-type drug delivery systems for a large number of proteins (Polk et al., 1994; Gombotz and Wee, 1998; Tonnesen and Karlsen, 2002). Positively charged proteins (e.g., TGF β1) can react with the carboxylic acid groups of the alginates, which may result in protein denaturation. To prevent loss of activity, additives such as polyacrylic acid are often used. The protection of the protein has been attributed to the shielding effect of the polyacrylic acid from the low molecular fragments of the alginates (Mumper et al., 1994). Amsden et al. (1999) reported that protein diffusion was highest in alginate gels prepared from alginates with a low guluronic acid fraction, which was associated with greater flexibility of the alginate backbone. In general, higher backbone flexibility results in higher solute diffusion. Tomida et al. (1993) reported that theophylline-loaded alginate gels showed zero-order release kinetics, stating that release could be controlled by changing coating thickness. Iannuccelli et al. (1996) found that gentamycin sulfate (GS) could selectively interact with the mannuronic residues of the alginates and did not compete with the calcium ions involved in gelation. They observed that calcium ions preferentially reacted with the polygluronic sequences, though the polymannuronic sequences could also play a small role during cross-linking. Thus, alginates rich in mannuronic acid residues were preferred for GS delivery systems given their higher GS binding capacity. Chan et al. (1998) reported that the release of drugs from alginate microspheres was affected by the presence of various additives, such as poly (vinylpyrrolidone) and ethylcellulose. The group reported that high-viscosity ethylcellulose reduced the drug release rate while poly (vinylpyrrolidone) was able to increase the flowability of the alginate microspheres. Takka and Acartürk (1999) reported that the alginate's molecular weight did not affect the release pattern of nicardipine HCl, a neutral molecule, but that the release rate was increased with the addition of various polymers such as Carbopol® 941, HPMC and Eudragit® RS 30. They also reported that when Ca^{2+} was used as a cross-linker, drug release was prolonged when compared with particles cross-linked with Ba^{2+} and Sr^{2+}. These findings could be used to modulate the release of drugs from the alginate microspheres, in which other polymers could be deliberately incorporated or the cross-linker could be changed. Imai et al. (2000) reported that the release of pindolol, a basic drug, was slow for gel beads prepared with the low-molecular-weight alginates. They proposed that drug release could be modulated by using alginates with different molecular weights. Kulkarni et al. (2000) developed polymeric sodium alginate microparticles by precipitating sodium alginate in methanol and subsequently cross-linking the particles with glutaraldehyde. The group reported that with an increase in cross-linking, there was a decrease in microsphere swelling, though the loading efficiency of nimesulide, a water-soluble drug, increased. Hodson et al. (1995) made tablets from blends of various drugs and alginate. They reported that water-soluble drugs were released at a faster rate in simulated gastric fluid than in the simulated intestinal fluid. However, drugs having poor water solubility showed the opposite effect.

16.18.2 Alginates as in situ Gelling Agents

Miyazaki et al. (2000) reported a novel method for the in situ gelation of sodium alginate for controlled oral delivery. The oral administration of the sodium alginate solution to rats was subsequently followed by a solution containing calcium ions. When the solutions reached the acidic stomach, the release of free Ca^{2+} promoted gelation. The group reported that there was an increase in the bioavailability of theophylline as compared to oral sustained-release formulations.

16.18.3 Alginates in Oral Delivery Systems

Dennis et al. (2002) filled hard gelatin capsules with a mixture of drug, alginate and a pH-independent polymer (HPMC). Upon ingestion, the capsules absorbed gastric fluid thereby initiating surface hydration of the polymer. The formation of a surface gel layer led to air entrapment and the capsules began to float. Over time, the gel layer eroded resulting in the movement of the gel-dissolution interface towards the core of the capsules. Eventually, the device lost its floatability and passed into the intestinal tract where the alginates dissolved (due to the basic environment in the intestine), thereby making the delivery system more porous leading to drug release (see http://patents1.ic.gc.ca/details?patent_number=2081070).

Alginates have good mucoadhesive properties and have been used in conjunction with chitosan in various biomedical applications because of this property. Miyazaki et al. (1994) developed alginate-chitosan tablets for the sublingual delivery of ketoprofen. Tennessen and Karlsen (2002) reported that alginate microparticles showed strong muco-adhesion towards the stomach mucosa. Coating of the particles with chitosan did not change their muco-adhesion property. Such examples demonstate that alginate-based systems can be developed for drug delivery in the stomach.

Mandel et al. (2000) reported that alginate-based formulations containing antacids and H_2-receptor blockers can be used in the treatment of heartburn and esophagitis by acting as a barrier against acid reflux. Katayama et al. (1999) developed a liquid preparation consisting of sodium alginate and ampicillin (an antibiotic) for the eradication of *Helicobacter pylori*. Once ingested, this preparation apparently spread on the stomach wall releasing the incorporated drug on the mucosa. Lastly, sodium alginate has been used to mask the bitter taste of drugs such as amiprilose hydrochloride.

16.18.4 Alginates as Encapsulating Agents

Encapsulation of cells and DNA within the alginate matrix has shown great promise in biomedical applications. Esquisabel et al. (1997) encapsulated bacillus Calmette-Guérin (BCG) (a vaccine against tuberculosis) within a calcium alginate matrix using an emulsification method for microencapsulation. They reported that the size of the particles depended on the viscosity of the oil used for emulsification. Smith (1994) developed an enteric delivery system for DNA by encapsulating DNA into a calcium alginate matrix. Similarly, Alexakis et al. (1995) encapsulated DNA in alginate microspheres, which were further coated with chitosan. The delivery system was orally administered to rats. The microparticles were recovered from the rat feces and were analyzed for the activity of the entrapped DNA molecules. The results indicated that the encapsulated DNA was biologically active and could be substantially recovered.

16.18.5 Alginates as Wound-healing Materials

Tonnesen and Karlsen (2002) also reported that alginates could be used as wound-dressing materials in the form of powder, films or fibers. The calcium alginate matrix promoted the exchange of calcium ions with sodium ions on the wound surface, which enhanced the blood-coagulation cascade, imparting a hemostatic property to the matrix.

16.18.6 Alginates in Ophthalmic Delivery Systems

Cohen et al. (1997) reported the in situ formation of an ophthalmic drug delivery system from alginates, which underwent gelation in the ocular sack without any external cross-linking agent. The extent of alginate gelation and release of an incorporated drug (pilocarpine) was dependent on the percent of glucoronic acid residues in the polymer backbone. The group also reported that alginates having a glucoronic acid content of >65% instantaneously gelled. In vitro results in rabbits indicated that the intra-ocular pressure-reducing effect of pilocarpine was increased to 10 h as compared to 3 h when the drug was administered in solution. In addition, the dissolution of the matrix in the dissolution medium was negligible for the first 12 h at 37°C.

16.19 SUMMARY

Although synthetic polymers are used more extensively in the field of drug delivery, biopolymers and their derivatives are rapidly gaining in importance. This is mainly due to their intrinsic properties that render them appealing. In general, they are non-carcinogenic, mucoadhesive, biocompatible and biodegradable. For example, chitosan has been used in oral and nasal delivery systems, where the mucoadhesive property of the polymer plays an important role. Furthermore, the properties of the biopolymers can be tailored via chemical modification to further expand their functionality. For example, sodium carboxymethylcellulose has been used as a viscosity-building agent and as a binder in pharmaceutical formulations but does not show pH-dependent swelling behavior. However, when it is esterified with acryloyl or methacryloyl chloride, it shows pH-dependent swelling and becomes insoluble in water (Pal et al., 2006a). Along with being biocompatible and non-carcinogenic, biopolymers are also ideal candidates for the development of matrices for tissue engineering and wound dressings. For example, Pal et al. (2006b) developed transparent starch hydrogels for use as a wound dressing. In tissue engineering, researchers are physically entrapping growth factors within biopolymer matrices for the development of specific cells and tissues.

In closing, the importance of biopolymers in the development of matrices for the controlled release of drugs and other bioactive compounds will continue to increase. Other than developments in the biomedical and pharmaceutical industries, it is likely that biopolymer usage will see rapid growth in the areas of cosmetology and nutraceutical delivery.

ACKNOWLEDGMENTS

The authors gratefully acknowledge funding from the Advanced Foods and Materials Network of Centres of Excellence (AFMNet) and the Natural Sciences and Engineering Research Council of Canada (NSERC) during the completion of the chapter.

References

Adhirajan, N., Shanmugasundaram, N., and Babu, M. (2007). Gelatin microspheres cross-linked with EDC as a drug delivery system for doxycyline: Development and characterization. *Journal of Microencapsulation* **24** (7), 659–671.

Alexakis, T., Boadi, D. K., Quong, D., Groboillot, A., Oneill, I., Poncelet, D., and Neufeld, R. J. (1995). Microencapsulation of DNA within Alginate Microspheres and Cross-Linked Chitosan Membranes for In-Vivo Application. *Applied Biochemistry and Biotechnology* **50** (1), 93–106.

Ali, J., Arora, S., Ahuja, A., Babbar, A. K., Sharma, R. K., Khar, R. K., and Baboota, S. (2007). Formulation and development of hydrodynamically balanced system for metformin: *In* vitro and in vivo evaluation. *European Journal of Pharmaceutics and Biopharmaceutics* **67** (1), 196–201.

Amsden, B. and Turner, N. (1999). Diffusion characteristics of calcium alginate gels. *Biotechnol Bioengineering* **65** (5), 605–610.

Anchisi, C., Meloni, M. C., and Maccioni, A. M. (2006). Chitosan beads loaded with essential oils in cosmetic formulations. *Journal of Cosmetic Science* **57** (3), 205–214.

Andersen, S. O., Peter, M. G., and Roepstorff, P. (1996). Cuticular sclerotization in insects. *Comparative Biochemistry Physiology B-Biochemistry and Molecular Biology* **113** (4), 689–705.

Bachelder, E. M., Beaudette, T. T., Broaders, K. E., Dashe, J., and Frechet, J. M. J. (2008). Acetal-derivatized dextran: An acid-responsive biodegradable material for therapeutic applications. *Journal of the American Chemical Society* **130** (32), 10494.

Berger, J., Reist, M., Mayer, J. M., Felt, O., Peppas, N. A., and Gurny, R. (2004). Structure and interactions in covalently and ionically cross-linked chitosan hydrogels for biomedical applications. *European Journal of Pharmaceutics and Biopharmaceutics* **57** (1), 19–34.

Bertone, M., Dini, V., Romanelli, P., Rizzello, F., and Romanelli, M. (2008). Objective analysis of heterologous collagen efficacy in hard-to-heal venous leg ulcers. *Wounds* **20** (9), 245–249.

Bigi, A., Cojazzi, G., Panzavolta, S., Roveri, N., and Rubini, K. (2002). Stabilization of gelatin films by cross-linking with genipin. *Biomaterials* **23** (24), 4827–4832.

Bin Choy, Y., Cheng, F., Choi, H., and Kim, K. (2008). Monodisperse gelatin microspheres as a drug delivery vehicle: Release profile and effect of cross-linking density. *Macromolecular Bioscience* **8** (8), 758–765.

Boateng, J. S., Matthews, K. H., Stevens, H. N. E., and Eccleston, G. M. (2008). Wound healing dressings and drug delivery systems: A review. *Journal of Pharmaceutical Sciences* **97** (8), 2892–2923.

Bohidar, H. B. and Jena, S. S. (1993). Kinetics of Sol-Gel Transition in Thermoreversible Gelation of Gelatin. *Journal of Chemical Physics* **98** (11), 8970–8977.

Bolliet, C., Bohn, M. C., and Spector, M. (2008). Non-viral delivery of the gene for glial cell line-derived neurotrophic factor to mesenchymal stem cells in vitro via a collagen scaffold. *Tissue Engineering Part C: Methods* **14** (3), 207–219.

Bonferoni, M. C., Sandri, G., Gavini, E., Rossi, S., Ferrari, F., and Caramella, C. (2007). Microparticle systems based on polymer-drug interaction for ocular delivery of ciprofloxacin – I. In vitro characterization. *Journal of Drug Delivery Science and Technology* **17** (1), 57–62.

Bonferoni, M. C., Sandri, G., Gavini, E., Rossi, S., Ferrari, F., and Caramella, C. (2007). Microparticle systems based on polymer-drug interaction for ocular delivery of ciprofloxacin – II. Precorneal residence times. *Journal of Drug Delivery Science and Technology* **17** (1), 63–68.

Brannon-Peppas, L. and Peppas, N. A. (1991). Equilibrium swelling behavior of dilute ionic hydrogels in electrolytic solutions. *J Control Release* **16** (3), 319–329.

Brannon-Peppas, L. and Peppas, N. A. (1991). Equilibrium swelling behavior of pH-sensitive hydrogels. *Chem Eng Sci* **46** (3), 715–722.

Brannon-Peppas, L. (1997). Polymers in Controlled Drug Delivery. *Medical Plastics and Biomaterials Magazine,* **Nov.** 34.

Bu, H. Z., Gukasyan, H. J., Goulet, L., Lou, X. J., Xiang, C., and Koudriakova, T. (2007). Ocular disposition, pharmacokinetics, efficacy and safety of nanoparticle-formulated ophthalmic drugs. *Current Drug Metabolism* **8** (2), 91–107.

Burzio, L. A. and Waite, J. H. (2000). Cross-linking in adhesive quinoproteins: Studies with model decapeptides. *Biochemistry* **39** (36), 11147–11153.

Butler, M. F., Ng, Y., and Pudney, P. D. A. (2003). Mechanism and kinetics of the cross-linking reaction between biopolymers containing primary amine groups and genipin. *J Polym Sci A1* **41** (24), 3941–3953.

Campoccia, D., Doherty, P., Radice, M., Brun, P., Abatangelo, G., and Williams, D. F. (1998). Semisynthetic resorbable materials from hyaluronan esterification. *Biomaterials* **19**, 2101–2127.

Carlsson, A., Karlström, G., Lindman, B., and Stenberg, O. (1988). Interaction between ethyl(hydroxyethyl)cellulose and sodium dodecyl sulphate in aqueous solution. *Colloid Polymer Science* **266**, 1031–1036.

Carlsson, A., Karlström, G., and Lindman, B. (1990). Thermal gelation of nonionic cellulose ethers and ionic surfactants in water. *Colloids Surface* **47**, 147–165.

Çaykara, T. and Kantoğlu, Ö. (2004). Thermal behavior and network structure of poly(N-vinyl-2-pyrrolidone-crotonic acid) hydrogels prepared by radiation-induced polymerization. *Polym Adv Technol* **15** (3), 134.

Chajra, H., Rousseau, C. F., Cortial, D., Ronziere, M. C., Herbage, D., Mallein-Gerin, F., and Freyria, A. M. (2008). Collagen-based biomaterials and cartilage engineering. Application to osteochondral defects. *Bio-Medical Materials And Engineering* **18**, S33–S45.

Chan, A. W., Whitney, R. A., and Neufeld, R. J. (2008). Kinetic controlled synthesis of pH-responsive network alginate. *Biomacromolecules* **9** (9), 2536–2545.

Chan, L. W. and Heng, P. W. S. (1998). Effects of poly (vinylpyrrolidone) and ethylcellulose on alginate microspheres prepared by emulsification. *Journal of Microencapsulation* **15**, 409–420.

Chase, M. R., Raina, K., Bruno, J., and Sugumaran, M. (2000). Purification, characterization and molecular cloning of prophenoloxidases from Sarcophaga bullata. *Insect Biochemistry and Molecular Biology* **30** (10), 953–967.

Chavanpatil, M. D., Khdair, A., and Panyam, J. (2006). Nanoparticles for cellular drug delivery: Mechanisms and factors influencing delivery. *Journal of Nanoscience And Nanotechnology* **6** (9-10), 2651–2663.

Chen, M. C., Tsai, H. W., Chang, Y., Lai, W. Y., Mi, F. L., Liu, C. T., Wong, H. S., and Sung, H. W. (2007). Rapidly self-expandable polymeric Stents with a shape-memory property. *Biomacromolecules* **8** (9), 2774–2780.

Christensen, F. N., Hansen, F. Y., and Bechgaard, H. (1982). Mathematical model for in vitro drug release from controlled release dosage forms applied to propoxyphene hydrochloride pellets. *J. Pharm. Sci.* **71**, 694–699.

Chung, S., Martins, A., Correlo, V. M., Oliveira, J. T., Marques, A. P., Neves, N. M., Mano, J. F., Sousa, R. A., and Reis, R. L. (2008). Dependence of cell proliferation on the morphology of starch based scaffolds for tissue engineering. *Tissue Engineering: Part A* **14** (5), 718.

Cohen, S., Lobel, E., Trevgoda, A., and Peled, Y. (1997). A novel in situ-forming ophthalmic drug delivery system from alginates undergoing gelation in the eye. *Journal of Controlled Release* **44** (2-3), 201–208.

Cravotto, G., Tagliapietra, S., Robaldo, B., and Trotta, M. (2005). Chemical modification of chitosan under high-intensity ultrasound. *Ultrasonics Sonochemistry* **12** (1-2), 95–98.

Dai, Y. N., Li, P., Zhang, J. P., Wang, A. O., and Wei, Q. (2008). Swelling characteristics, drug delivery properties of nifedipine-loaded pH sensitive alginate-chitosan hydrogel beads. *Journal of Biomedical Materials Research* **86b** (2), 493–500.

Dai, Y. N., Li, P., Zhang, J. P., Wang, A. Q., and Wei, Q. (2008). A novel pH sensitive N-succinyl chitosan/alginate hydrogel bead for nifedipine delivery. *Biopharmaceutics & Drug Disposition* **29** (3), 173–184.

DeFail, A. J., Edington, H. D., Matthews, S., Lee, W. C. C., and Marra, K. G. (2006). Controlled release of bioactive doxorubicin from microspheres embedded within gelatin scaffolds. *Journal of Biomedical Materials Research Part A* **79A** (4), 954–962.

Deng, L. D., Zhai, Y. L., Lin, X. N., Jin, F. M., He, X. H., and Dong, A. J. (2008). Investigation on properties of re-dispersible cationic hydrogel nanoparticles. *European Polymer Journal* **44** (4), 978–986.

Dill, K. A. and Bromberg, S. (2003). Molecular Driving Forces. Garland Science, London.

Doi, K., Ikeda, T., Marui, A., Kushibiki, T., Arai, Y., Hirose, K., Soga, Y., Iwakura, A., Ueyama, K., Yamahara, K., Itoh, H., Nishimura, K., Tabata, Y., and Komeda, M. (2007). Enhanced angiogenesis by gelatin hydrogels incorporating basic fibroblast growth factor in rabbit model of hind limb ischemia. *Heart and Vessels* **22** (2), 104–108.

Dong, Z. F., Wang, Q., and Du, Y. M. (2006). Alginate/gelatin blend films and their properties for drug controlled release. *Journal of Membrane Science* **280** (1-2), 37–44.

Dorski, C. M., Doyle, F. J., and Peppas, N. A. (1997). Preparation and characterization of glucose-sensitive P(MAA-g-EG) hydrogels. *Polym. Mater. Sci. Eng. Proc* **76**, 281–282.

Draget, K. I., SkjakBraek, G., Christensen, B. E., Gaserod, O., and Smidsrod, O. (1996). Swelling and partial solubilization of alginic acid gel beads in acidic buffer. *Carbohydrate Polymers* **29** (3), 209–215.

Duan, X. and Sheardown, H. (2005). Cross-linking of collagen with dendrimers. *Journal Of Biomedical Materials Research Part A* **75A** (3), 510–518.

Duan, X. D., McLaughlin, C., Griffith, M., and Sheardown, H. (2007). Biofunctionalization of collagen for improved biological response: Scaffolds for corneal tissue engineering. *Biomaterials* **28** (1), 78–88.

Dumitriu, S. and Chornet, E. (1998). Inclusion and release of proteins from polysaccharide based polyion complexes. *Advanced Drug Delivery Reviews* **31** (3), 223–246.

Edgar, K. J. (2007). Cellulose esters in drug delivery. *Cellulose* **14**, 49–64.

Ehtezazi, T., Washington, C., and Melia, C. D. (2000). First order release rate from porous PLA microspheres with limited exit holes on the exterior surface. *Journal of Controlled Release* **66** (1), 27–38.

Enderle, J., Blanchard, S., and Bronzino, J. (2005). Biomaterials: properties, types and applications. Introduction to Biomedical Engineering. Academic Press, San Diego, California.

Esquisabel, A., Hernandez, R. M., Igartua, M., Gascon, A. R., Calvo, B., and Pedraz, J. L. (1997). Production of BCG alginate-PLL microcapsules by emulsification internal gelation. *Journal of Microencapsulation* **14** (5), 627–638.

Falk, B., Garramone, S., and Shivkumar, S. (2004). Diffusion coefficient of paracetamol in a chitosan hydrogel. *Materials Letters* **58** (26), 3261–3265.

Farooqui, N., Myung, D., Koh, W., Masek, M., Dalal, R., Carrasco, M. R., Noolandi, J., Frank, C. W., and Ta, C. N. (2007). Histological processing of pH-sensitive hydrogels used in corneal implant applications. *Journal of Histotechnology* **30** (3), 157–163.

Ferreira, B. M., Sousa, R. A., and Reis, R. L. (2008). Development of novel carrageenan scaffolds for tissue engineering using rapid prototyping. *Tissue Engineering: Part A* **14** (5), 837–838.

Fick, A. E. (1855). Über Diffusion. *Ann Der Phys* **94**, 59–86.

Figueiro, S. D., Macedo, A. A. M., Melo, M. R. S., Freitas, A. L. P., Moreira, R. A., de Oliveira, R. S., Goes, J. C., and Sombra, A. S. B. (2006). On the dielectric behaviour of collagen-algal sulfated polysaccharide blends: Effect of glutaraldehyde cross-linking. *Biophysical Chemistry* **120** (2), 154–159.

Flory, P. J. and Rehner, J. J. (1943). Statistical Mechanics of Cross-Linked Polymer Networks II. Swelling. *J Chem Phys* **11** (11), 521–526.

Flory, P. J. and Rehner, J. J. (1943). Statistical Mechanics of Cross-Linked Polymer Networks I. Rubberlike Elasticity. *J Chem Phys* **11** (11), 512–520.

Foldvari, M., Oguejiofor, C., Afridi, S., Kudel, T., and Wilson, T. (1998). Liposome encapsulated prostaglandin E_1 in erectile dysfunction: correlation between in vitro delivery through foreskin and efficacy in patients. *Urology* **52** (5), 838–843.

Fournier, R. L. (1998). Basic Transport Phenomena in Biomedical Engineering. Taylor and Francis.

Friess, W. and Schlapp, M. (2006). Sterilization of gentamicin containing collagen/PLGA microparticle composites. *European Journal of Pharmaceutics And Biopharmaceutics* **63** (2), 176–187.

Fundueanu, G., Constantin, M., and Ascenzi, P. (2008). Preparation and characterization of pH- and temperature-sensitive pullulan microspheres for controlled release of drugs. *Biomaterials* **29** (18), 2767–2775.

Gazzaniga, A., Palugan, L., Foppoli, A., and Sangalli, M. E. (2008). Oral pulsatile delivery systems based on swellable hydrophilic polymers. *European Journal of Pharmaceutics and Biopharmaceutics* **68** (1), 11–18.

Gennaro, A. L. (1990). Remington's Pharmaceutical Sciences. *Mack Publishing Company, Pennsylvania.*

Ghosh, S. and Jassal, M. (2002). Use of polysaccharide fibres for modem wound dressings. *Indian Journal of Fibre & Textile Research* **27** (4), 434–450.

Girod, S., Boissière, M., Longchambon, K., Begu, S., Pétheil, C. T., and Devoisselle, J. M. (2004). Polyelectrolyte complex formation between iota-carrageenan and poly(-lysine) in dilute aqueous solutions: a spectroscopic and conformational study. *Carbohydrate Polymers* **55** (1), 37–45.

Gombotz, W. R. and Wee, S. F. (1998). Protein release from alginate matrices. *Advanced Drug Delivery Reviews* **31** (3), 267–285.

Groening, R., Cloer, C., Georgarakis, M., and Mueller, R. S. (2007). Compressed collagen sponges as gastroretentive dosage forms: In vitro and in vivo studies. *European Journal of Pharmaceutical Sciences* **30** (1), 1–6.

Gunasekaran, S., Ko, S., and Xiao, L. (2007). Use of whey proteins for encapsulation and controlled delivery applications. *Journal of Food Engineering* **83** (1), 31–40.

Guo, B. L., Yuan, J. F., Yao, L., and Gao, Q. Y. (2007). Preparation and release profiles of pH/temperature-responsive carboxymethyl chitosan/P (2-(dimethylamino) ethyl methacrylate) semi-IPN amphoteric hydrogel. *Colloid and Polymer Science* **285** (6), 665–671.

Gupta, P., Vermani, K., and Garg, S. (2002). Hydrogels: from controlled release to pH-responsive drug delivery. *Drug Discov Today* **7** (10), 569–579.

Gutowska, A., Seok Bark, J., Chan Kwon, I., Han Bae, Y., Cha, Y., and Wan Kim, S. (1997). Squeezing hydrogels for controlled oral drug delivery. *J Control Release* **48** (2-3), 141–148.

Hariprasad, S. M., Shah, G. K., Chi, J., and Prince, R. A. (2005). Determination of aqueous and vitreous concentration of moxifloxacin 0.5% after delivery via a dissolvable corneal collagen shield device. *Journal of Cataract And Refractive Surgery* **31** (11), 2142–2146.

Hasegawa, T., Kawazome, A., Yanagimoto, G., Hayashi, T., Seki, T., Akimoto, M., Todo, H., and Sugibayashi, K. (2007). Analysis of skin disposition of flurbiprofen after topical application using dual agar gel discs-inserted rats. *Biological & Pharmaceutical Bulletin* **30** (11), 2135–2140.

Hayashi, K., Kubo, T., Doi, K., Tabata, Y., and Akagawa, Y. (2007). Development of new drug delivery system for implant bone augmentation using a basic fibroblast growth factor-gelatin hydrogel complex. *Dental Materials Journal* **26** (2), 170–177.

He, H. Y., Cao, X., and Lee, L. J. (2004). Design of a novel hydrogel-based intelligent system for controlled drug release. *Journal of Controlled Release* **95** (3), 391–402.

Hench, L. L. and Jones, J. R. (2005). Biomaterials, artificial organs and tissue engineering. CRC Press, New York.

Higuchi, T. (1963). Mechanisms of sustained action mediation. Theoretical analysis of rate of release of solid drugs dispersed in solid matrices. *J Pharm Sci* **52**, 1145–1149.

Hodsdon, A. C., Mitchell, J. R., Davies, M. C., and Melia, C. D. (1995). Structure and behaviour in hydrophilic matrix sustained release dosage forms: 3. The influence of pH on the sustained-release performance and internal gel structure of sodium alginate matrices. *Journal of Controlled Release* **33** (1), 143–152.

Hoffman, A. S. (2002). Hydrogels for biomedical applications. *Advanced Drug Delivery Reviews* **54** (1), 3–12.

Hu, S. H., Liu, T. Y., Tsai, C. H., and Chen, S. Y. (2007). Preparation and characterization of magnetic ferroscaffolds for tissue engineering. *Journal of Magnetism and Magnetic Materials* **310** (2), 2871–2873.

Iannuccelli, V., Coppi, G., and Cameroni, R. (1996). Biodegradable intraoperative system for bone infection treatment .1. The drug/polymer interaction. *International Journal of Pharmaceutics* **143**, 195–201.

Ikada Y. (2006). Tissue Engineering: Fundamentals and Applications, Academic Press, 1 edition.

Imai, T., Kawasaki, C., Nishiyama, T., and Otagiri, M. (2000). Comparison of the pharmaceutical properties of sustained-release gel beads prepared by alginate having different molecular size with commercial sustained-release tablet. *Pharmazie* **55** (3), 218–222.

Isiklan, N., Inal, M., and Yigitoglu, M. (2008). Synthesis and characterization of poly(N-Vinyl-2-pyrrolidone) grafted sodium alginate hydrogel beads for the controlled release of indomethacin. *Journal Of Applied Polymer Science* **110** (1), 481–493.

Jayakumar, R., Reis, R. L., and Mano, J. F. (2006). Phosphorous containing chitosan beads for controlled oral drug delivery. *Journal of Bioactive and Compatible Polymers* **21** (4), 327–340.

Jayakumar, R., Reis, R. L., and Mano, J. F. (2007). Synthesis and characterization of pH-sensitive thiol-containing chitosan beads for controlled drug delivery applications. *Drug Delivery* **14** (1), 9–17.

Jayakumar, R., Selvamurugan, N., Nair, S. V., Tokura, S., and Tamura, H. (2008). Preparative methods of phosphorylated chitin and chitosan – An overview. *International Journal of Biological Macromolecules* **43** (3), 221–225.

Jeong, S. I., Lee, Y. M., and Shin, H. (2008). Tissue engineering using a cyclic strain bioreactor and gelatin/PLCL scaffolds. *Macromolecular Research* **16** (6), 567–569.

Ju, Y. M., Yu, B. Z., Koob, T. J., Moussy, Y., and Moussy, F. (2008). A novel porous collagen scaffold around an implantable biosensor for improving biocompatibility. I. In vitro/in vivo stability of the scaffold and in vitro sensitivity of the glucose sensor with scaffold. *Journal of Biomedical Materials Research. Part A* **87a** (1), 136–146.

Juntanon, K., Niamlang, S., Rujiravanit, R., and Sirivat, A. (2008). Electrically controlled release of sulfosalicylic acid from cross-linked, poly(vinyl alcohol) hydrogel. *International Journal of Pharmaceutics* **356** (1-2), 1–11.

Kao, M. C., Matsuno-Yagi, A., and Yagi, T. (2004). Subunit proximity in the H+-translocating NADH-quinone oxidoreductase probed by zero-length cross-linking. *Biochemistry* **43** (12), 3750–3755.

Kashyap, N., Kumar, N., and Kumar, M. N. V. R. (2005). Hydrogels for Pharmaceutical and Biomedical applications. *Critical Reviews in Therapeutic Drug Carrier Systems* **22** (2), 107–149.

Katayama, H., Nishimura, T., Ochi, S., Tsuruta, Y., Yamazaka, Y., Shibata, K., and Yoshitomi, H. (1999). Sustained release liquid preparation using sodium alginate for eradication of Helicobacter pyroli. *Biological and Pharmaceutical Bulletin* **22**, 55–60.

Katchalsky, A. and Michaeli, I. (1955). Polyelectrolyte gels in salt solutions. *J Polym Sci* **15** (79), 69–86.

Kavanagh, C. A., Rochev, Y. A., Gallagher, W. A., Dawson, K. A., and Keenan, A. K. (2004). Local drug delivery in restenosis injury: thermoresponsive co-polymers as potential drug delivery systems. *Pharmacology & Therapeutics* **102** (1), 1–15.

Khor, E. and Lim, L. Y. (2003). Implantable applications of chitin and chitosan. *Biomaterials* **24** (13), 2339–2349.

Khor, E. (1997). Methods for the treatment of collagenous tissues for bioprostheses. *Biomaterials* **18** (2), 95–105.

Kim, K. J., Lee, S. B., and Han, N. W. (1994). Kinetics of crosslinking reaction of PVA membrane with glutaraldehyde. *Korean Journal of Chemical Engineering* **11** (1), 41–47.

Kleinmann, G., Larson, S., Hunter, B., Stevens, S., Mamalis, N., and Olson, R. J. (2007). Collagen shields as a drug delivery system for the fourth-generation fluoroquinolones. *Ophthalmologica* **221** (1), 51–56.

Kobayashi, H., Minatoguchi, S., Bao, N., Yasuda, S., Misao, Y., Uno, Y., Takemura, G., Fujiwara, T., Tabata, Y., and Fujiwara, H. (2006). New Drug Delivery System using an Erythropoietin Gelatin Hydrogel Sheet Selectively Protects the Heart against Myocardial Infarction through Angiogenesis. *Journal of Cardiac Failure* **12** (8), S167. Supplement 1.

Koob, T. J. and Hernandez, D. J. (2002). Material properties of polymerized NDGA–collagen composite fibers: development of biologically based tendon constructs. *Biomaterials* **23** (1), 203–212.

Korsmever, R. W. and Peppas, N. A. (1983). Macromolecular and modeling aspects of swelling-controlled systems. Controlled Release Delivery Systems, Roseman, T. J. and Mansdorf, S. Z. (eds.). Marcel Dekker, New York.

Kramer, K., Kanost, M., Hopkins, T., Jiang, H., Zhu, Y., Xu, R., Kerwin, J., and Turecek, F. (2001). Oxidative conjugation of catechols with proteins in insect skeletal systems. *Tetrahedron* **57** (2), 385–392.

Krucinska, I., Komisarczyk, A., Chrzanowski, M., and Paluch, D. (2007). Producing wound dressing materials from chitin derivatives by forming nonwovens directly from polymer solution. *Fibres & Textiles in Eastern Europe* **15** (5-6), 73–76.

Kulkarni, A. R., Soppimath, K. S., Aralaguooi, M. I., Aminabhavi, T. M., Rudzinski, W. E. (2000). Preparation of Cross-Linked Sodium Alginate Microparticles Using Glutaraldehyde in Methanol. Drug Development and Industrial Pharmacy, http://0-www.informaworld.com.innopac.lib.ryerson.ca/smpp/title~content=t713597245~db=all~tab=issueslist~branches=26-v2626 (10), 1121–1124.

Kurisawa, M. and Yui, N. (1998). Dual-stimuli-responsive drug release from interpenetrating polymer network-structured hydrogels of gelatin and dextran. *J Control Release* **54** (2), 191–200.

Kurisawa, M. and Yui, N. (1998). Gelatin/dextran intelligent hydrogels for drug delivery: dual-stimuli-responsive degradation in relation to miscibility in interpenetrating polymer networks. *Macromol Chem Phys* **199** (8), 1547–1554.

Lang, Y. Y., Li, S. M., Pan, W. S., and Zheng, L. Y. (2006). Thermo- and pH-sensitive drug delivery from hydrogels constructed using block copolymers of poly(N-isopropylacrylamide) and Guar gum. *Journal of Drug Delivery Science and Technology* **16** (1), 65–69.

Lee, D.W., Shirley, S.A., Lockey, R.F., Mohapatra, S.S. Thiolated chitosan nanoparticles enhance anti-inflammatory effects of intranasally delivered theophylline. Respiratory Research7: Art. No. 112, 2006 (page numbers not for citation). [http://respiratory-research.com/content/pdf/1465-9921-7-112.pdf]

Lee, M. H., Boettiger, D., and Composto, R. J. (2008). Biomimetic carbohydrate substrates of tunable properties using immobilized dextran hydrogels. *Biomacromolecules* **9** (9), 2315–2321.

Leong, K. W. and Langer, R. (1988). Polymeric controlled drug delivery. *Adv Drug Deliv Rev* **1** (3), 199–233.

Levine, D. S., Raisys, V. A., and Ainardi, V. (1987). Coating of oral beclomethasone dipropionate capsules with cellulose acetate phthalate enhances delivery of topically active anti-inflammatory drug to the terminal ileum. *Gastroenterology* **92**, 1037–1044.

Li, J., Wang, B. C., and Liu, P. (2008). Possibility of active targeting to tumor by local hyperthermia with temperature-sensitive nanoparticles. *Medical hypotheses* **71** (2), 249–251.

Li, J., Wang, B. C., Wang, Y. Z., Liu, P., and Qiao, W. L. (2008). Preparation and characterization of thermosensitive nanoparticles for targeted drug delivery. *Journal of Macromolecular Science.* Part A. *Pure & Applied Chemistry* **45** (10), 833–838.

Liang, H. C., Chang, W. H., Lin, K. J., and Sung, H. W. (2003). Genipin-cross-linked gelatin microspheres as a drug carrier for intramuscular administration: In vitro and in vivo studies. *Journal of Biomedical Materials Research* **65a** (2), 271–282.

Liang, S., Zhang, L., and Xu, J. (2007). Morphology and permeability of cellulose/chitin blend membranes. *Journal of Membrane Science* **287** (1), 19–28.

Lih, E., Joung, Y. K., Bae, J. W., and Park, K. D. (2008). An in situ gel-forming heparin-conjugated PLGA-PEG-PLGA copolymer. *Journal of Bioactive and Compatible Polymers* **23** (5), 444–457.

Linnes, M. P., Ratner, B. D., and Giachelli, C. M. (2007). A fibrinogen-based precision microporous scaffold for tissue engineering. *Biomaterials* **28** (35), 5298–5306.

Liu, J., Chan, S. Y., and Ho, P. C. (2002). Polymer-coated microparticles for the sustained release of nitrofurantoin. *Journal of Pharmacy and Pharmacology* **54**, 1205–1212.

Liu, J., Meisner, D., Kwong, E., Wu, X. Y., and Johnston, M. R. (2007). A novel trans-lymphatic drug delivery system: Implantable gelatin sponge impregnated with PLGA-paclitaxel microspheres. *Biomaterials* **28** (21), 3236–3244.

Liu, L. S., Won, Y. J., Cooke, P. H., Coffin, D. R., Fishman, M. L., Hicks, K. B., and Ma, P. X. (2004). Pectin/poly(lactide-co-glycolide) composite matrices for biomedical applications. *Biomaterials* **25** (16), 3201–3210.

Liu, W. G., Griffith, M., and Li, F. F. (2008). Alginate microsphere-collagen composite hydrogel for ocular drug delivery and implantation. *Journal of Materials Science* **19** (11), 3365–3371.

Long, Z., Xu, J., and Pan, J. (2007). Immobilization of Serratia marcescens Lipase and Catalytic Resolution of Trans-3-(4′-methoxyphenyl)glycidic Acid Methyl Ester. *Chinese Journal of Catalysis* **28** (2), 175–179.

Lu, B., Wen, R., Yang, H., and He, Y. J. (2007). Sustained-release tablets of indomethacin-loaded microcapsules: Preparation, in vitro and in vivo characterization. *International Journal of Pharmaceutics* **333** (1-2), 87–94.

Makhija, S. N. and Vavia, P. R. (2003). Controlled porosity osmotic pump-based controlled release systems of pseudoephedrine. I. Cellulose acetate as a semipermeable membrane. *Journal of Controlled Release* **89**, 5–18.

Malm, C. J., Emerson, J., and Hiatt, G. D. (1951). Cellulose acetate phthalate as an enteric coating material. *Journal of American Pharmaceutical Association* **40**, 520–525.

Malm, C. J. and Fordyce, C. R. (1940). Cellulose esters of dibasic organic acids. *Industrial & Engineering Chemistry Research* **32**, 405–408.

Mandel, K. G., Daggy, B. P., Brodie, D. A., and Jacoby, H. I. (2000). Review article: alginate-raft formulations in the treatment of heartburn and acid reflux. *Alimentary Pharmacology & Therapeutics* **14** (6), 669–690.

Markland, P., Zhang, Y. H., Amidon, G. L., and Yang, V. C. (1999). A pH- and ionic strength-responsive polypeptide hydrogel: Synthesis, characterization, and preliminary protein release studies. *Journal of Biomedical Materials Research* **47** (4), 595–602.

Martin, A., Swarbrick, J., and Cammarata, A. (1991). Physical Pharmacy. Varghese Publishing house, Bombay.

Mathew, S. T., Devi, S. G., and Sandhya, K. V. (2007). Formulation and Evaluation of Ketorolac Tromethamine-loaded Albumin Microspheres for Potential Intramuscular Administration. *AAPS PharmSciTech* **8** (1) Article 14, DOI: 10.1208/pt0801014.

Mayo, L., Quaglia, F., Borzacchiello, A., Ambrosio, L., and La Rotonda, M. I. (2008). A novel poloxamers/hyaluronic acid in situ forming hydrogel for drug delivery: Rheological, mucoadhesive and in vitro release properties. *European Journal of Pharmaceutics and Biopharmaceutics* **70** (1), 199–206.

Mayo-Pedrosa, M., Cachafeiro-Andrade, N., Alvarez-Lorenzo, C., Martinez-Pacheco, R., and Concheiro, A. (2008). In situ photopolymerization-coated pellets for pH-dependent drug delivery. *European Polymer Journal* **44** (8), 2629–2638.

McDowell, L. M., Burzio, L. A., Waite, J. H., and Schaefer, J. (1999). Rotational echo double resonance detection of cross-links formed in mussel byssus under high-flow stress. *Journal of Biological Chemistry* **274** (29), 20293–20295.

Mi, F., Sung, H., and Shyu, S. (2005). Characterization of ring-opening polymerization of genipin and pH-dependent cross-linking reactions between chitosan and genipin. *J Polym Sci* **43** (10), 1985–2000. A1.

Mi, F., Sung, H., and Shyu, S. (2000). Synthesis and characterization of a novel chitosan-based network prepared using naturally occurring cross-linker. *J Polym Sci* **38** (15), 2804–2814. A1.

Miyazaki, S., Kubo, W., and Attwood, D. (2000). Oral sustained delivery of theophylline using in-situ gelation of sodium alginate. *Journal of Controlled Release* **67** (2-3), 275–280.

Miyazaki, S., Nakayama, A., Oda, M., Takada, M., and Attwood, D. (1994). Chitosan and sodium alginate based bioadhesive tablets for intraoral drug delivery. *Biological and Pharmaceutical Bulletin* **17**, 745–747.

Mockel, J. E. and Lippold., B. C. (1993). Zero-order drug release from hydrocolloid matrices. *Pharmaceutical Research* **10** (7), 1066–1070.

Moffat, K. L. and Marra, K. G. (2004). Biodegradable poly-(ethylene glycol) hydrogels cross-linked with genipin for tissue engineering applications. *Journal of Biomedical Materials Research* **71b** (1), 181–187.

Montisci, M. J., Giovannuci, G., Duchene, D., and Ponchel, G. (2001). Covalent coupling of asparagus pea and tomato lectins to poly(lactide) microspheres. *International Journal of Pharmaceutics* **215** (1-2), 153–161.

Moussy, Y., Guegan, E., Davis, T., and Koob, T. J. (2007). Transport Characteristics of a Novel Local Drug Delivery System Using Nordihydroguaiaretic Acid (NDGA)-Polymerized Collagen Fibers. *Biotechnology Progresss* **23**, 990–994.

Mukherjee, D. and Banthia, A. K. (2006). Preparation of adrenochrome hydrogel patch, gel, ointment, and the comparison of their blood coagulating and wound healing capability. *Materials and Manufacturing Processes* **21** (3), 297–301.

Mumper, R. J., Hoffman, A. S., Puolakkainen, P. A., Bouchard, L. S., and Gombotz, W. R. (1994). Calcium-Alginate Beads For The Oral Delivery Of Transforming Growth Factor-Beta (1) (Tgf-Beta (1)) – Stabilization Of Tgf-Beta (1) By The Addition Of Polyacrylic-Acid Within Acid-Treated Beads. *Journal of Controlled Release* **30** (3), 241–251.

Muzzarelli, R. A. A. (1973). Natural Chelating Polymers. Pergamon Press, New York.

Nagy, I. Z., Ohta, M., and Kitani, K. (1989). Effect of centrophenoxine and BCE-001 treatment on lateral diffusion of proteins in the hepatocyte plasma membrane as revealed by fluorescence recovery after photobleaching in rat liver smears. *Experimental Gerontology* **24** (4), 317–330.

Nesti, L. J., Li, W. J., Shanti, R. M., Jiang, Y. J., Jackson, W., Freedman, B. A., Kuklo, T. R., Giuliani, J. R., and Tuan, R. S. (2008). Intervertebral disc tissue engineering using a novel hyaluronic acid-nanofibrous scaffold (HANFS) amalgam. *Tissue Engineering: Part A* **14** (9), 1527–1537.

Ofner, C. M., III and Bubnis, W. A. (1996). Chemical Swelling Evaluations of Amino Group Cross-linking in Gelatin and Modified Gelatin Matrices. *Pharm Res* **13** (12), 1821–1827.

Ofokansi, K. C., Adikwu, M. U., and Okore, V. C. (2007). Preparation and evaluation of mucin-gelatin mucoadhesive microspheres for rectal delivery of ceftriaxone sodium. *Drug Development and Industrial Pharmacy* **33** (6), 691–700.

Oladiran, G. S. and Batchelor, H. K. (2007). Determination of ibuprofen solubility in wax: A comparison of microscopic, thermal and release rate techniques. *European Journal of Pharmaceutics and Biopharmaceutics* **67** (1), 106–111.

Pal, K., Banthia, A. K., and Majumdar, D. K. (2007). Biomedical evaluation of polyvinyl alcohol–gelatin esterified hydrogel for wound dressing. *Journal of Materials Science: Materials in Medicine* **18** (9), 1889–1894.

Pal, K., Banthia, A. K., and Majumdar, D. K. (2006). Characterizations of the prepared corn- starch based hydrogel membranes. *Journal of Applied Biomaterials and Biomechanics* **4**, 38–44.

Pal, K., Banthia, A. K., and Majumdar, D. K. (2007). Preparation and characterization of polyvinyl alcohol-gelatin hydrogel membranes for biomedical applications. *AAPS PharmSciTech* **8** (1), 21–24.

Pal, K., Banthia, A. K., and Majumdar, D. K. (2006). Preparation of Novel pH Sensitive Hydrogels of Carboxymethyl Cellulose Acrylates: A Comparative Study. *Materials & Manufacturing Processes* **21**, 877–882.

Palmieri, G. F., Bonacucina, G., Martino, P. D., and Martelli, S. (2002). Gastro-resistant microspheres containing ketoprofen. *Journal of Microencapsulation* **19**, 111–119.

Park, J. B. and Lake, R. S. (1990). Biomaterials – An Introduction. Plenum, New York.

Park, S. J. and Kim, C. H. (2007). Chitosan for tissue engineering. *Tissue Engineering and Regenerative Medicine* **4** (4), 471–477.

Pawde, S. M. and Deshmukh, K. (2008). Characterization of polyvinyl alcohol/gelatin blend hydrogel films for biomedical applications. *Journal of Applied Polymer Science* **109** (5), 3431–3437.

Peng, X. H., Zhang, L. N., and Kennedy, J. F. (2006). Release behavior of microspheres from cross-linked N-methylated chitosan encapsulated ofloxacin. *Carbohydrate Polymers* **65** (3), 288–295.

Peppas, N. A., Bures, P., Leobandung, W., and Ichikawa, H. (2000). Hydrogels in pharmaceutical formulations. *European Journal of Pharmaceutics and Biopharmaceutics* **50** (1), 27–46.

Peppas, N. A., Hilt, J. Z., Khademhosseini, A., and Langer, R. (2006). Hydrogels in biology and medicine: From molecular principles to bionanotechnology. *Advanced Materials* **18** (11), 1345–1360.

Peppas, N. A. and Merrill, E. W. (1977). Cross-linked poly (vinyl alcohol) hydrogels as swollen elastic networks. *J Appl Polym Sci* **21** (7), 1763–1770.

Peppas, N. A. (1987). Hydrogels in Medicine and Pharmacy. CRC Press, Boca Raton, FL. Vols. I–III.

Petite, H., Frei, V., Huc, A., and Herbage, D. (1994). Use of diphenylphosphorylazide for cross-linking collagen-based biomaterials. *Journal Of Biomedical Materials Research* **28** (2), 159–165.

Pettenazzo, E., Thiene, G., Gatti, A. M., Pasquino, E., Talenti, E., Noera, G., Bottio, T., and Valente, M. (2001). Is the tricuspid position suitable for testing replacement bioprosthetic valves in the sheep model? *The Journal Of Heart Valve Disease* **10** (4), 513–519.

Pica, K., Tchao, R., and Ofner, C. M. (2006). Gelatin-methotrexate conjugate microspheres as a potential drug delivery system. *Journal of Pharmaceutical Sciences* **95** (9), 1896–1908.

Pierucci, A. P. T. R., Andrade, L. R., Baptista, E. B., Volpato, N. M., and Rocha-Leao, M. H. M. (2006). New microencapsulation system for ascorbic acid using pea protein concentrate as coat protector. *Journal of Microencapsulation* **23** (6), 654–662.

Polk, A., Amsden, B., Deyao, K., Peng, T., and Goosen, M. F. A. (1994). Controlled-Release of Albumin from Chitosan-Alginate Microcapsules. *Journal of Pharmaceutical Sciences* **83** (2), 178–185.

Pongpaibul, Y. and Whitworth, C. W. (1986). Preparation and in vitro dissolution characteristics of propranolol microcapsules. *International Journal of Pharmaceutics* **33**, 243–248.

Prabu, P., Dharmaraj, N., Aryal, S., Lee, B. M., Ramesh, V., and Kim, H. Y. (2006). Preparation and drug release activity of scaffolds containing collagen and poly (caprolactone). *Journal of Biomedical Materials Research Part A* **79A** (1), 153–158.

Prestwich, G. D., Marecak, D. M., Marecak, J. F., Vercruysse, K. P., and Ziebell, M. R. (1998). Controlled chemical modification of hyaluronic acid. *Journal of Controlled Release* **53**, 93–103.

Puttipipatkhachorn, S., Nunthanid, J., Yamamoto, K., and Peck, G. E. (2001). Drug physical state and drug-polymer interaction on drug release from chitosan matrix films. *Journal of Controlled Release* **75** (1-2), 143–153.

Rajvaidya, S., Bajpai, R., and Bajpai, A. K. (2006). Effect of gamma irradiation on the interpenetrating networks of gelatin and polyacrylonitrile: Aspect of cross-linking using microhardness and cross-link density measurements. *J Appl Polym Sci* **101** (4), 2581–2586.

Rambo, C. R., Recouvreux, D. O. S., Carminatti, C. A., Pitlovanciv, A. K., Antonio, R. V., and Porto, L. M. (2008). Template assisted synthesis of porous nanofibrous cellulose membranes for tissue engineering. *Materials Science & Engineering. C, Biomimetic Materials, Sensors and Systems* **28** (4), 549–554.

Ravi Kumar MNV (2000). A review of chitin and chitosan applications. *Reactive and Functional Polymers* **46** (1), 1–27.

Reverchon, E. and Antonacci, A. (2006). Chitosan microparticles production by supercritical fluid processing. *Industrial & Engineering Chemistry Research* **45** (16), 5722–5728.

Rokhade, A. P., Agnihotri, S. A., Patil, S. A., Mallikarjuna, N. N., Kulkarni, P. V., and Aminabhavi, T. M. (2006). Semi-interpenetrating polymer network microspheres of gelatin and sodium carboxymethyl cellulose for controlled release of ketorolac tromethamine. *Carbohydrate Polymers* **65** (3), 243–252.

Rokhade, A. P., Shelke, N. B., Patil, S. A., and Aminabhavi, T. M. (2007). Novel interpenetrating polymer network microspheres of chitosan and methylcellulose for controlled release of theophylline. *Carbohydrate Polymers* **69** (4), 678–687.

Ruel-Gariepy, E. and Leroux, J. C. (2004). In situ-forming hydrogels – review of temperature-sensitive systems. *European Journal of Pharmaceutics And Biopharmaceutics* **58** (2), 409–426.

Sacks, M. S., Hamamoto, H., Connolly, J. M., Gorman, R. C., Gorman, J. H., and Levy, R. J. (2007). In vivo biomechanical assessment of triglycidylamine cross-linked pericardium. *Biomaterials* **28** (35), 5390–5398.

Sadeghi, M. and Hosseinzadeh, H. (2008). Dilution of microbicide gels with vaginal fluid and semen simulants: Effect on rheological properties and coating flow. *Journal Of Pharmaceutical Sciences* **97** (2), 1030–1038.

Sadeghi, M. and Hosseinzadeh, H. (2008). Synthesis of starch-poly(sodium acrylate-co-acrylamide) superabsorbent hydrogel with salt and pH-Responsiveness properties as a drug delivery system. *Journal of Bioactive and Compatible Polymers* **23** (4), 381–404.

Saito, H., Taguchi, T., Aoki, H., Murabayashi, S., Mitamura, Y., Tanaka, J., and Tateishi, T. (2007). pH-responsive cross-linkers swelling behavior of collagen gels prepared by novel based on naturally derived di- or tricarboxylic acids. *Acta Biomaterialia* **3** (1), 89–94.

Saito, K., Fujieda, T., and Yoshioka, H. (2006). Feasibility of simple chitosan sheet as drug delivery carrier. *European Journal of Pharmaceutics and Biopharmaceutics* **64** (2), 161–166.

Sankalia, M. G., Mashru, R. C., Sankalia, J. M., and Sutariya, V. B. (2007). Reversed chitosan-alginate polyelectrolyte complex for stability improvement of alpha-amylase: Optimization and physicochemical characterization. *European Journal of Pharmaceutics and Biopharmaceutics* **65** (2), 215–232.

Saravanan, M., Dhanaraju, M. D., Sridhar, S. K., Ramachandran, S., Sam, S. K. G., Anand, P., Bhaskar, K., and Rao, G. S. (2004). Preparation, Characterization and in vitro release kinetics of Ibuprofen polystyrene microspheres. *Indian Journal of Pharmaceutical Sciences* **66** (3), 287–292.

Sarmento, B., Ribeiro, A. J., Veiga, F., Ferreira, D. C., and Neufeld, R. J. (2007). Insulin-loaded nanoparticles are prepared by alginate ionotropic pre-gelation followed by chitosan polyelectrolyte complexation. *Journal of Nanoscience and Nanotechnology* **7** (8), 2833–2841.

Sarojini, H., Medepalli, K., Terry, D. A., Alphenaar, B. W., and Wang, E. (2007). Localized delivery of DNA to the cells by viral collagen-loaded silica colloidal crystals. *Biotechniques* **43** (2), 213–221.

Satish, C. S., Satish, K. P., and Shivakumar, H. G. (2006). Hydrogels as controlled drug delivery systems: Synthesis, cross-linking, water and drug transport mechanism. *Indian Journal of Pharmaceutical Sciences* **68** (2), 133–140.

Satoh, T., Kakimoto, S., Kano, H., Nakatani, M., Shinkai, S., and Nagasaki, T. (2007). In vitro gene delivery to HepG2 cells using galactosylated 6-amino-6-deoxychitosan as a DNA carrier. *Carbohydrate Research* **342** (11), 1427–1433.

Saylor, D. M., Kim, C., Patwardhan, D. V., and Warren, J. A. (2007). Diffuse-interface theory for structure formation and release behavior in controlled drug release systems. *Acta Biomaterialia* **3** (6), 851–864.

Scherlund, M., Brodin, A., and Malmsten, M. (2000). Nonionic cellulose ethers as potential drug delivery systems for periodontal anesthesia. *Journal of Colloid Interface Science* **229**, 365–374.

Seki, T., Kanbayashi, H., Chono, S., Tabata, Y., and Morimoto, K. (2007). Effects of a sperminated gelatin on the nasal absorption of insulin. *International Journal of Pharmaceutics* **338** (1-2), 213–218.

Sha, J. and Kaiming, Y. (2007). Nanoparticle-Mediated Drug Delivery and Gene Therapy. *Biotechnology Progress* **23** (1), 32–41.

Shang, J., Chen, X., and Shao, Z. Z. (2007). The electric-field-sensitive hydrogels. *Hua Xue Jin Zhan* **19** (9), 1393–1399.

Shi, X. W., Du, Y. M., Sun, L. P., Zhang, B. Z., and Dou, A. (2006). Polyelectrolyte complex beads composed of water-soluble chitosan/alginate: Characterization and their protein release behavior. *Journal of Applied Polymer Science* **100** (6), 4614–4622.

Silva, C., Silva, C. J., Zille, A., Guebitz, G. M., and Paulo, A. C. (2007). Laccase immobilization on enzymatically functionalized polyamide 6,6 fibres. *Enzyme and Microbial Technology* **41** (6-7), 867–875.

Siti-Ismail, N., Bishop, A. E., Polak, J. M., and Mantalaris, A. (2008). The benefit of human embryonic stem cell encapsulation for prolonged feeder-free maintenance. *Biomaterials* **29** (29), 3946–3952.

Sivadas, N., O'Rourke, D., Tobin, A., Buckley, V., Ramtoola, Z., Kelly, J. G., Hickey, A. J., and Cryan, S. A. (2008). A comparative study of a range of polymeric microspheres as potential carriers for the inhalation of proteins. *International Journal of Pharmaceutics* **358** (1-2), 159–167.

Sloan, K. B. (1989). Prodrugs for dermal delivery. *Advanced Drug Delivery Reviews* **3** (1), 67–101.

Smith, T. J. (1994). Calcium alginate hydrogel as a matrix for enteric delivery of nucleic acids. *Biopharm* **4**, 54–55.

Sosa, M. A. G., Fazely, F., Koch, J. A., Vercellotti, S. V., and Ruprecht, R. M. (1991). N-Carboxymethylchitosan-N, O-sulfate as an anti-HIV-1 agent. *Biochemical and Biophysical Research Communications* **174** (2), 489–496.

Souto-Maior, J. P. A., Reis, A. V., Pedreiro, L. N., and Cavalcanti, O. A. (2008). Phosphated pectin application in the development of films for drug delivery systems. Evaluation of permeability properties and swelling. *Revista Brasileira De Ciências Farmacêuticas* **44** (2), 203–213.

Sprockel, O. L. and Price, J. C. (1989). Evaluation of sustained release aqueous suspensions containing microencapsulated drug-resin complexes. *Drug Development and Industrial Pharmacy* **15**, 1275–1287.

Sriamornsak, P., Nunthanid, J., Luangtana-anan, M., Weerapol, Y., and Puttipipatkhachorn, S. (2008). Alginate-based pellets prepared by extrusion/spheronization: Effect of the amount and type of sodium alginate and calcium salts. *European Journal of Pharmaceutics and Biopharmaceutics* **69** (1), 274–284.

Sripriya, R., Kumar, M. S., Ahmed, M. R., and Sehgal, P. K. (2007). Collagen bilayer dressing with ciprofloxacin, an effective system for infected wound healing. *Journal of Biomaterials Science-Polymer Edition* **18** (3), 335–358.

Strauss, G. and Gibson, S. A. (2004). Plant phenolics as cross-linkers of gelatin gels and gelatin-based coacervates for use as food ingredients. *Food Hydrocolloids* **18** (1), 81–89.

Subrahmanyam, C.V.S. (2006). Diffusion. Textbook of Physical Pharmaceutics, New Delhi.

Sun, G. M., Zhang, X. Z., and Chu, C. C. (2007). Formulation and characterization of chitosan-based hydrogel films having both temperature and pH sensitivity. *Journal of Materials Science-Materials in Medicine* **18** (8), 1563–1577.

Sung, H. W., Huang, D. M., Chang, W. H., Huang, L. L. H., Tsai, C. C., and Liang, I. L. (1999). Gelatin-derived bioadhesives for closing skin wounds: An in vivo study. *Journal of Biomaterials Science-Polymer Edition* **10** (7), 751–771.

Sutter, M., Siepmann, J., Hennink, W. E., and Jiskoot, W. (2007). Recombinant gelatin hydrogels for the sustained release of proteins. *Journal of Controlled Release* **119** (3), 301–312.

Suyama, K. and Tsunooka, M. (1998). Effects of Cross-linking by Quinones on Dyeing of Irradiated Films of Copolymers Bearing Photobase Generating Groups. *Journal of Applied Polymer Science* **68**, 1177–1184.

Takayanagi, T. and Motomizu, S. (2006). Chitosan as Cationic Polyelectrolyte for the Modification of Electroosmotic Flow and Its Utilization for the Separation of Inorganic Anions by Capillary Zone Electrophoresis. *Analytical Sciences* **22** (9), 1241–1244.

Takezawa, T., Takeuchi, T., Nitani, A., Takayama, Y., Kinooka, M., Taya, M., and Enosawa, S. (2007). Collagen vitrigel membrane useful for paracrine assays in vitro and drug delivery systems in vivo. *Journal of Biotechnology* **131** (1), 76–83.

Takka, S. and Acartürk, F. (1999). Calcium alginate microparticles for oral administration: III. The effect of cross-link agents and various additive polymers on drug release and drug entrapment efficiency. *Pharmazie* **54** (2), 137–139.

Taleb, M. F. A. (2008). Radiation synthesis of polyampholytic and reversible pH-Responsive hydrogel and its application as drug delivery system. *Polymer Bulletin* **61** (3), 341–351.

Taravella, M. J., Balentine, J., Young, D. A., and Stepp, P. (1999). Collagen shield delivery of ofloxacin to the human eye. *Journal Of Cataract And Refractive Surgery* **25** (4), 562–565.

Tate, M. C., Shear, D. A., Hoffman, S. W., Stein, D. G., and LaPlaca, M. C. (2001). Biocompatibility of methylcellulose-based constructs designed for intracerebral gelation following experimental traumatic brain injury. *Biomaterials* **22**, 1113–1123.

Tebbe, D., Thull, R., and Gbureck, U. (2007). Correlation between heparin release and polymerization degree of organically modified silica xerogels from 3-methacryloxypropylpolysilsesquioxane. *Acta Biomaterialia* **3** (6), 829–837.

Tenjarla, S. N., Kasina, R., Puranajoti, P., Omar, M. S., and Harris, W. T. (1999). Synthesis and evaluation of N-acetylprolinate esters — novel skin penetration enhancers. *International Journal of Pharmaceutics* **192** (2), 147–158.

Theeuwes, F. (1975). Elementary osmotic pump. *Journal of Pharmaceutical Sciences* **64**, 1987–1991.

Thu, B., Bruheim, P., Espevik, T., Smidsrod, O., Shiong, P. S., and Braek, G. S. (1996). Alginate polycation microcapsules I. Interaction between alginate and polycation. *Biomaterials* **17**, 1031–1040.

Tomida, H., Nakamura, C., Yoshitomi, H., and Kiryu, S. (1993). Preparation of Theophylline-Loaded Calcium Alginate Gel Capsules and Evaluation of Their Drug-Release Characteristics. *Chemical & Pharmaceutical Bulletin* **41**, 2161–2165.

Tomihata, K. and Ikada, Y. (1997). Cross-linking of Hyaluronic Acid with Glutaraldehyde. *Journal of Polymer Science: Part A: Polymer Chemistry* **35**, 3553–3559.

Tomita, S., Sato, K., and Anzai, J. I. (2008). pH-sensitive thin films composed of poly(methacrylic acid) and carboxyl-terminated dendrimer. *Sensor Letters* **6** (1), 250–252.

Tongwen, X. and Binglin, H. (2000). A mechanism on the drug release into a perfect sink from a coated planar matrix with a super-saturation loading in the core. *International Journal of Pharmaceutics* **197** (1-2), 23–24.

Tonnesen, H. H. and Karlsen J. (2002). Alginate in Drug Delivery Systems. Drug Development and Industrial Pharmacy, http://0-www.informaworld.com.innopac.lib.ryerson.ca/smpp/title~content=t713597245~db=all~tab=issueslist~branches=28-v2828 (6), 621–630.

Trimukhe, K. D. and Varma, A. J. (2008). Complexation of heavy metals by cross-linked chitin and its deacetylated derivatives. *Carbohydrate Polymers* **71** (1), 66–73.

Tseng, C. L. and Lin, F. H. (2008). Preparation of gelatin nanoparticles with EGFR selection ability via biotinylated-EGF conjugation for lung cancer targeting. *Biomedical Engineering-Applications Basis Communications* **20** (3), 161–169.

Tseng, C. L., Wang, T. W., Dong, C. C., Wu, S. Y. H., Young, T. H., Shieh, M. J., Lou, P. J., and Lin, F. H. (2007). Development of gelatin nanoparticles with biotinylated EGF conjugation for lung cancer targeting. *Biomaterials* **28** (27), 3996–4005.

Verestiuc, L., Nastasescu, O., Barbu, E., Sarvaiya, I., Green, K. L., and Tsibouklis, J. (2006). Functionalized chitosan/NIPAM (HEMA) hybrid polymer networks as inserts for ocular drug delivery: Synthesis, in vitro assessment, and in vivo evaluation. *Journal Of Biomedical Materials Research Part A* **77A** (4), 726–735.

Vertommen, M. A. M. E., Cornelissen, H. J. L., Dietz, C. H. J. T., Hoogenboom, R., Kemmere, M. F., and Keurentjes, J. T. F. (2008). Pore-covered thermoresponsive membranes for repeated on-demand drug release. *Journal Of Membrane Science* **322** (1), 243–248.

Vinas, C. R., Breen, A., Damodaran, G., Ritter, T., O'Brien, T., and Pandit, A. (2007). A cross-linked collagen scaffold as a vehicle for gene delivery. *Tissue Engineering* **13** (7), 1635–1636.

Waite, J. H. (1990). Marine Adhesive Proteins – Natural Composite Thermosets. *International Journal of Biological Macromolecules* **12** (2), 139–144.

Waite, J. H. (1990). The phylogeny and phemical diversity of quinone-tanned glues and varnishes. *Comp. Biochem. Physiol* **97B** (I), 19–29.

Walt, D. R. and Agayn, V. I. (1994). The chemistry of enzyme and protein immobilization with glutaraldehyde. *Trends In Analytical Chemistry* **13** (70), 425–430.

Wang, B., Xu, X. D., Wang, Z. C., Cheng, S. X., Zhang, X. Z., and Zhu, R. X. (2008). Synthesis and properties of pH and temperature sensitive P(NIPAAm-co-DMAEMA) hydrogels. *Colloids And Surfaces* **64** (1), 34–41.

Wang, J., Tabata, Y., and Morimoto, K. (2006). Aminated gelatin microspheres as a nasal delivery system for peptide drugs: Evaluation of in vitro release and in vivo insulin absorption in rats. *Journal of Controlled Release* **113** (1), 31–37.

Whateley, T. L. (1992). Microencapsulation of drugs. Harwood Academic Publishers, Amsterdam.

Wu, D. Q., Wang, T., Lu, B., Xu, X. D., Cheng, S. X., Jiang, X. J., Zhang, X. Z., and Zhuo, R. X. (2008). Fabrication of supramolecular hydrogels for drug delivery and stem cell encapsulation. *Langmuir* **24** (18), 10306–10312.

Wu, L. and Brazel, C. S. (2008). Modifying the release of proxyphylline from PVA hydrogels using surface cross-linking. *International Journal of Pharmaceutics* **349** (1-2), 144–151.

Wu, L. Q., McDermott, M. K., Zhu, C., Ghodssi, R., and Payne, G. F. (2006). Mimicking Biological Phenol Reaction Cascades to Confer Mechanical Function. *Advanced Functional Materials* **16**, 1967–1974.

Yamamoto, A. (2007). Study on the colon specific delivery of prednisolone using chitosan capsules. *Yakugaku Zasshi-Journal of the Pharmaceutical Society of Japan* **127** (4), 621–630.

Young, R. J. and Lovell, P. A. (1991). An Introduction to Polymers. Chapman and Hall, London.

Young, S., Wong, M., Tabata, Y., and Mikos, A. G. (2005). Gelatin as a delivery vehicle for the controlled release of bioactive molecules. *Journal of Controlled Release* **109** (1-3), 256–274.

Yu, H. Q., Lu, J., and Xiao, C. B. (2007). Preparation and properties of novel hydrogels from oxidized konjac glucomannan cross-linked chitosan for in vitro drug delivery. *Macromolecular Bioscience* **7** (9-10), 1100–1111.

Yu, S. H., Wu, Y. B., Mi, F. L., and Shyu, S. S. (2008). Polysaccharide-based artificial extracellular matrix: Preparation and characterization of three-dimensional macroporous chitosan, and heparin composite scaffold. *Journal Of Applied Polymer Science* **109** (6), 3639–3644.

Yuan, Y., Chesnutt, B. M., Utturkar, G., Haggard, W. O., Yang, Y., Ong, J. L., and Bumgardner, J. D. (2007). The effect of cross-linking of chitosan microspheres with genipin on protein release. *Carbohydrate Polymers* **68** (3), 561–567.

Zhang, C., Ping, Q., Zhang, H., and Shen, J. (2003). Preparation of N-alkyl-O-sulfate chitosan derivatives and micellar solubilization of taxol. *Carbohydrate Polymers* **54** (2), 137–141.

Zhang, H., Gu, C. H., Wu, H., Fan, L., Li, F., Yang, F., and Yang, Q. (2007). Immobilization of derivatized dextran nanoparticles on konjac glucomannan/chitosan film as a novel wound dressing. *BioFactors* **30** (4), 227–240.

Zhou, Y. S., Yang, D. Z., Ma, G. P., Tan, H. L., Jin, Y., and Nie, J. (2008). A pH-sensitive water-soluble N-carboxyethyl chitosan/poly(hydroxyethyl methacrylate) hydrogel as a potential drug sustained release matrix prepared by photopolymerization technique. *Polymers For Advanced Technologies* **19** (8), 1133–1141.

Zhu, Y. and Chan-Park, M. B. (2007). Density quantification of collagen grafted on biodegradable polyester: Its application to esophageal smooth muscle cell. *Analytical Biochemistry* **363** (1), 119–127.

Zikakis, J.P. (1984). Chitin, Chitosan and Related Enzymes, Academic Press, Orlando.

CHAPTER

17

Amyloid Fibrils – Self-Assembling Proteins

Victoria A. Hughes and Dave E. Dunstan

Department of Chemical and Biomolecular Engineering, University of Melbourne, Vic., 3010, Australia

17.1 INTRODUCTION TO PROTEIN MISFOLDING AND FIBRIL FORMATION

The design of new materials through self assembly of biomolecules is at the center of various fields in science, technology and industry. Conventionally formed from covalent molecules including co-polymers and dendrimers, the materials can generate a wide range of structures and properties.

It is important to distinguish between aggregation and self assembly – aggregation is a self-association reaction generally resulting in the formation of amorphous, aggregates that are heterogeneous in structure. Self assembly involves spontaneous organization of molecules under thermodynamic equilibrium, towards a state of free energy via non-covalent interactions such as hydrogen bonding, electrostatic, hydrophobic and van der Waals (Pauling, 1960; Graveland-Bikker and de Kruif, 2006). The ability of the material to self assemble and the ability to tailor-make the final product with several desirable properties makes amyloid fibrils a very interesting class of materials. It is becoming increasingly difficult to use traditional methods for the design of more complex combinations of properties and structures and researchers have turned their attention to nature's 'bottom up' approach starting from the molecular level.

Molecular self assembly is ubiquitous in biological systems and underlies the formation of various complex biological systems (Whitesides et al., 1991). Proteins, peptides and DNA have been used to fabricate molecular biomaterials. One important example of protein self assembly is amyloid fibril formation (Dobson, 1999, 2004a, 2004b). Amyloid fibrils, most commonly known for their connection to neurodegenerative disease, are fibrils rich in β-sheet with a highly ordered cross-β core structure (Serpell, 2000; Serpell and Smith, 2000) conferring strength and stability. Generally accepted as being a generic property of proteins and polypeptides, amyloid formation can be controlled by careful choice of chemical and physical environment of the protein or polypeptide.

Kasapis, Norton, and Ubbink: Modern Biopolymer Science
ISBN: 978-0-12-374195-0

17.2 AMYLOID FORMATION, NATURE AND DISEASE

Protein aggregation is of critical importance in a wide variety of biomedical situations ranging from conformational disorders associated with neurodegenerative disease, such as Alzheimer disease, to the production, stability and delivery of protein drugs and food.

Amyloid was first recognized in 1854 as a pathological starch-like proteinaceous substance deposited between cells in various tissue and organs of the body in a wide range of clinical situations (Virchow, 1854a, 1854b). The β-sheet structure of the amyloid fibrils was first suggested in the 1980s by Glenner, principally due to improvements in electron microscopy and biochemical techniques (Glenner and Wong, 1984). Although each discipline has its own specific angle of interest it is the synergy between theoretical studies, biophysical experiments, cell biology, medicine and physical chemistry that is now beginning to provide common principles that rationalize the sequence from the native state of the protein to the misfolded amyloid state to the formation of amyloid fibrils.

Proteins that are normally soluble are transformed into insoluble fibrillar structures. The process involves a conformational change and the self assembly into fibrils. Amyloid precursor proteins do not demonstrate any sequence or structural homology, yet the mature fibrils appear to share similar highly organized multimolecular morphology which in itself has implications for understanding protein folding. There are several properties exhibited by all amyloid fibrils:

1. All amyloid fibrils can be stained with the diaaxo dye Congo red and exhibit apple green birefringence when viewed under polarized light (Klunk et al., 1999)
2. Under electron microscopy, amyloid fibrils exhibit uniform fibrils about 6–10 nm in diameter. The fibrils are straight, unbranched and have a smooth surface
3. X-ray diffraction patterns are dominated by a relatively sharp and intense 4.7 Å meridional and weaker and more diffuse 10-Å equatorial reflections which indicate a particular type of β-sheet (Serpell, 2000).

Much research has centered on trying to develop an understanding of the molecular level processes that result in fibril formation, the precise details of the molecular mechanism of amyloid fibril formation is still largely unknown. Amyloid fibrils are very large in terms of molecular mass and are insoluble and cannot in general be crystallized (Serpell et al., 1999). Low-resolution models have been generated from existing techniques such as cryoelectron microscopy and solid-state NMR. A number of groups are now concentrating on growing microcrystals of amyloid-forming peptides and are suggesting more detailed models based on X-ray diffraction (Meersman and Dobson, 2006).

It is now generally accepted that amyloid-like fibrils are defined by quasi-crystalline core region of the polypeptide rich in a highly extended β-strand conformation in which the backbone amide groups form an extensive network of hydrogen bonds that run parallel to the fibril long axis and perpendicular to the polypeptide chain. The amino acid side chains will affect the structure and stability of the precursor protein but will not influence the stability of the fibril. The typical length of polypeptide involved in the formation of each strand making up the core structure is between 5 and 12 amino acids. X-ray diffraction on microcrystals suggests that water is excluded from the core (Sawaya et al., 2007); a conclusion that is supported by fiber diffraction data from hydrated and dehydrated fibrils which showed similar patterns (Squires et al., 2006). Any portion of the polypeptide chain that is not involved in the core structure is suggested to be highly flexible and solvent-exposed, chemical modification of such area should not alter the structure or properties of the fibril. Several authors have found that proteins that are not currently associated with any disease can be manipulated to form amyloid-like fibrils (Guijarro et al., 1998; Chiti et al., 1999; Damaschun et al., 1999). Especially surprising was the finding that α-helical proteins such as myoglobin and apocytochrome c can form fibrils. These observations have led Dobson and co-authors to propose that amyloid fibril formation is a generic property of proteins (Dobson, 1999a; Fandrich et al., 2001); an argument further supported by the ability of seemingly unrelated proteins to form mixed fibrils (TF, 2001). It is therefore postulated that there is a common underlying mechanism of formation of amyloid fibrils, an idea supported by the fact that the intermolecular bonds that stabilize the fibrils involve the peptide backbone, which is common in all proteins and peptides.

17.3 WHY IS THERE SUCH A GREAT INTEREST IN AMYLOID FIBRILS?

17.3.1 Disease

About 20 human diseases, systemic or localized to various organs, are associated with fibril formation. The strict definition for amyloidosis diseases stipulates that these conditions must result in extracellular deposition of amyloid fibrils thus describe Alzheimer's disease, senile systemic amyloidosis, the spongiform encephalopathy and several prion diseases (GR et al., 1988; Bessen et al., 1997; Horwich and Weissman, 1997; TF et al., 2001; WH et al., 2002; Xing and Higuchi, 2002; Robinson et al., 2004). Conditions such as Parkinson and Huntington diseases are related pathologies but are not in the strictest sense classed as amyloidoses, as the deposition is limited to intracytoplasmic and intranuclear deposits (Goedert et al., 1998; Panktatz and Foroud, 2004). Twenty-five proteins have been found in proteinaceous deposits in various conformational diseases as summarized in Table 17.1 (Xing and Higuchi, 2002; Westermark et al., 2005).

One of the main proteins studied in vitro is Aβ due to its association with Alzheimer's disease (AD). AD is a progressive neurodegenerative disorder affecting a large proportion of the aging population. The major protein component of the characteristic extracellular amyloid plaque is the β-amyloid peptide consisting of 30–42 amino acid residues. Aβ is produced as a result of sequential proteolysis of APP by secretase

TABLE 17.1 Summary of Proteins have been Found in Proteinaceous Deposits in Various Conformational Diseases (Zerovnik et al., 2002)

Disease	Protein component	
Alzheimers	Tau, Aβ42	Neurofibrillary tangles
Picks disease	Tau	Pick bodies
Progressive supranuclear palsy	Tau, heat shock proteins	Neurofibrillary tangles
Demetia with Lewy bodies	α-synuclein	Lewy bodies Dystrophic neruite
Parkinsons	α-synuclein, crstallins	Neurofilaments
Huntingtons	Polyglutamine of huntingtin	Intracellular inclusion
Spincerebeller ataxias	Expanded Glu repeats of ataxin	Intracellular inclusion
TSE	prion protein, Cathepsin B	Endosome-like organelles

during cellular metabolism. AB peptide is produced by β secretase at site N-terminal to the start of the transmembrane domain and γ-secretase within the transmembrane domain. A second pathway involving a γ-secretase results in cleavage of APP protein in the center of the Aβ peptide region and therefore precludes the formation and deposition of the Aβ peptide. APP is a single transmembrane protein expressed in various cells including neurons. Native Aβ is produced in vivo and is found in the plasma and cerebrospinal fluids, importantly it is soluble and is not deposited. It is thought to have a role as an antioxidant (Kontush, 2001; Kontush et al., 2001).

The pathology of AD is characterized by senile plaques (dystrophic neurites surrounding a central amyloid core), intracellular neurofibrillary tangles composed of phosphorylated Tau protein in the brain cortex together with loss of synapse and neuronal death (Xing and Higuchi, 2002). In 99% of AD cases these is no mutation in the PS1 or PS2 genes. Mutations in three proteins have been described leading to early onset of AD. Rare familial AD is associated with a series of mutations on the APP gene, these mutations involve the residues where enzymatic cleavage occurs. This results in an over-production of Aβ and based on these observations the Aβ cascade hypothesis was formulated. This proposed that the progressive cerebral accumulation of Aβ initiated a complex multicellular cascade that results in neurotransmitter alterations and impairment in memory and cognitive functions. The PS1 gene on chromosome 14 and ps2 gene on chromosome 1 are associated with early-onset AD. It is thought that Presnilin proteins either alter the trafficking of APP or they may be the γ-secretases responsible for cleavage of APP. In AD, The neural toxicity is thought to arise from the extraneural plaques that are created by the Aβ amyloid fibrils, yet some studies indicate soluble Aβ oligomers, also known as ADDLs, Aβ-derived diffusible ligands may also be toxic (Catalano et al., 2006). It is still matter of much debate as to whether the fibrillar aggregates are the side product of some pathology or the main cause of disease. Senile plaques identical to those seen in AD are characteristic of cognitive decline in individuals with Down syndrome (Dickinson, 1997). The chromosomal abnormality of most DS patients is a triplication of chromosome 21, the site of the APP gene (Lemere et al., 1996). Lewy bodies have also been detected in a significant number of DS patients (Lippa et al., 2000).

Senile amyloidosis is a form of amyloidosis in which the incidence and severity of amyloid deposition increases with age without any particular diseases or predisposing conditions (Cornwell et al., 1995).

Tau protein normal function is to stabilize microtubules and has numerous sites for phophorylation. Hyperphosphorylations cause the development of insoluble filaments that deposit in the cell body (Geddes, 2005).

Parkinson disease is the second most common neurodegenerative disease, several proteins are implicated in the disease including α-synuclein, synphilin and parkin (Zhang et al., 2000). α-synuclein is a small, soluble, acidic 14 kDa protein, that is naturally unfolded and becomes helical on binding to synaptic vesicles. In vivo the function is thought to regulate synaptic vesicles and neurotransmitter release. A hallmark of Parkinson disease is the presence of Lewy bodies, of which α-synuclein is the main component (Campbell et al., 2000). Both α-synuclein and synphilin are required for Lewy body formation. Protein aggregates with Lewy bodies are composed of not only α-synuclein but also Tau and Aβ, proteins related to other neurodegenerative diseases. Missense mutations in the α-synuclein gene have been identified in instances of early-onset Parkinson's and cause an increase in the propensity to aggregate (Lashuel et al., 2002). Genetic predisposition and environmental factors are known to increase the risk of Parkinson's, but the etiology is still unknown. The cytotoxicity of the α-SYN is also still unknown, but some authors suggest that the soluble oligomers may be more toxic and harmful to the cell and the formation of filaments and Lewy bodies may be cytoprotective means of sequestering large amount of unfolded material in a defined volume (Lashuel et al., 2002).

Much in vitro work has been carried out on β_2-microglobulin because if its role in dialysis-related amyloidosis (DRA) (Drueke, 2000). Deposits of β_2M amyloid are found in the carpal tunnel and are found in many patients with end-stage renal disease and undergo regular dialysis treatment (Jadoul, 1998; Goto et al., 2005; Sumi et al., 2005; Yamamoto and Gejyo, 2005; Calabrese and Miranker, 2007). Protein misfolding is a result of mutations in gene coding resulting in the incorrect amount of protein being secreted in cystic fibrosis (Dobson, 1999). A growing body of evidence suggests that β-cell failure in type 2 diabetes correlates with the formation of pancreatic islet amyloid deposits, indicating that islet amyloid may have an important role in β-cell loss in this disease. Islet amyloid polypeptide (IAPP; amylin), the major component of islet amyloid, is co-secreted with insulin from β-cells. In type 2 diabetes, this peptide aggregates to form amyloid fibrils that are toxic to β-cells (Marzban et al., 2003; Porat et al., 2004).

Prion diseases are a range of transmissible spongiform encephalopathies and are closely related to a range of progressive disorders in lower mammals (Prusiner et al., 1983). Scrapie was the first of these diseases recognized and affects sheep. Human variants include Gerstmann-Straussler-Scheinker disease, familial fatal insomnia, and Creutzfeldt-Jacob disease (Xing and Higuchi, 2002). Prion encephalopathies can be both inherited due to mis-sense mutations and the likely cause of prion disease is an infectious particle termed a prion, which is devoid of nucleic acid and seems to be composed entirely of prion protein (Bessenet al., 1997; Xing and Higuchi, 2002). There are several examples of tumors that contain deposits of amyloid (Meenaghan et al., 1972) as confirmed by conventional techniques.

17.3.2 Toxicity

Several experiments relating to the toxicity of the mature fibrils, soluble precursors and intermediate protofilaments have been conducted. Although amyloid is known to be toxic, there is considerable debate as to whether it is the amyloid itself (Lorenzo and Yankner, 1994), the oligomeric precursors or the molten globule state (Ceru and Zerovnik, 2008) that are the most toxic. Pastor et al. (2008) suggest that the amyloid prefibrillar aggregates exert their cytotoxic effect through a common cell death mechanism related to a particular quaternary structure regardless of polypeptide sequence, length and amino acid chirality. Zamotin et al. (2006) in a separate study saw no cell death until characteristic ThT fluorescence indicates β-sheet formation. It has been suggested that nearly all proteins have the ability to form amyloid fibrils under certain conditions, which in itself has implications for understanding protein folding.

17.4 AMYLOID FIBRILS IN NATURE

Amyloid fibrils are found in nature having a function seemingly unrelated to disease. Curli are proteins of 1–10 μM in length and with a width of 4–12 nm, consisting of CSGA protein found on the surface of *E. coli* (Chapman et al., 2002). The proteins are important for adhesion and development of biofilms. The proteins are known to have amyloid structure. Not just limited to *E. coli*, amyloid-like adhesions have also been described on the surface of bacteria and fungi. Chaplin is a hydrophobic protein secreted by Gram-positive *Streptomyces*, important in the formation of aerial hyphae at the air–water interface (Elliot et al., 2003). Hydrophobins secreted from fungi are also important at the air–water interface promoting absorption (Wosten and de Vocht, 2000; Mackay et al., 2001). Fibrillar structures of unknown composition have also been found in river biofilms and microbial aggregates from activated sludge waste water treatment (Liss et al., 1996). Vimentin and microtubules are examples of biological materials that must self assemble in order to function correctly. The shell-like caspid of viruses are formed from polypeptides, templated by DNA, that then fold and self assemble.

17.5 PROTEIN FOLDING AND MISFOLDING IN THE CELL

Proteins are synthesized on ribosomes from genetic information encoded in the cellular DNA. The protein folding can be co-translational where the nascent chain remains attached to the ribosome, in the cytoplasm unattached from the ribosome or in specific compartments such as the endoplasmic reticulum. These processes are believed to be universal, although the exact details will depend on the environment of the protein. The proteins may not be completely folded so residues that are buried in the native folded structure may be accessible to solvents, or due to the crowded nature of the other cell protein and/or lipid membranes leading to inappropriate interactions (Shtilerman et al., 2002; Despa et al., 2005). The use of molecular chaperones has been nature's way of preventing such behavior (Dobson, 2004). It is believed that the molecular chaperones increase the efficiency of the overall folding process by reducing the probability of aggregation, however some are known that accelerate the slow steps within the process (Treweek et al., 2003; Thorn et al., 2005; Zhou et al., 2006).

17.6 AMYLOID FORMATION AND BIOTECHNOLOGY

17.6.1 Pharmaceutical

The understanding of protein folding and misfolding is becoming of increasing importance in the pharmaceutical industry, particularly in the development of therapeutic proteins. Understanding protein conformation and production protocols on the therapeutic proteins is of vital importance. Therapeutic proteins are manufactured in a variety of ways including extraction from blood and tissue, expression in cell lines of plants and animals. The production process may involve multiple steps including exposure to other proteins, surfactant, salt solutions and shear differences associated with purification and exposure to different interfaces. Each of these situations is known to induce fibril formation in vitro. Maintaining proteins at low pH is vital during the production and purification of protein-based drugs such as insulin (Smith et al., 2007). The storage of proteins is also an important factor in determining the shelf life of these materials.

Much in vitro work has been carried out on β2-microglobulin (β2M) because if its role in dialysis-related amyloidosis (DRA). Deposits of β2M amyloid are found in the carpal tunnel in many patients with end-stage renal disease who undergo regular dialysis treatment (Kock, 1992; Goto et al., 2005; Yamamoto and Gejyo, 2005).

17.6.2 Biotechnology

The aggregation of proteins into insoluble aggregates is a major problem in the processing of proteins on both the large and small scale. As discussed earlier the propensity for proteins and polypeptides to form amyloid fibrils is governed by increased temperature, pH, addition of metal ions and flow. Some if not all are involved in bioprocessing techniques. Several food proteins, including whey protein, soy protein, and egg white proteins aggregate and form amyloid fibrils under these conditions (Akkermans et al., 2006; Bolder et al., 2006a, 2006b, 2007a, 2007b). The potential benefit of amyloid fibrils from food is discussed below.

17.7 FIBRIL FORMATION PATHWAYS

Protein aggregation occurs when the stabilizing forces of the native state protein are perturbed, resulting in a non-specific coalescence of the disordered conformation driven by interactions between solvent-exposed hydrophobic groups usually buried within the native structure. Several authors have proposed that proteins that have not been associated with any disease can form amyloid-like fibrils (Doi, 1993; Dobson, 1999, 2004a, 2004b). This is especially surprising for myoglobin and apo-cytochrome c, which are α-helical proteins. This has led Dobson and coauthors to propose that amyloid fibril formation is a generic property of all proteins and peptides (Dobson, 2004a). In 1968 Levinal proposed that protein folding was a result of specific pathways, now much evidence supports the protein-folding landscape theory (Dobson, 1999, 2004a, 2004b). This model depicts the folding process as the approach to and then the descent into a funnel-shaped free energy landscape, the native state lies at the bottom of the funnel with minimal free energy. There are multiple pathways to the protein folding and misfolding and several models have been proposed to explain the mechanism of amyloid fibril assembly.

17.7.1 Nucleation-Dependent Polymerization

Jarrett and Lansbury (1992) suggest that protein aggregation follows a nucleation–polymerization mechanism. The kinetics of this pathway are controlled by two key parameters: nucleation rate and elongation rate. The rate of amyloid formation is in general directed by the ability to overcome an energetically unfavorable nucleation step. Formation of the nucleus and elongation are dictated by concentration of protein and conformational change. Factors affecting these parameters will alter the rate of fibril formation.

The nucleus formation phase is rate-limiting and requires a series of steps, however once the nucleus is formed the association of monomer becomes thermodynamically favorable resulting in a rapid extension of amyloid fibrils. This model has been used to explain the conversion of prion proteins cellular form to the prion protein pathogenic form. The model can also explain the kinetics of Aβ and APO AII fibril formation (Xing and Higuchi, 2002).

17.7.2 Nucleation Conformational Model

A second model is the nucleation conformational conversion mechanism proposed by Serio et al. (2000). This model recognizes the fact that protein aggregation is accompanied by a conformational transition of the disordered polypeptide into a more stable tertiary structure. However, in the absence of high-resolution structures for mature fibrils or intermediates, the connection between the mechanism of amyloid formation and protein folding remains tenuous. The nucleated conformation conversion model states that oligomers lacking a fibril competent conformation accumulate and associate into a nucleus where conformational conversion takes place as the rate-determining step. Molten globules appear to be crucial intermediates in forming amyloid nucleus. When those complexes undergo a conformational change on association with nuclei, rapid assembly follows.

17.7.3 Off Pathway Model

The off pathway folding model suggests that in the initial refolding step, an amyloidogenic intermediate forms in a parallel reaction. The step is irreversible in contrast to the normal folding phase where monomer and dimer are in equilibrium. Nucleus formation follows the initial partitioning of fibril competent and non-competent conformations. Filament formation then takes place, followed by filament elongation by end to end addition of intermediate. Fibrils form by lateral end to end association of the filaments.

17.7.4 Partially Folded Intermediates as a Prerequisite to Fibril Formation

Correct protein folding is of fundamental importance for the quality of pharmaceuticals, the food industry and biochemical reactions (Uversky et al., 1992, 2001; Dobson, 2004; Uversky and Fink, 2004). The relationship between conformational and structural properties of proteins and the intermolecular interactions that give rise to aggregation has been the basis of many studies (Uversky et al., 1992, 2001; Uversky and Fink, 2004). Molten globules exhibit a compact globular shape and still retain a certain amount of secondary structure, but lack a defined tertiary structure. Several authors report data that indicate that the first critical step in the fibrillogenesis is the partial unfolding of the proteins (Uversky et al., 1992, 1997; Uversky and Ptitsyn, 1996; Uversky and Fink, 2002; Grudzielanek et al., 2005). It is difficult to trap and characterize such intermediates under physiological conditions as they are only transiently populated on the fibrillization pathway and evidence of such intermediates comes from indirect evidence. Proteins can exist as natively folded such as insulin (Grudzielanek et al., 2005), transtherythrin, β2-microgloblin (Armen and Daggett, 2005) or natively unfolded such as α-synuclein (Uversky et al., 2001, 2002; Luk et al., 2007), amylin and amyloid β-protein (Maji et al., 2005), both groups have been shown to form partially folded intermediates as a prerequisite to fibril formation. The conformational change associated with the partial denaturing of the native state does not destabilize the non-covalent interactions between groups within the protein, resulting in only partial unfolding. The high propensity for the partially unfolded state to aggregate is in part due to exposure of buried hydrophobic residues and hydrogen bond acceptors and donors, which remain buried in the native state. Any aggregates formed from the native state are thought to bind reversibly through a weakly bonded system due to the lack of hydrophobic residues and hydrogen bond acceptors and donors. Partial unfolding from the natively folded protein can be influenced by denaturants such as urea and guanidinium chloride (Uversky and Ptitsyn, 1996), increased concentration of such denaturants form highly unfolded states that have a low propensity to aggregate due to disruption of the non-covalent interactions (Chiti et al., 1999). The propensity of a protein to aggregate is also solvent dependent. In the presence of alcohols and TFE acylphophotase greater than 35% v:v the protein did not aggregate, the unfavorable aggregation may be due to an increase in intramolecular hydrogen bonding (Chiti et al., 1999).

17.8 ANALYTICAL TECHNIQUES TO STUDY AMYLOID FORMATION

The main theories on the mechanism of fibril formation include the nucleation-dependent model and the conformational-dependent theory. This section describes the analytical techniques currently available to study the main processes in the fibril formation pathway: unfolding, protofilament formation and fibril formation.

17.8.1 Fluorescence

One method for identifying amyloid fibrils is by staining with Congo red, under polarized light. Amyloid samples exhibit green birefringence when stained with Congo red. Thioflavin T is a fluorescent dye widely used to study amyloid formation. Excitation at 450 nm produces fluorescence at 482 nm. Intrinsic fluorescence uses the absorption and emission properties of tyrosine and tryptophan residues.

17.8.2 X-ray Diffraction

Wide angle X-ray scattering (WAXS) is usually performed on dried samples. Stalks may also be dried in the presence of a magnetic field to improve alignment. Small-angle X-ray scattering (SAXS) can provide information on the secondary structure over larger length scales (5–100 nm).

17.8.3 Dynamic and Static Light Scattering

Analysis of the intensity autocorrelation function of scattered light from a particle undergoing translational or rotational diffusion enables the determination of a diffusion coefficient. And hence, from the Stoke–Einstein equation the effective hydrodynamic radius is determined. Static light scattering can be used to obtain the molecular weight of peptide aggregates and also to provide indication of the particle shape, from measurements of the angular dependence of the scattered intensity. Light scattering has been used quite widely to monitor the formation of amyloid protofilaments and fibrils.

17.8.4 Circular Dichroism

The secondary protein structure changes during amyloid fibril formation can be characterized using CD, the method is based on fingerprint feature of the far UV region (190–250 nm). Secondary structure composition is from the peptide bond region, below 240 nm, which is due to a weak but broad n–π^* transition at around 220 nm and a more intense π–π^* absorption at around 190 nm. Characteristic minima are 208 and 222 nm α-helix, 216–220 nm β-sheet and 195 nm random coil. The spectra are generally analyzed using algorithms from databases based on X-ray crystal structure (Kelly et al., 2005). Tertiary structure fingerprint region occurs between 290–320 nm from aromatic amino acid residues.

17.8.5 FTIR Spectroscopy

The amide I band of the FTIR spectrum at 1620–1640 cm is associated with B sheets. In comparison to coils or helices this band is shifted to lower frequency. A further side band at 1680–1690 cm has been ascribed to hydrogen bonding. Although ab initio calculations suggest that it may be due to the vibrational coupling of residues the narrow intense band observed for some peptides at 1610–1620 cm is ascribed to an anti-parallel aggregated structure. The amide II band around 1550 cm mainly results from the N–H bending vibrations (in D_2O, the hydrogen atom of the N–H bond exchanges with a deuterium atom) and as a consequence the amide II band is shifted by approximately 100 cm to 1450 cm. Since a full quantum chemical analysis of the vibrational modes of peptides or proteins is not usually possible. The analysis of FTIR data relies on 'fingerprinting' or peak fitting methods. FTIR measurements can be extended to study linear dichroism on aligned samples with isotope labeling (polarized FTIR) and to vibrational circular dichroism.

17.8.6 NMR Spectroscopy

Solid-state NMR spectroscopy has provided much detail on the structure of amyloid fibrils, various high-resolution experiments employing magic angle spinning can be performed using C- or N-labeled peptides. Homonuclear and heteronuclear 2D and 3D spectroscopy may also be performed to give information on interatomic distance and torsion angles for isotopically labeled peptides in the dry-state. Hydrogen/deuterium exchange techniques have also been employed to probe structural change during amyloid formation. Exchange rates provide information on the structural rearrangements of sub-segments of a protein or peptide during folding, unfolding, or fibrillization. This method has also been used to provide a 3D structure for AB42 (Tycko, 2003, 2006a, 2006b; Dyson and Wright, 2005; vander Wal et al., 2007).

17.8.7 AFM and EM

These methods allow the direct imaging of peptide fibrils. Atomic force microscopy (AFM), which is a scanning probe microscopy technique, provides images at the surface. To date, the method has mainly been used to image fibrils from solutions dried onto planar solid substrates such as mica. Electron microscopy includes transmission electron microscopy (TEM) of dried films on holey grids (these are usually stained with heavy-metal-containing compounds to enhance contrast) or scanning electron microscopy (SEM) of surfaces. Cryo-techniques are often preferred as a method to avoid artefacts caused by slow drying (McAllister et al., 2005).

17.9 TECHNIQUES FOR STUDYING AMYLOID FIBRIL FORMATION

Understanding each stage of the fibril-forming process will help understanding of how to inhibit fibril formation or promote fibril formation for use in nanotechnology and bioprossing.

17.9.1 Soluble Intermediates: Protein Denaturation, Folding and Partially Folded Intermediates (Molten Globule State)

Correct protein folding is of fundamental importance for the quality of pharmaceutics, food industry and biochemical reactions (Uversky et al., 1992, 2001; Dobson, 2004a; Uversky and Fink, 2004). The relationship between conformational and structural properties of proteins and the intermolecular interactions that give rise to aggregation has been the basis of many studies. A common observation is that fibrillization starts from an intermediate state which is either partially unfolded or partially folded. In vivo partial unfolding may happen as a result of lowered protein stability due to mutation, change in local pH at membranes, oxidative or heat stress. Partial folding may occur as a result of exposure to hydrophobic chemicals such as pesticides. Other factors that cannot be ruled out are exposure to shear flow (Hill et al., 2006) or hydrophobic surfaces (Sluzky et al., 1991). Mutations that destabilize the native protein structure increase the concentration of partially folded conformations and accelerate the rate of fibril formation due to an increase in concentration of the partially folded conformation. Differences of just one amino acid in a sequence can dramatically alter the propensity of a protein to fibrilize. Turkey lysozyme has one critical Asp residue (D101) replaced by Gly compared to chicken lysozyme, reducing the propensity to aggregated, due to inefficient nicking and fragmentation which is needed for the lysozyme fibrillization (Mishra et al., 2007). Mutants of human lysozyme, I56T, F57I, W64R, lower the stability of the natively folded protein making molten globular states more accessible under physiological conditions, the familiar mutations are associated with non-neuropathic systemic amyloidosis in vivo (Mishra et al., 2007).

Surface-active globular proteins are destabilized by interactions with lipid surfaces (Sluzky et al., 1991; Sharp et al., 2002). The partial unfolding of insulin in the presence of hydrophobic surfaces, is thought to be a result of the less stable monomer rather then the dimers and hexamers, whose conformation is stabilized due to the hydrophobic regions of the structure being involved in the formation of the dimer or hexamer (Sluzky et al., 1991). The use of UV absorption spectroscopy and quasi electric light scattering shows that aggregates grown at the surfaces were larger than those found in solution without surfaces, suggesting surfaces are important in controlling the unfolding and the aggregation in protein solutions (Sharp et al., 2002). Lipid-coated surfaces have also been found to increase the kinetics of unfolding of insulin, particularly surfaces that were electrostatically repulsive, which is thought to be a result of the increase in the free energy of adsorption (Sharp et al., 2002). This is particularly interesting as in vivo proteins will be in contact with lipid interfaces such as lipid membranes and amyloid isolated from patients have a high lipid protein. DNA-induced destabilization of the prion protein leads to increases in fibrillization with DNA acting as a catalyst to the protein unfolding (Nandi et al., 2002). Specific binding of ligands can significantly reduce the amyloidogenicity of a protein by stabilization of the native protein structure (Chiti et al., 2001). The native tetramer state of TTR was found to be stabilized by anions including chloride minimizing fibril formation.

Techniques for studying the monomeric protein and protein unfolding generally utilize low concentrations of samples to maintain a molecule in the monomeric or native form. Many investigations to study the effects of unfolding on the fibril formation use TFE, a solvent known to stabilize folded proteins (Chiti et al., 1999).

To explain the molecular basis of amyloid fibril formation, it has been suggested that fibrillization can occur when the rigid native structure of a protein is destabilized, forming a partially unfolded conformation or intermediate (Dobson, 1999, 2004a, 2004b). The formation of the non-native partially unfolded conformation is required to form amyloid fibrils. Several authors suggest that the change in conformation or formation of an intermediate may result in an amyloidogenic conformation. Such is the case for β-2-microglobulin which has been shown to unfold with two distinct intermediates. The first I1, is native like and has a large amount of secondary structure while the second intermediate I2 is substantially disordered, the two different morphologies of the fibrils are related to the intermediate from which they form, which itself can be controlled by changing the pH (Armen and Daggett, 2005). Similar observations have been seen for immunoglobulin light chains (Khurana et al., 2001). The formation of the beta sheet, characteristic of amyloid fibrils, needs considerable conformational change from the rigid, tightly packed native protein, due to the constraints of the tertiary structure (Dobson, 1999, 2004a, 2004b). Such is not the case for the natively unfolded proteins such as α-synuclein, amylin and amyloid β-protein, whose secondary structure is ill-defined and fibrillization starts with partial folding and stabilization of the partially folded conformation (NK, 2004; Goers et al., 2002). Partial refolding of Aβ has been detected in the early stages of fibrillization (Hope et al., 2004). The fibrillization of α-syn has been shown to significantly increase if the partially refolded intermediate is stabilized by low pH (Uversky et al., 2001), high temperatures (Uversky et al., 2001), pesticides (Uversky et al., 2002) and metal ions (Uversky et al., 2002). For natively unfolded proteins, oxidation of methionine or nitration of tyrosine stabilize the unfolded form and considerably slow the fibrillization process (Uversky et al., 2002, 2005). It is an interesting observation that conditions favoring the aggregation-prone conformation of the protein accelerate both the nucleation phase and elongation phase, suggesting the molten globule structure is likely involved in both the formation of a nucleus or seed and formation of protofibrils (Uversky and Fink, 2004).

A consensus has emerged that the different modes of in vivo and in vitro aggregation are linked by the implication of destabilized protein molecules that, by overcoming free energy barriers, undergo non-native self assembly as a competing pathway to native functional folding (Dobson, 1999). Aggregation and folding of partially denatured intermediates are thought to underlie some common principles that minimize the free energy of the destabilized protein state. Some suggestions include saturation of hydrogen bonds, the reduction in solvents accessible surface area and the burial of hydrophobic residues (Thirmalai et al., 2003; Grudzielanek et al., 2005). It is concluded that amyloid formation in vitro can be initiated by destabilization of the native protein state, yet retention of non-covalent bonds is favorable (Chiti et al., 1999). The in vitro fibrillization of beta Lg, induced by urea, is governed by a critical balance between unfolded native protein and destabilization of intermolecular interactions.

Fluorescence is an extremely powerful technique to study protein folding, dynamics, assembly and interactions. The power of fluorescence spectroscopy lies in its broad applicability, sensitivity and ease of use (Lindgren et al., 1995, 2005). Extrinsic fluorescent dyes are particularly useful for detecting and studying the folding and unfolding of proteins, and can provide information on the molten globule intermediates. A free rotation of fluorescent moieties in solution is inversely proportional to relative molecular mass, processes that affect the movement of a fluorescently tagged object, e.g. association into larger complexes, can be effectively monitored using fluorescence polarization. Free rotation of fluorophore linked proteins will be hindered during the course of assembly as monomers associate to form larger intermediates, ultimately fibrils.

The rotational dynamics of insulin in the early stages of protein aggregation have been monitored using covalent labeling of the protein, changes in the conformation of the protein during the formation of the self-assembling nuclei were monitored using changes in the fluorescence anisotropy of the dye (Grudzielanek et al., 2005). Changes in the anisotropy of non-covalently linked fluorophores have been used to study the proteins. It is also suggested that polarization and anisotropy measurements are compatible with HTPS assays for drug discovery based on fibril inhibition assays (Luk et al., 2007). The binding of ThT to HETs showed an increase in the anisotropy as the fibril formation process proceeded, due to a reduction in Brownian tumbling as the ThT binds to the fibrils (Sabate and Saupe, 2007). Molten globule states have been shown to have a high affinity for hydrophobic fluorescent probes, including ANS (Uversky et al., 1996) and more recently Nile red has proved to be a reliable alternative for steady-state fluorimetry and microscopy. Nile red and the rotor molecule DCJV are suggested as fluorophores to detect the early stages of aggregation (Lindgren et al., 2005).

Almost all proteins have natural fluorophores Trp, Tyr and Phe, and are powerful tools that allow us to study conformational changes associated with protein folding without perturbing the system. Tyrosine and typtophan are the most commonly occurring fluorophores in proteins. The unfolding of proteins can be monitored by changes in the Typ fluorescence due to its polarity-sensitive emission. The changes manifest as both intensity and wavelength shifts. Typically a Typ buried within a protein has an emission around 335 nm whereas a fully solvent accessible type has emission at 355 nm.

The emission of Typ can also be quenched by the presence of adjacent disulfide bonds as is the case for TYP35 in immunoglobulin light chain, unfolding of the protein at pH 2 or in the presence of 8 M urea, decreasing the quenching and an increase in intensity is seen. Differences in the wavelengths of the emission, at pH 2 the emission is blue shifted (345 nm) compared to the emission of the native state, with urea the emission maxima is red shifted (355 nm). At pH 2 , the emission suggests the Trp is buried – the structure is partially folded suggesting two denatured states, one unfolded at pH 5 and one partially unfolded at pH 2, the latter corresponding to the amyloidogenic, fibril-forming/competent structure.

Monitoring the intrinsic Tyr fluorescence of α-synuclein during fibril formation shows changes in both the intensity and shape of the emission spectrum, due to a change in the pka of the phenolic hydroxyl group of Tyr from 10 in the ground state to 4–5 in the first excited singlet state. The excitation of Tyr can lead to proton transfer to nearby glutamate or aspirate residues, leading to a decrease in the Tyr fluorescence but an increase in the tyrosinate fluorescence, shown as a bathochromic shift from 305 nm to 340 nm. The decrease in tyrosinate fluorescence and corresponding increase in Tyr parallels the assembly into fibrils (Zhu et al., 2003; Dusa et al., 2006; Munishkina and Fink, 2007). The fluorescence of the Tyr phenol ring can be quenched by exposure to hydrated peptide carbonyl groups and through hydrogen bond formation with peptide carbonyls or with carboxylate side chains of aspartic acid or glutamic acid. Tyrosine is generally insensitive to its environment, Typ mutations are constructed to give more information on the local protein environment during folding/unfolding. Y39w and y125 mutants of synuclein show that tryptophans are exposed in the monomeric unfolded α-syn. During the fibril formation process there is a blue shift in emission, suggesting that Trp 39 becomes partially buried in the core of the fibril. At acidic Ph, Y39w and y125 show emission at 345 and 345.5, indicating hydrophobic collapse of α-syn, suggested to be a result of loss of negative charges in the c terminal region (Zhu et al., 2003). Although phenyl alanine has both a low extinction coefficient and low quantum yield, both absorption and fluorescence measurements can prove complicated. The presence of phenyl alanine within a protein should not be ignored, particularly if it is in close proximity to Tyr or Trp. The close proximity can influence the quantum yield of Tyr or Trp by energy transfer (Munishkina and Fink, 2007). Lifetime studies can give complementary information by use of lifetime changes or changes in the anisotropy of the different residues.

Site-directed spin labeling in conjugation with electron paramagnetic spectroscopy can be used to study protein assemblies of any molecular weight. Spin label mobility and accessibility can be used to study protein unfolding (Margittai and Langen, 2006). A protein shows characteristic high-mobility spectra after GdnHCl induced folding, the spectra has sharp, narrow-spaced lines, compared to broad lines with increased separation for natively folded proteins. If the protein is partially, completely unfolded or folded, differences in accessibility to paramagnetic colliders can give information.

Near-UV-CD can give information on the tertiary structure changes that occur during the amyloid formation process. The near-UV wavelength bands are a result of interactions between aromatic side chains in the folded conformation, some contribution to the band may arise from disulfide bridges. Any events affecting the packing around the local probes will be mirrored in the near-UV spectra. The use of far UV-CD will give information on the secondary structure of proteins.

In principle the folding pathway of a protein could be fully characterized using IR spectroscopy. However, the direct use of IR spectroscopy is limited by 'spectral congestion' of proteins and denaturants (Sagle et al., 2004); significant time resolution is also needed to detect rapidly interconverting intermediates. One area of the spectra that is free from protein and denaturant absorption is 1800–2700 cm^{-1}, this transparent window allows sufficiently strong absorptions to be observed directly. A method using C-D bonds, which absorb at 2100 cm^{-1} has been developed, the structural, spectral and temporal resolution allows the detection and characterization of equilibrium unfolding intermediates (Sagle et al., 2004).

Sedimentation assays can give much detail on the structure of the intermediates, evidence of small spherical globular aggregates, termed seeds, sedimented at 50 000s (10^{10} Da) and 30 s (10^6 Da), imaging with AFM then shows these seeds associating and acting as nuclei (Serpell, 2000; Serpell et al., 2000).

SDS PAGE can be run in the presence of both reducing (β-mercaptoethanol) and non-reducing agents to yield information on the effects of hydrolysis on fibril formation and to examine fragmentation patterns. The techniques are applicable to all stages of the fibrillization pathways. It has proved particularly useful for examining the effects of hydrolysis on hen egg white lysozyme and understanding of fibril formation from both fragments and full-length protein, with the author suggesting that a combination of both may hold the key to understanding amyloidogenesis of proteins (Mishra et al., 2007). High-throughput screening strategies will be key for identifying inhibitors of misfolding and fibrillization. Techniques for identifying the intermediates in the process such as EM, filter traps and sedimentation are not compatible with HTS. Most commonly used are optical and fluorescence methods.

17.9.2 Protein Aggregation - Protofilaments and Oligomers

The formation of different oligomers such as protofilaments and protoaggregates, represents the second stage in the aggregation process. Pathways of fibril formation, fibril morphology and the stability of protofilament intermediates are strongly influenced by experimental conditions, the elongation rates in the overall fibril formation mechanism are dependent strongly on the stability of these intermediates (Tattum et al., 2006).

Aβ-derived diffusable ligands is the term used to describe oligomers/protofibrils found in vivo, they are found an order of magnitude higher in the brain and cerebrospinal fluid of AD patients compared to age-matched controls (Teplow et al., 2006). Several recent reports suggest that the formation of the protofibrillar structures is responsible for the toxicity in the senile plaques and Lewy bodies, rather than the amyloid fibrils. It is also suggested that oligomers of misfolded proteins are a common factor in amyloidogenesis in addition to the fibrillar structure (Lindgren et al., 2005). Prefibrillar structures could also serve as a common intermediate between amyloidosis and other misfolding diseases, including Parkinson's and prion diseases (Lindgren et al., 2005).

Experimental evidence indicates that early aggregates such as the soluble oligomers and protofilaments have a critical role in promoting pathological effects in disorders (Small et al., 2001; Caughey and Lansbury, 2003; Zamotin et al., 2006). Although the cytotoxic nature of the protofilaments remains unclear, they may interact with cell membranes and trigger responses that ultimately lead to cell death, such as oxidative stress (Turnbull et al., 2003) and the increase in free Ca^{2+} levels.

Some evidence for non-linear protofilaments is known, and suggested as a second generic structure to the linear protofibrils (Mantas et al., 2003). The filament intermediates of APOCII have been shown to form ribbon-type structures. Calcitonin forms ribbon-type protofibrils that assemble and form hollow tube-type structures by twisting back upon itself so that the edges can interact (Lashuel et al., 2002).

Variability of different synthetic Aβ batchs to form the protofilament intermediates was dependent on the initial secondary structure of the peptide and/or the oligomerization state. It has been suggested that slight differences in the synthesis, purification or lypophilization may alter significantly the peptide conformation and fibrillogenic properties (Soto et al., 1995). The kinetics of fibrillization have been studied using numerous techniques including turbidity, sedimentation and the thioflavin T fluorescence binding assay. These techniques only give information on the high-molecular-weight aggregates or the disappearance of the soluble protein (Walsh et al., 1997).

Protofilaments have been described as the subunits of mature fibrils, two or more protofilaments make up a fibril, often arranged in a helical manner, with a periodic left hand twist (Gosal et al., 2002, 2004). Transerthyrin protofilaments are beta sheet helix with four sheets composed of beta strands running perpendicular to the axis, with a 15^0 twist around a central axis with one beta strand and hydrogen bonded region (Serpell et al., 2000). The study of the intermediates in the fibril formation is complicated by several effects including high concentration of proteins which cause monomer self association causing an increase in the conformational complexicity (Teplow et al., 2006). They are further complicated by a mixture of covalently associated oligomers which are metastable and detectable in the experimental time. The use of spectroscopic techniques such as CD, FTIR and NMR is limited in studying the non-covalently self-associated oligomers as is SDS-PAGE. A combination of PICUP and SDS-page has been applied to determine the oligomerization states of aβ 1–40, 42 (Teplow et al., 2006). This methods freezes the different equilibrium states using covalent stabilization.

A combination of proteolysis and mass spectrometry is a technique that has been used to study conformational changes in proteins with a high propensity to aggregate. IMS is described as an in vacuo technique where the molecules of differing sizes are separated in a helium gas matrix in a drift tube, much the same way as molecules are separated using SDS-PAGE, but without any chemical bias.

17.9.2.1 *Sedimentation Velocity Assays*

Options for the separation and structural characterization of amyloid fibrils and their solution behavior are limited by extreme size and insoluble nature of the fibrils. The oligomeric states can be distinguished by molecular weights and sedimentation co-efficient. Preparative ultracentrifuge has been used extensively as a consistent way of separating mature fibrils from the monomers and protofilaments (Mok and Howlett, 2006).

Analytical centrifugation is a technique that couples centrifugation with an optical system to determine the concentration distribution of a sample in solution as it sediments (Mok and Howlett, 2006). It can provide a robust and precise method for determining sedimentation of amyloid fibrils under solution conditions. It has several advances over methods such as SEC and SDS PAGE as the analysis does not involve interference from denaturing solution, matrices or influence from surfaces, all known to affect fibril formation (Mok and Howlett, 2006). Sedimentation has increased resolution compared to diffusion-based assays such as light scattering and size exclusion chromatography (SEC). For particles of constant friction co-efficient, sedimentation rates depend on a $^2/_3$ power of the molar mass, diffusion-based methods depend on the Stokes radius and separation is based on $^1/_3$ power of the mass (Mok and Howlett, 2006). For preparative sedimentation, in general short low speed around 14 000–16 000 g is sufficient to pellet fibrils, whereas higher speeds are needed to sediment the monomers and protofilaments.

17.9.2.2 *Quasielastic Light Scattering*

The technique of quasielastic light scattering is particularly well suited for resolving individual stages in polymerization processes and examining kinetics, and has been applied to study all phases of fibrillogenesis and determining rates of fibril formation and elongation. The use of QLS in several independent reports suggests that small oligomers or seeds formed at early stages of the Aβ fibrillization process may be micelle-like structures with a diameter of 140 Å (Serpell, 2000). The formation of Aβ(1-40) micelle-like structure is above the critical micelle concentration of 0.1 Mm, QLS determined the hydrodynamic radius was to be 7 nm, the same hydrodynamic radius was found for micelles formed from Aβ (10–39), these micelles are believed to act as the nuclei for elongation and can explain the concentration-independent kinetics at concentrations greater than C^* (Yong et al., 2002). Small-angle neutron scattering was used to elucidate details of the structure of the micelle (Yong et al., 2002).

17.9.2.3 *Imaging*

The presence of protofilaments is most commonly detected by AFM imaging or electron microscopy. The protofilaments were first described in 1997 and since then much research has centered on understanding the formation and the convergent assembly into mature fibrils. The protofibrils are generally less than 150 nm, and are flexible and narrow (Teplow, 2006; Teplow et al., 2006). Several examples of protofilaments formed in vitro show a beaded structure, giving significant support to the idea that the protofilaments are formed from monomer self association. Time-lapse AFM studies using in situ AFM of Aβ1-42 upon hydrophilic mica and hydrophobic graphite, yield information on the structures and dimensions of the intermediates but also on the effects of surfaces on such structures; 50–60 Å globular aggregates coalescing over time were seen on the hydrophilic support. Narrow elongated sheets orientate in parallel assemblies in the direction of the carbon atoms in the graphite support (height 180 Å and 10–12 Å), suggesting a single β-sheet. More commonly ex situ time lapse AFM is used, which can give dimensional information on the height, periodicity and contour length (Gosal et al., 2002, 2004).

17.9.2.4 *Spectroscopy*

Fluorescence-based assays have also proved to be useful information on the formation of protofibrils. Nile red and ANS are fluorescent dyes that have been shown to have a higher affinity for protofibrillar species than native species. The covalent attachment of a fluorescent probe to a protein and investigation of a number of fluorescence properties such as intensity, anisotropy, lifetime and spectral distribution proved to be a sensitive indicator for the early stages of fibrillization of α-syn (Thirunavukkuarasu et al., 2008). FCS-derived methods allow efficient analysis of protein aggregation in neurodegenerative diseases such as prion diseases at the molecular level (Giese et al., 2005).

17.10 DETECTION OF AMYLOID FIBRILS

Much research has centered on elucidating the structure of amyloid fibrils. The heterogeneity, non-crystalline properties and high molecular weight limit the use of traditional techniques such as crystal X-ray diffraction and multidimensional NMR. Nonetheless, several techniques have been developed to give detail in the structure and nature of the fibrils.

It is generally accepted that amyloid fibrils have a cross β-structure (Serpell, 2000). X-ray diffraction from the amyloid fibrils has led to the conclusion that there is a generic cross beta type structure. The cross beta X-ray diffraction pattern has two distinct characteristic signals, a sharp reflection at 4.7 Å along the same direction as the fiber which corresponds to the spacing along the fiber axis (Blake and Serpell, 1996) and is accepted as representing the distance between hydrogen bonded β-strands in the beta sheet. A second broader equatorial reflection 10–11 Å perpendicular to the fiber direction is associated with side chain interactions. Higher-order synchrotron X-ray diffraction revealed weaker bands and reflections including second order of the 4.7–4.8 spacing at 2.4 Å. The X-ray pattern of TTR- Met 30 variant has been analyzed in detail and is in good agreement with other ex vivo and synthetic amyloid structures, In this model the protofilaments that make up the fibrils that are composed of a number of β sheets running parallel to the axis of the protofilaments, with their component β-strands closely perpendicular to this axis (Blake and Serpell, 1996). Such a model has also been revealed using cryo-electron microscopy (Serpell and Smith, 2000). The regular orientation of the strands and sheets with respect to the fibril axis may explain the alignments of the fibrils in a magnetic field. X-ray diffraction and electron microscopy on aggregates of several amyloidogenic proteins suggest a similar cross β amyloid fibril structure, with a common cross-β structure running perpendicular to that of the fibril axis. Amyloid fibers appear to be composed of 2–6 filaments, each filament having a diameter of 50–60 Å. Alternative structures for amyloid fibrils have been suggested such as the β-helix and concentric cylinder (Kishimoto et al., 2004), and suggested that the typical X-ray reflection arising from the fibril inter-sheet spacing, which is attributed to cross-β structure, may be an artefact from fibril dehydration before the X-ray structure analysis, this is in contrast to a number of other workers whose evidence supports the cross-beta structure for fibrils including the recent work by Squires et al. who saw little difference between hydrated and dehydrated fibrils of TTR105–115 and lysozyme (Squires et al., 2006). X-ray scattering has also proved useful for helping to understanding the hydration of fibrils (Squires et al., 2006), this is particularly relevant for understanding the sample preparation for AFM. Both the 1D and 2D scattering patterns showed excellent agreement between hydrated and solution fibrils, providing strong evidence that the core structure of the fibrils is independent of the hydration level. The hydration levels were found to have little effect on the inter-sheet or inter-strand reflections. The conditions used here have important connotations for other methods currently used to characterize fibrils. The X-ray pattern for solution and dehydrated samples shows that solution conditions used in spectroscopy and other biophysical methods do not affect the fibrils structure, conversely the use of 'dry' AFM techniques should not significantly alter the fibril structure for imaging techniques.

Imaging techniques have proved extremely useful in the elucidation of fibril structure and formation mechanism. Electron microscopy can reveal details of the length and degree of branching associated with fibrils. In addition to electron microscopy, AFM has proved a very useful tool in fibril formation studies. Although the resolution of AFM is not significantly improved compared to electron microscopy, dimensional information such as height, periodicity and contour length can be easily extracted (Hatters et al., 2003). The secondary protein structure changes during amyloid fibril formation can be characterized using CD, the method is based on fingerprint feature of the far UV region (190–250 nm). Characteristic minima are 208 and 222 nm α-helix, 216–220 nm β-sheet and 195 nm random coil. The spectra are generally analyzed using algorithms from databases based on X-ray crystal structure.

Amyloid fibrils all show characteristic optical behavior. Congo red apple green birefringence in cross polarized light is a classic optical technique for the determination of amyloid fibrils (Glenner and Wong, 1984). The binding of Congo red is considered specific to the amyloid fibrils but has recently been questioned due to in vitro investigations that show binding to partially folded conformations. One of the most commonly used dyes is thioflavin T. The thioflavin T molecule consists of a pair of benzothiazole and diaminobenzene rings connected with a C–C bond. ThT undergoes a strong increase in fluorescence at around 480 nm when excited at 450 nm and has become characteristic of the binding to amyloid fibrils. Widespread use of ThT is not accompanied by widespread understanding of the binding mechanism. Several models for ThT binding are known in the literature, several groups have proposed that a torsional non-radiative relaxation may limit the fluorescence emission. When bound to the fibrils rotation around the C–C bond would be inhibited, leading to an increase in fluorescence quantum yield. ThT emission has also been suggested as arising from excimer formation in a number of systems including DNA, amyloid fibrils and cyclodextrans. Further debated is the nature of binding of the ThT to the fibril structure. Khurana et al. propose that ThT binds to fibrils via the formation of micelles, from an ethanolic ThT solution (Khurana et al., 2005). Groenning et al. (2007) suggest that excimer formation within cavities of the fibrils (and cyclodextrans) results in increase in fluorescence, a view mildly in conjunction with Krebs et al. who suggest that ThT may bind along the long fibril axis in the channels formed by side chain residues of amino acids in the β-sheet suggested from a study of insulin spherulites at pH 2 with an unknown HCl concentration (Krebs et al., 2005). Yet, Groenning et al., in a separate study, suggest that the binding of ThT may be between protofilaments as they associate into fibrils (Groenning et al., 2007). Some studies with κ-casein are emerging that suggest ThT can bind with prefibrillar species (Thorn et al., 2005, 2008). These studies neither agree on the binding sites of ThT or the nature of the molecular binding. Table 17.2 summarizes the structure and function of fluorescent dyes commonly used to study amyloid fibril formation.

Fluorescent analogs of both Congo red and thioflavin T have been synthesized in an attempt to increase specific binding to fibrils, quantum yield and blood–brain barrier permeability, for in vivo detection of fibrils (Klunk et al., 2001). One group of dyes, closely related to cyanine dyes, is of particular interest for in vivo imaging due to the red shifted absorption and emission properties of the dyes. This area of the spectrum is of particular interest due to the low absorption and emission of endogenous chromophores in biological tissue (Na et al., 2000), currently such dyes have only been demonstrated as in vitro dyes for fibril quantification. Luminescent conjugated polymers are also an exciting alternative for the 'decoration' of amyloid fibrils in vitro and in ex vivo samples. They are particularly interesting due to the ability to fine due to the emission properties which are directly related to the length of the Π-conjugated polymers.

Nile red microscopy-detected calcitonin fibrils at early time points which were not detected using UV spectroscopy and steady-state fluorescence demule. Although Nile red is traditionally used, exposure of hydrophobic sites and study membrane interactions has been used to study surface properties of tubulin dimers and to differentiate between post-translational methylation of an enzyme. Nile red is currently used due to the stability at low pH commonly used in in vitro fibril formation studies and as discussed above the absorption and emission are red shifted relative to ThT, reducing interferences from scatter.

Solid state NMR has proved to be a very useful technique for studying the structure of fibrils, although the samples are non-crystalline, high-quality NMR spectra are obtained due to the well-defined molecular structure and a high signal-to-noise ratio is easily achieved due to the ability to get high concentrations of sample by lipophilization or centrifugation (Tycko, 2003, 2006b). Amyloid samples can also be prepared with selective or uniform isotropic labeling in large quantities. 13C NMR chemical shifts, prepared with either uniform or site-specific 13C, is used for determination of β-strand location in ordered or non-ordered segments. NMR has also proved useful in describing details of parallel and anti-parallel β sheets and the arrangement of such sheets within the fibrils. The exact methods are beyond the scope of this review and the reader is directed to Tycko (2003, 2006a, 2006b, 2006c) and references therein for more detailed information on the subject. The exchange of protons in a protein, particularly the amide proteins, depends on solvent accessibility, hydrogen bonding and spatial arrangement of amino acids.

The solvent accessibility of different morphologies along the fibril formation pathway has also been studied using H/D exchange couple with analysis by mass spectrometry (Yamaguchi et al.,

TABLE 17.2 Structure and Function of Dyes Commonly used to Study Amyloid Fibril Formation. (Hawe et al., 2000)

Dye	Solubility	Application	λ Ex/nm	λ Em/nm
ANS 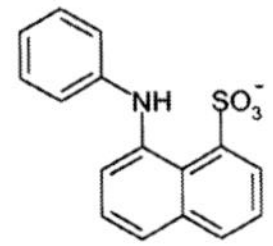	Aqueous Ethanol	Surface hydrophobicity unfolding/folding aggregation conformation	350–380	505
Bis-ANS 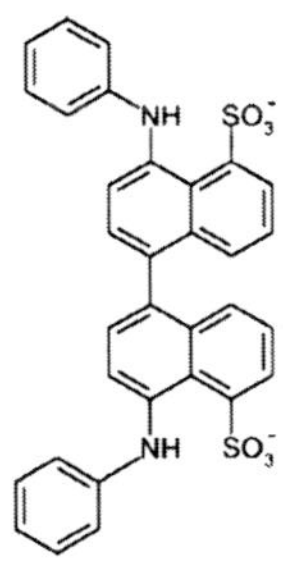	Aqueous Ethanol	Surface hydrophobicity unfolding/folding aggregation conformation	385–400	515
Nile Red 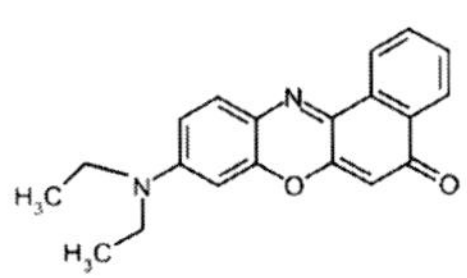	Ethanol DMSO	Surface hydrophobicity unfolding/folding aggregation conformation	540–580	660

(Continued)

TABLE 17.2 Structure and Function of Dyes Commonly used to Study Amyloid Fibril Formation. (Hawe et al., 2000)—*Cont'd*

Dye	Solubility	Application	λ Ex/nm	λ Em/nm
Thioflavin T 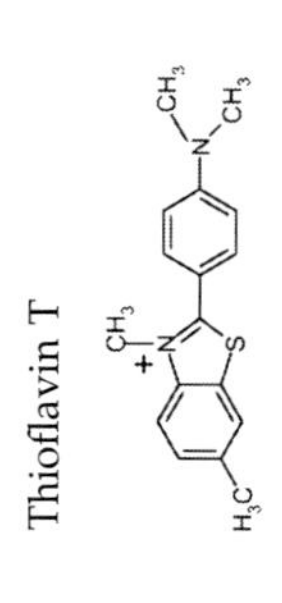	Aqueous	Fibrillation	450	480–490
DCVJ 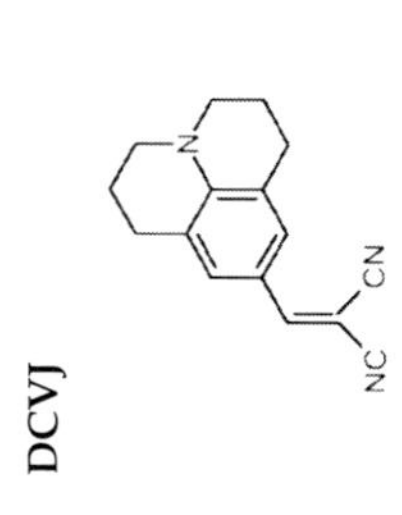	Ethanol DMSO	Micro-viscosity of protein environment rigidity	450	480–505

2004). It has proved a useful tool for insight into structural rearrangement of the HET's prion protein, regarding the solvent accessibility of the c terminal domain of the aggregating protein. Although the method seems sensitive and yields important information, the instrumentation is complex and would pose a difficult technique to use routinely.

The formation and changes to the beta sheet associated with the formation of fibrils can be monitored by an increase in the Amide 1 band at 1615–1640 nm. The fibrillization of BSA was monitored as a change in the intensity of the 1620 nm peak as function of time shows the characteristic sigmoid curve. Broadening of the same peak is assumed to suggest an increase in random coil. An investigation into the fibrillization of acylphosphatase showed the presence of two amide bands at 1685 and 1613 cm^{-1}(the amide 1 region) at early time points and suggests that the beta sheet is formed by hydrogen bonding between aggregates (Chiti et al., 1999).

The formation and changes to the beta sheet associated with the formation of fibrils can be monitored by changes in Raman spectra. The development of a new technique, drop coated deposition Raman coupled with hydrogen deuterium exchange can be used to obtain high-quality Raman spectra, the fibrillization of insulin was confirmed using these methods, by the presence of the spectral signatures and the presence of a previously unreported band at 3300 cm^{-1}.

Combinations of steady-state birefringence and centrifugal filtration have been used to develop quantitative methods for determining the fibril concentration in WPI, whey protein isolate. Samples were assumed to be aligned in the 5 s^{-1} steady-state shear rate, and no damage to fibrils was seen, the steady-state birefringence is related to the length and concentration of fibrils. The results also showed good correlation with ThT binding assay in which the concentration of bound ThT is proportional to the total fibril length, in a solution of excess ThT, the change in fluorescence intensity is commonly used as a relative measure of the total length concentration (Bolder et al., 2006).

17.11 ALTERNATIVE MODELS TO THE CROSS *β* STRUCTURE

17.11.1 *β*-helical

These models are based on a *β*-helical fold of globular proteins with a triangular rather than circular projection along the fibril axis (Jenkins and Pickersgill, 2001). Studies based on hydrogen deuterium exchange and proline scans have proposed the *β*-helix for full-length A*β* due to the exchange protection along substantial lengths of the peptide back bone as a result of hydrogen bonding.

17.11.2 Nanotubes

Several peptides have been shown to form amyloid fibrils which resemble nanotubes (Perutz et al., 2002; Lu et al., 2003; Ipsen and Otte, 2007). Successive turns in poly-l-glutamate are linked by hydrogen bonds between both the main chain and side chain amides, and side chains point alternately into and out of the cylinder. Self-assembling nanotubes with positively charged surfaces of very different inner and outer curvature could offer a unique and easily accessible scaffold for nanotechnology.

17.11.3 Non-linear Structure: Spherulites and Capsids

Spherulite formation, determined by a distinct Maltese cross pattern when viewed by cross polarizers is commonly observed during polymer and metal crystallization. Both bovine insulin and *β*-Lg have been shown to form spherulites (Krebs et al., 2005; Bolder et al., 2006; Rogers et al., 2006; Domike and Donald, 2007). The virus core shell has also been determined to be amyloid in nature, thought to arise from an equilibrium between a complete caspid and monomers and is highly dependent on temperature and salt concentration (Kegel and van der Schoot, 2004, 2006; Zandi et al., 2006).

17.12 FIBRIL FORMATION KINETICS

Aggregation depends on two sets of determinants that are described on the basis of kinetic and thermodynamic aggregation measurements (Figure 17.1). The physio-chemical environment including pH, temperature, ionic strength and presence and concentration of solvents and analytes affects the propensity of proteins and peptides to aggregate. It is suggested that all proteins and peptides can form fibrils under certain conditions in vitro. The fibrils show ultrastructure and dye-binding characteristics 'identical' to fibrils formed from disease-causing proteins. In most cases the fibrils were obtained from solution conditions where the native conformation was destabilized by pH, temperature, ionic strength or the addition of alcohols, salts or metal ions. This has led to the idea that the ability of fibrils to form in vitro is dependent on the conformational stability of the protein, defined as the difference in free energy between the folded and unfolded protein conformations (Dobson, 1999, 2004). It is also worth considering the intrinsic properties of the protein or polypeptide such as content and distribution of the residues with high or low hydrophobicity, charge, polarity, aromaticity and the propensity to form *β*-sheet.

17.13 CONDITIONS THAT PROMOTE FIBRIL FORMATION

The stability of a protein is determined by the free energy difference between the unfolded and folded state. The kinetics and thermodynamics of the fibril formation process are complex. To

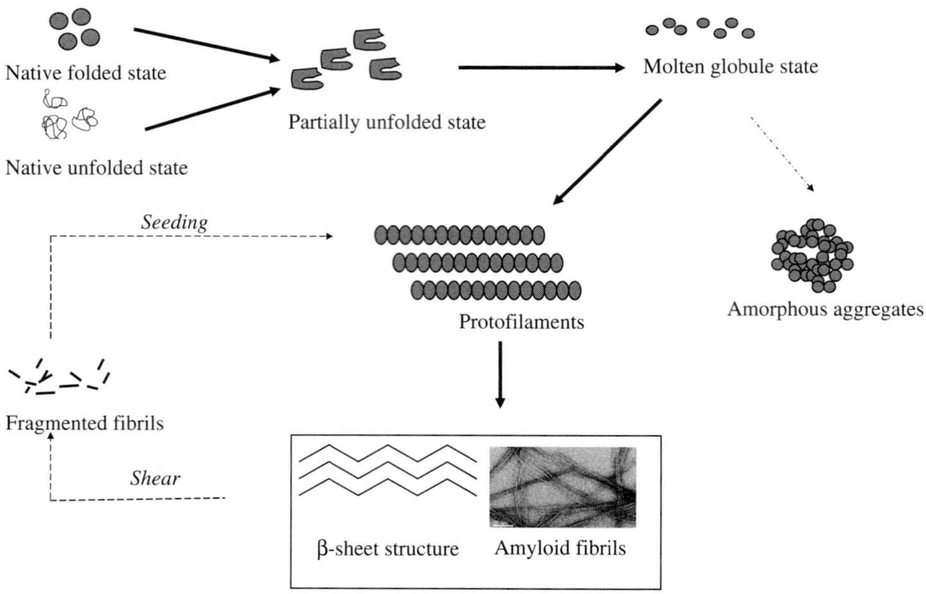

FIGURE 17.1 Suggested pathway to amyloid fibril formation with the solid black lines indicating processes that can be controlled using shear, temperature, pH. Ionic strength and chemical methods.

induce fibril formation in vitro strategies to destabilize the native state of the protein are involved, reducing the free energy of unfolding, favoring the population of partially unfolded states or increasing the flexibility. As discussed earlier, protein aggregation first involved the partial unfolding of the peptide or a change conformational stability. Secondly the unfolded peptides must have the ability to approach each other to a suitable distance for efficient intermolecular interactions, this is related to colloidal stability.

The kinetics and energy barrier of the two processes may greatly affect the ability of protein to aggregate. A small energy difference between the natively folded or partially folded state or a low kinetic barrier will readily allow the formation of aggregation-prone conformers. A low energy barrier for colloidal stability would also be advantageous for the formation of van der Waals or chemical interactions. Charge–charge interactions on the protein surface are also important for stabilizing proteins, modulation of the ion pairing on a protein surface by site-directed mutagenesis or chemical modification can optimize the protein surface for aggregation. The change in charge groups on the surface of ovalbumin and changes to the aggregation were attributed to a change in the rate of denaturation and not the colloidal forces. Changes to the surface groups of the protein leads to a drop in the activation of energy related to the endothermic denaturation step. The ionic network on the surfaces of the proteins and the geometry of the ion pairs and the location of the surface will play a role in the unfolding. Changes to the ionic network will change the activation energy to unfolding and will therefore affect the ability to form fibrils. Changes in the solution condition will affect the electrostatic interactions of the protein.

Using Aβ as a model system and proteins at the critical concentration, inclubation at 37°C for two weeks, all the samples at all types and concentration used in the study formed fibrils. The propensity to aggregation was increased and the C_c was decreased with the strongest effect for LiCl and $MgCl_2$. The salts used in the study were found to affect both the lag phase and growth phase of freshly prepared Aβ sample, the salt-induced fibrils show more bundle of fibrils compared to their control, suggesting stronger self assembly. The results are consistent with previous finding on α-synuclein, insulin, and β2-microglobulin. The formation of fibrils from microglobulin (b2) is inhibited by higher concentration of salt and a pathway to non-fibrillar aggregates is favored, this is a result of the higher salt concentrations causing a partial refolding of the peptide.

The effect of the addition of salts has a major effect on the formation of fibrils, salt addition does not just cause a shift in the equilibrium between the native unfolded state and the single aggregated state but can also influence the growth rate and the morphology of the fibril for some but not all of the proteins studied, the structure of α-syn does not change on salt addition. Changes to solvent conditions can have dramatic effects on the stability of native proteins and the stability of intermediates. Water is a polar molecule and can interact with charges in the surface of proteins, disfavoring exposure of hydrophobic side chains. Hydrophobic residues point towards the center, exposing polar and charged groups on the surface as this is more entropically favorable. The hydrophobic side chains are known to provide efficient protection of backbone hydrogen bonds and guide the protein-folding pathways by decreasing the dielectric co-efficient of surrounding water and enhancing Coulombic interactions between charged groups. Water will also compete with the amino acids for hydrogen bond formation. The strength of electrostatic interactions is proportional to the reciprocal of dielectric constant. Controlling ionic strength and dielectric constant can change the energetics of protein stability greatly and affect fibrillization. TFE is less polar than water and increases the solubility of the hydrophobic groups leading to unfolding of the protein, the use of TFE has also been shown to induce aggregation in globular proteins.

17.13.1 Temperature

In many manufacturing processes involved in the biotechnology fields such as food technology thermal treatments are involved. Knowing the structures before and after heat treatment is important for allergenic properties. Electron microscopy has shown that ovalbumin and bovine serum albumin form a fibrous gel of linear aggregates upon heating. BSA irreversibly forms intermolecular β-sheet on heating above 70^0C (Pearce et al., 2007).

17.13.2 Chemical

Any substance that is added to the in vitro assay that favors the partial unfolding of the proteins has been shown to significantly enhance fibril formation, supporting the hypothesis that any protein can form fibrils under certain conditions (Dobson, 2004a, 2004b). Fibrils formed from acylphosphatase in a mixture of water and 2,2,2,-trifluoroethanol is due to the TFE increasing the solubility of the hydrophobic groups while decreasing the solubility of peptide groups, structures that have exposed hydrophobic groups and buried peptide groups have been favored (Schmittschmitt and Scholtz, 2003).

The concentration of GuHCl can have a dramatic effect on the fibrillization due to the degree to which a natively folded protein unfolds. The solubility of globular proteins is determined by the amino acid content, the pKa values of the ionizable residues and environmental factors such as temperature and pH and the presence of co-solvents (Schmittschmitt and Scholtz, 2003). Proteins are generally less soluble near their isoelectric point (pI) where the net charge is zero (Schmittschmitt and Scholtz, 2003). Protein fibril formation correlated with the pH dependence of the protein solubility rather than the protein conformational stability for RNAase and small acidic peptide pI 3.5. Insulin is one protein that is known to exhibit a range of aggregation behaviors depending on the pH of its surrounding. Insulin has a pI of 5.5 and fibrillization at this pH yields an amorphous gel, fibrillization in acetic acid pH < 2 yields linear fibrils with a decreased lag time compared to fibrillization using pH 2.6 (0.1% HCl). The differences in the aggregation are due to the effect of pH on the native structure. On decreasing the pH from 5.5 to 2.6 to <1 a change from hexamer, dimer to monomer is seen (van der Linden and Venema, 2007). The propensity of Aβ to aggregate increases at a pH lower than 7.4, the aggregates formed at pH 8 do not resemble those of 7.4 and cannot be easily converted to fibrils when the pH is changed to 7.4 (Wood et al., 1996).

Derivatives of cholesterol-containing aldehyde groups are generated as a result of antibody-catalyzed ozone production during the inflammation processes associated with atherosclerosis (Zhang et al., 2004), the presence of such by-products can enhance the fibrillization of Aβ dramatically. The mechanism is thought to involve covalently modified Aβ peptides, via condensation of peptide amino groups resulting in a more hydrophobic peptide which can act as potent nuclei for fibrillization (Zhang et al., 2004).

The growing consensus is that lipid involvement is important in amyloid formation (Knight and Miranker, 2004; Kawasaki et al., 2006; Takahashi et al., 2006; Griffin et al., 2008). Many amyloidogenic proteins are surface active and the interaction of protein surfaces can have a profound effect. Proteins can reversibly bind to negatively charged lipid surfaces and the interactions promote fibril formation. There is also significant evidence of proteins inserting into membranes and forming stable α-helical conformations that inhibit fibril formation. The propensity of Aβ for example to form beta sheets is increased due to electrostatic interactions. The effect of phospholipids on amyloid fibril formation by human apolipoprotein (apo) C-II using phosphatidylcholine derivatives comprising acyl chains of up to 14 carbon atoms increased the rate of apoC-II fibril formation in an acyl-chain-length- and concentration-dependent fashion, while high micellar concentrations of phospholipids completely inhibited amyloid formation. At lower concentrations of soluble phospholipid complexes, fibril formation by apoC-II was only partially inhibited (Griffin et al., 2008).

17.13.3 Mutations

It is interesting to note that about half of all sequence alterations in genetic disorders are missense mutations that change a single amino acid in the chain (Krawczak et al., 2000). Mutations can change the intrinsic properties of a protein. One such outcome can be the shift in equilibrium between natively folded and misfolded protein. Mutations to hydrophobic regions of acylphosphatase were shown to significantly alter the propensity to aggregate and adopt the β-sheet conformation. It is from such studies that the changes in the aggregation rate were correlated to changes in the hydrophobicity and β-sheet aggregation propensity. The regions are distinct from regions that control protein misfolding. Several simple dipeptides have been shown to form nanotubes in aqueous solution ascribed to the ability to form π–π stacking interactions, although substitution of F by L does not prevent amyloid formation, although it is hindered if F is replaced by A, highlighting the importance of hydrophobicity in fibril formation. Introducing aromatic residues in a separate study reduced the propensity to aggregate but was correlated to a reduction in hydrophobicity and intrinsic β-sheet rather than specific interaction between the aromatic groups (Bemporad et al., 2006).

17.13.4 Shear

The role of shear flow on amyloid formation of β-lg has been investigated using fluorescence spectroscopy and AFM. Shear-induced formation of spherical aggregates was observed in a couette cell, these aggregates could act as seeds for subsequent fibrillization by incubation. The fibrillization was significantly enhanced relative to the unseeded control solution. Preformed fibrils were degraded at into shorter fibrils during prolonged shear or at high shear rates (Hill et al., 2006). Similar breakage of β-lg fibrils was seen under flow conditions, resulting in the formation of rod-like particles (Castelletto and Hamley, 2007). Such results have significant implications for understanding the mechanism of fibrillization in vivo, physiological shear rates in vivo are significant, and range widely from 100 to 8000 s^{-1} in blood vessels and the exocellular matrix (Charm and Kurland, 1972).

17.13.5 Interfaces

Proteins are known to change their structure, activity and self assembly in the presence of various interfaces. Studying the effects of collagen and lipid layers is important as they represent in vivo surfaces. Many different surfaces interact with biological fluids in vivo including cell membranes, cornea, bone and menisci at joints. It is also important to consider the synthetic surfaces used in constructing biomaterials for medical devices and implants. Other surfaces of non-biological origin, such as those involved in treatment include dialysis membranes and invasive surgical instruments such as needles and catheters are all surfaces to which proteins interact and can be absorbed. There are several factors that need to be addressed when discussing surfaces such as surface potential, hydrophobicity, heterogeneity and topology. When considering the absorption of proteins to the surfaces, wettability and surface roughness should also be considered.

There is significant evidence to suggest that absorption or interaction of proteins with surfaces enhances rate of protein aggregation (Sharp et al., 2002; Grudzielanek et al., 2007). Membrane surface can easily increase the concentration of a precursor molecule via electrostatically driven association, resulting in a surface with a much higher concentration than in bulk solution. In addition membrane binding can influence the conformational state of the native protein (Campbell et al., 2000; Lashuel et al., 2002). Together with an increased surface concentration the effect of surfaces will have a dramatic influence on the kinetics of fibril formation. Insulin aggregation in the presence of hydrophobic poly(tetrafluoroethylene) can lead to surface-enhanced nucleation, increasing the propensity of proteins to aggregation and fibril formation. The importance in understanding such behavior is essential for the understanding of the clinical syndrome injection localized amyloidosis and problems caused during the commercial production, storage and delivery of insulin.

Nucleation of β-microglobulin fibrillization has also shown to be enhanced in the presence of different nanoparticles such as copolymers, cerium oxide particles, quantum dots and carbon nanotubes, the understanding of the processes here will have considerable implication in the design and use of the fashionable nano drug delivery and biosensors which are currently under investigation. The studies show that external surfaces can significantly influence the initiation and growth of fibrils. Nayak et al. have recently conducted an investigation into several different synthetic polymers and the effect on the aggregation of insulin (Nayak et al., 2008). Hydrophobic or rough surfaces were shown to have the greatest effect. The decrease in the lag time associated with the presence of hydrophobic and rough surfaces suggests an increased rate of nucleation which may result from proteins absorbed on the surfaces acting as a reactive center, increasing the local protein concentration, a second explanation is an absorption-induced or hydrophobic-induced conformational change leading to a faster unfolding of the native state.

The results of such studies give insight into the nucleation process itself, but also highlight the need for careful consideration of polymers used in instruments that are in contact with biological fluids to avoid inducing fibril formation. Such results also suggest that the choice of an appropriate surface has potential as a platform on which to control and enhance the formation of new protein assemblies.

The use of enzymatic hydrolysis to improve the functional properties of food has attracted immense interest and attention from academics and industry (Otte et al., 2005; Ipsen and Otte, 2007). The use of proteases to induce structural changes and cause the self assembly of highly ordered functional nanostructures for uses in food areas and also non-food areas such as nanotechnology is under investigation.

17.14 TAKING LESSONS FROM NATURE

When properly controlled, protein aggregation often results in novel materials with many uses. The promising physical properties of fibrils are exceptional in the realm of biologically derived molecules. In nature the self assembly of material does not require a large expenditure of energy, and nanoscale structures are formed through a series of optimized processes that utilize the tendency of a system to minimize its overall free energy thereby minimizing activation energies. An in-depth understanding and the ability to control the thermodynamics is needed for the production of the nanoscale structures in food and bioprocessing technologies. Thus using a system of self assemblies in nature is beneficial.

There are several reasons why amyloid fibrils are candidates for novel materials in nanotechnology and biomedicine. As discussed above the self assembly of amyloid fibrils can be controlled by a variety of chemical and physical parameters. Amyloid fibrils also have other desirable features including strength, topography of the fibrils and the potential to design and incorporate functionality. With significant evidence suggesting that the toxicity is from the soluble oligomers rather than the fibrils themselves, the potential use of amyloid fibrils in biotechnology is immense.

17.14.1 Strength and Stability

Environmental conditions have a dramatic impact on fibril formation and careful control of these properties in combination with de novo designed peptide or biogenetically engineered peptides opens up the possibility of endless smart materials. α-helix-based materials tend to be more reversible than β-strand units primarily due to the β-strand being stabilized by hydrogen bonding and hydrophobic interactions. β-sheet structure has extensive inter-peptide and/or inter-chain hydrogen bonding leading to an increased rigidity and irreversibility. AFM nanoindentation is a suitable measure of the elasticity of fibrils, and can provide quantitative measurements on the local elasticity.

Amyloid fibrils are resistant to protease activity and dehydration. AFM studies of insulin fibrils have recently been found to have a strength comparable to that of steel and a Young's modulus of 3.3 Gpa. Table 17.3 compares the Young's radius for amyloid fibrils and naturally occurring material. The stability of the fibril of interest and processing conditions and uses of such fibrils is also of importance. The stability of fibrils to dissociation into the oligomers could yield problems such as toxicity or decrease in use, inactivity or adverse effects. The methods of destabilization can also yield information on the nature of assembly and binding within the fibrils, for example mature fibrils. TTR105-115 is unaffected by high hydrostatic pressures which are known to have no effect on hydrogen bonding. Under DMSO which is a known strong competitor of hydrogen bonding the complete dissociation further concludes that one of the major forces in the fibril of TTR fragment is hydrogen bonding.

TABLE 17.3 Comparison of the Young's Radius for Natural and Synthetic Materials

	Young's Radius/ GPa
Spider silk	1–1.5 (Smith et al., 2006)
Vinementin	0.3–0.9 (Guzman et al., 2006)
Microtubule	0.8 (Graveland-Bikker et al., 2006)
Insulin fibrils	0.2–0.8 (Smith et al., 2006)
Steel	0.6–1.8 (Smith et al., 2006)
Poly(ethylene terephthalate) (PET)	2.7 (Zhao et al., 2008)

17.14.2 Topography and Surface Interactions

The ability to vary fibril surface shape and area can be used to determine interactions of the fibrils with the environment. Fibrils from differing polypeptide chains can vary in size and can have a variety of structures including straight fibrils, twisted fibrils (Hatters et al., 2003), ribbon-like fibrils (Griffin et al., 2008), laterally associating fibrils and spherulites (Krebs et al., 2005). Differing growth conditions can be used to control the self-assembly process. Hydrophobins secreted from fungi are also important at the air–water interface, promoting absorption and may offer advantages in biotechnology for applications involving the reversal of surface polarity (Wosten and de Vocht, 2000; Mackay et al., 2001).

A change in peptide chirality may be used to change the handedness of the fibril twist around the protofilament axis. The ability of amyloid fibrils to form from more than one peptide to produce mixed or doped fibrils (MacPhee and Dobson, 2000) also leads to the possibility to incorporate functionality for surface interactions with cells and non-protein materials such as fluorescent polymers and dyes. Covalently attaching extrinsic fluorophores to residues in the protein has been shown to have little affect on aggregation properties (MacPhee and Dobson, 2000).

Careful choice of amino acid sequence within the peptide can determine core or surface properties. Amino acids targeted to be on the surface can also in theory be used to introduce functionality via covalent linkages to material such as bioactive groups, enzymes or non-protein groups. As described above the interactions with surfaces can increase the propensity to aggregate and direct the fibril morphology. In short amyloid fibrils as self-assembling material meet two criteria: (1) the ability to self-assemble and (2) they exhibit desirable properties for the particular function, if not present naturally, de novo and genetically engineered peptides are a real possibility.

As a final conclusion to the usefulness and immense potential of amyloid fibrils some examples of innovations of amyloid fibrils will be described.

17.15 NANOTUBES AND NANOWIRES

Metalized nano structures based on a genetically engineered yeast protein Sup 35 have demonstrated Ohmic conductance. The introduction of cysteine residues in the chain allowed surface modification with gold colloids, coupled with photographic techniques to produce the conducting cold nanowires. Hollow fibril nanotubes can be filled with silver ions that can be reduced to silver atoms. Enzymatic digestion of the peptide outer shell will produce silver nanowires of uniform dimensions. The formation of electroactive luminescent self-assembled nanowires from co-assembly of an amyloidogenic protein and a semiconducting oligoelectrolyte has significant implications for designing and assembling conducting nanomaterials and electroactive material (Herland et al., 2005; Channon et al., 2008). Coupled with the ability to co-assemble more than one type of amyloidogenic protein within a fibril suggests endless possibilities for smart materials.

Being comparable in size to biological scaffolds peptide nanowire networks have been explored for materials for tissue engineering. A self-assembling peptide nanowire network has been shown to support sufficient axon regeneration to return functional vision to hamsters in which the optic track has been severed. Amyloid fibrils have been applied to support cells and influence cell morphology and behavior (Gras et al., 2008). Food-based proteins have also been found to form nanotubes under partial hydrolysis from proteases (Ipsen and Otte, 2007).

Fibrils can be formed from seeding of the same or differing peptides, thereby generating mixed fibrils comprising two independent species, providing further evidence that the protein aggregation is a generic property of all proteins (MacPhee and Dobson, 2000), this finding also has significant implications for understanding the mechanism of disease. It is also possible to generate a protein conjugate from two proteins or peptides with differing propensities to aggregate. Molecular wires produced in this way by mimicking nature's nanowires (Leys and Scrutton, 2004), have been investigated using a porphyrin-binding protein on an amyloid scaffold composed of two covalently fused proteins with differing propensities to aggregate (Baldwin et al., 2006).

Peptide–PEG conjugates have been shown to exhibit pH-dependent fibrillization, with the propensity to aggregate being governed by the site of Peg conjugation. Conjugation at the terminus of the peptide resulted in ordered well-distributed fibrils, whereas conjugation to the center of the peptide resulted in an intractable material insoluble in aqueous solutions. Similar effects have been seen for hydrocarbon-based conjugates and polyferrocenylsilane-modified peptide chains. For the hydrocarbon-based conjugates often called peptide amphiphiles, the key feature of the self assembly is the formation of the β-sheet in the core region of the peptide closest to the alkyl chain, preventing hydrogen bonding in the first four amino acids results in the formation of spherical micelles (Ulijn and Smith, 2008). The formation of rods by peptide amphiphile is driven by a combination of the hydrophobic interaction between the alkyl chain and the hydrogen bonding of the peptides .The percentage mass of conjugates has also been seen to affect both the propensity to aggregate and the nature of the structure formed. The effects of conjugation highlight the importance of amino acid side chains in the fibril formation process.

17.16 FIBRILLAR GELS

Several food proteins including whey protein, soy protein and egg white proteins can aggregate under certain conditions (Ipsen, 1994; Otte et al., 1999, 2005; Ipsen and Otte, 2007). The whey protein fibril is of particular interest due to its potential use in functional foods (van der Linden, 2006; van der Linden and Venema, 2007). Whey proteins are known for their functional and nutritional properties, and the gelation of whey proteins is an important functional property used in the food industry.

Currently protein aggregate gels are used as gelling agents and as encapsulation agents in the production of food. Gelling agents are very attractive as they can then be used to produce foods with high amounts of proteins due to the high protein to weight ratio, these have the potential to suppress satiety and oppose obesity, others suggest it as a way to reduce sarcopenia in the elderly. Currently amorphous gels are attracting attention for food gelling agents. The potential uses of fibrils and fibril-based gels as alternative gelling agents are beginning to emerge (Bolder et al., 2006; Foegeding, 2006; van der Linden, 2006). It has been well established that globular proteins can form gels varying in appearance from opaque to clear and have high or low water-holding capacity. Both functions are related to the protein network making up the gel.

The formation of fibril-based gels, yields transparent or translucent gels that have a high water-holding capacity. The formation of the fibril-based gels can occur at low ionic strength and pI < pH < PI, the rigidity and structure of fibrils depending on the Ph and ionic strength used (Foegeding, 2006). It is apparent from the literature that the formation of food-based fibrils relies heavily on changes in pH and ionic strength. Fine-stranded aggregates have been found at neutral and low pH without addition of salts, addition of NaCl shifts the network from fine stranded to particulate (Foegeding, 2006). Recently studies have shown that the fibrils formed from ovalbumin in conditions relevant to food processing may be amyloid like, showing the characteristic fluorescent dye responses and X-ray diffraction patterns (Pearce et al., 2007). Other proteins including the whey proteins, α-lg, β-lg and BSA, and soy proteins are also known to form fibrils under conditions of high temperature and low Ph. Seeding and stirring (Bolder et al., 2006, 2007; Hill et al., 2006) have been utilized to accelerate the fibrils' formation of whey protein fibrillization as an alternative to heating or increasing the conversion to improve efficiency.

Formation of fibril-based gels involving cold set gelation has recently emerged. The process involves heating the protein at a concentration less than the critical gelation concentration at a temperature above the protein's denaturing temperature. Cooling and the addition of mono- or polyvalent salts induce the gel formation. The cold set gels have been found to have a fine-stranded structure, increased water-holding capacity and higher gel strength. The increased water capacity of fibril-based gels offers an alternative to air and foam in low-calorie products, although gels from different sources can yield gels with differing properties such as texture. Examples are common in the food industry and more recently in biomaterials. Heat-induced gelation, aggregation and gelation occur at the same time, whereas in cold gelation the two processes happen in separate processes. This means that one can take advantage of the two-step process and control or alter the properties of the soluble aggregates before formation of the gel.

Seeding experiments are known to induce and favor fibril formation. The formation of β-lg gels from preformed fibrils results in the use of lower concentrations of the protein and would result in a more efficient method of producing the gels in industry. The 'recycling' of fibrils from the same or different processes is possible and could improve efficiency further, since recent evidence suggests that non-native protein can induce fibril formation (Bolder et al., 2006). The 'cross pollination' of fibril formation may lead to a variety of fibrils with differing properties such as strength and water capacity. This also has implications in other areas and supports the hypothesis that amyloid formation is a generic property of all proteins and peptides (MacPhee and Dobson, 2000).

The use of enzymatic hydrolysis to improve the functional properties of food has attracted immense interest and attention from academics and industry (Otte et al., 2005; Ipsen and Otte, 2007). The use of proteases to induce structural changes and cause the self assembly of highly ordered functional nanostructures for uses in food areas and also non-food areas, such as nanotechnology and pharmaceuticals. Partial hydrolysis of α-lactalbumin by *Bacillus lincheniformis* delivers building blocks that can self assemble into filaments at neutral pH (Otte et al., 1999, 2005).

The key to the utilization of the nanofibrils in the food or biotech industry is stabilization on the fibrils and a complete understanding of the toxicity. α-lg is thermally stable at temperature consistent with the pasteurization of milk, yet higher temperature causes a transition to random aggregates. Freeze drying and redistribution do not affect the structures (Ipsen and Otte, 2007). The fibrils produced are irreversible upon dilution and robustness against other treatments including changes in pH and temperature. Fibrils prepared from β-Lg at pH 2 do not dissolve or dissociate when the pH is changed to 7 (Bolder et al., 2006). Understanding how the fibril gel behaves in flow will also be an important consideration in processing. In flow, the fibrils within the gel will align with the direction of flow, this suggests that more fibrils will be needed per unit of volume to interact to the same extent as a gel at rest, this will have implications for processes such as bottling, where under flow a fibril solution does not show gelation but does once at rest. It has also been noted by several groups that there is apparent breakage of the fibrils during shear flow. How this affects the gel strength may be two-fold, a reduction in fibrils may lead to a decrease in gel strength and water capacity due to a reduction in fibrils. Conversely the fragments of fibrils are known to induce fibrils formation as preformed seeds for many food proteins, being particularly useful for scaling up the process (Veerman et al., 2005; Akkermans et al., 2006; Hill et al., 2006; Castelletto and Hamley, 2007).

17.17 FUTURE INNOVATIONS?

The production of polymer nanofibrils using electrospinning can produce nanofibrils in the size range 10–1000 nm with varying mechanical, electrical and thermal properties. The nanofibrils have a large surface to area ratio governing the use as material to produce tissue templates, protective clothing and medical prostheses (Weiss et al., 2006). The applicability of electrospraying polymers for the food and agricultural systems is low due to the limited number of food biopolymers, applicability to forming protein-based amyloid fibrils remains to be seen.

Currently the use of nanotechnology-based products in the food industry is increasing rapidly. One suggested application of nanotechnology is for the production of edible coatings (Ukai et al., 1976; Valverde et al., 2005; Zhao, 2007) for food to prolong shelf life by reducing exposure to harmful environmental effect. Could amyloid fibrils derived from food-based proteins be an alternative to the current material used? Based on recent research (MacPhee and Dobson, 2000; Baldwin et al., 2006; Channon et al., 2008; Gras et al., 2008) the possibility of functionalizing the fibril or fibril surface with other ingredients such as antioxidant, nutrients and flavors may be possible.

Electron spun fibers are currently used in several ways in the food industry including the food packaging, building elements within food and for bacterial scaffolding but the use is limited by the few number of biopolymers, the possibility of using amyloid fibrils may help these applications.

Fibrillar structures of unknown composition have also been found in river biofilms and microbial aggregates from activated sludge waste water treatment (Liss et al., 1996). The usefulness of such compositions recovered from the Floc and sludge for nanotechnology has not yet been noted.

Table 17.4 summarizes some suggested potential uses of food-based nanotubes derived from α-lactalbumin (Otte et al., 2005; Ipsen and Otte, 2007).

17.18 CONCLUSIONS

Amyloids have potential for nanotechnology applications including roles in catalysis, electronics, plastic, supporting cells and maybe even in therapies. The key to the uses is an understanding of the formation mechanism and toxicity related to the fibrils.

TABLE 17.4 Potential Future Uses of Amyloid Fibrils

Potential uses (Graveland-Bikker and de Kruif, 2006; Doi, 1993)	Structure-related function
Viscosifer	High aspect ratio and stiffness
Gelation	High stiffness at low concentration
	Transparent gel
	Reversible gel formation
	Controlled disassembly
Encapsulation	Controlled flavour release
	Targeted drug release
	Modification of internal and external surfaces
Tissue engineering	Provides scaffolds for cell cultures
Nanotemplates	Selective metal deposition

Molecular self assembly is ubiquitous in chemical and biological systems, formed by the association of several individual entities into a coherent organization without external intrusions through specific interactions such as covalent, hydrogen bonding, hydrophobic and electrostatic interactions. Polypeptides have gained much attention in the area as biosensors, molecular photodiodes and transistors. Amyloid fibrils are a well-investigated example of self-assembled polypeptides and have been shown to be very versatile material in various fields including molecular electronics, drug delivery and tissue engineering. Several simple dipeptides have been shown to form nanotubes in aqueous solution ascribed to the ability to form π–π stacking interactions, although substitution of F by L does not prevent amyloid formation, although it is hindered if F is replaced by A, highlighting the importance of hydrophobicity in fibril formation. The ability to specifically engineer polypeptide and protein sequences means the ability to design smart material with specific chemical and biological features, which is expected to result in significant biotechnological advances in electronics and tissue engineering. The design of smart peptide-based materials is emerging. A material is deemed 'smart' if it can respond to external cues and can be defined at the structural or functional level. Particular interest is in smart materials whose response is reversible. With respect to protein- and peptide-based materials, α-helix-based material tends to be more reversible than β-strand units primarily due to the β-strand being stabilized by hydrogen bonding and hydrophobic interactions. β-sheet structure has extensive inter-peptide and or inter-chain hydrogen bonding leading to an increased rigidity and irreversibility. The ability to specifically engineer polypeptide and protein sequences means the ability to design smart material with specific chemical and biological features, which is expected to result in significant biotechnological advances in electronics and tissue engineering. We conclude that amyloid-based fibrils are excellent candidates for use in biotechnology, nanotechnology and the food industry.

References

Akkermans, C., Venema, P., Rogers, S. S., van der Goot, A. J., Boom, R. M., and van der Linden, E. (2006). *Food Biophysics* **1**, 144–150.

Armen, R. S. and Daggett, V. (2005). *Biochemistry* **44**, 16098–16107.

Baldwin, A. J., Bader, R., Christodoulou, J., MacPhee, C. E., Dobson, C. M., and Barker, P. D. (2006). *J. Am. Chem. Soc.* **128**, 2162–2163.

Bemporad, F., Calloni, G., Campioni, S., Plakoutsi, G., Taddei, N., and Chiti, F. (2006). *Acc. Chem. Res.* **39**, 620–627.

Bessen, R., GJ, R., and C. B (1997). *J. Biol. Chem.* **272**, 15. 227.

Blake, C. and Serpell, L. (1996). *Structure* **4**, 989–998.

Bolder, S. G., Hendrickx, H., Sagis, L. M. C., and van der Linden, E. (2006). *Journal of Agricultural and Food Chemistry* **54**, 4229–4234.

Bolder, S. G., Hendrickx, H., Sagis, L. M. C., and van der Linden, E. (2006). *Applied Rheology* **16**, 258–264.

Bolder, S. G., Sagis, L. M. C., Venema, P., and van der Linden, E. (2007). *Journal of Agricultural and Food Chemistry* **55**, 5661–5669.

Bolder, S. G., Vasbinder, A. J., Sagis, L. M. C., and van der Linden, E. (2007). *International Dairy Journal* **17**, 846–853.

Calabrese, M. F. and Miranker, A. D. (2007). *Journal of Molecular Biology* **367**, 1–7.

Campbell, B. C. V., Li, Q. X., Culvenor, J. G., Jäkälä, P., Cappai, R., Beyreuther, K., Masters, C. L., and McLean, C. A. (2000). *Neurobiology of Disease* **7**, 192–200.

Castelletto, V. and Hamley, I. W. (2007). *Biomacromolecules* **8**, 77–83.

Catalano, S., Dodson, E., Henze, D., Joyce, J., Krafft, G., and Kinney, G. (2006). *Current Topics in Medicinal Chemistry* **6**, 597–608.

Caughey, B. and Lansbury, P. T. (2003). *Annual Review of Neuroscience* **26**, 267–298.

Ceru, S. and Zerovnik, E. (2008). *FEBS Letters* **582**, 203–209.

Channon, K. J., Devlin, G. L., Magennis, S. W., Finlayson, C. E., Tickler, A. K., Silva, C., and MacPhee, C. E. (2008). *J. Am. Chem. Soc.* **130**, 5487–5491.

Chapman, M. R., Robinson, L. S., Pinkner, J. S., Roth, R., Heuser, J., Hammar, M., Normark, S., and Hultgren, S. J. (2002). *Science* **295**, 851–855.

Charm, S. E., and Kurland, G. S. (1972). In CardioVascular Fluid Dynamics, Vol. 2., Academic Press, London.

Chiti, F., Webster, P., Taddei, N., Clark, A., Stefani, M., Ramponi, G., and Dobson, C. M. (1999). *Proceedings of the National Academy of Sciences* **96**, 3590–3594.

Chiti, F., Taddei, N., Stefani, M., Dobson, C. M., and Ramponi, G. (2001). *Protein Sci* **10**, 879–886.

Damaschun, G., Damaschun, H., Fabian, H., Gast, K., Krober, R., Wieske, M., and Zirwer (1999). *Proteins* **39**, 204–211.

Despa, F., Orgill, D. P., and Lee, R. C. (2005). *Annals of the New York Academy of Sciences* **1066**, 54–66.

Dialysis-related amyloidosis (2005). *From molecular mechanisms to therapies* **1753**, 1–3.

Dialysis-related amyloidosis (2005). *From molecular mechanisms to therapies* **1753**, 4–10.

Dickinson, D. W. (1997). *J.Neuropathol. Exp. Neurol* **56**, 321–339.

Dobson, C. (1999). *Trends in Biochem Sci.* **24**, 329.

Dobson, C. M. (1999). *Trends in Biochemical Sciences* **24**, 329–332.

Dobson, C. M. (2004). Methods Investigating Protein Folding, Misfolding and Nonnative States. *Experimental and Theoretical Methods* **34**, 4–14.

Dobson, C. M. (2004). *Protein Misfolding and Human Disease and Developmental Biology of the Retina* **15**, 3–16.

Doi, E. (1993). *Trends in Food Science & Technology* **4**, 1–5.

Domike, K. R. and Donald, A. M. (2007). *Biomacromolecules* **8**, 3930–3937.

Drueke, T. B. (2000). *Nephrol. Dial. Transplant* **15**, 17–24.

Dusa, A., Kaylor, J., Edridge, S., Bodner, N., Hong, D. P., and Fink, A. L. (2006). *Biochemistry* **45**, 2752–2760.

Dyson, H. J. and Wright, P. E. (2005). Nuclear Magnetic Resonance of Biological Macromolecules. *Part C*, 299.

Elliot, M. A., Karoonuthaisiri, N., Huang, J. Q., Bibb, M. J., Cohen, S. N., Kao, C. M., and Buttner, M. J. (2003). *Genes & Development* **17**, 1727–1740.

FÃndrich, M., Fletcher, M. A., and Dobson, C. M. (2001). *Nature* **410**, 165–166.

Foegeding, E. A. (2006). *Food Biophysics* **1**, 41–50.

Geddes, J. W. (2005). *Experimental Neurology* **192**, 244–250.

Giese, A., Bader, B., Bieschke, J., Schaffar, G., Odoy, S., Kahle, P. J., Haass, C., and Kretzschmar, H. (2005). *Biochemical and Biophysical Research Communications* **333**, 1202–1210.

Glenner, G. G. and Wong, C. W. (1984). *Biochemical and Biophysical Research Communications* **120**, 885–890.

Goedert, M., Spillantini, M. G., and Davies, S. W. (1998). *Current Opinion in Neurobiology* **8**, 619–632.

Goers, J., Permyakov, S. E., Permyakov, E. A., Uversky, V. N., and Fink, A. L. (2002). *Biochemistry* **41**, 12546–12551.

Gosal, W. S., Clark, A. H., Pudney, P. D. A., and Ross-Murphy, S. B. (2002). *Langmuir* **18**, 7174–7181.

Gosal, W. S., Clark, A. H., and Ross-Murphy, S. B. (2004). *Biomacromolecules* **5**, 2408–2419.

Goto, Y., Bellotti, V., and Esposito, R. Biochimica et Biophysica Acta (BBA) – Proteins & Proteomics

G. R, M. MP, G. DF, K. L and P. SB, *J. Biol. Chem.* 263 (1988) 4950.

Gras, S. L., Tickler, A. K., Squires, A. M., Devlin, G. L., Horton, M. A., Dobson, C. M., and MacPhee, C. E. (2008). *Biomaterials* **29**, 1553–1562.

Graveland-Bikker, J. F. and de Kruif, C. G. (2006). *Trends in Food Science & Technology Functionality in Complex Foods* **17**, 196–203.

Graveland-Bikker, J. F., Schaap, I. A. T., Schmidt, C. F., and deKruif, C. G. (2006). *Nano Lett.* **6**, 616–621.

Griffin, M. D. W., Mok, M. L. Y., Wilson, L. M., Pham, C. L. L., Waddington, L. J., Perugini, M. A., and Howlett, G. J. (2008). *Journal of Molecular Biology* **375**, 240–256.

Groenning, M., Norrman, M., Flink, J. M., van de Weert, M., Bukrinsky, J. T., Schluckebier, G., and Frokjaer, S. (2007). *Journal of Structural Biology* **159**, 483–497.

Groenning, M., Olsen, L., van de Weert, M., Flink, J. M., Frokjaer, S., and Jørgensen, F. S. (2007). *Journal of Structural Biology* **158**, 358–369.

Grudzielanek, S., Jansen, R., and Winter, R. (2005). *Journal of Molecular Biology* **351**, 879–894.

Grudzielanek, S., Smirnovas, V., and Winter, R. (2007). *Chemistry and Physics of Lipids* **149**, 28–39.

Guijarro, J. I., Sunde, M., Jones, J. A., Campbell, I. D., and Dobson, C. M. (1998). *Proceedings of the National Academy of Sciences* **95**, 4224–4228.

Guzmán, C., Jeney, S., Kreplak, L., Kasas, S., Kulik, A. J., Aebi, U., and Forró, L. (2006). *Journal of Molecular Biology* **360**, 623–630.

Hatters, D. M., MacRaild, C. A., Daniels, R., Gosal, W. S., Thomson, N. H., Jones, J. A., Davis, J. J., MacPhee, C. E., Dobson, C. M., and Howlett, G. J. (2003). *Biophys. J.* **85**, 3979–3990.

Hawe, A., Sutter, M., and Jiskoot W. *Pharmaceutical Research* **25**(200), 1487–1499.

Herland, A., Björk, P., Nilsson, K. P. R., Olsson, J. D. M., Åsberg, P., Konradsson, P., Hammarström, P., and Inganäs, O. (2005). *Advanced Materials* **17**, 1466–1471.

Hill, E. K., Krebs, B., Goodall, D. G., Howlett, G. J., and Dunstan, D. E. (2006). *Biomacromolecules* **7**, 10–13.

Hope, A. D., Myhre, R., Kachergus, J., Lincoln, S., Bisceglio, G., Hulihan, M., and Farrer, M. J. (2004). *Neuroscience Letters* **367**, 97–100.

Horwich, A. L. and Weissman, J. S. (1997). *Cell* **89**, 499–510.

Ipsen, R. (1994). *Carbohydrate Polymers* **25**, 225–226.

Ipsen, R. and Otte, J. (2007). *Biotechnology Advances* **25**, 602–605.

Jadoul, M. (1998). *Nephrol. Dial. Transplant* **13**, 61–64.

Jarrett, J. T. and Lansbury, P. T. (1992). *Biochemistry* **31**, 12345–12352.

Jenkins, J. and Pickersgill, R. (2001). *Progress in Biophysics and Molecular Biology* **77**, 111–175.

Kawasaki, T., Asaoka, K., Mihara, H., and Okahata, Y. (2006). *Journal of Colloid and Interface Science* **294**, 295–303.

Kegel, W. K. and van der Schoot, P. (2004). *Biophys. J* **86**, 3905–3913.

Kegel, W. K. and van der Schoot, P. (2006). *Biophys. J.* **91**, 1501–1512.

Kelly, S. M., Jess, T. J., and Price, N. C. (2005). *Biochimica et Biophysica Acta (BBA) – Proteins & Proteomics* **1751**, 119–139.

Khurana, R., Gillespie, J. R., Talapatra, A., Minert, L. J., Ionescu-Zanetti, C., Millett, I., and Fink, A. L. (2001). *Biochemistry* **40**, 3525–3535.

Khurana, R., Coleman, C., Ionescu-Zanetti, C., Carter, S. A., Krishna, V., Grover, R. K., Roy, R., and Singh, S. (2005). *Journal of Structural Biology* **151**, 229–238.

Kishimoto, A., Hasegawa, K., Suzuki, H., Taguchi, H., Namba, K., and Yoshida, M. (2004). *Biochemical and Biophysical Research Communications* **315**, 739–745.

Klunk, W. E., Jacob, R. F., and Mason, R. P. (1999). *Analytical Biochemistry* **266**, 66–76.

Klunk, W. E., Wang, Y., Huang, G.f., Debnath, M. L., Holt, D. P., and Mathis, C. A. (2001). *Life Sciences* **69**, 1471–1484.

Knight, J. D. and Miranker, A. D. (2004). *Journal of Molecular Biology* **341**, 1175–1187.

Kock, K. M. (1992). *Kidney international* **41**, 1416–1429.

Kontush, A. (2001). *Free Radical Biology and Medicine* **31**, 1120–1131.

Kontush, A., Berndt, C., Weber, W., Akopyan, V., Arlt, S., Schippling, S., and Beisiegel, U. (2001). *Free Radical Biology and Medicine* **30**, 119–128.

Krawczak, M., Ball, Edward V, Fenton, Iain, Stenson, Peter D., Abeysinghe, Shaun, Thomas, Nick, and Cooper, David N. (2000). *Human Mutation* **15**, 45–51.

Krebs, M. R. H., Bromley, E. H. C., and Donald, A. M. (2005). *Journal of Structural Biology* **149**, 30–37.

Krebs, M. R. H., Bromley, E. H. C., Rogers, S. S., and Donald, A. M. (2005). *Biophys. J* **88**, 2013–2021.

Lashuel, H. A., Petre, B. M., Wall, J., Simon, M., Nowak, R. J., Walz, T., and Lansbury, P. T. (2002). *Journal of Molecular Biology* **322**, 1089–1102.

Lemere, C., Blusztajn, J., Yamaguchi, H., Wisniewski, T., Saido, T., and Selkoe, D. (1996). *Neurobiol. Dis* **3**, 16.

Leys, D. and Scrutton, N. S. (2004). *Current Opinion in Structural Biology* **14**, 642–647.

Lindgren, M., Svensson, M., Freskgard, P. O., Carlsson, U., Jonasson, P., Martensson, L. G., and Jonsson, B. H. (1995). *Biophys. J.* **69**, 202–213.

Lindgren, M., Sorgjerd, K., and Hammarstrom, P. (2005). *Biophys. J.* **88**, 4200–4212.

Lippa, C. F., Luise Schmidt, V. M. Y. L. M., John, Q., Trojanowski., G. G., Cornwell, K. H., 3rd, Johnson, and Westermark, P. (1995). *J Clin Pathol* **48**, 984–989.

Liss, S. N., Droppo, I. G., Flannigan, D. T., and Leppard, G. G. (1996). *Environmental Science & Technology* **30**, 680–686.

Lorenzo, A. and Yankner, B. (1994). *Proc. Nat. Acad. Sci.* **91** (25), 12. USA 243.

Lu, K., Jacob, J., Thiyagarajan, P., Conticello, V. P., and Lynn, D. G. (2003). *J. Am. Chem. Soc.* **125**, 6391–6393.

Luk, K. C., Hyde, E. G., Trojanowski, J. Q., and Lee, V. M. Y. (2007). *Biochemistry* **46**, 12522–12529.

Mackay, J. P., Matthews, J. M., Winefield, R. D., Mackay, L. G., Haverkamp, R. G., and Templeton, M. D. (2001). *Structure* **9**, 83–91.

MacPhee, C. E. and Dobson, C. M. (2000). *J. Am. Chem. Soc.* **122**, 12707–12713.

Maji, S. K., Amsden, J. J., Rothschild, K. J., Condron, M. M., and Teplow, D. B. (2005). *Biochemistry* **44**, 13365–13376.

Margittai, M. and Langen, R. (2006). in. *In* Amyloid, Prions, and Other Protein Aggregates, Part C, 122–139, Wetzel, I. K.a. R. (ed.). Academic Press.

Marzban, L., Park, K., and Verchere, C. B. (2003). *Experimental Gerontology* **38**, 347–351.

McAllister, C., Karymov, M. A., Kawano, Y., Lushnikov, A. Y., Mikheikin, A., Uversky, V. N., and Lyubchenko, Y. L. (2005). *Journal of Molecular Biology* **354**, 1028–1042.

Meenaghan, M. A., Appel, B. N., and Greene, G. W. (1972). Oral Surgery, Oral Medicine. *Oral Pathology* **34**, 908–919.

Meersman, F. and Dobson, C. M. (2006). *Biochimica et Biophysica Acta (BBA) – Proteins & Proteomics* **1764**, 452–460.

Mishra, R., Sörgjerd, K., Nyström, S., Nordigården, A., Yu, Y. C., and Hammarström, P. (2007). *Journal of Molecular Biology* **366**, 1029–1044.

Mok, Y.h.t.g. F. and Howlett, G. J. (2006). *In* Amyloid, Prions, and Other Protein Aggregates, Part C, 199–217, Wetzel, I. K.a. R. (ed.). Academic Press.

Munishkina, L. A. and Fink, A. L. (2007). *Biochimica et Biophysica Acta (BBA) – Biomembranes Amyloidogenic Protein – Membrane Interaction* **1768**, 1862–1885.

Na, R., Stender, I. M., Ma, L., and W.H.C (2000). *Skin Research and Technology* **6**, 112–117.

Nandi, P. K., Leclerc, E., Nicole, J. C., and Takahashi, M. (2002). *Journal of Molecular Biology* **322**, 153–161.

Nayak, A., Dutta, A. K., and Belfort, G. (2008). *Biochemical and Biophysical Research Communications* **369**, 303–307.

Otte, J., Schumacher, E., Ipsen, R., Ju, Z. Y., and Qvist, K. B. (1999). *International Dairy Journal* **9**, 801–812.

Otte, J., Ipsen, R., Bauer, R., Bjerrum, M. J., and Waninge, R. (2005). *International Dairy Journal* **15**, 219–229.

Pankratz, N. and Foroud, T. (2004). *NeuroRX Biomarkers and Surrogates* **1**, 235–242.

Pastor, M. T., Kümmerer, N., Schubert, V., Esteras-Chopo, A., Dotti, C. G., López de la Paz, M., and Serrano, L. (2008). *Journal of Molecular Biology* **375**, 695–707.

Pauling, L. (1960). Nature of the Chemical Bond and the Molecular Crystals: An Introduction to Model Chemistry. 3rd ed, Cornell University, Ithaca, NY. ed.

Pearce, F. G., Mackintosh, S. H., and Gerrard, J. A. (2007). *J. Agric. Food Chem.* **55**, 318–322.

Perutz, M. F., Finch, J. T., Berriman, J., and Lesk, A. (2002). *Proceedings of the National Academy of Sciences* **99**, 5591–5595.

Porat, Y., Mazor, Y., Efrat, S., and Gazit, E. (2004). *Biochemistry* **43**, 14454–14462.

Prusiner, S., McKinley, M., Bowman, K., Bendheim, P., and Bolton, D. (1983). *Cell* **35**, 349.

Robinson, S. R., Dobson, C., and Lyons, J. (2004). *Neurobiology of Aging: Challenging Views of Alzheimer's Disease–Round II* **25**, 629–637.

Rogers, S. S., Krebs, M. R. H., Bromley, E. H. C., van der Linden, E., and Donald, A. M. (2006). *Biophys. J.* **90**, 1043–1054.

Sabaté, R. and Saupe, S. J. (2007). *Biochemical and Biophysical Research Communications* **360**, 135–138.

Sagle, L. B., Zimmermann, J. r., Dawson, P. E., and Romesberg, F. E. (2004). *Journal of Americal chemical society* **126**, 3384–3385.

Sawaya, M. R., Sambashivan, S., Nelson, R., Ivanova, M. I., Sievers, S. A., Apostol, M. I., Thompson, M. J., Balbirnie, M., Wiltzius, J. J. W., McFarlane, H. T., Madsen, A. O., Riekel, C., and Eisenberg, D. (2007). *Nature* **447**, 453–457.

Schmittschmitt, J. P. and Scholtz, J. M. (2003). *Protein Sci* **12**, 2374–2378.

Serio, T. R., Gashikar, A. G., Kowal, A. S., Sawicki, G. J., Moslehi, J. J., Serpell, L., Arnsdorf, M. F., and Lindquist, S. L. (2000). *Science* **289**, 1317–1321.

Serpell, L. C., Fraser, P. E., and Sunde, M. (1999). *In* Amyloid, Prions, and Other Protein Aggregates, 526–536, Wetzel, R. (ed.). Academic Press.

Serpell, L. C. (2000). *Biochimica et Biophysica Acta (BBA) – Molecular Basis of Disease* **1502**, 16–30.

Serpell, L. C. and Smith, J. M. (2000). *Journal of Molecular Biology* **299**, 225–231.

Serpell, L. C., Sunde, M., Benson, M. D., Tennent, G. A., Pepys, M. B., and Fraser, P. E. (2000). *Journal of Molecular Biology* **300**, 1033–1039.

Sharp, J. S., Forrest, J. A., and Jones, R. A. L. (2002). *Biochemistry* **41**, 15810–15819.

Shtilerman, M. D., Ding, T. T., and Lansbury, P. T. (2002). *Biochemistry* **41**, 3855–3860.

Sluzky, V., Tamada, J., Klibanov, A., and Langer, R. (1991). *Proceedings of the National Academy of Sciences* **88**, 9377–9381.

Small, D., SS, M., and JC, B. (2001). *Nat. Rev. Neurosci* **2**, 595.

Smith, J. F., Knowles, T. P. J., Dobson, C. M., MacPhee, C. E., and Welland, M. E. (2006). *Proceedings of the National Academy of Sciences* **103**, 15806–15811.

Smith, M. I., Sharp, J. S., and Roberts, C. J. (2007). *Biophys. J.* **93**, 2143–2151.

Soto, C., Castaño, E. M., Asok Kumar, R., Beavis, R. C., and Frangione, B. (1995). *Neuroscience Letters* **200**, 105–108.

Squires, A. M., Devlin, G. L., Gras, S. L., Tickler, A. K., MacPhee, C. E., and Dobson, C. M. (2006). *J. Am. Chem. Soc.* **128**, 11738–11739.

Sumi, Y., Hayashi, Y., Hattori, H., and Ueda, M. (2005). *International Journal of Oral and Maxillofacial Surgery* **34**, 696–698.

Takahashi, T., Yamashita, H., Nagano, Y., Nakamura, T., Kohriyama, T., and Matsumoto, M. (2006). *FEBS Letters* **580**, 4479–4484.

Tattum, M. H., Cohen-Krausz, S., Khalili-Shirazi, A., Jackson, G. S., Orlova, E. V., Collinge, J., Clarke, A. R., and Saibil, H. R. (2006). *Journal of Molecular Biology* **357**, 975–985.

Teplow, D. B. (2006). *In* Amyloid, Prions, and Other Protein Aggregates, Part C, 20–33, Wetzel, I. K.a. R. (ed.). Academic Press.

Teplow, D. B., Lazo, N. D., Bitan, G., Bernstein, S., Wyttenbach, T., Bowers, M. T., Baumketner, A., Shea, J. E., Urbanc, B., Cruz, L., Borreguero, J., and Stanley, H. E. (2006). *Acc. Chem. Res.* **39**, 635–645.

T. F, F. G, D.u. P, B. M and S. M, Adv. Prot. Chem. 57 (2001) 171.

Thirumalai, D., Klimov, D., and Dima, R. (2003). *Current Opinion in Structural Biology* **13**, 146–159.

Thirunavukkuarasu, S., Jares-Erijman, E. A., and Jovin, T. M. (2008). *Journal of Molecular Biology* **378**, 1064–1073.

Thorn, D. C., Meehan, S., Sunde, M., Rekas, A., Gras, S. L., MacPhee, C. E., Dobson, C. M., Wilson, M. R., and Carver, J. A. (2005). *Biochemistry* **44**, 17027–17036.

Thorn, D. C., Ecroyd, H., Sunde, M., Poon, S., and Carver, J. A. (2008). *Biochemistry* **47**, 3926–3936.

Treweek, T. M., Morris, A. M., and Carver, J. A. (2003). *Australian Journal of Chemistry* **56**, 357–367.

Turnbull, S., Tabner, B. J., Brown, D. R., and Allsop, D. (2003). *Neuroscience Letters* **336**, 159–162.

Tycko, R. (2003). *Progress in Nuclear Magnetic Resonance Spectroscopy* **42**, 53–68.

Tycko, R. (2006). *In* Amyloid, Prions, and Other Protein Aggregates, Part C, 103–122, Wetzel, I. K.a. R. (ed.). Academic Press.

Tycko, R. (2006). *Quarterly Reviews of Biophysics* **39**, 1–55.

Tycko, R. (2006). Nanomedicine: Nanotechnology. *Biology and Medicine* **2**, 271.

Ukai, Y. N., Tsutsumi, T., and Marakami, K. Vachon C, D'Aprano, U.S. 1976.

Ulijn, R. V. and Smith, A. M. (2008). *Chemical Society Reviews* **37**, 664–675.

Uversky, V. N. and Fink, A. L. (2002). *FEBS Letters* **515**, 79–83.

Uversky, V. N. and Fink, A. L. (2004). *Biochimica et Biophysica Acta (BBA) – Proteins & Proteomics* **1698**, 131–153.

Uversky, V. N. and Ptitsyn, O. B. (1996). *Folding and Design* **1**, 117–122.

Uversky, V. N. and Ptitsyn, O. B. (1996). *Journal of Molecular Biology* **255**, 215–228.

Uversky, V. N., Semisotnov, G. V., Pain, R. H., and Ptitsyn, O. B. (1992). *FEBS Letters* **314**, 89–92.

Uversky, V. N., Winter, S., and Löber, G. (1996). *Biophysical Chemistry* **60**, 79–88.

Uversky, V. N., Narizhneva, N. V., Ivanova, T. V., Kirkitadze, M. D., and Tomashevski, A. Y. (1997). *FEBS Letters* **410**, 280–284.

Uversky, V. N., Li, J., and Fink, A. L. (2001). *J. Biol. Chem.* **276**, 10737–10744.

Uversky, V. N., Li, J., Bower, K., and Fink, A. L. (2002). *NeuroToxicology* **23**, 527–536.

Uversky, V. N., Yamin, G., Souillac, P. O., Goers, J., Glaser, C. B., and Fink, A. L. (2002). *FEBS Letters* **517**, 239–244.

Uversky, V. N., Yamin, G., Munishkina, L. A., Karymov, M. A., Millett, I. S., Doniach, S., Lyubchenko, Y. L., and Fink, A. L. (2005). *Molecular Brain esearch Molecular Mechanisms in Parkinson's Disease* **134**, 84–102.

Valverde, J., Valero, D., Martýnez-Romero, D., Guillen, F., Castillo, S., and Serrano, M. (2005). *J AgricFood Chem 53*, 7807–7813.

van der Wel, P. C. A., Lewandowski, J. R., and Griffin., R. G. (2007). *J. Am. Chem. Soc.* **129**, 5117–5130.

van der Linden, E. and Venema, P. (2007). *Current Opinion in Colloid & Interface Science* **12**, 158–165.

van der Linden, E. (2006). *Worlds Poultry Science Journal* **62**, 439–442.

Veerman, C., Sagis, L. M. C., Venema, P., and van der Linden, E. (2005). *Journal of Rheology* **49**, 355–368.

Virchow, R. (1854). *Virchows Arch. Pathol. Anat Physiol.* **6**, 268–271.

Virchow, R. (1854). *Virchows Arch. Pathol. Anat Physiol* **6**, 135–137.

Walsh, D., Lomakin, A., Benedek, G., Condron, M., and Teplow, D. (1997). *J. Biol. Chem.* **272**, 22. 364.

Weiss, J., Takhistov, P., and McClements, D. J. (2006). *Journal of Food Science* **71**, R107–R116.

Westermark, P., Benson, M. D., Buxbaum, J. N., Cohen, A. S., Frangione, B., Ikeda, S. I., Masters, C. L., Merlini, G., Saraiva, M. J., and Sipe, J. D. (2005). *Amyloid* **12**, 1–4.

W. H, M. MD, G. V, S. S and S. A, Proc. Natl. Acad. Sci. USA 99(6) (2002) 3563.

Whitesides, G. M. (1991). J. P. Mathias and C. T. and Seto. *Science* **254**, 1312–1319.

Wood, S. J., Maleeff, B., Hart, T., and Wetzel, R. (1996). *Journal of Molecular Biology* **256**, 870–877.

Wosten, H. A. B. and de Vocht, M. L. (2000). *Biochimica Et Biophysica Acta-Reviews on Biomembranes* **1469**, 79–86.

Xing, Y. and Higuchi, K. (2002). *Mechanisms of Ageing and Development* **123**, 1625–1636.

Yamaguchi, K.i., Katou, H., Hoshino, M., Hasegawa, K., Naiki, H., and Goto, Y. (2004). *Journal of Molecular Biology* **338**, 559–571.

Yamamoto, S., and Gejyo, F. Biochimica et Biophysica Acta (BBA) – Proteins & Proteomics

Yong, W., Lomakin, A., Kirkitadze, M. D., Teplow, D. B., Chen, S. H., and Benedek, G. B. (2002). *Proceedings of the National Academy of Sciences* **99**, 150–154.

Zamotin, V., Gharibyan, A., Gibanova, N. V., Lavrikova, M. A., Dolgikh, D. A., Kirpichnikov, M. P.,

Kostanyan, I. A., and Morozova-Roche, L. A. (2006). *FEBS Letters* **580**, 2451–2457.

Zandi, R., van der Schoot, P., Reguera, D., Kegel, W., and Reiss, H. (2006). *Biophys. J.* **90**, 1939–1948.

Zerovnik, E., Pompe-Novak, M., Skarabot, M., Ravnikar, M., Musevic, I., and Turk, V. (2002). *Biochimica et Biophysica Acta (BBA) – Protein Structure and Molecular Enzymology* **1594**, 1–5.

Zhang, Y., Dawson, V. L., and Dawson, T. M. (2000). *Neurobiology of Disease* **7**, 240–250.

Zhang, Q., Powers, E. T., Nieva, J., Huff, M. E., Dendle, M. A., Bieschke, J., Glabe, C. G., Eschenmoser, A., Wentworth, P., Jr., Lerner, R. A., and Kelly, J. W. (2004). *Proceedings of the National Academy of Sciences* **101**, 4752–4757.

Zhao, R., Torley, P., and Halley, P. J. (2008). *J.Material science* **43**, 3058–3071.

Zhao, D. L. Y. (2007). *Comprehensive Food Science and Food Safety Reviews* **6**, 60–75.

Zhou, W., Zhu, M., Wilson, M. A., Petsko, G. A., and Fink, A. L. (2006). *Journal of Molecular Biology* **356**, 1036–1048.

Zhu, M., Li, J., and Fink, A. L. (2003). *J. Biol. Chem.* **278**, 40186–40197.

Z Mantas Malisauskas, a., Vladimir, J., Jass, W., Noppe, C. M., Dobson, and Morozova-Roche, L. A. (2003). *Journal of Molecular Biology* **330**, 879–890.

CHAPTER

18

Hydrocolloids and Medicinal Chemistry Applications

Liam M. Grover and Alan M. Smith
School of Chemical Engineering, University of Birmingham, Edgbaston, UK

18.1 DRUG DELIVERY

The techniques for delivering substances into the human body have been widely researched by mankind for literally thousands of years both from a medicinal and recreational perspective. Generally speaking the approach is determined by the nature of the formulation that is to be delivered and therefore rational design and forethought is required to produce effective medicines that are fit for purpose. Biopolymers in particular are widely used within pharmaceutical products, traditionally as excipients such as binders, fillers, thickeners and disintegrants. Indeed, glance through the pharmaceutical handbook of excipients and you will find a large quantity of biopolymers listed for pharmaceutical use. The description of excipients, however, is gradually shifting towards functionally active materials rather than the traditional definition of inactive or inert ingredients. Biopolymers are utilized in the majority of drug dosage forms including oral, ocular, nasal sprays, topical formulations (gels and ointments), pulmonary delivery, and even parenteral delivery. These traditional dosage forms are the subject of much research to improve stability, drug efficiency, to reduce costs, increase patent compliance and improve therapeutic performance. The wide variety of biopolymers available with diverse compositions and properties, natural origin along with their ease of production make them a popular research focus within pharmaceutics. Apart from numerous applications in traditional dosage forms, biopolymers have recently been used for the controlled delivery of biological material such as proteins, peptides, and vaccines. To discuss all of the biopolymers that have been investigated for use as drug delivery systems would require considerably more space than this chapter allows, therefore attention is focused on biopolymers that have been investigated to provide modified-release properties and the use of biopolymers in biopharmaceutical formulations.

18.1.1 Oral Delivery

The major challenge in relation to optimal delivery of drugs lies in achieving the required dose at the required time in the required place. This may relate to a clinical need for accuracy of dosing in the case of drugs which have

Kasapis, Norton, and Ubbink: Modern Biopolymer Science
ISBN: 978-0-12-374195-0

potentially dangerous side effects if taken in excess, or to situations in which a consistent level of the drug in the body is required (such as with hormone replacement therapy) or simply to improve patient compliance when the treatment regimen is inconvenient. To achieve uptake of the drug via the oral route, disintegration of the dosage form is required followed by dissolution of the released drug. In the case of drug formulations intended for oral delivery it is usually the case that an immediate-release formulation is delivered to reach or possibly exceed the therapeutic window. The problem with such an approach is that once the drug has been fully metabolized the therapeutic effect is lost and a repeat dose becomes necessary, a situation that is avoided with sustained delivery as illustrated in Figure 18.1. Moreover, certain drugs are unsuitable for immediate release in the stomach and require delayed-release formulations to target the drug further down the gastrointestinal tract (GIT). The physiological variability in the GIT therefore needs to be considered when designing oral drug delivery systems, especially when formulating drugs with a narrow therapeutic index. Fluctuations in pH and enzyme concentrations can vary in the fed and fasted state, which can cause changes in the solubility of certain drugs which then alters the rate of absorption. The naturally fluctuating environment of the GIT not only has effects on drug molecules but also on the excipients used to formulate the dosage form.

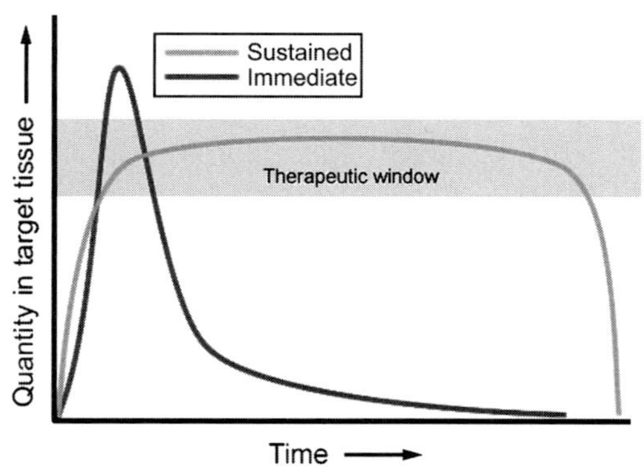

FIGURE 18.1 Simplified illustration of drug release profiles in relation to dosage form highlighting the difference in therapeutic activity of an immediate-release and a sustained-release system.

18.1.1.1 Tablets

Standard uncoated immediate-release tablet formulations that are prepared using methods such as direct compression of powder blends or compression following wet or dry granulation tend to include biopolymers such as starch or microcrystalline cellulose (MCC) as disintegrants or binders. Additionally, super swelling materials can be incorporated at low levels (2–4% w/w) as superdisintegrants. Materials such as sodium starch glycolate (trademarked Explotab® or Primojel®) and crosscaramelose sodium (cross-linked sodium carboxymethyl cellulose), which can swell to between 8–12 times their original size on contact with water, disrupt the integrity of tablet matrices resulting in disintegration. These simple tablets are designed to dissolve in the stomach to deliver drugs locally in the GIT or for systemic effects. To achieve modified-release from tablet matrices there are several approaches that can be used to either sustain or delay the release of the drug depending on what drug release profile is required to achieve the optimum therapeutic effect. One method of controlling release from tablet matrices is by the addition of polysaccharides such as hydroxypropyl methylcellulose (HPMC). When a tablet (containing HPMC) encounters the aqueous environment of the stomach, hydration of the HPMC occurs, forming a gel layer surrounding the dosage form which creates a barrier to diffusion. Release of the drug then occurs by diffusion through the gel layer and as the gel layer gradually erodes within the aqueous dissolution medium. The rate of drug release is therefore controlled by the rate of drug diffusion through the swollen HPMC gel layer. Increasing the proportion of HPMC in the formulation reduces diffusion of the drug and delays the erosion of the tablet matrix due to an increased swelling volume and a resulting larger gel layer. The rate of polymer swelling and dissolution can be increased by using lower-viscosity HPMC, which subsequently increases drug release rate. Xanthan is another polysaccharide that has shown to retard drug release, however, the rate of release is dependent on the ionic strength (Talukdar and Kinget, 1995) of the release media. It has also been reported when galactomannans (Sinha et al., 2004) and more recently konjac glucomannan are incorporated with xanthan in tablet matrices release rate is reduced further. Incorporation of gluco- or galactomannans within xanthan tablet matrices also provides the potential for colon-targeted delivery. The swollen gel layer formed by synergistic interactions of xanthan and gluco/galactomannans which retards release in the upper GIT is degraded in the colon as the mannans are hydrolyzed by β-manannases secreted by colonic micro-flora increasing drug release rate (Liu et al., 2007; Fan et al., 2008).

18.1.1.2 *Pharmaceutical Capsules*

The majority of two-piece hard capsules used for pharmaceuticals are made from gelatin and were first produced in 1834 by Francois Mothes who dipped a mold into a solution of gelatin and allowed it to dry before removing the resultant capsule. The dip molding principle of manufacture has remained relatively unchanged since. Today manufacture of the capsules is carried out using automated machinery, where stainless steel pins (for the body and the cap) are dipped into an aqueous gelatin solution (25–30% w/w), which is maintained at about 50–55°C. As the pins are withdrawn, they are rotated to distribute the gelatin evenly and cooled to below the gelation temperature. The capsules are then allowed to dry at controlled humidity and then stripped from the pins, trimmed and assembled. Usually capsules are prepared with a self-locking mechanism or can be hermetically sealed once filled with the pharmaceutical contents.

Gelatin is still the material of choice for capsule production due to excellent film-forming properties and rapid dissolution in gastric fluid; pharmaceutical companies, however, have been forced to develop capsules prepared from non-animal sources for numerous reasons including a rise in vegetarianism, religious objections to animal-derived products and instability with drugs that are hygroscopic or contain reactive aldehyde groups. There have been several alternatives to the gelatin capsule formulated (Yamamoto et al., 1995; Ogura, 1998; Cade et al., 2003) that are on the market, most notably Shionogi Quali-V® and Capsugel V-Caps®, both these capsules are based on low-viscosity HPMC as the film-forming material and a gelling agent (Quali-V use κ-carrageenan, V-caps use gellan gum) to provide structure to the HPMC during production. Both gelatin and HPMC capsules are designed to dissolve and release their contents in the stomach, however, for certain drugs that are unstable or reactive at gastric pH and/or dissolution of the capsule is required further down the GIT. The most successful method currently employed to achieve this is spray coating with synthetic-acid-insoluble methacrylate polymers either on the exterior surface of the capsule shell or the loaded drug itself, however recently research groups have focused on developing capsules with bulk enteric properties built into the capsule shell. Indeed, this has already been achieved in one-piece soft gel capsules used for oil-based preparations. Enteric Softgel capsules developed by Banner under the tradename Entericare™ claim enteric properties by blending solutions of polymers used for enteric coating (polymethacrylate, Cellulose acetate Phthalate or Shellac) with gelatin solutions, however no two-piece hard capsule with enteric properties has been developed to date. Several attempts to form gelatin capsules by cross-linking with aldehydes have resulted in mixed success (Brown et al., 1998; Pina and Sousa, 2002; Marchais, 2003).

More recently capsule films prepared from tertiary mixtures of HPMC/gellan/alginate have shown good film-forming properties and acid-delayed rupture at gastric pH due to the presence of alginate and gellan (Smith and Perrie, 2007).

18.1.1.3 *Oral Liquids*

A wide variety of biopolymers are used in the formulation of oral pharmaceutical liquids, whether to thicken solutions, suspend dispersions or to stabilize emulsions. Functionality, however, is not only restricted to providing structure to oral liquids. Responsive characteristics desirable in directed drug delivery and applications of biopolymers in which triggered or environmentally induced changes in state utilized in addressing a clinical problem are frequently incorporated into oral liquid dosage forms. For instance, gel-forming biopolymer systems such as alginate and gellan have a rapid sol–gel transition on exposure to acid which is a particularly attractive property for oral liquids due to the low pH of the stomach. The most well-known example of a product that utilizes this form of physiological responsive sol–gel transition therapeutically is Gaviscon®, which has been commercially available for almost 40 years. Here, alginate is used to prevent gastric reflux in combination with bicarbonates as an oral liquid buffered at neutral pH. When swallowed and contact is made with H^+ ions in the gastric fluid the alginate is rapidly cross-linked forming a gel 'raft' on the surface of the gastric fluid, preventing acid reflux. This has been an extremely successful product with several formulations available in the product range, the majority of which are based on alginate. A more recent example of acid gelation in the stomach utilizes gellan gum to increase the gastric residence time of the antibiotic clathromysin resulting in improved eradication of the pathogen *H. pylori* (Rajinikanth and Mishra, 2008). In this study the authors describe the two main prerequisites for the in situ gelling system; optimum viscosity for ease of swallowing and gelling capacity and rapid sol–gel transition due to ionic interaction. These prerequisites can relate to several polysaccharide systems and therefore provide many options for formulation. Another attractive property of solutions of polysaccharides is their ability to adhere to physiological tissue which has been investigated in oral liquids for targeted delivery of drugs to poorly accessible target sites such as the esophagus, where transit time is less than 16 seconds (Batchelor, 2005). This bioadhesion or more specifically mucoadhesion can significantly increase the bioavailability of drugs formulated as an oral liquid or enable formulations to act as a protective barrier, coating physiological lesions. Interactions of polysaccharide with the mucus layer and other mechanisms of mucoadhesion will be discussed later in this chapter.

18.1.1.4 *Alternative Oral Delivery Systems*

Demand for alternative oral delivery systems is particularly strong in the pediatric and geriatric markets, where non-compliance is high due to patient difficulty in swallowing conventional tablets or capsules. To address this problem there have been many developments utilizing the physical properties of hydrocolloids. One such dosage form is orally disintegrating tablets (ODT) of which there are several patented commercially available formulations which include among others Zydis (gelatin), OraSolv (Starch, HPMC), DuraSolve (MCC), FLASHDOSE® (polymaltodextrins, polydextrose) and WOWTAB (ethylcellulose). These dosage forms are prepared by different technologies but all incorporate biopolymers within the matrix to provide different functional properties. Of these ODT formulations the Zydis® dosage form was the first to gain FDA approval and has the most preparations that are currently on the market. The Zydis® system consists of drug entrapped in a lyophilized preparation of gelatin and mannitol and dissolves almost instantly (2–3 seconds) on contact with saliva removing the need for water. The dispersed or dissolved drug and excipients are then swallowed with saliva. It has been claimed that bioavailability can be increased as a consequence of dissolution and subsequent absorption through the buccal and esophageal mucosa (Seager, 1998). These systems are particularly useful for hydrophobic drugs accommodating up to 400 mg, however, they are limited to ~ 60 mg of hydrophilic drugs. Other major drawbacks of lyophilized preparations include the high cost of processing and poor mechanical properties which cannot be accommodated in standard blister packs. Orasolve, DuraSolve and WOWTAB are manufactured using conventional tableting techniques, such as direct compression of powder blends (at a reduced compression force) or compression following wet or dry granulation, which dramatically reduce production costs compared with lyophilization. These directly compressed tablets have the advantage of obtaining a high dosage of both hydrophobic and hydrophilic drugs and have good mechanical properties, which however significantly increase disintegration time. The fast disintegrating action of directly compressed ODT is facilitated by combinations of superdisintergants, effervescing agents and highly water-soluble excipients and mechanical properties are controlled by addition of polysaccharides such as MCC, starch or ethylcellulose. Successful formulation of these products is therefore a delicate balance of adequate structural properties suitable for processing and rapid disintegration on exposure to saliva. Other variants of ODTs have also been developed, using technology platforms such as molding and flossing (Dobetti, 2001). Demand for inventive formulations that utilize melt-in-the-mouth properties continues to grow due to their convenience in administration. This is one area where there appears to be potential to apply technology developed for food applications to pharmaceutical products. A recent example by Agoub et al. (2007) revealed melt-in-the-mouth behavior in synergistic xanthan-konjac gels prepared at pH 3.5 and below. The suggested application in this particular study relates to where 'melt-in-the-mouth' characteristics are important for product quality, and where moderate acidity is acceptable or necessary (e.g. fruit jellies). Therefore it seems reasonable to argue that technologies such as this could potentially be adapted for pharmaceutical orally disintegrating dosage forms. Additionally, dosage forms designed to disintegrate and dissolve in the mouth require considerations not generally given to standard oral dosage forms such as mouth feel and taste-masking. Mouth feel is improved in ODTs by the addition of effervescing excipients such as sodium bicarbonate and taste-masking achieved by incorporation of flavors, acidity regulators, artificial sweeteners such as aspartame allied with techniques such as polymer coating applied to bitter-tasting pharmaceutical actives prior to tablet formulation. For a detailed review on taste-masking oral pharmaceuticals refer to Sohi et al. (2004).

There are already numerous oral formulations on the market that are alternatives to classic tablets and capsules which include lozenges, chewable tablets, chewing gums and buccal films that either use or have the potential to use the multifunctional properties of biopolymers. One particularly inventive dosage form is Clarosip® (Grunenthal Ltd.), which contains the antibiotic clarithromycin. This product is in the form of a drinking straw which is loaded with granules formulated using carrageenan, HPMC and methacrylic acid ethylacrylate copolymer to provide enhanced mouth feel, taste-masking and enteric properties. The drinking straw contains one dose which is administered using the straw to sip the patient's favorite drink (carbonated drinks are preferred as this helps taste-masking and improves mouth feel). The granules are then swallowed along with the drink. Each straw contains a gauge to inform the patient when the full dose has been taken. It is simple but innovative technologies such as this that have the potential to significantly improve patient compliance and therefore therapeutic efficacy.

18.1.2 Ocular Delivery

The normal response of the eye from the insertion of foreign material is the blink reflex and tear production, both of which serve to eliminate low-viscosity solutions more rapidly than higher-viscosity solutions. As a consequence of this, common eye drops incorporate viscosity-enhancing polymers to increase contact time and bioavailability, however, due to high rates of shear produced when blinking, rheological properties of the formulations also need to be considered. The application of shear-thinning polymers in particular provides reduced resistance to blinking, resulting in greater patient acceptance. Although lacrimal clearance of viscous eye drops occurs at a slower rate than Newtonian solutions, a further increase in residence time can be achieved by utilizing biopolymers with a rapid sol–gel transition in situ. This approach was first developed by Pramoda et al. (1979) who exploited the pH-dependent sol–gel transitions of xanthan-locust bean gum (LBG) mixtures which exist as a liquid formulation below pH 3.5 but once applied to physiological pH of the eye (~pH 7) gelation is induced achieving increased residence time. Application of acidic liquids to the eye may be uncomfortable to the patient and perhaps this technology would be better applied to a less-sensitive target site such as the nasal cavity where the average baseline pH ~6.3 (Washington, 2000) which would still facilitate the formation of a synergistic xanthan-LBG gel.

A more practical approach to in situ gelation within the eye utilizes the composition of tear fluid. Tears are composed of a complex mixture of enzymes (lysozyme and lactoferrin), antibodies, organic acids, vitamins, glucose, cholesterol and electrolytes including Na^+, K^+, and Ca^{2+} (Edman, 2002). This presence of cations in tear fluid has provided pharmaceutical scientists with the opportunity to utilize the ionotropic gel-forming properties of polysaccharides, such as deacetylated gellan gum, for an increased residence time and subsequently an increased duration of therapy. Indeed, ophthalmic formulations are the most frequently encountered examples of deacetylated gellan gum in current pharmaceutical use. One such example is Timoptic®, an ophthalmic solution (that contains the active ingredient timolol malate, to treat glaucoma) and utilizes deacetylated gellan gum to provide the capacity for the formulation to thicken on application to the surface of the eye. This thickening is a result of a suppression of negatively charged carboxyl groups of gellan gum polymer chains by cations present in the tear fluid (Table 18.1) prior to which the solution viscosity is sufficiently low to enable simple dispensing. The initial application of the formulation provokes the tear production providing the cations for gelation, and as more tears are produced (because of the increase in viscosity of foreign material in the eye) the gel is strengthened. The therapeutic effect of the inclusion of gellan gum has been shown to an extent that once-daily administration of a formulation containing gellan gum is equivalent to a twice-daily administration of a standard timolol ophthalmic solution (Shedden et al., 2001). This mechanism of in situ gelation by

TABLE 18.1 Ionic Content of Tear Fluid Adapted from Edman (2002)

Electrolyte	Concentration mMol/L
Calcium	0.57
Sodium	~140
Potassium	15–29
Chloride	120–135
Bicarbonate	26

interaction with tear fluid has also been demonstrated in xanthan-based formulations where sol–gel transition is induced by the presence of lysozyme (Bawa et al., 1999). Interactions between ionic polysaccharides and ocular mucins have been studied with increasing interest recently with the aim of developing mucoadhesive systems incorporating a range of ophthalmic drugs (Ceulemans et al., 2002; Maltese et al., 2006; Sandri et al., 2006). To appreciate the potential use of mucoadhesive polymers in drug delivery it is important to understand the nature of mucoadhesion and how the physical and chemical properties of biopolymers are ideally suited for this purpose.

18.1.3 Mucoadhesion

Mucoadhesive biopolymers are of great interest to pharmaceutical science, as they can increase the residence time of the drug at the target site providing more time for absorption, ultimately increasing bioavailability. This can be vitally important where there is a constant flow of fluid or particulate matter as in the GIT. In a recent review by Zhang et al. (2008) the authors state that in pharmaceutical terms the object of mucoadhesion can be defined as attachment of drug delivery applications to mucus or mucus membranes and mucoadhesives described as natural or synthetic materials used in drug delivery systems that lead to mucoadhesion. Furthermore, the authors also describe the physicochemical properties a polymer should possess to be mucoadhesive. These properties include hydrophilicity, numerous hydrogen-bonding functional groups, viscoelastic properties when hydrated, and for pharmaceutical applications they should be easily combined with drugs to provide sufficient control over drug release.

Several mechanisms of mucoadhesion have been proposed which describe a number of polymer interactions with mucins, which are the main constituent of mucus. Mucins are glycoproteins that are highly glycosylated consisting of ~80% carbohydrates as oligosaccharide chains based on 5–15 monomers, primarily N-acetylgalactosamine, N-acetylglucosamine, fucose, galactose, and sialic acid, there are also traces of mannose and sulfonic acid (Bansil and Turner, 2006). Due to a high prevalence of sialic acid and sulfonic acid terminating the oligosaccharide side chains the mucin has a net negative charge (Gandhi and Robinson, 1994), which provides the opportunity for electrostatic attraction. Moreover there are regions of the mucin that are free from oligosaccharide side chains providing domains for hydrophobic interactions also the size of the mucins (up to 50 MDa) and provide the possibility of physical entanglement. In depth analysis of these mechanisms is beyond the scope of this chapter, for a more detailed account of these mechanisms the author points to articles by Smart (2005) and Sudhakar et al. (2006).

Chitosan in particular has well-known mucoadhesive properties and has been investigated widely for use in drug delivery systems. The mechanism of mucoadhesion in chitosan is thought to be due to electrostatic interactions between the positively charged amino groups of chitosan and the negatively charged sialic groups of the mucins. These interactions are strong at acidic pH where the charge density of the chitosan is high (depending on the degree of deacetylation) (Harding, 2006). Formulation of mucoadhesive chitosan systems has been developed for a wide range of target delivery sites, for example chitosan-based mucoadhesive tablets have been developed for oral, sublingual, and buccal delivery. Liquid formulations in the form of eye drops for mucoadhesive ocular delivery are another example (Felt et al., 1999). The use of alginate in a mucoadhesive system has been shown to be particularly useful when targeting poorly accessible target sites such as the esophagus. In another study by Batchelor et al. (2002) 2% solutions of sodium alginate (high G, medium molecular weight) were shown to retard transit time through the esophagus for up to 30 min when measured in vitro using porcine esophageal tissue. The incorporation of drugs into such a system would enable localized therapy on the esophageal epithelium. Measurements of esophageal retention using in vivo techniques such as gamma scintigraphy have not always correlated with predicted in vitro measurements as demonstrated by McCargar et al. (2001), this is probably due to the design of in vitro apparatus that had not taken into account peristalsis, posture position or saliva flow. In vitro measurements have since been improved by using whole ex vivo esophagus tubes, periodic washing with artificial saliva and simulated periodic peristaltic waves performed by roller along the length of the esophageal tube. This method was used by Richardson et al. (2005) in demonstrating esophageal mucosa retention time of sodium alginate, which was prolonged for up to ~60 min when dispersed in glycerol.

Mucoadhesion of polysaccharides relies solely on non-covalent interactions, therefore only weak levels of adhesion are achieved which can be problematic for certain drugs where sustained release is required. To overcome this problem mucoadhesive polysaccharides such as alginate and chitosan can be synthetically thiolated to produce thiolomers, which enable the formation of disulfide bonds between the mucoadhesive thiolomer and cysteine-rich subdomains of mucin. Thiolated polymers can improve mucoadhesion to vastly different extents. For example, alginate-cysteine thiolomer has been shown to achieve a four-fold increase in mucoadhesion, whereas for chitosan-iminothiolane thiolomer mucoadhesion is increased 250-fold in comparison with polymers prior to thiolation. The array of biopolymers with potential thiolation sites will surely lead to a greater number of biopolymer-based thiolomers for applications in mucoadhesive drug delivery systems.

This section has provided a brief overview of how biopolymers are used in a variety of traditional dosage forms and how developments within pharmaceutics are requiring more functional excipients for targeted drug delivery to improve efficiency and patient compliance. The remainder of this chapter will focus on utilization of biopolymers in the development of drug delivery systems for biopharmaceuticals and as tissue engineering scaffolds.

18.1.4 Medicine of the Future

The desire for personalized medicine is at the forefront of medical and pharmaceutical research. Advances in pharmacogenomics have released the potential to design-directed therapeutics formulated and prescribed based on the knowledge of patient's genetics, environment including genetic predispositions or treatment-related facts, such as a patient's particular drug response. In principle, this has the potential to revolutionize treatments, reducing long-term costs due to individually tailored treatments and improved patient compliance. Technical aspects of pharmacogenomics are beyond the scope of this chapter, however biopolymers can play an important role in delivering personalized medicines which are already being exploited most notably in tissue engineering. It is in this area of medical research where the use of biopolymers can have a great impact in the future.

18.1.4.1 The Development of Biopharmaceuticals

Originally biopharmaceuticals were restricted to growth hormones and insulin. However, since the elucidation of the human genome, genetic disorders have become potentially treatable using peptides, proteins, and nucleic acids driven by pharmacogenomics and pharmacoproteomics which in turn have provided pharmaceutical scientists with a great challenge to formulate these delicate molecules into effective drugs.

Proteins, peptides, genes and live vaccines are limited to parenteral delivery and to highlight the difficulty of developing a viable non-parenteral biopharmaceutical production in 2006 Pfizer produced the first commercially available form of inhaled insulin trademarked Exubera® and in early 2008 the product was removed from the market as it did not meet patient needs and financial expectations.

Oral delivery has historically been the preferred route of drug administration for patients, as it offers a convenient pain-free treatment. To deliver biopharmaceuticals orally, however, is an extremely tough challenge due to first-pass metabolism effects, poor bioavailability and susceptibility to extreme fluctuations in pH and abundance of digestive enzymes in the GIT. However, the variety of physiological conditions within the GIT provides potential to utilize the properties of biopolymers to create stable formulations that are physiologically responsive. The development of suitable biopharmaceutical formulations using polysaccharide delivery systems is an attractive proposition due to mild processing conditions and low toxicity (formulation using synthetic materials can often involve organic solvents, high shear mixing or high temperatures). Additionally, due to the high development costs of biotech drugs there is great interest in reformulation as patents begin to expire on some of the early biopharmaceutical products in an attempt to keep one step ahead of a growing number of biogeneric companies creating their own version of off-patent products.

Physiological responsive hydrogels such as chitosan and alginate have attracted increasing attention for the delivery of biopharmaceuticals especially with respect to oral drug delivery due to favorable properties such as mucoadhesion and pH sensitivity, along with the ability to form microspheres and nanospheres. Indeed a large amount of research published in the area of oral protein delivery using polysaccharides has focused on alginate microspheres as the vehicle of choice due to the mild cross-linking conditions having a minimal effect on protein denaturation. Microspheres can be prepared using several techniques to achieve the desired size. Simple extrusion of polymer into a solution of cross-linker is probably the simplest technique used. Other frequently used techniques involve single or double emulsion systems to form liquid polymer droplets in the aqueous dispersed phase that can be cross-linked by addition of a cross-linker with the particle size governed by the size of the polymer droplets formed in the emulsion which can be reduced by using high shear mixing.

The immobilization of proteins by Ca^{2+} cross-linking of continuous phase of alginate can provide a stabilizing effect for entrapped proteins and the pH responsive sol–gel transition has also been investigated to transit proteins through the harsh acidic environment of the stomach, releasing the protein in the intestine. This method of delivery has been investigated for the entrapment and oral delivery of live bacteria as probiotics (Lee and Heo, 2000) or as vaccines (Dobakhti et al., 2006). Although alginates have been shown as promising vehicles for oral delivery there are issues such as drug leaching, imbibing of acid, and competitive inhibition of carboxyl groups by some positively charged proteins affecting stability (Espevik et al., 1993). To overcome these problems alginate formulations have been tailored further by coating with positively charged polymers such as poly-L-lysine and chitosan to create multilayer microspheres (Anal et al., 2003), also blending with other biopolymers such as pectin (Liu and Krishnan, 1999) and chitosan to produce polyelecrolyte complexes (PECs), which affects gel network complexity and pore size, parameters that have the potential to be manipulated to achieve the desired effect. Alginate-chitosan PECs, for example, can be produced with a range of drug release profiles by varying parameters such as polymer ratio, molecular weight, the degree of acetylation of the chitosan and the G:M ratio of the alginate.

It is known that chitosan can enhance drug absorption by opening the intercellular tight junctions of the gastrointestinal tract which combined with mucoadhesive properties and pH sensitivity make it a popular choice as a coating material for particulate drug delivery systems. For example, chitosan has been investigated as a coating material for drug delivery systems such as liposomes for oral delivery of the peptide calcitonin and was found to possess excellent retention and penetrative property into the intestinal mucosa of rats (Takeuchi et al., 2005). Chitosan has also been used as a mucoadhesive coat on peptide-loaded DL-lactide/glycolide copolymer (PLGA) nanospheres for pulmonary delivery, which resulted in prolonged mucoadhesion for sustained drug release at the absorption site (Yamamoto et al., 2005).

The versatility and intrinsic properties of biopolymers provide ideal vehicles for directed drug delivery and will continue to assist with the progress of future medical practices not only in the controlled release of therapeutic compounds but also in developing regenerative medicine. Indeed, the controlled release of bioactive proteins in modern medicine is not confined to delivery in vivo but also to the rapidly expanding area of tissue engineering in vitro. An important aspect of tissue engineering is to provide suitable delivery systems for growth factors that can stimulate desired cell responses both in vitro and in vivo, to accelerate tissue regeneration. Polysaccharide microspheres have been explored for the controlled release of growth factors within three-dimensional tissue engineering scaffolds, to augment cellular proliferation and extracellular matrix formation. Lee et al. (2004a) presented results suggesting that transforming growth factor (TGF-β1) which controls cell proliferation, and differentiation loaded into chitosan microspheres then incorporated into cell scaffolds seeded with chondrocytes, have the potential to enhance cartilage formation. In another study vascular endothelial growth factor (VEGF), essential for angiogenesis, has shown high encapsulation efficiency in alginate microspheres followed by sustained zero-order release by diffusion over 3 weeks (Elcin et al., 2001). The in vitro delivery of growth factors is destined to provoke even greater interest in the future as the discipline of tissue engineering develops. The remainder of this chapter will focus on the functional role of biopolymers in recent tissue engineering research and potential future directions.

18.2 TISSUE ENGINEERING

The mean life expectancy of the developed world rose significantly in the 20th century due to major advances in medical technology. As a consequence, the average age of the population has increased, which has resulted in a greater and growing demand for novel technologies that are aimed at the replacement of diseased and damaged tissues. Although the gold standard for the replacement of the majority of tissues is autograft (harvested from the patient) (Gamradt and Lieberman, 2003), there exist significant problems with donor site morbidity (Arrington et al., 1996) and lack of availability (Sandor et al., 2003). Tissues derived from cadavers (allograft) and animals (xenograft) address both the harvesting and availability issues, however, there is a significant risk of implant rejection following implantation as a result of an immunogenic reaction or disease transmission (Patience et al., 1997). One potential solution to avoid immunogenic response and availability issues would be to harvest the patient's cells and culture them ex vivo to produce tissues for eventual implantation. This approach to tissue replacement was originally termed tissue engineering in 1993 by Langer and Vacanti (Langer and Vacanti, 1993) and has since been the focus of a significant and high-profile global research effort.

Cells, to some extent, may organize themselves into crude tissue-like structures in a simple culture system (L'Heureux et al., 2006;

Neagu et al., 2006). It is now widely recognized, however, that in order to reproduce functional tissues in vitro it is necessary to culture the cell populations on substrates that provide relevant mechanical, chemical and biological cues to direct tissue formation (Stevens, 2008). Such structures are typically referred to as scaffolds and have been formed from a range of ceramic (Habraken et al., 2007), metallic (Witte et al., 2008) and polymeric materials (synthetic and biologically derived) (Jagur-Grodzinski, 2006). Hydrogels are becoming widely used as scaffolds for the replacement of both hard and soft tissues due to their biological compatibility in a wide range of applications, which is attributed to their hydrophilic nature and the capability for a relatively high rate of molecular diffusion throughout their structures (Baroli, 2007).

Such is the widespread application of hydrogels in the field of tissue engineering, that an entire book could be dedicated to their use in a plethora of contrasting tissue types. As a consequence, this part of the chapter focuses on the current state-of-the-art in the area, with an emphasis on how cells interact with and can modify hydrogel matrices. This is of particular importance since one of the principal advantages of hydrogel-based materials in this application is that they provide an environment more akin to human tissue than other synthetic materials traditionally used to support cell growth (Abbott, 2003). A range of processing technologies will be discussed that can be used to microstructurally modify hydrogel-based materials to counteract mass transport limitations. Finally, a range of currently available engineered tissue structures will be discussed in addition to an important emerging area – the development of complex tissue interfaces using hydrocolloid derived structures.

18.2.1 Cell Adhesion

The adhesion of cells to polymeric matrices is an important consideration when selecting a hydrogel for use as a tissue engineering scaffold. Hydrogel materials derived from mammalian extracellular matrix (ECM) components such as collagen, fibrin and chitosan typically allow for cell adhesion (Imai et al., 2000; Urech et al., 2005; Hoemann et al., 2007). Materials derived from other sources, such as alginate, however, do not readily interact with mammalian cells (Lansdown and Payne, 1994). Hydrogels that are used in tissue engineering can therefore be defined either as cell adherent or non-cell adherent. Each class of material has its advantages. Cell-adherent hydrogels, for example, can typically be restructured in situ by a cell population, which may secrete enzymes that are capable of resorbing the matrix and can concurrently produce their own ECM (Urech et al., 2005). In contrast, alginate has been shown to preserve the phenotype of cells and maintain dimensional stability with time. In cartilage replacement, for example, it is imperative that implanted cells maintain a chondrocytic phenotype in order that they synthesize a cartilage-like matrix. Bonaventure et al. (1994) demonstrated that dedifferentiated chondrocytes encapsulated in an alginate matrix expressed cartilage-specific genes and allowed the formation of a cartilaginous tissue within 15 days.

18.2.2 Cell-Adhesive Hydrogels

The cell-adhesive hydrogels that are most frequently used in the fabrication of tissue engineering scaffolds include fibrin, collagen and chitosan. The adhesion of cells to the surface of these hydrogels is mediated through the attachment of specific cell adhesion molecules present on the surface of the cell membrane to cell adhesion proteins adsorbed to the material surfaces. These membrane-bound proteins known as integrins are heterodimers, with an α and β group. There have been 18 α-units and eight β-units described in the literature, which allow for a certain amount of selectivity to the cell adhesion process. Depending on the α–β

sub-unit combination, the resulting integrins (24 different combinations are currently known) can bind to specific proteins, including fibronectin, laminin, vitronectin and collagen, and ligands such as the RGD sequence (arginine–glycine–aspartic acid) which may be absorbed or attached onto a range of different substrates.

Integrins do not simply mediate adhesion, but are thought to be largely responsible for transmitting stresses to the cell cytoskeleton, which in turn enables the cell populations to respond to a range of mechanical stimuli. Integrins are transmembrane proteins and attach not only to ligands external to the cell, but also to the actin filaments present on the interior of the cell (Figure 18.2). It is the attachment of cells in this manner, which determines the morphology, function and shape of cells and allows them to both migrate and reorder their local microenvironment. After binding to the cell adhesion ligands has occurred, closely clustered integrins form dot-like complexes known as focal complexes. Focal complexes can be broken down rapidly to enable migration through an adhesive substrate in a 'hand-over-hand' manner. Eventually, the focal complexes evolve into focal adhesions, which bind bundles of actin fibers and enable a very strong adhesion to the substrate. Focal adhesions may then further evolve to form fibrillar adhesions, which are thought to be involved in the organization of the pericellular matrix.

The attachment of cells to the surface of materials enables the migration of cells through the hydrogel, which is essential in the production of a structure that mimics the geometry and organization of the replaced tissue. Furthermore, the adhesion allows the cell to monitor its local geometry to some extent and re-organize the fibrous structure accordingly. A major drawback with both fibrin and collagen gels is that the cells compact and contract the gel via strong adhesions resulting in significant shrinkage following cell attachment and this must be taken into account when using these materials as scaffolds.

18.2.2.1 Collagen

There are 20 different forms of collagen found in the human body. The most abundant form of

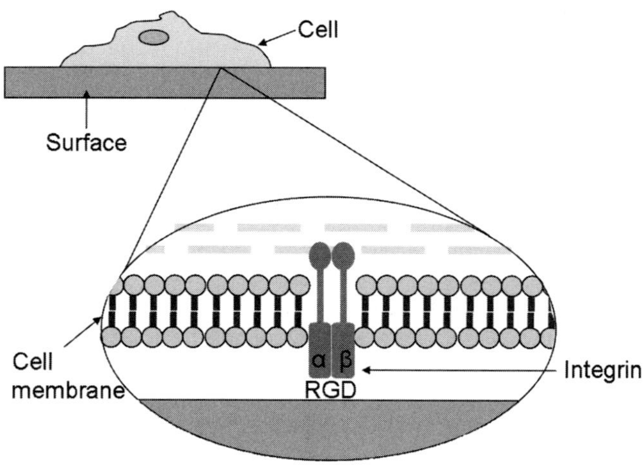

FIGURE 18.2 A schematic diagram of an adhesion between a 2D surface and a single cell. Integrins present within the membrane of the cells interact with proteins or ligands, such as the RGD sequence, that are readily absorbed onto the surface of the material. These initial adhesions may subsequently develop into focal adhesions, fibrillar adhesions or 3D matrix adhesions. The attachment of the heterodimeric integrin to the actin cytoskeleton of the cell is the mechanism by which stresses and strains are detected by the cell and can result in significant changes in cell phenotype.

collagen found in many different tissues throughout the body is type I collagen. It is typically well accepted on implantation in the body and therefore has been used in a wide range of different medical applications, including: dermal replacement (Otto et al., 1995), localized drug delivery (Chan et al., 2008), bone graft replacement (Leupold et al., 2006). Collagen is typically harvested from rat tail tendons or from bovine cartilage by a process of acid digestion. The collagen is then stored in acidic conditions until required and may be gelled by returning the pH value of the colloid to physiological pH. The collagen monomers self-assemble to form collagen fibers, which subsequently become physically entangled to form a weak gel. By controlling either collagen concentration or pH value during the gelling process, it is possible to manipulate both the fiber diameter and pore size of the resulting gels (Pedersen and Swartz, 2005). The weakness of the gel is attributed to the fact that gelling occurs only as a consequence of physical chain entanglement. It is possible to enhance strength by cross-linking the gel with glutaraldehyde, but this process is toxic to the cell population encapsulated within the gel structure. The mechanical properties exhibited by the gel improve on contraction by the cell population, however, this process can take many weeks, which makes clinical application unfeasible. Recent work has focused on uniaxially unconfined compaction of the forming collagen gels. Excess water is removed from the gel, resulting in densification and therefore a rapid improvement in mechanical properties and less dramatic shrinkage following cell seeding. Numerous recent advances in the use of the technique have seen the incorporation of aligned pores within the collagen structure (Nazhat et al., 2007), embossing of the surface with defined textures (Brown et al., 2005) and the fabrication of three-dimensional tissues by manipulating the collagen gels following multiple compactions (Aou Neel et al., 2006). Importantly collagen gels can support the growth of a range of different cell types, including kidney cells (Reid et al., 1993), smooth muscle cells (Song et al., 2000), fibroblasts (Feng et al., 2003a, 2003b), endothelial cells (Nakamura et al., 2008), and cardiomyocytes (Feng et al., 2003b). Collagenase enzymes secreted by most cell types enable the modification of the gel structure by cell populations to provide favorable niches for the maintenance of their phenotype, and facilitating migration through the gel structure.

Collagen has been used widely in the in vitro production of skin replacements. Skin was first tissue engineered in the mid 1970s by the co-culture of fibroblasts and keratinocyte cells (Rheinwald and Green, 1975), with the intention of subsequently applying the co-culture to the surface of the damaged skin as a replacement for autograft tissue. Although this approach was later successful in the treatment of relatively superficial wounds, where only an epidermal replacement was required, in some cases it is necessary to also provide a replacement for the dermis that can be incorporated into the body with no significant deleterious side effects. Hydrogels of collagen and hyaluronic acid have been widely investigated for this application, although clinical success has been limited due to the extensive contraction of the engineered skin-like material and also localized skin blistering. Numerous workers have attempted to reduce contraction and thereby enhance clinical success by chemically or physically modifying the hydrogel structure. Although some success has been reported, to date there are only three products commercially available which comprise bovine collagen (Apligraf, Orcel and Permaderm) (MacNeil, 2007).

18.2.2.2 Fibrin

Fibrin gels are used widely as supports to study the contractile ability of a range of cells (Carr, 2003) and to determine the propensity of both chemicals and materials to encourage angiogenesis (the formation of new blood vessels) (Maier et al., 2007). In the human body, fibrin is essential to enable the creation of a seal following significant tissue damage. In order to form a gelled fibrin matrix, fibrinogen is combined with thrombin and $CaCl_2$. The use of three different precursors to the formation of the gel means that there is considerable scope to modify the mechanical and chemical properties exhibited by the gel (Pedersen and Swartz, 2005), which typically exhibits shear and storage moduli of at least three orders of magnitude greater than those exhibited by collagen-based gels (Table 18.2). The resulting matrix is a densely gelled fibrin structure through which cells capable of secreting matrix metalloproteinases (MMPs) can migrate. Cells are typically seeded directly onto the surface of the gel and can then migrate through and modify its structure. This approach has been used to tissue engineer a range of soft tissues including ligaments (Hankemeier et al., 2007), tendons (Baar, 2007) and muscles (Huang et al., 2005).

18.2.2.3 *Chitosan*

Chitosan can be used in hydrogel form for the replacement of a range of tissues and as a tissue engineering scaffold. Similarly to collagen, chitosan is soluble in a weak acid and can be induced to gel by adjusting the pH value of the hydrocolloid to neutral or physiological pH, though with chemical modification it is possible to gel chitosan in response to the application of UV radiation (Amsden et al., 2007) or a temperature change (Chenite et al., 2000). The main use for chitosan-based polysaccharides in tissue engineering is as a scaffold for the formation of new cartilage (Suh and Matthew, 2000). Due to its cationic nature, chitosan can form complexes with glycosaminoglycans and can immobilize chondroitin sulphate, thus mimicking the structure of a cartilaginous ECM. It has been shown that chitosan-based hydrogels can therefore effectively support cultures of articular chondrocytes maintaining their chondrocytic phenotype.

TABLE 18.2 Summary of Literature Values for a Selection of Mechanical Properties Exhibited by Collagen, and Fibrin Hydrogels (Values from Pedersen and Swartz, 2005)

Gel → Mechanical Property ↓	Collagen	Fibrin
Shear storage modulus (Pa)	0.15–50	150–520
Shear loss modulus (Pa)	0.02–8	30
Tensile modulus (kPa)	1–33	31–112

18.2.3 Non-cell Adhesive Hydrogels

Alginate is the most often investigated hydrogel derived from natural sources that is used in tissue engineering that is non-adherent to cells (Augst et al., 2006). One of the first uses for alginate in cell-based therapies was as an encapsulation medium for pancreatic islets in order to facilitate immunoisolation (Lim and Moss, 1981). More recently, alginate has been investigated for use in a wide range of other medical applications, including: the delivery of a range of pharmaceuticals and biopharmaceuticals (Peters et al., 1998), as a wound dressing (Matthew et al., 1995) and as a carrier of cells for direct implantation or for engineering tissues ex vivo. Alginate has a number of advantageous properties, which make it favorable for use as a tissue engineering scaffold. Its mild gelation, with a range of different cations, means that cells can be encapsulated with little risk to their viability and they can remain vital in encapsulated form for a considerable period of time (at the time of writing the author has maintained fibroblast cultures in an encapsulated state for more than 150 days; Figure 18.3). The biggest risk to cell viability is during the encapsulation process, when in alginate hydrocolloids of $\geq$2wt% the shear forces are sufficient to cause cell death (Figure 18.4).

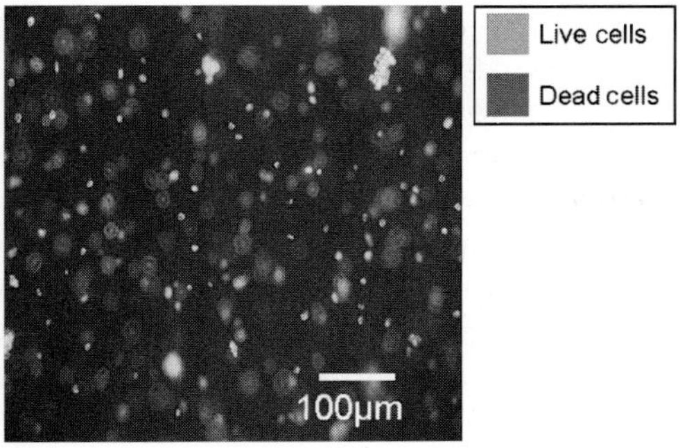

FIGURE 18.3 3T3 fibroblast cells seeded in an alginate bead (diameter 6 mm) and stained with a LIVE/DEAD assay. The cells were cultured for periods of up to 150 days following seeding. Live cells are stained in green and dead cells in red. Even after 150 days in culture the majority of the cells were alive.

A significant advantage that is associated with the use of alginate is that its mechanical properties can be relatively easily tailored to a given application. The mechanical properties that are exhibited by alginate can be influenced by changes in molecular weight distribution (M:G ratio), cross-linking cation density, gelation temperature and alginate concentration. The high degree of control that can be exercised over the mechanical properties that are exhibited by alginate-based hydrogels is a significant advantage in the control of cell differentiation. It is now well established that cellular phenotype and to some extent function are heavily influenced by local fluctuations in elastic modulus. Relatively high moduli, for example, are thought to favor the differentiation of mesenchymal stem cells to differentiate to cells exhibiting an osteoblastic phenotype and gels of comparatively low modulus have been shown to favor the differentiation of marrow stromal cells to a fibroblastic phenotype.

Interestingly, when cell–cell adhesions are stronger than cell–matrix adhesions, as is the case with alginate, cell aggregation tends to

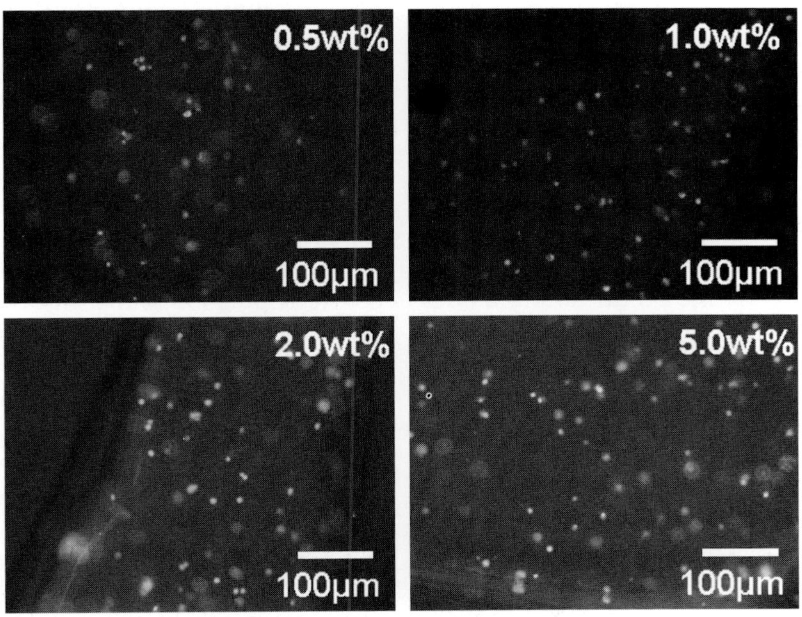

FIGURE 18.4 LIVE/DEAD assay of 3T3 fibroblast cells encapsulated in 0.5, 1.0, 2.0 and 5.0wt% alginate and cultured for one day. The cells stained in green are considered to be alive and those in red are considered dead. An increase in gel concentration resulted in a concomitant increase in the incidence of cell death, this effect has previously been attributed to an increase in the shear stress required to homogenize the cell dispersion with an increase in alginate concentration.

occur, which many researchers have exploited in order to form the precursors to many tissues or to form functional components of tissues. Hepatocytes and fibroblasts, for example, have been aggregated following 3D culture in an agarose gel (Dean et al., 2007). A significant problem with the formation of spheroids within a hydrogel matrix is poor mass transport to the center of the spheroid, which can cause cell cytotoxicity at its center.

While the relative dimensional stability of alginate is attractive for use and many applications and certain cell types (chondrocytes for example) have been shown to maintain their phenotype for prolonged periods of time in culture (Bonaventure et al., 1994), alginate hydrogels have been widely modified in an attempt to induce cell attachment. As previously outlined, cell adhesion takes place as a result of interactions between cell adhesion molecules such as fibronectin, the biopolymer and integrins present in the membranes of adherent cells (Figure 18.2). Workers have attempted to enhance cell adhesion by the incorporation of fibronectin, but have found that the cells tend to interact with the hydrogel in a non-specific and non-uniform manner. The incorporation of RGD sequences, fibronectin-derived cell adhesion complexes, yielded a significantly more homogeneous cell distribution within the hydrogel matrix (Augst et al., 2006) and this modification is now widely used in the production of tissues using alginate as a scaffold.

Alginate has also been used widely in the delivery of chondrocytes (Grimmer et al., 2004), osteoblasts (Lee et al., 2004b) and bone marrow stromal cells with the intention of regenerating diseased or damaged bone and cartilage. It has been demonstrated that the delivery of osteoblasts derived from calvaria and chondrocytes can significantly enhance the rate of bone formation in vivo, although it has also been shown that in order to achieve good bone regeneration it is essential that the alginate gel can to some extent degrade within the body. The likely reason for this is that cells encapsulated in alginate matrices are known to be metabolically inhibited while encapsulated, which would minimize or even prevent the formation of new ECM by the cell population. Once released from encapsulation, the cells encapsulated within the gels were shown to return to their normal metabolic state. Alginate has also been used in the production of scaffolds that seek to augment osteochondral defects, which are complex interfaces formed between bone and cartilage (Gelinsky et al., 2007). The incorporation of a calcium phosphate component into the scaffold to form a layered biphasic structure has been shown to successfully support the cultivation of chondrocytes and osteoblasts in vitro, however, to date no data exist to prove that this approach would succeed in vivo.

Recent work in the author's group has sought to produce a bilayered skin replacement using an alginate hydrogel seeded with fibroblasts as the dermal component and a population of keratinocytes cultured on the surface of the hydrogel to form the epidermal component (Figure 18.5). The encapsulation of the fibroblast cells within the alginate matrix inhibited cell proliferation, meaning that there was no need for mitotic inhibition of the fibroblast population and furthermore a stratified layer of keratinocytes has been shown to be formed on the upper surface (Figure 18.6). The weak adhesion of the keratinocyte population to the surface of the unmodified alginate means that it is possible that the stratified layer could be transferred onto the surface of superficial epidermal injuries. The concurrent application of the alginate structure may also effectively 'seal' the wound, lowering the chances of future infection. Indeed, due to its hemostatic nature alginate is already used widely as a wound dressing (Matthew et al., 1995).

18.2.4 Mechanical Conditioning

Through their adhesion to materials and the ECM, cells are constantly monitoring and

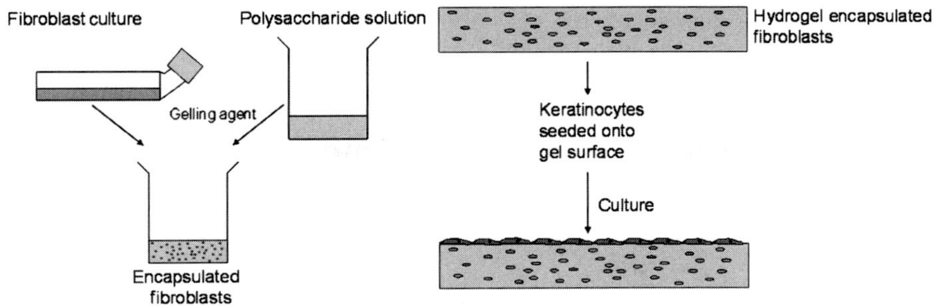

FIGURE 18.5 A schematic showing a method for the fabrication of a full-thickness dermal substitute comprising fibroblasts encapsulated in an alginate matrix and a keratinocyte layer.

responding to external mechanical stimuli by migrating, differentiating or laying down ECM. Consequently, bulk tissues are able to respond with time to the application of varying levels of load, remodeling their structures accordingly. Two of the most obvious and extreme examples of mechanical-conditioning-induced structural changes are the significant loss of bone mass in astronauts in microgravity or the significant gain in bone mass in the playing arm of a tennis player. One of the major problems that is associated with the in vitro production of tissues is that in the absence of mechanical conditioning the tissues will bear little structural or mechanical similarity to the tissue that they are designed to replace. The structures of tissues are directly influenced by cells, which modify the ECM to create their own niches. To further complicate matters, cellular phenotype is significantly affected by the mechanical properties that are exhibited by the ECM. To generate tissues that bear a resemblance to those in vivo, therefore, it is essential to provide the correct mechanical and chemical cues to control cell phenotype while allowing the cells to modify their own environments to generate new tissues representative of those found in the body. To generate tissues that are able to function in vivo

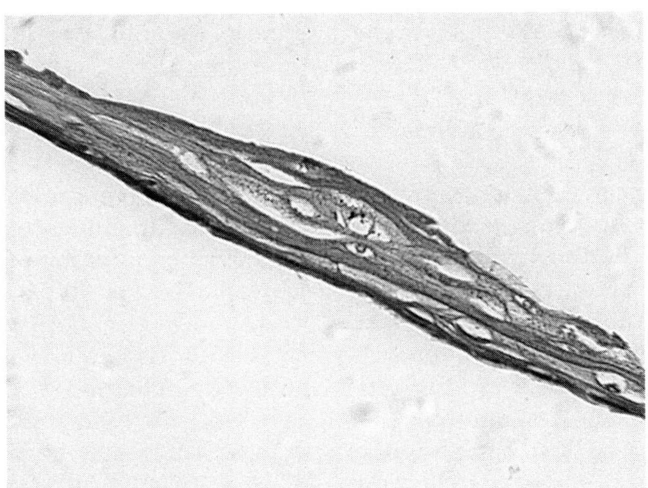

FIGURE 18.6 A histological section through a layer of stratified keratinocytes grown on the surface of an alginate hydrogel, the section was stained using hemotoxylin and eosin.

researchers have developed a range of complex bioreactors in order to define loads and thereby condition the resulting tissue such that it will be able to resist the load to which it is exposed in vivo. The type of reactor used is largely dependent on the type of tissue that is to be produced. Tendons and ligaments, for example, are exposed to uniaxial tensile loads, which stimulate the cells encapsulated within the scaffold to align in-line with the principal stress axis and produce a highly conserved matrix (Berry et al., 2003; Baar, 2007). In contrast, blood vessels are exposed to pulsatile flows and are therefore stimulated in a reactor which simulates the flow of blood (Solan et al., 2001). It is very important to consider, however, that upon initial seeding the matrix will bear little resemblance to that in vivo and therefore the way in which the load is transmitted to the seeded cells will be very different to the situation in mature tissues. When cells are seeded at relatively low densities in the hydrogel matrix the mechanical properties exhibited by the hydrogel will dominate and it will therefore support the largest part of the load and shield the cells from both stress and strain. The situation will be particularly complex in a non-affine gel (such as collagen) where stress and strain distributions are highly complex and difficult to model computationally (Pedersen and Swartz, 2005).

18.2.4.1 *Cell Morphology*

The morphologies of cells that are cultured in a 3D cell-adherent hydrogel are very different to those cultured on a 2D surface. Fibroblast cells that are cultured on surfaces, for example, spread and exhibit very prominent cellular extensions. In comparison, when cultured in a 3D cell-adherent gel the cells tend to take on a spindle-like, stellate or dendritic morphology. Cell seeding density and then application of loads to cell-seeded gels are also well documented to have a marked influence on cell morphology and the organization of cells within the structure, which tend to align with the principal axis of stress. In addition to the stress and strain concentrations in tissues the morphology of the cell population can result in the formation of different classes of adhesions between the cells and the ECM. It is thought that the chemical complexity of explanted tissue, for example, is essential to the formation of 3D adhesions rather than simple focal or fibrillar cell adhesions. Matrix stiffness has also been implicated as having a significant influence on the rate of proliferation of cells, with stiffer materials allowing more rapid cell proliferation than softer materials.

18.2.4.2 *Cell Migration*

It has also been widely reported that many cell types will migrate towards stiffer regions of a substrate. One group has previously immobilized fibronectin-coated beads in an optical trap and demonstrated that cells tend to strengthen their integrin-mediated adhesion proportionally to the force required to restrain the bead. This property has been employed by one group in order to guide the formation of neuronal cells through a hydrogel-based tube (Luo et al., 2001). Other authors have reported that although cells tend to move along stiffness gradients toward the stiffest part of a material, the rate at which the cells move is considerably faster on a soft substrate than on a hard substrate.

As well as influencing cell morphology, the mechanical environment to which a cell is exposed has an influence on cell phenotype. This is of obvious importance in tissue engineering, particularly since the cells responsible for the formation of bone, cartilage and other commonly engineered tissues are each derived from the same pluripotent cell source, the mesenchymal stem cell. The differentiation of mesenchymal stem cells to form bone has been shown to occur more readily on substrates of relatively high modulus (GPa) such as ceramics and more compliant substrates have been shown to support the formation of fibroblasts and chondrocytes. Although it is a complex set of environmental stimuli, both chemical and mechanical, that are known to stimulate the formation of these different cell types and not mechanotransduction alone, providing a suitable mechanical environment to encourage the formation of the desired cell type is obviously an important consideration in the design of a suitable scaffold material.

18.2.5 Microengineering of Hydrogels

In the body tissues are permeated by a network of blood vessels which serve to supply cells encapsulated within the ECM with nutrients and oxygen and also to remove their metabolic waste products. The absence of blood vessels within the majority of tissue engineering scaffolds means that the size of a tissue that can be engineered in vitro is hindered by mass transport limitations. There are two ways in which these mass transport problems can be solved: (1) 'microgels' may be formed, which when laden with cells can be self assembled to form tissues; (2) hydrogel monoliths can be precisely structured to form a network of capillary-like channels, which enable mass transport or the precise delivery of small quantities of nutrients or active molecules to the cell population. The production of both kinds of structure has been made possible by the increased availability of microfabrication facilities, which enable the production of precisely defined polymeric or silicon based molds.

18.2.5.1 Microgels

The fabrication of microgels of controlled morphology formed by a process of soft-lithography from alginate has previously been reported by Qiu et al. (2007). They demonstrated that it was possible to generate alginate particles of defined morphology down to the order of 10 μm and suggested that the increased surface area to volume ratio of the gels would negate mass-transport problems. Such processing methodologies have enabled researchers to take a new 'bottom-up' approach to the development of tissue-engineered structures, allowing more control over tissue structure than ever before. By directing the assembly of controlled morphology particles containing cells capable of forming different tissues, it is possible to effectively design complex tissue interfaces at the micro-level. Recent work by the Khademhosseini group (Du et al., 2008), for example, has demonstrated that it is possible to direct the assembly of similar gel-based structures into structurally complex multi-tissue constructs by exploiting the tendency of liquid–liquid systems to minimize their surface energy. The rapid progress in this area is extremely exciting as it will enable the production of ever more complex structures, opening up the possibility of forming extremely complex tissues with precisely defined biological and mechanical properties in vitro.

18.2.5.2 Microfluidic Scaffolds

Microchanneled hydrogel structures have also recently been reported in the literature. Choi et al. (2007) formed a precisely microchanneled alginate structure using soft-lithography. Using a micromachined silicon surface, it is possible to form channeled structures with very precisely defined structures with feature sizes down to 30 μm by gelling the alginate on the surface of a silicon wafer (Figure 18.7). The presence of the aligned pore structure throughout the hydrogel monolith would have helped to address mass transport issues, but importantly such channels allow the chemical environment of the encapsulated cells to be precisely controlled by means of the delivery of precise quantities of growth factors and other stimuli. In common with the manufacture of microgels, this approach to some extent would enable the production of complex tissue interfaces in vitro. In addition, it is possible that this technology could allow us to develop

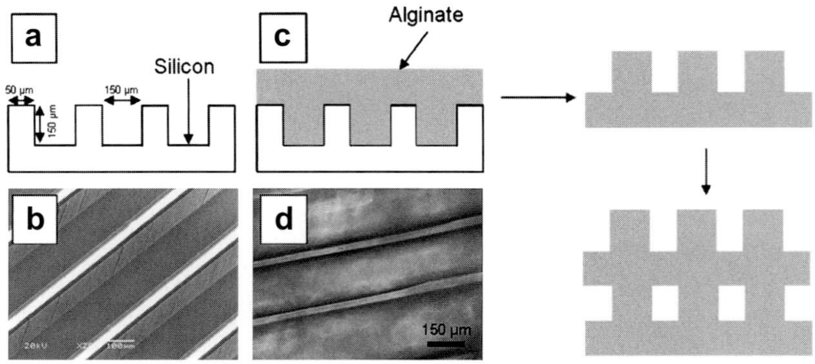

FIGURE 18.7 The fabrication of a microchanneled structure using a micromachined silicon surface as a mold (A, B and C). The alginate structures cast onto the surface showed good feature reproduction (D).

high-throughput assays to evaluate cell and tissue response to a range of chemical stimuli. Developing such technologies is becoming more important in a range of different industrial sectors as the use of animal testing to determine the biological response to a range of both chemicals and drugs is becoming increasingly taboo.

18.3 FUTURE HORIZONS

Biopolymer-derived hydrogels are now used widely in the delivery of drugs and are finding increased use in the fabrication of tissue-engineered structures. Their biological compatibility in a range of applications and exciting recent developments in microscale processing technologies means that we can exercise unprecedented control over their microstructures and regional variations in chemical and physical environments. As the requirement for tissue replacements increases and with the development of new biopharmaceutical products, their widespread application in regenerative medicine in the next decade or so is set to increase significantly.

ACKNOWLEDGMENTS

Dr. L M Grover would like to acknowledge Miss Nicola Hunt and Miss Pauliena Bohari for contributing their data to the chapter.

References

Abbott, A. (2003). Cell culture: Biology's new dimension. *Nature* **424**, 870–872.

Abou Neel, E. A., Cheema, U., Knowles, J. C., Brown, R. A., and Nazhat, S. N. (2006). Use of multiple unconfined compression for control of collagen gel scaffold density and mechanical properties. *Soft Matter* **2**, 986–992.

Agoub, A. A., Smith, A. M., Giannouli, P., Richardson, R. K., and Morris, E. R. (2007). "Melt-in-the-mouth" gels from mixtures of xanthan and konjac glucomannan under acidic conditions: A rheological and calorimetric study of the mechanism of synergistic gelation. *Carbohydrate Polymers* **69**, 713–724.

Amsden, B. G., Sukarto, A., Knight, D. K., and Shapka, S. N. (2007). Methacrylated glycol chitosan as a photopolymerizable biomaterial. *Biomacromolecules* **8**, 3758–3766.

Anal, A. K., Bhopatkar, D., Tokura, S., Tamura, H., and Stevens, W. F. (2003). Chitosan-alginate multilayer beads for gastric passage and controlled intestinal release of protein. *Drug Development and Industrial Pharmacy* **29**, 713–724.

Arrington, E. D., Smith, W. J., Chambers, H. G., Bucknell, A. L., and Davino, N. A. (1996). Complications of iliac crest bone graft harvesting. *Clinical Orthopaedics and Related Research*, 300–309.

Augst, A. D., Kong, H. J., and Mooney, D. J. (2006). Alginate hydrogels as biomaterials. *Macromolecular Bioscience* **6**, 623–633.

Baar, K. (2007). Uniaxial stretch increases collagen synthesis in fibrin-based engineered tendons. *Tissue Engineering* **13**, 1773.

Bansil, R. and Turner, B. S. (2006). Mucin structure, aggregation, physiological functions and biomedical applications. *Current Opinion in Colloid & Interface Science* **11**, 164–170.

Baroli, B. (2007). Hydrogels for tissue engineering and delivery of tissue-inducing substances. *J Pharm Sci* **96**.

Batchelor, H. (2005). Bioadhesive dosage forms for esophageal drug delivery. *Pharm Res.* **22**, 175–181.

Batchelor, H. K., Banning, D., Dettmar, P. W., Hampson, F. C., Jolliffe, I. G., and Craig, D. Q. M. (2002). An in vitro mucosal model for prediction of the bioadhesion of alginate solutions to the oesophagus. *International Journal of Pharmaceutics* **238**, 123–132.

Bawa, R., Hall, R. E., Kabra, B. P., Teague J. E., Cagle, G. D., Markwardt, K. L., and Shah, M. V. (1999). Gelling ophthalmic compositions containing. United States Patent 6261547.

Berry, C. C., Shelton, J. C., Bader, D. L., and Lee, D. A. (2003). Influence of external uniaxial cyclic strain on oriented fibroblast-seeded collagen gels. *Tissue Engineering* **9**, 613–624.

Bonaventure, J., Kadhom, N., Cohensolal, L., Ng, K. H., Bourguignon, J., Lasselin, C., and Freisinger, P. (1994). Reexpression of Cartilage-Specific Genes by Dedifferentiated Human Articular Chondrocytes Cultured in Alginate Beads. *Experimental Cell Research* **212**, 97–104.

Brown, J., Madit, N., Cole, E. T., Wilding, I. R., and Cade, D. (1998). The effect of cross-linking on the in vivo disintegration of hard gelatin capsules. *Pharmaceutical Research* **15**, 1026–1030.

Brown, R. A., Wiseman, M., Chuo, C. B., Cheema, U., and Nazhat, S. N. (2005). Ultrarapid engineering of biomimetic materials and tissues: Fabrication of nano- and microstructures by plastic compression. *Advanced Functional Materials* **15**, 1762–1770.

Cade, D., Scott, R., and He, X. (2003) Polymer film compositions for capsules. United States Patent 6,517,865.

Carr, M. E. (2003). Development of platelet contractile force as a research and clinical measure of platelet function. *Cell Biochemistry and Biophysics* **38**, 55–78.

Ceulemans, J., Vinckier, I., and Ludwig, A. (2002). The use of xanthan gum in an ophthalmic liquid dosage form: Rheological characterization of the interaction with mucin. *Journal of Pharmaceutical Sciences* **91**, 1117–1127.

Chan, O. C. M., So, K. F., and Chan, B. P. (2008). Fabrication of nano-fibrous collagen microspheres for protein delivery and effects of photochemical crosslinking on release kinetics. *J Control Release* **129**.

Chenite, A., Chaput, C., Wang, D., Combes, C., Buschmann, M. D., Hoemann, C. D., Leroux, J. C., Atkinson, B. L., Binette, F., and Selmani, A. (2000). Novel injectable neutral solutions of chitosan form biodegradable gels in situ. *Biomaterials* **21**, 2155–2161.

Choi, N. W., Cabodi, M., Held, B., Gleghorn, J. P., Bonassar, L. J., and Stroock, A. D. (2007). Microfluidic scaffolds for tissue engineering. *Nature Materials* **6**, 908–915.

Dean, D. M., Napolitano, A. P., Youssef, J., and Morgan, J. R. (2007). Rods, tori, and honeycombs: the directed self-assembly of microtissues with prescribed microscale geometries. *Faseb Journal* **21**, 4005–4012.

Dobakhti, F., Rahimi, F., Dehpour, A. R., Taghikhani, M., Ajdary, S., Rafiei, S., and Rafiee-Tehrani, M. (2006). Stabilizing effects of calcium alginate microspheres on mycobacterium bovis BCG intended for oral vaccination. *Journal of Microencapsulation* **23**, 844–854.

Dobetti, L. (2001). Fast-Melting Tablets: Developments and Technologies. *Pharmaceutical Technology*, 44–50.

Du, Y., Lo, E., Ali, S., and Khademhosseini, A. (2008). Directed assembly of cell-laden microgels for fabrication of 3D tissue constructs. *Proc Natl Acad Sci U S A* **105**.

Edman, P. (2002). Biopharmaceuticals of Ocular Drug Delivery (Pharmacology & Toxicology).

Elcin, Y. M., Dixit, V., and Gitnick, T. (2001). Extensive in vivo angiogenesis following controlled release of human vascular endothelial cell growth factor: Implications for tissue engineering and wound healing. *Artificial Organs* **25**, 558–565.

Espevik, T., Otterlei, M., Skjakbraek, G., Ryan, L., Wright, S. D., and Sundan, A. (1993). The Involvement of Cd14 in Stimulation of Cytokine Production by Uronic-Acid Polymers. *European Journal of Immunology* **23**, 255–261.

Fan, H. Y., Wang, K., Liu, M. M., and He, Z. M. (2008). In vitro evaluations of konjac glucomannan and xanthan gum mixture as the sustained release material of matrix tablet. *Carbohydrate Polymers* **73**, 241–247.

Felt, O., Furrer, P., Mayer, J. M., Plazonnet, B., Buri, P., and Gurny, R. (1999). Topical use of chitosan in ophthalmology: tolerance assessment and evaluation of precorneal retention. *International Journal of Pharmaceutics* **180**, 185–193.

Feng, Z., Yamato, M., Akutsu, T., Nakamura, T., Okano, T., and Umezu, M. (2003a). Investigation on the mechanical properties of contracted collagen gels as a scaffold for tissue engineering. *Artificial Organs* **27**, 84–91.

Feng, Z., Matsumoto, T., and Nakamura, T. (2003b). Measurements of the mechanical properties of contracted collagen gels populated with rat fibroblasts or cardiomyocytes. *J Artif Organs* **6**.

Gamradt, S. C. and Lieberman, J. R. (2003). Bone graft for revision hip arthroplasty. *Clinical Orthopaedics and Related Research*, 183–194.

Gandhi, R. B. and Robinson, J. R. (1994). Oral Cavity As A Site for Bioadhesive Drug-Delivery. *Advanced Drug Delivery Reviews* **13**, 43–74.

Gelinsky, M., Eckert, M., and Despang, F. (2007). Biphasic, but monolithic, scaffolds for the therapy of osteochondral defects. *International Journal of Materials Research* **98**, 749–755.

Grimmer, J. F., Gunnlaugsson, C. B., Alsberg, E., Murphy, H. S., Kong, H. J., Mooney, D. J., and Weatherly, R. A. (2004). Tracheal reconstruction using tissue-engineered cartilage. *Archives of Otolaryngology-Head & Neck Surgery* **130**, 1191–1196.

Habraken, W. J. E. M., Wolke, J. G. C., and Jansen, J. A. (2007). Ceramic composites as matrices and scaffolds for drug delivery in tissue engineering. *Advanced Drug Delivery Reviews* **59**, 234–248.

Hankemeier, S., van Griensven, M., Ezechieli, M., Barkhausen, T., Austin, M., Jagodzinski, M., Meller, R., Bosch, U., Krettek, C., and Zeichen, J. (2007). Tissue engineering of tendons and ligaments by human bone marrow stromal cells in a liquid fibrin matrix in immunodeficient rats: Results of a histologic study. *Archives of Orthopaedic and Trauma Surgery* **127**, 815–821.

Harding, S. E. (2006). Trends in mucoadhesive analysis. *Trends in Food Science & Technology* **17**, 255–262.

Hoemann, C. D., Chenite, A., Sun, J., Hurtig, M., Serreqi, A., Lu, Z., Rossomacha, E., and Buschmann, M. D. (2007). Cytocompatible gel formation of chitosan-glycerol phosphate solutions supplemented with hydroxyl ethyl cellulose is due to the presence of glyoxal. *Journal of Biomedical Materials Research Part A* **83A**, 521–529.

Huang, Y. C., Dennis, R. G., Larkin, L., and Baar, K. (2005). Rapid formation of functional muscle in vitro using fibrin gels. *Journal of Applied Physiology* **98**, 706–713.

Imai, K., Sato, T., and Senoo, H. (2000). Adhesion between cells and extracellular matrix with special reference to hepatic stellate cell adhesion to three-dimensional collagen fibers. *Cell Structure and Function* **25**, 329–336.

Jagur-Grodzinski, J. (2006). Polymers for tissue engineering, medical devices, and regenerative medicine. Concise general review of recent studies. *Polymers for Advanced Technologies* **17**, 395–418.

L'Heureux, N., Dusserre, N., Konig, G., Victor, B., Keire, P., Wight, T. N., Chronos, N. A. F., Kyles, A. E., Gregory, C. R., Hoyt, G., Robbins, R. C., and McAllister, T. N. (2006). Human tissue-engineered blood vessels for adult arterial revascularization. *Nature Medicine* **12**, 361–365.

Langer, R. and Vacanti, J. P. (1993). Tissue Engineering. *Science* **260**, 920–926.

Lansdown, A. B. and Payne, M. J. (1994). An evaluation of the local reaction and biodegradation of calcium sodium alginate (Kaltostat) following subcutaneous implantation in the rat. *J R Coll Surg Edinb* **39**.

Lee, J. E., Kim, S. E., Kwon, I. C., Ahn, H. J., Cho, H., Lee, S. H., Kim, H. J., Seong, S. C., and Lee, M. C. (2004a). Effects of a chitosan scaffold containing TGF-beta 1 encapsulated chitosan microspheres on in vitro chondrocyte culture. *Artificial Organs* **28**, 829–839.

Lee, K. Y., Alsberg, E., Hsiong, S., Comisar, W., Linderman, J., Ziff, R., and Mooney, D. (2004b). Nanoscale adhesion ligand organization regulates osteoblast proliferation and differentiation. *Nano Letters* **4**, 1501–1506.

Lee, K. Y. and Heo, T. R. (2000). Survival of Bifidobacterium longum immobilized in calcium alginate beads in simulated gastric juices and bile salt solution. *Applied and Environmental Microbiology* **66**, 869–873.

Leupold, J. A., Barfield, W. R., An, Y. H., and Hartsock, L. A. (2006). A comparison of ProOsteon, DBX, and Collagraft in a rabbit model. *Journal of Biomedical Materials Research Part B-Applied Biomaterials* **79B**, 292–297.

Lim, F. and Moss, R. D. (1981). Microencapsulation of Living Cells and Tissues. *Journal of Pharmaceutical Sciences* **70**, 351–354.

Liu, M. M., Fan, J. Y., Wang, K., and He, Z. (2007). Synthesis, characterization, and evaluation of phosphated cross-linked konjac glucomannan hydrogels for colon-targeted drug delivery. *Drug Delivery* **15**, 397–402.

Liu, P. and Krishnan, T. R. (1999). Alginate-pectin-poly-L-lysine particulate as a potential controlled release formulation. *Journal of Pharmacy and Pharmacology* **51**, 141–149.

Luo, Y., Dalton, P. D., and Shoichet, M. S. (2001). Investigating the properties of novel poly(2-hydroxyethyl methacrylate-co-methyl methacrylate) hydrogel hollow fiber membranes. *Chemistry of Materials* **13**, 4087–4093.

MacNeil, S. (2007). Progress and opportunities for tissue-engineered skin. *Nature* **445**, 874–880.

Maier, A. K. B., Kociok, N., Zahn, G., Vossmeyer, D., Stragies, R., Muether, P. S., and Joussen, A. M. (2007). Modulation of hypoxia-induced neovascularization by JSM6427, an integrin alpha 5 beta 1 inhibiting molecule. *Current Eye Research* **32**, 801–812.

Maltese, A., Borzacchiello, A., Mayol, L., Bucolo, C., Maugeri, F., Nicolais, L., and Ambrosio, L. (2006). Novel polysaccharides-based viscoelastic formulations for ophthalmic surgery: Rheological characterization. *Biomaterials* **27**, 5134–5142.

Marchais, H., Cayzeele, G., Legendre, J. Y., Skiba, M., and Arnaud, P. (2003). Cross-linking of hard gelatin carbamazepine capsules: effect of dissolution conditions on in vitro drug release. *Eur J Pharm Sci.* **19**, 129–132.

Matthew, I. R., Browne, R. M., Frame, J. W., and Millar, B. G. (1995). Subperiosteal Behavior of Alginate and Cellulose Wound Dressing Materials. *Biomaterials* **16**, 275–278.

McCargar, L., Crail, D., Dansereau, R., Myers, W., and Lane, M. (2001). The in-vitro porcine adhesion model is not predictive of the esophageal transit of risedronate tablets in humans. *International Journal of Pharmaceutics* **222**, 191–197.

Nakamura, M., Mie, M., Mihara, H., Nakamura, M., and Kobatake, E. (2008). Construction of multi-functional extracellular matrix proteins that promote tube formation of endothelial cells. *Biomaterials* **29**, 2977–2986.

Nazhat, S. N., bou Neel, E. A., Kidane, A., Ahmed, I., Hope, C., Kershaw, M., Lee, P. D., Stride, E., Saffari, N., Knowles, J. C., and Brown, R. A. (2007). Controlled microchannelling in dense collagen scaffolds by soluble phosphate glass fibers. *Biomacromolecules* **8**, 543–551.

Neagu, A., Kosztin, I., Jakab, K., Barz, B., Neagu, M., Jamison, R., and Forgacs, G. (2006). Computational modeling of tissue self-assembly. *Modern Physics Letters B* **20**, 1217–1231.

Ogura, T., et al. (1998). HPMC capsules: an alternative to gelatin. *Pharm. Technol. Europe* **10**, 32–42.

Otto, W. R., Nanchahal, J., Lu, Q. L., Boddy, N., and Dover, R. (1995). Survival of Allogeneic Cells in Cultured Organotypic Skin-Grafts. *Plastic and Reconstructive Surgery* **96**, 166–176.

Patience, C., Takeuchi, Y., and Weiss, R. A. (1997). Infection of human cells by an endogenous retrovirus of pigs. *Nature Medicine* **3**, 282–286.

Pedersen, J. A. and Swartz, M. A. (2005). Mechanobiology in the third dimension. *Annals of Biomedical Engineering* **33**, 1469–1490.

Peters, M. C., Isenberg, B. C., Rowley, J. A., and Mooney, D. J. (1998). Release from alginate enhances the biological activity of vascular endothelial growth factor. *Journal of Biomaterials Science-Polymer Edition* **9**, 1267–1278.

Pina, M. E. and Sousa, A. T. (2002). Application of Hydro-alcoholic Solutions of Formaldehyde in Preparation of Acetylsalicylic Acid Gastro-Resistant Capsules. *Drug Development & Industrial Pharmacy* **443**.

Pramoda, M. K. and Lin, S. (1979). Mixed xanthan gum and locust beam gum. US Patent Number 4136173.

Qiu, C., Chen, M., Yan, H., and Wu, H. K. (2007). Generation of uniformly sized alginate microparticles for cell encapsulation by using a soft-lithography approach. *Advanced Materials* **19**, 1603.

Rajinikanth, P. S. and Mishra, B. (2008). Floating in situ gelling system for stomach site-specific delivery of clarithromycin to eradicate H-pylori. *Journal of Controlled Release* **125**, 33–41.

Reid, G. G., Gorham, S. D., and Lackie, J. M. (1993). The Attachment, Spreading and Growth of Baby Hamster-Kidney Cells on Collagen, Chemically Modified Collagen and Collagen-Composite Substrata. *Journal of Materials Science-Materials in Medicine* **4**, 201–209.

Rheinwald, J. G. and Green, H. (1975). Formation of A Keratinizing Epithelium in Culture by A Cloned Cell Line Derived from A Teratoma. *Cell* **6**, 317–330.

Richardson, J. C., Dettmar, P. W., Hampson, F. C., and Melia, C. D. (2005). Oesophageal bioadhesion of sodium alginate suspensions 2. Suspension behaviour on oesophageal mucosa. *European Journal of Pharmaceutical Sciences* **24**, 107–114.

Sandor, G. K. B., Rittenberg, B. N., Clokie, C. M. L., and Caminiti, M. F. (2003). Clinical success in harvesting autogenous bone using a minimally invasive trephine. *Journal of Oral and Maxillofacial Surgery* **61**, 164–168.

Sandri, G., Bonferoni, M. C., Chetoni, P., Rossi, S., Ferrari, F., Ronchi, C., and Caramella, C. (2006). Ophthalmic delivery systems based on drug-polymer-polymer ionic ternary interaction: In vitro and in vivo characterization. *European Journal of Pharmaceutics and Biopharmaceutics* **62**, 59–69.

Seager, H. (1998). Drug-delivery products and the Zydis fast-dissolving dosage form. *Journal of Pharmacy and Pharmacology* **50**, 375–382.

Shedden, A. H., Laurence, J., Barrish, A., and Olah, T. V. (2001). Plasma timolol concentrations of timolol maleate: timolol gel-forming solution (TIMOPTIC-XE) once daily versus timolol maleate ophthalmic solution twice daily. *Doc Ophthalmol* 103.

Sinha, V. R., Mittal, B. R., Bhutani, K. K., and Kumria, R. (2004). Colonic drug delivery of 5-fluorouracil: an in vitro evaluation. *International Journal of Pharmaceutics* **269**, 101–108.

Smith, A. M. and Perrie, Y. (2007). Polymer Film Formulations for the Preparation of Enteric Pharmaceutical Capsules. Filed 29/03/07 GP 0706178.1 (Patent pending).

Smart, J. D. (2005). The basics and underlying mechanisms of mucoadhesion. *Advanced Drug Delivery Reviews* **57**, 1556–1568.

Sohi, H., Sultana, Y., and Khar, R. K. (2004). Taste masking technologies in oral pharmaceuticals: Recent developments and approaches. *Drug Development and Industrial Pharmacy* **30**, 429–448.

Solan, A., Prabhakar, V., and Niklason, L. (2001). Engineered vessels: Importance of the extracellular matrix. *Transplantation Proceedings* **33**, 66–68.

Song, J., Rolfe, B. E., Hayward, I. P., Campbell, G. R., and Campbell, J. H. (2000). Effects of collagen gel configuration on behavior of vascular smooth muscle cells in vitro: Association with vascular morphogenesis. *In Vitro Cellular & Developmental Biology-Animal* **36**, 600–610.

Stevens, M. M. (2008). Biomaterials for bone tissue engineering. *Materials Today* **11**, 18–25.

Sudhakar, Y., Kuotsu, K., and Bandyopadhyay, A. K. (2006). Buccal bioadhesive drug delivery – A promising option for orally less efficient drugs. *Journal of Controlled Release* **114**, 15–40.

Suh, J. K. F. and Matthew, H. W. T. (2000). Application of chitosan-based polysaccharide biomaterials in cartilage tissue engineering: a review. *Biomaterials* **21**, 2589–2598.

Takeuchi, H., Matsui, Y., Sugihara, H., Yamamoto, H., and Kawashima, Y. (2005). Effectiveness of submicron-sized, chitosan-coated liposomes in oral administration of peptide drugs. *International Journal of Pharmaceutics* **303**, 160–170.

Talukdar, M. M. and Kinget, R. (1995). Swelling and Drug-Release Behavior of Xanthan Gum Matrix Tablets. *International Journal of Pharmaceutics* **120**, 63–72.

Urech, L., Bittermann, A. G., Hubbell, J. A., and Hall, H. (2005). Mechanical properties, proteolytic degradability and biological modifications affect angiogenic process extension into native and modified fibrin matrices in vitro. *Biomaterials* **26**, 1369–1379.

Washington, N., Steele, R. J., Jackson, S. J., Bush, D., Mason, J., Gill, D. A., Pitt, K., and Rawlins, D. A. (2000). Determination of baseline human nasal pH and the effect of intranasally administered buffers. *Int J Pharm* **5**, 139–146.

Witte, F., Calliess, T., and Windhagen, H. (2008). Biodegradable synthetic implant materials. Clinical applications and immunological aspects. *Orthopade* **37**, 125–130.

Yamamoto, T., Matsuura, S., and Aka, K. (1995). Capsule shell. *United States Patent* 5,431,917.

Yamamoto, H., Kuno, Y., Sugimoto, S., Takeuchi, H., and Kawashima, Y. (2005). Surface-modified PLGA nanosphere with chitosan improved pulmonary delivery of calcitonin by mucoadhesion and opening of the intercellular tight junctions. *Journal of Controlled Release* **102**, 373–381.

Zhang, L., Russell, D., Conway, B. R., and Batchelor, H. (2008). Strategies and therapeutic opportunities for the delivery of drugs to the esophagus. *Critical Reviews in Therapeutic Drug Carrier Systems* **25**, 259–304.

Index